RAPID EXCAVATION and TUNNELING CONFERENCE 2017 PROCEEDINGS

**EDITED BY COLIN A. LAWRENCE
and ANTHONY DEL VESCOVO**

PUBLISHED BY THE

SOCIETY FOR MINING, METALLURGY & EXPLORATION

Society for Mining, Metallurgy & Exploration (SME)
12999 East Adam Aircraft Circle
Englewood Colorado 80112
(303) 948-4200 / (800) 763-3132
www.smenet.org

The Society for Mining, Metallurgy & Exploration (SME) is a professional society whose more than 15,000 members represent professionals serving the minerals industry in more than 100 countries. SME members include engineers, geologists, metallurgists, educators, students, and researchers. SME advances the worldwide mining and underground construction community through information exchange and professional development.

ISBN 978-0-87335-451-6
Ebook 978-0-87335-452-3

On the Cover: Photo of "Bertha," the 17.45-meter EPB for the SR99 Bored Tunnel in Seattle.

Contents

Part 6: Conventional Tunneling

Part 7: Large-Span Tunnel Cavern

Part 8: Future Tunneling

Part 14: Stations and Cross Passages

Part 15: Grouting and Ground Modification II

Preface

The Rapid Excavation and Tunneling Conference (RETC) is the largest and most prominent conference on a tunneler's calendar and is held regularly in the United States. Over the years, its has been developed into a very successful format that attracts the entire underground industry. Delegates attend not only from North America but from all over the world. In recognition of its international attendance, participation in the RETC International Committee includes 16 members.

The RETC Executive Committee extends a warm welcome to all delegates attending the 2017 RETC in San Diego. Bill Mariucci is the conference chair this year and is ably supported by his vice chair, Victor Romero, and the other Executive Committee members. The committee has been competently assisted by SME staff and their efforts are recognized for making all the necessary arrangements and logistics in support of the conference. The committee has also hand-selected the session chairs and meticulously chosen the authors with the aim of maintaining the high standards for which this conference is renowned. We also acknowledge the hard work and spare time that has been dedicated by many people behind the scenes to allow this conference to take place. The conference would not be possible without the significant support of our sponsors for which we are very grateful.

The three days of conference presentations include 114 papers divided into 19 sessions running on four tracks. As you might expect, the papers cover the full breadth and depth of the tunneling business ranging across all markets in the industry. A faithful attempt has been made at grouping the presentations into sessions that have a common theme or interest to delegates. With so many papers from both diverse and broad cross sections of our industry, this has proved to be quite a feat. We trust that you will find the sessions interesting and relevant.

The presentation topics support the view that our industry continues to be buoyant across the United States. Many major projects are currently in the planning stage, under final design, in or having completed construction, or have been in operation for many years and are now in need of rehabilitation. You will see tunnels for many end uses that demonstrate the versatility in application for underground project solutions. Several projects are pushing the envelope and are considered state of the practice for tunneling, where others continue to complete critical components of much larger programs that will provide future resiliency, sustainability, and opportunities for growth in the various municipalities. With the aging infrastructure of our many historic towns and cities, the potential for going underground to address these issues has never been greater. This fuels the continued challenge to our industry for being ever faster, and cheaper, while maintaining or improving safety and quality. We do this by continuously looking for innovation and improvement in everything we do. It certainly is an exciting time to be involved in, or to join, the tunneling industry. And to be successful, we must continue to grow to meet the demands of our business. Toward that end, we are happy to announce that we now have a robust Young Tunnel Professionals community, members of which you may meet at the conference. Investment in our young professionals has never been so important to our future with the current climate of the tunneling industry.

Please seize the opportunity of this RETC to network and meet other members of this very special industry. In doing so, you'll meet new friends and business acquaintances who will extend the friendship of our close and very special underground community. Enjoy the conference.

Colin A. Lawrence
Anthony Del Vescovo

Executive Committee

Session Chairs

Jeff Brandt
Traylor Bros., Inc.

Kurt Braun
L-7 Services LLC

Pierre Ciuffarin
Frontier-Kemper Constructors, Inc.

Joe Clare
Mott MacDonald

Thomas Costabile
Skanska

Adam Curry
Moretrench

Dave Dorfman
Schiavone Construction Co. LLC

Greg Emslie
McMillen Jacobs Associates

Geoff Fairclough
Schiavone Construction Co. LLC

Andrew Finney
CH2M Hill

Shaun Firth
CH2M Hill

Erica Fredrickson
Traylor Bros., Inc.

Rick Gomez
Gomez International, Inc.

Gregory Hauser
Dragados USA

Josh Jonasen
Traylor Bros., Inc.

Peter Kottke
Kiewit

Nate Long
Jay Dee Contractors, Inc.

Steve Maggipinto
Schiavone Construction Co. LLC

Robert Marshall
Frontier-Kemper Constructors, Inc.

Justin McCain
Tutor Perini

Daniel McMaster
Mott MacDonald

A.G. Mekkaoui
Jay Dee Contractors, Inc.

Bianca Messina
Skanska

Dwight Metcalf
Kiewit

Jack Nakagawa
Tutor Perini

Cody Painter
WSP | Parsons Brinckerhoff

Frank Perrone
Mott MacDonald

Mark Peterman
Kiewit

Peter Procter
Mott MacDonald

Gregory Rogoff
McMillen Jacobs Associates

David Smith
Parsons Brinckerhoff

David Sowers
Washington State Department of Transportation

Mike Stolkin
J.F. Shea Co. Inc.

Richard Taylor
Traylor Bros., Inc.

Matt Trotter
Kiewit

Darren VonPlaten
Traylor Bros., Inc.

Moussa Wone
DC Water and Sewer Authority

James Wonneberg
McMillen Jacobs Associates

International Committee

Pressure Face TBM I

Chairs

Jeff Brandt
Traylor Bros., Inc.

Cody Painter
WSP | Parsons Brinckerhoff

Accomplishing Extraordinary Tasks as a Machine Supplier for Metro Doha

Karin Bäppler ▪ Herrenknecht AG, Germany
Dirk Schrader ▪ Herrenknecht AG, Germany

With the construction of the Metro Doha in the State of Qatar a vision is accomplished to establish a modern, safe, efficient and integrated public transportation network. Upon final completion of multiple phases, the network will comprise four metro lines with a length of 216km and 100 stations. Phase 1 comprises 112km of underground tunnels (56km twin tube tunnels) and a total of 21 Earth Pressure Balance Tunnel Boring (EPB) machines were deployed by the four different Joint Ventures. The project was initiated, managed and supervised by the owner Transport Authority Qatar Rail. Phase 2 of the Doha Metro will start upon the completion of Phase 1 in 2019 (Qatar Vision 2030). The owner placed full faith in the manufacturer's abilities to deliver equipment of high quality, state of the art standard within the given time frame. This paper focuses on the manufacturer's scope of delivery and on the supplier's high standards of quality in production, logistics and services. The report is completed by a chapter with remarks on the manufacturer's analysis of TBM availability on the individual tunnelling sections that completed tunnelling works.

INTRODUCTION

In emerging regions of the world the growth of the cities is continuing unabated. To maintain and increase the mobility of citizens in urban cities, more extensive and efficient networks of infrastructure are required. The most common and long-term design approach is to build such infrastructure beneath the surface. Midsize cities and established industrial nations are following this trend and are upgrading or supplementing their existing infrastructure networks to be well prepared for the future. The Middle East region is emphasizing the need for mobility and new very efficient infrastructure systems are going to be built.

Qatar Rail was given the mandate to design and develop a new rail network and thus also the Doha Metro system of which a large part is underground. Qatar Rail's mission is to provide modern, reliable, attractive, safe and sustainable integrated railway services. Its objectives are to deliver the metro system in time and to the desired quality in a safe and cost effective manner. The focus of this report is on accomplishing extraordinary tasks as a machine manufacturer for the construction of the three metro lines of Doha Metro, the Green Line, Red Line and Gold Line.

Herrenknecht became the exclusive TBM supplier and delivered 21 Earth Pressure Balance (EPB) machines for Phase 1 of the Metro Doha. Herrenknecht group brands also delivered comprehensive additional equipment such as 57km of tunnel, shaft and overland conveyor belts, 474 segment moulds, 31 multi-service vehicles, rolling stock, navigation systems and surface monitoring. In 2012 a regional subsidiary was founded in Qatar with the main objective of supporting the client, consultants and contractors directly and as closely as possible. Herrenknecht technically supported the construction companies during the time of tender preparation and after contract award. Such support was further extended with the start of tunnelling works as follows but not limited to:

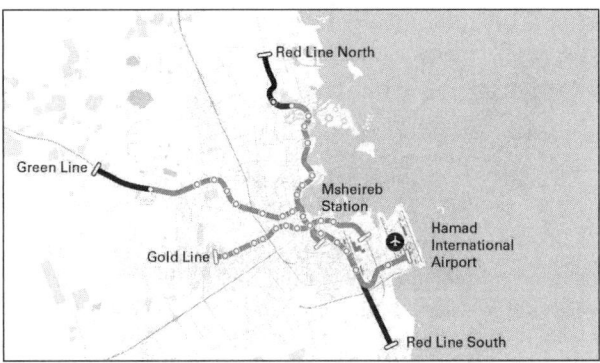

Figure 1. Doha Metro network linking Doha's main destinations

- Provision of technical advice and expertise for the TBM assembly, tunnelling, transfer and dismantling works
- Provision of specialized technicians (engineers, mechanics, electricians, operators, welders)
- Provision and set-up of TBM spare & wear parts and logistics
- Provision of special equipment for TBM assembly, transfer and dismantling
- Supply and refurbishment of excavation tools and the support of all group brands
- TBM lubricants
- Working visas
- Import and export of equipment, spare & wear parts

DETAILED OVERVIEW OF THE TUNNELLING SECTIONS FOR THE METRO GREEN LINE, RED LINE, AND GOLD LINE

Phase 1 of Doha Metro comprises the construction of three new metro lines: Green Line, Red Line (North, South) and Gold Line. The metro lines comprise parallel twin-tube tunnels excavated and lined by 21 EPB machines with shield diameters ranging between 7.05 and 7.11 meters. The tunnel lining consists of steel fiber reinforced concrete elements with an internal diameter of 6.17 meters.

The Red Line starts at Al Wakra in the south and ends in Lusail in the north. The 55.42km Red Line, also known as the Coast Line, will connect the Hamad International Airport with the City Center. The Red Line is separated into two lots, Red North (22.79km) from Msheireb to Qatar University and Red South (32.63km) from Hamad International Airport to Msheireb. The 33.68km Green Line connects Al Mansoura in the east with Al Riffa in the west and is also known as the Education Line because it passes through Education City. The 23.32km east-west Gold Line extends from Ras Bu Abboud to Al Aziziyah.

The geological conditions in the project area are characterized by Simsima Limestone, Midra Shale and Rus Formation. The Simsima Limestone is composed of weak to moderately strong weathered dolomitic and chalky limestone with fissures and cavities. The unconfined compressive rock strengths vary in general between 20–50MPa with some up to 120MPa. The Midra Shale is a weak to moderately weak, slightly weathered and well cemented mudstone. The Rus Formation is a soft limestone,

Figure 2. Typical geological conditions at TBM tunnel face (picture from Red Line South)

dolomitic and chalky limestone with gypsum with predicted strengths of 5 to 15MPa. Karstic features are found throughout the country and were taken into consideration when designing the TBMs. Due to the given gradient of the metro alignment the tunnels faced all three geological formations described above. Based on the subsurface conditions in the project area the main issues for the design of the TBMs have been:

- Heterogeneous tunnelling conditions along the tunnel drives
- Tunnelling at depths between 11 to 45 meters beneath the ground water table
- High hydrostatic pressures of up to 4.5bar
- Probability of facing karstic features with possible high water inflows
- Ground water classified as very aggressive with high salt and sulphate content

In order to reduce possible risks due to extreme variable geological conditions an extensive ground investigation program was executed in advance of the tunnelling works. This program comprised geotechnical field investigations, drilling campaigns with borehole logging, pumping tests and a ground water quality study to characterize the geotechnical and hydrogeological conditions along and around the planned metro network and underground metro stations. To identify possible changes in geology during tunnelling operation the EPB Shields and cutting wheels were prepared to install the Bore-Tunnelling Electrical Ahead Monitoring (BEAM) system to identify these changes about 20 meters ahead of the tunnel face.

The Doha Metro contract was split in five design and build civil contracts (four sections of Tunnel & Stations and one Major Station). The details about the respective tunnelling lots are described in the following sections with focus on the TBM works. Special technical and logistical challenges are highlighted in the following chapter that is related to the simultaneous operation of 21 EPB TBMs.

Doha Metro, Green Line

The design and build contract was awarded to the Joint Venture of Porr (Austria), Saudi Binladin Group (KSA) and HBK (Qatar).

For the 33.38km of tunnel construction, Herrenknecht supplied six EPB Shields (Ø7.05m), tunnel, shaft and overland belt conveyors, 16 multi-service vehicles, segment moulds, cooling and grouting plants and comprehensive tunnel-surface monitoring.

Figure 3. Assembly of the first two TBMs (S-846 and S-847) of in total four TBMs at Al Messila Station

Four TBMs were assembled at Al Messila Station and two at Though Shaft. Along the tunnel drives all TBMs had several intermediate breakthroughs. The machines excavated in depths of 11 to 35 meters and were designed for a maximum operation pressure of 4.5bar. The first two TBMs (S-846 and S-847) started excavation in September and October 2014 for the twin tube tunnels from Al Messila Station toward Msheireb Station. Each machine excavated and lined a section of about 4.1km with a total of four intermediate breakthroughs. At the end of November 2015 and the beginning of December 2015 the two machines celebrated their final breakthroughs. Best weekly performances of up to 216 meters were achieved.

The two TBMs (S-844 and S-845) that started tunnelling in January 2015 from Al Messila Station in opposite direction for 6.1km toward Education City Station had their final breakthroughs in March 2016. Along this tunnel section from Al Messila Station to Education City Station the two machines had each three intermediate breakthroughs and a maximum performance of 212.8 meters per week was managed. The 6.5km long twin tube drives (S-848 and S-849) from Tough Shaft to Education City Station were characterized by one intermediate breakthrough. Weekly performances of up to 283.2 meters were achieved and both EPB TBMs finished their drives end of February 2016.

Doha Metro, Red Line South

The design and build contract was awarded to the Joint Venture of Qatar Diar-Vinci Construction (Qatar), GS E&C (Korea) and Al Darwish Engineering (Qatar).

For the 32.63km tunnel, Herrenknecht supplied five EPB Shields (Ø7.05m) and logistic rolling stock.

The scope of the Red Line South contract comprises the design and construction of the underground works beneath central Doha between the Msheireb Underground Station and the New Doha International Airport (Hamad International Airport) including five underground stations. The machines operated at depths of 11 to 35 meters below surface and were designed for a maximum operation pressure of 4.5bar.

Two of the five TBMs were assembled at M10/11 Switchbox and the other three at Umm Ghuwailina. Along the tunnel drives all TBMs had several intermediate breakthroughs. The two TBMs (S-860 and S-862) that were assembled at M10/11 Switchbox

Table 1. Overview about the tunnelling sections of the six EPB Shields used for the underground sections of Doha Metro, Green Line

No.	Breakthrough	TBM	Station	Tunnel Length (m)
1	Partial	S-846	Switchbox M30/M31	624.0
2	Partial	S-847	Switchbox M30/M31	627.2
3	Partial	S-846	Hamad Hospital Station	640.0
4	Partial	S-847	Hamad Hospital Station	652.8
5	Partial	S-846	White Palace Station	403.2
6	Partial	S-845	Al-Qadeem Station	2,270.4
7	Partial	S-847	White Palace Station	398.4
8	Partial	S-844	Al-Qadeem Station	2,270.4
9	Partial	S-848	Evacuation Shaft 04	3,694.4
10	Partial	S-845	Al Shaqab Station	939.2
11	Partial	S-849	Evacuation Shaft 04	3,694.4
12	Partial	S-844	Al Shaqab Station	948.8
13	Partial	S-846	Al Bidda Station	1,670.4
14	Partial	S-847	Al Bidda Station	1,648.0
15	**Final**	**S-846**	**Msheireb Station**	**745.6**
16	**Final**	**S-847**	**Msheireb Station**	**766.4**
17	Partial	S-845	Qatar National Library Station	1,606.4
18	Partial	S-844	Qatar National Library Station	1,588.8
19	**Final**	**S-848**	**Education City Station**	**2,760.0**
20	**Final**	**S-849**	**Education City Station**	**2,769.6**
21	**Final**	**S-845**	**Education City Station**	**1,340.0**
22	**Final**	**S-844**	**Education City Station**	**1,336.0**

excavated each a first section of about 3.7km toward Umm Ghuwailina before they started with the excavation of the second 5.5km section toward the Doha International Airport. Tunnelling of TBM S-862 was successfully completed on September 25, 2016 which also marked the total completion of 112km tunnelling works for Doha Metro Phase 1.

The other three TBMs (S-861, S-863 and S-864) started operation from Umm Ghuwailina. Two machines excavated the first section of about 2.5km toward Msheireb with the second section toward Ras Bu Abboud of 2.8km. The third machine (S-864) excavated a section of 4.3km (first 1.3km section toward M40/104 Switchbox and the second and third sections toward Msheireb of about 1.5km each). Tunnelling works are completed and TBM dismantling is ongoing.

Doha Metro, Red Line North

The design and build contract was awarded to the Joint Venture of Salini-Impregilo (Italy), SK E&C (Korea) and Galfar Al Misnad (Qatar).

For the 22.79km of tunnel, Herrenknecht supplied four EPB machines (Ø7.05m) and segment moulds.

The tunnel alignment is 11 to 35 meters below surface and all machines were configured to operate with maximum pressures of 4.5bar. Two machines (S-865 and S-867) were used for the twin tube tunnels of about 3km (1st section) and 4.2km (2nd section). The two EPBs were assembled and launched from the bottom of Al Qassar Station.

Table 2. Overview about the tunnelling sections of the five EPB Shields used for the underground sections of Doha Metro, Red Line South

No.	Breakthrough	TBM	Station	Tunnel Length (m)
1	Partial - 1th Drive	S-863	Al Doha Al Jadeda Station	704.0
2	Partial - 1th Drive	S-861	Al Doha Al Jadeda Station	707.2
3	Partial - 1th Drive	S-860	Al Matar Station	2,761.6
4	Partial - 1th Drive	S-862	Al Matar Station	2,768.0
5	Final - 1th Drive	S-864	Switchbox M40/104	1,222.4
6	Final - 1th Drive	S-863	Msheireb Station	2,337.6
7	Final - 1th Drive	S-861	Msheireb Station	2,331.2
8	Final - 1th Drive	S-860	Umm Ghuwailina Station	916.4
9	Final - 1th Drive	S-862	Umm Ghuwailina Station	913.6
10	Partial - 2nd Drive	S-863	Switchbox M40/104	810.0
11	Final - 2nd Drive	S-864	Msheireb Station	1,500.8
12	**Final**	**S-861**	**Ras Bu Abboud Station**	**2,790.0**
13	**Final**	**S-863**	**Ras Bu Abboud Station**	**1,920.0**
14	**Final**	**S-864**	**Msheireb Station**	**1,502.4**
15	**Final**	**S-860**	**Hamad International Airport**	**5,381.0**
16	**Final**	**S-860**	**Hamad International Airport**	**5,435.0**

Two other EPBs (S-866 and S-868) excavated the first 2.3km long drive between Corniche and Msheireb and then the 1.8km long twin tube tunnel section between Corniche and Doha Exhibition and Convention Centre (DECC). Along the tunnel drives all TBMs had several intermediate breakthroughs.

On February 16, 2015 the tunnel section and TBM S-868 between Corniche to Al Diwan was flooded as a result of water ingress through the TBM screw conveyor of about 400 l/sec. The TBM S-868 had already excavated 1,100m and had almost reached the deepest level of the alignment between the stations (tunnel invert 35m below GL). The JV immediately started an extensive pumping scheme and the installation of a cut-off bore pile wall with additional wells in front of the TBM. The inflow of water was stopped and the TBM was recovered. It was found out that a big boulder had blocked the screw conveyor discharge gate. The TBM was fully refurbished inside the tunnel with the replacement of all electrical components, including the control cabin, in only three months so that the TBM was able to restart on May 16, 2015 and successfully completed the remaining tunnel section without any further problems.

Doha Metro, Gold Line

The design and build contract was awarded to the Joint Venture of Aktor (Greece), Yapi Merkezi (Turkey), Larsen & Toubro (India), STFA (Greece) and Al Jaber (Qatar).

For the 23.32km of tunnel construction, Herrenknecht supplied six EPB machines (Ø7.05m), tunnel, shaft, and overland belt conveyors, 15 multi-service vehicles, segment moulds, cooling and grouting plants and comprehensive tunnel-surface monitoring. Herrenknecht also set up a Joint Venture with Commodore for the production and supply of the steel fiber reinforced tunnel segments. The same JV also supplied the tunnel segments for 30km of the STEP tunnel in Abu Dhabi.

Two machines (S-920 and S-921) were assembled at Ras Bus Abboud (Airport City Station), each excavating a section of about 4.4km toward Msheireb Station with two partial breakthroughs. Another four TBMs started excavation at Al Sudan Station. Two machines excavated each a section of 3.5km toward Msheireb and two EPBs

Table 3. Overview about the tunnelling sections of the four EPB Shields used for the underground sections of Doha Metro, Red Line North

No.	Breakthrough	TBM	Station	Tunnel Length (m)
1	Partial - 1th Drive	S-866	Al Bidda Station	1,605.0
2	Final - 1th Drive	S-867	DECC Station	2,977.5
3	Final - 1th Drive	S-865	DECC Station	2,971.5
4	Final - 1th Drive	S-866	Msheireb Station	712.5
5	Partial - 1th Drive	S-868	Al Bidda Station	1,598.9
6	Partial - 2nd Drive	S-867	Katara Station	1,231.5
7	Partial - 2nd Drive	S-865	Katara Station	1,236.0
8	Final - 1th Drive	S-868	Msheireb Station	796.5
9	Partial - 2nd Drive	S-866	West Bay South Station	747.0
10	Partial - 2nd Drive	S-867	Legtaifiya Station	889.5
11	Partial - 2nd Drive	S-865	Legtaifiya Station	877.5
12	Partial - 2nd Drive	S-868	West Bay South Station	769.5
13	**Final**	**S-866**	**DECC Station**	**1,083.0**
14	**Final**	**S-868**	**DECC Station**	**1,068.0**
15	**Final**	**S-865**	**Trough Golf Course**	**2,097.0**
16	**Final**	**S-867**	**Trough Golf Course**	**2,125.5**

Figure 4. Gold Line, Station Souq Waqif

excavated and lined the twin tubes of 3.7km in opposite direction toward Al Aziziyah Station. Along the tunnel drives all TBMs had several intermediate breakthroughs.

The tunnel boring machines operated at depths of 13 to 35 meters and were designed to operate with maximum operation pressures of up to 4.5bar.

SPECIFIC PROJECT REQUIREMENTS FOR THE METRO GREEN LINE, RED LINE, AND GOLD LINE

The Metro Doha project is a key infrastructure project with the target of systematically building an extensive underground local public transport artery in only a few years. A key element in the success of such a large- scale infrastructure project is the manufacturing and delivery of the TBMs. A common factor for the construction of the three new Metro Lines in Doha was that all 21 TBMs were purchased from the same supplier. This had the advantage to reduce interfaces between different construction components to a minimum and to guarantee a smooth construction process.

Table 4. Overview about the tunnelling sections of the six EPB Shields used for the underground sections of Doha Metro, Gold Line

No.	Breakthrough	TBM	Station	Tunnel Length (m)
1	Partial	S-922	Al Joaan Station	1,261.5
2	Partial	S-923	Al Joaan Station	1,267.5
3	Partial	S-920	Qatar National Museum Station	2,224.5
4	Partial	S-921	Qatar National Museum Station	2,221.5
5	Partial	S-925	Al Waab Station	1,489.5
6	Partial	S-924	Al Waab Station	1,489.5
7	Partial	S-922	Al Sadd Station	654.0
8	Partial	S-923	Al Sadd Station	669.0
9	Partial	S-923	Bin Mahmoud Station	879.0
10	Partial	S-920	Souq Waquif Station	1,186.5
11	Partial	S-921	Souq Waquif Station	1,174.5
12	Partial	S-925	Sport City Station	1,615.5
13	Partial	S-922	Bin Mahmoud Station	892.5
14	Partial	S-924	Sport City Station	1,612.5
15	**Final**	**S-925**	**Al Aziziyah Station**	**610.0**
16	**Final**	**S-920**	**Msheireb Station**	**954.0**
17	**Final**	**S-923**	**Msheireb Station**	**736.0**
18	**Final**	**S-924**	**Al Aziziyah Station**	**620.0**
19	**Final**	**S-922**	**Msheireb Station**	**743.5**
20	**Final**	**S-921**	**Msheireb Station**	**977.8**

The requirement from the owner, Qatar Rail, was the commitment of all partners involved in the construction to build a modern and efficient public transportation network that will be operational in time for the FIFA World Cup 2022. The German TBM manufacturer Herrenknecht was the exclusive TBM supplier for the underground sections of the Metro Doha project and delivered 21 EPB machines that will be described in more detail in the following.

TBM Design and Scope of Delivery of TBM Supplier

All 21 tunnel boring machines that were used for numerous tunnelling sections for the Green Line, Red Line South, Red Line North, and Gold Line are EPB machines with diameters between 7.05 to 7.11 meters.

The design of the cutting wheel of the EPB Shields for all four contracts was similar with opening ratios of 32% to 38%. The cutting wheels were equipped with 17-inch disc cutters, buckets and cutting knives to deal with the prevailing limestone formations of differing constitution.

The twin tube tunnels of Metro Doha are constructed at maximum depths of 45 meters beneath the ground water table. The machines are designed with operating pressures of 4 and 4.5bar according to the specific tunnelling sections. The predicted variable geological conditions, the probability of facing karstic features, cavities and local high water inflows were considered during tunnel excavation. The EPB Shields can be operated in open mode, compressed air mode and in closed mode with active face support. The machines were designed to perform—if required—radial inclined injection drillings through the shield for crown injections and horizontal injection drillings into the face area.

Figure 5. Cutting wheel design of Metro Doha EPB shield

As the project site is in the city area, the safety and protection of the residents is of utmost importance. This required a settlement controlled TBM operation. The chosen tunnelling technology enables to safely control ground settlements through an adequate face pressure management and the immediate support of the excavated ground with a lining that consists of reinforced concrete elements. As the annular gap constitutes a structural risk with regard to the bedding of the lining and subsidence at the surface it is essential to efficiently backfill the annular gap between the outside of the segmental lining and the excavated surface of the ground. The backfilling of the annular gap stabilizes the segmental ring and preserves the natural state of stress in the surrounding soil so that settlement can be controlled. The backfill protects also the segments and the shield tail from ground water inflows. The backfill material is injected through grout lines incorporated in the tailskin at the rear of the shield structure. All 21 machines used a two-component grout type to backfill the annular gap. The two component grout consists of an A-component (cement based grout with bentonite with a retarder and stabilizer) and a B-component. The B-component is an accelerator that is typically sodium silicate.

The advantages of the two component grout are:

- Quicker stabilization of the ring
- Shorter setting times
- Optimized working times
- The processing properties and the pumpability of the two component grout are not affected by downtimes or advancement speeds.

The first machine for Doha Metro was manufactured in Schwanau and was inspected by HE Jassim Saif Ahmed Al Sulaiti, Qatar's Minister of Transport, in February 2014. Already in August of the same year tunnelling started for the first lot of Doha Metro on the Red Line North. Only one year later all 21 machines were in action.

In addition to supplying all 21 EPB machines, the scope of delivery of the TBM supplier for this mega project also included special solutions to support the construction companies in realizing multiple tunnelling operations smoothly and reliably. The support of the Herrenknecht Group was able to cut construction process interfaces by delivering several major packages from its group companies with navigation systems from VMT GmbH, belt conveyors from H+E, multi-service vehicles from Techni-Métal Systèmes and moulds from Herrenknecht Formwork.

In 2012 Herrenknecht opened a branch office for the tender period to support the client, the Joint Ventures, consultants and partners involved in the project. In 2013, prior to the start of the tunnelling works, a subsidiary was founded in Doha (Herrenknecht Tunnelling Doha LLC, HQA). This subsidiary provided ongoing comprehensive technical support during all TBM tunnelling phases including planning, assembly, transfer, tunnelling and dismantling on a 24/7 basis. A warehouse was opened in 2014 for the storage of spare & wear parts plus the refurbishment of cutter discs. HQA refurbished approximately 1,600 cutters for all JV partners during the project. Special equipment for the TBM transfers was planned and delivered. During the tunnelling period up to 100 Herrenknecht specialists including project managers and project engineers worked on all TBMs. Several student trainees from universities in Germany also worked for HQA. The support of all Herrenknecht group brands (TMS, H+E, VMT, MSD) was also organized and supported through HQA. With such a set-up, Herrenknecht worked very closely with all contractors, partners and the client.

The scope of delivery of the comprehensive additional equipment comprised the supply of 31 multi-service vehicles (MSVs) from Techni-Métal Systèmes of which 16 MSVs were supplied for the Green Line and 15 MSVs to operate on the contract of Gold Line. In addition to the MSVs the supply contract included a permanent field service technician and a consignment stock of spare parts to cover both projects.

The six TBMs of the Green Line used navigation and information technology from VTM GmbH. The navigation system is used to determine and calculate all necessary data and information for navigating the TBM along the tunnel alignment. The information technology comprises the integrated risk and information system for the tunnel, geomonitoring and deformation monitoring. The machines on the Red Line North and the Gold Line also used the navigation technology of VMT GmbH.

Herrenknecht Formwork was able to deliver a total of 474 high-quality and high-precision segment moulds for the Metro Doha Phase 1 in Qatar. For the Red Line North 126 moulds for stationary production were delivered. The contract for the Green line comprised 180 moulds for stationary production and handling equipment and the contract for the Gold Line included 168 moulds for stationary production with handling equipment and ring design. Thus Herrenknecht Formwork could contribute with its delivery of moulds and segment handling equipment to the production of approximately 350,000 segments for the three contracts Green Line, Gold Line and Red Line North.

H+E was able to support the Gold Line and Green Line with tunnel conveyors, stockpile conveyors and cross or transfer conveyors. For the Green Line they used a total 37,400m of tunnel conveyors, 250m (100m and 150m) of stockpile conveyors and 103m (47m and 56m) cross or transfer conveyors.

For the contract Red Line South muck cars were used for muck transport from the tunnels and the Joint Venture of Red Line North used their conveying equipment from their earlier project STEP in Abu Dhabi.

Figure 6. Additional equipment such as conveyor belts, e.g., for Doha Green Line (picture at Al Messila Station)

Demand on TBM Production and Logistics for the Machine Manufacturer

For the large-scale construction project of Metro Doha three new metro lines were built comprising twin tube tunnels of 55km in length. The overall project was divided into four civil engineering lots to optimize construction operations and to safeguard the completion of the overall construction program. A critical aspect in such large-scale construction projects is the purchasing and delivery of the necessary equipment. To realize the project in time a total of 21 TBMs were needed to fulfil the expectations.

All TBMs were required to be delivered in a time range between February 2014 and March 2015 that led to the question if the available machine manufacturers had the capacity and the possibility to deliver the machines in the given period of time. Herrenknecht was able to do so and could also guarantee to manufacture and deliver all 21 machines including back up system and associated equipment within the set time. The delivery period for all machines for the four metro contracts comprised less than one year. Seventeen out of 21 machines were manufactured, assembled and factory tested at the Herrenknecht factory in Germany. The four machines for the Red Line North contract were manufactured, assembled and factory tested in the Herrenknecht factory in China. All machines were delivered on site on time.

1. Apart from being able to produce and deliver such a great number of TBMs, an effective and efficient logistics is also one of the key success factors for such large-scale construction projects and is also relevant during the construction phase. The logistics have to ensure that all components will be in the right quantity and at the right time at the point of demand. In order to avoid delays to the project, the logistics need to be planned very carefully, in advance and monitored on a daily basis to ensure high reliability of all construction activities.

Complex regulations made the import of 21 TBMs to Qatar an enormous task for the machine manufacturer. Coordinating the timing of thousands of parts weighing up to 64 tons each called for high expertise in transport logistics. Delivery to job sites with pinpoint accuracy in the designated assembly time slot requires exact planning and coordination with the Joint Ventures on site. In the peak phase of assembly, a loaded ship carrying TBM components left Antwerp for Qatar every two weeks.

TBM AVAILABILITY AND PERFORMANCE

The tunnelling sections of the Green Line, Red Line South and North and Gold Line comprised tunnelling sections of between 624 meters and 3,694 meters. The logistics of the diverse tunnelling drives was handled with mucking out via conveyor belts for the Green Line, Gold Line and Red Line North and with muck cars for the Red Line South. According to the contract the machine supplier warranted that each machine achieves a machine availability of 90%. The machine availability is calculated as follows:

$$\text{Machine Availability} = \frac{(T_t - T_d \times 100\%)}{T_t}$$

with T_t as the available working time in minutes and T_d the downtime in minutes due to TBM breakdown or repair which are solely due to a default by the supplier. Maintenance of the TBM forms part of the available working time in minutes.

The TBM availability for all construction lots was 97% overall and for the Green and Golden Line mainly in the range of 99% and thus far above the warranty (Figure 7).

Figure 7 shows the TBM availability along the single tunnelling sections with a more specific subdivision into the main activities illustrated with Figure 8 such as advance, ring build, pipe extension, logistics and delays from the manufacturer as well as delays from the JVs.

CONCLUSION

The Metro Doha project is one of the current mega projects with the aim of building a new and large-scale inner city infrastructure system underground within the given time frame and with maximum safety. Herrenknecht accomplished extraordinary tasks as

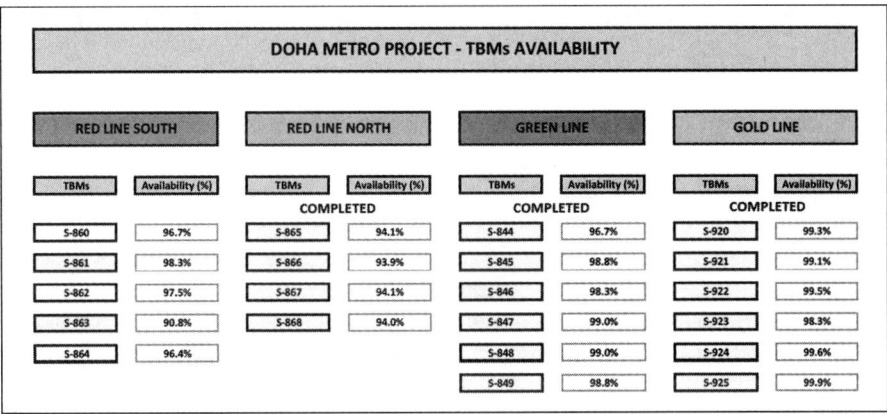

Figure 7. Overview of TBM availabilities for the overall tunnelling sections of each machine and each JV

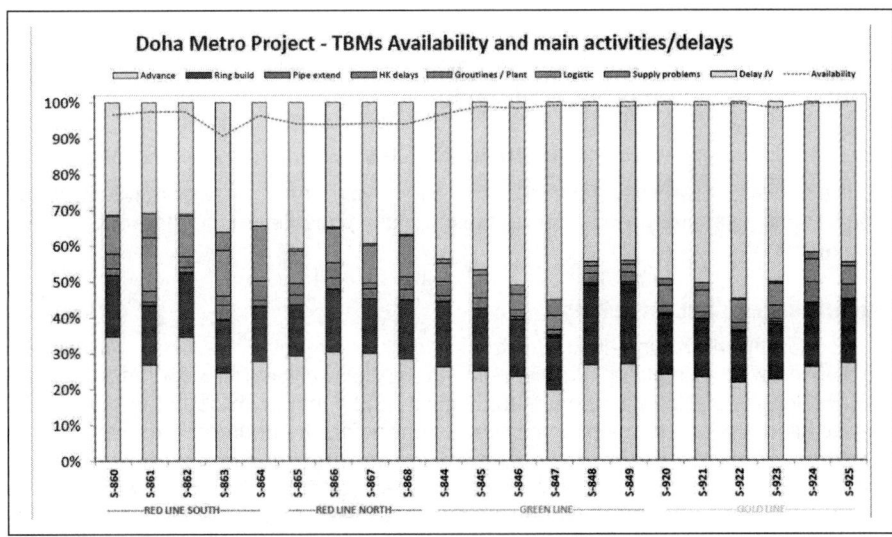

Figure 8. Overview of TBM availabilities subdivided in main activities including delays

a machine manufacturer for the Metro Doha by delivering state of the art technology with 21 TBMs and tailored solutions for this large-scale project reliably and on time. The demand for the contracted companies was to realize high performance tunnel construction with minimum impact to the population and the environment. This was achieved with mechanized tunnelling technology that has established itself in recent years as pioneering technology emphasized by planning security and handing over the completed construction on time. Nevertheless, such mega projects can only be successfully implemented if all parties involved in the construction activities including the owner, designers, consultants, contractors and suppliers work hand-in-hand. This is equivalent to the statement made by Dr. Markus Demmler, Senior Director Qatar Integrated Railway Project: "Managing 21 TBMs working smoothly beneath a metropolis like Doha is only possible with 100% commitment from all partners involved."

Innovations and Efficiency in Urban Tunnelling—A Case Study of the Eglinton Crosstown LRT in Toronto, Ontario

Dan Ifrim ▪ Hatch
Andre Solecki ▪ Hatch
Iqbal Hassan ▪ Metrolinx
Paul Cott ▪ Smith & Long

ABSTRACT

The Eglinton Crosstown LRT project with 10km of twin tunnels 6.5m DIA in EPB conditions is the largest infrastructure project initiated in Canadian history and is a perfect example of challenging working conditions in urban tunnelling. Working space constraints, hauling of muck, water discharge, noise and vibration limits along with "just in time" delivery of segmental lining and other materials posed a real challenge to the tunnel construction logistics.

This paper discusses the project challenges and the design and contractor's innovative solutions with focus on tunnel logistics, project specific conditions restrictions and, the impact on TBM productivity and efficiency.

INTRODUCTION

From a social perspective, a bored tunnel for subway, road, utility or railway through urban areas is generally the preferred option. Many projects around the world enforce this statement.

Challenges in Urban Tunnelling

Challenges in urban tunnelling are related to environmental protection, real estate, logistics as well as political and financial implications of dealing with them. Dealing with challenges starts in the conceptual design and continue through the preliminary and detailed design when solutions are sought before tendering the project. Not all problems find the best answer in design; during construction the Contractor may find alternate resolutions to challenges and cost savings through innovative solutions. Some of the challenges may find answers in alternate procurement such as design-built or P3 or early Contractor involvement or Owner procured equipment or materials.

Urban tunnelling may involve small diameter, tunnelling such as microtunnels, to large diameter tunnels such as metro size or larger. The intent of this paper is to discuss the challenges related to metro size tunnels with focus on the Eglinton LRT Project in Toronto, Ontario, Canada.

Managing challenges starts in the conceptual design and continues through the preliminary and detailed design when solutions are sought before tendering a project. During construction the Contractor may find alternate resolutions to the solutions found during design challenges and cost savings through innovative solutions and means and methods are finalized.

URBAN TUNNELLING CASE STUDY

Project Description

The Eglinton Crosstown Light Rail Transit Project is part of an $8.4 billion (2010$ CAD) investment from the Ontario provincial government to expand transit in Toronto, Canada. The project is being implemented by Metrolinx, an agency of the Government of Ontario, to provide 19 kilometers of dedicated LRT service connecting the City of Toronto from west to east along Eglinton Avenue. The Crosstown will be the first LRT constructed in Toronto and will link regional and local transit, reduce travel time along the route by half and provide over a thousand jobs during construction.

Two tunnelling contracts, the West Contract (from Keele to Yonge) and the East Contract (from Yonge to Brentcliffe) are currently under construction to provide operational service by the year 2021. The underground portion of the LRT alignment is 10 kilometers in length and will be excavated in six tunnel drives grouped into three segments of twin tunnels.

Tunnel excavation from the west and east contracts commenced in 2013 and 2015 respectively. Tunnels have been excavated using Earth Pressure Balance (EPB) Tunnel Boring Machines (TBMs) through a series of glacial tills, interglacial and glacial deposits.

As part of the Eglinton Crosstown project, nine cut-and-cover & three mined stations will be constructed under separate contracts once the tunnel excavation is complete."

Project Challenges

All urban tunnelling challenges are related to tight working space, narrow right of way and interference with existing utilities and this project was no different. Challenges with narrow right of way, surrounding structures and utilities together with solutions to counter these challenges are analysed below.

Surrounding Infrastructure

Urban tunnelling poses unique challenges for a number of reasons one of them being that the surrounding infrastructure already exists. The risk of constructing tunnels adjacent to existing buildings, under existing roads, crossing existing highways, utilities or other tunnels needs to be mitigated and reduced to acceptable levels. In most of the cases the existing infrastructure needed to be protected and kept in operation during all phases of the project.

Twin Tunnels on Narrow ROW

Narrow right of way (ROW) challenges were related to protecting existing structures and ensure no obstructions will impede the tunnelling process. On the Eglinton Crosstown, the horizontal separation from edge of property line to tunnel is less than 1 meter.

Comprehensive settlement monitoring plan was adopted, to ensure a quick response to eventual settlements. Over 1400 buildings spanned the tunnelling zone of influence, all of which needed to be analyzed. Data collection and interpretation of buildings in the zone of influence and identification of possible obstructions was carried out; existing soil anchors protruding the proposed tunnel path were identified during design and the tunnel alignment design was modified to mitigate any potential obstruction impacts during tunnelling.

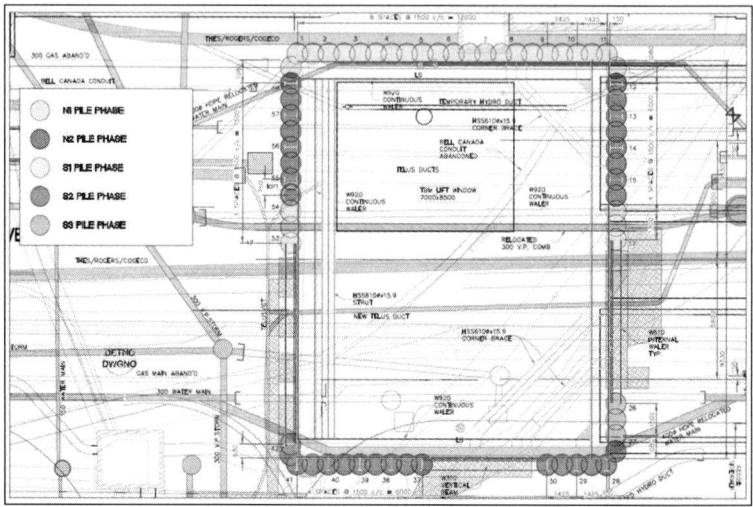

Figure 1. ES-3 existing subsurface conditions—shaft support planning

Utilities

Limited records of old existing utilities created additional challenges and generated unplanned cost and schedule extension. Some level of inaccuracy is always expected; however inaccuracies at the location of extraction shaft 3 (ES-3) were beyond expectations.

A number of undocumented utilities were found upon excavation and consequently shaft temporary support and excavation with typical mechanical equipment needed to be halted. A number of methods were employed to document existing utilities to enable shaft temporary support; however that was not possible along the entire shaft perimeter. The process was complemented by hand excavation and erection of temporary supports on ongoing basis. Casing and supporting of existing utilities and services were employed where relocation was not deemed possible.

Use of Jet-Grouted Head Walls

Jet grout blocks were the preferred method of excavation support for station headwalls since they reduced the number of required utility relocations and can reach deep installation between existing utilities. Secant piles were used a Keele, Caledonia and Dufferin stations since station design had already advanced to 30% and the station designers did not accept jet grout SOE.

Multiple brick sewers were identified along the tunnel alignment and two sections were identified as sensitive by a CCTV inspection. A slip lining was installed in the sensitive sections of sewer in advance of tunnelling. Jet grouting was adopted in several locations in lieu of the secant piles to reduce the impact on existing utilities relocation.

Obstructions to Tunnelling

As the Eglinton tunnel alignment is located in a congested and developed area, the possibility of a man-made obstruction to tunnelling was considered a major risk. Consequently, the design included a desktop investigation and historical archive search to identified potential obstructions. Where known obstruction were found,

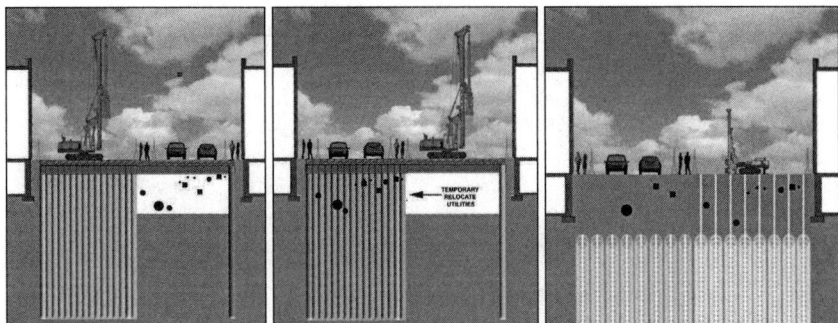

Figure 2. Secant piles vs. jet grouting support planning

the alignment needed to be re-designed or the obstruction needed to be removed in advance of tunnelling.

One significant obstruction to tunnelling was support of excavation installed for the Eglinton West Subway which was cancelled in 1995 after construction had started. No records of the status of work upon cancellation were available, so a site investigation was required to determine what had been installed. The test pit investigation confirmed 21 steel piles which impeded the tunnel alignment and it was decided to remove these prior to tunnelling.

Significant obstructions were present at the Spadina subway line as a result of the box structure and associated support of excavation. The obstructions contributed to the division of tunnel drives as discussed in the following section.

Crossing Existing Subway Lines

Challenges specific to Eglinton LRT project are related to the interchange station at Allen Road and Eglinton Avenue for interchange with University-Spadina Subway Line and interchange station at Yonge Street and Eglinton Avenue for interchange with Yonge Subway Line. The construction of the interchange stations includes replacement and relocation of existing utilities in the area.

The process started at the design and planning phase and included the removal and relaunching of the West tunnel TBMs at Allen Road, construction of the respective shafts and transportation of the TBMs above ground at Allen Road. The process continued with the temporary storage of the West tunnels TBMs in ground beneath Yonge Street and construction of an exit shaft for the removal of the East tunnels TBMs. All these extreme solutions were part of both Designer and Contractors efforts to mitigate the specifics constrains at the optimum cost.

Crossing University-Spadina Subway Line

During the design phase for the project the vertical alignment around Allen Road/ Eglinton West Station was identified as a challenge due to the number of constraints in the vicinity including the University-Spadina Subway box structure and associated abandoned support of excavation, an 1830mm sewer and a 1980mm sewer installed beneath Eglinton Avenue. The following bullets describe the rationale for specifying the extraction, transport and re-launch of the TBMs at Allen Road and over the Toronto Transit Commission (TTC) existing subway box structure:

Figure 3. Allen Road and Eglinton Avenue

- Tunnelling beneath the TTC box structure would require pre-support such as compensation grouting or mud jacking and would need approval from the TTC.
- The tunnel alignment would be required to be lowered further than the current design to mine beneath the TTC structure in order to any avoid obstruction with the TTC box structure and abandoned support of excavation. This would increase the future station excavation depth and consequently increase construction costs.
- Since tunnelling is a continuous linear construction process, mining the entire alignment from Black Creek to Yonge (6.4 km in length) would extend the construction schedule in order to complete the inverts, walkways, cross passages and other finishing works. Staging tunnelling in two separate drives allows turning over the first portion of tunnels between Black Creek to Allen earlier than staging tunnelling from one location.
- The distance from Black Creek to Yonge is approximately 6.4 km. Tunnelling this distance without any significant refurbishment posed a risk to tunnelling. Having the TBMs re-surface at Allen Road would allow them to be fully refurbished at approximately the mid-point of the tunnel alignment, reducing the risk of significant machine downtime.

TBM Move

To reduce the schedule impact of extracting and relaunching the TBMs, the contractor decided to lift and transport the TBMs in one piece instead of disassembling, transporting and reassembling the machines as originally envisioned.

Each TBM, weighted approximately 430 tonnes each, was transferred from ES-1 to LS-2 above ground over the existing Spadina Subway line. The transfer was done overnight with minimal road closures in one weekend as discussed in more detail by Liebno et. al. (2016).

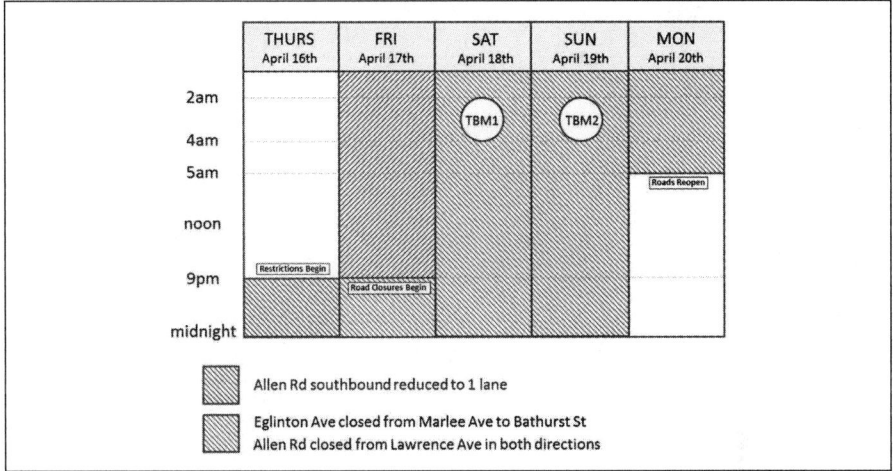

Figure 4. TBM move schedule

Figure 5. TBM move at Allen Road

The move was orchestrated by the contractor and a specialized heavy lift subcontractor Mammoet and consisted of mobilizing a hydraulic strand jack gantry crane and a hydraulic self-propelled modular trailer. This TBM move was an innovative solution to the urbanized challenges and ultimately reduced the TBM assembly and disassembly schedule and provided less construction impact to the public. A schedule for the move was provided to public (Figure 4).

Crossing Yonge Subway Line

Similarly to crossing the University-Spadina Subway line, it was undesirable to tunnel through due to complexity of the Yonge subway station; therefore the tunnel drives were configured to terminate at each side of the Yonge Subway station. As a result, extraction shafts were required on either side of the intersection of Yonge and Eglinton. This location was the most comprehensively developed intersection on the alignment.

Given the project procurement structure and timing the final exit shaft of the tunnels both on the West and the East contract were not ready in time for TBMs retrieval.

On the West Contract the TBMs were advanced within the Yonge station limits and parked there until the station excavation advances far enough to allow for TBM removal.

On the East Contract the TBMs were advanced within the Exit Shaft limits and parked there until shaft excavation advanced far enough to allow for TBMs removal. During the parked period the TBMs were subject to regular monitoring to ensure the TBM remained sealed to prevent ground loss.

Table 1. Working site area

Area (m²)	Launch Shaft 1	Launch Shaft 2	Launch Shaft 3
Site	12,000	7,600	17,353
Shaft	1,200	1,780	2,057
Surface (incl. decking)	10,800	7,200[1]	17,353
Office	270	240[4]	300
ubstation	300	260	542
Grout Plant	330	175	450
Water Treatment	60	60	60
Muck Pit (Surface) & Excavator Work Space	600+400	360+200[2]	467
Muck Pit (Shaft)	NA	167	NA
Contaminated Soil Pit & Excavator Work Space	160+100	185+200[2]	410
Crane Pad	1×190	2×190	1×190
Segment Storage	1,200	700	2,150
Segment Heating	200	140	408
General Storage	1,500	800	216
Mechanical Shop	300	300	218
% Open Space[3]	50%	44%	45%
Offsite Storage	—	14,000	—
Offsite Owner Offices	450	450	600
Offsite Owner Parking	—	1,000	1,000

Table 2. Tunnelling duration and production data

Tunnel Drive	Drive 1	Drive 2	Drive 3	Drive 4	Drive 5	Drive 6
Tunnel Length (m)	3546.7	3546.5	2864.3	2863.3	3281	3262
Duration (days)	347	303	204	210	309	281
Tunnelling duration (days)	221	208	172	172	161	156
Average production over duration (m/day)	10.2	11.7	14.0	13.6	10.6	11.6
Average production while mining (m/day)	16.05	17.05	16.65	16.65	20.38	20.92
Best day production (m)	33	35	33	32	40	47.5
Best Week (m)	128	138	128	125	148	158
Best Month (m)	426	411	486	408	553	485

Settlement Monitoring

In urban environments controlling settlement is essential to protect the roadway, utilities and adjacent structures. Monitoring instrumentation was installed along the tunnel alignment and instruments were monitored at a frequency determined by the TBM's proximity. Review and alert levels corresponding to volume losses of 0.5% and 1% respectively were set. In general, settlement performance far exceeded the predicted review level. On the west tunnels drive, the average settlement was less than 5mm and corresponded to and average volume loss of 0.17%. The exceptional control of ground loss on the Eglinton Crosstown greatly mitigated impacts and damaged to adjacent structures along the narrow tunnel right-of-way (Solecki et al 2016).

Challenges to Tunnelling Logistics

Small Working Area Footprint

Launch and extraction shafts locations and footprint were restricted by available land.

Launch Shaft 1 (LS-1) was located on a public park and within two lanes of Eglinton Ave. Road traffic continued to use two lanes beside the shaft. Extraction Shaft 1 (ES-1) was constructed under Eglinton Ave from a public parking lot. Decking was installed over half of the shaft for road traffic to continue one lane in each direction.

Launch Shaft 2 was the most restricted launch site. LS-2 was initially a public parking lot and required realigning an expressway ramp to provide working space. LS-2 was similarly located entirely under Eglinton Ave with decking over half the shaft providing two lanes of traffic and decking on the other half of the shaft within the site providing 780m^2 of additional area for storage and equipment.

Launch Shaft 3 was the largest launch site and was constructed under two lanes of Eglinton Ave and a large hillside boulevard. The extraction shafts for the Drives 3 and 4 (ES-2) and Drives 5 and 6 (ES-3) were located at the most heavily developed area on the alignment at Yonge Street and Eglinton Ave. ES-2 was planned to be constructed at the location of a former bus terminal and ES-2 was constructed under the Eglinton roadway.

Launch shaft #2 (LS-2) was more restrictive in size than LS-1 and LS-3 due to location, conditions and required TBM launching in a umbilical configuration. An initial mining phase included completion of the TBM assembly and installation of the tunnel continuous conveyor. From Table 2 we could understand the complexity of the launching process at each drive correlation the duration required for TBM assembly, and initial mining with the total duration.

Efficiency of launching was clearly affected by the location constraints; however daily productions were within the same range.

Innovations were required particularly at LS-2 for logistics due to the constrained working space. Innovations included installing decking over the shaft within the site as previously discussed. The Contractors employed two trains for each drive operated from the launch shaft and used a 'California Switch' to allow passing prior to entry into the tunnel. At LS-2, two lift windows were left in the decking to provide the crane access to the materials trains for each tunnel. And a third opening was used by the muck removal crane. The decking accommodated the ventilation equipment and segment storage as well. The contractor also used stacked construction trailers to fit within the reduced footprint of construction site.

Decking as Solution to Traffic Management

Eglinton Ave is an arterial road in Toronto so maintaining two-way traffic flow at all times was a requirement of the tunnelling contracts. Since the launch and extraction shafts and emergency exit buildings were constructed partially or wholly under Eglinton Avenue, the Designer and Contractors developed and implemented rigorous plans for traffic control and management in relation with the work schedule at these locations. At LS-1 and LS-3 open lanes of traffic were maintained around the shaft. At ES-1, LS-2 and ES-3 road decking was required to provide lanes for two-way traffic.

Staging plans were developed and implemented maintain two-way traffic work was performed. Street closures were only adopted for limited times and only at off peak times (weekend nights).

Multiple Handling of Materials

As a result of constrained site space, particularly at LS-2, just-in-time delivery of materials and off-site storage yards and warehouses were required to maintain a flexible supply of materials.

Figure 6. Decking of working shaft at LS-2

Ahead of starting tunneling, the following logistics measures were implemented:

- A trucks staging area was made on the incoming route of the mucking and tunnel precast trucks. With the City of Toronto approval, the contractor built a two way staging zone on the trucks route in nearby proximity to site, within radio communication with the site.

Figure 7. Continuous conveyor at LS-3

- A temporary storage yard was set up, again, on the tunnel precast truck route and within 4 km from the site, to act as an overflow/buffer mean for the tunnel precast delivery and other materials (rail, tunnel conveyor frames, tunnel conveyor belts, etc.)

- Muck removal from site with the trucks was timely segregated from tunnel precast delivery-the muck removal trucking started 5 am finishing 4:00 pm (80 to 140 trucks/day), and the tunnel precast receiving started 4:00 pm finishing 9:00 pm (15–20 trucks/day).

- All warehousing configuration on site was built on two levels, to reduce the surface footprint and allow more site storage and vehicles maneuvering.

- A section of the open area of the shaft was decked within the site, engineered to withstand the storage of additional 30 precast tunnel rings.

Continuous Conveyors

The muck resulted by the tunnel excavation need to be removed out of the tunnels. Both tunnel contractors for the West and East contracts installed innovative setups to remove the muck by continuous conveyors. The continuous conveyors were delivering the muck from the TBM straight to surface located muck pits at two of the three working sites (Keele/Eglinton and Leslie/Eglinton).

At Launch Shaft 2 there was insufficient space for a conveyor to transport material up to the surface muck pit so the Contractor adopted a hybrid solution by installing a two chamber muck pit in the shaft from where the muck was extracted by a crane with a clam to accommodate for the small foot print.

Figure 8. In shaft muck pit (LS-2) vs. portal muck pit (LS-3)

The continuous conveyors were designed and manufactured by the specifics of each working sites. The benefits of the continuous conveyors were becoming visible with the length of the tunnel drives; the trains were strictly utilized to transport the materials and consumables for the TBM and tunnel advance including the segmental lining. These include greater TBM advance availability, less down time and simplified logistics when compared to rail-based muck removal by rail cars.

Muck Handling and Treatment

Although the TBMs were using only bio-degradable additives at times the muck was more fluid than required for immediate transportation.

In other instances while mining through jet-grouted walls the muck did not meet the criteria for disposal at regular disposal sites and required transportation by cisterns at treatment locations and disposal at special approved disposal sites.

The contractors adopted ingenious logistics and at times incurred additional cost to maintain the planned schedule.

As the removal of the muck from the shaft was done by clamming, the shaft muck pit was staged, creating an isolated discharge for each tunnel. That allowed not only segregation of soils between the tunnels, but also ensure productive mining for one TBM while the other one was mining through the jet grout walls. Also on the surface a secondary "contaminated" muck pit was built, as the high PH soil of the jet grout walls was stored in this isolated pit for trucking. Muck removal was staged (talk about shaft muck pit and double handling).

Interventions at Secant Pile and Jet Grout Headwalls

Secant pile and jet grout headwalls were used to perform interventions to avoid work in hyperbaric conditions. Yamashita et al. (2015) presented the contractors innovative processes developed to control flowing ground during interventions.

At secant pile headwalls, the TBM was advanced so the head was in full contact with the wall. External dewatering was used to lower the ground water and chemical grout was used to seal the TBM to the treated ground to prevent flowing water and soil towards the depressurized head.

In the jet grout headwalls the cutterhead was advanced to the midpoint of the wall and chemical grout was used to seal around the machine. The jet grout permitted interventions to be completed without dewatering the soil.

CONCLUSIONS

Urban tunnelling is different and most of times is a difficult and risky endeavour. Money invested in heavy construction impacts upon other community interests and services and the politics are more complex than in rural areas. Urban tunnelling projects demand a client with the ability to select and combine competent specialised engineers to get the best quality product for the taxpayer.

Experienced Engineers, Consultants and Contractors, employing state of the art TBMs, systems and technologies, performing a comprehensive risk mitigation and management together with an innovative approach to tackle specific challenges can result in a very successful project such as the Eglinton Crosstown LRT.

REFERENCES

Liebno, D., Yamashita, J., Nishikokura, O., and Sheehan, M. "Mechanized Tunnelling Driving Toronto's Big Move Eglinton Scarborough Crosstown Tunnel Construction (ECLC1-15)," WTC 2016, San Francisco, USA.

Noah Johnson, The Robbins Company, Urban EPB Tunneling in Limited Space: A Case Study of the San Francisco Central Subway Project

Solecki, A., Taghavi, A., and Hassan, I. "Redefining Settlement Control Industry Standards with Modern Mechanized EPB Tunnelling: Eglinton Crosstown LRT Case Study," WTC 2016, San Francisco, USA.

Yamashita, J., Nishikokura, O., Sheehan, M., and Stewart, C., TBM Cuttinghead Interventions at Eglinton-Scarborough Crosstown Tunnel Construction (ECLC1-15), RETC 2015, SME, New Orleans, USA, 962–970.

Wear of Cutting Tools on an EPB TBM Tunneling Through Glacial Soils

Lisa Mori ▪ Jay Dee Contractors, Inc.
Ehsan Alavi ▪ Jay Dee Contractors, Inc.
Brian Hagan ▪ Jay Dee Contractors, Inc.
Michael A. DiPonio ▪ Jay Dee Contractors, Inc.

ABSTRACT

Prediction of tool life and wear for soft ground pressurized TBMs is a challenging task for contractors, machine manufactures, and designers. The issue of cutterhead inspection and maintenance in Earth Pressure Balance (EPB) tunneling can be a dangerous, time consuming, and costly process, particularly when the ground is unstable. In this case, cutterhead inspection and tool maintenance are performed under pressurized conditions or "hyperbaric interventions." Severe primary and secondary wear on the cutters has been seen in some of the projects in glacial soils including Seattle area. This paper discusses the experience gained regarding cutting tool wear from an EPB TBM that mined through 8.85 km of glacial soils in Seattle, WA over the course of two separate tunneling projects.

INTRODUCTION

During the last 15 years, several tunneling projects in Seattle including the Brightwater tunnels, Beacon Hill tunnels, Mercer Street Tunnel, Henderson Way Tunnel and Sound Transit's University Link and Northgate Link tunnels were excavated by using pressurized shielded machines.

Despite the fast growth in the use of soft ground Pressurized TBMs in the tunneling industry, prediction of tool wear prior to excavation and monitoring and maintenance of tools during the excavation is a challenging task for contractors, machine manufactures and designers. The issue of cutterhead inspection and maintenance in Earth Pressure Balance (EPB) and Slurry TBMs can be a dangerous, time consuming, and costly process particularly when the ground is unstable. In this case, cutterhead inspection and tool maintenance are performed under pressurized conditions or "hyperbaric interventions." Severe primary and secondary wear on the cutters has been seen in some of the projects in glacial soils. As an example, mining on the two central tunnels of the Brightwater Project (BT2 and BT3) was halted in May 2009 due to severe wear damage to the cutterhead on both machines (Tunneling Journal 2011). Although the BT2 machine was repaired and finished its run, the BT3 machine was stranded.

This paper will discuss in detail the severe primary and secondary wear on the cutters of an EPB TBM that tunneled through 8.85 km over the course of two projects in Seattle, WA. In addition, some of the experience and remedies that were utilized to address the cutterhead wear and improve the life of the cutters along these two projects are discussed. In the next section an overview of these projects as well as their subsurface geotechnical properties are presented.

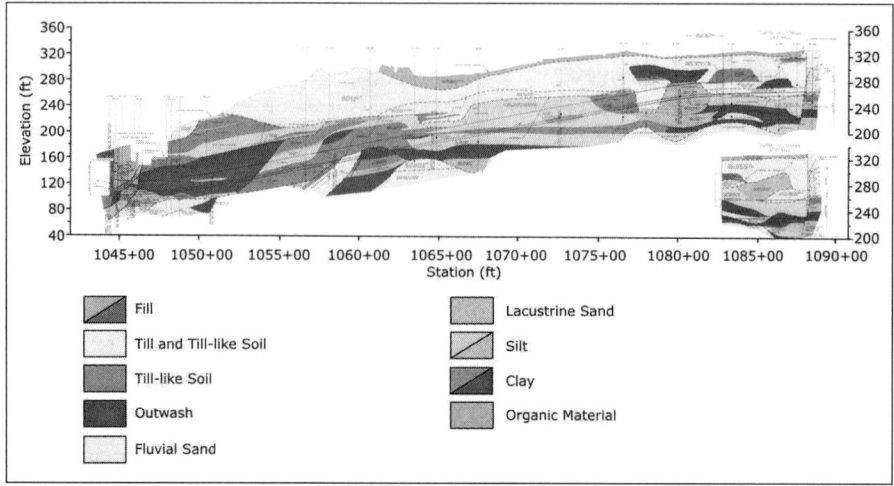

Figure 1. U230 University Link Light Rail Extension geological profile

University Link Light Rail (U230) Tunnel Project

The University Link Project consists of approximately 5.1 km of twin Light Rail Transit tunnel commencing from the north end of the Pine Street Stub Tunnel in downtown Seattle to the University of Washington Station near Husky Stadium. Jay Dee Contractors, Inc., Frank Coluccio Construction Company, and Michels Corporation have formed a Joint Venture (JCM U-LINK, JV) to construct the portion of the tunnels from Capitol Hill Station (CHS) to the Pine Street Stub Tunnel (PSST) which is called Contract U-230. This contract includes the construction of twin tunnels which are each 1.2 km in length.

Geologic description of this project can be divided to fluvial deposits, glacial deposits and lacustrine and glaciolacustrine deposits, all of which have been glacially overridden. Figure 1 shows the geological profile of the project.

N125 Northgate Link Light Rail Extension Project

The Northgate Link Extension will extend service north from the University of Washington to the University District, Roosevelt, and Northgate neighborhoods by 2021, and is expected to cost approximately $2.1 Billion. Most of this 6.9 km extension will be underground, and the N125 contract includes the construction of 5.6 km of twin EPB tunnels. Also included are the excavations of the Maple Leaf Portal (MLP) where the light rail will transition from tunnels to elevated guide-way and two large underground station boxes, one for the University District Station (UDS) and one for the Roosevelt Station (RVS).

The geology of the N125 project is very similar to that of University Link in that the tunnels are constructed through glacially overridden fluvial and lacustrine deposits. The two major differences are that N125 will encounter a large amount of glacial till and till-like deposits and that N125 will encounter much more coarse-grained soils overall than University Link. Figure 2 shows the geological profile of the project.

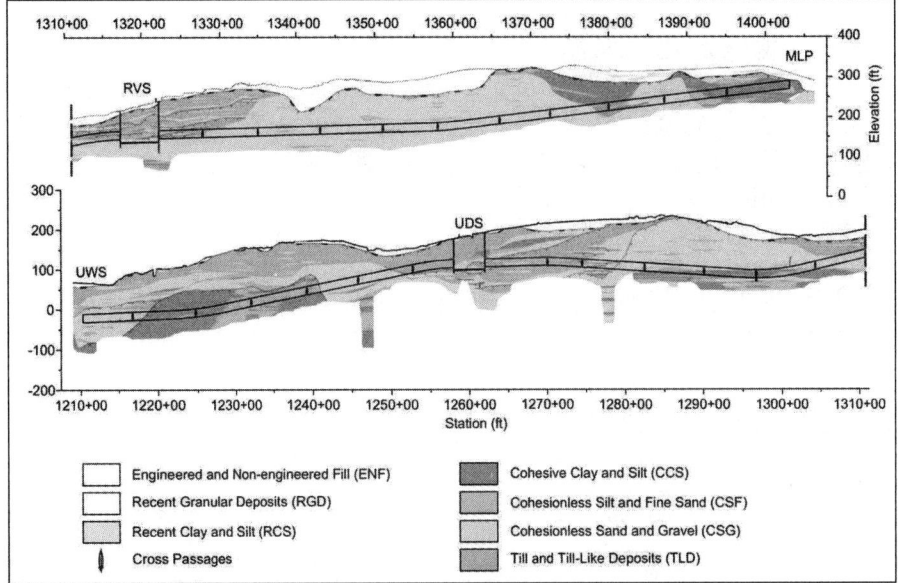

Figure 2. Northgate Link Light Rail Extension geological profile

U230 and N125 Tunnel Boring Machines

A new HitZ TBM was fabricated by Hitachi Zosen for use by JCM on the Sound Transit U230 project. The TBM fabrication was completed in 2011. The Sound Transit U230 Project consisted of twin-bored (northbound and southbound) tunnels each with an approximate length of 1200 meters from Capitol Hill Station to Pine Street Stub Tunnel in Downtown Seattle. The tunnels had an outside diameter of 6.27 m and an inside diameter of 5.74 m. The HitZ TBM completed both tunnels, starting in 2011 and finishing in 2012 (Figure 3).

Figure 3. Original machine in factory

JCM decided to refurbish this TBM and use it for the excavation of one of the twin tunnels in N125 project (see Figure 4 and Table 1).

CUTTING TOOL WEAR MEASUREMENT METHODS

Scrapers, precutters, and disc cutters were utilized in different sections along the alignment of these two projects. A unique identification and measuring system was developed to monitor the cutter wear. The scrapers were numbered from the center to the periphery of the cutterhead, for each of the two sides of the spoke, left and right (see Figure 5). E.g., scraper 53R is the third scraper on the right side of spoke 5. The wear of scrapers was defined by the remaining length of the scraper protruding from

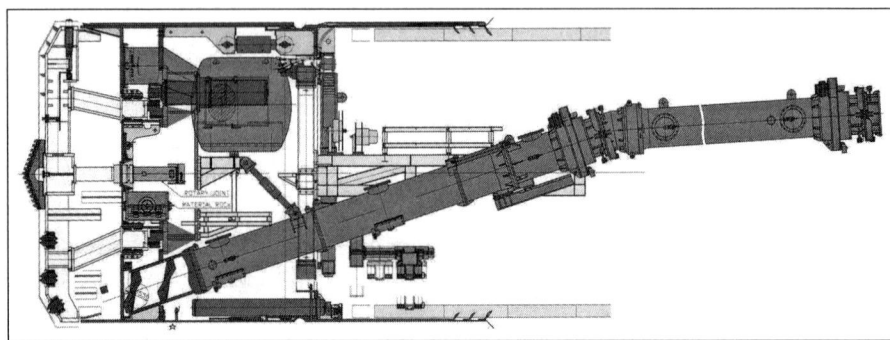

Figure 4. Side view of the refurbished Northgate Link TBM design

Table 1. Specifications for the Refurbished North Link HitZ TBM

Hitachi Zosen		
Excavation Diameter	With Soft Ground Tools	6,640 mm
Cutterhead	Type	Bidirectional, mixed ground
	Opening ratio	45%
	Cutterhead drive	Electric motors with VFD
	Cutterhead power	720 kW (8×90 kW)
	Cutterhead speed	0~2.2 rpm
Torque	Cutterhead working torque	2970 kN-m
Thrust	Trust jack stroke	2,300 mm
	Maximum thrust	40,000 kN
Electrical	Primary voltage	13,800V
	Protection	Class 1, Div 2
Conveyors	Screw conveyor diameter	800 mm Shafted
	Screw conveyor type	Two stage, periphery drives, end discharge, and shaft style auger with replaceable wear protection on flights and casing
	Speed	1.0~18.3 rpm
	Torque	80 kNm
	Back-up conveyor belt width	750 mm
Weights	TBM weight (approx.)	320 tonnes
	Back-up weight (approx.)	200 tonnes

the spoke subtracted from the original length (see Figure 6a). The measurement was taken at three points along the scraper, in the middle and on each side (see Figure 6b). The same unique system of numbering was developed for precutters (see Figure 5). The wear of pre-cutters was defined by the remaining length of the carbides sub-tracted from their original length (see Figure 6c).

Two different types of precutters were used on the U230 project: low precutters and high precutters. The main difference between the low and high precutters was their welded position. Low precutters had a height of 110 mm and were welded into the cutterhead spoke while high precutters had the same initial height but were welded into the interchangeable boxes (overall height from the spoke was 130 mm for high precutters). Based on the experience gained through the U230 project, it was decided to use the same height for all the precutters. The overall height of the precutters was increased from 130 mm (from the spoke) in U230 to 170 mm (from the spoke) in N125 project.

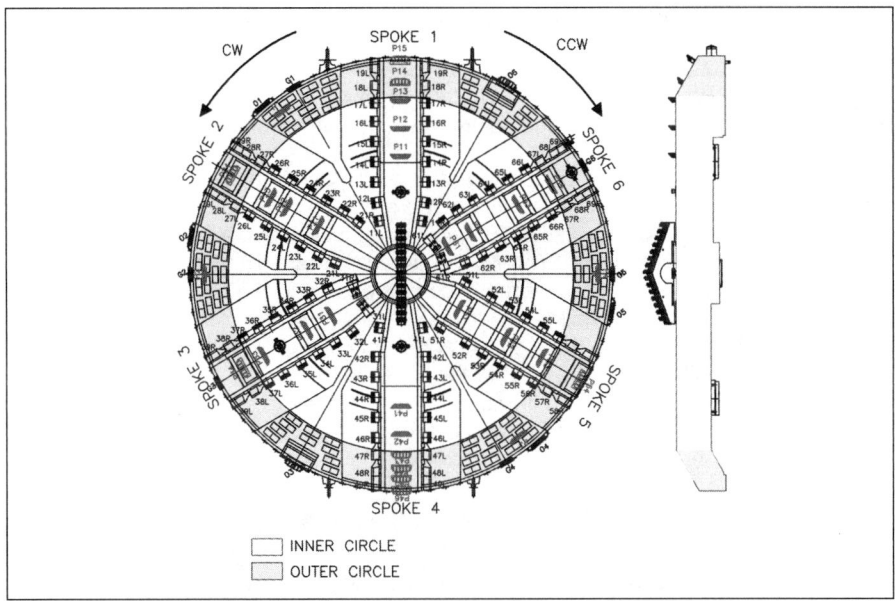

Figure 5. Numbering of scrapers, precutters, gauge cutters, and overcutters on the cutterhead

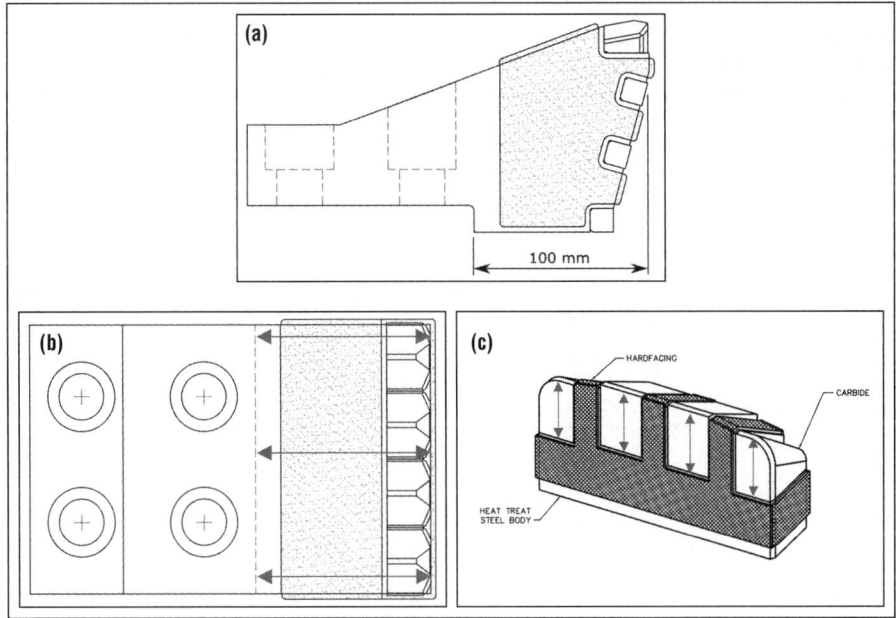

Figure 6. Side view of scraper with 100 mm protruding from spoke (a). Wear measurement locations for precutters (b) and scrapers (c) depicted with red arrows.

The cutterhead can also be divided into two areas: the inner circle and the outer circle (see Figure 5). The cutterhead structure of the inner circle is flat and perpendicular to the tunneling direction. The cutterhead structure of the outer circle is tilted at an angle of 60 degrees to the tunneling direction. These two areas and the resulting difference in cutter orientation has also been considered in the wear analysis.

The wear measurement was performed after each tunnel section for both projects. For the U230 project the wear was measured after the NB tunnel finished, then the cutterhead was refurbished and the TBM was relaunched for the SB tunnel. The wear was measured again after the SB tunnel finished the run. The wear measurements of the NB and SB tunnel can be directly compared, since the same TBM was used and they mined through basically the same geology. For the N125 project the wear was measured in each of the three shafts after each of the three tunnel sections were excavated. In the RVS and UDS shafts, the cutterhead was then refurbished and the TBM was relaunched for the next tunnel section. After the tunnel section from UDS to UWS the cutterhead was also refurbished, because the TBM was reused to excavate the SB section from UDS to UWS.

RESULTS AND DISCUSSION

This section of the paper presents the analysis of the wear data that was collected during the U230 and N125 projects. The analysis mainly focuses on the wear of scrapers and precutters and does not include wear of gauge cutters, overcutters, or disc cutters.

U230 Wear Measurement Results

Figure 7 shows the wear of scrapers on the inner and outer circle for the NB and SB U230 tunnels plotted against the traveled distance of the scrapers. The traveled distance is calculated from circumference of the circle along which the scraper travels multiplied by the total number of cutterhead rotations in each tunnel. The filled symbols stand for the NB scraper wear data and indicate that the wear increases with traveled distance, at least up to the transition zone between inner and outer circle. The wear of scrapers on the outer circle seem to decrease slightly with traveled distance and therefore their distance from the front of the cutterhead. The unfilled symbols stand for the SB scraper wear data and indicate that the wear of scrapers on the inner circle increases with traveled distance and that the wear of scrapers on the outer circle decreases with traveled distance.

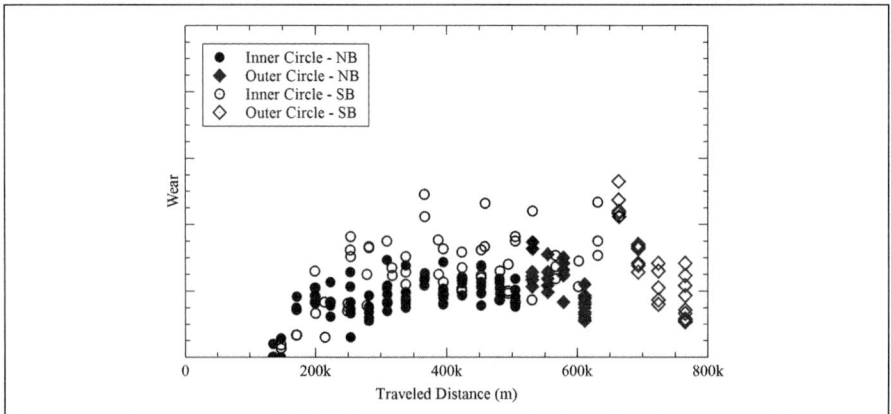

Figure 7. Wear of scrapers plotted against traveled distance for the NB and SB tunnel of the U230 project

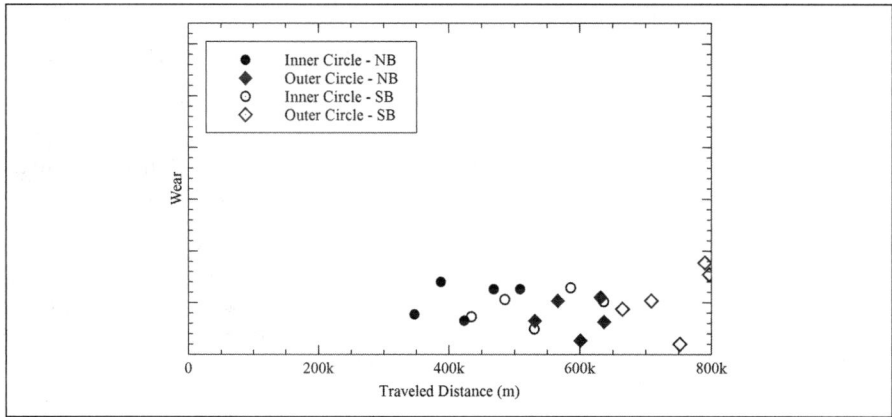

Figure 8. Wear of low precutters plotted against traveled distance for the NB and SB tunnel of the U230 project

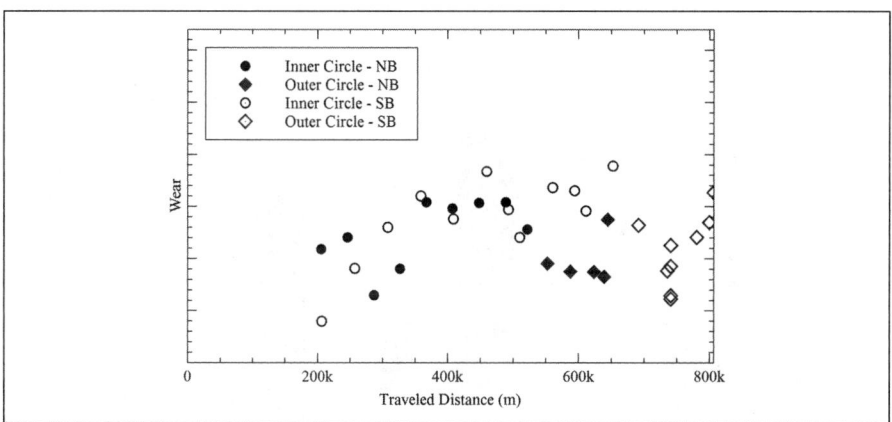

Figure 9. Wear of high precutters plotted against traveled distance for the NB and SB tunnel of the U230 project

Figure 8 shows the wear of low precutters on the inner and outer circle of the NB and SB U230 tunnels plotted against the traveled distance. Filled symbols stand for NB wear data and unfilled symbols stand for SB wear data. The data does not indicate that the wear is correlated to the traveled distance. However, in contrast to the scraper wear data the highest wear was measured on the precutters on the outer circle that had the highest traveled distance.

Figure 9 shows the wear of high precutters on the inner and outer circle of the NB and SB U230 tunnels plotted against the traveled distance. The filled symbols stand for the NB wear data while the unfilled symbols stand for the SB wear data. The data indicates that for both NB and SB tunnels the wear increases with traveled distance up to the transition zone between inner and outer circle and then decrease. However, the precutters on the outer circle with the highest traveled distance have wear of similar magnitude than the cutters at the transition zone.

N125 Wear Measurement Results

The soil type TLD encountered on the N125 project contained an excessive amount of gravel, cobbles, and boulders. An increased amount of damage to the cutterhead structure and cutters was experienced in tunnel sections that contained a high amount of soil type TLD. Additional cutterhead interventions were necessary to replace damaged cutters before the planned refurbishments in the shafts. Two of the interventions between RVS and UDS needed to be performed as hyperbaric interventions due to high water pressures at their locations. The cutterhead chamber was entered under pressures of 1 bar to 1.5 bar. 41 scrapers were replaced during the first hyperbaric intervention and 37 scrapers were replaced during the second hyperbaric intervention. Majority of the damage was caused due to encountering several large size boulders that basically introduced significant load to the cutters and broke off the carbide inserts and/or bolts holding the scrapers into the cutterhead spokes. The wear of the cutters was not measured during these interventions.

Figure 10a shows the number of cutters changed during and at the end of each of the three tunnel sections: MLP-RVS, RVS-UDS, and UDS-UWS. The number includes both cutters that were changed in the shaft during the standard cutterhead

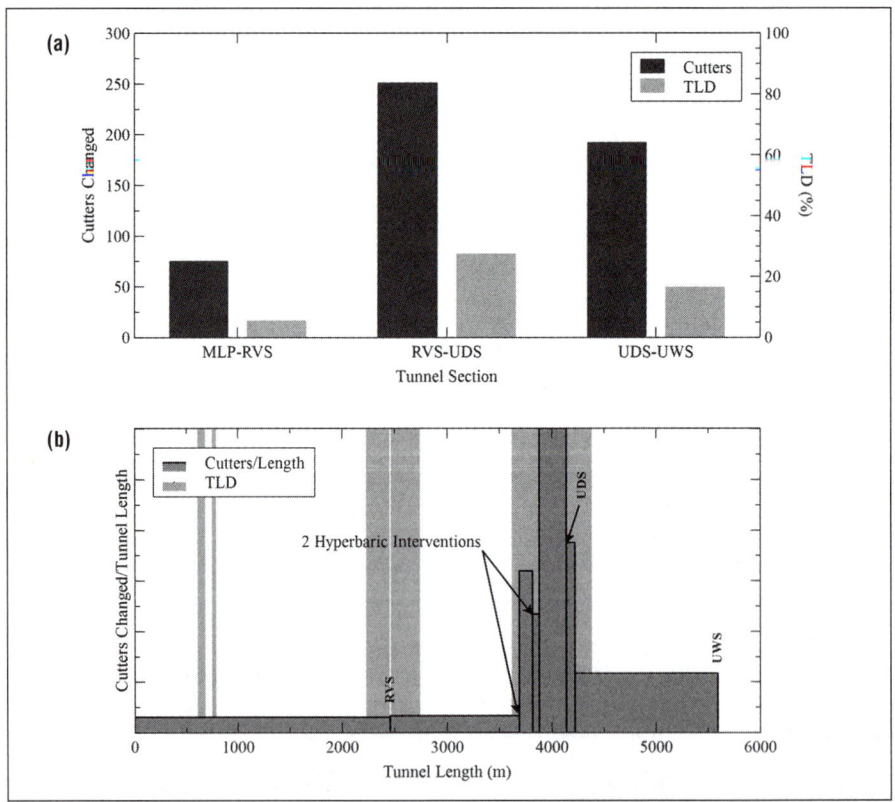

Figure 10. Number of cutters changed during and after each of the three N125 tunnel sections and the percentage of TLD present in each tunnel section (a). The ratio of number of cutters changed to length of tunnel after which they were replaced plotted against the tunnel length (b). The area of each bar gives the number of cutters changed and the width of the bar gives the length of the tunnel that was excavated before they were changed. Areas with soil type TLD in the tunnel cross section are highlighted in purple.

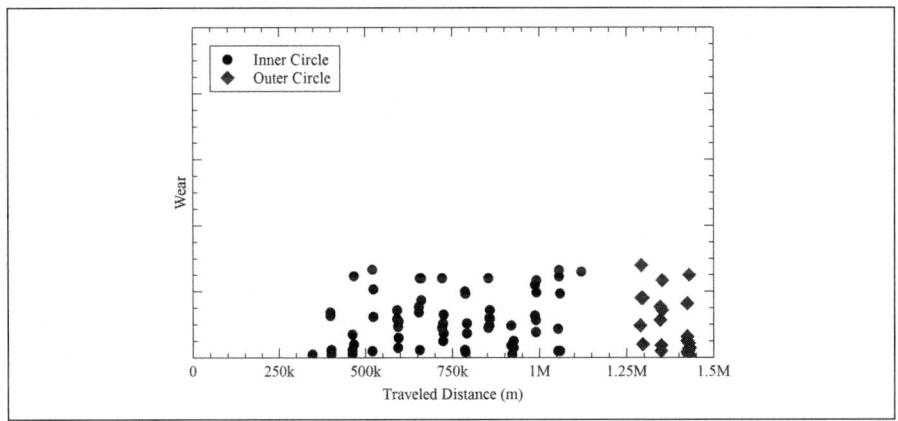

Figure 11. Wear of scrapers plotted against traveled distance for the N125 tunnel section from MLP to RVS

refurbishments and cutters that were changed during interventions. Figure 10a also displays the percentage of soil type TLD for each tunnel section. The number of cutters changed seems to increase with the percentage of TLD. In the tunnel section from MLP to RVS with the lowest amount of TLD, no intermediate intervention was necessary and only the standard cutterhead refurbishment was performed at RVS. In the tunnel section from RVS to UDS with the highest amount of TLD, two hyperbaric and one atmospheric cutterhead intervention were performed in addition to the standard cutterhead refurbishment at UDS. In the tunnel section from UDS to UWS with the second highest amount of TLD, one atmospheric cutterhead intervention was performed in addition to the standard cutterhead refurbishment.

Figure 10b shows the ratio of number of cutters changed to length of the excavated tunnel (before changing the cutters) plotted against the tunnel length. The width of each of the green bars shows the length of tunnel that was mined before the cutters were changed. The area of each green bar gives the number of cutters changed during either one of the standard refurbishments or one of the additional cutterhead interventions. Areas along the tunnel alignment that contain soil type TLD are highlighted in purple on Figure 10b. The figure shows that the ratio of cutters to tunneled length increased significantly in the area around UDS shaft where four cutterhead interventions were performed and where a high percentage of soil type TLD is present.

Figure 11 shows the wear of scrapers, measured after the MLP to RVS tunnel section, plotted against the traveled distance of each scraper. The traveled distance is calculated from circumference of the circle along which the scraper travels multiplied by the total number of cutterhead rotations in each tunnel section. The traveled distance seems to have no apparent influence on the scraper wear. However, the wear seems to be highest closest to the transition from the inner to outer circle.

Figure 12 shows the wear of precutters, measured after the MLP to RVS tunnel section, plotted against the travelled distance of each precutter. For the precutters it looks like the wear increases with the traveled up to the transition from the inner to the outer circle. The wear is highest at the transition from the inner to outer circle and then decreases with the traveled distance. The main reason for decrease of wear in the outer circle area despite the higher travel distance is due to significant increase in the number of cutters in this area compared to the inner circle area. In other words, there are redundant cutters that travel through the same radius or the area of coverage of each cutter has been reduced significantly.

Figure 13 shows the wear of scrapers sorted by their installation location. RVS notation refers to the scrapers that were installed during the standard refurbishment in the RVS shaft. R 2430 and R 2512 indicate that the scrapers were installed during the hyperbaric interventions performed at rings 2430 and 2512. R 2554 indicates that the scrapers were installed during the atmospheric intervention at ring 2554. The filled symbols stand for scrapers on the inner circle while the unfilled symbols stand for scrapers on the outer circle. There seems to be no correlation between the traveled distance and the wear for any scrapers installed during an intervention. The wear of scrapers on the inner circle that were installed at RVS seems to increase with the traveled distance. The gap in wear data of scrapers installed at RVS indicates that the location of scrapers that were replaced during intervention was concentrated around the change from inner to outer circle. There seems to be no correlation between traveled distance and

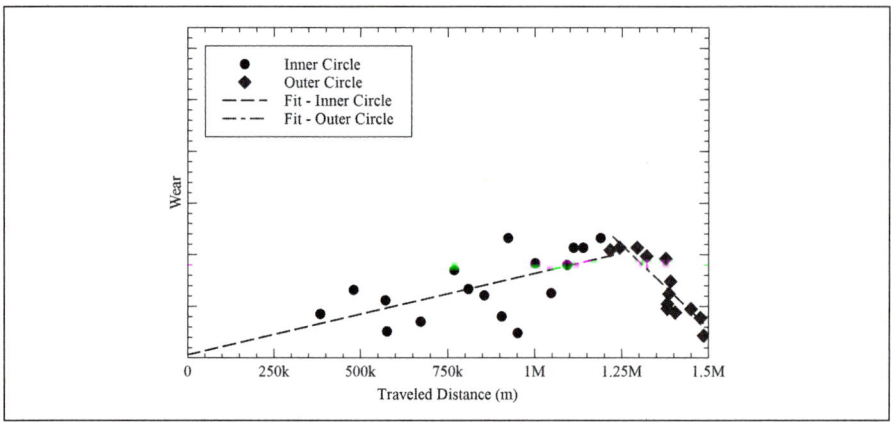

Figure 12. Wear of precutters plotted against traveled distance for the N125 tunnel section from MLP to RVS

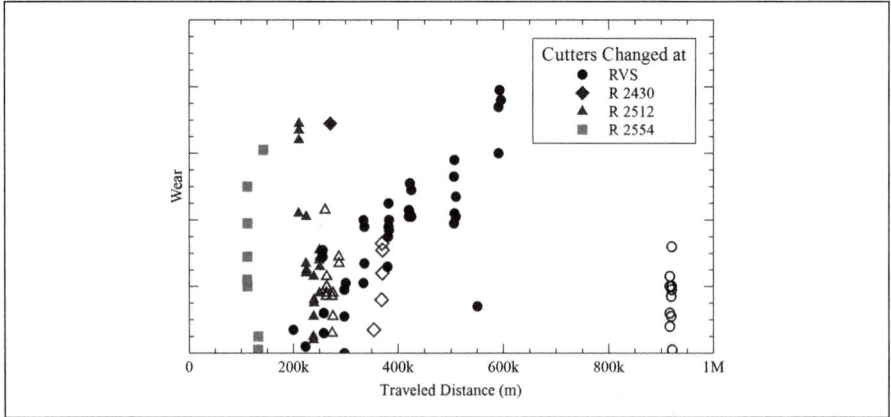

Figure 13. Wear of scrapers plotted against traveled distance for the N125 tunnel section from RVS to UDS. The different symbols indicate the location when the scrapers were installed. The filled symbols stand for scrapers on the inner circle and the unfilled symbol stand for scrapers on the outer circle.

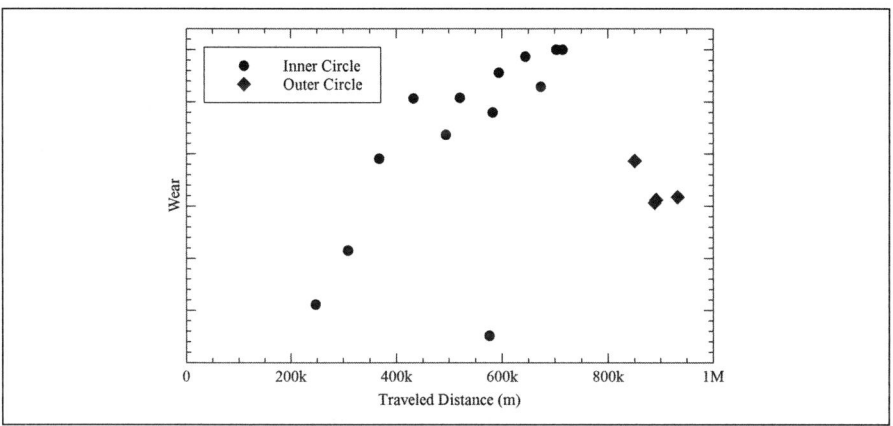

Figure 14. Wear of precutters plotted against traveled distance for the N125 tunnel section from RVS to UDS

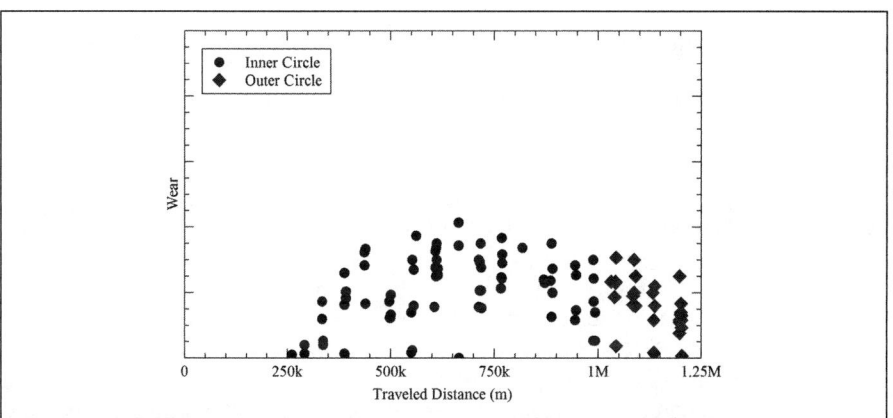

Figure 15. Wear of scrapers plotted against traveled distance for the N125 tunnel section from UDS to UWS

wear of scrapers on the outer circle that were installed at RVS, however the wear is lower than the wear of the closest scrapers on the inner circle. This again could be another explanation that the increase in frequency of precutters on the outer circle was helpful in reducing the wear of scrapers in this zone.

Figure 14 shows the wear of precutters measured after the RVS to UDS tunnel section. Several of the precutters were replaced by disc cutters during the three intervention on this tunnel section and are therefore not included in the analysis. The wear of precutters on the inner circle seems to increase with the traveled distance with one outlier. The wear of precutters on the outer circle is lower than the wear of the closest precutters on the inner circle.

Figure 15 shows the wear of scrapers after the tunnel section between UDS and UWS plotted against the traveled distance. The wear of scrapers on the inner circle seems to increase with the traveled distance up to a certain point, after which the wear decreases and continues to decrease on the outer circle. The highest wear is not at the transition zone from inner to outer circle like for the other two tunnel sections, but is somewhere on the inner circle.

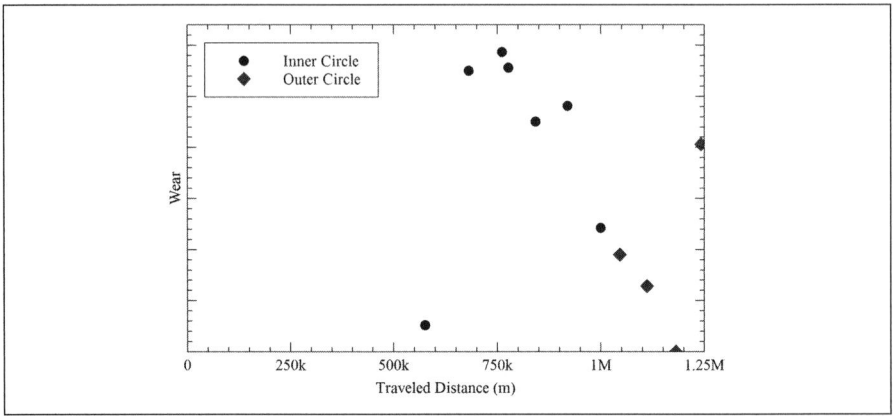

Figure 16. Wear of precutters plotted against traveled distance for the N125 tunnel section from UDS to UWS

Figure 16 shows the wear of the precutters measured after the tunnel section between UDS and UWS plotted against the traveled distance. Due to the small amount of available data points, no clear correlation can be seen from the data, but the wear on the outer circle seems to be generally lower than on the inner circle.

CONCLUSION

From the analysis of the cutting tool wear data collected during the U230 and N125 projects the following conclusions can be drawn:

- An abrasive soil type containing cobbles and boulders can lead to significant damage to the cutting tools even after a short distance of tunneling. The number of cutters changed during and after each of the N125 tunnel section directly correlates with the amount of abrasive TLD soil type.

- The distance that a cutting tool travels has an influence on the amount of wear, as can be expected. For the flat part of the cutterhead the amount of wear generally increases with the traveled distance. However, this is not the case for the cutting tools on the tilted part of the cutterhead, which leads to the next conclusion.

- The shape of the cutterhead structure and the resulting orientation of the cutting tools also has an influence on the amount of wear. The farer back the cutting tool is from the front of the cutterhead the smaller the wear. Furthermore, the amount of wear is generally highest at the transition zone from the flat to the tilter area of the cutterhead structure.

- The amount of damage caused by breakage of carbide inserts in general is much higher than the damage caused by abrasive soils.

Impact of Conditioned Soil Parameters on Tool Wear in Soft Ground Tunneling

Mansour Hedayatzadeh ▪ Politecnico di Torino University
Jamal Rostami ▪ The Pennsylvania State University
Daniele Peila ▪ Politecnico di Torino University

INTRODUCTION

The use of mechanized shielded tunnel boring machines (TBMs), has grown rapidly in various soft ground tunneling projects. In the recent years, Earth Pressure Balance (EPB) shields have been successfully used in many tunneling projects in urban areas while the range of soil types that it can handle has expanded. One of the critical issues for successful EPB tunneling is the use of proper soil conditioning. The effects of soil abrasiveness on primary and secondary wear of tools and machine components is one of the most important aspects of soft ground tunneling, estimation and assessment of, costs, and schedule of a project. Among the parameters that can affect tool wear, soil conditioning is the most critical one that can be controlled by the operators. To assess the impact of the soil conditioning on tool wear, a new test apparatus with new propeller has been developed. Specific test procedure has been developed and many tests have been conducted. Review of the initial test results shows that soil conditioning reduces wear of tool and other machine parts. More tests are underway to evaluate the effect of soil conditioning on machine torque requirements through a parametric study.

There are four main forms of wear as discussed by Rabinowicz (1995). Abrasive wear, corrosive wear, surface fatigue wear and erosion wear. Abrasion wear is most related to interaction of metals and hard materials such as soil constituent minerals. It occurs when a rough hard surface or a soft surface containing hard particles, slides on a softer surface and ploughs a series of grooves in softer material. The material form the grooves is displaced in the form of wear particles. In mechanized tunneling, abrasion is the most suitable term to define tool wear. It should be noted that the abrasion mechanism in soft ground tunneling is deferent than hard rock TBM. In soft ground tunneling the cutter head, chamber and screw conveyor of EPB are in contact with soil under pressure or load. In this situation the soil properties have modified by pressure and create the interface between metals and soil these two parameters including internal friction and interface friction can describe the wear phenomenon and motion of wear including: sliding and shearing (rolling).

Some of the recent publications have focused on wear in mechanized soft ground tunneling. Rostami et al. (2012), Peila et al. (2012), Barzegari et al. (2013), Jakobsen et al. (2013) and Alavi Gharahbagh et al. (2014) have developed various laboratory apparatus to evaluate the wear and torque of soft ground TBMs. These tests can also include soil conditioning. The result of these tests have shown the possibility of exploring the impact of various soil conditioning parameters on wear of cutterhead and cutting chamber as well as cutterhead torque. In order to examine the impact of earth pressure and surcharge loading on tool wear and behavior of conditioned soil, a new testing device was designed to simulate the conditions of cutter head, pressure chamber and modified soil in the pressurized face of an EPB machine (Hedayatzadeh et

 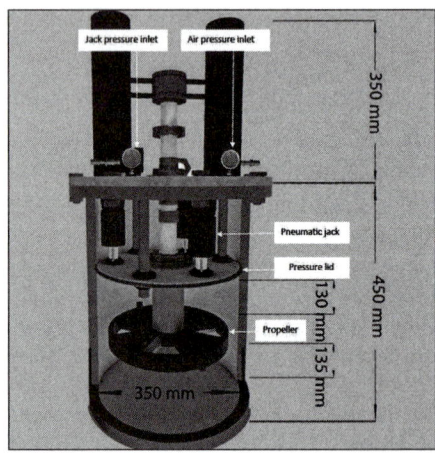

Figure 1. Picture and schematic drawing of the soil abrasion testing device

al., 2015). The device has been used to test different soil conditioning, water content, ambient pressures and the results show the feasibility of using the proposed device for evaluation of wear and effectiveness of soil conditioning parameters.

TEST DEVICE AND TESTING PROCEDURE

A soil abrasion testing device and procedure has been developed and used for quantifying soil abrasion as discussed by Rostami et al. (2012) and Alavi Gharabagh et al. (2014). This unit was subsequently modified to allow for application of direct loading on the soil to mimic surcharge loading for this study. The device consists of a cylindrical chamber, 350 mm in diameter and 450 mm in length which is charged with selected soil under desired conditions (i.e. dry, moist, saturated, conditioned etc.). The chamber dimensions were chosen especially for soils that is containing large gravel and for simulating the in-situ conditions of the soil as closely as possible. The rotation speed of the propeller in most tests were set at 60 rpm. There are plans underway to modify the unit to reduce the rotational speed to near zero to allow evaluation of soil rheology. The test device and the testing system was designed to allow for control of the ambient pressure through application of the compressed air from the top lid. The modifications involved applying pressure on the soil using a push plate pressed on the top of the soil sample by expansion of three pneumatic jacks. This arrangement simulates the working conditions in the chamber of EPB tunneling machines where the soil particles are pressed against the tools and abrade various surfaces at high contact pressures.

The loading pressure can be controlled by jack pressure range from 3 to 11 kPa load. Figure 1 shows the general view of testing device. In order to take into account both type of wear that can happen in soft ground excavation, an especially propeller was designed featuring two sets of blades, including surface and lateral blades. The blades were made by aluminum (a tensile strength of the material: 440MPa; elastic modulus: 3.3 GPa). Four pins (arms) were designed to mix the conditioned soil during the test. The pins could create a movement in the soil to expose the blades to fresh batch of soil. The dimensions and pictures of the propeller are shown in Figure 2.

In order to assess the capabilities of the modified soil abrasion testing device, a testing program was designed and several tests were performed. These tests focused

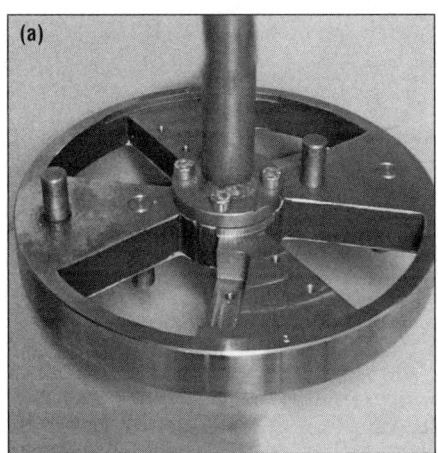
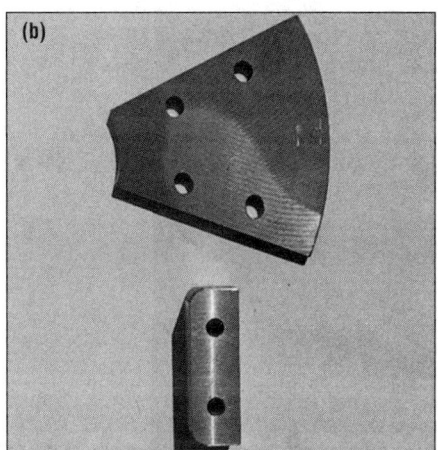

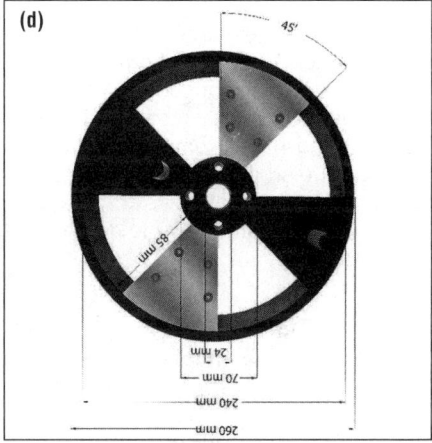

Figure 2. Picture of (a) front, (b) surface and lateral blades, (c) back, and (d) dimensions of the propeller

on evaluating the effect of conditioned soil parameters and ambient / surcharge load pressure on tool wear. For this purpose the soil samples were dried in the air and subsequently mixed with proper amount of water to control the water content of original soil samples. The quarzitic sand samples tested without altering the samples original grain size distribution. The soil at given moisture or water content (WC) was then mixed with the desired soil conditioning with thoroughly measured amount of surfactant to adjust Concentration ratio (Cf), Foam Injection Ratio (FIR). After preparation of the conditioned soil, a slump test was performed to capture the optimal conditioning parameters, based on procedure that was suggested by Peila et al. (2009). The soil mass needed for each test, was almost 40 kg and filled the chamber to the height of about 300 mm from the bottom. The surface and lateral blade were weighed separately before each test using a high-precision scale with a resolution of up to of 0.0001 g. After weighing blades were installed on the propeller and the propeller was positioned about 150 mm from the bottom and the lid is secured by using a set of twelve bolts.

Control of the plate load on the soil, the pneumatic jack pressure was selected. Together with air pressure in the chamber that represented the ambient (pore) pressure, they

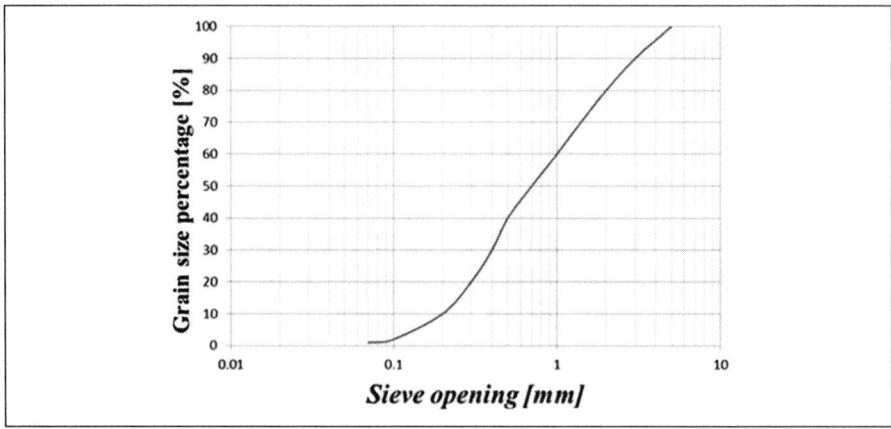

Figure 3. Sieve curve and characteristics of studied soil

constituted the testing condition. The range the pore pressure was from 2 to 6 bar and plate load pressure from 3 to 11 kPa. After completion of the set up, the tests commenced and continued for 30 minutes in constant speed of 60 rpm. The propeller torque was directly measured by instrumented arms attached to the chamber. All sensory and measurement systems were calibrated before tests on regular basis. After each test, the propeller blades were washed and dried for measurement of the weight loss. Test variables for this study included applied load on the plate, water content, FIR, FER and CF to determine the impact of each parameter on the wear of the blades.

TEST RESULTS

To investigate the effect of above mentioned parameters, several tests were carried out on silica sand. The grain size distribution curve of the sand is shown in Figure 3. The main constituent minerals include 97.1% Quartz and 2.4% kaolinite. USCS class of the soil is SP (poorly graded sand). Friction angle is 40°, Specific gravity (Gs) of the sample is 2.67, and D60 of 1, D10 of 0.2 mm, and Uniformity coefficient is 5. The tests were conducted using two conditioning agents including foam A (stabilized surfactant) and foam B (non stabilized surfactant).

Comparison of weight loss versus normal load applied on the soil allows to assess the impact of earth pressure (or depth of tunnel) on wear. Several tests were conducted for various values of ground pressure, each using different conditioning parameters. The related results are summarized in Figure 4. Figure 4 shows that the weight loss increases with increasing the amount of normal load on the soil. As expected, there is a reduction in the volume of pore spaces between soil particles and more solid–solid contact, greater solid–solid friction, and soil mass–metal blades contact stress, leading to higher wear.

Figure 4 also indicates that the torque also increases with increasing applied load. The described trends can be observed with all the conditioning sets. Figures 5 and 6 show the examples of the measured torque diagram by varying the applied pressure for two set of tests. The results prove that it is possible to decrease torque value from 11 to 3 kPa pressure that exerted by soil on propeller. In order to determine the effect of soil conditioning on wear, the soil was conditioned by different conditioning sets (FIR, FER and WC). Before each test, a slump test was performed to capture the optimal

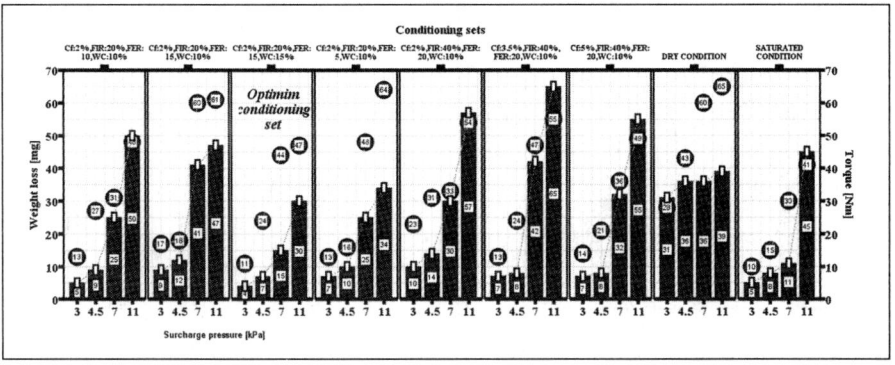

Figure 4. Comparative four axis relationships between weight loss and torque at different load pressures for different soil conditioning

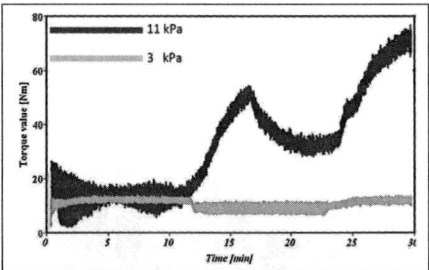

Figure 5. Comparison of torque requirement in different tests as a function of time for 3 kPa and 11 kPa (FIR: 20%, FER: 15, WC: 15%, Air pressure: 3 bar), for foaming agent A

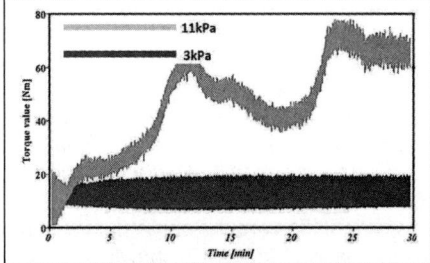

Figure 6. Comparison of torque requirement in different tests as a function of time for 3 and 11 kPa (FIR: 20%, FER: 15, WC: 10%, Air pressure: 3 bar)

conditioning parameters, based on procedure that was suggested by Peila et al. (2009). The optimum soil conditioning was observed to be FIR=20%, FER=15 and WC=15. Figure 7 shows the amount of weight loss and torque with two different types of foaming agents. The results clearly points to superior performance of stabilized foam versus the non-stabilized foam, since using stabilized foam allows the bubbles to stay intact over longer time and can bear more load.

Figure 8 illustrates the weight loss and measured torque by using different FIR. As it can be expected, the increasing FIR would enhance lubrication between soil particles and reduce the solid–solid

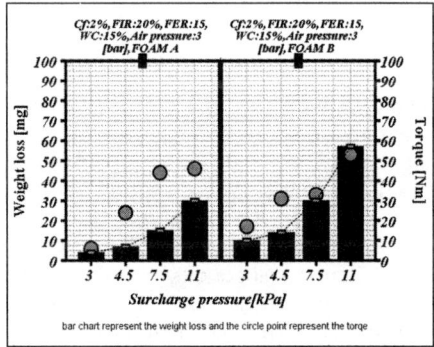

Figure 7. Comparison of weight loss and torque at different loading pressures for different soil conditioning types (Foam A, Stabilize, and Foam B unstabilized)

interactions. More investigation was done to evaluate the impacts of WC, FER, and Cf on tool wear. Primary results show the high interdependence of wear on the water content of soil together with the FIR. The combined effects can be seen in Figure 9. Figure 10 shows the measured of weight loss as a function of FER.

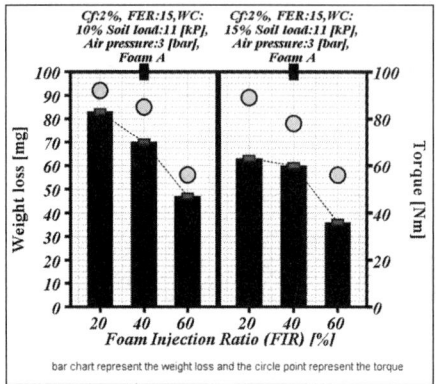

Figure 8. Comparison of weight loss and torque at different FIR for different soil conditioning

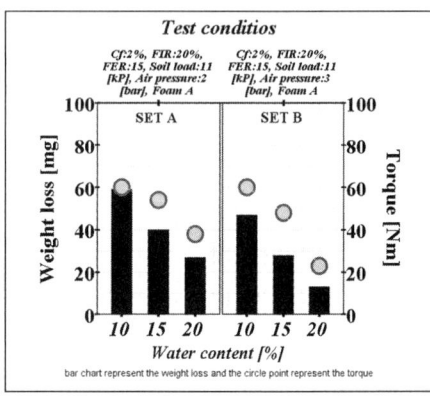

Figure 9. Comparison of weight loss and torque with different water content

The overall trend is that with increase in FER, the weight loss is the same for low load pressure, but decreased in high load pressure. Further testing needs to be carried out to verify this trend. Figure 11 shows the variation of weight loss with Cf. The results shows that for Cf=2 and 3.5, the weight loss are the same but for Cf=5, it decrease. Figure 12 shows the worn propeller and blades after several tests. This shows the level of wear that can be inflicted on the test pieces even on a short period of time in a laboratory setting that can simulate accelerated wear on various machine parts.

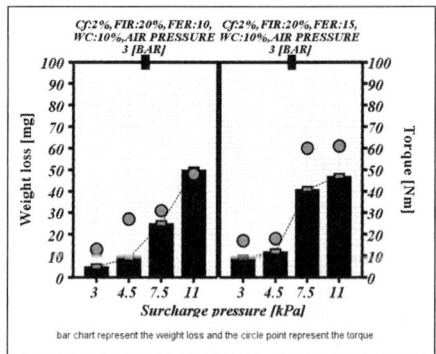

Figure 10. Comparison of weight loss and torque for different FER

CONCLUSION

The proposed soil abrasion and wear testing device with an improved propeller and test methodology have allowed to observe the impacts of various parameters including the ambient and surcharge pressures, water content, and soil conditioning parameters on wear and torque requirements of a soft ground tunneling machines. The testing unit is able to simulate the condition of various parts in the cutting chamber of EPB tunneling machine. The results of preliminary testing on a silica sand sample with different soil conditioning under changing ambient and soil pressures show that the load or earth pressure is has a major impact on wear and torque requirement of the machine. The results show that use of suitable soil conditioning can substantially reduce both wear and friction of the soil that is linked with the torque on the cutterhead. Also, the use of stabilized foam has proved to reduce wear and torque, especially if long life cycles and circulation in the pressure chamber is expected. Further studies are underway to characterize the effect of various soil conditioning parameters on wear and torque requirements in various soil types.

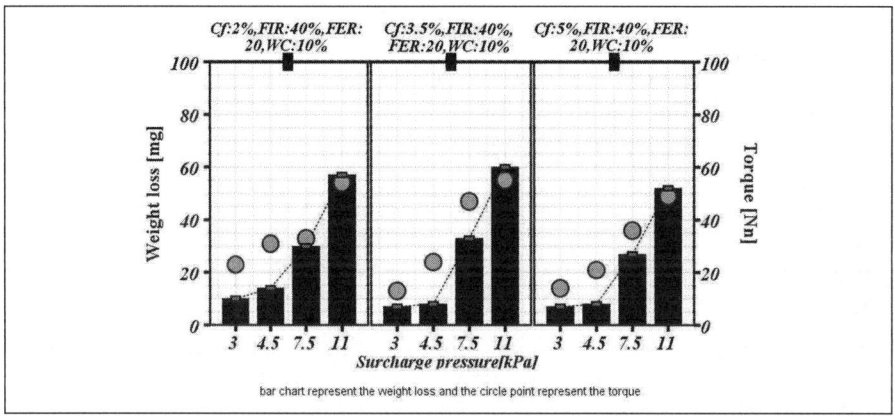

Figure 11. Comparison of weight loss and torque for different surfactant consternation factor (Cf)

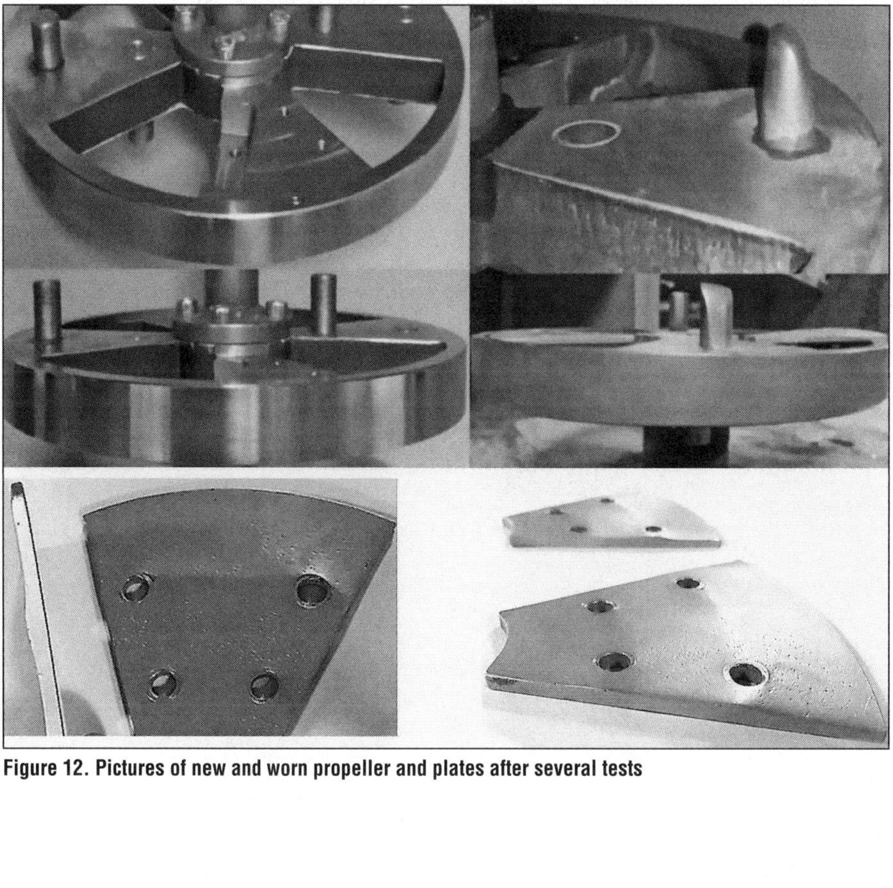

Figure 12. Pictures of new and worn propeller and plates after several tests

REFERENCES

Alavi Gharahbagh E., Rostami J., Talebi K.: "Experimental study of the effect of conditioning on abrasive wear and torque requirement of full face tunneling machines," Tunnelling and Underground Space Technology (2014), 41, 127–136.

Barzegari G., Uromeihy A., Zhao J.: "A newly developed soil abrasion testing method for tunneling using shield machines," Quarterly Journal of Engineering Geology and Hydrogeology (2013), DOI: 10.1144/qjegh2012-039.

Hedayatzadeh M., Rostami J., Peila D.: "Impact of conditioned soil parameters on tool wear in soft ground tunneling," (2015), World Tunneling Congress, Dubrovnik, Croatia.

Jacobsen P.D., Langmaack L., Dahl F.,Breivik T.: "Development of the Soft Ground Abrasion Tester (SGAT) to predict TBM tool wear, torque and thrust," Tunneling and Underground Space Technology (2013), DOI: http://dx.doi.org/10.1016/j.

Peila D., Oggeri C., Borio L.: "Using the slump test to assess the behavior of conditioned soil for EPB tunneling," Environmental & Engineering Geoscience, (2009), Vol. XV, No. 3, pp. 167–174, DOI: 10.2113/gseegeosci.15.3.16.

Peila D., Picchio A., Chieregato A., Barbero M., Dal Negro E., Boscaro A.: "Test Procedure for Assessing the Influence of Soil Conditioning for EPB Tunneling on the Tool Wear," (2012),World Tunneling Congress, Bangkok, Thailand.

Rabinowitz E. 1995. Friction and wear of materials. New York: Wiley. 315p.

Rostami J., Alavi Gharahbagh E., Palomino A.M., Mosleh M.: "Development of soil abrasivity testing for soft ground tunneling using shield machines," Tunneling and Underground Space Technology Journal (2012), 28, 245–256.

Removal of Interfering Tiebacks Using SEM in Advance of TBM Mining on the Regional Connector Project

Christoffer Brodbaek ▪ Mott MacDonald
Derek Penrice ▪ Mott MacDonald
Eren Kusdogan ▪ Regional Connector Constructors (Skanska-Traylor JV)
Christophe Bragard ▪ Regional Connector Constructors (Skanska-Traylor JV)

ABSTRACT

The Regional Connector Transit Corridor (RCTC) comprises a 1.9-mile underground light-rail project woven through downtown Los Angeles (LA). This $1-billion design-build project's reference design indicated the need for a large and intrusive open excavation, approximately 100-feet long and 30-feet wide, to remove existing tiebacks within the public right-of-way and in conflict with the RCTC bored tunnels. To minimize surface impacts, the design-build team, Regional Connector Constructor (RCC), proposed an alternative tieback removal concept comprising a vertical shaft and mined adit. This paper describes the use of Building Information Modeling as a powerful design tool to optimize the location, size and alignment of the concept and demonstrates its feasibility to the Project Owner, LA Metro, and describes the shaft and adit construction in a dense urban setting and the tieback removal process.

INTRODUCTION

The Regional Connector Transit Corridor (RCTC) is a significant expansion of the Los Angeles County Metropolitan Transportation Authority (Metro) system. This design-build project is owned and operated by Metro with Regional Connector Constructor (RCC)—a Skanska/Traylor Brothers joint venture—as the design-builder and Mott MacDonald as lead design engineer. The 1.9-mile underground light-rail project consists of a 1.1-mile twin bored tunnel, 0.5-miles of cut and cover tunnel, a cavern to be constructed by sequential excavation methods and three underground stations, also to be constructed by cut and cover methods. The bored tunnels are mined with a 21.6-foot diameter Earth Pressure Balance (EPB) Tunnel Boring Machine (TBM). Along the alignment various obstructions from the construction of adjacent building basements are known to exist. In most areas, the tunnel alignment clears existing obstructions, but in one location, at the Bank of America Building located at S Flower St and 3rd St downtown LA a conflict with existing tiebacks was unavoidable and removal of temporary tiebacks was necessary. The conflicting tiebacks were located up to 57 feet below existing grade and impacts 80 feet of the tunnel alignment. The tunnel alignment and location where tieback removal is required are shown in Figure 1.

LAYOUT AND ARRANGEMENT OF TIEBACK REMOVAL SHAFT AND ADIT

The Owner's reference design, provided in the Request for Proposal (RFP), featured a 30 × 100-foot large and 57-foot deep open excavation to remove the interfering tiebacks. During detailed design RCC recognized that the majority of the tiebacks were located in the lower half of the right track tunnel and could be removed from an approximately 24-foot diameter shaft and a small 11-foot diameter horseshoe shaped adit, resulting in significantly less excavation and backfill and substantially reduced traffic and utility impacts at street level in comparison with the RFP concept. The

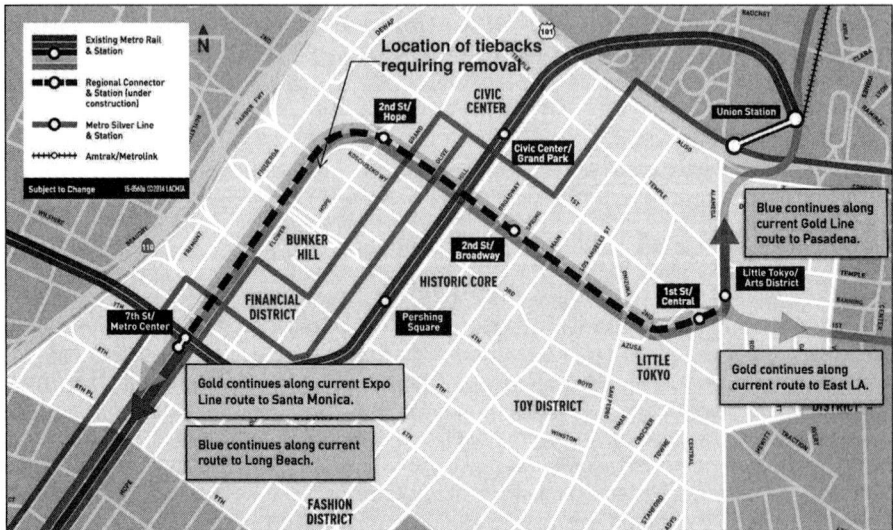

Figure 1. Tunnel alignment and location of existing tieback requiring removal

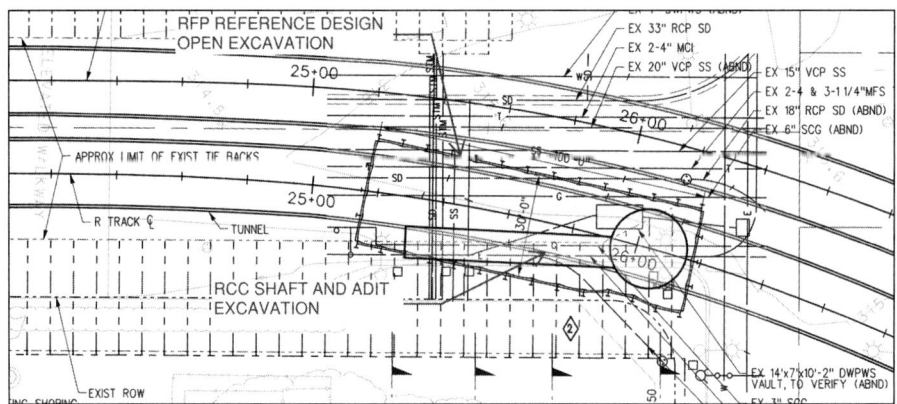

Figure 2. RFP reference design with shaft and adit alternate overlaid in yellow

reference RFP design with the shaft and adit excavation alternate overlaid are shown in Figure 2.

The general arrangement of the tieback removal shaft and adit is shown in Figure 3. The tiebacks have a seven-foot nominal spacing in plan corresponding to the soldier pile spacing of the Bank of America Support of Excavation (SOE).

The optimal location of the shaft and adit was determined utilizing a three-dimensional Building Information Modeling (BIM) system, which includes the geometry of adjacent buildings, SOE and tiebacks, the final TBM tunnels and underground utilities. The BIM model was based on detailed information available in design drawings of the Bank of America building's basement, constructed in the early 1970s. To confirm the location of the tieback removal shaft and adit, a 30-foot deep pothole was excavated during the design phase to confirm tieback locations and to validate the tieback removal shaft and adit's size and location. Because as-built conditions may differ from the design drawings, the tieback removal shaft and adit design was kept flexible by using

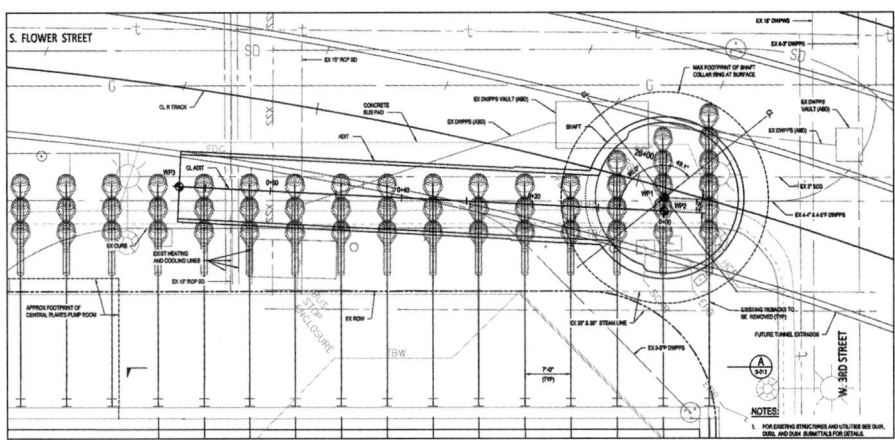

Figure 3. Layout of shaft and adit for removal of tiebacks

shotcrete liners at the interfering tieback levels. This makes pocket excavations and adit enlargements feasible to accommodate misalignment and relocation of tiebacks that may have occurred during construction of the Bank of America Building.

GROUND CONDITIONS

The ground conditions at the tieback removal site consist 30 feet of Artificial Fill and about five to eight feet of sandy Young Alluvium underlain by Fernando Formation.

The Artificial Fill is a variable mix of soils and includes sand, clay, shale and construction debris. Young Alluvium is a silty to poorly graded sand with varying fines content. The Fernando Formation is an extremely weak to very weak massive clayey siltstone to silty claystone with an unconfined compressive strength in the range 25 to 300 psi. Such soils and weak rock may be excavated with small excavation equipment without blasting or pre-treatment. The Fernando Formation contains some discontinuities, but is predominately massive at the tieback removal shaft and adit site. Perched groundwater generally exists within the lower part of the fill and alluvial deposits, due to the relatively low permeability of the underlying Fernando Formation. A Regional groundwater level is present within the Fernando Formation, and shaft and adit inflows from the Fernando Formation is limited to drips and slow seeps occurring along joints, shears, and bedding planes. The Geotechnical Baseline Report (GBR) indicated that inflow during excavating of the RFP reference design 30 × 100-foot open excavation should not exceed 100 gallons per minute, this equates to maximum anticipated inflow to the shaft of approximately 30 gallons per minute. The excavation in the Fernando Formation was essentially dry, but inflow from the alluvium was present during excavation, and was limited to approximately one gallon per minute at most, which is substantially less than the maximum inflows indicated in the GBR.

The underground works for the project are classified as "Potentially Gassy" by the State of California, Department of Industrial Relations, Division of Occupational Safety and Health Administration (CAL/OSHA). Per the GBR methane and hydrogen sulfide gases may be encountered during mining of the adit and shaft excavation. These gasses are expected to exist and seep through pore spaces and discontinuities, and will be generated from off-gassing of groundwater that flows into the excavation. The concentration, pressure, and volume of these gasses are expected to be sufficiently low that the inflow and off-gassing of these gasses can be mitigated within the shaft and

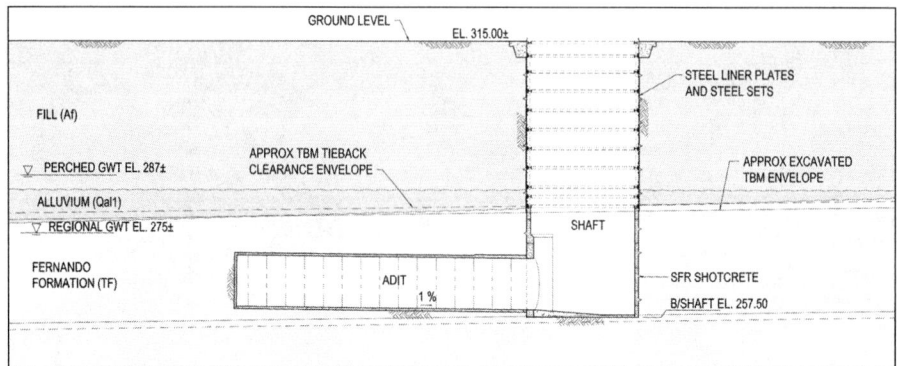

Figure 4. Shaft and adit profile and ground conditions

adit excavations through adequate ventilation and proper shotcrete application. Gas monitoring was performed daily per CAL/OSHA requirements. Gas measurements were essentially zero during excavation, and if any gas was registered it was well below action threshold values. Geologic stratigraphy and groundwater conditions are shown in Figure 4 together with the shaft and adit layout in profile.

SHAFT AND ADIT DESIGN

As the purpose of the shaft and adit construction is to clear the TBM tunneling path for steel obstructions, no steel could be left in place from the shaft and adit construction within the TBM tunnel excavation area—this is a prime motivation for using shotcrete as much as possible in the shaft and adit. During feasibility evaluation of the shaft and adit; RCC conducted a construction risk assessment, which helped the design of the shaft and adit identify required toolbox measures construction should be prepared to implement in response to adverse construction conditions. A list of identified construction risks and proposed mitigation strategies are listed in Table 1. One of the larger risks identified is not being able to access the tiebacks. Using BIM technology, the shaft and adit were strategically located to confirm that all tiebacks can be accessed and cut from the shaft and adit. Tiebacks outside of the shaft and adit would require removal from small pocket excavations in the side walls of the shaft and adit. This is possible since the tiebacks primarily are in the Fernando Formation and shotcrete is available at the site for rapid deployment and ground stabilization.

To minimize excavation and impact to traffic the tieback removal shaft footprint and adit diameter were kept as small as possible to clear all conflicting tiebacks within the TBM clearance envelope.

Shaft Design

Initially it was planned to construct the shaft with steel fiber reinforced (SFR) shotcrete over the full shaft depth. Concerns about insufficient ground standup time in the artificial fill and alluvium to allow application of shotcrete led to a preference for steel liner plates and steel ribs for the shaft in these formations whilst maintaining a shotcrete liner in the Fernando Formation. However, as the contact between the alluvium and Fernando Formation is at the TBM tunnel crown, the last round of liner plates will be in the TBM clearance envelope. Consequently, the last round of liner plates will have to be removed during backfilling of the shaft prior to TBM mining. The shaft layout, tieback locations and the TBM tunnel are shown in Figure 5.

Table 1. Construction risks and mitigation strategies

Construction Risk	Mitigation Strategy
Unstable ground during shaft excavation results in building/utility damage etc.	Permeation grout around shaft perimeter. Reduced unsupported wall height. Metal sheets
Higher than anticipated groundwater elevation.	Grouting or dewatering around shaft perimeter. Reduced unsupported wall height.
Ground conditions different than suggested by profile – alluvium in crown of adit.	Observe excavated materials in shaft. Probe drill prior to mining of adit. Grouting as necessary.
Adit instability/inundation from perched groundwater in alluvium over adit.	Drill drain holes from shaft. Allow seasonal perched water to drain to sump in advance of mining.
Shaft/adit excavation does not recover all expected tiebacks.	Each tieback expected to be removed will be identified by location, i.e., pile line and elevation, and will be recorded during excavation. Locally enlarge adit to find expected tiebacks – pocket excavation. Face intervention at the TBM.
Bank of America building in BIM model at incorrect location and/or as-built data are incorrect.	Confirm building corners with survey and/or obtain additional as-built information.
Tiebacks not installed per as-builts – length/inclination.	Review actual locations/installations of tiebacks within shaft relative to expected positions.
Connectivity of annular backfill grout with building basement due to cutting of tiebacks.	Potential risk regardless of whether shaft and adit are constructed. RCC to control pressures and monitor grout volumes for potential communication.

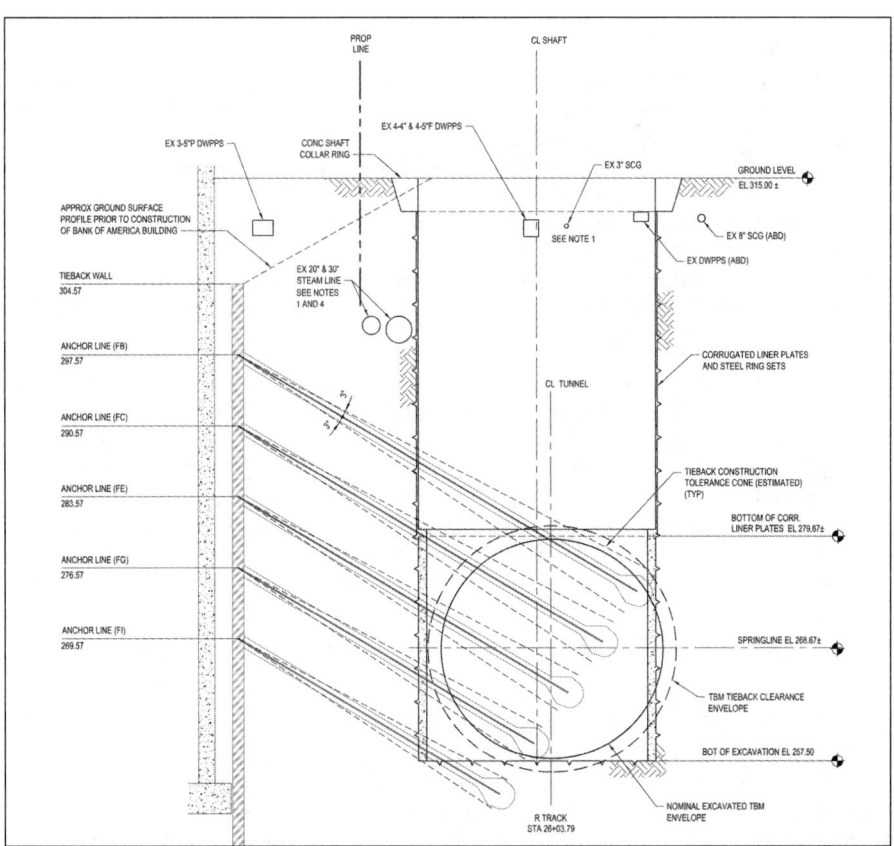

Figure 5. Shaft layout and tieback location

The shaft was designed applying drained long-term ground properties and apparent earth pressures, which were compared to axi-symmetrical earth pressures for circular shafts. Due to the relatively rigid nature of the shaft the apparent earth pressures, which are greater than axi-symmetric earth pressures, were applied for structural design of the shaft. Surcharge loads from construction equipment were included as non-uniform shaft pressures around the shaft, introducing additional bending moments in the shaft liner.

The shaft has a nominal external diameter of 24 feet and features:

- 3.33-foot deep concrete collar at street level, which provides support for the hanging steel sets and utilities.
- 24-inch corrugated steel liner plates with steel sets at nominal four-foot spacing in the upper 35 feet of shaft.
- A 10-inch thick SFR shotcrete liner, except at the adit opening where the liner was strengthened by locally increasing the shotcrete liner thickness to 18 inches, to account for increased tension and compression liner forces around the adit opening.

Section similar to the one shown in Figure 5 were developed in BIM and were imperative for the design, construction planning and confirmation that each tieback could be cut from the shaft and adit over the entire TBM clearance envelope.

Adit Design

The adit is a horseshoe shaped short tunnel, sized to provide adequate access to remove tiebacks interfering with the future TBM tunnel. To minimize potential surface settlements, and expedite construction, it was decided to perform pocket excavations to cut tiebacks across the entire TBM tunnel clearance envelope rather than enlarge the adit cross section. Pocket excavations to clear tiebacks are practical since they take place in the stable Fernando Formation. The general shape of the adit, location of tiebacks and indication of pocket excavation requirements for tieback removal is indicated in Figure 6. Similar sections were cut at each tieback in the BIM model to confirm access in the adit to clear the tiebacks and identify areas where pocket excavations would be required.

The structural verification of the adit liner and estimates of anticipated settlements were analyzed in a two-dimensional finite element analysis. For design purposes it was assumed the adit would be mined full face with a maximum four-foot advance and fiberglass polymer lattice girders placed for profile control. Fiberglass polymer lattice girders were used to allow for future TBM mining trough the adit. The four-foot advance length resulted in 60 percent ground relaxation, which was included in the finite element analysis by reducing the ground pressures to 40 percent around the adit perimeter prior to activation of the liner. The adit design was based on the construction sequence:

1. Excavate four-foot advance – full face assumed for design
2. Apply two-inch shotcrete initial liner
3. Install lattice girder
4. Apply four-inch shotcrete to final six-inch liner thickness.

The liner was verified at each stage of construction assuming a nominal one-day cycle time per advance to account for shotcrete strength gain with time. The maximum

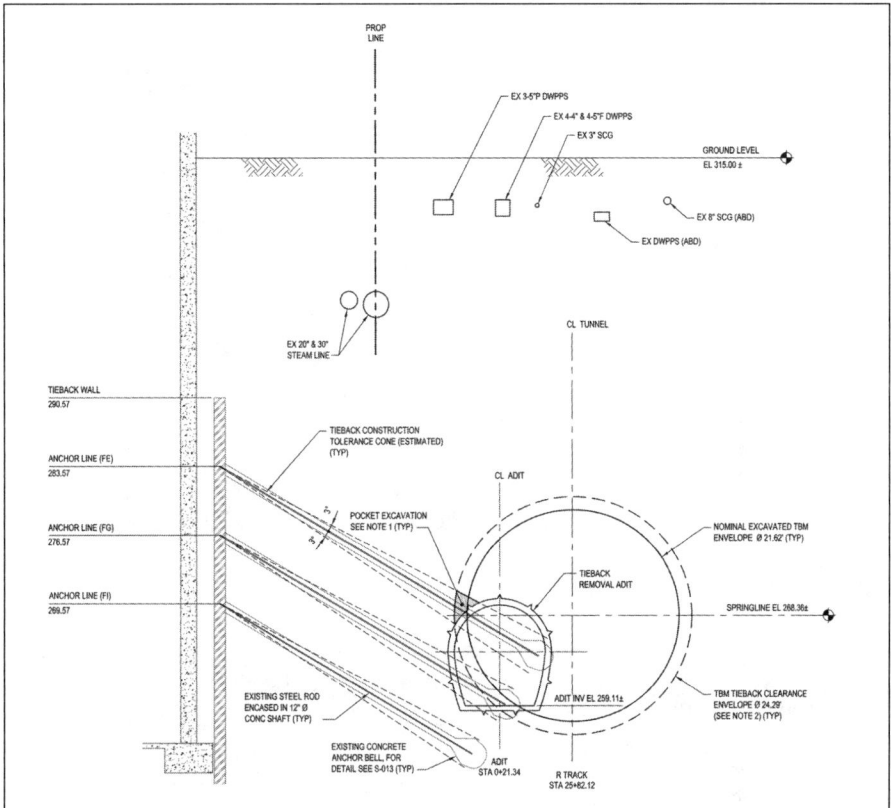

Figure 6. Adit layout and tieback locations

accumulated ground displacement from mining the adit was estimated in the finite element analysis to be 0.24-inch at the crown and 0.17-inch at the springline. This indicates relatively small ground deformations from adit excavation, which meet the project settlement criteria of maximum 0.5-inch settlement of existing structures and utilities. Accumulated deformations around the adit opening are shown in Figure 7.

Shotcrete Requirements and Testing

Both the shotcrete shaft and the adit were designed for minimum 5,000 psi 28-day compressive strength and a residual flexural strength of 290 psi. Panel tests were shot prior to construction as part of the mix design confirmation and approval. The minimum and average compression strength achieved from testing are listed in Table 2 together with specified minimum strength at 1, 3, 7 and 28 days. Flexural strength tests were performed per ASTM C 1609. The panel tests showed flexural strength in the range 371 to 389 psi, with an average of 382 psi, which was well above the 290 psi minimum residual flexural strength.

Tieback Removal Program

The existing tiebacks were installed for temporary support only and are no longer in service; they can therefore be removed without compromising the adjacent Bank of America building. To determine whether a tieback potentially could conflict with TBM

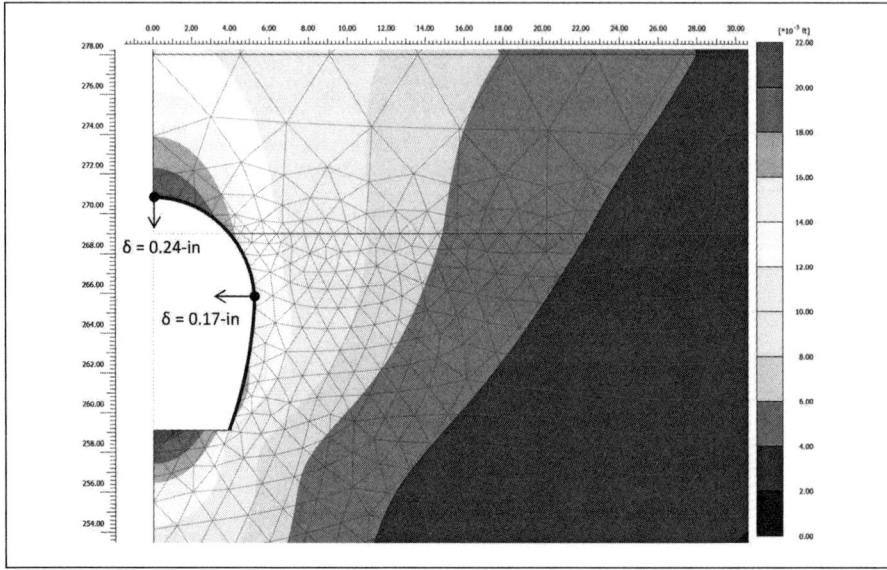

Figure 7. Accumulated ground deformations during mining

Table 2. Compression strength test results of shotcrete

Shotcrete Age (days)	Compression Strength, f'$_c$ (psi)		
	Minimum per Specification	Minimum of Tests	Average of Tests
1	1500	2532	3941
3	3200	3268	4740
7	4000	4981	5911
28	5000	5793	7037

mining, a 3-degree tieback installation tolerance cone was included in the BIM assessment, as indicated in Figure 5 and Figure 6. This was a major assumption for the design and planning of the shaft and adit and would only be validated during excavation and tieback removal. Additional mitigation would be required if the construction tolerances exceeded the 3 degrees assumed. The primary mitigation strategy was to perform pocket excavations together with local ground

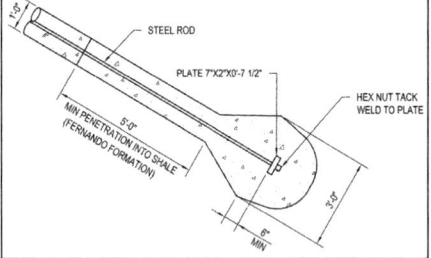

Figure 8. Tieback anchor geometry

stabilization if needed. The BIM model was used to identify locations where pocket excavations are required to cut the tiebacks. This was very valuable during construction, as it helped locating the tiebacks outside of the shaft and adit side walls.

According to as-built information, the tiebacks consist of a 1-inch to 1 ⅜-inch steel rods encased in concrete, and shown with a belled anchor tip—a typical tieback is shown in Figure 8.

The anticipated tieback removal program and tieback removal status are shown in Figure 9, which illustrates the Bank of America SOE soldier pile wall in elevation and the tieback location at the SOE wall. During design each tieback was categorized reflecting the reason for its removal as follows:

- Tieback within TBM clearance envelope. Removal required.
- Tieback within TBM clearance envelope. Removal required. Pocket excavations likely required.
- Tieback within TBM clearance envelope, but steel appears to be outside of tieback clearance envelope – may conflict based on tieback construction tolerance cone.
- Tieback within shaft excavation. Removal required for shaft access.
- Tieback removal may be required if encountered in adit excavation.
- Tieback removal not required. Tieback is outside of TBM clearance envelope.

A total of 30 tiebacks were identified as conflicting with the TBM tunnel excavation and would require removal. During construction the tieback removal schedule was used to maintain a log of tieback removal progress, facilitating construction feedback and documentation throughout the tieback removal process. The removed tiebacks are indicated in Figure 9 and are discussed in more detail in the next section of this paper.

CONSTRUCTION AND TIEBACK REMOVAL

Shaft Construction

Construction of the shaft started by installation of the first two rounds of liner plates, the concrete collar and preparations for support of active utilities including a gas line and two electric duct banks located below the concrete collar. The utilities were supported in-place across the shaft by hanging them off the collar. Subsequent excavation and liner plates were installed in four-foot advances through the fill and alluvium above the perched groundwater table, which was encountered approximately 33 feet below street level. The last two rounds of steel liner plates were installed in two-foot advances as they took place below the perched water table in alluvium sands.

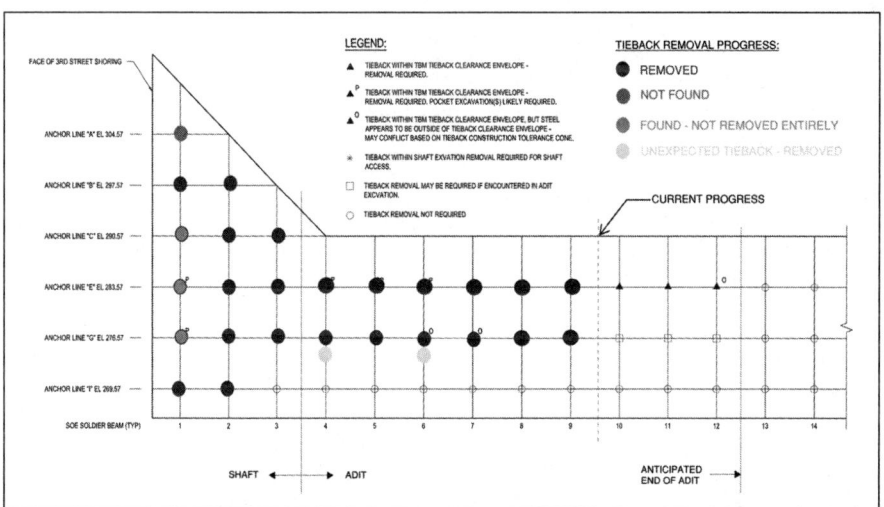

Figure 9. Tieback removal program. Anchors shown at SOE soldier pile wall.

Figure 10. Liner plates installed and shotcreted lower part shaft under construction

Figure 11. Shaft breakout for adit excavation

The alluvium turned out to be slightly deeper than the geologic profile indicated at the shaft location. The GBR indicated that the alluvium below the groundwater table may be running indicating a potential short standup time for vertical cuts. RCC decided to apply a fast setting shotcrete to overcome the apparent seepage at the bottom of the alluvium layer and running ground potential, rather than adding another round of steel liner plates and steel sets, which would require removal during backfilling of the shaft, since they would be in the TBM clearance envelope. The fast setting shotcrete worked well in the slightly seeping alluvium in combination with additional weep holes in the liner and limiting the advance length in this transition zone to a three-foot vertical cut. Once in the stable Fernando Formation excavation continued to final invert in advances of five feet.

Shaft excavation took place utilizing a small excavator. The steel liner plates were installed at a rate of 2 feet advance per day, this included grouting behind liner plates and steel set installation. The shotcrete liner shaft was advanced at a rate of 5 feet per day in the Fernando Formation.

A picture of the essentially completed shaft is shown in Figure 10. The shaft breakout for mining of the adit was performed upon shaft completion as shown in Figure 11. The exposed Fernando Formation is observed in the background in this picture, prior to application of flashcrete. It indicates the Fernando Formation is massive with stable excavation conditions for the subsequent mining of the adit.

Tieback Removal in Shaft

As shown in Figure 9 the first three soldier beam tiebacks are located in the shaft. The tiebacks for the second and third soldier beam were at their anticpated locations and were readily cut. However, three of the tiebacks at the first soldier beam were angled up to 13 degrees in plan from their design orientation, which is appreciably greater than the three-degree tieback installation tolerance assumed for shaft layout. These tiebacks were well outside of the shaft and deeper pocket excavations were required to clear the TBM path. To maintain construction progressing the three tiebacks were partially cut and surveyed in place, and the remaining part of the tiebacks removed during backfilling of the shaft.

The upper tieback of the first soldier beam was not found during excavation, it may not have been installed or it may have been removed during construction of the Bank of America building. The per design location indicates this tieback would not interfere with the TBM mining as it was located above the TBM tunnel.

The anchor line "C" tieback, see Figure 9, was located in the alluvium sand and Fernando Formation transition zone, which makes pocket excavation at this level more challenging. This tieback was removed with two small pocket excavations in combination with local groundwater drainage using well-points. The two other tiebacks requiring removal during backfilling were located in the Fernando Formation, which has better standup time for access and were cut without drainage and partial excavations.

Adit Construction

Although the adit was designed for full face excavation with maximum four-foot advance lengths, the actual mining of the adit was staged into effectively two drifts per four-foot advance:

1. Excavate above springline and apply two-inch flashcrete
2. Excavate below springline, apply two-inch flashcrete plus four-inch shotcrete for final liner above and below springline.

A small excavator was used for the adit excavation. No measurable groundwater was encountered in the adit and the fractures in the Fernando Formation were essentially dry, which made the excavation progress effectively with a four-foot advance rate per day.

Where tiebacks were encountered they were cut utilizing a torch, once initial flashcrete had been placed on exposed ground. If pocket excavation were required to remove a tieback, the full six-inch liner would be placed to stabilize the ground during pocket excavation and tieback removal. Two unexpected tiebacks were encountered in the adit at Anchor Line "G" indicated with green in Figure 9, both tiebacks were cut and removed. A picture of the adit during construction is shown in Figure 12. An encountered tieback at the adit face is shown in Figure 13.

The locations of tiebacks in the adit compared well with the per design tieback location implemented in the BIM model. The total adit length mined to remove all interfering tiebacks was 60 feet as planned per the BIM model. Subsequently the shaft and adit were backfilled with low strength cellular concrete.

Backfilling and Shaft Removal

Upon completion of excavation, the shaft and adit was backfilled with a flowable, low density Cement Low Strength Material (CLSM). The CLSM has an in-place density of

Figure 12. Excavation of adit   Figure 13. Tieback encountered at the adit face

27 pcf with a minimum 100 psi unconfined compressive strength, which will be easily excavated during TBM mining. Grout tubes were installed along the adit crown to
contact grout any gaps and voids between the CLSM and adit liner occurring during
placomont of tho oollular oonorctc. Thc ahaft ia backfillad in atagea, allowiiiy ieiiiuval
of liner plates in conflict with the TBM alignment and allowing for pocket excavations to
retrieve portions of tiebacks beyond the shaft perimeter. During completion of backfilling the upper eight feet of the steel liner plates and the utility support was removed
and roadway restored in accordance with City of Los Angeles requirements.

Engineering Support During Construction

To facilitate construction and engineering input during shaft and adit excavation RCC
developed templates for Daily Meeting Reports (DMR) and Required Excavation
Support Sheets (RESS). The DMR and RESS were filled out during daily progress
meetings and included ground deformation and convergence instrumentation data,
gas monitoring, groundwater flow, geologic observations, construction contingencies
applied, shotcrete test data, general observations and status of tieback removal.

Both the DMR and RESS's were effective mechanisms to document excavation and
excavation support changes requiring acceptance from the Engineer of Record to
confirm proposed modifications during construction. Examples include a) relocation
of steel sets to clear utilities at the top of shaft, b) shotcrete not meeting specified
strength at a one location in the shaft and c) engineers review of proposed pocket
excavation methods. In each instance the Engineer was able to provide construction
feedback in a timely fashion, permitting construction to progress on a day to day basis
without delays. The direct communication and construction documentation the DMR
and RESS forms provided turned out to be an effective way to keep all parties, including Metro and Metro's representatives, up to date on construction progress and status.

CONCLUSION

The tieback removal process required to clear the future TBM tunnel path was successfully constructed and executed. The use of advanced BIM technology to strategically locate and size the shaft and adit was beneficial for the design, planning and execution of the works. In addition, the BIM model provided visualizations of locations where pocket excavations would be required to clear the TBM tunnel path and confirmation that all conflicting tiebacks would be accessible from the shaft and adit. The use of BIM also facilitated the construction approval process with Metro and the City of Los Angeles as it provided them with detailed analyses and visualization of tieback access from the shaft and adit.

The majority of the anticipated tiebacks within the TBM clearance envelope were located and cut generally in good agreement with the model predictions, although some of the tiebacks deviated more than the original three-degree installation tolerance allowed for in the planning and design phase. Such tiebacks were removed from additional local pocket excavations. On that basis the use of BIM as an optimization tool for the shaft and adit design and construction has been successfully demonstrated on this project.

The reduced footprint of the excavation at street level resulted in substantial less excavation, backfilling and impact to traffic and reduced construction traffic through downtown LA. The utilization of common construction methods allowed the shaft and adit construction to progress continuously, while tiebacks were removed as they were encountered during the excavation.

ACKNOWLEDGMENTS

The authors would like to thank Metro for providing permission to publish this paper. We are greatly appreciative for everyone contributing to the design and construction of the tieback removal shaft and adit including LA Metro, City of Los Angeles Department of Engineering, Bender Consulting on dewatering and KOA Corporation on traffic planning. In particular, we would like to thank the onsite personnel for their continuous feedback during construction. Additional thanks to Carlos Herranz Calvo with Mott MacDonald for providing the construction pictures utilized in this paper.

Port Mann Water Supply Tunnel: Lessons for the Future

Gregg W. Davidson ▪ McMillen Jacobs Associates
Frank Huber ▪ Metro Vancouver
Murray D. Gant ▪ Metro Vancouver

ABSTRACT

After more than a decade of planning, design, and construction, the Port Mann Water Supply Tunnel was successfully brought into service in early 2017. The tunnel increases Metro Vancouver's (MV) capacity to meet water demand while increasing system reliability, and represents the first of several tunnel projects planned to improve seismic capacity of the system following a Maximum Credible Earthquake event. The contracting and management issues associated with delivering the project have been as complex and timeconsuming as some of the design and construction challenges. As well as explaining how the design and construction challenges were successfully addressed, the paper presents how Metro Vancouver approached contractor procurement through prequalification and negotiated proposal phases, presents the risk management approach and project team structure, and describes how the lessons learned on the project can be applied to future tunnel projects in the region.

INTRODUCTION

Metro Vancouver provides a reliable source of safe, high-quality drinking water to approximately 2.5 million people in its 18 member municipalities, one electoral area, and one treaty First Nation. This service includes acquiring and maintaining the water supply, operation of treatment facilities to ensure quality, and delivery of potable water to these local government members. Water is collected from three mountainous watersheds—Capilano, Seymour, and Coquitlam—and is delivered by an extensive system of 25 reservoirs, 19 pumping stations, and over 500 km (300 mi) of transmission mains.

The existing Port Mann Main–Fraser River crossing was constructed in 1974. It consists of a 1,200 mm (48 in.) diameter welded steel pipe, approximately 1,000 m (3,300 ft) in length, crossing the Fraser River just downstream of the Port Mann Bridge. The crossing is a primary water supply link to municipalities south of the Fraser River. A location map is shown in Figure 1. This crossing was damaged by riverbed scour in May 1997, causing temporary but significant water supply problems to several municipalities south of the Fraser River during the summer of 1997. As it was recognized that a new crossing would be required in the future to address capacity increases and seismic deficiencies, the repair, completed in 1998, consisted of replacing only the damaged section of the water main and providing a protective riprap apron to secure it against future scour and undermining. This temporary repair was intended to provide an acceptable level of service for about one to two decades, during which time the long-term options for a new crossing would be fully explored and a suitable course of action developed.

After extensive study, the Port Mann Water Supply Tunnel was selected as the long-term solution to replace the existing water main crossing. The new water main is sized to accommodate future growth in the region, is located below the depth of riverbed scour, and is designed to remain operational following a major earthquake with a return period of 10,000 years.

PROJECT OVERVIEW

During the mid-1990s, Metro Vancouver initiated a seismic review and retrofit program for its key facilities, and developed its own seismic design standard, which required critical infrastructure such as marine crossings to remain operational after a 1:10,000-year event. As part of this review process, it was identified that many of the critical service delivery marine crossings would not survive even a moderate earthquake, such as an event with a 475-year return period.

Prior to assessing the long-term options for a new crossing, MV conducted a seismic vulnerability assessment on the existing Port Mann crossing in 2001. The proposed crossing was assessed for damage potential when subjected to seismic events with 100-year and 475-year return periods; and the Maximum Credible Earthquake (MCE) event, with a return period of 10,000 years. Although no damage potential was identified for the 100-year earthquake, the study concluded that the crossing would fail at several locations during a 475-year or MCE event. It was determined that the subsoil would liquefy, causing permanent ground deformation along both riverbanks towards the river. These conclusions are generally consistent with observed damage to pipelines during past earthquakes in places such as California and Japan.

MV undertook preliminary design of a new crossing of the Fraser River in 2002–2004. This work, which was based on limited geotechnical information, indicated that a new crossing would likely need to be trenchless and at least 30 m (100 ft) beneath the river bottom to avoid future river scour (1:500-year event) and withstand an MCE seismic event. As part of the study, several different trenchless methods and different pipe diameters were evaluated. Trenchless methods considered were horizontal directional drilling (HDD), microtunneling, and tunneling using a pressurized face tunnel boring machine (TBM). A welded steel pipe installed in a tunnel, constructed using a pressurized face TBM, was eventually selected as the optimal water main crossing replacement solution.

Detailed design of the project was awarded to the Fraser River Tunnel Group (Ausenco, McMillen Jacobs Associates, and Golder Associates) in late 2006 and was completed in early 2010. An extensive geotechnical exploration program, combined with significant seismic analyses, including geotechnical and structural modeling, was undertaken to complete the design. Environmental, archaeological, social, and public impact assessments were also carried out during the detailed design phase. Additional geotechnical

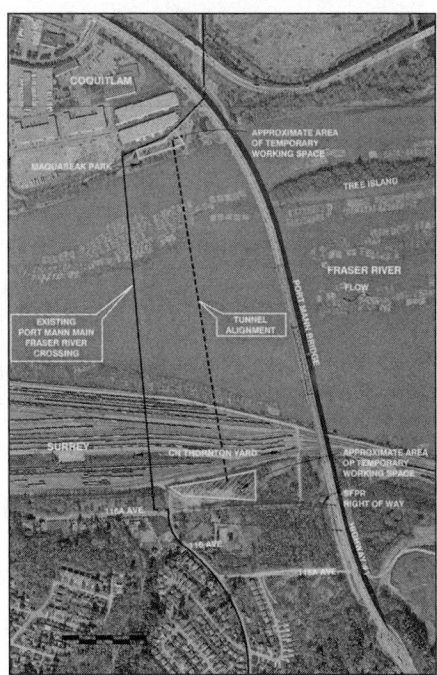

Figure 1. Port Mann water supply tunnel location plan: Fraser River crossing

explorations were carried out in 2010 at both the south and north shafts in order to provide data beyond the full depth of the proposed shaft bottom.

The final project configuration consists of a 3.5 m (11.6 ft) outside diameter bored and segmentally lined tunnel mined between two deep vertical shafts. The tunnel connects to the existing Port Mann Main on the north and south banks of the Fraser River. Two new valve chambers were constructed integral to the concrete shaft lining to control water flows through the tunnel pipeline. The tunnel is approximately 1,000 m (3,300 ft) long and was sized to install a 2,135 mm (84 in.) diameter full penetration butt-welded steel water main. The tie-in piping required to connect the new tunnel to the existing water main consists of a 1,220 mm (48 in.) diameter butt-welded steel pipe. Provisions were included in the design to allow for a future twin 1,525 mm (60 in.) diameter water main to connect into the tunnel when required because of increased demand. Shaft construction was facilitated by installation of unreinforced concrete slurry walls prior to excavating the interior soils and constructing a cast-in-place reinforced concrete lining. Muck handling, materials stockpiling, office facilities, and labor parking were located at the TBM launch shaft on the south side of the river. The steel water main was installed in the tunnel via the south shaft.

Procurement of the tunnel contractor commenced in 2010, and the contract was awarded to a Joint Venture of McNally Inc. and Aecon Constructors in mid-2011. The project was completed in late 2016, with the water main going into service in early 2017.

SUBSURFACE CONDITIONS

Geomorphology

The Project site is within the lower reaches of the Fraser River Valley, about 35 km (22 mi) east (upstream) of the mouth of the Fraser River in Georgia Strait. The valley is bounded by the Coast Mountains to the north and Cascade Mountains to the southeast. The topography at the project site ranges from approximately El. +5 m (+16 ft) on the north river bank at the North Shaft to El. −19 m (62 ft) near the center of the main river channel. The south river bank near the south shaft is approximately El. +4 m (+13 ft).

Tectonic Setting and Seismicity

The Project is located in a seismically active area. Site seismicity results from the thrusting (subducting) of the Juan de Fuca Plate beneath the Continental North American Plate. The offshore plate tectonic setup has resulted in shallow crustal earthquakes occurring within the Continental Plate, deep intraplate earthquakes occurring in the subducting plate, and interplate earthquakes occurring at the contact between the plates.

Over the past several decades, intraplate earthquakes have occurred at regular intervals—Campbell River (M7.3, 1946), Olympia (M7.1, 1949), Seattle/Tacoma (M6.5, 1965), and Nisqually (M6.8, 2001). A sitespecific seismic hazard study completed for the Project area indicated intraplate and interplate earthquakes are the dominant seismic risks at the site, with the intraplate earthquakes controlling the seismic risk for periods less than about 1 second and both intraplate and interplate earthquakes contributing similar seismic risk for periods longer than 1 second. The controlling earthquake scenarios included a M7.25 intraplate earthquake occurring at a distance of 55 km (35 mi) and an M8.8 interplate subduction earthquake occurring at a distance of 160 km (100 mi) from the site.

Regional Geology

The sediments that fill the Fraser River Valley were deposited during glacial, interglacial, and postglacial periods over the last 120,000 years. The Project tunnel and shafts were excavated through these sediments, as well as through fill deposits that mantle both river banks. The thickness of the interglacial, glacial, and postglacial sediment layers varies along the tunnel alignment because of uneven deposition and localized erosion. Based on results of the geotechnical exploration program, seven Tunnel Soil Groups (TSGs) were defined to represent soils expected within the shaft and tunnel excavations and are described below.

- TSG0 – Silty Clay to Clayey Silt: Clays and silts deposited in proglacial lakes during glacial advance. This unit is characterized by artesian conditions.

- TSG1 – Silty Sand, Sand and Gravel, and Silty Clay: A till-like, very dense or hard, poorly sorted, heterogeneous mixture of clay, silt, sand, gravel, cobbles, and boulders. Like TSG0, it also has an artesian piezometric level.

- TSG2 – Silty Clay: Overlies TSG1 and typically consists of soft to very stiff silty clay to clayey silt with varying plasticity. Cobbles were encountered infrequently throughout this soil group.

- TSG3 – Gravel: A relatively thin and flat-lying gravel bed forms the upper contact of TSG2 along almost the entire tunnel alignment. The gravel layer is compact to dense, and contains variable amounts of sand.

- TSG4 – Sand with Gravel, and Silt Interlayers: Clean sand containing a trace to some gravel. The sands are very loose to very dense, although they are typically compact. Scattered within TSG4 are layers and lenses of silts and clays. Cobbles were also encountered infrequently.

- TSG5 – Peat, Sand, and Silt: Loose to compact (or firm to stiff) interbedded sand, silt, and amorphous and fibrous peat with wood fragments.

- TSG6 – Fill: Fill deposits that mantle alluvial deposits along both the north and south Fraser River shorelines. The fill is a heterogeneous mixture of compact to dense, sand, gravel, cobbles, and organics.

The tunnel was driven through TSG1 and TSG2, whereas the shafts were excavated through TSG1 to TSG6 (see Figure 2).

Groundwater

Hydrogeological conditions at the Project area are controlled by the combined influences of local topography, complex geology, and Fraser River tidal cycles. The upland areas that rise more than 100 m (330 ft) above the Fraser River, both north and south of the Project area, are regional-scale topographic features that are also regionally significant groundwater recharge areas. The water table below the upland areas is elevated approximately 30 to 50 m (100 to 165 ft) above the Fraser River level, and groundwater generally flows toward the Fraser River. Groundwater flow is physically constrained by the presence of relatively low permeability sediments below the Fraser River floodplain, resulting in zones of highly pressurized groundwater. This pressure induces a partial upward groundwater flow gradient.

Piezometric levels above the ground elevation were recorded at various depths in the till-like sediments and in the underlying silt and clay deposits. These levels are indicative of artesian conditions at the south shaft site and to a lesser degree at the north shaft site.

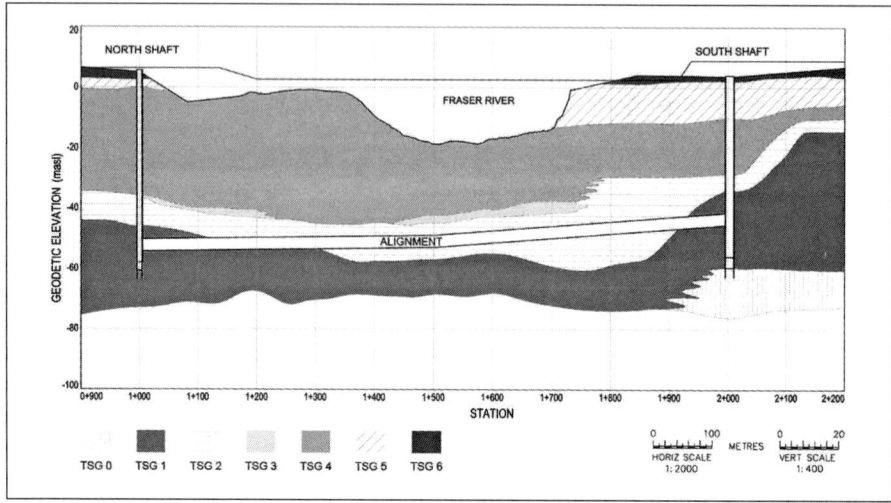

Figure 2. Tunnel geological profile

KEY DESIGN REQUIREMENTS

At the highest level, the key design requirements include:

- Crossing must remain operational at full capacity following a Maximum Credible Earthquake (MCE) with a return period of 10,000 years and a peak horizontal ground acceleration of 0.7g.
- Crossing must be located below future river scour level associated with a 1 in 500-year return period flood event.
- Crossing must be sized to accommodate future growth in water demand.

Seismic

The most challenging design requirement was seismic loading conditions. As part of its ongoing seismic resilience program, MV intends to be able to provide potable water to the entire region at winter demand levels shortly after a major earthquake. To meet this objective, system modeling has indicated that five marine crossings, including Port Mann, must remain operational at full capacity after an MCE event. In the case of Port Mann, site-specific ground response analyses were undertaken to quantify the magnitude and pattern of ground deformations resulting from the design earthquake scenarios. Of particular interest were the profiles of permanent ground deformations caused by earthquake-induced soil liquefaction occurring towards the Fraser River and the resulting interaction with the shafts and tunnel.

The ground response analyses were undertaken using 2D numerical methods of analyses (FLAC) that included a geotechnical model that extended approximately 1.5 km (0.9 mi) across the river, including both river banks, and about 200 m (650 ft) on either side of the shoreline. The model's lower boundary was extended to firm ground (glacial till) encountered at depths varying from 35 m (115 ft) on the south bank to in excess of 50 m (165 ft) on the north bank. The free-field permanent ground displacements obtained from the FLAC analyses were applied to the shafts and the tunnel in the SAP2000 structural model via constant displacements applied to the end of nonlinear soil springs. These free-field displacements were very large, exceeding 6 m (19.6 ft) at both the top of the south and north shafts. The free-field displacements used in the analyses were taken at the end of the shaking for the MCE ground motions considered. As a result, the permanent shaft and piping design allows over 2 m (6.6 ft) of differential displacement due to soil liquefaction between the north valve chamber and tunnel below. Both the concrete shaft and steel water pipe are designed to yield during the MCE, but to remain functional after the event.

Scour

During detailed design, hydrologic analyses were carried out to determine design river scour depth. Factors affecting river scour include general scour due to occurrence of extreme floods, periodic scour due to formation of bedforms at the river bottom, long-term degradation caused by river dredging, and local scour associated with existing structures in the river. Predicted design scour levels were estimated by combining all of the factors for the 1 in 200 and 1 in 500 year return period flood events. Analyses results indicated a maximum design scour level of El. −31 m (−102 ft) at the river's deepest part for the 1 in 500 year return period flood event—approximately 12 m (39 ft) below current river bottom elevation.

Future Demand

The Port Mann Water Supply Tunnel was designed for a design life of 100 years. The 2,135 mm (84 in.) diameter water main in the tunnel was sized to meet future growth in demand from communities south of the Fraser River, and it more than doubles the capacity of the existing 1,220 mm (48 in.) diameter crossing. Valve chambers located at the top of each shaft include piping and valves that connect into the existing 1,220 mm (48 in.) diameter land section as well as piping and valves for a 1,525 mm (60 in.) diameter future connection to meet growth requirements.

CONTRACTING APPROACH

The Project presented several unique conditions and risks, including:

- Soft soils capable of liquefying during an earthquake
- A deep tunnel alignment requiring deep slurry wall construction and tunneling through soft ground under high groundwater pressures
- Concrete tremie slabs at the base of each shaft
- Ground improvement at the shaft break-out and break-in locations using slurry wall panel methods
- Tunneling utilizing sophisticated, pressurized-face TBM equipment under the Fraser River
- Potential to carry out TBM inspection and maintenance under compressed air pressure over 6 bar
- Environmentally sensitive construction sites

These factors translated into stringent requirements and a complex design needed to meet, in particular, the post-MCE operating requirements. As a result, Project specifications were more prescriptive than would normally be the case. Some Project components have extremely tight tolerances. These unique constraints and conditions required specialized construction techniques and equipment, and proven experience on behalf of the contractor's workforce to execute the works. To meet these challenges, an experienced tunneling contractor, together with experienced specialist subcontractors, was required. To ensure the construction team selected for this unique and complex project could competently complete the work, MV decided to modify its procurement process, from its more common low bid approach.

The construction contract for the Port Mann Water Supply Tunnel was procured using a two-stage process involving a Request for Qualification (RFQ) followed by a Request for Proposal (RFP). The intent was to short-list up to five qualified contractor/ subcontractor teams during the RFQ stage, then submit an RFP to each of the short-listed teams. This approach allowed for increased flexibility in evaluating and selecting the best qualified team to perform the high-risk and specialized work associated with this project.

The RFQ/RFP strategy differs from a more conventional two-stage contracting approach, whereby contractors are prequalified based on their past experience and performance only, followed by an Invitation-to-Tender (ITT), after which the work is awarded to the lowest compliant bidder. MV saw several disadvantages in using the ITT approach: the requirement to award to the lowest price bidder, no ability to negotiate, and bidders can change their teams between the RFQ and ITT stages. This can also occur for the RFQ/RFP approach; however, the likelihood is reduced, since the proponent is required to submit methodology statements during the RFQ and RFP stages. These statements are included as part of the evaluation criteria in short-listing (during RFQ stage) and selecting the successful contractor (during RFP stage). The contractors are required to hold their teams from the RFQ to the RFP stage, or accept the risk of not being accepted if they change their teams at the RFP stage. Another advantage to an RFP approach is that it allows more flexibility in negotiating terms and conditions of the contract. By allowing this flexibility, it mitigates the concern over disqualifying a suitably qualified contractor if that contractor disagrees with a certain (perhaps minor) condition of the contract.

The RFP required submission of more detailed technical information, as well as a detailed cost proposal. A point rating system was used to score both technical and financial criteria for each proposal. MV then entered into negotiations with the proponent having the highest overall ranking. This allowed both parties (MV and contractor) to explore opportunities for risk reduction, methodology improvements, risk sharing, and contractual enhancements. As well, it provided the opportunity for the parties to get to know each other better, and at an early stage begin building trust, an effective team, and a positive working relationship. In the case of the Port Mann Water Supply Tunnel Project, the negotiations took approximately three months but resulted in more effective risk sharing and construction methodologies.

RISK MANAGEMENT/RISK SHARING

Public utilities such as MV are risk averse organizations. Therefore, it was a difficult decision to move forward with such a complex and challenging tunneling project, but it had been clearly demonstrated that a tunnel was the only approach that could meet the stringent design criteria for the new crossing.

The approach from the outset was to understand the risks associated with the Project and then reduce, eliminate, or share them fairly. It was desirable that the party or parties in the best position to deal with any residual risk should take it on. During the detailed design process, a comprehensive quantitative risk assessment and management analysis was undertaken. This involved development of a draft list of project risks (risk register) covering all aspects and phases of the Project. Several workshops were then held throughout the Project duration with industry and project-specific experts to refine the register—adding and deleting risks, providing clarity around each specific risk, developing probabilities for each risk, and estimating potential cost impacts. Mitigation strategies were assessed for each risk, leading to further revisions to the risk register. In the case of Port Mann, a large number of risks were identified during late detailed design. A probabilistic risk analysis was then undertaken via a Monte Carlo simulation. This resulted in a probability distribution curve for project cost and schedule. MV selected an 80th percentile target value (20% chance of cost-exceedance) for budget purposes, which translated into an appropriate contingency value to include in the project budget based on the identified risks. Where possible, ownership of these risks was assigned to the party that could best manage them. For example, procurement risks were taken on by the owner, and delays in construction equipment delivery were assigned to the contractor.

As it turned out, this risk assessment process did an excellent job of identifying which risks could be mitigated and what those mitigation strategies should be, which risks could not be mitigated and estimating an appropriate contingency for those, and assisting with decisions around which party should take on each risk and how that should be reflected in the contract documents. However, it is important to note that a risk register and related probabilistic analysis is only one component of an effective risk management program. For the Port Mann project, a number of additional strategies were implemented that contributed to the successful completion of this project, including:

- Involving knowledgeable designers and construction managers—a team with extensive experience in underground design and construction
- Establishing an independent Technical Review Board of industry leaders during design and construction that reported directly to the owner
- Using a Geotechnical Baseline Report (GBR) to clearly define the boundary for establishing changed site conditions
- Establishing a Dispute Resolution Board (DRB) to resolve serious disputes early on
- Using a partnering approach to develop effective relationships throughout the entire Project team
- Using an alternate procurement method to select a highly qualified construction contractor

During the entire project, the DRB was used only once, to help address an issue around a unique change in tax law. All other disputes were resolved by the Project team.

MANAGEMENT STRUCTURE

The Project management structure was carefully selected to ensure adequate checks and balances were included when addressing the complex design and construction issues that were anticipated to arise during construction. The overall project responsibility rested with the owner, Metro Vancouver. Head office MV representatives included a project sponsor from the senior management team and a lead senior engineer. MV site staff included two engineers, one of whom was the owner's point of contact between head office and the site team. The detailed design team (Ausenco, McMillen Jacobs Associates, Golder Associates) provided engineering services during construction in the form of full-time and part-time site staff, depending on the phase of the construction work.

A construction management firm, Hatch Infrastructure (formerly Hatch Mott MacDonald), was contracted to oversee the entire construction phase, with all communication between design team and contractor being handled directly by the Hatch management team on site. This team included staff with tunneling expertise that effectively liaised with both the design team and the contractor. Hatch also reviewed the designer's cost estimate and constructability aspects of the Project before contract award and then participated with the rest of the Project team in the contractor procurement process.

In addition, MV established a construction-phase Technical Review Board (TRB) that consisted of internationally recognized engineering experts. MV used this Board to obtain independent professional advice when challenging issues that arose during the work. As a result of the TRB's independence and the separation from the design team and the construction managers, better decisions were made, which contributed to the success of the Project. Occasionally, some major decisions had to be made directly by MV staff. These were supported by input from the TRB.

DESIGN CHALLENGES

The key design requirements, described above, and the Project location in an active seismic zone presented design challenges and special design considerations that the team had to develop for the various work elements. For the north and south shafts, the lateral spreading anticipated to occur during a seismic event was of the order of several meters. This necessitated development of a specific, yet innovative, design approach to accommodate these ground movements while also ensuring that the facility remained operational. The design approach for the shafts is summarized below:

- Unreinforced slurry wall panels for shafts up to 60 m (200 ft) deep in challenging ground. The shafts needed sufficient embedment in TSG1 in order to remain stable during and after an earthquake.
- The slurry walls formed the initial lining of each shaft. Once the shaft was excavated, a final, cast-inplace lining was formed within each shaft. This final lining was designed to provide the long-term resistance to earth and water pressures as well as the resistance to seismic design pressures. Because of the design sensitivity to structural stiffness of the final lining, even small changes suggested by the contractor had to be reviewed by the design team. Significant changes to the design that would result in increased stiffness of the final liner could not be considered.
- Coupled vs. decoupled initial and final linings—north shaft final lining needed to be decoupled from the slurry wall panels. This was done by installing a slip layer between the two linings.
- Extremely tight diametric tolerances between initial and final shaft linings. Extensive industry research was required to develop an attainable vertical tolerance on the slurry wall panels. The verticality tolerances were 0.4% and within typical practice limits for slurry wall panel construction.

Regarding the South Shaft, artesian groundwater pressures was a significant component of design.

- Groundwater was a key consideration in shaft design. The high piezometric levels affected slurry wall depth and raised concerns about bottom stability and slurry wall construction. Dewatering was considered to reduce the artesian water pressure in TSG1, but was not considered feasible due to the large estimated volumes of water (4,000 liters/minute per well), issues related to discharge (permits, treatment, and location), and concerns over ground settlement caused by dewatering.

- Calculations showed that it was possible for the slurry pressure to exceed the artesian water pressure at the bottom of the slurry wall, provided a heavy slurry was used and a small starter berm at the ground surface was constructed. For the South Shaft, specific gravity (SG) of the slurry on the order of 1.1 to 1.15, combined with a slurry level of about 1.5 m above the existing ground surface was estimated to provide a slurry pressure about 10% greater than the water pressure at the shaft. A starter berm at the North Shaft was not necessary, although the higher SG slurry was required.

Developments during the engineering analyses on the ground behavior during the MCE led to a greater understanding of the extent of soil liquefaction and the degree of lateral soil displacement at depth. This resulted in the tunnel vertical alignment being lowered significantly below the liquefiable soils to ensure that the structural shafts were able to remain functional after the MCE event. For the tunnel itself, the primary design issues related to the carrier pipe performance. These issues included:

- Development of backfill grout around pipe for confinement in seismic events: Provided required strength while exhibiting properties required for large-scale pumping and backfilling operations.
- Grout ports: Designed and fabricated to be fully watertight (i.e., zero infiltration and exfiltration).
- Seismic requirements: Compressive and tensile behavior within tunnel pipe due to ground deformations during an MCE event resulted in special welding requirements, which were developed and included in the Contract Documents.

CONSTRUCTION CHALLENGES

Each element of the Project presented specific challenges during construction. The site team, ably led by the McNally Aecon JV, successfully addressed and resolved each of these challenges, ultimately leading to the Port Mann Water Supply Tunnel being brought into service in early 2017. For the shafts, the construction challenges comprised:

- Concrete overpour control: Strict design tolerance for permanent lining dimensions required a lot of milling of the slurry wall surfaces. Higher than designed concrete strength made removal of overpour difficult (as well as placement of secondary panels difficult).
- Placement of the shaft tremie slab went smoothly. However, preparation work of the keyways and removal of large chunks of concrete overpour were extremely time-consuming.
- Development of TBM break-in and break-out blocks at depths of 60 m (200 ft): Ground replacement techniques were proposed by the contractor and consisted of constructing break-in and break-out blocks using hydromilling equipment and backfilling with low-strength concrete.

The tunnel excavation and lining process experienced several challenges, including:

- TBM launch was carried out through a reach of bouldery ground, presenting groundwater control issues together with accelerated mechanical wear/ breakdowns in the tunnel drive's early stage.

- Due to the limited work space in the South Shaft, the TBM launch sequence had to be undertaken in smaller increments than is typical, essentially assembling and launching one gantry car at a time.

- Constructability and logistical challenges with TBM internal confines under high hydrostatic pressure conditions: Limited work space, difficult access for repairs, hyperbaric intervention.

- Maintaining the tunnel alignment was challenging when trying to maintain required face pressures in mixed ground (clay/till interface).

- Tail void grouting was attempted but stopped because of line plugging issues. As a result, annular grouting of the segmental lining was carried out through the segments. This was highly successful with a fully grouted and dry tunnel evident on completion of tunnel excavation.

- Over-excavation: Difficulty maintaining high face pressures in mixed ground led to instances of higher muck weights than expected.

- Several interventions were required to repair mechanical failures, unblock the screw conveyor, remove boulders from the cutterhead, remove blockages of cutterhead openings and replace worn tools. All interventions were completed under free air. However, one intervention towards the end of the drive required a ground freezing operation because of a combination of poorer ground conditions and mechanical issues that did not allow safe depressurization of the chamber. Alternatives to hyperbaric interventions were not explicitly defined in the Contract Documents; however, acceptable approaches were advanced through the contractor's ingenuity.

- Break-out into the North Shaft was complicated by the need to balance high hydrostatic pressures. Tight alignment tolerance at tunnel eye was required due to the extensive array of structural reinforcement. Several alignment checks utilizing gyroscopic equipment were required to accurately confirm the TBM position prior to breakthrough.

Following on from completion of the tunnel excavation, installation of the 2,135 mm (84-in.) carrier pipe commenced (see Figure 3). Challenges associated with this operation included:

- While the cellular backfill operation was successfully carried out, development of an acceptable grout mix design and construction measures to control heat of hydration proved challenging. Removal of standing water prior to placement was critical to quality of backfill placed.

- Vertical alignment had to be adjusted to fit within the as-built tunnel alignment while still meeting minimum clearances and grades.

- Pipe ovality: During steel pipe installation, it was observed that several pipes had "squatted," typically shortening in the vertical direction. This led to the establishment of a circularity monitoring program, and daily pipe circularity measurements for select pipes. Stulling was installed where necessary to prevent further deformation. Upon completion of the backfill program, it was observed that several squatted pipes had slightly regained circularity, which was thought to be due to hydrostatic pressures from the backfill acting at the pipe springline, and counteracting the squat.

- Horizontal alignment: During steel pipe installation, work area restrictions caused by welding and other equipment resulted in the steel pipe alignment survey being completed after pipe welding was complete. The survey results determined that the pipe horizontal alignment exceeded the alignment tolerance in three locations. Analysis concluded that the additional horizontal eccentricity did not appreciably affect the pipe stability for the design level earthquake load condition. The pipe alignment was accepted as is.

Figure 3. Pipe installation photo

- Likely due to alignment difficulties with the north shaft bottom elbow, the north shaft steel pipe length vertical alignment tolerance was exceeded. To address this, the shaft pipe was surveyed, then a miter cut was made to bring the following pipe piece (the top elbow) back into alignment.

KEY LESSONS LEARNED

The Port Mann Water Supply Tunnel provided several challenges, from planning through design development to contractor procurement and construction. These challenges were all successfully overcome, providing Metro Vancouver, as well as the designers and contractors involved, with valuable lessons for the design and construction of future complex tunneled marine crossing projects. A summary of the key lessons learned, and the benefits afforded, is presented below:

- Carry out a comprehensive geotechnical exploration program. Avoid having to go back and carry out additional borings by adopting this approach early in the design development phase.

- Best value procurement of the tunnel contractor is the preferred way forward. MV intends to continue with RFQ/RFP procurement approach for future tunnel projects. Every effort should be taken to encourage skilled, qualified, experienced contractor teams to bid the work.

- Ensure that the procurement process fully considers skills and qualifications of specialist subcontractors. Consider having subcontractor representation at negotiations.

- Encourage contractor innovation and flexibility in selection of means and methods while still maintaining the functional requirements of the project. Ensure the design allows provision of a TBM that addresses key constructability issues, not just minimum geometry requirements.

- Transition zones (geology) are the most challenging. Mixed ground, especially with boulders is the most demanding on excavation equipment. Characterizing these zones prior to construction should be a priority. Contingency measures should be well developed for addressing potential issues preconstruction (such as interventions, blockages, and breakdowns).

- Success of cellular concrete backfill is a function of a good mix design and adequate care and precautions during placement to remove water, maintain temperatures, and monitor volumes.

- Establish tunnel pipeline alignment tolerances that are reasonably achievable. This was critical because of the high seismically induced compression and tension strains in sections of the alignment. When sections of the pipeline fell out of tolerance, the design team had to spend considerable time during construction re-evaluating the original seismic analysis of the pipeline.

- Make sure that contractor-designed elements (break-out blocks, temporary work to construct valve chambers, etc.) do not adversely affect the performance of the final works.

- Strive for straightforward, practical design solutions by spending the right amount of time early in the design phase when seismic behavior is being analyzed. This lesson has already been adopted on a subsequent MV marine crossing where a very comprehensive analysis of the lateral ground deformations was performed early in the detailed design, resulting in more manageable displacements for structural design element.

- Wherever possible, avoid the use of nonstandard design elements (in this case the "slip sheet" membrane between initial and final South Shaft linings). Use conventional steel and concrete materials to the full extent possible.

- During the Port Mann contract negotiations, significant agreement was attained on equitable sharing of geotechnical risk. This should be considered for all tunneling projects.

- Continue and expand on benefits achieved from the use of a technical review board and collective partnering approach such as those that were seen on the Port Mann Water Supply Tunnel Project.

ACKNOWLEDGMENTS

The authors would like to acknowledge the many contributions of Ausenco, McMillen Jacobs, Golder Associates, Hatch Mott MacDonald, and their subconsultants, the McNally Aecon JV, as well as Metro Vancouver staff in the design and construction of this project.

Instrumentation and Monitoring

Chairs

Gregory Rogoff
McMillen Jacobs Associates

Moussa Wone
DC Water and Sewer Authority

Material Flow Reconciliation: Risk Management for Pressurized Face Tunneling

Ulf G. Gwildis ▪ CDM Smith
John E. Newby ▪ CDM Smith

ABSTRACT

Overexcavation during pressurized face tunneling is a risk factor with respect to unplanned surface deformation and damage to existing installations above the tunnel. Risk management typically includes surface and subsurface deformation monitoring of critical areas identified during planning and design. However, continuously reconciling material flows during the mining process is a risk management tool that allows recognizing and correcting causal trends before effects are observed. This paper describes the reconciliation methods for different TBM types, provides project examples from Seattle area tunnels, and discusses risk management implementation approaches involving the owner, designer, contractor, and third party asset holders.

INTRODUCTION

For estimating the surface impact of tunneling operations in urban areas several analytical approaches exist, e.g., geometric analysis (Cording and Hansmire, 1975), closed-form solution (Bobet, 2001), and numerical modeling. All these approaches require an estimate of the ground losses that will occur beyond the geometrically required excavation volume. The tunnel design has to make the assumption that the ground loss can be limited to a certain percentage—e.g., 0.5% based on recent TBM performances or a higher number using a more conservative approach—when specifying a pressurized face tunnel boring machine (TBM) to be operated in closed mode and tail void grouting. The calculated surface deformations then allow determining the impact on existing structures above the tunnel and pre-tunneling construction needs such as ground improvement or reinforcement of foundations. In urban areas the tunnel design typically specifies the installation of instrumentation such as multipoint extensometers (MPBX), inclinometers, surface, structure, and utility settlement points for monitoring deformations. Contractual definition of action limits for the measurements of each instrument requires the contractor to take mitigating action if a threshold has been reached.

However, at most tunnel projects the spacing of the instruments exceeds the length of a mining-and-ring-building step by the order of a magnitude or more; also, installation of instrumentation may not even be considered for tunnel sections identified as low risk regarding deformation impact. Another factor limiting the value of surface deformation monitoring is the time delay between the tunnel excavation and causally linked effects observed at the ground surface such as settlement troughs or—in cases of significant overexcavation—sinkholes. This time delay varies between hours and months depending on the depth of the excavation below ground surface (bgs), the composition of the overburden, and other factors (Figures 1 and 2).

Operating the TBM at sufficiently high face support pressures in soils with no standup time under atmospheric conditions is an essential requirement for controlling the excavation process and according monitoring is standard practice. For example, during

Figure 1. Sinkhole in sandy overburden few hours after TBM overexcavation at about 50 m (150 ft) bgs

Source: Sound Transit
Figure 2. Chimney in overburden with till and clay layers several months after TBM overexcavation at about 50 m (150 ft) bgs

Earth Pressure Balance TBM excavations, the time graphs of the bulkhead pressure sensors help identifying instances where insufficient mining pressure below static pressure was applied, which is the case when the pressure signature shows increasing values when the advance is stopped for the ring-building step (Alavi et al., 2016). However, ensuring mining pressures above static pressure is a necessary but not a sufficient requirement for minimizing the risk of overexcavation and associated risk of deformation damage to existing infrastructure.

Monitoring of the various material flows of the tunnel excavation process—tunnel muck or backflow slurry, soil conditioners and/or bentonite slurry, tail void grout, and secondary grout—and reconciling the data with the geometrical dimensions of the TBM and the tunnel liner—diameter of the cutterhead including gauge cutters, shield diameter extrados, and tunnel liner diameter extrados—is an additional risk management tool that goes beyond the tracking of mining pressures and allows addressing the limitations of a tunnel project's deformation monitoring program.

RECONCILIATION FOR EARTH PRESSURE BALANCE TBM (EPBM)

The general approach is to compare the excavation volume of each excavation step or 'push' (tunnel face area A_f times advance length L) converted into weight via the average in-situ density of soil materials in the tunnel face (δ_f) with the measured weight of the excavated muck (M_{bs}) during that push under consideration of soil conditioner (M_{sc}) and bentonite quantities (M_b) added, which results in the reconciliation value RV_1 (Equation 1). As the TBM advances incrementally, the balance of filling the tail void with grout is later added for the same ring location by amending Equation 1 by the difference of the quantity of tail void grout injected (M_{tvg}) minus the theoretical weight needed under consideration of the space between excavated diameter and the area of the tunnel liner extrados (A_{tl}) and the tail void grout density δ_{tvg} (Equation 2). The resulting reconciliation values RV_2 is then used for determining if additional measures such as secondary grouting are required for compensating for potential overexcavation.

$$RV1 = \frac{\text{Mbs} * 100}{(\text{Af} * \text{L} * \delta\text{f}) + \text{Msc} + \text{Mbi}} \qquad [\%] \qquad (1)$$

$$RV2 = \frac{\left(\text{Mbs} + \text{Mtvg} - \left((\text{Af} - \text{Atl}) * \text{L} * \delta\text{tvg}\right)\right) * 100}{(\text{Af} * \text{L} * \delta\text{f}) + \text{Msc} + \text{Mbi}} \qquad [\%] \qquad (2)$$

Reconciliation can be performed either by weight or by volume, requiring the densities of the various materials involved in the process. While the densities of added materials are usually well defined, the in-situ density of the excavated soil volume needs to be estimated based on soil laboratory test data or geotechnical baseline values in conjunction with a realistic assessment of the composition of the tunnel face, or based on muck density measurements under consideration of a bulking factor. Muck samples taken during a push for purposes such as checking the adequacy of soil conditioning can also be a valuable data source for verifying face conditions, especially in variable geology (Gwildis et al., 2009).

Irrespective which approach is chosen, accurate recordings of the chainage from start to end of a push are required that are linked via time stamp to weight (or volume) measurements of the muck extruded from the conveyor screw. Furthermore, time-stamped recordings of the quantity of added conditioners, bentonite, tail void grout, and secondary grout are needed for each ring location.

A system sketch of EPBM material flow data that provides the input to the data processing spreadsheet for the approach illustrated by Equation 1 is shown by Figure 3.

During the TBM drive, the tunnel contractor may modify the earth pressure distribution in the excavation chamber, e.g., by purging any air volumes originating from soil conditioning and building up in the plenum at the crown elevation as indicated by apparent densities determined from the bulkhead pressure sensors and their relative elevations (Mori et al., 2016). The contractor may also drive the TBM in a manner that the excavation chamber filling is intentionally compacted at the end of a push, e.g., in preparation of a weekend stop, by halting conveyor screw extrusion of muck while continuing to advance the TBM. In order to account for these variations in tunneling procedure, the reconciliation sheet should include columns showing reconciliation values for lengths of two or three neighboring rings, allowing e.g., to recognize compensating effects of pre or post underexcavation events.

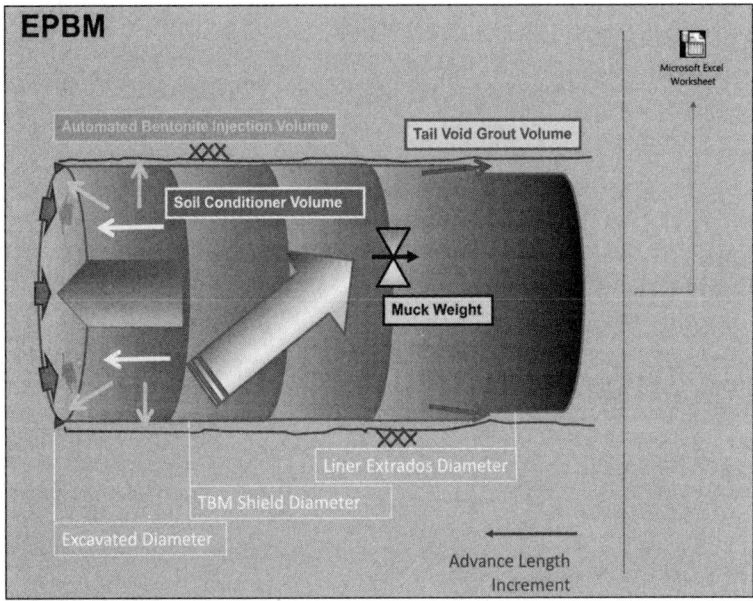

Figure 3. EPBM material flow system graph

RECONCILIATION FOR SLURRY TBM (STBM) AND HYBRIDS— ADDITIONAL CONSIDERATIONS

An STBM uses a slurry circuit for transporting the excavated soils from the excavation chamber of the TBM to a separation plant at the ground surface. The dry weight of the excavated material (M_d) can be determined by comparing flow rate and specific weight of the bentonite slurry that is pumped to the excavation chamber (Q_{sa}, γ_{sa}) with the material-loaded backflow slurry to the separation plant (Q_{sr}, γ_{sr}) (Bochon et al., 1999). The general relationship is shown by Equation 3 (specific weight of water, γ_w).

$$Md = \sum_{t=0}^{n}(\frac{\gamma sr - \gamma w}{1 - \frac{\gamma w}{\gamma sr}} * Qsr) - \sum_{t=0}^{n}(\frac{\gamma sa - \gamma w}{1 - \frac{\gamma w}{\gamma sa}} * Qsa) \quad \text{[kN]} \qquad (3)$$

The TBM data acquisition system output typically provides cumulatively increasing weight values as the TBM advances during a push. Completely emptying the excavation chamber from the material excavated during each push is important for obtaining accurate results. Evidently, measuring the weight of the solids after separation from the slurry at the plant could be another approach that should result in the same value.

After having determined the dry weight of the excavated material, the reconciliation for STBMs generally follows the procedure based on weight described for EPBMs after modifying Equations 1 and 2 for the specific input parameters of STBMs. Same as for EPBMs, verifying the in-situ density estimates of the soils constituting the tunnel face by taking muck samples in regular intervals is recommended. During a STBM operation, the sampling is done at the discharge of the separation plant and may require some reconstituting if there are various discharge points, e.g., at centrifuges and shakers. A system sketch of STBM material flow data sources is provided with Figure 4.

If hybrid TBM types are utilized in granular soils with little fines content that include material flow processing elements such as separation plants or slurryfiers inside the

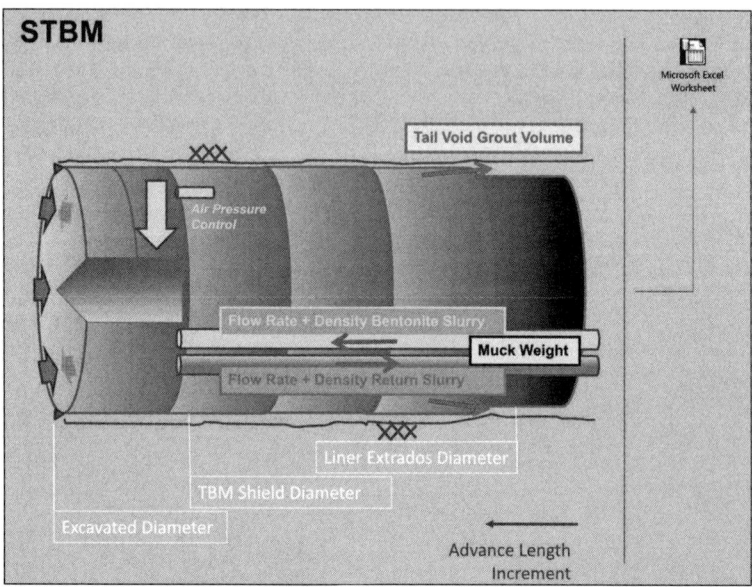

Figure 4. STBM material flow system graph

shield (Maidl et al., 2016), further customizing of the reconciliation based on the procedures described herein is required.

SEATTLE AREA TUNNEL PROJECT EXAMPLES

Over the past three decades more than a dozen TBM-driven tunnels with excavation diameters between 3.7 m (12 ft) and 17.5 m (57.5 ft) have been or are being constructed in the Seattle metropolitan area with its unique geological setting of interlayered glacial, glaciomarine, and non-glacial deposits (Gwildis et al., 2014). The geotechnical tunneling conditions—characterized by frequent and sometimes rapid transitions between full face and mixed face conditions ranging from overconsolidated, hard, cohesive soils with some stand-up time under atmospheric conditions to fast raveling or flowing granular soils, combined with coarse components such as cobbles and boulders found frequently at certain types of geologic deposits and boundaries—have proven to provide for challenging TBM mining and contributed to some spectacular ground surface impact events. Seattle area experiences have helped to recognize the importance of monitoring and tightly controlling TBM material flows for mitigating the risk of infrastructure damage in a highly developed urban area. The following project examples illustrate an evolutionary process of increasing focus on TBM material flow data, initially used for post-impact event forensic investigations and later for more comprehensive monitoring efforts during TBM mining advance by the tunnel contractor and also the owner and, depending on the risk level and the risk impact level, affected third-party asset holders.

Beacon Hill Light Rail Tunnel

The Beacon Hill Project included 1,310 m (4,300 ft) of twin tunnels and an underground train station in glacial and lacustrine, glacially overconsolidated deposits. The tunnels were completed in 2008 utilizing a 6.5 m (21.2 ft) diameter EPBM that was equipped with two conveyor belt scales for muck weight measurements. 18 months after completion of tunneling, the discovery of a <1 m (<3 ft) wide hole on a property

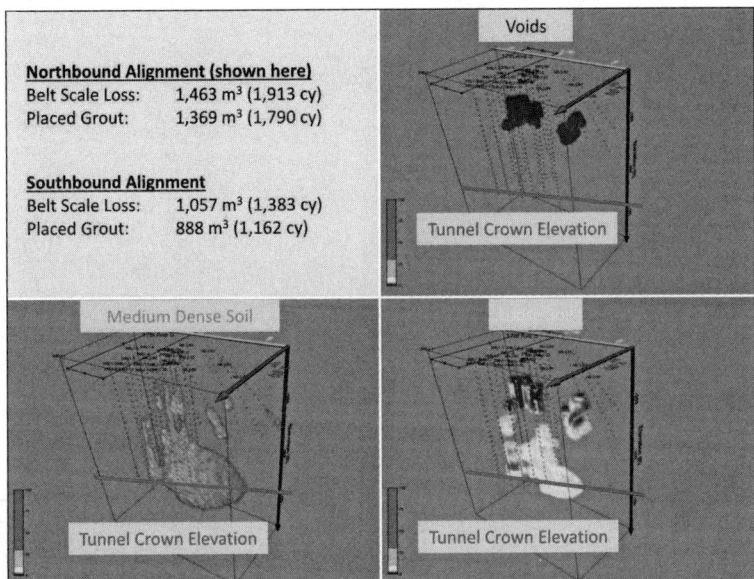

Source: Sound Transit

Figure 5. Post-event comparison of belt weight data and surface grouting volumes

above the tunnel alignment (Figure 2) triggered a multi-phase ground loss exploration and grouting program for remediating the situation (Figure 5). Post-construction review of the TBM belt scale data, considered erroneous at the time of construction, resulted in an interpreted ground loss volume of 2,520 m³ (3,296 cy) which correlated within 2% of the injected grout volumes plus delineated disturbed soil volumes (Robinson et al., 2012). This project served as an early demonstration of the value of conveyor belt scales as a tool for muck flow monitoring for subsequent tunnel projects in the area.

Brightwater Effluent Conveyance Tunnel

Construction of the Brightwater Conveyance System between 2006 and 2011 included four tunnel drives and one tunnel completion drive in glacially overconsolidated gla-cial and non-glacial deposits as well as an alluvial valley filling utilizing 2 STBMs and 2 PBMs excavating a total distance of 20.3 km (12.6 mi.). The operational parameters of the TBMs that were provided continuously to the owner's data base and visualiza-tion system included in all cases the excavation weights per ring, derived from con-veyor belt scales of the EPBMs and slurry circuit flow meter and density sensors of the STBMs. The construction management teams and the geotechnical consultant to the owner used the data for evaluating the surface impact risk for each tunnel align-ment section as well as for forensic analyses in cases where surface deformation was suspected as potentially tunneling-related.

An example of a forensic investigation conducted during construction is a sinkhole of about 6 m (20 ft) in diameter that had opened overnight where a residence's drive-way had been. The operational parameters and data of the STBM indicated that the slurry yield values had been increased but overexcavation was allowed to continue over a distance of 4 to 5 rings (Figure 6). Because the overburden at this location consisted of granular soils with only thin layers of fine-grained soils interbedded, the ground loss volume moved quickly upwards, breaching a perched aquifer base layer

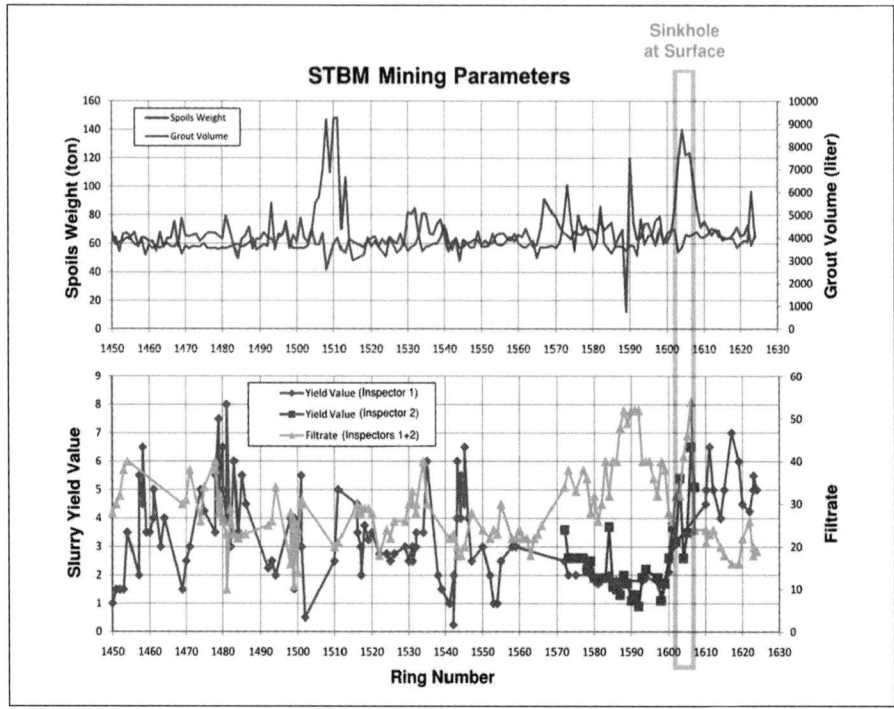

Figure 6. Forensic analysis of STBM parameters of tunnel section with sinkhole occurrence

as documented by the change of hydrostatic head of a nearby monitoring well, and then creating a sinkhole congruent with the ground loss pattern at the tunnel elevation (Figure 1).

Recent and Ongoing Seattle Area Tunnel Construction Projects

Tunnel construction projects that have been completed since 2013 or are still on-going are the Sound Transit tunnels for light-rail system extensions of the University-Link (U220: 3,475 m (11,400 ft) of twin tunnels; U230: 1,183 m (3,880 ft) of twin tunnels) and the Northgate Link (N125: 5617 m (18,430 ft) of twin tunnels) and WSDOT's Alaskan Way Viaduct Replacement Tunnel for SR 99 (2,825 m (9,270 ft)) undercrossing the Seattle waterfront and downtown areas. In all cases EPBMs were utilized with excavation diameters ranging from 6.5 m/6.6 m (21.2 ft/21.6 ft) for the light rail tunnels to 17.5 m (57.5 ft) for the double-deck highway tunnel. These tunnels were advanced predominantly in variable, glacially overconsolidated, glacial and non-glacial deposits. The alignments of these tunnels run generally in North-South orientation, i.e., parallel to the direction of the repeated glacial advances and associated erosion patterns, resulting in less variation of tunnel face conditions than experienced during tunneling along the East-West oriented alignments of the previously referenced projects.

The generally longer reaches of relatively uniform face conditions in fine-grained lacustrine deposits, granular till and outwash deposits, or mixed face conditions allow for better trend analyses. Relative minima and maxima of the reconciliation values determined for each ring are more likely related to specifics of the TBM operation than just an expression of abruptly changing face conditions. At the locations of extensometers

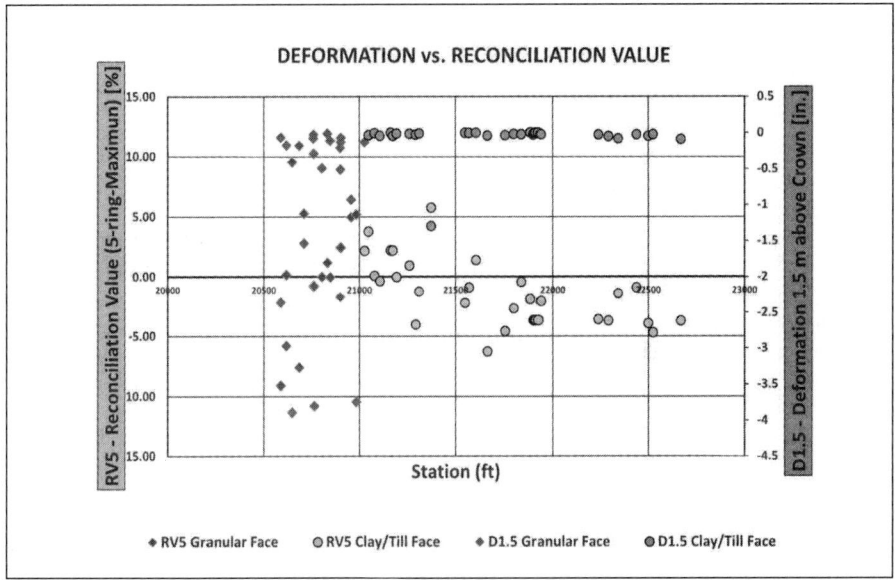

Figure 7. Correlation of reconciliation results and extensometer movement

measuring the tunneling-induced deformations directly (1.5 m (5 ft)) above the crown, correlating the measured deformations with the reconciliation values allows an understanding of the fluctuation range for each set of face and crown soil conditions as well as an understanding of how well the excavation process is being controlled (Figure 7).

DISCUSSION AND RECOMMENDATIONS

Material flow reconciliation values summarize for each mining advance step the many variables of the interaction of the TBM operation process and the geotechnical conditions encountered, the latter being characterized in the Seattle area by highly variable glacial geology. This results in a high data variance and requires scrutiny when determining the average in-situ soil density of the tunnel face as a calculation input parameter. The following approaches can be considered:

1. Using a density value based on the geotechnical baseline information (ideally verified by muck samples taken in regular intervals);

2. Using a density values derived from muck sample descriptions and density testing under consideration of a bulking factor;

3. Using a density value back-calculated and extrapolated from past reconciliation values set to zero (which assumes that excavation of the past rings did not include significant over- or underexcavation);

4. Or, a combination of the above.

Figure 8 illustrates an example of material flow reconciliation on a per-ring basis performed by independent parties for the same tunnel section in variable glacial geology, using. in addition to slight procedural variations. different approaches for estimating the in-situ density of the tunnel face as input parameter. The graph illustrates the general fuzziness of the reconciliation values as evident by the data variance; however, the graph also provides clear information regarding locations where the risk of overexcavation having occurred may be considered as increased.

In soil types with some stand-up time, a ring where reconciliation values indicate over-excavation followed by one where reconciliation values indicate underexcavation may have evened out the mass balance. If that is not the case or if overexcavation trends extend over neighboring rings, a more thorough investigation and consideration of corrective action such as secondary grouting may be in order. Reconciliation values should always be interpreted in the context of geotechnical data and other TBM opera-tional data, primary among them the TBM bulkhead pressure sensor data, whose timegraphs indicate if the TBM is operated at higher or lower pressures than required for face support, and whose gradient over the height of the face ('apparent density') indicates if phenomena such as air bubble build-up from soil conditioners used in EPBM mining are a factor. Another question to look into is if bentonite injected dur-ing extended EPBM stops for maintaining face pressure does or does not need to be included in the reconciliation calculations.

It is our conclusion that despite the fuzziness of the results, TBM material flow rec-onciliation has evolved as an essential risk management tool for pressurized-face tunneling projects in urban areas that is indispensable for tracking, assessing, and pro-actively responding to the risk of surface impact damage. The focus of this tool is on trend analysis and this tool must be supplemented with other tools such as monitor-ing of critical TBM operational parameters and deformation measurements at eleva-tions between the tunnel crown and the ground surface.

During the design phase, management of the risk of overexcavation should include specifying the equipment and data collection requirements for providing the basis for allowing material flow reconciliation. The construction contract should be set up that the contractor is required to provide the owner with real-time TBM operational data that include excavated material weights and to perform reconciliation on a per-ring basis.

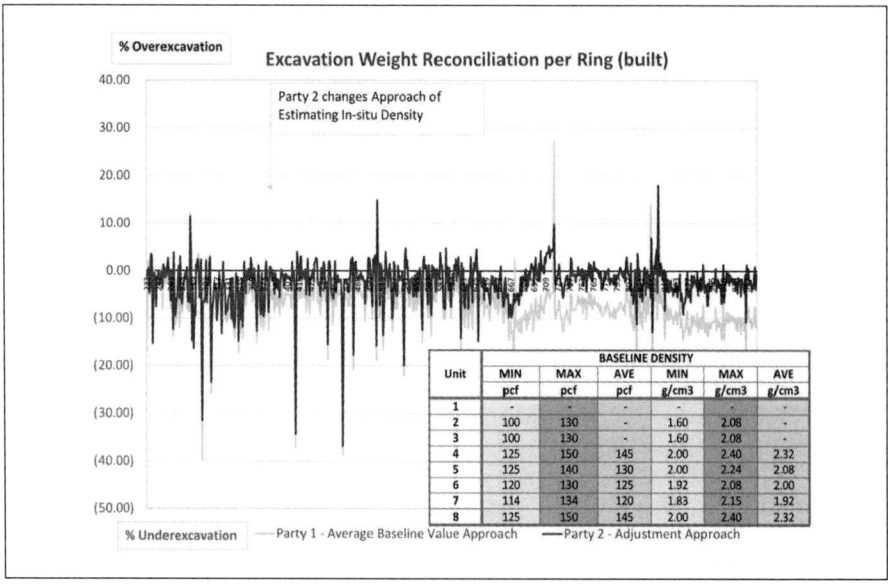

Unit	MIN	MAX	AVE	BASELINE DENSITY MIN	MAX	AVE
	pcf	pcf	pcf	g/cm3	g/cm3	g/cm3
1	-	-	-	-	-	-
2	100	130	-	1.60	2.08	-
3	100	130	-	1.60	2.08	-
4	125	150	145	2.00	2.40	2.32
5	125	140	130	2.00	2.24	2.08
6	120	130	125	1.92	2.08	2.00
7	114	134	120	1.83	2.15	1.92
8	125	150	145	2.00	2.40	2.32

Figure 8. Comparison of two independent reconciliation calculations using different approaches for estimating average in-situ densities of the tunnel face

During construction, management of the risk of overexcavation should include continuous material flow reconciliation by the contractor, who forwards the calculations and results to the owner. In an urban area with an increased risk of high-value impact, the owner and third party asset holders may consider performing their own independent material flow reconciliation. A mechanism to jointly compare and assess results and data trends such as regular meetings is recommended.

In addition, the owner and third party asset holders may consider using risk projection tools for the tunnel sections ahead that consider factors such as face conditions, overburden conditions, and also the contractor's performance up to that point. These risk projection tools assist in outlining high-risk areas, thereby managing expectations and minimizing the likelihood of surprises.

REFERENCES

Alavi, E., Frank, G., Hagan, B., LaVassar, C., Mori, L., Dugan, D., and Capka, R. 2016. Comparison between calculated and actual face pressures in EPB TBMs—Case studies in Seattle, WA. *Proc. WTC 2016.*

Bobet, A. 2001. Analytical solutions for shallow tunnels in saturated ground. *Journal of Engineering Mechanics.* 127(12):1258–1266.

Bochon, A., Rescamps, Y., and Chantron, L. 1999. La détection des anomalies d'excavation au tunnelier a pression de boue: Mèthode mise au point sur le chantier EOLE. *AFTES 1999.*

Cording, E.J., and Hansmire, W.H. 1975. Displacements around soft ground tunnels. *Gen. Report of Session IV,* 5th *Panamerican Conf. on Soils Mech. and Foundation Eng.* 571–632.

Gwildis, U., Maday, L., and Newby, J. 2009. Actual vs. baseline tracking during TBM tunneling in highly variable geology. *Proc. NAT 2014.* 250–262.

Gwildis, U., Robinson, R., Sage, R., and Sowers, D. 2014. Geotechnical planning advances for TBM projects in glacial geology. *Proc. NAT 2014.* 707–717.

Maidl, U., Comulada, M., Turolla Maia, C.H., and Di Dio Pierri, J.C. 2016. Shield tunneling in pure sands. *Proc. WTC 2016.*

Mori, L., Mooney, M., Alavi, E., Frank, G., DiPonio, M. 2016. Assessment of EPB soil conditioning on two TBMs by using apparent density. *Proc. WTC 2016.*

Robinson, R., Sage, R., Clark, R., Cording, E., Raleigh, P., and Wiggins, C. 2012. Conveyor belt weigh scale measurements, face pressures, and related ground losses in EPBM tunneling. *Proc. NAT 2012.* 65–72.

Managing Ground Control with Earth Pressure Balance Tunneling on the Alaskan Way Viaduct Replacement Project

Edward J. Cording ▪ University of Illinois at Urbana-Champaign
Jack T. Nakagawa ▪ Tutor Perini Civil West
Justin J. McCain ▪ Tutor Perini Civil West
Anthony F. Stirbys ▪ Tutor Perini Civil West
David Sowers ▪ Washington State Department of Transportation
Jorge Vazquez ▪ Dragados USA
Cody Z. Painter ▪ WSP | Parsons Brinckerhoff

INTRODUCTION

Washington State Route 99 (SR99) extends along Seattle's Elliott Bay waterfront on the two-level Alaskan Way Viaduct, a reinforced concrete structure built in the 1950s. The structure will be replaced by cut and cover approach structures and a single, 1.8 mile (2.8 km) long, 57.4-ft (17.5-m) tunnel accommodating a double deck structure with two-over-two traffic lanes and a breakdown lane, longitudinal ventilation ducts, and emergency passenger egress. The design-build contractor, Seattle Tunnel Partners (STP), a joint venture of Dragados USA and Tutor Perini, selected a Hitachi-Zosen earth pressure balance machine (EPBM).

The first 1,500 feet of TBM advance, designated as the South End Settlement Mitigation Plan (SESMP) section, was located beneath the project construction yard and immediately adjacent to the Alaskan Way Viaduct. The section was used to gain depth to the tunnel crown, from 15 to 70 feet, and to monitor ground movements and develop control procedures prior to tunneling under the Viaduct and structures in Pioneer Square. A row of drilled shafts, both secant piles and spaced drilled shafts, were placed between the Viaduct foundations and the tunnel to protect the viaduct against ground movements. In this section, the tunnel advance began with the crown in hydraulic fill placed early in the last century on loose alluvial soils of former tidal flats, and in underlying glacial sands and till below 30-foot depth.

Results of monitoring of ground movements and TBM performance in the first 1,000 feet of the SESMP section in November, 2013 was summarized by Cording et al, 2015. Tunneling was halted on December 6, 2013 and during 2014 and 2015 an access shaft was installed, the TBM was advanced into the shaft and the main bearing was removed and replaced, among other modifications. In early 2016, the reassembled TBM was advanced through the last 500 feet of the SESMP section, including two test sections in which vertical and lateral ground movements were monitored that confirmed that ground movements were being controlled and surface settlements prevented. During this period, the capabilities of the STP teams of managers, TBM engineers, operators, quality control, safety, and geotechnical monitoring, and WSDOT construction management were drawn together to fully implement a coordinated program of monitoring of ground and TBM performance, and control, review, and adjustment of TBM operation parameters.

On April 29, 2016, the TBM began its advance beneath pile foundations of the Alaskan Way Viaduct located 15 to 30 feet above the tunnel crown in glacial lacustrine clays and outwash sands. Advance continued beneath brick bearing wall structures in

82

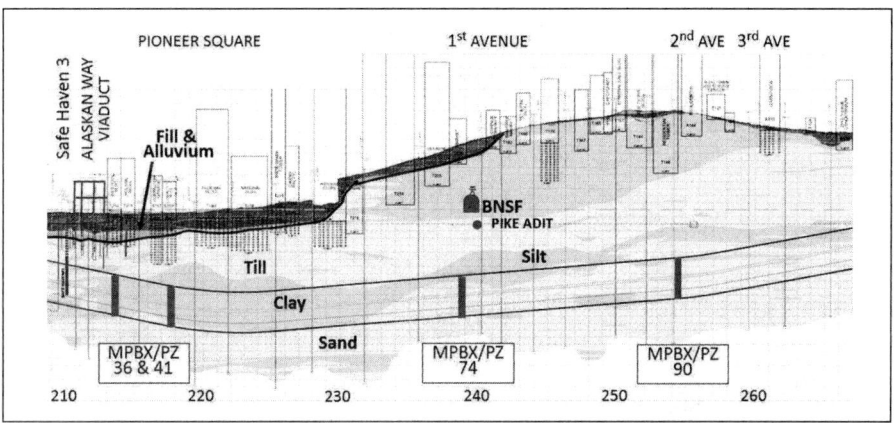

Figure 1. Profile in central 5,000 feet of tunnel showing geology and key instrumentation sections

Pioneer Square with the tunnel face largely in very hard lacustrine clays with a continuous pro-glacial outwash sand aquifer at invert level. Advance continued to the North, beneath First Avenue, the BNSF tunnel, and buildings of downtown Seattle, through mixed ground consisting of dense glacial till and outwash sands, and lacustrine silt and clay (Figure 1).

Over the entire tunnel alignment, STP established an extensive program of instrumentation. The data was available to project participants on the web-based GIS software, GeoScope, provided and maintained by SolData (now known as SolData/Sixense). Structures in the potential influence zone of the tunnel were monitored with automatic structure monitoring points (ASMPs) as well as level surveys. Level lines consisting of liquid levels, ASMP prisms, or tiltmeters were placed within sensitive structures located above the tunnel alignment in Pioneer Square, and within the Burlington Northern tunnel, and the Pike St. Adit.

Ground movements and groundwater pressures were monitored using multiple position extensometers with vibrating wire piezometers placed in boreholes spaced an average of 50 feet apart along the entire alignment. The primary goal was to identify and evaluate the location and causes of ground movement and changes in groundwater pressure around the advancing tunnel boring machine. Particular focus was placed on relating data from this unprecedented array of deep extensometers and piezometers to key TBM operating parameters controlling ground movement such as face pressures, bentonite injection and pressures in the steering gap, and grouting around the segmental lining at the tail of the shield.

TUNNELING COORDINATION

The AWVR project has developed an integrated approach to managing the dynamic tunneling operation through team collaboration and open lines of communication. Figure 2 illustrates the interactions.

Daily, at the beginning of the shift, the STP team and the Construction Monitoring Task Force (CMTF) meet review the previous day's tunnel operations and the plans for the day. The CMTF is chaired by the Contractor's tunnel manager and includes the Contractor's project managers, TBM engineers engaged in quality control, maintenance, supply and operations, the geotechnical monitoring team, safety engineer,

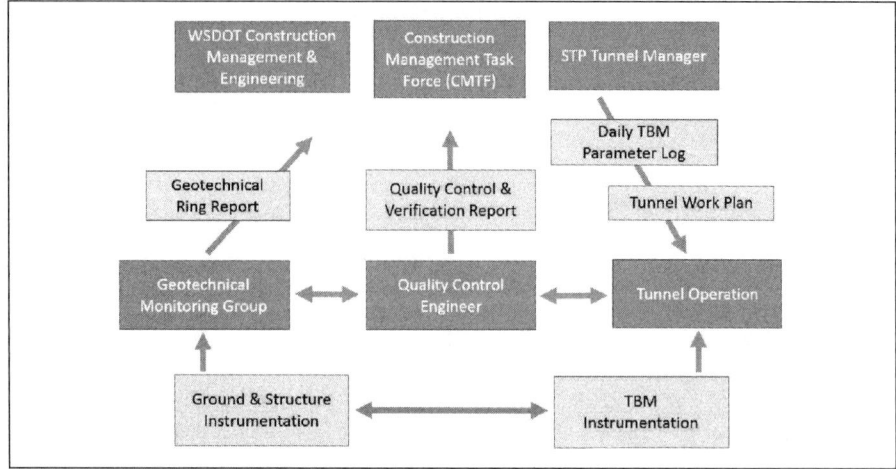

Figure 2. Tunnel control interaction chart

and Washington State Department of Transportation (WSDOT) engineering and construction managers and chief inspector. The CMTF is responsible for planning, implementing procedures, and making recommendations to control ground movement. WSDOT provides technical input and quality verification, or audits, of STP's quality plan. The CMTF reviews the previous day's tunneling activities that are summarized in the Quality Control and Verification report and reviews any changes in the Daily TBM Parameter Log (Figure 3). The Log prescribes targets and ranges for key operating parameters and is provided to TBM operations. The tunnel manager, in directing the TBM operation, notifies CMTF participants of issues that develop in real time and the changes or mitigations that may be initiated, and solicits input.

A tunnel work plan and a Daily TBM Parameter Log, which prescribe targets and ranges for key operating parameters, is provided to TBM Operations (Figure 3). The CMTF reviews the Daily TBM Parameter Log and any recommended adjustments.

Throughout each shift, the TBM data is evaluated by Quality Control and TBM Operations teams and summarized in automated reports following completion of each 2-meter advance. The results for the previous 24 hours are summarized in the Quality Control and Verification report presented in the at the daily meetings. On each shift, a geotechnical engineer from the Geotechnical Monitoring team monitors instrumentation that includes deep ground movements and groundwater pressures, TBM pressures and manual and automated survey displacements of structures, and provides updates to Quality Control and TBM Operations staff in the tunnel. The instrumentation data within 200 feet of the TBM is summarized in a Geotechnical Ring Report emailed to project staff and management after each shove. The results are presented the next day in the project and CMTF meetings. Geotechnical Ring Reports include the following:

- Ground conditions, description of muck, and muck density
- Geological profile showing TBM and location of multiple position extensometers/piezometers
- Continuous time plots of displacement of Automatic Structure Monitoring points (ASMPs) on structure exteriors, displacements of level lines within structures, and survey leveling at the ground surface and on structures

Daily TBM Parameter Log		Report #:	135	revision #:	1
		Start Station:	25838	Current Ring #:	977
		Date From:	12/9/16	Time From:	5:00 AM
		Date To:	12/10/16	Time To:	4:59 AM

TBM PARAMETERS	Target Advance Speed(mm/min):	40	Target Total CHD Force (kN):		45,000		
	Target CHD Torque (%):	40%	Target Total Thrust Force (kN):		155,000		
	Target CHD Rotation Speed (rev/min):	1.0	Penetration Rate (mm/rev):		40		
WORKING PRESSURE	**TOP Target Earth Pressure (bar):** (average of sensors 11 &12)	**3.2**	Green Range	3	to	3.5	
	Keep System Settings (bar): (based on average of sensors 11 &12)	Lower	3	Upper	3.05		
GROUT PRESSURE	TOP Target Grout Line Pressure (bar):	5.3	Green Range	5	to	5.8	
	MID. TOP Target Grout Line Pressure (bar):	6.1		5.8		6.6	
	MID. BOT. Target Grout Line Pressure (bar):	7.1		6.8		7.6	
	BOTTOM Target Grout Line Pressure (bar):	7.5		7.2		8.0	
GROUT VOLUME	Grout Volume Target (m3):	26	Green Range	22.1	to	31.2	
GAP	TOP Target Gap Line Pressure (bar):	4.0	Green Range	N/A	to	8.0	
	Gap Volume Target (m3):	4.14	Green Range	N/A	to	5.4	
	Gap Keep Settings (bar): (based on gap sensors #6 and #3)	Lower	2.80	Upper	2.85		
SOIL CONDITIONING	Planned Ground Conditioning Recipe:	Recipe 1 SLF (3.0%) plus Rehosoil 211 (0.5%)					
	FER Target: 12	FER Range: 10 to 14	FIR Target: 85%	FIR Range: 40% to 100%			
	Slump Range: 5	to 7	Concentration Target: 3.0%	C(%) Range 3% to 5%			
	Target. Belt Scale Weight (ton):	GBR Density: 1.85 Bank m³: 478	884	Comments:	Density/Slump tests to be carried out at least once per shift. Except when consistency is too wet that it will be once per ring		

Figure 3. Daily TBM Parameter Log, 12/9/2016 to 12/10/2016

- Continuous time plots of extensometer (MPBX) displacements, focusing on the deep anchor displacements occurring as the TBM approaches and passes the extensometer location
- Continuous time plots of dynamic groundwater pressure changes measured by piezometers around the advancing TBM, correlated with TBM face (chamber bulkhead) pressures of the conditioned muck
- Time plots of pressures in the bentonitic fluid-filled steering gap around the shield exterior, measured by earth pressure cells on the arch and crown of the shield

The aggregate of these operational procedures, descriptive plans and open lines of communication among contractor management, tunnel operations, and owner alike, have contributed to effective coordination of the extensive machine and geotechnical instrumentation data reviewed in real time by the various project teams, and contributed to consistent TBM operation and control of ground movement. The open channel review and discussions among project teams, as well as the daily review in project and CMTF meetings, has created an environment in which information is shared and information-based decisions are timely executed.

TBM CONTROL

Figure 4 is a profile and section of the AWVR EPBM showing instrumentation used to monitor pressures and volumes for correlation with ground movements and groundwater pressures. Pressures are continuously maintained on and around the TBM from the face to the tail of the shield, both during and between advances of the TBM,

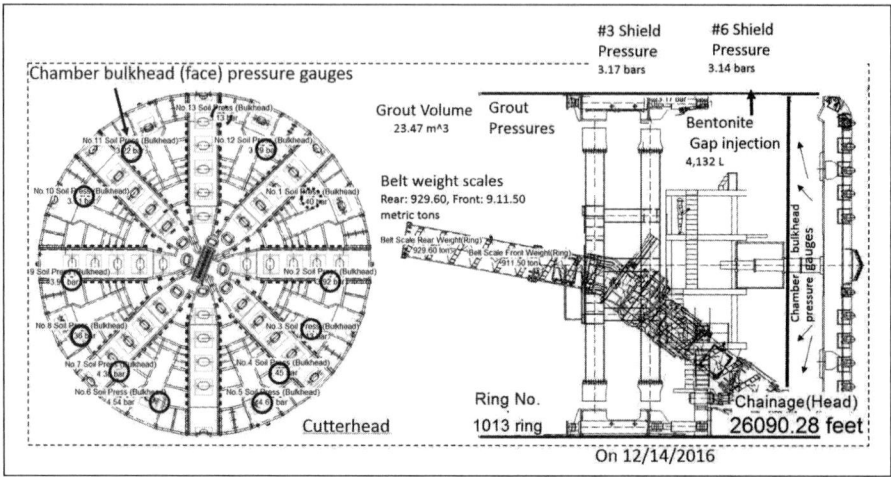

Figure 4. GeoScope readout of TBM functions

24 hours a day, seven days a week. Pressurized face shields are often described as preventing ground loss by pressurizing the face and continuously grouting under pressure through the tail as the shield is advanced. However, pressurized face shields also pressurize and fill the overcut gap around the shield body with conditioned muck or bentonite injection and prevent ground loss into the shield gap that would otherwise require special measures, such as compensation grouting to reduce settlements to acceptable levels. Thus, ground loss is minimized by providing a pressurized envelope around the entire shield: the face, the shield body, and the tail, until the tail grout around the segmental concrete tunnel lining sets. Filling and pressurizing the overcut gap around the shield as well as pressurizing the face and filling the tail gap has resulted in the excellent performance of modern pressurized machines in minimizing ground loss.

The AWVR EPBM has thirteen pressure cells mounted on the bulkhead of the chamber wall and is also equipped with six earth pressure cells on the arch and crown of the shield. Pressures near the pump in the bentonite injection lines to the shield body are also monitored. Figure 4 illustrates the machine functions that are monitored with SolData's GeoScope program which also provides the electronic readout of ground movement and piezometric pressures.

Face Pressure Control

The earth pressure gauges in the upper chamber of the TBM are used as the reference for control of the TBM pressures. In a full face of permeable, potentially flowing ground, balancing the water pressures in the upper face is the critical condition for minimizing ground loss. Figure 4 shows the location of pressure gauges on the chamber bulkhead, located behind the cutterhead. The differential water pressure between upper and lower pressure gauges (45 feet apart vertically) is 1.3 bars. The higher density of the conditioned muck produces a differential pressure in the range of 1.6 to 2.2 bars for material with specific gravity ranging from 1.2 to 1.7. Thus, face pressure balanced with water pressure at the top pressure gauge of the TBM would result in a pressure at the bottom gauge that is 0.3 to 0.9 bars above the water pressure.

For tunneling beneath Pioneer Square, with clay in the face and a sand aquifer at the invert, the critical condition was to balance pressure at the invert. In this case, the lower face pressure gauges were monitored and maintained above the measured piezometric pressures in the deep sand aquifer.

During the TBM advance, the upper face pressure is maintained 0.8 bars above the static groundwater pressure, to balance active earth pressures and provide a safety factor for variation in pressures. Face pressures during the advance were typically controlled within +/- 0.2 bars during tunneling beyond Safe Haven 3 (Figure 5). In the figure, pressure differentials between upper and lower gauges located 40 feet apart indicate that specific gravity of the muck was 1.2 during the shove and increased to 1.4 over the weekend. In this span of drive, with TBM face in predominantly granular material, conditioner had a Foam Injection Ratio (FIR, actual foam volume/volume of ground in place) of a typical 80% to 90%, and a Foam Expansion Ratio (FER, foam volume/foam solution volume) of 11% to 13%.

Piezometers placed with extensometers in boreholes at an average of 50-foot centers along the alignment provide a means of evaluating the dynamic groundwater pressures that are influenced by TBM operation pressures. Consistent control of face pressures during the advance is aimed at preventing the face and shield pressures from dropping below dynamic elevated ground water pressures in permeable soils.

Face pressures between advances were maintained by automatic bentonite injection into the chamber keep system. Between the 2-meter (6.56 ft.) advances, the targets for the upper earth pressure cells at the crown on the shield perimeter were set 0.2 to 0.15 bars below the target pressures during shove.

A vent line was extended into the top of the upper chamber, and was checked by venting throughout the advance to prevent accumulation of air in the chamber.

Reconciliation of muck weight and volume was carried out continuously during the shove. Two belt weight scales were used to compare muck weights with calculated weights based on the theoretical volume excavated and in place density of the soil. The muck weight vs advance was plotted continuously throughout the shove as a reference for the TBM engineers and operator to minimize over- excavation. Reconciliation of weight and volume is prepared and summarized for each ring advance, adjusting for the weight of material injected into the chamber and screw. Target range for excavated weight was +2% to −4% the theoretical weight. Typically, the weights fell within this range.

Shield Gap Pressure and Volume Control

With pressurized face TBMs, properly conditioned muck, particularly in fine grained soils with adequate space around the cutterhead, can flow from the cutterhead to the shield body and fill the overcut gap, preventing ground loss. However, it has also been observed that in sandy, gravelly soils the conditioned muck may not readily flow around the cutterhead and fill the gap. Injection of a bentonite mixture provides a consistent way to fill and pressurize the gap. This practice was developed to handle the larger ground losses on large diameter TBMs and is also important for limiting ground loss in smaller diameter EPBMs, particularly in granular materials in low cover situations. On the AWVR project, bentonite filling of the overcut was specified. STP developed the following procedure, which became standard practice:

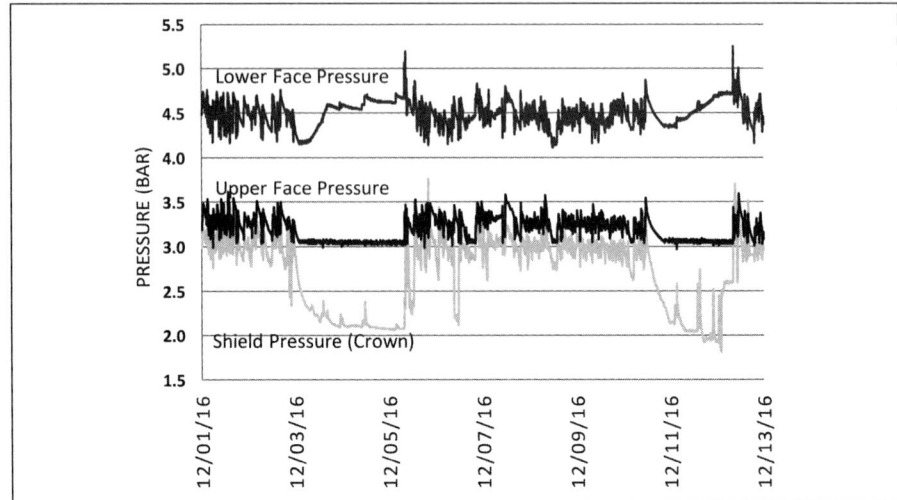

Figure 5. Time plot of upper and lower face pressures and shield pressures, 12/1/2016 to 12/13/2016

- Maintain shield fluid pressures, as measured by minimum pressure on the earth pressure gauge on the shield body. Target pressure is maintained between a typical minimum and maximum of 0.4 bar below and 1.3 bar above reference face pressure.

- Inject volume of bentonite to fill 30mm overcut: theoretical volume per ring is 4.14 m^3, with a target injection of 5.4 m^3.

- Pump pressure for bentonite injection set at a maximum of 6 bar, but is typically adjusted throughout the push by operators based on dynamic changes to face/shield pressure.

- Bentonite concentration injected is 7.0%. Where sandy soils have been encountered near the crown of the shield, polymer has been injected along with bentonite.

- Between pushes, bentonite is automatically injected into the shield gap to maintain minimum shield pressure described above.

Tail Grout

Monitoring of extensometer displacements throughout the drive has shown that the continuous injection of grout to fill the gap between the shield and the tunnel lining has prevented ground loss at the tail, and in some cases, caused a small heave.

Tail grout pressures are typically 2 to 2.5 bar above face pressure, however this can be adjusted by parameter log change, with larger changes occurring when ground conditions change (e.g. transition from predominantly clay to predominantly granular in-situ condition). Theoretical grout volume is 26.0 m^3 per ring, but target grout injection is determined by pressure.

Secondary Grouting

Grouting through secondary grout ports pre-installed in ring segments is performed once per week, and additional secondary grouting is performed if determined necessary by TBM management team/CMTF on review of excavation reconciliation and instrumentation data.

GROUND BEHAVIOR

Ground conditions in the 5,100 feet of the alignment tunneled from April 28, 2016 to December 15, 2016, are illustrated in Figures 1 and 6. The figure shows the extensometers/piezometers (MPBX/PZ) at four locations that were selected to illustrate observed ground behavior in clays and in sands, in the lower cover section beneath Pioneer Square and in the high (~200-foot) cover section north of Pioneer Square.

Multiple position borehole extensometers (MPBX) were installed at an average spacing of 50 feet along the entire tunnel alignment. The MPBXs have rods anchored at different depths below the surface with the deepest anchor located 5 feet above the tunnel crown to evaluate the location of any ground movement occurring ahead of, over, or behind the TBM shield. Vibrating wire piezometers (PZ) were installed at the bottom of each of the MPBX borings to measure groundwater pressures: both ambient conditions and the changes generated by the pressure wave around the advancing TBM.

The primary goal was to identify and evaluate the location and causes of ground movements around the advancing tunnel boring machine. Particular focus was placed on relating data from this unprecedented array of deep extensometers and piezometers to key TBM operating parameters controlling ground movement such as face pressures, bentonite injection and pressures in the steering gap, and grouting around the segmental lining at the tail.

Tunneling Beneath Alaskan Way Viaduct and Pioneer Square

The TBM advanced from Safe Haven 3 at Station 210+00 on April 29, 2016, passing 15 to 30 feet beneath pile foundations for four Alaskan Way Viaduct piers, then crossed Yesler Way, and advanced beneath century-old brick bearing wall structures in Pioneer Square where ground elevation was +16 feet, and tunnel depth 70 to 110 feet. Much of area was originally tide flats which were filled with sand displaced during regrades early in the early 1900s. At tunnel depth are glacially overridden soils; mixed sands and lacustrine clays at the Alaskan Way Viaduct, and hard lacustrine clays beneath Pioneer Square buildings. At and above the tunnel crown are lacustrine clays as well as till (material generally described as dense to very dense sand and gravel and silty clay, and cohesionless sand and gravel layers).

At and below the tunnel invert is a major continuous sand and gravel aquifer. TBM face pressures in the lower portion of the cutterhead chamber were maintained above the ground water pressures in the aquifer. A piezometer at Columbia Street, in the continuous sand formation below the tunnel invert, responded to face pressures when the TBM cutterhead was at two locations in Pioneer Square, indicating that the cutterhead had intersected the lower sand and gravel aquifer at those locations.

During tunneling beneath the Alaskan Way Viaduct and buildings in Pioneer Square, there was no discernable effect of tunneling on settlement of structures or the ground surface (as measured by the ASMP points on the structures, the liquid level lines within the structures, and the survey leveling of ground surface and extensometer heads). ASMPs typically recorded twice-daily variations in vertical displacement influenced by tidal changes that were on the order of ±0.05 inches.

Figure 6 is a plan showing extensometer/piezometer (MPBX/PZ) locations in Pioneer Square. The response of most of the piezometers located 5 feet above crown indicated that they were in clay soils, except for two of the piezometers, where piezometric levels tracking with the face pressures and shield body pressures provided evidence of permeable lenses (as shown in Figure 7 for MPBX/PZ 36).

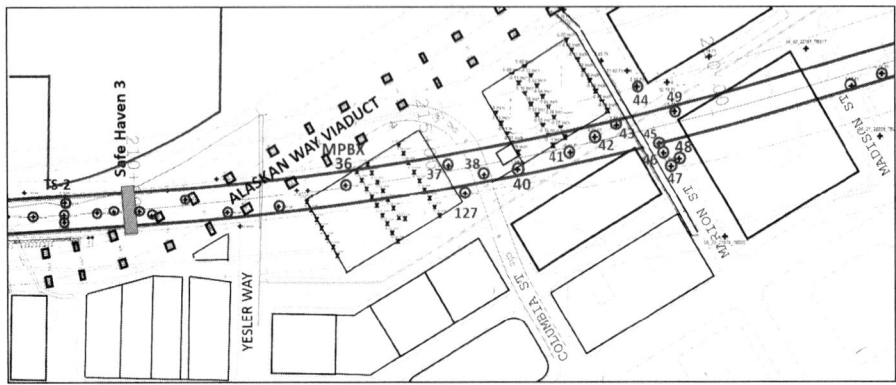

Figure 6. Plan view: tunnel layout, structures, buildings, and key instrumentation

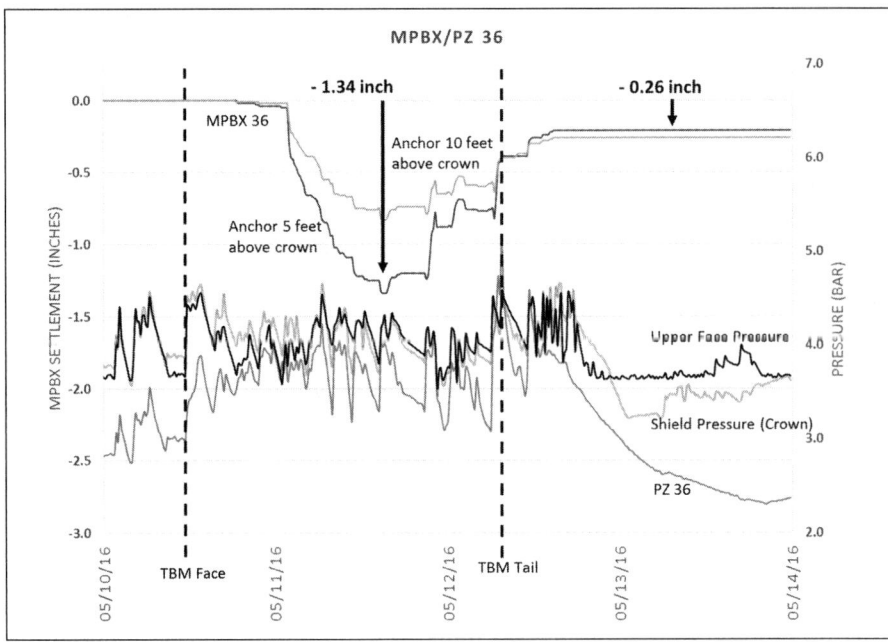

Figure 7. MPBX/PZ 36 displacements with TBM face and shield pressure during TBM passage

Case 1: Pioneer Square, Sand at Tunnel Crown

Figure 7 illustrates the type of response that occurs when sand lenses are in the tunnel crown. Piezometer PZ 36 responded to and tracked with face and shield body pressures. Face pressures throughout the shove were maintained above the dynamic groundwater pressure.

The deep anchors, 5 and 10 feet above the crown, showed a small, sharp settlement beginning 20 feet behind the face. Settlements flattened out over the shield at 1.3 inches for the anchor 5 feet above the crown. Heave occurring near the back of the shield and at the tail resulted in recovery of most of the settlement. Total settlement after the shield had passed was 0.2 inches. There was some increase in face

Table 1. Instrumentation Test Section 3 settlement and volume loss, tunnel depth: 90–100 feet

MPBX Number	Deep Anchor Settlement, 5 ft Above Shield, in.	Half Width of Assumed Trough, 5 ft Above Shield, ft	Volume Loss, 5 ft Above Shield		Percent Volume Loss 5 ft Above Shield
			ft^3/ft	m^3/m	
36	0.2	56	0.9	0.08	0.034%
37	0.01	56	0.05	0.05	0.002%
38	0	56	0	0	0

pressure as settlements leveled out and an increase in bentonite volume injected as the rebound occurred near the tail of the shield.

Table 1 shows total movement of deep MPBX anchors during/due to tunneling, and volume loss.

Case 2, Pioneer Square, Clay at Tunnel Crown

Inclinometers placed ahead of the TBM recorded lateral displacements toward the cutterhead that were typically 0.1 inches when the TBM approached within 10 ft. of the inclinometer. However, the deep extensometer anchors located 5 feet above the tunnel showed no settlement ahead of the face and negligible, but measurable settlement over the body of the TBM shield, typically in the range of 0.02 to 0.05 inches.

Figure 8 (MPBX/PZ 41) illustrates the type of response that typically occurred throughout Pioneer Square when clay was in the tunnel crown. Piezometric pressure in PZ 41 dropped 1 bar below ambient over the front portion of the shield and recovered over the back half of the shield and tail. At the same time, the deep anchor, 5 feet above the shield, displaced downward 0.04 inches and recovered to 0.02 inches behind the shield.

Unlike piezometers in permeable soils, the piezometers in the clay do not track with the face or shield pressures as the TBM passes. Clays do not drain freely, so the piezometers respond to stress changes occurring in the clay due to volumetric and shear stresses generated by the difference between overburden/at-rest pressures and the pressures applied at the face and around the perimeter of the shield. In Pioneer Square, the groundwater table was located near the surface causing the differential between the overburden pressure and the upper face/shield body pressures required to balance groundwater and active pressures was relatively small: overburden pressure was typically 5 to 6 bars, and upper face pressure was 4 to 4.5 bars. Settlement was minimized as a result of this low pressure differential, the high stiffness of the soil, and the pressurization and filling of the annular spaces around the shield and tail.

Table 2 shows total movement of deep MPBX anchor and volume loss during/due to tunneling.

Tunneling North of Pioneer Square Through Downtown Seattle

As the tunnel proceeded north of Pioneer Square up First Avenue, ground elevation increased to 160 feet and cover over the tunnel increased to 200 feet. Throughout, groundwater levels remained flat at elevation +5 to +10, near tide level in Puget Sound. With the increasing cover, the differential pressure between overburden pressure and the face and body pressures increased (12 bars overburden vs 3 bars face pressure).

The pattern of deep anchor settlement and piezometric pressures were similar to those observed during tunneling beneath Pioneer Square. The higher differential between the overburden pressures and TBM pressures resulted in some increase in

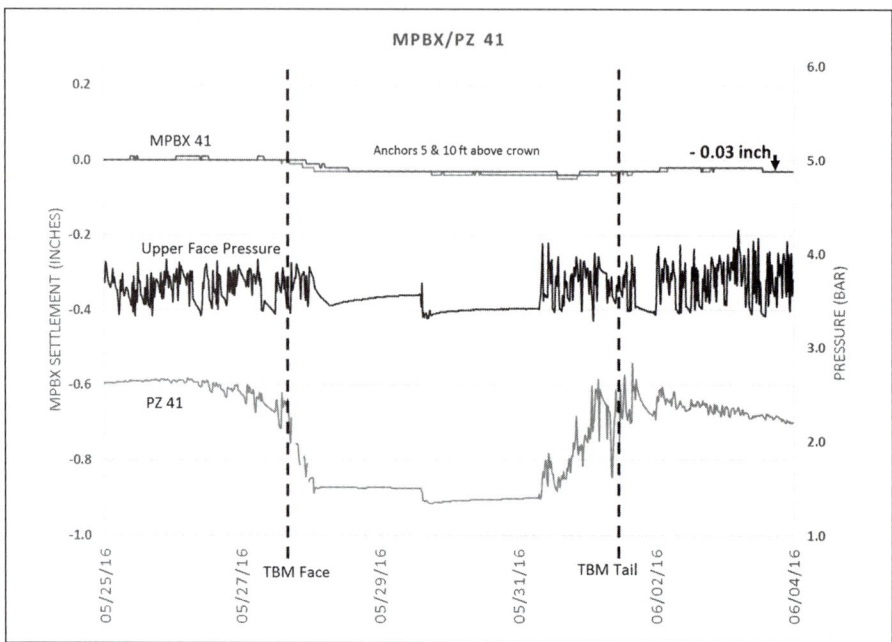

Figure 8. MPBX/PZ 41 displacements with TBM face and shield pressure during TBM passage

Table 2. Instrumentation Test Section 4 settlement and volume loss, tunnel depth: 100–110 feet

MPBX Number	Deep Anchor Settlement, 5 ft Above Shield, in	Half Width of Assumed Trough, 5 ft Above Shield, ft	Volume Loss, 5 ft Above Shield		Percent Volume Loss 5 ft Above Shield
			ft³/ft	m³/m	
40	0	56	0	0	0%
41	0.03	56	0.15	0.015	0.006%
42	0.01	56	0.05	0.05	0.002%
43	0.02	56	0.1	0.01	0.004%

deep anchor settlement to totals in the range of 0.2 to 0.6 inches, well below maximum settlement criteria because of the high stiffness of the ground and the complete pressurization and filling of annular spaces around the shield body and tail. Throughout, bentonite volumes injected into the shield steering gap during the advances were close to the target of 4.14 m³ per advance.

High permeability sand and gravel occupied an increasing portion of the tunnel face. Lacustrine clays continued to be present in the upper face of the TBM throughout much of the section to the vicinity of Station 245+00 to 250+00, where soils at the tunnel crown transitioned to silts and outwash sands and till-like soils (see Figure 1).

Case 3. Clay in Crown, High Cover, Approaching BNSF Tunnel

Presented are the differential settlements of the MPBX/PZ anchors and the piezometric pressures as the TBM approached the BNSF tunnel (Figures 9 through 11) Figure 9 is a plan and Figure 10 is a profile showing the five MPBX/PZs located prior to the BNSF tunnel, and two MPBX/PZs beyond the BNSF. From the deep anchor

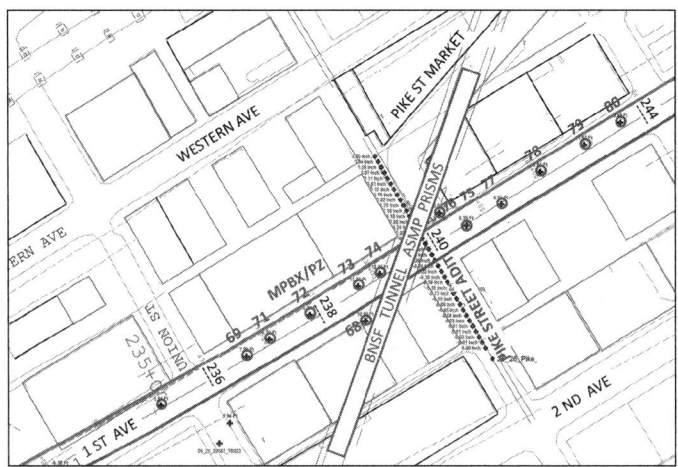

Figure 9. Plan view, BNSF Test Section with key structures and instrumentation

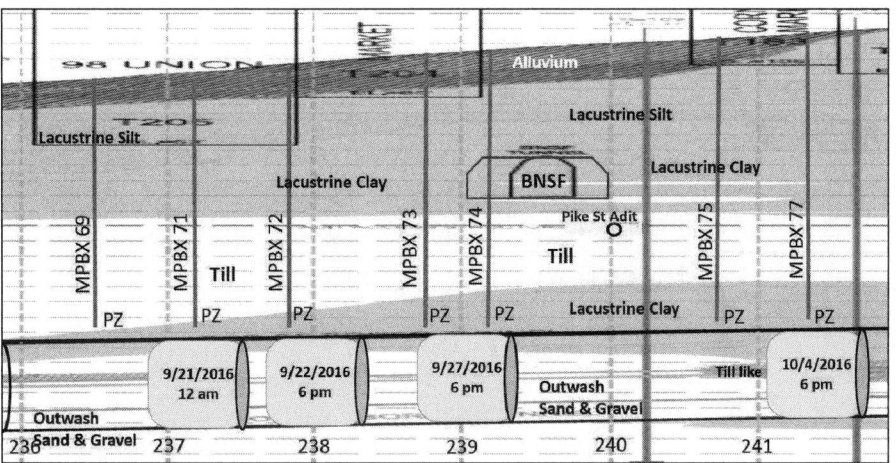

Figure 10. Profile, BNSF Test Section with key structures and instrumentation

displacements, estimates were made and reported of the settlements of the BNSF tunnel, which were anticipated to be below maximum settlement criteria and below any damage levels.

The deep anchors located 5 feet above the TBM had differential settlements (with respect to the extensometer head) of 0.1 to 0.5 inches. Surface settlements were estimated to be on the order of 0.1 inches and within the range of the variations in settlement of the structures above the tunnel, based on ASMP measurements at the surface. This value is added to the differential displacements of the extensometer anchors to obtain the total (Table 3).

A small amount of settlement occurred beginning up to 30 feet ahead of the face, as shown for MPBX/PZ 74 in Figure 11. The drop in piezometric pressure in the clay soils above the crown of the tunnel also began ahead of the face, with recovery of pressure over the shield.

Table 3. BNSF Test Section settlement and volume loss, tunnel depth: 180–200 feet

		Deep Anchor, 5 ft Above Crown		Estimated Volume Loss	
	Station	Differential Settlement, in.	Total Settlement, in.	Volume, ft³/ft	% Volume
MPBX 69	236+69	0.4	0.5	2.3	0.09%
MPBX 71	237+13	0.2	0.3	1.4	0.05%
MPBX 72	237+88	0.1	0.2	0.9	0.035%
MPBX 73	238+75	0.35	0.45	2.1	0.08%
MPBX 74	239+16	0.37	0.47	2.2	0.08%
MPBX 75	240+73	0.34	0.44	2.1	0.08%
MPBX 77	241+37	0.5	0.6	2.7	0.11%
BNSF Tunnel	239+00 to 240+00	(85 feet above TBM)	0.2	2.2	0.08%
Pike St Adit	240+00	(70 feet above TBM)	0.22	2.2	0.08%

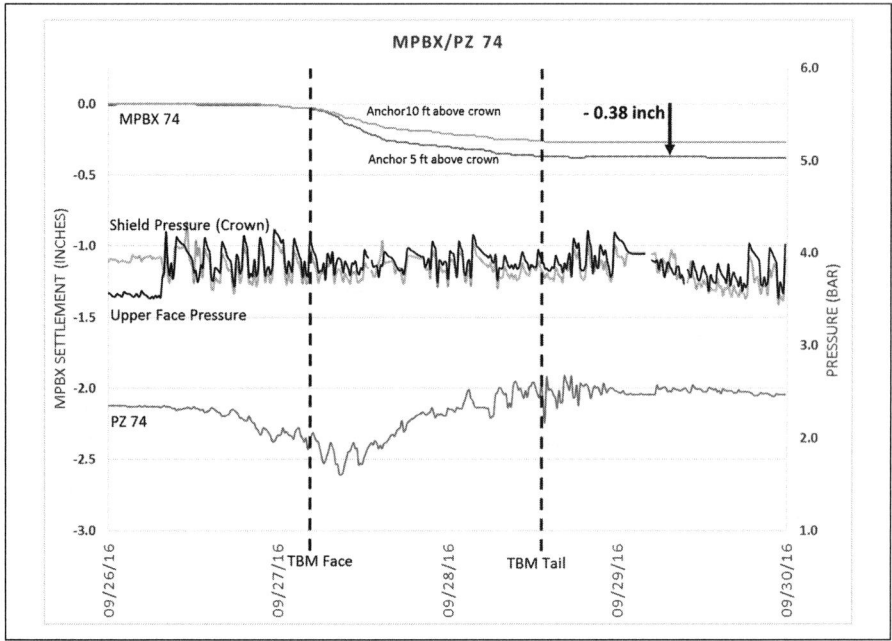

Figure 11. MPBX/PZ 74 displacements with TBM face and shield pressure during TBM pass

The piezometric pressure dropped in an undrained response to the total stress changes in the ground due to shear and volumetric stresses imposed during TBM advance.

Case 3, Continued: BNSF Tunnel Displacement

The actual settlements of the BNSF tunnel were close to predicted values, within contract-prescribed settlement criteria, and well below damage levels.

Figure 12 shows the vertical displacement and locations of the ASMP prisms in the BNSF tunnel for the half of the settlement profile west of the TBM centerline. Measurements are shown for October 4, 2016 after the TBM passed beneath the BNSF tunnel. Prisms are located on the walls and on the double track ties. Similar measurements were made in the BNSF tunnel to the east of the TBM tunnel centerline.

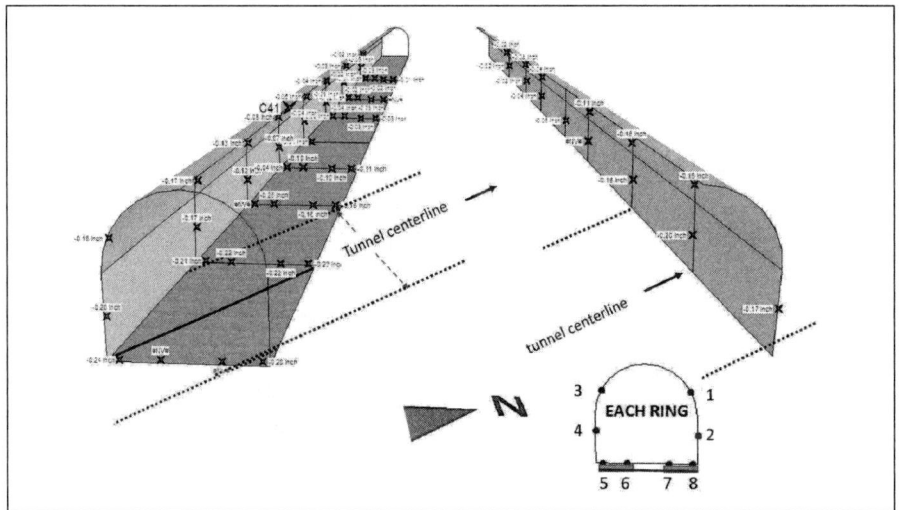

Figure 12. Vertical displacement measured by ASMP prisms in BNSF West Section (from GeoScope)

The TBM tunnel passes at a skewed angle beneath the BNSF tunnel, and Figure 13 shows both the displacements with distance along the BNSF tunnel and with perpendicular distance from the TBM tunnel axis. Small settlements of 0.01 to 0.02 inches were measured in the BNSF tunnel when the TBM had approached within approximately 2 diameters of the BNSF tunnel, and increased to 0.1 inches as the TBM cutterhead began passing beneath the BNSF tunnel (with the settlement trough skewed eastward due to the angle of TBM approach at ~40 degrees to the BNSF axis). After the TBM passed the BNSF tunnel, total settlements reached 0.2 inches at the TBM centerline. The vertical displacements are exaggerated in the figure in order to show the sequence of these small displacements. The precision of the ASMP instrumentation has permitted showing the shape of a classic settlement trough even though settlement values are small.

Prior to advancing beneath the BNSF tunnel, deep extensometer displacements were used to determine expected settlements at the BNSF tunnel. A settlement of 0.2 inches was estimated, assuming a 40-degree angle of draw from TBM springline to the BNSF to determine the half width of the settlement trough which is equal to 2.5 i (the inflection point in a Gaussian-shaped trough), and assuming no volume change in the trough. Actual volumes and settlements were as predicted, indicating that there was very little volume change between the TBM tunnel and the BNSF tunnel, 85 feet above (Table 3).

Case 3, Continued: Pike St. Adit

The Pike St. Adit, a 9.5 ft. concrete tunnel housing a number of smaller sewer pipes, is immediately beyond the BNSF crossing and is 70 feet above the crown of the AWVR tunnel. An instrument array consisting of a series of connected tilt beams (essentially a horizontal in-place inclinometer) was placed in the adit, extending 175 feet each direction from the centerline of the TBM tunnel. Settlements are measured assuming the points at the ends of the array are fixed. The measurements in the Pike St. Adit provided precise information on the shape and volume of the settlement trough above the TBM, even though the settlement magnitudes were small (Figure 14).

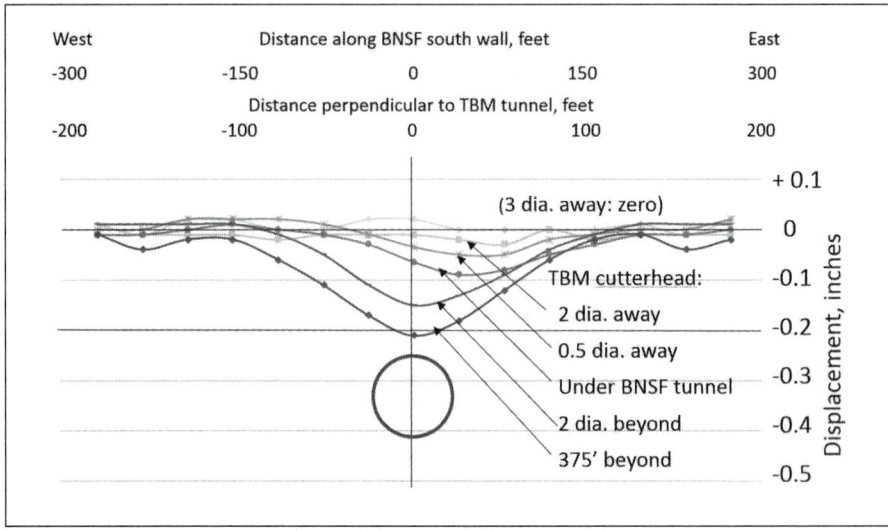

Figure 13. Settlement trough measured by ASMP prisms in BNSF tunnel

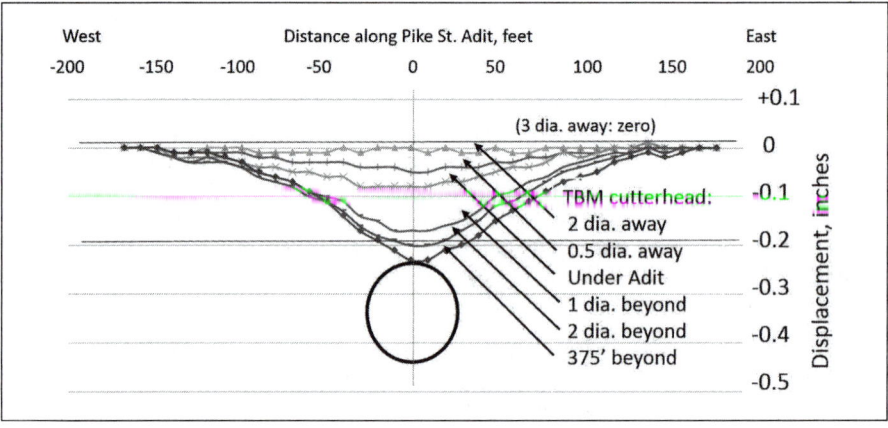

Figure 14. Settlement trough measured with tilt beam in Pike Street Adit

Small settlements (on the order of 0.01 inches) were detected in the adit above the centerline of the TBM tunnel when the cutterhead approached within approximately two TBM diameters of the adit. After the TBM passed, maximum settlement at tunnel centerline reached 0.22 inches, and the settlement trough had a classical Gaussian-shaped distribution above TBM tunnel. The volume of the settlement trough was 2.2 ft³/ft advance (0.08 percent of tunnel volume), approximately the same as the BNSF tunnel.

Figure 15 shows the relationship between angular distortion, lateral strain, and damage levels in masonry structures. It is also used for describing cracking in other types of walls. The points near the zero axes represent the angular distortion and lateral strain at the BNSF tunnel and Pike Street Adit, based on the slope of the settlement trough. The distortions are far below the "negligible" boundary and below the values estimated in pre-construction analyses for negligible damage to the BNSF tunnel.

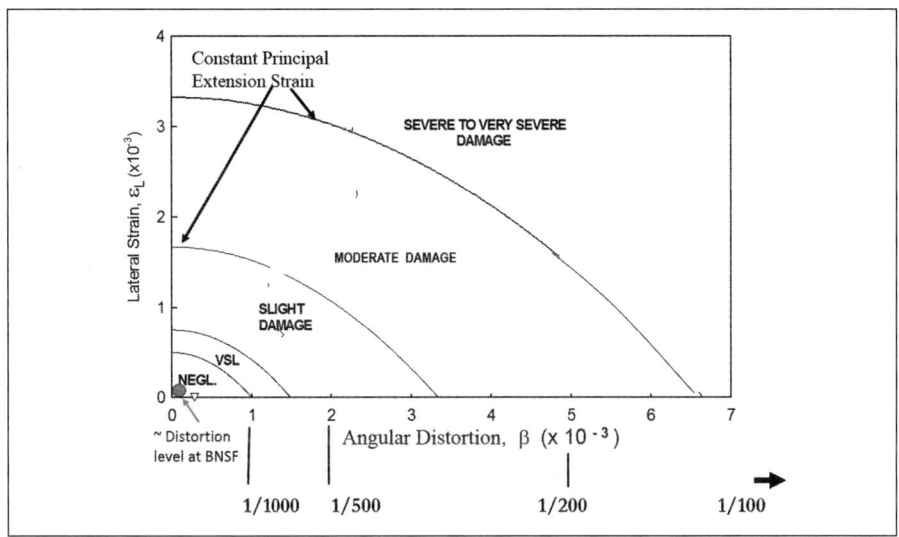

Source: Cording et al. 2010, modified after Boscardin & Cording 1989.

Figure 15. Relationship between angular distortion, lateral strain, and damage levels

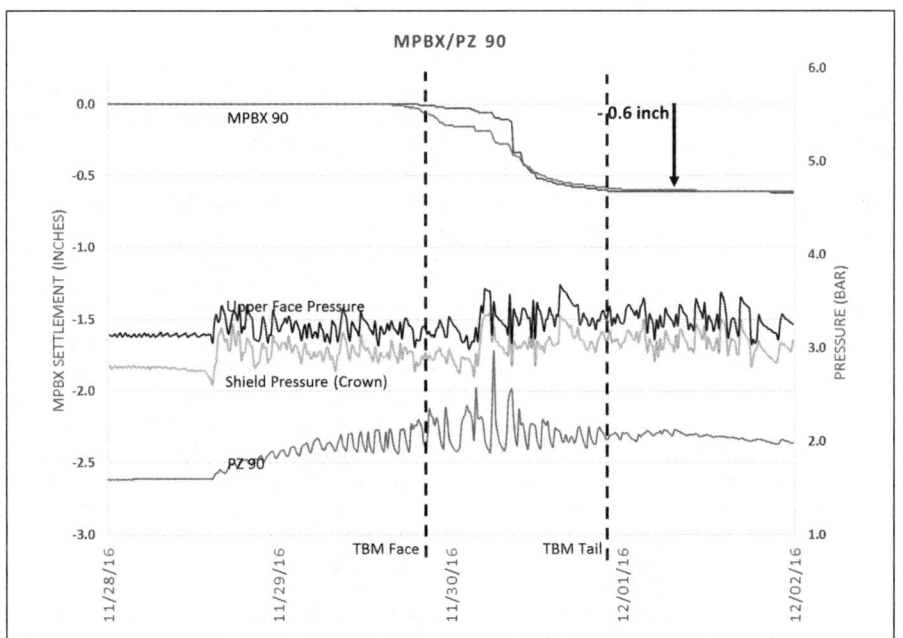

Figure 16. MPBX/PZ 90 displacements with TBM face and shield pressure during TBM pass

Case 4. Sand in Tunnel Crown

Settlements and piezometric levels at MPBX/PZ 90, in mixed soils, including sands at the tunnel crown, are shown in Figure 16. Groundwater pressure 5 feet above the shield began increasing 120 feet ahead of the TBM, from 1.6 bars to 2.0 bars as the cutterhead passed and continued to increase over the shield, to values temporarily

Table 4. Station 254+70: settlement and volume loss, tunnel depth: 200 feet

MPBX	Station	Differential Settlement	Total Settlement	Volume, ft³/ft	% Volume
90	254+70	0.6 inch	~0.7 inch	2.9	0.11%

approaching within 0.5 bars of the pressure on the shield body. Deep anchor differential settlement of 0.6 inches occurred over the central portion of the shield in the same range as those measured with clayey soils in the crown as in Case 3. Throughout this reach, as more sandy soils were encountered in the crown, polymer was injected into the shield gap with the bentonite in order to increase viscosity and provide additional support to the sands above the TBM.

Table 4 shows total displacement of deep MPBX anchor during/due to tunneling, and estimated volume loss.

CONCLUSIONS

On the Alaskan Way Viaduct Replacement project, the aggregate of operational procedures, use of agreed target criteria for key TBM operating parameters, open lines of communication among contractor management, tunnel operations, and owner have contributed to effective coordination of the extensive machine and geotechnical instrumentation data reviewed in real time by the various project teams, and contributed to consistent TBM operation and control of ground movement.

Throughout the drive beneath the Alaskan Way Viaduct, Pioneer Square, and structures in downtown Seattle, a continuous pressure envelope was maintained from the face to the tail of the shield, both during and between advances of the TBM. Upper face pressures required to balance groundwater pressures and active earth pressures at the face also served as the reference for the pressures that were maintained in the 30 mm shield gap filled with bentonite. After advances of the EPBM, as bentonite was automatically injected into the chamber behind the cutterhead and into the shield gap, a cake was formed that maintained the TBM pressures above the groundwater pressures. Deep extensometer anchors and piezometers monitored displacements and groundwater pressures immediately above the TBM. Piezometric pressures in sands increased as the TBM approached, tracking below face pressures. In contrast, in clay soils, piezometric pressures dropped as the cutterhead passed in response to undrained stress changes due to the difference between the overburden pressure and face/shield pressures. The smaller ratio of overburden to TBM pressure in Pioneer Square resulted in settlements immediately above the shield of less than 0.05 inches in the very hard clay soils, but north of Pioneer Square, the increased overburden pressure resulted in settlements in the range of 0.2 to 0.7 inches immediately above the shield, far below project settlement criteria. Throughout the drive, surface settlements were in the range of 0 to 0.1 inches, less than recorded variations due to environmental effects and below any damage levels.

The percent volume of ground loss 5 feet above the shield ranged from 0.0% to 0.03% in Pioneer Square and from 0.03% to 0.11% at depths of 200 feet. Compared to both the modeled 0.30% for this TBM drive and the current state-of-the-practice criteria sometimes applied to large diameter TBMs of 0.50% (which is based on the assumption that percent volume loss values used in the design of a large diameter TBM is proportional to percent volume loss values used for smaller diameter machines), the low volume loss measured on the Alaskan Way Viaduct Replacement Project shows ideal TBM performance.

DC Clean Rivers Project: Geotechnical Instrumentation Programs for Protecting Critical Infrastructures in the Nation's Capital

Lei Fu ▪ AECOM
Peter Kottke ▪ McMillen Jacobs Associates
Brad Murray ▪ McMillen Jacobs Associates
Stephen Njoloma ▪ McMillen Jacobs Associates
Rafael Castro ▪ JCK Underground
Moussa Wone ▪ DC Water

ABSTRACT

The District of Columbia Water and Sewer Authority (DC Water) Long Term Control Plan (LTCP) (also known as the DC Clean Rivers (DCCR) project) consists of about 14 miles (23 km) of large-diameter tunnels and deep shafts constructed in soils. A total of 15 shafts with diameters up to 149 feet (45 m) and depths up to 193 feet (59 m); three large-diameter tunnels have been constructed in three contract divisions (Divisions A, H, and P) and procurement of a fourth tunnel project (Division J) is underway. The projects of the LTCP are located in the crowded urban environments and extensive geotechnical instrumentation programs were installed to monitor performance of shaft and tunnel construction for the protection of the critical infrastructure, including DC Metro, bridges, water lines, sewers, etc. Experience gained from each project allowed the protection of structures and geotechnical instrumentation requirements to be better adapted for the follow-on projects. This paper discusses how the structure protection criteria were set and tied to the instrumentation monitoring requirements in the RFP documents of the completed divisions and the implementation of the instrumentation programs during design and construction phases. Lessons learned are also included.

PROJECT INTRODUCTION

DC Water is currently implementing its Long Term Control Plan (LTCP) for Combined Sewer Overflows (CSOs) through the DC Clean Rivers Project (DCCR). The DCCR includes a 14 mile-long (23 km) tunnel system that will store and convey combined sewer and wet weather flows to DC Water's Blue Plains Advanced Wastewater Treatment Plant. There are 13 contract divisions in the DCCR. Figure 1 shows the locations of major tunnel contract divisions, including Division A: Blue Plains Tunnel (BPT); Division H: Anacostia River Tunnel (ART); Division J: Northeast Boundary Tunnel (NEBT); and Division P: First Street Tunnel (FST). The BPT, ART, and NEBT have a finished internal diameter (ID) of 23 feet (7 m); the FST has a finished ID of 20 feet (6.1 m). The construction of the BPT, ART, and FST tunnels and shafts were completed in August 2015, November 2016, and December 2015, respectively. These projects were delivered using design-build contracts. In this paper, the contractor means the design-builder. A total of 15 shafts with diameters up to 149 feet and depths up to 193 feet were constructed in Divisions A, H, and P. These shafts serve as tunnel boring machine (TBM) launching or receiving shafts, sewer drop shafts, ventilation shafts, dewatering shafts, or screening shafts. This paper focuses on the geotechnical instrumentation programs of the completed major contract divisions.

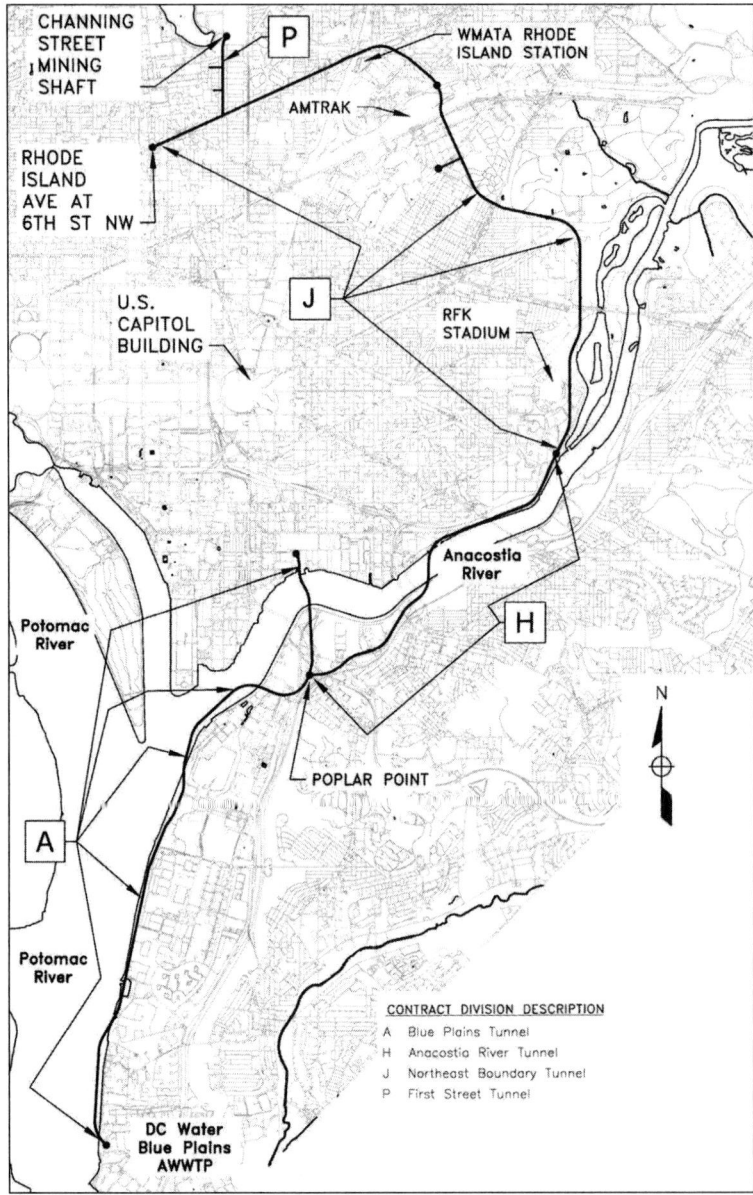

Figure 1. Major contract divisions in the DCCR Project

REGIONAL GEOLOGY

The project site is located at the western edge of the Atlantic Coastal Plain Physiographic Province and just to the east of the Piedmont Physiographic Province. Figure 2 shows a geotechnical profile of the primary tunnel system. The subsurface materials in the project area include the following geologic units from the top down (from youngest to oldest):

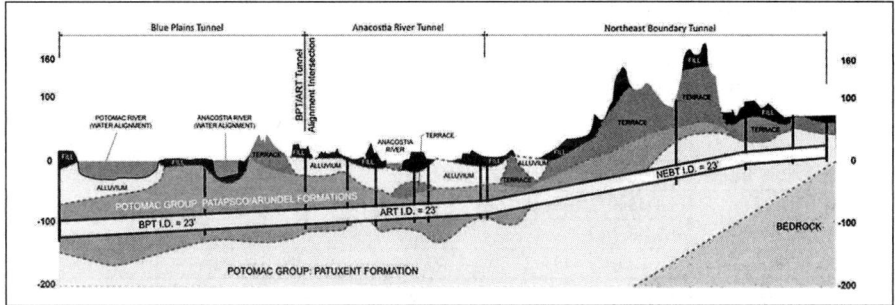

Figure 2. Geotechnical profile along the DCCR Project

- Recent Fill consists of a heterogeneous mixture of clay, silt, sand, and gravel-sized particles and can include cobbles and boulders.

- Quaternary-age Alluvium consists of interbedded deposits of clay, silt, sand, and gravel, with cobbles and boulders and varying amounts of organic materials.

- Cretaceous-age Patapsco/Arundel Formation (undivided) of the Potomac Group consists of predominantly thick sequences of silt and clay, locally with sand and gravel interbeds. The overall thickness of the Patapsco/Arundel Formation varies, but can be over 120 feet (36.6 m).

- Cretaceous-age Patuxent Formation of the Potomac Group is more coarse-grained overall, characterized predominantly by silty and clayey sand and gravel to relatively clean sand and gravel layers, locally with silt and clay interbeds.

- Bedrock consists of metamorphic rocks, generally in the form of dark grayish green amphibolites (e.g., tremolites and actinolites) along with lesser extents of schists and gneiss.

TUNNEL AND SHAFT CONSTRUCTION METHODS

The large diameter tunnels in the DCCR, including BPT, ART, and FST, were excavated using earth pressure balance tunnel boring machines (EPB TBMs). Precast concrete segments were used as the tunnel liner.

The shafts in Divisions A, H, and P were constructed using slurry diaphragm walls (D-walls) with excavation occurring "in the wet" and "in the dry, and ground freezing (see Table 1). For the dry excavations—including BPT-DS, BPT-SS, CS-DS, CSO-019NS, CSO-019SS, MS-DS, CSO 005-DS, CSOO 007DS, and CSO 018-DS, groundwater depressurization by pumping was performed to lower the groundwater level below the bottom of shaft excavation. For the wet excavations—including JBAB-DS, PP-JS, and MPS-DS—the top portion of the shafts were conventionally excavated to the normal groundwater level. The remaining excavation was performed with the shaft filled with water to provide the necessary base stability. The bottom slab of the shaft was constructed by the tremie concrete method. The shafts were dewatered and the final liner was installed in the dry. In Division P, AS-DS, VSDS, and PS-DS were constructed using the ground freezing method for ground support and bottom stability.

Table 1. Summary of shafts installed in the DCCR

Contract Division	Shaft	Shaft ID, ft (m)	Shaft Depth, ft (m)	D-wall Depth, ft (m)	Shaft SOE	Excavation Method (dry or wet)	Groundwater Depressurization (yes or no)
A	BPT-DS	149 (45.4)	193 (58.9)	198.5 (60.5)	D-walls	Dry	Yes
	BPT-SS	76 (23.2)	156 (47.6)	180 (54.9)	D-walls	Dry	Yes
	JBAB-DS	55 (16.8)	143 (43.6)	148 (45.1)	D-walls	Wet	No
	PP-DS	60 (18.3)	135 (41.2)	139 (42.4)	D-walls	Wet	No
	MPS-DS	60 (18.3)	120 (36.6)	124 (37.8)	D-walls	Wet	No
H	CSO 019-NS	75 (22.9)	105 (32.0)	135 (41.2)	D-walls	Dry	Yes
	CSO 019-SS	75 (22.9)	99 (30.2)	135 (41.2)	D-walls	Dry	Yes
	MS-DS	63.5 (19.4)	105 (32.0)	120 (36.6)	D-walls	Dry	Yes
	CSO 018-DS	32 (9.8)	102 (31.1)	136 (41.5)	D-walls	Dry	Yes
	CSO 007-DS	16 (4.9)	115 (35.1)	128 (39.0)	D-walls	Dry	Yes
	CSO 005-DS	15 (4.6)	107 (32.6)	128 (39.0)	D-walls	Dry	Yes
P	CS-MS	55 (16.8)	159 (48.5)	144 (43.9)	D-walls	Dry	Yes
	AS-DS	20 (6.1)	96 (29.3)	—	Frozen ground	Dry	No
	VS-DS	23 (7.0)	86 (26.2)	—	Frozen ground	Dry	No
	PS-S	22.5 (6.9)	90 (27.4)	—	Frozen ground	Dry	No

PROTECTION OF EXISTING STRUCTURES

The DCCR is located in the crowded Washington DC environment. Protection of various existing structures, including buildings, underground utilities, and roadways, is crucial for the success of the project. The owner, DC Water, put a great deal of effort during the bidding document preparation on this aspect of the project. In each contract division, an inventory was prepared for the existing facilities that were within the zone of influence (ZOI) estimated to be caused by shaft and/or tunnel construction. For shaft or other open-cut excavations, the ZOI corresponds to a zone that extends from the edge of excavation pit to a point where the estimated ground movement is minimal (i.e., less than 0.1 inches). For tunnels, the ZOI extends from the tunnel centerline to the point of near zero displacement at the edge of the settlement trough. For Washington Metropolitan Area Transit Authority (WMATA) structures, the WMATA definition of the ZOI was adopted. A preliminary construction impact analysis was performed by DC Water for each of existing structures, using analytical, empirical and/or numerical methods. Based on the results of the screening analysis, mandatory performance requirements limiting the construction-induced ground, structure, and utility movements were set in the contract documents. For structural protection purposes, existing structures were classified and assigned a tier, each with specific action by contractor. For example, in Division H, there were five tiers:

- Tier 1A. Ground movement mitigation or structure strengthening required prior to construction for Tier 1A structures to achieve structure protection criteria or structural design criteria/codes.
- Tier 1B. Tier 1B structures may experience damage at anticipated levels of ground movement that can be repaired post-construction. Enhanced monitoring and post-construction inspection and repair (if needed) of these structures were considered the preferred protocol for addressing potential impacts.

- Tier 2. Tier 2 structures are expected to stay within the structure protection criteria or structural design criteria/codes at the anticipated levels of ground movement.

- Tier 2X. Tier 2X Structures are Tier 2 structures that are of significant importance; therefore, special scrutiny is required for the structures. The potential cost of impacts is the justification for requiring the contractor to perform a detailed assessment using the actual construction means and methods and advanced analysis methods (i.e., three-dimensional numerical modeling). If the detailed analyses by the contractor show impacts above damage criteria, the structure was then re-classified as a Tier 1 structure, which is further classified as either a Tier 1A or Tier 1 B structure.

- Tier 3. Tier 3 structures are structures that have not been classified. The contractor is responsible for the protection, modification, or relocation of existing utilities and facilities required to accommodate the contractor's means and methods that have not been anticipated by the owner. If a Tier 3 Structures is reclassified as a Tier 1A or 1B structure, complete a detailed assessment on the structure.

Division A had three tiers (Tiers 1, 2a and Tier 2b). Tier 1 structures were structures that were expected to be impacted by construction; Tier 2a and T2b were not. The differences between Tier 2a and T2b were movement limits. The Division A tier classification were fairly simple to use, but more were added for Divisions H, J, and P. Specifically, Tier 1B, 2X, and 3 were added. The benefit of having additional tiers includes more options for addressing construction impacts in the case of Tier 1B, enhanced risk management of critical infrastructure in the case of Tier 2X, and ability to allocate more responsibility to contractor but also give contractor flexibility in assigning protection requirements based on their means and methods in the case of Tier 3.

The actual construction impacts on the existing structure, ground, and groundwater conditions were monitored by the geotechnical instrumentation programs discussed in the following sections and structural inspections performed before, during, and after construction.

The movement limits for ground, structures, and utilities were developed in terms of action and maximum values/levels based on types of structures using protection criteria or structural design criteria/codes. "Action" level is a geotechnical instrumentation reading that triggers a set of review and mitigation actions to ensure that the "Maximum" level is not exceeded. "Maximum" level is a maximum permissible geotechnical instrumentation reading that corresponds to a potential temporary work stoppage to prevent damage to structures.

GEOTECHNICAL INSTRUMENTATION PROGRAMS

A geotechnical instrumentation plan depicting minimum requirements for number and type of instruments to be used was included in the request for proposal (RFP) documents. As part of the detailed design, the contractor finalized the layout and provided additional geotechnical instruments based on the contractor's means and methods and structural protection needs. The geotechnical instrumentation monitored and documented how the construction activities affected the surrounding ground, groundwater, and existing structures. This included evaluating whether the contractor's methods and procedures met the project requirements, the effectiveness of corrective changes where performance did not meet requirements. Instrumentation was also used to confirm the anticipated ground behavior and design assumptions and provided required

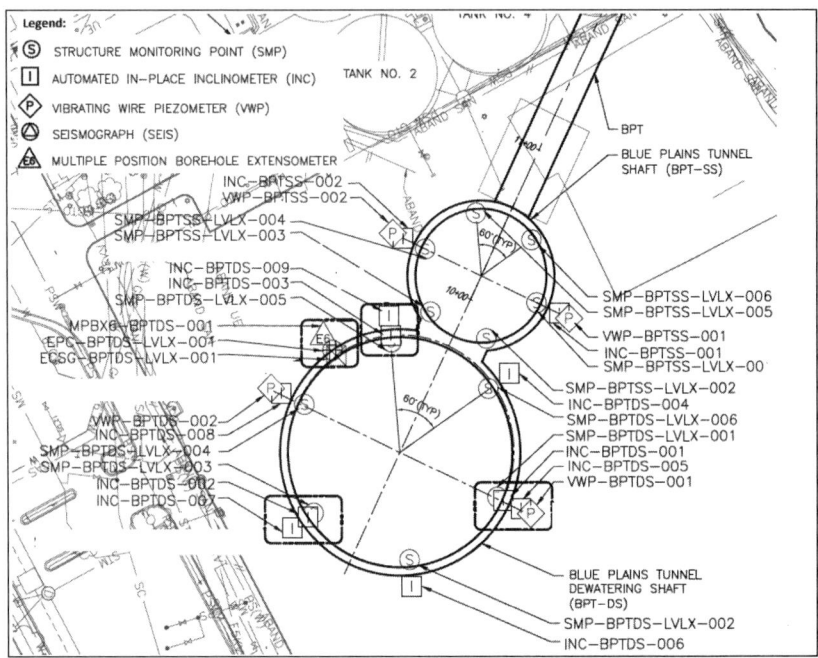

Figure 3. Geotechnical instrumentation installed at the BPT-DS and the BPT-SS site

information to third parties. As an example, Figure 3 shows the instruments installed at the BPT-DS and the BPT-SS site. Table 2 includes a list of the instruments installed in the DCCR.

Ground Instrumentation

Ground instrumentation was installed in or on the ground surface to monitor construction impacts on the ground and groundwater conditions. Ground survey points and extensometers were installed along the tunnel alignments and in the area influenced by open-cut excavations. Multiple position borehole extensometers were installed at different depths above the tunnel crown to measure the tunneling induced ground movements at different distances above the tunnel crown. The measured movements were used to confirm if the ground loss values and distribution of ground movements over depth were as expected or, for the tunnels, if allowable volume loss limitations were met. The extensometers were automated instruments. However, manual readout of the instrument was used to confirm the automated readings. When needed, the reference head of the extensometer was surveyed and the head survey data was used to convert the extensometer readings to ground movements (heave or settlement, rather than relative movement) at the locations of the extensometer sensors.

Ground monitoring arrays consisting of ground survey points spread out at regular intervals along the tunnel alignment were used to measure the shape of settlement trough, which is the distribution of the ground surface settlements in a profile perpendicular to the tunnel alignment. In Division A, typically there were five survey points in one array located at offsets of –70 feet (–21 m), –30 feet (–9 m), 0, 30 feet (9 m), and 70 feet (21 m) from the tunnel centerline.

Table 2. Summary of geotechnical instruments installed in the DCCR

Monitoring Purpose		Instruments
Ground	Ground movement	Optical survey point (manual, automated), extensometer (MPBX, SPBX), inclinometer (automated, manual), Borros anchor
	Groundwater level	Monitoring well, vibrating wire piezometer
	Horizontal earth pressure	Earth pressure cell
Structure		Optical survey point (manual, automated), tiltmeter
Utility		Utility monitoring point (manual, automated), crack gage
Support of excavation system	Movement	Optical survey point (manual, automated), in-wall inclinometer
	Concrete strain	Strain gauge (concrete, rebar)
Noise and vibration		Geophone, seismograph

Inclinometers (manual and automated types) measured the full depth profiles of lateral ground movements adjacent to shaft or other open-cut excavations. Inclinometer data was analyzed by assuming bottom fixity which requires the inclinometer casing to extend beyond the displacement zone caused by excavation. Lack of bottom fixity was noticed in some inclinometers installed adjacent to shafts. In Division A, automated in-place inclinometers (IPIs) were initially used around shafts, but the readings from them were found to be questionable, with possible data drifting and unexplained data jumps. The possible causes included vibrations from the construction equipment and accumulated errors from very long inclinometer strings with up to 20 sensors. After an in-depth investigation into the IPI data issue, the decision was made to allow the IPIs to remain on the project. However, the readings from this kind of instrument were deemed only as an indicator of potential ground movements. If IPI readings were above or approached the movement limits or in question, the IPI sensors were removed and manual inclinometer surveys were conducted to verify the IPI data. A similar approach was also used in Division H. Additionally, in Division A, during shaft excavations, manual inclinometer surveys were required to be performed at the beginning of a shaft excavation, and at 25%, 50%, 75%, and 100% of shaft excavation depths. In Division P, three shafts were built using the ground freezing method. Ice was formed in some inclinometer casings installed in the frozen ground, which led to erroneous readings and in some cases inability to pass the inclinometer probe through the casing. The solution was to de-ice using hot water or other hot fluids.

Structural Instrumentation

Structural instrumentation was installed on existing structures, including buildings, bridges, near surface structures, and pavements that were in the potential ZOI of the tunnels, shafts, or other open-cut excavations. Typical structural instruments included structural monitoring or survey points, tiltmeters, and crack gauges. For points located on the exposed surface of structures, the seasonal changes (i.e., the reading increases or decreases with the ambient temperature) were clearly observed.

Utility Instrumentation

Utility instrumentation was installed on existing utilities (waterline, sewer pipe, gas line, etc.) that maybe impacted by tunnel, shaft, or other open-cut excavations. Typical utility instrumentation included utility monitoring points and crack gages. Utility monitoring points were typically installed on the utilities less than approximately 20 feet (6 m) deep. To ensure a secure attachment, the fiberglass rod of the utility monitoring point was required to be epoxy-glued to the outside surface of the utility pipe. For utilities deeper than 20 feet (6 m) or under groundwater levels, it was sometimes difficult to keep the boring hole open to install the casing of the utility monitoring point. An alternative was to install a Borros anchor at the bottom depth of the utility pipe to measure the ground movement at the utility location.

Support of Excavation System Instrumentation

As indicated in Table 1, most large diameter shafts were built using concrete diaphragm wall systems. The inclinometers with their casings being cast into the diaphragm walls were used to measure the lateral deformation of the diaphragm walls. The support of excavation systems (i.e., soldier piles and lagging and secant pile walls) for other open-cut excavations were monitored for the lateral deformation using optical survey points mounted to the inside face of the support of excavation systems.

TBM Performance

The contractor was required to monitor TBM performance and provide the monitoring data to the owner, including TBM operating performance parameters (advance rate, cutterhead torque, rotation speed, applied pressure, etc.), guidance data (segment ring interval, tunnel centerline, offsets, etc.), and excavated volumes.

Geotechnical Instrumentation Management System

In each contract division, the contractor was required to provide and maintain a geotechnical instrumentation management system (GIMS). The GIMS was an integrated, internet-accessible system that receives, stores, transfers, displays, and reports data from geotechnical instrumentation and selected TBM operating parameters. The system issued alerts for the readings exceeding threshold values, i.e., action or maximum levels. In addition, the owner purchased and maintains a separate tunneling monitoring system, the Tunneling Process Control (TPC) by Babendererde Engineers. Each contractor was required to provide the GIMS data to the TPC. An advantage to interface real time GIMS data with TBM data was evident in relating the deep ground movement measured by extensometers to the position of the TBM. While the movement limits established by the contract were for the total ground movement, it was important to determine when in relation to the TBM the ground movement occurred. The contractors would then use the movement relative to the TBM position to make necessary changes.

GEOTECHNICAL INSTRUMENTATION SYSTEM IMPLEMENTATION

Implementation of the GIMS in a contract division involved the contractor and the owner's construction manager (CCM), and the owner's program consultant organization (PCO). The contractor was responsible for installing and maintaining instruments, collecting data, preparing data reports, and operating the instrumentation website. The CCM inspected the contractor's instrumentation work. The PCO oversaw the implementation of the GIMS. Figure 4 illustrates the data flow in the GIMS. The data reports (daily, weekly, or monthly) summarized data gathered during a reporting period, and discussed the status of instruments (installation, malfunction, damage, and repair) and the status of the readings compared with threshold limits.

All instrumentation-related issues—including malfunctioning instruments, damaged instruments, installation issues, questionable data, alerts, and resolutions to the alerts—were summarized in an issues log that is included in the data reports. The instrumentation issues, installation schedule, data interpretation, and other related topics were discussed in weekly instrumentation meetings that were attended by the

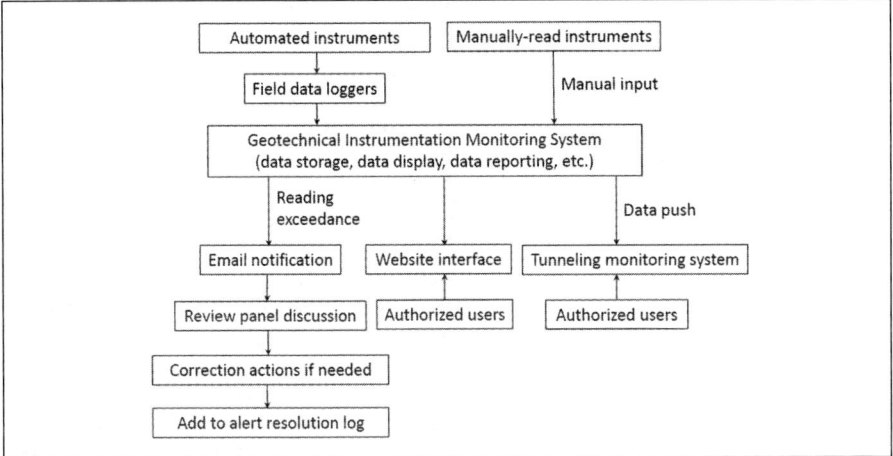

Figure 4. A typical geotechnical instrumentation data flow chart

contractor, the instrumentation engineer, the CCM, the engineer of record (EOR), and the PCO.

The GIMS automatically issued alerts by comparing monitoring data with action and maximum levels. If a reading exceeded the action level or the maximum level, an action level (amber) alert or a maximum level (red) alert was issued, respectively. The contractor's response to an alert included the following steps:

- Manual reading to confirm automated reading
- Inspection to determine if damage occurred
- Re-evaluation of threshold limits
- Evaluation of construction performance
- Adjust means and methods

LESSONS AND RECOMMENDATIONS

Geotechnical instrumentation programs and data management for the DCCR project were a critical component of the project execution and played a key role in the success of the projects with regard to the protection of existing facilities, construction safety, and construction process adjustment. Below are some lessons and recommendations for similar applications:

- Install instrumentation to cover all ground and all important existing structures and have reasonable redundant instruments. Instrumentation monitors construction impacts and protects the owner and the contractor from potential spurious claims.
- For a large project like the DCCR, with the extensive amount of instruments installed and data collected each day, various instrumentation issues should be expected. Issues logs, response level plans, and weekly instrumentation meetings are very useful tools for resolving the issues in a timely manner. Having the contractor provide an initial response to instrumentation alerts within 24 hours has served the DCCR well. The alerts should be discussed in depth in weekly meetings. The resolution to an alert should be documented

in an issues log that is reviewed in regular progress or instrumentation meetings.

- Automated extensometers should have manual readout in case that the automated readings needs to be confirmed by manual data. The manual readout should allow manual readings to be taken without disturbing the electronic readout. In addition, extensometers should include a setup for surveying extensometer reference head movements. The measured head movements can be used to estimate absolute ground movements at extensometer sensor locations.

- The fiberglass rod of the utility monitoring point needs to be securely attached (i.e., epoxy glued) to the outside surface of the utility pipe to avoid potential separation of the rod from the utility.

- The benefit of the automated inclinometer or IPI includes continuous data that can be obtained and used in places that have restricted access during active construction. However, for IPI's installed around deep shafts, readings can be affected by construction vibrations, accumulated errors, electric noises, etc., which could lead to unstable or questionable data. Manual inclinometers are proven to be more reliable. Where IPI's are used, it is a good practice to take both manual and automated baseline readings. In case there are questions about the IPI readings, manual surveys can be performed to verify the automated readings.

- Inclinometers need to extend beyond the displacement zone caused by adjacent excavation so that absolute deformation data can be calculated by assuming base fixity. For deep shaft excavations, the displacement zone could be evaluated using analytical, empirical, or numerical methods.

- When an inclinometer casing is installed in the freezing ground, ice can form in the casing, leading to survey errors. Solutions include de-icing using hot water or tiling the casing with brine. However, de-icing could be time consuming when monitoring frequency is high (e.g., weekly).

A Smart Disc Cutter Monitoring System Using Cutter Instrumentation Technology

Kamyar Mosavat ▪ The Robbins Company

ABSTRACT

Current disc cutter instrumentation technology is designed to be a conveniently mounted instrumentation package that monitors individual cutter rpm, wear, temperature, and vibration. A data logger service receives the cutter information wirelessly using low-power radio technology and displays cutter conditions in real time. With cutter instrumentation, the operator continuously monitors cutter conditions, which results in higher efficiency, lower incidence of down time, and prevents unexpected ring wear-related damage from causing further damage to bearings and hubs. Cutter instrumentation technology has been tested on Robbins' rock machines and results from previous and recent projects are presented. Design improvements for longer lifetime and increased reliability are discussed.

WHY CUTTER INSTRUMENTATION?

In mechanized tunneling continuous information from the excavation face is essential. The ultimate goals of cutter instrumentation are to monitor real-time individual cutter operation, acquire more realistic cutterhead thrust force values, and gain a better knowledge of the geology in front of the cutterhead. Analysis of this information can provide in-depth knowledge of machine excavation. Information about cutter operation has direct and indirect advantages: It helps better predict and monitor cutter usage rates, and it can reduce the cost of unplanned cutter or ring replacement, which can result in a better planning of inventory, manpower, and cutter rebuild requirements. Another merit of cutter instrumentation is to maintain assembly health by monitoring individual cutter operation. An instrumentation system can notify an operator of uneven or harsh ring wear and makes it possible to prevent unnecessary seal or bearing changes. Additionally, it can prevent cutterhead damage caused by a late cutter change (see Figure 1).

HISTORY

Instrumentation at Robbins has about 25 years of history, which can be divided into two period: the time before 2007 and after 2007. In the late 1980s a cutter instrumentation system was developed in-house and installed on a few projects such as Svartisen

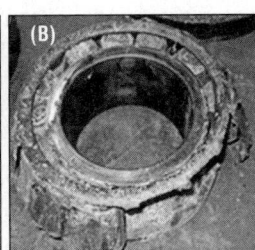
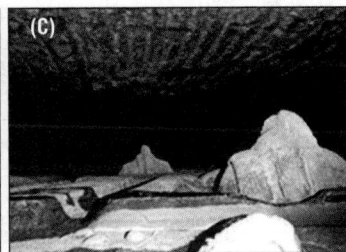

Figure 1. (A) Wiped out cutter wear; (B) damaged bearing; (C) face of excavation

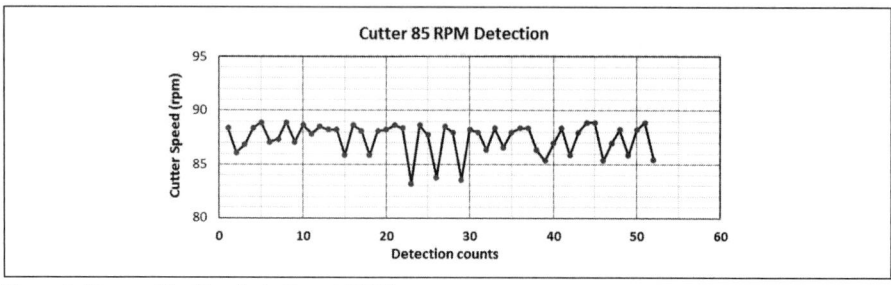

Figure 2. Niagara Machine first attempt (2008)

in Norway. The cutter monitoring system provided two main functions, consisting of cutter load measurement and cutter rpm. A magnetic sensor was installed along with a strain sensor in the cutter housing and a small permanent magnet was pressed into a hole in each cutter hub. The strain measurements and cutter rpm on individual cutter housings were read, amplified, converted to a digital signal, and transmitted to a computer where they were recorded. This information was logged as history of the rock formation type and homogeneity. Load and wear could be combined to better predict usage rates versus penetration to plan inventory and cutter rebuild requirements. The actual cutter instrument on the cutter housing was wired to a junction box and then to the data acquisition computer (DAQ) through hydraulic hoses. A slip ring was used to transfer data from the DAQ to the operator cabin, which couldn't accommodate continuous data collection. However, a fair amount of data was recorded to determine cutter force in different materials. In the 1980s and 1990s several projects implemented the systems. In 1993, Robbins mobile miners were instrumented with strain gages and DAQ and monitored cutting action and forces. The issue of interrupted and short time data collection remained the major obstacle until early 2000s when the idea of wireless transmission was developed with available technologies. The new wireless configuration was successfully tested later in 2007.

During an experiment in 2007, successful radio communication was established with a radio transducer and a receiver. Further in 2008, in an attempt to read cutter rotation, one instrument was built and installed in the Niagara TBM—a 14.4 m diameter open type, hard rock machine and the largest in the world of its type. The instrument was installed in the cutter housing and mounted to the wedge bolt. The system survived for a short period of time and provided spotty cutter speed data (See Figure 2). In 2009, an attempt at resolving the data readings problem led to the design of a new instrument prototype. In the same year five of those instruments were welded to cutter housings of a 12.4 m diameter open type hard rock machine for China's Jinping-II Hydropower project. This effort improved the system and outlined the outstanding issues. The welded mounting was not convenient for field staff, instrumentation enclosure worn out quickly, and the readings were spotty. As a result, the wedge bolt mounting option was selected as the most reliable design that maintained a good life for the enclosure. Furthermore a stainless steel sleeve was used to cover the plastic enclosure to prevent abrasive material from damaging the instrument. Between 2010 and 2012 the Niagara TBM, as well as three more open-type TBMs for Malaysia's Pahang-Selangor Water Tunnel, were equipped with the new prototype. Cutterhead speed was entered manually to calculate the cutter wear on the Niagara TBM (See Figure 3a) and later acquired from the PLC in Malaysia (See Figure 3b). Longer survivability of the instrument along with better radio communication was a turning point in the prototype design. The cutter instrumentation issues were alleviated to a certain extent and the enclosure survived longer than previous prototypes. However some issued remained.

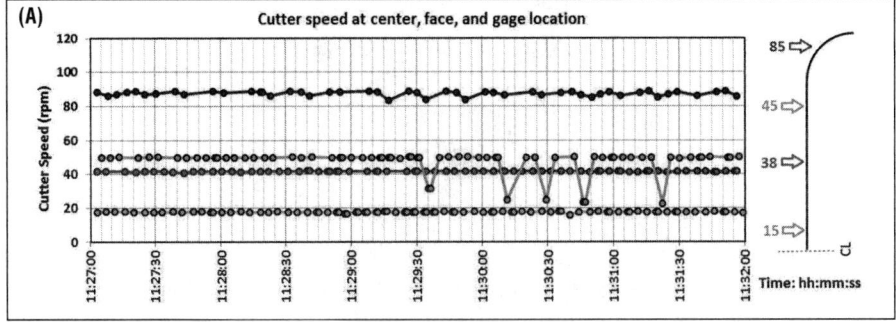

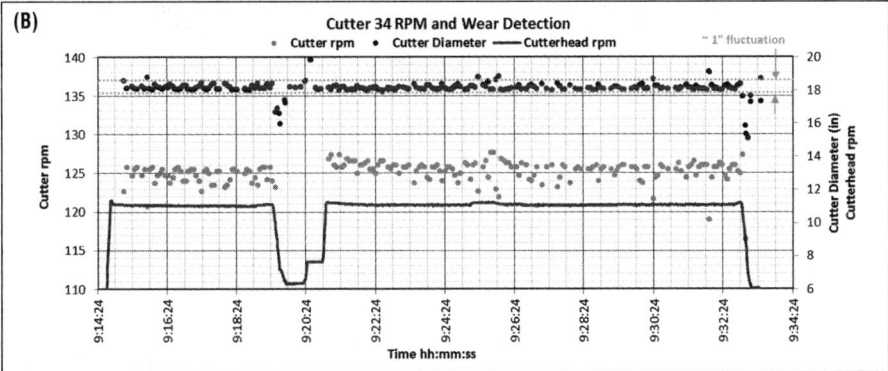

Figure 3. (A) Niagara second attempt (2010); (B) Malaysia Pahang-Selangor (2012)

Cutter speed measurements in Niagara were clearly correlated with the locations showing a decrease in speed from the gage toward the center. In Malaysia the cutter speed was fluctuating within 5 rpm and not enough data was available to analyze and smooth the data. Wear results were fluctuating within a 25mm (~1in) boundary. Although the wear measurements were not satisfactory, it was helpful to demonstrate that cutter speed was in the correct range. A computer screen was developed as an interface to control the operation (see Figure 4).

In 2014 after some design improvements on SmartCutter, 10 instruments were installed on the cutter housing wedge bolt and transmitted data for less than a month at Norway's Røssåga project (1.5 hours data is shown in Figure 5). Battery capacity was insufficient to provide longer data collection. Cutter data such as speed, wear, temperature, peak vibrations, instrument status, and battery status was displayed on the operator cabin interface. The magnets that were installed in the cutter hub were used to detect rotation. As the hub wore out the magnet occasionally fell out, which was also an issue during previous projects, and at Røssåga a few magnets failed during mining and caused issues with cutter speed. Wear values were not compared to the job site measurements to confirm the accuracy.

In 2016, a longer test was undertaken at the AMR project in India after making some critical technical improvements on the new prototype called SmartCutter to assure the reliability of the system. The previous battery was replaced with a higher capacity version. The magnet installation process was changed by using a special potting technique to secure the magnet from falling out due to wear or vibration induced breakage. Sealant was used on the enclosure that protects the instrumentation board and

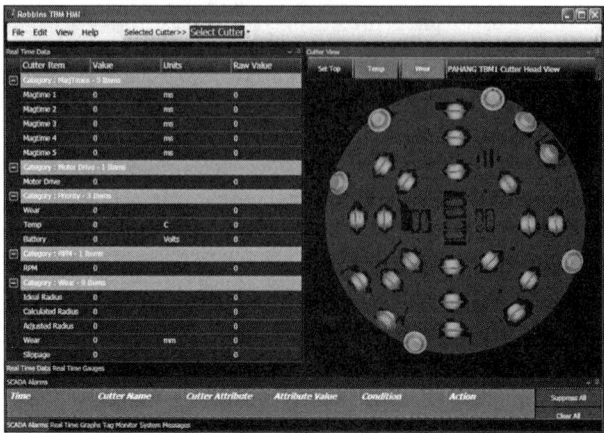

Figure 4. Røssåga TBM, Norway—SmartCutter data analysis (2014)

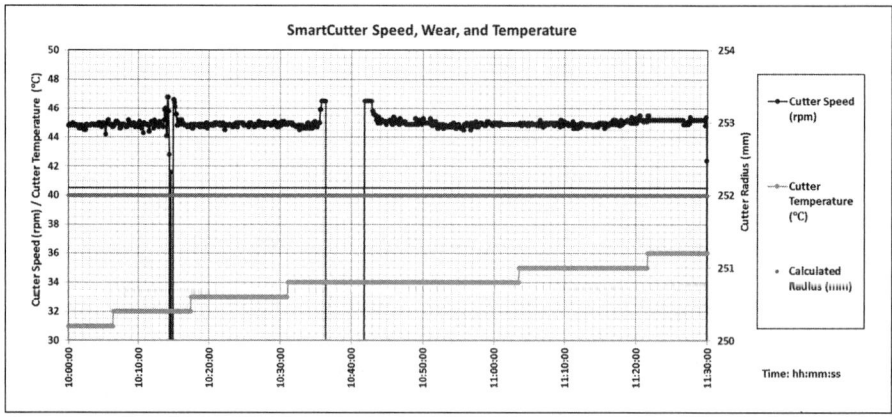

Figure 5. Malaysia Pahang-Selangor operator SmartCutter display

battery from moist, mud, and water infiltration. Five instrument units were installed on the cutterhead and two receivers were mounted in the shield. A field service person was assigned to commission the system and monitor its operation.

AMR Double Shield Hard Rock TBM

The AMR cutterhead diameter is 10.0 meters with a total of 70 cutter positions. Five instruments (SC1 to SC5) were initially installed at the gage area in position 60 to 64. Throughout July, August, and September 2016, the location of these cutters changed from the gage area to the face area as the cutters were moved inward to positions where the allowable cutter wear is greater. To take an in-depth look into the wear patterns and data evaluation, two sets of data analysis were designed: a micro analysis that investigated short report timing and a macro analysis that provided overall trends throughout a cutter's wear life.

Figure 6a shows the AMR SmartCutter system diagram. Each instrumentation box was installed on the wedge bolt of the cutter housing. Two gateways were installed on the structure of the machine conveyor (See Figure 6b). Two gateways ensured the communication link was maintained at all times. In the event of a loss of communication the

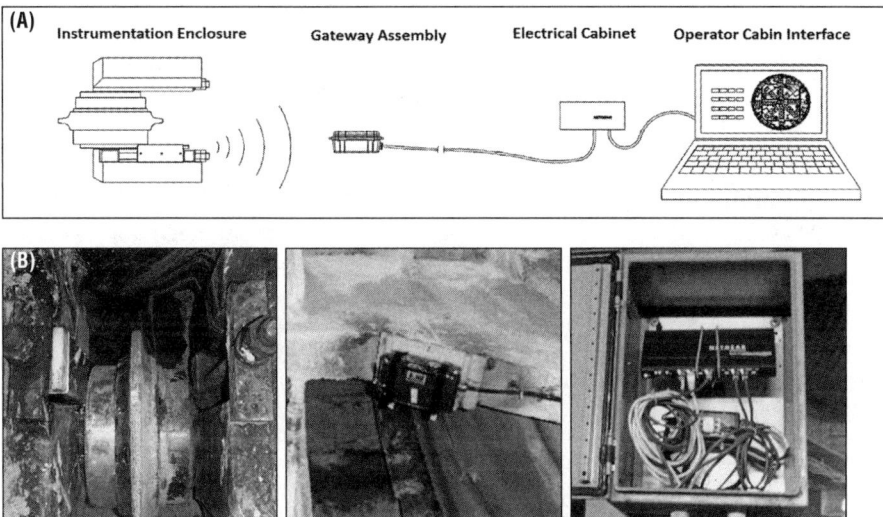

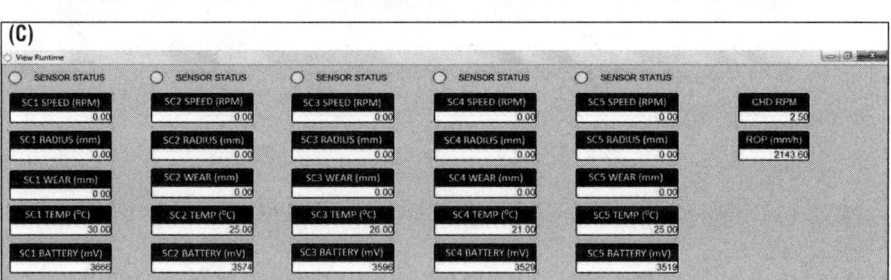

Figure 6. AMR project (2016) (A) system diagram; (B) equipment installation; (C) test monitoring platform

operator was alerted on the monitor with a red alarm (See Figure 6c). The instrumentation battery capacity was increased beyond the normal cutter wear life, meaning that instrumentation could operate throughout one or more cutter changes. Additionally, the battery capacity and status was continuously reported and displayed on each instrument at the operator screen.

AMR Speed and Wear Analysis

The magnetic sensor inside the instrument enclosure senses the time of each cutter revolution in milliseconds and reports the cutter speed (ω_{DC}). Knowing the disc cutter radius, the cutter's distance from the cutterhead center line, and also the cutterhead speed (ω_{CHD}), one can derive the cutter speed. Now using the same correlation and knowing the ω_{DC}, the cutter radius, hence its wear, can be calculated. Defining a reasonable sampling and radio data transfer rate is critical to generating meaningful data. At the same time a data filtering algorithm is required for representative and accurate wear calculation.

Figure 7 shows eight minutes of unfiltered data for a single cutter. In this plot anticipated cutter speed at a certain amount of wear and instant ω_{CHD} are presented. ω_{DC} of both a brand new disc and a worn out disc are shown in orange and black dashed lines, respectively. The blue line represents the anticipated ω_{DC} at 20mm wear. The ω_{DC} reading is shown in green dots. This value is calculated from reported cutter

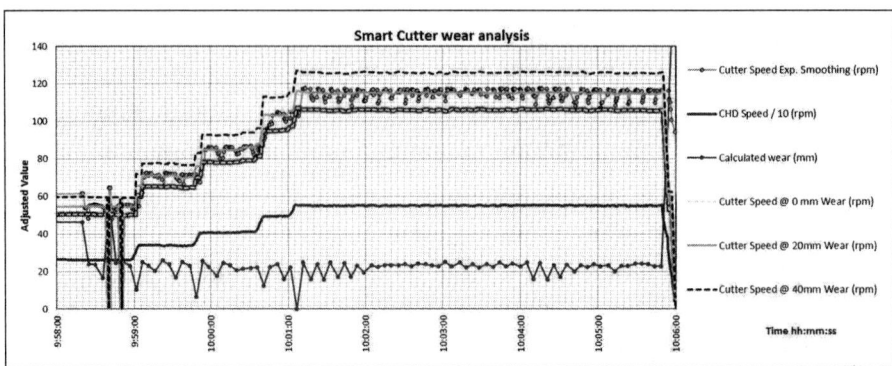

Figure 7. Eight minutes of unfiltered SmartCutter data from AMR (micro analysis)

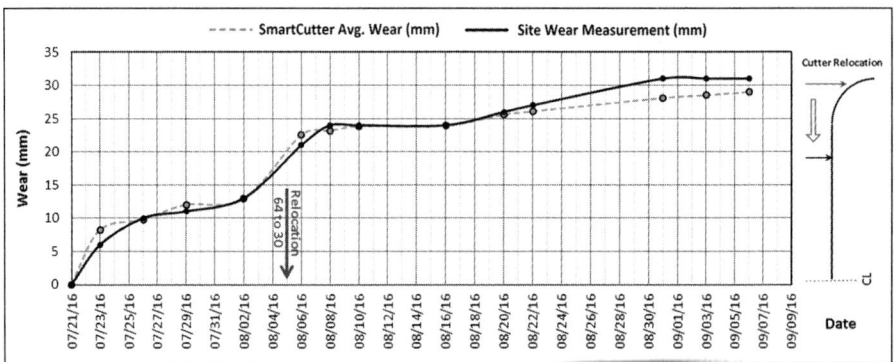

Figure 8. SmartCutter wear reading in comparison to job site measurements for SC1

complete revolution time. In a normal cutter operation ω_{DC} is expected to remain between anticipated minimum and maximum anticipated cutter speed. From this figure one can determine that the highest speed that has the majority of the data represents the normal speed in which the cutter is rolling without interruption. If there are any hiccups where less than true rolling occurs, a reduction in ω_{DC} is to be expected. The cutter wear is calculated from ω_{DC} and is plotted in the red line. ω_{DC} perfectly correlates with the changes in ω_{CHD} at every step between 9:59 to 10:01 as it was expected.

Because cutter radius from the cutterhead center is about 20 times larger than a 20 inch cutter radius at the last gage location, a small change in ω_{CHD} can have large influence in the wear calculation. Even the second decimal of the ω_{CH1D} number have influence. Although wear values showed the same trends as the jobsite measurements, an offset was observed initially. It was believed that the gap was inherited from the ω_{CHD} degree of accuracy at AMR TBM. With applying 1.03 factor in ω_{CHD} wear, the offset was resolved. Using an encoder on the drive in this particular project can increase ω_{CHD} accuracy.

Figures 8 and 9 shows the wear results of all five SmartCutters after increasing the ω_{CHD} by 1.03. The Cutter relocation is also displayed on these plots. This macro analysis shows a very close correlation of SmartCutter average wear values and the actual field measurements on the cutters, especially within the bold increases in wear (i.e. Figures 8 and 9c).

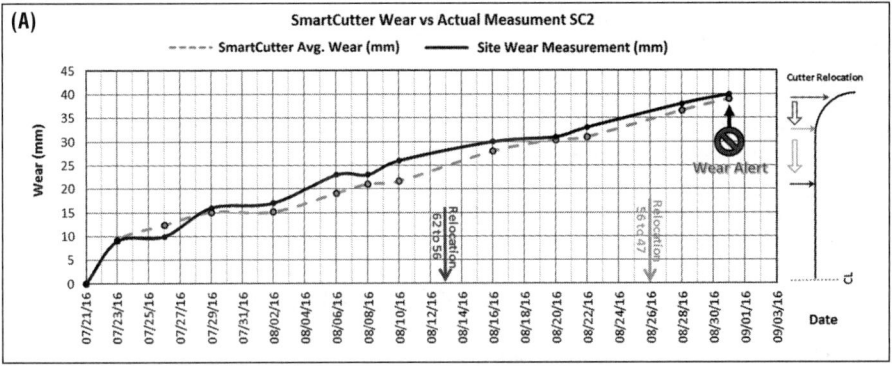

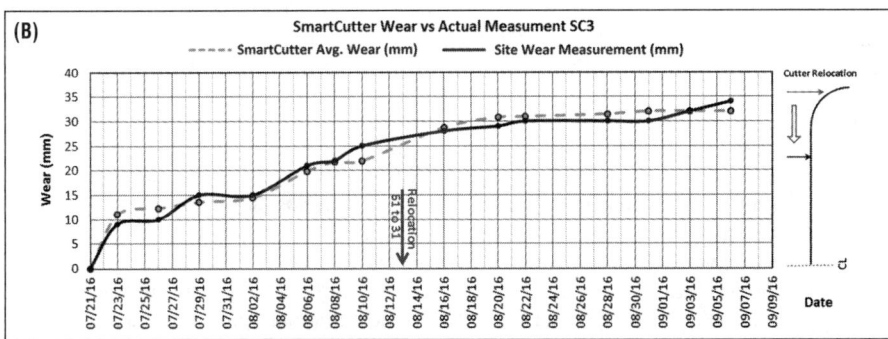

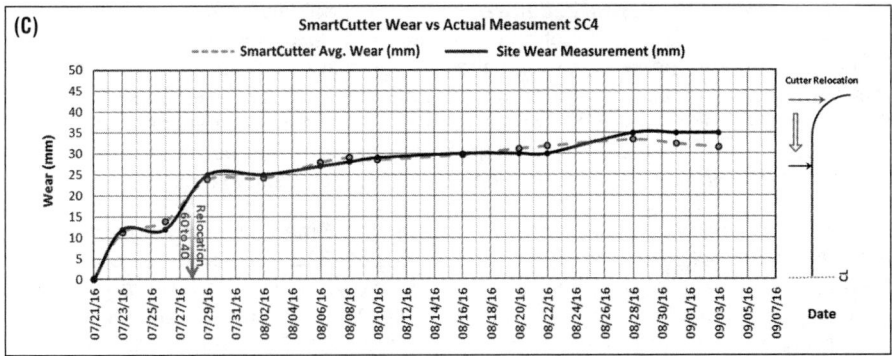

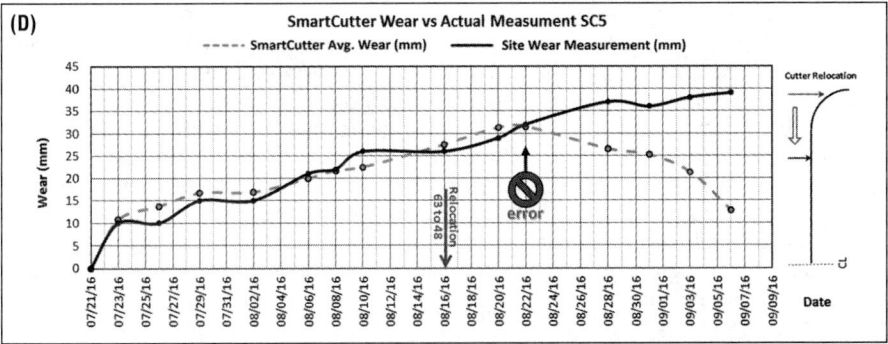

Figure 9. SmartCutter wear reading in comparison to job site measurements for SC2 to SC5

Figure 9d shows an error, which means although sensor status was correct the rotation was not measured quite satisfactorily. This is the result of interruption in cutter rolling (i.e. an uneven cutter ring wear), which caused longer revolution time and consequently lower wear values. In such conditions an error is displayed to alert the operator regarding ω_{DC} and the program will only post the maximum previous wear. The latest cutter disc radius is used to plot and display the anticipated ω_{DC} and the operator can review ω_{DC} timeline and locate when the problem starts (in this case 8/22/16). Figure 9a shows an alert that informs the operator that the cutter is close to the maximum wear. Operators can set certain wear limits for each cutter in the program alert setting. In many cases alerts can prevent unexpected cutter ring wear-related issues from causing further damage to bearings and hubs.

AMR Temperature and Vibration Analysis

The cutter instrumentation system constantly monitors the cutter assembly's temperature and displays it on the operator's screen. The main purpose of recording this parameter is to scan the cutter assembly function and identify abnormal ring temperature. In highly abrasive geologies or high strength rock material, disc cutter ring temperature will ramp up. This can cause higher wear and, if it passes the lubrication boiling temperature, it can cause bearing damage. In extreme cases it can result in seal failure due to rapid increase in internal pressure. Board temperature is also shown to confirm that in the current design the electrical board and battery is kept at a low enough temperature to maintain reliable performance. Additionally, this information can be coupled with other cutter parameters and provide an insight into cutter operation. Figure 10a shows the temperature of a brand new cutter throughout several mining periods during one day. During cutterhead standstill natural ground temperature can be observed (i.e. time 3:00 to 5:00 and 9:00 to 12:00). This plot shows that the temperature of the brand new cutter was high in the beginning (125°C) and decreases gradually. One possible explanation is that the cutter experiences high stress at the brand new cutter rounded tip corner at the gage area and as it wears for a few millimeters, the stress reduces and temperature drops under 100°C. Figure 10b also confirms it by correlating wear and temperature in macro analysis within successive days of excavation. In this plot the blue dash line shows the wear. Temperature is shown with a maximum, minimum line, and average values in green dashed line.

The wireless radio technology that is used in the SmartCutter configuration provides some limited bandwidth for data transmission. The accelerometer used in the SmartCutter reports acceleration in the X, Y, and Z direction as well as the peak acceleration. The time domain analysis of these data can provide some insight into cutting action and the geology in front of the machine. However, for distinguishing ground characteristics, frequency domain analysis will be essential, which requires an enormous amount of data to be transmitted and processed. Excavation propagates a certain range of vibration that is influenced by cutting action, cutter assembly mechanism, disc cutter ring condition, geology, and other machine noises. The hidden dimension is the power and frequency of all individual parameters. For instance, one can claim that geology-related vibration frequencies vary with material type or changing conditions within identical geologies. Figure 11 shows a time domain macro analysis on average of peak acceleration perpendicular to cutting surface within a homogeneous geology. It is very clear from this plot that as the cutter wears out the acceleration drops down gradually. Although frequency domain analysis is believed to reveal useful information on geology and cutting pattern, it is not included in the current SmartCutter package.

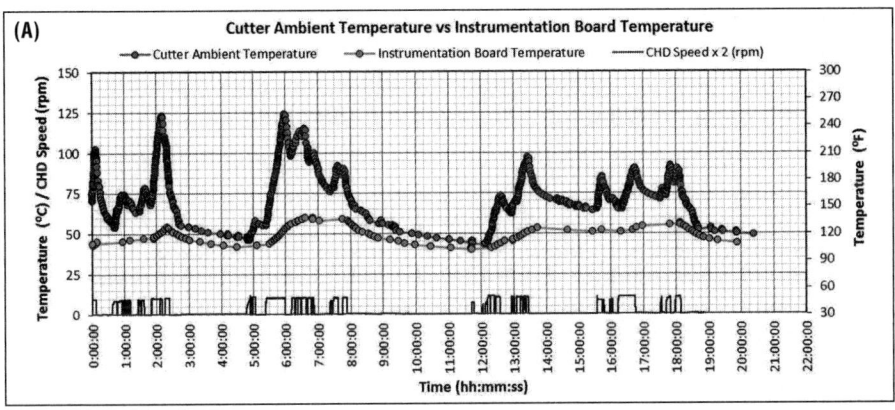

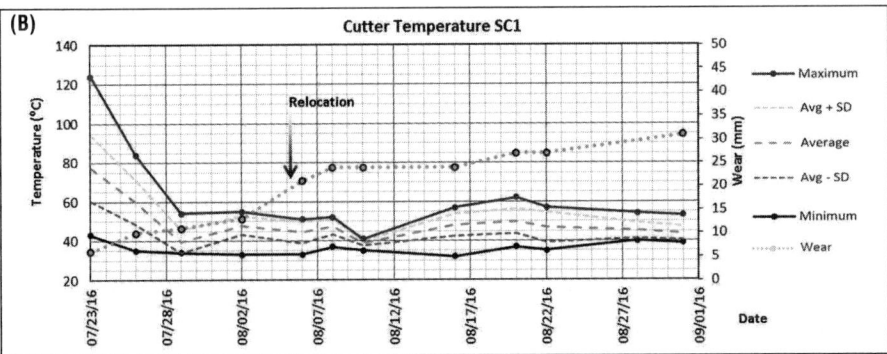

Figure 10. SC1 cutter temperature (A) during one day of mining; (B) during continuous days of mining

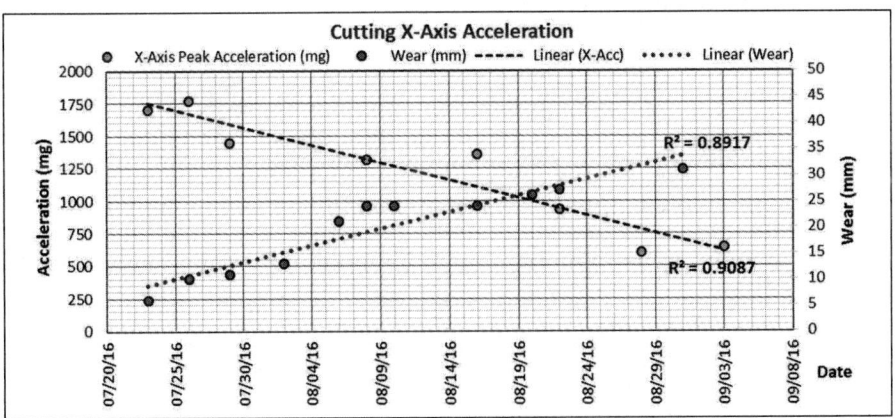

Figure 11. SC1 peak vibration in comparison with cutter wear within continuous days of operation

Future Development

The SmartCutter system has been successfully tested on several projects and the results from the AMR project have been discussed in this paper. Cutter instrumentation for other Robbins products is presently under development.

SmartCutter Version I is currently capable of measuring and recording cutter speed, wear, temperature, and vibration. A separate user-oriented screen with alerting system is designated to control the operation. The alert system can take a step forward and communicate with the PLC to reduce the TBM thrust, in case of a continuous abnormal cutter force report or identified higher strength ground through vibration. In some geologies SmartCutter can be programmed to control the cutterhead speed and thrust to achieve highest cutter production.

Lighter instrumentation, ease of installation, and longer life will lead to the SmartCutter version II design consideration. Cutter positioning sensors and strain gages are additional instruments under the design and testing phase for the next generation. As discussed earlier in this paper, about a quarter century ago Robbins tested cutters with strain gages on several projects. The results of the future testing and historical data will be presented in a separate paper. An alternative communication system is also under development to provide communication in pressurized conditions.

PART

Shafts

Chairs

Dwight Metcalf
Kiewit

Thomas Costabile
Skanska

Dugway Storage Tunnel Ground Freezing at the Shaft DST-1: A Focus Study on the Successful Application of Ground Freezing Around an Open Shaft Excavation

Jim Kabat ▪ Salini-Impregilo Healy JV
Roberto Bono ▪ Salini-Impregilo Healy JV
Giacomo Pini ▪ Salini-Impregilo Healy JV
Ryan Sullivan ▪ Northeast Ohio Regional Sewer District

ABSTRACT

The Dugway Storage Tunnel (DST) is a combined sewer overflow deep tunnel project for the Northeast Ohio Regionals Sewer District (NEORSD) in Cleveland, Ohio and will provide additional storage of combined sewer flows during wet weather events reducing the number of combined sewer discharges into the environment. The tunnel alignment is approximately 2.8 miles long, excavated with single shield hard rock TBM 27 ft. diameter. The precast concrete segments with steel fibers are the finished internal lining of the tunnel with diameter 24 ft. Depths to tunnel invert generally range from 180 to 250 ft. below ground surface. The project will have a total of six shafts with internal lined diameter between 16 to 50 ft. and four adit connections between shafts and tunnel.

The excavation of the DST-1 Access shaft, the project's main shaft for the tunnel operations, required specialized ground control methods to allow for the safe excavation of the shaft along with maintaining the stability of the existing Easterly Interceptor Sewer, the NEORSD's primary sewage conduit to their Easterly Treatment Plant. This paper focuses on the Contractor's selected method of ground freezing for the control of these ground conditions and its effectiveness.

INTRODUCTION

The principal component of the Dugway Storage Tunnel (DST) project is a 24-feet finished diameter CSO storage tunnel. The tunnel, shafts, and appurtenant facilities will be used for the collection and storage of the combined sewage and storm water runoff from the northeastern part of Cleveland, Ohio and for conveyance to the NEORSD's Easterly Wastewater Treatment Plant for treatment.

Dugway Storage Tunnel (DST)

The Dugway Storage Tunnel project (DST) includes the construction of the following elements (Figure 1):

- Main tunnel excavated with single shield 27-ft diameter hard rock TBM. Total length of the tunnel 14,840 feet (approximately 2.8 miles)—24 feet internal lined diameter.
- Six deep shafts between 180 and 250 feet with internal lined diameter between 16 to 50 feet
- Four adit connections between shafts and main tunnel.

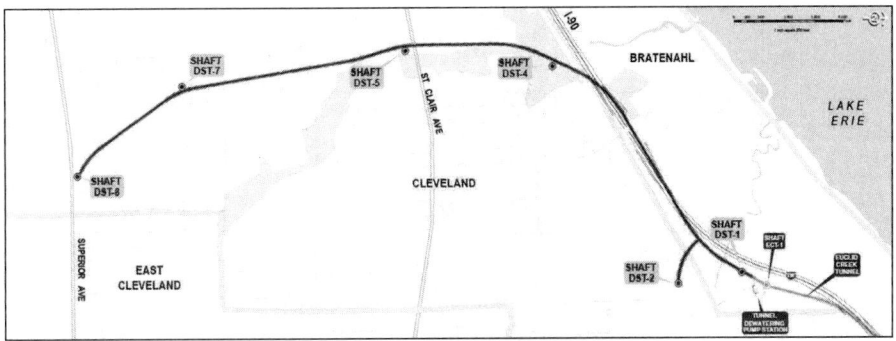

Source: NEORSD

Figure 1. General layout of the Dugway Storage Tunnel Project

- Near surface structures (Diversion Structures, Gate Structures, Control Vaults, Ventilation Vaults, Drop Manholes)
- Site/civil restoration
- Modifications to existing regulatory structures

The project area is mainly older residential (pre-1950s), interspersed with commercial properties and urban parks.

SOIL EXCAVATION OF SHAFT DST1

The construction of the 48 feet diameter DST-1 shaft through the overburden was successfully completed to El. 512 (a depth of 96 feet) on July 22, 2016. The shaft was excavated by conventional methods using an excavator loading ten cubic yard muck boxes and a 200-ton crane hoisting the boxes to the surface. Two-flange liner plates and W12×72 steel ribs were installed for lateral ground support as the excavation proceeded downward. The space between the excavation and liner plates was filled with cementitious grout.

To support the ground encountered during the excavation of the shaft between El. 542 and El. 525 feet, a ground freezing system was installed. This paper discusses the shaft DST-1 shaft site soil conditions, the ground freezing method applied to control the ground conditions between El. 542 and El. 525 feet and the effect of the frozen ground on the liner plate shaft support during both the establishment and maintenance of the freeze wall. It will describe the monitoring required as the freeze wall formed and then thawed to verify the stability of the shaft support system during the entire process.

The successful results allowed the shaft excavation from EL. 542 to EL. 512 feet to be completed safely and without ground loss and ground water infiltration.

GEOTECHNICAL SCENARIO

The Dugway Storage Tunnel DST-1 shaft was constructed through fill material, glacial till and lacustrine deposits overlaying Chagrin Shale bedrock. Specifically, the excavation consists of 96 feet through soil and 119 feet in shale rock for a total depth of 215 feet Based on the Geotechnical Baseline Report (GBR), the subsurface was divided into five engineering classes of soil material and the Chagrin Shale as shown in Figure 2 and described in the following sections.

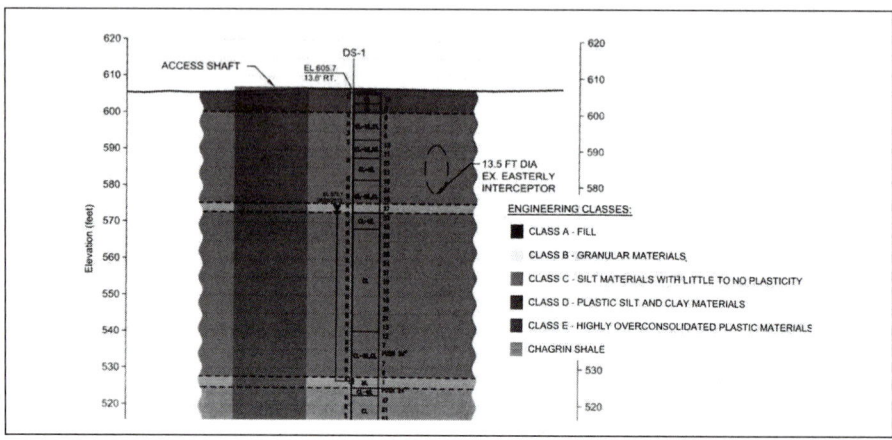

Source: NEORSD
Figure 2. Geological section of the Shaft DST-1

This section also shows the NEORSD Easterly Interceptor located within 50 feet of the DST-1 shaft. Special consideration was needed in the planning, excavation and ground freezing to insure the stability of this critical conduit to the Easterly Wastewater Treatment Plant. The type of materials involved were classified as:

- Fill material is (**Class A**) found in the upper 20 feet of the shaft and includes variable material such as topsoil, various backfill and building materials.

- Granular materials (**Class B**) is comprised of sand, gravelly sand, silty sand and clayey sand. Per the geological profile provided within the GBR, this type of material was not anticipated in the DST-1 shaft excavation. However, additional geotechnical investigation performed during construction by the JV Salini-Impregilo Healy, demonstrated the presence of this class of soil at the shaft location.

- Silt Materials with little to no plasticity (**Class C**) are comprised of silt, sandy silt, gravelly silt and clayey silt. These materials do not show a plastic behavior and exhibit a rapid dilatant behavior when tested using the field dilatancy test.

- Plastic Silt and Clay material (**Class D**) are comprised of clay, silty clay and clayey silt. The plasticity index of this class is higher than 4, so they have a non-dilatant behavior.

- Highly over-consolidated plastic material, (**Class E**) encountered during the subsurface exploration are comprised of clay, silty clay and clayey silt. Due to the over-consolidated nature, this class of soil given has a high N value and based on USCS is classified as *hard* material (B. Giurgola, 2016).

GROUND FREEZING

As the excavation of the shaft had reached a depth of approximately 64 feet (El. 542), unstable ground conditions with a water and soil inflow developed in the south-west side of the base of the shaft. The flow of water and soil was approximately 10–15 gpm. To stabilize these conditions and as a precaution due to the proximity of the Easterly sewer, loose soil material was dumped into the shaft to raise the base to El. 550 and the shaft was flooded with water to Elev. 570 ft. (equalizing the external ground water level).

Two different methods were attempted to control ground water pressure and to continue the safe excavation of the DST-1 Shaft, but neither proved to be successful. At this point it was determined that a ground freezing system would provide the highest probability of success in these difficult ground conditions. Three (3) additional boreholes were drilled around the outer perimeter of the DST-1 Shaft to determine the properties of the soil to design the freeze system properly.

Moretrench America Corporation was the company selected to design, install and maintain the ground freezing system. The ground freezing system utilized by the subcontractor was the Poetsch process, a closed-circuit system using brine. This process relies on large primary refrigeration plants to chill a secondary coolant down to negative 30 degrees Celsius, which is continuously circulated through a distribution manifold at the ground surface and refrigeration pipes which are embedded in the ground. The entire brine circulation is a close and sealed system. At no time is brine exposed to the soil and the entire process is non-toxic, non-flammable, and nonexplosive. This system created a frozen ground column around each freeze pipe with thickness between three and six feet. The refrigeration plant is powered by commercial electricity with appropriate safety controls. The Poetsch process has been employed successfully on ground freezing projects for over one hundred years, in shafts, excavations and tunnels in similar ground to the what was anticipated on this project. It is a proven and reliable means of stabilizing the earth and controlling ground water.

Design Model and Loads

The detailed freezing method developed by Moretrench included the ground thermal design, drawings and parameters to analyze the shaft area. Also, laboratory tests conducted on soil samples retrieved with Shelby tube during the supplementary boring program. The frozen earth wall was designed only as a mean of stabilizing the soil and controlling the groundwater. The liner plates, ribs and grout remained the primary shaft support system when the excavation proceeded. The thermal analysis was performed using a time dependent finite element program called TEMP/W developed by GeoSlope, Calgary, Alberta.

Figure 3 shows the anticipated soil temperature contours after 30 days of ground freezing. This profile was updated with the actual position of the freeze pipes obtained through inclinometer surveys after drilling was completed. It was later updated with actual brine temperature measured daily from the site.

With the freeze wall established next to an open structure, there was a significant concern that as the water within the soil freezes, it will expand volumetrically approximately nine percent. This expansion can create additional pressure on the rib and liner plate shaft support system already in place. To evaluate these pressures, Moretrench tested the volumetric expansion on the soils when frozen in the laboratory before the operations on site began. The results of these tests were incorporated into a PLAXIS analysis that yielded a theoretical increase in the pressures on the ribs and liner plates.

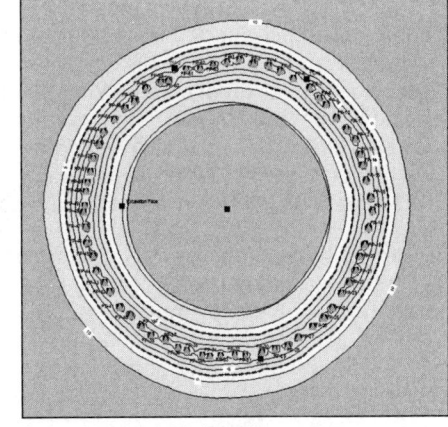

Source: Moretrench Final Report

Figure 3. Temperature contours, Day 30

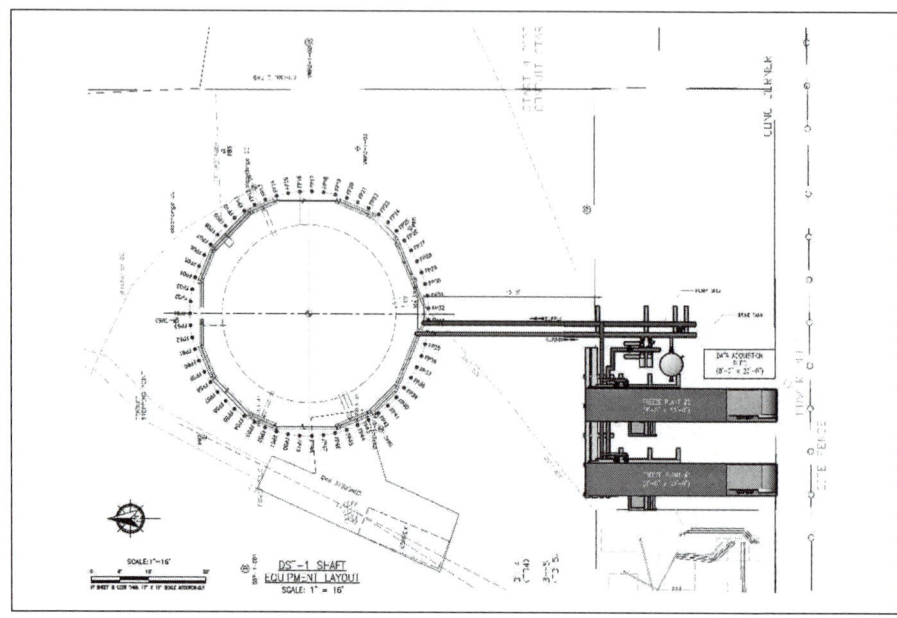

Source: Moretrench Final Report
Figure 4. Equipment layout

FK Engineering (SIH's engineer of record for support of excavation) assisted the study on the effect of the freeze wall pressure on ribs and liner plates.

To monitor closely the actual effect of this expansion, it was decided to install strain gauges on six of the shaft ribs on both north and south side and 2 load cells behind the shaft liner plates. Also, because the computed value had shown an increase in the loads on the ribs between elevation 551 and 542 feet, beyond the safe working loads for these ribs, the internal flange of the three ribs were reinforced with ¾" rolled steel plates before the full freeze wall was established.

The freeze pipes were installed to a depth of 100 feet placing them approximately ten (10) feet into the underlying Chagrin shale. This would effectively establish a vertical cut-off wall from the surface into the shale through the very low permeable soils overlying the shale (J. Sopko, 2016). The design ground temperature for thermal model had to consider the fact that cementation grouting had taken place in the some of the soil prior to the decision to freeze the ground. Temperature probes installed in the ground indicated that in the areas where the grouting had taken place, the ground temperature was higher than anticipated. Temperatures were measured at 20 degrees Celsius in grouted ground when normal temperatures were measured to be 13 degrees Celsius.

To confirm the complete water cut-off due to the freeze wall, four open pipes piezometers were installed and were monitored through automatic sensors to record any variations. Three of the piezometers were install on the outside perimeter of the wall and one inside the shaft. Additionally, three temperature monitoring pipes were installed three feet from the center of the freeze wall to record the ground temperature at ten feet vertical interval.

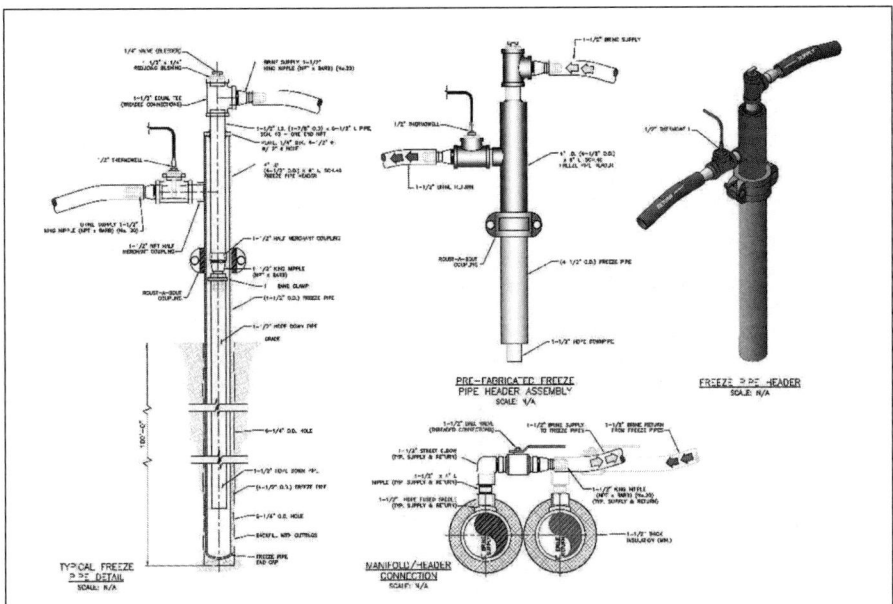

Source: Moretrench Final Report
Figure 5. Freeze pipe details

Technical Details of the Freeze System (and Brine)

The freeze was accomplished using two electrically driven 350 HP refrigeration plants which are custom built for the sole purpose of ground freezing. The refrigeration plants chilled and circulated calcium chloride brine of approximately 1.28 specific gravity though the closed pressure tested manifold system designed to ensure balanced circulation of the coolant throughout all the refrigeration pipes (J. Sopko, 2016). The system operated at flow rates of approximately 20 gpm through each pipe, consistent with the computed heat transfer coefficient used in the thermal analysis. Figure 4 shows the equipment layout next to the DST-1 shaft.

Details of the freeze pipes and recirculation connectors are provided in Figure 5.

Sequence of the Operations and Durations

The first operation to be completed on site was to drill the holes using mud rotary drilling equipment for all the 63 freeze pipes to design elevation (606 to 506 feet). Verticality test were performed in every freeze pipe to maintain consistent spacing between the freeze pipes and ensure overlap of the freeze columns at full depth. Two additional freeze pipes had to be drilled and added in locations where the spacing at the bottom of the pipes had exceeded the tolerance. All freeze pipes were then pressure tested to confirm that they were sealed.

Temperature monitoring probes were installed in the location where the freeze pipes had deviated over four feet from each other (Figure 4).

Once all the pipes were installed and tested, inner PVC pipes of 1 ½" were installed and connected with the header pipe (see Figure 5 and Figure 6 for more detail).

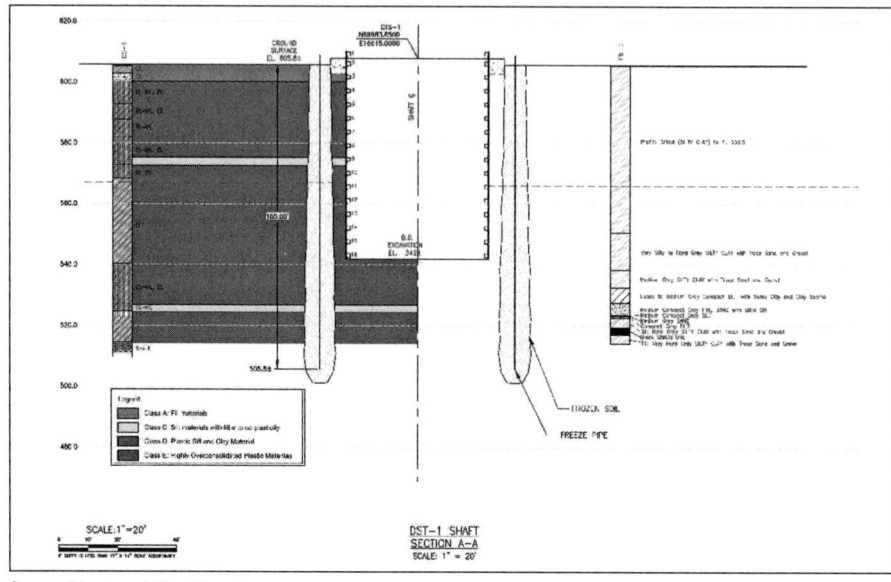

Source: Moretrench Final Report
Figure 6. Shaft section

Table 1. Main operations during freezing

Date	Main Operations
4/12/16–5/17/16	Drilling and installing freeze pipes and temperature monitoring pipes
5/18/16–5/24/16	Set-up and Installation of main supply pipoo and plants
5/25/2016	Activated the freeze plants and started supplying chill brine to the pipes
5/25/16–6/27/16	Started the freeze system and time required to form the 3 feet thick wall
6/27/2016	Freeze system switched to maintenance mode
6/28/16–7/22/16	System turned into maintenance mode as excavation progressed downward
7/22/2016	Rock excavation started (using blasting) and system was turned off

The header pipe was then connected to the brine supply and return lines from the freeze plants. Once all the connections were completed and tested, the system was filled with calcium chloride brine chilled to−30°C and recirculated through the pipes. Refer to Table 1 for a full list of the main freeze operations that took place on site.

Monitoring Procedures

The monitoring procedure during freezing operations were required to be extremely accurate and precise to detect very small changes in the system parameters and the ground temperatures. The primary data points to be analyzed were:

- the supply and return temperatures from the manifold to the freeze plants,
- the flow rate from both plants,
- the supply pressure and return pressure
- the temperature of each freeze monitoring point set at 10 feet vertical intervals in the temperature monitoring pipes.

The brine supply pressure (21 psi) and return pressure (11 psi) and brine flow (330gpm) were also monitored to detect possible leaks or system failure.

The freeze pipes and plants supply and return temperature were recorded and monitored to analyze the ground temperature variations over time. The supply temperature of the brine decreased in the first weeks from 26.5 °C to −30 °C during the initial system set-up. The return temperature decreased from −23°C to −28°C at slower rate as expected due to the energy needed to the lower the temperature in the ground as the operations progressed forward. Three temperature monitoring pipes were installed around the outer perimeter of the shaft at three feet from the center of the freeze pipes. Temperature probes were installed in these pipes at ten feet vertical intervals to the full depth of 100 feet. The temperature was recorded continuously at each of the probe location. This data was plotted into temperature profiles as the freeze wall formed. A slight variation in the temperature profile would indicate locations and elevations where the freeze wall may not be building up to the design wall thickness. In this situation, the freezing time would have to be increased to assure the establishment of the full three-feet thick design freeze wall in every location.

However, as mentioned earlier, increasing the freeze time presented an additional problem. Because of the proximity of the freeze wall to the existing rib and liner plate shaft, increasing the freeze time would likely increase the wall thickness and could force the failure of the shaft support system. To monitor the effect, that the growing freeze wall was having on the shaft ribs and liner plates, strain gauges were placed on the internal and external flanges and the web of Rib No. 8, 9, 10, 15, 16, & 18 and load cells placed behind the liner plates above Rib No. 8 (refer to Figure 6 for location of the ribs) With these instruments in place, the increase in the load on the ribs were measured as the freeze wall increased in sized and this increase was then compared to the computed increased values. Additional monitoring of two existing inclinometers approximated 50 feet from the outer perimeter of the shaft took place to measure the horizontal displacement on the ground due to the freeze wall formation. This was correlated to the increase in load to the shaft wall. Figure 7 shows an aerial view of the shaft and freeze pipes.

The other relevant source of information during the monitoring procedure of the shaft conditions, were the three (3) open standpipe piezometers installed around the outside of the shaft and the one installed inside the shaft. An increase in the level of the water in these piezometers normally would indicate the closure and full formation of the freeze wall.

Figure 7. Aerial view of the shaft and freeze pipes

Monitoring Results and Effects of the Establishment of the Freeze Wall

The temperature profiles in Figure 8 shows a generally consistent downward temperature trend around the entire perimeter of the shaft. The strain gauge readings in Figure 9 shows an increase in the stress on the ribs during the establishment of the freeze within the expected range. The inclinometer readings in Figure 10 had indicated a horizontal ground deflection of 0.35 inches toward the shaft before the freeze was established and a deflection of 0.25 inches away from the shaft after the freeze was in place. The open standpipe piezometers had no significant change in elevation.

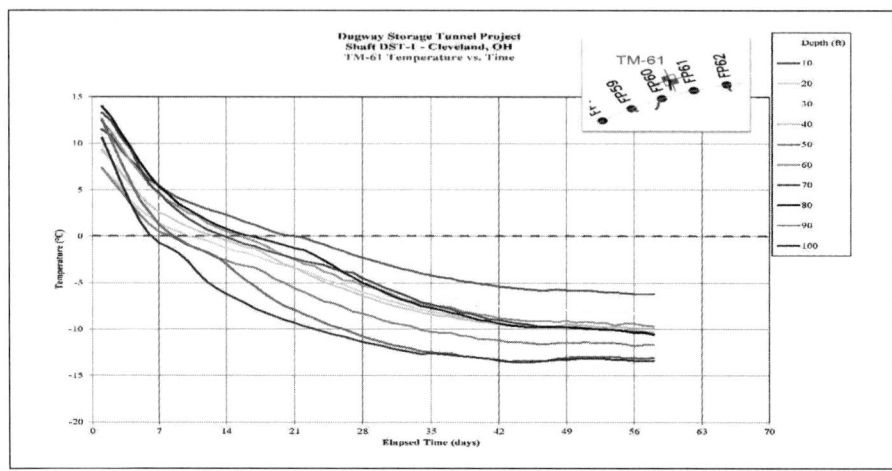

Source: Moretrench Final Report
Figure 8. Temperature monitoring

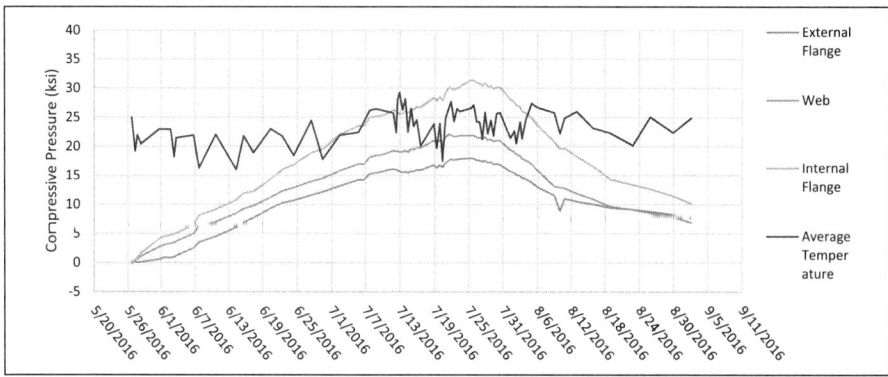

Figure 9. Strain gauges on Rib 8 South Station

On the June 20, 2016, the temperatures recorded on the temperature monitoring pipes were below –7 degrees Celsius at every elevation, and all return temperatures at every location around the shaft showed no differential in temperature. The outside piezometer did not show any ground water level variation during the previous three weeks. Based on this data, it was decided to proceed with the drawdown of the water in the shaft.

The drawdown was performed in three stages over three days. In the first stage the water in the shaft was drawn down ten feet at a rate of 330 gal/min while, monitoring closely any variation of temperature at all locations and the ground water level in the open standpipe piezometers. These reading remained stable as the water was drawn down. The water level in the open pipe piezometer inside the shaft was decreasing with the level of the water in the shaft, as expected, due to the communication between the permeable layer at the bottom of the shaft and the piezometer.

The same procedure was followed for the second and third stages of the shaft water drawdown. As the pumping operation were completed, all the monitoring indicated that the ground was stable, and the piezometer readings collected continuously were not detecting any changes in the water level around the shaft.

At this point the freeze system was switched to maintenance mode (injection brine initially −27 °C then increased to −25°C) with the purpose of slowing down the growth of the freeze wall thus reducing further loading of the shaft support system. As seen in Figure 8, there was a general decrease of temperature at different depths in Temperature Pipe No. 61 as the freeze wall was established. However, a general leveling trend in the temperatures occurred as the maintenance mode took effect on Day 35.

As the freeze system was switched into maintenance mode the strain gauges and load cells to the shaft did not show an immediate leveling off or decrease in the load on the shaft support. To relieve some of the pressure on the ribs, the blocking between the rib and liner plate was systematically removed and replaced. Figure 9 show change in compressive pressure on the ribs from the start of the freeze, to the maintenance, to the completion of the excavation and thawing of the freeze wall.

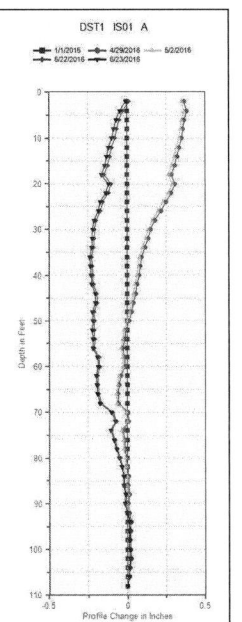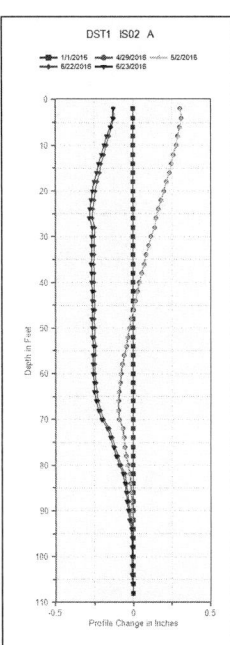

Figure 10. Inclinometer reading IS-1-01 and IS-1-02 (direction towards the shaft)

An additional consideration was made during the excavation of the remaining soft ground section of the shaft. Because it was understood that the shear strength of ground that had been frozen and then thawed would be reduced, the spacing of the remaining shaft ribs was changed from 4.5 feet to 3.0 feet on center.

One month later, on July 22, 2016, the soft ground excavation was completed, the concrete collar at the rock soil interface was placed and the freeze system switched off. It is estimated that it took approximately two month after the completion of the excavation for the freeze wall thaw.

CONCLUSIONS

Freezing the ground did provide the expected results with no site settlement around the shaft, no boils or intrusion of fines into the shaft and no drop in the ground water level. The method was costly and time consuming, taking 45 days to install and 35 days for the freeze wall to form. It is, however, an assured method for controlling difficult ground and groundwater conditions in deep shaft excavations. Further, installing a freeze system around an open excavated shaft proved to be feasible, when proper instrumentation and monitoring is applied.

In the future, additional temperature monitoring pipes should be installed to guarantee the best control of the temperature around the shaft and consequently confirm the thickness of the freezing wall with more certainty.

The success of this activity at the Dugway Storage Tunnel DST-1 Shaft can be attributed to the cooperation between the Client Northeast Ohio Regional Sewer District, the Main Contractor Salini-Impregilo Healy JV and the subcontractor Moretrench America Corp.

REFERENCES

Giurgola, B., (2016) Salini-Impregilo Ground Water Control DST-1—Final Report.

Sopko, J., (2016) Morethench Dugway Storage Tunnel—Final Report.

Design of the Hemphill Deep Pump Station Shafts Using Blind Bore Drilling Techniques

Yong Wu ▪ Stantec Consulting Services Inc. (Part of JP2/PRAD, Stantec, and Chester Engineers JV)
Tao Jiang ▪ Stantec Consulting Services Inc. (Part of JP2/PRAD, Stantec, and Chester Engineers JV)
Julian Prada ▪ Stantec Consulting Services Inc. (Part of JP2/PRAD, Stantec, and Chester Engineers JV)
Brian Jones ▪ City of Atlanta, Department of Watershed Management

ABSTRACT

The Hemphill Pump Station, part of the City of Atlanta Raw Water Supply Program, consists of five shafts each equipped with a deep submersible pump. The shafts are approximately 420 feet deep, 9.5 feet in excavated diameter, and will be constructed from the ground surface using blind bore drilling techniques primarily in hard and abrasive metamorphic rocks. Welded steel liners will be lowered into the shafts and grouted in place. A grouting program is implemented before commencing shaft excavations to reduce the rock mass permeability and to prevent potential drilling fluid losses during shaft development. Plaxis3D is used to model the junctions between the shafts and adjacent tunnel to ensure the stability of the pillars between the shafts.

PROJECT BACKGROUND

The current water supply program operated by the City of Atlanta (COA) Department of Watershed Management (DWM) consists of four aged raw water pipelines, one of which dates back to the early 1890s. Based on previous assessments completed by DWM, the entire system is at or will soon reach its recommended useful life. As such, the COA acquired the Bellwood Quarry in 2006 with the long-term goal of converting the quarry to a raw water storage facility with a volume of approximately 2.4 billion gallons. The water storage facility will greatly enhance the reliability of the drinking water supply to the greater Atlanta metropolitan area.

The Chattahoochee River is the source of water supply to the quarry facility. The facility will be operated in an "offline" mode, with raw water being stored in the quarry before being withdrawn for treatment at the Hemphill Water Treatment Plant (HWTP) and/or Chattahoochee Water Treatment Plant (CWTP). The offline operating mode will include routine withdrawal of water from the quarry and replenishment with Chattahoochee River water.

A conveyance system is required to connect the quarry, HWTP, CWTP, and the Chattahoochee River. The conveyance system includes an approximately 4.5-mile-long TBM tunnel with a finished diameter of 10 feet and multiple shafts and connecting adits. The overall project has been divided into two design packages. The first package, termed the Phase 1, connects the quarry to the HWTP, which includes a TBM tunnel of approximately one mile long and four shafts and associated connecting adits at the quarry site. Quarry highwall stabilization is also part of the Phase 1 package. The stabilization measures include scaling, rock bolting, and installation of a drape system up to 300 feet high. The second package, termed the Phase 1 Extension, connects the HWTP to the CWTP and the Chattahoochee River, which includes a TBM tunnel of approximately 3.5 miles long, five shafts and associated connecting adits at

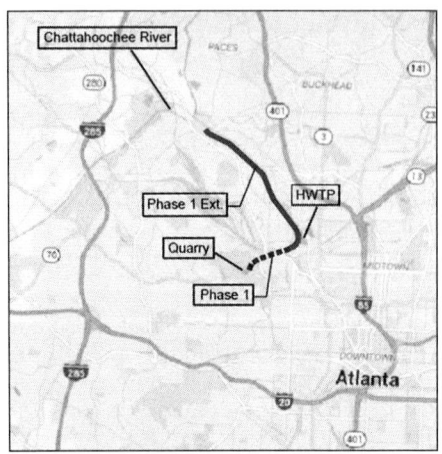

Figure 1. Project location

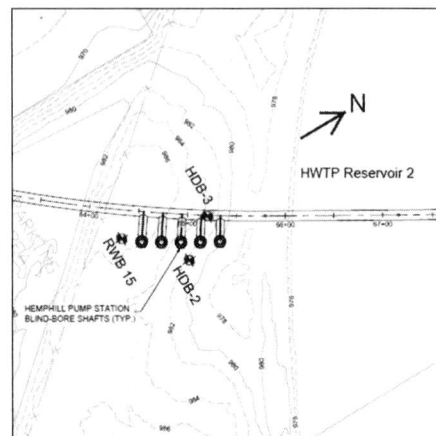

Figure 2. HWTP site plan

the HWTP site, and one shaft at the end of the TBM tunnel near the CWTP and the Chattahoochee River.

The project location is shown in Figure 1, which is generally in the Northwest part of downtown Atlanta, Georgia. As shown in Figure 2, five shafts are planned at the HWTP site to provide connections to the TBM tunnel for water access. The center-to-center spacing between adjacent shafts is approximately 20 feet. The shafts are offset approximately 30 feet to the right (east) of the TBM tunnel and are between 100 and 200 feet south of the HWTP Reservoir 2. Each shaft is approximately 420 feet deep with a minimum excavated diameter of 9.5 feet in the bedrock. Steel liners with an inside diameter of 76 inches will be installed in the shafts. Connections between the shafts and the TBM tunnel are through the adits. The adits will have a modified horseshoe-shaped excavation face, approximately 10 feet both in width and in maximum height. The adits will be lined with cast-in-place concrete. The finished inside diameter will be eight feet.

Blind bore drilling techniques from the ground surface will be used to excavate the five shafts at the HWTP site. The adits will be excavated using controlled drilling and blasting techniques. This paper focuses on the design of the blind bore shafts as well as the stability of the pillars between the adits.

GEOLOGIC SETTINGS

Regional Geologic Setting

The project is located in the Piedmont Physiographic Province (McConnell and Abrams, 1984). The geology of the Piedmont in the greater Atlanta area generally consists of medium-grade metamorphic rock with granitic intrusions. These crystalline rocks are some of the oldest rocks in the Southeastern United States. They were generally formed before and during the building of the Appalachian Mountains. Since their origin, the rocks have undergone a complex history of metamorphism, weathering, and deformation.

Various structural features are present in the rocks, including folds, fractures, and lineaments. The high pressures and temperatures at great depths resulted in a full range of deformational styles, ranging from medium-grade metamorphism, through fully-welded ductile shearing and mylonite formation, to brittle fracturing with rocks that commonly contain hydrothermally deposited minerals. At shallower depths, structures like exfoliation fractures were formed in the rocks due to erosion of overburden and unloading. The exfoliation fractures mainly occur along the foliation "planes" and tend to be open and act as conduits for water movement through the rock mass.

Lineaments, which are surface topographic expressions of underlying rock mass or crustal structure, occur throughout the Piedmont. The lineaments are often controlled by weathering associated with discontinuities in the rock. In many cases, the lineaments represent fracture zones in the underlying bedrock. At depth, the fracture zones are typically cemented with weathered minerals. At shallower depths, erosion of these weathering minerals often results in zones of broken, water-bearing rocks and topographic features such as valleys and draws.

Soil to Rock Transition

The ground conditions at the project site can be divided into a soil zone, an underlying transition zone, and a bedrock zone. The transition zone and, to some degree, the components of the soil zone, are derived from the underlying bedrock as a consequence of weathering. Since weathering is facilitated by fractures, joints, and rock compositions, contacts between these zones are anticipated to be highly irregular.

The soil zone mainly consists of thoroughly degraded residual soils, although layers and lenses of rock and partially weathered rock can occur locally within the soil zone. The transition zone primarily consists of partially weathered rock; however, layers or lenses of soils and/or fresh rock can be present. The rock mass strength of the transition zone is typically much lower as compared to the bedrock zone due to the presence of abundant weathered joints and reduced intact strength of partially weathered rock.

The bedrock zone is dominated by fresh rock. The HWTP site resides in the Clairmont Melange geologic unit, which includes variable gneisses, schists, and granites. The Clairmont Melange has a poorly to welldeveloped foliation which is typically low angle and undulatory. Lithology within this rock unit is extremely contorted, with quite variable foliations over short distances. In general, random fractures are abundant in the unit, while through-going joint sets are scarce and not well-developed.

Groundwater

In the Atlanta area, the primary groundwater source is infiltration from the ground surface into the overlying soil zone. Precipitation consistently recharges the soil zone. The transition zone typically contains abundant open fractures and can become a major storage source for groundwater. The bedrock zone has fewer open fractures with depth than the transition zone. However, large fractures with the ability of producing large volumes of water do exist in the bedrock. Within the bedrock, groundwater is typically conveyed along foliation joints, fractures, veins, and other geologic features that have been enhanced by differential weathering.

GROUND CONDITIONS AT HWTP SITE

As shown in Figure 2, three deep borings (HDB-2, HDB-3, and RWB-15) were drilled in the vicinity of the blind bore shaft locations. Based on the information collected from these borings, the expected subsurface profile and groundwater level are presented in Table 1.

Highly weathered/fractured rocks were encountered in the borings, particularly in RWB-15. Distribution of Rock Quality Designation (RQD) measured from the recovered rock cores is presented in Figure 3, which indicates that approximately 72% of the encountered rocks are good to excellent, 18% are fair, and 10% are poor to very poor. Rock Mass

Table 1. Expected subsurface profile at HWTP blind bore shafts

Profile	Top Elevation, ft.
Ground Surface	983±4
Groundwater Level	947±5
Top of Transition Zone	947±8
Top Bedrock Zone	936±8

Rating (RMR) distribution, developed from the RQD along with intact rock strength, joint spacing, joint condition, and packer testing results, is presented in Figure 4. This figure indicates that approximately 30% of the rock mass is characterized as "very good," 65% of the rock mass is characterized as "good," and 5% of the rock mass is characterized as "poor" to "fair." However, the actual RMR rating at the HWTP site is anticipated to be somewhat lower than what Figure 4 indicates due to the following reasons: (1) the intact rock strength used in the RMR evaluation was obtained from rock cores recovered from other borings. The tested rock samples are less weathered that the rocks encountered in the HWTP borings; and (2) no packer testing was performed in RWB-15 due to borehole instability concerns. It is anticipated that the ratings for the groundwater would be lower had the packer results from RWB-15 been available and included in the RMR evaluation.

PRE-EXCAVATION GROUTING

A suite of borehole geophysical tests were performed in HDB-2 and HDB-3. The geophysical tests were not performed in RWB-15 due to borehole instability concerns as indicated by the highly fractured rock core samples recovered from this boring. As shown in Figure 5, analysis of the geophysical data from HDB-2 and HDB-3 indicates one low-angle foliation joint set and two high-angle non-foliation joint sets. The geophysical data also indicate numerous fractures within the identified joint sets, which contain apertures ranging from 0.25 to 5 inches. These open fractures could act as water communication conduits during the blind bore shaft sinking operations. Consequently, a pre-excavation grouting program was developed and carried out at the HWTP site.

The pre-excavation grouting program was designed to address several concerns that were raised during the design. First, construction of the shafts could potentially pose significant risks to the HWTP reservoir as the shafts are located in close proximity of the reservoir. The City required that all risks associated with inadvertent dewatering of the HWTP reservoir due to construction of any aspect of the project be kept

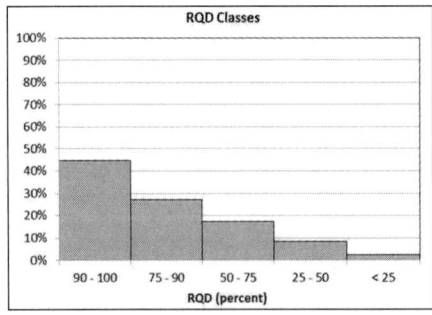

Figure 3. RQD distribution

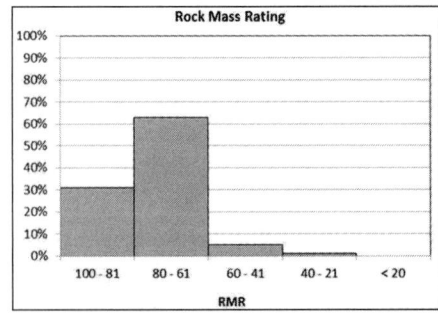

Figure 4. RMR rating

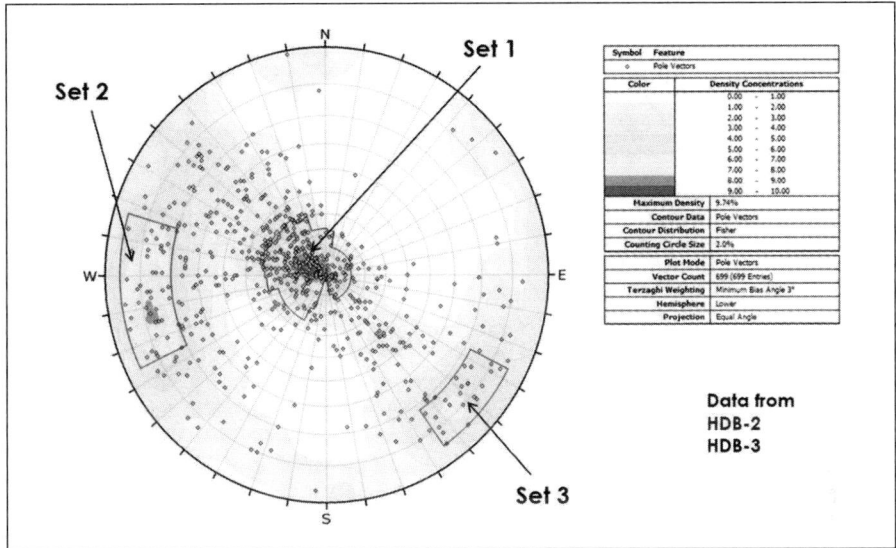

Symbol	Feature	
o	Pole Vectors	

Color	Density Concentrations	
	0.00 - 1.00	
	1.00 - 2.00	
	2.00 - 3.00	
	3.00 - 4.00	
	4.00 - 5.00	
	5.00 - 6.00	
	6.00 - 7.00	
	7.00 - 8.00	
	8.00 - 9.00	
	9.00 - 10.00	

Maximum Density	9.74%
Contour Data	Pole Vectors
Contour Distribution	Fisher
Counting Circle Size	2.0%
Plot Mode	Pole Vectors
Vector Count	699 (699 Entries)
Terzaghi Weighting	Minimum Bias Angle 3°
Hemisphere	Lower
Projection	Equal Angle

Data from
HDB-2
HDB-3

Figure 5. Stereoplot of measured joints at HWTP site

to an absolute minimum. Second, since blind bore drilling techniques will be used to construct the shafts, cuttings will be removed from the shafts by reverse circulations. Therefore, the shafts need to be filled with drilling fluid and it is critical to prevent drilling fluid loss during shaft sinking operations. In addition, the shaft excavation will be concurrent with the TBM tunnel mining. Some shafts will be completed before the TBM arrives at the HWTP site, and some of the shafts will be drilled after the TBM has mined through the area. A potential for drilling fluid loss and TBM tunnel flooding exists. Considering all the risks and the highly fractured rocks encountered in the borings, a pre-excavation grouting program was deemed necessary. The main purpose of the grouting program is to lower the rock mass permeability such that potential risks for loss of the drilling fluid as well as water communication between the shafts and the HWTP reservoir can be reduced. In addition, properties of the rock within the grouted zone will be improved through consolidation of the rock mass, thus to facilitate shaft sinking and adit excavation operations.

In order to enhance the potential for intersecting the identified joint features as indicated from the geotechnical investigation and analysis, inclined grout holes were selected. The holes were oriented 10° off vertical at a bearing of 260°. Note that due to site restrictions, some of the grout holes were kept vertical. A typical cross section of the pre-excavation grouting is shown in Figure 6.

The grouting program consisted of 42 primary holes and 44 secondary holes (71 inclined and 15 vertical) with a staggered spacing of eight feet between the primary and secondary holes. The area of the target grouted zone is approximately 6,000 square feet. Grouting was generally performed from bottom up in 20-foot stages. The grouting work was completed by Hayward Baker, a subcontractor of the Atkinson/Technique Joint Venture, in October 2016. Analysis of the grout data indicates that favorable results have been achieved and the pre-excavation grouting program has fulfilled the design intent.

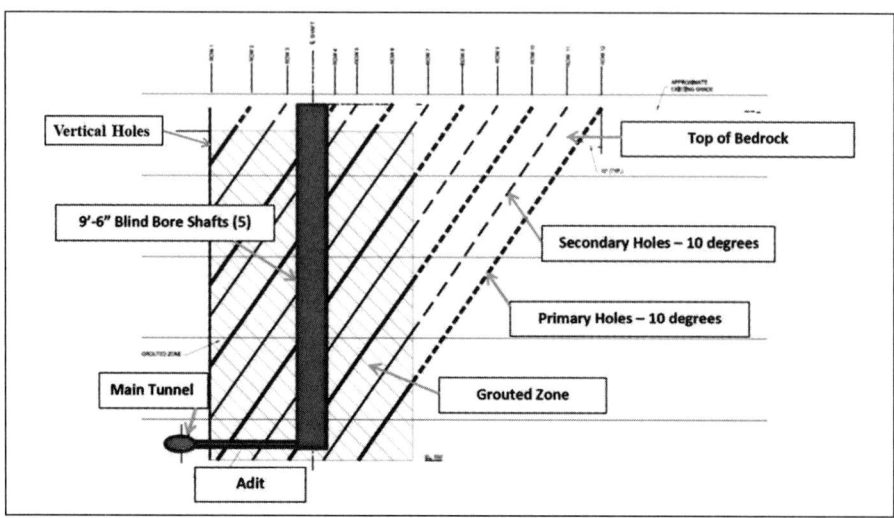

Figure 6. Pre-excavation grouting from ground surface

BLIND BORE SHAFTS

The five shafts at the HWTP site will be constructed using blind bore drilling techniques from the ground surface. Following completion of the pre-excavation grouting, permanent steel casing will be installed down to the top of the bedrock at each shaft location. The steel casings have an inside diameter of 11 feet with a wall thickness of ¾ inches. A two-pass approach will be used to excavate the overburden and the transition zone. An 8-foot-diameter shaft will be drilled down to refusal first. Once this is completed and the top of rock elevation is verified, a core barrel with over cutter arms will be used to cut the excavation out to 12 feet. Bentonite slurry will be used to stabilize the excavation during both drilling processes. After the excavations have been completed, the 11-foot-diameter steel casing will be lowered down to the bedrock and the annulus will be backfilled with grout. The casings will be installed by A.H. Beck Foundation Company, Inc., who is subcontracted with the Atkinson/Technique Joint Venue.

Upon completion of the surface casing installation, blind bore drilling of the shafts in rock will begin. The shaft drilling rig has a rotary table which provides the torque or turning action for the reamer. Throughout the entire shaft development, both the shaft and the hollow drill string are filled with water to create two independent columns of water. The water column inside the drill string is made much lighter by injecting compressed air. The heavier water column inside the shaft thus pushes down and across the bottom of the shaft. The water is then forced through a small opening on the reamer body and displaces the lighter water in the drill string to create upward flow (reverse circulation). The reverse circulation generates tremendous vacuum at the reamer opening and removes the cuttings from the face. Maintaining a constant water level in the shaft during the entire drilling operation is critical. In addition to cuttings removal, the water also provides outward pressure on the shaft wall to improve the shaft stability (Fiscor, 2009). The returned water from the shaft is collected in a surface pond adjacent to the shaft site, and the water is re-circulated to the drilling operation after the cuttings have settled out.

Keeping verticality of the shafts within the specified tolerance is critical for the pumps to function as designed. With well-controlled blind bore drilling, accuracies of 0.25% are attainable. Higher accuracies up to 0.1% have been achieved by drilling a small diameter pilot hole and using it as guidance during the blind bore drilling (Keeble, 1990). The pilot holes for this project will be directionally advanced utilizing an optical technique that allows for continuous monitoring of the deviation. Upon completion of the directional drilling, an optical survey will be performed to verify that the pilot holes meet the required verticality. Once the verticality is verified, a 9.5-foot-diameter reamer will be used to ream the shaft to its full diameter. The reamer is designed to follow the pilot holes to ensure the shaft verticality.

Upon completion of the blind bore drilling, a steel liner with an inside diameter of 76 inches will be installed in the shaft and grouted in place. The steel liner will be fabricated in 40-foot-long sections. The sections will be lowered down into the shaft in sequence and connected together using full penetration welds during the installation. All welds will be ultrasonically tested.

It is anticipated that backfilling the excavation below the steel liner and the liner annulus will take place continuously but in multiple stages. Fixed length steel lines with threaded connections will be utilized as the tremie pipes. Grout placement for each stage will start with the end of the tremie pipes being embedded into the grout of the previous stage. The tremie pipes will not be moved during the grout placement. The pumps utilized for pumping the grout will be capable of overcoming the additional pressure head associated with the tremie pipes being embedded more than 15 feet. During the grouting operation, the grout elevation will be monitored to ensure that the rising of the grout around the steel liner will be fairly uniform. The entire shaft will be kept full of water during the steel liner placement and grouting process. Thus, the external grout pressures are offset by the water pressure inside the steel liner. Dewatering of the shaft is not allowed until the grout is cured and reaches a compressive strength of at least 1,500 pounds per square inch (psi).

Due to the strict requirements of the vertical turbine pumps, the steel liner is required to be installed such that its center is within one inch of the shaft centerline at the surface as shown on the Contract Drawings. Inclination of the liner in the shaft is required to be no greater than two inches per 100 feet section and a cumulative total of four inches for the overall depth of the shaft. Upon completion of the steel liner installation, downhole survey will be performed to verify that these tolerances have been satisfied.

STEEL LINER DESIGN

During normal operations, the shafts at the HWTP site are filled with water and therefore, the steel liners are not in danger as the internal water pressure would offset the external pressure. However, conditions may occur that the shafts need to be dewatered; therefore, the steel liners are required to be capable of withstanding full external pressure without internal balancing pressure. The critical failure mode in this case is buckling.

Buckling of the steel liner occurs at a critical circumferential stress at which the liner becomes unstable and fails. The groundwater level at the HWTP site is approximately 370 feet above the bottom of the liner; thus a hydrostatic pressure of 240 psi (equivalent to a 370-foot water head multiplied by a factor of safety of 1.5) was used for computing the minimum required steel liner wall thickness.

Analytical methods have been developed for determining the critical buckling pressure for a cylindrical steel liner. It is noted that different analytical methods produce different results; thus, it is prudent to use several methods in a design. For this project, analytical methods developed by Jacobsen (1974), Vaughan (1956), and Bulson (1985) were used to compute the required steel liner wall thickness in order to withstand an external pressure of 240 psi. Equations for these analytical methods can be found in USACE (1997) and Hunt et al. (1995).

The analytical method developed by Jacobsen assumes that the buckling failure involves formation of a single lobe parallel to the axis of the liner. Determination of the critical external buckling pressure for a cylindrical steel liner with a given diameter and wall thickness requires solution of three simultaneous nonlinear equations with three unknowns. The method developed by Vaughan is not based on the single lobe assumption; instead, it is based on distortion of the liner represented by a number of waves. In the analysis using the Bulson's method, the backfill grout between the steel liner and the surrounding rock was treated as a "rigid" filter. The surrounding rock provides restraint to deformation of the steel liner in proportion to its compressibility. The critical external buckling pressure of the steel liner is dependent on the stiffness of the liner itself and the surrounding rock. For the purpose of the liner buckling analysis, the surrounding rock mass was conservatively assumed to be poor.

All calculations were implemented in a Mathcad application. Results of the analyses indicate that a minimum wall thickness of ½ inches is required for the 76-inch inside diameter steel liner.

PILLAR STABILITY ANALYSIS

The five blind bore shafts are connected to the main TBM tunnel through five short adits. The pillars between adjacent adits are only about nine feet in thickness. Given the in-situ stress level at the adit depth, which is about 420 feet below the ground surface, and the surrounding rock conditions encountered in the borings, the stability of the thin pillars is a big concern in the design, particularly if overbreak occurs during the adit excavation. As such, a three-dimensional finite element model, using the computer program Plaxis3D, was developed to investigate stress concentrations and pillar stabilities. Computed stress distributions in the vertical and horizontal directions around the junctions are shown in Figures 7 and 8, respectively.

The computed stresses in the pillars are lower than the strength of the surrounding rock mass; therefore, with proper excavation methodology and timely initial support, the pillars are anticipated to be stable. In the Contract Documents, the sequence of the adit excavation is specified such that concurrent excavation of adjacent adits is

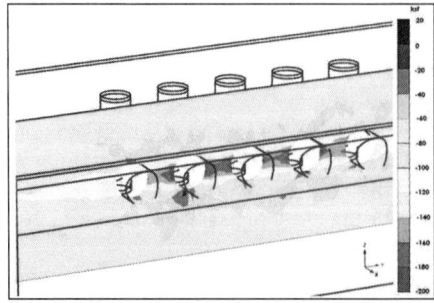

Figure 7. Vertical stress distribution

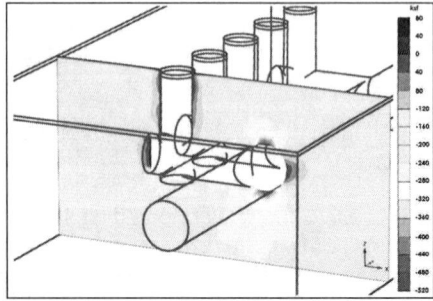

Figure 8. Horizontal stress distribution

not allowed. The maximum excavation length for each round is limited up to four feet and initial support, consisting of steel ribs with hardwood blocking and welded wire fabric lagging, is required to be installed prior to excavating the next round. In order to minimize overbreaks, the adits will be excavated using controlled blasting with line drilling around the periphery of the excavation.

CONCLUSIONS

The City of Atlanta is building resiliency into its water infrastructure by implementing the raw water supply program and converting the former Bellwood Quarry into a water storage facility. The pump station shafts at the HWTP site are key components of the project. It will be the first blind bore shaft construction in the Atlanta region.

This paper presents various design considerations given to the shafts and connecting adits. The preexcavation grouting has been completed. Installation of the permanent steel casings in the overburden and transition zones is currently ongoing. The blind bore shaft sinking operation is anticipated to start in early 2017.

ACKNOWLEDGMENTS

The authors would like to thank the City of Atlanta Department of Watershed Management for making this project possible and Stantec's tunnel designers Adam Bedell, Don Del Nero, Steve Fradkin, Konner Horton, and Rick Ponti, who have also made significant contributions to the project.

REFERENCES

Fiscor, S. (Editor-in-Chief). 2009. Blind Shaft Development—SDI Breaks New Ground in Shaft Development and Ground Stabilization. Coal Age Magazine, February, pp. 37–40.

Hunt, S.W., Heuer, R.E., and Safdar, A.G. 1995. Casing Collapse at the CT-8 Dropshaft in Milwaukee. 1995 RETC Proceedings, pp. 197–218.

Keeble, S. 1990. Alternative Methods of Shaft Sinking through Water-Bearing Strata. The Mining Engineer, December, pp. 207–214.

McConnel, K.I. and Abrams, C.E. 1984. Geology of the Greater Atlanta Region. Department of Natural Resources, Environmental Protection Division, Georgia Geologic Survey. Bulletin 96.

U.S. Army Corps of Engineers (USACE), 1997. EM 1110-2-2901 Engineering and Design Tunnels and Shafts in Rock. May 30.

Hecla Mining Lucky Friday No. 4 Shaft Challenges and Possibilities

George Sturgis ▪ Hecla Limited
David Berberick ▪ Hecla Limited
William Strickland ▪ Hecla Limited
Matthew Swanson ▪ Cementation
Eunhye Kim ▪ Colorado School of Mines
Gabriel Walton ▪ Colorado School of Mines

ABSTRACT

Hecla Mining Company has completed sinking the No.4 Shaft. The 5.5-m (18-ft) diameter, concrete lined shaft is collared at the 4940 level of the mine and has a final mine level depth of 8620, or 2,923-m (9,590-ft) below the surface. The project has overcome numerous engineering and technical challenges including changing and squeezing ground conditions, complex logistics, and high heat conditions. Once completed, the shaft project will position the Lucky Friday Mine to potentially access over 78 million ounces of silver resources; nearly half of what has already been mined in the 70-year history of the Lucky Friday.

INTRODUCTION

Hecla's Lucky Friday Mine is a deep, underground producer of silver, lead and zinc in the Coeur d'Alene mining district of northern Idaho (Figure 1). Primary access into the mine is via the Silver Shaft, a 5.5-m (18-ft) diameter, concrete lined shaft sunk to an original depth of 1,890-m (6,200-ft). Current production is from the Gold Hunter vein system, approximately 1,525-m (5,000-ft) northwest of the Silver Shaft.

Lucky Friday's #4-Shaft is a 5.5-m (18-ft) finished diameter internal shaft, or winze, being sunk to the north of the Gold Hunter vein system, from the mine's 4760 level, 1,743-m (5,720-ft) below the surface, to the 8620 level, 2,923-m (9,590-ft) below the surface. The shaft is designed to provide access to the deep, high-grade zones of the Gold Hunter in order to increase the mine's production and operational life. Design and engineering began in 2006, with construction starting in the fourth quarter of 2009.

Three production levels with shaft stations are planned at the mine's 6500, 7500, and 8300 levels. Twenty-four meters (80-ft) below each production level is a skip-loading pocket. Production muck will be dumped into 4.3-m (14-ft) diameter bins on the production level, then fed via chutes, vibratory feeders and conveyor to skip loading flasks at the shaft.

GEOLOGY

The Gold Hunter orebody is hosted in the Precambrian Wallace formation of the Belt Supergroup. The Wallace formation consists of thinly bedded argillites with a roughly east-west strike and a near vertical dip (Figure 2). These bedding/foliation planes are typically coated with talcy minerals. The rock mass has an estimated rock mass uniaxial compressive strength of 48-mPa (7,000 psi) (Pakalnis, 2008) and is highly anisotropic, being distinctly weaker when loaded parallel to foliation The maximum in-situ stress strikes roughly northwest-southeast with a ratio of maximum horizontal to vertical stress of about 1.5 to 1.

Prior to excavation of #4-Shaft, it was recognized that the primary failure mode of the ground would be buckling and opening of the beds when excavating parallel to bedding. Bored raises had shown buckling of the north and south walls in the past (Figure 3).

Figure 1. Location of the Lucky Friday Mine

INITIAL GROUND SUPPORT DESIGN AND EXCAVATION

Preconstruction design called for 2.4-m (8-ft), 22-mm (#7) Dywidag or rebar resin-grouted bolts installed on 1.2-m (4-ft) centers with welded wire mesh or chain link mesh during excavation, followed by a nominal 300-mm (1-ft) thick, unreinforced, concrete liner poured in 6-m (20-ft) sections, within approximately 9-m (30-ft) of the advancing face.

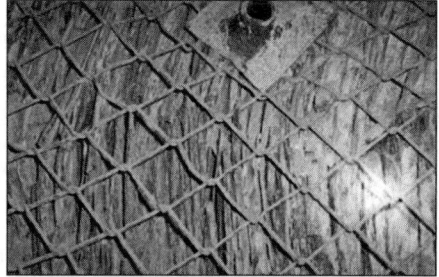

Figure 2. Typical vertical bedding (foliation) of less than 15mm in the argillites of the Wallace formation

There had been early discussions of the potential need to change the excavation shape to control the buckling of the argillite beds (Board, 2011). However, the stresses for the first 450-m (1,500-ft) were below the compressive strength of the rock, possibly due to being in the stress shadow of the mined out stopes south of the shaft, and the ground proved to be stable. Consequently, ground support was reduced to 1.8-m (6-ft) SplitSets on nominal 1.5-m (5-ft) spacing, with welded wire mesh or chain link. The concrete liner was kept within approximately 12-m (40-ft) of the bench. Virtually no ground water was encountered during sinking. As evidence of the lack of significant time-dependent deformation, the shaft remained stable for 16-months of inactivity, December 2011, through March 2013, while the mine's Silver Shaft was rehabilitated.

Figure 3. Buckling of argillite beds in north and south walls of a 1.8-m (6-ft) diameter bored raise

SHAFT LINER ISSUES APPEAR

The 6500 shaft station, 2,274-m (7,460-ft) below the ground surface, was cut in early January 2014, without incident, although bowing of the hanging rods between the brow and the sill was noted as the shaft was sunk below the level. The first report of cracking in the liner was made on January 31, with a large crack running horizontally around the liner 1.8-m (6-ft) noted above the brow. A vertical crack with 6-mm (¼-in) offset appeared in the north wall at the same elevation in early February, with spalling

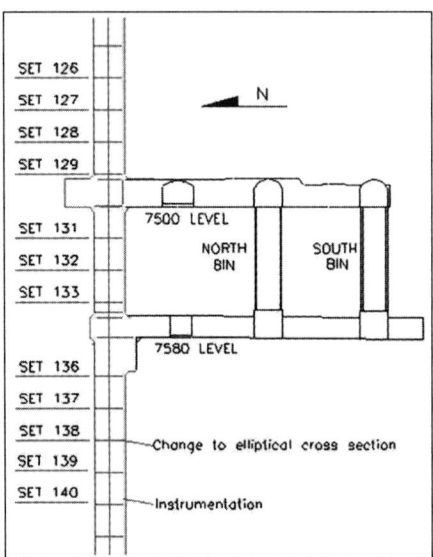

Figure 4. Vertical section of the shaft at the 7500 and 7580 mine levels

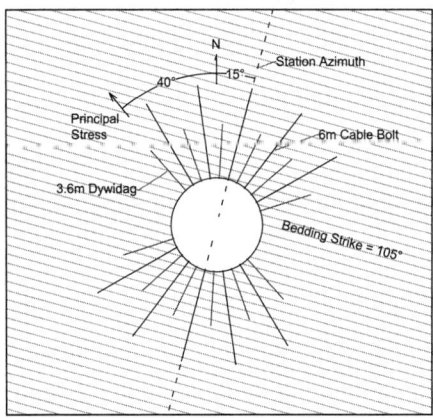

Figure 5. Ground support installation beginning at the 7580 level

and breakout to a hanging rod reported soon after.

Believing this fracturing to be caused by opening and buckling of the argillite beds as had been seen in the bored raises, shaft ground support was increased to include cable bolting prior to placing concrete. Grouted, 18-mm (0.7-in), bulbed cables, with 6-m (20-ft) minimum embedment were placed in rows spaced 1.5-m (5-ft) apart vertically. Each row consisted of five cables in the north wall and five cables in the south, spaced 1.2-m (4-ft) apart horizontally, to provide deep anchorage and restraint to bedding separation. This bolting pattern was continued until the shaft was 12-m (40-ft) below the 6580 loading pocket. No difficulties with ground control were noted between the cessation of cable bolting below the 6580 and the approach to the 7500 level.

Cable bolting resumed as the shaft bottom came within 12-m (40-ft) of the 7500 production level, (Figure 4), approximately 2,580-m (8,460-ft) below the ground surface, on July 22, 2014. The concrete liner was completed to the brow of the 7500 shaft station in early August of that year, and excavation of the level began. By October 28th, the shaft had reached the brow of the 7580 level, and severe, vertical fracturing of the north wall of the concrete liner above the 7500 brow developed. Shaft ground support was increased to five 6-m (20-ft) cable bolts per row on the north and south walls spaced on a 1.2m × 1.2m (4ft × 4ft) pattern, with 3.6-m (12-ft) Dywidags placed midway, or "five spotted," between them (Figure 5).

Cracking of the shaft liner between the 7500 and 7580 levels began in early November with some simple flaking and spalling at set 132, approximately midway between the two levels. At that time, the decision was made to install shaft instrumentation to monitor rock mass radial deformations as well as tangential and radial stresses induced in the concrete liner. A study also began on alternative shaft excavation shapes to improve stability. Damage to the existing concrete liner continued and expanded, ultimately stretching from 30-m (100-ft) above the 7500 level to 12-m (40-ft) below the 7580 level, an extent of nearly 75 meters (245-ft). This area eventually required installation of steel liner plate for reinforcement and, in some areas, actual demolition and replacement of concrete.

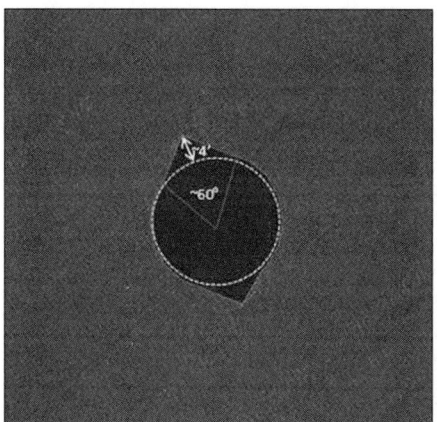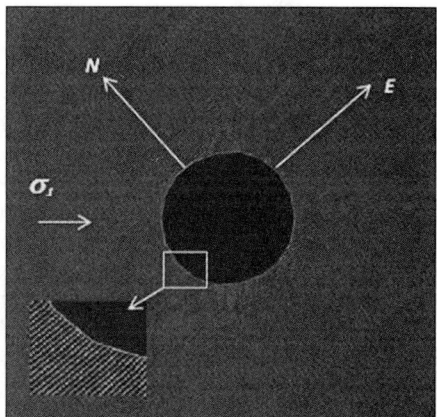

Figure 6. Circular shaft shape used in the model. The inset shows the model has a series of thinly spaced beds with an east-west strike (left). Notched (1.2.m) shaft cross-section used in the model (right).

CHANGE IN SHAFT EXCAVATION SHAPE

Traditionally, timbered shafts in the district were mined with "the long horizontal axis oriented normal to the bedding of the country rock. Experience has shown that such a position is best able to resist rock pressure at depth" (McWilliams, 1960). After consideration, it was decided to change the excavation to a more elliptical shape with the long axis perpendicular to bedding, similar to the orientation of a timbered shaft, believing this would provide a more stable opening and reduce point loads on the concrete shaft lining. To examine the impact of such a non-circular shape, some simple numerical simulations of both circular and notched shaft cross sections were run in UDEC, a "discontinuum" type program developed by Itasca Consulting Group. The models simulated a series of thinly-spaced beds oriented east-west, with the major principal stress oriented N40°W and assumed to be 1.5 times the vertical (gravity) stress, or about 100-mPa (14,500-psi) (Figure 6).

The simulation, as expected, showed the bedding tends to slip and buckle at the north and south sides of the shaft where the beds are more-or-less tangent to the wall in the circular case (Figure 7 left). The results showed an elliptical-shaped movement zone that is about 1.8-m (6-ft) deep on the north and south sides of the shaft (Figure 7 right). The maximum movement was shown to be directly perpendicular to bedding at opposite sides of the shaft. Based on this modelling, the decision was made to change the excavated cross section of the shaft to an elliptical shape, long axis perpendicular to bedding, with the final dimensions determined by the capabilities of the equipment being used to mine the shaft.

INSTRUMENTATION DATA ANALYSIS

To understand the nature of the observed ground movements, several extensometers were installed in circular shape and in elliptical shape. A total of one East-oriented extensometer, six West-oriented extensometers, ten North-oriented extensometers, and eleven South-oriented extensometers were identified as viable for analysis. Given the symmetry of the shaft geometry with respect to in-situ stress and geological structure, diametrically opposed sets of extensometers were analyzed together (i.e. East with West, North with South).

In the estimation of the depth of inelastic deformation at each N-S extensometer location, the deepest pair of anchors showing significant relative extension (dilatancy) are considered to bound the depth of the yield zone. For example, in Figure 8, the data for EXT #45 indicate a depth of yield between 6 m and 8 m. The data from EXT #45 show another interesting feature observable in many of the extensometer data sets: the apparent contraction of the collar relative to the toe following liner installation. This contraction either corresponds to movement of the collar away from the excavation and lining, or the movement of the toe (10 m depth) towards the excavation at a rate greater than that of the collar. Since there is no physical mechanism to explain movement of the collar away from the excavation, it appears likely that the ground from 10 m to 1 m behind the excavation wall is moving very slightly (i.e. ~2 mm) towards the excavation, while the ground at the collar is being held approximately in place by the liner. Although these levels of deformation are small and do not necessarily imply

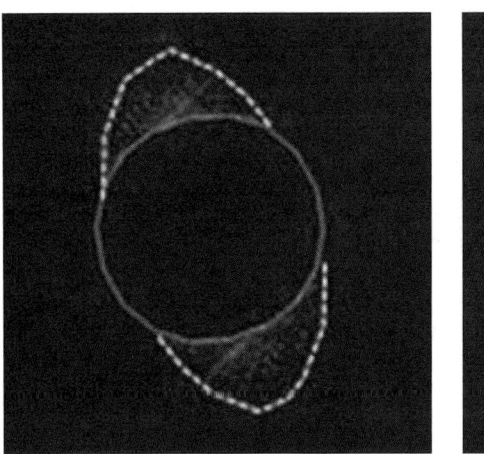 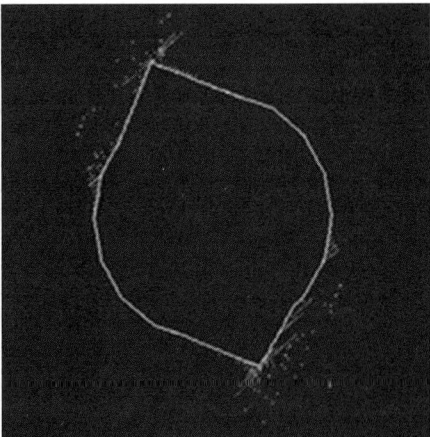

Figure 7. Region of shear displacement on bedding, approximately 1.8-m (6-ft) deep, for the circular shaft cross section (left). Dashed, white line shows the elliptic-shaped failure zone and Failure region as shown by shearing bedding planes is largely eliminated by simply mining out the potential failed region before it is formed (right).

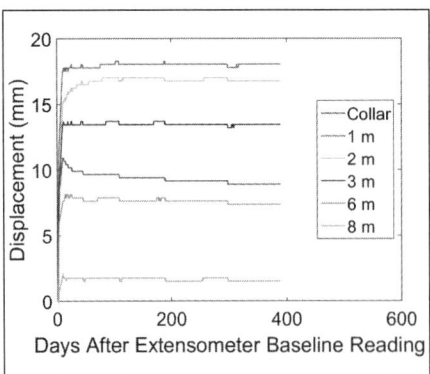 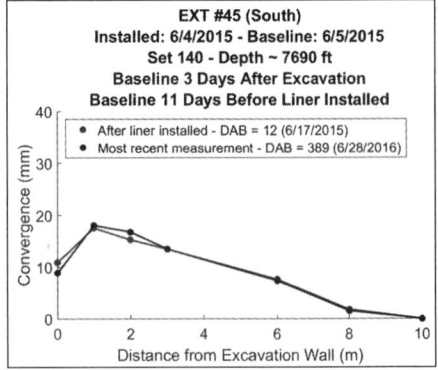

Figure 8. Extensometer 45 data for all anchors as a function of time (left) and measured displacements at two key times (right). Note the change in strain gradient around 8 m depth.

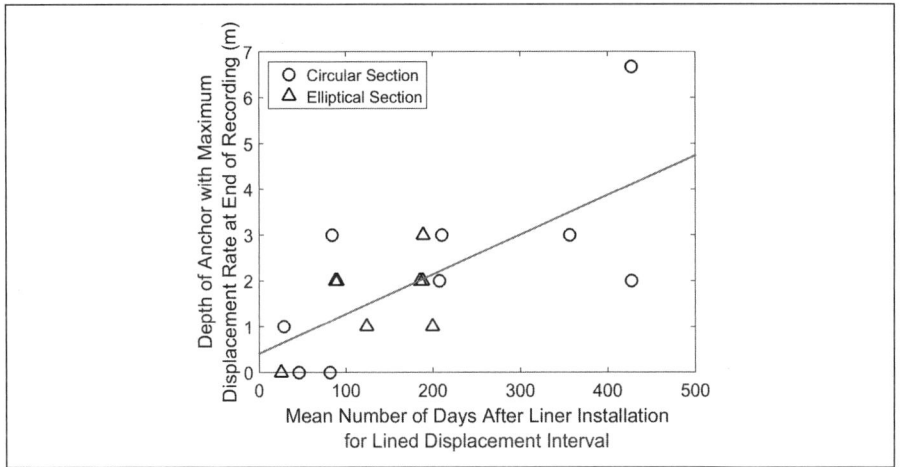

Figure 9. Changes in dilatant zone depth as a function of time after liner installation

ground yield at depths greater than 10 m, this observation has important implications for future instrument design (lengths greater than 10 m should be considered) and for interpretation of extensometer results (measured collar displacements may be lower than the true absolute displacement of the collar in cases where the extensometer toe has moved).

Also shown in Figure 8 is that the extensometer anchor with the most displacement for EXT #45 from installation (3 days after excavation) to present is 1 m deep. The most plausible explanation is that the zone of greatest dilatancy initiated at the excavation boundary migrated to 1 m behind the excavation wall within the first three days following excavation. This trend of the highest dilatancy zone moving deeper into the rockmass over time is consistent with observations made more broadly across the complete set of extensometer data, particularly following liner installation (see Figure 9). This phenomenon is likely due to the stiff concrete liner limits displacement at the ground-liner contact (i.e. the extensometer collar). Because the shallow region around the excavation, which has already dilated, is effectively much softer than the beds further into the rockmass, and this deeper zone has lost significant confinement due to shallow dilatancy, the deeper zone can easily dilate while re-compacting the shallow dilatant zone (i.e. inducing closure of previously dilated bedding); see Figure 10. Although a similar trend can be seen in the extensometer data for unlined ground, the liner was installed shortly after extensometer installation in most cases, leaving limited data from which to draw definitive conclusions.

An examination of all N-S oriented extensometers found that although the patterns of displacement and dilatancy within the yield zone are time dependent, the maximum depth of yield generally tended to remain consistent at any given location throughout time. It was also found that similar to the trend in maximum depth of dilatancy (shown in the Figure 9), the apparent depth of yield was similarly insensitive to excavation shape (see Figure 11). In both cases, yield depths on the order of ~6 m are typical. The more extreme yield depths tended to be observed on extensometers located between levels, which corresponds to the areas with the most liner issues, or near the bottom of the shaft with increased stress and reduced liner thickness. Despite showing the same depths of yield, the circular and elliptical excavations showed significant

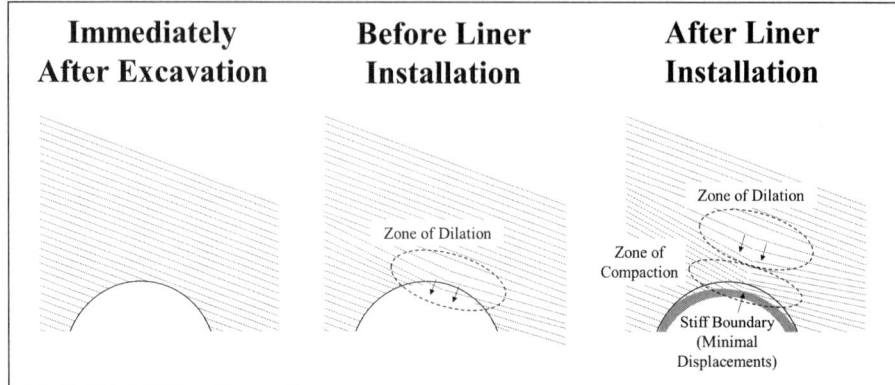

Figure 10. Schematic illustrating the mechanism for progressive deepening of the zone of dilation beyond the shaft following liner installation

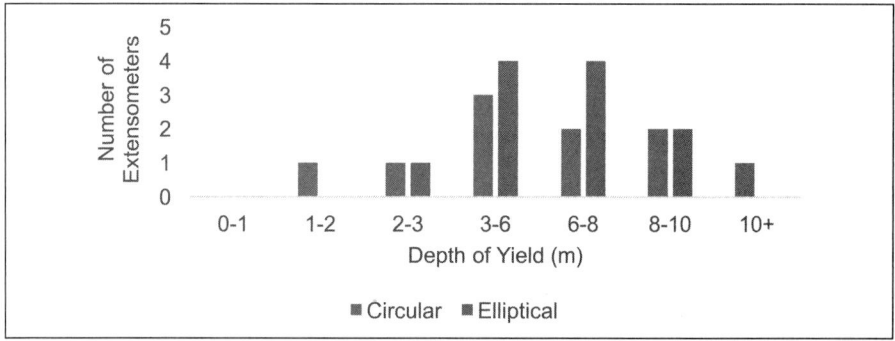

Figure 11. Histogram showing distribution of depths of yield identified from extensometer data

differences in the magnitudes of displacements recorded, with the latter being almost an order of magnitude lower than the former for both the unlined and lined cases.

FURNISHING

The shaft has been furnished with steel buntons and guides on 6.1-m (20-ft) vertical spacing as shown in Figure 12. When the deformation was first observed, planning was initiated as how best to furnish the shaft given this permanent distortion of the shaft liner. While it was known that deformation existed, the extent was not well defined. Prior to the completion of sinking, an opportunity to better define the deformation was proposed by Cementation USA. This plan involved high definition surveying of the shaft using a 3D laser scanner to measure the variances in the shaft liner to a level never before utilized in conjunction with furnishing. Because the shaft was not sunk to full depth, the scan was completed in two phases. The first phase provided baseline data as there remained time to modify the permanent shaft steel. After the completion of sinking, the remainder of the shaft as scanned. Key areas of the phase one were rescanned, including areas where additional deformation may have occurred.

The chosen equipment for the scans were the Leica ScanStation P20 and P40 models, used on the first and second scans, respectively. These portable units are capable of scanning with a point density in excess of 430,000 points per sqm (40,000 per

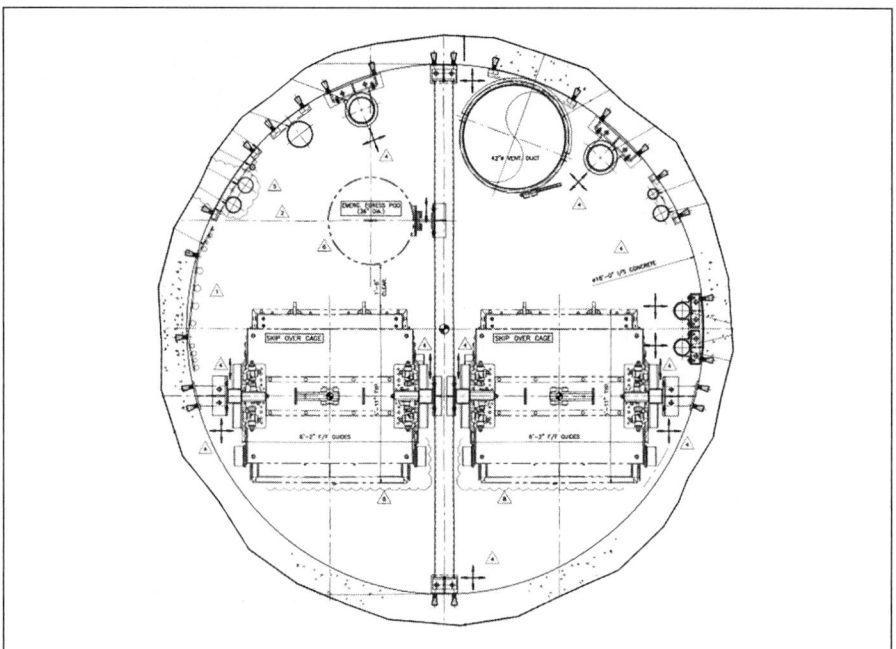

Figure 12. Typical shaft furnishing plan view

square foot) within the 5.5-m (18-ft) shaft diameter. This provided sufficient accuracy and resolution for this analysis and recording of defects on the face of the shaft liner. The methodology for scanning the shaft involved placing the scanner unit on an aluminum arm that was mounted to the shaft wall using two bolts connected to existing shaft steel inserts. A separate scan was taken in each 6.1-m (20-ft) shaft set to ensure consistent accuracy of data within each set. Control targets were placed in each scan and left in place for the adjacent scans, allowing all of the individual scans to be tied together with a high level of precision. The product of each scan was a point cloud, all of which were integrated and registered into AutoCAD files in which dimensional analysis could be completed. Dimensions were verified between the north and south concrete inserts as well as the east and west inserts to model how the designed shaft steel would fit at each set (see Figure 13).

The dimensions from the scan model data were exported to an Excel file and collectively analyzed relative to a common shaft centerline. This produced a deviation-from-design value for each of the four insert locations in each set. Analyzing this data as a whole made clear the extent of the shaft liner deformation, with the worst-case horizontal deviation exceeding 120-mm (5-in) in one location. A 61-m (200-ft) vertical stretch of the shaft above 7500 Level, which had liner plates installed previously to contain concrete spalling resulting from the ground stress, was tight enough to encroach on minimum conveyance clearances. By design, the conveyances require a minimum 60-mm (2.5-in) of clearance from any part of the conveyance to the shaft wall. This tightest area of the shaft reduced this clearance to far less than the minimum in several locations. Detailed analysis of the deformation made clear that it was not possible to install the shaft steel as it had been designed and partially fabricated at that point. Simply, the steel would not fit as designed and could not be installed plumb.

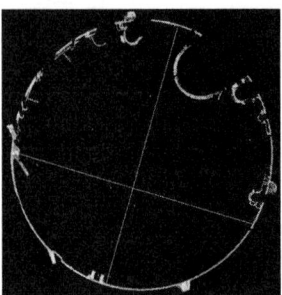

Figure 13. Dimensions were typically pulled from the midpoint of the top two inserts (left). Typical shaft set scan data with dimension pulled between inserts (right).

It also became clear that special provisions would need to be made in the tightest area of the shaft to maintain the required conveyance clearances from the shaft liner.

Liner deformation throughout the shaft required the steel to come slightly out of plumb to maintain adequate clearance from the shaft wall or level steel installations. In order to prevent sudden changes in guide alignment, the shaft side lines used as control for shaft liner pours and steel installation were adjusted. Existing positioning brackets on the lines were reset to best fit the concrete in the section of shaft in which they were installed. The end result of this exercise was side lines that were not plumb but instead followed the shaft liner in a smooth, gradual manner, allowing for smooth conveyance travel and adequate clearance throughout the shaft. In order for the guides to properly follow the control and maintain alignment despite uneven concrete, sets of wall brackets with varying dimensions were designed and ordered. These various wall brackets would account for shaft wall deviation while allowing the guides to maintain their alignment within the shaft barrel. Three additional wall bracket styles were designed and procured to supplement the brackets of standard design: A long "wide style" bracket, a short "tight style A" bracket, and a shorter "tight style B" bracket. Collectively, these three supplemental styles of brackets in addition to the standard style would cover liner deviation up to 45-mm (1.75-in) wide of design and up to 80-mm (3.13-in) tight of design, covering the expected deviation throughout the shaft.

Throughout furnishing, the various wall bracket styles proved invaluable to the installation of the shaft steel. Out of the 184 steel sets installed within shaft liner sets, 161 of them required at least one of the modified bracket styles. Without these brackets, the time required to modify and fabricate custom brackets in these sets would have exponentially exceeded the cost of procuring the additional modified brackets ahead of time. It should also be noted that there were significantly fewer modified brackets used in the final 50 sets of the shaft, where the elliptical concrete liner was installed. This change further verified the performance of the elliptical pour design in resisting deformation in the axis of the primary ground stress, allowing the shaft liner to better hold its design dimension.

The tightest portion of the shaft above the 7500 Level required special considerations beyond what was used in the rest of the shaft to maintain minimum conveyance clearance. The installed liner plates and deformed concrete in this area were tight enough in some locations to actually enter the design conveyance compartment. With the skips already designed and fabricated, only minimal gains could be made without complete re-fabrication. The scroll wheels, which serve to activate the dumping action of the skips, extend from the side of the skip near the corner with this tightest clearance. These wheels were reduced in diameter from 200-mm to 150-mm (8-in to 6-in), which gained approximately 19-mm (0.75-in) of additional clearance from the liner. To gain the remaining required clearance, it was decided to gradually shift the conveyance compartments towards the center of the shaft through this deformed section. This shift would increase conveyance clearances throughout the full tight zone and required no additional time beyond what was already scheduled for steel installation in these sets. The compartment shift was accomplished by transitioning to narrower steel buntons through this section, allowing the conveyance compartments to shift closer to each other by 100-mm (4-in), increasing clearance from the shaft wall by 50-mm (2-in) on each side. The transitions into and out of the narrow bunton zone were designed to be smooth and gradual to limit any additional wear on the conveyance rollers and slippers as well as to avoid overstressing any of the shaft steel due to resulting lateral loading.

Based on preliminary alignment verification in the shaft, the alignment of the guide strings appears to be excellent throughout the length of the shaft. While fine tuning of a handful of splices remains, the vast majority are within specification due to these modified wall brackets and attention to detail during installation. The narrow bunton zone through the tightest portion of the shaft also served its design intent. As each set of steel was installed through this area, measurements were taken from the guides to the tightest clearance areas to ensure that adequate clearance had been achieved. All sets were verified to have proper clearance, eliminating the need for any notching of the liner or liner plates. The guide splices through the transitions into and out of this zone show no measurable difference in alignment due to the gradual nature of the transitions. Final testing will be completed with the production conveyances once installed, but no issues are anticipated based on the verifications and testing performed to date.

An additional training and testing supplement that assisted with these unique furnishing conditions was the construction of a shaft furnishing training center on surface, as shown in Figure 14. This mock-up of a typical shaft steel set was constructed by pouring an exact replica of the shaft liner A-ring in a small excavation, complete with bolt inserts for the shaft steel. Wall brackets were installed in the ring, and the steel was constructed within it as a training exercise. All steel was full scale, with the exception of the guides, which were shorter than a typical set for logistical reasons. The original intention of the center was to train crews on the proper and safe installation of steel, a purpose that was fulfilled. However, the training center also came to serve as a test center for modifications and improvements, such as determining the best design for steel installation jigs, concrete and liner plate notching, and installation of modified wall brackets.

Figure 14. Surface training facility

LOGISTICS

The laydown area available for use underground was severely limited. This required just-in-time delivery for temporary and permanent equipment and supplies throughout the life of the project. Of particular note was the 19,270-m^3 (25,200-yd^3) of concrete and 4,820-m^3 (6,300-yd^3) of shotcrete required of the project. The self-consolidating concrete (SCC) mix used in the shaft is supplied by a surface batch plant and delivered through a 150-mm (6-in) inside diameter borehole slickline to the mine's 4900 level where it is loaded into transmixers and transported to #4-Shaft collar. Concrete is then loaded into buckets and lowered to the shaft forms. Sampling follows American Concrete Institute (ACI) standards, with one sample set taken for every 38 cubic meters (50-yd^3) at the pour site by ACI certified technicians. Samples are cured and tested in a certified, on-site lab. The 28-day compressive strength of the SCC mix has been averaging over 50-mPa (7,400-psi).

HEAT

The design criteria included virgin rock temperatures of 65°C (150°F). To achieve a workplace temperature of 29°C (85°F), a centralized refrigeration plant was installed on the 4900 Level to accommodate up to six refrigeration units; three units were installed during shaft construction. The 972-ton centralized refrigeration plant supplies 3,600-lpm (950-gpm) of chilled water in a closed circuit arrangement to an 865-ton Bulk Air Cooler (BAC) positioned on 5900 Level. During the project the three refrigeration units were fully loaded by the 5900 Level BAC and the associated thermal loss in the distribution system. As mining operations move laterally and deeper the cooling capacity can be expanded by adding up to 3 additional refrigeration units to the centralized plant, increasing the total capacity to 1,944-tons.

CONCLUSIONS

Over 284-m (930-ft) of shaft was mined using the elliptical cross section, with no evidence of cracking or other damage to the concrete liner. Pressures and deformations seen in the instrumentation at set 140, 2,640-m (8,668-ft) below surface, have not exceeded the liner strength. The 8300 production level, more than 2,830-m (9,290-ft) below the ground surface, is complete with no damage to the shaft liner. Further instrumentation has been installed to confirm the performance of the elliptical excavation between the 8300 and 8380 shaft stations, where the liner will be subjected to conditions similar to those that caused deformation and fracturing of the liner between the 7500 and 7580 levels.

Since 1891 Hecla has produced 400 million ounces of silver, with 160 million ounces coming from the Lucky Friday since operations started in 1942. 4 Shaft allows access over the next twenty years to Proven and Probable Reserves of 79 million ounces, and Measured and Indicated Resources of 128 million ounces. The 7500 level provides an exploration platform from which additional reserves will be identified and extracted.

REFERENCES

Board, M., Itasca Denver (2011), "Geotechnical Trip Report to #4 Shaft Project, Lucky Friday/Gold Hunter Mine," Technical Memorandum 1917.

Kim, E. and Walton, G. (2016), "Report of No. 4 Hecla Shaft, Mullan, Idaho," Phase 1 Report.

McWilliams, J.R., and E.G. Erickson (1960), "Methods and Costs of Shaft Sinking in the Coeur d'Alene District, Shoshone County, ID," U.S. Bureau of Mines IC-7961.

Pakalnis, R., Pakalnis and Associates (2008), "Pre-Development Report (Geotechnical) to #4 Shaft—Hecla Limited—Lucky Friday," Report No. LFM-2/08.

Strickland, B., Board, M., Sturgis, G. & Berberick, D. (2016), "Elliptical Shaft Excavation in Response to Depth Induced Ground Pressure," SME Annual Meeting, paper 16-084.

An Overview of the SR99 TBM Access Shaft, Seattle, Washington

Phillip A. Burgmeier ▪ Brierley Associates
Jacob Mitchell ▪ Brierley Associates
Gregory Hauser ▪ Seattle Tunnel Partners Joint Venture

ABSTRACT

The SR99 bored tunnel will replace the seismically vulnerable, double-deck Alaskan Way Viaduct. "Bertha," at 57.3-ft in diameter, was, at the time, the largest EPB-TBM in the world. The TBM launched in July 2013 and stopped after mining approximately 1000-ft when it was discovered that the outer seals had been damaged. An access shaft was installed in front of the TBM to facilitate removal and repair of the cutter head assembly. Access shaft design and construction overcame difficulties including buried structures, archaeological concerns, limited site access, complex geology, adverse groundwater conditions, and proximity to the adjacent viaduct that remained in service. The project required installation of deep secant piles up to 10-ft diameter, multi-stage dewatering, structural and hydraulic grouting, structural and geotechnical modeling, and real-time instrumentation. The TBM successfully mined into the access shaft in early 2015 and the 2000-tonne cutter head was removed from the shaft via modular lift tower supported on the shaft.

INTRODUCTION

The City of Seattle, Washington USA has two major North-South highways carrying traffic through the City. Interstate Route 5 through the center of the city between Lake Washington and Puget Sound, and State Route 99, the Alaska Way Viaduct along the West side of the City following the Puget Sound shoreline.

The Alaska Way Viaduct (AWV) was built in the 1950s and was nearing the end of its useful life in 2001 when it was impacted by the 6.8 magnitude Nisqually earthquake. This event, coupled with age, made the AWV unsound and mandated its replacement. The Washington Department of Transportation (WSDOT), owner of the AWV, along with the City of Seattle and King County, reviewed options for a replacement. In 2009 they agreed on a tunnel beneath the City to replace the Viaduct. The new tunnel was conceived to be a single bore with two lanes in each direction, a shoulder on one side, and emergency egress for each direction of traffic.

The Project was advertised as a Design Build Project and proposals were received in December 2010. Notice to Proceed for Administration and Design was issued to Seattle Tunnel Partners (STP), a Joint Venture of Dragados USA (DUSA) and Tutor Perini Corporation (TPC), on February 7, 2011.

The TBM was furnished by the Hitachi Zosen Corporation of Osaka, Japan. The machine design was completed in March 2012, and the manufacture, assembly in Osaka, and testing were completed by January 2013. The TBM was shipped to the Project site via ship in large assemblies by April 2013, was launched in June 2013, and with various stops and re-starts, the tunnel was advanced about 1,000-feet by early December 2013.

Early morning on December 4, the TBM encountered a steel obstruction that caused damage to the cutter head. The TBM continued to mine ahead after hitting the obstruction but quickly lost performance and the ability to advance and, by December 7, the TBM was halted to assess the damage. During January 2014, crews entered the cutter head under compressed air to remove debris and examine the cutter head. By early February of 2014, it became apparent that the main bearing seals had failed and the decision was made to design an access shaft to allow for replacement of the main bearing seals. The TBM had mined about 1,025-lf.

ACCESS SHAFT DESIGN

In March 2014, Brierley Associates developed conceptual designs for an access shaft to assist STP with recovery of the TBM cutter head. The shaft location was subjected to site constraints including the presence of settlement mitigation piles (SESMPs) along each side of the tunnel alignment, the proximity of the Alaskan Way Viaduct pile foundations, and the presence of the large TBM. With the TBM inside of the shaft, a 62-ft by 38-ft clear zone was required in front of the machine to remove the cutter head. Given these constraints, a 3D model was developed to explore potential concepts for shaft support to about 110-ft depth. These existing site conditions and constraints were included in the 3D model via importing as-built survey information to ensure that the conceptual designs were constructible.

Access shaft concepts explored the possibility of installing a shoring system around the cutter head, followed by excavation to expose and remove the cutter head. Early concepts included a roughly square-shaped secant pile shaft and a racetrack type layout with a single line of bracing over the TBM; however, both of these concepts were abandoned given the amount of steel bracing that would be required to support a square shaft, and bracing conflicts given the size and position of the TBM within an oblong shaft.

A roughly circular-shaped secant pile shaft that derived support primarily through ring compression without internal bracing was selected for the design. The access shaft location was shifted north of the TBM so that a complete ring of secant piles could be installed to depths below the TBM, creating a roughly cylindrical shape through which the TBM would mine. This approach provided a groundwater cutoff underneath the TBM.

A roughly-circular shaft was desired in order to minimize the development of moments in the compression ring, but the SESMPs conflicted with a continuous circular secant pile ring. The compression ring's continuity was interrupted at the SESMP locations where newly installed secant pile shafts would be required to bear directly on the existing heavily-reinforced concrete SESMPs. In addition, the 3.3-ft SESMP piles were installed on roughly 4-ft centers parallel to the tunnel alignment which meant that there was not enough room between SESMP piles to install tangent piles and would therefore leave soil zones present within the line of ring thrust. The layout shown in Figure 1 was developed to utilize new secant piles installed tight to existing SESMP piles, followed by grouting to replace the soil remaining in the interstitial spaces between new and existing piles, thereby creating a continuous thrust ring. Figure 2 shows the structural model that was created using as-built information and illustrating the real geometry of the shaft comprised of secant and tangent piles.

A mix of 3.3-ft (1.0-m), 5-ft (1.5-m), 8-ft (2.5-m), and 10-ft (3.0-m) diameter unreinforced secant piles were used to develop the final shaft plan geometry. The South Wall, through which the TBM would mine, was approximately 18-ft thick, and the effective wall thickness at northern portion of the shaft ring was approximately 6-ft. The

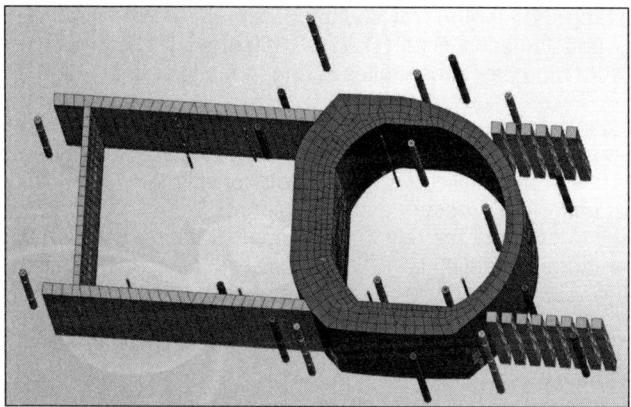

Figure 1. Finite element mesh of access shaft and "bathtub" along with dewatering system

secant pile ring was designed to accommodate an excavation depth of about 116-ft and required a secant pile vertical deviation tolerance of 0.6%.

To improve shaft stability and to support the weight of the TBM after break-in, a massive reinforced concrete TBM cradle was installed at the bottom of the shaft. The 5 to 10-ft thick cradle was designed to support the dead weight of the machine following break-in.

The contract documents indicated that the TBM was currently located within a silty to sandy artesian aquifer indicating that water inflow and material loss could be encountered during break-in. To mitigate this risk, a "bathtub" which extended south from the access shaft piles along the line of the SESMP piles and past the TBM skin to the previously constructed precast tunnel liners was installed as a cut off to reduce dewatering requirements. The bathtub design called for jet grout between SESMP piles and across the top of the precast tunnel liners. Additional chemical grouting was later installed to seal underneath the tunnel liners.

Mining through the shaft wall would create a 58-ft diameter hole within a roughly circular shaft of 80-ft inside diameter. Finite element modeling was used to evaluate the stresses within the shaft concrete prior to and following TBM penetration. The model suggested that the secant piles adjacent to the opening could develop unacceptable bending stresses within plain concrete. Therefore, eight double W24×104 structural steel beams were installed into secant piles on each side of the TBM penetration.

The use of a shoring system consisting solely of an unreinforced secant pile compression ring at depths in the 100-ft

Figure 2. Structural model with as-built secant pile geometry

range was first reported for the New Irvington Tunnel Vargas Shaft in Fremont, CA by Lindquist, E.S. and Jameson, R (2011). The SR99 project is believed to be the first that incorporated 10-ft diameter secant piles in ring formation.

Groundwater control generally followed the Observational Approach as described by Carl T. Terzaghi and presented in Peck, R.B., (1969). The finite element model developed for this project utilized a multi-stage construction analysis procedure and a sequential stress–pore pressure analysis method. Transient groundwater analyses were performed to evaluate the results of pump testing and to calibrate the finite element model stratigraphy and hydraulic conductivity against field measurements as the excavation advanced. This approach relied on abundant groundwater drawdown data collected in real-time near the shaft, the results of additional construction-phase test borings, and manual groundwater level readings at existing monitoring wells located away from the shaft.

The groundwater model was utilized during the construction phase to investigate suspected leaks through the shaft wall associated with the interstitial grouting, to account for unanticipated subsurface conditions which required real-time design of dewatering elements, and to assist in the evaluation of excavation base stability in the very-fine sand / silt that was present at the bottom of the excavation. Utilizing the Observational Approach to dewatering allowed the development of a calibrated groundwater model to supplement engineering calculations for potential excavation base stability, internal soil piping, and global soil plug failure modes.

ACCESS SHAFT CONSTRUCTION

Malcolm Drilling was contracted by Seattle Tunnel Partners to install the access shaft secant piles and jet grouting. Construction began in early May 2014 with the installation of the jet grout cutoff wall above the TBM and interstitial grouting between the SESMP piles for the bathtub area. Grouting work was completed by mid-May and was followed by installation of the secant piles.

Secant pile installation began using a Bauer BG50 drill rig to install the 3.3-ft piles. Subsurface conditions necessitated the use of steel casing and a flooded shaft to ensure hole stability. The first 3.3-ft pile installation revealed a discrepancy between SESMP as-built drawings and the conditions that were encountered. This discrepancy was located only along the western line of SESMP piles and meant that a redesign of the compression ring was required. Installation of 3.3-ft piles continued along the eastern side of the access shaft while the western side SESMP piles were exposed and surveyed. Redesign of the western side of the compression ring was completed within a week without construction delay. The design-build team replaced all 6.6-ft piles in the original design with a combination of 3.3-ft, 5-ft, 8-ft, and 10-ft piles during this redesign.

Installation of 5-ft piles using the BG 50 was phased in with 3.3-ft pile installations and the installation of these smaller diameter piles continued until mobilization of the equipment to install the 8-ft and 10-ft shafts was complete. Two Liebherr 885s and one Liebherr 895 equipped with grabs were used in conjunction with casing oscillators and a casing rotator for the 3.0 m shafts. Generally, one of the 885s was used as a service crane. Work proceeded 24 hours a day, 7 days a week using two 12 hour shifts.

Additional challenges were encountered with respect to the as-built conditions of the SESMP piles. As the SESMP piles were not originally designed to be part of the access shaft structure, the verticality tolerances were not as tight as those for the new secant piles, which resulted in most adjacent shafts conflicting with SESMP piles at depth. The design team and construction teams coordinated in the field to resolve conflicts as they arose. Installation of all secant piles was completed in early September 2014. A total of 84 piles were installed to a depth of 131-ft, including (40) 10-ft secant piles and (14) 8-ft secant piles.

After secant pile installation was complete, grouting of the interstitial spaces between SESMP piles was attempted via jet grouting. Due to the presence of cobbles and the existing SESMP piles, jet grouting was ineffective. An alternative method was chosen where the interstitial spaces were flushed via air and water using a Foremost DR24 drill rig and then tremie grouted through the drill string.

During secant pile installation, instrumentation consisting of fourteen piezometers and one multipoint borehole extensometer was installed in and around the shaft walls. Instrumentation and drilled shaft installation occurred simultaneously resulting in, during the peak of work at-grade, a total of five holes—two large diameter secant piles with the Liebherr cranes, one small diameter shaft with the BG 50, and two geotechnical borings—being advanced concurrently within a 200-ft by 100-ft yard.

During piezometer installation, it was discovered that the assumed clay plug at the bottom of the access shaft was predominantly silt rather than clay. Laboratory testing of the soil samples from the geotechnical borings confirmed that the silt was non-plastic. Piezometer installations confirmed the presence of isolated aquifers under pressure. The presence of highly erodible silt increased the risk of shaft bottom instability and resulted in the design-build team revisiting the shaft excavation method. Two viable options were considered: excavate the access shaft in the wet, or redesign the dewatering system. The dewatering system was redesigned with passive elements originally designed to control localized instability repurposed as part of an active system that would reduce pore pressures in the shaft plug.

In preparation for the dewatering system redesign, an additional piezometer was installed at a depth of 215-ft in the deep aquifer located underneath the shaft. This piezometer confirmed the existence of artesian conditions below the shaft plug and reaffirmed the team's decision to install additional dewatering measures given the subsurface conditions that were encountered. The Observational Approach to dewatering was again chosen and specific design goals were set for the dewatering system and additional dewatering measures were installed as required based on field measurements. The design goals set for this system were to reduce the water pressure in the deep aquifer to prevent heave of the access shaft soil plug, and to reduce the upward pressure gradient within the silt unit that composed the access shaft soil plug to control the potential for soil piping and local heave. To meet these goals, four deep dewatering wells were installed into the deep aquifer, three dewatering wells were installed into the silt strata to the north of the shaft, two of the dewatering wells in the bathtub were deepened to penetrate the silt strata south of the shaft, and five dewatering wells were installed in the shaft interior penetrating the silt strata. Redesign, construction, testing, and evaluation of the new dewatering system proceeded concurrently in order to reduce schedule impacts.

Once the dewatering wells were installed and the design goals set forth by the team were met, excavation of the access shaft began in mid-October 2014. Shortly after starting, excavation was halted due to an archaeological find. After a short hiatus, excavation was allowed to resume. Approximately 50-ft into the excavation, it was discovered that portions of the interstitial grouting program had been ineffective. Excavation was halted while the construction team came up with a plan to effect remedial repairs of the interstitial spaces ahead of the excavation. Figure 3 shows the Access Shaft partially excavated during this time.

The 3.3-ft piles that were located inboard of the SESMP piles were removed in order to gain direct access to the interstitial spaces. Removal of the 3.3-ft piles was accomplished via chipping from a crane lowered rig as shown in Figure 4. After the removal of the 3.3-ft piles, shotcrete was placed to ensure continuity of the compression ring to the excavation level and a sequential excavation plan was put in place which integrated excavation, removal of the 3.3-ft piles below grade, and placement of shotcrete. After reaching 90-ft in depth, the design team made the determination that the excavation plan where the shotcrete trailed the excavation was not sufficient to prevent potential failure of the compression ring and a new plan was devised where the interstitial spaces were remediated ahead of the excavation. This work consisted of using a small articulating drill at multiple vertical angles and directions to inject grout under pressure into the interstitial spaces from inside the excavation. This method was successful and excavation resumed.

When the excavation reached approximately 105-ft, the internal piezometers reported that the hydraulic gradients in the silt plug no longer met design limitations. Additional

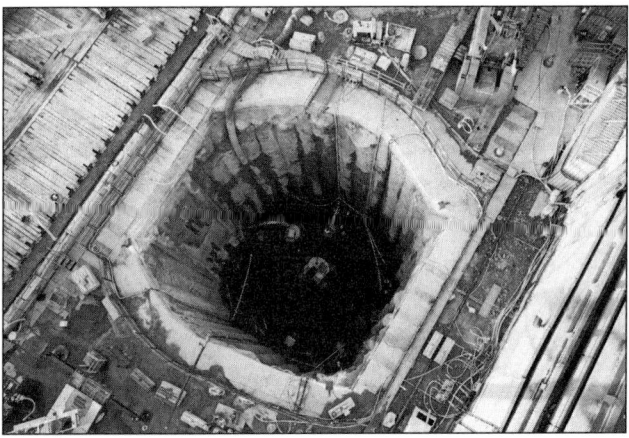

Figure 3. Excavation of access shaft

Figure 4. Removal of 3.3-ft diameter secant piles prior to interstitial grouting

Figure 5. TBM penetration

dewatering elements were installed in the form of well points installed from the excavation base. The well points were sufficient to relieve the excess hydraulic gradient and allow excavation to continue.

Excavation was completed in late January 2015; however, piezometers located in the bathtub indicated that a significant amount of water was still present above the tunnel invert. Therefore, horizontal drains were installed from inside of the access shaft under the TBM to further reduce the water level in the bathtub such that material migration through the annulus after the TBM entered the shaft would not occur.

TBM BREAK-IN

Mining into the shaft began in mid-February 2015 and shaft break-in occurred on February 19th. Survey prisms were located on the shaft walls around the TBM penetration at 3, 6, 9, and 12 o'clock positions as well as at the center point. As expected, the largest shaft wall deformation, about 1-inch, occurred at the center point just prior to collapse of the secant pile wall over the area of shaft penetration. Survey prisms located on the MLT foundation cap beam and inclinometers located within the secant piles generally registered less than 0.25-in. Figure 5 shows the TBM in the shaft after successful break-in.

MODULAR LIFT TOWER

The TBM's Cutter Drive Unit (CDU) was removed from the shaft using a Modular Lifting Tower (MLT) from heavy lifting company Mammoet shown in Figure 6. Brierley provided the foundation design to support the MLT and it was decided that, due to the large loads the pick would impose, deep foundation elements were necessary. The final design incorporated the previously installed SESMP piles, the access shaft secant piles, and an additional ten drilled shafts installed to the north of the access shaft.

All deep foundation elements were capped with two 274-ft long cap beams located to either side of the access shaft with a span of 74.5-ft between the two cap beam centerlines, which matched the centerline spacing of the MLT skids. The MLT was supported by two pairs of skid beams centered on each cap beam. Early design meetings between Brierley and Mammoet indicated that the original skid beam pair design could result in large torsion across the access shaft clear span. In order to provide

Figure 6. Modular lift tower

relatively uniform bearing on the cap beam, Mammoet redesigned the MLT's skid beam system to rely on hydraulics to equalize pressure across the skid beam pairs. The cap beam final design was a heavily-reinforced 5-ft thick concrete cap to support about 130 kips per foot of vertical load along with a nominal horizontal load.

After disassembly of the TBM skin and removal from the shaft, the CDU was removed on March 30, 2015. The CDU weight was approximately 1,900-tons. This load was approximately 80% of the maximum design load for the cap beam and for the MLT.

The CDU was lifted from inside the shaft over approximately 18 hours. The CDU's weight was slowly transferred until the majority of the weight was on the strands and then the final connections between the TBM body and the CDU were removed. The CDU was then raised out of the access shaft and rotated 90 degrees before being placed on bearing blocks located directly south of the access shaft.

The cap beam was surveyed during the CDU pick in order to measure real-time deformation. Survey prisms were located at the third points of the section of the cap beam on top of the access shaft. During the pick, up to roughly 0.3-in cap beam lateral deformation was recorded and up to 0.1-in of vertical deformation was recorded. These cap beam deformations were slightly lower than the estimated 0.5-in; however, the 0.5-in deformation estimate was based on a pick weight of 2,200-tonne.

TBM REPAIRS

Once the cutter head and the CDU were on the surface as a single unit, they were separated to allow replacement of the main bearing seals. Both the inner and the outer seals and the main bearing itself were replaced. This work required the removal of the bearing block and the inner cylinder. These components were stored on site. The center pipe was also removed and replaced after the main bearing and outer seal system were installed.

The repairs required careful measurements and placement of each of the components to ensure that the final product was properly aligned with the cutter column and the main bearing and cutter drive unit itself. During this time, additional gussets and stiffener plates were installed to support the new seal system, including stiffener plates on the bearing block and the forward shell sections that were removed and the lower sections that were still attached to the main body of the TBM still in the shaft.

As the seals and main bearing were being replaced, modifications to the cutter head to allow better muck flow and replacement of all of the cutters was being completed. Additional face cutters were welded to the cutter head and hard facing placed to provide additional protection of the cutter head face.

All of the repairs were completed and the cutter drive unit was lowered back into the shaft and attached to the TBM main body by late August. Through September, careful measurements and delicate movements of the CDU were carried out to ensure that the CDU was properly positioned within the TBM body, the forward shield pieces were replaced and welded around the CDU, and the CDU was welded to the TBM main body in the shaft. Final positioning, welding, and final system testing all occurred in late 2015. After all systems were confirmed to function properly, backfilling of the shaft commenced in December followed by operational testing prior to break-out.

The operational testing was successful and the TBM advanced through the shaft backfill material and contacted the north face of the Access Shaft where it remained through the end of 2015. Break out from the shaft started right after the first of the year in 2016 and proceeded smoothly until the TBM was clear of the shaft wall. The TBM then advanced to just West of where the tunnel alignment crossed under the existing Alaskan Way Viaduct where a planned cutterhead inspection was performed to confirm that the tooling was ready for the critical crossing underneath the existing Viaduct.

Due to concerns about potential settlement of the Viaduct while tunneling proceeded below, WSDOT made the decision to close the Viaduct to traffic while the TBM advanced under the foundations of the Viaduct. On Friday, April 29, 2016, the TBM began working around the clock and successfully tunneled below the Viaduct without negative impact to the existing Alaskan Way Viaduct. Since crossing below the Viaduct there have been several stops for cutterhead inspections and cutters have been changed using the free air process from the spokes of the cutterhead. As of this writing, as of December 1, 2016 the TBM has advanced to STA 255 + 47, and installed Ring #929 of 1,426 total Rings. The TBM is currently below Second Ave and Lenora Street, heading NE and more than 65% of the way to the hole through at the North Access Shaft. The current schedule anticipates TBM hole through by June 20, 2017 and Substantial Completion of the Project by Nov. 21, 2018.

CONCLUSION

The SR99 TBM Access Shaft was successfully constructed and excavated in 2014. Consisting of a mixture of existing reinforced drilled shafts, newly installed secant piles, jet grouting, and a dewatering system across three aquifers, the TBM Access Shaft utilized state of the art construction methods through challenging geology including installation of 10-ft diameter secant piles to depths of 131-ft. The Access Shaft allowed the successful retrieval and repair of the 1,900 ton, 57.3-ft diameter tunnel boring machine's cutter drive unit, and a successful relaunch of the TBM through the North wall of the Access shaft.

REFERENCES

Lindquist, E.S. and Jameson, R. 2011. Secant pile shoring—developments in design and construction. DFI 36th Annual Conference on Deep Foundations.

Peck, R.B (1969). Advantages and limitations of the observational method in applied soil mechanics, Geotechnique, 19, No. 1, pp. 171–187.

Gas Control in a Vertical Boring Machine Advanced Shaft in Shale

Ryan P. Sullivan ▪ Northeast Ohio Regional Sewer District
Michael J. Schafer ▪ MWH/MM Jv
Michael A. Piepenburg ▪ MWH/MM Jv
Phil Kassouf ▪ Triad Engineering

INTRODUCTION

Gas inflows into Cleveland, Ohio tunnels are well-documented and historically addressed with ventilation. Until recently, shafts were thought open and accessible enough to avoid gas buildup. Dugway Storage Tunnel (DST) shaft DST-7 was completed using a vertical boring machine (VBM). During shaft excavation, gas inflow rates exceeded the volume controllable with the VBM ventilation system and the VBM could not be advanced. Nearly thirty working days of delay occurred before gas levels dissipated enough to allow safe shaft entry and completion of the excavation. A case history of the shaft excavation with recommendations regarding dealing with gas impacts on future VBM-advanced shafts is provided.

PROJECT OVERVIEW

The Northeast Ohio Regional Sewer District's (NEORSD) Project Clean Lakes is an on-going EPA-mandated 25-year-long construction program designed to reduce combined sewer overflows (CSOs) from the Cleveland, Ohio metropolitan area into Lake Erie. This project will reduce CSOs from 4.5 billion gallons per year to less than 0.5 billion gallons per year. The three-billion-dollar program consists of the construction of: large-diameter storage/conveyance tunnels; shafts at key CSO pick-up points; underground pump stations; wastewater treatment plant capacity and treatment improvements; and green infrastructure. Of the six large-diameter tunnels included as part of the program, the Euclid Creek Tunnel (ECT) was the first and was completed in August 2015. The second tunnel, the Dugway Storage Tunnel (DST) is now under construction with a project completion date of mid-2019. Design of the DST project was performed between 2013 and 2015 by a MWH/MM, joint venture between Montgomery Watson Harza and Hatch Mott MacDonald (now Mott MacDonald).

The DST project is located below the predominately residential neighborhoods of northeast Cleveland. Lake Erie, with a surface elevation of 571 feet is located less than one-half mile north of the downstream end of the tunnel; the DST-1 shaft. The project alignment crosses below the plains that slope from the lake's shoreline up towards the southeast. Ground surface elevations above the tunnel range from 605 feet above mean sea level (NAVD 1988) at the DST-1 shaft to 653 feet at the DST-8 shaft, the tunnel's upstream terminus.

The Dugway Storage Tunnel is a 24-foot finished-diameter tunnel, excavated through the Chagrin Shale at depths of 180 to 235 feet. The tunnel is lined with gasketed steel-fiber-reinforced precast-concrete segments. As shown on Figure 1, the tunnel starts at Shaft DST-1, a 49-foot-diameter shaft that will be used to launch the tunnel boring machine (TBM) and connect to the adjacent Euclid Creek Tunnel. From DST-1, the tunnel extends nearly 14,500 feet upstream to Shaft DST-8; a 56-foot-diameter shaft

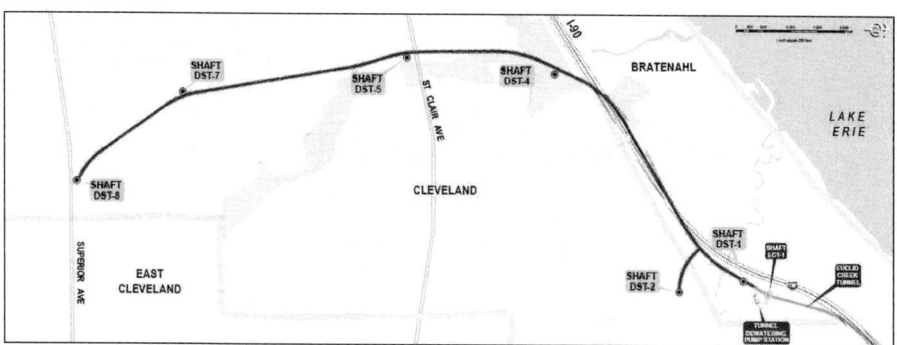

Figure 1. Dugway storage tunnel site map

Table 1. DST shaft dimensions

Shaft Designation	Approx. Mat'l Thicknesses (ft)		Total Shaft Depth (ft)	Diameter (ft)	
	Fill/Soil	Rock		Excavated	Finished
DST-1	93	122	215	45.5 (soil), 49.0 (rock)	40.0
DST-2	76	131	207	32.0 (soil), 28.5 (rock)	25.0
DST-4	115	73	188	26.0 (soil), 24.5 (rock)	21.0
DST-5	98	106	204	22.0 (in soil)	16.0
DST-7	19	200	219	19.25 (in rock)	
DST-8	9.5	239.5	249	56.0 (soil & rock)	50.0

that will be used for retrieval of the TBM, dropping of tunnel flow, and relief of surge pressures.

Four other shafts, designated DST-2, DST-4, DST-5, and DST-7, are also shown on Figure 1. As shown on Table 1 these shafts range from 19.25 to 56 feet in excavated diameter. The four shafts were located to relieve existing over-capacity sewers and capture flows from future CSO collection systems. All of the DST shafts will be lined with reinforced cast-in-place concrete. Baffle drop structures will be constructed in all shafts except DST-1. Eleven additional structures, all <50 feet deep, will serve as diversion structures, drop manholes, gate structures or gate control structures.

The DST Project was awarded to the Salini-Impregilo-Healy JV (SIH) in February 2015. SIH completed the DST-1 shaft and are currently constructing the DST tunnel. Cleveland-based subcontractor Triad Engineering (Triad), as a subcontractor to SIH, excavated the DST-2, DST-4, DST-5, and DST-7 shafts. SIH subcontractor Marra Services Inc. (Marra) excavated the DST-8 shaft. Concrete lining of the shafts and construction of baffle drop structures was performed by Cleveland-based Northstar Ohio Inc.

SITE GEOLOGY

All of the DST project excavations and shafts were advanced through fill and soil to depths as great as 115 feet. Excavation of the Chagrin Shale bedrock in these same shafts ranged from 75 to 240 feet below grade. A generalized geologic profile along the DST alignment is shown in Figure 2.

A layer of fill extending from the ground surface to depths ranging between 3.0 and 40.5 feet below ground surface is present along the DST alignment and consists of clay or sand mixed with a wide range of organic materials and debris. The underlying glacial-derived soils along the DST alignment are comprised predominantly of plastic silt and clay and highly over-consolidated plastic clay, with seams and layers of low-plasticity/non-plastic silt and sand.

As shown in Figure 2, the topography of the soil/rock contact varies and is a result of Pleistocene-age glaciation that carved out a series of valleys in the buried bedrock surface. The apparent lowest point in the bedrock topography above the DST alignment is between shafts DST-2 and DST-4. The tunnel was designed to be a sufficient distance below this low point for stability and ease of construction.

The Chagrin Shale is the one of three members of the Ohio Shale and was initially deposited in deep (oxygen poor) marine basins during the late Devonian Period (~365 million years ago). The Chagrin Shale is an olive gray to blue gray colored thin-bedded shale with randomly spaced interbeds of siltstone and sandy siltstone. Locally, up to 30% of the rockmass may be comprised of siltstone interbeds. The shale is comprised predominantly of clay (illite or illite/smectite) with subordinate amounts of silt.

Bedding planes in the shale are typically spaced <¼ inches apart, oriented within five degrees of horizontal, and dip gently towards the southeast. When allowed to dry, core samples of the shale tend to part along bedding planes into thin "poker-chip" like discs. The siltstone/sandstone interbeds are generally between $1/16$ and 4 inches thick and spaced between 1 and 24 inches apart. Thicker interbeds up to 18 inches thick have been recorded but are rare.

Three sets of joints have been observed within the Chagrin Shale but are not commonly encountered or observed. As a result, over 80% of the RQD values recorded during the DST subsurface exploratory program were rated as excellent (90 to 100%). Joint spacing is estimated to average approximately 12 feet.

Baseline groundwater levels at the six DST shafts were stated in the project Geotechnical Baseline Report (GBR) and are within six to twelve feet of the ground surface. Based on water pressure (packer) tests performed on the shale during the DST subsurface exploration program, the shale typically exhibits a mass hydraulic conductivity measuring orders of magnitude less permeable than 1×10^{-5} cm/sec. "Flush" flows initially measuring between 50 and 200 gpm (before tapering off) have

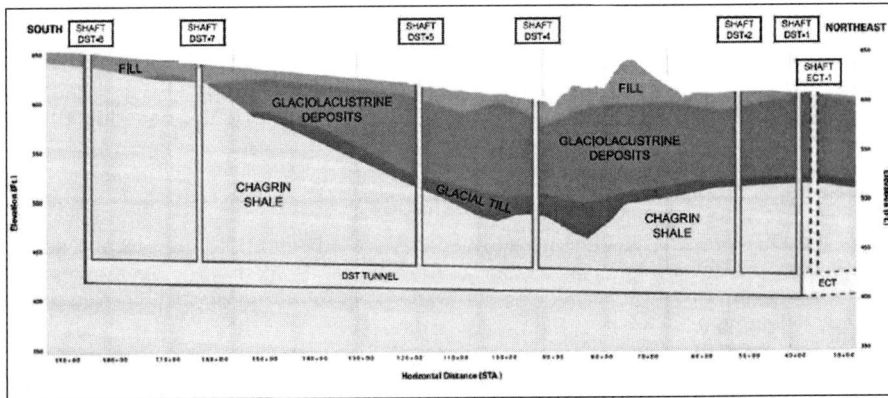

Figure 2. Generalized geologic profile along the DST alignment

been encountered in the Chagrin Shale at other locations in the Cleveland area. No flush flows were encountered during the excavation of the six DST shafts.

The Ohio Shale members are known to be significant gas producing formations. Natural gas inflows into Cleveland-area tunnels excavated through the Chagrin Shale have been documented for over 100 years, beginning with a fatal explosion in the early 1900s during the excavation of water intake tunnels below Lake Erie. Recent NEORSD tunnels, including the Mill Creek Tunnel Contract experienced extended delays due to natural gas inflows. Naturally occurring gas in the Ohio Shale in the Cleveland area is between 60 and 90% methane with the remaining balance comprised of ethane, propane, butane, and carbon dioxide. Hydrogen sulphide is also present in the Chagrin Shale.

Presently, while gas can be measured at the top of the borehole corresponding to depths drilled, this information is not typically indicative of when or where gas will be encountered in the tunnel and shaft construction. A comparison of the rock conditions and gas inflow locations recorded on the DST boring logs with the location of the gas inflows recorded during the shaft excavations did not find any correlation between gas inflows observed in the shafts and the percentage of siltstone interbeds within the shale, the rockmass RQD or the presence (or absence) of recordable gas inflows in the boring logs. No correlation was observed between the shafts regarding the rockmass condition or quality nor could a gas-bearing layer or horizons be delineated between the shafts even after completion of excavation.

DESIGN OVERVIEW

All of the DST shafts except DST-4 were excavated between June 2015 and November 2016. Once the shaft invert was advanced through the soil and into the Chagrin Shale bedrock, the shaft excavation contractors used any one or a combination of the following excavation methods:

- Mechanical excavation (hoe rams, jack hammers, and excavators)
- Chemical rock splitters (in combination with an excavator)
- Vertical Boring Machine (VBM)
- Drill and Blast Excavation

The DST Contract Documents did not preclude any shaft rock excavation methods. Historically all four excavation methods have been successfully used on other NEORSD projects in the Cleveland area.

The DST Contract Documents provided the Contractor with historical information regarding gas inflow on previous NEORSD projects as well as requirements for gas monitoring and ventilation at the following locations:

- The DST GBR outlined gas inflow history on six recent (1975 to present) NEORSD underground projects built in the Cleveland area. The GBR also referred to the Division 1 Safety Specification for the ventilation requirements for gases encountered during excavation.
- The Safety Specification referenced OSHA regulations and in particular the requirements in 29 CFR Part 1982.000, Subpart S entitled "Underground Construction" which provides details regarding ventilation systems for tunnel and shaft work. In the event that gas is encountered, the DST specifications require the Contractor to demonstrate that his ventilation levels exceed

the OSHA requirements before any gas-delay related claim will be considered. For the 19.25-foot excavated diameter DST-7 shaft, the Contractor was required to provide a minimum flow velocity of 30 cfm of fresh air to the shaft bottom per application of the code to his methods and equipment.

- The Safety Specification required monitoring of the shaft atmosphere with a combustible gas analyzer (CGA) for oxygen, methane, and other flammable and toxic gases. This specification also sets the alarm levels for oxygen at <19.5% or >22% and methane and other flammable gases at >10% of the lower explosive limit (LEL). The CGA and other gas meters were required to be connected to de-energize electrical circuits on the excavation machine when gas concentrations exceed acceptable limits as defined by OSHA and/ or when combustible gas concentrations exceeded a designated threshold (not more than 10% of the LEL). A portable gas meter was to be provided as backup to the equipment mounted units.

DST Contract Bid Item No. 5 financially compensates the Contractor for up to 30 days of standby time in the event that tunnel mining operations are shut down due to gas encountered when mining. However, this bid item was written specifically for tunnel excavation, not shaft excavation. Although gas had been observed in shafts prior to the DST Contract, no gas-related delays or delay claims had been made for shafts excavated in the Chagrin Shale. It is the authors' opinion that this is the result of shafts being accessible to more and larger ventilation equipment and more open to the atmosphere than a tunnel. In addition, a shaft is excavated over a short length at a finite location while a tunnel horizontally traverses miles with a greater probability of encountering gas.

VERTICAL BORING MACHINE (VBM)

Triad has successfully used their Vertical Boring Machine on numerous Cleveland area shafts in the Chagrin Shale and elsewhere in the United States. Without modifications, the VBM can excavate shafts between 18 and 24 feet in diameter. The VBM consists of: a rotating cone-shaped cutterhead equipped with carbide tipped "bullet-shaped" teeth; a central screw conveyor which lifts the excavated material up to a set of crane maneuvered muck buckets for out-of-shaft transport and disposal; a jacking system which braces the VBM off of the shaft walls; and a thrusting system to propel the machine downwards. The VBM is raised and lowered into the shaft with a crane that also supports the shaft excavation activities. Access to the equipment is from a lower work deck positioned about 8 feet above the leading edge of the cone and an upper work deck positioned about 13 feet above the lower work deck.

As part of the VBM excavation effort at each of the DST shafts, a ventilation system comprised of fans and scrubbers was located on the ground surface adjacent to the shaft being excavated and fresh air sent to the shaft bottom through a 30-inch diameter metal fanline hanging along the shaft wall. Figure 3a shows the fanline and the top of the VBM as it begins to excavate the DST-2 shaft. Upon completion of the shaft excavation, the VBM was secured to a crane, a labor crew working off of a suspended platform supported by a second crane removed the sections of the fanline, and the VBM lifted out of the shaft. The fanline was then re-installed along the shaft wall and a labor crew working off of a suspended work platform began installation of rock dowels and mesh along the shaft walls. Figure 3b shows a view of the shaft fanline, the suspended work platform being lowered into the DST-2 shaft and the installed rock dowels and mesh.

Figure 3. (a) Top of the VBM in the DST-2 shaft; (b) rock dowel and mesh-lined VBM-excavated shaft (DST-7) and suspended work platform viewed from above

A gas monitor was secured to the VBM and was set to cut-off the VBM power supply if oxygen max/min levels or >10% LEL is recorded. During the advancement of the VBM excavated shafts, a hand-held four-gas monitor was carried by the VBM operator.

DST-2 AND DST-7 SHAFT EXCAVATION SEQUENCE

The DST-2 shaft was excavated with the VBM between November and December 2015. Gas was encountered during the shaft excavation and the inflows were observed by the authors bubbling out of the shaft wall through small (<1 inch long and <½ inch wide) horizontally-oriented openings. The gas levels during the DST-2 shaft excavation period did not exceed the 10% LEL limit and no gas-related VBM shutdowns occurred.

Upon completion of shaft DST-2, all equipment was mobilized to the DST-7 shaft and excavation into the shale at a depth of 25 feet below ground surface (elevation 612 feet) began on January 4, 2016. By February 2, 2016 the shaft invert had been advanced uneventfully down to a depth of 162 feet (elevation 475.2 feet). Ventilation into the shaft was provided with two surface-based 40 horsepower (hp) fans and a 30-inch diameter metal fan line.

On February 3, 2016 the VBM crew was evacuated from the DST-7 shaft and the VBM shut down due to gas levels exceeding 10% LEL. Between February 3rd and 17th, the shaft heading was advanced to a depth of 177.7 feet (el. 459. 5) with periodic gas-related shut-downs and crew evacuations. On February 16th, Triad and SIH provided documentation of the gas level readings in excess of 10% LEL and formally notified the NEORSD of a differing site condition (DSC) at the DST-7 shaft.

A third 40hp fan was connected to the existing ventilation system on February 17th and the fans were set so that they could either transfer fresh air from the surface to the shaft bottom or evacuate gassy/dusty air from the shaft bottom to the surface and allow fresh air to enter through the mouth of the shaft. Shortly afterwards, NEORSD and Triad personnel using hand-held anemometers recorded air flows across the shaft that exceeded the specified 30 cfm ventilation requirement for the entire shaft and documented that even with this level of air flow, the gas levels in the shaft remained above 10% LEL. Based on the field observations and the provided readings, the NEORSD responded to the DSC claim letter with a work order pre-authorization which (per the DST Contract) authorized the Contractor to proceed with the additional work up to a pre-determined value and to document on a time-and-material basis any costs associated with this claim.

Between February 19th and 28th, shaft advancement progress remained slow and sporadic, initially due to wet and sticky shale and then due to the presence of gas which limited access for both repairs and excavation. Triad also brought a 750 cfm air compressor and six (four 36-inch-long and two 18-inch-long) venturi air movers to the site and secured the venturi at various locations on the lower deck above the VBM cutterhead and in the lifting auger chamber as shown in Figure 4. A seventh venturi was added at a later date. Triad then spent time adjusting the location of the various size venturi to determine which combination helped generate the greatest amount of air movement and turbulence within the cutterhead cone.

The shaft advancement progress between February 3rd and April 8th and the gas level readings (as % LEL) recorded daily by the Contractor is shown on Figure 5. Up to 0.8ppm of hydrogen sulphide was encountered sporadically in the shaft during this same interval. Incidents of hydrogen sulphide readings are not shown on Figure 5.

Venturi Air Movers

Figure 4. Venturi air blowers located above the bottom of the excavation (left) and in the VBM lifting auger chamber (right)

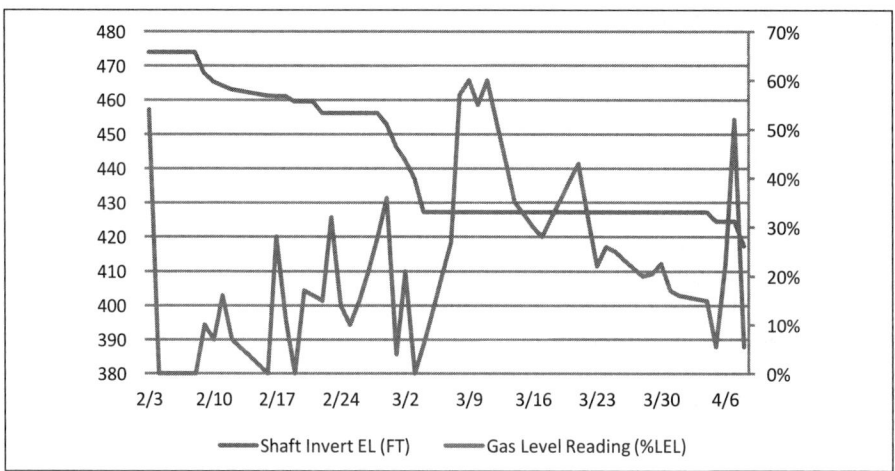

Figure 5. Gas level readings and shaft excavation progress between February 3 and April 8, 2016

Although still sporadic, the shaft excavation advance rate improved after February 29th possibly due to the combination of the ventilation system adjustments and the venturi usage. By the end of shift on March 4th the invert was 209 feet below the ground surface, about 11 feet above the shaft's final design depth of elevation 417.68 feet.

On March 7th, with the fans and venture in full operation mode, , gas levels up to 62% LEL were recorded in the shaft by Triad with a resultant stoppage of all underground work through April 4th. Gas levels in the shaft were confirmed by independent readings collected by the NEORSD's on-site inspector using a third-party gas meter. During this time period, the NEORSD, design team and Triad met frequently to monitor the gas levels and venting progress and determine a course of action to complete the shaft excavation work. Various options discussed between the NEORSD, the design team and Triad included:

- Drilling vent holes around the DST-7 shaft to attempt intersecting the gas outside of the shaft and allowing for accelerated gas dissipation in the shaft. However, as discussed previously. based on field observations and past experience, the gas in the Chagrin Shale is not predictable based on geology and any discrete vent hole has a greater chance of missing a gas-bearing zone, feature or fissure then encountering one, requiring very tight spacing of holes. Additionally, to reach the gas inflows around the shaft, each vent hole would have to be drilled to a depth of 200 feet.

- Increasing the amount of air movement in the shaft bottom to further dilute and dissipate gas inside the shaft. The venturi air movers used around the VBM were adequate to mix the air and produced flows between 600 and 1,100 cfm. The air flow in the shaft at these speeds was violent and greatly reduced the Triad shaft crew's production. Bringing more fresh air from the surface was considered to be a more viable and practical option to dissipate the gas inflows.

- Flooding the bottom ten feet of the shaft with water to reduce the likelihood of sparks igniting the gas. The VBM cutterhead is capable of functioning below water; however, wet shale becomes slick and sticky and cannot be reliably lifted by the VBM's auger.

- Terminating the shaft at the current elevation and starting the 60.5-foot long adit connecting the DST-7 shaft to the Dugway Storage Tunnel at an elevation 11 feet higher than planned. The elevation change would increase the adit gradient from 0.09% to 33.9%. Although the adit and the proposed gradient change could be physically constructed, the option was abandoned due to system-wide hydraulic constraints.

- Abandoning the DST-7 shaft and constructing a new shaft at a nearby location with the VBM. This option did not merit much discussion as the available project space was limited and the site was scheduled to be turned over to another contractor in July 2016. Furthermore, there was no guarantee that the relocated shaft would not encounter the same gas levels.

- Remove the VBM from the DST-7 shaft, relocate and complete the DST-5 shaft excavation, and return the VBM to the DST-7 site after gas levels had dropped to permissible levels. Removal of the VBM as positioned would require cutting torches, a situation incompatible with gas levels >10% LEL. This idea would be reconsidered once gas levels in the shaft fell beneath acceptable levels for work with cutting torches.

- Leave the VBM in the DST-7 shaft, wait until the gas levels dropped, and resume excavation at a later date.

The NEORSD opted to wait until the gas levels dropped below permissible levels. Triad temporarily closed the site except to make daily gas readings to document conditions and redistributed their labor forces to other on-going projects. Triad also made further adjustments and repairs to the ventilation system to reduce air volume and pressure losses.

Through the end of March, weekly discussions were held to review the status of the gas levels and re-visit the aforementioned options. As shown on Figure 5, during the last weeks of March 2016, the overall gas levels appeared to be slowly dropping. By the beginning of April, the NEORSD and Triad agreed that if gas levels continued this trend, Triad would cut the VBM out of the shaft by mid-April I and vacate the DST-7 site until gas levels were further reduced.

The recorded gas level on April 5th was 5.5% LEL and the VBM, after nearly a month of inactivity, advanced the shaft bottom an additional three feet. Gas levels >10% LEL encountered on April 6th and 7th precluded excavation. On April 8th, excavation resumed with recorded gas levels <5.5% LEL and the shaft invert was advanced to its final depth.

Upon completion of the shaft, the VBM was removed within a week and the project team was able to observe the sources of the gas inflows. The gas inflow observed by the authors in the DST-7 shaft occurred along the northern quadrant of the shaft at elevation 437 feet through a 15-foot long horizontally-oriented ½-inch wide opening located beneath a <1-inch thick siltstone interbed. A cloud of gas or water vapor was readily visible extending outwards from the shaft wall.

Gas inflows into the DST-7 shaft were never observed to fully dissipate. For several weeks after the VBM was removed, a gassy odor, smelling similar to propane or butane, was noticed coming out of the shaft mouth by the authors and Triad crew. In December 2016, nearly 10 months after gas was first encountered in the DST-7 shaft, the Northstar crew completing the shaft's concrete liner reported encountering gas emitting from a <50 square foot section of exposed Chagrin Shale in the bottom of the shaft. Using a ventilation system with a configuration and capacity similar to that used to initially support the VBM operation, the gas levels in the shaft bottom were recorded at <10% LEL. On December 13, 2016, within minutes after a fan electrical failure, the gas levels in the shaft bottom rose to 35% LEL and the Northstar crew evacuated. Once the ventilation system was repaired and the system re-activated, gas levels dropped to below permissible levels, work resumed and the final section of concrete lining installed without further incident.

Upon completion of the DST-7 shaft excavation, the Owner and Contractor met to resolve the costs related to the February 16, 2016 SIH/Triad DSC claim and subsequent NEORSD-issued work order pre-authorizations. Because Triad was able to divert most of the DST-7 shaft labor crew to other on-going NEORSD projects, the majority of the incurred costs were related to equipment (VBM, crane) idled by the standby time and the use of a larger air compressor to operate the venturi air movers. A mutually acceptable compensation package for 53 calendar days of gas-related excavation delay was reached during the claim settlement negotiation.

INCIDENTS OF GAS INFLOWS IN THE DST-5 AND DST-8 SHAFTS

The VBM was mobilized to the DST-5 site and began excavating the 19.5-foot excavated diameter shaft beginning on July 7, 2016. Shaft excavation progress was good until a gas inflow on July 14th exceeded the 10% LEL threshold and triggered an automatic VBM shut-down. At this point, the work was halted and the crew evacuated from the shaft. Triad promptly brought the seven venturi and larger capacity air compressor to the site and placed these items in the same manner as used at the DST-7 shaft. Ventilation occurred overnight and the readings taken the morning of July 15th indicated that gas levels were below the permissible levels to allow shaft access. The crew removed the venturi, turned off the compressor, and completed the shaft excavation to the target depth of 204.8 feet (el. 411.6) by July 19th. In this case, it is not clear if the prompt actions taken by Triad helped reduce the gas inflow levels or if the encountered gas pocket was small and bled off quickly.

On August 30, 2016, during the excavation of the 56-foot-excavated-diameter shaft DST-8, a gas pocket was encountered at a depth of 165 feet (elevation 488 feet) while an excavator-mounted hoe ram (approximate weight = 45,000 lbs) was mechanically breaking up the Chagrin Shale in the shaft invert. According to Marra's foreman and excavator operator, the bottom of the shaft heaved and lifted the excavator upwards 6 to 12-inches, and then settled back to the original position, with numerous resulting cracks in the shale. Gas levels when the ground was lifted are reported to have reached 40% LEL, but as verified by SIH's on-site safety officer, dissipated rapidly once the ventilation system flow rate across the shaft bottom was increased to 41,000 cfm. The release of the gas occurred at a depth about 25 feet above a gas inflow recorded in the geotechnical boring advanced at this site. After this event, the DST-8 Access Shaft excavation was completed without further incident.

CONCLUSIONS

As the excavation of five of the six DST shafts through the Chagrin Shale has been completed, the following conclusions related to shaft excavation in the Chagrin Shale have been drawn by the Owner, design team, and Contractor:

- The VBM successfully completed the DST-2, DST-5 and DST-7 shaft excavation and remains a competitively viable means to advance a shaft in the Chagrin Shale. NEORSD Construction Supervisors and Program Managers indicate that there will not be any changes to the specifications to restrict the use of the VBM for shaft excavation.

- There was little room to secure more than seven venturi air movers and a 27-inch and 30-inch diameter airline to the VBM and shaft walls and still have room for the VBM and mucking operation. It did not appear that Triad could have reasonably achieved the delivery of additional air to the working VBM face or generated more turbulence to help lift the inflowing gas up and into the air mass exiting the shaft mouth.

- The occurrence of gas and the resulting LEL levels encountered in the Chagrin Shale are unpredictable. If all reasonable ventilation improvements have been made, the best option may be to delay construction activities until gas levels subside. Depending on depth of the gas inflow, larger diameter vent holes could be drilled on a case by case basis.

- The Contract Documents, including the use of a GBR clearly delineated the baseline conditions and the terms of what items would or would not be accepted as part of a claim and how the items would be paid. With the GBR in hand, the Owner and Contractor were able to cooperatively focus on the resolution of the problem at hand.

- The ground behavior and conditions observed during the shaft excavation through the Chagrin Shale were consistent with the baseline conditions established in the GBR.

- At this time, field delineation of gas bearing layers in the Chagrin Shale is not possible, however, over the years, it appears that more shafts have not encountered gas stoppages versus those that have and the chance of encountering a gas stoppage is considered to be a lower risk that can still be successfully dealt with the enforcement of a safety-conscious culture, ventilation and clear contract language.

ACKNOWLEDGMENTS

The authors would like to thank the following individuals for time spent in discussion on this topic or review of the manuscript: *NEORSD*: Kellie Rotunno and Doug Gabriel, *MWH/MM JV*: Michael Vitale, Rory Ball, and Barry Doyle, *Triad Engineering*: Cliff Kassouf and Tom Clingan. Thanks are also extended to Richard Raynak (MWH/MM Jv) for preparation of the figures.

Large Diameter and Deep Shafts: Unique Design and Construction Challenges

Raymond Blanchard ▪ CH2M Inc.
Edgardo Ross ▪ CH2M Inc.
Harald Leiendecker ▪ CH2M Inc.
Rodolfo Aradas ▪ CH2M Inc.

ABSTRACT

Over the past five to ten years, there has been an increasing number of underground infrastructure projects worldwide using different geometries incorporating large diameter, deep shafts into their configuration. Example projects Ch2m Inc. has direct design and construction experience with include shafts for:

- Blue Plains Dewatering shafts (one of five shafts) for District of Columbia Water and Sewer Authority CSO Clean Rivers Project in Washington DC
- Matanza-Riachuelo Catchment Sanitation Program; La Plata River Outfall.

The design, construction and critical instrumentation monitoring of such shafts brings unique challenges. These challenges can be dealt with in a number of ways, including the use of 3D structural models with a better understanding of soil structure interaction incorporated into the analyses of all loads on the shafts, with resulting savings in material and more flexibility in construction sequencing. The authors draw upon their global experience of design and construction of these large diameter and deep shafts to identify some key unique challenges, and to suggest particular ways in which they can be overcome during design and/or construction.

INTRODUCTION

Over the past five to ten years, there has been an increasing number of modern underground infrastructure projects worldwide using different geometries incorporating large diameter, deep shafts into their configuration. These structures are commonly associated to conveyance tunnels, either as part of water or sewerage projects, as they serve the functional requirement of being head and surge chambers as well as facilitating the logistics of tunnel launching and operation.

The design, construction and critical instrumentation of such shafts brings unique challenges as the large depths bring a significant increase of external ground and water pressures that stresses the structure (typically diaphragm walls), condition the excavation sequence, and requires a detailed analysis of the execution of openings.

The authors draw upon their global experience of design and construction of these large diameter and deep shafts to identify some key unique challenges, and to suggest particular ways in which they can be overcome during design and/or construction. Example projects Ch2m Inc. has direct design and construction experience with include shafts for:

- Blue Plains Dewatering shafts (one of five shafts) for District of Columbia Water and Sewer Authority CSO Clean Rivers Project in Washington DC. The Blue Plains project is part of a large scheme to reduce the discharge of

171

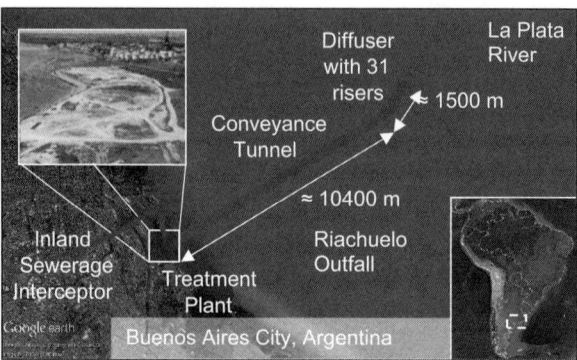

Figure 1. Project location and layout of the works

combined sewer overflows into the local water ways in the District of Columbia and surrounding areas which is referred to as the Long Term Control Plan (LTCP). The system is comprised of four large diameter tunnels through the soft ground of the District and several large, deep shafts.

- Matanza-Riachuelo Catchment Sanitation Program; La Plata River Outfall, The Riachuelo Sanitation Program, currently being implemented for the Riachuelo Catchment in Buenos Aires, Argentina, comprises the execution of 16.6 km of in land conveyance tunnels, followed by 12 km of outfall into the La Plata River, considered to be one of the largest estuaries in the world. The outfall will convey a maximum flow of $27m^3/s$ inside a 4300mm diameter tunnel.

Figure 1 shows a schematic layout of the works.

CURRENT APPROACHES

The design of large and deep shafts includes several structural and geotechnical challenges and demand the need of multiple analytical approaches. Of prime importance is the need to use soil structure interaction modelling to represent the impact of construction sequence and the variability of geotechnical conditions at depth in the development of external earth pressure loads; in this respect, 3D FEM modelling becomes very important when dealing with multi-cell structures of irregular geometries. Obtaining accurate earth pressure loads, results in savings in material and more flexibility in construction sequencing.

However, soil structure interaction modelling gives predominance to the representation of ground properties through a portfolio of constitutive models but tends to simplify the representation of the structural elements; for instance, it is common to simulate an entire diaphragm wall by a continuous element of constant thickness, neglecting the horizontal discontinuity between panels and the loss of verticality amongst them.

Therefore, it is becoming normal practice to also develop an independent structural FEM model, which receives external pressure loads from the soil model, but focuses in the determination of internal forces (axial, shear and flexural loads) accounting for more detailed representation of features such as openings, jointing and stress concentration at the base slab level.

These models need to be further complemented by more detailed analysis (sometimes using hand calculations) around specific features such as reinforcement needs

Figure 2. Large 139 ft and 81 ft diameter by 195 ft deep shafts at Blue Plains Project, USA

to take shear stresses in multiple panels, in particular when they form "Y" panel connections between cells.

The resulting forces are complemented with hand calculations to complete the design and the reinforcement specially for the case of horizontal forces to take into consideration the eccentricity due to deviations in the vertical direction, and stress concentration at joints and openings.

Finally, it is crucial to highlight the importance of sound geotechnical investigations and the installation of a good monitoring equipment to support the development of the models, not only during the design stage but also during construction and post construction stages to provide a good degree of safety during the operational life of the structure.

LARGE DIAMETER SHAFTS—UNIQUE CHALLENGES

Key Unique Design Challenges and Solutions Blue Plains Dewatering shafts

Base Slabs

Large diameter base slabs, especially in deep shafts, invariably result in extensive water pressure loads under the bases of the slabs that are taken into account during the design process. For large diameters, as shown in Figure 2, a typical flat slab of uniform thickness will result in a thick base slab with extensive reinforcement, particularly in the top face. However, the use of dome shaped base slabs with tension rings to take the resulting radial forces in hoop tension significantly reduces both the thickness and amount of reinforcement required.

In addition, the placement of reinforcement in the radial and circumferential (ring) directions, with a small square grid in the center, also significantly reduces the quantity of reinforcement by approximately 25% to 30% compared to typical flat slab rectangular grid patterns as shown in Figure 3. The radial and circumferential layout allows for greater spacing of the reinforcement than is required at the center of the base slab as the internal moments reduce at increasing distance from the center.

Skin Friction at Temporary Supports

Potential uplift and flotation of the entire shaft due to large water pressure loads under the base of a large shaft resulting from high ground water elevations are critical design

Figure 3. Radial and circumferential (ring) reinforcement layout

issues. For diaphragm wall shafts in soil, the designer should consider a reduction of skin friction due to the presence of bentonite cake. The friction capacity of the soil/ diaphragm wall interface is impacted by the construction method itself. First, the slurry trench phase results in a reduction of initial lateral stresses (see following section). Second, the bentonite support forms a filter cake on the trench walls, which might not be displaced during concreting of the panel. Especially in granular soils, such caking of bentonite can be observed during the excavation of the shafts.

Reduction of Earth Pressures on Circular Structures

The accurate assessment of the earth pressure distribution is essential to the design of large and deep shafts in soils such as the Blue Plains DS/SS shaft. Construction of diaphragm wall panels causes considerable stress changes in over-consolidated soil deposits and generally can induce ground movement reducing the at-rest earth pressure.

Recommendations in the literature, in particular industry guidelines developed in the UK (Gaba, 2003), suggest that realistic bending moments for diaphragm walls can be obtained assuming a lateral earth pressure coefficient of unity ($K_0 = 1.0$) prior to excavation even for over consolidated clays with coefficient of earth pressure at-rest K0 well above 1.0. Further reductions, which would result in earth pressure coefficients below unity, need to be justified by detailed analysis, like the finite element calculations presented here. For the diaphragm wall shaft, two different construction processes are of particular importance. The trench excavation procedure and simultaneous slurry support, which stabilize the trench, and the concreting procedure, which takes place once the reinforcing cage has been installed in the slurry-filled trench. In order to estimate reliably earth pressure redistribution, numerical calculations can be made using the finite element method (FEM).

FEM Computations

Soil structure interaction software such as Plaxis3D can be utilized to simulate staged construction as shown in Figure 4. The construction processes of the slurry wall panel are modelled in multiple steps.

Modelling of Wet Concrete Pressure

Commonly used in modelling wet concrete pressures for slurry wall trenches is an approach described by Lings 1994. Lings showed that for the three cases investigated,

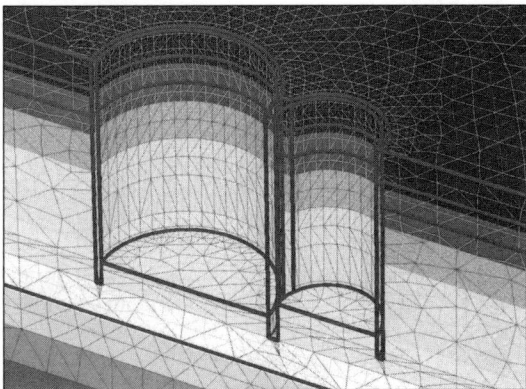

Figure 4. Generalized FEM computation model view (Plaxis 3D)

a bi-linear function depending on the trench's depth is capable to describe an envelope of the real pressure distribution originated by the fresh concrete. The lateral pressure exerted by wet concrete placed under bentonite in a diaphragm wall panel follows a bi-linear pressure envelope, in which full fluid concrete pressures apply only above a critical depth (h_{crit} at approx. 20% of the trench depth). Below this depth, pressures increases with depth following the slope of the hydrostatic bentonite line. The bilinear envelope covering the fresh concrete pressure curves can be approximated as follows:

$$pc = \begin{cases} \gamma_{conc} \cdot z, z < h_{crit} \\ \gamma_{bent} \cdot (z \cdot h_{crit}) + (\gamma_{conc} \cdot h_{crit}), z > h_{crit} \end{cases}$$

Within the numerical model we can simulate the concrete placement using the bilinear function curve. Effective lateral stresses can then be extracted from the computations and plotted over construction staging.

Overcoming Design Control Challenges by Monitoring

An advanced method was used on the Blue Plains Project shafts to monitor and assess stresses and strains in diaphragm wall panels. The use of a 3D soil-structural model provided a good understanding of deformation and internal forces and was used to compute the predicted stresses and strains in a diaphragm wall, especially around the large openings in the shaft wall. Instrumentation in form of rebar strain gauges, concrete strain gauges and earth pressure cells was then used to measure actual wall deformations and concrete strains for comparison with the predicted values.

Threshold and alert trigger strain limit values for various concrete strengths and reinforcement were provided for the strain gauge instruments located on the horizontal, inside-face, reinforcement at different elevations. Strain trigger alerts values were provided for each instrument at various elevations and shown in Green, Amber and Red Zones according to their severity. Alert level severity takes into consideration the concrete strain in the elastic, inelastic and failure zones for various anticipated equivalent concrete compressive strengths. Finally, several potential remedial action plans were developed to be implemented, based on the "trigger" point monitoring values and the ability to accommodate the unexpected. In the event, there was no need to implement any of these remedial action plans.

Key Unique Design Challenges and Solutions Matanza-Riachuelo Catchment Sanitation Program

Functional Requirements and Geometry

The shaft was designed as a multi-cell cast in situ diaphragm wall structure to resist all ground and water external loads during the construction stage of the works, followed by the execution of an internal lining to withstand the external and internal hydrostatic pressures and provide water tightness during the operational stage.

The geometry consists of four intersected cells with a variable diameter that varies between 14m to 16.50m and a depth of approximately 58m as shown in Figure 5. Bored diaphragm walls with a section of 1.20m × 2.80m are executed using a Hydrofraise with a tolerance in terms of loss of verticality of 0.50%.

With an earth pressure of 410 kN/m^2 and a hydrostatic pressure of 520 kN/m^2 at the bottom of the slab, the ultimate horizontal hoop forces reach 5900 kN/m on the walls of Cell 2 and 3, 7100 kN/m on the walls of Cell 1 and 4, and 9400 kN/m on the central walls. The maximum vertical bending moments are 900 kNm/m on the central cells and 1500 kNm/m in in Cell 1 and 4.

Geotechnical Conditions

The geology of the area is characterized by a sequence of glacial and inter-glacial sedimentary deposits comprising primarily sand and clays. The four main geological layers encountered here are: 1) Postpampeano Formation, composed by soft to very soft, normally consolidated and pre-consolidated silts and clays. 2) Puelche Formation, a semi confined aquifer, characterized by dense sand deposits of high permeability. 3) "Blue" Clays, a layer of plastic, pre-consolidated and very stiff clays of low permeability. 4) Paraná Formation, characterized by dense silty and clayey sand deposits of high permeability. The water table level is between +3.87m and −0.63m. Table 1 shows the soil parameters and the Figure 6 shows a schematic representation of the geological formations relevant for the design of the structure.

Layout and Shape of the Structure

The overall shape of the structure imposes the first and most important boundary condition as far as the structural design is concerned. Two extreme configurations can be typically defined: rectangular shafts favor the availability of interior space and avoids the complexities of materializing openings, at the expense of large vertical

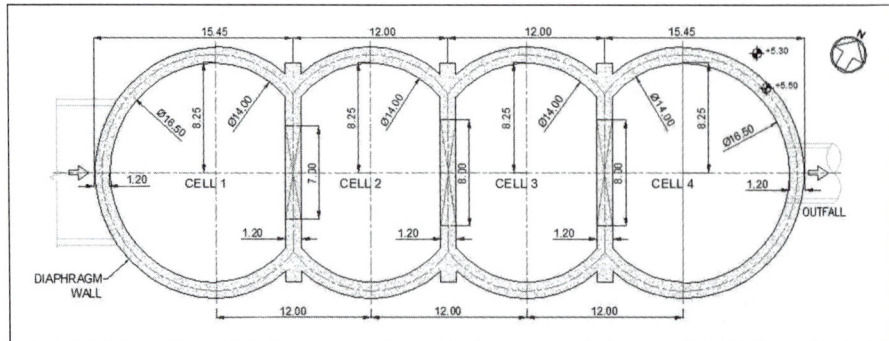

Figure 5. Shaft layout

Table 1. Soil parameters

Description	γ (kN/m^3)	C_u (kN/m^2)	c' (kN/m^2)	ø' (°)
Silty sands of loose compacity	18	35	0	27.5
Clays of soft consistency	16	25	0	25.5
Clays of firm consistency	19	90	10	28
Sands of very dense compacity (Puelchense)	20	—	0	37.5
"Blue" clays of very firm consistency	17	120	15	29
Sands of very dense compacity (Paranaense)	20	—	0	40

flexural loads that require intermediate propping elements. On the other hand, perfect circular shafts exhibit a more efficient structural behavior with predominant hoop forces, but space and openings become more difficult to materialize. In this project, the operational set up for the TBM implied the need for large size openings between cells which, if a perfect circular shape was adopted, it would have led to a dimension much greater and costly than originally envisaged; this alongside with other factors such as specification requirements to maintain the number of cells and constructability aspects in relation to the materialization of the "Y" junctions between cells, led to the design of a compartments with variable diameters. One of the consequences of the variable diameter shape is the modification of the relative axial and flexural stiffness in horizontal direction, causing to change its behavior in terms of load and stress distribution. In the portion of cell with a larger radius the vertical flexural behavior takes more importance whilst in the sector with a smaller radius the horizontal axial behavior takes more importance.

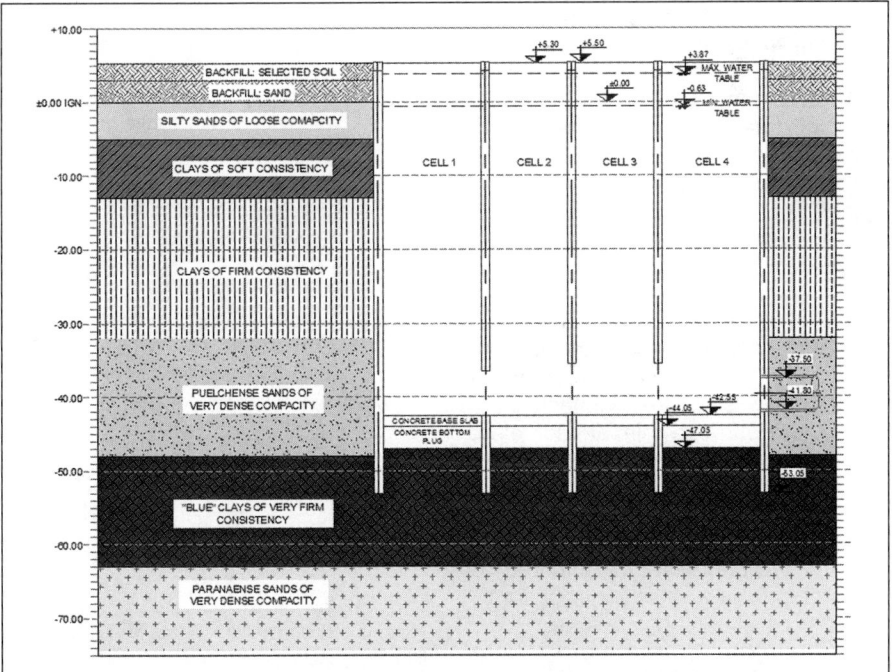

Figure 6. Geological profile

Joints

The variable diameter shape and the loss of verticality, that significantly decreases the available section to take compression between panel joints, reduces the compression and flexural capacity, increasing the eccentricity of the internal loads.

In the "Y" panels as shown in Figure 7, the verifications are dependent on the reinforcement cage and joint arrangement. In this case because of the construction limitations, the use of three cages was necessary. The addition of a special reinforcement in the central cage prevents a shear plane parallel to the central wall without reinforcement. Having the shear plane perpendicular to the lateral panel adds the necessary compression to avoid the necessity of reinforcement in that plane.

Loss of Verticality

The combination of higher hydrostatic and earth pressures, and the loss of verticality that reduces the contact area between panels become critical in the design of the thickness of diaphragm walls, particularly for large and deep shafts as shown in Figure 8.

For Lote 3 project, the consideration of the maximum loss of verticality between panels, leaves a remaining available section of 55% of the wall thickness at the bottom of the concrete plug and 50% at the bottom of the panel. With a concrete of f´c = 35 MPa and hoop stresses that reach 14 MPa, the compression concrete capacity is close to its limit.

The loss of verticality and the subsequent reduction of the contact area between joints was not included in the FEM model, but dealt with in the final calculation taking into consideration the eccentricity of the axial load between the panel and the joint area, and adding the resulting flexural moment at each elevation to the original flexural moment from the FEM model. Then every joint and every panel was analyzed with interaction diagrams in order to determine the needs for the final reinforcement calculation.

One of the common assumptions in the FEM model is that the diaphragm wall is a continuous structure. A check of the eccentricity on every joint at each elevation have been made to verify that the joints are completely under compression.

A more detailed FEM model have been made in SAP2000 to analyze the behavior of the panels and the joints.

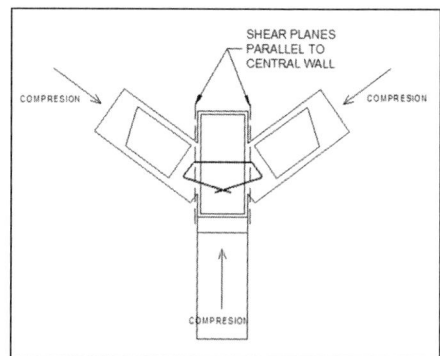

Figure 7. Typical layout of a Y panel connection showing transverse reinforcement

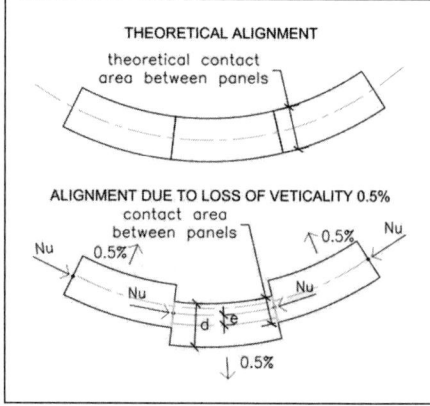

Figure 8. Eccentricity due to loss of verticality between panels

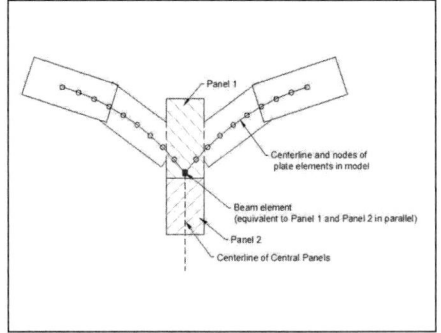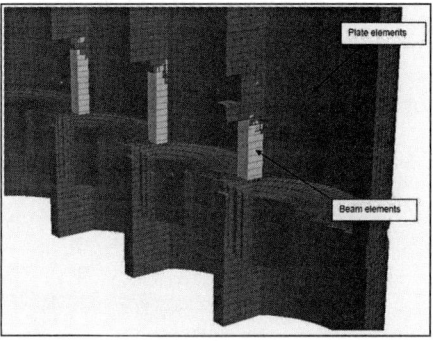

Figure 9. Verification of openings

Openings

The large openings in the panels between cells as shown in Figure 9 leave on either side of the central walls a very limited space to take compression coming from the perimeter walls. The impossibility to complete, not even partially, the inner lining before finishing the construction of the tunnel, because of schedule and operational limitations, leads to the need to take large internal forces on the remaining section of panels on both sides of the opening. Several mitigation options were discussed, such as the provision of reinforcement as steel plates, but were discarded at the light of complexity of installation under, schedule and number of necessary anchors crossing a highly reinforced section. Therefore it become critical to verify those remaining sections, which are composed by two panels (a center and a "Y"panel), for shear and flexural forces. The existence of joints and the discontinuity of reinforcement from one cage to another does not allow to consider the section as monolithic for the verifications. An alternative FEM model with an equivalent section of the panels working in parallel has been made solely to assess the forces taken by those sections, and then a calculation of the shear and flexural moment distribution between the panels and cages has been made taking into consideration the different relative stiffness.

REFERENCES

Gaba, A.R. et al. "Embedded retaining walls—guidance for economic design" Construction Industry Research and Information Association, London 2003.

Laechler, A.,"Bedeutung herstellungsbedingter Enfluesse auf das Trag- und Verformungsverhalten von Schlitzwaenden, Institut fuer Geotechnik der Universitaet Stuttgart, 2009.

Lings, M.L, Ng, C.W.W., Nash, D.F.T "The lateral pressure of wet concrete in diaphragm wall panels cast under bentonite," Proc. Of Institution of Civil engineers, Geotechnical engineering, 1994, 107, pp 163–172.

René Schäfer, Theodor Triantafyllidis,"Spannungs- und Porenwasserdruckentwicklung infolge Schlitzwand-Herstellung in weichen bindigen Böden" Bautechnik, Volume 83, Issue 3, pp 186–201.

Loreck, C., Theodor Triantafyllidis,"Berücksichtigung des Frischbetondrucks bei der FE-Simulation der Schlitzwandherstellung " Bautechnik, Volume 84, Issue 9, pp 646–655, September, 2007.

PART

Grouting and Ground Modification I

Chairs

Daniel McMaster
Mott MacDonald

Erica Fredrickson
Traylor Bros., Inc.

Cellular Backfill—A Review of Some of the Basics

Raymond Henn ▪ RW Henn LLC
David Crouthamel ▪ McMillen Jacobs Associates

ABSTRACT

The requirement to backfill the void or annular space between the excavation and the final lining system, associated with tunnel and shaft construction, has existed for many years. In the past the primary backfill materials used were concrete, sanded grouts, neat cement grouts or blown-in sand/pea gravel. However, starting approximately 20 years ago the use of cellular cement products has come into wide use as the backfill material of choice. This is due to its ability to flow and be pumped long distances and its relativity lower cost. However, as the use of cellular backfill has increased so have problems associated with its placement and ability to meet design goals.

To help better understand and mitigate some of these problems this paper discusses approaches that designers, general contractors and specialty contractors should consider when utilizing cellular backfill.

INTRODUCTION

The need to fill the void space or annulus created between an excavated tunnel or shaft and the final lining system has long been a requirement in underground construction. Most of those types of voids are created as a result of employing the "two pass" tunneling lining method. Filling of this void is called backfilling. For many years the materials used as backfill were primarily reinforced and non-reinforced concrete, sanded grout, neat cement grout, or blown-in sand/pea gravel. However, in the past 20 years or so the use of cellular grouts/cellular concrete as the backfill material has greatly increased. Most cellular grout is made of a mixture of water, Portland cement (with or without the addition of fly ash), a foaming agent and admixtures such as a water-reducer or superplasticizer.

It is the authors' preference to use the term "cellular grout" as opposed to "cellular concrete" since by ACI Concrete Terminology an ACI Standard (ACI CT-13), concrete contains aggregates. Normally cellular grout does not contain sand (finite aggregate), but almost never contain course aggregates.

We believe there are several reasons for this increased use of cellular grout as backfill, such as:

- The increase utilization of the various trenchless tunneling methods. Tunnels constructed using trenchless method tends to be smaller in size, excavated diameter or cross sectional area. Making man-entry for placing backfill from inside the secondary liner, utilizing pre-installed grout posts, more difficult if not impossible. It is these workspace restrictions that are the drivers as to when to utilize grout ports installed during the pipe/liner fabrication versus grout delivery pipes installed within the void/annulus space.
- The ability of cellular backfills to be pumped and flow longer distances.

- The unit weight of cellular grouts can be controlled by mix design and generally can range from 560 Kg/m^3 (35 lbs/CF) to 1040 Kg/m^3 (65 lbs/CF).

Keeping in mind that plain concrete has a unit weight of approximately 2240 Kg/m^3 (140 lbs/CF). Using cellular grouts with a lower unit weight helps reduce material costs as well as reduce buoyancy (floatation) forces and loads on the final tunnel lining systems. The authors, who specialize in the design and construction of grouting programs used for tunneling and underground construction application, have noticed that within the past several years there have been problems associated with cellular backfill grouting programs in general.

We would like to focus on two conditions which we believe can have the biggest negative impact on the quality of the in-place backfill and cause the most serious problems. These are flowing and standing water in the void space being backfilled. These conditions need to be addressed during design, when writing the specifications and in the field during preplacement/placement of the backfill. Allowing cellular grout and other types of cement based backfills to be placed in flowing water should never be allowed. However, with proper design and preplacement/placement methods, backfilling can be done to displace standing water and minimize damage to the cellular backfill during its placement.

Flowing water can be best controlled by employing techniques such as: installation of panning in combination with dewatering piping and/or a "French drain" system. Standing water is best handled by first removing the water and then filling in the low spots in the invert with "dental" concrete prior to installing the carrier pipe or other final lining systems. Allowing backfill to be placed into flowing or standing water can result in having a total absence of, or at the very least, poor quality in-place backfill behind the liner, particularly in the invert and up to the spring line areas of the tunnel. These conditions, if left uncorrected, can lead to carrier pipe failure (buckling) or liner distress which may materialize either during construction or at some point in the future after the project has been put into service.

Any and all of these conditions can result in a defective final product, time consuming and expensive rework, scheduling delays, disputes, and even a catastrophic failure of the liner system.

It is important to understand that "backfill grouting" and "contact grouting" are two different processes (methods) performed for different reasons. The textbook *AUA Guidelines for Backfilling and Contact Grouting of Tunnels and Shafts* (Henn, 2003) offers definitions of these two grouting processes. This paper deals only with cellular grout backfills.

Design Criteria for Cellular Backfill

The design criteria of the cellular backfill will vary depending on the design and use of the carrier pipe or other types of final liner. It will also vary based on amount of ground load, external hydrostatic head and ground load sharing with the initial support, such as steel sets and lagging, steel casing, exposed rock, shotcrete, pre-cast concrete segments, etc. The highest risk pipe is large diameter steel pipe with external hydrostatic loads. Large diameter steel pipe becomes more flexible and sufficient restraint behind the pipe and maintaining suitable circularity is more critical. The pipe must be sufficiently restrained uniformly around the pipe.

Table 1. Recommended cellular backfill performance criteria

Performance Criteria	Range of Values	Comment
Strength	1–5+ MPa (145–725 psi)	Dependent on external structural loads
Max heat of hydration	43–72°C (109–162°F)	Dependent on structural tolerance
Shrinkage	0.03–0.05% Max	Dependent on structural tolerance
Foamed density	$\geq$1040 kg/m^3 (65 PCF) or $\leq$1040 kg/m^3	Presence of water or no water

Backfill with insufficient strength or highly variable strength, such as stratified layering of weak and stronger zones, can lead to the pipe to go out of round or buckle due to inadequate confinement.

Confinement of the pipe is also provided by limiting the amount of gap development behind the pipe. The minimum performance criteria which should be specified for the cellular backfill are presented in Table 1 which are typical for steel pipes. For other pipe materials, such as fiberglass pipe, Reinforced Concrete Cylinder Pipe and others the values will vary depending on material selection and project requirements.

In addition to the requirements described in Table 1, corrosion protection of steel pipe in aggressive groundwater, including saline or high sulfide conditions by substitution of up to half of the cement with fly ash. The use of fly ash reduces the permeability of the cellular backfill, as well as the rate of development of heat of hydration.

Effects of Water Present During Placement on Cellular Backfill

Water can either be flowing or standing or both within the annular space being back-filled. Fresh cellular backfill, as it is introduced into water, can be weakened by dilution as it moves through the water or if flowing water comes into contact with the backfill the cement can be completely washed out. Dilution can cause the cellular backfill to become stratified in density and strength within the annular space. An example of this is shown in Figure 1, where the backfill, after removing a damaged steel pipe, is not uniform in density or strength. The variability of strength is typically much less than the target strength requirements.

If the compressive strength requirements are not met, the pipe will not be restrained sufficiently to withstand external hydrostatic pressures without buckling or damage. Likewise, if the cellular backfill strength is too low, long term ground loads may not be reliably carried by the final lining without the lining distorting. The plain steel pipe is not designed to carry hydrostatic or ground loads on its own, the composite effect of sufficient cellular backfill strength and steel pipe are designed to carry these forces. If the pipe is not properly restrained, the pipe can deform, loose circularity and the factor of safety reduced relative to buckling.

Figure 1. Stratified cellular backfill of variable strength and density due to the presence and damage of water

Figure 2. Erection of intermediate bulkheads within the tunnel alignment

The cause of damage during placement of the cellular backfill is due to the presence of water and the travel distance of the cellular backfill through the water. The travel distance can be reduced by creating closed cells or bulkheads along the length of the pipe. Figure 2 shows the erection of an intermediate bulkhead, to create a cell along the length of the tunnel. The use of intermediate bulkheads limits the distance that the backfill has to travel along the length of the tunnel and helps preserve the integrity of the cellular backfills strength. Figure 3 shows a very well designed, constructed, and organized portal bulkhead. In Figure 4 a section view of a tunnel shows intermediate and portal bulkheads and the backfill injection piping installed through the bulkhead and in the annulus between the excavated tunnel and the carrier pipe. If flowing water is present, panning should be placed securely on the tunnel wall or

Figure 3. Portal bulkhead

segments to shield the water from the cellular backfill. The panning needs to drain into dewatering piping and or a "French drain" system. Likewise, if standing water is present, every effort should be made to evacuate the water prior to placing the cellular backfill to limit the exposure to water. Every effort should be made to minimize the exposure of the cellular backfill to standing or flowing water. If standing water is present, the mix design should have a wet density greater than water as indicated in Table 1.

Limiting Gap Development Behind Pipe

Another critical component to the performance of the placed cellular backfill is the limitation of a gap immediately behind the wall of the pipe. This assumed gap allows deformation of the steel lining in response to external pressure, forming a lobe whose

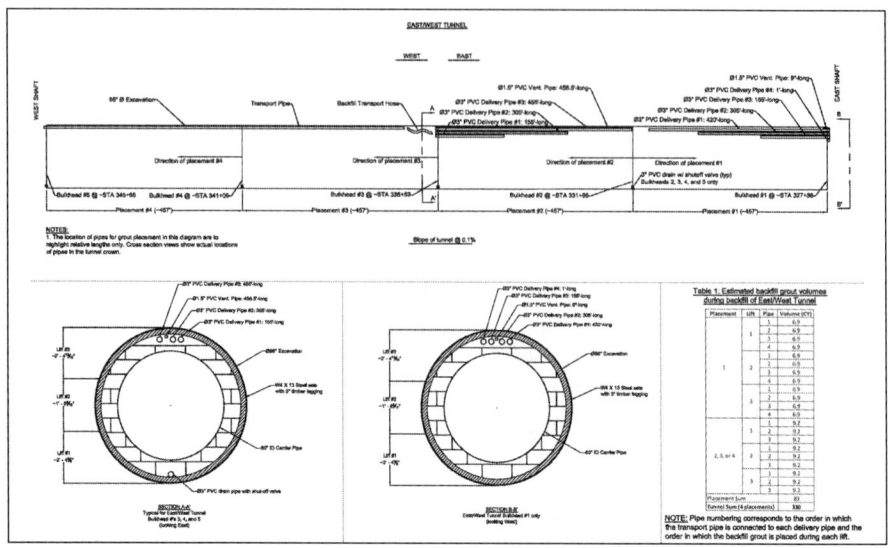

Figure 4. Section view of tunnel

stability or factor of safety is determined (Figure 5). Two methods of analysis are Amstutz, (1970) and Jacobsen, (1974) which are based on prevention of single lobe bucking of steel pipes and apply a factor of safety of 1.5 against buckling. Both methods for plain cylinder stool pipes assume a minimum gap exists between the outside surface of the steel lining and the surrounding backfill. The magnitude of this assumed gap accounts

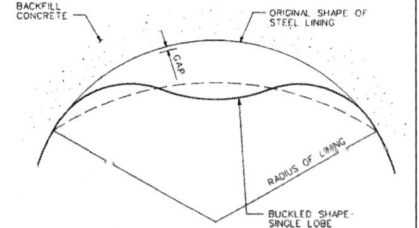

Figure 5. Single lobe buckling principal

for backfill shrinkage and thermal deformations in the steel lining due to heat of hydration. To meet this requirement of minimized gap, thermal rise due to heat of hydration of the cellular backfill and shrinkage must be limited to prevent enlargement of the gap. An excessive gap growth, if not treated, can result in reduced factors of safety against buckling due to external pressures.

Mitigation of gap development should be based on specifying maximum shrinkage and heat of hydration performance requirements which should be determined based on final lining design. The specification should specify the maximum temperature rise of the placed backfill as a performance requirement. The mix designs should be tested in accordance with ASTM C186, Test Method for Heat of Hydration of Hydraulic Cement.

Heat of hydration (HoH) can be limited a number of ways. The most reliable methods is replacing cement content with fly ash in the mix design. Another method is to restrict or limit lift heights during placement. Another procedure, which is less effective, is using chilled water as part of the mix water, or filling the carrier pipe with water.

Testing for heat of hydration should be performed in a fashion which mimics the volume of mass cellular placement in the field. Figure 6 shows a 1 × 1 meter square block with a height thickness of 375 mm as an equivalent lift height.

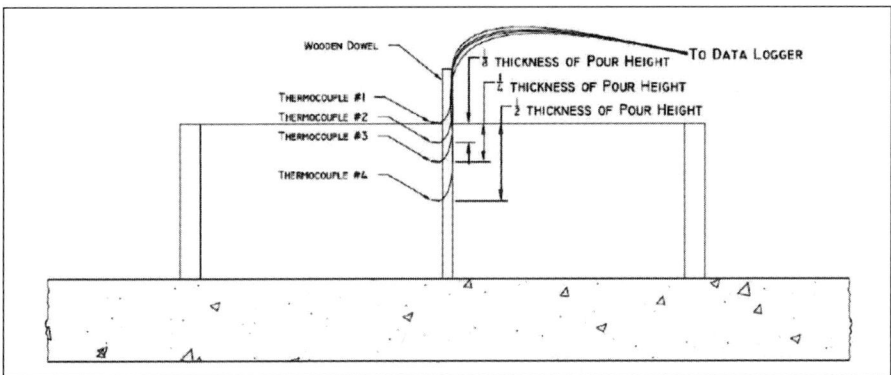

Figure 6. Heat of hydration testing arrangement

Heat of hydration should be addressed first and foremost in the mix design and lift height. Tests have shown that using a 100 percent cement and water mix designs will result in excessive temperatures. Figure 7 (top) shows a 425 mm lift height test with a mix design using 100 percent cement, 0 percent fly ash in the mix. The test shows an approximate temperature rise of 99.7 °C. Figure 7 (bottom) shows a 375 mm lift height test with a mix design using 50/50 percent cement and fly ash. The test shows a reduced HoH of 64 °C. peak value which meets a specified target of no more than 72°C.

While it can be difficult to predict the in situ maximum core temperatures, other factors will have an impact on the placed field HoH temperatures including:

- Use of fans and cooling water on the inside surface of the pipe to keep pipe/concrete temperatures down.
- Surface temperatures of previous placed lifts.
- The effects of heat sink through either rock or precast concrete tunnel segments.
- Humidity of the curing environment and the amount of water present.
- Location of lift behind the pipe as the highest lift towards the crown produces more heat than the invert lift.

If a large gap behind the steel pipe does form, as evidenced by excessive measured heat rise and sounding of the pipe, skin grouting procedures would have to be conducted. The process of skin grouting requires the drilling and taping of holes through the steel lining, which then have to be sealed. This process should avoided by using all available tools to limit the HoH in the first place and ensure that the gap growth is limited.

RECOMMENDATIONS

Various possible deficiencies have been identified in the placement of cellular backfill for tunnel and shaft and the performance of various lining designs. The design of these linings is based on performance requirements of specified strength and gap limitations. These performance requirements should be highlighted in the design documents along with anticipated methodologies to meet these requirements for the installed backfill product. It is recommended that provisions should be made in the contract documents that the following requirements be made:

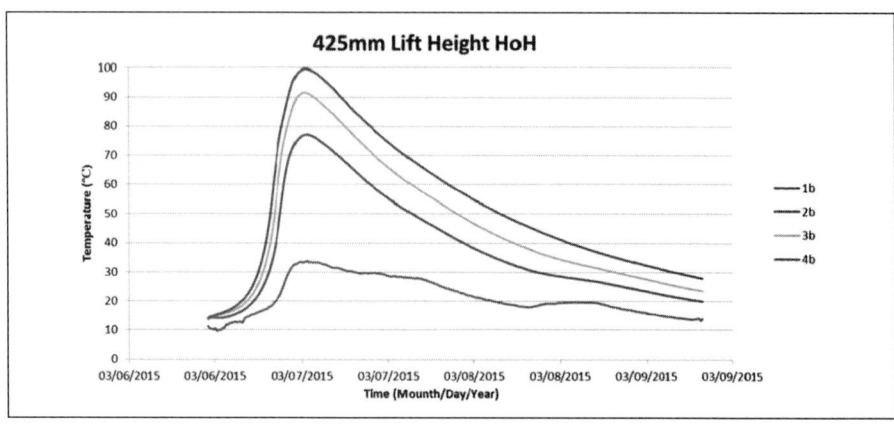

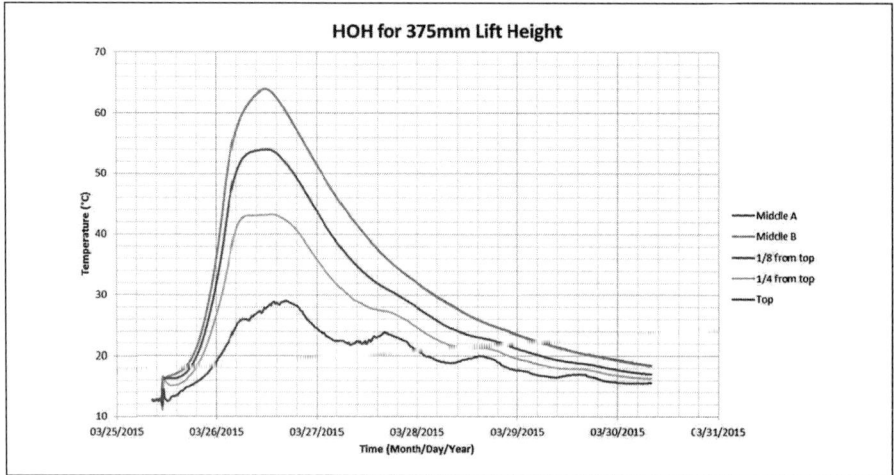

Figure 7. Comparison of lift height and flyash content

- Placement in flowing water should never be allowed. If flowing water is present some means of controlling it, such as panning, dewatering piping or a French drain system should be used.
- In the case of standing water the low spots in the invert should be filled in with concrete prior to placing the carrier pipe.
- In the presence of standing water, wet density of cellular backfill should be greater than water throughout the ± tolerance range of its wet density. This will be necessary to displace the water.
- In the presence of standing water, placement cells should be created to limit the longitudinal travel distance of the cellular backfill through water. The spacing of these cells will depend on the ability or inability to remove the water prior to placing a lift and the amount of standing water. Spacing can range from as little as a few hundred feet to three to four thousand feet if the water depth is reduced and the backfill wet density is significantly greater than water.
- Heat of hydration rise should be specified and methods to control it including modified mix designs using upwards to 50% fly ash and restricting lift

heights. Testing procedures which mimic field conditions should be required to ensure that the selected mix designs and installation procedures will reliably meet the temperature restrictions under in situ placement conditions.

REFERENCES

American Concrete Institute, January 2013, Publication ACI CT-13, ACI Concrete Terminology.

Amstutz, E., 1970, "Buckling of Pressure Shaft and Tunnel Linings," *Water and Power Dam Construction*, Nov., pp. 391–399.

Henn, R.W., et al. 2003, AUA Guidelines for Backfilling and Contact Grouting of Tunnels and Shaft, ASCE Press/Thomas Telford, pages 3–12.

Jacobsen, S., 1974, "Buckling of Circular Rings and Cylindrical Tubes Under External Pressure," *Water Power*, Dec., pp. 400–407.

Copenhagen Cityringen Project: Big Data to Manage Quality Control in Megaprojects

Livia Cicinelli ▪ Seli Tunneling Denmark ApS (Salini Impregilo Group)
Valerio Violo ▪ Seli Tunneling Denmark ApS (Salini Impregilo Group)
Frank Stahl ▪ Tunnelsoft
Thomas Gronbach ▪ Dropbox

ABSTRACT

As Operators in the tunneling sector we are well aware of the importance of the Quality in our everyday work. Historically, "to do things right" was the informal definition of goals, but is has emerged and established as an integrated and systematic contractual requirement. We all became familiar with the words auditing, documentation, traceability (work orders, parts, working hours, etc), identificability, configuration, non-conformity management, and learned the great benefits and improved visibility around our project. Implementing the contractual requirements is challenging in terms of organizing work, resources and managing the vast reporting requirements resulting.

This session shows how visionary project development, strong planning, software design, field testing and execution (the paramount PDCA, Plan, Do, Check, Act of the Quality Assurance Management Systems) is making this task more streamlined and more efficient. We will outline how an implementation aimed for the project users and workers involved rather than fulfilling mere regulatory and contractual requirements boosts project success. We will highlight the software/service our system has been integrated with: TPC Tunneling Process Control, provided by Tunnelsoft, a branch of Babendererde Engineers and the cloud service leader provider (Dropbox).

INTRODUCTION

The Cityringen is an extension of the existing metro line. The 15.5 km long twin tunnels will form a new circular line with 21 new structures (17 + 4 crossovers). When finished it will interchange with the existing metro, rail and bus infrastructure at key locations around town, taking 90% of Copenhagen citizen at less the 600m from a metro station (Figure 1).

The Contract has been awarded in April 2011 to the Copenhagen Metro Team (CMT), a joint venture between the Italian companies Salini Impregilo, Tecnimont and Seli SpA.

Tunnels are being excavated with 4 Seli-Kawasaki EPB TBMs, two of which manufactured in Denmark by Seli Tunneling Denmark ApS, which is also operating them. At the moment this article is written nearly 90% of the tunnels have been completed and over 50% handed over for the electromechanical works.

Seli Tunnelling Denmark is a Danish company constituted in 2012 as a subsidiary of Seli SpA for this project and subsequently acquired by the Salini Impregilo Group in 2015.

The first two TBMs have been launched in Nørrebroparken shaft (NØP) towards Sønder Boulevard (SBV) respectively in July and December 2013. TBMs 3 and 4 have

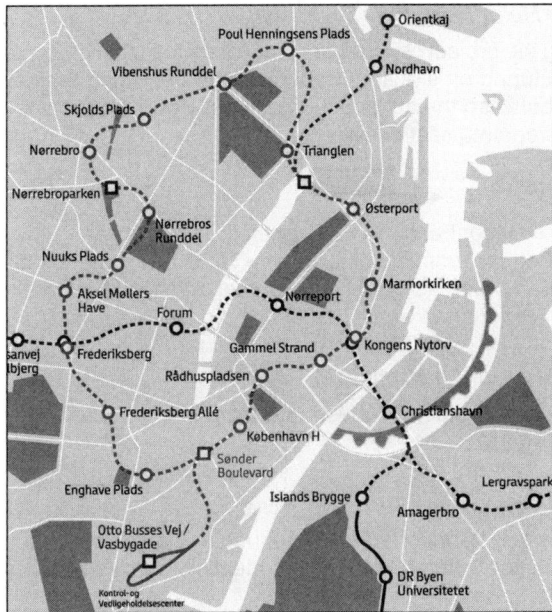

Figure 1. Cityringen general layout

been manufactured and tested in a Copenhagen facility by Seli Tunneling Denmark, before being transported to the Tømmergraven site (south side of the city), where they were re-assembled and launched in June and August 2014.

Tunnel excavation will be completed between December 2016 and February 2017.

The four TBMs are EPBs, identical and built to the following specification:

- Cutter head Diameter: 5.84 m.
- Shield Diameter: 5.74 m un-tapered.
- Shield length: 10.5 m-double articulation.
- Lining Diameter: External 5.5 m/ Internal 4.9 m.
- Minimum Radius: Horizontal 180 m /vertical 625 m.
- Maximum gradient: 6%.

PROJECT PROCESSES

Description of the Actors

The actors involved in the project are multiple and each one of them owns a part of the process as far as quality is concerned. Therefore collaboration and communication is at the base of a successful project.

Metroselskabet, Project Owner

In a design build project the client is directly involved, thru its representative, in the construction process and is the recipient of the inspection reports for segment lining, analysis report of mining parameters, settlements etc. The project owner is the final instance when tracing the reporting chain from bottom to top.

CMT, Copenhagen Metro Team, Main Contractor

CMT is executing the project construction. It acts as the main contractor and is responsible for manufacturing on site and producing and delivering the segment lining. It is an intermediate between the construction company and the project owner and is also responsible of all communication with the client.

Seli Tunneling Denmark, Sub-contractor

Seli Denmark is a subcontractor to CMT, even though it is part of the same group. Seli Denmark is building the tunnels, installing the lining and generating quality document flow for reporting on progress, quality measures, and delivery of parts. Among others, main responsibility of tunnel excavation and lining QC lies with Seli Denmark.

Mobilbaustoffe, Segment Manufacture

Mobilbaustoffe is the company that is manufacturing the segment lining in its facility in North-east Germany.

DSV, Port Authorities

The segment lining is transported to Copenhagen via ship, stored at the harbor and transported to site according to the TBM production.

The wide range of organizations and the number of people involved, requires the collaboration task to be designed for its users in systems that allow easy exchange of documents among them and convenient interfaces with systems. Processes that rely strongly for its success on documents require a well-defined content based collaboration approach.

Segment Production and Journey from Factory to Tunnels

The into graphic of Figure 2 shows, in a simplified way, the journey of the prefabricated elements from the factory in Northern Germany, to Copenhagen, the sites and, eventually, the tunnels. From the document point of view the transportation process is complex and collaboration is very important.

Delivery at Sassnitz Harbour

CMT is responsible for sending in advance a 2 months sailing schedule look ahead to DSV, in order to plan all the shipments and to inform Mobil Baustoffe accordingly. After the production and the database with the segment data is created, the segments are delivered to Sassnitz harbour.

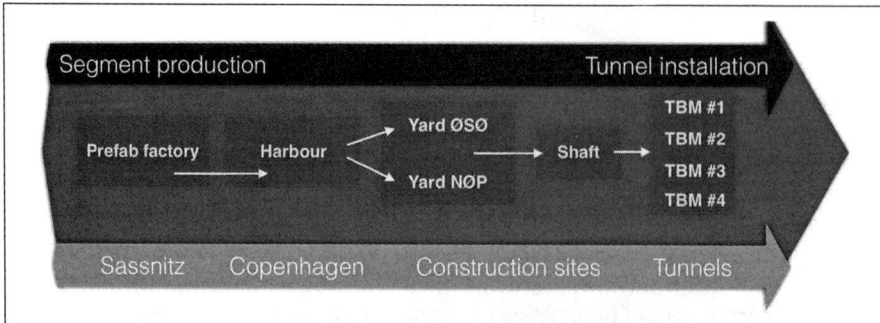

Figure 2. Segment lining journey from the factory to the tunnel

Together with the FOB (Free on Board – Incoterms 2010) are issued two documents:

1. A Shipping list handed over to CMT by Mobil Baustoffe, and signed by CMT Quality Control representative.
2. Load pack list is handed over to DSV by CMT. This document provides the list elements loaded on the vessel.

After the segments loading process, just before the shipment, CMT Quality Control inspector carries out a second inspection: each inspected ring is marked on the side by with green/yellow/red paint, which corresponds to ok to load/repairs needed/ring rejected. Only green marked rings shall be transported. Upon completion of the loading, cargo manifest is issued by DSV to CMT to confirm that all loads listed in the pack list have been effectively loaded on the vessel and are transported to Copenhagen Malmö Port (CMP).

DSV Shipment

A delivery note is issued by DSV when the cargo has been delivered to Copenhagen port (CMP).

After the segments unloading process, CMT Quality Control inspector carries out a third inspection. Another checklist is issued and co-signed by CMT, MBS representatives and DSV.

Each inspected ring is remarked on the side by with green/yellow/red paint. Only double green marked rings shall be transported to sites.

Storage and Transport

Segments could be temporarily stored at the CMP (Copenhagen Malmö Port) or brought directly to the sites (NØP-ØSØ) by DSV trucks.

Delivery at Site

During delivery on site an inspection by CMT takes place to ensure proper unloading, handling and temporary storage. Another checklist is issued.

The Quality Department representatives of Seli Denmark under the responsibility of their Technical Office Manager perform joint inspection with CMT responsible site personnel for the acceptance of the segments on site (Acceptance at Yard—done with TPC scanning). Each time there is a new lot of TBM concrete segments delivered on site by the Carrier (DSV), CMT notifies verbally or by e-mail Seli Denmark responsible personnel.

At this point more than 10 people from 5–8 organizations were involved in establishing reporting combined with a complexity of working in-office as well as mobile and the product has been already inspected three different times.

The Number of the Project

The tunnel lining consists of concrete segments, 6 "stones" forming a ring 1,4 m long. Segments are precast at a factory in Germany, shipped to Copenhagen Harbor where they are kept in storage until needed, then transported to each one of the 2 TBM jobsites, where they are inspected, transported inside the shaft, moved to the excavation front inside the tunnel and eventually installed. We are talking 31.000 m × 6/1,4 m

=133.000 elements, each one to be identified and traced from the moment of casting to the moment of hand over to the client.

Besides the documents involved in the segment lining production and transport, in the day-to-day life of the project, the different Quality System involved require their quota of document production: many different employees and actors take part of the construction process, from the TBM pilot that has to mine keeping the TBM in the alignment and operate it within the design parameters, to the QC engineers that verify the parameters and create the relevant reports according to the various procedures. The same operatives are also the recipients of the working instructions and procedures, of the design output such as face pressure to be applied, muck weight and all relevant alerts and alarms. Having fours machines operating simultaneously leads to a very high number of control documents and reporting.

The four TBMs are operated each by a shift of 11 employees, supervised by two tunnel superintendents per site and one site manager per site. The flow of information from the main contractor down to the company technical office and, from here, to the mining crew has been particularly well .determined and documented.

It is also very important mentioning the parties in the quality processes:

Each TBM is currently excavating, per average, 16 rings per day. This gives a total of approximately 60 to 70 rings per day. Required reporting, for each one of these rings, foresees:

1. Muck weight report (alarm/alert);
2. Excavation report;
3. Shift report with statistical incidental of stoppages, mining etc,
4. Ring erection report that includes keystone position and/or damages
5. Mechanical and electrical shift reports;
6. Ring inspection;
7. Ring joint client/contractor inspection;
8. Ring repair report;
9. Ring condition report.

All above reporting, for us up to 400 units in a good day only for mining operation from (1) to (4), is the normal in tunnel construction. What we had in mind at the beginning of the project was to reduce the amount of the paper-reporting, making it automatic as much as possible, to let it "happen in the background," but at the same time having the information readily available for collaboration, back analysis, and improvement of working methodologies.

Moreover, points from (6) to (9) are paramount since they essentially close all open NCRs, determining the handover of a certain tunnel stretch to the client with all implication that that has in term of payments, milestones and guarantees.

In addition to the above points, the tunnel crews, both mining crews and repair crews, are the recipient of design documents, in particular:

a. TBM mining parameters;
b. Procedures;

c. Special instructions;

d. Break thru or other special passage drawings

Content based collaboration by Dropbox provides instant access to all documents, integration in various systems via an open API (Application Programming Interface), fastest synchronization of document content in the field with Dropbox and an easy to use user interface for all roles ready to use.

The Adopted Solution: Mining Operation

The Planning of the System started already in the mobilization phase, we worked very closely with the developers, we analyzed our workflow and they tailored the process to our needs. It was a long process, but we are all very pleased with the results.

The first decision was not to have a server for office, company needs, but to rely instead entirely on a cloud based system such as Dropbox. This substantially reduced costs for server maintenance and made collaboration easier.

The second step was to give as much access as we could to cloud based company services, to do so we implemented a only-softcopy-distribution policy: all procedures, design drawings and parameters transmittal forms are distributed online and made available on a Dropbox dedicated folder. All managers have smartphones and/or tablets that can access the shared folders and save them for offline use.

On the fours TBM we set a basic Internet connection. This has been possible by simply using the optic fiber data cable, that is used underground to transmit TBM data. The connection is used to receive information end, especially, to transmit the reports mentioned at point (5): in this case the online reporting is particularly useful since, in case of break down, spare parts are ordered to the warehouse including a photograph of the broken part and the warehouse barcode, for faster delivery.

For Tunnel Data processing we adopted the software TPC. The software in/output have been developed together with the supplier and tailored on our procedures and, therefore, on our quality system: point (1), (3) and are automatically generated by the system and made available to the technical office and QC engineers. The reports mentioned at point (2) and (4) are the only still manually filled in by the TBM operator and the TBM foreman. Operators are instructed to fill in as many information as they have, including commentaries and information that would be otherwise lost, relevant to time losses or problems occurred during the shift and that are also used to improve the TPC records of events (a QC engineer enters all information that are written on the report into the online TPC report).

The Adopted Solution: Segment-Lining Handling

Our target was complete traceability from the production factory in Germany to one of the four TBM and to the final position in the tunnels under Copenhagen.

This was a particularly challenging task, also in consideration of the long journey that segment lining takes from production to site, as previously described.

The system is actually quite simple; a database and a few scanners and cloud storage form it. The use of the cloud-database for storing information about the tunnel lining wouldn't be so special per se; what we think is innovative is that it is used as part of the handover protocol, and that the client accepts the electronic signature.

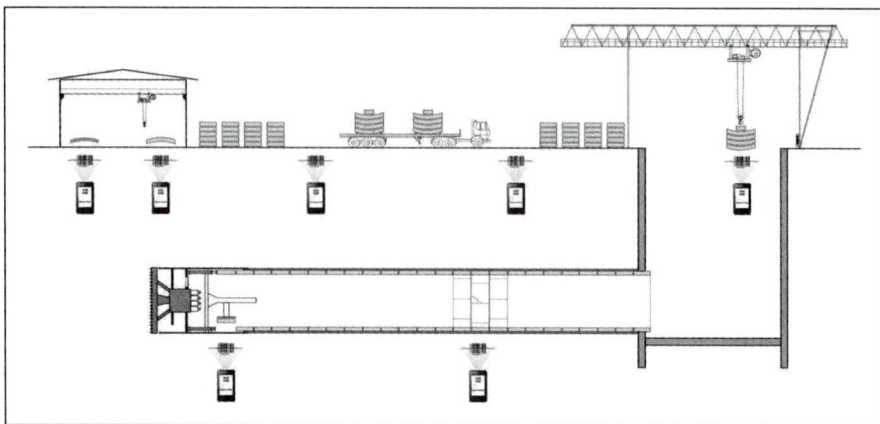

Figure 3. Different scanning section for segment lining

The tunnel elements each bear a tag with a barcode that can be scanned. We opted to use iPods for the scanning, because they are small, easy to handle and have a decent camera. Moreover, they're relatively inexpensive, compared to industrial bar code readers, and cannot be stolen thanks to the iOS peculiarity of "find my iPhone" that makes them unusable if stolen. The segment lining manufacturer sends lists of the segments that are input in the database: each "stone" is associated with a unique number; a set of six stones forms a package with associated a unique number. The whole package is completely scanned at the first inspection as they enter site, during the inspection that determines whether the ring can be accepted or not. Our first scan will have a unique reference to the paper documentation the segment is travelling with from the factory that, being outside our company jurisdiction, kept their own quality processes.

At the arrival the segments are scanned for damages and can be:

- Accepted;
- Marked for repair;
- Rejected.

In case of rejection they're painted red and re-used for temporary purposes such as sliding into the stations or as dummy ring for TBM launching. Those requiring repair, are stored in a separate area (quarantine) until ready to be re-inspected and, eventually, accepted.

After this first step, as long as the package is kept together, only one stone needs to be scanned and this is very important because all operative that will scan one segments of the package from this moment on, will track the whole ring. In a segment in a package has to be substituted, a new package must be formed to keep the process streamlined and limit further scanning.

As you can see in the flowchart (Figure 2) and the drawing (Figure 3) there are different points of scanning stations: at any given moment of the flow it is possible to take pictures, signal different type of damage, write comments, and produce reports. Once installed inside the Tunnel the System univocally assigns to the package a position corresponding to the sequential Ring number. A series of inspections is then performed; starting from the moment the ring exits the TBM, to map the defects situation, type, and position status.

At any time it is possible to extract statistical information to show amount and distribution of the defects which can be easily put in relation to the alignment and the position of the TBM: this is of course extremely important at the moment of planning the interventions to repair the defects, to organize the allocation of personnel and material, time schedules and deliveries. It can also be very useful from the Quality point of view to help identifying the root causes of recurring damages, for example we might notice they happen in certain specific geometrical condition or during a shift in particular: this way the production team can put in place preventive action, for example giving the TBM operator new instruction or retrain the segment erector operator if necessary.

SEGMENT DAMAGE REPORTING AND REPAIR

Inspection

As said all history of each segment lining stone is followed by a scan that generates an entry into our segment-tracking database.

Once a section of tunnel is completed, or partially completed, the QC engineer carries out a first inspection. The inspection is carried out with an iPod with the segment-tracking app installed; a measuring device, to measure crack width (only cracks wider than 0.2mm must be repaired) and a flashlight. We decided to be, during the first inspection, very conservative: in case of doubts whether a repair is needed, according to the contract specification, we recorded the damage. This caused our first inspection/statistic to be more populated than what we had to repair.

Figure 4 shows a screenshot of the application on one iPod used for the inspection. The inspection may also include photographs of the actual damage.

Once the inspection is completed, the system generates a condition report (Figure 5) which is automatically saved on a Dropbox folder for further access, and that is sent to the main contractor and, by the QA of the main contractor, to the project owner. This condition report is the basis on which the joint client/contractor inspection is carried out.

The joint inspection goes thru the same tunnel section and may include further observation or may, as it is in most cases, reduce the one inserted during the first inspection declassifying damages. When the inspection is completed and electronically signed by the client representative and the company QC engineer we proceed to the repair.

Repairs are also recorded in the application, by mean of a change of status for a given ring. However for repair we also produce a paper report that is signed and stored in the technical office.

After all repair are completed, we proceed with the two lasts inspections with the project owner and with the main contractor (CMT).

We have therefore four inspections that are all digitally signed and that do not produce any paper document, being the only paper document those filled in by the repair crew in the tunnel.

The above-described process offers the following advantages given by the cloud use:

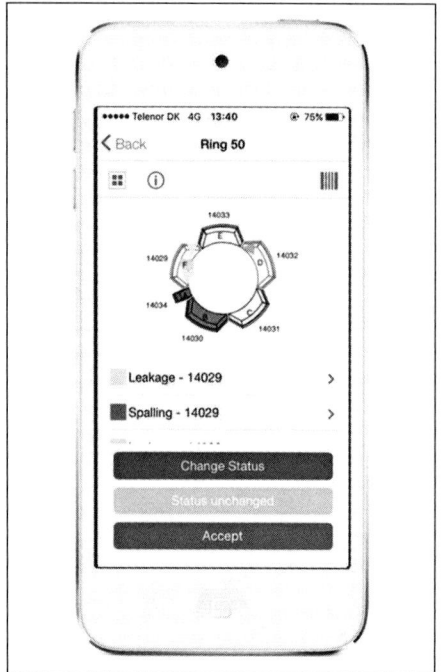

Figure 4. Mobile application

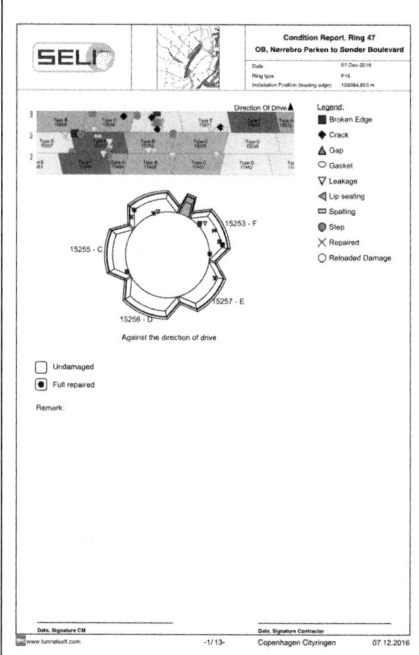

Figure 5. Detailed condition report

- No paper, or nearly no paper, document;
- Signature are easy to trace since are electronically stored;
- Documents are easily shared and it is very easy to search for a certain stretch having to open just an app and not a paper folder;

Moreover the process is very fast: areas with limited damages or repairs, can be walked thru in matter of minutes or hours, repairs can be inspected and approved by the contractor and the client just with a tap.

Once the repair is approved within the app, a report similar to the one in Figure 5 can be generated, with the difference that the final report includes all the inspection history plus the repair process. A normal condition report for one ring can be made of hundreds of pages. However, being the document digital, search within the different report is easy and fast.

Statistical Analysis

Having all data recorded in a database is common practice for what concerns TBM mining. Back analysis on excavated stretches aims to improve TBM performances and behaviour in the machine interaction with the ground and the environment.

However managing the complete segment acceptance, tracking and repair process is something that hasn't been done too often so far. The incredible amount of data and its immediate accessibility for analysis and collaboration makes extremely easy to determine statistical distribution of the segment damages and facilitates the creation of handover programs that can be aggressive and at the same time realistic.

Table 1. Recorded versus repaired damages

	Nørrebroparken to Sønder Boulevard			Nørrebroparken to Østersøgade	
	Recorded	Repaired		Recorded	Repaired
Cracks	319	242		48	29
Spalling	1183	749		859	670
Broken Edges	478	306		66	47
Leakages	1352	869		54	44

With reference to the Figure 1, once the first drive from Nørrebro Parken to København H has been completed, we had to dismantle the TBM, proceed backward from the arrival to the starting portal removing all utilities and, eventually, carrying out the tunnel repair to handover the tunnel.

It is important to mention that, as shown in Table 1, the amount of repair in the first drive was at the beginning higher than what expected and higher than what we experienced for the following stretches. Moreover while we were handing over the tunnel sections, the main contractor was literally following us with the tunnel invert casting and another subcontractor was behind them to take charge of the first EM installations. All this put an enormous pressure on our teams and without a very efficient quality system and fast inspection method we would have delayed the other works. In average we handed over a tunnel section every three weeks.

As Table 1 is showing, the number of necessary repairs on the drives we are currently finishing is dramatically lower: we have already allocated for each tunnel stretch, between consecutive stations, one week, after utilities removal, for final handover. As mentioned, Table 1 shows the total amount of damages registered during the first inspection and the actual number of the repairs that, per the specification, had to be carried out. Considering the length of the drives, 4825m versus 3094m, we notice that the incidence of the leakages dropped from 0.18 unit/m to 0.01 unit/m: i.e. one order of magnitude less.

Another important consideration is the number of repaired broken edges vs initially recorded: approximately 40% of damages recorded during the first inspection was actually within the tolerances and was easily written off during repair and, lastly, during hand over.

In our opinion the first inspection is paramount to the correct definition of the schedule and the resources that will be employed during the repair and handover process: a too "optimistic" inspection could potentially lead to allocate limited resources and thus missing the handover milestone. On the other hand a too pessimistic first inspection could take us to "overshooting" the necessary resources.

Figure 6 is an example on how "big data," when easily accessible and usable, can give back important information and help back-analysis. The figure shows the distribution of the leakages in the project. The biggest circles are in the first stretch and represent 700 leakages. These were mainly due to the learning curve and to the TBM launching that, for the limited dimension of the launching shaft, were particularly complex and did not allow a smooth and continuous operation of the machine. The number of leakages diminished with the following stretches until disappearing almost completely: the smaller circles in the most recent stretches represent 0 to 5 leakages.

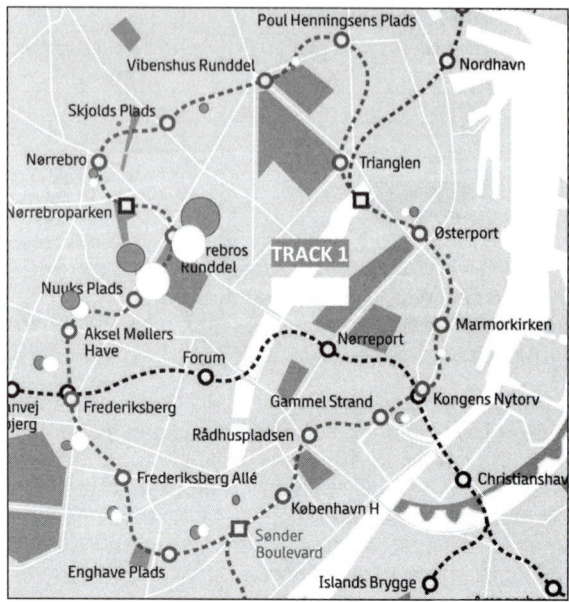

Figure 6. Cluster distribution of leakages

OUTLOOK

A primary goal is a regularly check all site workflows in matter of effectiveness and quality. The aim of these reviews is to optimize the (IT) process and to possibly introduce new software functionalities. In addition to that, all parties sitting together to define feature requests, to develop even more requirement specific software solutions for the construction workflow.

In the following, a few outlooks for possible optimizations are given. Some are already currently being integrated and some other will be studied in the next future.

Automated Delivery of Frequent Reports via Email

Daily routine requires distribution of different reports. Currently, the TPC System performs the distribution every morning at 7 am via Email with a PDF attachment. In an attempt to reduce email traffic and centralizing the reporting effort, daily PDF reports will be stored in Dropbox automatically via an API integration between TPC and Dropbox. Users only need to check their Dropbox folders if they would like to review any report. Reminders, if a report hasn't been reviewed could be triggered via automated Dropbox notifications.

Safety Related Reports (Less Frequent)

Safety related reports, that need to be available and reviewed when an incident occurs require a different procedure. In the event of a safety incident Tunnelsoft triggers a notification via Dropbox or email with THE LINK to the appropriate document, instead of an attachment to an email. No attachment operation entails multiple benefits such as

- Most importantly: No malware exchange over email
- One central file—diminish number of versions

- Revoke access to a file immediately any time
- Fully mobile enabled
- Post Data Operation

Data Archiving

After project completion, the TPC data will need to be archived for multiple years based on current regulatory requirements. Over time the TPC software and its database are stored on a USB hard drive. Some risks and downside of the USB archive is that it is not available for users that required instant access. Also the USB drive can get lost or damaged. Additionally, the USB standard has further developed with new computers only allowing USB 3 type C plugs creating incompatibility with earlier versions and possible access limitations.

Archiving TPC and its data in Dropbox could be an alternative or additional backup solution. Dropbox core business expertise, providing scalable, highly secure and accessible storage solution built for users can circumvent any issues arising with USB hard drive archiving and offer an option for evolving the TPC solution.

Copenhagen Cityringen Project: Complex Passage and Obstacle Removal Under Existing Metro Station

Valerio Violo ▪ Seli Tunneling Denmark ApS (Salini Impregilo Group)
Antonio Raschillà ▪ Seli Tunneling Denmark ApS (Salini Impregilo Group)

ABSTRACT

The Cityringen project foresees the construction of a new circular metro line in and around the city center of Copenhagen. In his 31 km drive the TBMs had to pass near or under existing infrastructure that, in some cases, were extremely sensitive in terms of allowable settlements and risk of disruption.

This article describes the passage under the existing and operating Frederiksberg metro station and Frederiksberg mall.

INTRODUCTION

The Cityringen is an extension of the existing metro line. The 15 km long twin tunnels will form a new circular line with 21 new stations (17 + 4 crossovers). When finished it will interchange with the existing metro, rail and bus infrastructure at key locations around town, taking 90% of Copenhagen citizen at less the 600m from a metro station (Figure 1).

The Contract has been awarded in April 2011 to the Copenhagen Metro Team (CMT), a joint venture between the Italian companies Salini Impregilo, Tecnimont and Seli SpA.

Tunnels are being excavated with 4 Seli-Kawasaki EPB TBMs, two of which manufactured in Denmark by Seli Tunneling Denmark ApS, which is also operating them. At the moment this article is written nearly 90% of the tunnels have been completed and over 50% handed over for the electromechanical works.

Seli Tunnelling Denmark is a Danish company constituted in 2012 as a subsidiary of Seli SpA for this project and subsequently acquired by the Salini Impregilo Group in 2015.

The first two TBMs have been launched in Nørrebroparken shaft (NØP) towards Sønder Boulevard (SBV) respectively in July and December 2013. TBMs 3 and 4 have been manufactured and tested in a Copenhagen facility by Seli Tunneling Denmark, before being transported to the Tømmergraven site (south side of the city), where they were re-assembled and launched in June and August 2014.

Tunnel excavation will be completed between December 2016 and February 2017.

The four TBMs are EPBs, identical and built to the following specification:

- Cutter head Diameter: 5.84 m.
- Shield Diameter: 5.74 m un-tapered.
- Shield length: 10.5 m-double articulation.
- Lining Diameter: External 5.5 m/ Internal 4.9 m.
- Minimum Radius: Horizontal 180 m /vertical 625 m.

- Maximum gradient: 6%.
- Maximum operative pressure: 5bars.

FREDERIKSBERG STATION AND MALL

Description of the Area

Approaching the station of Frederiksberg the two TBM drives pass underneath a Mall and the existing operating metro station.

Figure 2 shows an aerial view of the area. The spaces are extremely limited: the rectangular with blue dashed line is the new Frederiksberg Station, while the one with orange dash line is the existing Frederiksberg station, which is adjacent to the *Frederiksberg Centret*, shopping mall. The two tunnel drives, white lines in the picture, pass underneath the shopping mall and the existing station that has been in operation during the all time, to then enter into the new station under construction.

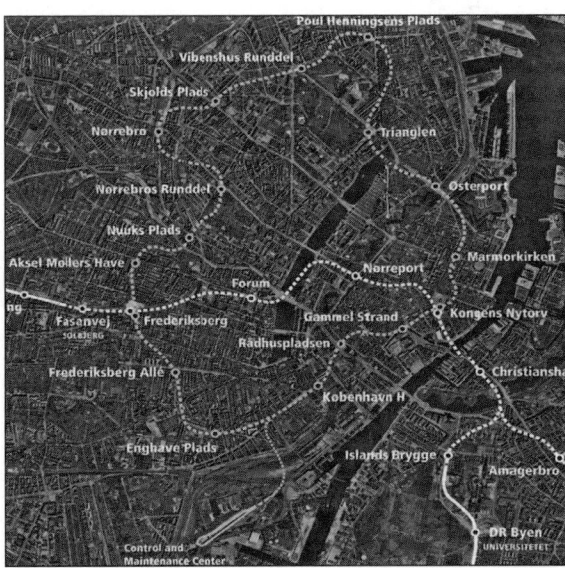

Figure 1. Cityringen general layout

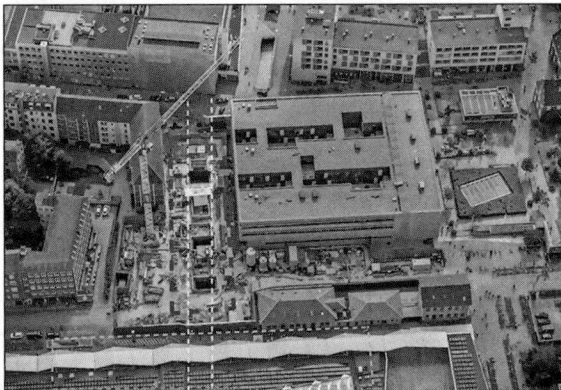

Figure 2. Aerial view of the area

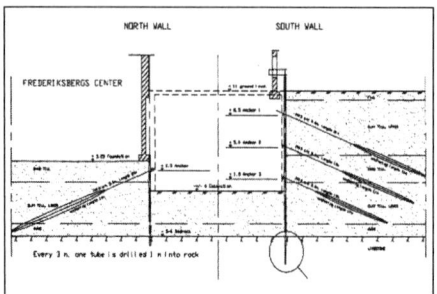 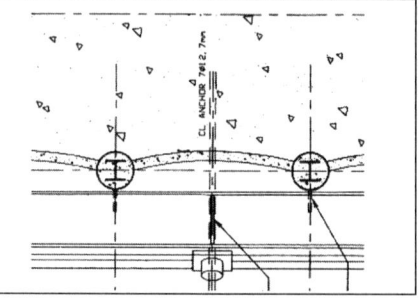

Figure 3. Available as build drawing shows the king posts (right) and the anchors (left)

During the construction of the first metro station, different temporary retaining method had been applied on the station walls, namely anchors and king posts. The TBMs drives interfere with both structures, on one side and the other of the existing station. While anchors were expected to be corroded and not to pose a threat to the TBM advance, king posts represented a serious problem: the king posts are a commonly adopted retaining structure in Scandinavia. In our case it consisted of steel permanent casings, approx. 200mm in diameter, reinforced with H100mm profiles, interconnected to form a wall-like structure. They are installed with a spacing of 1000mm. According to the available as-built drawings (Figure 3) the elements have a length of approx. 20m. It was clear that, to allow TBM passage, king posts had to be removed.

As far as the anchors were concerned, their poor conditions have also been confirmed by a partial pre-excavation that exposed the top part. However, even though the anchors were expected to be a minor risk for TBM advance, this risk had also to be mitigated.

The Risk Assessment

Approaching the station the TBMs encounter first the anchor zone, then a first line of king posts and finally, while passing under the station in operation and immediately prior to break thru, the second line of king posts.

The TBM drive is located in mixed ground, with the lower part of the tunnel face in UCL (Upper Copenhagen Limestone) and gravel/silty till (quaternary deposits) above spring line.

All these structures were temporary retention measures for the station construction, and have been since decommissioned.

It has been assessed that the anchors were going to be, probably, corroded enough not to stand a threat for the TBM. However the risk of blocking the cutter head and/or the screw conveyor and having to enter the chamber to manually cut them has been considered not negligible. While the cutting operation could have been carried out under hyperbaric conditions, we decided that it was safer to proceed with the empty chamber: in this way in case of a screw conveyor blockage it would have been possible to empty the chamber to remove the anchors.

As far as the king posts are concerned the second line, closer to the break thru, has been removed from the surface since the area was relatively easily accessible from the construction site. The first line was instead inaccessible and had to be removed

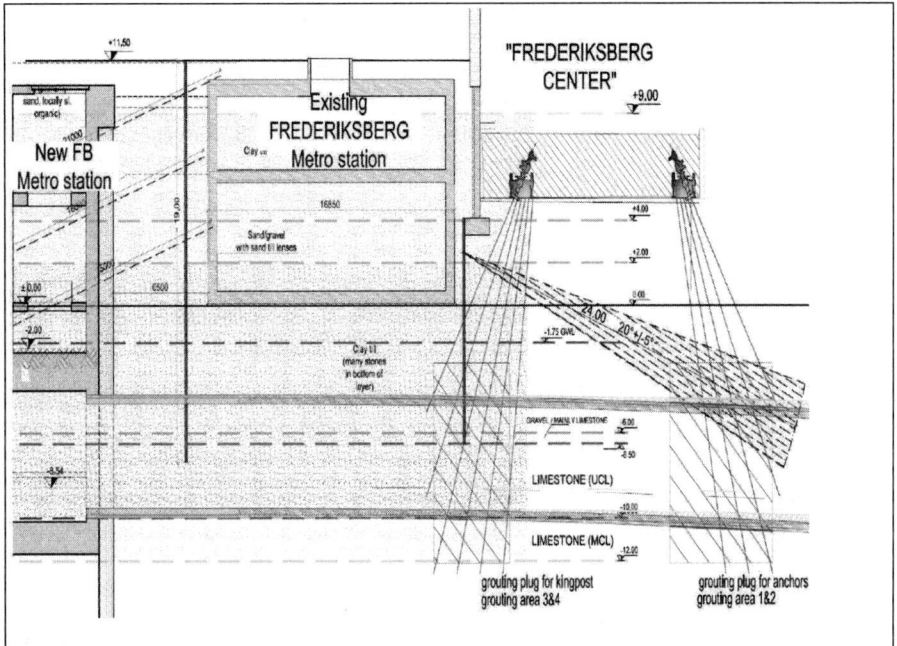

Figure 4. TBM drives, from right to left, and the consolidated plugs 1–2 and 3–4

from inside the tunnel. To this extent, a special cutting procedure from inside the excavation chamber, had to be developed.

The Designed Mitigation Measures

In order to mitigate the risk connected with the TBM passage and the removal of the obstructions, it has been decided to proceed as follows:

- The accessible king posts have been removed from surface with a dedicated drilling rig and a 300mm casing coring;
- To mitigate the risk involved with an obstruction removal while mining thru the area of the anchors, it has been opted to create a consolidated plug. The TBM would approach the area with the chamber half empty and, in case of blockage due to anchors, it would have been possible to enter in the excavation chamber, in atmospheric conditions, to remove it;
- The same consolidated plug has been used for the removal of the king posts from the TBM, for which a special procedure has been studied and applied;
- Eventually a compensation grouting has been put in place to accompany the TBM mining under the exiting station, in consideration of the soft ground above spring line and the cover limited to 3m (Figure 5). The compensation grouting has been carried out from the station, during the construction phases, due to unavailability of space on surface;
- Prior to TBM break thru into the station and in the area where the second line of king posts has been removed, a ground improvement has been carried out, approximately 4m long and 6m wide, in order to mitigate the risk of water ingression during break thru operation.

THE CONSTRUCTION PHASE

Removal of King Posts

The accessible king posts, closer to the break thru, have been removed with a Casagrande drilling machine that carried out a coring, with a 300mm casing, on top of the king posts. The head of the king posts have been exposed with a pre-excavation, then the coring extracted them from the ground.

Grout Plug for King Posts and Anchors

The grouting operations on the mall side of the station, before passing underneath the existing station, have been complex and required the execution of drilling and grouting from inside the basement of the shopping mall (Frederiksberg Centret).

In order to carry out the drilling work, the machine had to be dismantled and transported, in parts of manageable dimensions, thru a service elevator in the basement of the shopping mall. Cooperation with the shopping mall has been paramount for the success of these operations that have been entirely carried out without any disruption to the businesses.

The machine chosen for the task was a Comacchio MC 400 that is, without the boom, only 3,1m long. The grouting technique foresaw the use of TAM (Tube á Manchettes) with valve every two meters. The intervention consisted in four grouted plugs: 2 plugs per TBM, one in the anchors area and one for the intervention on the king posts. The TBM would have:

- to enter the first block mining with lower pressure and partially empty chamber;
- after exiting and re-establishing face pressure, bore a non-consolidatod mixed ground area;
- reach the second plug where once more the pressure can be decreased and the chamber partially emptied;
- advance until the king posts where it was foreseen to work under atmospheric pressure.
- once removed the king posts, the TBM would resume mining in mixed ground, under Frederiksberg station and up to the break thru point approximately 50m ahead

Compensation Grouting from the Station

The final 30m from the king post removal area, consolidated, to the station had to be bored in mixed ground and with a low cover under the existing station. The combination of quaternary deposit at the top and the low cover, that required a limited face pressure, required an additional mitigation measure. This consisted compensation grouting carried out with sub-horizontal drilling from the station. Figure 5 and Figure 6 show the compensation grouting lay out. The drilling from the station took place from the top of the false tunnel for the TBM entrance, during construction phase, with a compact machine similar to the one used to carry out the ground improvement from inside the mall basement.

Due to proximity with the existing Metro Station, the working procedures feature an intensified monitoring interface. All the phases of ground improvement have been designed to mitigate the risks of disruption to the operating metro station, from the

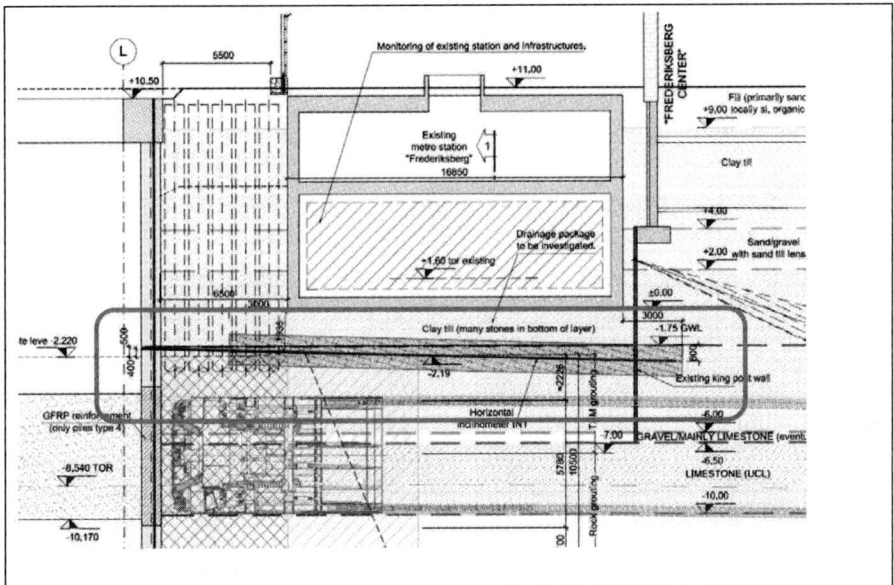

Figure 5. Compensation grouting from the station in construction

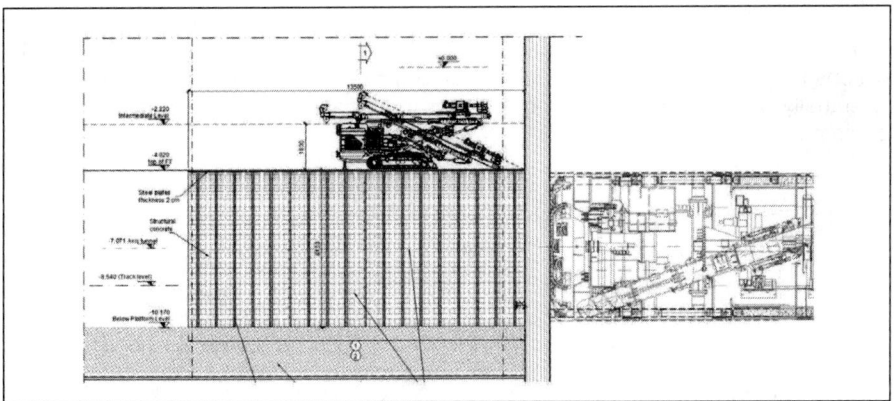

Figure 6. Drilling position on top of break through false tunnel

grouting to pre-condition the soil beneath the existing station (contact grouting, first phase), to the post-grouting (second phase) for compensating the possible volume losses induced by the TBM passage while breaking-thru the new station. Additional monitoring stations had been added to the metro station technical rooms.

The Procedure for Removal of King Posts from the Tunnel: Torch Cutting

Once the TBM reached the king posts, it was necessary to enter the excavation chamber, expose the steel structure and cut it with a safe technique.

It was assessed that the fastest way was to cut the king posts with torch and to remove them in smaller parts. This procedure required different OHS considerations that, already at the design phase, have been addressed:

- We installed a ventilator with an exhaust line of 300mm. The fan was powerful enough to extract the fumes and to push them, thru the small section, to the central part of the back up where a second exhaust line took the fumes in the tunnel up to the portal;

- An electric winch system, plus a monorail thru the airlock has been installed to evacuate the steel parts;

- A pump has been placed in the excavation chamber with a dedicated line thru the bulkhead, in order to manage possible water inflows.

All the above applied if working in atmospheric conditions and with the crown relatively stable, relying in the consolidated plug. However to mitigate the risk of non efficient grout plug and to possibly allow cutting operation in hyperbaric condition, a plan B has been studied and put in place.

The Procedure for Removal of King Posts from the Tunnel: Hydro-Blast

To be able to operate in hyperbaric condition, in case the working environment conditions did not allow an extensive enough campaign of torch cutting, we were ready to proceed with hydro-blasting cutting.

This procedure involved bringing into the back up the hydro-blast equipment, fixing the cutting nozzle on the cutter head and slowly rotating the cutter head clockwise and anti-clockwise to perform the cutting. The operation could be carried out without direct supervision: the crew could wait, under pressure, in the safe haven of the airlock and access the excavation chamber during pauses of the cut, to assess the process. To fit the diesel equipment and the sand tank on the tunnel service train, several modifications had to be put in place, such as removing wheels and modifying the chassis. Eventually a special frame that could fit the train flat car has been installed on the equipment.

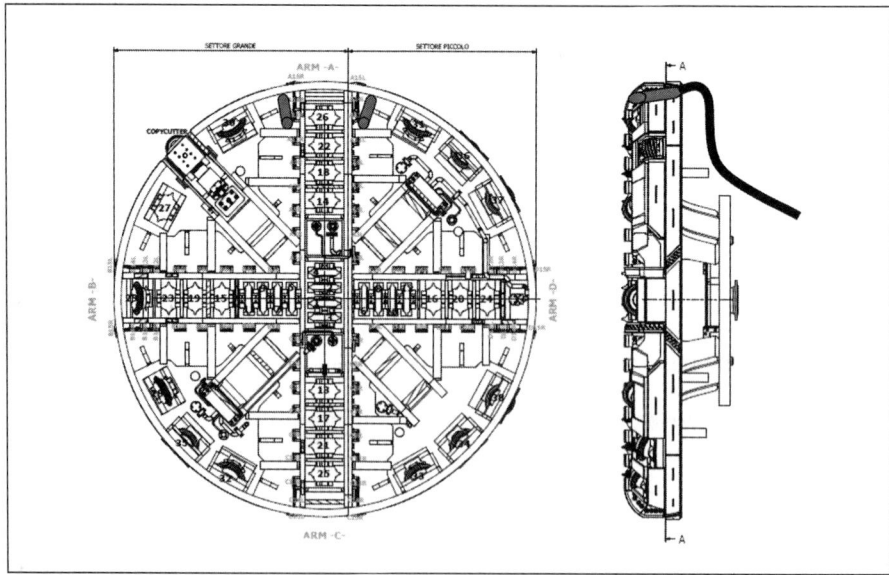

Figure 7. Nozzle installation on the cutter head

Figure 7 shows the position of the nozzle on the cutter head.

To assess the efficiency of the methodology, a full-scale test had been carried out with a mock up of the king post embedded in concrete that gave a very satisfactory result.

Safety Considerations

Figure 8 shows the sight of the king post pile as it appeared from inside the excavation chamber. As it can be seen, the space is extremely confined and accessing the king post to expose it and prepare it for the cutting is not an easy task.

In consideration of the particular working condition during the cutting operation, we tailored a dedicated training and safety indoctrination program. In particular we focused on the confined space (which is the same as per hyperbaric or cutter head maintenance) adding a special focus to the air quality monitoring:

- General information about the possible contaminants during torch cutting and/or sand blasting;
- Possible health effects and short-term health effects;
- Monitoring plan and action plan in case of air-borne contaminants detection;
- Mitigation measures on TBM;
- First aid;
- Dedicated PPE and manual handling of steel parts in the excavation chamber.

As an additional precaution we also monitored contaminants such as benzene and chlorinated compounds, which are gases known to be often found in Copenhagen ground water.

An engineer from the technical office or OHS department has supervised all works.

Since the rear part of the TBM was not under a consolidated ground there was the risk of water coming from the rear and running above the shield, to the cutter head. In order to mitigate this further risk we designed and installed nozzles to inject waterproofing foam. These nozzles were kept at the head, ready to be used in case of non-negligible water ingression from above the shield.

THE TBM PASSAGE

Cutting Operation

In August 2014 the first TBM approached the area of interest.

As defined by the procedure, the TBM advance rate has been decreased at the entrance of the consolidated plug. At the chainage where the anchors might have be found the excavation chamber has been partially emptied in order to allow opening the bulkhead door in case of cutter head blockage.

As foreseen, the anchors did not represent a problem and the TBM passed thru the area without any stoppages or problem.

Once passed the anchors area, the excavation chamber has been filled again to bore between the two consolidated blocks. When inside the second plug, the excavation chamber has been emptied again. Approaching the point where the as built was indicating the presence of the king posts, the advance speed has been reduced and, at the king posts known chainage, the TBM stopped.

On August 16th the first TBM reached the king posts. Cutting operation started early in the morning, immediately after having exposed the king post by washing away the concrete and the soil they were embedded in. For the first two king posts, cutting lasted only 6 hours; after the cutting was completed the first TBM had been able to resume mining. It is worth mentioning that, during the cutting operation, water started seeping from over the shield and into the excavation chamber. The quantity was easily controlled by mean of the installed pumps, but an increase in the flow required the injection of sealing foams thru the ports in the shield which helped reducing the flow.

Figure 8. King post from inside the excavation chamber

According to the design the passage underneath the station was very delicate: the narrow spaces did not allow an extensive ground improvement campaign and the mixed ground at the cutter head lead the designer to envisage a volume loss of 0.5%, which was also above the limit that could be handled by the compensation grouting. It was therefore extremely important to limit the volume loss and to operate the TBM in the best possible way.

The TBM resumed mining on August 17th and it only had 32 meter to go to get to the station. Particular care had been placed in alerting the crews and making them aware of the conditions, during "awareness" meeting carried out with the crews where the operatives had been informed of the conditions/risks and required actions weeks in advance of the operation. In fact the control of the mining parameters, in particular of the face pressure, has been extremely good during the entire delicate passage: during this phase it hadn't been necessary to inject to compensate settlement, in fact only a minimum of post grouting had been carried out to make sure that the soil wasn't de-tensioned.

Monitoring During TBM Passage

During TBM advance from the consolidated plug, underneath Frederiksberg station and toward the break thru in the new Frederiksberg station, an extensive campaign of monitoring has been carried out. The campaign consisted of:

- Real time TBM parameters monitoring, with special focus on pressure, torque, apparent density in the chamber and backfilling parameters;
- Building monitoring, from inside the Metro Station;
- Deep instrument monitoring, extensometers and inclinometers including sub-horizontal inclinometer from the new station;
- Ground water table monitoring.

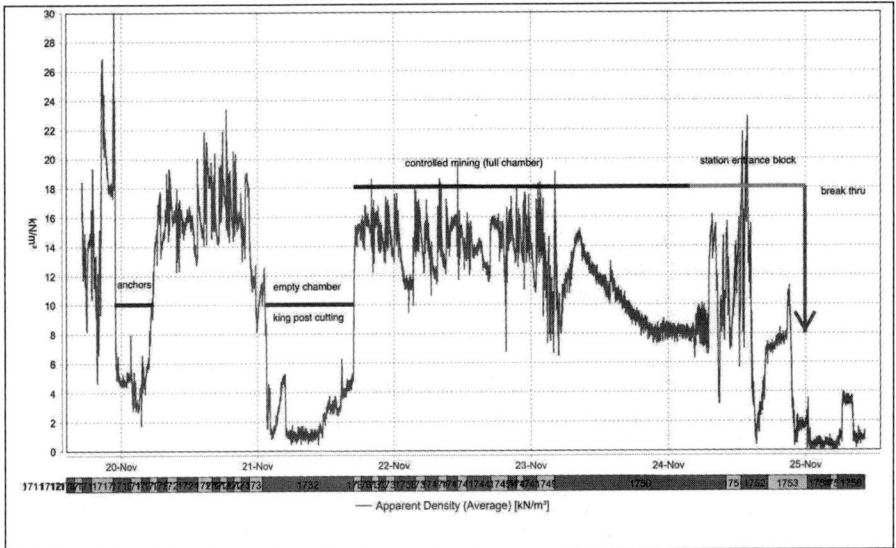

Figure 9. Apparent density during the passage

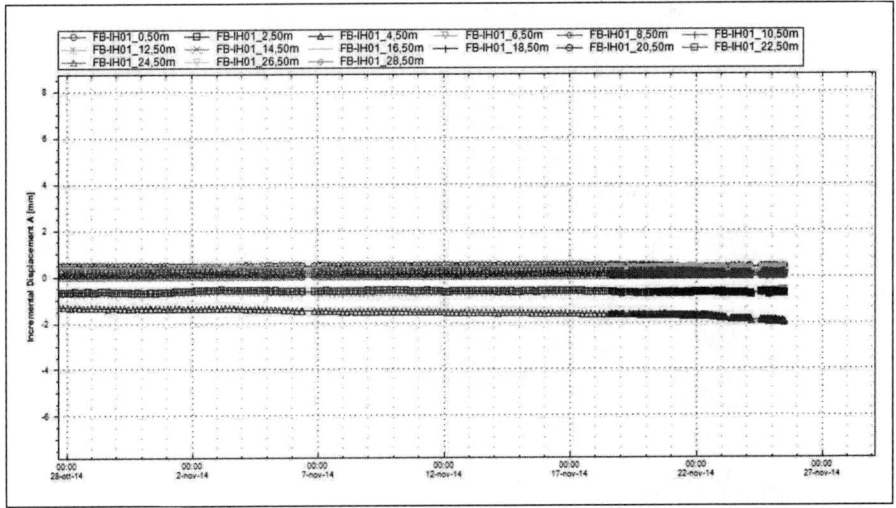

Figure 10. Horizontal inclinometers show limited settlements

The crews managed the mining parameters very well; the pressure management of TBM #1 from the king post area to the break thru has been within the maximum allowed oscillation of 0.3bar. Pressure has been always kept in the required ranges and the same can be said for backfilling (pressure and volumes).

The graph in Figure 9 is also very interesting. It shows the apparent density that, in EPB mining, is an indication of the degree of filling of the excavation chamber: apparent density is often used to distinguish between true EPB mining, where the excavation chamber is filled with muck over a certain minimum density, from mining in unsafe conditions, with the chamber partially filled with the compressed air released from the soil conditioning foam. In the graph it is possible to follow the TBM advancement, noting the passage thru the first consolidated plug, when the chamber is partially emptied; between the two plugs, chamber full, and into the second plug where the king post have been cut. Finally from the second plug to the break thru, while passing under the existing metro station, the TBM mined in controlled and safe conditions with full chamber.

The control of the mining parameters is well reflected from the monitoring results: the sub-horizontal inclinometers from the station, above the TBM passage, did show only negligible settlements (Figure 10). Volume loss has been minimal also confirmed by the limited post-grouting.

CONCLUSIONS

Overall it is unavoidable that during such challenging project constructions, some passages are particularly difficult in term of interference with existing structures. However these difficulties can and have to be overcome with good planning and design. It is of the outmost importance the collaboration of all the parties involved: from the project owner, to the designer, from the TBM tunnelling management to the operative crew (TBM foremen and TBM pilots) that is actually building the tunnel.

Both TBMs completed the difficult passage in less then a week: the first in August 2014 and the second in November of the same year; the quick and successful passages are the result of carefully designed and programmed operations that involved different disciplines which culminated in the successful tunnel construction.

Grouting and Ground Modification—Copenhagen Cityringen Project: Compensation and Jet Grouting as Mitigation Measures for TBM Operation Under Historical Building

Antonio Raschillà ▪ Seli Tunneling Denmark (Salini Impregilo Group)
Valerio Violo ▪ Seli Tunneling Denmark (Salini Impregilo Group)
G. Kafantaris ▪ Copenhagen Metro Team JV (Salini Impregilo Group)

ABSTRACT

The Cityringen project will form a new circular line in the city centre and will consist of 21 new stations (17 + 4 transfer station) and approx. 31 Km of EPB twin-bored tunnels.

The new tunnels passed under the historical building of Magasin du Nord, McDonald's and two pubs in the heart of the city with a minimum cover of about 4 m.

To minimize the movements, a complex system of compensation grouting and jet grouting soil treatment was performed. A detailed action plan between the parties involved allowed coping with the critical TBMs passage successfully.

INTRODUCTION

The Cityringen is an extension of the existing metro line underground Copenhagen and will consist of 21 new structures (17 + 4 transfer station) and 15.5 kilometers twin tunnels loop around the city. The 15.5 km City Circle Line will form a new circular track inside the city centre and it will intersect the existing M1 and M2 lines at Kongens Nytorv and Frederiksberg stations, and suburban train services at København H, Østerport and Nørrebro. It will extend the Metro network to the Nørrebro, Østerbro areas and the central station at København H (Figure 1). The two new lines M3 and M4 will run on the same tracks, but the M3 will be a circle route running around the entire line, whereas the M4 will drive from the station to the north side of the city.

The €1.5bn design build contract has been awarded by client Metroselskabet in January 2011 to the Copenhagen Metro Team (CMT), a joint venture between the Italian companies Salini, Tecnimont Civil construction and Seli SpA. Since bankruptcy of Seli SpA and Tecnimont, Salini Impregilo owns now 100% of the JV shares. Seli Tunneling Denmark ApS, a subcontractor to CMT, was initially constituted as a Seli SpA subsidiary and, following the economic difficulties of Seli SpA, has been acquired in 2015 by Salini Impregilo. Seli Denmark is in charge of the tunnels excavation and lining and owns the four 5.84m diameter Seli/Kawasaki EPBMs.

Ground conditions are at times challenging, most of the alignment runs through Limestone with occasional inclusions of flint bands, but it occasionally dips in and out of a mixed face to include sand, clays and tills. Invert depth is up to 45m under a high water table and requires machines capable of withstanding the resulting 4.5bar operational pressures. The north drives, from the station of Nørrebro to Vibenhus Rundell, are entirely in soft ground, sand and glacial deposits.

TRACK AND TBMs MAIN FEATURES

The tunnels are being excavated with 4 Seli-Kawasaki EPB TBMs (Figure 2), two of which manufactured in Denmark by Seli Tunneling Denmark ApS, which is also operating them. TBMs 1 and 2, were launched in July and December 2013 from Nørrebroparken shaft towards Sounder Boulevard. Here the machines were disassembled and moved to Øster Søgade shaft to mine towards København Central Station (KH).

TBM 3 and 4 have been built and tested in Denmark in 2013 and launched from Tommergraven shaft to København Central Station in June and August 2014. Once in KH machines were disassembled and moved back in Nørrebroparken shaft to complete the last stretch to ØSØ.

The four TBMs are identical. Table 1 summarizes the main technical features.

PASSAGE UNDER HISTORICAL BUILDING

One of the main challenges of the Cityringen is the passage under the historical building of Magasin du Nord, Mc Donald's and two old pubs in the heart of the city, with a minimum cover of about 3.7 m.

TBMs 1 and 2, were reassembled and launched from Øster Søgade shaft respectively in October 2015 and January 2016 to complete the stretch Øster Søgade – KH by the beginning of 2017 (Figure 3). At the moment this paper is written, 90% of the tunnels have been completed.
In July 2016 the first TBM arrived at Kongens Nytorv station ready to cope with the most critical passage of the whole project (Figure 4).

The drive of the machines from the station of Kongens Nytorv is complex since launching. In the first 40m the TBMs have to pass under the so-called transfer tunnel: this

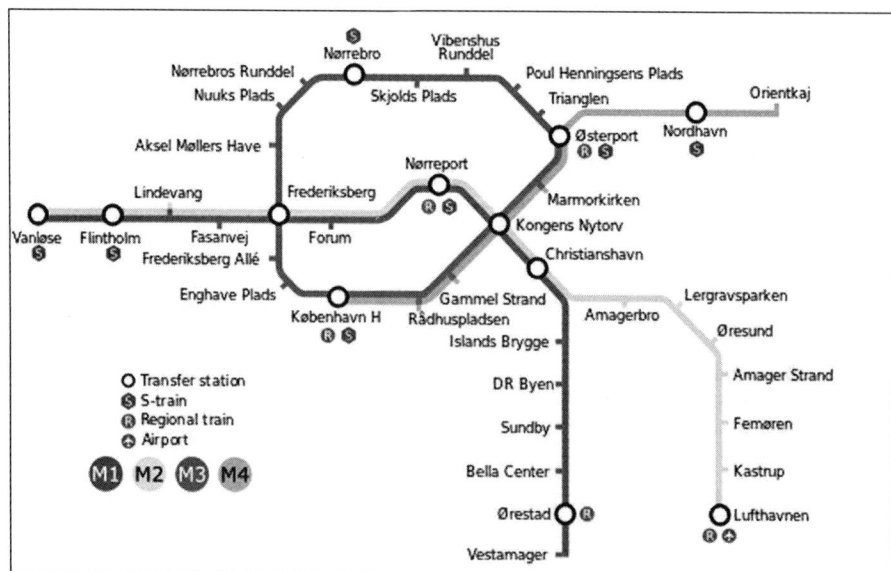

Figure 1. Copenhagen metro network

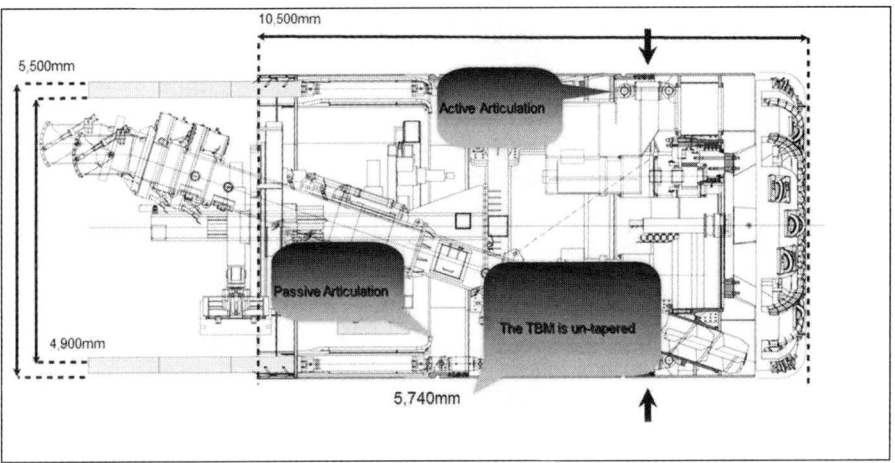

Figure 2. Seli-Kawasaki TBMs shields configuration

Table 1. TBMs main features

Cutterhead diameter	5,84 mt
Shield diameter	5,74 mt untapered
Shield length	10,50 mt
Ring configuration	Universal 5 + key
Lining diameter	Outer 5,50 mt / Inner 4,90 mt
Minimum radius	Horizontal 180 mt / Vertical 625 mt
Maximum gradient	6%
Maximum operative pressure	5 bars
Number of cutters	38 (17–18" peripheral)
Maximum load per cutter	267 kN

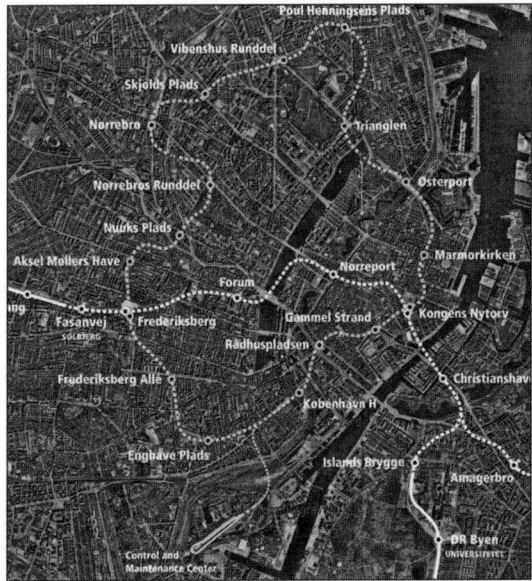

Figure 3. Cityringen layout

Figure 4. Kongens Nytorv station and compensation grouting shaft

is the connection between the existing metro and the new one and, at the time of the passage, its roof slab and retaining walls were in place.

The tunnels of the existing, and operating, metro line are right below, being the clearance only 150 cm. The existing tunnels, differently from the one we just completed, have been built in limestone.

Next the TBM drives pass underneath the two old pubs to then dive under the basements of Magasin du Nord (MdN) and Mc Donalds (McD) with a minimum clearance of 3.7m from the deepest foundation.

The first 70m after the exit from Kongens Nytorv station, before reaching the competent limestone, the tunnel alignment is in soft ground, a mix of sand, sand-till and clay.

A complex system of compensation grouting and jet grouting soil treatment was put in place, in order to mitigate the risks for the buildings and to minimize the settlements.

This paper will describe the design and construction phases of the soil improvements execution, the passage of the two TBMs and the detailed action plan between the parties involved which allowed facing the critical TBMs passage successfully.

DESIGN AND MONITORING

Pre-Design Phase and Key Risks

During the early days of the project, in 2011/2012, the review of the available information (Buildings Database) for this critical passage brought to light the first important risk to be assessed and addressed: the indication of presence of steel micro-piles hammered to unregistered depths below the foundation pillars of Magasin du Nord as temporary underpinning for the lowering of its basement (Figure 5). The potential consequences of clashing with a micro-pile or more, in soft ground and with such low overburden, with the associated risks of a hyperbaric intervention and/or a long stoppage under the building for their removal, made the pursuit of a definitive solution to the problem a must. Additional investigations failed to provide further input; in fact even more uncertainty was added by unsuccessful attempts to simply locate micropiles indicated by the available design drawings. In agreement with the Client, Metroselskabet,

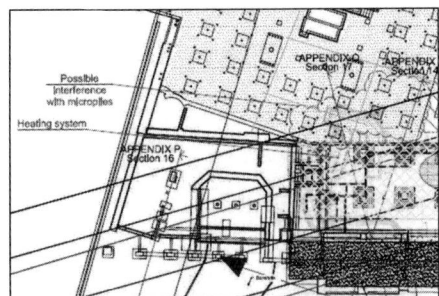

Figure 5. Magasin du Nord foundation interference

the risk was addressed by a modification (lowering) of the track alignment, introducing a very aggressive slope of 6%, immediately after the exit of the station. The probability to encounter micro-piles was in this way mitigated and made negligible. As a second result, lowering the TBM alignment shortened the distance to be bored in soft ground, prior to reaching the Limestone bedrock.

Other important risks identified during the pre-design phase:

- The presence of large granite boulders in the quaternary deposits, which presence has been recognized during the excavation of the existing metro station;
- Unregistered and/or potentially not decommissioned structures and services under the buildings' basements;
- The complex structural system of Magasin du Nord which is made of different blocks of different ages (from 1857 to early 20th C.) and complex typologies inter-connected one with each other. Moreover the available information about beams, columns and/or connections of the different elements was extremely sparse;
- The poor condition of the old pubs that required, already during the construction of the first metro, the underpinning of one of the corner, to cope with the important displacement.

Preliminary Design Phase

The effects of the tunneling activities on the overlaying structures (Magasin du Nord and McDonald's) were studied by Lombardi Ingegneria S.r.l through an integrated 3-D geotechnical and structural FEM analysis (Figure 6). A fully coupled soil-structure interaction was considered and the entire excavation process was reproduced in a 3D model where the buildings were adequately modelled. In this way a risk assessment analysis was performed, evaluating directly the effects of the excavation in terms of settlements and variations of the internal actions as far as the excavation process proceeds.

The input necessary to feed into the FE model with regards to the structures affected by the TBM, was provided by a detailed buildings' condition review, performed by a specialized subcontractor and supervised by CMT and their consultants. Despite a reasonably good knowledge of the structure acquired in terms of geometric properties (position and dimensions of structural members), due to unavoidable limitations during the execution of the survey in the busiest mall of Copenhagen, some uncertainties remained regarding:

- Material properties
- Properties of connections and structural joints
- Geometry of many structural elements

The evaluation of different scenarios within the FE model, in terms of volume loss (0.5% and 1.0%) and elasticity modulus for the sand resulted in respective profiles of displacements, angular distortions and actions for the buildings overlaying the tunnels.

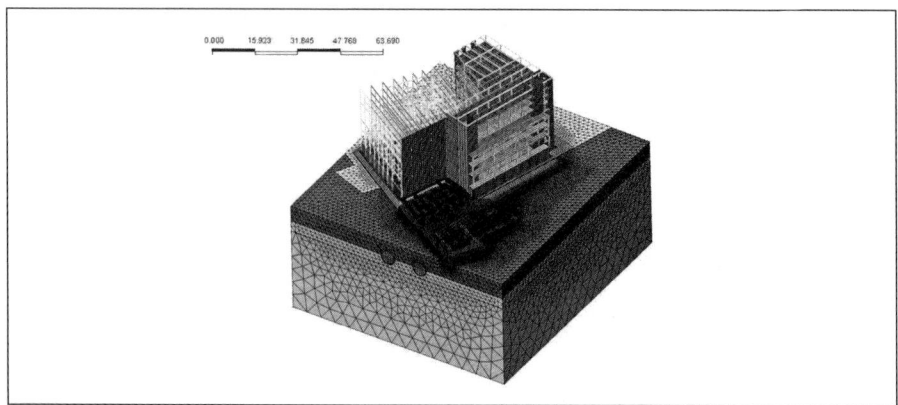

Figure 6. Magasin du Nord 3-D FEM analysis

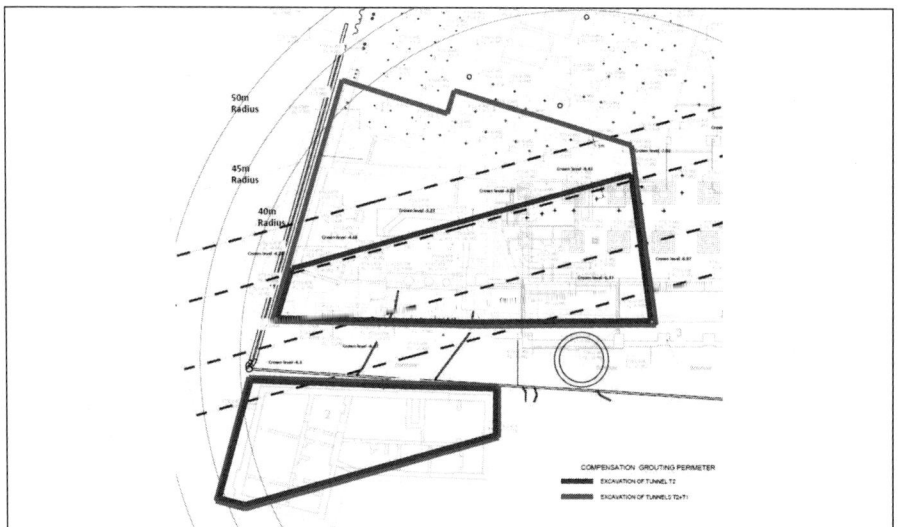

Figure 7. Compensation grouting area of influence

On the basis of the results of the analyses for the different scenarios, the parts of the buildings where mitigation measures were required were thus defined (Figure 7).

Two sets of threshold values (alert/alarm) were defined with regards to the initiation of mitigation activities/measures:

- Type A—For the cast-iron pillars of MdN and the foundations for the McDonalds and the pubs: An alert distortion limit of 1/2000 and an alert displacement of 2 mm and alarm distortion limit of 1/1000 and an alarm displacement of 3.5mm;

- Type B—For all other foundations: An alert distortion limit of 1/2000 and an alert displacement of 2.5mm and alarm distortion limit of 1/1000 and an alarm displacement of 5mm

Considering that more than 80% of the buildings' foundations fall under Type A, the respective threshold values for mitigation measures application were applied to all foundations and Type B limits were eventually discarded. The previous values refer to the activation of mitigation measures and not the start of functional damage, which of course corresponded to a much higher displacements.

The completion of the FEM analysis and definition of the predicted effects of the tunneling works was followed by a feasibility study of different mitigation measures, for the limitation of the buildings' displacements:

- Permeation grouting
- Compensation grouting
- Jet grouting
- Ground freezing
- Underpinning/structural interventions

The following evaluation criteria were applied and an overview of pros and cons was produced for each technique to be finally assessed:

- Compatibility with the site geological and hydrogeological conditions
- Effectiveness to counteract the impact of the tunneling activities on the buildings
- Suitability to tackle pre-identified risks (micropiles, boulders, unknown underground structures, buildings' heterogeneity)
- Compatibility with risks associated with the TBM operation (face pressure loss/blow out, breakdown/stoppage, etc)
- Compliance with the time schedules of the station and the TBM works
- Eventual expropriation requirements (Magasin and McDonalds basements, realization of additional shafts, etc)
- Implementation cost

The feasibility study was concluded with the adoption of compensation grouting, executed from a shaft installed between MdN and the pubs, addressing potential movements of all four buildings, in combination with jet grouting underpinning of the foundation pillars of McDonald's.

Detailed Design Phase

The development of the detailed design, for both jet grouting and compensation grouting, was undertaken with the full involvement of CMT's selected subcontractor for the ground improvementworks, Keller Grundbau GmbH, and the contribution and supervision by the experienced geotechnical consultant RD Geotech.

With regards to jet grouting under McDonald's basement, the minimum performance requirements for the columns were defined by Lombardi Ingegneria S.r.l and confirmed by a site trial executed by Keller:

- E=800 MPa
- UCS=2MPa
- Diameter=1.50m

The detailed design comprised 86 columns, 6.5m to 8.5m long, reaching the limestone level (Figure 8). The execution took place between September 2015 and January 2016 by Keller and the basement was refurbished and delivered back to its Owner in accordance with the agreed expropriation window.

The detailed design of compensation grouting, on the basis of locating a 6.5m internal diameter shaft between MdN and the pubs, foresaw the installation of 43 steel TAMs (max length 42m) for grouting during the TBM passage and 12 additional TAMs for ground improvement during the preconditioning phase (Figure 9). A maximum horizontality error of 1% was defined for the drillings and a target heave of 1mm–2mm was specified with regards to the preconditioning phase injections.

Keller started with the drilling and TAM installation works in August 2015 and completed the preconditioning injections in April 2016, well before the first TBM transit. The high presence of granite boulders in the quaternary soil, as highlighted during the pre-design phase, resulted in the need to re-drill or even abandon the original location for an important number of TAMs. Staged checks of the drillings' horizontality led to backfilling and re-drilling up to 7 times for some holes. In other cases, provided that the TAM could be installed and be operational and the geometrical impact (proximity to foundations or the tunnel) was tolerable, the 1% horizontality error was slightly compromised.

As a result of the foregoing, the pre-conditioning execution did not follow the completion of installation of the TAMs but instead, Keller were instructed to alternate between drilling and preconditioning in order to ensure compliance with the overall time schedule.

The design of the compensation grouting for the TBM transit was based on the following assumptions:

- 1% maximum volume loss
- 10m daily TBM advance rate
- 50% of TBM shield in soft ground under MdN
- 10% grouting efficiency

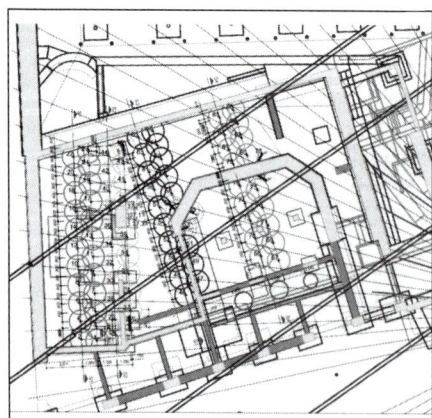

Figure 8. Jet-grouting under McDonald's basement

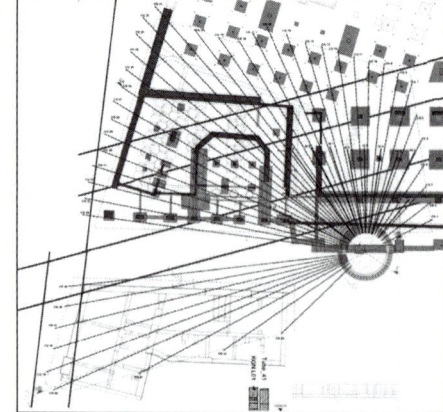

Figure 9. Compensation grouting TAM installation

For a flow rate of 10L/min, an injection module with 4 packers operational at all times would provide sufficient capacity during the TBM transit.

Design Injection Plans (D.I.P) prepared prior to the TBM transit, as sets of provisional injections targeting to maintain any settlements within the threshold values in case a movement trend would be detected, complied with the assumptions above and with the restricted zones for injection agreed with SELI Denmark (Figure 10).

Structural Monitoring

The real time control of displacements on the structural elements of the buildings affected by the TBM passage was primarily assigned to a liquid levelling (LL) sensor system provided by GeTec, acting as a subcontractor of Keller. All structural elements of the four buildings within the zone of influence were equipped with sensors and the respective measurements were transmitted to the Cityringen project database, KRONOS, visible to CMT, Seli Denmark and to the project owner Metroselskabet (Figure 11).

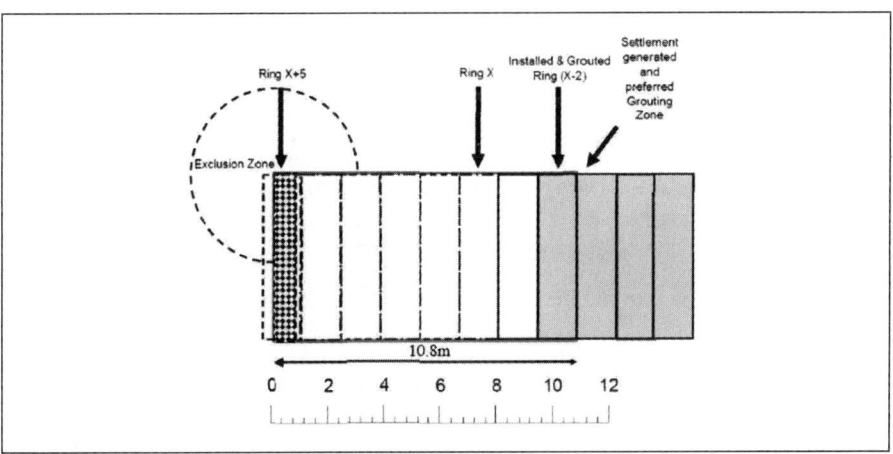

Figure 10. Grouting zone during TBM passage

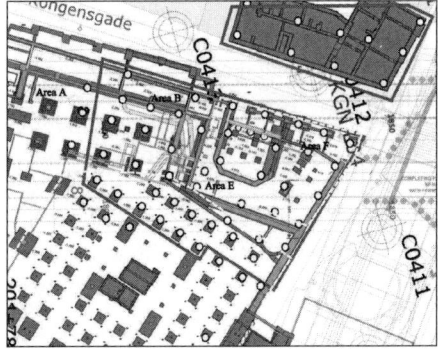

Figure 11. Sensors for the buildings displacement control

Precise levelling performed by the monitoring crews of CMT, on the same structural elements, intended to confirm the validity of the LL system, quantify the contribution of external influences (e.g., temperature) but also provide backup in case of system malfunction.

Finally, CMT measured the facades of the buildings with automated systems.

TBMs PASSAGE

Actions Taken Before the Passage

We had two important phases, before and during the tunnel excavation. And everything has been carefully planned for both.

Before the start of excavation from Kongens Nytorv station meetings with foremen and pilots were held to analyze the situation and the issues that could possibly arise from mining operations. During those meetings, that we called "awareness campaigns," we decided that it was of utter importance to share with the crews all the information about the passage that we were about to cope with, including a detailed yet easy to understand risk assessment.

We discussed the expected geology, all risks related to it and to the adjacent structures and the possible different behavior of the TBM under the buildings; possibility of the screw and/or cutter head blockages due to boulders etc.

Plan and section drawings have been posted in the TBM control cabin showing the actual position of the machine, meter by meter for the whole 70 m critical stretch from launching from Kongens Nytorv to the safer geology, in limestone.

The main excavation parameters, such as pressure, extracted weight and backfilling, were connected to the TPC (tunneling process control software) and broadcast to Seli Denmark managers and supervisors, with relevant alert for sensitive data such as face pressure and muck weight. The alerts were also sent via mobile app and smartphone.

The TPC system is also normally connected to the Kronos system, the monitoring software used by CMT: in this way Client, designer and partners were able to monitor the advance of the TBM on real time.

For the excavated weight control, besides the two usual scales, double scales have been implemented and the filling degree of the wagons was compared to the TBM speed. To do this, sensors were installed on the back up gantry, next to the discharge point, in order to measure the level of the muck in the muck car box related to the actual TBM advance and an alarm sensor has been set in the control cabin to alert the operator in case of excavation over the 10% compared to the theoretical value.

TBM Mechanical Enhancement

To mitigate some of the risk connected to the TBM operation, we decided to carry out some mechanical improvements.

While mining in soft ground and, especially, at the contact between different geotechnical formation, we experienced the tendency of the TBM to slow down and to suffer increased thrust forces and, generally, friction. To mitigate this behavior we enhanced the tail shield towing system and the bentonite injection system.

- **Tail Shield Towing System.** It was redesigned to work up to 36.000 kN, by adding a mechanical towing system and by re-designing and re-building the hydraulic system in order to be able to operate up to 400 bars. In addition to this the shield was reinforced with 50 mm circular shaped plates.

- **Bentonite System.** Bentonite has proven positive effect in reducing the friction during the advance; to maintain face pressure during short or long-term stoppages and to stabilize the steering gap around the shield, i.e the void created by the cutter head over-cut. In order to enhance the bentonite system, we added six additional injection points in the middle and tail shield to be able to inject in the over cut to achieve the best efficiency of the above-mentioned positive effect of Bentonite injection. In particular the use of bentonite as a friction reducing measure, proved to be extremely useful.

Actions Taken During the Passage

The coordination between the TBM excavation performance, the execution of the compensation grouting and a real-time monitoring system for the control of the buildings response and application of prompt and adequate mitigation measures for any abnormal displacements, was paramount during the TBM passage. The clear definition of the roles and responsibility among the involved parties and the synergy among the organization, the processes and the communication during the tunneling operation was also essential.

In order to enhance collaboration and communication we decided to adopt the most traditional way.

A site office has been set up near to the Kongens station where CMT, Seli Tunneling Denmark and Keller technicians were present during the passage of the 2 TBMs, on 24 hours 7 days per week, to follow the smooth running of the operations: nothing better than sharing the same office to collaborate and to exchange information in a quick and effective way. Also the project owner was often on site, gathering information as the TBM proceeded and coordinating with the involved third parties, in particular the pub owners and the Magasin Du Nord management.

Each involved company set up two level of support:

- On site. An on site technician/engineer in constant contact with the operative personnel. For example in the case of Seli Denmark the on site engineer granted an immediate access to the TBM status beyond what the automated system and the mobile application is able to indicate. The on site engineer was in contact with the mining crew thru radio and was always able to determine what particular action was on progress at any given moment. This proved to be particularly valuable for monitoring personnel that could know in advance how long a stoppage was going to last, whether was for normal ring build operation or for a mechanical problem, thus giving a longer time span for possible compensation injection;

- On call. For all those decisions that could not be taken on site, each organization provided an on call senior engineer or manager. Who was on call was constantly connected to the machines thanks to the mobile application and/ or remote access software and, in case of call, could quickly understand the situation and take decisions accordingly.

Monitoring During the Passage

According to the above on site/on call presence, each company provided its own level of monitoring support.

CMT was in charge of the real time review and analysis of the monitoring system recordings. It was also in charge of the compensation grouting injections review and the compliance with the pre-defined action and emergency plan.

Seli Tunneling Denmark was responsible for the TBM excavation monitoring and the coordination of the communication flow between the tunneling crews and the monitoring crews.

Keller oversaw the compensation grouting and liquid level monitoring system and the design injection plan adaptation.

The monitoring personnel communicated and cooperated with Seli Tunneling DK during TBM critical strokes advance, giving a prompt feedback of how the adjacent building were reacting to TBM behavior and to consequently coordinating with the grouting operations.

The TBM passage for both machines was highly successful in terms of recorded settlements which did not exceed 0.8mm, thus remaining within the predefined threshold values and not necessitating any compensation grouting injections. Limited post grouting, following the TBM passage, recovered the very low displacements incurred.

The following two examples show how the accuracy of the monitoring, both at TBM level and at displacement level, allowed us an extremely detailed follow up of all operation and, consequently, a very informed decision taking process

Figure 12 compares the back-filling two-component pressure with the movements of the ground, recorded under the corner of the first building encountered during the excavation.

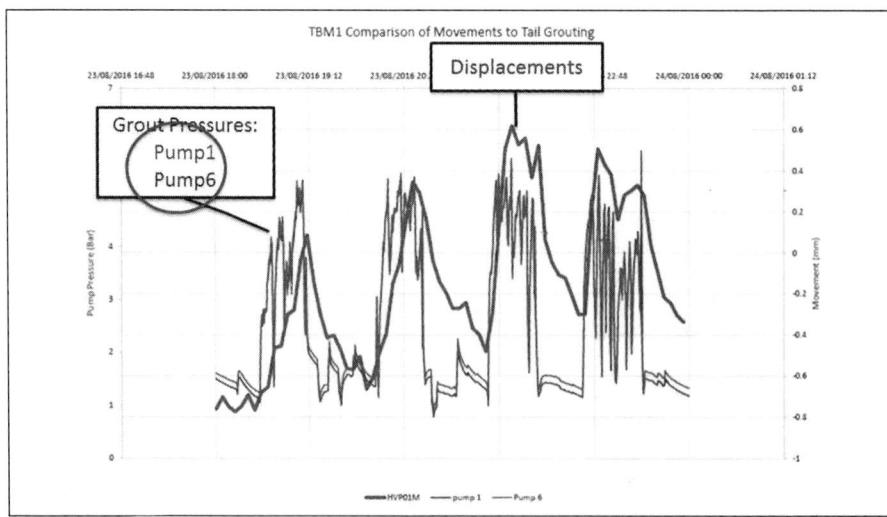

Figure 12. Grout pressure vs. ground displacements

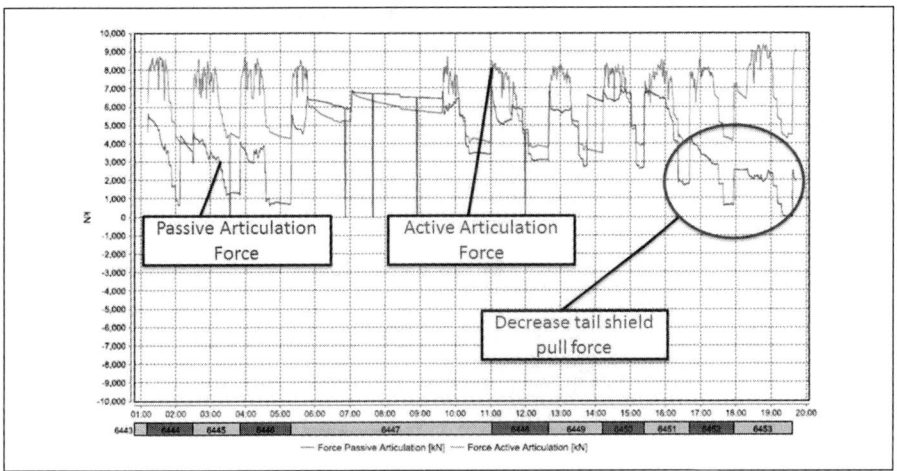

Figure 13. Articulation force vs. geology

It is clear and interesting how the influence of the backfilling injections is immediately visible on the building displacements. During each stroke, the grout injection pressures were directly influencing the ground heaves. Once the pressures were reaching the peak, the displacements were increasing too. We could follow up on production just looking at the building behavior.

Monitoring the correlation between TBM parameters and building behavior gave us the possibility to promptly react and change the TBM influence accordingly, if necessary.

Another very interesting correlation is shown in Figure 13. Here the thrust force on the tail shield, or better the pulling force on the tail shield articulation system, is correlated to the geology. The decrease of the pull force indicated that also the tail shield was, at the point, safely into the limestone bedrock.

CONCLUSIONS

The excavation of the critical 70m stretch lasted 3 days per machine and it took place without any disruption: both TBMs bored in the extremely unfavorable conditions with an average advance rate of 22.3m per day. Both TBM raced to the next station, Gammel Strand, completing the 439m stretch in only 25 days, keeping a very good average daily production of 17.6m.

The maximum settlement recorded was less than1 mm and compensation grouting injection during the TBM operations was eventually not necessary.

While the design set at a conservative 1.0% of volume loss, the recorded volume losses didn't exceed 0.1%.

The whole design process lasted three years. It has been complex and involved different disciplines,actors and negotiations with the project owner. It required delicate interfaces with third parties, both during expropriation process and during the actual works and monitoring process. This long undertaking culminated in the three uneventful days of the actual TBM passage. Overall a good example of how accurate design study before and careful execution, with clear distribution of responsibilities after, allowed to complete in a very short time one of the most delicate and challenging passage of the project.

The Crossing Under the Alaskan Way Viaduct

Enrique Fernandez ▪ Dragados USA
Gregory M. Hauser ▪ Dragados USA
Francisco Gonzalez ▪ Dragados USA
Carlos Herranz ▪ Mott MacDonald
Andrew Herten ▪ HNTB

SUMMARY

The 2001 Nisqually earthquake caused damages to the Alaskan Way Viaduct, (SR-99) in Seattle (WA). A bored tunnel was the solution to replace it and for that, the largest TBM ever built was manufactured in Japan. The approach for the first 1,000 feet of tunnel, running parallel to the viaduct, was to create a protected area by secant piles, jet grouting and more allowing the mega TBM to excavate in and under non engineered fill materials before crossing under the viaduct with minimal cover. This paper will describe the measures put in place to protect the Viaduct and the results of those measures.

SHORT HISTORY OF THE PROJECT

On February 28th, 2001, the Nisqually earthquake shook the city of Seattle with an intensity of 6.7 on the Richter scale. The existing Alaskan Way Viaduct which runs along the Seattle shore was severely damaged. From the three alternatives considered: repairing the viaduct, build a cut and cover tunnel or a bored tunnel, the bored tunnel was the selected option to replace it. This solution allowed building the new infrastructure while the viaduct continued in operation after some remedial works were implemented. The double deck concept from the viaduct was transferred to the bored tunnel solution, thus a single large bore concept was developed.

In December 2010, WSDOT (Washington State Department of Transportation) awarded the SR 99 Bored Tunnel Alternative Design-Build Project to Seattle Tunnel Partners (STP), a Dragados USA and Tutor Perini Corp joint venture. The design team

Figure 1. Tunnel cross section

was led by HNTB and Mott MacDonald was selected by WSDOT to provide construction management services.

The new tunnel is 9,270 feet long and runs entirely below the city on a south - north alignment. In order to minimize the ramps at the portal structures, a shallow cover of the tunnel at these areas was a must on the design. A very detailed analysis of ground treatments and preventive protections at the launching portal were developed and finally applied on site. Considering the minimal overburden of 13 feet at the launching portal as well as the poor unconsolidated residual soils encountered at the beginning of the tunnel, the ground improvement included three safe havens, an anti-buoyancy concrete slab, tangent piles walls and a massive jet grouting treatment to improve the ground over the tunnel crown. Also a careful monitoring plan was performed to control the tunnel excavation and its impact on buildings and other structures, like the viaduct, along the alignment.

The third safe haven, built with mortar piles and jet grouting was the last protected area prior to the crossing under the viaduct, where the TBM was thoroughly checked and fitted with new cutting tools.

DESIGN APPROACH FOR THE VIADUCT CROSSING

A key innovation during the project proposal and design stage was STP's settlement assessment. WSDOT's RFP based its analysis on a constant volume loss of 0.5% throughout the alignment and STP challenged this assumption based on previous TBM tunneling experiences and the extensive use of numerical modeling on specific areas, where preliminary assessment showed displacements beyond allowable limits. The result was a reduction in the number of structures to be protected and a cutting edge analysis of those requiring specific mitigation measures, as it is the case of the viaduct.

Assuming a constant volume loss as an input for settlement assessment and mitigation measures design can be conservative in tunnels with variable overburden and varying ratios of total overburden vs overburden in over consolidated ground. On the other hand, the approach can be unconservative in portal areas, where the learning curve in the operation of the machine and the variability of the shallower soil layers can lead to higher displacements.

STP's approach to settlement control in the project was that the primary protection of existing structures relied on the proper operation of the TBM through face pressure, shield injection and tail grouting control. On top of this, in the crossing under the viaduct, a combination of ground improvement measures and structural strengthening was proposed to reduce the tunneling induced displacements and its impact on the structure.

The geo-structural mitigation measure consisted on a series of micropile walls formed by two lines of 7" diameter, 0.453" wall thickness, API N80 pipes spaced 1'-6" on centers between lines and 3'-3" on centers in longitudinal direction. The geometrical definition of the walls was a great challenge due to existing utilities and the proximity of the tunnel to the viaduct's deep foundations, which consists on a combination of steel and concrete piles and reinforced micropiles. This led to the definition of 22 groups of micropiles, with a maximum length of 138'. Micropiles within the same group were tied with 1'-6" thick reinforced concrete pile cap beams.

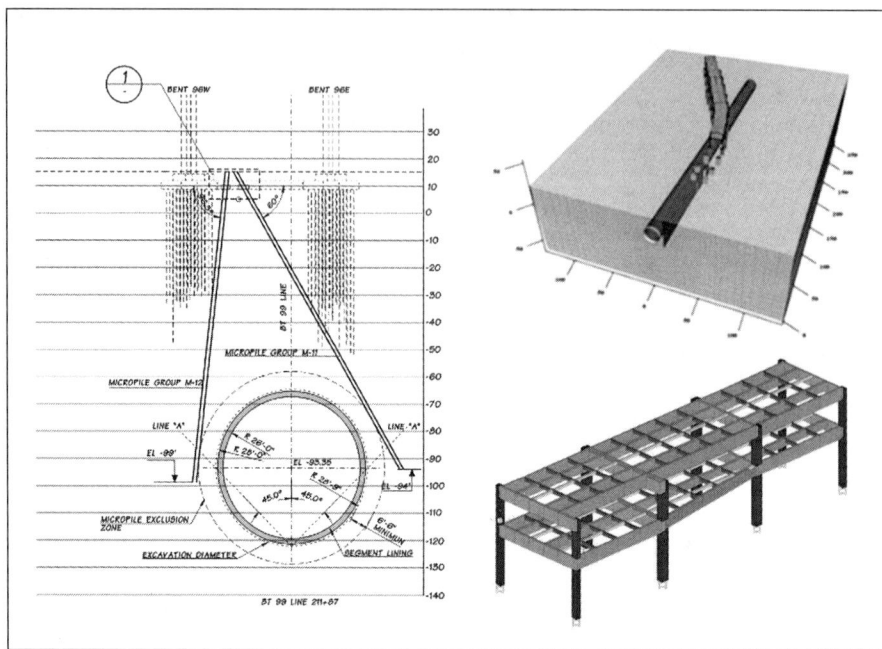

Figure 2. Micropile walls layout, FLAC3D and SAP2000 models

Table 1. FLAC3D model results—Expected maximum EPB-induced displacements at AWV foundations

Total Displacement (inches)	Vertical		Horizontal Longitudinal		Horizontal Transverse	
	1.0		0.2		0.5	
Differential Displacement (inches)	Transverse Direction			Longitudinal Direction		
	Vertical	Horizontal Longitudinal	Horizontal Transverse	Vertical	Horizontal Longitudinal	Horizontal Transverse
	0.7	0.15	0.35	0.7	0.15	0.35

The analysis of the crossing was performed with the use of a multi-staged three-dimensional FLAC3D model to simulate the TBM excavation under the viaduct and to assess the effectiveness of the micropiles in reducing the tunnel induced settlements. A small strain soil constitutive model was implemented to account for a variable stiffness of the ground dependent on the shear strain.

The expected maximum settlement at surface level in the crossing ranged between 0.7 and 1.4 inches with volume losses between 0.25 and 0.4%, depending on the section within the crossing.

The differential displacements in the viaduct's foundations obtained (Table 1) were used in a step-by-step structural SAP2000 model to design the viaduct's shear and flexural FRP strengthening, taking also into account the loads and displacements that the AWV had suffered to date (including those from the 2001 Nisqually earthquake).

EPB EXCAVATION CONCEPT

The SR 99 Bored Tunnel alignment runs along a complex, highly variable geological environment, comprised of a heterogeneous mixture of glacial and non-glacial or inter-glacial deposits below groundwater table. These soils feature clays, silts, sands, gravels, cobbles and boulders along the geotechnical profile. Therefore, the tunnel face is not in homogeneous soil conditions at any location. This is a fact, but does not mean that it is a problem and the excavation is progressing in homogeneous mix face conditions, without sudden change in the composition of these mixed materials.

Based on the existing geotechnical information, an EPB (Earth Pressure Balance) machine was clearly the most recommended and extended solution to excavate these anticipated soils as well as to operate in mixed face (mix soils) conditions. To develop the best machine possible for this project, the most reliable manufacturers in the world were involved. Finally the Hitachi Zosen proposal was selected and an EPB 57'4" diameter was selected to excavate the tunnel. The Japanese manufacturer built the machine fitted with proven technologies. The most relevant is the free air intervention feature to inspect and replace the cutting tools, which is the most innovative solution on western tunnels despite it has been applied in Japan for more than 15 years.

The EPB concept is to maintain the face stability by compensating the earth and water pressure from the ground with the thrust obtained by pushing the TBM against the segmental lining. The equilibrium between the soil entering the cutterhead plenum and the soil extracted through the screw conveyor is the key to keep the plenum fully filled and consequently avoid the risk of loosen soil above the cutterhead entering the chamber. In non-cohesive soils, this fact can lead to ground settlement issues or even sinkholes. On the other side, entering in the chamber more soil than extracted can easily create soil plugs and obstructions which would require hyperbaric interventions to clean the chamber.

The second critical area to avoid settlements is the TBM shield where there is a gap between excavation diameter and shield diameter. To compensate this gap, bentonite is injected through ports around the shield at the required pressure. The third one is the gap between excavation diameter and outer lining diameter which is also grouted through the tail shield by pressure with two component grout. With these basic concepts of control of muck material and proper grouting of the gaps, the risk of settlement is minimized and the crossing under critical structures, like the viaduct, can be done in a safe way.

The tunnel lining precast segmental rings, 9+1 universal shape, has an outer diameter of 55' 9" and are 6' 6" long and 2' thick. The annular gap between the outer diameter of the lining and the ground, as well as its backfilling has been specially designed to mitigate surface settlements. This theoretical behavior has been proved in previous experiences, especially in Madrid M-30 By-Pass tunnels where 2 different 49' diameter TBMs were used in parallel drives, each one with different gap, and the obtained subsidence data proved the relevance of the gap in the settlements produced. The one with bigger gap, Herrenknecht had 8.5", drove its tunnel first generating higher

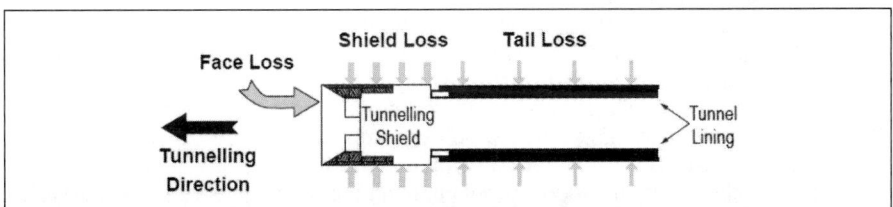

Figure 3. Sources of volume loss due to tunnel excavation with an EPB machine

settlement than the second TBM with smaller gap, Mitsubishi had 7", which goes against the normal results when both tunnels are dug with similar TBMs. Generally, the second drive produces higher subsidence due to the ground relaxation during the first drive. In the SR99 TBM, the gap between the ring outer diameter and the excavation is limited to 8" to minimize this fact. In addition, the gap between the excavation and the shield is only 0.6" in radius due to the active articulation shield configuration.

This methodology also simplified the preventive measures that would normally be taken in advance of the TBM excavation to protect the structures along the alignment that might be sensitive to or that could be affected by the tunnel excavation.

LESSONS LEARNED DURING TBM DRIVE FROM THE PORTAL TO SH3

The launch of the TBM in June 2013 began very shallow, especially for such a large TBM, with only approximately 13 feet of cover over the crown of the machine. The concern was what affect the TBM and tunnel excavation would have on existing structures and especially the Alaskan Way Viaduct, which would remain in service during the entire tunnel drive. To protect the Viaduct and other structures, STP installed a system of settlement mitigation measures, (the South End Settlement Mitigation Plan or SESMP), that consisted of reinforced concrete secant piles along both sides of the tunnel drive, jet grouting of the fill soils that comprised the upper half of the tunnel face at launch and a reinforced concrete buoyancy slab, tied into the secant piles on each side to prevent the tunnel from floating out of the ground during the early shallow launch phase. All these measures proved successful and there was little or no impact to the Viaduct or any of the existing buildings along the route up to Safe Haven #3.

Along with the SESMP, a vast array of monitoring devices and settlement points were established to monitor the ground settlement, movement of structures and ground water pore pressures as the TBM advanced north and dove below the manmade fills into the various layers of Glacial soils consisting of sand and gravel, tills and till like material, clay and silts, cobbles and some boulders. These monitoring devices included Near Surface Settlement Points (NSSP), Automated Structure Monitor Points (ASMP) on the adjacent viaduct and nearby buildings, Manual Structure Monitor Points (MSMP), and Utility Settlement Points (USP). Instrumentation has also been installed to monitor ground movements and pore water pressures immediately around the advancing TBM, and correlate the information with tunneling conditions. For this purpose, a series of Multiple Position Borehole Extensometers (MPBX) with Piezometers (PZ) have been located at an average spacing of 50 feet along the entire tunnel alignment. Inclinometers (INCL) were also installed adjacent to and ahead of the advancing TBM to monitor both lateral displacements at the side of the tunnel and longitudinal displacements into the face of the TBM. Figure 4 is a plan view of all the instrumentation installed between Station 205+00 and 210+00.

Monitoring of these devices continued throughout all phases of tunnel excavation, both the Automated Monitoring points and the manual points were read regularly, (the automated points read every hour) and the results analyzed and reviewed by the production supervisors to ensure that there were no threats to the existing structures. For all of the points monitored, there were no impacts to the Viaduct or to any of the other existing structures, movements generally in the −0.25 inch range were all that occurred. All of this monitoring data was and continues to be reviewed daily by the Construction Monitoring Task Force which is comprised of the Owner, WSDOT, the City of Seattle, WSDOT Consultants, JV Consultants, the Monitoring Group engineers lead by Tony Stirbys and the Tunnel production managers, engineers and supervisors.

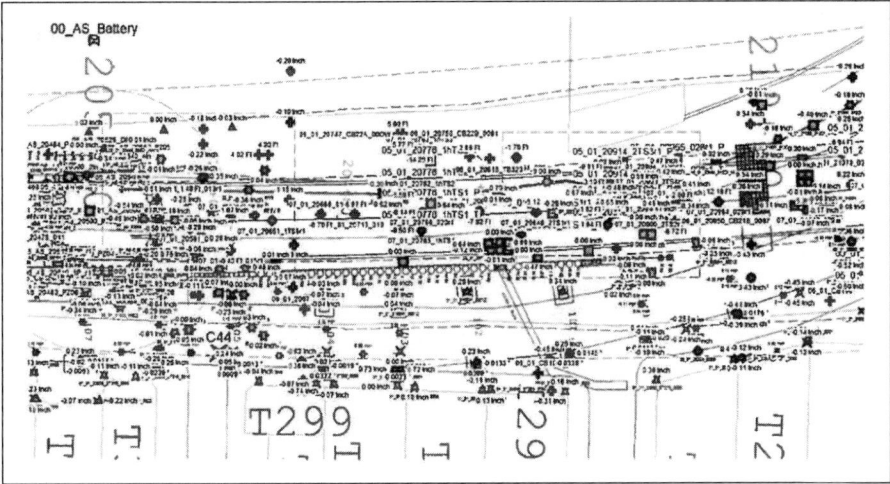

Figure 4. Plan view of instrumentation

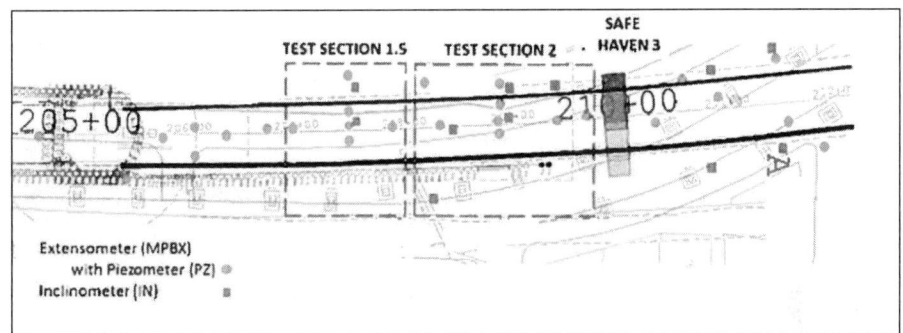

Figure 5. Plan view of monitoring between access shaft and Safe Haven 3

The above procedures of monitoring existing structures and reviewing the results continued for the initial 1,000 foot initial drive, the TBM breakdown and ensuing repairs including the installation of the Access Shaft and re-launch of the TBM in the final approach to the Viaduct and into Safe Haven #3. In order to have very precise and detailed documentation to confirm that the TBM could be operated and could perform properly without damaging either the Viaduct or the other existing structures, two test sections with additional monitoring devices were set up prior to the TBM entering Safe Haven #3. These were Test Sections 1.5 and Test Section 2, and plan views of both are shown in Figure 5. Figure 6 shows the Geotechnical Profile from the Access Shaft to Safe Haven #3 and the extensometer (MPBX) locations and anchor depths. From the access shaft to Safe Haven 3, the depth to the crown of the TBM increases from 50 to 70 feet. The extensometers (MPBXs) along the tunnel centerline have anchors to measure ground settlements 5 feet above the tunnel crown and at several shallower depths between the tunnel and the ground surface. A piezometer is located at the bottom of each extensometer boring to measure ground water pressures; both ambient conditions and the changes generated by the pressure wave around the advancing TBM.

There is another paper being presented at this conference that deals specifically with the monitoring and soil reaction to the TBM, SR 99 Tunnel Project Monitoring: A Dynamic and Integrated Approach, and if anyone wants more detail of the monitoring program they should refer to that paper. For our purposes here, let it suffice to say that the monitoring did show that the TBM could perform while tunneling under existing structures without any adverse impact to those structures. The Surface and Structure settlements were generally in the 0 to −0.25 inch range with no detrimental impact to any of the structures.

Based on this monitoring program and the results, it was agreed by all involved that the TBM was capable of advancing below the Viaduct and Downtown Seattle without adverse effects to the existing structures. WSDOT gave STP permission to continue below the Viaduct and this began on Friday, April 29, 2016.

GROUND IMPROVEMENT UNDER THE VIADUCT

During the months leading up to the actual crossing under the Viaduct, there was a great deal of discussion regarding ground improvement thru this area. At one location the footings for the Viaduct would be as close as 15 feet to the TBM and the new tunnel. There was great concern for the stability of the Viaduct and what impact the Tunnel would have on that stability.

There was also concern for what impact to the Viaduct any ground improvement activities might have on the Viaduct stability. There was a layer of cobbles that has caused settlement to the Viaduct while installing the Pin piles between the AWV Bents and footings and the Tunnel thru this area and great concern that any ground modifications might cause more harm than good. There was also the realization that after crossing the Viaduct and moving into the Downtown areas that there was virtually no further surface access to the tunnel, if the TBM was not capable of mining below the Viaduct, it would not be capable of mining the remainder of the tunnel.

After a careful review of the monitoring data and the results achieved, surface settlement of 0 to 0.25 inches, it was decided to not install any further ground modifications below the Viaduct. The TBM had been repaired and all indications showed that it

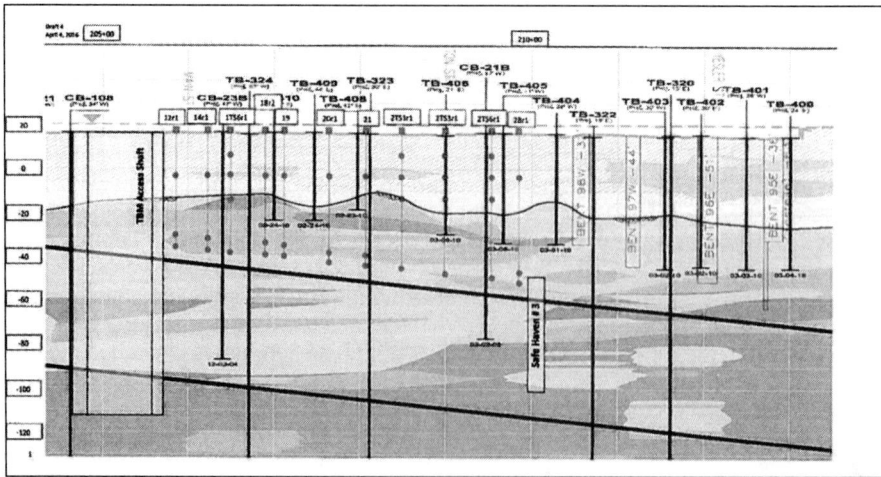

Figure 6. Geotechnical profile from access shaft to Safe Haven 3

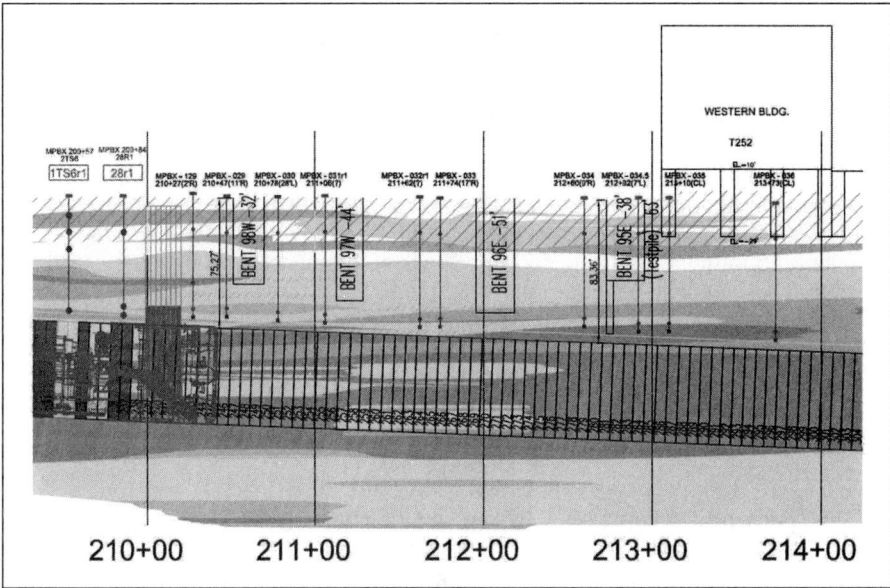

Figure 7. Vertical section

was functioning better than planned. The settlement monitoring showed much less impact to the existing structures than originally anticipated. It was decided to allow the TBM and the JV crews to show what they were capable of and WSDOT and the JV all agreed to forgo any further ground modifications and to begin tunneling below the Viaduct.

THE CROSSING

The AWV poor structural conditions forced STP to extreme cautions operating the TBM in this stretch of the tunnel drive. Pile footings were 15 feet above the tunnel crown. Micropile barriers were as close as 5 feet to the TBM perimeter. Isotropic ground wasn't expected through the tunnel face. Raveling sands and silts were present above and within the TBM face. Till and till like deposits, layers of clay and boulders were likely to be encountered on this always saturated soil. For reference, Figure 7 shows the Geologic Profile below the Viaduct and depicts the location of the Viaduct footings in relation to the Tunnel.

Every single foot to mine was therefore a challenge and minor ground loss could potentially induct some differential settlement that might compromise the Viaduct's structural stability.

STP's approach was nothing different that operating the TBM in strict EPB mode, no more, no less, and that is what was done. Three different calculations were made to determine the target face pressure, and the more conservative was chosen. Keeping the TBM chamber full of muck at all times was paramount, and as such, 2 each motorized 8" purge lines were fitted to the TBM bulkhead during the SH3 stoppage. A Polymer+Bentonite system was ready to continuously inject a viscose fluid into the TBM overcut. This gap backfill and much more importantly, the grout backfill around the concrete rings (so called tail grout) was injected following pressure criteria, rather than theoretical volume, which we all know is a regular and dangerous practice. Secondary grouting was feasible should the theoretical minimum volume not have been achieved.

QC, QA, and TBM Operators were thoroughly briefed in the importance of operating the TBM as a pure EPB. Some features were fitted to the TBM cabin screens to help these men on their key roles. As examples, a screen showing the face pressure distribution top to bottom was fitted. This screen allowed them to visualize and keep a linear, therefore hydrostatic pressure, through the chamber height, preventing any air bubble to form in the upper section. Similarly, a couple of views were created in the operator screens to have "on-real-time" measures of the excavated material, compared with the theoretical. Should the deviation be higher than 1%, a red alarm would pop up on the cabin screens, advising the driver and QC of potential over-excavation. Additionally, the QC engineers undertook weight and volume reconciliation twice per ring. As such, every shove was interrupted always at half of the advance to sample the muck on the conveyor belt.

In summary, EPB mode was controlled and monitored by,

i. **Face pressure control,** by keeping the chamber full of muck. Plotted pressure lines in the screens and purge valves.

ii. **Shield pressure control,** by continuously injecting a pressurized viscose bentonite+polymer fluid.

iii. **Weight reconciliation,** by fitting on-real-time weight comparison screens and sampling the excavated material to perform weight and volume reconciliation twice per ring.

iv. **Tail backfill,** by injecting bi-component grout following pressure criteria.

v. **Briefing the crews,** by providing bespoke training on the risks we were facing and the tools in place to mitigate those to the whole mining crews.

Figure 8. Crossing plan view

On April 29th, 2016, the challenge started. This industry knows the continuous operation is good practice to mitigate settlement, and agreement was made to operate 24/7 mining under the AWV. Traffic was suspended by WSDOT as a precaution and it was envisaged to resume by 15th May.

On May 11th, the more critical section had been completed and vehicle traffic did resume. 62 rings, 403 feet had been successfully mined in 12 days, an average of 33.6 feet a day or 5.2 rings a day.

ANALYSIS OF THE GROUND SETTLEMENT RESULTS

The information obtained from the monitoring instruments was compared with the EPB control readings (face, shield and tail pressures) during construction. The plan view below shows the crossing layout.

As mentioned in the description of the design approach, the WSDOT Technical Requirements (TR) called for 1" transversal and ½" longitudinal maximum differential settlements in the bents. The structure was partially strengthened and ground improvement in the form of micropile barriers were put in place prior to tunneling. Phased FEA models were carried out considering the historical settlement the structure incurred during the last years, and the tunnel potentially induced settlements assuming a 0.2% ground loss. With those outputs the actual structure was then checked, and it was considered that if the tunneling settlements were within the predicted values, the viaduct would continue being operative.

The maximum permitted tunnel induced settlements are shown in Tables 2 and 3 (see expected values). Rather than a generic ground loss estimate, the FEA held STP to an extremely small differential settlement between adjacent columns. The oblique tunnel-viaduct crossing made the operation even more challenging.

The automated and manual survey data installed in the bents provided real-time data which were collected and analyzed. In the ground itself, the MPBXs and inclinometers were providing accurate readings in real time as well.

Table 2. Absolute and relative movements, transversal direction

	Bent 95			Bent 96			Bent 97		
(Inches)	95E	Differential	95W	96E	Differential	96W	97E	Differential	97W
Expected	−0.88	−0.63	−0.25	−0.90	−0.51	−0.39	−0.38	0.55	−0.93
Actual	0.22	0.04	0.18	0.16	−0.06	0.22	0.16	−0.02	0.18

Table 3. Absolute and relative movements, longitudinal direction

	Bent 95/96 E			Bent 96/97 E		
(Inches)	95E	Differential	95W	96E	Differential	96W
Expected	−0.88	0.02	−0.90	−0.90	−0.52	−0.38
Actual	0.22	0.06	0.16	0.16	0.00	0.16

	Bent 95/96 W			Bent 96/97 W		
(Inches)	95E	Differential	95W	96E	Differential	96W
Expected	−0.25	0.14	−0.39	−0.39	0.54	−0.93
Actual	0.18	−0.04	0.22	0.22	0.04	0.18

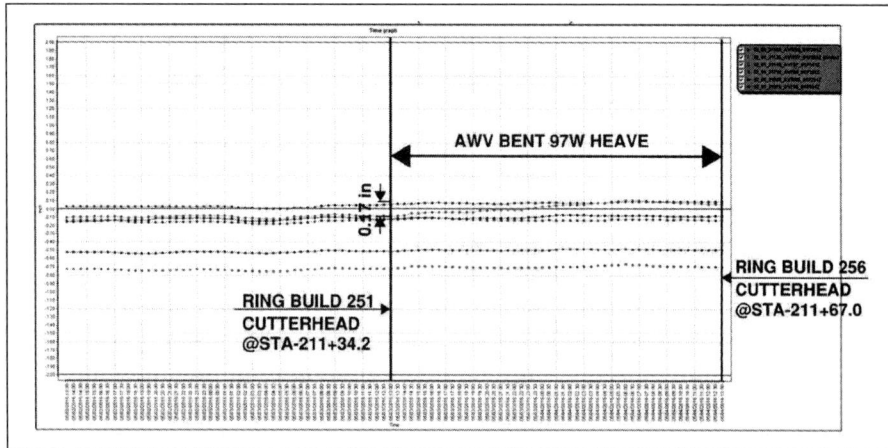

Figure 9. Displacements on viaduct's bents

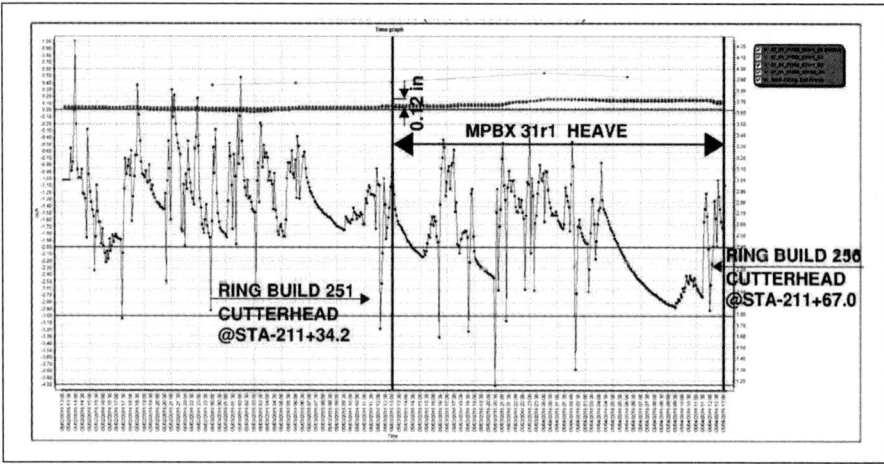

Figure 10. Displacements on MPBXs

It should be noted that several events caused displacements in the area before the TBM excavation (such as deferred displacements caused by the 2001 Nisqually earthquake). In bents with previous movements, those are in general significantly higher than the TBM induced displacements.

Because of the earthquake damages to each of the bents and the historical settlement in the area, bents 95, 96 and 97 were the most critical.

Tunnel induced vertical settlements had been assessed at approximately 1" in some of the columns of these bents (FEA models), which combined with the induced settlement in the two adjacent columns, resulted in 0.5"–0.6" differential for each of these pairs. With those figures, there was no room for errors.

Tables 2 and 3 include the tunneling induced settlement for both the expected and the actual values in the three critical bents, 95, 96 and 97. The numbers shown as "expected" in the table, are the limits that our designer and the owner considered "upper limits" to safely re-open the viaduct.

The tables are self-explanatory. None of the columns settled the expected inch, rather, all of them heaved between one and two hundreds of an inch. On the same note, differential settlements, expected in the 0.55" region, were actually one order of magnitude less, in the five hundreds of an inch range.

Figures 9 and 10 show an example of the monitoring data registered during the crossing. The light green series on Figure 9 corresponds to Bent 97 West, showing a heave of approximately 0.17 inches. The displacement occurred while the TBM advanced under the bent.

Figure 10 shows a similar response in depth, with the MPBX closer to Bent 97 West experiencing a similar heave between the same TBM locations.

Both figures show how the injection through the shield caused a response similar to compensation grouting, with upwards displacements on the surrounding ground that got transmitted to the structure to be protected.

The heave induced in Bent 97 West suggests not only an effective "passive" control of ground displacements through face pressure, shield bentonite injection and tail two component grouting to fill the corresponding gaps, but the ability to actively induce displacements on the ground through pressured injections, like compensation grouting would provide.

The displacements induced in the viaduct during the crossing are one order of magnitude lower than design predictions. The chamber full of pressurized muck, the bentonite-polymer pressurized overcut and the pressurized grout/tail gap injection did their job. The EPB theoretical mode is feasible in the real world.

CONCLUSIONS

There is room for improvement in design to better predict current capacities of the EPB tunneling industry in controlling ground displacements, a key feature in urban areas. It should be kept in mind though that risk management is essential in underground design and construction, and the confidence in predictions must be evaluated by the Owner, Contractor and Designer considering specific ground conditions, the sensitivity of existing structures and potential impact in case of damage.

The successful crossing under the Alaskan Way Viaduct with a very large TBM in the conditions described in the paper is the result of the joint effort between Owner, Contractor, Designer and Consultants, working together with a single mission, deliver this outstanding infrastructure to the City of Seattle while minimizing the construction impact.

As tunnelers, we have the challenge to open new infrastructures under congested areas without disturbing the neighborhood. Current tunneling technology allows us to do this.

PART

Hard Rock TBM

Chairs

Mike Stolkin
J.F. Shea Co., Inc.

Geoff Fairclough
Schiavone Construction Co. LLC

Rockbursts in TBM Tunnels—Analysis and Countermeasures

Gary Peach ▪ Multiconsult AS
William B. Dobbs ▪ MWH
Bruce Ashcroft ▪ Multiconsult AS

ABSTRACT

Twin 8.5 m diameter, 10.5 km long parallel headrace tunnels under overburdens of up to 1,870 m with high horizontal stresses are being excavated for the Neelum Jhelum Hydroelectric Project in northeastern Pakistan using two main beam gripper Tunnel Boring Machines (TBMs). Rockbursts presented a significant danger to personnel and equipment. Rockburst characteristics were compiled on both TBMs to create a database for future reference. Correlation of rockburst characteristics improved the ability to predict the likelihood, location, severity and number of rockburst events. This in turn allowed for optimization of rockburst countermeasures including horizontal and vertical relief holes, full ring steel support spacing, shotcrete application, over cutters and alignment change.

INTRODUCTION

The Canadian Rockburst Support Handbook defines rockbursts as "damage to an excavation that occurs in a sudden or violent manner and is associated with a seismic event." (Kaiser et al., 1996). This definition, widely used in the mining industry, was deemed too restrictive for the Neelum Jhelum project. The term "rockburst" was expanded to refer to any release of detectable seismic energy linked with TBM excavation. A rockburst did not necessarily damage the tunnel excavation or the TBM. It may have just been a noise. A broader definition proved more meaningful because even small rockbursts can affect TBM operations, and small events can be followod by much larger ones.

Rockbursts can be self-initiated or remotely triggered. Self-initiated rockbursts occur when tangential stresses near the excavation boundary exceed the rock mass strength, and failure proceeds in an unstable or violent manner (Kaiser et al., 1996). Once the stresses exceed the strength of the rock mass, it can violently and suddenly fracture.

Remotely triggered rockbursts are caused by large seismic events which occur away from the tunnel. They can cause dynamic stress releases near the tunnel opening and in turn lead to rockburst damage. Figure 1 shows different rockburst damage mechanisms.

The rockbursts described in this paper belong to the first category, also known as strainbursts. Fracturing and dilation occur when the stresses near the opening exceed the rock mass strength triggering a violent release of excess energy. The primary source of energy causing fracturing and dilation comes from the strain energy stored in the rock around the opening (Kaiser & McCreath, 1993).

Rockbursts were first encountered in deep mines and much of the rockburst literature comes from the mining industry. Traditional mining methods such as drill and blast do not require high concentrations of plant and equipment present near the excavation

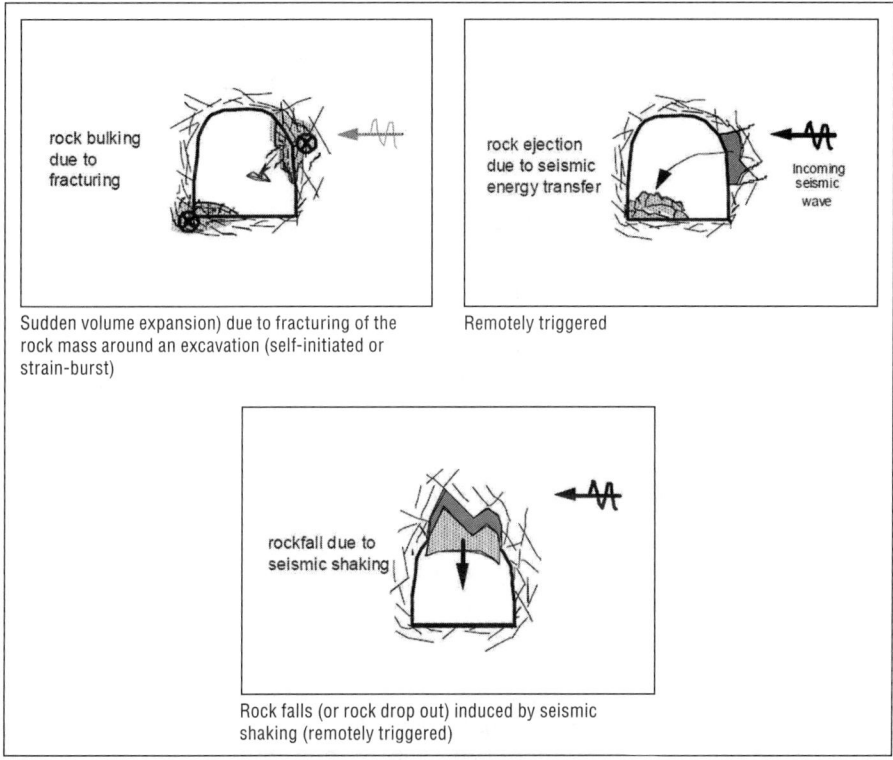

Figure 1. Rockburst damage mechanisms (Kaiser et al., 1996)

face. This contrasts with TBM excavation which continually concentrates complex equipment and manpower close to the excavation face.

PROJECT DESCRIPTION

The project is located in the Muzaffarabad District of Azad Jammu & Kashmir (AJK), in north-eastern Pakistan within the Himalayan foothill zone known as the Sub-Himalayan Range. The terrain is rugged with ground elevations that range from 600 to 3200 meters above sea level.

The project is a run-of-river scheme, employing 28.6 kilometers of headrace and 3.6 km of tailrace tunnels that bypass a major loop in the river system, transferring waters from the Neelum River into the Jhelum River, for a total static head gain of 420 m (Figure 2). The headrace tunnels include both single bore (31%) and twin bores (69%). The tailrace tunnel consists of a single tunnel.

Design capacity of the waterway system is 283 m^3/s. The project will have an installed capacity of 969 MW, generated by four Francis turbines located in an underground powerhouse.

Tunneling commenced in 2008 using conventional drill & blast techniques. It soon became clear that excavating a 13.5-kilometer-long section of the headrace twin tunnels (under high hilly overburden that precluded construction of additional access adits) would take too long.

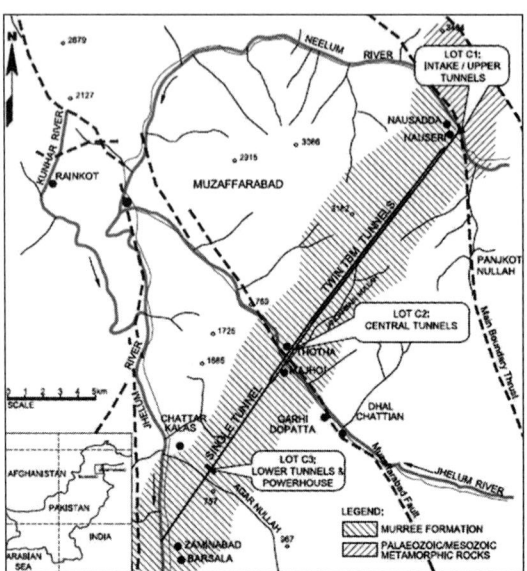

Figure 2. Project layout showing TBM Twin tunnels (in bold), major faults (dashed) and simplified alignment geology

The contract was amended to incorporate two 8.5-meter diameter main beam gripper hard rock TBMs to excavate approximately 10.5 km of twin headrace tunnels (Figure 2). The gripper design offered flexibility for the expected conditions: possible squeezing ground given the relatively weak rock mass and overburdens up to 1870 meters, and the potential for rockbursts in the stronger beds.

GEOLOGICAL SETTINGS

The entire project was excavated in the sedimentary rocks of the Murree Formation, which is of Eocene to Miocene age. The Murree Formation consists of interbedded strata of sandstone, siltstone and mudstone that have been tightly folded and thrust with generally steep bedding dips and strikes sub-perpendicular to the tunnel alignment. The rocks types have a significant calcareous content (20–30%).

The TBM tunnels are being driven through a zone bounded by two major Himalayan faults that trend sub-perpendicular to the tunnels: the Main Boundary Thrust, and the subsidiary Muzaffarabad reverse/thrust fault.

Lithologies

Siltstones & Silty Sandstones

A suite of dark reddish-brown siltstones that grade to silty sandstones constitutes the most commonly occurring rock type on the project, with about 70% of the TBM tunnels excavated in this unit. Typical Uniaxial Compressive Strengths (UCS) are 50–70 MPa.

Mudstones

Mudstones, which also are invariably dark reddish-brown, represent the weakest rocks in the Murree sequence, with UCSs in the 30 to 40 MPa range,

Sandstones

These are the strongest members of the Murree sequence. They are usually grey, with sharply defined contacts. About 21% of the TBM tunnels to date have been excavated in this material. Bedding thicknesses vary from a few meters to over 50. Of 30 samples tested recently, 77% fractured in the 130–170 MPa range, with the remaining 23% exhibiting higher strengths, up to 230 MPa.

Groundwater

Fractured sandstone beds form the principal aquifers in the Murree Formation. Approximately 20% of the TBM tunnels have been excavated in sandstone, yet the tunnels are mainly dry, indicating limited connection to recharge sources.

In-Situ Stresses

In-situ stress measurements to assess hydro jacking potential began where twin headrace tunnels cross under the Jhelum River and at the underground power-house (Figure 2). Hydro jacking test results varied widely, due to the anisotropic rock sequence, with large variations in stiffness between sandstones and mudrocks. Most results clustered loosely around a horizontal-to-vertical stress ratio (k) of unity. High horizontal stresses might have been expected parallel to the tunnel axis, given, this is the major principal stress orientation indicated by the World Stress Map.

Over-coring tests in sandstone beds in the TBM tunnels found a tectonically altered zone of high stresses (k up to 2.9) with the major principal stress oriented sub-horizontally and sub-perpendicular to the tunnel azimuth.

ROCKBURSTS

No rockbursts occurred in the first 2.3 km of both TBM tunnel drives, after which they gradually increased in frequency and severity. By November 2014, with 4.7 km and 4.3 km of the tunnels excavated in the left and right tunnels respectively, regular rockbursts warranted systematic recording. Rockburst events were categorized by magnitude, from "noise only" to "major rockburst." The system aimed to correlate timing and distribution of rockbursts and facilitate selection of mitigation measures at the TBM.

The most severe rockburst occurred on 31 May 2015 on the trailing TBM. This event, equivalent to a Magnitude 2.4 earthquake on the Richter scale, severely damaged the TBM and the recovery took 7 months. An unusual combination of local geological conditions, including a bedding strike parallel to tunnel direction, and unrecognized massive sandstone beds in the sidewalls masked by siltstone, contributed to its severity. The event made clear the need for better ground investigation to identify rockburst-prone conditions ahead of each TBM to allow the TBM management to implement rockburst countermeasures. This would boost tunnel personnel confidence and morale.

Rockbursts in sandstone beds accounted for 76.6%, with the remainder occurring in siltstone units. The more powerful rockbursts originated within sandstone beds.

The initial center-to-center spacing between the tunnels was 33 m. It was acknowledged that increasing the spacing between the two tunnels should reduce rock stresses in the pillar and thus reduce the risk of rockbursts. The spacing was increased gradually to 55.5 m by 'stepping out' the leading TBM over a distance of 700 m.

Figure 3 shows the cumulative rockbursts for the lead and trailing TBMs.

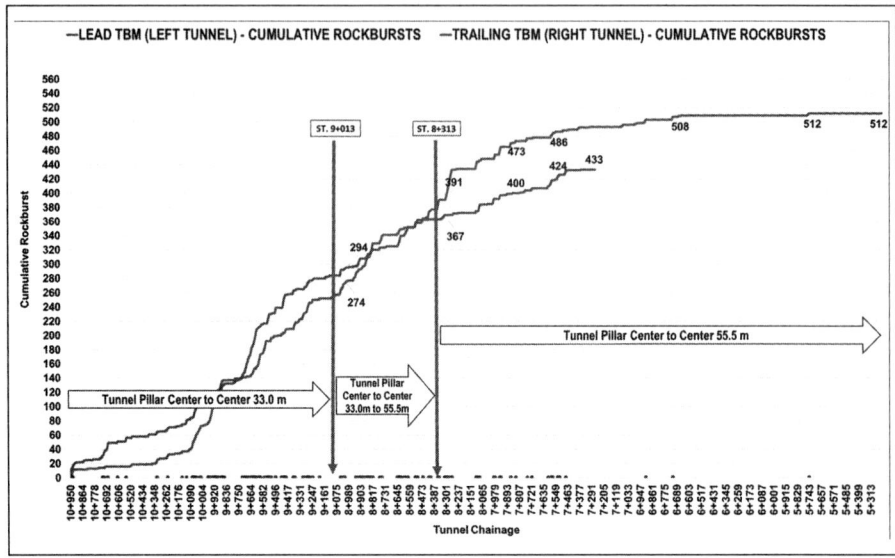

Figure 3. Rockbursts recorded between November 2014 and November 2016

The chart shows that before increasing the tunnel spacing, the frequency of rockbursts at both TBMs was similar, with a significant increase after 6 kilometers had been excavated. The rockburst activity encountered on the two TBMs was similar while the center-to-center distance between the two tunnels was gradually increasing. After the step-out, rockburst activity on the trailing TBM dropped significantly.

Rockburst activity on the leading TBM virtually ended after 9 km of TBM drive.

The rockbursts have been categorized as follows:

Category 1: Noise only: a slight popping sound is heard no damage to the support or ejection of rock

Category 2: Noise and weak rockburst: a popping sound is heard and there may be some light damage to the support and surrounding rock

Category 3: Noise and medium rockburst: loud popping sounds are heard and there may be splitting, spalling or shallow slabbing to the support and surrounding rock

Category 4: Noise and major rockburst: loud sound similar to an explosion, violent ejection of rock into the tunnel and severe damage to the installed support.

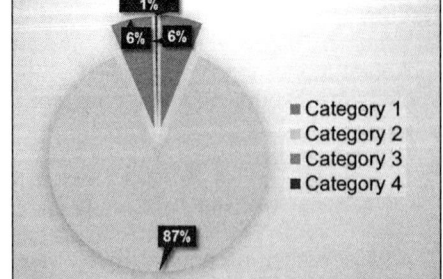

Figure 4. Rockburst distribution by category

Figure 4 shows that the majority of rockbursts are classified as Category 2.

ROCKBURSTS DURING TBM ACTIVITY

TBMs follow an established activity sequence. Rockbursts were recorded during the following activities:

- Excavation, i.e., the TBM is advancing with cutter head rotating with the gripper shoes pushing against the tunnel periphery and the thrust rams pushing the cutter head forward
- Installation of initial rock support at the L1 zone (just behind the shield), i.e., once the TBM has advanced a full stroke, necessary rock support is installed at the L1 zone
- Other, e.g., regular maintenance or repair work being carried out on the TBM

Figures 5 shows the activities underway when rockbursts were encountered on the TBMs.

Most rockbursts (64.6%) on both TBMs occurred when the machines were thrusting and cutting through the rock mass. This is not surprising as the rockbursts recorded were strainbursts, where tangential stresses exceed the rock mass strength and triggered rock mass fracturing and dilation.

Just under a third (30.5%) of the overall number of recorded rockbursts on both TBMs occurred during rock support installation at the L1 zone, which exposed more workers to danger.

After the step-out of the leading TBM, the number of rockbursts encountered on the trailing TBM during excavation increased from 57% to 79%. The proportion of

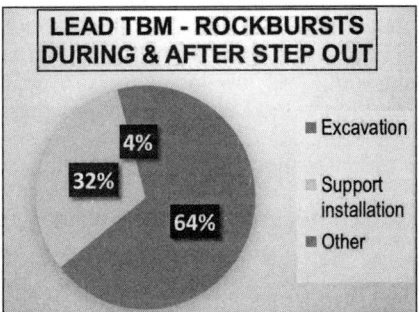

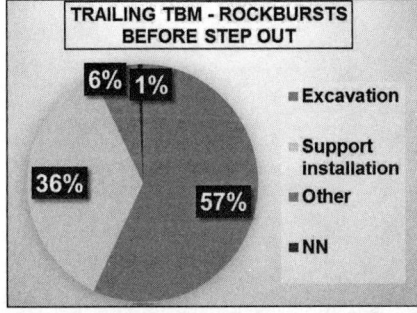
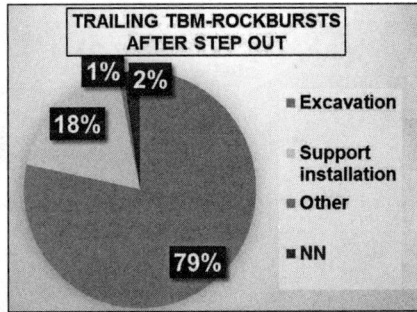

Figure 5. Frequency of rockbursts depending on the activity of the leading TBM before and after the "step out"

rockbursts occurring during support installation dropped from 36% to 18%. Reduced rockburst activity while workers are installing rock supports created safer working conditions.

A small portion of rockburst activity (4.9%) was recorded when activities other than excavation or support installation took place, such as during maintenance periods.

ROCKBURST DISTRIBUTION BY TBM LOCATION

Rockbursts were encountered at the following locations on the TBMs: (1) Ahead of or at the TBM face (2) TBM shield (3) L1 zone and (4) Behind the L1 zone

Figure 6 shows the different locations on the TBM.

Overall over 80% of the rockbursts on both TBMs occurred at either the TBM shield or the L1 zone. Safety considerations make the L1 zone the least desirable location to experience rockbursts.

However, after the step out, the trailing TBM experienced fewer rockbursts at the shield and more at the tunnel face. This is a safer location for rockbursts to occur on the TBM since the shield protects the workers from the face. After the step-out, the trailing TBM recorded almost the same proportion of rockbursts at the L1 zone as before (roughly a third). Although this is a hazardous location for workers, the trailing TBM also experienced 79% of its rockbursts during excavation (Figure 6), reducing exposure of workers at L1 zone during this activity.

Over a third (40.2%) of the rockburst activity was recorded on the TBM shield. The shield is 6 m long. This means that more than half of the rockburst activity (including that at the TBM face) occurred less than one diameter behind the excavated face.

The results also show that overall only 5.2% of the rockbursts occurred behind the L1 zone. On the trailing TBM, the proportion of rockbursts occurring in this section was halved after the step-out. This low incidence proved beneficial because the 60-meter section between the L1 and L2 zones contains densely arranged electrical and hydraulic equipment that blocks access to the tunnel sidewalls and crown. Any repairs in this area must be postponed until the L2 zone reaches the damaged area.

Figure 7 shows the rockburst distribution by location on the leading and trailing TBMs.

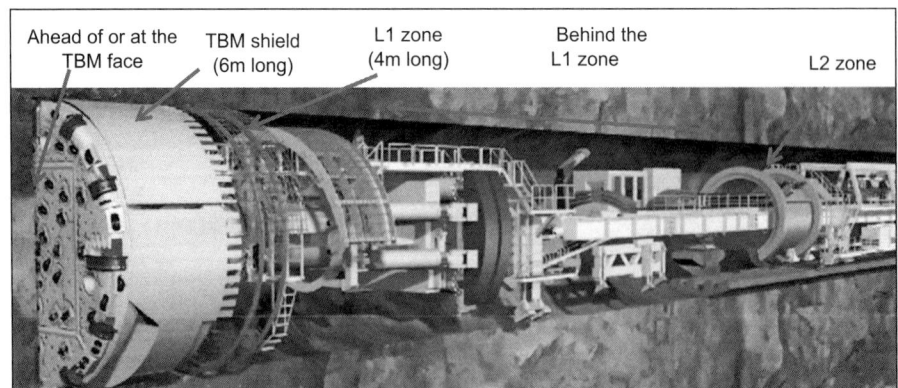

Figure 6. Schematic of gripper TBM (indicative only), Courtesy of Herrenknecht

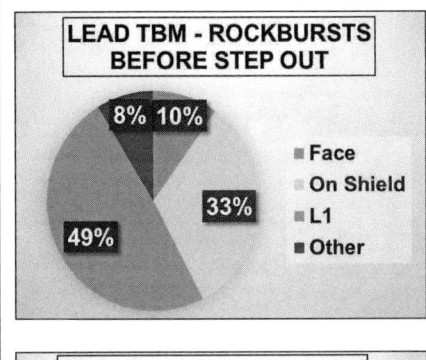
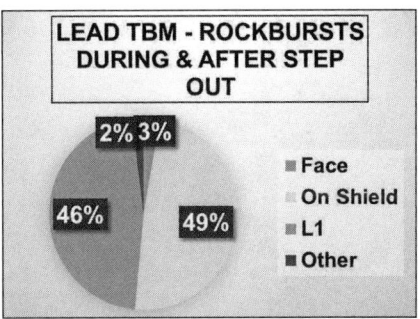
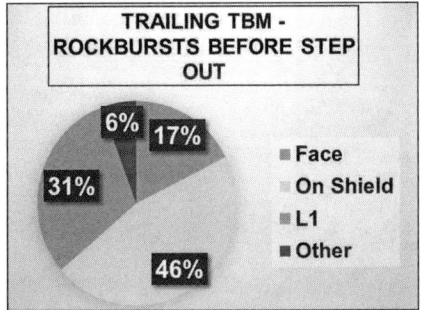
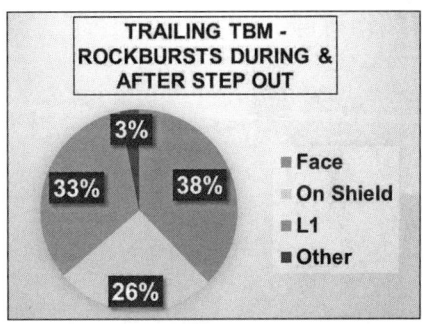

Figure 7. Frequency of rockbursts by location before and after the "step out"

ROCKBURST FREQUENCY ADJACENT TO AND WITHIN THE SANDSTONE BEDS

Probing ahead of the face was performed routinely to detect the presence and thickness of sandstone beds. Rockbursts were recorded with the cutterhead at the following positions:

- A 2-meter siltstone entry zone before the sandstone bed
- The first quarter of the sandstone bed
- The second quarter of the sandstone bed
- The third quarter of the sandstone bed
- The fourth quarter of the sandstone bed
- In the 2-meter siltstone exit zone after the sandstone bed

Figure 8 shows the numbers of rockbursts according to the position of the TBM cutter heads within the sandstone bed where the rockbursts occurred, before and after the step out in alignment.

The leading TBM encountered more rockbursts in the 2-meter contact entry zone and the first quarter of the sandstone bed than the trailing TBM. When this hard, brittle rock mass was encountered, any strain-burst activity was released before the TBM cutterhead had reached a quarter of the way through that sandstone bed.

By contrast, the trailing TBM encountered fewer rockbursts in the 2-meter contact entry zone and the first quarter of the sandstone bed and more in the 2-meter contact exit zone and the last quarter of the sandstone bed both before and after the step-out. After the step out, the trailing TBM experienced greater rockburst activity

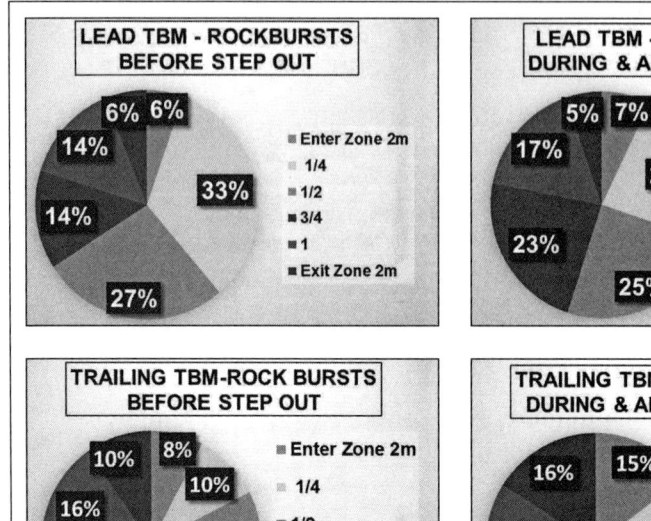
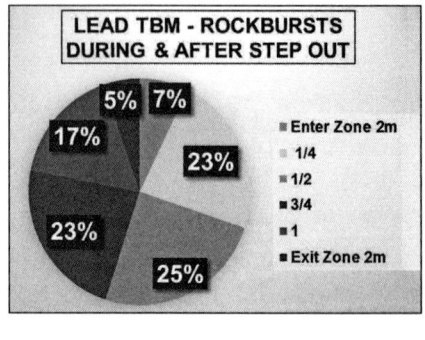
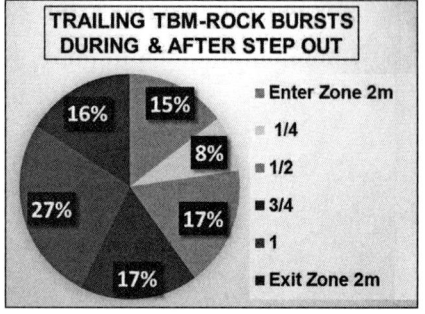

Figure 8. Frequency of rockbursts adjacent to and within sandstone beds both TBMs

in the 2-meter contact entry zone and the 2-meter contact exit zone (31%). The TBM personnel were aware of the increased instances of rockburst activity as they excavated through these contact zones and were able to apply countermeasures when appropriate.

Overall the results shown an evenly distribution as to where the rockbursts occurred within the sandstone bed.

ROCKBURST MITIGATION MEASURES

Some rockbursts were expected on the project, although regular rockburst activity was not encountered until about half way into the TBM drives. At that time the extent of the phenomenon and the 'tectonic zone' of high sub-horizontal principal stresses ('k' up to 2.9) had not been identified. Counter measures developed used the probe and rock bolt drills and shotcrete equipment available on the TBMs (Figure 9).

Longitudinal Relief Holes

Drilling of longitudinal stress relief holes ahead of the tunnel face can fracture the rock mass, thereby releasing stress and reducing rockburst potential. Holes can be drilled with the probe drill and should be closely spaced enough so that the rock between cracks or fractures to relieve the stress. The holes can be concentrated in highly stressed parts of the rock mass. On this project, the holes were drilled using the probe drill with 64-mm drill bits at a lookout angle of 10°. Visible rock mass fracturing between the holes, was rare but popping and cracking in the rock indicated the gradual stress relief.

Figure 9. Drilling of horizontal and vertical relief holes, and shotcrete application in L1 zone

Radial Relief Holes

Radial stress relief holes can reduce the likelihood of rockbursts by shifting the tangential stress peaks away from the excavated perimeter. The holes must be large enough and closely spaced enough so the rock between the holes cracks and breaks. This creates a stress-relieved zone around the excavation perimeter. Fewer holes are required in fractured rock. The holes were drilled 3.85 meters deep using a 64 mm drill bit in the upper 100° of the tunnel. Again, visible rock mass fracturing between the holes, was rare but popping and cracking in the rock indicated stress relief (Figure. 9).

Horizontal Side Wall Probe

A significant contributor to the 31 May 2015 severe rockburst was a local change in strike of the rock strata from perpendicular to the tunnel alignment to parallel. This hid the rockburst-prone sandstone beds behind siltstone beds. In order to detect future hidden sandstone beds, side probe holes were drilled at 5-meter intervals on both sides of the excavated tunnel at tunnel axis height. This activity began on all TBM tunnels in siltstone after the severe rockburst of 31 May 2015.

Installation of Shotcrete at the L1 Zone

Reinforcement of the rock mass begins with installation of rock bolts and wire mesh, used routinely on the TBM. Steel fiber-reinforced shotcrete can provide additional functionality, since the substantial post-peak strength of the reinforced shotcrete can contribute significantly to the energy absorbing capability or toughness of a support system (Kaiser et al. 1996).

The TBMs used on this project allow shotcrete application from three areas; the L1 zone as initial support, 20 meters further back where the invert is sprayed, and

60 meters behind the face at the L2 zone where the permanent support installation is finalized. The L2 zone contains a spray robot beam with one lifting beam mounted per side. The TBMs were designed to have most of the shotcrete applied by two robots in the specially shielded area at the L2 zone (Figure 9).

Where rock conditions require less support, it is preferable to apply most of the shotcrete at the L2 zone to allow quicker installation of initial rock support and faster resumption of excavation. However, 94% of rockbursts were detected at the front 10 m of the TBM. Therefore, to effectively mitigate rockburst, most of the shotcrete had to be applied as soon as possible, i.e., at the L1 zone. Once the TBMs encountered regular rockbursts, up to 62% of the shotcrete was installed at the L1 zone.

Shotcrete is a versatile material that can be installed as a stiff element, as a closed-ring shotcrete lining or as a yielding element, e.g., isolated shotcrete panels. Certain elements (e.g., mesh reinforced shotcrete) may transition from stiff to ductile behavior at large displacements and thus form very tough retaining elements. Shotcrete acts primarily as a retaining element, but also helps to strengthen the rock mass by preventing raveling at the surface and subsequent excessive rock mass bulking (Kaiser et al., 1996).

Shotcrete installation at the tunnel axis is particularly important, due to the high horizontal in-situ stresses present there, to cover support elements and prevent damage from the gripper shoes.

Full Ring Steel Support

The original purpose of full ring steel supports was to support the tunnel at faults, large overbreak areas, and soft and squeezing ground. These supports are time consuming to install and can be installed at spacing's ranging from 0.9 to 1.6 meters. The spacing directly influenced the daily advance rate. Initially these supports had been installed in large overbreak areas adjacent to sandstone beds. As the tunnel advanced, rockbursts commenced and Category 4 events caused major equipment damage and serve damage to rockbolts, mesh, mining straps. The full ring steel supports, however, remained mostly intact even when dislodged. These elements remained rigid but certainly prevented more extensive damage to rock supports and equipment and most importantly provided a degree of protection to TBM personnel.

TBM Overcutters

The excavation diameter of a TBM is the cutter head width plus the protrusion of the gauge cutters. As the cutterhead turns and excavates the tunnel, the gauge cutters wear faster, leading to a reduction in the tunnel diameter. Regular replacement of gauge cutters keeps the cutterhead diameter close to the original design. The remainder of the TBM including the back-up gantries is designed to fit within this diameter with a clearance envelope.

Overcutting was achieved by extending the cutters located on the cutterhead periphery using shims to increase the effective width of the cut and by replacing the gauge cutters with larger diameter cutter wheels (from 17 to 18 inches). Both methods were employed to increase the tunnel diameter by 100 mm.

Change of Alignment

The initial center-to-center spacing between the tunnels was 33 meters. It was anticipated that increasing the spacing between the two tunnels to 55.5 m would reduce stresses in the rock pillar and thus mitigate rockburst potential. The spacing was increased gradually by 'stepping out' one of the TBMs in a tunnel widened by the use of over cutters. The over cutters allowed an increased clearance from support to TBM equipment and allowed for a smaller turning radius for the 190-meter-long TBM. Figure 3 shows that the trailing TBM experienced fewer rockbursts after the change of alignment.

Heavy Rockburst Support

The combination of overcutters, full ring steel support, shotcrete in L1 followed by application of a further shotcrete layer up to a total of 350 mm allowed installation of a much heavier/stronger rock support. This support became integral in dealing with the Category 1, 2 & 3 rockbursts and significantly reduced the impact of Category 4 rockbursts. This support provided safety for the TBM personnel and more protection to the main TBM components located in the front 100 meters of the TBMs.

ROCKBURST SCENARIOS COUNTERMEASURES

Countermeasures for TBM Activities

Table 1 summarizes the countermeasures that are appropriate to mitigate the occurrence of rockbursts during various tunneling activities.

Countermeasures for TBM Locations

Table 2 summarizes countermeasures that are appropriate for rockbursts depending on where they occur on the TBM.

Table 1. Summary of rockburst countermeasures by activity

Activity on the TBM	Appropriate Countermeasures
Excavation	• Installation of longitudinal relief holes
Installation of rock support	• Installation of longitudinal relief holes • Installation of radial relief holes • Application of shotcrete at the L1 zone • Horizontal sidewall probe • Installation of overcutters
Other	• Application of shotcrete at the L1 zone

Table 2. Summary of rockburst countermeasures by location

Location on the TBM	Appropriate Countermeasures
Face	• Installation of longitudinal relief holes
Shield	• Installation of longitudinal relief holes
L1 zone	• Application of shotcrete at the L1 zone • Installation of longitudinal relief holes • Installation of radial relief holes • Horizontal sidewall probe • Installation of overcutters
Other	• Application of shotcrete at the L1 zone • Installation of overcutters

Countermeasures for TBM Position Adjacent to or Within Sandstone Bed

The following countermeasures are appropriate to mitigate the occurrence of rock-bursts that occur within the first quarter of the sandstone bed:

- Installation of longitudinal relief holes
- Installation of radial relief holes
- Application of shotcrete at the L1 zone
- Installation of overcutters

CONCLUSIONS

No single technique can or should be relied upon for rockburst mitigation; a multi-faceted approach should be pursued. The rate of rockbursts detected was relatively uniform for a large part of the tunnel drive as shown in Figure 3. When excavating through highly stressed ground any stored energy must be released which is why the countermeasures outlined in this paper were applied rigorously.

The TBM could not be economically modified mid-way through its drive to implement other countermeasures (such as installing dynamic supports or longitudinal steel slats). This demonstrates the importance of selecting countermeasures to protect TBM personnel during TBM design.

The following were concluded:

- 'Stepping out' the lead TBM clearly reduced the rockburst activity on the trailing machine.
- After the step out, the rockburst activity on the trailing TBM during excavation increased but the proportion of rockbursts occurring during support installation halved.
- The step-out had little effect on the frequency of rockburst activity at the L1 zone.
- After the step-out, the trailing TBM experienced a significantly larger proportion of rockburst activity in the siltstone contact zones coming into or out the sandstone bed.
- Effective Horizontal and vertical relief holes (popping and / or cracking noise) on the lead TBM were applied on the trailing TBM and these countermeasures reduced overall the effects of rockbursts, thus improving progress.
- Most rockbursts occurred when the TBMs were excavating.
- On the trailing TBM, the proportion of rockbursts occurring during excavation increased to 79%, and during support installation the proportion of rockbursts decreased from 36 to 18%.
- The proportion of rockbursts ahead of or at the TBM face (least hazardous) on the trailing TBM increased to 38% after the step out.
- Distribution of rockbursts (number and category) in sandstone and siltstone allowed full ring steel support spacing to be adjusted and thus improve progress.
- On the trailing TBM fewer rockbursts were encountered in the 2-metre contact entry zone and the first quarter of the sandstone bed.
- On the trailing TBM, more rockbursts were encountered in the last quarter of the sandstone bed and the 2-meter contact exit zone before the step-out than

on the leading TBM and an even higher proportion of rockburst activity was recorded in these two sections on the trailing TBM after the step-out.

REFERENCES

Kaiser P.K. & McCreath D.R. (1993), "Rock Mechanics Considerations for Drilled or Bored Excavations in Hard Rock," *Tunnelling and Underground Space Technology*, 9 (4), 425–437.

Kaiser P.K., McCreath, D.R. & Tannant, D.D. (1996), *Canadian Rockburst Support Handbook*, Mining Re-search, Directorate, Sudbury, Canada

Jehle (2011) *Roof Support and Ground Control Equipment.*

Heidbach, O., Tingay, M., Barth, A., Reinecker, J., Kurfeß, D., and Müller, B. 2008. The World Stress Map database release; doi:10.1594/GFZ.WSM.

Large-Diameter 20-Inch Disc Cutters: A Comparison of Tool Life and Performance on Hard Rock TBMs

Stephen Smading ▪ The Robbins Company

ABSTRACT

Optimization of disc cutter life and penetration rate in hard rock can be one of the biggest predictors of project success. With hard rock TBMs being used today in ever more difficult conditions and longer tunnels, the question of which type of disc cutter to be used becomes critical. At one such project in Northeastern China, varying disc cutter tool steels and sizes were put to the test on a total of nine different 8.5m diameter hard rock, Main Beam TBMs from various manufacturers. The TBMs excavated sections of a vast water tunnel in similar granitic geology.

This paper will look at the development of 20-inch disc cutters and the case for large diameter cutters, using the most recent example in China as a focus area of study. Varying advance rates, cutter life, tool steels, and challenges excavating the rock will be discussed. The paper will conclude with recommendations for optimal cutter life in TBMs destined to bore long tunnels in hard rock.

HISTORY

When the author began his career in 1980, TBMs featured 14 and 15.5 inch diameter cutters. The 15.5 inch cutter was an improvement over the 14 inch with a larger bearing set and its corresponding increase in thrust capacity. Shortly thereafter the 15.5 inch cutter was expanded to 17 inches by mounting a larger diameter disc on the same bearing set. While the thrust capacity of the cutter hadn't changed, the premise was that the 17 inch disc, with increased sacrificial material, would increase the mean time between cutter changes. This proved to be the case and the 17 inch cutter became the de facto standard even after the 19 inch cutter was developed in the late 1980s for the Svartisen Hydroelectric Project in Norway.

The TBMs employed at Svartisen were the first "high performance" TBMs. These HP TBMs were designed with both increased thrust and cutterhead power compared with earlier machines. While the 19 inch cutter featured increased thrust capacity, the material of the disc, the chrome/moly/nickel steel historically used, was not up to the task. It was not until the introduction of tool steel, and later modified tool steel, that the benefits of increased thrust capacity were able to be fully realized. While the benefits of 19 inch cutters were clear to the contractors actually using them, it would be at least fifteen years before they were generally accepted by the industry.

INCREASING CUTTER SIZE

There are two distinct benefits to be realized by employing larger cutters; higher thrust capacity and longer wear life.

Higher thrust capacity enables efficient boring in harder formations. To efficiently cut rock, the thrust force applied to an individual cutter must overcome the penetration resistance of the rock and initiate chip formation. Once the critical pressure has been achieved, penetration increases rapidly with a relatively small increase in cutter load. The critical pressure increases with rock strength and it is primarily for this reason that larger cutters have been developed to bore in harder rock. This principle is illustrated in Figure 1.

Longer wear life is the result of an increased volume of sacrificial material in the larger diameter disc ring. The migration from 19 inch to 20 inch on the same bearing core is based on the same principle as the improvement made to the 15.5 inch cutter by installing a larger diameter 17 inch disc back in the 1980s. The relative wear volumes of 17, 19, & 20 inch discs are compared in Figure 2.

The cross section of a 17 inch and a 19 inch cutter are effectively identical when considering just the sacrificial portion or "blade" of the disc. 30mm wear has been assumed for the 17 & 19 inch sizes in Figure 2, and the relative wear volumes are given. The 19 inch disc has 12% more wear volume than the 17 inch disc. The 20 inch disc however, has substantially more wear volume (60%) when compared to the 19 inch. The tip of the 20 inch disc has been extended by 13mm on the radius compared to a 19 inch disc while the rest of the disc profile remains unchanged.

The same principle of extended tip discs can be applied to the 17 inch cutter and when geology permits, numerous TBM operators have chosen an 18 inch disc or even

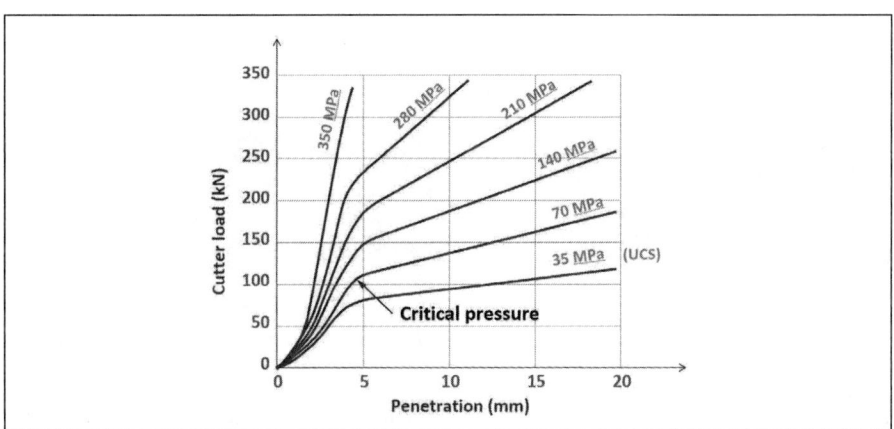

Figure 1. Cutter load vs. penetration rate in rock based on rock strength (MPa UCS)

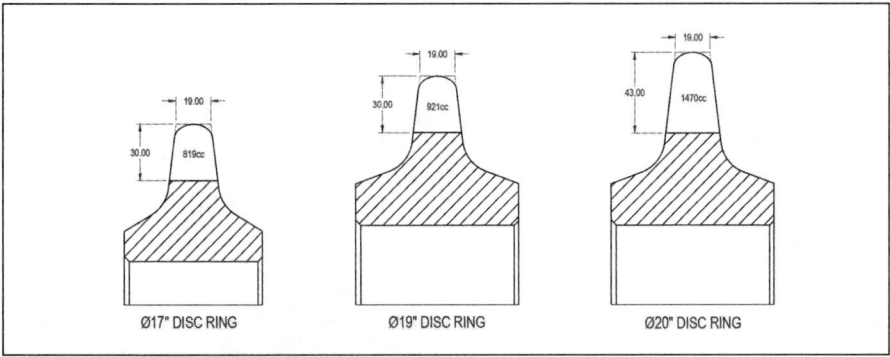

Figure 2. Relative wear volumes of 17-inch, 19-inch, and 20-inch disc rings

a 19 inch disc mounted on a 17 inch bearing core in order to take advantage of the added sacrificial material. This can be an effective solution in pressurized face tunneling to extend the time between cutterhead interventions.

LIAONING NOW PROJECT

One of the longest tunnels in recent history is Northeastern China's Liaoning NOW Water transfer project, measuring 120km in length. The government-commissioned tunnel, for irrigation and drinking water, has been divided into nine lots, designated T1 through T9 (for Tunnel No. 1 to 9, etc.). Each lot, except for T7, was excavated by TBM. Lot T7 is utilizing drill and blast. Lots T1 and T2 utilized new Main Beam machines from another manufacturer. Contractor China Sinohydro Bureaus 3 & 4, responsible for lots T3 and T4 respectively, elected new Robbins Main Beam TBMs, 8.53m in diameter. Similarly, T5 contractor Shanxi Hydraulic Engineer Construction Bureau ordered an 8.53m Robbins Main Beam. Chinese equipment supplier NHI contracted with Robbins to supply Main Beam machines of the same diameter for T6 and T8, and a rebuilt Robbins machine at 8.03m was provided for lot T9. All eight machines were ordered with Robbins continuous conveyors for muck removal (see Table 1 for summary).

Each of the eight TBMs excavating the Liaoning NOW project bored two consecutive tunnels ranging from 5 to 10 km long, totaling about 15 km each. The difficult and long tunnels pass through mainly granite, granite gneiss, and schist geology of varying abrasivity, and this geology was similar for all the drives. Mountainous terrain including valleys and rivers requires versatile ground support. Cover varies widely, from as little as 97 m to as high as 590 m at T6. Despite their nearly identical designs, some of the TBMs were fitted with 19-inch disc cutters and some with 20-inch cutters at the request of the various contractors. As such the project afforded a unique opportunity to test the relative efficiency of the two cutter sizes in similar geology (see Figure 3).

Three TBMs on the project utilized 19-inch cutters. Five TBMs utilized 20-inch cutters. The large number of machines on a single project presented a unique opportunity to compare not only 19 inch and 20 inch cutters but also to compare discs

Figure 3. Cutterhead installed with US-manufactured 20-inch cutters

Table 1. Liaoning NOW project summary

Lot No.	Cutter Size	TBM Type	Diameter	Length 1st Drive	Length 2nd Drive	Contractor
T1	19	Other Manufacturer	8.53m	7347	10016	China Sinohydro Bureau 14
T2	19	Other Manufacturer	8.53m	11453	8868	China Sinohydro Bureau 15
T3	20	Robbins	8.53m	8833	6507	China Sinohydro Bureau 3
T4	20	Robbins	8.53m	7275	8206	China Sinohydro Bureau 4
T5	20	Robbins	8.53m	7412	5270	Shanxi Hydraulic Engineer Construction Bureau
T6	20	Robbins/NHI	8.53m	6406	3650	Ministry of Rail Bureau 18 (MOR 18)
T8	20	Robbins/NHI	8.53m	5234	3354	CRTG
T9	19	Robbins (used)	8.03m	7495	2668	China Sinohydro Bureau 6

manufactured by different suppliers. Data on disc cutter consumption was not made available for the entirety of each tunnel but we were able to obtain sufficient data that some interesting trends were observed. Table 2 illustrates the distances over which disc cutter consumption data were obtained. Most data came from the first drives of each contract.

19-INCH VS. 20-INCH CUTTERS

One would expect average cubic meter rates per disc cutter consumed to be roughly proportional to the additional volume of sacrificial material in a 20 inch cutter when compared to a 19 inch cutter. I-n addition one would expect that monthly advance rates might be somewhat higher as a result of less frequent cutter changes. These data lend support to both premises when comparing average cubic meter rates or when comparing the monthly advance rates from all eight tunnels.

Advance Rates

Figure 4 summarizes the advance rates of all eight TBMs. The first two columns show the monthly advance averages for the portions of each contract where disc cutter data were made available. The 19 inch cutters averaged 446m/month compared to 554 m/month for the 20 inch cutters. In this project, the machines with 20 inch discs had on average a 24% better monthly advance rate than those with 19 inch cutters.

Table 2. Data available for disc usage

Lot No.	1st Drive, m	2nd Drive, m
T1	7324	0
T2	4807	0
T3	8833	6507
T4	7275	8206
T5	4007	0
T6	3722	0
T8	1955	0
T9	6435	0

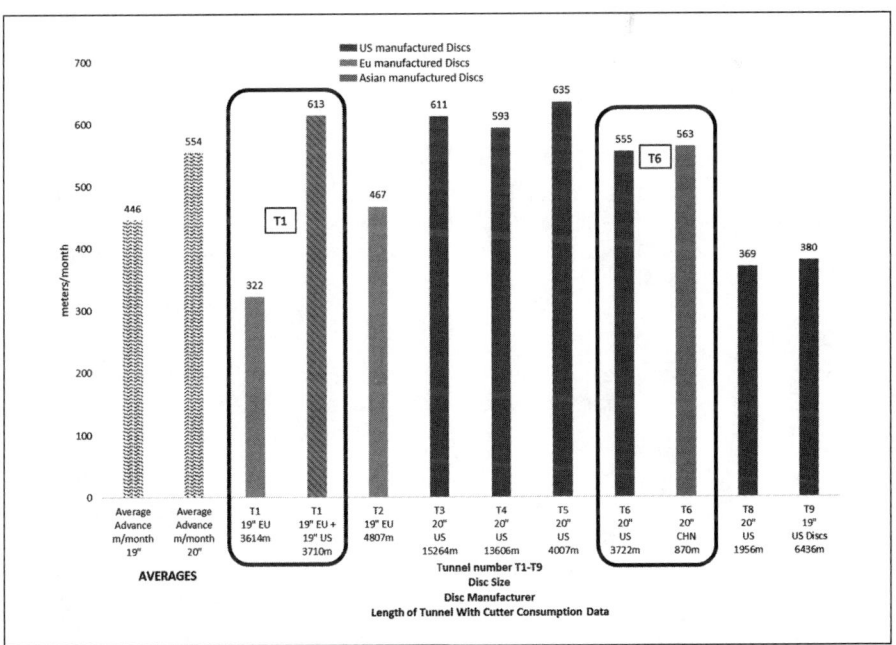

Figure 4. Average monthly advance based on disc cutter size

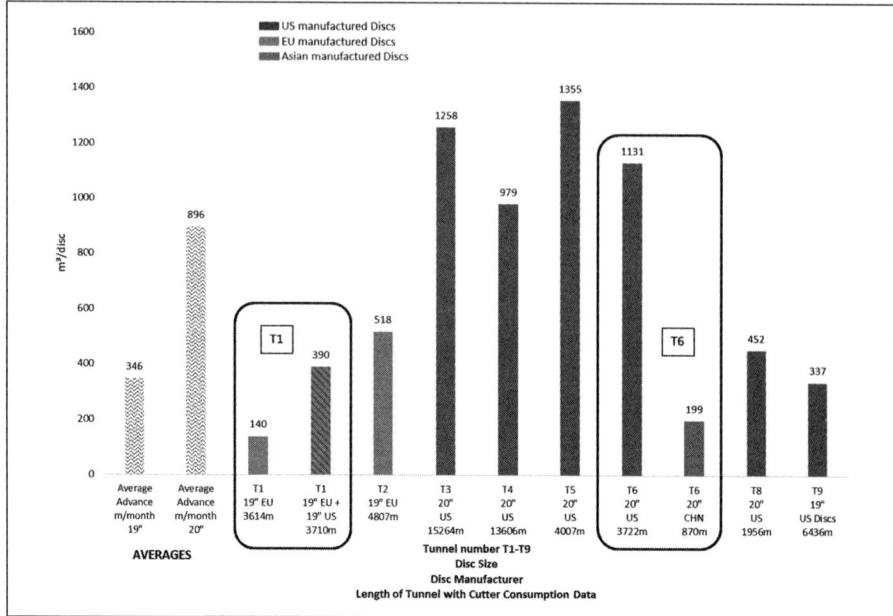

Figure 5. Cubic meters consumed per cutter disc

Cubic Meter Rates

Figure 5 is a summary of cubic meters per cutter disc consumed for all eight TBMs. Cubic meters per cutter disc, the best measure of cutter life and also a good indicator of total cutter cost, are also telling. The averages for all eight machines are shown in the first two columns of Figure 5. The average for 19 inch cutters is 346 cubic meters bored per cutter and for 20 inch cutters, the number is more than twice as high at 896 cubic meters bored per cutter.

DISCS MATTER

During the execution of this project there were two separate opportunities to compare the relative merits of different manufacturer's discs against each other; those manufactured in the US, vs. those manufactured in both Europe and Asia. This occurred on the T1 and T6 contracts.

Contract T1

The most striking difference compares the performance of the T1 TBM on its first drive using 19 inch cutters procured by the contractor from a European supplier with 19 inch cutters made in the US.

At T1, the contractor bored the first 3614m of the tunnel with cutters manufactured in the EU. For the final 3710m of the first drive, the contractor switched to a mixture of US sourced cutters and EU cutters. The US cutters were used in the transition area of the cutter profile. This is the area where the face transitions from being flat to turning out to cut the gage. It is well understood that the cutters in this area are the most highly loaded of all the cutters on the cutterhead.

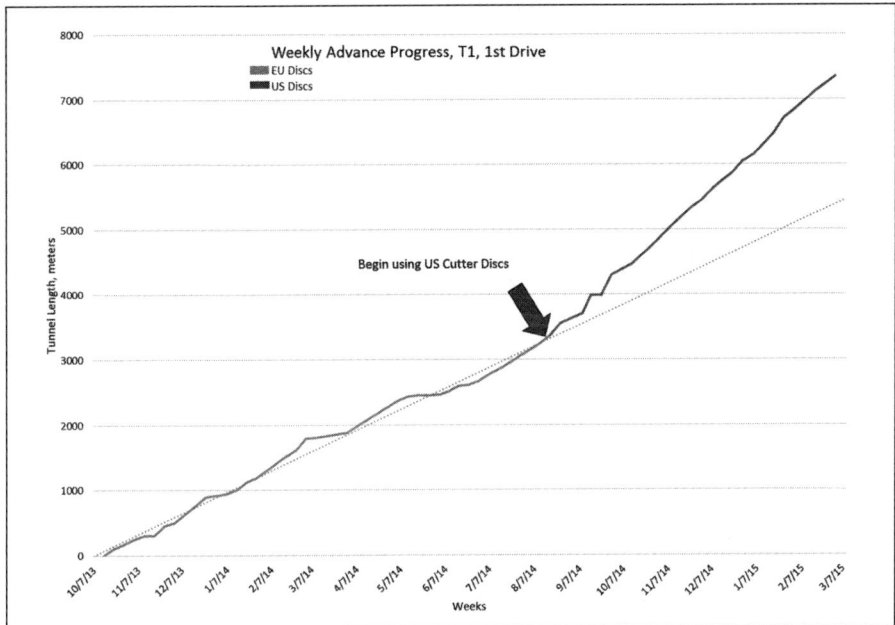

Figure 6. Relative advance rates over the length of the T1 drive

The performance was markedly better during the second half of the first drive with 543 discs consumed vs. 1479 discs consumed in the first half. Of the 543 discs, 225 were US discs and 318 were EU discs. This equates to a performance increase from 140m³/disc to 390m³/disc by deploying US discs in the transition area. Refer back to Figure 5 where the two bar graphs for T1 are enclosed with a black border and labeled T1. The EU discs are colored green and the mixed EU & US discs are shown in blue and green. In addition to an improved cubic meter/disc rate, the monthly advance rate improved significantly on T1 after the changeover. This is clearly illustrated in Figure 4 where the two bar graphs for T1 are enclosed with a black border and labeled T1 showing 322 meters/month for the first 3614m and 613m/month for the final 3710m of T1s first drive. Figure 6 shows the relative advance rates of both sections over the length of the first drive.

Contract T6

At T6, the contractor bored the first 3722m of the tunnel with cutters manufactured in the US. Then the contractor switched completely from the US cutters to Asian manufactured cutters. Data were only available for 870m after the changeover to Asian discs, but it was long enough to observe two phenomena.

First, referring back to Figure 4, there is no significant difference in monthly advance rates seen in the two bar graphs enclosed in black and labeled T6. With US cutters, the average was 555m/month and slightly higher at 563m/month with the Asian discs installed. The advance rates comparing both types of discs are also illustrated in Figure 7 where no significant difference in the slopes of the data are apparent.

The important difference between the US discs and the Asian discs is readily apparent in Figure 5 where the cubic meters per cutter disc consumed are compared. Refer the two bar graphs enclosed with a black border and labeled T6. During the first 3722m

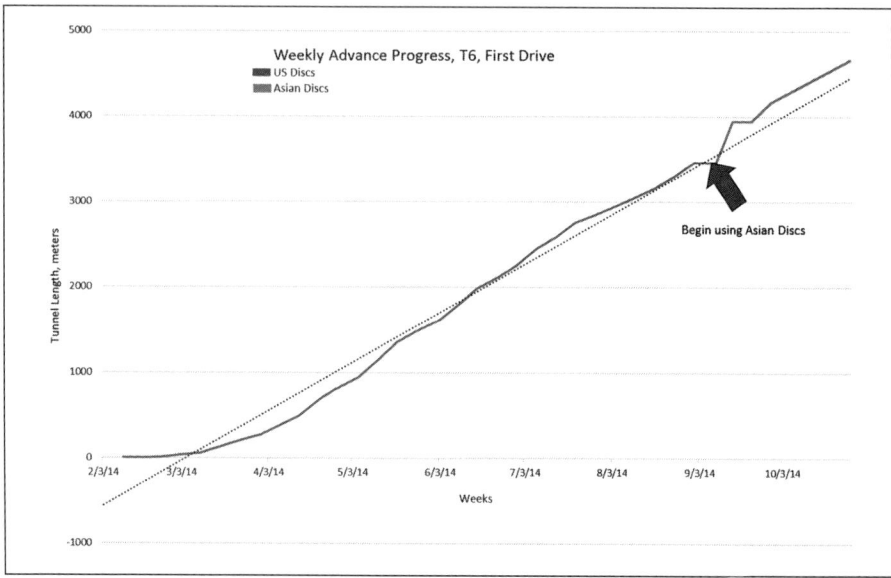

Figure 7. Advance rates on the T6 drive

of T6, US discs were used and the average rate of rock excavated per disc consumed was 1131m^3/disc. Compare that the performance of the Asian discs at 199m^3/disc. Excavation performance of the discs was reduced by a factor of 5.7. This represents a significant cost increase both in discs consumed (even when the discs are substantially cheaper) and in down time for cutter changes.

CONCLUSION

Contractors should carefully consider whether the use of larger discs will provide an economic benefit. The choice of 20 inch cutters over 19 inch cutters (coupled with an aggressive cutterhead management program) can provide longer time between cutter changes and longer overall life between rebuilds.

Contractors will also benefit from carefully considering the technology applied by each manufacturer to their disc cutters. Any competent manufacturer can make a disc cutter but the proof of the quality of the disc will not be apparent until the steel meets the rock. Nearly all disc cutter manufacturers now offer a tool steel disc ring and most have similar composition. It is, however, less the composition of the disc ring than it is the subsequent processing that makes the difference in performance. Less expensive disc cutters will not be economical in hard rock when considering the total cost over the duration of a project and this becomes more and more significant as the rock becomes ever more challenging.

Successful Excavation of Mexico City's Emisor Poniente II Wastewater Tunnel—Use of a Dual-Mode, Crossover TBM in Challenging Geology

Roberto Gonzalez ▪ Robbins Mexico
Martino Scialpi ▪ The Robbins Company

ABSTRACT

In July 2015, the launch of a dual mode, Crossover type TBM marked the start of Mexico City's next challenging wastewater project: the Túnel Emisor Poniente (TEP II). The 5.5 km long tunnel travels below a mountain at depths of 170 m as well as a section just 8 m below residential buildings, and the geology is equally varied. Ground consists of andesite and dacite with bands of tuff and fault zones, as well as a section of soft ground at the tunnel terminus.

This paper will detail the unique 8.7 m diameter Crossover TBM designed for the challenging conditions, and the successful excavation of the machine through fault zones, soft ground, and more. Strategies for excavation and advance rates, and downtimes will be analyzed. As the machine can be converted from hard rock mode to EPB mode in the tunnel, the authors will also look at the conversion process and how both modes worked to excavate in widely varying geological conditions.

INTRODUCTION AND HISTORY

In the last 100 years, Mexico City has sunk by nearly 12 m. As a result, the city's buildings, main streets, sewage systems, etc. have been extensively damaged. In addition, the city historically faces serious problems of flooding during the raining season. In 2006 there was a high risk that major floods might occur in the city and suburbs, affecting a population of 4 million—six districts within the Federal District and three municipalities of the State of Mexico—by flooding an area of 217 square km. The areas with greatest risk of flooding are the historic downtown area and the Mexico City Airport and surrounding areas.

In 2007, Mexican President Felipe Calderon labeled this situation a "National Emergency" and designated it as a top priority of the National Infrastructure Program.

Two main actions were proposed:

1. Repair, maintenance and recovery of the slope of the Túnel Emisor Central, the main sewage system of the city.
2. The construction of the Túnel Emisor Oriente.
3. The construction of the Túnel Emisor Poniente II.

Background

The history of Mexico City is inextricably linked to the issue of its geographic location. The Metropolitan Area of the Valley of Mexico is built on a closed basin, which originally formed a lake system consisting of five large lakes: Texcoco, Xaltocan, Zumpango, Xochimilco and Chalco. Tenochtitlan, the ancient capital of the Mexica civilization, covered an estimated 8 to 13.5 square km, situated on the western side of the shallow Lake Texcoco.

The city was connected to the mainland by causeways leading north, south, and west of the city. These causeways were interrupted by bridges that allowed canoes and other traffic to pass freely. The bridges could be pulled away if necessary to defend the city. The city was interlaced with a series of canals, so that all sections of the city could be visited either on foot or via canoe.

After the Conquest, the Spanish rebuilt and renamed the city. The valley contained five original lakes called Lake Zumpango, Lake Xaltoca, Lake Xochimilco, Lake Chalco, and the largest, Texcoco, covered about 1,500 square kilometers of the valley floor. However, as the Spaniards expanded Mexico City, they began to drain the lake waters to "control flooding."

In the rainy season, these lakes were converted into just one lake of two thousand square kilometers. This condition explains the periodic floods that since the founding of Tenochtitlan inhabitants have faced and the resulting need to build major drainage works to control and evacuate wastewater and rainwater.

The idea of open drainage canals first came about after a flood of the colonial city in 1555. The first canal, known as Nochistongo, was built in 1605 to drain the waters of Lake Zumpango north through Huehuetoca, which would also divert waters from the Cuautitlán River away from the lakes and toward the Tula River. Another canal, which would be dubbed the "Grand Canal" was built parallel to the Nochistongo, ending in Tequixquiac. The Grand Canal consists of one main canal, which measures 6.5 meters in diameter and 50 km long, and three secondary canals, built between 1856 and 1867. The canal was completed officially in 1894 although work continued thereafter. Despite the Grand Canal's drainage capacity, it did not solve the problem of flooding in the city. From the beginning of the 20th century, Mexico City began to sink rapidly and large pumping facilities needed to be installed in the Grand Canal, which before had drained the valley purely with gravity. Currently, and despite its age, the Grand Canal can still carry 150 m^3/s out of the valley, but this is significantly less than what it could carry as late as 1975 because continued sinking of the city (by as much as seven meters) weakens the system of water collectors and pumps and is altering the canal's slope.

As a result of the decreased capacity, another tunnel, called the Emisor Central, was built to carry wastewater in the 1970s. Although it was considered the most important drainage tunnel in the country, it has been damaged by overwork and corrosion of its 6 m diameter walls. Because of the lack of maintenance, there has been a gradual decrease in this tunnel's ability to carry water.

The System Today

Today the capacity of the drainage system in the metropolitan area is insufficient and has serious problems. Just comparing the capacity it had in 1975 with what it has now shows a significant decrease in efficiency: a 30% lower ability to convey wastewater with nearly twice the population. This decrease is mainly due to steady sinking of the City of Mexico, caused by overexploitation of aquifers of the Mexican valley.

The Emisor Central, designed to transfer rainwater in storm peaks, has operated for 15 years past its design capacity, and has been in continuous use without maintenance. In addition the sewer is transferring untreated or "black" water, and this has caused accelerated wear in the upper section of the tunnel. Although the Emisor Central is the tunnel upon which the security of the eviction of wastewater and storm water of the valley falls, it must close during the dry season months for repair and maintenance each year. This raises the urgent need for an alternative tunnel with the ability to maintain system operation throughout the year.

In order to solve the problem of the drainage system it was deemed necessary to build a new deep tunnel system: the Túnel Emisor Oriente, 62 kilometers and seven meters finished diameter, and the Túnel Emisor Poniente II, a 5.5 km long tunnel with 7 m finished diameter (see Figure 1).

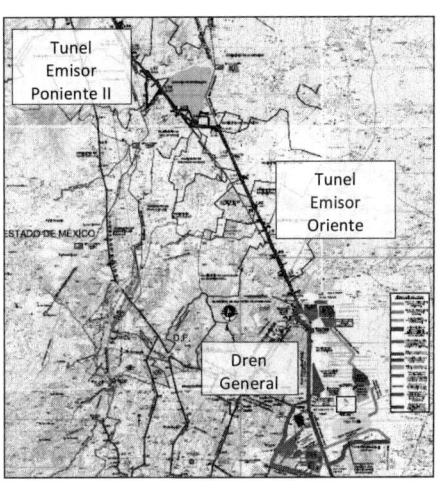

Figure 1. Map of the future wastewater tunnel network

TÚNEL EMISOR PONIENTE II (TEP II)

The Emisor Poniente II tunnel has three main purposes:

1. Expand the capacity of drainage, which will reduce the risk of flooding in the west side of the city.
2. Reduce the overexploitation of aquifers, which exacerbates the sinking of the northwest area.
3. Water treatment of the wastewater to promote its reuse in agriculture, instead of using sewage water for agriculture.

While the Emisor Oriente tunnel (TEO) has been written about in previous papers, we will mention it here with regards to its role in the overall system. TEO is currently being built by the federal government, with a trust between the Government of the State of Mexico, Mexico and Hidalgo, with an initial investment for construction of 9,600 million pesos. The tunnel starts at port interceptor tunnel No. 2, the "River of the Remedies" and ends in the town of Atotonilco in Hidalgo (output Portal), where the area's first wastewater treatment plant is under construction. It passes through the municipalities of Ecatepec de Morelos, Atenco, Tonatitla, Nextlalpan, Jaltenco Zumpango, Huehuetoca Atotonilco Tequixquiac and Hidalgo. It will have a capacity of 112 m³/sec of wastewater. Currently the drainage system of the valley of Mexico has a displacement capacity of 195 m³/sec, but with the commissioning of the TEO and TEP II, it will have a total of 345 m³/sec.

The TEP II will connect in an open channel to the Emisor Poniente I, and the combined flow will be transferred to the same treatment plant in the Municipality of Atotonilco, in the state of Hidalgo. The plant will be responsible for water reuse for agricultural irrigation, and will be the second largest plant of its kind in the world.

In order to excavate TEP II, a consortium of Aldesem, Proacon, and RECSA was chosen by the national water and irrigation management authority of the Mexican Government (CONAGUA). The consortium sought an alternative to a typical tunnel boring machine due to complex geological conditions, and chose an 8.7 m diameter dual mode, Crossover-type TBM manufactured by The Robbins Company. The machine is capable of excavating in both a pressurized EPB mode and a non-pressurized, hard rock mode.

Project Challenges

Geological Conditions

The ground conditions of the pipeline are mainly andesite rock with a compressive strength of 1500–2,500 km/cm^2; however, the tunnel also passes through two faulted areas with tuff and sand in contact zones, and the last 900 meters are in softer ground consisting of tuff and alluvial lake clays.

In order to deal with the abrasive, hard rock conditions, the TBM was designed to utilize 20-inch diameter disc cutters—the first time these large cutters have been used in Mexico. The machine was planned to be launched in non-pressurized hard rock mode and converted to EPB mode in the last 900 m—a section that also happens to be in a residential area with low cover just 1.5 times the machine diameter. The TBM bore also faced some local stigma and a lack of experienced personnel for the conditions—the last time that a TBM was able to bore through rock efficiently in Mexico was a 4.5 m diameter Robbins Double Shield that worked on a clean water tunnel back in 1998. This machine would need to be able to bore both hard rock and soft ground in an efficient manner (see Figure 2).

Limited Site Space

Due to existing infrastructure, housing, and a nearby water tank the launch site for TEP II was fairly small, which posed a huge challenge for logistics. The entire jobsite had to fit into a space of less than 10,000 m^2. The machine was assembled using Onsite First Time Assembly (OFTA), a method that involves initial assembly of TBM components on location, and can save both time and money that is then passed on to the contractor.

Major parts were refurbished and customized for majority hard rock conditions, as the machine had most recently bored a tunnel in softer rock in Laos. Many components were new including the shields, the cutterhead, main drive gearboxes, bull gear, and main bearing. Sub-Systems were factory-tested and shipped to the jobsite for initial assembly, where there was limited area for staging. Components had to be lowered into an 11 m wide pit. The machine was over 100 m long, but the starting chamber was only 50 m in length and required assembly of the back-up system in two stages.

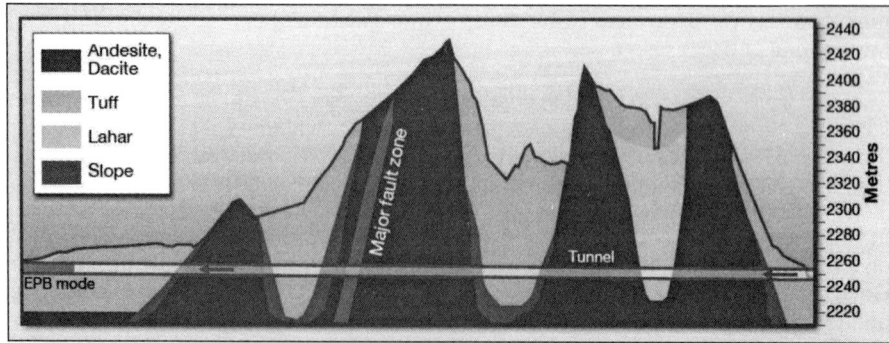

Figure 2. Project geology and corresponding machine mode

Figure 3. Cutterhead installation in a small launch shaft

Overall, OFTA took about 12 weeks—a process estimated to have saved at least 60 days as compared to a similarly-sized machine assembled in a shop (see Figure 3).

TBM DESIGN CONSIDERATIONS

Adaptable Cutterheads

The custom-designed Crossover TBM was engineered with a robust, back-loading cutterhead to tackle variable conditions. High pressure, tungsten carbide knife bits can be interchanged with 20-inch diameter carbide disc cutters depending on the ground conditions. Specialized wear detection bits lose pressure at specified wear points to notify crews of a needed cutting tool change.

The opening ratio in the cutterhead can be changed by removing bolted plates for the larger opening EPB configuration.

Twenty-five injection ports spaced around the periphery of the machine are used for injection of various additives depending on ground conditions, and for probe drilling. Additives such as Bentonite are used to condition the muck for removal by belt conveyor.

Drilling Equipment

The TBM is equipped with two type of drills. A canopy drill is able to install pipes and form the umbrella, to be able support the ground above the machine in particular. This procedure is used on the contact areas or where the face is not self-supporting, mainly when boring through tuff and sand.

The second drill is a probe drill, for ground investigation in front of the TBM, and grouting ahead of the TBM in areas where the geology could be too fractured or non-self-supporting.

Continuous Conveyors for Limited Space

Muck from the TBM is deposited from the screw or machine conveyor to a fabric belt conveyor mounted on the trailing gear, which transfers to a Robbins side-mounted continuous conveyor. The continuous conveyor carries the muck to the launch shaft. The slope for the horizontal belt of 11.7% eliminated the need for a vertical conveyor. Once at the surface, a radial stacker deposits muck in a kidney-shaped pile for temporary storage.

Two-Stage Main Drive Reducers

The TBM torque is able to be increased when needed to excavate through soil and fault zones using two-stage gear reducers. In rock mode, torque can go up to 5,200 kNm and the cutterhead is able to go as fast as 6 rpm. In EPB mode the torque is able to go up to 9,917 kNm at 3.6 rpm, with a breakout torque of 14,875 kNm. The change is able to be made during tunneling with only the turn of a lever on each reducer.

Active Articulation

The articulation system was included in the design for EPB mode. Active articulation engages articulation cylinders between the front and rear shields to steer the machine independently of the thrust cylinders.

Bulkhead Closure Gates

The TBM has the capacity to close the gates in the cutting chamber, in order to avoid the entrance of running ground and water to the TBM for emergency, during the excavation in rock mode.

Special Antiroll System

The skew system will be used while boring in rock mode. This system allows the TBM to correct the roll by applying a perpendicular force to the thrust cylinders and thus correcting the roll of the TBM.

Two-Part Grout System

This is the second time that the two component grout system has been utilized in Mexico City, and the system has proven itself to work in Mexico's difficult geology. A two-part A + B grouting system (grout plus aooclerant) hardens quickly and fills the annular gap while minimizing settlement.

Conversion Between Modes

The TBM started off tunneling in hard rock mode. Conversion to EPB mode requires several steps to take place. Removable plates must be installed in the cutterhead first, in order to create a larger opening ratio, from 6% to 25%. Disc cutters are then changed out for knife-edge bits and scrapers, two types of soft ground tooling. Next a rotary union is installed to inject additives and foam in front of the cutterhead in order to make a good mixture of the material. Once the opening ratio has been changed to be more like an EPB, the belt conveyor must also be removed and the screw conveyor installed to take material from the bottom of the mixing chamber. With this diameter of machine both the screw and belt cannot be installed at once. The process can be done in about eight weeks (see Figures 4–5).

LAUNCH AND EXCAVATION

The Crossover machine was launched in August 2015 in a hard rock configuration and mounted with 20-inch diameter disc cutters—a risky move given that the first sections of tunnel were in softer soils before the TBM hit more solid rock. There was some worry that this could clog the cutterhead as it might be sticky, but this thankfully did not happen. In fact, the machine's advance rates picked up quickly, with project records set in December, and again in January after the machine achieved a best day of 42.8 m and a best week of 185.1 m.

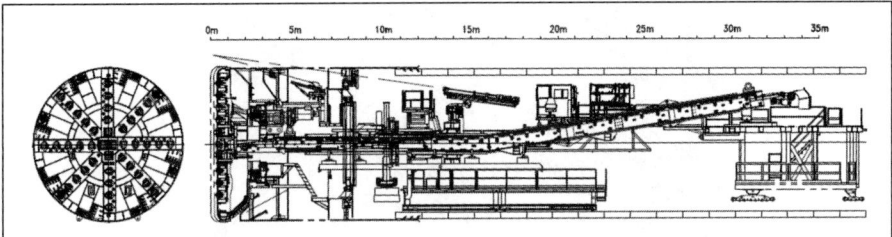

Figure 4. TBM in rock mode

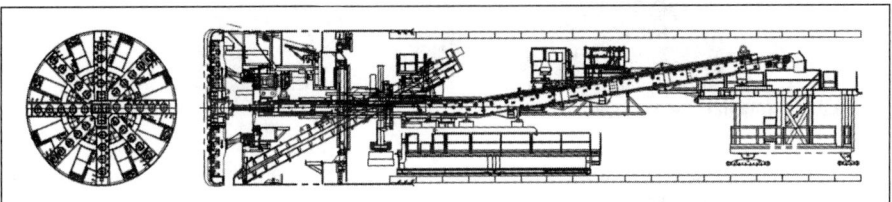

Figure 5. TBM in EPB mode

Early in 2016 the TBM hit the first of several contact zones, a 30 m wide fault of frac-tured and blocky rock. While the excavation through the contact zone was slow going, progress picked up again in the more competent andesite rock. After an intermediate breakthrough in March 2016 into an 80 m deep shaft followed by inspection and main-tenance, the TBM continued on (see Figure 6).

By June 2016, the TBM was flying through fairly competent rock and had achieved two national records for TBM advance—57 m in one day and 702.2 m in one month. Continuous conveyors certainly contributed to the result, as the system at TEP II can remove 600 to 900 tons of material per hour, and has been able to match the speed of the TBM.

The next challenge for the machine was the major fault zone known as Falla Norte de Barrientos. More water was encountered than expected—about three liters per second--as well as soil that required cleaning out of the tunnel invert. The TBM was in rock mode so there was some spillage from the conveyor and a lot of cleanup. But, they were able to get through the zone in about 30 segments' length and advance rates went up again. Once out of the fault zone the ground was primarily fractured andesite.

CURRENT CHALLENGES

The core samples for the last 900 meter long section in soft ground weren't taken from the precise axis of the tunnel, due to houses and private property in the way. A large cavern of about 50 m^3 was encountered at only 14 meters' coverage in Autumn 2016, with high risk of ground settlement and damage to the houses above. The excavation was immediately stopped to analyze the situation; the cavern was filled with special Polyurethane Epoxies, then pea gravel and grout, to consolidate the empty space.

The conversion has also been a learning process. There was no special shafts con-structed to be able to change the TBM from boring in Rock Mode to EPB. Several modifications had to be done to the cutterhead and bulkhead, and the ground at the face had to be stabilized for the workers to enter the cutting chamber and work on the

Figure 6. Intermediate breakthrough in March 2016

conversion. Due to space the screw conveyor and TBM belt conveyor weren't able to be installed at the same time. It was very challenging to be able to assemble a large component such as the screw conveyor inside the TBM.

Once the machine is through its last section of soft ground, disassembly will present a further challenge. The narrow exit site is flanked by equally narrow roads and nearby houses. The machine cannot be backed through the tunnel and must be disassembled and removed from the exit site using a 200 metric ton capacity gantry crane.

CONCLUSIONS

Crossover TBMs will start to be applied more and more in tunnels all over the world, as tunnels are trending deeper and more complex. TBMs must be able to adapt to bore in different and changing conditions, but an accurate geological study is needed to maximize effectiveness. Geological studies can detect features like fault zones and caverns ahead of time so that mitigation strategies can be planned from the start.

The machine at TEP II was, nevertheless, able to break country-wide records and was by far the most efficient way to excavate a tunnel. Good tunnel management and logistics reduce downtime, while a cooperative manner by all parties involved has enabled focusing on tunnel goals.

REFERENCES

Asociación Mexicana de Ingeniería y Obras Subterráneas. Trascendencia futura, estado de arte y desarrollo histórico de las obras subterráneas en Mexico.

Gonzalez Obregón, Luis. Memoria Historica, Tecnica y Administrativa del desgue del Valle de Mexico 1449/1900. Oficina Impresora de Estampillas. Mexico, 1902.

High Cover TBM Tunneling in the Andes Mountains— A Comparative Study of Two Challenging Tunnel Projects in Chile

Carlos Lang ▪ The Robbins Company
Mark Belli ▪ The Robbins Company
Pablo Salazar ▪ The Robbins Company

ABSTRACT

The Andes Mountain range is among the youngest and most complex in the world, geologically speaking. Tunneling projects, particularly for hydroelectric and water transfer schemes, are not new to the range but their past history has met with mixed success. Two new projects utilizing very different tunnel boring machines and excavation strategies are now providing a testing ground for modern underground construction equipment in the Chilean Andes.

This paper will analyze two projects: the Alto Maipo and Los Condores Hydroelectric Projects, located approximately 100 km apart in the Andes Mountains. The two strategies being employed will be analyzed in detail, as one project is using an open-type Main Beam TBM plus extensive ground support, while the other is utilizing a Double Shield TBM and segmental lining. The authors will look at TBM performance and ground conditions encountered in the two tunnels and what effects the TBM selection and ground support strategy may have had on each tunneling operation.

ALTO MAIPO HYDROELECTRIC PROJECT

The Alto Maipo Hydroelectric Project for AES GENER and Luksic group is intended to channel the water of the main river that supplies water to the Santiago, the capital city of Chile. The project is located 70 km southeast of Santiago de Chile, in the municipal district of San Jose de Alto Maipo in the Metropolitan Area of Santiago at an altitude of 1025 to 2550 m.

The Inter-American Development Bank (IDB) signed a USD $195 million loan agreement with Alto Maipo SPa, owned by AES Gener (60 percent) and Antofagasta Minerals (40 percent), as part of a financing package of more than $1.2 billion for the Alto Maipo Hydroelectric Project. The project is co-financed by the International Finance Corporation, the U.S. Overseas Private Investment Corporation and six commercial banks. The total project cost will be roughly $2 billion, 60 percent in the form of debt and the remaining 40 percent in equity.

In Santiago there are only two main rivers: one of them, the Mapocho River, is 80% fed by Maipo river waters. The project aims to channel the water of the main tributary streams of the Maipo River into a 70 km long tunnel. The project includes, among other things, five intakes; 67 km of tunnels excavated both by Drill & Blast and TBM with 800–1000 m of average overburden; two caverns—Alfalfal II and Las Lajas—with a combined capacity of 531 MW; and around 17 km of high voltage transmission lines.

The contract was awarded to two different joint ventures: Constructura Nuevo Maipo (CNM) and Strabag-Voigh Hydro. Constructora Nuevo Maipo S.p.A (CNM), a Joint

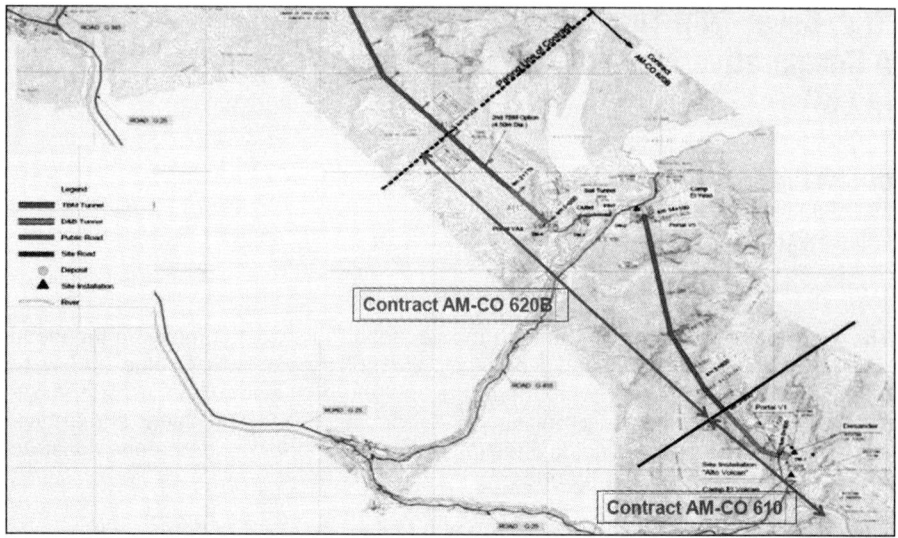

Figure 1. Map of project contracts

Venture between the companies HTCC-Hochtief Construccion Chilena Ltda—an affiliate company of the German firm Hochtief Solutions A.G.—and the Italian CMC-Cooperativa Muratori & Cementisti di Ravenna (CMC), is responsible for the execution of the works under the Contract AM-CO 610/620 at a cost of over USD $280 million.

Figure 1 shows the portion of the project awarded to CNM including the following scope of works.

AM-CO610 Scope:

- Alto Volcan System
- El Volcan Tunnel System (upstream portion of El Volcan tunnel)

AM-CO620B Scope:

- El Volcan tunnel system (downstream portion of El Volcan tunnel)
- El Yeso Headrace System
- Alfalfal II System (Alfalfal II Headrace Tunnel, upstream section)

The project schedule foresaw the employment of one open-type hard rock TBM for boring part of the El Volcán Tunnel and part of the Alfalfal II Tunnel.

After being assembled at the portal next to the site installation called "El Yeso," the plan was to have the TBM bore one 7 km section of the El Volcán Tunnel and, as an option, another 1,500 m beyond that point.

At the end of boring the El Volcán Tunnel, the TBM would be disassembled and transported through the bored and lined tunnel back to the portal "El Yeso," and from there the machine would be moved to the portal of the Alfalfal II Tunnel.

The first section of the Alfalfal II Tunnel from station km 0 + 000 to km 3 + 250 was planned to be excavated by conventional methods—drill & blast. The second section from station km 3 + 250 to km 6 + 250 would be bored with the above TBM from

El Volcán Tunnel. The machine would be moved along the conventional bored tunnel section up to the underground face.

Disassembly and removal of the machine from the tunnel at the end of boring was planned to commence through the bored and lined tunnel back to the portal.

In January 2015 Robbins supplied a 4.13 m open gripper type hard rock TBM with back-up System and a Robbins Continuous Conveyor System for muck transport along the bored tunnel to the portal area (see Figure 2).

The TBM with the back-up system and tunnel conveyor system were meant for boring El Volcán headrace tunnel and part of the Alfalfa II Tunnel, allowing quick

Figure 2. Robbins first open gripper type Main Beam TBM inside El Volcán Tunnel

access directly behind the cutterhead for the installation of extensive rock support measures such as rock bolts, steel mesh, ring beams and shotcrete.

Main Technical Features of the TBM for Rock Support

The open type TBM supplied for excavation in hard rock was furnished with a hard rock cutterhead dressed with 17" Wedgelock™ back-loading cutters with 267 kN nominal load capacity each. For immediate support of the bored tunnel tube closest to the face, support shields on the roof and at the sides were provided as well as a vertical front support. Finger shields were attached to the roof support shield.

For primary rock support in the so-called L1 section, extending from the rear end of the support shields to the level of the TBM gripper assembly, two roof drills were provided as well as one probe drill, a ring beam erector, work platforms and stands and shotcrete application (see Figure 3).

The two (2) roof drills were assembled on a ring shaped support structure to allow for rock bolting service at the tunnel wall and at 106° on either side from the top crown position.

The probe drill was initially designed with a lookout angle of 5 degrees to the tunnel wall to allow for a conical pattern of drilled holes just outside the TBM bore excavation, so that the perimeter of the bored tunnel could be pre-grouted if necessary. The drill rig was attached to a ring shaped support carriage to allow for service around the circumference of the bored tunnel face and in dedicated spots of the face. The ring structure comprised a crown mounted segment which could rotate through 360 degrees. It was conceived to have one rig assembled at all times and to add another rig in case of need, which would be stored outside of the tunnel.

Under the protection of the support shields the ring beam erector allowed for installation of a full circle ring beam set or beam segments. A full circle ring beam was meant be pre-assembled in the erector and horizontally displaced for installation at the tunnel wall. Work platforms allowed the ground support to be installed under the personal protection of the roof shield finger support system.

As shown in Figure 4 and in lieu of the above traditional "finger shield," the roof shield was designed to incorporate the Robbins-McNally "slat type" ground support system. The roof shield featured 35 roof "fingers" to protect the workers from rock falling in the TBM area. Each finger was installed in a pocket made of a structural angle bar and could be totally or partially removed when McNally rock supports were needed. The free pockets could then be used to install slats or rebar as needed for the McNally system.

One (1) additional roof drill and a shotcrete robot were supplied for secondary rock support in the so-called L2 working section, which follows the above L1 working section. The additional roof drill was assembled on a ring type positioner and allowed for roof bolting along the circumference of the bored tunnel wall except for the section covered by the invert segment.

With the robot operated via a remote control, two successive 150 mm thick layers of shotcrete could be applied for a total thickness of 300 mm along the circumference of the bored tunnel, except for the section covered by the invert segment.

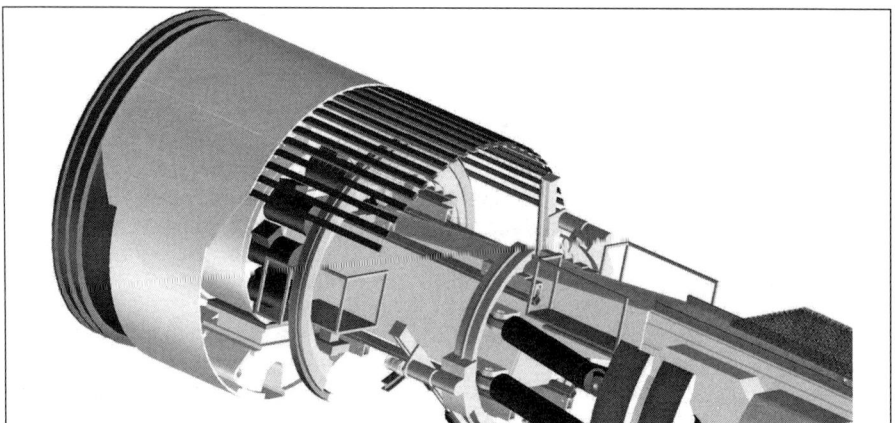

Figure 3. L1–working section ground support

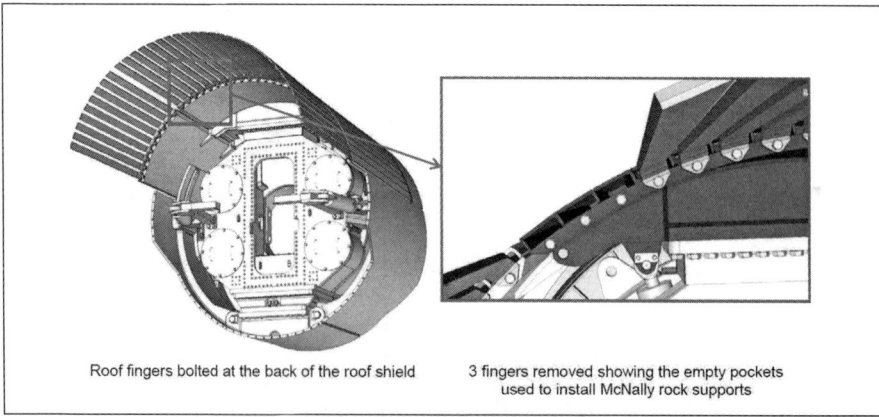

Roof fingers bolted at the back of the roof shield   3 fingers removed showing the empty pockets
used to install McNally rock supports

Figure 4. Finger shield removal and McNally System installation

Upon request by the project owner, a second open gripper type Main Beam TBM is being delivered at the time of preparation of this paper.

Geology

The project was to be bored under a relative high overburden of up to 1,500 m at El Volcan tunnel in variable geology of the so-called "Abanico Formation," ranging from competent andesite to andesitic breccias with some sandstone and tuff sections and some fault zones. In general, these rocks showed a good or fair geotechnical rock quality; however, in some highly altered tuffaceous sections the geotechnical rock quality was poor.

In December 2014, excavation of the El Volcan Tunnel started with conventional drill & blast methods from the El Yeso portal. The excavation of the tunnel started in soil conditions (colluvium and sliding deposits) for a stretch of 30.55 m, to continue with 130.23 m of drill and blast excavation up to chainage 13+968.40 where the TBM would boring.

The Robbins TBM-1 started to bore on 15th June 2015, and during the first 15 days the TBM advanced 106.33 m along the tunnel. Rock mass presented good to fair quality with RMR values between 50 and 58 in the first meters of excavation with the TBM. Rock mass was composed of andesitic breccias with signs of alteration. The crown area was more fractured and it was observed that there were sub-horizontal joints that generated wedges and planar roofs.

During tunnel boring the most commonly anticipated support type was *rock support type I*, consisting of welded wire mesh and resin end-anchored rock bolts 2.4 m long and 22 mm in diameter, positioned to ±60 ° and ±25 ° at 2 m intervals. Less commonly used would be *rock support type II*, consisting of 7 cm of shotcrete, welded wire mesh and resin end-anchored rock bolts positioned to ±60 ° and ±25 ° at 2 m intervals. *Rock support type III* would consist of resin end-anchored rock bolts 2.4 to 3.9 m long and 22 mm in diameter, positioned to ±60 ° and ±25 ° as well as spot bolts at 0° every 1 m, along with welded wire mesh and 10 cm of plain shotcrete distributed through 165° at the crown. Figures 5 and 6 show the rock mass in a stretch with rock class II with sub-horizontal joints and RMR=51.

Figures 7 and 8 show the support increased to type III when rock mass became worse (RMR approx. 40), due to two sub-horizontal joint systems at the roof that generated significant over-break.

Figure 5. Planar roof and sub-horizontal joints   Figure 6. RMR=51

Figure 7. Sub-horizontal joints, RMR=42

Figure 8. 15 July 2015, support type III

Figure 9. TBM excavation, overbreak with RMR = 32

Figure 10. TBM excavation, overbreak with RMR = 32

Figure 11. Ch. 13+862.31 and Ch. 13+861.50, support type IV

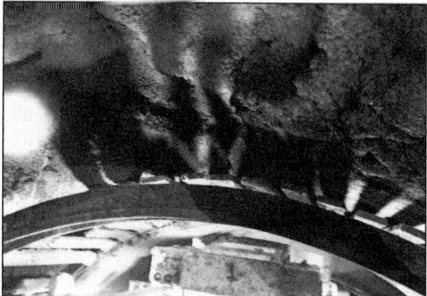

Figure 12. Ch. 13+862.31 and Ch. 13+861.50, support type IV

Figures 9 and 10 show intense scaling after each TBM stroke when the rock mass continued getting worse, with more fractures detected and RMR ranging from 32 to 34. Three joints/faults systems were detected—two sub-horizontals and one vertical in the crown—that caused continuous over-break.

At a certain point, *support type IV* was required (Figures 11 and 12), which required yielding steel set installation due to the presence of a sub-horizontal fault/joint (dipping slightly upwards).

On 31st October 2015, the excavation works restarted with RMR ranging from 55 (type II) to 70 (type I), and with ground composed of andesitic breccia and sub-horizontal faults and joints that generated wedges and planar roofs.

On 8th December 2015 a 3 m long material detachment between the L1 and L2 area of the TBM was detected and loose material fell into the mesh, making it necessary to strengthen the support. The use of McNally elements as steel straps placed between rock bolts installed transversally and longitudinally around the affected area was instructed by the geologists, despite this support type not being part of the approved initial support design (see Figure 13).

Figure 14 shows the right wall and crown highly fractured with support type III installed.

McNally slats were installed in the top 150° with steel arches as well as 5 cm of plain shotcrete distributed in 150° at the crown as a result of the rock mass quality steadily worsening with highly fractured rock and a lot of material falling during excavation.

It is important to highlight that previous to this happening, scaling and use of support was what was used to control the over-break in the crown. However, this called for manual scaling and manual mucking of the unstable blocks and loose material, which ended up causing repeated damage to the TBM equipment and creating a potentially hazardous situation for the workers. The McNally System (specially designed for rock bursting conditions) combined with light arches solved both of these problems.

Exploration drillings were also carried out to detect the presence of water. A number of 30 m long consolidation grout injections up to 50 bar pressure were instructed by the owner. Even with these measures the rock conditions continued to be very poor with

Figure 13. 8th December 2015, McNally System as additional support due to material detachment

Figure 14. Advance 14th December 2015, Ch13+293

fractured rock, clay infillings and water inflows occurring due to the failure of the grouting works.

The sub-horizontal fault that had been initially detected continued, and faulted material got worse with crushed rock and significant clayey and weathered material. The presence of water washed out the clay, causing more over-break, first at the crown and over the machine, then later at the sides. These conditions required the installation of McNally systems and yielding steel arches in order to hold the loose rock again in place.

Figure 15. Initial support including McNally slats and water inflow conditions

Due to the excess load (dead weight of loose rock) over the steel sets, the stretch was further reinforced with steel sets fixed with self-drilling bolts at the crown and resin end-anchored bolts at the sides and floor. During the drilling of the bolts it was detected that the fault presented a thickness of approximately 5.0 m, considering that rock appeared 1 m over the invert and 2 m from the crown.

In order to control the area, chemical grouting was required to fill the cavities produced by the combination of clay infillings and water inflow. This chemical grouting was carried out during the 14th to 24th of January 2016 (see Figures 15 and 16).

Figure 16. Initial support including McNally slats and water inflow conditions

In a second step, a strong framework made up of mesh and shotcrete was prepared to consolidate and seal the area. Additionally, extra support was installed close to the steel set profiles as probe drilling was prescribed and this required partial cuts of some steel sets to properly execute the test.

Conclusions for Alto Maipo

During the excavation of the El Volcan tunnel from the El Yeso portal, the presence of sub-horizontal joints and faults was continuously detected, which to a large extent affected and determined the excavation rates and the rock support requirements. These sub-horizontal structures along the excavation through the Abanico Formation had never been outlined in the Geological Baseline Report.

Standard ground support measures were applied to handle these sub-horizontal joints in the initial drill and blast excavation, but this became a serious problem for a small diameter (4,13 m) open-type TBM that in principle had been selected for good rock conditions. Under difficult ground conditions, the TBM operation demanded continuous manual scaling and manual mucking in order to remove loose material caused by wedges formed from these sub-horizontal joints in the tunnel crown.

The situation became even worse when these sub-horizontal joints turned into a sub-horizontal fault (shear fault) with an estimated thickness of 5 m, combined with high water inflow. This was managed with the application of all types of grouting injections.

The open-type main beam TBM provided by Robbins allowed quick access directly behind the cutterhead for the installation of rock support with rock bolts, steel mesh, ring beams, shotcrete and the McNally System, making this machine the most reliable and suitable choice for competent to slightly fractured rock. When the percentage of very poor rock mass quality is fairly small the machine is the superior method, as slow progress in bad ground can be compensated by the high excavation rates potentially reached under fair and good ground conditions. It may also be the preferable choice in heavily squeezing ground due to the short length of the machine front section (shielded area of about 5 m), since the ground may not deform fast enough to settle on the machine.

However, the TBM may not be the best choice when boring under extensive bad ground conditions with sub-horizontal joints and faults.

As a result of the lessons learned during the excavation under the above mentioned difficult ground conditions some improvements were implemented on the TBM:

- An extension of the side supports and the vertical front support to better face extreme squeezing ground conditions was carried out by means of the attachment of 3 × A572 GR50 steel plates (full penetration weld) to the existing side and vertical front support which extended the overall dimensions by 606 mm along each side and by 1264 mm for the vertical front support. The roof support length remained the same.
- A new probe drill setup was installed, aiming for a better lookout angle of 3 degrees to the roof. This included modification of the roof, side and vertical front supports by installing probe drill tubes and installation of a moveable drill ring that could travel 1500 mm and through two positions: a "working position" (only when probe drilling) at the aft end, and the "parking position" at the rear end, when no probe drill works were being executed. Roof drill travel was not affected, remaining with the original 2000 mm stroke.
- A mechanical system for removal of large blocks in the invert area was installed, including a handling sled with 0,5 m^3 capacity. The system was meant to be pulled under the bridge area by a winch system similar to the one used for pulling the disc cutters or steel ring.
- A newly designed dewatering system was installed to cope with 25 lps industrial water (dirty water) from the invert area and 100 lps infiltration water (clean water).
- New storage areas on the TBM backup system were installed for basic rock support materials such as steel sets, cement bags, rock bolts and others.
- A new, more user-friendly and versatile shotcrete robot setup was installed to reduce the shotcrete rebound.

As a result of these new features and modifications four (4) additional decks were added to the current backup system, including all associated changes and additions to conveying frames, piping, ventilation ducts and lines, cabling and hosing.

LOS CONDORES HYDROELECTRIC POWER PLANT

"Clementina" is the name given by Endesa Chile to the 4.56 m (15.0 ft) diameter Double Shield TBM built specifically by The Robbins Company for the 150 MW hydroelectric power plant of Los Cóndores, which involves the excavation of a 12 km (7.5 mi) intake tunnel for a design flow rate of 25 m³/s.

The hydropower plant is located in the upper basin of the Maule River, 360 km south of Santiago, Chile, where water flow regulation will be made by means of the existing water reservoir of the Laguna El Maule. The project expects an average annual energy output of 642 GWh (48% load factor) and requires a total investment of US$662 million. Operations are planned to start at the end of 2018.

The project key features are the water-intake facilities located at the reservoir Laguna del Maule at 2,500 m altitude, a 2×220 kV high voltage transmission line running for 87 km to connect the power plant to the substation facility, the 12 km long headrace tunnel, a 470 m-long vertical pressure shaft lined in concrete, and the turbine cavern.

The contractor responsible for the works, Ferrovial Agroman, met Robbins representatives in Solon, Ohio, USA in August 2015 to complete Factory Acceptance Testing of the TBM for the job. The Double Shield TBM was shipped to site in September 2015 to bore two sections of the tunnel lined with 4+1 concrete segments of 1,200 mm average length and OD 4,200 mm / ID 3,700 mm. The first tunnel section measures 6.0 km (3.7 mi) and the second 4.4 km (2.7 mi).

TBM site assembly started in mid-October 2015 onto a steel cradle provided by Ferrovial at "Plataforma Lo Aguirre," a common laydown area for muck discharge and logistics for the two tunnels. Two months later, on January 2015, the TBM assembled onto the steel cradle was lifted and loaded onto a MAMMOET 27 m long × 3 m wide heavy lift 16-axle hydraulic modular motorized trailer specially designed for transport operations with heavy loads requiring a high bending moment (see Figure 17).

In a second step the entire assembly was transported through the 1,0 km long pre-excavated Drill & Blasting (D&B) access tunnel up to the launching chamber. Lashing material to prevent sliding and/or tipping of the load was used to secure the TBM and cradle onto Mammoet trailer as well as anti-slip material between all steel contact areas to promote friction (see Figure 18).

The TBM was launched in February 2016. Currently the TBM has bored more than 1.2 km of the access tunnel—the first drive (downstream) of the headrace tunnel.

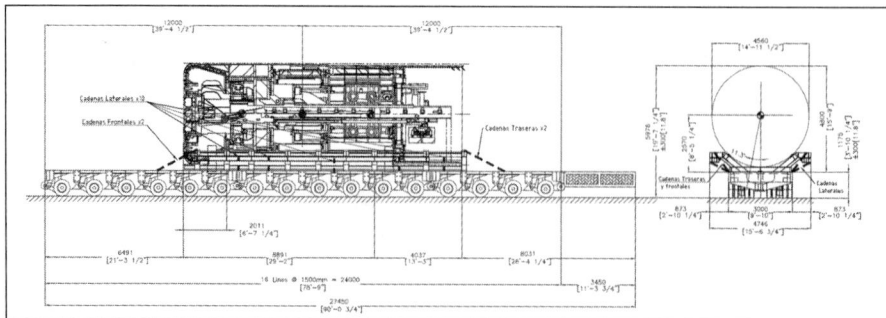

Figure 17. Mammoet heavy lift 16 axle hydraulic modular motorized trailer used to transport the TBM up to the tunnel launching chamber in Los Condores project

The tunnel is being bored at a maximum depth of 500 m below rock consisting of sandstone, tuff, and pyroclastic breccias. The rock was tested at strengths up to 100 MPa UCS, with at least two fault zones expected during tunneling.

Figure 18. TBM being moved into launching chamber

Robbins Double Shield TBM by the Name of "Clementina"

The Double Shield TBM "Clementina" provided by Robbins included many standard elements and several new features that were included to help the customer cope with whatever conditions may arise. As standard, the TBM was built with a 2.65 m diameter main bearing for the 4.56 m outer diameter, as well as four variable frequency drive (VFD) electric motors that provide power to the cutterhead. Together with a maximum cutterhead thrust of over 22,000kN at 450 bar, the maximum cutterhead torque is over 1,800 kNm.

Among the specific features adopted for the TBM was the inclusion of gear reducers on the motors. By reprogramming the gearboxes on the VFM motors, extreme breakout torque could be increased.

The back-loading cutterhead is dressed with 30 × 17" Robbins wedge lock disc cutters and has a 19 mm to 12 mm wear capacity on the gauge cutters, depending on the cutter position. To be better prepared to cope with fractured and squeezing conditions, the cutterhead is also equipped with an overcutter facility. Once engaged, the overcutters can increase the bore dimension by 25 mm and 50 mm on the radius, corresponding to a tunnel increase of 50 mm or 100 mm on the diameter. The 50 mm overcutter setting on the radius is engaged in extreme squeezing ground and seriously reduces the risk of the shielded TBM becoming trapped. The overcutting facility can be adjusted through the crown and must be non-centric to the normal cut diameter to keep the invert of the extended cutterhead diameter on the same line. To achieve this, the cutterhead and the bearing unit is lifted onto a new axis.

Another special feature of "Clementina" is the system for probing and pre-excavation drilling should it be needed. Probing to >40 m ahead of the face is a systematic requirement.

The drill ports also allow for forepoling in an array over the crown and, in potentially squeezing conditions, lubrication of the shield can be injected through the same shield ports. Control and cleaning of muck is possible on the machines by closing off the muck chute in the excavation chamber with a horizontal guillotine door. The system will be used in the event that the cutterhead fills with material.

Drills are mounted on a ring beam providing for a permanent drill installation immediately behind the segment erector. From this forward location, the drill can install 16 drill holes on a 7 degree outlook angle through the drill ports around the forward shield (4 holes) and the gripper shield (15 holes) and through two horizontal holes in the cutterhead.

Lastly, "Clementina" was specifically designed for disassembly and removal from a blind heading since there will be no possibility of breakthrough at end of each drive, neither at the upstream site of the new intake facilities nor at the downstream junction with the 470 m deep high-pressure shaft into the underground powerhouse. The TBM, therefore, will be dismantled within the finished tunnel and the components, including the sections of the cutterhead, retrieved back through the completed and segmentally lined tunnel. The machines has been designed and manufactured specifically to achieve this final operation with efficiency.

Challenges During Excavation

Highly variable and rapidly changing ground conditions were encountered during the excavation of the first 1 km of the headrace tunnel downstream, including mixed-face conditions and unexpected and constant water inflow that presented great challenges and decreased the advance rates dramatically. These events, still ongoing at the time of paper writing, may trigger potential delays in the future if they are not addressed with a detailed understanding of the unfavorable conditions and the problems involved in TBM tunneling under mixed-face ground.

From the viewpoint of the rock-machine interaction, some mitigation measurements such as the selection of the correct TBM type, modifications and improvements on the TBM, conditioning of ground and finally the optimization of TBM operation are to be taken.

Muck Chute Failure Under Highly Variable and Rapidly Changing Ground Conditions

Mixed ground conditions with very hard andesite (up to 150–200 MPa) on one side of the tunnel face and a very fine hydrothermal sand deposit from rhyolite on the other side were encountered. Combine this with a larger than normal rotary unit to accommodate the many cutterhead ports, and the reduced capacity muck chute is what contributed to the failure and forced the TBM to stop for three weeks until a new muck chute was provided and installed.

This issue typically occurs in blocky ground when a boulder becomes caught between the muck chute and the rotating cutterhead. When the muck chute becomes blocked, the rock and material will usually overspill the chute. However a blockage can also occur in sticky clays or loose material (sand), especially when the muck chute spoil window is not big and the impact pad is not steep enough for fine material.

The real issue to overcome is to prevent boulders from getting caught between the cutterhead and muck chute. The operator must be aware of the ground conditions and the chance of boulders entering the cutterhead, and capable of recognizing the sign of a boulder getting caught (basically a large jolt of the machine as the boulder gets caught between the muck chute and the cutterhead). The operator must then stop the cutterhead as soon as possible to prevent it from being sheared off (see Figure 19).

As it is more a matter of keeping the cutterhead clean and ensuring the correct hardware is in place, Robbins and Ferrovial decided to revise the cutterhead inspection protocol and have the personnel in charge of the tasks ensure the cutterhead is being cleaned thoroughly, including checking the bucket lips and grill bars more often.

If blocky ground is not the root cause for the chute failure, then increasing the discharge capacity of the muck chute may be the right decision to make. A tradeoff between the rotary union capacity to fit as many injection ports as practically possible and a chute

with a steeper front loading plate to better handle the actual ground conditions
(soft ground) was made, even when that
entailed downsizing the rotary union and
associated manifold and plumbing. This
therefore affected the capacity to inject
foam, water and other additives. A revision of the general assembly with a new
belt length and modified take up frame
was also required.

High Water Inflow Rates

Another no less important challenge has
been the steady increase in water inflow
up to a maximum of 3,500 l/min. The
presence of water exacerbated the issue

Figure 19. Blocky material (andesite with fine sand from weathered rhyolite) inside the cutterhead

of face stability by washing away the small particles of weathered rhyolite that allowed
progressively large sections of the face to become unstable and added fluidity to in-
situ ground and excavated muck. This fluidized muck made it more difficult to control
the flow into the head, the telescopic shield and into the rear section in the ring building area.

Small particles of sand dragged by the pressurized water inflow remained accumulated underneath the main thrust cylinders and caused continuous damage to them
during the re-gripping operations. Water inflow also made it harder for the customer to
keep the annulus grout from washing out.

Depending on conditions, i.e., water quantities, pressures and flow rates, some methods have been tried:

- Add anti-dilution additives to the grout
- Inject sodium silicate into the grout at the injection nozzle to achieve a very
  rapid set without negative effects of too much water in the annulus. Dosages
  to be tried and evaluated outside of the tunnel first.
- Selectively drain the annulus (de-pressurize); i.e., locally pull the water out
  and replace with grout
- Selectively drain the nearby rock mass
- Do more pre-excavation grouting in front of or around the TBM

In all cases, it is necessary to keep the annular grouting program around the segmental liner "tight" to the tail shield since lots of water is often released from the rock
mass as soon as the cutterhead passes. These modifications are still in progress
and a more detailed report of the most successful options will be provided during the
conference presentation.

FINAL CONCLUSIONS

Whether an open type, Main Beam TBM or a shielded hard rock TBM, the Andes is a
mountain range that presents challenges. While both projects utilize different methodologies—McNally slats at Alto Maipo contain ground in combination with other forms
of support, while concrete segments will line the entire Los Condores Tunnel—both
are successful methods. Shielded TBMs can be outfitted with a number of additions including multi-speed gearboxes and shield lubrication to avoid the problem of

machine shields becoming stuck in fault zones. As such, both shielded and open-type machines are capable of performing well in fractured rock. In both cases common methods, including continuous probe drilling and grouting, can inform operators of conditions ahead of the TBM in order to make adjustments accordingly, grout off water, etc. And in both cases well-trained operators and a team with agreed-upon protocols for bad ground conditions is key to project success. As both machines navigate the challenging Andes Mountain range we will continue to analyze and further refine our opinion of the two excavation strategies.

PART  **6**

Conventional Tunneling

Chairs

Mark Peterman
Kiewit

Dave Dorfman
Schiavone Construction Co. LLC

John Hart Generating Station Replacement Project—Underground Works: Project Update and Challenges Encountered

Matt Kendall ▪ Frontier-Kemper Constructors, ULC

ABSTRACT

The Frontier-Kemper/ASL Joint Venture was formed to excavate the underground workings of the John Hart Generating Station Replacement Project in Campbell River, British Columbia. The existing surface 126-MW John Hart Generating Station originally completed in 1947 on the Campbell River, represents approximately 17% of the total generating capacity on Vancouver Island. The underground replacement project will create a more reliable, seismically robust, and environmentally friendly facility with an increased installed capacity of 132 MW. InPower BC General Partnership is a special-purpose vehicle created by SNC-Lavalin Capital Inc., and contracted by British Columbia Hydro and Power Authority (BCH) to design, build, finance and maintain the replacement power station under a 19 ¾ year public-private partnership (PPP) with BCH. The underground excavation works portion of the project is 77% complete as of the end of November 2016 and is scheduled to achieve substantial completion on or ahead of schedule by July 2017.

INTRODUCTION

The John Hart Generating Station Replacement Project (JHGSR) is located on Vancouver Island in Campbell River, British Columbia, Canada. It is positioned on a long-term lease concession from the Province of British Columbia in the Elk Falls Provincial Park along tho Campbell River. The original surface generating station completed construction and was commissioned in 1947. That plant is still in operation today and will continue to be through construction of the new underground station. Part of the project is to demolish the old power generating station after the replacement plant and facilities are commissioned and running. The dam was built on the Campbell River and thus created the John Hart Reservoir upstream to the surface powerhouse. This reservoir feeds the intake to three each 1.4km surface penstocks. The first kilometer being of wood stave construction then transitioning to steel penstock for the steeper gradient down to the surge towers and then on the powerhouse. The current station has produced electricity for about 70 years and has seen operating expenditures rise and turbine output decline due to age. The current station is exposed to tremors and earthquakes due to its location on the Ring of Fire. Given its construction and current conditions there are a lot of concerns over survivability of the station even with a minor magnitude earthquake. Additionally the Campbell River and Quinsam River tributary are prime habitat for a variety a salmon species. During unit(s) unplanned downtime

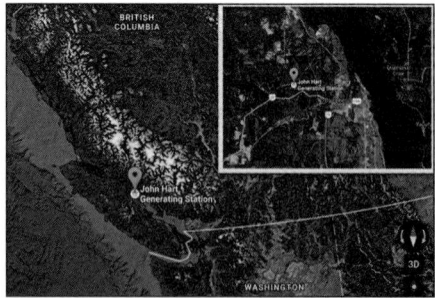

Figure 1. Plan view showing the location of the John Hart GSR Project on Vancouver Island, BC

there exists a significant delay to releasing overflow at the dam spillway to keep adequate water downstream in the river for the fish viability. All of these factors led to BCH's decision to opt for the design and underground powerhouse option presented by the winning team of SNC-Lavalin, Aecon, and Frontier-Kemper Constructors.

The principal feature of the replacement project is the underground powerhouse. The powerhouse excavation has an approximate volume of 64,000m³, is 94m long, 23m wide at the crown and 40m tall at its deepest point. For comparison, this powerhouse is approximately one football field long and at its highest point meets that of a 13-story building. Another innovated feature of this powerhouse is the bypass valve chamber. In the event of a unit outage (planned or unplanned), the bypass system allows for continuous water discharge out of the tailrace tunnel to the Campbell River benefitting the aquatic species living in the river. The replacement station also contains +2.0km of combined power and tailrace tunnels, 1km of other access tunnels, inlet and outlet manifolds, surge chamber, gate chamber, surge shaft, and intake shaft. All of this has to be excavated and constructed within the tight confines of the environmentally sensitive park concession and during ongoing operations of the current surface station which has many vibration sensitive features. In all, more than 300,000m³ of in situ rock will be excavated from this underground complex. Once the new, underground station is commissioned, the surface station will be decommissioned, and demolition will commence. Legacy environmental issues will be remediated and surface reclamation completed. InPower BC will then run the new station for approximately 15 years; then hand over the facility to BCH for the subsequent 80 years-rounding out the 100 year design life. The focus of this paper will be on the underground excavation aspects and challenges encountered thus far and how they were successfully met to bring the project in on schedule.

STAKEHOLDERS

- Client: British Columbia Hydro and Power Authority ("BC Hydro")
- Build Consortium: InPower BC General Partnership - a special-purpose vehicle created by SNC-Lavalin Capital Inc., and contracted by British Columbia Hydro and Power Authority (BCH) to design, build, and finance and maintain the aforementioned station under a 19 ¾ year public-private partnership with BCH.
- Design-Builder: SNC-Lavalin Incorporated
- Civil Contractor: ASL JV—a joint venture formed between Aecon Constructors, a division of Aecon Construction Group Incorporate and SNC-Lavalin Constructors (Pacific) Incorporated. Aecon is the managing partner.
- Civil Subcontractor: FK/ASL JV—a joint venture between Frontier-Kemper Constructors, ULC (a Tutor Perini company) and ASL JV. Frontier-Kemper Constructors, ULC is the managing partner. ASL JV essentially provided surface services support.

SCOPE OF WORK

FK/ASL JV's scope of work is more than 300,000 m³ of underground excavation (including its support) and concrete tunnel inverts. This includes all tunneling of varying cross sectional areas, mass excavation (Powerhouse) and two shafts. All of this work is being completed with a relatively standardized fleet of modern rubber tire mining equipment, conventional shaft sinking gear, and Frontier-Kemper's in-house raise boring equipment.

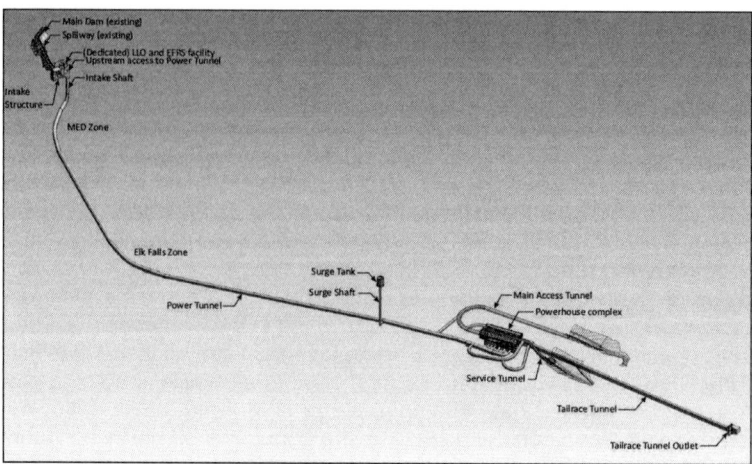

Figure 2. Orthogonal view of the project as bid

EXCAVATION SUMMARY

Access Tunnels

The project has roughly 1km of access tunnels; primarily with two cross sections, 6.0m w × 6.2m h and 9.0m w × 6.2m h. The main access tunnel (MAT), roughly 400m long, is 9.0m w × 6.2m h and declines down from the portal on the north side of the existing pen stocks to the Service Bay of the Powerhouse. This tunnel will be discussed further in the Challenges Encountered section of the paper. The other access tunnels with a 6.0m w × 6.2m h section starts with the Service Tunnel (L40) which declines down on the south side of the existing pen stocks to crown of the powerhouse. As the name implies this tunnel will eventually house the majority of services (piping, conduits, feeders, and ventilation) for the operation of the new underground powerhouse. Off of the L40 Service Tunnel, two more access tunnels are developed. The first being Ramp B which spirals down to the upstream (U/S) side of the powerhouse. Off of this ramp is Ramp C which drives down to the downstream (D/S) side of the powerhouse. Ramp C intersects the D/S Surge Chamber which then transitions to the Tailrace Tunnel. Apart from the muckbays being the same cross section the last is the access tunnel to the Gate Gallery which is located above the Surge Chamber off of the MAT.

Power Tunnel

Including the U/S intake manifold where the water bifurcates to the three units, the Power Tunnel extends 1,580m from the eastern boundary of the powerhouse to the intake shaft to the west and is the critical path for FK/ASL's scope of work. The cross section of the Power Tunnel is 8.1m w × 8.3m h D shape and is mined full face. The tunnel has various different gradients. From the intake manifold it inclines positively at 12%, levels off to +0.50% for the majority of the way, then steepens to +15.25% for the final +140m to the Intake Shaft under the Middle Earthen Dam (MED). The power tunnel has two zones: the MED and the Non-MED zone. The MED zone has a different ground support regime, a steel fiber reinforced final lining along the entire perimeter (including the invert) and a thicker invert concrete slab. This is required by the designers given the proximity to the Middle Earthen Dam and concerns over pore pressure in the dam after "water-up" and long term MED stability.

The purpose of this tunnel is to convey water from the new John Hart Reservoir Intake via the intake shaft to the three turbines in the underground powerhouse which are design to produce 132MW. Proceeding east to west the Power Tunnel intersects Ramp D which provides western access to the MAT and Service Bay of the Powerhouse during excavation. It then ties in with the U/S Surge Shaft and continues west were it will eventually tie in with the Intake Shaft lower elbow. Spaced approximately every 252m are muck bays to support tunnel face development. Just outby of each of the muckbays are slashes to temporarily house portable mine refuge chambers. As the new muckbays are excavated and put into service, the portable refuge chambers are skidded forward accordingly.

Tail Race Tunnel

The tail race tunnel, with a cross section of 6.5m w × 10.7m h D shape, starts just east of the Ramp C intersection. The purpose of this tunnel is to convey the water that has collected in the downstream Surge Chamber of the powerhouse and carry it to the tail race exit portal where the water will be re-introduced to the Campbell River just north of the old surface powerhouse location. The tunnel is 534m long with a −0.1% slope, including a +16m transition from the Surge Chamber, and given its dimensions combined with the equipment fleet selection for the job, it is being excavated top heading and bench. Due to the top heading and bench methodology the tunnel is mined in three reaches. The top heading is excavated 6.5m w × 6.7m h with rounds taken ordinarily either 5m or 6m in depth depending on ground behavior and jumbo availability. The bench dimension is 6.5m w × 4.0m h and is more productive than the top heading; since the crown is already supported and with the extra free face no burn cut is required. As with the other tunnels, muck bays are located on an average spacing of 252m. In the case of the tail race tunnel, two will be required except only in the bench portion on the invert elevation.

Originally this tunnel was supposed to be excavated from the exit portal towards the powerhouse surge chamber. This would have been slightly more beneficial from a drainage stand point during face development and would have less congestion on the access ramps C, B, and L40 plus had less ventilation demands. Drawbacks would have been the satellite nature of the portal causing more logistically issues for flow of equipment, materials, and supervision. Due to some unforeseen infrastructure issues around the old surface powerhouse it was later deemed not feasible. It was decided to mine the tunnel instead from Ramp C. Ultimately the pros and cons were a wash and the impact so far has been tolerable.

Surge Chamber

This is a large excavation in its own right. The surge chamber has dimensions of 9.0m w × 14.7m × 92.5m long and the height tappers towards the west when alongside of the bypass outlet tunnels. For the same reasons listed for the tail race tunnel, the surge chamber was mined top heading and bench. The purpose of this heading is twofold. First it collects all of the water exiting the turbines via the three outlet draft tubes and/or any water being bypassed via the outlet bypass tunnels. The water then transitions to the tail race where it gravity drains to the exit portal back to the Campbell River. It will also contain concrete guides for each of the six outlet tunnel gates.

GATE GALLERY

As with the Surge Chamber this is also part of the downstream excavation north of the powerhouse. It is smaller in dimension than the Surge Chamber at 6.0m w × 12.5m h avg × 71.5m long. It is positioned in plan off center to the south and above the Surge

Chamber. As the name implies it will house the gates for the three outlet draft tube tunnels and three outlet bypass tunnels. For the same reasons listed for the tail race tunnel and the surge chamber, the Gate Gallery was mined top heading and bench. There are three smaller man way tunnels coming off of the gate gallery to provide secondary access and egress to/from the powerhouse. Two of the small 2.5m w × 2.5m h man way tunnels connect the powerhouse to the gate gallery. The southeast manway at the invert off the gate gallery and the southwest manway near the crown of the gate gallery. There is a third manway which connects the gate gallery to the Main Access Tunnel for secondary egress. These smaller manways were mined with jacklegs and stopers and mucked with the smaller ST 710 LHD.

Lastly between the Gate Gallery bench invert and the crown of the Surge Chamber will be six narrow slots to guide the gates between the two chambers. On average the slots will be 5.15vm in height. The slots vary in length between 5.4m up to 7.2m and all of the slots will be 2.7m wide typical. Due to the poor ground conditions on the west end and the narrow rock pillars not only between the slots themselves, but between the two chambers we've chosen to mechanically excavate. We will drill off and fire a small wagon wheel drop raise. With relief holes already drilled along the perimeter and body of the slot we will then mechanically break the rock with a hoe ram mounted on the 200 LC tracked excavator with the broken material passing the drop raise to be collected and removed with a remote LHD in the Surge Chamber. These slots will be retreated from the Gate Gallery and advanced in remote mucking from the Surge Chamber. They will proceed east to west and we will apply shotcrete and install rock bolts off of an adjustable work deck suspended by a 50mt rubber tire crane.

POWERHOUSE (MASS EXCAVATION)

The powerhouse is one of the key features for this project even though not on the CP in FK/ASL JV's SOW. For the Prime and Civil Contractor it is the principal feature of the project and vital that FK/ASL execute the excavation per the design and as quickly as possible due to all of the follow on works that take place after the handover of the excavation and area. This is a large excavation at roughly 64,000m^3. It is 94m long × 23m w (at the crown) × 40m h. The crown is elliptical and is 10m tall. Below the crown the powerhouse walls step in 2m to establish a shelf for the main overhead crane concrete plinths. The plinths were designed to be 2m wide and 1m tall and to be installed by the civil contractor in parallel with ongoing benching activities. From there the remaining powerhouse is 19m wide. In plan stepping down the powerhouse has a series of stepped reductions in the plan perimeter to different elevations to account for the concrete works and different final chamber requirements.

The crown was accessed via the L40 (Service Tunnel) and once the powerhouse crown boundary was reached the 6.0m w × 6.2m h access tunnel flared out to a crown pilot tunnel 8m w × 8m h. The pilot tunnel was driven in a sufficient distance then the north and south sidewall drifts were slashed out to the full width of the powerhouse and driven concurrently with the pilot tunnel but staggered. The support requirements were 100mm of polyfiber reinforced shotcrete followed by alternating rows of 6m and 4m dual corrosion protected bolts (DCPs). In this case the prime recommended the use of DSI's CT-Bolt. The bolt has a point anchor shell for immediate active support and then is post grouted after the blasting front is at least 8m advanced.

Once the crown was fully developed and supported the crews stepped in 4m total and proceeded to line drill and take the first blind bench 7m deep. An internal ramp was extended from the Service Tunnel into the powerhouse to access the invert of the bench. After the bench was completely mucked out after a series of blasts the ribs and

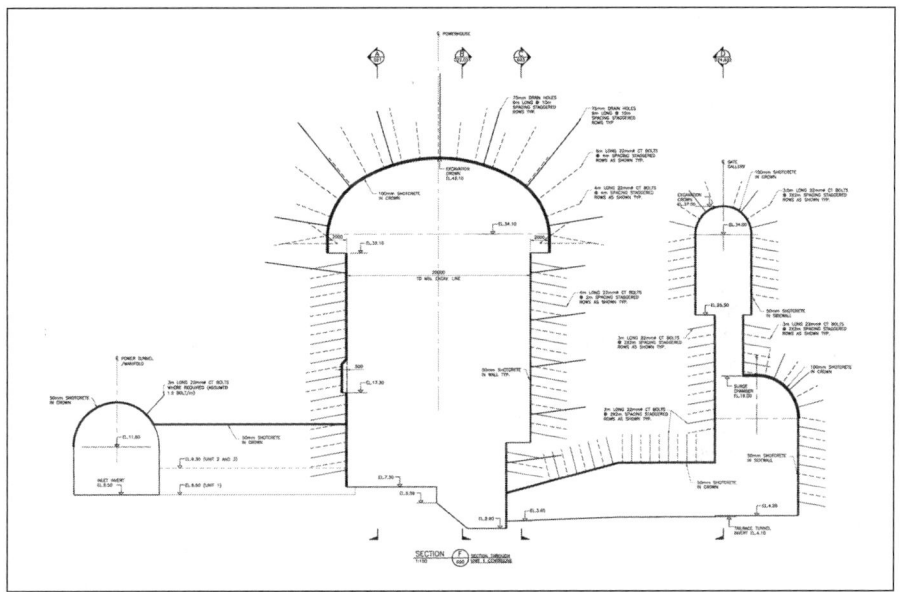

Figure 3. Section view of the powerhouse, surge chamber, gate gallery, power tunnel, inlet, and draft tube as bid

end walls were supported. At that time the drill jumbos were positioned perpendicular to the long axis of powerhouse to drill off the 2m × 2m pillars of rock remaining in the crown on the side wall drifts to create the rock shelf for the aforementioned concrete plinths. Anchors were then drilled and installed in the walls to secure the rebar cages to for the plinth pours. While the concrete forming and pouring were taking place by the civil contractor on the opposite side of the powerhouse in a staggered position FK/ASL were drilling and installing large anchors in the crown of the powerhouse for the future false ceiling. Once one side finished the positions were swapped in an effort to reduce cohabitation interference in the crown. Well that was the theory anyways. There of course were incidents of cohabitation interference but they were dealt with as amicably as possible.

A planned 3m × 3m glory hole was key to efficient muck transfer for the powerhouse. The original plan was get Ramp C down to the Surge Chamber as quickly as possible, turn west and develop just beyond draft tube #2 for some tail room. The crews would then back up and fully develop draft tube #3 and punch into the low point in the powerhouse about 3m. Next from the crest of bench #2 access via the Service Tunnel the crews were to drill and shoot a 3m × 3m square drop raise aligned with the north wall of the powerhouse intersecting the extension of draft tube #3. The glory hole was designed to be shot bottom up in three separate blast similar to that of a VCR (vertical crater retreat). Not only would this be more efficient for the cycle, this would then get the bulk of the mucking equipment out of the powerhouse affording less congestion and interference in the benches themselves and keep the trucking out of the powerhouse. The added benefit for crews in subsequent benches would be a nice free face to shoot to. For the first blind bench the crews had to develop a sinking cut first to create this area. For reasons to be discussed later, the arrival of the glory hole to the power house operation was delayed. This will be discussed further in the paper in the Challenges Encounter section. Once the glory hole was in place then all operational access for drilling, charging, and ground support would come from the Service Tunnel via internal powerhouse ramping for benches one, two, and three. After that and once the Service Bay had been developed subsequent to the Main Access Tunnel arrival to the west the crews would advance in flat and recover the internal ramp from the east, stepped in yet again and began benching down towards the inlet tunnels with more internal ramping all the time tipping shot muck down the glory hole.

INTAKE SHAFT

Developed from the bottom of a partially pre-excavated rectangular intake structure east of the existing John Hart Reservoir Dam. The shaft had a finished inside diameter of 6.5m and was 62m deep. During conventional sinking the shaft was serviced by two cranes. The HC 165 crawler crane was for heavier picks such as the shaft jumbo, work deck, muck buckets, and shotcrete buckets. The second rubber tire crane was exclusively for man travel. The drilling was performed with a 32 ft four boom Watson Shaft jumbo with Canon drifters. CAT 305.5 tracked excavators mucked to 6 yd^3 muck buckets with chain bridles. Shotcreting was manually applied with a hand nozzle off the single stage work deck and wall bolting with jacklegs.

SURGE SHAFT

The surge shaft is located on the south side of the Power Tunnel and is 200m west from the Ramp D intersection with the Power Tunnel. It was 105vm with a finished inside diameter of 4.3m and is designed to replace the three surface surge tower over the existing penstocks. It is supposed to handle any upstream surge from the powerhouse.

The shaft was excavated using Frontier-Kemper's Ingersoll-Rand 211 raise bore drill (RBM). The 13¾-in. diameter pilot hole was drilled top down from the surface location and was sub-drilled approximately 1m below grade of the underground access drift off the Power Tunnel. The pilot hole was probed and confirmed its location to within millimeters of intended location. The access drift then pushed through. Subsequently an invert bench was excavated in the access tunnel to serve as a long term rock trap for the surge shaft. The reamer was then assembled underground in the access drift. The reamer was stabbed and reaming commenced. The first 100m had an average penetration rate of 5.26vm per work day (19 days). The last 5m took 4 days due to the very hard nature of the rock at the collar and the desire to be cautious and not risk breaking of any big wedges or potentially shear the stem and dropping the head. Once the RBM was picked off the collar and demobilized the HC-165 crawler was brought over and positioned. Off of a two stage work deck the shaft was inspected and ground support prescribed in 25m reaches. For the majority of the shaft only spot bolting with double corrosion protected (DCP) rock bolts was required.

EQUIPMENT FLEET

The following considerations were weighed when evaluating and selecting the appropriate fleet. Taken into account were the following: tunnel cross sections, lengths of tunnel and volume of mass excavations, gradients, compatibility and versatility of said equipment, upfront CapEx, and potential resale of the equipment at completion.

A rigorous selection process took place for equipment selection. All major vendors were invited to participate along with a couple of used equipment vendors as well. A lot of effort was expended during this process to ensure the proposals were evenly comparable to guarantee the fairest possible evaluation. Many considerations factored into the final decision such as: plant loading to ensure delivery to our schedule, purchase price, bundled options and discounts, levels of automation and integration,

Table 1. Major equipment selected

Equipment Type	Manufacturer	Model	Fleet Size
Drill Jumbos	Atlas Copco	E2C	03
LHD	Atlas Copco	ST 14	04
Mine Truck	Atlas Copco	MT42	04
Shotcrete Sprayer	Normet	8100VC	02
Shotcrete Transmixers	Normet	LF600	03
Charging/High Lift	Normet	9905 BT	01
Charging/High Lift	Normet	9915 BA	01
Scissor Deck	Normet	MF 540	01
Ring/LH Drill (used)	Davis	DK 60	01
Telehandler (used)	Skytrak	10042	03
Tracked Excavator (used)	John Deere	200 LC	01
Mechanical Scaler (used)	Oldenburg	SV 26	01
LHD (used)	Atlas Copco	ST 710	01

estimated EOE, service, training, proximity to critical and spare parts, consignment parts options, sales support during bidding, etc. Ultimately FK management selected Atlas Copco for jumbo drills, LHDs, and mine trucks. All of the Atlas Copco E2C drill jumbos came with their proprietary Advance Boom Control (ABC) Regular system which provided a good level of drilling automation. Normet was selected for explosive charging vehicles, shotcrete sprayers, shotcrete transmixers, and scissor deck. It's important to note the two charging vehicles were fairly utilitarian as they could double for alternate high lifting capacity, and one of them was equipped with a hydraulic manipulating arm to install lattice girders. FK also opted for the hydro scaling option on the shotcrete sprayers as well to be a primary tool for scaling and rock washing prior to shotcrete application. There were some other ancillary mobile equipment as well such as Skytrak telehandlers, tracked excavator, ring drill for grout drilling, bench drills for the powerhouse, man carrier tractors and RTVs, and eventually a backup mechanical scaler. See Table 1 for major fleet.

Those familiar with the operating limits of most of this equipment may ask why a three boom jumbo or a larger LHD were not selected. The biggest factors why FK steered away from that was the varying tunnel cross sections to excavated, the impact of theoretical efficiency gains over length of the larger tunnels versus the increased capital cost, the thought the crews could achieve similar productivities with two smaller LHDs versus that of one larger LHD, and ability to quickly salvage at end of the job.

With the selected equipment and varied means & methods the crews could mine all the varying tunnel cross sections and not be overly exposed if a piece of equipment had downtime-planned or unplanned. In the larger diameter tunnels (8.1m and 9.0m widths) the crews muck with two ST 14 LHDs. After the blast the crews fast muck back to the nearest muck bay thus allowing the men to get back and face support as quickly as possible. Subsequently and in parallel to face support, drilling and charging the crews can muck haul to surface. The muck haul to surface is now taken off the critical path of the face cycle itself. In the smaller diameter tunnels, the crews muck with just one ST 14. The 4th unit is on standby for planned preventative maintenance or unplanned breakdowns. With respect to the drill jumbos FK wanted to bolt using the face jumbo but also have the ability to take 5m face rounds at a minimum depending on ground conditions. In the smaller diameter tunnels bolting with 5m fixed feed slides was not feasible due to the operating kinematics of the jumbo so FK opted for 18ft/10ft splitfeeds. This way after the in cycle polyfiber reinforced shotcrete (PFRS) reached

J2 (≥0.5MPa) the driller could bolt then roll over, extend the feeds and start drilling the face. In the case of the power tunnel given its width and height and the requirements in the specifications for a fan of five each 30m long probe holes every 22.5m along the tunnel alignment FK opted for the fixed 5m feeds as there was sufficient clearance to turn 90 degrees and bolt. Additionally be bolted on Atlas Copco's Rod Handling System (RHS) to both fixed feeds to facilitate the probe hole fans. That was not an option with the splitfeeds. The RHS carousel can hold up to 10 each 8 foot drill steels for this plus the 5m steel already on the slide.

PROJECT CHALLENGES

Delay to Start Date

Surface trench development to access and expose the bedrock for portal develop-ment is the scope of the civil contractor. The as bid design concept was for an open cut with slopes laid back 2:1. Once on site and some bed rock confirmation holes were drilled it was discovered the rock to hard pan interface had changed and was much more variable than previously thought. As such the lay back option was out as it would exceed the client's concession boundaries in the Provincial Park. Next a sheet pile option was looked at but not deemed feasible due to boulders. Finally a shotcrete soil nail wall was settled on. Unfortunately these design delays and actual excavation works delayed the portal delivery for the L40 Service tunnel by 2 ½ months. Likewise the L20 Main Access Tunnel portal was delayed almost 3 months. Frontier-Kemper/ASL also met the challenge of rock encountered at higher elevation in the portal cut by excavating approximately 50m more main access tunnel where it was foreseen originally to be surface excavated. The portal face was moved back in the case of the MAT. The combined effect of course was a delay to all of the downstream milestones and works.

Buried Valley in the Main Access Tunnel

Once the L20 Main Access Tunnel started it went very well. It was the best rock with the most favorable jointing orientation job to date. Half barrels from the perimeter holes could be seen on the tunnel walls almost all the way. The face was 145m inby of the portal and the crews were on the last round of the first muckbay. The crew drilled off both faces and fired them at the same time. Once mucking commenced the crew noted more water than usual. As they proceeded further it became worse. Then mucking was halted and piping reviewed to make sure there were no plumbing issues. None were discovered. Then the call came over the radio that they thought they had a problem. When supervision & management arrived the water was steadily rising out of the shot muck and creeping up the ramp. It was obvious FK/ASL had hit something but the magnitude wouldn't be realized until later. The face pump was quickly over run and the crew had to evacuate the water in the drill water line and convert it into a discharge line and with a second electric pump to increase the output capacity. Even that wasn't enough to arrest the rising water. It was decided that one of the 8HP pumps should be switched out with a contingency 15HP pump and eventually the crew was able to stall the rise of the water and slowly start to gain ground. At peak the inrush was averaging 500gpm.

After drawing the water down the crews started to muck on the face. Eventually they were able to expose the feature. The feature discovered was an oblong shaped win-dow in the basalt face roughly 20 to 30cm thick by 150cm tall by 200cm wide plus or minus. In that window was sand, cobbles and 40,000psi granite boulders the size of rugby balls. Eventually the subsurface water feature drained off over a few days and stabilized at an average of 40gpm.

This was an obvious game changer and this sat on FK/ASL's project critical path. At that time FK/ASL was geared up for a traditional drill & blast job with rubber tire mining equipment. Nobody had any forehand knowledge of the potential for a soft ground mixed face tunneling scenario nor did FK at the time have all the appropriate equipment on site to deal with it. Fortunately Frontier-Kemper Constructors has on staff experienced soft ground sequential excavation method (S.E.M.) personnel and also a fleet of equipment for that method of tunneling. Frontier-Kemper Constructors immediately began probing the extents of the feature and immediately mobilized specialized S.E.M. equipment including Cat 328 Tunnel excavator and a grout mixing/injecting plant. In parallel the civil contractor mobilized an exploratory core driller to test the feature as well. Working concurrently with SNC's designers, soft ground tunneling design drawings were produced to allow the quick release of a lattice girder order from DSI. Similarly, FK reacted quickly with a canopy pipe umbrella system for pre-support all the while not completely sure what lay ahead. At the time FK liaised with Dywidag-Systems International (DSI) who was supplying all of the pipe canopy materials to other FK projects in the United States and fortunately for us was actually being loaded into sea containers for delivery. FK were able to divert the loads and were able to get

that shipment rerouted to Campbell, BC instead. Within 6 ½ weeks of the event taking place on May 14th 2015 FK had all the materials and equipment on site to being soft tunneling through this feature. The experts ultimately postulated that a glacier had receded and deposited this material and this eventually became what is known as a buried valley. During topheading SEM development the crews encountered some tree trunks and carbon dating aged the material to be greater than 50,000 years old.

Figure 4. Photo showing the buried valley discovery in the MAT on 14-May-2015

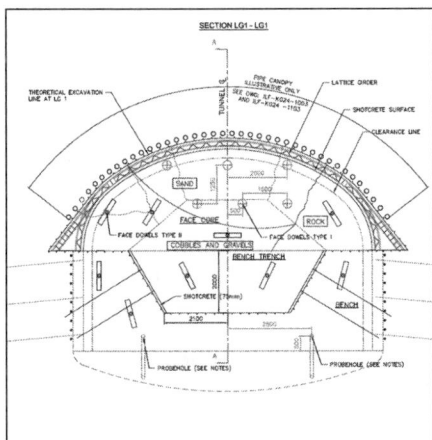

Figure 5. Section view (face) of the excavation sequence and ground support for a pipe canopy in the MAT. Note mixed face condition.

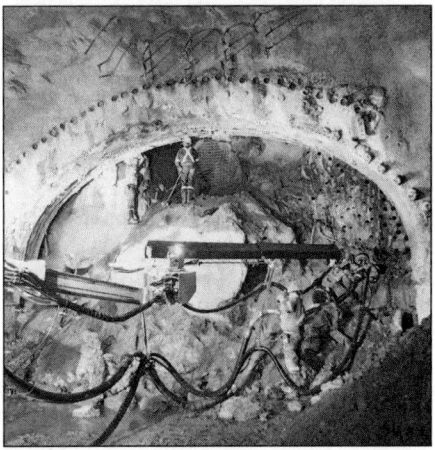

Figure 6. Actual photo showing mixed face conditions in the MAT per Figure 5. Note partial triple canopy.

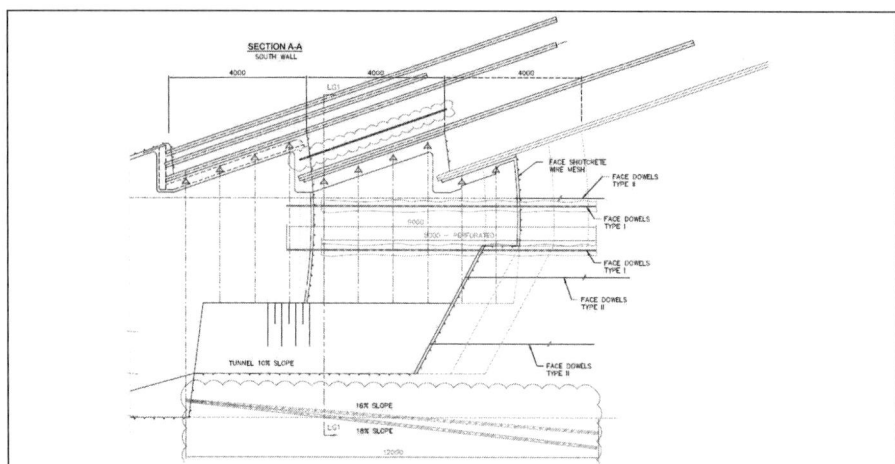

Figure 7. Section view (long) of the excavation sequence and ground support for a pipe canopy in the MAT

During the soft tunnel mobilization period the civil contractor brought in drillers to attempt to define the extents of this zone. Additionally FK/ASL were tasked to also perform some longhole probes with our jumbos from the crown of the powerhouse and crown of the topheading of the surge chamber. At the same time the civil contractor hired an engineering consulting company to perform the design on the SEM tunneling temporary works. In parallel this consultant liaised with the prime to incorporate certain components of the temporary works into 100 year life permanent design.

Eventually it was determined that the zone along the original tunnel alignment was ±54m wide and open to the north, in elevation, and with some extension south and sub tunnel grade. The plan was to mine SEM with the topheading all the way through the zone then back up and take a flat bench with the option of full ring closure in the invert if heaving occurred. Frontier-Kemper Constructors has established soft ground tunneling experience and was instrumental in bringing solutions to the project team to overcome this challenge. To build off of that and bolster the project team in the field FK/ASL decided to bring in two operators/trainers and one senior engineer with significant experience in SEM tunneling from Beton-und Monierbau Ges.m.b.H (BeMo) in Austria whom they worked with before on other major projects. This was done to not distract or take away resources from the other important ongoing tunnel activities elsewhere in the project. In parallel FK/ASL management and BeMo offered more input for the design and meet with ground support vendors to round out the necessary materials to complete the temporary works for the Civil Contractors hired tunnel engineers. DSI was already slated to supply the pipe canopy materials to another project which was routed to John Hart. It was decided to stay with DSI. DSI provided the AT-139 Threaded 5 ½" OD Ported Canopy Pipe, starter tubes & bits, hallow bar self-drilling spiles, all courses of CP 130/8/11 lattice girders, packers and all bolt on drill accessories for the E2C jumbo. Due to the quantity of pipe canopy to be installed FK/ASL later transitioned to DSI's squeeze connection pipe canopy, the hydraulic squeezing unit, and pipe canopy autoloader. This proved to be more efficient and provided less exposure to men in the man basket.

The SEM excavation started off slowly. The crews had to be trained up since none of the local miners on the project had ever work in soft ground tunnels previously. It was compounded by the extreme mixed face conditions at the onset. The wedge of basalt in the face was thicker west to east, so for the first two full top heading canopies the face had sand, cobbles and gravel, boulders, basalt rock, and in places even ancient tree trunks. This impacted pipe canopy installation speeds. Grouting the mixed conditions was difficult and had a fair bit of trial and error and eventually a micro fine was used and a mix dialed in. The civil contractors engineering consultant directed grouting and the retirement criteria be it pressure or volume.

After half way through the third topheading canopy FK/ASL received a stop work order (SWO) from the civil contractor and were instructed to mine a bypass tunnel to the south. The thought being the bypass tunnel could steepen up and just get under the feature but stay in basalt. Given the estimated narrow nose pillar over our heads and the potential that the crews could break into the feature again and risk a potential inrush it was agreed to continue with a modified SEM canopy through this bypass zone until safety through. Once through the zone and confident in adequate head cover FK/ASL returned to normal drill & blast operations. From the time the window was discovered and the project had returned to drill & blast operations 363 days had transpired. Almost one full year.

Schedule Pressures in the Power House

In conjunction with the service & main access tunnel portal delays and the unplanned buried valley discovery in L20 Main Access Tunnel there were massive implications to the project design, excavation plan, ventilation plan, haulage profiles and overall job schedule. The job was turned up on its head and the project team had to react and adjust to account for these complications. A bypass plan was developed to accommodate a prolonged delay in the arrival of the L20 MAT to the service bay level of the Powerhouse. This main access tunnel was essential for FK/ASL in the completion of the powerhouse as the higher access tunnel from L40 would get beyond reach of the sinking benches in the powerhouse. Something drastic had to be done to keep the power house going all the while with an eye to the main power tunnel critical path and the respective milestones. See the as bid 3d rendering and compare to the as constructed 3D rendering in Figure 8.

Originally Ramp B was designed to intersect the power tunnel manifold just west of the power tunnel rock trap (see Figure 8) with Ramp C taking off Ramp B as required to access the surge chamber and tail race tunnel. The L20 Main Access Tunnel was to be mined down grade with an approximate 200 degree turn to access the power house at the service bay level. Coming off that 200 degree turn tangentially was Ramp D to provide access for power tunnel development as that was the critical path. The power tunnel inlet manifold and inlet tunnels to the powerhouse were to be mined opportunistically via Ramp D and/or Ramp B. Prior to the Main Access Tunnel buried valley discovery Ramp B was redesigned to a 360 degree spiral to account for the L40 Service Tunnel redesign. Once the crews hit the buried valley and the implications realized it was decided to re-sequence the entire project for underground development (see Figure 9). At a risk assessment workshop it was agreed to assume negotiating the buried valley with SEM techniques could take up to one year. Thus a connector tunnel was mined to the west off the base of Ramp B to access the powerhouse inlet manifold. This was now the new critical path for FK/ASL.

The previously mentioned issues made the efficient excavation of the powerhouse significantly more challenging given lack of access. With the MAT delayed that meant a corresponding delayed arrival to the Service Bay on the west side of the powerhouse. Mining through the powerhouse inlet manifold, up Ramp D, then turning east

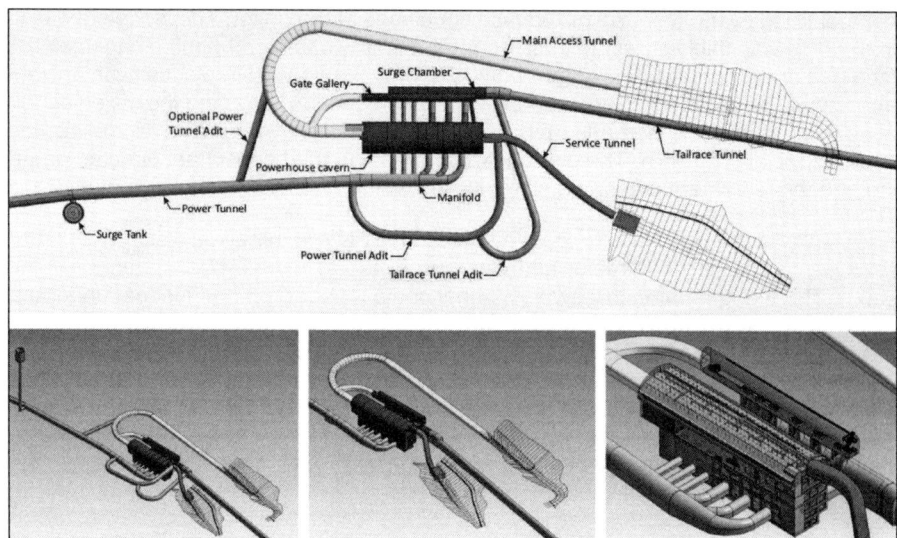

Figure 8. Plan and orthogonal views of the Powerhouse and surrounding areas before encountering the buried valley in the MAT

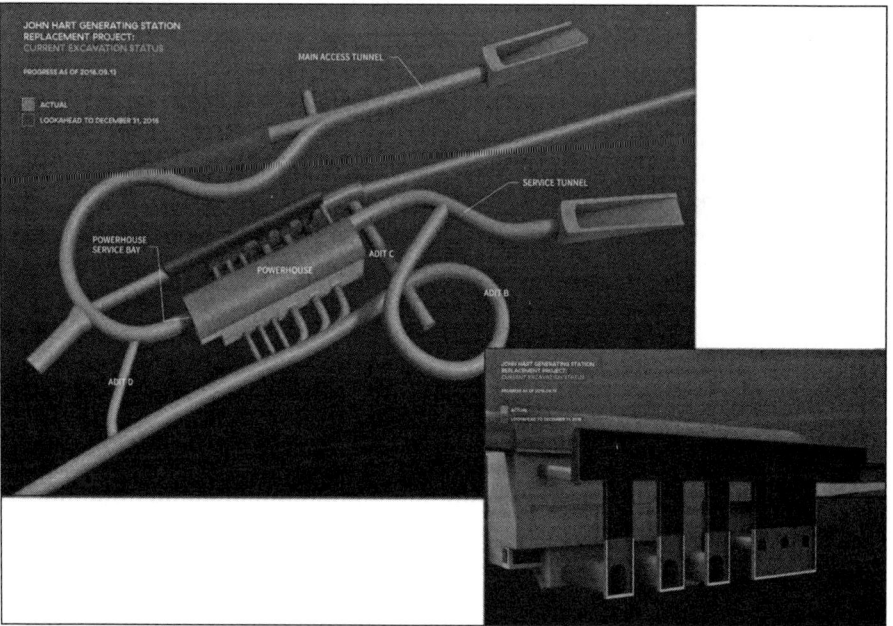

Figure 9. Orthogonal views of the Powerhouse and surrounding areas after encountering the buried valley in the MAT

to the Service would take longer thus causing a delay in the powerhouse. This meant more internal powerhouse ramping from the east was required to keeping blind benching down to the service bay elevation. Once the crews leveled off on the service bay it was decided to punch out of the powerhouse on the MAT alignment to hasten the connection with the crews developing up from the west mentioned above. The knock on effect of this strategy however delayed the gloryhole development since where the collar location over draft tube #3 as designed was not accessible due to the internal eastern ramping. Once the MAT was connected at Ramp D the crews could then access the powerhouse from the west and recover the eastside internal ramp. After that was done access with the L40 Service Tunnel was gone but crews could then mine the gloryhole though from one bench lower than originally anticipated. After the gloryhole was commissioned benching progressed downward from the west via the MAT.

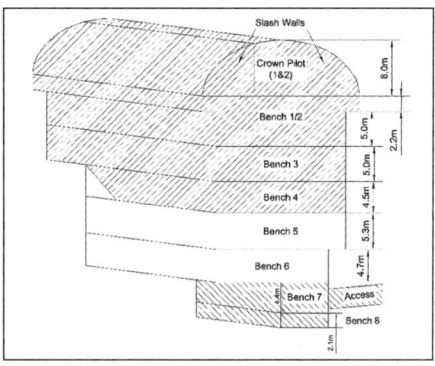

Figure 10. Powerhouse ortho view

Something had to be done to try an overcome these compounding delays to the powerhouse. During an internal FK/ASL planning meeting a technique was created to help chip away at the delay. The solution was concurrent benching with a crown pillar (see Figure 10). By this time draft tube #2 and #1 (#3 being used draw point for gloryhole) had been driven up to the north powerhouse boundary as well as all of the inlet tunnels. Concurrent with bench 4 being mining vertically bench #7 was mined horizontally via the #1 & #2 draft tubes with a 10m crown pillar overhead (benches 5 & 6). After bench #7 was completed, and smaller bench #8 was taken blind. Benches 7 & 8 were then backfilled with shot muck so that subsidence would be minimal when bench 6 was eventually taken. This concurrent work saved about a month on the powerhouse schedule and proved to be an excellent innovative idea from the Frontier Kemper on site management and supervisory team. Next benches 5 & 6 were excavated and supported to their maximum extent via the western internal ramp. At that point the inlet tunnel #3 (farthest east) was used to ramp up and recover the western internal ramp.

CONCLUSION

The John Hart Generating Station Replacement Project is a very high profile PPP in Canada and the first of its kind with British Columbia Hydro and Power Authority. The geology on the project thus far has not been as anticipated and has provided the project team with difficulties and corresponding challenges. Together with the prime and civil contractor the civil subcontractor FK/ASL JV has leveraged its experience and tunneling know how to find solutions to surmount these challenges and has been successful. This has allowed the project to recover from the delays created by these issues and has put the underground excavation portion of the project back on schedule. As such the balance of the underground excavation is on track to meet the milestones laid out in the project agreement.

In closing I'd like to share with the readers a quote by Herbert C. Hoover that has stuck with me during my career of which certain aspects of this great project has reminded me of.

"The mining engineer's works, on the other hand, depend at all times on many elements which, from the nature of things, must remain unknown. No mine is laid bare to study and resolve in advance. We have to deal with conditions buried in the earth. Especially in metal mines we cannot know, when our works are initiated, what the size, mineralization, or surroundings of the ore bodies will be. We must plunge into them and learn, - and repent."

Sewer Tunnel Excavated Under and Adjacent to Treacherous Terrain, Including Landfills, Oil Refinery, Crowded Streets, and Significantly Contaminated Material and Utilities

Russell Vakharia ▪ Los Angeles County Sanitation District
Rosann Parachuelles ▪ Los Angeles County Sanitation District

CASE STUDY

The following is a case study of the 2.5 miles long, 12-foot diameter sewer tunnel undertaken by the Sanitation Districts of Los Angeles County in two phases between 2009 and 2015.

Background of the Sanitation Districts

The Sanitation Districts of Los Angeles County (District) are a federation of public agencies that serves the wastewater and solid waste needs of over 5.1 million people in 78 cities in Los Angeles County. This includes the design, construction, operation and maintenance of 11 wastewater treatments plants, 36 pumping plants, landfills, recycling centers, gas-to-energy facilities, refuse-to-energy facilities, 1,320 miles of main trunk sewer lines, and other facilities.

Project

The project, entitled Joint Outfall "C" Phases 1 and 2 (hereinafter referred to as 'JOC'), was comprised of over 2.5 miles of 9.5 foot outside diameter reinforced concrete pipe, to provide relief to the existing sewer system. The line was installed primarily via tunneling, with a relatively small open cut section. The project also included 7 underground structures, some of which were used to tie the new line into an existing parallel sewer.

Tunnel Alignment

The tunnel excavated for this project was approximately 12 feet in diameter rib and lag tunnel, with multiple horizontal and vertical curves.

Figure 1. Sewer alignment

The tunnel was excavating using an open faced tunnel machine with a a ribs and lagging tunnel support system.

The cover over the excavated tunnel was generally between 25 and 35 feet. The ground was comprised primarily of sandy soil with some silt. The borings showed that groundwater was below the tunnel invert, although a couple of borings showed water just 2–3 feet below invert.

The first half of the tunnel alignment, including tunnel pits, ran under busy streets adjacent to numerous homes, schools and commercial businesses. The raised numerous challenges regarding traffic control, community relations and safety. The second half of the alignment ran adjacent to landfills, under multiple active railroad lines, an oil refinery and numerous oil and other utility lines. This treacherous alignment posed numerous challenges regarding ground settlement of key facilities, management of contaminated material and numerous interfaces with third party agencies.

Figure 2. Rib and lag tunnel

Tunnel Machine

The contract documents did not specify the particular type of tunneling machine to be used. The contractor proposed and used an open face tunnel machine with articulated steering and laser guidance system. The rib and lagging tunnel support system was fully erected in the tail shield and then expanded into place by a high capacity tunnel ring expander system. Steel spacers were then welded into place between the expanded rib flanges at the 4 and 8 o'clock positions. The tunnel machine shield had a 36 inch brow that extended beyond the lower part of the machine, allowing the tunnel face to be excavated while providing continuous support to the tunnel crown. The District excavator arm did not extend out beyond the extended brow.

One of the cities that the tunneling was performed under, was concerned that the open face tunnel machine that was proposed by the contractor, may not be able to control ground settlement while tunneling through sandy material under busy streets. There were discussions regarding requiring pre-tunnel grouting of the tunnel alignment or use of a pressurized closed face tunnel machine. Both these options would have had resulted in several million dollars in additional costs and several months of delay to the project, which had already begun. District and City staff met with the contractor and the District agreed to provide settlement monitoring data to allay the City's concerns.

In fact, the open faced tunnel machine worked very well. Care was taken not to excavate beyond the buried hood of the machine. The District monitored work at the heading with a combination of an inspector and video camera.

Surface settlement of no more than 0.02 foot was observed throughout the entire project, with the exception of a small section where tunneling passed an old slough overlain with fill, which is discussed in greater detail below.

Had a closed faced, pressurized face machine been specified for this project, it would have added millions to the cost of the project and months to the project schedule. The open face machine was also relatively simple and easy to repair and made it easier to remove objects encountered during mining. Given the lack of groundwater in the tunnel alignment, decent ground conditions and safe mining practices, the use of an open faced machine resulted in a major cost and time savings to the project.

Ground Settlement Monitoring

Ground settlement monitoring was performed with a combination of extensometers and ground surface monitoring points.

Multi-point borehole extensometers were installed every 200 feet over the alignment, with each instrument having two points, the lower one being 5 feet above the tunnel crown and the upper point 7 feet below the ground surface. Baseline readings were taken prior the tunnel machine reaching within 100 feet of the point. Data from each point was transmitted every 15 minutes using wireless data loggers that were tied into a WiFi system. This allowed the contractor and owner to simultaneously monitor ground settlement and to receive electronic alerts if action levels were exceeded. The extensometer lower threshold limit was 0.5 inches at 5 feet above the tunnel crown, except when passing under railroad tracks, when it was 0.25 inches. The upper threshold limit was 1 inch at 5 feet above the tunnel crown, except under the railroad right of way. The typical pattern at each extensometer, was some small movement in the lower point as the machine passed, which was then reversed as the rib/lagging set was expanded into place, resulting in negligible net movement. The upper point of most extensometers showed no movement even as the machine passed.

Community Outreach

The first phase of the project included a 1,000 foot long open cut section and 3 big underground structures constructed on a busy street with houses, schools and numerous businesses on each side. Large equipment, pieces of pipe, excavated spoils and other materials were staged on the street. Extensive traffic control was set up to divert traffic around work areas. The following efforts were made to minimize the impacts to the community:

- Prior to the start of construction, the District Project Manager and Design staff attended local community meetings and made presentations with pictures to show how the work would look like and what the potential impacts to the community were
- Construction work hours were adjusted to accommodate schools
- Signage was erected to direct customers to certain businesses
- The District hand delivered letters to all homes and businesses in the area informing them of the project and providing them the Project Manager's telephone number as well as an after hours emergency number
- When complaints were received, they were responded to promptly. The Project Manager would personally respond to phone calls and discuss issues with members of the community. When actions could be taken to allay community concerns, this was done. When corrective action was not possible, this was explained to community members, who appreciated prompt responses and explanations.
- Project staff attended the District's quarterly community outreach meetings to provide updates on the project and answer questions.

Challenges Working with Third Parties

One of the major challenges of this project was working with the numerous third parties involved. This included, but was not limited to:

- **Refinery and Oil Facility Work:** Tunneling was undertaken under a refinery and oil facility. The tunnel receiving pit was also in the oil facility. In order to construct the receiving pit and the underground structure to tie the new sewer into the existing, a large number of oil lines had to be relocated.

  All work in the oil facilities had to be performed under extremely stringent safety standards and work rules. The oil facilities were concerned that tunneling and equipment loads may damage the numerous underground oil and jet fuel lines. The contractor was therefore required to perform detailed settlement analysis for tunneling and provide detailed equipment and lift plans to show that the underground lines would not be damaged. The contractor was also required to plate the area to distribute the equipment and material loads.

- **Oil Companies:** Numerous oil and jet fuel lines were in the area that was tunneled. Oil companies often imposed stringent requirements for tunneling near or under their lines. Some lines had been sold multiple times to different companies which resulted in poor as-built drawings and confusion from oil company field representations regarding identification of lines and their nature. Oil company rules and requirements were often inconsistent and constant adjustments had to be made in the field.

- **Railroads:** The tunneling went under several major freight railroads, including the Alameda Corridor, which is a major transportation corridor from the Ports of Los Angeles and Long Beach. The railroads reviewed all tunnel support and settlement monitoring information.

- **AQMD:** Southern California has some of the nation's most stringent air quality rules, including how contaminated material is to be stored and handled.

Community Issues—Tunnel Pit Adjacent to Mall

One of the tunnel pits on the project was shown on the plans as being near one of the entrances to a local mall. There were discussions with the mall owner to reduce the impact to business by constructing a new entrance in another location, add signage and take other measures. However, given the slowdown in the economy at the time, the mall owner objected strenuously to all the options and called the City to voice his displeasure. The City informed the District that it could not place the tunnel pit in the location shown, particularly given the large amount of materials and equipment that would be placed on the surface near the pit area. The City stated that, given the current economic climate, it could not afford the impacts to businesses.

The District and contractor met to discuss potential alternatives. After several discussions, a plan was developed to eliminate that tunnel pit and adjacent structures and to tunnel on a curve around that location. This option had not been previously considered during the design phase of the project, because the new alignment sent the tunnel just under and very close to multiple oil lines. The new alignment also required an existing

Figure 3. Oil line relocation

utility line to be relocated at considerable cost. The existing and new alignment are depicted in Figure 4.

Liner Plate Tunnel

A 100-foot section of the connection between two junction structures (JS 1 and 2 depicted in Figure 4) ran under a busy street, with utility lines above and below, making it impossible to construct with the specified open cut method. Relocating the utilities would have taken an inordinate amount of time and delayed the entire project.

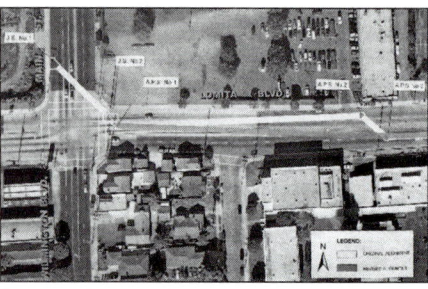

Figure 4. Revised alignment

The solution the District and contractor agreed upon was to use a liner plate tunnel with a small loader to remove spoils. Pressure was kept on the face using while the liner plate was assembled by hand.

CalOSHA Tunnel Classification, Re-Classification, and Tunnel Machine Retrofit

The tunnel for Phase 1 was initially classified as Potentially Gassy with Special Conditions. Towards the end of the Phase 1 tunnel mining, the tunnel machine encountered some refined gasoline that had leaked from a pipeline above the alignment many years ago. OSHA was notified and they reclassified the tunnel as Gassy with Special Conditions. This required the tunnel machine and related equipment to be retrofitted to be "permissible." This included things like: adding fans and increasing fan duct size to increase air flow, replacing electrical powered equipment and tools with air-powered ones, installing spark arrestors, installing explosion proof electrical equipment and wiring, sealing off conduits, adding heat exchangers to cool the oil for the larger permissible motors, adding LEL sensors and automatic shutoff of the tunnel machine if the LEL levels rose. Since permissible equipment and wiring is generally larger, this required major modifications to the tunnel machine. All this work had to be performed on a tunnel machine that was over 1,500 feet from the shaft. The retrofit work took approximately 8 weeks to perform and cost over $1 Million. Other actions taken included, increasing the air flow to the tunnel face, having an industrial hygienist evaluate the conditions in the tunnel and prepare an action plan for air monitoring after the resumption of mining. The tunnel miners also had to take Hazwoper training classes. When mining resumed, the face was sprayed with an agent to reduce airborne particulates of the contaminants.

Phase 2 of the project was also initially classified by CalOSHA as Potentially Gassy with Special Conditions. However, the contract documents for Phase 2 required the contractor to bid the project assuming a Gassy classification, including permissible tunnel equipment and worker training. The contract documents further stated that if the tunnel was reclassified during construction, the contractor would be granted up to a 10 day non-compensable time extension for any delays resulting from OSHA inspections or other work related to the reclassification. The Phase 2 tunnel was in fact reclassified during construction after contaminated material was encountered. Given the wording in the contract documents, the contractor began the project using permissible equipment and was ready to implement other measures such as increasing air flow, all of which was accomplished in a little over 10 days, resulting in no impact to the project.

CalOSHA also required escape shafts or chambers to be placed every 5,000 feet. The contractor installed steel casings with stairs, as escape shafts, to meet this requirement. This was done at the contractor's cost since the contract documents required the project to be bid as a Gassy tunnel.

Importance of Constructability Reviews and Incorporating Lessons Learned

The issue above, in which the contract documents were changed between Phases 1 and 2 of the project, is just one of many such examples. Both the design and construction management of this project were performed by experienced District staff. This allowed for feedback from the field regarding constructability issues. Other examples of this included:

- Including approved traffic control plans in contract documents instead of having the contractor submit them. The contractor was free to submit an alternative at their own risk.
- Utility lines were relocated prior to contract bid, to reduce the possibility that delays to the relocation would impact the project
- The contraction scope and potential methods were reviewed with Review construction with refinery representatives to determine their concerns about loads on existing underground lines. Language was included in the contract documents for the contractor to submit detailed lift plans with structural calculation prior to performing certain work.

Encountering Water and Oil Saturated Squeezing Ground

While tunneling under an oil refinery in an area that decades ago had been a slough, overlain with fill, dark colored water began seeping through the rib and lagging support system into the tunnel.

The ground also appeared to be squeezing in, pushing through some of the lagging window boards and also resulting in cracks to some of the lagging. Although air sensors did not indicate contaminant levels harmful to humans, mining was stopped and CalOSHA was contacted.

The source of the water appeared to be an unlined temporary water storage area adjacent to the tunnel alignment. Although soil borings performed prior to contract bid had indicated no groundwater in the area, these borings had been done multiple years earlier when the area had been dry. However, just prior to the tunnel machine's approach, this area had been filled with water that had been used to water test some oil storage tanks and some rain water had been pumped in.

Unlike most of the other areas of the tunnel alignment, this short section of the old slough had a higher clay content.

Figure 5. Seepage into tunnel

The water seepage into the tunnel almost stopped completely within a couple of days after beginning.

To reinforce the tunnel and prevent any chance for additional seepage of water, the contractor inserted ¾" thick rolled steel liners into the tunnel, expanded them against the existing ribs and lagging and then grouted the small annular space before continuing mining.

Freight Train Hitting Piece of Pipe

Although the following incident had no direct bearing on the outcome of the project, it did cause quite a stir.

One day, the Project Manager received a call from the job site inspector that a 40,000 pound piece of pipe had been hit by a train while being transported to the site and it was unclear as to the nature of the damage to the train.

When the Project Manager arrived at the site shortly thereafter, he saw the piece of pipe lying on the track apron and the train at a halt. By that time, the area was swarming with police, members of the railroad and the contractor's personnel.

Evidently, the truck trailer carrying the pipe had been hit by the train while crossing the tracks.

Luckily, the train had not derailed.

Inefficiency Claim Resulting from Working Out of Sequence

The Contract Documents led the contractor to believe that the City would allow one traffic control plan to be implemented over a large portion of Phase 1 of the project, which was built under a street with significant housing, schools and businesses on both sides. However, when traffic control plans were submitted by the contractor, the local City required the work to be performed in smaller phases, with each phase having to be restored prior to work beginning on the next phase. This delayed approval of the traffic control plans. This problem was

Figure 6. Water source (tunnel alignment shown with arrow)

Figure 7. (a) Steel liner prior to insertion; (b) steel liner inserted in tunnel

compounded by the fact that the traffic control plans were approved in a different order than the contractor had intended to work. This resulted not only in a delay to the start of work, but also to the contractor working out of sequence, resulting in inefficiencies.

Following completion of the work, the contractor submitted a delay and inefficiency claim for traffic control impacts. The impacts included, among other things: delays to the start of work, equipment having to remain on site for a longer

Figure 8. Pipe on tracks

duration than planned, equipment having to be mobilized and demobilized multiple times, and other related impacts. The District evaluated the contractor's claim using a variety of methods, including a review of the monthly CPM schedules, as well as plotting and overlaying the contractor's planned equipment usage with actual equipment usage. The analyses showed that the contractor did indeed have to keep certain equipment on the project longer than planned and incurred additional mobilization and demobilization costs.

Closing

This challenging project, performed using an open faced tunnel machine with a rib and lagging tunnel support system, was completed successfully despite treacherous work conditions.

Design and Construction of the Capitol Connector Pedestrian Tunnel

Andrew M. Stone ▪ Stone + Howorth
John Jacoby ▪ Stone + Howorth
Matt Over ▪ McMillen Jacobs Associates
Joe Schrank ▪ McMillen Jacobs Associates

ABSTRACT

The Tennessee Capitol sits atop a limestone precipice in the heart of downtown Nashville and is home to the state senate and legislative chambers. Legislators currently access the building via the Motlow Tunnel from their offices in Legislative Plaza. With the relocation of their offices to the Cordell Hull Building, a new access for legislators to the State Capitol is required. Work for the utility and pedestrian tunnel involves extending the existing elevator shafts 50 feet down from the terminus of the Motlow Tunnel to a new access tunnel from the newly renovated Cordell Hull Building. The schedule of the entire rehabilitation project was driven by the need to relocate the legislators prior to the 2018 legislative session. This timeline required very aggressive design and construction schedules. This paper discusses design and construction challenges associated with the tunnel mining and elevator shaft extensions within the confines of this historic structure.

PROJECT BACKGROUND

Built in the 1970s, Legislative Plaza is currently home to the legislative offices for all state representatives and is located south of the Tennessee State Capitol Building. Currently, access to the State Capitol is provided via the Motlow Tunnel (circa 1950s) to two elevators, which connect to the main floor of the State Capitol. Because the legislative offices were in need of renovation and modernization, the State of Tennessee elected to move them to the Cordell Hull Building, which sits on 5th Avenue to the east. In order to provide the same level of access to the State Capitol, the idea of the Capitol Connector Pedestrian Tunnel (CCPT) was developed.

In July 2015, Stone and Howorth in conjunction with McMillen Jacobs Associates began the planning and design of the fast tracked CCPT. The project consists of a mined 435 linear foot (LF) tunnel to provide direct access to the voting floor in the State Capitol for legislators from their offices in the renovated Cordell Hull Building (Figure 1). The project also includes extension of the existing elevator shafts and creation of an emergency egress stair to the new tunnel approximately 50 feet below the State Capitol.

From the beginning, the project has presented unique challenges: the fact that the proposed alignment passes adjacent to several historic structures (such as the Tennessee State Capitol, a statue of President Andrew Jackson, the tomb of President James K. Polk, and the John Sevier Building); the need for the design and construction team to work around the legislative schedule; and the difficulties of working in an existing marble-lined tunnel to advance the shafts and meet the project timelines. This paper discusses each of these challenges and how the team worked to solve them.

CCPT Overview

The typical tunnel excavation cross section for this project is 17.3 feet wide by 15.5 feet high to provide an 8- to 12-foot-wide by 11.5-foot-tall walkway with utility chases on both sides. Three vestibules with an excavation cross section of 25.5 feet wide by 20 feet long by 17.5 feet high are included along the tunnel and will have hung dome vaulted ceilings. The tunnel terminates in an elevator lobby, which is 14- to 23-feet wide and 23-feet long. The new tunnel will connect into the Motlow Tunnel via an extension of the shafts and services of the existing elevators, which currently connect the Motlow Tunnel to the Capitol Building. An egress staircase will be constructed in a new shaft from the Motlow Tunnel down to the new elevator lobby.

Project Schedule

The fast track schedule was the result of the late addition of the tunnel to the overall Cordell Hull Renovation Project and the need to have all facilities open for the 2018 Legislative Session. When the design team mobilized, construction was slated to commence in the fall of 2015 with the demolition of the Central Services Building. From the outset, the team had only seven months to generate the Geotechnical Data Report (GDR) and Geotechnical Baseline Report (GBR), and to design the tunnel and its components.

Other challenges related to the need for collection of geotechnical data and the eventual construction to work around the legislative schedule. Fieldwork could only occur during nonlegislative periods, and all data collection had to be coordinated with the State Capitol staff, including security. The GDR and GBR were completed in the fall of 2015 while the 30% design of the tunnel was underway. Final (100% Construction) drawings for the CCPT were produced at the end of March 2016 with the tunnel contractor mobilizing to the site at the start of the second quarter of 2016.

GEOTECHNICAL INVESTIGATION

Geotechnical Exploration Program

The geotechnical investigation consisted of an initial geotechnical exploration program performed within the footprint of the proposed construction from August 12 through August 18, 2015. This program included the drilling of four borings with the installation of a piezometer in each, and the collection of representative soil and rock samples for subsequent laboratory testing. In addition, packer testing was performed within each of the four borings at two different elevations. The boring locations are shown in Figure 1.

Because of the low rock quality designation (RQD) values in the boring near the tunnel portal, a secondary geotechnical exploration was performed on January 18, 2016. This included the drilling of one boring to a depth of 40.5 feet, and the installation of a piezometer upon the completion of drilling.

Geologically, the site is located in the northwest corner of the Outer Central Basin of Tennessee. The underlying subsurface stratigraphy consists of two main categories:

- Artificial fills consisting mainly of organic, medium stiff, moist topsoil, brown medium stiff, moist silty clay, and concrete and asphalt pavement over an aggregate subbase.

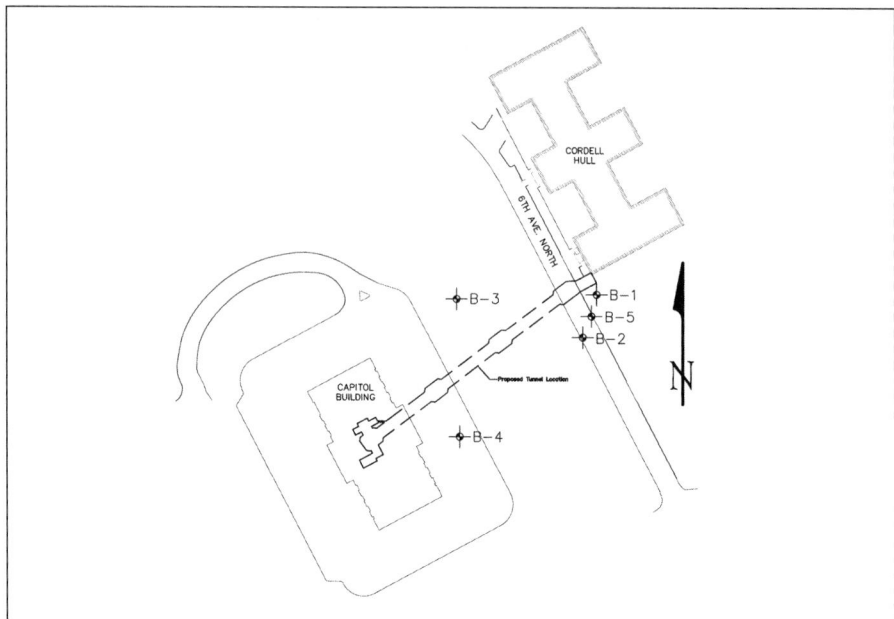

Figure 1. General location map and boring plan for the CCPT

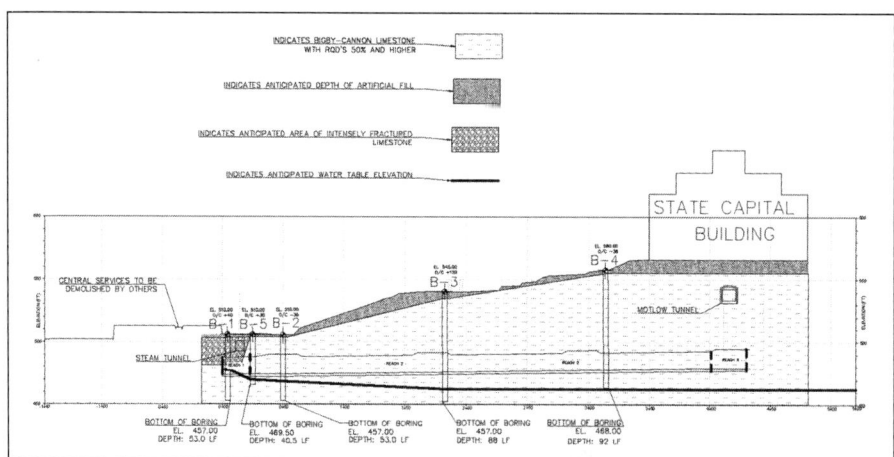

Figure 2. Geological conditions along the tunnel alignment

- Bedrock (Bigby-Cannon Formation) consisting of strong, lightly weathered, blue-grey limestone. This bedrock constituted the bulk of the material excavated during construction (Figure 2).

The site consists of moderately steep to steep slopes (18%–31%) with level areas at the top and toe. There are no known major faults traversing the site since the nearest major fault is the New Madrid fault located in Memphis, Tennessee, approximately 200 miles away.

Table 1. UCS with Young's Modulus (ASTM D7012 Method C)

Boring	Elevation (ft)	Bulk Density (lb/ft³)	UCS (psi)	Young's Modulus (psi)
B-1	471.5	168.40	12,776	7,400,000
B-2	472.3	168.30	7,714	4,100,000
B-3	471.4	164.00	13,507	7,500,000
B-4	474.4	168.40	16,763	7,800,000
Average Below Tunnel Zone			*12,690*	*6,700,000*
B-1*	485*	167.90*	4,710*	4,400,000*
B-2	488.2	164.20	16,741	7,100,000
B-3	488.3	166.40	12,075	8,200,000
B-4	491.3	168.7	12,213	10,600,000
Average within Tunnel Zone			*11,435*	*7,575,000*
B-1	494	168.30	12,961	8,100,000
B-2	499.7	168.40	8,525	4,400,000
B-3	500.8	167.80	19,551	9,000,000
B-4	506.8	168.80	19,564	6,700,000
Average Above Tunnel Zone			*15,150*	*7,050,000*

*Laboratory sample failure occurred because of a pre-existing discontinuity, resulting in a lower compressive strength in comparison to other laboratory samples taken.

Laboratory Testing

Amec Foster Wheeler (AMECFW) of San Diego was commissioned to perform laboratory testing aimed at evaluating index and strength characteristics of selected rock core samples from the borings. The core samples collected from each of the borings represented the material below the invert, above the crown, and within the zone of the proposed CCPT. The rock testing included Unconfined Compressive Strength (UCS) with Young's Modulus, Brazilian Tensile Strength, Point Load Strength Index, and Cerchar Abrasion Test. The results of the UCS testing are presented in Table 1.

Geotechnical Considerations

Based on the results of the GDR, the tunnel was divided into three reaches. Each reach is described in detail in the GBR and is summarized below.

Reach 1 (Station 0+00 to Station 0+26). Reach 1 consists of the Bigby-Cannon Limestone. It extends 26 feet from the portal near the Cordell Hull Building towards the State Capitol Building and ends prior to Vestibule #1. The ground surface slopes slightly from east to west along the alignment, providing a consistent 16± feet of cover. Moderately weathered, intensely fractured Bigby-Cannon limestone was anticipated from the crown of the proposed tunnel to just below ground surface where the upper approximately 1.5 feet below ground surface are composed of concrete or asphalt paving and a crushed stone subbase. Discontinuities within the upper 20 feet below the ground surface of this reach were primarily horizontal, closely spaced, and included frequent rubble zones and 1- to 3-inch clay infillings within the near surface weathered zones.

Reach 1 had RQD values of between 20% and 100%, and these values were anticipated to be lowest closest to the portal, increasing as tunneling progresses toward Station 0+26. Within the approximate tunnel zone, RQDs were between 20% and 76% near the portal and increased to between 84% and 100% at boring B-2. Low RQD values within the upper portion of the tunnel as well as the shallow ground cover above the tunnel required additional initial support techniques in this region.

Reach 2 (Station 0+26 to Station 4+01). Reach 2 extends 375 feet from Station 0+26 to the start of the elevator lobby below the State Capitol Building at Station 4+01. This reach consists of the bulk of the tunnel excavation including the three vestibules. Cover over the tunnel in this reach ranges from approximately 16 feet at Station 0+26 to approximately 60 feet because the ground slopes steeply upward toward the State Capitol Building. Ground cover consists almost entirely of limestone bedrock with the exception of artificial fills that consist of asphalt paving and grassed areas ranging from 1.5 to 3.0 feet in depth. Several sensitive structures lie within this reach, which required surface monitoring throughout the tunneling process to ensure that they were unaffected by the vibrations caused during construction.

The Bigby-Cannon limestone within Reach 2 is strong, occasionally fractured, with little to no weathering. RQD values within this reach ranged between 76% and 100%, and were typically greater than 92%. Within the approximate tunnel zone, RQDs were between 86% and 100%. It was anticipated that an unsupported section of tunnel for this reach would exhibit stable behavior with the potential for isolated block failures.

Reach 3 (Station 4+00.97 to Station 4+32). Reach 3, which consists primarily of Bigby-Cannon Limestone, extends 32 feet from Station 4+01 to the shaft below the State Capitol Building at Station 4+32 and includes the elevator lobby area as well as the three shafts. This entire reach of tunnel is directly below the State Capitol Building. Ground cover is thick in this reach of the tunnel at approximately 60 feet.

The Bigby-Cannon limestone within Reach 3 is very strong, occasionally fractured, with little to no weathering. RQD values within this reach ranged between 76% and 100%, and were typically greater than 92%. Within the approximate tunnel zone, RQDs were between 92% and 100%. It was anticipated that an unsupported section of tunnel for this reach would exhibit stable behavior with the potential for isolated block failures.

ALIGNMENT CONSIDERATIONS

Alignment of the new tunnel system with the existing elevator shafts was the primary objective of the CCPT design. Some of the key constraints that drove the geometry of the alignment were: limiting the vertical extension of the existing shafts without destabilizing the existing Motlow Tunnel; providing a means of emergency egress besides the portal or elevators, via the existing Motlow Tunnel; and the fact that the Motlow Tunnel terminates at the elevator shafts, forcing any additional connections to be on the south side of the existing elevators. Refer to Figure 3 for a 3D rendering of the shafts, cavern, and connecting tunnels.

Since the rock in Nashville is typically of a high quality and usually requires blasting, the vertical shaft construction, using either the Motlow Tunnel or new tunnel to remove the muck, was identified as the most impactful to the construction schedule and needed to be as limited as possible.

Additionally, there were three other constraints: first, the portal elevation was dictated by a connection to a subterranean level of an existing building; second, the Americans with Disabilities Act (ADA) maximum allowable slopes for walkways limited the grade of the tunnel; and third, if the new tunnel excavation resulted in such shallow excavation below the Motlow Tunnel that it required additional support measures, any construction savings in excavation would be lost to structural considerations. Ultimately, the ADA requirements governed without destabilizing the Motlow Tunnel or impacting the planned portal location, and a grade of 2.5% between vestibules and a 1% grade in the vestibules were established.

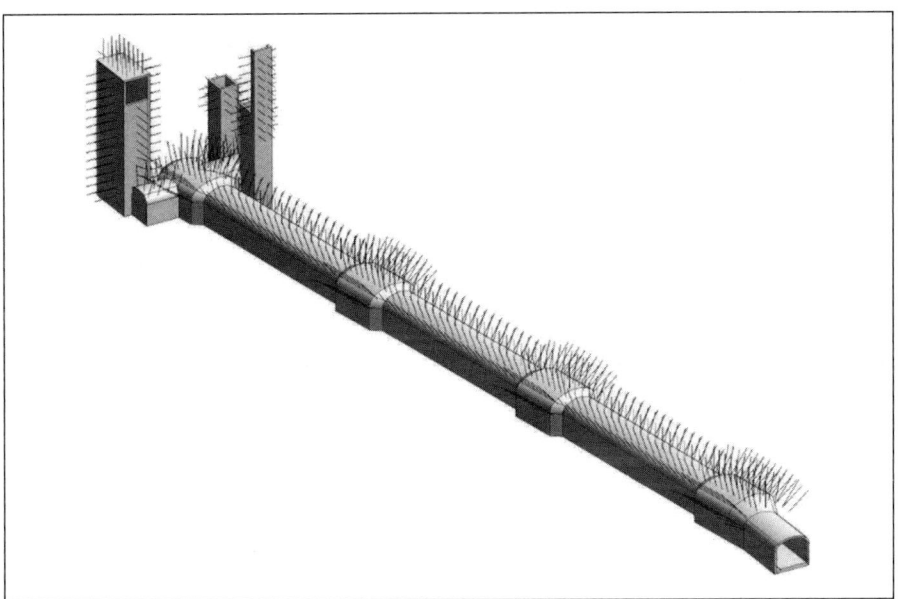

Figure 3. Capitol connector pedestrian tunnel 3D rendering

The shape and terminus of the CCPT meant emergency egress could not be attained by means of a cross passage and exit via a parallel structure. Moreover, the elevators do not qualify as a means of emergency egress. Therefore, the need for an egress stairwell shaft was apparent early during the design.

Initially, the egress shaft was sited to connect all the way to the Capitol Building. In its current state, the Motlow Tunnel has no emergency egress except at the portal. The initial plan was to provide a redundant egress means by constructing the stairwell shaft to tie into the Motlow Tunnel as well as the Capitol Building. However, after careful discussion with the State Fire Marshal, it was determined that the tunnels (Motlow and CCPT) could mutually serve as the alternate means of emergency egress for each other. Therefore, the emergency egress stairwell shaft is technically a winze that connects the tunnels, without a surface expression.

The last challenge to placement of the emergency egress was that the Motlow Tunnel does not extend past the existing elevators (running approximately north from portal to the elevators). Therefore, any new shaft construction had to be to the south of the existing elevators, which dictated the connection being on the opposite side of the elevator lobby.

The last alignment consideration was how to provide proper ventilation, electric, audio visual, and plumbing to the new tunnel. Ultimately, the most efficient choice was to tie into the Capitol Building systems. This was achieved by installing a utility chase in the southeast corner (back face) of the existing east elevator shaft, which will extend the full height of the existing and extended elevator shafts. Extension of utilities along the back wall of an elevator shaft ultimately requires wrapping the utilities around the perimeter of the shaft to connect to the tunnel interior, which would result in a widening of the shaft to accommodate the ductwork and mounting hardware. Rather than enlarge the shaft (which would have reduced the critical pillar geometry) or extending the tunnel excavations further, a small-diameter bore (24 inches) was selected that could connect the back of the utility chase to the north sidewall of the typical tunnel. This allows all utilities to be brought down from the Capitol and hidden in the interstitial space behind the internal build-out and the tunnel lining intrados.

DESIGN CONSIDERATIONS

The CCPT design process required balancing and understanding several design objectives. The key design requirements included extension of existing elevator shafts to connect with new construction, the new construction to be a fully waterproofed system, and protection of a series of overlying and adjacent sensitive structures. Notwithstanding the external constraints, the geometry of the tunnel terminus involves a cavern with three connections and three shafts in close proximity.

For the standard tunnel section, the initial support consisted of 6 inches of steel fiber reinforced microsilica shotcrete and 10-foot-long double corrosion-protected (DCP) grouted rock bolts. This initial support was supplemented as needed for other features of the tunnel, which are described in the following sections. Also discussed are the structural and geotechnical engineering approaches required to meet the design requirements. Figure 4 presents the tunnel plan, and Figure 5 presents the tunnel profile.

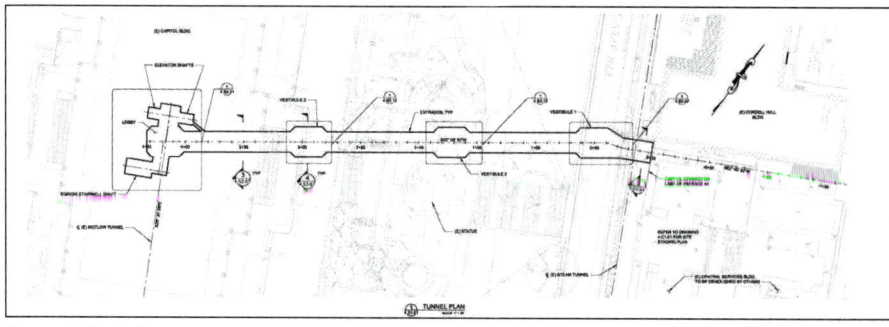

Figure 4. Capitol connector pedestrian tunnel plan

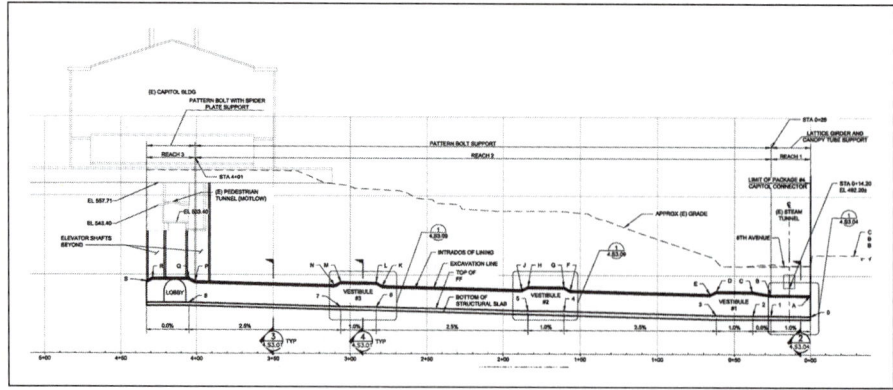

Figure 5. Capitol connector pedestrian tunnel profile

Tunnel Turn Under

The turn under for the CCPT passes through an existing reinforced concrete wall (from the former Central Services Building) and almost immediately traverses below 6th Avenue. Directly behind the existing wall, there was a short region of backfill before competent limestone was encountered. Below 6th Avenue is a steam tunnel (original construction 1973), approximately 5 feet above the crown of the tunnel. Because of the very sensitive nature of the utilities, steam tunnel, and overlying road, grouted canopy spiles on 18-inch centers were installed in a semicircular array from quarter-arch to quarter-arch combined with lattice girders on 4-foot centers to provide a two-way load resisting system to support the excavation. The lattice girders were installed with 3 inches of shotcrete to the exterior, meeting American Concrete Institute (ACI) standards for minimum clear cover and additionally serving as the permanent support underneath 6th Avenue.

Main Tunnel Features

There are several unique features to this relatively short tunnel. First, the tunnel has intermediate vestibules, areas that are approximately 2 feet taller and 8 feet wider than the typical section. The vestibules are located approximately every 100 feet along the tunnel and are approximately 20 feet long, excluding the tapered connections into and out of the typical section. The vestibules serve an operational purpose, in compliance with ADA specifications, by breaking up the 2.5% typical grade in the tunnel with intermediate 1% regions.

The vestibules required a change in the ground support to longer and closer spaced DCP grouted rock bolts (13 feet long instead of 10 feet, and 4 feet on center spacing instead of 5 feet). Because of the change in the excavated shape (to a larger radius), which increased the compression in the permanent lining crown, the permanent lining did not need to be increased from the 12-inch thickness in the typical section.

Second, the tunnel has a pivot (not a curve), to provide both a perpendicular turn under at the portal and the desired architectural shape at the elevator terminus. The tunnel pivot was placed in the transition region leading into Vestibule #1, where the geometry makes the connection between the typical tunnel sections that are on different bearing angles. This asymmetric section is the only region of the tunnel without a vertical slope. The pivot is required to position the last vestibule in the system (referred to as the elevator lobby) in a way that aligns the extended shafts with existing elevator shafts above as well as positioning the egress stairwell such that it connects appropriately to the Motlow Tunnel. Architecturally, the pivot also serves to break up the visual sight lines within the system.

Third, the tunnel is fully tanked. A spray-on waterproofing membrane was selected as the main waterproofing element for the tunnel for two principal reasons. First, since admixtures can only reduce the shotcrete permeability, a membrane was preferred. Second, the tunnel has a high level of geometric variance and many junctions. Because of the difficult installation of a sheet membrane in these conditions, especially at junctions, the flexibility of the spray-on membrane to adapt to the final shape of the system made it the preferred solution.

Elevator Lobby

At the terminus of the tunnel is an enlarged cavern with two near-perpendicular connection tunnels (referred to as "adits" in Figure 3), and three shafts. The principal design challenges in this area were the irregularities to the geometry and the coordination of the new shafts with existing structures.

The geometric constraints at the tunnel terminus led to three key design decisions. First, the elevator lobby, the main tunnel, and both adits all have either different radii or crown heights, which required 3D analysis to determine the stresses in the structural system and surrounding rock mass. Second, the connection of the shafts and the crowns of the adits form shallow troughs, which required a drainage system on the lining extrados to prevent ponding of groundwater. Finally, the lobby is supported principally by pillars that are exposed on three faces, which needed to be checked for stability in the initial condition and designed with an appropriate confinement in the permanent condition.

Structurally, this is the only region in the tunnel where the lining spans two ways. The geometric transitions in the elevator lobby often induce stress concentrations, simply because the load path to points of support is indirect. To mitigate these stress concentrations without thickening the lining, welded wire fabric (WWF) was specified and the permanent lining was connected to the permanent rock bolts using spider plates (rebar welded to nuts that threads onto the protrusion of the rock bolt past the face plate). The geotechnical design in the elevator lobby (which has the same geometry as a vestibule) required the same rock bolt length as the vestibules, but at a tighter spacing because of the discontinuity in the sidewalls at the connecting tunnels.

At each of the shaft breakouts, the connecting tunnel crow radii intersect to form a concave surface that could entrap percolating groundwater. Because the internal waterproofing system is a spray-on membrane, it is possible that these surfaces could intercept gravity drainage and lead to ponding. These locations are also where header beams are required to support the adjacent arch and overlying shaft wall, therefore, a design that would relieve pressure on the extrados of the lining at these locations was appropriate. The design approach was to install a geotextile membrane and strip drain that would move water to the sides of the shaft breakouts, where it would be free to continue draining via gravity through the rock mass. To protect the strip drain from the shotcrete installation, a protective, light gauge, galvanized metal sheet encases the geotextile and drain system.

The rock pillars provide support to the majority of the elevator lobby, the elevator shafts, the egress stairwell shaft, and the connecting tunnels, and have a much higher vertical effective stress than the rock mass anywhere else in the system. The pillars are typically exposed on three sides, and at a minimum, were 15 feet tall when initial support was installed. To protect these pillars in the short term as well as under permanent conditions, a stability analysis was conducted that established minimum confinement and pillar dimensions. The minimum pillar dimension was 5 feet in width based on Hoek-Brown rock mass analysis (Hoek and Brown, 1997), and the initial lining design (6 inches of fiber reinforced shotcrete) incorporated WWF to confine the pillars.

CONSTRUCTION CONSIDERATIONS

The Cordell Hull Building Renovation Project was procured as a Construction Manager at Risk (CMAR) project delivery method. The Nashville office of Skanska USA Building Inc. of Parsippany, New Jersey (Skanska) was selected and hired by the state in July 2015.

The tunnel and shafts were a standalone contract package within the Cordell Hull Building Renovation Project. Skanska obtained multiple bids for the tunnel construction and recommended a tunnel contractor based on qualifications and price. Clark Construction Group, LLC of Bethesda, Maryland and its subsidiary, Guy F. Atkinson Construction, LLC of Golden, Colorado were selected.

Excavation of Reach 1 was completed by roadheader because of the proximity to the Cordell Hull Building, steam tunnel, and 6th Avenue. Excavation proceeded slowly because of the rock quality and learning curves associated with new operations and the fast mobilization. Some gravel fill was encountered on the north side of the portal at the corner of the Cordell Hull Building. Water was also an issue near the portal.

Blasting started at the beginning of Reach 2. The first six blasts were 4-foot advances, allowing all the parties involved to become comfortable with blasting and the blasting protocols to be fine-tuned. Blast advances (rounds) were limited primarily based on peak particle velocity (PPV) limits and concern about potential damage to nearby structures. From Station 0+59 to Station 3+54, the blast lengths generally varied from 6 to 8 feet long with occasional longer (10-foot-long) rounds.

As the excavation approached the Capitol Building, concern about the PPVs required the contractor to limit blast lengths to 4 and 5 feet. However, as the excavation continued under the building, the PPVs generally decreased, allowing the contractor to use 6-foot-long rounds to excavate the elevator lobby.

The shafts were initially excavated from the Motlow Tunnel with hand-operated equipment (jackleg drills and splitters) until the new tunnel, elevator lobby, and adits were excavated and access was possible from below. This excavation process was slow, and extended until all three shafts were excavated to a depth of 17 feet below the Motlow Tunnel invert. The muck was removed from the top of the shafts through the Motlow Tunnel until raises were blasted from below. Once the raises were excavated, the miners accessed the shafts from below and blasted from the top down, pushing the muck into the raises and removing it through the new tunnel. Final lining of the CCPT is anticipated to be completed in early April 2017.

ACKNOWLEDGMENTS

The authors thank the State of Tennessee office of General Services, Centric Architecture, Skanska and Clark/Atkinson for their support in the publication of this paper.

REFERENCES

ASTM D7012-14. 2014. *Standard Test Methods for Compressive Strength and Elastic Moduli of Intact Rock Core Specimens under Varying States of Stress and Temperatures*. West Conshohocken, PA: ASTM International.

Hoek, E. and Brown, E.T. 1997. Practical estimates of rock mass strength. *Intnl. J. Rock Mech. & Mining Sci. & Geomechanics Abstracts*. 34(8): 1165–1186.

Geotechnical Considerations for the ORBEEC Drumanard Tunnel

David Neil ▪ WSP | Parsons Brinkerhoff

ABSTRACT

The ORBEEC alignment required crossing the historic Drumanard property to complete the connection of I-265 between Kentucky and Indiana. The Drumanard property is listed on the National Historic Register and surface disturbance of the property is not allowed. The most viable option to overcome this obstacle required a pair of shallow three lane tunnels each approximately 1700 ft. long for the Kentucky HWY 841 approach to the proposed bridge. In order to determine the feasibility of a tunnel a limited geotechnical boring program was performed. The program included groundwater monitoring, core sampling and a small geophysical mapping program which was completed during the 2004 to 2007 time frame. These data were used for the prefeasibility study and a comprehensive test drift and underground coring program was designed and bid with the information to be used to form the geotechnical basis for a final design. For various reasons the test drift plan was abandoned and in 2011 a series of three horizontal core holes, one in each tunnel crown and one in the pillar was completed by S&ME. This data was incorporated into a comprehensive Geologic Baseline Report (GBR) and Geologic Data Report (GDR) prepared by Golder Associates. The GBR and GDR are used in the tunneling industry to present a baseline of geologic conditions a contractor may expect to encounter on a given project and used for input into planning, designing, and performing the work. Included in the GBR are the hydrogeologic conditions, the physical characteristics of the rock mass, the classification of the rock lithologic units to be excavated and the ground support requirements for initial excavation, the stress within the units, and the instrumentation requirements to monitor the reaction of the rock mass to the excavation process.

INTRODUCTION

The East End Crossing of the Ohio River Bridges Project (EEORB) alignment was determined by the alignment of I-265 in Indiana and Kentucky during the 1980s. As traffic levels increased the need to complete the connection of I-265 and I-65 in Indiana via a new bridge over the Ohio River was no longer an option but a requirement to accommodate area population and economic growth. Environmental activist opposition to the project centered about the crossing of the historic Drumanard property and its position on the National Register of Historic Places. With the possibility of bridging or a cut and cover tunnel excluded by the Registry, a tunnel became the only viable option to cross the property.

The Drumanard Tunnel is the twin tube access to the new east end bridge. The South Portals are located at the intersection of SR 841 and US HWY 42 and progress northward some 1,700 lf to allow traffic to cross over a second bridge at Harrods Creek and onto the new EEORB. The tunnels were constructed by a Sequential Excavation Method (SEM) using drill and blast. Construction was under the overarching Public Private Partnership (P3) between the Indiana Finance Authority (IFA) and Walsh Vinci Balfour Beatty (WVB) and the developer is a Joint Venture between Walsh and Vinci (WVC). At completion ownership of the tunnel and the 841 extension of I-265 reverts to KYTC at final acceptance.

As a part of the feasibility of tunneling under the property in the 2004–2007 time frame Hatch Mott McDonald as a sub to designer H.W. Lochner prepared a Geologic Baseline Report (GBR) and Geologic Data Report (GDR) to establish the baseline conditions expected for tunneling. As a part of this investigation HMM proposed an exploratory tunnel located in the crown of the Northbound bore to supplement the geologic information required for design due to the lack of access to the surface of the Drumanard property. The proposed exploratory tunnel was bid in 2007 but the program was not awarded due to cost. As the EEORB project feasibility progressed a supplemental GBR/GDR was prepared by Golder Associates using additional geologic information and data from three 1,500 foot horizontal bores by S&ME along with additional surface bores and a geophysical investigation to determine top of rock after KYTC acquired ownership of the property.

GEOLOGIC BASELINE REPORT

The GBR and GDR are used in the tunneling industry to establish contractual responsibility for risk associated with unforeseen site conditions. The GBR provides the basis for initial support design and instrumentation to monitor conditions during the work. This data and information is to assist potential bidders in evaluating the requirement for excavating and supporting the ground. Risks associated with the data provided in these two documents are allocated to the contractor and risks associated with more severe conditions are allocated to the owner.

The GBR is typically organized into several sections including the anticipated ground conditions, potential subsurface hazards, and construction considerations. The document provides a range of intact rock properties, a record of the borings and geophysical investigations, baseline groundwater monitoring, and environmental restrictions or requirements. This information is used by the designer to prepare his concept designs and is normally supplemented by additional geotechnical information as required as the project moves to full design.

Anticipated Ground Conditions

The tunnels are located in the Bluegrass Physiographic Province of central Kentucky and bordered on the north by the Ohio River and Harrods Creek. The principal formations include the Louisville Limestone, a grey fossiliferous, soluble limestone with shale inclusions, the Waldron Shale, a greenish grey dolomitic shale with severe degradational properties and breaks down when exposed to air and moisture. The Waldron overlies the Laurel Dolomite which is a stable massive grey dolomite with minor limestone. In the area of the tunnel the Waldron is in the tunnel invert at the South Portal rising at about a 10% grade to form the crown of the tunnel at the North Portal. The Waldron averages about 12 ft. thick throughout the tunnel alignment. The Louisville Limestone is noted for pinnacled karstic features with multiple solution enlarged joints and bedding planes and sinkholes filled with fat clays. Perched water is common at the contact with between the limestone and shale. The Ohio River Valley has a high horizontal stress field that also had to be accounted for in the design.

The portals were established at a bench level some 90 feet below the surface of HWY 42 at the south end and only some 22 feet below the surface at the North Portal. The tunnels were divided into three reaches in the Golder Supplemental GBR/GDR: Reach 1 from the South Portal approximately 450 feet to a point where the Waldron rises to the crown; Reach 2 approximately 600 feet to where the Waldron rises to become the dominate rock in the tunnel crown and; Reach 3 where the Waldron composes all of the crown rock and is the area of the shallowest cover and is additionally

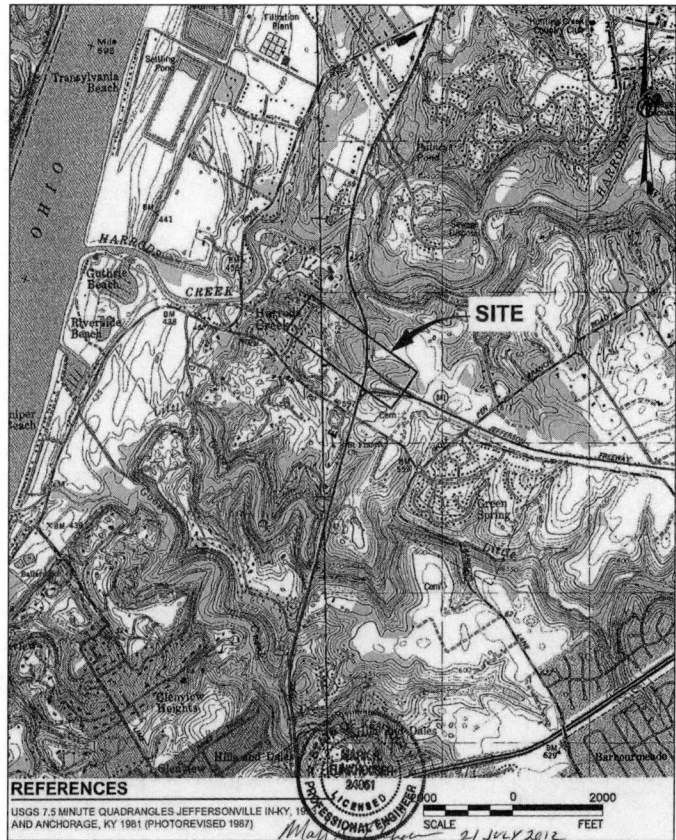

Figure 1. Site location map. Louisville–Southern Indiana Ohio River bridges project—Twin highway tunnels.

the point where the Drumanard Creek passes directly over the centerline of the tunnel alignment. The tunnels were excavated using the sequential excavation method to a dimension of 55'wide × 35' high to accommodate 2 lanes of traffic with a potential for a third lane and are separated by a 40 foot pillar. Two cross passages were excavated in the pillar on 500 foot centers for excapeways.

Baseline Intact Rock Properties

Cores from the HMM and Golder, and S&ME drilling programs were tested to provide the baseline rock properties. The data is presented in max min to provide a range of values for the Developers designer and geotech to use. Along with the rock physical properties all of the rock mass properties are described and evaluated for their impact on the excavation.

Golder evaluated the RMR value when evaluating the rock mass rating along with the Norwegian Q system. The RMR rating was considered the most simple and accurate for this geologic environment and was used by the GBR to define the rock mass. The RMR results were then used to define the initial ground support requirements for each of the three reaches. Reach 3 had the additional factor of low ground cover with little heavily pinnacled Louisville Limestone remaining above the Waldron where

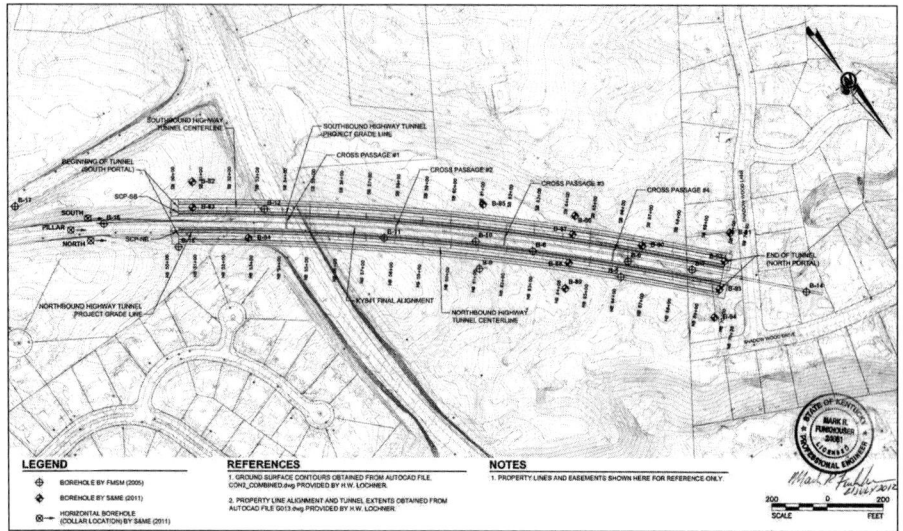

Figure 2. Boring location map

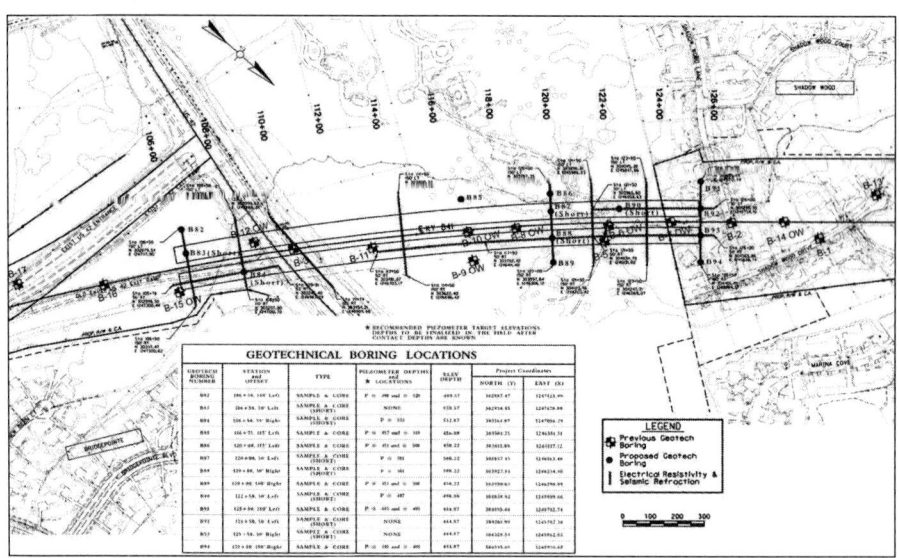

Figure 3. Tunnel geotech testing coordinates

a pressure arch could not form and the dead weight of the shale must be supported from within the tunnel.

S&ME horizontal bores were completed in 2011/2012 in each bore and in the pillar. The bore alignment was mapped using the Devico tool and NQ cores were recovered for testing.

On completion of the horizontal drilling program the core was preserved at the KYTC facility in Frankfort for viewing by potential bidders.

Table 1. Baseline range for intact rock properties (Golder GBR August 2012)

		Measured		Baseline Range	
		Min	Max	From	To
Uniaxial Compressive Strength, psi	Shale	1,924	19,298	1,800	20,000
	Limestone	1,638	38,829	1,500	40,000
	Dolomite	975	28,640	800	30,000
Point Load Index I_{50}, psi	Shale	24	1,792	20	2,000
	Limestone	130	1,871	20	2,000
	Dolomite	31	1,775	20	2,000
Brazilian Split Tensile Strength, psi	Shale	150	2,635	100	3,000
	Limestone	414	4,413	400	5,000
	Dolomite	651	2,996	600	4,000
Peak Slope Index from Punch Penetration Test	Shale	61	61	60	120
	Limestone	37	168	30	180
	Dolomite	96	165	80	200
CERCHAR Abrasiveness Index	Shale	0.32	0.6	0.25	0.75
	Limestone	0.7	2.5	0.5	2.5
	Dolomite	0.5	1.4	0.5	2.5
Slake Durability Index	Shale	15	99	10	99

Table 2. Data summary and baseline values of RQD and fracture spacing

RQD	Formation	No.	Min	10%	20%	50%	80%	90%	95%	Max	From	To
						Percentile					Baseline Range	
Reach 1	Limestone	186	0	85	94	100	100	100	100	100	0	100
	Shale	5	94			99				100		
	Dolomite	13	93	99	100	100	100	100	100	100		
Reach 2	Limestone	131	38	95	97	100	100	100	100	100	0	100
	Shale	83	0	43	73	100	100	100	100	100		
	Dolomite	10	98			100				100		
Reach 3	Limestone	59	18	59	71	90	98	100	100	100	0	100
	Shale	247	0	19	48	93	100	100	100	100		
	Dolomite	86	57	96	98	100	100	100	100	100		

Fracture Spacing, ft	Formation	No.	Min	10%	20%	50%	80%	90%	95%	Max	From	To
						Percentile					Baseline Range	
Reach 1	Limestone	148	0.35	1.21	2.50	7.95	10	10	10	10.2	0.1	10
	Shale	5	2.50			10				10		
	Dolomite	13	1.10	1.48	1.80	5	10	10	10	10		
Reach 2	Limestone	91	0.78	1.98	2.53	6	10	10	10	12	0.1	10
	Shale	51	0.42	1.43	2.50	10	10	10	10	10		
	Dolomite	10	1.46			3.70				10		
Reach 3	Limestone	57	0.17	0.47	0.63	1	1.6667	3.5	5	5	0.1	10
	Shale	161	0.17	0.69	1.00	5	10	10	10	12		

1. RQD and fracture spacing data shown are from measurements taken immediately following extraction of the core.
2. Fracture spacing is calculated for each core run by dividing the length of the run by the number of fractures observed in the core run. For those core runs where no fracture was observed, the fracture spacing is conservatively assumed to equal the length of the core run.

Construction Considerations

In addition to the description of the physical properties of the units in the rock mass, the GBR also describes the groundwater conditions and hydraulic conductivity as fairly low, on the order of 1×10^{-6} cm/s higher rates of 1×10^{-3} cm/s will occur along rock discontinuities. Additional conditions described include In-situ stresses, Discontinuities, and adverse geologic features. In the case of the Drumanard tunnels these include the karst dissolution features and voids primarily in the pinnacled Louisville Limestone and the shallow cover on the north end where the Drumanard Creek has eroded the limestone surface and the Waldron Shale has little cover. To monitor potential subsidence the Developers Geotech put in a series of MPBX and UMPBX instruments along with a series of surface prisms and settlement monitoring points. Each extensiometer was a three point system and all were remotely monitored and alarmed when movement exceeded a certain point. Trends were graphed and recorded by both the developer and the on site geologist and verified by IFA's geotech.

Figure 4. S&ME horizontal bores Devico tool

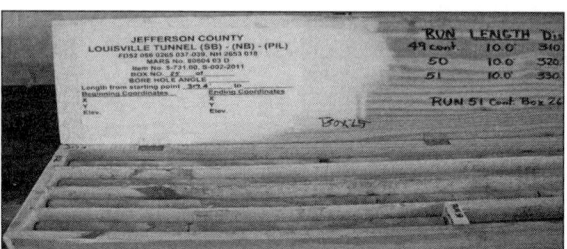

Figure 5. Core from horizontal drilling program

Figure 6. Horizontal drilling

Figure 7. Mining the top heading

Figure 8. Southbound excavation complete

Initial ground support was provided by installing 12' swellex roof bolts on 6' centers. This provided adequate support where the main roof was the massive Louisville Limestone. The Waldron shale was covered with a flashcoat of shotcrete 1" thick within 8 hours where ever it was exposed to control slaking. The Developer chose to scale with a hydraulic hoe ram mounted to an excavator which contributed to excessive overbreak when the shale became the dominate rock in the crown of the tunnels.

Construction of the Drumanard Tunnel

All of the GBR and GDR data along with supplemental information required by the Developers Geotech firm, Lachel Associates, was incorporated into it's the design of the initial ground support and excavation sequence. The sequential excavation method (SEM) was selected as the most efficient using drill and blast as the excavation method. SEM requires a constant and consistent monitoring of the face during excavation by an experienced geologist to determine the ground support requirements round by round. At Drumanard, a 4 step excavation sequence was selected where the top center heading was mined first to a 18' × 24' dimension, then the right wing

Figure 9. Southbound tunnel complete

Figure 10. Portals complete

was kept back 45' and the left wing 45' behind the right. After the entire top heading was mined out and supported the remaining bench was taken in one lift.

A series of small falls in the Southbound tunnel during excavation of the center heading in the Class 3 ground and it's associated shallow cover was observed and recorded by the instrumentation. Shortly after the Southbound tunnel center drift holed through, excessive movement was recorded on the settlement points just behind the North Portal at the centerline of the Southbound tunnel. Shortly after that a fall occurred some 100' behind the last unexcavated side drift. The Developer's Geotechnical consultant developed a contingency support plan which was implemented using steel sets and a laminated shotcrete girder with 15' resin bolts on 4' centers to control the ground and complete the excavation. The void above the sets was filled with concrete and the sets and lagging became a permanent part of the initial support. As a part of this plan the Northbound drift initial support was reinforced with a laminated shotcrete beam using 15' resin bolts on 4' centers.

CONCLUSION

The Drumanard tunnel excavation was completed in September of 2015 and the installation of the final liner completed in June of 2016. The tunnels are presently being fitted out with the M&P equipment including Jet fans, Fire Life Safety and Fire Deluge systems and the SCADA control system to be monitored by Trimarc. The facts and recommendations contained within the GBR and GDR were confirmed by the excavation.

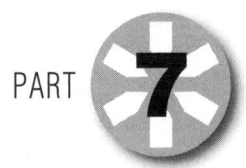

Large-Span Tunnel Cavern

Chairs

Darren VonPlaten
Traylor Bros., Inc.

Peter Kottke
Kiewit

Design and Construction of Indianapolis Pump Station Cavern

Verya Nasri ▪ AECOM
Alex Varas ▪ AECOM
Michael Miller ▪ Citizens Energy Group
Jose Castillo ▪ Southland Contracting

ABSTRACT

This paper presents the design and construction of a deep rock conveyance and storage tunnel and pump station cavern that was built about 70 m below ground surface for the City of Indianapolis. The TBM tunnel ran nearly 13 km in limestone with a finished inside diameter of 5.5 m. It provided a minimum storage volume of 300 million liters of untreated excess wet weather overflow. A pump station was planned for dewatering of the tunnel and discharging the CSO flow to a surface advanced wastewater treatment facility with a firm capacity of 340 million liters per day. The pumps were located within an 18.3 m wide and 24.4 m high deep mined rock cavern. The project also had several deep shafts including the TBM launch/screen & grit and retrieval shafts, the pump room cavern access/discharge and equipment shafts with inside diameters ranging from 4.9 to 13.4 m.

INTRODUCTION

The Deep Rock Tunnel Connector (DRTC) project is a major component of the City of Indianapolis Long-Term Control Plan (LTCP) to address combined sewer overflows (CSOs) from outfalls along the White River. The DRTC is the southernmost portion of a deep large diameter conveyance and storage tunnel system to provide overflow relief during wet weather events. The City entered into a Federal Consent Decree in 2006, which required development of the LTCP. In 2011, the City sold their water and wastewater assets to Citizens Energy Group (Citizens), a public charitable trust. Along with the assets, Citizens became party to the Consent Decree, and subsequently inherited the responsibility of complying with the Consent Decree, including construction of the LTCP projects. In 2012 construction of the DRTC project commenced under Citizens' ownership.

The DRTC project includes approximately 13 km of 6.1 m excavated diameter deep rock tunnel, three drop shafts, two major shafts: a launch shaft at the south end at the Southport Advanced Wastewater Treatment (AWT) facility, and a retrieval shaft at the north end. Also included in the DRTC project, constructed under a separate contract, is the tunnel dewatering pump station at the south end near the existing Southport AWT facility (Nasri 2011 and 2012). AECOM was retained by the City of Indianapolis, Department of Public Works to provide detailed design of the project. A joint venture of J.F. Shea Co., Inc. and Kiewit Corporation (SK-JV) built the tunnel and a joint venture of Oscar Renda Contracting and Southland Contracting (RS-JV) is currently finalizing the construction of the pump station cavern.

The pumps are located within an excavated pump room cavern approximately 82 m below grade. Flow from the pump station is discharged to the Southport AWT facility headworks. The DRTC Pump Station (PS) consists of above ground grit and screening

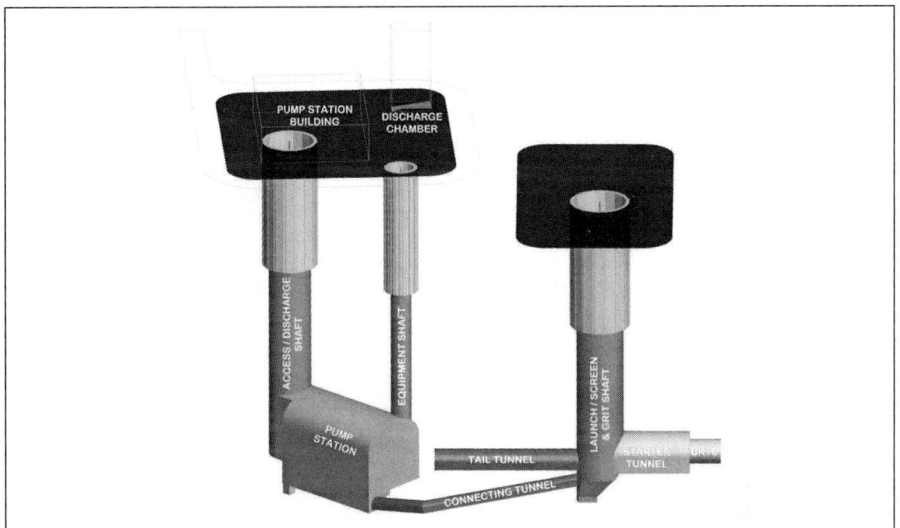

Figure 1. DRTC and pump station isometric view

removal equipment, a pump station building, below ground pump room, a 10.7 m diameter access/discharge shaft, and a 4.9 m diameter equipment shaft (Figure 1).

The span of the cavern is about 18.3 m, its height is around 24.4 m and its length is approximately 30.5 m. The crown arch consists of three circular arc segments. The crown arch geometry was determined to create a balance between the rock excavation volume and the opening stability. Also, it was ensured that the rock cover above the crown consists of sound limestone with a thickness of at least one third of the cavern span.

The pump room houses the main tunnel pumps and other electrical and mechanical equipment. An overhead gantry crane is provided inside the pump room for lifting the pumps and other equipment, and transporting them to a drop zone within the pump room from which they are lifted through the equipment shaft up to the surface.

The main tunnel was excavated by a single main beam rock TBM and supported mainly with rock dowels. The tunnel launch and retrieval shafts were excavated through 33 and 23 m overburden, respectively, by conventional soil excavation methods and approximately 37 and 43 m, respectively, to the tunnel invert elevation of rock by drill and blast method. The shaft excavation was supported within the overburden soil with a slurry wall and within the rock by rock dowels and shotcrete. Within the overburden, the slurry wall acts as the permanent wall for the shaft. Beneath the overburden, a 0.6 m cast in place concrete lining was installed as a permanent support for the shaft. The pump room was excavated by drill and blast method and supported by permanent rock bolts and shotcrete. Construction of the drop shafts was performed by advancing an over-size steel casing through the overburden and rock to the required depths.

GEOLOGICAL SETTING

The DRTC project site is predominantly located within the broad outwash valley of the ancestral White River, which was formed during the Wisconsinan glacial period. The soils consist primarily of unconsolidated glacial and glaciofluvial sediments. The overburden along the DRTC alignment consists mainly of sand and gravel with a few discontinuous till layers, which comprises the extensive outwash deposits filling the White River Valley. Throughout most of the county, the average sediment thickness is 30 m. However, overburden thickness can vary from less than 5 m to more than 90 m.

The presence of sub-rounded cobbles and boulders immediately above the shale bedrock and within the outwash was noted during drilling the DRTC borings. The cobbles and boulders tend to be made up of hard rock such as granite and gneiss from the Canadian Shield, transported by glacial action and deposited as either ice-rafted boulders in the sands and gravels, or embedded within the till. The depths and locations at which the boulders and cobbles were encountered are generally random in nature, although they were more frequently encountered immediately above the top of bedrock. The boulder sizes are ranging from about 0.3 to 1.5 m in diameter, and occasional boulders as large as about 2.4 m in diameter have been reported.

Devonian and Silurian carbonate rock underlie most of Indianapolis and the White River valleys where the DRTC project is located. The shallowest bedrock formation along the alignment is predominantly New Albany shale of Devonian-Mississippian geologic age. The shale of the New Albany Formation has eroded to a surface of relatively gentle relief and the low hydraulic conductivity of the shale allows the unit to act as a groundwater-confining layer where it overlies the Devonian and Silurian carbonates.

Underlying carbonate formations encountered in boreholes at depths of up to 90 m below grade consist of, in descending order, North Vernon limestone, Vernon Fork limestone, Geneva dolomite and Wabash dolomite. The DRTC tunnel horizon is located within the Vernon Fork Formation and Geneva Formation.

Groundwater in the vicinity of the DRTC alignment predominantly flows through an outwash aquifer system associated with the White River Valley. The extent of the confining layers is limited and the outwash aquifer is generally unconfined within the project area. The bedrock aquifer is confined by the New Albany shale along the majority of the DRTC alignment. The outwash aquifers and the bedrock aquifer can locally be in direct hydraulic connection in areas where the New Albany shale is absent. Both the outwash and the bedrock aquifer are productive enough to provide adequate quantities of groundwater for domestic, industrial, and municipal production wells. Productive bedrock aquifer zones are reportedly in areas where the New Albany shale is absent.

The sets of joints and faults associated with large scale sedimentary rock basins form a distinct, orthogonal structure and result in the formation of sub-vertical normal faulting, vertical jointing, and localized strike–slip faulting that is oriented relative to the shape of the basin. Regional and local drainage patterns are often controlled by the bedrock structure and are reflected to the surface in the form of angular, linear segments or lineaments, typically along the river and stream flow paths.

SUBSURFACE INVESTIGATIONS

The detailed design geotechnical investigation included 29 vertical and 10 inclined deep soil and rock borings and 12 shallow overburden soil borings along the tunnel alignment. The program also included water pressure testing in the bedrock, field slug testing, installation of wells and piezometers, laboratory soil and rock index property and strength testing, laboratory groundwater chemical testing, and field gas measurement. The engineering and index properties of the overburden soils and rocks were evaluated based on results of laboratory and in-situ tests, empirical correlations, past experiences with similar soils and rocks, and engineering judgment.

The laboratory unconfined compressive strength (UCS) tests of intact rock samples from the DRTC project borings show a wide range of strength values between 20 and 190 MPa. The range of values is primarily due to the composition and type of rock, but may also be influenced by the direction of bedding and/or fractures and the size and location of vugs or other features with respect to the test loading direction.

The strike and dip direction data obtained from Hanson Quarry and the adjacent underground mine revealed two distinct, continuous, vertical to sub-vertical, generally rough joint sets, with one horizontal to sub-horizontal, discontinuous, rough to smooth bedding set. Near vertical fractures were encountered in several of the borings completed for this project.

The condition of the rock mass within the upper 6 m of the bedrock is considered slightly weathered with open and stained joints and bedding planes. Partial or complete loss of drilling fluid was experienced at top of the rock in some borings. This observation indicates high permeability zones can be expected immediately below the rock surface.

Borehole water pressure (or packer) tests were performed to characterize the hydrogeologic conditions of the project area. In general, the higher permeability values are associated with features such as open joints, and horizontal or vertical fractures. The permeability values from the discontinuities in the Geneva dolomite are generally higher than those from the Vernon Fork limestone. This can be attributed partly to the fact that the Geneva dolomite is vuggier than the Vernon Fork limestone, and that some of the vugs are interconnected (Figure 2).

Seven samples from the shale, dolomite and limestone formations were tested in the laboratory to characterize their durability. The Slake Durability Index (SDI) for the tested limestone/dolomite exceed 98%, and all of the test results for limestone and dolomite fall into "Type I" and therefore dolomite and limestone from the North Vernon, Vernon Fork, Geneva, and Wabash units are considered to be highly durable.

Cerchar Abrasivity Index (CAI) testing was conducted. The CAI values range between 0.1 and 2.9, which represents the rock types anticipated at the tunnel horizon throughout the alignment. Approximately 56% of the data indicates very low to low abrasiveness, 31% has medium abrasiveness, and 3%, or one location only, exhibited high abrasiveness.

Seven samples selected from five borehole locations throughout the alignment were tested to determine Drilling Rate Index (DRI), Bit Wear Index (BWI), and Cutter Life Index (CLI) by SINTEF laboratory from Norway. The DRI ranged between 66 and 82, (high to very high); BWI ranged between 7 and 13 (extremely low to very low); and CLI ranged between 41 and 108 (very high to extremely high).

Observation wells or piezometers were installed as part of the geotechnical investigation program to monitor groundwater levels. A nest of wells and piezometers were installed at several locations to evaluate the potential connectivity between the outwash and bedrock aquifers. The monitoring data indicates that the outwash and bedrock aquifers have a strong hydraulic connection in areas where the shale is absent or thins out, or where the shale is fractured. In areas where the shale is relatively thick

 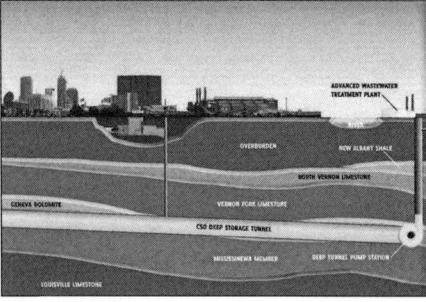

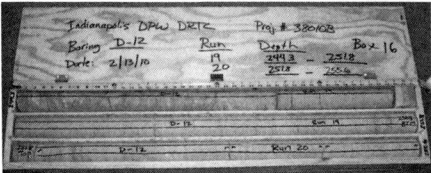 

Figure 2. Hanson quarry, geological profile, limestone and dolomite cores

and intact, there is a significantly weaker hydraulic connection between the outwash and bedrock aquifers.

Naturally occurring crude oil, methane, and hydrogen sulfide were identified in the DRTC borings. Crude oil stains were also observed on isolated locations on the exposed rock faces in Hanson Quarry and the nearby underground mines. Samples of the rock where oil staining was noted were chemically tested in the laboratory. The test results indicate that the petroleum encountered in bedrocks is naturally occurring. The amount of crude oil encountered was not significant relative to the volume of rock removed as core. The crude oil was principally observed as a coating on hairline fractures or as small droplets in vugs in the rock. Crude oil was encountered in most borings, and appears to be a regional feature. Interviews with operators at Hanson Quarry suggested that naturally occurring crude oil is encountered throughout the quarry, but quantities are so small as to not adversely affect quarry operations or require special handling.

DESIGN AND CONSTRUCTION APPROACH

The methodology adopted for designing initial ground support for DRTC project involved the following three steps: Ground conditions at different locations along the alignment were characterized based on the geological investigations, The initial ground support elements were designed based on the empirical Norwegian Geotechnical Institute Q-system, The adequacy of the initial support elements was verified using three different methods: Continuum Analysis (PHASE2), Jointed Rock Mass Analysis (UDEC), and Block Stability Analysis (UNWEDGE). These verifications confirmed the initial support complies with performance requirements. Selecting analysis methods depended on rock conditions including the frequency of joints and discontinuities.

The ground condition was characterized based on the geological and geotechnical investigations. A subsurface drilling program and field survey at a nearby quarry and underground mine were conducted to identify geological structures, such as possible rock joints, and their characteristics including orientation and spacing. This

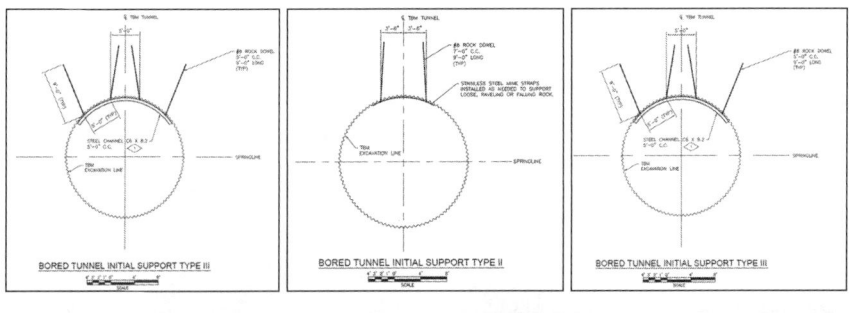

Figure 3. DRTC TBM tunnel initial support, main beam rock TBM, and excavated tunnel

geotechnical investigation provided the mechanical properties for the intact rock, rock mass, and discontinuities (joint properties).

Depending on the ground conditions encountered during excavation, three initial ground support types were used to support the DRTC tunnel. Initial ground support Type I with spot bolting was considered for excellent to good rock qualities based on the Rock Quality Index. Fully cement-grouted rock dowels (with no pre-stressing) were used as a part of the tunnel initial rock support system. Rock dowels consisted of No. 25, Grade 520, all-thread steel bars. The designed dowel length was 2.7 m for the 6.1 m excavated tunnel diameter. Systematic rock dowels were installed above the tunnel spring-line at 2.7 × 2.7 m for Type II and 1.8 × 1.8 m for Type III rock support (Figure 3). Additional channel sections, wire mesh, and mining straps were used where necessary as protection against falling rock. The geotechnical data indicated the groundwater infiltration was not high enough to affect the design of the initial support system. As a result of advanced design, high quality construction and a favorable geology, all the production rate world records for this size of tunnel diameter were broken (maximum 124.9 meters per day, maximum 515.1 meters per week, and maximum 1754 meters per month).

The pump room principal load-carrying component is the rock mass surrounding the cavern; and the ground support elements largely serve to maintain the integrity of the surrounding rock mass. The elements of the initial ground support in the cavern include rock bolts, dowels, and reinforced shotcrete. The use of tensioned reinforcement (bolts) or untensioned reinforcement (dowels) integrated with a fiber reinforced shotcrete layer forms a composite system with the rock mass. The shotcrete layer also plays an important role in bridging the gap between adjacent bolts or dowels and preventing progressive raveling of small pieces of rock that are not confined by the reinforcement.

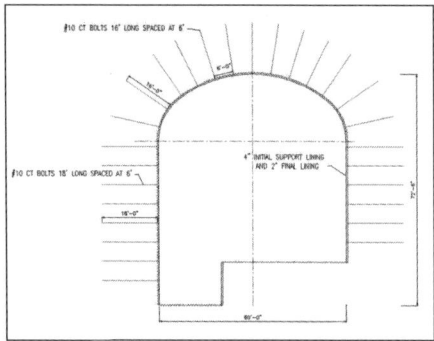 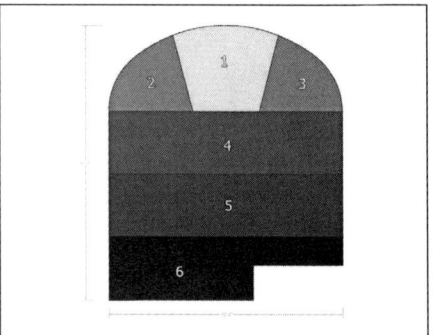

Figure 4. Pump station support system, excavation sequence, and construction photo

The primary support elements for the pump room cavern included 6 1 m long, 13.6 tons tensioned No. 32 bolts at 2,1 × 2.1 m pattern for crown and shoulders, and passive dowels with 2.1 × 4.3 m spacing for sidewalls combined with 0.15 m of fiber reinforced shotcrete (Figure 4).

One of the characteristics of the pump room cavern is its relatively large size of the span and height. Large cavern sections require multiple drill and blast drifts in excavation stage. The design of drift sizes and shapes were governed by stability, ground settlement, and vibration concerns. The excavation sequence for the pump room is shown on Figure 4. The excavation of the cavern included 6 drifts; drifts 1, 2, and 3 constituted the top heading excavation and drifts 4, 5, and 6 represented the bench excavation. The center-out excavation sequence adopted for the top heading. This allowed for a continuous dissipation and redistribution of the induced stresses away from the excavation profile and furthermore facilitated the gradual formation of rock arch over the crown.

During construction, the launch shaft accommodated the launching operations of the TBM and facilitated muck removal. After construction was completed, the launch shaft served as the screen & grit shaft. The retrieval shaft used to retrieve the TBM at the end of the TBM drive, and to accommodate TBM launching operations for future tunnel construction. The launch shaft and the retrieval shaft were both excavated through overburden, shale and limestone.

Key considerations in selecting the appropriate construction methods included preventing groundwater drawdown and providing support of excavation. Slurry wall was selected to provide temporary excavation support and permanent lining of the launch and retrieval shafts through the overburden soils. The slurry wall was embedded 1.5 m

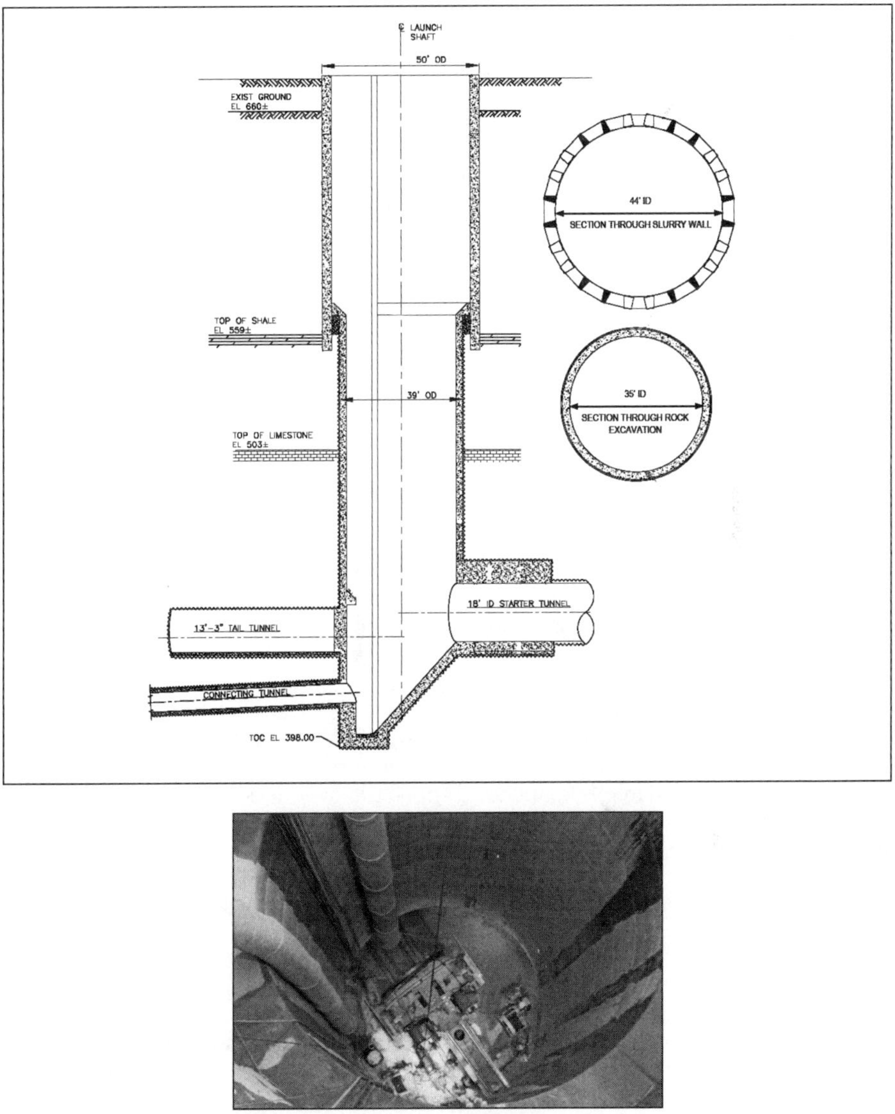

Figure 5. Launch shaft profile, sections, and construction photo

into the shale layer. The slurry wall panels were arranged to form a circular shaft. The entire "ring" was broken into overlapping straight chords, arranged as sets of primary and secondary panels. A ring beam was positioned near the slurry wall toe and rock interface.

The excavation in rock was carried out using the drill and blast method. The initial support for the portion of each shaft excavated through shale and limestone consisted of rock dowels and 0.1 m of fiber reinforced shotcrete. No final lining was constructed inside the slurry wall portion of the shafts, as the slurry walls act as permanent walls in the final condition. A 0.6 m thick cast-in-place concrete final lining was installed

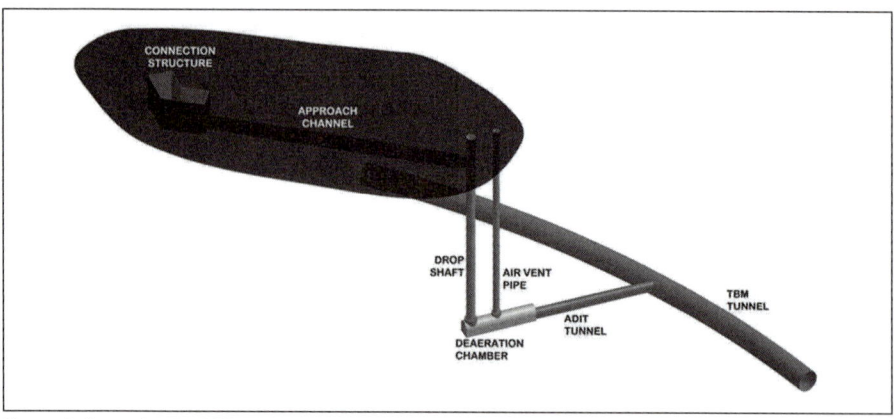

Figure 6. Example of a CSO connection structure

on the portion of the shaft constructed within the rock layers of shale and limestone (Figure 5).

Three vortex type drop shafts of 2.1, 1.7, and 1.4 m ID serve as collection points for the proposed CSO system. These shafts are paired with a 1.8, 1.4, and 1.1 m ID, respectively, air vent pipe that is connected to the drop shaft by an air vent connector pipe (Figure 6). The drop shafts and air vent pipes extend through overburden, shale and limestone from the ground surface down to the crown of the deaeration chambers. The deaeration chambers are horseshoe shaped tunnels connected to the DRTC tunnel through an adit. To build the drop shafts a steel casing with an inside diameter larger than the outside diameter of the drop shaft lining and outfitted with carbide teeth was used to rotate and cut through the overburden and rock layers. This provided temporary ground support to allow the shaft to be built. A minimum 0.23 m thick cast-in place concrete final lining was used for the larger shafts and the precast concrete pipe segments were used for the final lining of the smaller shafts and for this later the annular space between the steel casing and the outside of the segments was filled with concrete.

The mining of the adit tunnel began at the DRTC tunnel and proceeded towards the deaeration chamber, in a horseshoe shape. The adit tunnel was constructed larger than required to facilitate the installation of a fiberglass pipe, centered in the adit and encased in concrete. The mining continued from the adit and into the deaeration chamber. The deaeration chamber initial support consisted of rock dowels and shotcrete. The final lining included cast-in-place reinforced concrete.

The requirements for groundwater control included probe hole drilling, water, and gas monitoring, grouting behind and ahead of the TBM advance and maintaining water level below the haulage track rail. The main volume of water from the tunnel, adits, deaeration chambers, and shafts was pumped from the launch shaft, pre-treated, and discharged to a specified location near the Southport AWT plant. Limited treatment of groundwater and discharge to a stream under the terms of a NPDES permit was required. The design required the protection of all potable residential wells and residential water supplies by limiting the maximum allowable drawdown in adjacent drinking water wells to 0.6 m.

Drilling a probe hole ahead of the TBM during the tunnel drive over one of the reaches of the alignment was required. This probe hole was for monitoring potential water inflow and increased quantities of dangerous gasses. Pre-excavation grouting ahead of the tunnel face was directed if inflows from a 0.05 m diameter probe hole exceeded 570 liters per minute for a period of 30 minutes. This was incorporated to enhance the safety of the tunnel drive, to reduce the potential for flush inflow, and to reduce potential adverse effects upon the rate of production.

CONCLUSIONS

This paper presents the final design and construction aspects of the deep rock conveyance and storage tunnel and pump room cavern in Indianapolis. The geological and hydrogeological setting and subsurface investigation program is discussed and the general aspects of the detailed design and construction approach for the excavation and initial and final supports of the tunnels, shafts, adits and a large pump room cavern is described.

REFERENCES

Nasri, V., and Morgan, J. (2011), Indianapolis Deep Rock Conveyance and Storage Tunnel. Rapid Excavation Tunneling Conference 2011, San Francisco, CA, June 19–22, 2011, pp. 402–413.

Nasri, V., Morgan, J., and Varas, A. (2012), Indianapolis Deep Rock Pump Station Cavern. North American Tunneling Conference 2012, Indianapolis, Indiana, June 20–23, 2012, pp. 435–443.

Nasri, V., Morgan, J., and Varas, A. (2012), Indianapolis Deep Rock Conveyance and Storage Tunnel and Pump Station Cavern. International Tunneling Association World Tunneling Congress 2012, Bangkok, Thailand, 18–23 May, 2012.

Admiralty Station, Hong Kong: Rock Excavation and Support Challenges to Accommodate MTR's Two New Lines

Harry Asche ▪ Aurecon Australasia
Mike Bezzano ▪ MTR Corporation Ltd
Scott Smith ▪ Aurecon Hong Kong
Mark Wiltshire ▪ Laing O'Rourke

ABSTRACT

Hong Kong's new South Island Line and Shatin to Central Link requires the expansion of Admiralty Station into Hong Kong's first 4 line interchange station. Already the busiest interchange station in Hong Kong, expanding this station without interrupting services involves major engineering, logistics and rock mechanics challenges.

The new platforms are housed in a shallow cavern and platform tunnels immediately adjacent to the existing platforms and under major roads and high rise buildings. To facilitate connection with the existing station, the existing Island Line platform tunnel needed to be underpinned and up to 24m of rock below it removed, while keeping trains running safely above.

Multi-party collaboration, interactive design and construction processes were adopted that allowed program optimization as construction progressed.

INTRODUCTION

Hong Kong is a rapidly developing world city, with significant population growth and limited opportunities for space. Hong Kong's MTR Corporation provides the key transportation service for Hong Kong. MTR is undertaking several projects aiming to provide enhanced public transport options to emerging regions. These include the South Island Line (SIL) and the Shatin Central Link (North South Line) (SCL(NSL)). The SIL connects the existing Island Line (ISL) and Tsuen Wan Line (TWL) from Admiralty Station (ADM) to the Southern District of Hong Kong. The SIL(E) is approximately 7km long, and is a medium capacity line with stations at South Horizons, Lei Tung, Wong Chuck Hang, Ocean Park and Admiralty (ADM), comprising underground and elevated structures. The SCL(NSL) will provide access to emerging regions on the Kowloon Peninsula. Refer to Figure 1 to show the proposed railway development on Hong Kong Island.

SOUTH ISLAND LINE (EAST) CONTRACT 901

The SIL is divided into a number of construction contracts. The works for Contract 901 include an extension east of the existing ADM station to accommodate the new SIL and SCL(NSL) lines, an interchange concourse, circulation areas and plant rooms. The integrated ADM station will be the first four line interchange in Hong Kong, providing a convenient interchange between the new SIL, and SCL(NSL) and the existing TWL and ISL lines.

In May 2011, Contract 901 *Admiralty Integrated Station and SCL Enabling Works* was awarded to Kier-Laing O'Rourke-Kaden Joint Venture. A key aspect of the works was

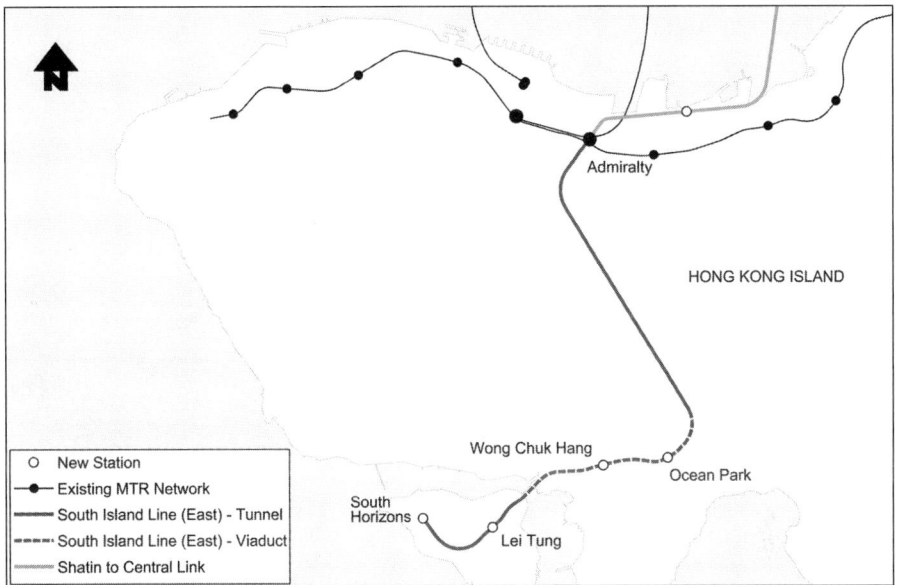

Figure 1. South Island Line (East)

the interfacing with the operating railway and the need to continue the full service of the existing lines without interruption, in what is one of the busiest interchange stations. In addition to the four new platforms that are under construction below and adjacent to the existing station, the contract includes an interchange concourse, relocation of existing passenger entrances and external landscaping.

The new interchange station box is excavated as a partial top down cut and cover box between the existing TWL and ISL finger platforms and is connected to the existing Admiralty Station to the west. The ADM station cavern will serve as the terminus of the SIL and is excavated to the south of the new station box. Two platform tunnels for the future SCL are also being excavated on each side of the ADM cavern. Several passenger adits connect the cavern with the platform tunnels providing fast transfer routes between the SIL and SCL lines. A plan of the Integrated Admiralty Station is shown on Figure 2.

Parties to the contract are shown in Table 1.

GEOLOGY, HYDROGEOLOGY

The project area is dominated by both granitic and volcanic rocks of the Jurassic-Cretaceous age. Kowloon Granite of the Lion Rock Suite is present at the northern fringe of the project area, where the ADM station is located; whereas volcanic rock of Ap Lei Chau Formation, which mainly consists of vitric tuff, is present to the south of the ADM station.

The granite intrusion is commonly associated with contact metamorphism and hydrothermal alteration. The tuffs, which have been intruded by granites, are likely to be altered / recrystallized by contact metamorphism.

Groundwater monitoring has been undertaken in boreholes for both the cavern and station box areas. This showed that the groundwater levels are generally a subdued

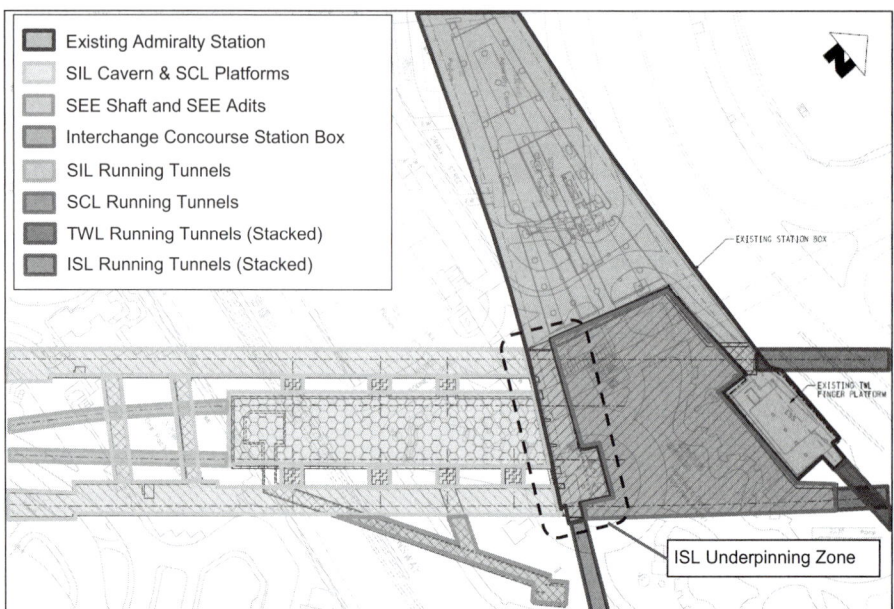

Existing Admiralty Station
SIL Cavern & SCL Platforms
SEE Shaft and SEE Adits
Interchange Concourse Station Box
SIL Running Tunnels
SCL Running Tunnels
TWL Running Tunnels (Stacked)
ISL Running Tunnels (Stacked)

EXISTING STATION BOX

EXISTING TWL FINGER PLATFORM

ISL Underpinning Zone

Figure 2. New works at Contract 901 in relation to the existing ADM Station

reflection of the surface topography. At the ADM cavern, groundwater levels vary by a few meters above rockhead in the granular sediment of the reclamation fill layers. Due to the large-scale excavations of the station box, access adits, platform tunnels and station cavern at the ADM site, groundwater drawdown was expected in the cavern area. Since

Table 1. Contract parties

Client	MTR Corporation
Detailed Design Consultant	Arup
Main Contractor	Kier-Laing O'Rourke-Kaden JV
Contractor's Designers	Benaim & Aurecon Hong Kong

large-scale dewatering has been carried out previously during earlier construction contracts in the Admiralty area, it was anticipated that any consolidation settlement due to groundwater drawdown had already occurred. Groundwater inflow was monitored during excavation works and ground treatment with grouting required where the inflow exceeded allowable limits. Groundwater levels are expected to recharge in the long-term case, affecting only the permanent lining design.

The NGI Tunneling Quality Index (1994), otherwise known as the Q-System, was adopted to quantitatively describe the rock mass quality through assessing available drillholes in the general vicinity of the project area. It was shown that the Q values in general ranged between 1–10, with mean values of 4.5 and 8.0 on the northern and southern side of the ADM cavern respectively.

CAVERNS FOR NEW STATION CONSTRUCTION

Cavern Geometry and Restrictions

The new interchange concourse is excavated as a cut and cover box between the existing TWL and ISL finger platforms. The ADM station cavern, excavated to the south of the new interchange concourse, will serve as the terminus of the SIL. Two

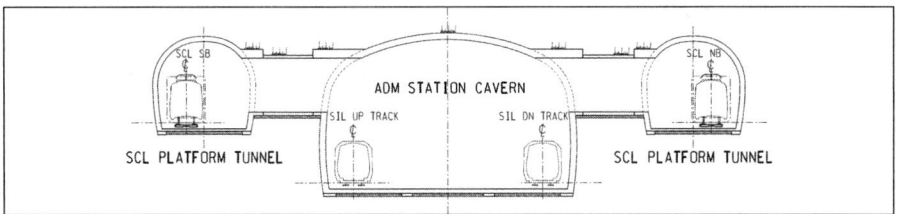

Figure 3. ADM station cavern and platform tunnels

platform tunnels for the future SCL(NSL) are also excavated on each side of the ADM cavern, with several passenger adits connecting the cavern with the platform tunnels providing fast transfer routes between the SIL and SCL(NSL) lines.

The cavern is oriented in a NE/SW direction and is situated underneath Harcourt Garden, Queensway, a major traffic corridor for east west traffic on Hong Kong Island, and in close proximity to the caisson foundations of the high-rise Pacific Place development.

The ADM station cavern is excavated in moderately and slightly decomposed granite with less than one span of rock cover above crown level. The cavern has a span of 26m and a height of 16m. A typical section of the ADM cavern and platform tunnels is shown in Figure 3.

Design of Caverns

Design Overview

The combination of cavern and platform tunnels with narrow pillars, together with the construction of passenger adits created a complex 3-dimensional excavation underground. This complex geometry drove a robust temporary support design methodology that included significant analysis over and above the conventional use of the NGI Q-Chart (Barton, 1994). The analyses for the temporary support of the cavern and tunnels included:

1. Empirical support design using the NGI Q-Chart (1994)
2. Semi-empirical and numerical analysis using methodologies described by Bischoff and Smart (1975) plus embedded beam modelling
3. Kinematic wedge analysis using UNWEDGE
4. Finite element analyses and finite difference analysis
5. Discontinuum methods (see below)

Design Check Using Discontinuum Methods

Given the complex geometry of the cavern, tunnels and adits, various forms of numerical modelling were used to finalize the temporary support design. Both UDEC and 3DEC were used. UDEC was the primary numerical modelling tool adopted for the temporary supports. UDEC is a two-dimensional distinct element method for discontinuum modeling which simulates the response of discontinuous media (such as a jointed rock mass) subject to either static or dynamic loading.

The UDEC modelling was used to:

1. Determine potential structural failure mechanisms of the rockmass around the excavation including the size of potential blocks that could form between the systematic rock dowels

2. Confirm suitability of the systematic rock dowel system including the size of the rock dowels (diameter and steel grade)

3. Estimate ground movement during the various stages of excavation

4. Confirm the impacts from the Pacific Place caissons at the south of the cavern; and

5. Test sensitivities of model inputs including:
   - Adoption of 10 different randomly generated rock mass models (within set boundaries)
   - Modification of the rock joint conditions
   - Increases and decreases of both horizontal stress and surcharges

3DEC is a 3D version of UDEC. Simplified 3DEC modelling was used to correlate any 3D impacts the junctions and adits had on the temporary support and overall tunnel stability.

Modelling necessarily involves simplification and a clear expression of the objective is the best way of choosing the modelling approach. The objective for the distinct element modelling, was to simulate the response of a fractured hard rock to excavation and rock bolting. Given this objective, and the fact that distinct element modelling is a heavy user of computer processing power, the following principles were applied:

- Only the rockmass immediately in the vicinity of the excavation, and subject to bolting, was modelled as a fractured rockmass.

- In 3D modelling, the elements were rigid only (not deformable). In 2D modelling, tests were made to compare between rigid and deformable elements. The difference in response between the two types of modelling is not significant for fractured hard rock.

- Multiple models were run of the same model with different pseudo-random seeds, to explore a range of responses created by the interaction between the jointing and the excavation.

For distinct element modelling to simulate fractured hard rock, it is obviously important that the model represents a realistic distribution of fractures. Many distinct element models are set up with quite unrealistic fracture distributions, typically with a regular spacing modified by a randomized offset. Priest (1993) suggests that the most appropriate distribution for joint spacing is the exponential distribution. One of the main advantages of the exponential distribution is that only a single statistical parameter is required, because the reciprocal of the average spacing is also the standard deviation.

However, using the exponential distribution in a distinct element model comes with some challenges. The production of random numbers that have an exponential distribution is straightforward because there is an analytical solution for the inverse cumulative distribution, so inverse transform method is easily implemented. A different problem arises from the fact that the speed of solution of distinct element models is inversely proportional to the size of the smallest element. The exponential distribution involves a larger number of very close points—one of the reasons that modelers routinely adopt a regular spacing in models, despite the fact that this is not seen in nature. Figure 4a and 4b shows the difference between regular spacing and the exponential distribution.

To avoid too many very small elements, the strategy is adopted that very close joints are merged (Figure 4c). The merged joints are then assigned reduced stiffness properties. The algorithm for generating joints is shown in the pseudo code below.

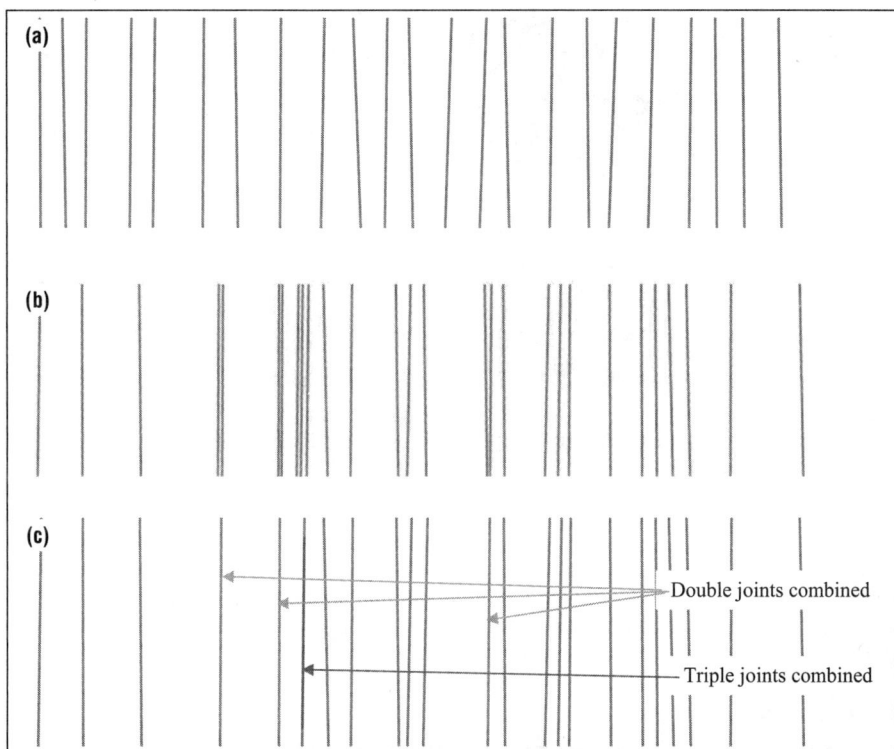

Figure 4. Schemes for generating joint spacing in distinct element models: (a) uniform spacing with added variation; (b) inverse exponential distribution—note some spaces are very small; (c) as b, except with combined joints modeled by property, to avoid very small elements

1. Set up model and generate the total model extent. Model comprises inner and outer model zones. Outer blocks are used to create the boundary conditions, inner block is to be jointed, excavated and bolted

2. Hide outer blocks and identify extents of inner block

3. Create excavation boundaries in inner block

4. For each joint set:
   a. Create first joint in inner block at furthest extreme corner
   b. Until the cumulative distance between the joints exceeds corner-corner distance; get new randomized separation between joint planes
      i. If separation is too small; then make same joint into a multiple joint
      ii. Else create new joint and add distance to cumulative sum

5. Remove all very small blocks

6. Unhide outer blocks and fix boundary blocks, define all block and joint properties

7. Create a "heavy layer" to simulate insitu horizontal stresses

8. Run model to equilibrium, remove heavy layer and re-equilibrate

9. Excavate, bolt, run model to equilibrium, report results.

 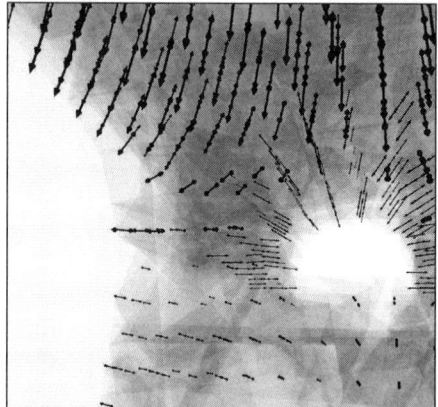

Figure 5. 3DEC model of cavern/platform access tunnel: (a) geometry and jointing and (b) rockbolts

Results of Analyses

A typical 3DEC model is shown in Figure 5. In the UDEC and the 3DEC models, bolts were deleted when shear/tensile strains reached limiting values. Bolt actions were extracted from the model and analyzed statistically. The assessment was made that a bolt pattern was acceptable if no more than 5% of bolts were at shear/tensile limits, and if the model retained stability in that case.

Construction

Due to an existing underground carpark within Harcourt Garden and program constraints from accessing the tunnels from the C&C box, all tunneling works were undertaken from the Supplementary Emergency Entrance (SEE) shaft and construction adit to the east of the Cavern. The SEE shaft ultimately acts as an entry/egress shaft for the operating station, and was constructed to its maximum size possible, and was within 3m of the existing ISL tunnel and immediately adjacent to the Harcourt Garden Car Park blocks.

The construction adits were excavated using drill and blast methods by full face blasting. A complex construction sequence was developed starting from the construction adit that allowed access to the cavern heading, and then benching in stages. The Cavern has been excavated developing three benches, the Crown with three faces and two lower benches with two faces (see Figure 6).

Probing provided the understanding of ground and groundwater conditions prior to the excavation from the excavation front, and was carried out in 25m lengths, maintaining a 10m overlap with the next round of probing. The probing results confirmed two transmissive geological features and pre-excavation grouting was carried out, with approximately 14,000 litres of grout being injected.

Given that multiple faces were concurrently blasted in the ADM cavern and

Figure 6. Excavation of Headings I, II, and III

SCL platform tunnels, the presence of transmissive geological features and narrow rock pillar (wide of <3m) between the cavern and platform tunnel created some specific challenges.

A shear zone was encountered initially 0.3m to 0.5m increasing to 2m to 3m moving toward the south. The mapped Q values of the face through the shear zone were low, between 0.5 and 0.75, and great care was taken to control any ground loss through this area. As the shear zone was shown to be water bearing, pre-excavation grouting was carried out. An additional design review was also undertaken to validate the design assumption and parameters adopted for the temporary supports in the shear zone. Additional temporary support by means of additional shotcrete to a maximum 390mm thick was designed to strengthen the rock pillar where the shear zone intersected the excavation.

UNDERPINNING AND SUPPORT OF THE ISL FINGER PLATFORM

Geometry and Restrictions

An area of high risk and complexity for Contract 901 was the underpinning of the Island Line (ISL) finger platform, to enable excavation of the new interchange concourse immediately below the existing platform. This platform needed to remain fully operational for the entire duration of the contract (refer Figure 2).

Excavation of the station box, for a width of 58m and depth of 24m immediately underneath the existing ISL finger platform and vent shaft structures, was required to create a critical section of the new ADM station, and had to be completed while maintaining railway operation and avoiding any incident affecting passenger services. This underpinning solution was a highly technical and challenging engineering feat never previously attempted.

Site location, program and overall sequence constraints demanded that access for this underpinning was possible only from the new Station Box side of the finger platform structure. In addition, the existing finger platform soffit slab, having originally been designed for transverse hogging between the side walls, was analyzed to be capable of spanning only 9m longitudinally without intermediate support.

Conventional full depth slot underpinning would have resulted in rock pillar supports with a height to width aspect ratio of approximately 5:1, and due to a multitude of constraints, only a very limited amount of advance geotechnical investigation was possible. Interpretation of this investigation data showed that extensive temporary rock pillar support would probably be required and, furthermore, that should the structure move outside of its operating parameters it would not be possible to move the structure back into its original position, and the railway would be affected.

An innovative staged support sequence was consequently devised, in which the pillar excavation and load from the finger platform would be transferred to alternating rock pillars with the use of temporary steel supports in 4.5m wide, 4.5m deep modules, as excavation progressed downwards (refer Figure 7). Rock pillar height to width aspect ratios could thereby be limited to 1:1 with associated reduction in temporary support and residual risk. This alternating sequence, however, necessitated an associated regime of finger platform load transfers between steel support columns, which required careful control and management, with a purpose built software system employed. In order to manage the risk to the lowest possible level the design was also based on a scenario whereby any one support could completely fail, shedding the

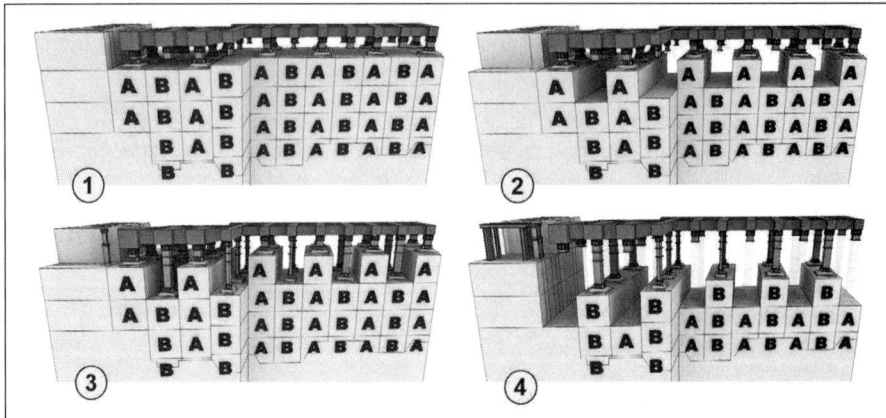

Figure 7. Staged sequence to remove rock beneath ISL finger platform

load to adjacent supports whilst maintaining a minimum factor of safety for the railway to continue to operate.

Rock Pillar Design Aspects

The rock pillar support design was developed to allow an observational approach, consistent with a conservatively feasible set of pre-interpreted rock joint orientations and conditions. Based on laboratory testing of joint conditions from the limited number of boreholes in the vicinity and in consideration of limitations concerning the practicality of observational assessment, two friction angles were considered in the design, average (37 degrees) and poor joint conditions (27 degrees).

A matrix of rock-bolt support options was developed, to define bolt spacing based on the joint orientations and conditions actually encountered.

The design adopted the use of high strength through-pillar bolts, and traditional rock bolts drilled from both sides of the rock pillar designed with maximum overlapping. The through bolts were to be plated each side of the pillar, to fully mobilize bolt capacity, maximize bolt spacing and avoid the need for grouting. The adoption of un-grouted bolts would enable their re-use as the excavation progressed.

Design Using Closed Form Solution

A simplified solution was developed to commence design. Assuming a worst-case joint (or combination of joints), the bolts were designed simply to apply a pressure in the horizontal direction. The increase in horizontal stress reduces the tendency for a pillar to slide along the joint plane.

Consider a pillar loaded vertically, with the worst joint in existence allowing the pillar to slide on the joint (see Figure 8).

Using Mohr's circle of stress, we get:

$$\sigma_n = \frac{1}{2}(\sigma_1 + \sigma_3) + \frac{1}{2}(\sigma_1 - \sigma_3)\cos(2\alpha) \tag{1}$$

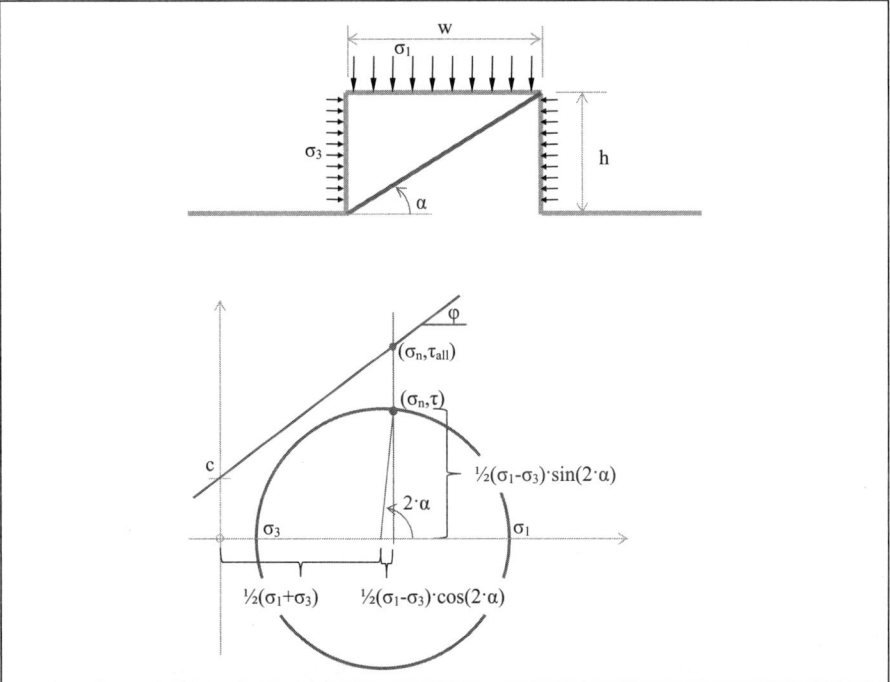

Figure 8. (a) Geometry; (b) derivation of conditions across joint

$$\tau = \frac{1}{2}(\sigma_1 - \sigma_3)\sin(2\alpha) \tag{2}$$

The Mohr-Coulomb equation of equilibrium for the joint is

$$\tau_{all} = \sigma_n \tan(\varphi) + c \tag{3}$$

Where c and ϕ are the cohesion and friction angle parameters for the joint.

The FoS is based on the shear strength:

$$FoS_\tau = \frac{\tau_{all}}{\tau} = \frac{\frac{1}{2}(\sigma_1 - \sigma_3)\sin(2\alpha)\tan(\varphi) + c}{\frac{1}{2}(\sigma_1 + \sigma_3) + \frac{1}{2}(\sigma_1 - \sigma_3)\cos(2\alpha)} \tag{4}$$

In this analysis, the rockbolts contribute the horizontal stress, and also contribute the cohesion c.

Design Optimization

The initial analysis was unrealistic in a number of ways; it conservatively assumed that the joints will exist in the worst orientation and that the weight of the pillar itself is treated simplistically as a uniform weight at the top.

The design methodology was subsequently updated to consider different joint orientations other than the worst case scenario. Figure 9 describes the final through-bolt design.

LOCATION	JOINT QUALITY	MINIMUM TEMPORARY ROCK SUPPORT			MAXIMUM UNSUPPORTED ADVANCE LENGTH (X) /DEPTH (Y)
		JOINT DIP ANGLE (°)	ROCKBOLT GRID (Z)	SHOTCRETE	
UPPER SLOT OF ROCK PILLARS	AVERAGE CONDITIONS ($\varphi = 37°$)	27–52 52–61 61–67 67–70 >70	1.5m 1.6m 1.7m 1.9m 2.1m	–	3.4m
	POOR CONDITIONS ($\varphi = 27°$)	27–52 52–61 61–67 67–70 >70	1.1m 1.2m 1.3m 1.6m 1.8m	–	2.3m
ROCK PILLARS	AVERAGE CONDITIONS ($\varphi = 37°$)	27–52 52–61 61–67 67–70 >70	1.1m 1.1m 1.2m 1.4m 1.5m	–	0.5m/2.8m* 1.3m/2.8m*
	POOR CONDITIONS ($\varphi = 27°$)	27–52 52–61 61–67 67–70 >70	0.8m 0.9m 1.0m 1.1m 1.3m	–	0.5m/1.9m* 1.3m/1.9m*

Figure 9. Through bolt spacing requirements for the 4.5 × 4.5m pillars

Design Check Using Discontinuum Methods

Although the closed form solution was expected to be conservative it did not take into consideration the impact of jointing in the third dimension. 3DEC was used in a similar way to described above for the cavern to confirm the suitability of the calculated bolting density by the closed form solutions. Figure 10 shows a typical 3DEC model.

The 3D modelling proved to be an important step in the design process as it both confirmed the suitability of the through bolts and identified the need for end bolts to prevent a block sliding failure mechanism at the front face of the pillars. Figure 11 demonstrates larger displacements at the front of the pillar. Sensitivity runs with higher pillars showed this failure mechanism to be a controlling design case.

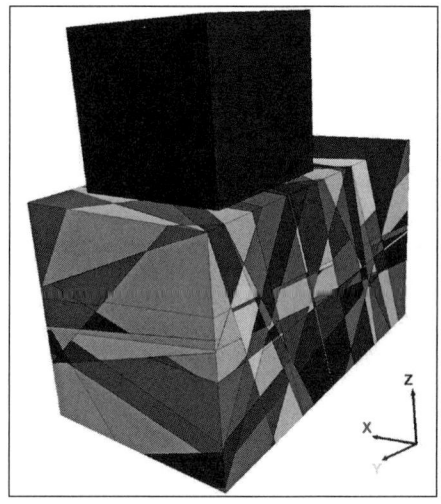

Figure 10. 3DEC model of pillar with applied load

Construction

Monitoring

Given the importance of maintaining safe continuous operations of the Island Line, a detailed instrumentation and monitoring regime was established to record the behavior of the existing Island Line finger platform and vent shaft structures, also the temporary underpinning structures and rock pillars. This regime included:

- Real time level and horizontal displacement monitoring inside the operational structure, using total stations and prisms (half hourly reporting)

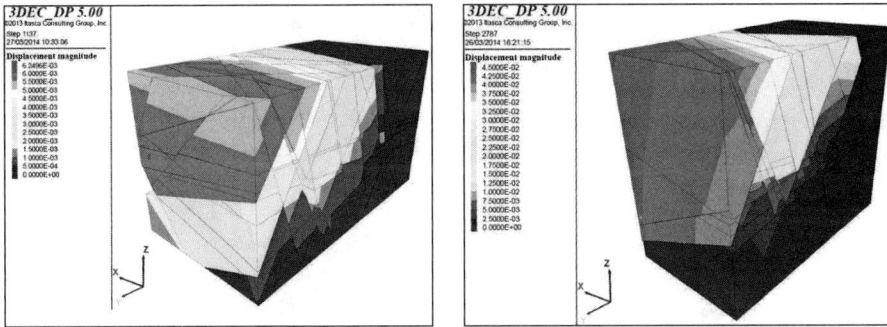

Figure 11. Large displacement at the nose of the pillars—4.5m and 6.75m cases

- Real time level monitoring inside the operational structure, using a bespoke vibration wire water levelling system (2 minute interval reporting)

- Real time 3-D movement sensors on the key interface between the finger platform structure and the running tunnels and existing station structure (2 minute interval reporting)

- Real time level and horizontal displacement monitoring of the rock pillars and finger platform & vent shaft from outside the structures, using total stations and prisms (half hourly reporting)

- Real time Pressure/Load sensors and Stroke displacement monitoring within the hydraulic jacking system (daily reporting but instant reporting during load transfer operations)

- Precise levelling of the finger platform and vent shaft from outside the structures (daily reporting)

- Crack width monitoring at the top of the rock pillars (reporting on an as-required basis)

Alert, Alarm and Action levels were set for each of the monitoring systems at 50%, 75% and 100% of the expected movements and jack loads, with the real time monitoring systems raising automatic text and e-mail warning messages at each level. A comprehensive management plan was implemented to deal with each respective level of warning raised. In addition, a daily review of the data and forthcoming construction activity was held between responsible people, to determine any action that may have been required.

Monitoring of the existing structure was established 3 months prior to the commencement of construction to establish a credible baseline and understand any impacts caused by station box excavation, temperature and other factors.

Construction

The excavation followed the original plan to the letter with the initial 4.5m × 4.5m 'A' and 'B' slots (see Figure 7) being used to excavate the rock using excavators mounted with hydraulic hammers. The 150+MPa rock, although jointed, proved very difficult to remove. To help with progress, the rock in each of the slots was pattern drilled in advance to assist excavation.

The through-pillar rock bolt concept led to downtime on those occasions when the rate of progress in adjacent slots varied. An alternative, "Grouted Bolt" solution was

Figure 12. Initial slot excavation beneath the existing ISL platform tunnel

developed to overcome this constraint, although it followed that bolts could no longer be re-used. A reduction in bolt spacing was also necessary since capacity became reliant on permissible rock / grout bond stress rather than bolt strength.

It soon became evident that excavation rates could be significantly improved if, somehow, the excavation module sizes could be enlarged to facilitate the employment of heavier (physically larger) plant.

Revised Pillar Excavation Sequence

A thorough review of the design and sequencing was carried out. It became evident that the embedded design redundancy had room to facilitate an alternative "double width slot—double width pier" excavation sequence, once the finger platform & vent shaft soffit slab support grillage had been constructed. Slots & pillars could thereby be doubled in width whilst still maintaining the one column loss and the minimum factor of safety.

Furthermore, actual geological data feedback to the designers facilitated further 3DEC analysis of 6.75m and 9.0m deep slots, which justified the increase in excavation module depths by 50% to 100% (slot heights were constrained by the already fabricated 2.25m underpinning column module lengths).

The re-engineered excavation module enlargements, associated alternative construction sequence (refer Figure 13) and employment of larger plant led to a significant increase in excavation rates.

Blasting of the Slots

In parallel to the assessment and implementation of the wider and deeper slots, a detailed impact assessment and construction methodology was developed for rockmass removal using drill and blast methods. Blasting had already been highly successful in improving the rock excavation rates from the adjacent station box with no impact to the operating railway.

Detailed liaison was undertaken with MTR Operations, Mines Division and the Buildings Department to enable approval for some limited blasting of the rock slots in the underpinning zone. It was decided to trial the blasting and 3 blasts were successfully undertaken. It was decided not to take this further forward as the excavation rates

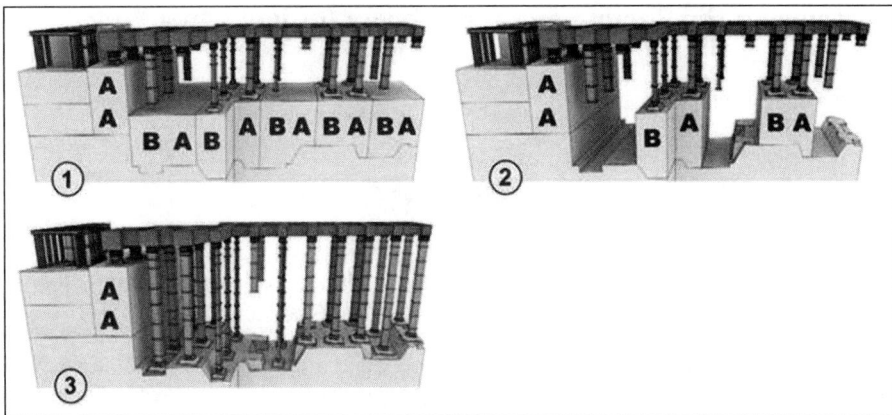

Figure 13. Revised pillar excavation sequence

from the wider and deeper slots was now advancing at a rate that was faster than the limited blasting could achieve.

CONCLUSIONS

Despite the huge scale and complexity of the works undertaken, there has been zero disruption, and no impact on the safety on the Tsuen Wan Line or Island Line train service throughout the whole construction period. This was explicitly a result of the remarkable levels of collaboration between all the parties involved, with innovative design, excellent construction supervision, rigorous planning and attention to detail on work permits and safety procedures. The safe completion of the work marks an outstanding achievement in completing work to customer satisfaction, and maintaining railway performance and the safest possible environment.

ACKNOWLEDGMENTS

The Authors thank and acknowledge the MTR Corporation and the contractor Kier-Laing O'Rourke-Kaden JV.

REFERENCES

Barton N R & Grimstad E, 1994. The Q-system Following Twenty Years of Application in NMT Support Selection, Geomechanik

Bischoff J A & Smart J D, 1977. A Method of Computing a Rock Reinforcement System Which is Structurally Equivalent to an Internal Support System; Proc. 16th Symposium on Rock Mechanics & Design Methods in Rock Mechanics

Grimstad, E., & Barton, N, 1994. The Q system following twenty years of application in NMT support selection. 43rd Geomechanic Colloquy, Salzburg. Felsbau, 6/94: 428:436.

Priest SD, 1993. Discontinuity analysis for rock engineering. Chapman and Hall, London.

Completing the Second Avenue Subway Project, New York

Jonalen Chua-Protacio ▪ Arup
Richard Giffen ▪ Arup

ABSTRACT

Phase 1 of the Second Avenue Subway Project is due for completion in December 2016. At a cost of $4.5 billion this phase will provide 4 new stations and 2 miles of new tunnels to extend the existing Q line to the upper east side of Manhattan. Future phases will extend north into Harlem and south to the Financial District.

This paper will focus on the challenges of completing and fitting out the tunnels, caverns and cut and cover structures previously constructed. The complexity of completing the platforms, mezzanines, public, back of house and ancillary structures together with interfacing with adjacent properties, utilities and third parties will be covered. In particular the lessons learned from the installation of track, escalators, elevators, architectural finishes and systems integration to a tight schedule while accommodating design changes will be covered. Particular focus will be paid to waterproofing methods used and remedial work done prior to completion.

INTRODUCTION

The Second Avenue Subway is the expansion of mass transit on the east side of Manhattan for the New York Metropolitan Transit Authority (MTA). The project was first conceived in the late 1920s with parts of the tunnel constructed in the early 1970s. However, due to the New York fiscal crisis in 1975, the project was halted and did not revive until the late 1990s with final design starting in 2006. The project consists of 8.5 miles of track, 16 new stations, one renovated station, and connections to existing subway lines. Due to its length and complexity, the Second Avenue Subway was subdivided into four phases. The first phase which is currently under construction consists of renovation of an existing station at 63rd Street, new mined cavern stations at 72nd and 86th Streets, and a new cut and cover station at 96th Street. This phase has an estimated cost of $4.45 Billion and is expected to have approximately 200,000 ridership.

Phase 1 was further divided into 10 different contracts to allow more contractors to participate thus increasing contractor competition. The split also allowed for smaller contracts to be awarded in a staggered fashion. The TBM tunnels were the first contract to be awarded in 2007 to spearhead the construction. The other contracts were awarded in the following years starting with 96th Street and ending with 86th Street.

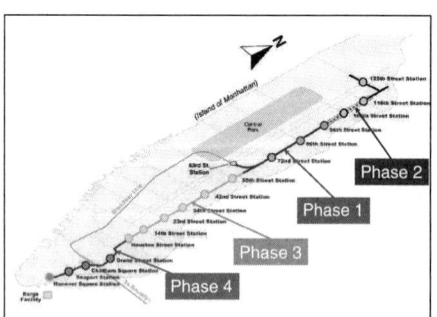

Figure 1. Project map

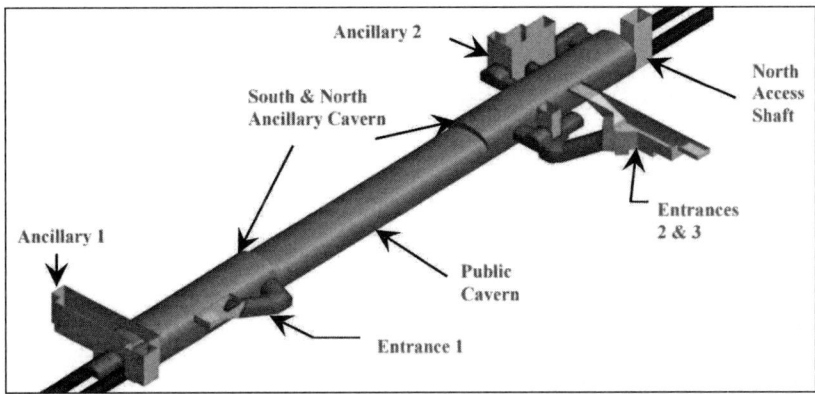

Figure 2. 86th Street Station rendering

CONSTRUCTION OVERVIEW

Construction began with the bored tunnels in 2007 followed by construction of the stations and systems packages. As of December 2016, the main structural work has been completed and the majority of the architectural and MEP work is installed. In order to achieve completion, Contractors have been working multiple shifts 7 days a week. As Phase 1 nears its finishing point, close coordination and interface amongst the design team, construction team, and the client becomes even more important to resolving critical field conditions. This involved real time coordination on site with various disciplines and construction teams.

As with any construction project, regardless of size or complexity, there are bound to be unforeseen field conditions. The Second Avenue Subway is no exception and due to the project size, there were many issues that had to be overcome. The following are some examples during the closing stages of the project where collaboration across multi-disciplines were key to a resolution.

North Access Shaft

The north access shaft is located at the north end of the 86th Street Station and was intended to serve as a construction access point for equipment, materials, etc. in and out of the station. Figure 2 depicts the schematic rendering of the station with the north access shaft highlighted in green which is sandwiched between the north ancillary cavern and the TBM tunnel to the north. The shaft is approximately 22 ft. wide by 32 ft. long on plan and extends from street level down to the track which is about 100 feet below grade. The shaft was constructed by blasting vertically in increments of 6 to 8 feet with rock bolts to support the rock. Rock bolts were installed at 5 to 6 foot on centers in each direction to ensure rock stability throughout the height of the shaft.

Once the shaft is no longer required, the station contractor would then install a segment of TBM liner within the shaft to connect the ancillary cavern to the existing northern TBM tunnel. However, due to the sequence of work, the shaft needed to remain accessible while services along the liner walls were required to be installed to meet milestones. Therefore, a TBM liner was not feasible and another construction method would be required to accommodate the schedule.

Meetings with the contractor were held to find a solution that would work for both the design and schedule aspect. A cast-in-place concrete box was deemed the simplest and quickest solution to the issue. The walls were to be constructed first and designed to cantilever from the invert slab so that services along the walls can be installed to meet schedule and the shaft would remain open to above for access. Once access is no longer needed, the roof would then be constructed to finish the box and backfilled to street level. Furthermore, in order to keep the services along the walls in a straight run, the inside surface of the CIP box was formed to be circular similar to the shape of the north TBM liner.

As one would expect, blasting for the shaft resulted in a rough surface with over break in some locations. The over break in the shaft was mainly on the north and east sides and ranged from a few inches to 8 feet in some areas. Figure 3 is an example of the over break scan across the shaft. The contractor preferred to pour additional concrete to make up for the over break in lieu of form-work and backfilling. Pouring the walls thicker would allow them to construct the walls faster and stay on schedule. To accommodate the contractor's means and method, the design team provided additional reinforcement in the walls for temperature and shrinkage on top of the reinforcement required for flexure. The final box configuration is shown on

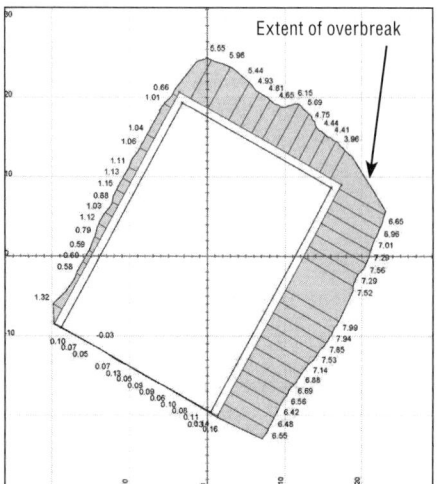

Figure 3. Overbreak plot

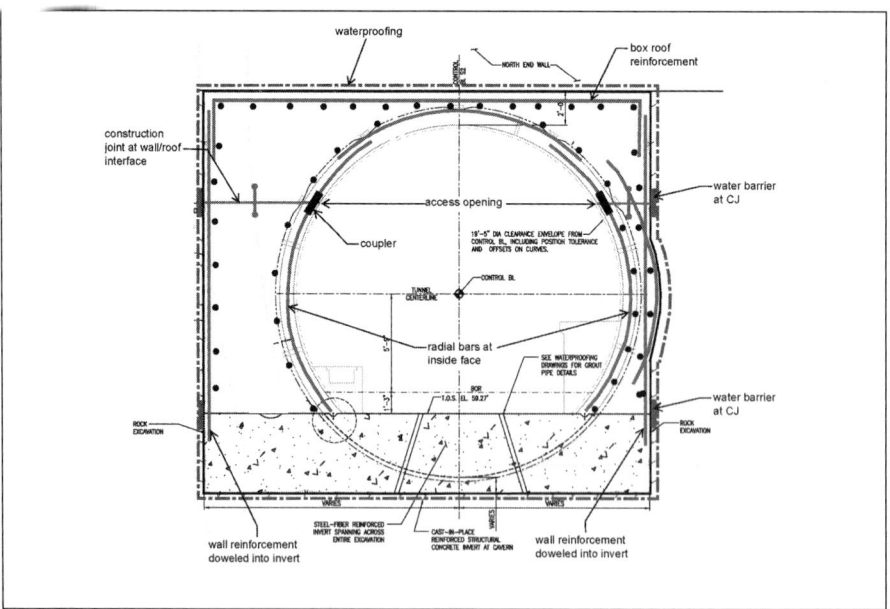

Figure 4. Reinforced tunnel box

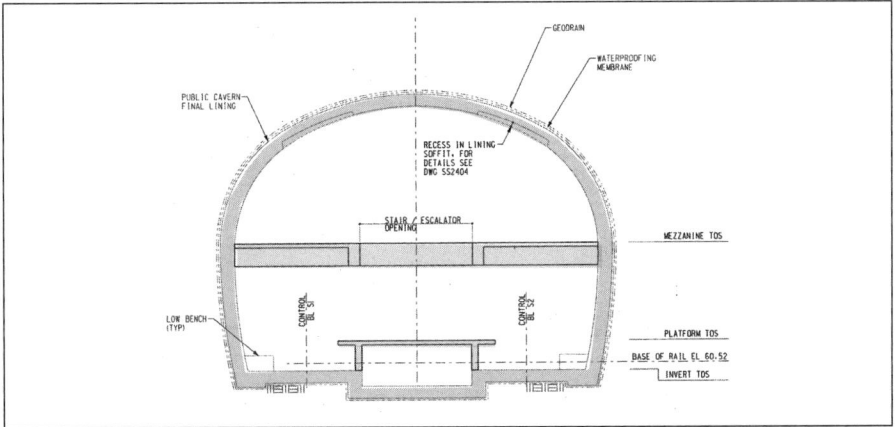

Figure 5. Typical cavern station section

Figure 4 which shows the reinforcement and construction joint between the wall and roof pours.

Misaligned Couplers

As mentioned earlier, Phase 1 was divided into 10 separate contracts. As part of this split, the station structure as shown in Figure 5 was constructed under 2 different contracts. The construction of the cavern liner was completed under one contract while the mezzanine, platform and finishes were completed under

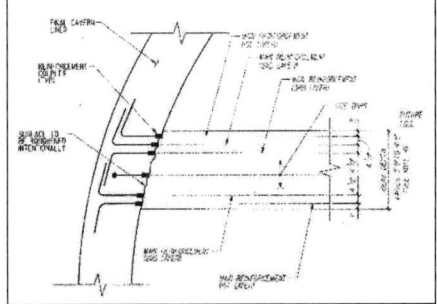

Figure 6. Typical mezzanine beam connection

another. Thus the connection of the mezzanine slab and beam to the cavern liner became a design issue to coordinate.

The cavern liner could have been blocked out to accommodate the mezzanine structure in the follow on contract. However, this would have been a challenge to locate and form the pockets in the densely reinforced liner. Another challenge would have been to coordinate the pockets with the cavern formwork which would have taken time and may end up not aligning correctly causing more issues. Therefore, reinforcement couplers were used in the liners to allow the follow on contract to thread the mezzanine rebar into as shown on Figure 6.

Most of the mezzanine beam connections aligned properly, however, there were a handful of locations where the couplers shifted position. The shift likely happened during the cavern pour where the dowels with couplers were not adequately secured in position and the weight of the wet concrete pushed the dowels out of place. These misalignments were discovered and the new position of the couplers were surveyed relative to where they should be. In some cases, the couplers were shifted far enough that the new location is outside of the beam perimeter making them unusable.

The contractor sub-consulted an engineer to redesign the connection with guidance and input from the design team. The design team was also responsible for reviewing the final connection design to determine whether it was adequate and behaves similar

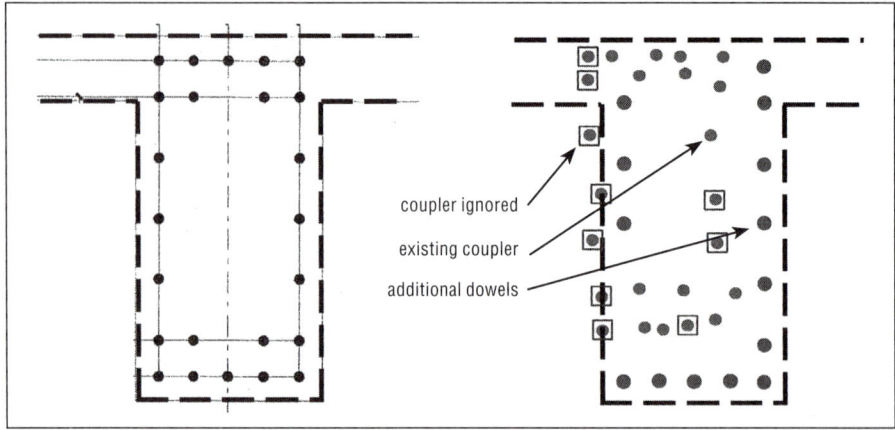

Figure 7a. Mezzanine beam rebar layout   Figure 7b. Example of as-built rebar

to the original connection. Although there were unusable couplers, the remaining couplers were still within the beam boundaries and were utilized as part of the connection. In addition to the existing couplers, new dowels were required to be drilled and fixed into the cavern liner to replace unusable couplers and reinstate capacity at the connection. As post-installed dowels are not as efficient as cast-in dowels or couplers, the total number of post-installed dowels were more than the number of unused couplers.

Figure 8. Roof of existing masonry building

Ancillary Wall Conflict with Existing Chimney

The project did not only have field issues underground but there are several above ground field issues that required coordination and redesign as well. The Ancillary 1 building at the 86th Street Station is one such example. This building consists of tunnel ventilation shafts, plumbing closets, mechanical rooms, as well as emergency egress stairs. It is located on a street corner adjacent to an existing 4-story masonry building. The footprint of the ancillary building was previously an existing masonry building which shared a party wall with the adjacent building, see Figure 8. The party wall consisted of two chimney structure which the buildings shared as well. During the demolition of the masonry building for the ancillary structure, the chimney for the demolished building should have been removed as part of the demolition. However, the flue was in very poor condition that removing one side would have meant removing the flue for the adjacent building, see Figure 9. Therefore, the chimney remained in place.

As the ancillary building was about to be constructed at street level, it was determined that the chimneys protruded too far out and would conflict with the new ancillary wall. The contractor provided survey information of how much the chimneys would encroach approximately 10 inches into the new wall. The design team then had the task of coming up with a solution in a short period so that construction schedule is not delayed. Coordination amongst mechanical, plumbing, and structural teams was necessary as each discipline would be affected by any revision to the wall.

The new wall was 12 inch thick and originally designed to span horizontally to intersecting walls that makes up the ventilation shafts. The wall thickness was reduced to 8 inches which alleviates four inches from the encroachment. In addition to reducing the wall thickness, the wall location was shifted 9 inches to clear the chimney structure and allow a small space for stay-in-place formwork. This shift had to be coordinated with each discipline as the program space in the ancillary building had to be reduced by the same 9 inches. Fortunately, the ventilation shafts had extra capacity to be slightly reduced. The plumbing chase had to be reconfigured in order to allow for the shift.

Figure 9. Existing flue in poor condition

Waterproofing

The new stations and tunnels for Phase 1 are typically lined with an external water-proofing barrier. During design a solution was adopted whereby mined sections would be lined with PVC and cut and cover sections would be lined with HDPE bonded to the outer surface. However after the construction contracts had been awarded, the client requested that all areas utilize PVC meaning that the design

Figure 10. Cut and cover wall waterproofing

for the cut and cover sections needed to be modified. The prime reason for doing this was that a PVC system as it is installed loose and broken into compartments, more easily incorporates the ability to later install grout to address any leaks. This design change was straightforward except for details relating to the cut and cover roof and interfaces. PVC on laid on the top of a roof surface is not typical and presents a number of challenges. In addition, on some sections the roof ties into sections of slurry wall which are expected to have some seepage. The detail at this interface proved problematic and contributed to significant areas of leakage within the roof areas. In addition the compartmentalization on the roof had to utilize tape which proved less effective at creating barriers than regular waterstop. The end result was that more grouting to the roof areas had to be installed than was initially expected causing an impact to follow on trades as many of the areas had been turned over for mechanical and electrical fit out.

Architectural Finishes

As construction nears the end, more focus has been placed on the aesthetics of the station finishes in preparation for the station opening. One of the key features of the public station is the architecturally exposed concrete finish of the station structure. The cavern liner within the public areas were intended to have a smooth and even

Figure 11. Mined cavern roof lining

Figure 12. Cavern liner prior to repair

Figure 13. Repaired cavern liner

finish. However, problems were encountered and in some areas the final appearance was uneven and did not meet the appearance approved during mockups, see Figure 12. After trials of utilizing different repair methods which included surface sand blasting, it was decided that a surface applied thin set mortar would be the best product to cover the blemished liner. The product used was Sikatop 123 Plus which is a cement based polymer which performs well for vertical and overhead applications. The repair provided the station a uniform and smooth finish as shown in Figure 13.

At isolated construction joints the Sikatop 123 Plus was found to be less appropriate as some of these joints experienced movement. Local cracking and spalling of the repairs were observed as shown in

Figure 14. Construction joint repair

Figure 14. Thus a separate repair was utilized at some joints which would allow for the movement and provide adequate protection and seal. Sikaflex 1a which is an elastomeric sealant was utilized in the joint and allow for 35% joint movement.

CONCLUSION

The design and construction of a project of the size of 2nd Avenue is challenging and to be successful requires the collaboration of all parties. Towards the end of the project as deadlines approach and pressures increase it becomes even more important that the design team are involved on a day to day basis with site activities and are brought in quickly to help resolve problems. On this project the design team were regularly on site and made a large contribution to the successful close out of the work and the particular challenges of the finishing trades. Completing this first phase of the Second Avenue Subway is an historical moment and all involved take pride in delivering the first new subway line in New York in over 60 years and a piece of infrastructure that will benefit generations to come.

REFERENCES

Giffen, R., Ezzeldin, K, & Towel, N., (2009), Second Avenue Subway, New York: Reconstructing Existing Buildings for New Station Entrances. ASCE Structures Congress 2009, Austin, TX.

Aksman, B., Grigson, R., Giffen, R., (2012), Second Avenue Subway, New York: Design of New Underground Stations. ASCE Structures Congress 2012, Chicago, IL, March 29–31.

Trabold, M., Giffen, R., Aksman, B., (2012), Second Avenue Subway, New York: Repair and Refurbishment of Existing Structures Adjacent to Deep Excavations. ASCE Structures Congress 2012, Chicago, IL, March 29–31.

Garavito-Bruhn, B., Napoli, A., Towel, N., (2012), Second Avenue Subway, New York: Refurbishment of an Existing Underground Station. ASCE Structures Congress 2012, Chicago, IL, March 29–31.

Voorwinde, M., Giffen, R.,(2012), Second Avenue Subway, New York : Design of a New Elevator Only Entrance. ASCE Structures Congress 2012, Chicago, IL, March 29–31.

Trabold, M., Giffen, R., (2014), Second Avenue Subway Project New York: Design and Construction of the 96th St Station. North American Tunneling Conference 2014, Los Angeles, CA, June 23–25.

Voorwinde, M., Giffen, R., & Powers, R. (2014). Second Avenue Subway Project: Design and Construction of 72nd Street Station and G3/G4 Cavern Final Linings. North American Tunneling 2014 Proceedings (pp. 301–309). Los Angeles, CA: Society for Mining, Metallurgy & Exploration Inc.

Voorwinde, M., Garavito-Bruhn, E., Dalton, L & Giffen, R. (2015). Second Avenue Subway Project: Design and Construction of Large Cavern Final Linings and Penetrations at 86th Street Station. SME Rapid Excavation and Tunneling Conference, New Orleans, LA.

Trabold, M., Giffen, R. and Lemus, P (2016). The Light at the End of a 40 Year Old Tunnel - Retrofitting of an Existing Tunnel as part of The New Second Ave Subway Project, New York. World Tunnel Congress, San Francisco.

Permanent Lining Design for Downtown Los Angeles Cavern

Justin Lianides ▪ Mott MacDonald
Carlos Herranz ▪ Mott MacDonald
Derek Penrice ▪ Mott MacDonald

ABSTRACT

Metro is constructing a lightrail corridor beneath Downtown Los Angeles—The Regional Connector Transit Corridor (RCTC). To provide operational flexibility, RCTC requires a track crossover adjacent to the proposed 2nd/Broadway Station. Rightofway constraints require that the crossover be constructed at relatively shallow depth using Sequential Excavation Methods (SEM), resulting in what is believed to be the largest tunneled crosssection in Los Angeles. The SEM cavern permanent lining must accommodate significant ground and groundwater loads, building surcharges, and seismic events with 2,500year return periods. Due to the complexity of the cavern geometry, dynamic timehistory analyses were needed to confirm the structure's seismic performance. This paper describes the numerical models required to demonstrate satisfactory performance of the permanent structure under long-term and seismic conditions.

INTRODUCTION

The Los Angeles County Metropolitan Transportation Authority's (Metro) highly prioritized Regional Connector Transit Corridor (RCTC) project will bring light rail subway through downtown Los Angeles and link multiple existing lines to improve public mobility within the City and County. A US$927-million design-build contract for the project was awarded in 2014 to Regional Connector Constructors (RCC), a joint venture of Skanska and Traylor Brothers. Mott MacDonald is RCC's principal designer. The project totals 3.1 km of new light rail infrastructure with 1.6 km of twin bored tunnels, 1.1 km of cut-and-cover tunnels, three cut-and-cover stations, 0.3 km of at-grade alignment, and a crossover cavern. The design of the RCTC tunnels and underground structures has been completed with a small portion of the project under additional design optimization efforts. The expected construction completion is in 2020. A location map showing the alignment, stations, and crossover cavern is shown in Figure 1.

A crossover structure is needed to permit rail trains to switch from one track to the other. A limited right-of-way at the crossover location, walled by existing basements and structures, made a cut-and-cover option infeasible. Thus, the project adopted one of its biggest design and construction challenges: a mined crossover cavern measuring 17.1-m wide by 11.0-m tall and 88 m in length, the overall design of which is described in Herranz et al. (2016). Once completed, the cavern will have the largest tunneled cross-section in Los Angeles.

The cavern is located 15 m below the centerline of 2nd Street. It will be constructed from a single portal at the temporary perimeter wall of the 2nd/Broadway Station excavation. For excavation stability, economy, and the protection of existing site infrastructure, sequential excavation methods (SEM) will be utilized for the full length of the cavern. Once the excavation is complete, permanent lining and internal structures will be built to accommodate crossover infrastructure, emergency egress, and an

358

overhead ventilation plenum. A rendered perspective from within the cavern showing the finished lining, rail infrastructure (excluding crossover), and plenum structure is shown in Figure 2.

GEOLOGIC AND GEOTECHNICAL CONDITIONS

RCTC is situated in the northern portion of the Los Angeles Basin, a major depression that has a thick accumulation of sediments dating back to the middle-Miocene Epoch. Locally, the cavern will be constructed in Pliocene-Series sedimentary rocks of the Fernando Formation (T_f). The formation is overlain by Holocene alluvial deposits of the Los Angeles River.

Geotechnical site explorations encountered massive, extremely to very weak, clayey siltstone with indistinct bedding at 6 m deep. The upper 6 m of this siltstone was identified as a weathered zone that transitions into more competent bedrock. The formation is overlain by approximately 3 to 5 m of coarse-grained alluvial deposits and 2 to 3 m of artificial fill. The groundwater regime consists of a perched groundwater table at 4 m deep and a regional groundwater table at 7 m deep. The siltstone has

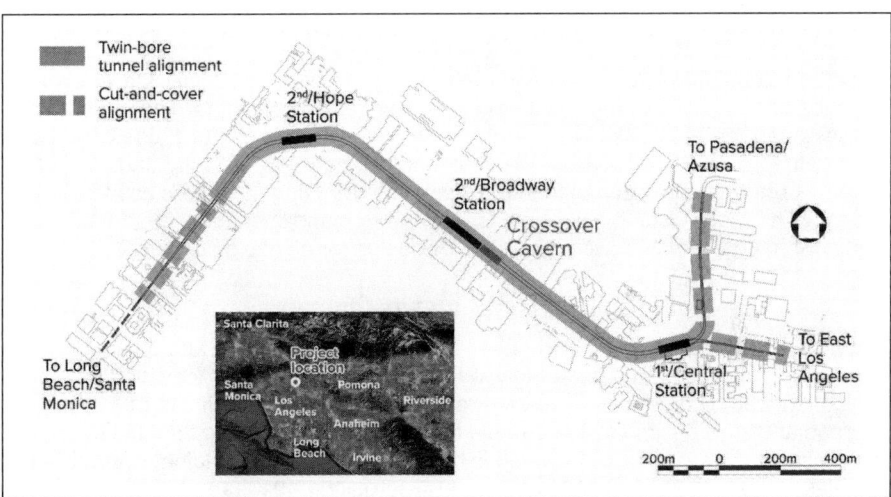

Figure 1. Location map of RCTC twin bore alignment with 2nd/Hope Station, 2nd/Broadway Station, 1st/Central Station, and the crossover cavern

Figure 2. Rendered perspective from within 2nd/Broadway Station towards the Crossover Cavern

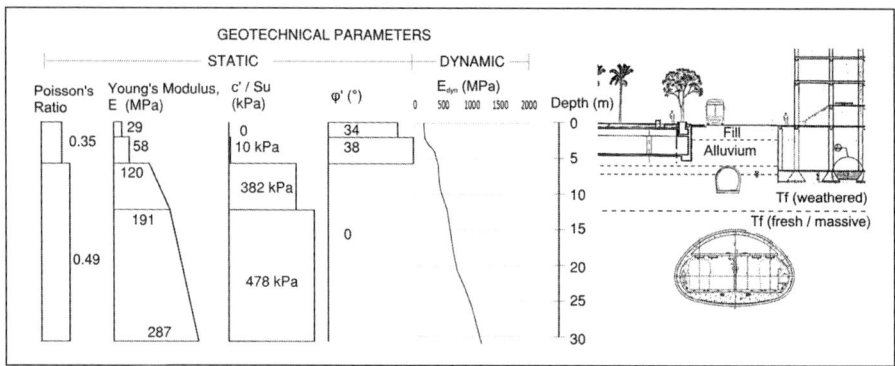

† E = Static Young's modulus, c' = static effective cohesion, φ' = static effective friction angle, Su = static undrained shear strength, E_{dyn} = small-strain Young's modulus

Figure 3. Site geotechnical profile

relatively low hydraulic conductivity, measured in the range of 4×10^{-7} to 2×10^{-4} cm/sec, with an average value of 1×10^{-6} cm/sec. Groundwater infiltration during excavation is expected to be limited to seeps, originating primarily from rock discontinuities.

Short-term-loading geotechnical parameters were selected for analyzing a relatively rapid loading of the T_f during SEM excavation and seismic events. The strength and stiffness parameters varied with depth as shown in Figure 3. For these conditions, the weathered and competent T_f was assumed to be predominantly undrained (i.e., develop pore pressures during loading), while the overlying soils were assumed to be drained. The Mohr-Coulomb strength failure criterion and linear elastic-perfectly plastic deformational behavior were selected for design for all geologic units.

EXISTING SITE INFRASTRUCTURE

The cavern will be built below portions of three adjacent buildings and directly below a large county storm drain. Accordingly, a focal point of the SEM excavation sequence evaluation was to determine if the Metro-mandated settlement limits of 13-mm and an angular distortion limit of 1/600 could be met along the full length of the cavern. At the critical cross-section, the cavern lies between the Higgins Building, a 10-story Los Angeles Historical-Cultural Monument that was built in 1910, and the10-story headquarters of the Los Angeles Police Department (LAPD) that was built in 2009. The Higgins building, with two basements levels, slightly overlies the south side of the cavern excavation. A one-level underground parking structure, adjacent to the LAPD headquarters, is just to the north of the cavern. The storm drain is a lightly reinforced concrete structure that measures 3×3 m and is located 5 m above the vertical centerline of the cavern. A cross-section showing the existing site constraints is provided in Figure 4.

SEM DESIGN OVERVIEW

The detailed design of the SEM cavern required extensive numerical modeling efforts due to the scale of the excavation and its interaction with existing site infrastructure. The design of the drift and heading sizes, sequencing, and temporary ground support required both two-dimensional (2D) and three-dimensional (3D) numerical models to evaluate the response of the ground, lining and overlying structures (Lianides

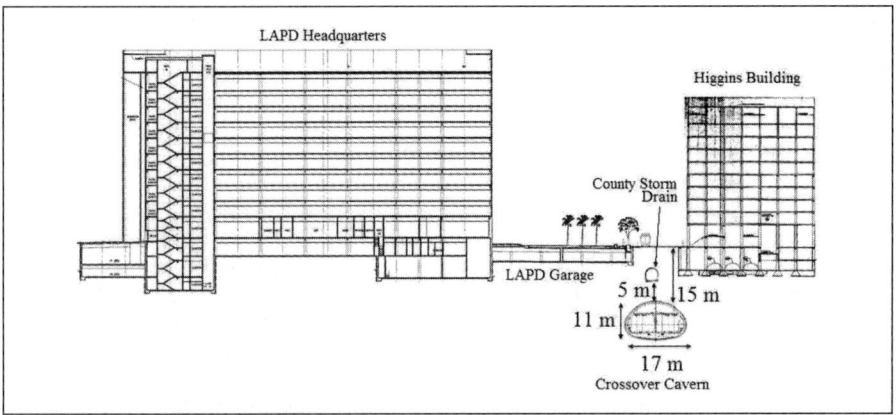

Figure 4. Site cross-section (facing east)

et al., 2016). The numerical models were generated in Itasca software *FLAC (*Fast Lagrangian Analysis of Continua), Version 8.0, and *FLAC 3D,* Version 5.0.

The models demonstrated that a two-drift sequence, with excavation advance lengths of 0.9 meters in top headings and 1.5 meters in benches, could meet Metro's settlement criteria. The main components of the temporary ground support include 30 cm of fiber-reinforced shotcrete (compressive strength of 35 MPa), lattice girders at 0.9 meters on-center, and local reinforcing steel at the corners of the temporary wall and drifts. Standup time will be improved in the break-out from the station with a pipe canopy system and as- needed face stabilization techniques including face wedge, fiberglass dowels, and drain holes.

PERMANENT LINING DESIGN AND ANALYSES

The Metro design criteria for the permanent lining design requires a 100-year service life, neglecting any beneficial contributions of the temporary ground support. At a minimum, the lining design requirements include ground loads, hydrostatic pressure, live loads within the tunnel and on surfaces above the tunnel, and seismic loads. To meet these requirements, a 28 MPa reinforced concrete lining was selected for analyses. Thicknesses varied from 460 mm along the perimeter arch to 1800 mm at the centerline of the invert.

The permanent lining will be cast against a pre-installed hydrocarbon-resistant gas-and-waterproofing membrane following SEM completion. The cast-in-place plenum structure will then be constructed as a 30-cm-thick slab supported by a 30-cm-thick center wall and two 42-degree inclined corbels. The wall will be discontinuous at the midpoint of the cavern to allow for rail crossover. Figure 5 provides a cross-section of the final structure arrangement.

Numerical Model

The numerical model for the permanent lining was based upon the final excavation stage of the SEM 2D FLAC model, where ground loads were initialized onto the lining with a full degradation of the temporary support. The permanent lining was connected to the grid with a slip-permissible interface assuming a friction angle of 10 degrees that was intended to correspond with irregularities in the temporary shotcrete support and the waterproofing membrane. The lining was modeled with elastic-perfectly plastic 'liner elements' using varying thicknesses and loading resistances that were based upon nominal reinforced capacities. The plenum slab and center wall were included with elastic 'beam elements' for dynamic analyses. The model measured 120-m-wide by 50-m-tall and the continua contained 46,870 zones that varied in area from 0.1 to 0.8 m² (see Figure 6).

The lining and interior plenum structure were evaluated under load and resistance factor design (LRFD) load combinations to meet long-term performance criteria. Different load factors were addressed in the model with individual loads proportioned to a final global load factor. In general, water loads controlled the static analyses of the permanent lining. Seismic analyses addressed two different seismic performance levels with a probabilistic design approach. The key design considerations incorporated in the seismic numerical models of the permanent lining are discussed in the following sections.

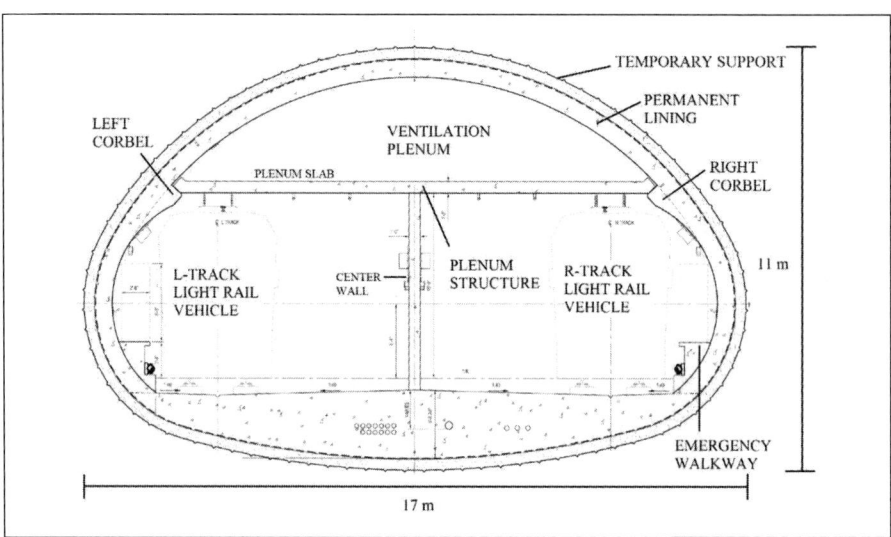

Figure 5. Cross-section of the permanent lining and plenum structure

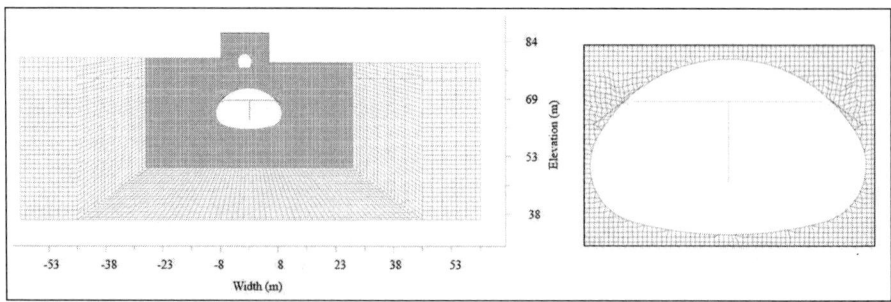

Figure 6. (Left) FLAC model for long-term analyses of permanent lining; (Right) focused view on the cavern lining and plenum structure

SEISMIC INPUT MOTIONS AND DESIGN CRITERIA

Local Seismicity

The Los Angeles basin is an area of high seismic potential due to influences of several strike-slip and blind thrust faults. Historically, the largest recorded event in the basin is the 1933 Long Beach earthquake, which had a moment magnitude (Mw) of 6.4. Areas adjacent to the basin have experienced more notable events, such as the 1971 San Fernando earthquake and the 1994 Northridge earthquake (both Mw of 6.7). These two earthquakes resulted in violent to extreme shaking and caused significant damage and collapses of structures.

There are no known active faults intersecting the RCTC alignment. However, Shaw et al. (2002) project a northwest striking section of the Puente Hills blind-thrust system to be approximately six kilometers below the southern portion of the RCTC alignment. This fault is believed to have caused the destructive 1987 Whittier Narrows Earthquake (6.0 Mw) and is estimated to have a maximum magnitude of 7.1 (M_w) (Shaw et al., 2002).

Design Earthquakes

Rigorous seismic analyses were required to demonstrate compliance of the permanent cavern lining with Metro Rail Design Criteria (MRDC). A probabilistic seismic design approach defined two design ground motion levels and performance levels. They are described in Table 1.

Horizontal Motions

Three horizontal ground motion time histories for each design earthquake were developed by spectral-matching acceleration response spectra of seed motions to target spectra. Target spectra corresponded to outcropping bedrock with a shear wave velocity of 560 m/sec, similar to that demonstrated by the Fernando Formation material at 45 m deeep. The three selected seed motions are summarized below:

- Event 1: 1989 Loma Prieta Earthquake, Mw=6.93, Reverse-Oblique fault mechanism
- Event 2: 1992 Cape Mendocino Earthquake, Mw=7.01, Reverse fault mechanism
- Event 3: 1995 Kobe Earthquake, Mw=6.90, Strike-slip fault mechanism

Once matched, the time histories for the ODE obtained peak ground accelerations (PGA) of 0.23 to 0.26 g and peak ground velocities (PGV) of 17 to 24 cm/sec. In similar context, the MDE PGAs ranged from 0.68 to 0.84 g and the PVGs ranged from 66 to 82 cm/sec. For these histories, acceleration response spectra of linear single degree-of-freedom systems are shown in Figure 7.

Table 1. Summary of design earthquake levels

Design Earthquake	Probability of Exceedance	Return Period (years)
Operational Design Earthquake (ODE)	50% in 100 years	144 (150 approx.)
Maximum Design Earthquake (MDE)	4% in 100 years	2,475 (2,500 approx.)

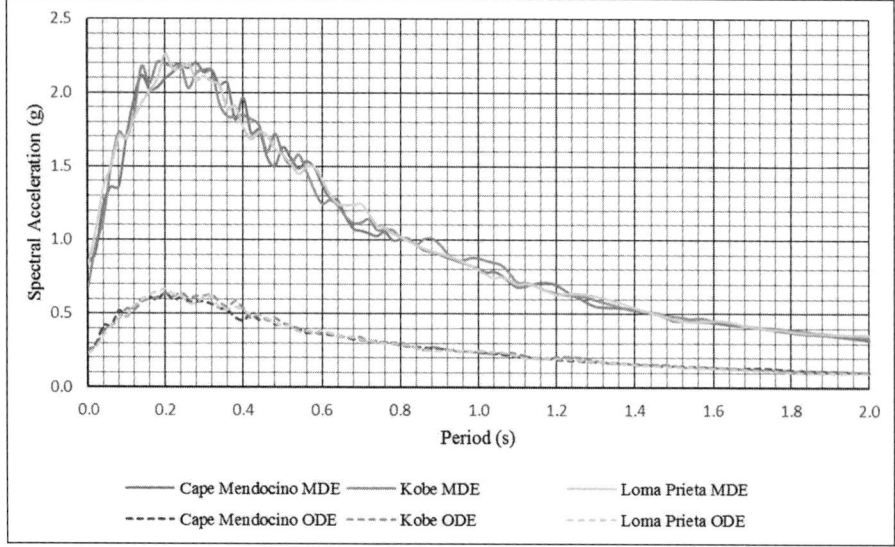

Figure 7. Acceleration response spectra (5% damping) of spectrally-matched motions

For comparative purposes, the upper 50 m of the site had an initial fundamental period of 0.44 seconds per the one dimensional (1D) simplified Rayleigh procedure (Dobry et al., 1976).

Vertical Motions

The presence of an underlying thrust fault may result in a significant vertical component of shear waves during a seismic event. As a minimum, MRDC (2013) required the cavern design to include additional structural loads that represented the effects of vertical ground motions. The vertical seismic load (EQ_{vert}) was to be combined with loads obtained from horizontal ground motions. For mined structures, EQ_{vert} was defined by the following relationship:

$$EQ_{vert} = \pm k_{sv}(DC + W_{LS})$$

where: k_{sv} is the vertical seismic coefficient, DC is the structural dead load, and W_{LS} is the weight of the loosened zone above the roof. To be consistent with these 2D continuum excavation models, W_{LS} was assumed to be equal to the static ground forces that acted on the lining.

The vertical seismic coefficient is defined as follows:

$$k_{sv} = \frac{2}{3}\frac{PHA}{g}$$

where: PHA is the peak horizontal ground acceleration and g is the acceleration of gravity. The ⅔ factor is consistent with horizontal to vertical acceleration ratios (V/H) measured near the Metro tunnels during the Northridge Earthquake (Hashash et al., 2001).

Initial design efforts considered whether the seismic analyses should use site-specific vertical time histories for the MDE and ODE events. However, without a set of established implementation procedures in the MRDC (2013), and after several discussions with Metro, Mott MacDonald elected to proceed with the defined EQ_{vert} loads for all seismic analyses.

Liquefaction

Liquefaction hazards were determined to not be present at the cavern site, due to the relatively dense alluvial soils posing a low liquefaction potential.

Seismic Performance Criteria

Metro Design Criteria requires the following load combination for both the MDE and ODE events:

$$U = 1.00 \cdot DC + 1.00 \cdot E + 1.00 \cdot WA + 1.00 \cdot LL + 1.00 \cdot EQ$$

where: U is the ultimate load, DC is dead concrete, E is earth, WA is water pressure, LL are live loads, EQ pertains to loads generated during the MDE or ODE event.

The seismic performance criteria, however, are different for the two design earthquakes. Following an ODE event, the cavern is required to be fully operational and that a post-earthquake inspection would indicate minimal to no damage. The permanent structures would behave elastically with no ductility demand. The performance criteria for an MDE event is to prevent collapse of the structure, allow some access to emergency vehicles, and limit damage to repairable levels.

SEISMIC ANALYSES FOR LINING DESIGN

MRDC Analysis Options

The deformed shape of the tunnel lining during shaking results in additional axial, bending, and shear loads. MRDC (2013) provided several options to determine these effects for circular bored tunnels, reinforced boxes, and station structures. The outlined procedures stated that in conditions of uniform soil, constant structure thickness, and where little interaction could occur with other structures, closed-form solutions were permissible for ovaling or racking analyses. These ground-structure interaction solutions are based on maximum expected seismic ground strain and are available in publications by Wang (1993) and Hashash et al. (2001).

Numerical methods were required for more complex scenarios that strayed from the simplified closed-form assumption set. The minimum numerical method consisted of applying peak seismic deformations to a 2D soil-structure interaction model with a solved static state. This is commonly known as the pseudo-static method. Those applied deformations could be obtained from 1D or 2D site response analyses. Where inertial effects of the structure were deemed significant or for non-rectangular stations, a nonlinear 2D dynamic time history model was required.

Selected Analysis Methods

The crossover cavern has limited applicability to the aforementioned analysis options for a variety of reasons. The cavern is not circular, nor rectangular; not lined with uniform thicknesses of concrete; situated in varied geotechnical conditions; and with proximity to the ground surface and existing infrastructure. All these complexities largely invalidated analytical solutions and left the design team with the option of using pseudo-static and dynamic methods. The pseudo-static approach simplifies the complex cyclic and inelastic response mechanism of the structure and does not

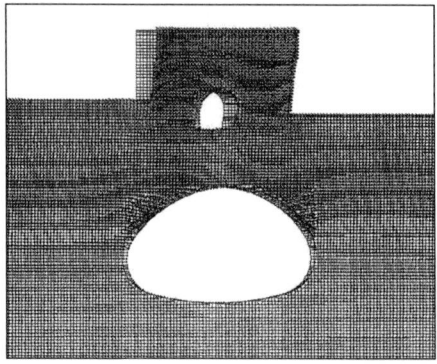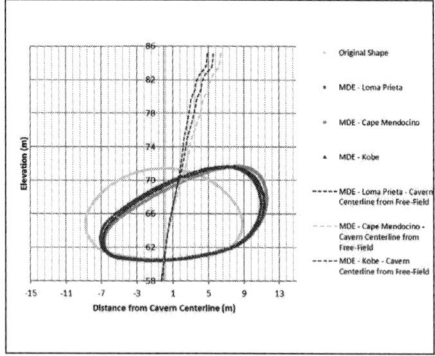

Figure 8. Pseudo-static deformations: (left) model displacement vectors; (right) deformed cavern shape alongside 1D free-field displacements (both scaled by a factor of 100)

account for nonlinear ground deformations. Thus, to further evaluate the lining design, the cavern was designed with both dynamic and pseudo-static numerical methods.

Cyclic Ground Stiffness

Cyclic ground stiffness was defined for implementation into seismic analyses prior to numerical analyses. Hysteretic shear-strain modulus reduction (G/G_{max}) and damping ratios were defined for each geologic layer, all were consistent with published curves. The fill unit used the Seed & Idriss (1970) 'Sand Mean' models. The alluvium unit used the Seed & Idriss (1970) 'Sand Mean' modulus degradation curve and the 'Sand Lower' damping curve. The Tf unit used the Darendeli & Stoke (2001) 'PI=20, OCR=5' curves.

Pseudo-Static Analyses

The SEM cavern was first analyzed with a pseudo-static approach for each of the MDE, ODE, and $\pm EQ_{vert}$ conditions. At the design section, a 1D soil column model was created with Shake 2000, a software for equivalent-linear site response analyses. Maximum instantaneous deformation profiles for each design earthquake were selected for analysis, which corresponded to the largest obtained relative displacement between the crown and invert of the cavern. The corresponding ground deformations were incrementally applied to the 2D FLAC model, up to one diameter in distance from the cavern, until the full deformational profile along those zones were reached. Elastic constitutive models were used, however with moduli based on G/G_{max} curves and shear strain levels. The stiffness of the lining was assumed to have a 50% reduction of the gross moment of inertia (I_g) to simulate lining cyclic ductility and some yielding under MDE distortions. Figure 8 shows a magnified displacement profiles of the MDE Cape Mendocino Event.

Dynamic Method

In collaboration with Itasca Consulting Group, Inc., the dynamic model was built upon the pseudo-static FLAC model after it was configured for dynamic analysis. Initial studies were directed toward confirmation of the grid for accurate wave transmission. Input velocity time histories had peak frequencies of up to 17 Hz. The corresponding wavelengths were greater than 10 times the length of the largest model zone area, which met minimum criteria recommended by FLAC (2016).

The model had relatively limited lateral extents and depth, thus boundary conditions for dynamic applications were used to model shaking behavior as if they were an infinite medium. The lateral extremities of the model were defined with free-field boundaries to locally force displacements to match 1D analyses. The base of the model was defined with a quiet boundary to absorb incoming reflected waves. In accordance with FLAC (2016) procedures, the base input motions were converted to shear-stress time histories when applied to the quiet boundary. A stiffness-proportional Rayleigh dampening of 0.1% was included in the dynamic models to reduce small-strain oscillations which could not be addressed by hysteretic relationships.

The response of Tf was studied from fully elastic behavior to elastic-perfectly plastic behavior conservatively assuming a dynamic shear strength, S_{ud}, and dynamic tensile strength, σ_{td}, equal to the following relationships:

$$S_{ud} = 1.0 \cdot S_u$$
$$\sigma_{td} = 0.17 \cdot S_u$$

Due to the size of the earthquakes, plastic deformations and tensile yielding of the Tf continuum developed during shaking with the aforementioned Tf dynamic strengths.

The plenum structure was fully incorporated into the model rather idealized with plenum structure loads on the lining. Because the permanent lining was modeled with 2D structural elements, the effective center wall height required the structure to be freely isolated within the center of the cavern. To unite the structure with the model, a cycling routine forced the base of the center wall to match displacements of the lining invert and prevent rotation. The overlying plenum slab was fixed to the top of the center wall and supported by inclined rollers at each corbel. At the corbels, the gap width between the permanent lining and plenum slab were continuously monitored during cycling. When the gap closed, an impact force was applied to both the slab and the corbel.

The modeled behavior of the structural elements was optimized for dynamic analyses. The stiffness of the permanent lining was reduced to approximately 30 to 35% of Ig for the cracked inertial values under static loading conditions. Furthermore, due to relatively large deformations during the MDE events, the lining was allowed to develop localized yielding and the structural rotations were later confirmed with moment-curvature analyses. The plenum structure attracted relatively large bending moments at the wall-to-slab junction, therefore a limited moment capacity was utilized to increase available ductility during model cycling. Finally, loads in the lining and plenum structure were compared with moment thrust interaction diagrams based on Caltrans (2013) expected concrete and steel strengths for MDE simulations and ACI 318 (2014) nominal strengths for ODE simulations.

The events were taken to 40-seconds in duration, which resulted in a 10-hour processing time for a Windows 7 64-bit machine with an Intel Xeon CPU E5-2687W processor (with hyper-threading enabled) and a model critical dynamic timestep of 1.6×10^{-6} sec. Data histories were recorded every 4000 steps in terms of forces, rotations, and displacements for all structural elements and nodes along the liner, plenum structure, and corbels. In addition, several key gridpoints were monitored for displacements to review ground response to the shaking and to compare deformations at the cavern to along the boundaries.

Summary of Model Behavior

A minimum of 14 dynamic simulations were required to demonstrate satisfactory performance of the permanent lining which included: three MDE + EQ_{vert}, three MDE

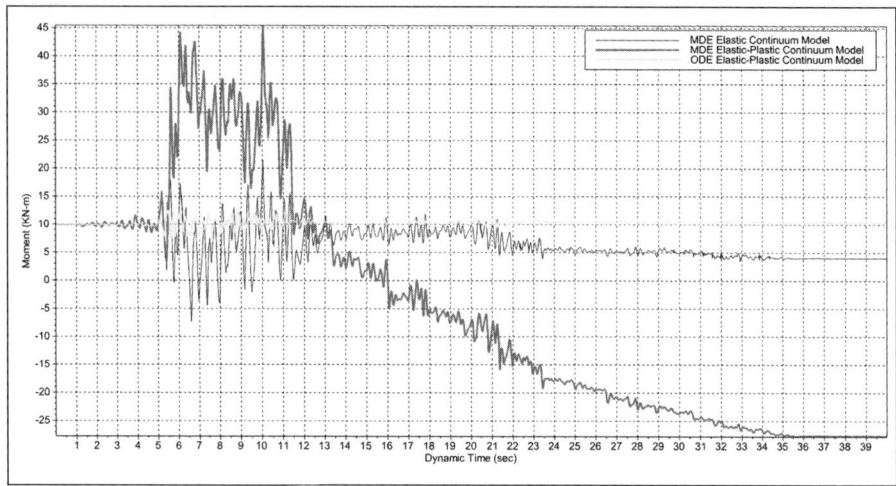

Figure 9. Dynamic moments during Cape Mendocino event (+EQ vert) at 45 degrees clockwise from the cavern crown

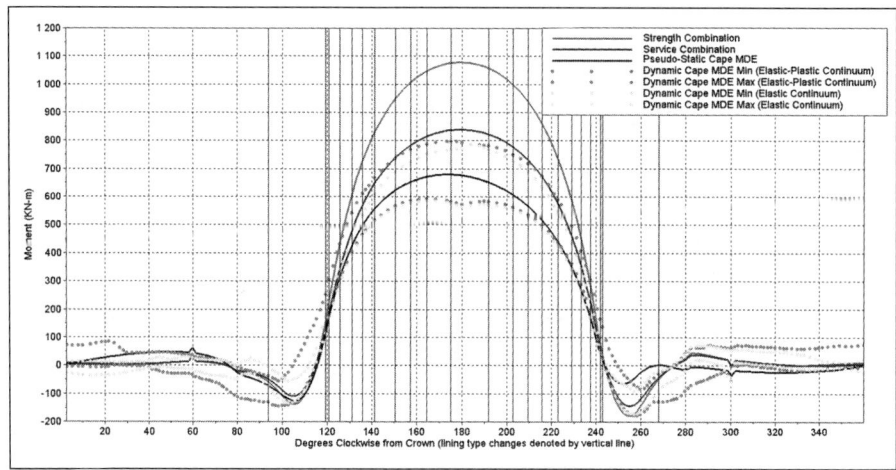

Figure 10. Comparison of moments along permanent lining for static, pseudo-static, and dynamic loading

$-EQ_{vert}$, three ODE $+ EQ_{vert}$, three ODE $- EQ_{vert}$, one temporary plus permanent lining model, and one permanent lining model with increased interface strength properties. Additionally, the process to create a standard dynamic base model took numerous iterations to properly capture the complexities of the interior structure shaking, cyclic ground behavior, ground plasticity, and the response of overlying structures and excavations. Continuous moment histories for elastic and elastic-perfectly plastic continuum models of the Cape Mendocino MDE and ODE dynamic simulations ($+EQ_{vert}$) are provided in Figure 9 for a 'liner element' at 45° clockwise from the crown.

The maximum and minimum loads obtained from each element history were compiled to provide a ranges developed during dynamic analyses. Figure 10 and Figure 11 provide a compilation of thrust and moment ranges for several load combination

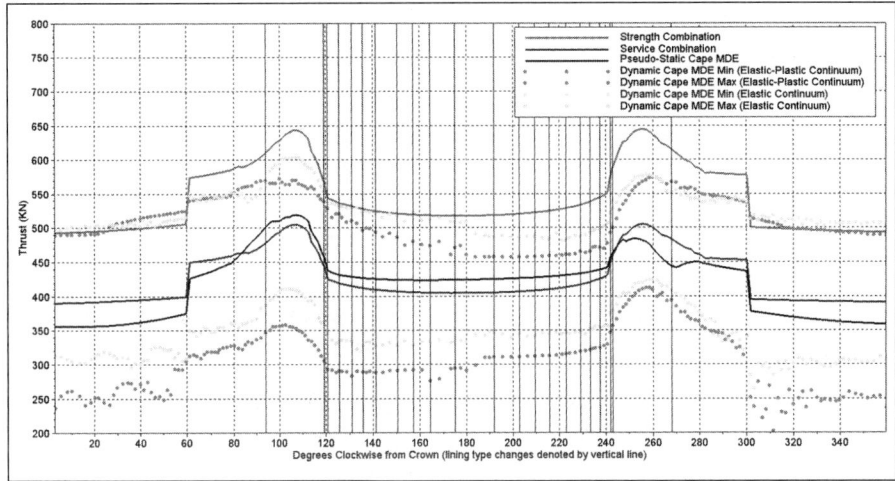

Figure 11. Comparison of thrusts along permanent lining for static, pseudo-static, and dynamic loading

scenarios and serves as a gage of overall model behavior. Specifically, the figures contain the following load combinations:

- Strength, where $U = 1.25\ DC + 1.35\ E + 1.25\ WA + 1.75\ LL$;
- Service, where $U = DC + E + WA + LL$);
- Pseudo-static $+EQ_{ver}$;
- Dynamic $+ EQ_{ver}$ for elastic continuum model; and
- Dynamic $+ EQ_{ver}$ for elastic-perfectly plastic continuum models ($S_{ud} = S_u$).

The figures demonstrate that the strength combination produced higher concrete flexural demands in the rigid invert and the dynamic elastic-plastic continuum model produced higher demands in the flexible lining near cavern springline.

CONCLUSIONS

The structural design of the crossover cavern permanent lining and interior structures required several static and dynamic load scenarios to demonstrate satisfactory long-term performance. Analyses methods addressed significant ground and groundwater loads, building surcharges, and seismic events with 2,500year return periods. Pseudo-static and dynamic time history analyses were performed due to the scale of the ground motions and complexity of the cavern structure. The analyses provided the following conclusions:

- The permanent lining's long-term performance was not controlled by one load combination along the perimeter; it fluctuated between several of them.
- Loads from dynamic analyses are sensitive to cyclic ground properties and will increase with continuum plasticity. Proper selection of dynamic strength parameters in the early stages will help target sensitivity studies in model trial runs.
- Dynamic analyses for MDE events increased flexural demands and produced localized yielding along the permanent lining. Expected concrete properties

and moment curvature analyses were utilized to confirm section capacity and ductility with the model loads.

ACKNOWLEDGMENTS

The authors would like to express their gratitude to Metro and RCC for authorizing the analyses and granting permission to publish the paper. Additional appreciation is given to Varun and Augusto Lucarelli of Itasca Consulting Group for their invaluable efforts in the development of the dynamic models.

REFERENCES

American Concrete Institute (ACI). 2014. *Building Code Requirements for Structural Concrete (ACI 318-14) and Commentary (ACI 318R-14).* Farmington Hills, MI: ACI.

California Department of Transportation (Caltrans). 2013. *Seismic Design Criteria.* Version 1.7. Sacramento, California: Department of Structure Construction.

Darendeli, M.B. 2001. *Development of a New Family of Normalized Modulus Reduction and Material Damping Curves.* Dissertation. University of Texas at Austin, p. 362.

Dobry, R., Oweis, I., and Urzua, A. 1976. Simplified procedures for estimating the fundamental period of a soil profile. *Bulletin of the Seismological Society of America.* 66(4):1,293–1,321.

FLAC Fast Lagrangian Analysis of Continua. 2016. *User's Guide.* Version 8.0. Minneapolis: Itasca Consulting Group, Inc.

Hashash, Y.M.A., Hook, J.J., Schmidt, B., Yao, J.I. 2001. Seismic Design and Analysis of Underground Structures. *Tunnelling and Underground Space Technology.* 16 (2001); 247–293.

Herranz, C., Z. Horvath, Lianides, J., and D. Penrice. 2016. SEM Crossover Cavern Design in Downtown LA. In *Proceedings for the 2016 Annual Conference. Tunnelling Association of Canada (TAC)*, Ottawa, Ontario, Canada, October 16–18.

Lianides, J., Z. Horvath, C. Herranz and D. Penrice. 2016. Numerical modeling for crossover cavern in downtown Los Angeles. In *Proceedings for the 4th Itasca Symposium on Applied Numerical Modeling in Geomechanics,* Lima, Perú, March 7–9. Minneapolis: Itasca International Inc. pp. 375–84.

Metro Rail Design Criteria (MRDC) Section 5. 2013. *Structural/Geotechnical.* Los Angeles: Los Angeles County Metropolitan Transportation Authority (Metro).

Seed, H.B. and Idriss, I.M. 1970. *Soil Moduli and Damping Factors for Dynamic Response Analyses.* Report No. EERC 70-10. University of California, Berkeley: Earthquake Engineering Research Center.

Shaw, J.H., Plesch, A., Dolan, J.F., Pratt T.L., and Fiore, P. 2002. Puente Hills Blind-Thrust System, Los Angeles, California. *Bulletin of the Seismological Society of America.* 92 (8): 2,946–2,960.

Wang, J. 1993. *Seismic Design of Tunnels, A Simple State-of-the-Art Design Approach.* Monograph 7. New York, New York: Parsons Brinckerhoff.

First Large-Diameter Hard Rock CSO Chamber in St. Louis

Dave Frierdich ▪ Metropolitan St. Louis Sewer District
Patricia Pride ▪ Metropolitan St. Louis Sewer District
Kevin Nelson ▪ Black & Veatch
Clay Haynes ▪ Black & Veatch

ABSTRACT

The Maline Creek Tunnel (MCT) will be the first large diameter chamber to store combined sewer overflows in St. Louis. The MCT is a key feature of Project Clear, Metropolitan St. Louis Sewer District's (MSD) Long Term Control Plan to address sanitary and combined sewer overflows to local streams and rivers. Project Clear's estimated cost is greater than $4 Billion dollars making it the largest public works project to date for the state of Missouri. The MCT includes the construction of a 40-ft diameter, 12.5 MGD submersible pump station, a 28-ft diameter × 2,700-ft long cavern, a 580-ft long × 6-ft lined connecting tunnel, three deaeration chambers, three intake structures, a shallow connector sewer constructed by pipe jacking, and 1,000 feet of 12-inch to 30-inch diameter near surface sewers. Bids were opened on March 10, 2016. The successful bidder was SAK/Goodwin JV with a bid of $82.3 M dollars. The Engineer's Estimate was $87.7 M dollars. The project was awarded in May 2016. The contractor mobilized on site in June 2016. The project is allocated 1567 calendar days to complete the construction.

BACKGROUND

St. Louis MSD serves approximately 1.3 million customers in St. Louis City and St. Louis County. Some of the oldest parts of MSD's sewer system were constructed in the 1850s and are combined sewers that convey both sanitary sewage and storm water in one conduit to treatment plants. During moderate to heavy storm events, storm water and wastewater discharges into local waterways from overflow points in the sewer system. These sewer overflow points act as relief valves when too much storm water enters the sewer system. Without these overflow points, many basements would experience backups and some streets would flood.

In 2007, the State of Missouri and the United States Environmental Protection Agency (USEPA) filed a lawsuit against MSD to reduce the sewer overflows. In August 2011, the Department of Justice filed a consent decree requiring MSD to spend approximately $4.7 billion dollars over the next 23 years to reduce the sewer overflows, increase water treatment capacity and provide other sewer system improvements. The Consent Decree between MSD, the USEPA and the Missouri Coalition of the Environment went into effect on April 27, 2012.

The MCT is located on the downstream end of the Maline Creek Watershed as indicated on Figure 1. The Consent Decree stipulated that the MCT would reduce the overflows at CSO BP 051 to four events or less, and 6 million gallons of untreated overflow volume in a typical year and that the MCT would reduce the overflows at CSO BP 052 to four events or less and 20 million gallons of untreated overflow volume in a typical year. The critical Consent Decree milestone for both CSO BP 051 and CSO BP 052 is 12/31/2020.

MAJOR PROJECT FEATURES

The following project features were designed to meet the aforementioned performance criteria for CSO BP 051 and CSO BP 052:

- 2,700 ft long, 28 ft diameter concrete-lined cavern
- 177 ft deep, 40 ft diameter concrete-lined shaft serving as a 12.5 million gallon per day (MG) pump station
- 580 ft long, 6 ft diameter connecting tunnel
- three deareation chambers
- three intake structures
- 213 ft of 24-inch diameter shallow connector sewer constructed by pipejacking
- 1,000 ft of 12-inch to 30-inch diameter near surface sewers

PREQUALIFICATION AND BID RESULTS

There was a very good response to the request for prequalification. Twelve domestic tunneling contractors and two European tunneling contractors were prequalified for the project. Six domestic tunneling contractors and one European tunneling contractor submitted bids. SAK/Goodwin Joint Venture submitted the successful bid at $82,828,282. The Engineer's estimate was $87,700,000.

The contractor was granted 1507 calendar days to substantially complete (SC) the project and an additional 60 calendar days beyond SC for final completion.

PUMP STATION (WORKING) SHAFT CONSTRUCTION

A Notice to Proceed was issued on May 13, 2016 and construction began on the working shaft in July 2016. Secant pile (SP) wall construction was required for the soil strata in the upper 40 ft plus or minus 2 ft. Nicholson was the subcontractor selected to construct the SP wall. Nicholson employed a Bauer BG-39 drill rig to install seventy eight 880 mm diameter unreinforced secant piles in a circular configuration. Some typical challenges with SP wall construction were encountered including protrusions, voids and windows.

After careful observation of the protrusions after exposure, it appeared that the protrusions emanated predominantly from the secondary piles. One plausible explanation was that some ground was lost during drilling of the secondary piles and concrete flowed into the voids during

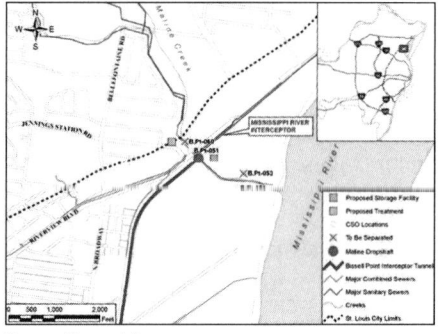

Source: Consent Decree
Figure 1. Maline Creek location map

Figure 2. Bauer BG-39 in center of photo

Figure 3. Drilling blast holes in the south half of the shaft footprint

concrete installation. The protrusions were removed from the interior of the SP wall with a hydraulic hoe ram mounted to a small backhoe during soil excavation.

Some voids developed in the secant piles above the protrusions. These voids were successfully repaired with shotcrete and welded wire fabric.

One secant pile was improperly located at the shaft collar resulting in a window. This pile was redrilled to remediate this problem. Nicholson elected to redrill two other piles as a precautionary measure even though these piles were within verticality tolerances according to instrumentation.

After these standard repair measures were implemented, the SP wall proved to be a very robust and nearly watertight support system. Only minor seepage was observed on the secant pile wall.

SAK/Goodwin is in the process of excavating the shaft bedrock at the time of this article's publication. For the upper 30 feet of rock, SAK/Goodwin has designed blast rounds that only excavate one-half of the shaft footprint in each blast. The contractor plans to start full shaft diameter shaft blasts at 30 feet below the top of rock.

DROP AND VENT SHAFT EXCAVATIONS

Kiewit Foundations Group (KFG) was the subcontractor selected to drill the drop and vent shafts on the project. KFG selected Ziegenfuss Drilling to drill pilot holes for the drop and vent shafts using a percussive down hole drill. Pilot drilling began in June 2016 and was completed in August 2016.

Table 1 shows the dimensional parameters for the shaft drilling work.

KFG used a Liebherr LB36 drill with soil augers to drill the upper section of the shafts that was comprised of soil materials. A polymer-based drilling additive was added to the water to create a viscous drilling fluid that supported the shaft walls until drilling progressed to bedrock and a steel casing was installed. A steel casing was installed after soil excavation was completed. A concrete mud slab was constructed in the shaft base by tremie methods. After the concrete slab had set for several hours the

Table 1. Drilled shaft data

Shaft Name	Casing Dia. in Soil (ft)	Excavated Dia. in Rock (ft)	Approx. Depth to Rock (ft)	Total Depth (ft)
COR* drop shaft	12	11	43	136
COR* chamber vent shaft	8	7	43	136
COR* cavern vent shaft	7	6	42	120
CHS† drop shaft	12	11	32	155
CHS† chamber vent shaft	8	7	32	155
CHS† cavern vent shaft	7	6	33	140
CSO 52 drop shaft	8	7	45	138
CSO 52 vent shaft	7	6	45	138
NI‡ shaft	11.5	N/A§	45	45

Notes:
* COR = chain of rocks
† CHS = Church Street
‡ NI = North Interceptor
§ N/A = not applicable

annular space between the soil and the steel casing was backfilled with grout by tremie methods.

After the steel casing was installed and grouted, KFG attached Berminghammer drill pipes and air swivel and a Center Rock, Inc. (CRI) cluster rock cutterhead to the LB36 to drill the bedrock. Cuttings were removed by compressed air supplied by seven 1450 cfm Sullair air compressors. CRI supplied four different cutterheads including 6 ft and 7 ft full diameter cutterheads and 9 ft and 11 ft diameter overreamers. The 11 ft diameter excavations were to be drilled in three passes. The 6 ft diameter cutterhead was to be used on the first pass; the 9 ft diameter cutterhead was to be used on the second pass and the 11 ft diameter cutterhead was to be used on the third pass. As of the time of publication, the two 11 ft diameter shafts have not been drilled.

The six ft and seven ft diameter cutterheads plugged with cuttings on several occasions requiring removal from the shaft excavation to unplug the cutterheads. Overall, drilling progress in the rock was favorable with an average excavation rate inclusive of all down time, plugs, maintenance, etc. of about 2 to 3 ft per hour.

DIVERSION STRUCTURE EXCAVATIONS

Three flow diversion structures will be constructed to convey combined sewer overflows to the MCT. The largest flow diversion structure (Chains of Rocks) is also the deepest diversion structure. The deepest section the Chain of Rocks structure is constructed in the upper section of the bedrock approximately 43 to 45 ft below grade. Nicholson was selected to install secant piles for the lower section of the Chain of Rocks diversion structure. The majority of the secant pile wall was non-circular requiring the installation of W24 × 104 beams in the secondary piles and walers and struts inside the SP wall. In addition, toe pins were installed in the base of the secondary piles. As of publication time, approximately 80 percent of the secondary piles have been installed. There were some problems with cross communication of concrete into adjacent piles that were corrected promptly.

Figure 4. Installing the steel casing

Figure 5. Center Rock Inc. cutterhead

Figure 6. Installing a beam in a secondary pile

The other excavations are shallower and will be supported by sheet piling. As of publication time, no work has been done on the shallower diversion structures.

SUMMARY

The MCT is the first large diameter hard rock CSO cavern to be constructed in the St. Louis, MO area. SAK/Goodwin JV was given a Notice to Proceed with construction on May 13, 2016. The working shaft, drop and vent shafts and the Chain of Rock diversion structure are being constructed now. The project is on schedule and no major problems have been encountered to date. The MCT is a key part of Project Clear's program to reduce water quality impacts to local streams and rivers in the St. Louis, MO area.

ACKNOWLEDGMENTS

Sincere thanks are due to the many team members including the Metropolitan St. Louis Sewer District (Owner), Jacobs Engineering Group Inc. (Design Engineer), SAK/ Goodwin JV (General Contractor), Ziegenfuss Drilling (Pilot Hole Drilling Subcontractor), Kiewit Foundations Group (Vent and Drop Shaft Drilling Subcontractor), Nicholson (Secant Pile Wall Subcontractor), and Black & Veatch, Kwame Building Group Inc., and AECOM (Construction Management Team).

REFERENCES

United States District Court for the Eastern District of Missouri Eastern Division, April 27, 2012, No. 4:07-CV-1120 (CEJ) Consent Decree United States of America and the State of Missouri (Plaintiffs) and Missouri Coalition for the Environment Foundation (Plaintiff/Intervenor) v. The Metropolitan St. Louis Sewer District (Defendant).

PART

Future Tunneling

Chairs

Andrew Finney
CH2M Hill

David Sowers
Washington State Department of Transportation

Overvaal Rail Tunnel: Securing the Economic Arteries of the Rainbow Nation

Jack Muir ▪ Aurecon Hong Kong
Hennie Gouws ▪ Aurecon Tshwane

ABSTRACT

South Africa's proposed new Overvaal rail tunnel, located on the Highveld escarpment will triple the rail capacity from the coalfields to Richard's Bay. The existing Overvaal rail tunnel was built in the mid-1970s through challenging conditions and was mapped and studied by Bieniawski to develop the Rock Mass Rating (RMR) System. Beneath the escarpment point, the tunnel exhibited squeezing conditions and continuing instability at the invert means that it needs regular maintenance.

A new tunnel has been designed to accommodate a double track tunnel. This paper outlines the unique design challenges for a rail tunnel in South Africa, from the unique geotechnical conditions to the challenges of retaining continuous operation of the existing tunnel during construction.

TRANSNET AND THE EXISTING TUNNEL

Transnet is a South African rail, port and pipeline company, headquartered in Johannesburg. It was formed as a limited company on April 1, 1990. A majority of the company's stock is owned by the Department of Public Enterprises of the South African government. The company was formed by restructuring into business units the operations of South African Railways, Harbours and other existing operations and products.

Transnet Freight Rail (TFR) is the business unit that manages the South African freight railway lines and was formerly known as Spoornet. One of the lines, which TFR manage is the Richards Bay coal line. It delivers coal from the Mpumalanga region to Richards Bay that houses the coal export facility.

The existing Overvaal Tunnel forms part of the Richards Bay coal line. It is located in Mpumalanga between Ermelo and Piet Retief. The tunnel was completed in 1976 and is situated on the only single track section of the line. Figure 1 shows the location of the existing tunnel.

A significant proportion of Transnet's freight traffic passes through the tunnel, which has been identified as a bottleneck for future rail traffic growth. Any incident closing this single track section of the line would have severe economic consequences. Therefore, the aim of the new Overvaal tunnel is to increase the capacity of the Richards Bay coal railway line with an additional set of tracks through this area. The tunnel is located approximately 30km south east of Ermelo and is approximately 3km long.

The existing Overvaal was completed and commissioned in 1976. As-built drawings, construction records and historic papers were collected and collated for use in the design of the new tunnel.

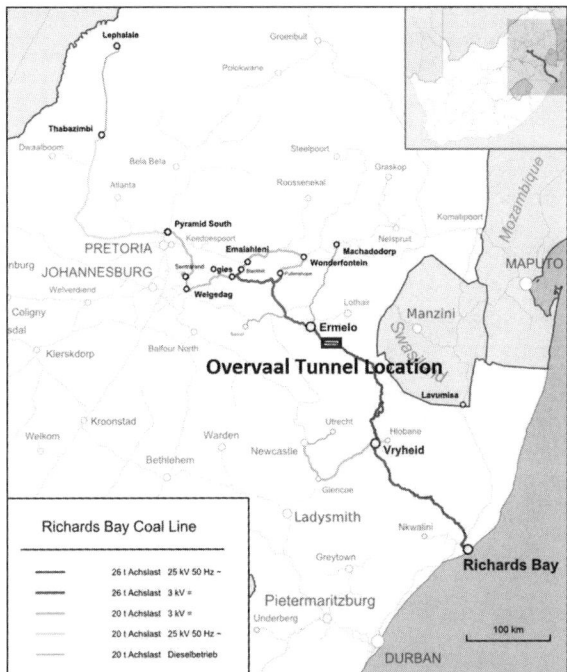

Figure 1. The location of the existing tunnel background (image courtesy of Wikimedia Commons)

The existing tunnel has a horse-shoe profile of approximately 5m width and 6m height. The first parts of the tunnel at both the eastern and western portals were built as cut and cover structures with the drill and blast excavation starting in competent rock with adequate cover. 8 No. cross passage stubs were excavated and lined from the southern wall of the tunnel at irregular intervals. An inclined ventilation shaft was located approximate midpoint of the existing tunnel and contains a roof collapse. Figure 2 shows a view of the tunnel from the western portal and a general view inside the existing tunnel.

The performance of the existing tunnel has been less than satisfactory over its 40-year period of service, the problems encountered include:

- High groundwater ingress
- Track slab deformation that has caused significant maintenance disruption
- Bowing of the sidewalls towards the eastern end of the tunnel evidenced by extensive wide cracks in the lining intrados

During construction of the tunnel, which took place in 1974 and 1975, conditions were found to be less than favourable than expected. Bieniawski was developing his geomechanics classification (Bieniawski 1974) and he carried out studies of in-situ measurement of ground displacement during construction (Bieniawski and Maschek 1975).

Investigations of historical records of the existing tunnel suggest that the tunnel was supported temporarily with rockbolts and mesh and in some locations, steel sets. Even in areas where tunnelling conditions were good, deterioration of the rock has been observed and wall bolting has been retrofitted through the permanent lining where the tunnel is deep, this has exasperated the groundwater ingress problem.

NEW TUNNEL REQUIREMENTS

Previous studies concluded that a new tunnel adjacent to the existing tunnel was the preferred option for removing the constraint of the single track existing tunnel. The new tunnel design had specific requirements which included:

- Double track tunnel with walkways on both sides of the tunnel and a physical barrier between the two tracks
- Barrier protected sanctuaries for material storage
- Stubs on the existing tunnel to be extended to the new tunnel to form cross passages
- Fire proof doors at the cross passages
- Drained monolithic concrete lining
- Finite element analysis of the existing tunnel and any other structures that may be vulnerable, and proposals for monitoring of vulnerable structures during construction and for a relevant period after construction.

Fire and life safety was considered for the new tunnel, given the tunnel would only be used for coal and less regularly for general freight. Fire and life safety performance required a risk based approach given the use of the tunnel, particularly the non-existence of passengers.

Mechanical and electrical services such as ventilation, power supply, tunnel lighting and communications were to be considered and performance requirements were based on review of South African and overseas practice. One unique design requirement was that the ventilation design of the new tunnel needed to provide ventilation of the existing tunnel with equivalent performance requirements.

GEOLOGY

The stratigraphy in the area surrounding the tunnel was identified as sandstones and siltstones of the Karoo Sequence. These sediments have been intruded by a massive dolerite sill through that the existing tunnel is excavated.

Fine grained, highly fractured dolerite is found at the contact with the Karoo sandstones and siltstone. The dolerite

Figure 2. A view of the tunnel from the western portal and a general view inside the existing tunnel (courtesy of Gary Davis)

transitions to widely jointed and coarse grained as depth increases. The coarse grained dolerite displays rapid weathering characteristics (Orr 1979). These rapid weathering characteristics provided challenges to the tunnel support. Figure 3 shows rapid weathering dolerite at surface exposures.

It is known that dolerite dykes occur within the sill and are expected to be intersected during tunnel excavation. A fault zone was also encountered during excavation of the existing tunnel, which caused problems in achieving stabilization of the tunnel and in controlling convergence. Unfavourable joint set orientations also led to significant overbreak in some areas and collapse of side walls in areas of high horizontal stress. This recorded behavior of the existing tunnel was obviously studied closely for the support design of the new tunnel.

Long sections of the existing tunnel were sourced from historical publications (Bienawski and Maschek, 1975). From this there were four lengths along the existing tunnel of reasonable quality rock, where tunnelling progressed at a good rate and where support was generally limited to rockbolting. The remaining areas involved greater difficulty in tunnelling, related to fault zones, high joint frequency and very poor joint infill. This phenomenon could be attributed to the presence of rapidly weathering dolerite, which is a problem that has been encountered in dolerite throughout South Africa in the Karoo supergroup (Orr 1979). Figure 4 shows a geological cross section that was produced in the book Engineering Geology of South Africa showing the expected tunnel conditions, the geomechanics classification done by Bieniawski and Mashek and the final lining installed in the existing tunnel.

Figure 3. Rapid weathering dolerite at surface exposures

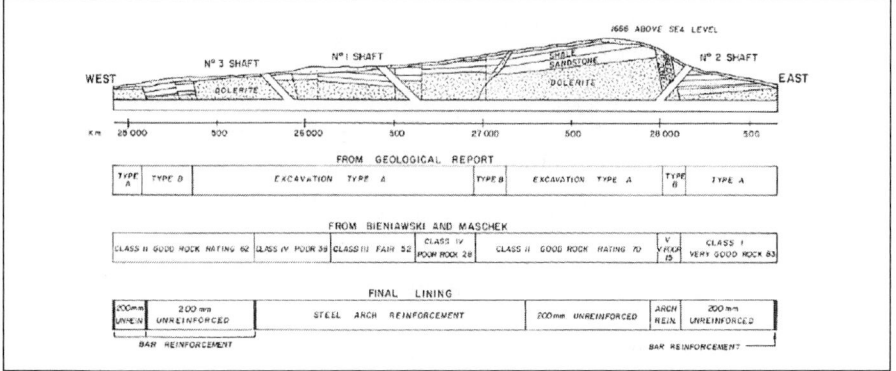

Figure 4. A geological cross section (from Engineering Geology of South Africa)

Descriptions of the rockmass quality of the existing tunnel are included in a number of publications (Brink 1983; Bienawski and Maschek 1975). In fact Bienawski used construction monitoring data in his development of the Rock Mass Rating (RMR) System.

PROJECT LIFECYCLE

Numerous previous studies had been undertaken prior to the detailed design phase of the new tunnel commenced. These spanned from prior to the existing tunnel being constructed such as ground information studies for the existing tunnel, to more recent pre-feasibility studies for the second Overvaal tunnel.

Track slab deformation and high groundwater ingress into the existing tunnel prompted studies into the cause and subsequent remedial measures. As part of these studies coring through the track slab and into the subgrade showed locations of very poor subgrade material that was causing movement of the track slab. After blasting mass concrete was cast in 2 to 3 layers to get the tunnel floor to the required level for the 300m track slab. Rapid weathering dolerite was hypothesized as contributing to the relative displacement of the track slab and detrimental effect to the track geometry. In some locations, settlement of 50mm of the track slab was encountered. The drained tunnel along with the pumping mechanism of the heavy axial loads were factors in the deterioration of the subgrade. Remedial measures included grouting the underside of the slab to replace the longitudinal distortion from the reduced bearing capacity of? the weathered dolerite. Grouting measures were only partially effective as grout holes started to perforate the track slab frequently and the rail shifted laterally to allow the grouting to be done. This caused further problems to the structure and delays to operation of the line.

When the two issues of poor performance and remedial measures for the slab, coupled with Transnet's strategic importance and direction for the line were considered, it was decided that the bottleneck of the Overvaal tunnel needed to be removed. Options included:

- New overland route meeting current geometric requirements
- Overland route making use of historical railway corridor prior to the existing tunnel
- Second tunnel adjacent to the existing tunnel

Ultimately the decision was made to go with the option of a second tunnel. Further engineering assessment was undertaken on the construction method that was going to be used for the detailed design of the tunnel. Drill and Blast construction and Tunnel Boring Machine (TBM) construction with a hard rock machine were considered and an FEL-3 level design undertaken and priced and programmed to allow decision for the method to be chosen to be taken forward. Table 1 outlines some relative advantages of each construction method considered at FEL-3. Figure 5 shows indicatively the final cross sections of the permanent lining for drill and blast and TBM options at FEL-3.

TUNNEL SUPPORT DESIGN

Based on previous studies, the design of the permanent tunnel lining was based off the assumption of a Sequential Excavation Method (SEM) primarily by drill and blast, with temporary support/ reinforcement provided to allow the casting of a cast insitu permanent concrete lining.

Table 1. Relative advantages of construction methodology for the new Overvaal tunnel

Drill and Blast Methodology	TBM Methodology
• Tunnelling could commence from both ends after completion of portal cuts • Relative high South African technology and capability in the South African mining industry • No major power requirements • Higher flexibility in construction methodology changes for encountering poor ground • Flexibility in permanent lining thickness based on actual ground conditions encountered	• Speed – tunneling rate faster and finished earlier • Lateral distance between tunnels shorter due to lower construction vibrations • Shorter cross passages • Less earthworks and excavation for portal cuts

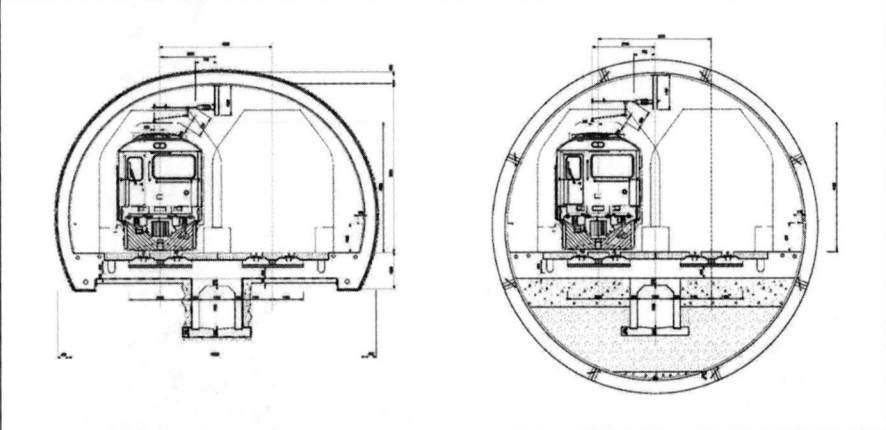

Figure 5. Indicative cross sections of the permanent lining for drill and blast and TBM options at FEL-3

Rapid weathering materials in a rock mass classification system, modern day shotcrete (versus the gunite they used in construction of the existing tunnel).

The existing tunnel construction and its behaviour in service provided essential information into the design of the support/reinforcement and lining of the new tunnel. There was significant differences between the expected and encountered ground conditions in construction of the existing tunnel. In particular the following characteristics of the existing tunnel were closely analysed to incorporate into the design.

Instability During Excavation

Evidence of instability during construction of the existing Overvaal tunnel was identified from the historic records. Overbreak was an issue in the existing tunnel and is expected to be caused by a sub-horizontal, shallow dipping joint set. This was encountered in approximately 1400m of the existing tunnel. The book Engineering Geology of South Africa has a particular chapter for tunnelling in dolerite which goes into much detail of the actual encountered ground condition and the following extract from the book gave particularly perturbing warnings to the ground conditions that could be expected in the new tunnel:

"Joint surveys were carried out on all of the limited rock outcrops in the vicinity of the tunnel. One low angled and three near-vertical joint sets were identified. It was recognised that in two of the sets the joints would be closely spaced in places. What was not revealed from the joint survey or the borehole cores

was the presents of large horizontal joints. These, in combination with closely spaced vertical joints, were to affect the overbreak conditions significantly in a number of places in the tunnel. The presence of these joints sets produced uncontrollable overbreak conditions during excavation of the tunnel. This led to an arbitration over changed conditions and a success-ful claim by the Contractor."

Figure 6 shows historical photos by Bieniawski that clearly shows the large horizontal and vertical joint surfaces leading to overbreak (top) and progressive collapse from high stress acting on columnar joints at the sidewalls (bottom).

Convergence During Construction

Convergence was measured during excavation of the existing Overvaal tunnel and this was documented in the CSIRO report on the monitoring undertaken by Bienawski and Maschek. There are two locations identified as considered the worst in terms of tunnel convergence. One measuring station measured movement of the tunnel walls of 31.5mm, which was not stabilised 5 months after excavation and was installed with a combination ot bolts and TH arches. Another station at the highly brecciated fault zone encountered convergence of the tunnel walls of 113.5mm this was not stabilised 6 months after excavation. A combination of bolts, TH sections and concrete pre-lining was used for the support in the existing tunnel. Rock falls were also encountered.

Figure 6. Images from construction of the existing tunnel (reproduced from Engineering Geology of South Africa, photo taken by R. Galliers)

The temporary support used in the existing tunnel can be assumed based on the CSIRO report by Bienawski and Maschek. This consisted of the following:

- Rockbolts—16mm dia bar, 1.6m long
- Gunite (Dry mix shotcrete)
- Wire mesh
- Steel arches TH type 25 kg/m^3

The permanent lining of the existing tunnel was drained and generally made up of a 200mm unreinforced concrete lining or a concrete lining including steel ribs in poor

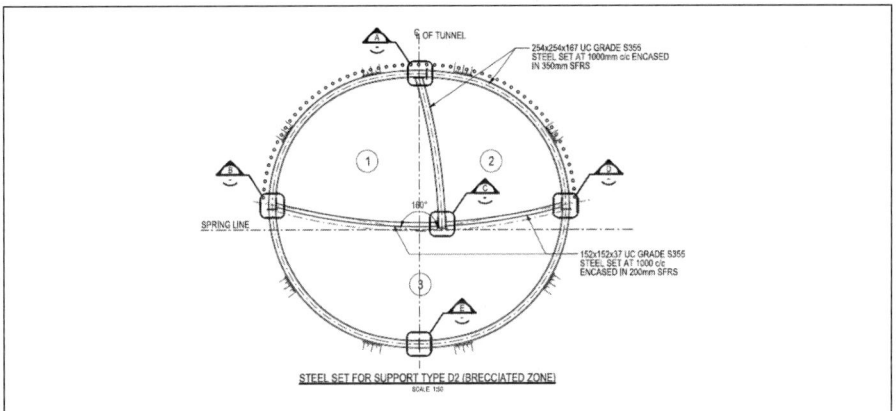

Figure 7. The temporary support requirements for the brecciated zone based on back calculated ground parameters from the existing tunnel

ground conditions. The walls sat on the invert and this restraint detail may have caused operational issues with sidewall movement.

Temporary Support

Generally in the good ground conditions expected along the majority of the tunnel alignment, a rock bolt and shotcrete temporary support solution was detailed. Failure modes were expected to be from discrete blocks from the discontinuities within the dolerite. Support would be specified based on the actual rock face conditions encountered and quality mapped by the Q-system. Locations where the Q-system was not considered adequate were the areas corresponding to the very large uncontrolled displacements in the existing tunnels and zones of rapid weathering dolerite. Finite element modelling was used for temporary support design in the areas of high wall convergence. Based on the understanding of the stress conditions and the exiting tunnel profile, back calculations were undertaken to iterate Generalized Hoek Brown parameters for the ground based on the convergence measurements in the existing tunnel. These yielded very poor ground parameters in no way being consistent with rock expected in these areas. High in-situ stress with the principal stress in adverse direction to the tunnel trend may have caused these large convergences in the rock expected and in-situ stress testing along the tunnel was lacking.

In these conditions the tunnel profile was changed to a more beneficial invert arch. This was to give a more effective stress distribution around the tunnel. Staged headings and benches were also specified to allow ground movement control and tunnel excavation lines were modified to allow convergences up to 60mm prior to permanent tunnel lining installation. Figure 7 illustrates the temporary support requirements for the brecciated zone based on back calculated ground parameters from the existing tunnel.

In areas of rapid weathering dolerite in the invert, based on experience from existing tunnel operations, it was decided to have an inspection, test, remove and backfill of select material solution. The material was considered too troublesome to deal with by pavement design methods. Rapid weathering dolerite is proposed to be defined by not just physical and visual methods but also petrographic (based primarily on olivine content) methods.

One factor in the design of the temporary support of the new tunnel while using existing tunnel historic data was quantifying how improvement in tunneling technology had improved over the last 40 years and how this may reduce adverse impacts in the new tunnel. One example of this is sprayed concrete. The existing tunnel used "gunite" that is by definition a dry mix type shotcrete. A wet mix shotcrete was specified by the designer to be used for the temporary support. This is generally readily available for modern tunneling methods and the rapid excavation of this to the excavation face after exposure may reduce the risk of the dolerite weathering and losing the strength of its mechanical properties.

Permanent Support

The current geological interpretation of the second tunnel alignment splits the tunnel into 4 different types of geotechnical zones and support requirements outlined in Table 2.

Generally when ground conditions allowed the drained, permanent lining was designed as unreinforced concrete with a kicker and permanent slab – Continuously Reinforced Concrete Pavement (CRCP) cast between the kicker (see Figure 5). Where ground was poorer and a lot of convergence was expected, mainly based off the experience

Table 2. Permanent lining design philosophy for expected ground zones

Geotechnical Zone	Approximate Length (m)	Geological Description	Permanent Lining Design Philosophy
Zone A	1350 (35%)	Overall rock mass conditions for tunnel are "Good," occasionally "Very Good " conditions might be encountered, or occasional "Fair" conditions. Generally similar conditions in crown area versus tunnel floor / invert.	This ground is expected to be equivalent to hard rock tunnelling with favourable in-situ stresses. The temporary support and permanent support will be designed independently of each other. Drill and Blast will be suitable for most if not all of this ground.
Zone B	633 (16%)	Typically "Good" to "Fair" rock mass conditions. Conditions in crown / upper tunnel area generally slightly poorer than in lower portion / invert.	The permanent lining design will be based on the rock quality based on face mapping and thickness specified using Norwegian Geotechnical Institute's (NGI) Rock Mass Quality or Q chart system (Barton et al, 1974; Grimstad and Barton, 1994). The permanent lining for these zones will be identified by the prefix AB.
Zone C	1663 (43%)	Generally "Fair" to "Poor," but conditions can be quite variable and in places "Good" or even "Very Poor" conditions are expected. Very broadly conditions in upper sections / crown are slightly better than the lower portion / invert.	The support in this ground is likely to be a combination of zones AB and D. The permanent lining specified in this ground will depend on the actual ground conditions encountered and will be specified on site depending on the actual behaviour of the ground. The Works Information documentation will specify the actual process used with the Contractor to specify this support.
Zone D	250 (6%)	Generally "Poor" to "Very Poor" rock mass conditions.	This ground is expected to be very difficult to tunnel through and there is likely to be instability when excavating the ground. Mechanical excavation will be required and rapid weathering dolerite is likely to change mechanical properties when exposed. Significant convergence may be encountered which will mean the permanent lining design needs to incorporate the temporary support measures. The permanent lining for this zone will be identified by the prefix D.

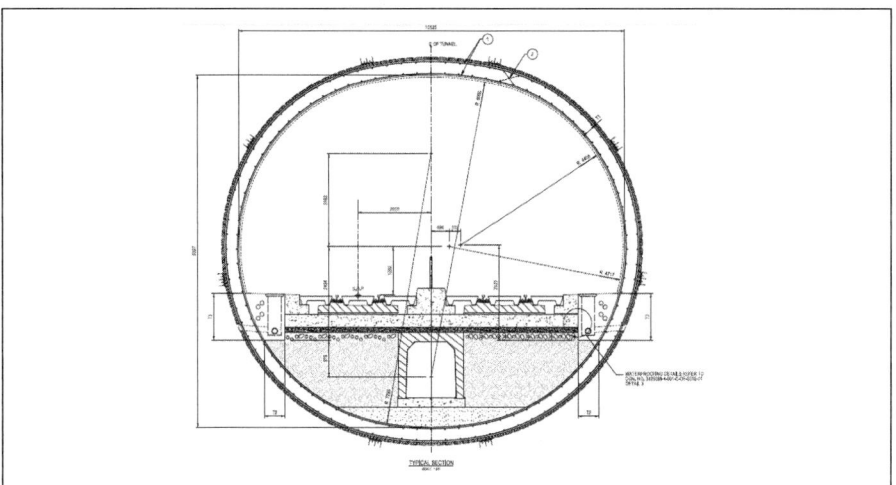

Figure 8. Permanent tunnel lining design in high stress breciatted zones using back calculated ground parameters

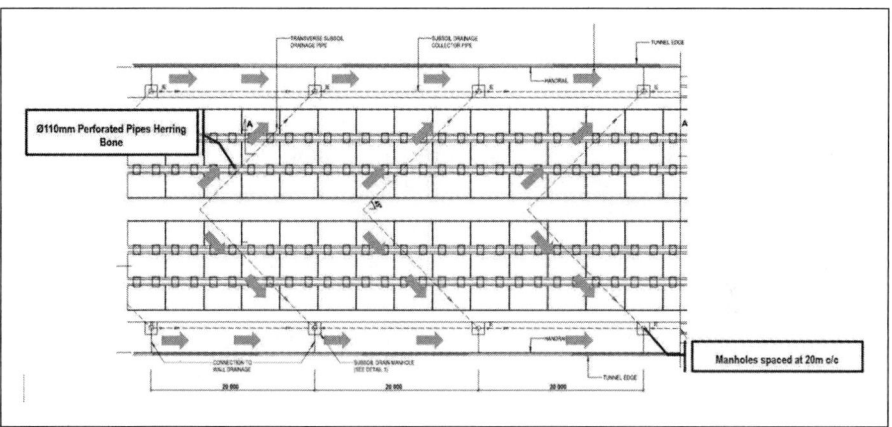

Figure 9. Herringbone under slab drainage system

of convergence in the existing tunnel, a curved invert was specified for the permanent lining. It was expected in the highly fractured fault zones that passive ground support such as girders and steel sets degraded that large ground pressures would come onto the lining, which would be more efficiently resisted by the permanent lining with an invert arch. This also doubled as a counter-measure for areas of rapid weathering dolerite given that more of the invert material would be removed and, although drained, the structure was continuous longitudinally, which was more likely to transfer loads across areas of poor ground in the invert (see Figure 8).

DRAINAGE

A drained permanent lining was specified in the project requirements. Issues that needed to be considered with regards to this was how this would affect the rapid weathering dolerite as anecdotally settlement of the rails in the existing slab could have been attributed to pumping action below the track slab. Some of the drainage

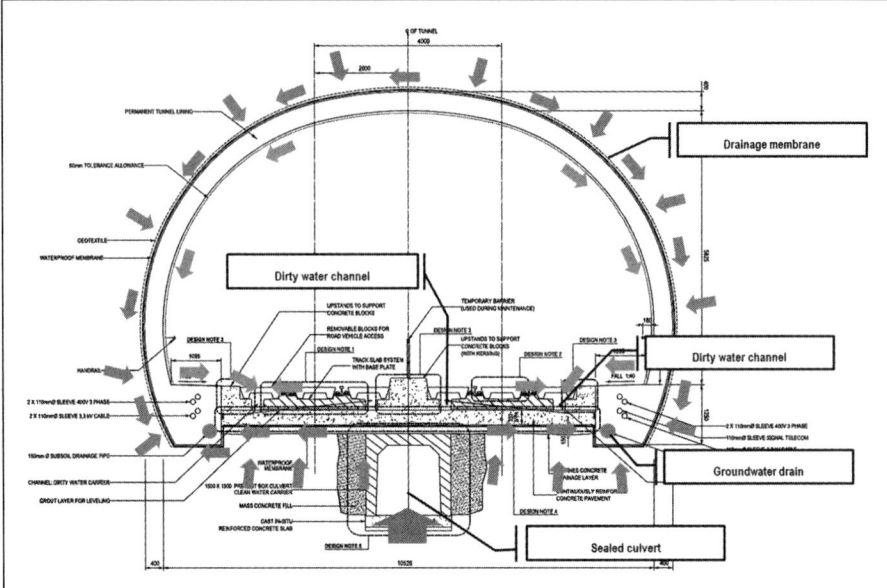

Figure 10. Concept of proposed drainage within the new tunnel

systems in the existing tunnel were visually clogged and this may have caused ground water built up and cracking and distortion of the walls. A herring bone system (see Figure 9) along with a no fines concrete system underneath tunnel floor (CRCP) create an extra redundancy with regards to the groundwater, thus preventing any potential pumping action below the CRCP.

The groundwater drainage system within the tunnel needed to deal with separate types of water within the tunnel:

- Water from the upstream portal catchment needed to be isolated environmentally and drained to the downstream side of the tunnel
- Water collected on the intrados of the tunnel from trains and or groundwater leaks was expected to be mixed with coal dust – this drained to evaporation ponds at the downstream end
- Groundwater was collected from the extrados of the lining and delivered though a system in the kicker with regular manholes – these manholes were design for ongoing maintenance to reduce the risk of clogging as experienced on the existing tunnel

These drainage systems are illustrated in Figure 10.

REFERENCES

Bienawski and Maschek. 1975. *The Civil Engineer.*

Brink, A.B.A. 1985. *Engineering Geology of South Africa*; p. 194–198.

Orr C., M. 1979. *Rapidly Weathering Dolerites.* The Civil Engineer.

California High-Speed Rail—Connecting and Transforming California—Design Considerations for Tunnels

Steve Dubnewych ▪ WSP | Parsons Brinckerhoff
Steve Klein ▪ WSP | Parsons Brinckerhoff
Ofelia Alcantara ▪ California High-Speed Rail Authority
Noopur Jain ▪ California High-Speed Rail Authority
Randy Anderson ▪ California High-Speed Rail Authority

ABSTRACT

Construction of the proposed California High-Speed Rail system will require extensive tunneling through mountain ranges in both the north and south regions of the State. Rail alignments under consideration could require between 45 to 50 miles of tunnels that range in length from several thousand feet to over 20 miles under a cover exceeding 2,000 feet at certain locations. Challenging geologic, hydrogeologic and seismic conditions are anticipated along potential tunnel routes. Geologic conditions range from unconsolidated alluvium to strong granitic rocks; tectonically sheared and deformed rock masses; high groundwater pressures and in situ stresses; and active earthquake faults. In addition, the mountain ranges exhibit environmentally sensitive areas that will have to be protected during construction. This paper discusses the design and construction challenges associated with the tunnels in this Statewide system.

BACKGROUND

The California High Speed Rail system is being planned, designed, constructed, and operated under the direction of the California High Speed Rail Authority (Authority). This will be the first high-speed rail system constructed in the U.S. The system will eventually include over 800 miles of rail service, with up to 24 stations.

High-Speed Rail is envisioned as a state-of-the-art, electrically powered, high-speed, steel-wheel-on-steel-rail technology, including innovative safety, signaling, and automated train-control systems. The trains will be capable of operating at speeds of over 200 miles per hour with an expected travel time between Los Angeles and San Francisco of under three hours.

This rail system will connect the major population centers of Sacramento, the San Francisco Bay Area, the San Joaquin Valley, Los Angeles, the Inland Empire, Orange County, and San Diego as shown on Figure 1. Phase 1 will provide service from San Francisco to Los Angeles and Phase 2 will extend service to Sacramento and San Diego. Because of the large size of the system, it has been broken into several project sections. Currently in Central Valley region about 119 miles of high-speed rail guideway is under construction. In Northern and Southern California regions, eight separate project sections are being evaluated, including preparation of an Environmental Impact Report/Environmental Impact Statement (EIR/EIS). According to the Authority's 2016 Business Plan the Silicon Valley to Central Valley (or "Valley to Valley") segment will be the initial operating segment, providing service between San Francisco and Bakersfield. Accomplishing this objective requires to San Jose to Merced segment

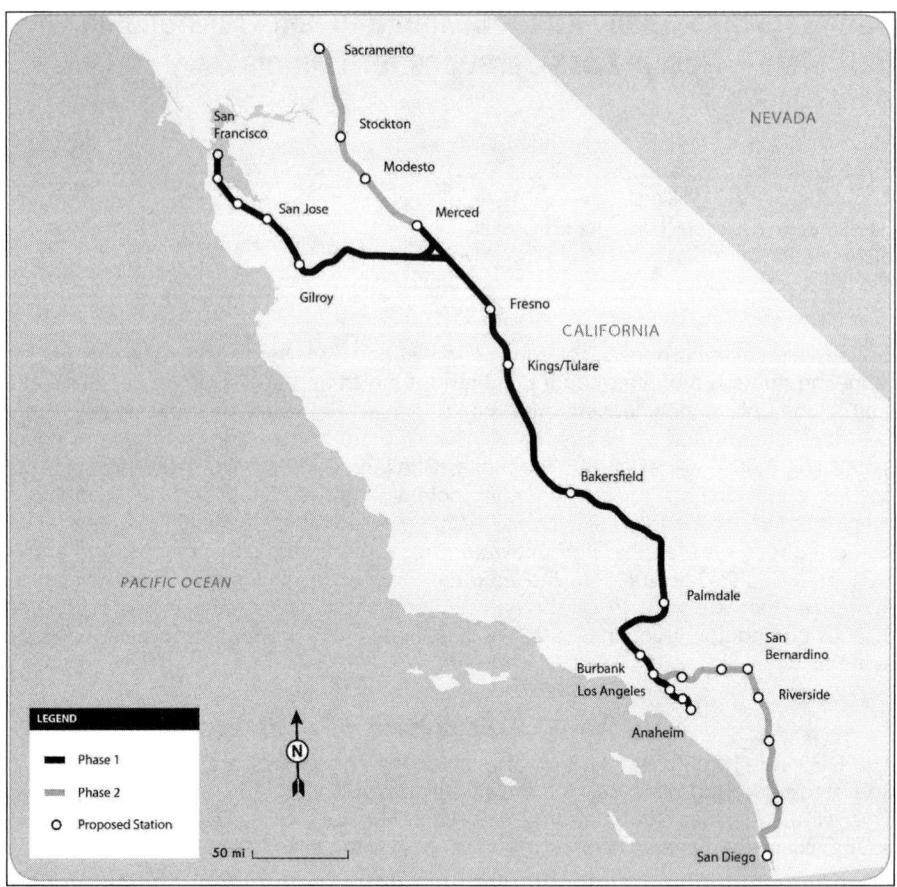

Figure 1. California high speed rail statewide system

to be constructed next. This segment is about 113 miles long and it also crosses the rugged California Coast Range.

Future sections of the project through the California Coast Range, Tehachapi Mountains and San Gabriel Mountains will involve significant tunnel reaches and, in some instances, unprecedented tunneling conditions. The geotechnical conditions identified along potential tunnel routes are very challenging including long tunnel sections, highly variable geotechnical conditions, groundwater pressures in excess of 50 bars, active fault crossings, and formations with a high potential for methane gas. This paper provides an over view of the expected tunneling conditions that need to be considered for alignment selection and design.

SAN JOSE TO MERCED PROJECT SECTION (CALIFORNIA COAST RANGE)

Alignment Overview

The San Jose to Merced project section extends south from San Jose along the Santa Clara Valley to the south side of Gilroy where it turns to the east and cuts through the California Coast Range at Pacheco Pass over to the San Joaquin Valley (Figure 2). The 24-mile section across Pacheco Pass will include about 15 miles of tunnels.

These will be the first significant tunnels constructed by the Authority for the high-speed rail program.

Environmental studies and conceptual engineering work is currently underway, including a comprehensive geotechnical investigation program. Several alternatives are being considered for the tunnel sections. Presently there will be three to four tunnels ranging from 1.7 to 10.5 miles in length. Depth of cover ranges from about 200 to over 1,000 feet. Twin tunnels in a single track tunnel configuration would be about 28 feet in diameter (ID).

Regional Geologic Setting

Pacheco Pass is in the Diablo Range, which is part of the California Coast Range. Formations present along the tunnel alignment include Franciscan Complex rocks and the Great Valley Sequence (referred to as the Panoche Formation). These formations were originally deposited in a subduction zone and were subsequently folded, faulted, and uplifted due to tectonic activity at a previous boundary between the Pacific and North American plates. The geologic structure produced by this severe tectonic activity results in the Panoche Formation flanking the Franciscan Complex on the west and east sides of the Diablo Range. Although similar in age and composition, there are some differences in the physical properties of the two formations, however both formations are pervasively fractured and sheared.

Two active northwest trending faults, the Calaveras and Ortigalita faults, are present at the west and east margins of the Diablo Range, respectively. These are right-lateral strike-slip faults. Other potentially active faults located just to the east of the Diablo Range are the O'Neill fault and the San Joaquin Blind Thrust faults.

The tunnel section on the west side of Pacheco Pass is expected entirely in the Panoche Formation and the other tunnels will be mainly in the Franciscan Complex rocks, with Panoche Formation on the east side of the Ortigalita fault.

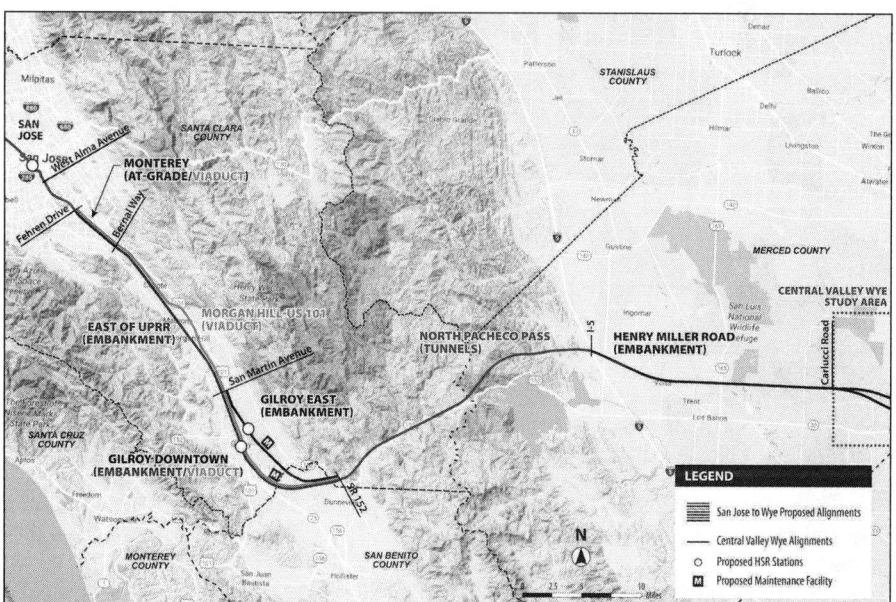

Figure 2. San Jose to Merced project section alignment details

Local Geologic Conditions

Available geologic data includes the results of 4 initial, deep core holes completed for the project (California-High Speed Rail, 2016) and investigations completed by the United States Bureau of Reclamation (USBR) for two tunnels near the high-speed rail alignment that were completed as part of the Central Valley Project in 1979 to 1983. As-built records from these tunnels, the Santa Clara Tunnel and Pacheco Tunnel Reach 2 provide additional insight into the geology and tunneling conditions in this area. Mobilization for additional geotechnical investigations were underway at the time of paper was written.

The Panoche Formation consists mainly of interbedded sandstone, siltstone, and shale. On the west side of the Ortigalita fault a significant conglomerate bed (over 60 feet thick) was encountered. The rock mass is generally weak to moderately strong and moderately to intensely fractured. Shear zones and sheared rock with clay infillings are present through the rock mass. Unconfined compressive strengths of intact rock samples generally range from about 3,000 to 15,000 psi.

Franciscan Complex rocks consist mainly of graywacke, a metamorphosed sandstone; sheared shale and claystone; and lesser amounts of greenstone, breccia, serpentinite, and chert. Portions of this formation are described as mélange which are described as a mixture of hard rock blocks or inclusions (typically graywacke, greenstone, breccia, and chert) embedded in a weak, pervasively sheared shale matrix. The weak matrix has a significant clay content and typically exhibits the strength of a stiff to hard clay, but softens significantly when exposed to water. The graywacke can be massive, strong rock with a strength in the range of about 3,000 to 25,000 psi, however, the behavior of the rock in a tunnel excavation is often dominated by the weak sheared shale matrix.

Expected Tunneling Conditions

The previous tunnels constructed by the USBR provides a fairly good indication of expected tunneling conditions in the Pacheco Pass area.

The Santa Clara Tunnel is approximately 5,100 feet and it was excavated with a 13.4-foot tunnel boring machine (TBM) through the Panoche Formation in 1981–82 (USBR, 1986a). Cover above the tunnel ranged from 65 to 500 with about 250 to 500 feet of cover over half the tunnel. Mainly thick bedded, moderately to widely fractured sandstone and very intensely to intensely fractured siltstone and shale was encountered. Variable thickness shear zones were also frequently encountered. The fully shielded TBM was advanced with thrust jacks pushing against a steel rib and timber lagging support system. Following tunnel excavation, a cast-in-pace concrete lining was constructed to provide a 9.7-foot finished ID tunnel.

Construction difficulties included face stability, and raveling and squeezing ground associated with the fractured and sheared rock conditions. Excessive caving in shear/fault zones resulted in overbreak. Groundwater inflows in 90% of the tunnel were described as dry to moist, with groundwater inflows less than about 2.5 gpm. Groundwater inflows in the remainder of the tunnel ranged from about 2.5 to 100 gpm.

The Pacheco Tunnel, Reach 2 was constructed from 1979 to 1983 (USBR, 1986b). It was approximately 5.3 miles long and excavated using drill-and-blast methods. Cover ranges 40 to almost 1,300 feet with most of the tunnel having about 600 feet of cover. The tunnel has a horse-shoe shape section, approximately 13 feet wide and 13 feet high, and it was supported with steel ribs and lagging and spiling installed to pre-reinforce the ground. In addition, a precast concrete invert was installed to provide lateral support for the tunnel sidewalls and to protect the invert from degradation. Geologic conditions encountered consisted of Franciscan Complex mélange and greywacke, with lessor amount of siltstone and chert. The rock mass was typically intensely to moderately fractured with frequent shear zones and sheared rock.

Construction difficulties were associated with the poor quality, sheared rock mass conditions. These conditions resulted in face instability, overbreak, and squeezing ground. In some areas jump sets had to be installed to stabilize the tunnel excavation and in other areas tunnel deformations were so large that the tunnel had to be remined and resupported. Groundwater inflows were generally low, less than 2.5 gpm over 90% of the tunnel. In the other areas groundwater inflows ranged from 20 to 200 gpm, however these inflows quickly dissipated. Methane gas was encountered in several portions of the tunnel and Cal/OSHA reclassified the tunnel as "Gassy" requiring tunneling operations halted temporarily until modifications could be completed to comply with the requirements of this classification.

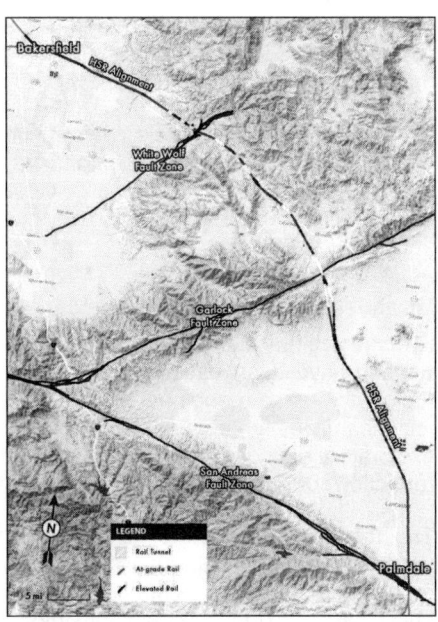

Tunneling Challenges

Tunneling at Pacheco Pass will be challenging due mainly to the poor rock mass conditions, consisting of weak, highly fractured and sheared rock. Tunneling methods will need to address ground instability, raveling conditions, overbreak, and squeezing ground. Groundwater inflows do not appear to be a major issue, although significant groundwater pressures could be encountered for the deeper alignments. In addition, special

Figure 3. Bakersfield to Palmdale project section alignment details

precautions to deal with methane gas will be required in the Franciscan Complex.

BAKERSFIELD TO PALMDALE PROJECT SECTION (TEHACHAPI MOUNTAINS)

Alignment Overview

The Bakersfield to Palmdale project section includes four alternatives and is approximately 80 miles in length, traversing valley, mountain, and high desert terrain, as well as urban, rural, and agricultural lands. From the north, this project section begins at the Bakersfield Station and travels south and southeast through the Tehachapi Mountains, then descends into the Antelope Valley where it terminates at the Palmdale Station.

Tunnels are needed through the mountains to meet operational requirements for grade and horizontal curvature and to limit or avoid the construction impacts on existing facilities. Up to nine tunnels have been identified along the HSR alignment through the Tehachapi Mountains as shown on Figure 3. The configuration of the nine tunnels consist of either single, double track tunnels or twin, single track tunnels, ranging in length from 0.3 to 2.5 miles and a total length of 8.8 or 9.7 miles depending on the final alignment chosen.

Regional Geologic Setting

The central portion of Bakersfield to Palmdale project section traverses the Tehachapi Mountains located at the southern tip of the Sierra Nevada geomorphic province. The Sierra Nevada province is a tilted fault block that separates the Great Valley on the west from the Basin and Range province to the east. The Sierra Nevada is approximately 400 miles long, with a steep eastern escarpment and a relatively gentle western slope. The structure of the Tehachapi Mountains is defined by the White Wolf and Garlock faults, as shown in Figure 3. The Tehachapi Mountains have been uplifted along the reverse White Wolf fault and shifted into an east-west trend by left-lateral movement along the Garlock fault. Bedrock units of the Tehachapi Mountains are primarily composed of granitic rock with isolated bands of metasedimentary rock.

Of the nine proposed tunnels in this section, three are anticipated to be in Tertiary aged sedimentary formations consisting of the Bealville fanglomerate and Bena gravel. The other six tunnels will be within granitic and metamorphic bedrock formations, including hornblende-biotite quartz diorite, quartz monzonite, hornblendee diorite, marble, and schist.

Local Geologic Conditions

Evaluation of the subsurface conditions to date for the Bakersfield to Palmdale Project Section have been based on available historic data, surficial mapping, accessible rock outcrops, and engineered cuts and limited field investigations. The limited geotechnical investigations completed to date consist of geologic mapping, drilling six exploratory borings using geotechnical rotary wash drilling methods and three seismic refraction surveys.

The Tehachapi Mountains subsection of the project section traverses rugged mountainous terrain of the Sierra Nevada. Elevation along this subsection of the alignment ranges from 850 feet to 3,800 feet above mean sea level. As the proposed high-speed rail alignment extends to the south into the Tehachapi Mountains, geologic conditions transition first to conglomerate and sandstone, then to igneous and metamorphic rock, dominated by quartz diorite and schist, gneiss, marble, and quartzite.

The granitic basement rocks (hornblende-biotite quartz diorite and quartz monzonite) observed at surface outcrops in the vicinity of tunnels are generally strong to extremely strong (>7,500 psi). These rocks become weaker when located in close proximity to major fault zones, particularly the White Wolf and Garlock faults. The degree of rock mass fracturing varies significantly and the units contain significant amount of abrasive material which will be encountered in more than half of the tunnels.

The schist and foliated portions of the hornblende-biotite quartz diorite, near areas of deformation, are likely to have an inherently wide range of strength characteristics because of anisotropies associated with foliation. Although generally heavily weathered, the units are noted as having a blocky to fractured rock mass structure.

The conglomerates (Bealville fanglomerate and Bena gravels) consist of fine-grained to coarse-grained materials including large boulders. The unit appears less cemented where coarser grained materials dominate within the matrix. Based on findings from borings, the unit becomes moderately to well-cemented at the proposed tunnel depths.

Expected Tunneling Conditions

A wide range of rock mass conditions will be encountered during the excavation of the Tehachapi Mountain Tunnels. Tunnels will be excavated through short lengths of weak sedimentary materials containing large hard rock boulders, and the longer hard rock tunnels will likely to contain zones of heavily fractured rock, weathering, and alteration.

Based on evaluations of surface outcrops a wide range of Geological Strength Index (GSI) values (Hoek, 2013) have been reported for each rock unit ranging from 20 to 86 (poor to very good conditions). The lower bound of GSI values could be anticipated for tunnels in vicinity of fault zones (California High-Speed Rail Authority, 2013).

An oil and gas field is present in the Central Valley to the southeast of Bakersfield and the Division of Oil, Gas, and Geothermal Resources (DOGGR) show oil and gas wells in vicinity of the proposed tunnels. The sedimentary source rocks that yield oil and gas underlie, and may be connected by fractures to the rock units that may be encountered during excavation of the tunnels. Encountering gas in a tunnel is mainly a safety issue during construction. In California the hazard is well recognized and the California Division of Occupational Safety and Healthy (Cal/OSHA) regulates tunnel construction to ensure that safe working conditions are maintained.

Currently there is no detailed groundwater information on any of the tunnels, however based on available information the three most northerly tunnels will likely be situated below the groundwater table. The issues and risks associated with construction in this condition such as water inflow, tunnel stability and tunnel dewatering will have to be addressed. It is anticipated all remaining tunnels located to the south will be above the groundwater table.

Tunneling Challenges

Tunnel design and construction for the Bakersfield to Palmdale project section will need to address the following issues:

Hazardous atmosphere—There is a potential for methane gas in sedimentary formations adjacent to oil/gas producing rock units. There is also documented active and abandoned oil wells within close proximity to the project boundaries that will need be avoided.

Boulder excavation—All the reaches of the tunnel within the conglomerate will contain boulders that may result in significant localized overexcavation/overbreak. Contractors will need to consider the impact of boulders on the excavation and ground support means and methods chosen.

Fault zones—Several active fault zones are mapped in close proximity to tunnel alignments. Rock quality near the fault zones are anticipated to be very poor and highly fractured.

Karst features—Karst features have been identified in the marble limestone units during site reconnaissance and will need to be carefully evaluated during design-phase geotechnical investigations.

PALMDALE TO BURBANK PROJECT SECTION (SAN GABRIEL MOUNTAINS)

Alignment Overview

The alternatives for the Palmdale to Burbank project section include three alignment alternatives as shown Figure 4. The approximately 38 to 44 mile section has three alignments under consideration including the SR14, E1, and E2 alignment and two station alternatives in Burbank. The project section extends through urban, rural, and undeveloped mountainous land. Each alignment alternative would involve areas of tunneling beneath the Angeles National Forest (ANF), including portions within the San Gabriel Mountains National Monument.

The SR14 alignment alternative is the most westerly alternative being considered and consists of 7 tunnels ranging from 0.5 to 9.5 miles in length. The easterly alignments E1 and E2 have 4 and 2 tunnels each respectively and tunnels up to 20 miles in length. Ground cover is also extremely variable ranging from one diameter of cover near the portals to in excess of 2,700 in case of the E2 alignment. Ground cover typically decreases to the west with average ground cover of less than 400 feet and over 1,000 feet for the SR14 and Eastern Alignment Alternatives, respectively.

Regional Geologic Setting

The San Gabriel Mountain range lies within the Transverse Ranges geomorphic province, which is an east-west trending series of steep mountains and valleys and is bounded on the north by the San Andreas Fault Zone, on the south and southwest by thrust and reverse faults of the Cucamonga-Sierra Madre fault complex, and on the east by faults of the San Jacinto Zone. Intense north–south compression has led to rapid increase in elevation of this geomorphic province and has also resulted in folding and faulting of petroleum-rich sedimentary rocks, making this area significant for oil production.

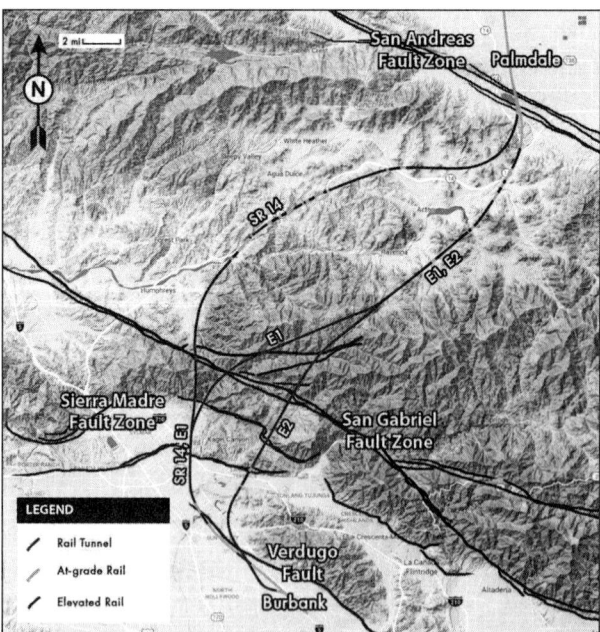

Figure 4. Palmdale to Burbank project section alignment details

The northern portion of the tunnel alignment alternatives extending to the Angeles National Forest, approach depths up to 2,700 feet beneath the crest of the San Gabriel Mountains. The alignments north of the San Gabriel Fault contain various igneous and metamorphic rocks and numerous rock formations with the predominant rock types being the San Gabriel Proterozoic anorthosite-gabbro complex and Cretaceous granodiorite.

South of the San Gabriel Fault are numerous thrust faults of the Sierra Madre Fault Zone (SMFZ) that expose igneous and metamorphic basement rocks belonging to the Cretaceous Josephine Mountain granodiorite formation and Cretaceous diorite gneiss that thrust over younger sedimentary rocks of the Saugus and Towsley formations.

South of the SMFZ, the alignments encounter alluvial deposits from the Big Tujunga, Little Tujunga, and Pacoima Canyons, and other smaller drainages fill this valley and the area around the Verdugo Mountains. Moving southwards the tunnels beneath the Verdugo Mountains, are expected to be in Tertiary sedimentary rocks belonging to the Topanga Group, Modelo and Cretaceceous granodiorite formations. The alignments emerge from tunnels through the Verdugo Mountains in the vicinity of the Verdugo Fault, a thrust fault that forms the front of the mountain range. After crossing the fault the tunnels are in Quartenary alluvial deposits.

Local Geologic Conditions

As part of the environmental evaluation of possible tunnel alignments from Palmdale to Burbank, the Authority embarked on a limited subsurface investigation of deep tunnel alignments beneath the western San Gabriel Mountains within the Angeles National Forest. The purpose of the investigation was to evaluate geotechnical and hydrogeological conditions of the deepest tunnel segments that might impact the feasibility of design and construction of tunnels beneath the mountains. The investigations included rock coring at six locations ranging in depth from 650 to 2,700 feet. Three of the core holes were inclined to pass through the San Gabriel fault and the Transmission Line fault to evaluate the width of fault gouge and sheared rock, and hydraulic conditions at intersections of the tunnels with faults. The two deepest core holes, 2,700 feet and 2,100 feet, were drilled to measure maximum in-situ water pressures and ground temperatures at the tunnel depths.

Igneous rocks are expected to be the predominant rock type encountered in the tunnels, accounting for a combined total of 45% of the length of SR14 tunnels and approximately two thirds of the length of the Eastern alignment alternatives (i.e., E1 and E2). The anorthosite complex has unconfined compression in-tact strengths that range from 700 to 29,000 psi, whereas the granodiorite strengths range from 700 to 22,000 psi. Sedimentary deposits are the next most abundant rock type, ranging from 11% to 28% of the length with metamorphic rocks being the least abundant except for in E2. Alluvial deposits are common to all of the alignments and range from 25% of the total for SR14 and decrease toward the east to 6% for E2. Alluvial deposits are relatively local in extent except toward the southern ends of the tunnel alignments where extensive deposits are present. Alluvial deposits are commonly associated with distinctive geomorphic features such as alluvial fans, floodplains and terraces. The alluvial deposits are typically sand, with gravel and cobbles.

Expected Tunneling Conditions

Geologic conditions along the tunnel sections are expected to be extremely variable ranging from massive very strong granitic and gneissic rocks to weak alluvial soils. A number of formation contacts, geologic contacts and fault crossings are anticipated for all of the alignment alternatives. The contacts could be highly weathered and likely conduits for groundwater flow. In addition, it should be anticipated that the rock mass will be hydrothermally altered due to the tectonic activity associated with the formation of the San Gabriel Mountain range. Crushed or sheared rock and high groundwater flow is also expected to be encountered within the three fault zones crossed by the tunnel alignment.

Detailed groundwater conditions are unknown; however, potential groundwater heads in excess of 50 bars (1,700 feet) could possibly be encountered beneath the Angeles National Forest. In most other areas, groundwater heads are expected to be significantly less, potentially in the range of 100 to 300 feet. Groundwater control will be important in water bearing soil/rock, especially where groundwater inflows in to the tunnel may result in significant loss of ground, tunnel instability, or unacceptable environmental impacts.

Tunneling Challenges

The entire study area crosses a complex geologic zone of structural folding and faulting. Some of the key geotechnical and construction considerations that affect tunnel design for the Palmdale to Burbank project section includes:

Long Tunnel Drives—Each of the alternative alignments has significant tunnels with between two and seven tunnel sections each and total tunnel lengths ranging from 22.3 to 27.3 miles. Individual tunnels range in length from 0.6 to 20 miles and each of alternative alignments has at least one tunnel that exceeds 9 miles in length. Maximum individual tunnel lengths for alignments SR14, E1 and E2 are 9.5, 13.6 and 20.7 miles, respectively.

High Groundwater Pressures—Groundwater pressures may exceed 50 bars along certain tunnel sections between Palmdale and Burbank. In most sections, the pressures are expected to be in the range of about 3 to 9 bars (100 to 300 feet of hydrostatic head). Issues that need to be considered include design of the lining system, impacts on wear of tunneling equipment and interventions for inspection and maintenance and the potential impacts of tunnel construction on the groundwater system

Variable Ground Conditions—For this project zones with geologic units exhibiting significantly different physical characteristics, such as soft ground (or alluvium) and rock, or weak rock and strong rock will be encountered within the same tunnel. Unstable ground conditions will be associated with cohesionless alluvium, as well as fault zones and fractured or weak, weathered/altered rock. The instability is magnified when these units are present and exposed to high groundwater pressures.

Active Fault Crossings—All of alignments cross the San Gabriel and San Madre fault zones. A fault chamber or special tunnel lining design will be required for a tunnel crossing an active fault to avoid tunnel rupture and possible collapse due to a large displacement from a seismic event.

DELIVERY APPROACH AND PROGRAM SCHEDULE

It is expected that the tunnels for this program will be delivered using a design-build approach, similar to the other construction packages previously awarded (CP1 to CP4). As indicated in the 2016 Business Plan, completion of the "Valley to Valley" segment will require the Pacheco Pass tunnels to be one of the next construction packages procured. Recently the Authority has retained a geotechnical investigation team to carry out explorations for the San Jose to Merced project section. These explorations will provide the data required for preparation of a Geotechnical Baseline Report to support the procurement process.

The tunnels in Southern California will trail the "Valley to Valley" segment by several years, however it is expected that geotechnical investigations will continue as required to support the planning and environmental process, and eventually procurement.

ACKNOWLEDGMENTS

Special thanks are due to others on the California High-Speed Rail program team including: P. Guptill (Kleinfelder), Jody Castle (EMI), J. Kan (ENGEO), C. Davis (WSPIPB), K. Fan (WSPIPB), R. Bhargava (WSPIPB), and P. Chilingar (WSPIPB) for their valuable contribution to this paper.

REFERENCES

California High-Speed Rail Authority, 2013. Bakersfield to Palmdale, Preliminary Geotechnical Design Report Tunnels.

California High-Speed Rail Authority, 2016, Connecting and Transforming California, 2016 Business Plan.

California High-Speed Rail Authority, 2016, San Jose to Merced Project Section, Draft Preliminary Geotechnical Data Report for Tunnel Subsection, September.

Hoek, E, 2013, Quantification of the Geological Strength Index Chart, 47th US Rock Mechanics/Geomechanics Symposium, San Francisco, CA, USA.

USBR, 1986a. Construction Geology—Santa Clara Tunnel. U.S. Department of the Interior, Bureau of Reclamation Mid-Pacific Region.

USBR, 1986b. Construction Geology—Pacheco Tunnel Reach 2. U.S. Department of the Interior, Bureau of Reclamation Mid-Pacific Region.

Design of Atlanta Raw Water Supply Program

Tao Jiang ▪ Stantec Consulting Services Inc. (Part of JP2/PRAD, Stantec, and Chester Engineers JV)
Don Del Nero ▪ Stantec Consulting Services Inc. (Part of JP2/PRAD, Stantec, and Chester Engineers JV)
Adam Bedell ▪ Stantec Consulting Services Inc. (Part of JP2/PRAD, Stantec, and Chester Engineers JV)
Brian Jones ▪ City of Atlanta Department of Watershed Management
Ade Abon ▪ City of Atlanta Department of Watershed Management

ABSTRACT

The City of Atlanta is converting an over-a-century-old quarry into a 2.4 billion-gallon raw water storage facility. The system consists of a deep hard rock tunnel, two deep pump stations, two dropshafts and multiple connecting adits and shafts. Bored with a TBM from tunnel portal at the quarry base, the tunnel is approximately 24,000ft long, 250ft to 450ft deep, 13ft in bore diameter, and is partially concrete lined with modified contact grouting to control water infiltration/exfiltration. CFD modeling is performed for the pressurized system. Various shaft construction techniques are used, including drill & blast, blind bore drilling and raisebore drilling.

PROJECT BACKGROUND

The current water supply program operated by the Department of Watershed Management (DWM) of City of Atlanta consists of four aged raw water pipelines, some of which goes back to 1890s. Based on previous assessments completed by the DWM, the entire water system is at or will soon reach its recommended useful life. As such, the City acquired the former Bellwood Quarry in 2006 with an intention to create a water storage facility with a volume of approximately 2.4 billion gallons to serve approximately 1.2 million people.

The quarry was used for mining granitic gneiss and crushed-stone aggregate production for almost a hundred years. As shown on Figure 1, it has almost vertical sides with benches in some areas. The averaged elevation of the top ground is about 860 ft. and at the base is about 540 ft. The proposed full pool level during storage is at elevation of 840 ft. The water to supply the quarry facility is taken from the existing water intake structure at the Chattahoochee River. The raw water is stored in the quarry before it is withdrawn for treatment at the Hemphill and or Chattahoochee water treatment plants. This offline operating mode includes routine withdrawals and replenishments.

As illustrated on Figure 2, the Hemphill Water Treatment Plant (HWTP) and the Chattahoochee Water Treatment Plant (CWTP) are in between the quarry and Chattahoochee River. A conveyance system, including a TBM tunnel and multiple shafts and adits, is required to connect these facilities above. The TBM is assembled at the quarry base and launched from the north side highwall through a tunnel portal.

The other major components of the project include two pump station shafts, one riser shaft and one dropshaft at the quarry, five blind bore pump station shafts at HWTP, and one construction shaft at CWTP, which will be converted into a dropshaft after the TBM construction. The system configuration at the quarry is demonstrated in the 3D rendering shown on Figure 3.

Figure 1. Bellwood Quarry

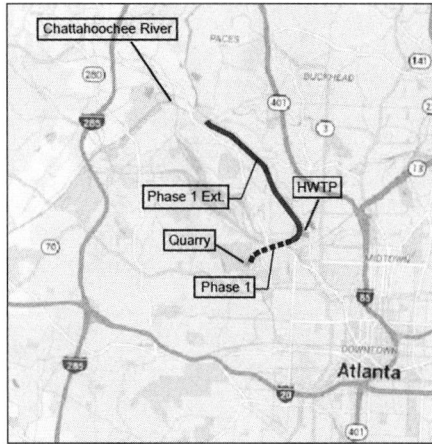

Figure 2. Project overview

GEOTECHNICAL INVESTIGATION

Local Geological Setting

The project is located in the Piedmont Physiographic Province. The Piedmont geology in the greater Atlanta area generally consists of medium-grade metamorphic rocks with granitic intrusions. These crystalline rocks are some of the oldest rocks in the Southeastern United States. They were generally formed before and during the building of the Appalachian Mountains. Since their origin, the rocks have undergone a complex history of metamorphism, weathering, and deformation.

Various structural features are present in the rocks, including folds, fractures, and lineaments. The high pressures and temperatures at great depths resulted in a full range of deformational styles, ranging from medium-grade metamorphism, through fully-welded ductile shearing and mylonite formation, to brittle fracturing with rocks that commonly contain hydrothermally deposited minerals. At shallower depths, structures like exfoliation fractures were formed in the rocks due to erosion of overburden and unloading. The exfoliation fractures mainly occur along the foliation "planes" and tend to be open and act as conduits for water movement through the rock mass.

Lineaments, which are surface topographic expressions of underlying rock mass or crustal structure, occur throughout the Piedmont. The lineaments are often controlled by weathering associated with discontinuities in the rock. In many cases, the lineaments represent fracture zones in the underlying bedrock. At depth, the fracture zones are typically cemented with weathered minerals. At shallower depths, erosion of these weathering minerals often results in zones of broken, water-bearing rocks and topographic features such as valleys and draws.

Groundwater

Groundwater occurs in all three zones of the subsurface, the soil zone, soil to bedrock transition zone, and the bedrock zone. The depth of the groundwater table varies significantly along the proposed tunnel alignments, ranging from less than 10 ft. to over 200 ft. The soil zone is generally considered to be a good producer of groundwater. The transition zone typically contains abundant open fractures and can become a major storage source for groundwater where its thickness is significant. The bedrock

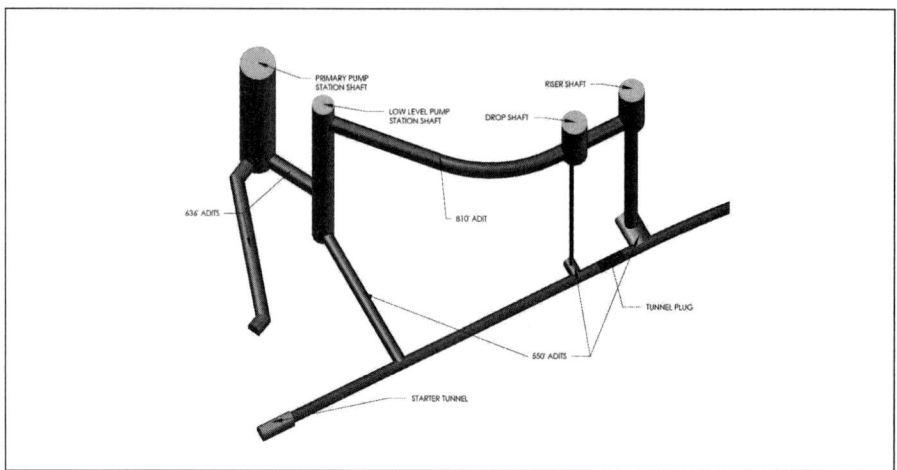

Figure 3. System configuration at the quarry

zone in the Piedmont generally has fewer open fractures with depth than the transition zone. However, large fractures with the ability of producing large volumes of water do exist in the bedrock. High-yield wells have been reported to produce sustained yields up to nearly 500 gallons per minute.

Potentiometric gradients may be steep in the Piedmont. Seasonal fluctuations in the water table are common in response to precipitations. Local observations of the water table rising and falling between 8 ft. to 14 ft. are common. Perennial streams are fed by bank seepage and upwelling groundwater along the course of their lengths.

Geotechnical Investigation Program

The geotechnical and hydrogeological field investigations for the Water Supply Program comprised 25 deep borings and 30 shallow borings. The deep borings were advanced along the tunnel alignment with the purpose of characterizing the bedrock conditions. The shallow borings were drilled at the locations of proposed shafts and surface structures and considered as supplemental borings with the purpose of characterizing the overburden soil conditions, including information on the transition from soil to rock. Drilling occurred in phases between August 2014 through August 2016 in concert with an evolving design.

Prior to initiation of the geotechnical investigation, all readily available, relevant geologic data was summarized and reviewed, and some field work was performed. Background information included lineament analysis, structural and tectonic geologic analyses including field mapping and hydrogeologic analysis. The geotechnical investigation was developed on the basis of information contained in these background reports. Triple-tube HQ coring was selected to obtain rock samples. In addition to coring, double-packer permeability testing was performed on all vertical boreholes. Once cores were extracted, they were logged and photographed.

Once drilling was complete, a suite of borehole geophysical tests were run in 21 of the deep borings. This provided the following information: optical and acoustic televiewer logs, full wave sonic logs, fluid temperature and conductivity logs, natural gamma logs, single point resistance logs, three arm caliper logs, and EM flowmeter flow measurement logs. These tests helped to further characterize the in-situ geologic conditions at depth while also providing hydrogeologic information and joint orientation data used to create stereoplots.

The depth of the tunnel (greater than 400 ft.in areas) warranted in-situ stress testing. Agapito and Associates conducted in-situ stress testing in three of the deep borings and attempted 12 tests, of which 7 were successful. They used the over-coring method as developed by Sigra, Pty of Brisbane, Australia. The purpose of this testing is to determine

Figure 4. Tunnel portal stabilization

the magnitude and direction of the secondary principle stress in the plane perpendicular to the borehole. The results were factored into tunnel excavation support design. Additional laboratory testing included tests to determine the intact properties of the rockmass. These tests include unit weight, unconfined compressive strength, Cerchar, Brazillian tensile strength, acoustic velocity, point load index strength, and abrasivity/drillability tests.

TUNNEL PORTAL STABILIZATION

The TBM is launched through a tunnel portal at the base of quarry. To secure the approximate 320 ft. tall quarry highwall above the portal, a drape system is designed to minimize the hazards of potential rockfalls (Figure 4). The system consists of TECCO 3 mm mesh from Geobrugg and rock dowels in the locations that are identified to have potential rock wedge failures. At the tunnel "eye," where the tunnel breaks into the ground from the portal, 20 ft. long spiles are installed along the tunnel crown to stabilize the transition area.

TUNNEL DESIGN

The tunnel is about 24,000 ft. long, 250 ft. to 450 ft. below ground surface, and is sloping up from the quarry to the dropshaft at CWTP with a slope of 0.2%. The tunnel will be partially concrete lined with an internal diameter of 10 ft. The service life of the final lining system is designed to be 100 years.

Tunnel Initial Ground Support

A two-pass tunnel support system is used in the design, which is common for Atlanta area hard rock tunnels. Excavation ground support will be installed immediately following the TBM excavation to stabilize the tunnel and provide a safe work area. The ground are categorized into three ground types (Types A, B and C) based on rock mass properties, and three excavation ground support types are developed, respectively. Type A support consists of two 5-ft long double corrosion protection dowels as both excavation support and permanent support, since most of Type A ground is not anticipated to be concrete lined. Type B support consists of four 5-ft long friction dowels with welded wire mesh, and Type C support consists of steel ribs with welded wire mesh as lagging. Both Type B and Type C ground will be concrete lined.

In the excavation ground support design, wedge analyses are performed to identify any potential rock wedges that may be formed during tunnel excavation. Stress analyses are conducted using finite element methods to identify potential overstressed areas or the areas that may exhibit plastic behavior. The excavation ground support is developed based on the potential failure modes identified in the analyses.

Tunnel Permanent Lining

After the excavation is completed, the tunnel will be partially concrete lined. Type B and C ground is required to be lined, while most of the Type A ground is anticipated remain unlined. The minimum lining thickness is designed to be 12 inches, not only for sustaining the design loads but for facilitating concrete placement with better quality. The double corrosion protection dowels installed in unlined tunnel sections in Type A ground is considered as part of the permanent lining system and will support the ground during the tunnel service life.

As the tunnel is part of the water storage facility, the permanent lining system not only needs to support all the external loads, including rock load and groundwater pressure, but also to sustain the internal water pressure, which is about 300 ft. head. The internal pressure will cause the concrete lining in tension, as such reinforcement is required for the lining in Type B and C ground as they provide less constrain to the lining. The transient pressures during filling the tunnel are also considered in the design.

Lining Section Selection Critria

It is determined in the design that Type B and C ground would receive reinforced concrete lining. For Type A ground, which is estimated to count about 70% of total tunnel length, the decision to line it or not depend on the field observation by the engineer and geologist with the following general rules:

- The structural stability of the rock mass to maintain the unlined tunnel opening throughout the design life;
- The adequacy of the installed excavation support elements to maintain that stability during the design life of the tunnel;
- Groundwater infiltration into the tunnel and its impact to operations and groundwater resources (if any) must be kept to within acceptable limits and locations,
- Skip length;
- Groundwater exfiltration into the rock mass beyond the tunnel and groundwater resources (if any) must be kept to within acceptable limits and locations; and
- The need to prevent adverse rock mass degradation during the 100-year design life of the tunnel.

Water Exfiltration

The tunnel concrete lining, whether unreinforced or reinforced, is considered permeable or semipermeable, given the fact that the concrete would crack for a number of reasons, such as shrinkage. The reinforcement without being pre-stressed may be able to reduce cracks or to redistribute the locations of cracks but would be difficult to completely prevent the lining from cracking. The water may migrate in (infiltration) or out (exfiltration) the tunnel through the cracks. As the tunnel will be full most of the time in the normal operation of the system, while the groundwater infiltration may mainly be a temporary construction issue during tunnel excavation, the water exfiltration will be a long term concern and should be addressed in the tunnel design.

The tunnel is a pressurized system and constantly under internal water pressure during storage mode. Under the pressure, the water in the tunnel may leak through the cracks in the lining and get into the features of surrounding rock mass. Particularly at some areas around CWTP, where the ground is below Elevation 840 ft. (the quarry full pool level), the water could seep through the ground surface and flood those areas. To prevent such event from occurring, a combined lining system is developed based on the understanding of local geology and the experience from previous projects. As illustrated in Figure 5, the system consists of tunnel concrete lining, modified contact grouting and grouted surrounding rock mass.

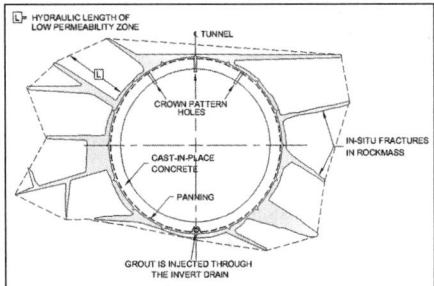

Figure 5. Water infiltration/exfiltration proof system

Modified Contact Grouting

As aforementioned, modified contract grouting is utilized to fill up the joints and voids behind the tunnel lining, and along with the tunnel lining, to form a waterproof system to reduce or even cutoff the water infiltration and/or exfiltration.

Modified contact grouting was developed on previous Atlanta tunnels to eliminate the cost and delays associated with making two grouting passes through the tunnel. Modified contact grouting differs from traditional contact grouting in that it combines contact grouting with consolidation grouting into a single operation. It is achieved in a single pass of labor, equipment and materials through the tunnel that the contact grouting to fill voids and consolidation grouting to fill rock joints thus to reduce ground-water inflow.

In the modified contact grouting program, the invert drain and panning serve as path-ways to deliver grout into the rock mass. By grouting the invert and crown in opposite directions, groundwater and grout are forced in the direction of the ungrouted tunnel. This systematically replaces water and air with grout down the tunnel. This also allows for grout to "push off" previously grouted areas and permeate the rock mass and concrete lining. This eliminates the need to drill precisely located grout holes that intersect known fracture locations in the rock.

Modified contact grouting requires injection pressures that exceed hydrostatic to force grout into the rock mass. By pumping grout with a smaller cement grain size at higher pressure, grout particles are pushed farther out into the rock mass through the same fractures that deliver groundwater to the tunnel. The result is that the tunnel lining and surrounding rock mass are grouted as far back into the rock mass as fracture aperture and cement particle size allow, as shown in Figure 4. This reduces the pressure gradient pulling the water into the tunnel, effectively lowering groundwater infiltration and exfiltration.

Tunnel Plug

As an important component of the tunnel hydraulic profile, a tunnel plug is designed in between the riser shaft and dropshaft at the quarry. The plug mainly serves two purposes: (1) during storage, the water level in the tunnel system can be maintained constantly at Elevation 840 ft. regardless the water level in the quarry, as such the operation depth of most of pumps is significantly reduced; (2) during maintenance, the water at the quarry can still be kept at full pool level while the tunnel is completely dewatered.

The plug is designed as a massive concrete pour, and the resistance against the water pressure from either side relies on the bonding between the plug and surrounding rock mass. Xypex admix is added in the concrete mix to help seal the cracks and thus to reduce the seepage through the plug.

Tunnel Hydraulic Transient Study

Given the complex hydraulic profile of the system, a computational fluid dynamic (CFD) modeling is performed to determine the transient pressure inside the tunnel during filling the system. The analysis indicates that the maximum transient pressure is approximately 27 psi, as an additional internal pressure, and the maximum vacuum pressure is about 16 psi, as an additional external pressure. The additional pressures are included as one of the load cases in the lining design.

SHAFT DESIGN

As aforementioned, the system consists of several shafts with different sizes, depths, and construction techniques to build. The pump station shafts at the quarry and drop/construction shaft at the CWTP will be built using conventional drill-and-blast methods from the top down. The tangential drop shaft and riser shaft at the quarry will be raisebored from the bottom up. The five pump shafts at HWTP will be drilled from the surface with the blind boring techniques. To the authors' knowledge, the five 9.5 ft. diameter blind bore shafts at HWTP are the largest and deepest shafts in the Piedmont geology to use this technique to date.

Blind Bore Shafts at HWTP

As shown on Figure 6, the five pump shafts are determined to be constructed using blind bore techniques since surface blasting is prohibited at HWTP due to adjacent reservoirs. Upon completion of the five 11-ft diameter steel casings in overburden, five 9.5-ft diameter 400-ft deep blind bore shafts will be drilled from the surface into rock through the steel casings in overburden.

Two blind bore rigs will be mobilized due to the schedule requirement. The rig has a rotary table to provide the torque or turning action for the reamer. Throughout the entire shaft development, both the shaft and the hollow drill string are filled with water to create two independent columns of water. The water column inside the drill string is made much lighter by injecting compressed air. The heavier water column inside the shaft thus pushes down and across the bottom of the shaft. The water is then forced through a small opening on the reamer body and displaces the lighter water in the drill string to create upward flow or reverse circulation. The reverse circulation generates tremendous vacuum at the reamer opening and removes the cuttings from the face. Maintaining a constant water level in the shaft during the entire drilling operation is critical. In addition to cutting removal, the water also provides outward pressure on the shaft wall to improve the shaft stability. The returned water from the shaft is collected in an adjacent settling pond, and the water can be re-circulated to the drilling operation after the cuttings have been settled out.

In order to meet the 0.25% verticality tolerance, a pilot hole is required for each shaft. The pilot hole will be directionally advanced utilizing an optical technique that allows continuous monitoring for the deviation. Once completed, an optical survey will be performed to verify that the pilot holes meet the required verticality.

Upon completion of the blind bore drilling, a 76" ID steel pipe casing with 1-inch wall thickness will be lowered into the shaft and grouted in place in wet. Each section of the steel pipe is 40-ft long and will be connected through welding. The welds will be ultrasonically tested. After the shaft construction, the vertical turbine type pumps will be installed inside the steel casings.

Pre-Excavation Grouting at HWTP

Figure 7 is the stereoplot that summarizes the findings from the geotechnical investigation around the blind bore shafts. From the plot, one low-angle foliation joint set and two high-angle non-foliation joint sets are identified. The geotechnical investigation also indicates numerous fractures within the identified joint sets, in which the apertures may range from 0.25 to 5 inches. These open fractures could act as water communication conduits between the adjacent reservoirs and the blind bore shafts and/ or between the blind bore shafts and surrounding rock mass. In order not to impact the reservoir and not to lose the water inside the blind bore shafts to surrounding rock mass, a pre-excavation grouting program is design at the HWTP site, as shown on Figure 8. In order to more efficiently intersect highangle features, the drill holes are oriented 10 degrees off vertical at a bearing of 260 degrees.

The pre-excavation grouting program was completed in October 2016. The total grout take is approximately 53,000 gallons, with roughly 68% of the grout being grouted in the primary holes.

Connection Between Blind Bore Shafts and Main Tunnel at HWTP

The five blind bore shafts are connected to the main tunnel through five short adits, respectively. The adits have a modified horseshoe shape with the excavated size of about 10 ft. × 1 0ft., the pillar in between the adits is only as thin as 8 ft. Given the in-situ stress level at the adit depth and the surrounding rock conditions, the stability of thin pillars becomes a big concern in the design, especially in the event of rock overbreaks. As such, given the complex configuration, a three-dimensional finite element model is created to investigate the stress concentrations and the pillar stability following the excavation. PLAXIS3D is used for the modeling. The modeling results are shown in Figures 9 and 10.

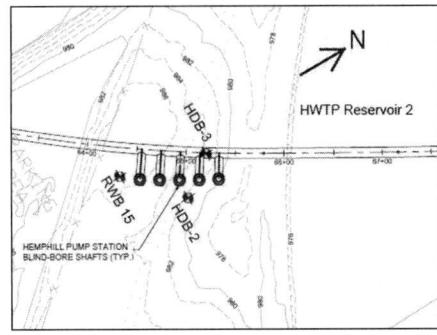

As anticipated, the model indicates that the most increased vertical stresses occur in the pillars while the horizontal stresses occur at the main tunnel crown. Although the stresses are dramatically

Figure 6. Blind bore shafts at HWTP

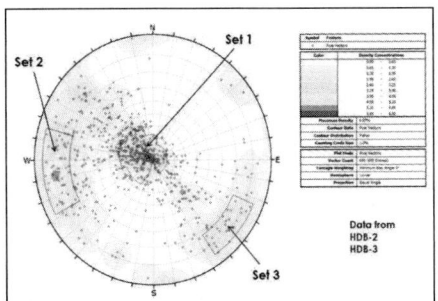

Figure 7. Stereoplot at HWTP

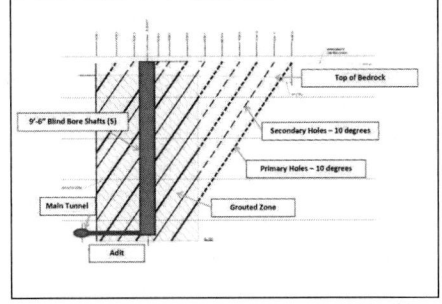

Figure 8. Pre-excavation program at HWTP

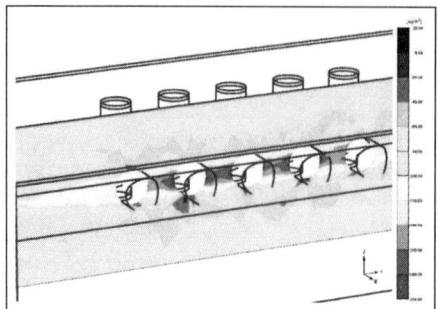

Figure 9. Vertical stress distribution

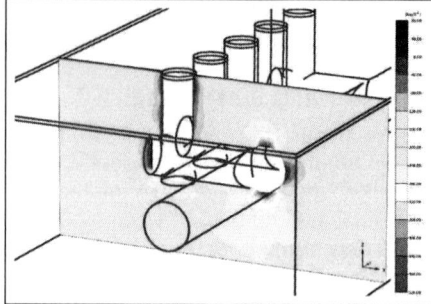

Figure 10. Horizontal stress distribution

increased, the magnitudes are still within the compressive strength of rock mass, the pillars are therefore considered stable. However, to ensure the safety of excavation and minimize the potential risks during construction, any two directly adjacent adits are not allowed to be concurrently excavated until the support is installed, and the mechanical rock excavation is specified to prevent overbreaks.

PROJECT DELIVERY METHOD

Due to the project schedule, which is primarily driven by the condition of the City's existing water infrastructure, the City considered alternative project delivery instead of traditional design-bid-build. The project schedule required a start date for construction of January 2016 and a substantial completion date of May 2019.

The City selected Construction Management at Risk (CMAR) as the project delivery method as the evaluation of schedule certainty, cost certainty, funding, team collaboration, quality, and City experience. This method entails a commitment by the CMAR for construction performance to deliver the project within a defined schedule and price, in this case by Guaranteed Maximum Price (GMP). The CMAR acts as consultant to the owner in the development and design phases, but as the legal equivalent of a general contractor during the construction phase.

CONCLUSIONS

The paper describes the Raw Water Supply Program that is being undertaken by City of Atlanta, and presents the design considerations and approaches for the tunnel and shafts. The project is now under way and will provide reliable water resource to the City once it is completed.

ACKNOWLEDGMENTS

The authors would like to thank the Department of Watershed Management of City of Atlanta for its commitment to make the project possible, also would like to thank PRAD Group and River to Tap, the teaming partners in the Joint Venture, who are responsible for the design of pipelines and pump stations, for their contributions to the project. The contributions and efforts from other members of Stantec's tunnel design team, including Julian Prada, PE, Yong Wu, PE, Ph.D., Steve Fradkin, PG, Rick Ponti, PG, and Konner Horton, EIT, are also greatly appreciated.

Annacis Island Wastewater Treatment Plant Tunneled Outfall System

John Newby ▪ CDM Smith Canada
Kapila Pathirage ▪ CDM Smith Canada
Ken Massé ▪ Metro Vancouver

ABSTRACT

Expansion to Metro Vancouver's Annacis Island Wastewater Treatment Plant (WWTP) in Delta, British Columbia will require a new 4.2 m ID tunneled effluent outfall system, including three shafts and multiple tunnel drives, extending into the Fraser River. Within Annacis Island, the conveyance tunnels will be located near the base of a 35 to 40 m alluvial sand deposit. The tunnel will terminate about 25 m beneath the river at a riser shaft connected to a near surface diffuser pipeline, risers, and ports. This paper addresses the overall project, which will begin construction in 2018, and challenges associated with the project.

INTRODUCTION

Existing Plant and Outfall

Metro Vancouver owns and operates five wastewater treatment plants in the lower mainland of British Columbia, Canada. The largest one, the Annacis Island Wastewater Treatment Plant (AIWWTP), is undergoing a major expansion, called the Stage 5 Expansion, and a new outfall is required to accommodate the plant's increase in hydraulic capacity.

The existing plant, located on Annacis Island in Delta, BC currently serves a population of one million people and discharges treated effluent into the Fraser River via an outfall that was built in the early 1970s. The river near the outfall experiences fluctuating depths due to tidal influences. At the design river level, the existing outfall can convey approximately 14.5 m^3/s (1,253 MLD); however, after the Stage 5 Expansion, an outfall with a capacity of 18.9 m^3/s (1,633 MLD) will be required. Metro Vancouver has decided to overbuild the new outfall to match the ultimate plant capacity which will be 25.3 m^3/s (2,176 MLD). Figure 1 is a map showing the location of the plant and the outfall path.

New Outfall System Design Requirements

The main objective of the project is to provide an outfall system that can convey 25.3 m^3/s without impacting the hydraulic gradeline of the treatment plant. There is a regulatory requirement for the new outfall, stipulated by the provincial Ministry of Environment, that the discharged effluent must achieve a minimum dilution ratio of 10:1 and meet water quality guidelines within 100 metres of the discharge point, as defined by the Municipal Wastewater Regulations. A target dilution ratio of 20:1 under slack water and low flows in the river was established by MV to help assure water quality guidelines are met.

A gravity outfall will meet the near-future capacity and dilution requirements; however, an effluent pump station will be required in the future as the plant's effluent flow increases and river levels increase due to sea level rise.

Seismic Resiliency

The Vancouver area is in a high seismic zone and seismic resiliency is important consideration for infrastructure design. MV's Seismic Design Criteria is based on the National Building Code of Canada 2010 (NBCC 2010) which requires new wastewater plants to be designed to a post-disaster performance level, capable of remaining operational with only minor repairable damage following a code-level seismic event. This event, the design earthquake, is defined as having an annual exceedance probability (AEP) of 1 in 2,475. The post-disaster perfor-

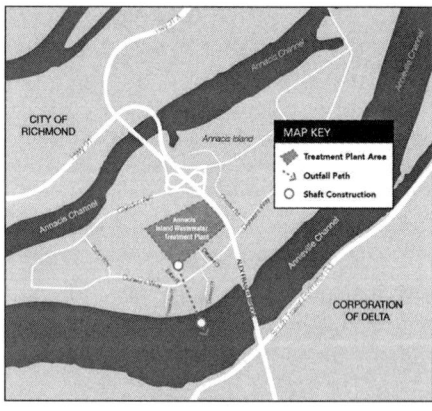

Metro Vancouver, 2016

Figure 1. Location map with Metro Vancouver region

mance level applies to all facilities for the new outfall, except those that are not critical to continuous operation.

Facilities being built as part of the Stage 5 Expansion will meet the Post Disaster performance objective; however, the existing plant infrastructure If the older infrastructure were to be rendered inoperable in an earthquake, the effluent would not be able to get out of the plant, so a bypass will be constructed to connect the new secondary clarifiers to the new outfall. This is called the "Post-disaster Bypass Conduit to the Outfall" (PDBCO), and it will be scoped into the contraction contract package with the outfall. Figure 2 shows the alignment of the new outfall as well as the PDBCO.

SITE CONDITIONS

Geologic Setting

The Annacis Island area has been subjected to three or more periods of glaciation. Prior to the first glaciation, the area comprised part of the seabed at the mouth of the Fraser River. As the glaciers retreated, fine-grained soils were initially deposited by the meltwater in a marine environment. These fine-grained soils were subsequently consolidated and a mixture of sand and gravel was deposited above the finer marine sediments. Overbank (floodplain) deposits comprising clayey silt to silty clay with varying amounts of organic material mantled the surface of Annacis Island prior the placement of sand fill prior to the 1970s to raise the land surface for future development of the island.

Surface and Subsurface Conditions

The land surface on Annacis Island is relatively level at El. 104 to 105[*] and the area surrounding the Annacis Island WWTP is primarily occupied by light-industrial facilities and warehouses. Water depths in the center of the Fraser River are typically 15 metres or more below Chart Datum (Local Low Water). Where sediment accumulates in the river bed, the Port of Vancouver performs dredging to maintain water depths in the navigation channel to a minimum of 10.9 metres below Chart Datum (equivalent to El. 87.51).

---

[*] Elevation datum for the Annacis Island WWTP is CVD28GVRD (Geodetic Datum) + 100 metres.

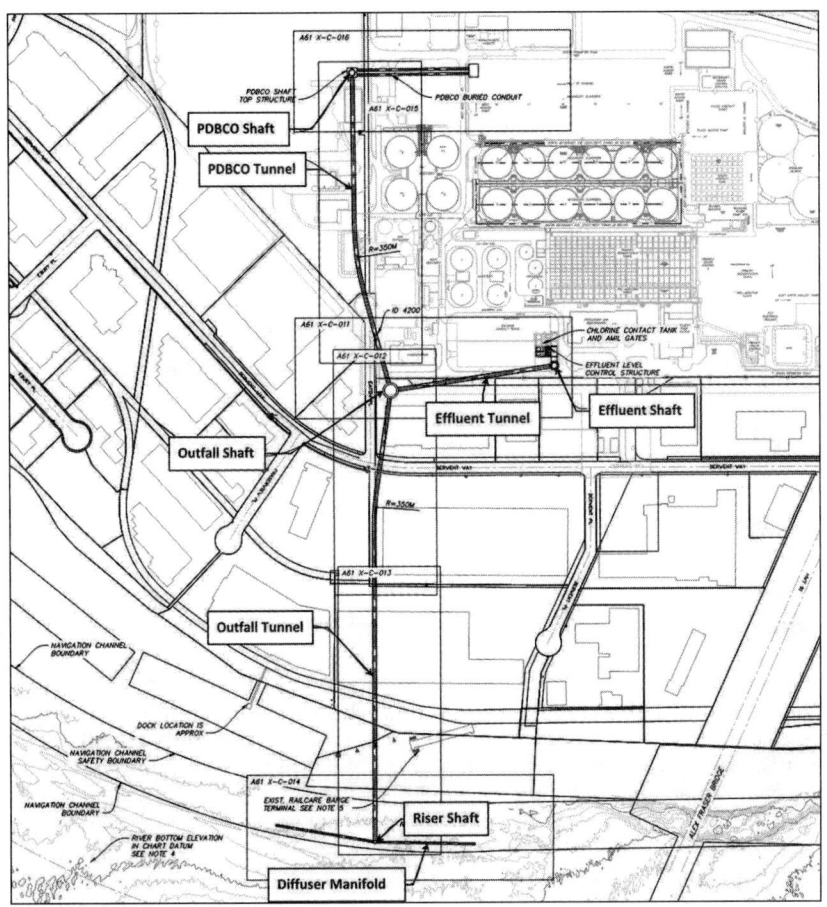

CDM Smith, 2016

Figure 2. Alignment map (with key project components)

Table 1. Soil units

Soil Unit	Thickness (m)		Geological Origin
	On–Land	Offshore	
Fill	2.5 to 5.5	—	Fill (primarily river sand)
Clayey silt/organic silt	0.5 to 3	—	Overbank sediments
Sand	30 to 45	27 to 30	Fraser River sand
Gravel to sandy gravel	—	—	Post-glacial channel erosion
Clayey silt/silty clay	>6 to >45	>13 to >30	Marine deposit
Clayey silt to silty clay	>22	>20	Glaciomarine deposits

Six soil units have been identified underlying the new outfall alignment as described in the order of depth encountered and geologic age in Table 1.

Groundwater levels on Annacis Island vary with the water level in the river, change in season, and amount of precipitation. Nominally, groundwater levels on land vary between El. 100 and 101 but can be a 1 to 2 m higher where influenced by high river tides, flood, or heavy precipitation.

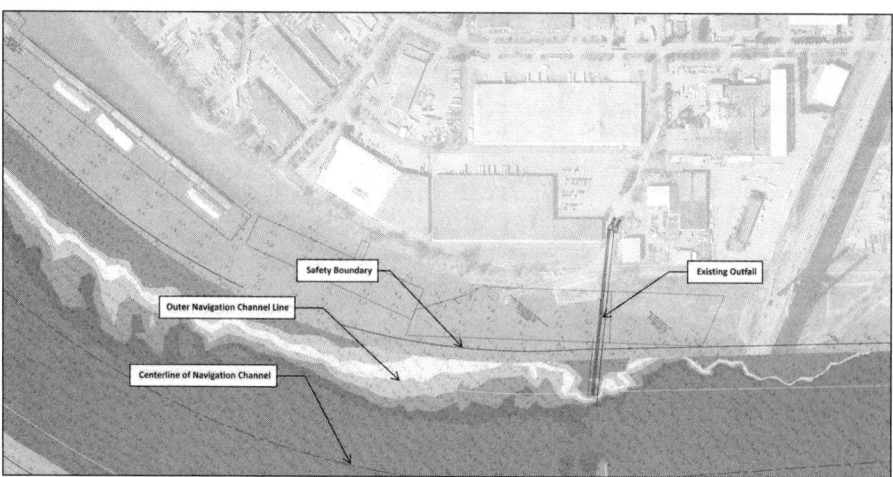

CDM Smith, 2016
Figure 3. Diffuser study area (with 2016 bathymetric survey overlay)

River Geomorphology

The Fraser River drains 232,000 km^2 of southern British Columbia, making it the largest river on the west coast of Canada. It has a snowmelt-dominated flow regime and sediment loads are significant during late spring and summer spring snowmelt runoff (freshet) when sand waves several metres high move through the river bottom off shore of the treatment plant. Construction of the Alex Fraser Bridge in 1984 created a constriction in river flow resulting in erosion of the river bottom downstream of the bridge creating an area of limited deposition of sand in the river bed. Figure 3 shows water depths in area of potential outfall diffuser locations outside the outer limit of the dredged navigation channel. Contour shading is at 0.5 m intervals with blue colors indicating greater that 10.5 m below Chart Datum. An area of deeper water with less sediment accumulation can be seen extending about 300 m downstream of the existing outfall; further downstream evidence of sand wave build-up is evident.

NEW OUTFALL SYSTEM DESCRIPTION

The new outfall system consists of tunneled conveyance from the treatment plant to the river, a river riser, and a diffuser pipeline buried in the river bed with multiple risers and ports extending above the river bed. Treated effluent is discharged from the plant into a drop shaft connected to the tunnel system and pass through another shaft which will serve as the wet well for the future effluent lift (pump) station. Key elements of the system are shown on the alignment map in Figure 2.

Outfall Shaft: This shaft will house the future effluent pump station receiving effluent from the tunnels connected to either the Effluent Shaft or the PDBCO Shaft. The final lining will be a minimum 16 m in internal diameter and the completed shaft will be approximately 39 m deep (top of base slab El. 86.7). It is assumed that the temporary excavation support system for each of the on-land shafts will utilize slurry walls. The Outfall Shaft will serve as the launch shaft for all the tunnel drives.

Effluent Shaft: This shaft will receive effluent from the existing output from the treatment plant. The final lining will have a minimum I.D. of 7.0 m and be the same depth as the Outfall Shaft. The Effluent Shaft will be designed to serve as a TBM receiving shaft for the Effluent tunnel.

PDBCO Shaft: This shaft will receive effluent from the newer (Stage V) portion of the treatment plant in the event older portions of plant are no longer functional following a design level earthquake. It could also serve as the effluent discharge point for future plant expansions. The dimensions and anticipated temporary excavation methods are the same as for the Effluent Shaft and it will serve as a TBM receiving shaft for the PDBCO Tunnel.

Effluent Tunnel: This tunnel segment will convey flow from the Effluent Shaft to the Outfall Shaft. The interior diameter will be a minimum of 4.2 m and its length is about 204 m. All tunnels are anticipated to be constructed using a pressurized face slurry pressure balance (SPB) or earth pressure balance (EPB) TBM with an initial precast concrete segmental tunnel liner. A secondary steel liner may be required for some distance outside each shaft to accommodate potential liquefaction-induced deformations.

PDBCO Tunnel: This tunnel segment will convey flow from the PDBCO Shaft to the Outfall Shaft. Its length will be about 385 m. The internal diameter and construction methods will be the same as for the Effluent Tunnel.

Outfall Tunnel: This tunnel segment will convey flow from the Outfall Shaft to the River Riser. Its length will be about 572 m. The internal diameter and construction methods will be the same as for the Effluent and PDBCO Tunnels.

Riser Shaft: This shaft will convey flow from the tunneled conveyance upward over 14 m to the river bed where it will connect to the Diffuser Manifold. It will include a tapered, curved connection between the 4.2 m diameter tunnel and the 3.8 m I.D vertical riser pipe. The temporary excavation support is anticipated to utilize a combined king pile and sheet pile cofferdam and be of sufficient size to serve as the final TBM receiving shaft with the TBM likely to be abandoned in-place.

Diffuser Manifold: This buried pipeline will distribute the effluent flow to as many as 48 risers and diffuser ports protruding through armor rock erosion protection at the river bed elevation. It will consist of two 120 m long sections with a minimum internal diameter of 2.5 m. Each port will be approximately 600 mm in diameter with flexible duck bill check valves located about 1 m above the river bed to avoid small sand waves expected to move through this area.

Other Components: There are other components to the project such as the connection to the Chorine Contact Tanks (CCTs) and a buried PDBCO conduit; however, they are not discussed in this paper.

DESIGN AND CONSTRUCTION CHALLENGES

Design Challenges

Several aspects of the design for the new outfall system presented challenges including limited hydraulic gradient to drive the effluent into the river under gravity conditions, providing the required post-disaster performance since the project site is susceptible to potential deep liquefaction, and physical and operational constraints on the diffuser location and configuration. Addressing these challenges drove many of the design decisions.

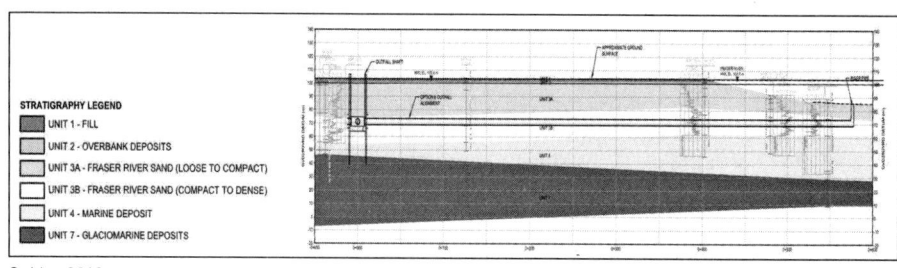

Golder, 2016
Figure 4. Outfall tunnel profile

On-Land Liquefaction-Induced Displacements: The upper 30 m of the Fraser River Sand deposit is comprised of loose to compact sand containing limited amounts of fine grained particles. This deposit has the potential to experience liquefaction-induced vertical and lateral displacements following a seismic event. Ground improvement consisting of vibro-replacement with stone columns to a depth of 30 m and a lateral extent of 25 m. beyond at-grade structure limits has been implemented at part of the Stage V treatment plant upgrades. This improvement is expected to reduce total and differential deformation such that the structures will meet post-disaster performance requirements.

It was judged impractical to perform ground improvement along the entire outfall conveyance route due not only cost, but also access restrictions and surface/building impacts. Also, it was uncertain that the anticipated reduction in vertical and lateral deformation would be adequate for the conveyance system to meet its post-disaster performance requirements. A design decision was made to deepen the tunnels and shafts such that they were in ground where liquefaction-induced displacements were limited and the structures could meet post-disaster performance requirements.

The lower portion of the Fraser River Sand deposit appears older and consists of compact to dense sand with significantly higher SPT blow counts and cone penetration resistance. Based on results of a preliminary liquefaction analysis, the tunnel invert was established at El. 69, as shown in Figure 4, and detailed liquefaction evaluation is underway to further assess the amount of potential deformation at the planned tunnel elevation. This analysis includes of two-dimensional (2D) ground response analysis (free-field) using the 2D finite difference computer code FLAC2D (Ver. 7.0) to assess the liquefaction potential of site soils and the resulting permanent ground deformation along the tunnel alignments. In addition, three-dimensional (3D) soil-structure interaction (SSI) analyses using the finite difference program FLAC3D (Ver. 5.0) are being performed to predicted the response of shaft and tunnel structures to the free-field ground displacements.

Preliminary, conservative results indicate that liquefaction-induced maximum permanent vertical displacements at the tunnel elevation could be 50 mm at the Effluent and PDBCO Shaft locations and 75 mm at the Outfall Shaft location. The maximum liquefaction-induced permanent lateral displacement at the tunnel depth for the on-land shaft locations is approximately 100 mm. The ability of a single pass liner to provide post-disaster performance given this amount of deformation is under further evaluation—a secondary steel liner may be required for some distance outside each shaft due to differential movement between the shafts and tunnels.

In-River Liquefaction-Induced Lateral Spreading: Slopes along the river bank are susceptible to much larger lateral liquefaction induced deformation often referred to as lateral spreading. Preliminary, conservative results indicate that liquefaction-induced maximum permanent displacements at the tunnel elevation could reach 200 mm vertical and 1.0 m lateral under free-field conditions. Lateral displacements at the river bed where the Diffuser Manifold will be located could experience maximum displacements of 0.5 m vertical and 1.8 m lateral. A decision was made to design the River Riser as a rigid structure keyed into deeper soils with internal piles to limit lateral deformations at the tunnel elevation. This also has benefits from a construction stand point as described later in this paper. The Diffuser Manifold will be designed with bendable sections that will allow it to deform without breaking so it can survive the anticipated lateral and vertical liquefaction-induced deformation at the river bed.

Hydraulic Gradient: Because the surface of Annacis Island is only slightly above Fraser River flood levels, there very little available hydraulic gradient (~2.4 m) to drive effluent flow between where it exits the treatment plant at the chlorine contact tanks and the design river flood level in the river. For the Annacis outfall, the treated effluent will primarily discharge into freshwater where momentum is the primary contribution to dilution within the river. Momentum is created by the discharge velocity at the diffuser ports. A goal of the diffuser design is to select small ports to achieve high discharge velocity while staying within available hydraulic head to discharge by gravity. To preserve as much hydraulic head as possible for dilution, head losses in the rest of the effluent conveyance had to be minimized. This is accomplished by using larger diameters for the tunnel, riser, and diffuser manifold, as well as limiting the number of corners and bends along the conveyance alignment. To meet discharge dilution requirements, the current design is optimized to use all available gravity head for Stage V flows and provides for a future effluent pumping station to add the hydraulic head necessary to achieve the required dilution for future expansions in plant capacity.

Diffuser Location and Configuration: Dilution r-equirements for the diffuser system dictated that it be in the deepest water possible, be 240 to 300 m in length with closely spaced diffuser ports, and be outside the dredged shipping channel. As can be seen in Figure 3, this essentially dictated the diffuser be located just outside the limits of the dredged channel and immediately downstream from the existing outfall. The river bed elevation at the diffuser location cannot be any deeper than the adjacent channel or it would quickly fill up with sediment, so this constrains the maximum water depth. To minimize hydraulic gradient losses due to bends in the conveyance alignment, as well as the overall length and cost of the conveyance to the river, a directly alignment from the Outfall Shaft to the River Riser was selected.

Construction Challenges

The construction associated with the new outfall system includes both on-land and in-river construction activities. Most the on-land construction activity is anticipated to take place within the AIWWTP and adjacent property being acquired for the Outfall Shaft and future pump station. The tunnels will pass beneath commercial properties, including crossing under one warehouse structure. The current design avoids construction activities in sensitive near-shore areas, however, construction will involve significant in-river activities related to installation of the riser shaft and the diffuser manifold, both of which will take place within the safety boundary bordering the Fraser River Navigation Channel. In-river activities will also be limited between March 1st and June 15th due to salmon migration in the Fraser River and may be constrained at other times due to fisheries or other environmental concerns.

On-Land Shafts: The current design assumes that temporary excavation support for the three on-land shafts (Outfall Shaft, Effluent Shaft, and PDBCO Shaft) will utilize unreinforced concrete slurry walls; however, final selection and design of the temporary excavation support system may be left for the selected contractor. Preliminary design of Outfall Shaft includes installation of a partial final lining to support three breakouts for launch of the three tunnel drives (see Figure 5). For both the Effluent and PDBCO Shafts, interior reinforcement may not be necessary to facilitate TBM break-in.

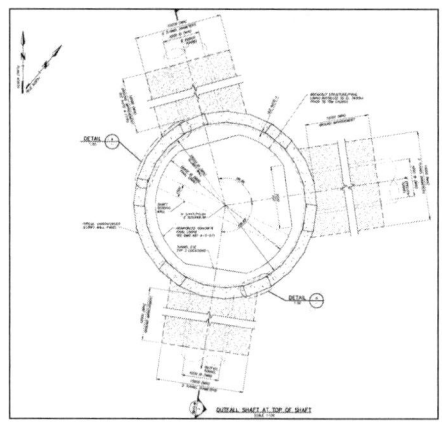

Subsurface conditions for the on-land shafts are anticipated to be similar, as shown on Figure 5. The excavation is anticipated to go through the Fill, Overbank Sediments, and Fraser River Sand deposits and bottom in the Marine Deposit. Preliminary analysis suggests that the shafts will have adequate base stability to be excavated in the dry, possibly with some ground improvement below the base slab. Monitoring will be required to confirm base stability and the contractor will be expected to have a contingency plan for tremie placement of the base slab in the wet.

Construction challenges associated with the shaft construction include maintaining the slurry-bentonite quality and level to prevent sloughing during excavation and achieving panel verticality within the required tolerances.

Tunnels: Although the tunnel drives are relatively short, it is anticipated that tunnel excavation will require a pressurized-face TBM due to the required minimum internal diameter to minimize hydraulic loss. The contractor may select either SPB or EPB TBM technology. The tunnel will use an approximately 254 mm-thick segmental concrete liner one-pass support system.

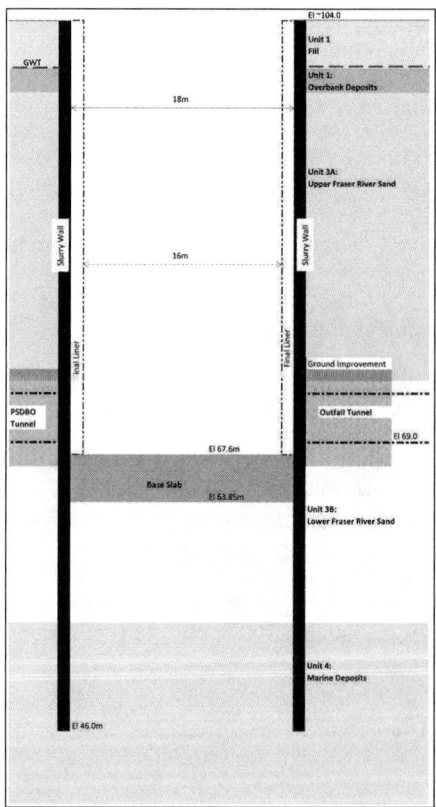

CDM Smith, 2016

Figure 5. Outfall shaft plan view and typical shaft section

The design anticipates that the first TBM drive will be launched from the Outfall Shaft to the Effluent Shaft. The TBM would be removed and re-assembled in the Outfall Shaft to begin the second drive to the PDBCO Shaft where it would be again removed.

The final drive will run from the Outfall Shaft to the Riser Shaft the TBM's steel shell would not be recovered and abandoned within the riser shaft.

Tunnel construction challenges include as multiple sequences of TBM assembly and removal, maintaining the required face pressure and ground conditioners to minimize abrasion and loss of ground in sand that will exhibit flowing behavior if not supported, and about 3 bars of ground water pressure.

River Riser: The riser shaft construction and its connection with the outfall tunnel are the most challenging tasks of this project. The project team has been involved in design of similar marine projects (and reviewed many others) where the tunnel to riser connection was completed in rock or firm ground. We are aware of no such connections completed in flowable sand. Therefore, the riser design assumes that a zone of 'controlled' ground will be completed within which the tunnel/riser connection can be completed.

Several temporary excavation support methods were evaluated for the Riser Shaft. The combined king and sheet pile wall cofferdam was selected because it has been successfully used by contractors to support similar marine cofferdams for bridge piers and foundations and utilizes relatively convention construction techniques. The cofferdam also has the benefit of protecting the interior work space from river currents and sediment as well as isolating construction activity impacts to the river environment. Figure 6 shows a plan and section views of the concept for the River Riser construction.

The basic concept includes installing the cofferdam from above the river surface to below the base of excavation; excavating soil within the cofferdam; installing internal piles extending into the underlying clay to provide permanent lateral resistance to lateral spreading; backfilling the excavation to near the river bed with sequence of a structural concrete based slab, lean mix concrete at the tunnel reception elevation, and another structural concrete slab with the riser pipe installed within the slab; removal of the cofferdam after completion of the riser structure.

The Outfall Tunnel would be completed later by driving the TBM into the lean mix portions of the riser shaft, secondary grouting through the TBM tail and shield to confirm lack of connectivity to the river, grouting of the pressure chamber, removal of the TBM equipment, and making an opening through the shield to complete the connection to the riser.

Diffuser Manifold: Construction of the diffuser will consist of dredging into the river bed, placement of bedding material; installation of pre-assembled, factory coated sections of the steel pipe manifold complete with the steel portion of the riser; backfilling to grade with native river sand material, and placement of armor rock generally as shows on Figure 7. Following installation, temporary blind flanges on the top of the riser will be removed and flexible risers with duck bill check valves installed.

Diffuser construction challenges include coordination with river navigation; dredging/ excavating the for the Diffuser Manifold (potentially in sections using temporary shoring on the land side to minimize the required dredging and backfill volume); and maintaining the grade during installation of the Diffuser Manifold.

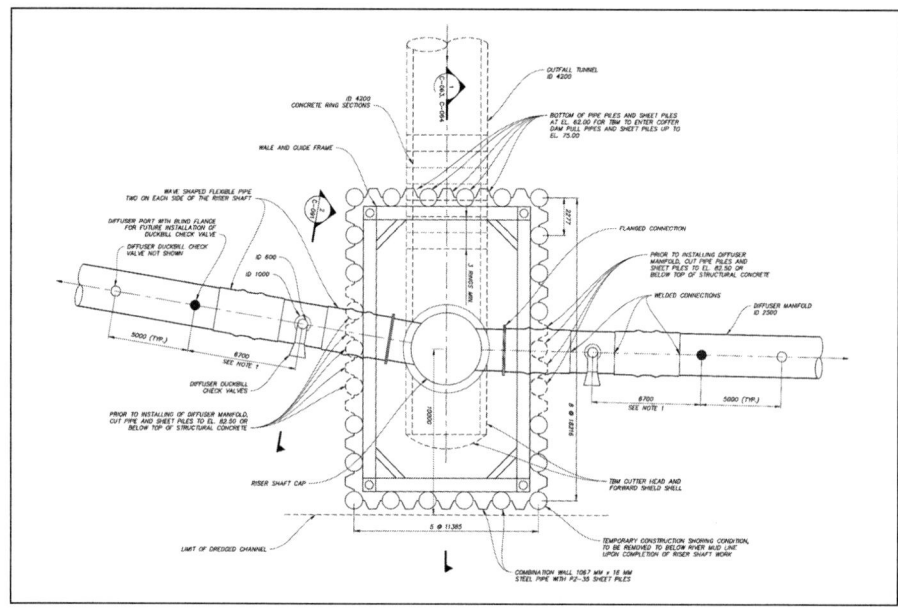

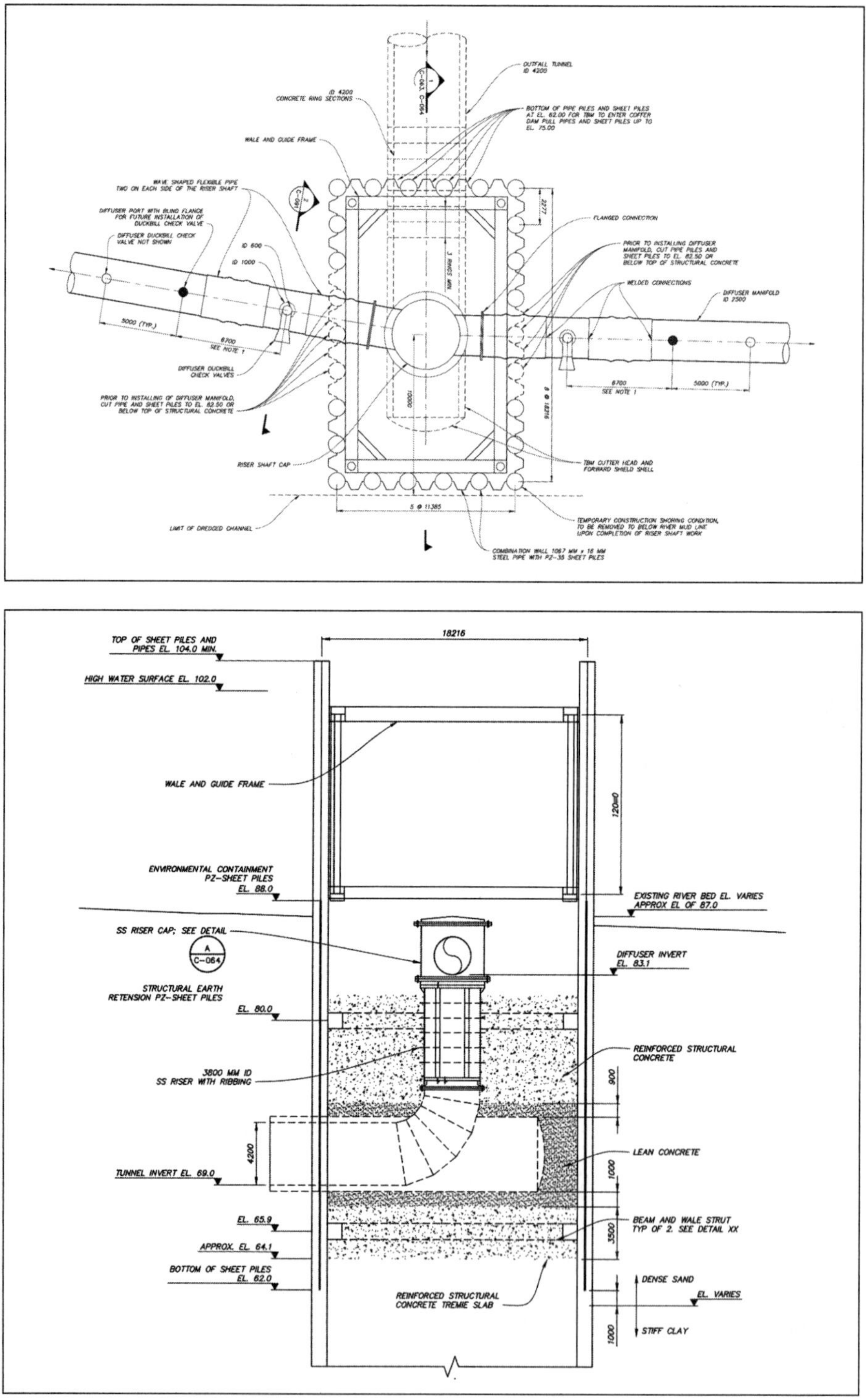

CDM Smith, 2016

Figure 6. River risers plan view and section

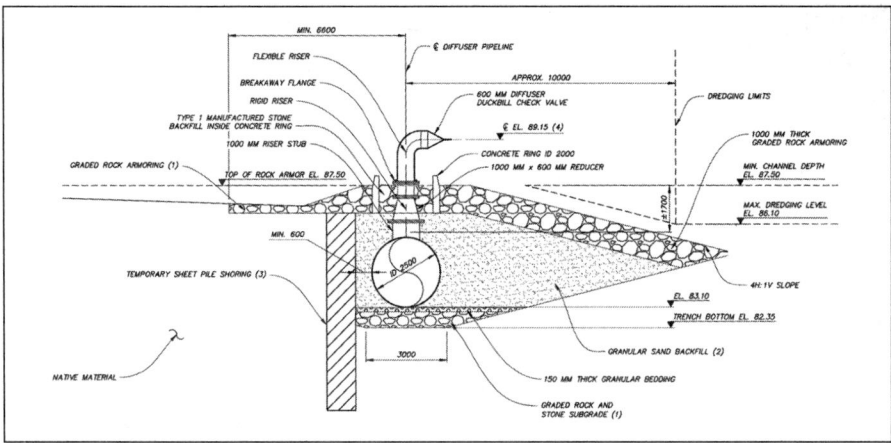

CDM Smith, 2016

Figure 7. Diffuser cross section

SCHEDULE

60% design will be completed in early in 2017 followed by a RFQ from interested contractors in mid-2017. Final design and permitting will be completed around November 2017 followed by issuance of an RFP to qualified contractors. Contract award and start of construction is anticipated by mid-2018.

Delivery of Design, Environmental Statement, Engineering, Construction Management of the UK's New High Speed Railway for the 21st Century

Colin Rawlings ▪ CH2M/High Speed Two (HS2) Ltd
Nita Rabadia ▪ High Speed Two (HS2) Ltd
Mark Howard ▪ High Speed Two (HS2) Ltd
Richard Sturt ▪ Arup
David Soper ▪ Birmingham Centre for Railway Research & Education
Alan Vardy ▪ Dundee Tunnel Research

ABSTRACT

High Speed Two (HS2) is a new $67billion (£56 billion) high speed railway which will form a "spine" of a reshaped rail network providing an engine for growth and connecting eight out of the top ten regions within Great Britain. There are about 64km of twin bore and 8km of twin cell cut-and-cover tunnels. Principles of Systems Engineering (International Council on Systems Engineering (INCOSE)) have been incorporated to drive processes and approaches to underground structure design & construction. In addition, HS2 Ltd established an efficiency challenge programme specifically tasked with generating savings for the project through updating and refining existing standards/ guidance documents in geotechnical engineering and underground construction. This paper describes aspects of the design, engineering & construction management and environmental aspects to ensure high quality and cost effective underground structures for this innovative high speed railway including HS2s approach to tunnel porous portal designs for the various HS2 underground structures.

INTRODUCTION

HS2 Ltd was established in January 2009 to develop proposals for a new high speed railway. CH2M is development partner for the preliminary design and the environmental statement which is the first to use the Infrastructure UK procurement route map. This is helping to boost efficiency and bring down the cost of UK rail infrastructure with early engagement with the industry (four major contractors—benefiting constructability, costing and programme assurance) using best skills & technologies, standardisation of designs, repeatable construction and off-site fabrication. The key design scope was broken down into four lots: civil & structural design services; railway system design services; environmental services; and land referencing services. Technical standards and specifications for all of these design aspects for the high speed railway have been set by the client company HS2 Ltd supported by the development partner.

Any project of this size is always going to have huge challenges. Principles of Systems Engineering have been used to drive processes and approaches to design development. This paper describes the management for the delivery and review of the designs (geotechnical & underground construction) and the environmental statement to ensure a high quality and cost effective new high speed rail network. Key environmental impacts of the project and the measures envisaged to manage & reduce any adverse effects locally (including public consultation) are being incorporated into the design. Other aspects being taken into account include sustainability, carbon footprint

420

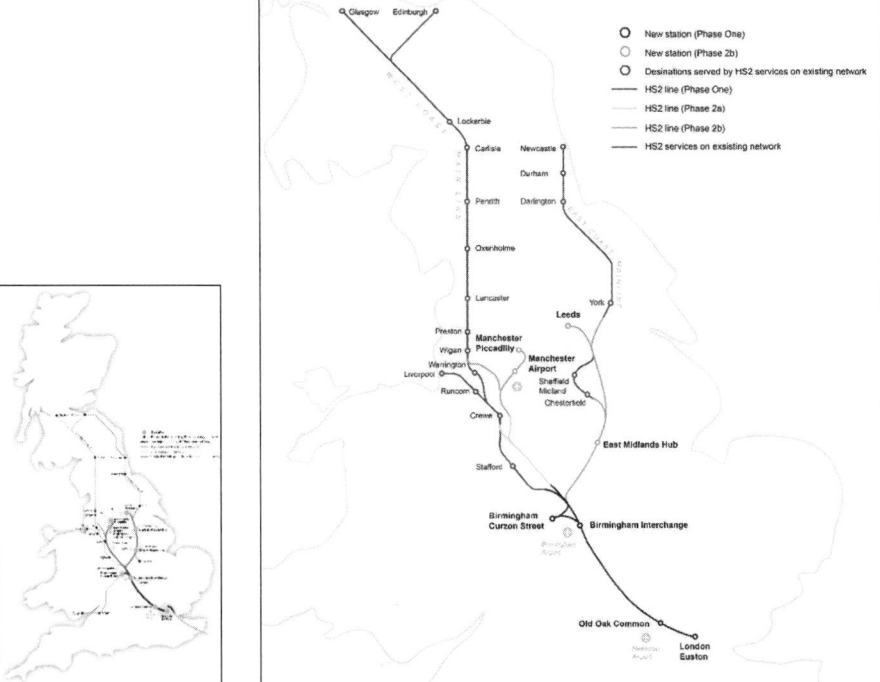

Figure 1. Proposed High Speed Two (HS2) rail network

and climate change. HS2 is also looking into areas of innovation in the train design. Key to achieving the real benefits will also be obtaining excellent interface between HS2 stations, current stations, and other transport modes.

HS2 is currently being developed in two phases: Phase One will run between London and Birmingham and tackle the overcrowding in the West Coast Main Line. Phase Two will do the same for the East Coast and Midland Main Lines Birmingham to Manchester and Leeds. Phase One currently comprises 293 bridge structures, 70 viaducts, four stations, two maintenance depots, six twin bore tunnels and five cut-and-cover tunnels along a route length of about 225km. Phase 2 currently comprises 179 bridge structures, 77 viaducts, four stations, four maintenance depots, seven three bore tunnels and two cut-and-cover tunnels along a route length of about 340km. Phase Two has been subdivided into Phase 2a Birmingham to Crewe (to prioritise this part of the route) and Phase 2b—the remainder. Tenders for the Phase 1 main works civil (design & build) contracts are to be returned by December 2016 and contract award is planned for summer 2017 with construction starting in Autumn 2018.

REQUIREMENTS FOR THE INTERIM PRELIMINARY DESIGN FOR GOVERNMENT APPROVALS (HYBRID BILL STAGE)

There are numerous engineering and technical challenges facing a project of this complexity from its initial conception to ultimate delivery and bringing into service (currently HS2 Phase 1 2026, Phase 2a 2027 & Phase 2b 2033—Figure 1). Of utmost importance at this stage of the project development cycle as described in this paper, is developing the scheme suitable for Government approval. This requires the resolution of a very specific and focused set of challenges, being: the development of

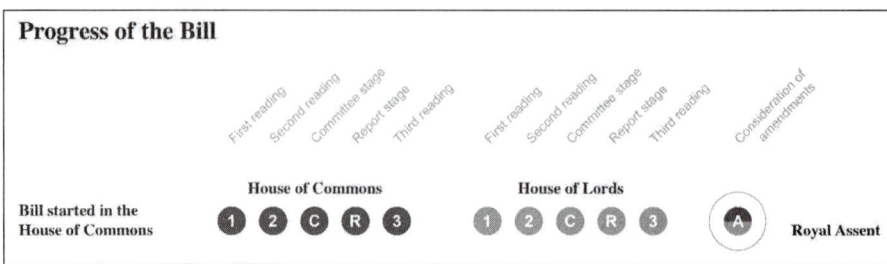

Figure 2. Progress of the hybrid bill through Parliament

engineering detail sufficient to demonstrate with confidence, that subject to a further detailed design process, a robust engineering solution exists; that significant impacts associated with constructing, operating and maintaining the future high speed railway have been identified, and their impacts assessed; the identification of land required for the purpose of construction, operation and future railway maintenance, including all associated activities both within and outside the immediate 'trace' of the railway; and that the cost and programme for implementation and bringing into service of the railway have been fully considered and identified. The stages in the progress of the hybrid Bill design through the Houses of Parliament are shown in Figure 2. The hybrid Bill has now gone through the House of Commons and is expected to go through the House of Lords in early 2017 with Royal Assent following shortly afterwards.

Environmental assessment has been key in route selection, design development, arrangements for construction and operation of the railway, and identifying measures to mitigate the project's environmental impacts. HS2 Ltd's aim has been twofold: to enable the nation to take full advantage of the opportunities and benefits offered by the project and to mitigate the potential adverse environmental impacts of the project as effectively as it reasonably can. The law requires that the Environmental Statement (ES) includes a description of the measures (Figure 3) envisaged to avoid, reduce and, if possible, remedy the significant adverse effects of the project and this has been achieved by avoiding adverse environmental effects, where reasonably practicable, through the design of the project. Where this has not been achieved HS2 Ltd has considered measures to reduce or abate such effects. Where despite efforts to avoid and reduce them, significant adverse environmental effects are predicted to occur, HS2 Ltd has proposed repair and compensation measures. In some cases, such measures may actually lead in the longer term to an overall improvement in the environment. The mitigation proposed depends on an assessment of the nature and severity of the adverse environmental effect and of the effectiveness and value for money of the mitigation measures under consideration.

Mitigation measures applied in the design of Phase 1 include developing the route to avoid adverse environmental effects (especially on residential properties, community facilities, public open spaces, businesses, farm buildings, sites of ecological and/or heritage importance and the wider landscape) where appropriate and reasonably practicable using tunnels and cuttings to reduce noise effects and provide visual screening for local communities. Other mitigation measures include: using earth mounding and planting to screen views and integrate the project into the local landscape; provide noise barriers (fence barriers or earth mounds) to reduce effects on communities and provide links across the route to maintain access for roads, public rights of way and properties and allow safe passage of wildlife. New habitats and other features of ecological value will be created to compensate for unavoidable habitat losses. A balance (as far as reasonably practicable) of cut and fill volumes so as to reduce the number

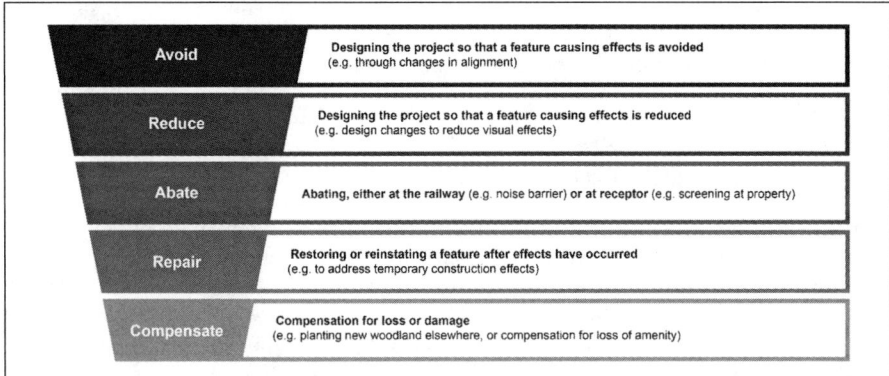

Figure 3. Approach to environmental mitigation

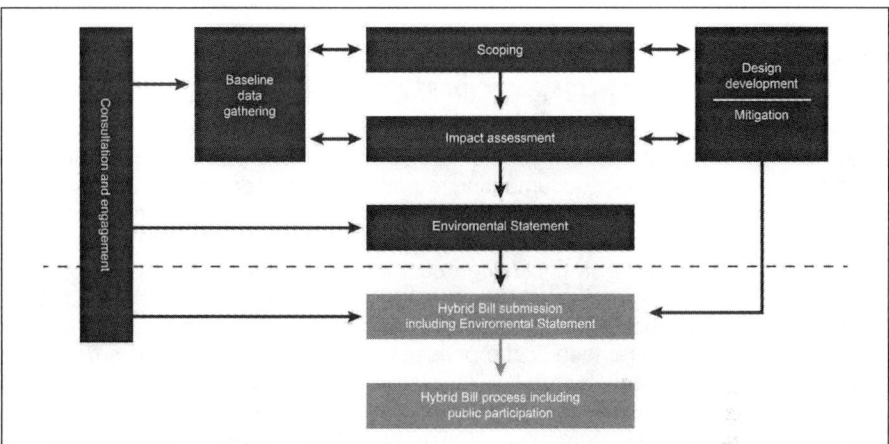

Figure 4. Application of Environmental Minimum Requirements (EMR) during design development

of heavy vehicles on local roads will be achieved. Flood plains/flood storage areas will be avoided or impacts reduced and balancing ponds to control surface water run-off will be provided.

In order to ensure that the environmental effects (Figure 4) of the Phase 1 project will not significantly exceed those assessed in the ES a set of controls known as Environmental Minimum Requirements (EMRs) will require HS2 Ltd and its design & build contractors to adopt measures to reduce the adverse environmental effects reported in the ES, provided that such measures are reasonably practicable and do not add unreasonable cost or delay to the construction or operation of this project. The EMRs include: general principles by which the Secretary of State commits that the environmental effects reported in the ES are not exceeded through application of the environmental mitigation assessed in the ES; the Code of Construction Practice & Environmental Memorandum (which is the framework for the HS2 Ltd and its contractors and stakeholders, such as the Environment Agency, Natural England and English Heritage work together to ensure that the design and construction of the Phase 1 is carried out with due regard for environmental considerations); a Planning Memorandum (which will set out an agreement between the Government and the local planning authorities relating to the processing of detailed planning approvals under the provisions of the Bill, including the design and appearance of stations, bridges, viaducts, ventilation shaft head houses, tunnel portals, noise barriers and earthworks); a Heritage Memorandum (which will set out a commitment to limit the impact on the historic environment) addressing the elements of the design and construction works that have a direct impact on heritage assets; and Undertakings and Assurances given during the passage of the Bill through Parliament.

If changes are required to the project due to circumstances that were not forseeable at the time of the ES, these will be assessed to ensure that they do not have significant adverse environmental effects greater than those identified in the ES. If such a change is found to have such effects, it will be subject to a separate development consent process and to environmental impact assessment in its own right. Changes which have occurred to the Phase 1 underground structures during the passage of the hybrid Bill through the House of Commons include: and extension to the Chiltern Tunnel (length now 15.8km) by 2.5km with an additional (intervention) shaft through the Chilterns Area of Outstanding Natural Beauty and a small extension of the Burton Green cut-and-Cover Tunnel.

PRINCIPLES OF SYSTEMS ENGINEERING AND APPLICATION ON HS2

Systems Engineering is an interdisciplinary approach to enable realization of successful systems. It focuses on; defining customer needs and the required functionality early on in the development cycle; documenting requirements; and design production & system validation—always considering the complete problem—(integration). At HS2 the principles behind the requirements management have been to ensure there are fully articulated user requirements (including operational and maintenance requirements) upfront to help minimise whole life cost and to make the right trade off decisions (quality, time, passenger requirements). Systems Engineering encompasses four major process groups which are: technical processes (stakeholder requirements definition, requirements analysis, architectural/civil/systems design, implementation, integration, verification, transition, validation, operation, maintenance, disposal & cross cutting technical methods); project processes (project planning, project assessment & control, decision management, risk management, configuration management, information management & measurement); agreement processes (acquisition & supply); and organizational project-enabling processes (life cycle model management, infrastructure management, project portfolio management, human resource management & quality management). An example Systems Engineering model is given in Figure 5.

HS2 has applied Systems Engineering in the development of the Interim Preliminary Design to; capture requirements and agree what is needed to gain acceptance; simplify the interfaces through integration and packaging of work; use simulation/testing to identify the options which deliver the outcomes before construction; introduce check points and obtain a better quality product; understand and minimize the impact of changes (effective change control & governance); and tailor Systems Engineering processes and tools to fit. This has been achieved by the use of the verification & validation process at various stages of design development including interdisciplinary design reviews and checks with all relevant parties. HS2 is working together with the Transportation Working Group (TWG) of the International Council on Systems Engineering (INCOSE) whose mission statement is to: promote the development and tailored application of Systems Engineering best practices to ground transportation systems with emphasis on roadways, rail, bus and metro. For HS2 the systems engineering approach has been applied to the civil, track, mechanical, electrical, train set and power requirements. The HS2 Sponsors (Department for Transport—DfT)

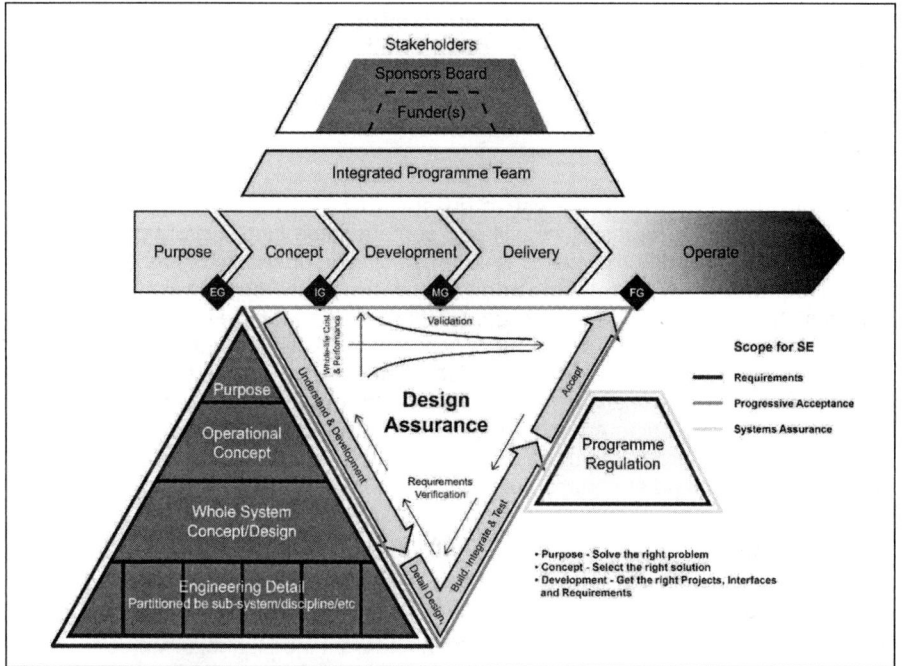

Figure 5. Example of systems engineering model approach on large multifunctional project

Requirements from the Development Agreement (between HS2 & DfT) have been captured in the HS2 Programme Requirements to align with the key HS2 project procurement route—civils followed by stations, rail systems and rolling stock to ensure the correct design decisions are taken to inform design development and minimise rework.

GEOLOGY, TOPOGRAPHY, GROUND & GROUNDWATER CONDITIONS ALONG THE HS2 ROUTE

The geology and topography of the proposed HS2 Phase One route varies since the solid geology along the route passes through a variety of geological formations comprising over-consolidated stiff clays, Chalk and other limestones, non-durable mudstones, sandstones and weak rocks. The youngest of these deposits occur in the south of the route in the Eocene London Basin. Northwards, the route passes through progressively older outcrops of basement strata from below the London Basin, passing through the Chalk outcrops of the Chilterns and the Cretaceous and Jurassic sands and clays in the vale past Aylesbury and beyond. Further north towards Birmingham, the Mercia Mudstone of the Knowle Basin, the Triassic older rocks of the Sherwood Sandstone and briefly Carboniferous rocks are encountered.

For Phase Two beyond Birmingham and towards Manchester the route crosses mostly the Cheshire Basin comprising Permo-Triassic rocks of mudstone, siltstone and sandstone of the Mercia Mudstone and Sherwood Sandstone Groups. Halite formations are also present locally. Similarly beyond Birmingham and towards Leeds the route crosses Triassic Mercia Mudstones, the Sherwood Sandstone Group, the Carboniferous Coal Measures (much of which has been mined from surface to great depth), Millstone Grit and Permian sandstones/ mudstones/dolomitic limestones. The structural geology is generally simple with broad basin structures, shallow synclines/ anticlines and few major faults. This solid geology is sporadically overlain by superficial deposits mostly comprising localised alluvial and river terrace sands and gravels, with more extensive glacial deposits of diamicton, fluvioglacial and glaciolacustrine materials further northwards. To the north of Manchester area some extensive peat deposits are also present.

Particular geohazards include existing quasi-stable slopes, compressible deposits, groundwater, soluble deposits, landfill and contaminated land & groundwater, abandoned mine workings (both shallow & deep), backfilled opencast mines and salt caverns. Groundwater levels are expected to be at the ground surface in the superficial deposits and much of the bedrock except within the Chalk where natural groundwater levels have been lowered by abstraction. At some locations underground construction is planned to be carried out within the saturated zone through groundwater source protection zones e.g., Chalk aquifers. A detailed programme of geotechnical investigations is currently being carried out along the HS2 Phase One route to determine more precisely the ground and groundwater conditions at the location of each of the proposed underground structures summarised in Table 1.

DETAILS OF UNDERGROUND STRUCTURES & GROUND CONDITIONS ALONG THE PHASE ONE & TWO HS2 ROUTE

Table 1 summarises the current status of the various underground structures along the HS2 route. All tunnels are designed as twin bored/mined tunnels/twin cell cut-and-cover tunnels.

CIVIL CONSTRUCTION CONTRACTS

Phase One has initially been subdivided into seven work packages (Figure 6) with bidders invited to tender for up to four packages and limited to two package awards. The work packages are as follows: Lot S1—Euston Tunnels & Approaches; Lot S2— Northolt Tunnel; Lot C1—Chiltern Tunnel and Colne Valley Viaduct; Lot C2—North Portal Chiltern Tunnel to Brackley; Lot C3—Brackley to Long Itchington Wood cut-and-cover Tunnel South Portal; Lot N1—Long Itchington Wood cut-and-cover Tunnel to the Delta Junction/Birmingham Spur; and Lot N2—Delta Junction to West Coast Main Line Tie-in. Following the invitation to tender, each tenderer will go through an evaluation process, with winning bidders planned to be announced in summer 2017, after Royal Assent. Later in the process successful firms will also be able to bid for further work—Phase 2a from Birmingham to Crewe.

A further four contract packages will be let for the four stations in Phase One, namely, Euston Lot S3, Old Oak Common Lot S4, Birmingham Interchange Lot N3 and Birmingham Curzon Street Lot N4. Construction for Phase 1 is expected to start in Autumn 2018.

TECHNICAL STANDARDS AND SPECIFICATIONS TO IMPROVE DELIVERY

HS2 Ltd's efficiency challenge programme was set up and specifically tasked with generating savings for the project through updating and refining existing standards/ guidance documents. Industry experts were involved from designers, contractors and professional organisations including the Institution of Civil Engineers, the Institution of Mechanical Engineers, the Royal Institute of British Architects and the Railway Industry Association. HS2 Ltd appointed the British Standards Institution (BSI) to undertake research into standardisation in the areas of civil engineering, buildings and railway

Table 1. Summary of the underground structures along the HS2 route

Underground Structure	Type	Approximate Length (m)	Expected Ground Conditions	Porous Portal (Hood)
PHASE 1				
Euston	TBM*	7,290	London Clay	No
Old Oak Common (OOC) Station	Station Box	9,30	London Clay	No
OOC Tunnel	Mined	320	London Clay	No
Victoria Road Crossover Box	Crossover Box	240	London Clay/Lambeth Group	No
Northolt	TBM*	13,420	As above	Yes
Chiltern	TBM*	15,840	Chalk	Yes
Wendover	C & C† Tunnel	1,420§	Chalk	Yes
Greatworth	C & C† Tunnel	2,100§	Glacial Till/ Oolite	Yes
Chipping Warden	C & C† Tunnel	2,490§	Marlstone/ Dyrham	Yes
Long Itchington Wood**	TBM*/SCL‡/C & C†	2,030	Mercia Mudstone	Yes
Burton Green	C & C† Tunnel	760§	Tile Hill Mudstone	Yes
Bromford	TBM*	2,810	Mercia Mudstone	Yes
PHASE 2				
Leeds				
Red Hill	C & C† Tunnel	200	Mercia Mudstone	—
Strelley	C & C† Tunnel	800	Glacial, Sherwood Sandstone	Yes
Hoyland	TBM*/ SCL‡	2,200	Coal Measures	Yes
Ardsley	TBM*/ SCL‡	1,200	Coal Measures	Yes
Woodlesford	TBM*/ SCL‡	1,100	—	—
Manchester				
Whitmore** (Phase 2a)	TBM*/ SCL‡/C & C†	1,000	Kidderminster	Yes
Madeley (Phase 2a)	TBM*/ SCL‡	700	Mercia Mudstone	Yes
Crewe	TBM*	3,600	Mercia Mudstone	Yes
Manchester	TBM*	12,800	Mercia Mudstone	Yes

*Tunnel Boring Machine; † Cut-and-Cover Tunnel; ‡ Sprayed Concrete Lining Techniques; § Phase 1 Cut-and-Cover includes porous portal length; ** Long Itchington Wood and Whitmore Tunnels are part cut-and-cover and part bored/ mined.

systems. The priority work identified by BSI included the: development and publication of an HS2 design specification covering railway systems, civil engineering and buildings; revision of BS8002: on earth-retaining structures, revised 2015, BS8004: on foundations, revised 2015 and BS8081: on ground anchorages, revised 2015 all to align with Eurocodes but to keep valuable non-contradictory complementary information; updating the Construction Industry Research & Information Association (CIRIA) document C580: on embedded retaining walls to align with Eurocodes and to include weak rocks; now C760 Guidance on Embedded Retaining Wall Design; and the production of new documents (BSI Publicly Available Specifications (PAS)) where standards do not currently exist and to comply with Eurocodes. These new standards include: BSI PAS 8810: April 2016—Tunnel Design—Design of Concrete Tunnel Linings— Code of Practice; BSI PAS 8812: January 2016—Temporary Works—Application of European Standards in Design—Guide; BSI PAS 8811: January 2017—Temporary Works—Major Infrastructure Client Procedures—Code of Practice; BSI PAS 8820: April 2016—Construction Materials—Alkali-activated Cementitious Material and Concrete—Specification. In addition to the impact of these standards upon efficiency, the new standards/documents provide sustainability and innovation in line with the UK Government's Construction 2025 Strategy.

DESIGN

General

The main factors governing the design of the HS2 underground structures can be found in Rawlings et al. (RETC 2015). This RETC 2015 paper described briefly the aerodynamic requirements of the underground structures and the following paragraphs cover the aerodynamic work in more detail, which was carried out by Arup, the Birmingham Centre for Railway Research & Education and Dundee Tunnel Research for HS2, in the preliminary design of the tunnel porous portals to achieve efficiencies in the HS2 tunnel preliminary design.

AERODYNAMIC DESIGN OF UNDEGROUND STRUCTURES

Production of Micro-Pressure Waves at Tunnel Portals

When a train enters or exits a tunnel, pressure waves are generated inside the tunnel. These waves propagate along the tunnel and are reflected at the portals. A micro-pressure wave (MPW) is a pulse of air pressure that is emitted from tunnel portals into the environment whenever these reflections occur. In general the MPWs are at frequencies below the audible range and of such small amplitude that they are rarely noticed. However, MPWs can take the form of so-called "sonic booms" for trains entering tunnels at high speed and can cause environmental impacts near the tunnel exit

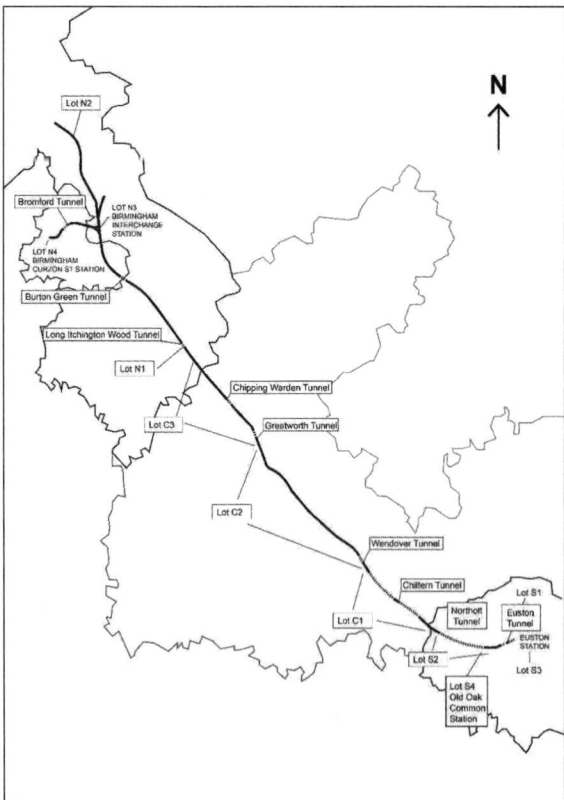

Figure 6. Geographical area subdivision of HS2 Phase 1 work packages

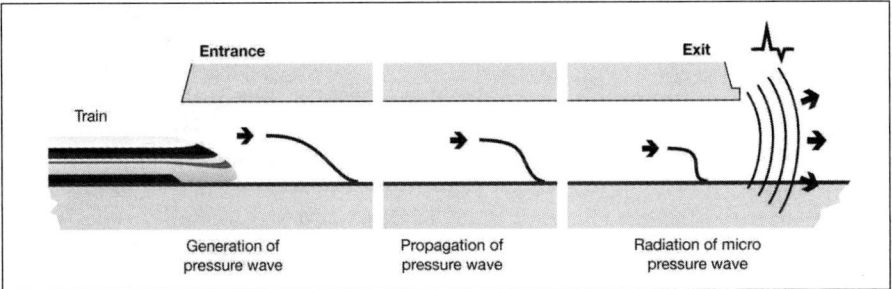

Figure 7. Micro-pressure wave generation

(Vardy 2008). The generation and transmission of MPWs take place in four phases (Figure 7) as follows: generation of a pressure wave inside the tunnel, typically caused by the nose of the train entering the tunnel; as the pressure wave propagates down the tunnel at the speed of sound, the gradient of the pressure wave may increase ("steepening") in long tunnels; when the wave reflects from the tunnel exit, a small proportion of the energy is radiated into the surrounding environment as a micro-pressure wave (the steeper the pressure wave, the greater the amplitude of the MPW); and the MPW is transmitted through the air outside the tunnel, reducing in amplitude with distance from the tunnel portal.

In long tunnels, the sound can occur before the train itself arrives at the tunnel exit because the speed of propagation of the pressure wave (speed of sound) is much greater than the speed of the train. For MPWs it is the pressure gradient that is significant because the amplitude of the MPW emitted outside the tunnel is approximately proportional to the gradient of the pressure wave inside the tunnel as it approaches the tunnel exit. The pressure gradient at the tunnel exit depends on the pressure gradient caused by entry of the train, and on any steepening or attenuation that occurs during propagation. Since the pressure gradient at the tunnel entrance varies with the train speed cubed, small increases of line speed can result in large increases of the amplitude of emitted MPWs.

Mitigation is commonly achieved by adding special entrance regions containing openings (porous portals) to delay the build-up of the pressure waves on train-entry, reducing their gradients and therefore leading to reduced amplitudes of MPWs emitted from the tunnel exit. The extent of mitigation required (the length of the porous portals) depends on several factors which include: line speed (at least a speed-cubed relationship); tunnel length (more MPW steepening occurs in longer tunnels); blockage ratio (i.e., cross-sectional area of train divided by the free cross-sectional area of the tunnel—the wave steepening effect increases with pressure amplitude); track type (ballast resists or reduces the wave steepening effect while slab-track does not); and proximity of receptors to the exit portal. "Receptors" being locations such as residential buildings where the negative effects of MPWs must be prevented.

Preliminary Design of HS2 Tunnel Entrance Porous Portals (Hoods)

The preliminary design of the HS2 porous portals was carried out using numerical modelling (software Thermo Tun—2014 & Arup TunX) which was validated by scale model testing (TRAIN—TRansient Aerodynamic INvestigation—Birmingham Centre for Railway Research & Education) to produce specific design parameters for each tunnel. The HS2 tunnels have relatively high blockage ratios as part of a cost-effective tunnel design, which further increases the need for mitigation of micro-pressure

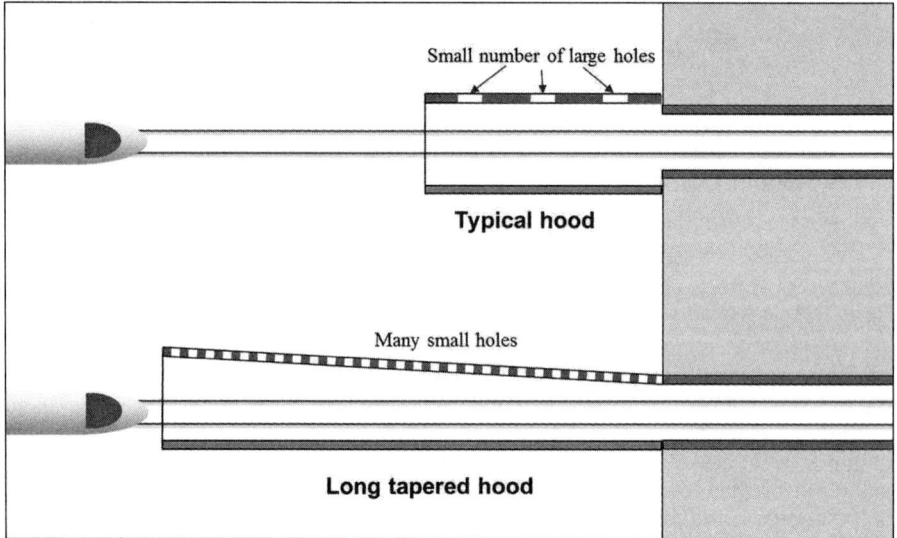

Figure 8. Conceptual illustration of differences between porous portal (hood) types (not to scale)

waves. All HS2 tunnels with an entry speed greater than about 200km/h (including shorter tunnels) have porous portals as part of an integrated aerodynamic design that addresses both aural comfort and micro-pressure wave issues. The maximum speed in the HS2 tunnels will be 360km/h. The HS2 generic porous portals' designs differ from typical porous portals (hoods) in several respects: (a) increased length, necessary because of the higher line speeds; (b) the cross-sectional area is tapered to match the area of the main tunnel, eliminating the typical design's step-change of area at the junction to the main tunnel; and (c) instead of a small number of large openings, the HS2 design has a large number of openings distributed continuously along the length of the porous portals (hood). These differences are illustrated schematically in Figure 8.

It is understood that no long tapered porous portals (hoods) have been constructed to date. However, it is considered that this design will have an important benefit in that the step change in area at the junction of the tunnel eye/start of the porous portal, which can be a strong potential initiator of MPWs, is avoided. The HS2 generic preliminary design can be varied in respect of the porous portal length and there is an optimum size and distribution of openings which minimise the gradient of the pressure wave caused by train entry. One of the aims of the scale model testing and associated analysis work was to identify this optimum to achieve efficiency in design.

Micro-Pressure Wave Acceptability Criteria

In order to design the tunnel porous portals for HS2, the acceptability criteria for micro-pressure waves (MPWs) arriving at receptors near the tunnel portals needed to be established. The key properties of a MPW that determine its audibility and other potential impacts are its amplitude (maximum pressure) and frequency content, measured at the location of the potential receptor. There are no internationally-agreed criteria. National Criteria have been developed in Japan, Germany and elsewhere, but so far none have been prescribed for use in the UK. In view of this the principle of having a two-part criterion (one to prevent loud noises near the tunnel entrance, and another that addresses potential impacts on residents) has been adopted for HS2.

The design target applied at the receptor locations has been determined with reference to measured data from High Speed One (HS1—the operating Channel Tunnel Rail Link), whose tunnels do not generate any audible or noticeable MPWs and has been taken to define a satisfactory condition.

Overall Design Methodology

The methodology for design of the HS2 tunnel porous portals (hoods) is shown in Figure 9. The aim of the methodology is to ensure that the MPWs experienced at receptors near HS2 tunnel portals will be no greater in amplitude than those deemed to be satisfactory arising from HS1 tunnels, taking into consideration the distance of receptors from each Phase 1 individual tunnel portal. From the provisional MPW amplitude limit at the receptors, the corresponding targets for pressure gradient inside the tunnels near the exit and near the entrance were calculated using methods set out by Vardy (2008) combined with one-dimensional analysis by specialist software.

The performance of the HS2 generic porous portal (hood) design was validated using scale model testing. The scale model test results were also used as a means of validating one-dimensional analysis of the porous portals (hoods) with ThermoTun (2014) and Arup's in-house program TunX. These programs were then used to analyse full-scale tunnel porous portals (hoods) of different lengths, optimise the design parameters such as opening size, and predict the maximum pressure gradient downstream of the entrance porous portal (hood). The porous portal length for each tunnel was then obtained by matching these predicted pressure gradients to the target pressure gradient for that tunnel entrance.

Scale Model Test Facility—Birmingham Centre for Railway Research & Education

Testing was carried out at the TRAIN (TRansient Aerodynamic INvestigation) rig operated by Birmingham Centre for Railway Research and Education, shown in Figures 10 & 11. This is a scale model facility (typically 1/25 scale) in which model trains can be fired down a 150m long track by a catapult system, with a maximum speed of 60–80m/s depending on the mass of the train model. Further description of the rig is provided by Baker et al, 2001. The TRAIN rig had a tunnel already installed (see

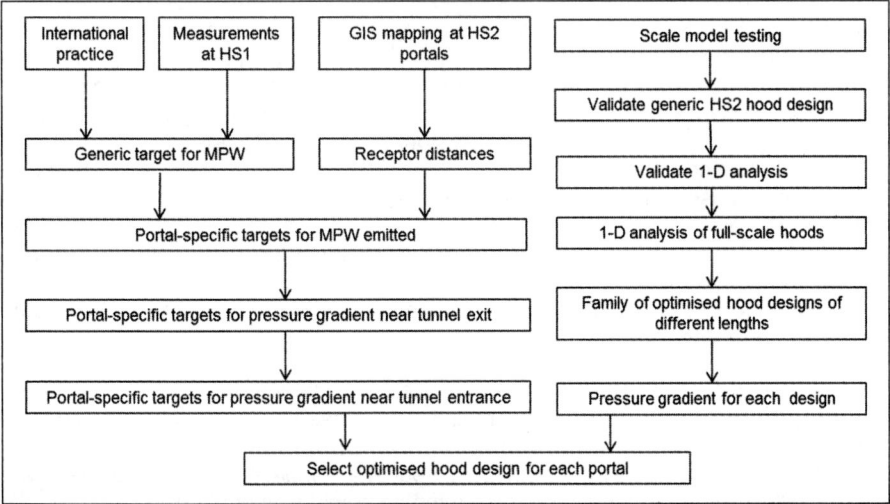

Figure 9. HS2 tunnel porous portal (hood) design methodology

Figure 10. TRAIN rig: (left) view toward end from which trains are fired; (right) tunnel

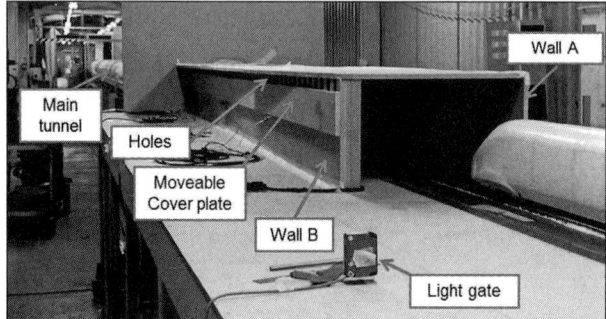

Figure 11. Scale model of porous portal (hood)

right-hand photo in Figure 10), with a cross-sectional area of 0.0736m^2 and a length of 24m (46m^2 and 600m full scale equivalent) respectively.

Scale Model Test Objectives and Setup

The objectives of scale model testing (Figures 10, 11 & 12) were to: validate the principles of the HS2 generic porous portal (hood) design, i.e., to confirm that the nose-entry pressure wave in HS2 tunnels will approximate the optimum ramp-shaped wave form without large peaks in the gradient; provide data from which the one-dimensional analysis can be validated, for a porous portal (hood) condition very similar to the actual HS2 design; and provide insight into the optimum size and distribution of openings, i.e., the size and distribution that minimises the pressure gradient. The train model was a 1/25th scale model of the Alstom AGV, the HS2 reference train. For all the tests presented here, the speed was 60m/s, the maximum practicable speed for the train model used. High-frequency piezoresistive miniaturised differential pressure transducers were mounted flush with the internal faces of both sides of the porous portal (hood) and the tunnel at full-scale equivalent distances of 25m, 50m, 75m, 100m, 125m and 200m from the entrance portal. A selection of the configurations tested is shown in Table 2. In all cases the length of the model porous portal (hood) was 100m (full scale equivalent).

The cross-sectional areas of the porous portals (hoods) and the total area of openings are presented as fractions of the main tunnel cross-sectional area, referred to as A_0: these fractions are then carried across to the full scale design. The openings are uniformly distributed along the length of the hood, i.e., for each test the cover plate was fixed parallel with the ground so that the opening height was uniform.

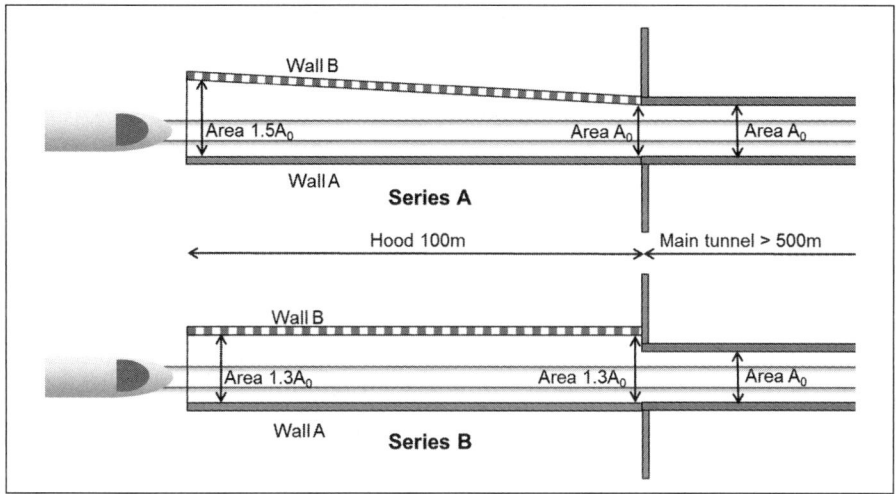

Figure 12. Scale model porous portal (hood): tapered and straight configurations in plan

Table 2. Scale model test configurations (A_0 means the cross-sectional area of the main tunnel)

	Porous Portal (Hood) Configuration	Area at Portal Entrance	Area at Junction to Main Tunnel	Total Area of Openings	Notes
Series A-0[*]	Tapered	$1.5A_0$	$1.0A_0$	$0.27A_0$	Openings smaller than optimum
Series A-1	Tapered	$1.5A_0$	$1.0A_0$	$0.41A_0$	Estimate of optimum opening size based on preliminary analysis
Series A-2	Tapered	$1.5A_0$	$1.0A_0$	$0.82A_0$	Openings larger than optimum
Series B	Parallel	$1.3A_0$	$1.3A_0$	$0.41A_0$	Effect of step-change of cross-sectional area

[*] Results of Series A-0 affected by vibration of scale model porous portal (hood)

Scale Model Test Results

The test results are reported in detail in Sturt et al. 2015.

One-Dimensional Numerical Analysis of the Scale Model Test Results and Comparison of Results

One-dimensional numerical analyses of all the tested conditions were performed, using the software ThermoTun (2014), which has been extensively validated by comparisons with full-scale and model-scale experiments, and also using Arup in-house software TunX. The purpose of this aspect of the study was to validate the use of the programs for the design of the porous portals (hoods) of the configurations tested here. Both programs gave acceptable agreement with experiments. A sample of the TunX analysis results is compared with scale model test results in Figure 13 and in Table 3. In the figure, the thinner lines are the analysis results while the thick continuous lines & the broken lines are test results. Pressures (above) and pressure gradients (below) are shown 200m (full scale equivalent) from the entrance portal. Results are given for Series A-1 (left), Series A-2 (middle) and Series B (right).

From the Series B results, it may be seen that the accuracy of the one-dimensional simulation reduces at the start of the pressure wave (initial nose entry) and during the final peak gradient (when the nose passes the step-change of cross-sectional area). At both these times, the governing phenomena are more three-dimensional in

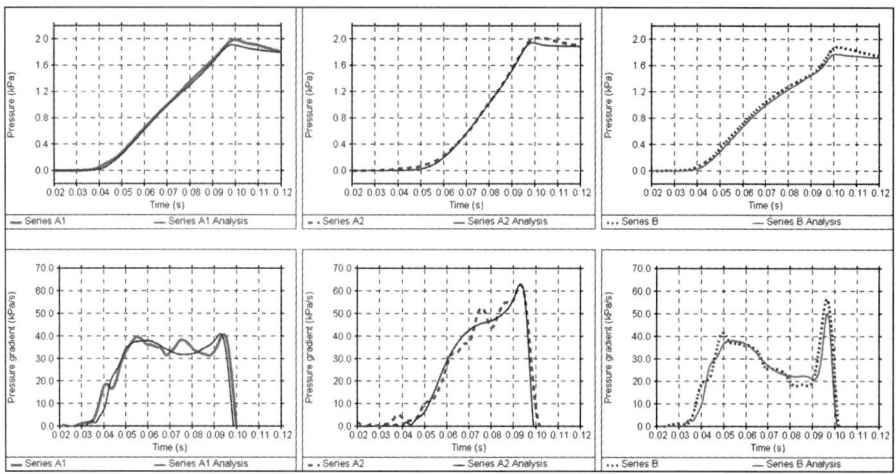

Figure 13. Results from analysis compared with scale model test

Table 3. Pressure gradients from scale model test and analysis

	Porous Portal (Hood) Configuration	Area at Entrance Portal	Area at Junction to Main Tunnel	Total Area of Openings	Maximum Pressure Gradient (kPa/s)	
					Scale Model	Analysis
Series A-0	Tapered	$1.5A_0$	$1.0A_0$	$0.27A_0$	(*)	45
Series A-1	Tapered	$1.5A_0$	$1.0A_0$	$0.41A_0$	41	41
Series A-2	Tapered	$1.5A_0$	$1.0A_0$	$0.82A_0$	63	63
Series B	Parallel	$1.3A_0$	$1.3A_0$	$0.41A_0$	56	50

* Results of Series A0 affected by vibration of scale model hood

nature. One-dimensional simulation methods can be applied with higher confidence to designs that successfully eliminate the influence of three-dimensional phenomena on the pressure gradient, such as Series A-1.

APPLICATION TO DESIGN OF HS2 POROUS PORTALS

The one-dimensional analysis methods illustrated have been used in the preliminary design of a group of full-scale porous portals (hoods) for the HS2 tunnels detailed in Table 1. The lengths of the porous portals vary but, in all cases, the cross-section is rectangular with area tapering from $1.5A_0$ at its entrance to A_0 at the junction with the main tunnel (where A_0 is the cross-sectional area of the main tunnel). The tapering area is achieved in the same manner as the Series A scale models tests (Figures 11 & 12), by setting one wall at an angle to the track.

The porous portal length for each tunnel was selected according to the process illustrated in Figure 9, and depends upon the line speed, the tunnel cross-sectional area, the tunnel length and the proximity of receptors to the exit portal. In order to satisfy the provisional targets for MPW amplitude at the receptor locations, each tunnel was assigned a maximum allowable predicted pressure gradient inside the tunnel just upstream of the exit. These were in the range 6–15kPa/s, with the higher values applying to tunnels with greater distances to the nearest receptor. These pressure gradients are less than half of the 40kPa/s suggested by Hieke et al (2009) as a threshold above which the resulting MPWs may have strong audible components. Design target pressure gradients near the entrance of each tunnel were then calculated such that, after

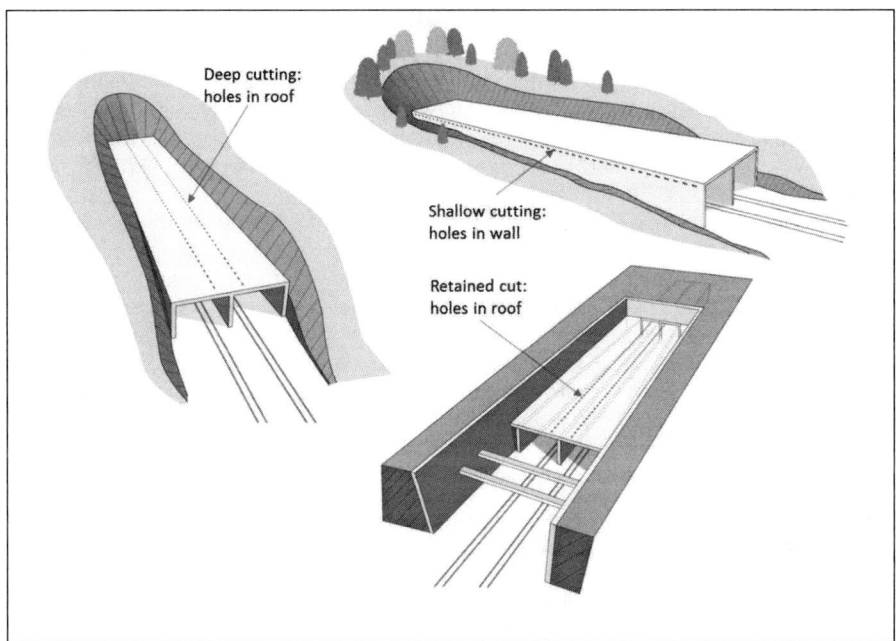

Figure 14. Sketches of HS2 porous portal (hoods) (not to scale)

wave steepening during propagation along the tunnel, the pressure gradient at the exit would not exceed the allowable value for that tunnel. As a conservative assumption for preliminary design purposes, only inertial steepening was assumed without any reduction associated with non-uniformities inside the tunnel such as cross passage. The porous portal length for each tunnel was then chosen so that the design target pressure gradient near the entrance of that particular tunnel (typically 3–8kPa/s) would not be exceeded. The 100m long hood used in this paper for reference purposes is a typical design outcome.

The openings for individual porous portals will be formed either in the wall (as in the scale model tests) or in the roof, according to the topography (see Figure 14).

The openings are of uniform height and their distribution along the porous portal (hood) is uniform. In the Series A scale model tests presented previously, the optimum total area of the openings is about 40% of the main tunnel area. Using one-dimensional analysis, this has been found to be appropriate irrespective of porous portal (hood) length. As a consequence, for the chosen uniformly distributed openings, the required height of the opening is inversely proportional to the hood length. HS2 is designed to be capable of two-way running on each track so porous portals are provided for both tracks at both ends of each tunnel (this is currently being reviewed by HS2). The two porous portals (hoods) are formed within a single structure with a non-porous central dividing wall as indicated in Figure 14. The entrance porous portal for the non-routine direction will serve as an exit porous portal for the routine direction. When a pressure wave reaches the tunnel exit, there is the possibility for its attenuation by the porous portal at the exit portal before emission of the micro-pressure wave into the environment. As a conservative assumption, however, they have been disregarded in the calculations of porous portal length for the HS2 tunnel entrances.

CONCLUSIONS

The High Speed Two (HS2) railway is currently one of the most substantial infrastructure projects in Europe. Despite other large and significant UK projects taking years to develop from concept to consent, HS2 by contrast has been developed and gained widespread support in a remarkably short timeframe. The development of the project has followed the environmental requirements and procedures as described in this paper for the hybrid Bill submission to the Houses of Parliament. For such a large project robust organisational methods and management procedures are required and a Systems Engineering approach has been adopted by HS2 including application to the economic design for tunnel porous portals. In addition CH2M is working in a fully integrated co-located client management team providing contract management, engineering management, project specification, project controls support and technical interface management. HS2 has also sponsored the updating of geotechnical and underground structure technical standards/documents as well as the creation of new technical standards to assist in achieving efficient designs.

ACKNOWLEDGMENTS

The authors wish to acknowledge the assistance of the HS2 Organisation in the preparation of this paper and for permission to publish. The authors also wish to gratefully acknowledge the work of the numerous consultants, contractors, their colleagues at HS2 and international experts involved in the underground structures' design. In this paper the opinions expressed by the authors do not necessarily reflect the opinions of HS2 Ltd, CH2M, Arup, Birmingham Centre for Railway Research & Education and Dundee Tunnel Research.

REFERENCES

Baker, C.J., Dalley, S., Johnson, T. & Quinn, A. & Wright, N. 2001. The slipstream and wake of a high-speed Train. In: Proc ImechE, Part F: Journal of Rail and Rapid Transit, 215, 83–99.

BS8002 2015. Code of Practice for Earth Retaining Structures, BSI, London, UK.

BS8004 2015. Code of Practice for Foundations, BSI, London, UK.

BS8081 2015. Code of Practice for Grouted Anchors, BSI, London, UK.

BSI PAS 8812 2016. Temporary Works—Application of European Standards in Design—Guide, BSI, London, UK.

BSI PAS 8810 2016. Tunnel Design—Design of Concrete Tunnel Linings—Code of Practice, BSI, London, UK.

BSI PAS 8820 2016. Construction Materials—Alkali-activated Cementitious Material and Concrete—Specification, BSI, London, UK.

BSI PAS 8811 2017. Temporary Works—Major Infrastructure Client Procedures—Code of Practice, BSI, London, UK.

CIRIA C760 2016. Guidance on Embedded Retaining Wall Design, Construction Industry Research & Information Association, London, UK.

Hieke,M, H-J Kaltenbach,H-J, Tielkes,T 2009. Prediction of micro-pressure wave emissions from high-speed railway tunnels. In: Proc 13th International Symposium on the Aerodynamics and Ventilation of Vehicle Tunnels.

INCOSE website: www.incose.org.

INCOSE TWG website www.incose.org/ChaptersGroups/WorkingGroups.

Infrastructure Procurement Route Map 2013. A Guide to Improving Delivery Capability. HM Treasury January 2013.

Rawlings, C., Carroll, J., Leggett, M., Harland, B., Gee, I., Portal Cabezuelo, V, & Jung, H 2015. High Speed Two (HS2)—General Overview of Project with Focus on Tunnelling Challenges. In: RETC, New Orleans.

Sturt, R., Baker, C.J.B., Soper, D., Vardy, A.E., Howard, M. & Rawlings, C. 2015. The design of HS2 tunnel entrance hoods to prevent sonic booms. In: Proc Railway Engineering-2015, Edinburgh, UK, 30 June–1 July.

ThermoTun (2014)—see the web site www.ThermoTun.com

UK Government—Industrial Strategy: government and industry in partnership—Construction 2025 www.gov.uk/...data/.../bis-13-955-construction-2025-industrial -strategy.pdf.

Vardy, A; 2008. Generation and alleviation of sonic booms from rail tunnels. In: Proceedings of the Institution of Civil Engineers, Engineering and Computational Mechanics, 161, September. p. 107–119.

Planning and Design of the New Ashbridges Bay Treatment Plant Outfall for the City of Toronto

Gary J.E. Kramer ▪ Hatch
Deborah Ross ▪ CH2M Canada
Fiona Duckett ▪ W.F. Baird and Associates
Justyna Kempa-Teper ▪ City of Toronto

ABSTRACT

The Ashbridges Bay Treatment Plant (ABTP) Outfall project by the City of Toronto involves construction of a new tunneled outfall that will convey treated effluent (water) from the ABTP into Lake Ontario. This new outfall will be built to allow cessation of operations of the existing outfall which is reaching the end of its service life and has limited hydraulic capacity.

Outfall construction will include mining a 3,500 m long, 7 m internal diameter tunnel through rock beneath the lakebed. The project will initiate at an 85 m deep, 14 m internal diameter onshore shaft, to be constructed adjacent to the shoreline and down-stream of the future UV Effluent Disinfection Facility. Treated effluent will flow by gravity from the plant through connecting conduits to the shaft and tunnel out into the lake. Vertical in-line risers will be constructed along the last 1,000 m of tunnel to connect the tunnel and convey flows to the lake. Tunnelling operations will be supported from the onsite shaft and the risers will be drilled from over-water barges. The investigations and design phase of the project work is planned through the end of 2017, tendering in Q1/Q2 of 2018 and construction starting in the third quarter of 2018 and extending to the end of 2023.

The paper presents the planning and design of the project including discussions of challenges such as optimization of the risers, meeting regulatory requirements, addressing underground construction in rock known to exhibit time dependent behavior and riser construction up to 48 m in length through 38 m deep soft ground lakebed deposits overlying the bedrock.

INTRODUCTION

The Ashbridges Bay Treatment Plant (ABTP) is located at 9 Leslie Street in Toronto, Ontario on the north shore of Lake Ontario (see Figure 1). The plant receives and treats wastewater from the trunk sewers serving the City of Toronto (City) and discharges treated effluent into Lake Ontario through an existing 1 km long outfall that was constructed and put into service in 1947. The City is planning to construct a new tunneled Outfall to replace the existing outfall which has insufficient capacity, does not meet current regulatory standards and is nearing the end of its service life.

The new Outfall will provide adequate capacity to convey treated effluent by gravity flow from the ABTP and disperse it into Lake Ontario. The dispersion of effluent into Lake Ontario will meet regulatory requirements based on Provincial Water Quality Objectives (PWQO), Ministry of the Environment and Climate Change (MOECC) conditions for approval and key public health water quality requirements at beaches and water treatment plant intakes. The new Outfall will be permitted to operate under an updated Environmental Compliance Approval (ECA) for the ABTP granted by MOECC.

PROJECT PLANNING

The need for a new Outfall at ABTP was first discussed in the mid-1980s. At the time, it was noted that the existing outfall and diffuser were not of sufficient size to handle both the present and future peak raw wastewater flows delivered by the ABTP sewer shed and therefore a new, larger outfall must be designed and constructed. The project area of the Proposed Ashbridges Bay Treatment Plant Outfall is shown in Figure 1. A conceptual design for a new outfall at the ABTP that called for effluent conveyance through three tunnel pipes was developed in 1986 (Gore and Storrie, 1986) and updated in 1995 (Gore and Storrie, 1995).

The Ontario Ministry of Environment and Climate Change (MOECC) issued a "Notice of Approval to Proceed for the Undertaking" (to the City of Toronto (the City)) for the three plant improvement projects identified in a 1997 Environmental Assessment (EA) Amendment, including a new outfall (this project), an Ultraviolet (UV) disinfection system (a separate project) and a new effluent pumping station (if required).

The City commenced implementation of the new outfall in 2010 with a cost estimate update (Hatch 2011) and budget planning that lead to a conceptual design update in 2015. The 2015 Conceptual Design (CH2M et al, 2015) was performed to review the previous outfall design and to identify and evaluate alternative outfall designs.

The new outfall design concept was developed to operate within the physical parameters of the study area while meeting the latest MOECC guidance for water quality, based on a Receiving Water Impact Assessment (RWIA) performed using the latest near-field and far-field modeling tools (CORMIX and MIKE 3) and water quality data.

Figure 1. Project area of proposed Ashbridges Bay treatment plant outfall

The conceptual design update called for a single large-diameter tunnelled pipe, which takes advantage of new, larger diameter tunnel boring equipment available today, reduced construction costs and hydraulic losses.

OUTFALL DESIGN REQUIREMENTS

Hatch, with sub-consultant partners CH2M and Baird are completing the preliminary design, detailed design and construction administration services for the proposed new ABTP Outfall in response to the City of Toronto Request for Proposal (RFP) No. 9117-15-766. This assignment (current project phase) was initiated in 2016 and will continue through the end of 2017, with tendering in Q1/Q2 of 2018, construction starting in Q3 of 2018 and extending to the end of 2023.

The Recommended Preliminary Design of the ABTP Outfall consists of the following major components:

- **On-shore Work Area:** along the shore of Lake Ontario, at ABTP, approximately 40,000 m^2.
- **Effluent Conduits:** approximately 80 m long, extending from the termination of upstream conduit sections from the Disinfection Project to the Outfall shaft.
- **Shaft:** 14 m Internal Diameter (ID) launch shaft on-shore at the ABTP site.
- **Tunnel:** Single 7 m ID tunnel constructed in bedrock and lined with pre-cast segmental lining, extending approximately 3,500 m straight out from the launch shaft beneath Lake Ontario.
- **Diffuser:** 50 risers (1 m ID) with diffuser ports (830 mm ID), constructed in-line with the tunnel at equal spacing along a 1,000 m length diffuser section (located 2500 m off-shore), extending vertically from the tunnel to the Lake bed.

The plan and profile for the proposed outfall design concept is presented in Figure 2.

Hydraulic Design Requirements

The new outfall will be designed to operate under gravity flow for the hydraulic design basis shown in Table 1 for the duration of its 100 year design life. In order for the outfall to operate under gravity flow, the available hydraulic head between the liquid level in the outfall shaft and the high Lake level is approximately 1.0 m. Outfall hydraulic elements were designed so that the total head loss along the tunnel and diffuser (risers and ports) does not exceed 1.0 m at the 3,923 MLD design peak flow.

Figure 2. Isometric view of proposed Ashbridges Bay treatment plant outfall

Table 1. Hydraulic design basis for the ABTPO

Description	Criteria
Average daily flow rate	818 ML/d
Peak instantaneous flow rate	3,923 ML/d
Design life	100-years
Design lake level elevation	75.76 m (IGLD1985)
Design shaft liquid level elevation	76.76 m (IGLD1985)
Maximum system head loss	1 m
Dispersion requirements	Provincial water quality objectives

Table 2. Approvals, permits, and consultation programs for the ABTPO

Authority/Agency	Anticipated Approvals/Permits
Toronto and Region Conservation Authority (TRCA)	• Development, Interference with Wetlands and Alterations to Shorelines and Watercourses Permit (site plan and stormwater management plan). • Excavated Material Disposal Site Agreement.
Aquatic Habitat Toronto (AHT)	• Not required.
Ports Toronto	• Harbour Master Authorization.
City of Toronto	• Sanitary/Storm Sewer Discharge Permit. • Site Plan Approval. • Building Permit and Occupancy Permit. • Noise Exemption Permit for Construction. • Application to Injure or Destroy Trees. • Toronto Public Utilities Coordinating Committee (TPUCC) Notification.
Utility Authorities	• Consent from various utility authorities
Ministry of the Environment and Climate Change (MOECC)	• Environmental Compliance Approval (Sewage). • Dredging/Disposal Approval. • Permit to Take Water (PTTW).
Ministry of Natural Resources and Forestry (MNRF)	• Work Permit (in-water construction timing window). • Crown (Water Lot) Easement. • Endangered Species Act Permit. • Blanket Drilling Licence.
Ministry of Labour (MOL)	• Notice of Project (NoP). • Notice of Tunnels, Shafts, Caissons and Cofferdams. • Notice of Diving Operations. • Confined Space Entry Permit.
Ministry of Tourism, Culture and Sport (MTCS)	• Archaeological Assessment Clearance.
Department of Fisheries and Oceans Canada (DFO)	• Project Authorization.
Environment and Climate Change Canada (ECCC)	• Migratory Bird Timing Window. • Mitigation Measures for Deleterious Substances.
Transport Canada	• Notices of Works.
Canadian Coast Guard	• Notice of Works.

The diffuser section is being designed to achieve uniform flow and velocity distribution through all diffuser ports under normal operation conditions. The diffuser system must also meet the dispersion requirements throughout the range of design flows to provide sufficient mixing. CFD modelling is being completed to optimize the diffuser design.

MEETING REGULATORY REQUIREMENTS

Various federal and provincial legislation and policies as well as municipal by-laws govern the planning, design, construction and operation of a new outfall. This section presents a high-level summary of the permitting and approval requirements for the ABTPO.

The outfall construction includes an on-shore and off-shore portion. The approvals, permits and consultation programs that are anticipated for the outfall construction and operation are summarized in Table 2. Consultation with the various authorities and agencies will be completed during the design phase to ensure that all necessary approvals and permits are obtained.

Effluent Quality Requirements

A Receiving Water Impact Assessment (RWIA) was completed to demonstrate that the proposed outfall design meets Ontario Ministry of Environment and Climate Change

Table 3. Provincial water quality objectives (PWQO)

Receiving Water Parameter	Plant Effluent Quality	PWQO
E. coli	675 counts/100 mL	100 counts/100 mL (geometric mean of at least five samples) at a designated beach
Total Phosphorus (TP)	1 mg/L	≤0.02 mg/L for ice-free period to avoid nuisance concentrations of algae
Temperature	18.5 °C spring 25.8 °C summer	The temperature at the edge of the mixing zone shall not be more than 10°C above the natural ambient water temperature.
Ammonia	18.7 mg/L spring 14.6 mg/L summer	Un-ionized Ammonia (UIA) ≤ 0.02 mg/L at edge of near-field mixing zone

(MOECC) regulatory requirements for surface water discharges (Provincial Water Quality Objectives or PWQO) and the conditions of the Environmental Assessment. The key parameters used in the assessment and their Provincial Water Quality Objectives (PWQO) are listed in Table 3. The RWIA used the latest ABTP effluent quality data to predict that the new outfall design will provide sufficient mixing based on both near-field and far-field mixing zone analysis.

A detailed data review was completed for the Conceptual Design Report (CDR) to define the background concentrations and physical conditions in Lake Ontario, near the project site. This data is considered representative of the current conditions. The ambient water quality was characterized in terms of TP, pH, ammonia, E.Coli, DO and TSS on a seasonal basis. It was shown that Lake Ontario is Policy 1 in the study area for the parameters evaluated, meaning water quality is better than the Provincial Water Quality Objectives (PWQO) for the parameters analyzed.

Physical parameters including bathymetry, lake levels, currents and temperature were also defined based on existing data and data collected for the project. Current speeds were less than 0.07 m/s 25% of the time. Temperatures were found to be isothermal in the spring, fall and winter and stratified in the summer. For the riser optimization, the CORMIX model was run for spring and summer conditions as these are more critical for mixing.

The dilutions required to meet the MOECC's Provincial Water Quality Objectives (PWQO) for the key parameters were calculated based on effluent and ambient water quality data. TP had the highest required dilution ratio, with a seasonally averaged value of 80, and it is therefore identified as the governing water quality criterion.

Numerical modelling was used to assess the impacts of the new outfall on the water quality in Lake Ontario at specific locations of interest (see Figure 1). The United States Environmental Protection Agency (USEPA) Cornell Mixing Zone Expert System (CORMIX) model was used to simulate the wastewater dispersion in the near-field and the Danish Hydraulics Institute's (DHI) MIKE3 model will be used to simulate far-field mixing.

The design constraints for the analysis included a constant exit velocity of 0.35 m/s (at 818 MLD). The average water depth where the risers are located is approximately 15 m, there is one port discharging per riser therefore the terms riser, and port are used interchangeably.

The CORMIX model was used in the riser optimization study, to optimize the number of ports/risers by comparing predicted dilutions for varying number of ports/risers to

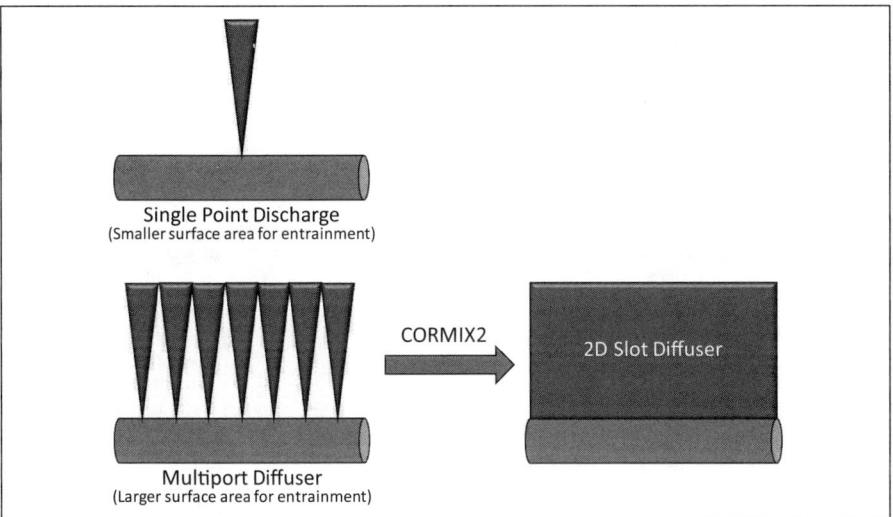

Figure 3. Single port vs. multiport diffuser and slot diffuser geometry

the required dilution for TP. Multiport diffuser systems are an efficient design to convey waste effluent away from shore in regions where lake currents generate conditions favorable for rapid dilution and dispersion. By distributing the flow to a number of jets along the diffuser, the total surface area for entrainment is increased compared to a single port, as illustrated in Figure 3. As a result, the diffuser system can achieve higher dilutions more rapidly compared to a single port/riser discharge.

The model predictions were considered at the edge of the NFR and at half the outfall pipe length, which is taken as a radius 1,250 m from the centre of the diffuser. The CORMIX2 model was used to predict dilution estimates for the multiport diffuser system. This model assumes a line source as shown in Figure 3. Recognizing that CORMIX2 tends to over-predict mixing, particularly for more widely spaced ports where plume merging is delayed, the predicted dilutions from CORMIX1 were considered.

Table 4 provides a summary of the predicted dilutions 100 m from the diffuser (estimated edge of NFR) and at the half pipe distance (1,250 m from the diffuser). Given the limited information available on the performance of widely spaced risers with respect to dilutions, a conservative approach was taken and the CORMIX1 results were considered. Acceptable mixing is predicted by both CORMIX1 and CORMIX2 at the half pipe distance for the 50 riser option and hence 50 risers were selected for the ABTPO.

REGIONAL GEOLOGY

Southern Ontario is characterized by a thin mantle of glacial soils overlying a thick sequence of Palaeozoic sedimentary rocks deposited directly on the Precambrian basement rock. These rocks span geologic time periods that include Cambrian, Ordovician, Silurian, Devonian, and Mississippian periods, which encompass a range of time from approximately 570 million years ago to 325 million years ago. Thurber Engineering (under Hatch) completed a geotechnical investigation between June and October 2016 to supplement the information gathered the 1986 Conceptual Design (Morton, 1988). A plan and profile view of the ABTPO showing interpreted geologic profile is shown in Figure 4.

Table 4. Comparison of CORMIX1 and CORMIX2 dilution estimates

Spring Conditions (Isothermal)						
No. Risers	Flow per Riser (m³/s)	Riser Spacing (m)	Initial Dilution (S) 100 m from Diffuser*		Initial Dilution (S) at Half Pipe (1250 m)	
			CORMIX2 (MPD)	CORMIX1 (SPD)	CORMIX2 (MPD)	CORMIX1 (SPD)
100	0.095	10	83:1	50:1	100:1	238:1
61	0.155	16	83:1	30:1	100:1	120:1
50	0.189	20	83:1	25:1	100:1	98:1
45	0.210	23	83:1	23:1	100:1	88:1
40	0.234	26	83:1	21:1	100:1	76:1
30	0.316	34	83:1	17:1	100:1	55:1
Summer Conditions (Stratified)						
No. Risers	Flow per Riser (m³/s)	Riser Spacing (m)	Initial Dilution (S) 100 m from Diffuser		Initial Dilution (S) at Half Pipe (1250 m)	
			CORMIX2 (MPD)	CORMIX1 (SPD)	CORMIX2 (MPD)	CORMIX1 (SPD)
100	0.095	10	76:1	24:1	90:1	175:1
61	0.155	16	76:1	16:1	90:1	93:1
50	0.189	20	76:1	13:1	90:1	76:1
45	0.210	23	76:1	12:1	90:1	63:1
40	0.234	26	76:1	11:1	90:1	57:1
30	0.316	34	76:1	9:1	90:1	40:1

Note: Dark shading indicates predicted dilution meets PWQO mixing requirements and light shading indicates predicted dilution does not meet PWQO mixing requirements.

Bedrock

The project area is underlain by the Upper Ordovician Georgian Bay Formation, described as a greenish to bluish grey non-calcareous shale. It is a fairly flat Formation, dipping a maximum of 1° to the south or southwest [9]. Lewis and Sly (1971) report that this Formation is up to 120m thick in the Toronto area. The formation is extends to Streetsville in the west and is bounded by the base of the Niagara Escarpment to Georgian Bay in the north.

The shale is interbedded with harder layers of limestone, siltstone and sandstone. These harder layers are often the site of bedding planes, and are generally less than 200mm thick. Regional, near vertical fractures zones striking just south of east and spaced 200 to 350m apart are present. These fracture zones are assumed be the result of isostatic crustal deflection following glacial retreat. The geological conditions in this region are characterized by two main features:

1. All rock formations are subjected to high in-situ horizontal stresses.
2. In some rock units, particularly shaly rocks, important time dependent deformation (often referred to as either squeezing or swelling) occurs subsequent to rock excavation.

The properties and behaviour of sedimentary shaly rock formations such as the Georgian Bay Shale in Southern Ontario in response to tunnel and shaft excavation can be generally characterized as follows:

- Generally, massive appearing rock with widely spaced or no vertical or inclined jointing (in the context of the scale of engineering works) and closely spaced sub-horizontal bedding planes;
- High in situ horizontal near surface stresses;

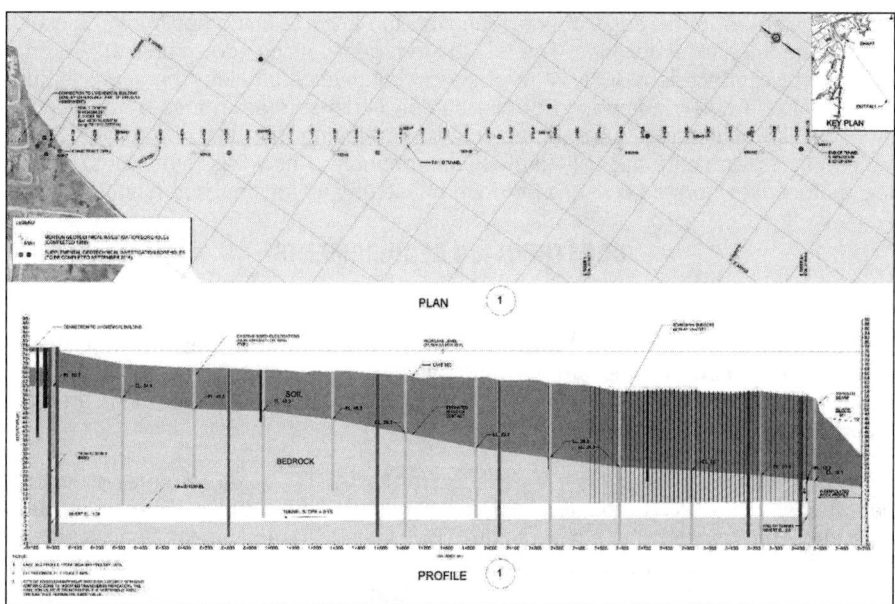

Figure 4. Plan and profile view of proposed ABTPO

- High susceptibility to slaking upon exposure (excavation or removal of overburden);
- Shales are considered to be highly fissile;
- Low intact rock tensile strength in the horizontal direction (order of magnitude 10% of UCS) and an order of magnitude less to virtually nil in the vertical direction;
- Relatively soft, weak, brittle failure in unconfined compression;
- Time dependent deformational behaviour (swelling and squeezing) that is highly stress dependent upon relief of initial stresses and exposure to wetting or humidity;
- Cross-anisotropic behaviour with respect to deformation, swelling and strength properties;
- Vertical Time Dependent Deformation rates greater than horizontal;
- Presence of gas and hydro-carbon (BTEX) compounds;
- High choride content in the pore fluid;
- High slake potential and
- Strength and stiffness reduction over time for exposed rock.

Overburden

Ashbridges Bay is a reclaimed lagoon east of the Don River with near-surface onshore materials consisting of landfill material deposited in the early part of the 20th century. Beneathe the surficial deposits, a layer of overburden ranging from 18m to 37m thick, consisting of mostly normal to unconsolidated sandy and silty lacustrine sediments, overlies bedrock consisting of Georgian Bay Formation shale. Groundwater levels in the overburden deposits are controlled by the elevation of Lake Ontario.

Moving offshore to the edge of the strandflat, the layer of fining sediments thickens and the bedrock level declines. The lake bottom gently slopes down until the Toronto Scarp, which provides roughly 40m of sudden bathymetric relief. The scarp feature appears to be solely composed of overburden sediments as it does not appear to manifest itself in the bedrock topography. A gently sloping strandflat, the Toronto Shelf, yields to a steep scarp approximately 3.5km offshore. Known as the Toronto Scarp, this feature likely formed during a low-lake phase as an erosional shore bluff feature.

CONSTRUCTION REQUIREMENTS

Tunnel Construction

The 2015 CDR provided the framework for the selection of tunnel diameter, tunnel excavation method and the tunnel lining system. The tunnel will be 3,450 m long, 7 m ID, mined through the sedimentary Georgian Bay Formation shale using an open faced TBM.

Two construction systems were considered for the ABTP Outfall; a single-pass pre-cast tunnel lining (PCTL) installed using a single shield TBM or a two-pass system construction process using a main beam TBM. The PCTL acts as the final lining and is installed as the tunnel construction advances. With a two-pass lining system, the initial ground support is installed as the tunnel is excavated to protect workers and equipment from overbreak and rock fallout until the final CIP concrete lining is installed; typically, once excavation is complete.

Due to the sub-aqueous nature of the tunneling, it is planned to incorporate specifications provisions for probing and grouting ahead of the TBM to address the potential for groundwater inflow and gas intrusion into the tunnel. The presence of poor rock at fault and shear zones, while infrequent, will also need to be addressed.

Time Dependent Deformations (TDDs) in Rock

The observed behavior of shales and shaly rocks in southern Ontario and elsewhere including Georgian Bay Shale has demonstrated an interesting combination of high in situ horizontal stresses and the potential for substantial time dependent deformations in response to tunnel excavation. Underground structures built in such rock formations have experienced various degrees of distress, particularly when time dependent deformations have not been explicitly considered during the design process.

Upon tunnel excavation, the in situ stresses within the rock mass redistribute around the tunnel opening and elastic (and possibly inelastic) deformations immediately manifest themselves as inward convergence into the tunnel excavation. In rock that exhibits TDD behavior, the convergence deformation continues over time and the amount of TDD that occurs is a function of the TDD potential of the rock, the magnitude of stress relief and stress levels remaining within the rock (see Figure 5).

Traditional practice in Southern Ontario area to address the TDD impacts has been to utilize a two pass lining system—initial support during tunnel mining followed by the delayed installation of a CIP tunnel lining. However, analysis using more recent developments that incorporate the stress dependency of TDDs (Lo and Hefny, 1996, Hawlader et al, 2005, Kramer and Moore, 2006) was carried out to determine if TDD will govern the choice of the tunnel lining system. It was found that, while the TDD behavior is anticipated to be more pronounced for a PCTL system (due to the short time lapse between excavation and support installation), the additional liner stresses

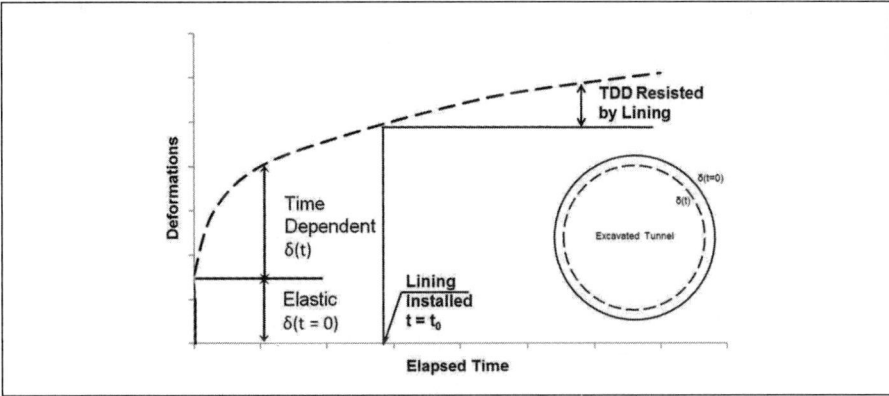

Figure 5. TDD convergence of tunnel excavation relative to lining installation

are within capacity of that system and hence not a differentiator between the two systems.

After comparing the benefits and challenges of a two-pass lining system with those of a single-pass system, a single-pass PCTL was chosen for this project on the basis of lower out-turn cost due to shorter schedule, acceptable TDD performance, superior durability and a safer working environment (control of gas inflow and fallout). The tunnel will be lined with a PCTL as the TBM advances. The TBM will likely be abandoned in place and all PCTL bolt pockets will be filled to provide a smooth lining surface once tunnel mining is complete.

Riser Construction

The Outfall diffuser section will be located approximately 2500–3500 m off-shore in Lake Ontario and will include 50 risers, in line with the tunnel, evenly spaced along the 1000 m diffuser length. The average riser depth (lakebed deposits plus bedrock depth) will be approximately 50 m (see Figure 6). The riser configuration will be as follows:

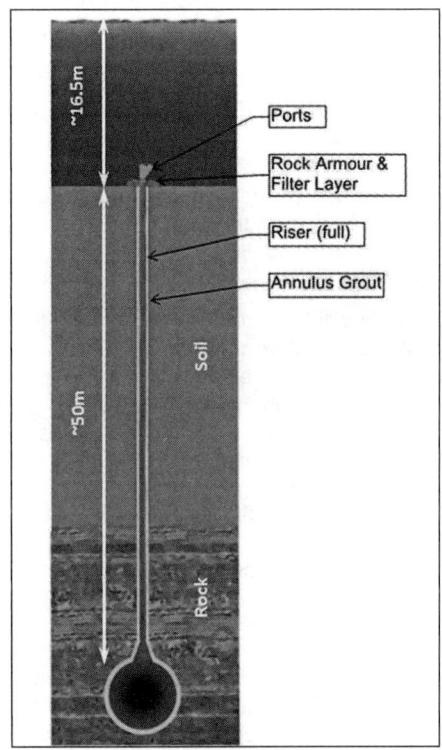

Figure 6. Typical riser configuration

- Stainless steel risers (1.0 m ID).
- One Fibre Reinforced Polymer (FRP) port per riser (830 mm ID).
- A high copper nickel coating will be applied to both the port and the upper flanged riser pipe to mitigate the migration of zebra mussels into the Outfall system.

The riser height, the depth of overburden, and the depth of rock vary as the riser's progress off-shore. Table 5 summarizes average dimensions for the ABTP Outfall Risers. The anticipated elevations, heights and dimensions of individual riser pipes will be established in Detailed Design. The depth and nature of the overburden lakebed deposits will make special provisions necessary for the excavation and installation of the risers in order to maintain hole stability and achieve a grout seal around the riser pipes.

Table 5. Estimated riser elevations, depths, and dimensions

Riser Dimension	Unit	Approx. Average Value
Riser Height	m	50
Estimated Riser Excavation Diameter (OD)	mm	1,500
Riser Diameter (ID)	mm	1,000
Port Diameter (ID)	mm	830
Lake Water Depth to Top of Riser	m	16.5
Estimated Depth of Overburden	m	36
Estimated Depth of Rock	m	14

Construction in the lake during the spring is prohibited by the Ministry of Natural Resources and Fisheries until the end of fish spawning season (July 1st) and is limited in the fall by deteriorating weather conditions. Typically, it is not productive to work in the lake after Oct. 31st. It is anticipated that three such marine construction timing windows will be required to complete marine construction activities. The project construction schedule will depend greatly on the first construction timing window. All necessary permits and approvals will be obtained prior to awarding the construction contract such that marine work can begin in the planned timing window. Discussions with the MNRF are ongoing to obtain permission to advance the construction timing window based on the results of fish habitat studies.

Shaft Construction

Effluent will flow from the upstream Disinfection Facility through conduits connected to the Outfall shaft. The completed shaft will convey treated effluent from the conduits to the Outfall tunnel. In addition, the shaft will serve as the main tunnel mining launch shaft during construction and as an access shaft during operation. The design of the main shaft will also include provisions to accommodate future conduits from the potential pump station and a satellite facility.

The proposed shaft is approximately 80 to 85 m deep and will be finished with a 14 m ID, CIP reinforced concrete final lining along its entire depth. This diameter is required to accommodate currently planned conduit connections and connections for a future east bypass conduit, potential future pump station and conduit from a future satellite treatment facility.

In the fill material, overburden soil and weathered rock zone, an excavated diameter of 17 m is required. The excavated diameter in competent rock is 16 m. The shaft plan and elevation configuration are shown in Figure 7 and Figure 8. The shaft will be excavated through approximately 19 m of fill and overburden soils and 61 m of shale bedrock of the Georgian Bay Formation. The fill and soil comprises 2.5 to 4.5 m of landfill material followed by 14.5 to 16.5 m of in-situ native soil and the bedrock comprises 1 m of mixed soil and shale bedrock intermediate material, and 63 m of shale bedrock of the Georgian Bay Formation.

Excavation methodologies will differ between the overburden and bedrock sections. Excavation will be performed in the fill and soil using conventional soil excavation equipment such as clamshells and excavators. Based on local experience, shaft

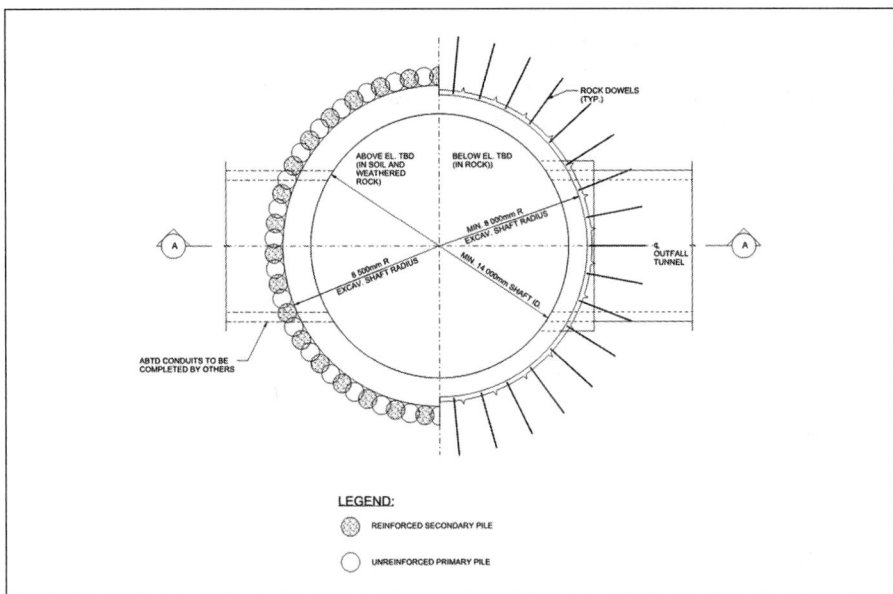

Figure 7. Plan view of proposed onshore shaft

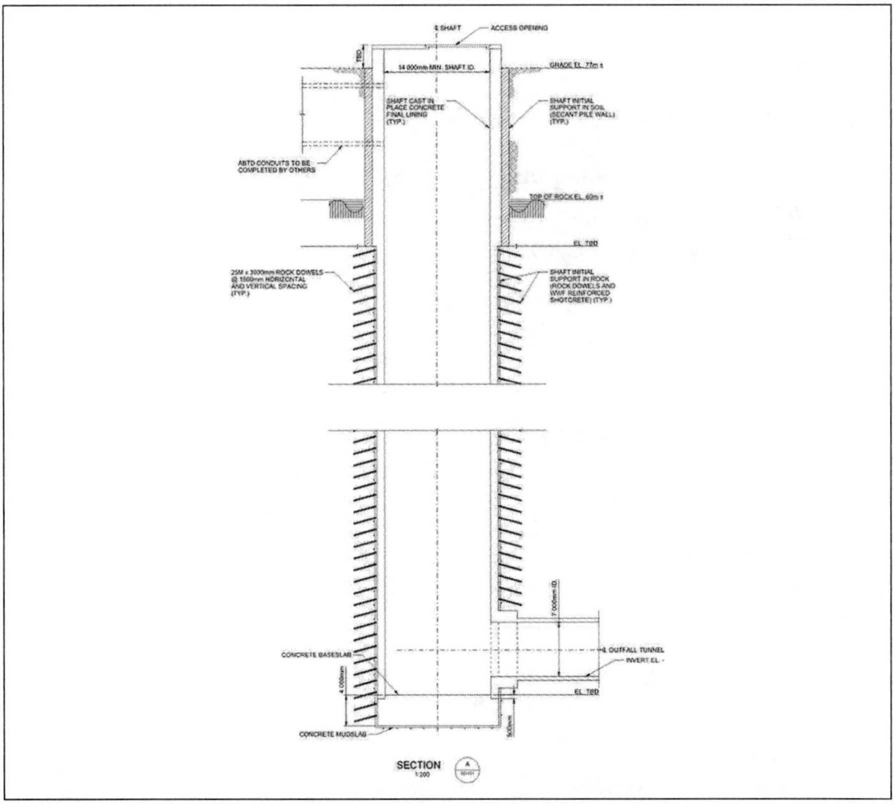

Figure 8. Elevation of proposed onshore shaft

sinking excavation within bedrock will likely be performed using hydraulic excavators equipped with percussive hammer (hoe ram) attachments.

Shaft initial support in the fill materials, overburden soils and weathered rock will be a sealed system consisting of a steel-reinforced, CIP concrete secant pile wall in a circular or combined semi-circular/quadrilateral configuration (see Figure 7). Alternatively, a slurry-wall system may be adopted. A sealed system is required due to the impracticality of active dewatering outside of the shaft excavation for groundwater control and soil stabilization in the fill material and overburden soil, resulting from inexhaustible groundwater recharge from Lake Ontario. Secant piles are planned to be 1,000 mm diameter with a 20% overlap between piles and advanced up to 5 m below the soil-rock interface to key into the rock, create a seal and penetrate into competent bedrock.

The inside diameter of the secant pile wall face should be 17 m (not including construction tolerance) allowing for a 500 mm rock bench width at the base of the secant pile wall. The initial support will be designed to incorporate the openings required to connect the effluent conduits to the shaft. The use of full-depth, temporary casing installed simultaneously while drilling will be required based on anticipated ground conditions consisting of saturated granular soils exhibiting flowing sands behavior.

Several operational design features are to be included in the shaft design to facilitate grit removal, inspection and maintenance, hydraulic optimization, and public safety. The bottom of the shaft below the elevation of the Outfall tunnel invert will be configured with a 0.5 m deep sump to serve as a grit trap to collect any accumulated solid material. A shaft cover is required to close the shaft and tunnel system against the external environment and is required for public safety. Pending confirmation by hydraulic modelling, the top of the shaft collar will be located 5 m above the mean water level of Lake Ontario and approximately 3 m above finished grade. The shaft cover will be located above the shaft collar.

PLANNED PROJECT IMPLEMENTATION

A bar schedule for the planned project implementation is provided in and shows that the planned period for prequalification and tendering will occur in late Q1 through Q3 2018 with a construction start in Q3 2018 (Figure 9). The project value (including design and contingencies) is estimated at approximately $340M (CAD).

Other aspects of the planned project implementation and preferred procurement strategy will involve preparing Contract Documents to incorporate the following:

- **Design Bid Build (DBB) Procurement:** Contract Documents will be prepared as a DBB contract in accordance with the Toronto RFP 9117-15-7166 terms and conditions.

- **Advanced Site Works Contract:** An "Advanced Works" contract package will be let to address schedule critical components before the main contract (e.g., tree removal, clearing and grubbing, erosion and sediment control installation).

- **Bidder Pre-Qualification:** Bidders will be pre-qualified to ensure that they are qualified to perform the work. Prequalification documents will be prepared using CCDC 11 in accordance with the Ontario General Contractors Association (2006) "A Guide to Prequalification of Contractors." Bidders will have to pre-qualify for both the marine work and the underground (tunnel and shaft) work and will be required to have COR Certification.

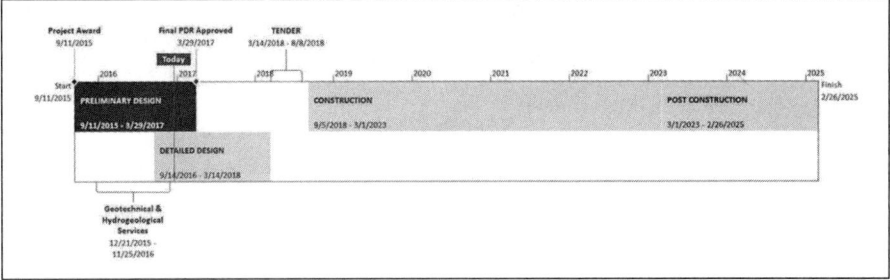

Figure 9. Planned project

- **Bid Form Format:** The Quantity and Price Schedule will be prepared to include both lump sum items and unit price items with pre-defined quantities. Methods for adopting the Toronto Project Tracking Portal System will be investigated during detailed design.

- **Provisional Bid Items:** Provisional bid items for site works, shaft, tunneling and marine construction will maintain flexibility with regards to changes during construction.

The following contract provisions for major underground infrastructure projects are also planned to be included in the Contract Documents:

- **Geotechnical Baseline Report:** will be incorporated in accordance with the ASCE Guidelines.

- **Differing Site Conditions Clause:** to establish the process by which a potential change of ground condition is discovered, investigated, evaluated and implemented.

- **Changes Clause:** to establish time limits for claim preparation and evaluation, profit allowances and overhead costs associated with time extensions and/or deducted work.

- **Escrow Bid Documents Clause:** to establish the contractor's assumptions, calculations and other information used to prepare their bid and to improve resolution of quantum disputes.

- **Disputes Resolution Board Clause:** to improve resolution of construction disputes by establishing the preferred disputes evaluation process, board member selection, update requirements during construction and cost sharing approach.

- **Partnering:** Provisions for a partnering agreement between the City and the successful contractor will be implemented, including partnering workshop, preparation of a project charter and issue resolution process, and periodic process of performance evaluation.

CONCLUDING STATEMENT

The ABTPO, when constructed, will significantly improve the water quality of the Toronto lakeshore and surrounding communities.

ACKNOWLEDGMENTS

The planning and design of the ABTPO is due to the contributions of many people from the City of Toronto, Hatch, CH2M and Baird and Thurber. The authors would particularly like to acknowledge the contributions of Alison Barlow, Vlad Petran and Nancy Afonso of the City, Kevin Waher, John Pehar and Lucie Clatworthy of Hatch, Emma Shen, Dan Olsen and Matthew Elliott of CH2M and Mike Fullarton of Baird.

REFERENCES

CH2M, Hatch and Baird (2015). Ashbridges Bay Treatment Plant Outfall Conceptual Design and Receiving Water Assessment.

Gore & Storrie Limited (1986). Main Treatment Plant Outfall Water Quality Study. Volume 1, Sections 1–11.

Gore & Storrie Limited (1995). Metro Works, Improving the Effectiveness of the Main Treatment Plant.

Hatch Mott MacDonald (HMM) (2011). New Outfall Cost Estimate Report.

Hawlader, B.C., Lo, K.Y. and Moore, I.D. (2005). Analysis of tunnels in shaly rock considering three-dimensional stress effects on swelling. Canadian Geotechnical Journal, 42:1–12.

Lo, K.Y. and Yuen, C.M. (1981). Design of Tunnel Lining in Rock for Long Term Time Effects, Canadian Geotechnical Journal, Volume 18, 24–39.

Lo, K.Y. and Hefny, A. (1996). Design of Tunnels in Rock with Long-Term Time dependent and Nonlinearly Stress-Dependent Deformation, Canadian Tunnelling, pp. 179–213.

Morton Geotech Inc. (1988). Geotechnical Report for Outfall Sewer at Main Treatment Plant, Contract D-9-87.

Planning of the San Francisco Public Utilities Commission's Channel Tunnel

R. John Caulfield ▪ Jacobs Engineering
Art Hamid ▪ Stantec/MWH Global
Manfred Wong ▪ San Francisco Public Utilities Commission

ABSTRACT

The proposed Channel Tunnel is a critical component of the San Francisco Public Utilities Commission's Central Bayside System Improvements Project (CBSIP), which is a key element of a $6.9-billion Sewer System Improvement Program to upgrade their aging and seismically vulnerable wastewater facilities. The tunnel will provide gravity conveyance and storage of combined sewage flows from the northeast sector of San Francisco to the Southeast Water Pollution Control Plant for treatment. The tunnel will be approximately 1.7 miles long and 24 feet in internal diameter. It will utilize a single-pass precast concrete segmental lining system. Tunneling challenges include excavating with pressurized face TBM technology through highly variable ground conditions including Franciscan Complex (rock and mélange), clayey/silty sands, stiff to hard clay (Old Bay Clay) and mixed-face conditions.

PROJECT BACKGROUND

The San Francisco Public Utilities Commission (SFPUC) is a department of the City and County of San Francisco that provides municipal power, drinking water and wastewater services to San Francisco.

San Francisco's combined sewer system collects both wastewater and stormwater in the same sewers, which is then pumped, treated to stringent state standards and discharged into the San Francisco Bay and Pacific Ocean. The sewer system includes tunnels, large storage/transport boxes, 1000 miles of collection sewers, 28 pump stations, and three wastewater treatment plants. After decades of service, much of the system is nearing the end of its useful life.

The existing system is also vulnerable to seismic damage. During the 1989 Loma Prieta earthquake, damage to SFPUC wastewater infrastructure was extensive. This included the 11,200-foot long, 66-inch diameter Channel Force Main, built in 1976, and a critical component of the conveyance system. The Channel Force Main carries 70 percent of the bayside dry weather sewage flow from the northern and central part of the city to the Southeast Water Pollution Control Plant for treatment. It is antiquated, not pile supported and not constructed to current seismic codes. There is also currently no conveyance redundancy for north/central bayside flows to the treatment plant. Figure 1 shows the SFPUC's wastewater collection system and the existing Channel Force Main.

This paper addresses the planning aspects of the proposed tunnel. Current planning is for a 24 foot internal diameter, 1.7 mile long gravity conveyance/storage tunnel with a 120 million gallon per day capacity lift station at its downstream end located on the south side of Islais Creek. It will be constructed in a highly congested urban and industrial area of San Francisco under challenging geotechnical conditions.

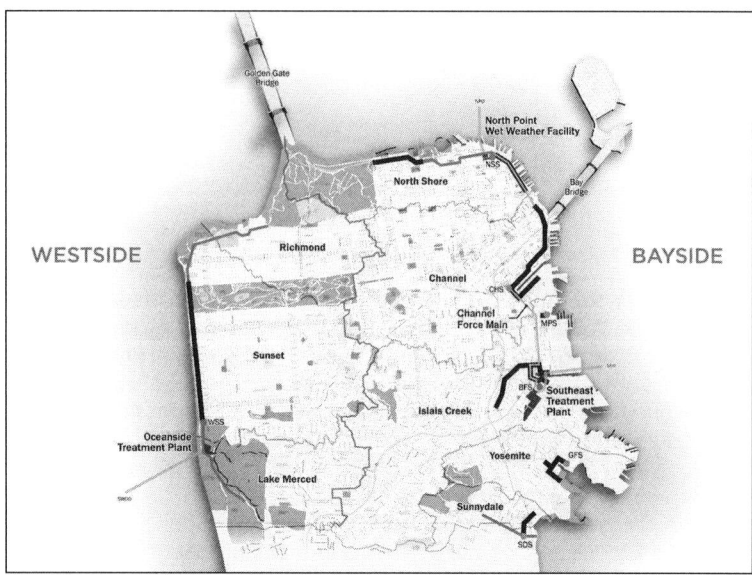

Figure 1. Schematic diagram of the SFPUC's wastewater distribution system

This project has completed the conceptual engineering planning phase that provides the 10 percent design for the project. The 35% tunnel design effort is currently underway.

GEOLOGIC CONDITIONS

The project site is located on the San Francisco peninsula within the Coast Ranges Geomorphic Province, with northwesterly trending ridges and valleys. Rugged hills consisting of Jurassic- to Cretaceous-aged bedrock are juxtaposed with low, flat-lying areas underlain by Quaternary sedimentary deposits. Bedrock underlying these sediments consists of consolidated rocks of the Franciscan Complex. This Franciscan bedrock consists of highly deformed, weathered, and intensely fractured sedimentary rocks of the Franciscan assemblage including mélange shales, sandstones, siltstones, mudstones, greywacke sandstone, serpentinite, and greenstone. Sediments overlying the Franciscan Complex include Old Bay Clays, Young Bay Muds, and Colma Sands with interbedded clays, and undifferentiated alluvium/colluvium. Historical development in the area resulted in placement of fill over substantial portions of modern estuaries, marshlands, and creek beds to reclaim land. A regional geologic map is shown in Figure 2 (Graymer et al., 2006).

The preliminary geotechnical investigation included the installation of open standpipe piezometers along the alignment to monitor groundwater levels. Piezometer readings indicate that the groundwater pressures are generally consistent at between 2 to 20 feet below the ground surface.

SEISMIC CHARACTERIZATION

Characterizing the potential seismic impacts on the facility and developing seismic design parameters will be critical activities in the final design process. The proposed Channel Tunnel is located in central coastal California which is a seismically active area. As shown in Figure 3, the proposed alignment does not cross any identified active faults but is situated between the San Andreas (7.4 mi to the southwest) and Hayward (11.1 mi to the northeast) faults, which are capable of generating large earthquakes (USGS, 2016). There are also numerous smaller active faults within 30 miles of the proposed alignment. A total of 15 earthquakes of M6.0 have occurred in the San Francisco Bay region between 1850 and the present. These include the 1906 M7.9 San Francisco earthquake and the 1989 M6.9 Loma Prieta earthquake.

The Channel Tunnel is expected to experience significant ground shaking during its specified design life of 100 years. As required by the SFPUC's general seismic design requirements, the project will be designed for ground motions that will have a 5% probability of exceedance in 50 years (975-year approximate return period). As part of the design studies, a probabilistic seismic hazard analysis (PSHA) for ground shaking along the alignment will performed. The purpose of this evaluation will be to estimate the levels of ground motions at a specified exceedance probability. Deterministic scenario ground motions will also be calculated and compared to the probabilistic ground motions. An evaluation of liquefaction potential will also be performed as part of the design.

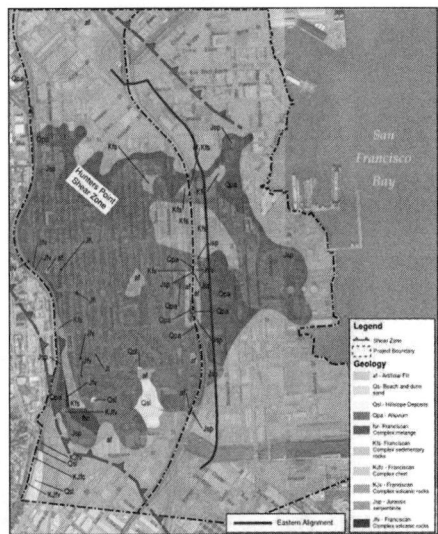

Source: Graymer et al. 2006.

Figure 2. Regional geologic map

Source: USGS, 2016.

Figure 3. Regional fault map

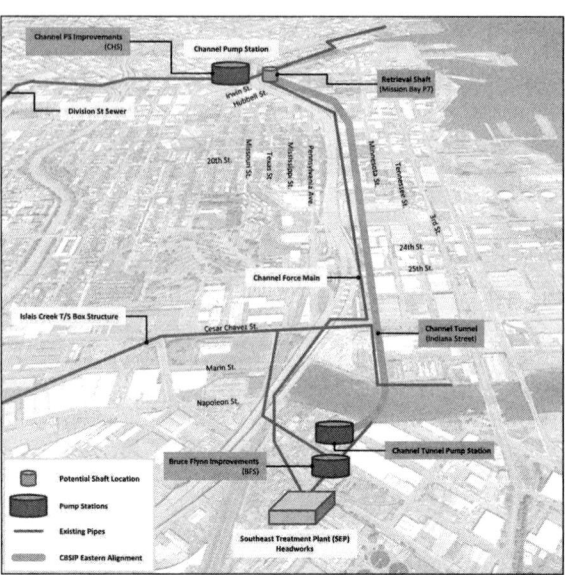

Figure 4. Proposed tunnel alignment

TUNNEL ALIGNMENT

Potential alignments for the Channel Tunnel were evaluated based on a number of different factors including anticipated geological conditions, tunnel and shaft construction methods, constructability and risk considerations, right-of-way acquisition, permitting requirements, operations considerations, hydraulic impacts, regulatory compliance, and project cost.

The preferred alignment (also called the Eastern Alignment) as depicted in Figure 4 was determined to provide the best overall value to the SFPUC. The tunnel alignment profile provides a gravity fed system to the downstream lift station that is approximately −130 feet deep to the tunnel invert at the upstream end and −148 feet at the downstream end. This 18 vertical foot difference provides an overall constant grade of 0.2 percent.

The preliminary alignment studies for the project identified some significant tunneling challenges that lead to the proposed depth. At shallower depths, the tunnel would pass through soft Young Bay Mud strata and multiple mixed-face zones that would be problematic to tunnel construction. The tunnel boring machine (TBM) steering and grade control, as well as control of face pressure were deemed to be significant issues. Potential settlement was also a consideration as the existing active Channel Force Main overlies the majority of the proposed tunnel alignment.

Additionally, building structures and infrastructure in the area were constructed on deep pile foundations to penetrate the very soft Young Bay Mud deposits. Avoiding these foundations with the proposed tunnel required going through them (and tunneling through the surrounding extremely soft soils) or avoiding them altogether. Avoidance options included; threading the tunnel through the piles and placing it within the extremely soft soils, removing the piles in the affected areas, or going deeper with the tunnel alignment. Based on an evaluation of these options and the current understanding of the project conditions, a deeper tunnel alignment was deemed to be the preferred alternative.

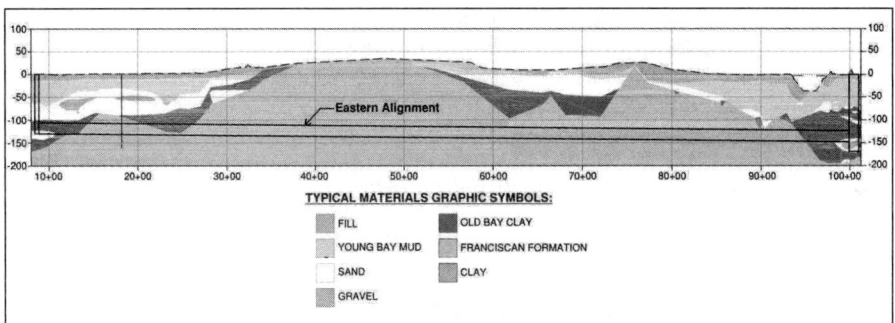

Figure 5. Generalized geologic profile along the tunnel alignment

While there are inherent risks and higher construction costs associated with the deeper alignment and associated shafts, these were weighed against the higher risks and associated mitigation costs of tunneling along a shallower alignment through very difficult geologic conditions and numerous obstructions.

TUNNELING CONDITIONS

Descriptions of the tunneling conditions are provided within this section. Tunnel construction for the Eastern Alignment will be performed in rock and highly weathered rock conditions; soil conditions; and some mixed-face conditions (i.e., rock and/or highly weathered rock and soil). Figure 5 provides a generalized geologic profile along the tunnel alignment.

The initial 400 to 500 linear feet of the tunnel drive from the launch shaft and under Islais Creek is anticipated to consist mostly of Old Bay Clays (sometimes referred to as Yerba Buena Mud). The Old Bay Clays consist of very stiff to hard, fat clay materials. They are over-consolidated and have typical undrained shear strengths in the range of 2000 to 4000 psf. They generally have good tunneling excavation characteristics and possess good standup time in the absence of face support.

From the north shore of Islais Creek and for the next 8000 to 8100 linear feet the tunnel drive is expected to encounter Franciscan Complex materials. The exception to this will be a relatively short 400 foot long section where a mixed face of Franciscan Complex and Old Bay Clays will be encountered. The rocks of the Franciscan Complex are highly variable in their degrees of fracturing, strength, hardness and weathering, ranging from soft to hard and from friable to moderately strong or massive. They have also been known to exhibit squeezing behavior. The rock types have been divided into the following categories:

- Mélange matrix,
- Greywacke sandstones, and
- Serpentinite (including greenstone)

The mélange is characterized by a pervasively sheared block-in-matrix fabric consisting of extremely weak to weak, dark gray to black clayey shale or blue green serpentinite matrix with abundant tectonic inclusions of greywacke. The block-in-matrix inclusions in the Franciscan Complex are known to range from gravel-size to over 100 feet in dimension. The greywacke is moderately strong to strong and typically occurs as smaller gravel to boulder-size pieces within the mélange matrix. The internal structure of the mélange is chaotic and depends on the degree and intensity of localized shearing and the size and distribution of the harder block inclusions. Bedrock strength is correspondingly variable, but is generally considered to be low in areas that display the greatest degree of shearing or a higher proportion of mélange matrix to block inclusions.

Greywacke sandstones found within the Franciscan Complex typically have wide ranging unconfined compressive strengths, from approximately 50 psi to over 9,000 psi. Calculated unconfined compressive strengths obtained from point load index testing range from nearly 100 psi to over 15,000 psi.

The serpentinite of the Franciscan Complex is slightly to moderately weathered, with altered infilling of discontinuities. The initial project geotechnical investigations also encountered moderately weathered, moderately strong to strong greenstone (metamorphosed basalt) beneath the surficial soils. This unit was highly fractured with localized shear zones that were completely weathered/altered and soft.

The majority of unconfined compressive strength tests performed during preliminary investigations returned values of less than 5,000 psi and typically less than 3,000 psi. Point load index tests performed on the greenstone had results ranging from 11 psi to 294 psi. Brazilian Tensile Strength tests indicated a wide range of tensile strengths from 20 to 1,080 psi, with a skew in the data towards the lower strengths.

Laboratory tests on collected samples indicate that much of the serpentinite contains levels of naturally occurring chrysotile asbestos.

The ground classifications indicate that a high percentage of the Franciscan Complex ground is poor to extremely poor. Q classification indicates that over 60 percent of the ground is "very poor" and RMR indicates that more than 55 percent of the ground is "poor," showing good correlation between both methods. The GSI values typically ranged from 15 ("very poor" and "disintegrated") to 60 ("blocky" and "good"), which suggests high variability in rock quality. It is anticipated that the excavation through the Franciscan Complex will present a number of challenges including potential squeezing behavior, highly variable ground conditions, abrasion of excavation tools, and the potential for blocking of the EPBM screw auger system and cutterhead.

The remaining approximately 500 feet of the tunnel excavation will encounter Old Bay Clays with some dense sand layers entrained within the lower part of the tunnel near the invert.

The entire tunnel alignment is located under the water table, potentially subject to approximately 3 to 4 bars (45 to 60 psi) of hydrostatic pressure. Further evaluation of the hydrogeologic characteristics of the subject formations will be part of the future project geotechnical investigations.

TUNNEL EXCAVATION AND LINING METHODS

Bored tunneling methods using a shielded pressurized face machine were considered to be the most suitable method for excavating the main tunnel drive based upon the current understanding of the geologic conditions. The machine will need to accommodate the wide range of rock, soil and mixed-face ground along the alignments as well as the groundwater conditions. A pressurized face shielded TBM will also facilitate the installation of the lining system anticipated.

The pressurized-face TBM will likely be an Earth Pressure Balance Machine (EPBM) given the recent tunneling success and contractor experience with these machines in the local geologic conditions. Recently completed EPBM excavated tunnels within the San Francisco Bay Area include the SFMTA Central Subway twin tunnels (ENR, 2015), the SFPUC Bay Tunnel (Caulfield et al., 2014), and the SFPUC Sunnydale Auxiliary Sewer Tunnel (Fippin et al., 2011).

A slurry shield TBM may also be viable, but space considerations for the extensive slurry surface plant involved may prove difficult at the TBM drive shaft site. However, the potential selection of a slurry shield TBM (STBM) will be provided within the contract documents to allow for contractor flexibility in means and methods.

Based upon the current understanding of the ground conditions, some of the machine capabilities and features would include:

- The ability to excavate in low strength mélange and serpentinite as well as blocks of hard sandstone.
- The ability to excavate in and restart after a stoppage in squeezing ground types.
- To handle the variable rock and soil conditions, it is anticipated that the cutterhead will be fitted with a combination of back-loaded disc cutters and drag picks as excavation tools.
- The ability to inject bentonite under pressure in the annular space between the shield and the excavated rock to help mitigate against squeezing ground conditions.
- The ability to grout ahead of the TBM face.
- Due to the ground conditions along the tunnel drive and an anticipated hydrostatic head of up 3 to 4 bars (45 to 60 psi), hyperbaric interventions may be required for machine maintenance and/or repairs.
- To avoid the risk of the head becoming jammed in squeezing ground, the machine would require variable speed drive motors, and be able to provide full torque at low rotation speeds.
- The EPBM main and tail shields would incorporate articulation to help facilitate line and grade adjustments, as well as negotiating curves.

The excavated tunnel and shaft muck will be screened for hazardous materials to identify the appropriate disposal location. It will be dispositioned for beneficial reuse wherever possible. However, it is anticipated that the majority of the muck will contain elevated levels of naturally occurring chrysotile asbestos. This will dictate special handling requirements and potential disposal as a classified hazardous waste. This will also require enhanced Personal Protective Equipment (PPE) for underground personnel during excavation, as well as increased air quality monitoring.

The tunnel lining will consist of precast concrete segments that will act as the initial support as well as the final lining system (i.e., a one-pass lining). The segments will be bolted and gasketed to limit water intrusion and the potential outflow of any effluent. The annular gap between the segments and the surrounding ground will be backfilled with a two part grout to limit potential settlement and provide good embedment for the lining system.

PROJECT SHAFTS

There are a number of proposed shafts for the project, including:

- An EPBM launch and main construction operations shaft which will also be used to house the permanent lift station at the south side of Islais Creek;
- An EPBM receiving shaft at the south side of Mission Creek, which will also be used as the permanent upstream drop shaft;
- A connection shaft near the downstream end of the tunnel to tie in an existing box sewer to provide wet weather operational flexibility;
- There are currently no intermediate vent shaft locations associated with the current alignment but they may be added during the design phase depending upon the results of hydraulic modeling and air entrainment studies.

TBM Launch and Lift Station Shaft

As depicted in Figure 6, the launch shaft for the EPBM will be constructed at a 1.72-acre site (2 Rankin Street) at the northwest corner of Rankin Street and Davidson Street. The site is located in an industrial area and is currently a vacant lot. Historically, it was occupied by a large fish packing plant that was developed sometime prior to 1938. That structure, which was supported on deep foundations, was demolished sometime around 2010. Some of the old foundation elements of the facility were left in-place and will need to be addressed during design and construction. With the large shaft size and amount of work activities centralized at 2 Rankin Street, the contractor's working space will also be very constrained, which will require efficient site planning and organization.

The internal diameter of the launching /lift station shaft is currently set at 110 feet, which is governed by the space needs of the permanent lift station structure. The shaft is anticipated to be excavated to a depth of approximately 183 feet.

The anticipated geological strata at the launching/lift station shaft site, based on preliminary site borings are as follows:

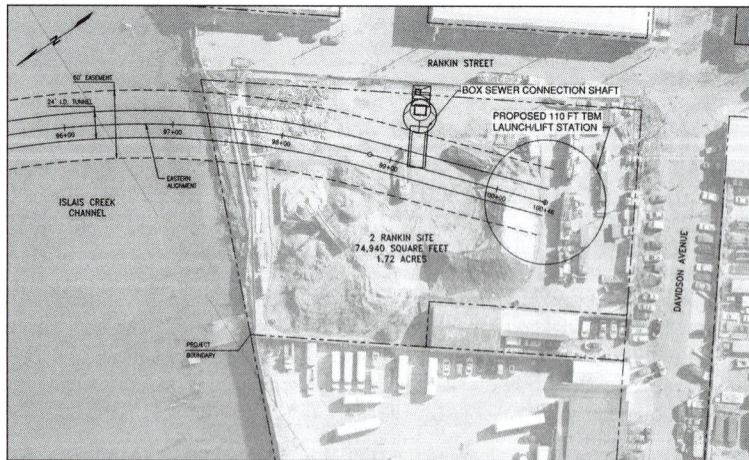

Figure 6. EPBM launch and lift station shaft

- 0 to 8 feet depth: Fill
- 8 to 93 feet depth: Young Bay Mud (CH)—very soft to stiff (increasing with depth)
- 93 to 99.5 feet depth: medium sand with silt and gravel (SP)—very dense
- 99.5 to 145 feet depth: Old Bay Clay (CH)—stiff
- 145 to 151 feet depth: fine to medium sand (SM)—very dense
- 151 to 164 feet depth: well graded gravel (GW)—very dense
- 164 to 169 feet depth: clayey sand with gravel (SC)—medium dense
- 169 to 184 feet depth: fat clay with silt (CH)—very stiff
- 184 to 191 feet depth: silty fine sand (SM)—medium dense
- 191 to 194 feet depth: well graded gravel with sand (GW)—very dense
- Below 194 feet depth: Franciscan Complex

The groundwater table in the vicinity is generally around 5 to 10 feet below the ground surface.

Receiving Shaft

The receiving shaft of the Eastern Alignment is to be located on a 1.96 acre empty lot referred to as Parcel P7 on the west side of Owens Street, just north of the University of California, San Francisco Medical Center.

The internal diameter of the receiving shaft is currently planned to be 45 feet, which will accommodate retrieval of the EPBM as well as the construction of permanent access, connections to the CHS and hydraulic drop structures. The shaft will be excavated to a depth of approximately 140 feet.

The anticipated geological strata at the receiving shaft site, based on preliminary site borings, are as follows:

- 0 to 19.5 feet depth: Fill
- 19.5 to 64 feet depth: YBC (CH)—very soft to soft
- 64 to 101 feet depth: sand (SP with some SM)—very dense
- 101 to 125.5 feet depth: OBC (CH)—very stiff
- 125.5 to 134 feet depth: sandy clay (CL)—very stiff
- 134 to 145 feet depth: sand with silt (SP-SM)—very dense
- 145 to 173 feet depth: OBC (CL-CH)—very stiff
- Below 173 feet depth: Franciscan Complex

The groundwater table is generally located about 8 feet below the ground surface.

Shaft Construction Methods

A number of different shaft construction methods were considered, including slurry diaphragm walls, ground freezing, secant piles, and caissons. Based upon the geologic information collected to date, as well as the size and depth of the shafts, slurry diaphragm walls are anticipated to be the preferred method of construction.

The slurry wall panels for the shafts would be keyed into the Franciscan Complex rock to provide a water cutoff system to allow them to be excavated in the dry with reduced dewatering efforts. Packer tests performed in the Franciscan Complex rock during the initial geotechnical investigation program had no water take which indicates very low permeability. The slurry wall embedment depth and shaft bottom stability would be confirmed following the detailed future geotechnical investigations during design. Additional grouting may be required at the toe of the slurry wall panels to improve the groundwater cutoff where zones of permeable features are encountered. Slurry wall trench stability may also be a concern for the construction of the panels, so deep soil mixing or other means of ground improvement may be utilized for strengthening the upper soft Young Bay Mud layers.

Water disposal will an important aspect of the construction. At the Rankin site, the only economical discharge point for collected groundwater will be the adjacent Islais Creek that discharges into San Francisco Bay. Because of this, stringent water disposal standards will be enforced on the project. The project contract documents will specify a water treatment facility that will sufficiently treat any large inflows into the shaft, particularly during the TBM break-out period.

Ground improvement by jet grouting will also be specified outside of the shafts to create a seal to mitigate the inflow of water and soil upon break-out/break-in of the machine through fiberglass rebar "soft eyes" located in the shaft walls.

PROJECT SCHEDULE

The final design of the Channel Tunnel is currently underway and is expected to be completed by the end of 2018. The project environmental documentation is being performed in parallel with the design and is also expected to be complete in 2018. The project is currently scheduled to go to construction in 2019.

SUMMARY AND CONCLUSIONS

The SFPUC's proposed 1.7 mile long, 24 foot diameter Channel Tunnel is a critical wastewater conveyance/storage facility for the City of San Francisco. Replacement of the existing antiquated force main pipeline with the tunnel is required to adequately address the project service requirements, and upgrade the existing seismically vulnerable wastewater facilities.

The geologic conditions are well suited for pressurized face Tunnel Boring Machines and recent local tunnels in the area have been successfully constructed using Earth Pressure Balance technology. Tunneling challenges include excavating through highly variable ground conditions including Franciscan Complex rock, stiff to hard clay (Old Bay Clay), sands and mixed-face conditions, all located under the groundwater table.

Based upon the current project information, the tunnel alignment is located at depths up to 130 to 148 feet to avoid potential conflicts with existing pile foundation obstructions and soft soils that would present very difficult tunneling conditions. This would require several deep, large diameter shafts which would be some of the deepest soft ground shafts ever constructed in North America. These will be further investigated during the design studies.

ACKNOWLEDGMENTS

The authors wish to acknowledge the support provided by the Jacobs Engineering, the San Francisco Public Utilities Commission, and Stantec/MWH Global, and for the preparation of this publication.

REFERENCES

Caulfield, R. John, Whitman E., Stevens J., Fleming K., and Wong J.I. 2014. Construction of the Bay Tunnel. *In Proceedings of the North American Tunneling Conference*, Los Angeles, California: SME.

ENR. San Francisco's $1.6 Billion Central Subway Stays on Course. November 23, 2015. www.enr.com/blogs/14-gold-rush/post/38027-san-franciscos-16-billion-central -subway-stays-on-course

Fippin R.L., R.J. Caulfield, M. Wong, W. Lee. 2011. Sunnydale CSO Tunnel—Dealing with Urban Infrastructure. *In Proceedings of the Rapid Excavation and Tunneling Conference*, San Francisco, California: SME.

Graymer, R.W. et al. 2006. Geologic Map of the San Francisco Bay Region. U.S. Geological Survey, Scientific Investigations Map 2918.

MWH/URS Joint Venture, April, 2015. Channel Tunnel Alignment Alternatives Analysis Technical Memorandum, Final Draft. Internal Report.

MWH/URS Joint Venture, May 5, 2015. Central Bayside System Improvement Program, Revised Preliminary Geotechnical Interpretive Technical Memorandum. Internal Report.

MWH, September 2016. Central Bayside System Improvement Program. Draft Interim Geotechnical Phase 2B Data Report (GDR) Addendum, Internal Report.

U.S. Geological Survey, 2016. Bay Area Faults.

PART **9**

Pressure Face TBM II

Chairs

James Wonneberg
McMillen Jacobs Associates

A.G. Mekkaoui
Jay Dee Contractors

EPB or Slurry TBM? Suffolk County, Long Island, NY, Outfall Replacement Tunnel

Michael S. Schultz ▪ CDM Smith
Greg Sanders ▪ CDM Smith
Mary Anne Taylor ▪ CDM Smith
John Donovan ▪ Suffolk County Department of Public Works

ABSTRACT

Suffolk County is planning to build a replacement outfall pipeline with a 4.6 km (2.8 mile) 3.3 m (10.8 ft) internal diameter soft ground tunnel. The Owner and designer had to consider significant regulatory and permitting requirements, maintaining the existing outfall during construction of the outfall replacement, potential risk issues, ground conditions and community acceptance during design. Major decisions included selecting the TBM method (slurry versus earth pressure balance machine), use of ground freezing versus secant piles for shaft construction, GBR baseline values, potentially variable ground conditions and working in and around environmentally sensitive areas. This project is expected to bid and be awarded in early 2017.

INTRODUCTION

Treated effluent from the Bergen Point Waste Water Treatment Plant (WWTP) is discharged through a 10.5 km (32,000-foot) long outfall that was constructed in 1977 and consists of 1,829 mm (72-inch) diameter prestressed concrete cylinder pipe (PCCP) and concrete coated steel pipe. The PCCP section of the outfall, starts at the WWTP effluent pump station and extends beneath the floor of the Great South Bay to the barrier island and then to the Atlantic Ocean out beyond the surf zone. Assessment by CDM Smith and an independent assessment of the pipeline concluded that the pipeline beneath the Great South Bay was potentially in a failed state and the design of a replacement outfall between the WWTP and the barrier island was required.

Replacement of the existing outfall requires the construction of approximately 100 meters (330 feet) of new near surface 1829 mm (72-inch) effluent pipeline and a 4.3 km (14,005 feet) new outfall tunnel with a minimum 3048 mm (120-inch) internal dimeter to replace the section of the existing outfall.

The near surface piping at the WWTP will convey the effluent from Effluent Ultraviolet Disinfection building to the tunnel launch shaft with a tie in to the Final Effluent Pump Station. The pipeline alignment will be located between an existing 15.24-m (50-foot) diameter water tank and the existing Final Effluent Pump Station in an open cut excavation.

The new tunnel will start at a shaft adjacent to the final effluent pump station and run beneath the mudline of the Great South Bay to the connection point near the existing outfall sample chamber, located at Gilgo State Park on the barrier island just north of Ocean Parkway. Figure 1 shows the proposed outfall alignment, existing outfall alignment and easement. The tunnel will be excavated in a southerly direction from the Launch Shaft located on Long Island to the Receiving Shaft on the barrier island. A slight up slope of 0.1 percent was selected as the gradient, since it is adequate for the

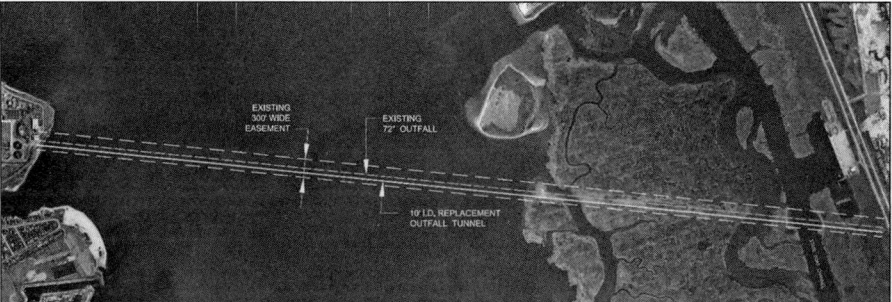

Figure 1. Tunnel alignment

long-term function of draining the tunnel. The invert of the tunnel at the Launch Shaft is set at elevation –100.0. At this elevation and slope, based on the anticipated tunnel diameter the minimum overburden between the tunnel spring-line and bottom of lowest dredge depth of the three boating channels is 5.1 diameters and the maximum hydrostatic pressure at invert is 3.5 bars.

The connection piping from the receiving shaft to the existing outfall will consist of an approximately 29.5 m (90-foot) section of 1829 mm (72-inch) dimeter near surface pipe installed in an open cut excavation. An 1829 mm (72-inch) Tee connection and a 13.1 m (40-foot) long pipe section are also included in this. To make the pipe connection a line tap and approximately 32.8 m (100 feet) of 1829 mm (72-inch) bypass line will be installed.

GROUND CONDITIONS

Soil Conditions

In general, the soil profile from the bottom of the bay below the mud line, elevation 0 to elevation –20, consists of loose sand and silt and soft to very soft clay and plastic silt. Below this loose to soft soil extending to about elevation –65 is predominately medium dense to dense silty sand, clayey sand and sandy silts. Within this general matrix there are isolated pockets of clayey silts, poorly graded sands and well graded sands typical of outwash deposits. Underlying this material and extending to below the tunnel invert the soil continues to exhibit a distribution of an outwash deposit however, the composition of the soil grain size is finer and generally the consistency of the sandy soil is medium dense to very dense and the clayey soils are stiff to hard.

As shown on Figure 2 the face conditions will often be composed of two or more soil groups. Depending on the overall composition of the soil at the face and the behavioral tendencies of the soil groups included in the composition changes in the tunneling ground behavior is anticipated over relatively short lengths of the tunnel drive. To maintain an efficient excavation and mucking process the tunnel boring machine to be selected by the Contractor must be capable of adjusting to changes in the ground considering the different soil compositions. These changes will likely require modifications to the conditioner mix frequently.

Groundwater

Groundwater levels are at or near the ground surface at the shafts and at sea level along the tunnel alignment. There are two groundwater aquifers generally recognized to exist in the project area; an upper glacial aquifer which includes the water

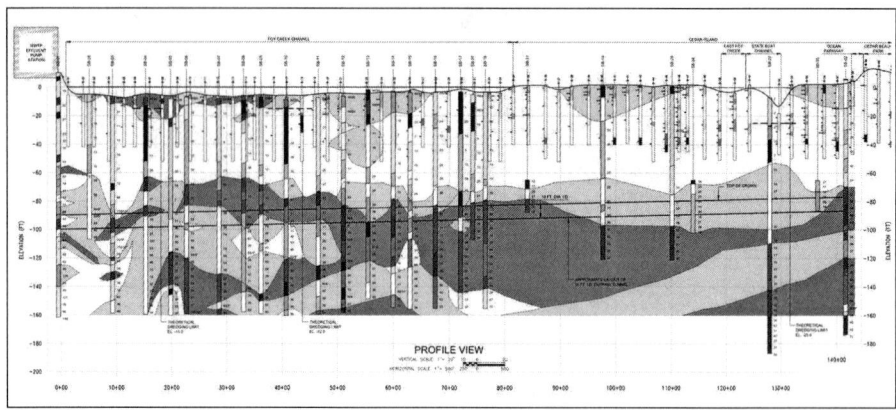

Figure 2. Tunnel profile

table throughout most of Long Island and the upper portion of the Magothy aquifer. Groundwater levels along the tunnel alignment are expected to be a result of the upper glacial aquifer. However, some piezometers at the receiving shaft site encountered artesian conditions resulting from the Magothy aquifer.

SITE RESTRICTIONS

Several factors associated with the tunnel site impose limitations on the proposed outfall replacement design and construction. One factor affecting the construction is that overall plant operations must be maintained while performing all contract work. This requirement includes the ability of the county to process over 30 million gallons of water per day every day during the construction efforts. This directly affects the launching shaft location which is bordered on the north by an existing 2743 mm (108-inch) effluent pipe and to the east by the Effluent Pump Station and the existing outfall. The conditions of the existing effluent pipe and the existing outfall are unknown and may be easily damaged. Both the pump station and the pipelines must remain in operation during construction and therefore stockpiling of materials and passage of equipment will not be allowed above the pipelines to reduce the risk of damage.

In addition to the outfall tunnel replacement project, the Bergen Point Waste Water Treatment Plant is currently undergoing a plant expansion consisting of several separate construction projects. Coordination with the other contracts will be critical and are expected to have an impact on muck removal, site access, and available work space. The scale of the other construction projects limit the available work area for the tunnel project to a narrow section along the south side of the plant, adjacent to the shoreline of the Great South Bay. A power drop for the launch shaft site will be provided by the owner near the north side of the waste water treatment plant (Figure 3). It will be the contractor's responsibility to determine the best location for routing the power to the launch shaft site and to coordinate installation of the power with the plant operations and at locations where it crosses the other contractors work areas.

To the south of the tunnel site work area, the shoreline of the Great South Bay is lined with riprap and disturbance of the riprap is strictly prohibited during construction. Water levels at all shaft locations are at or very near the ground surface and are anticipated to vary in accordance with the water level in the Great South Bay.

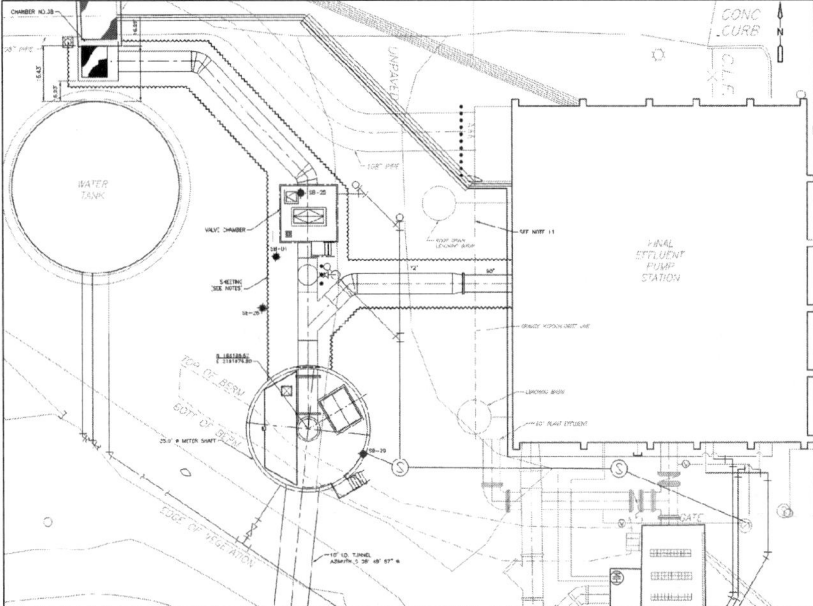

Figure 3. Launch shaft site

Although the receiving shaft site is not as congested as the launching shaft site, it has its own unique challenges. The receiving shaft is located on the barrier island near a protected wetland area. The existing outfall pipe is located to the east and like the launch shaft site, stockpiling of material and the passage of construction equipment will be prohibited above the pipe. A power drop will not be available at the receiving shaft site and the contractor is expected to provide portable power units if required. The proximity of the protected wetland, limits the northern edge of the work area and wildlife and insects, including mosquitos, are anticipated to be an issue during the summer months.

TBM SELECTION

During design, the selection of the TBM type (EPB versus Slurry machine) was made primarily based on the anticipated ground conditions and muck handling considerations.

As shown in Table 1 the Suffolk Outfall Replacement Tunnel will be in a soft ground mixed face conditions. Mixed face conditions are defined in the contract documents when two or more of the Soil Groups occur at the tunnel face. Variations in the soils at the tunnel face can be expected to be encountered along the tunnel alignment and the tunnel face is anticipated to rapidly transition from non-plastic to high plastic materials (silts to clays) over a relatively short tunnel length. However, some of the variations at the tunnel face will include a rapid transition to sands and one section is expected to contain a clean sand and gravel material as part of a mixed-face condition. Cobbles, boulders and other obstructions are also considered to be possible within the tunnel alignment, especially, at locations where the tunnel face is transiting from the fine-grained soil groups (silts or clays) to coarse grained soils (sand and gravel).

Table 1. Soil composition percentage by tunnel reach

Tunnel Reach	Reach Station Range (ft)	Percentage of Soil Composition Anticipated Within Tunnel Reach			
		Gravel	Sand	Silts	Clays
1	0+17.5 to 95+00	0 to 5	3 to 8	40 to 45	50 to 55
2	95+00 to 140+90	0 to 2	3 to 8	87 to 92	3 to 8

Although the summary of conditions would indicate that for a large percentage of the tunnel length where the finer grained soils are anticipated (Tunnel Reach 1), the conditions might be considered more favorable for use of an EPB machine. However, the anticipated mixed faced conditions along the tunnel alignment are expected to result in greater challenges for efficient use of soil conditioning when using an EPB machine. Although, EPB machines generally have a lower initial capital cost than slurry machines, the use of additional soil conditioning additives may increase the total cost to the project to a point that the higher equipment cost of a slurry machine is negated. An EPB machine equipped with the ability to respond to these changes in face conditions quickly will help to efficiently manage the muck excavation process and could possibly reduce the amount of soil conditioner that is needed.

The contract documents require that the TBM be operated in closed mode at all times and face pressure shall be maintained to mitigate risk for potential frac-out conditions by maintaining a slightly positive outward pressure. The baseline value for water head pressure at the tunnel invert along the entire tunnel profile has been set at 3.5 bars. Groundwater conditions in the northern portion of the tunnel alignment are likely representative of the upper glacial aquifer and are expected to be lower than the baseline value. However, a rapid increase in the water head at the tunnel face should be anticipated if tunnel alignment enters a soil stratum influenced by the Magothy aquifer and should be considered during machine selection.

Traditionally, Slurry TBMs have been considered better suited for compensating with rapid pressure changes and have been recommended for water pressure over 2 bar, though EPBs with higher-pressure limits of up to 7 bars have been used on previous projects such as the Harbor Siphons project in New York City and the Brightwater Conveyance Project in Seattle, Washington. It should be noted that for the central tunnel contract of the Brightwater project, an EPB machine was used to complete a tunnel with a water pressure of up to 7 bars that was not completed by a slurry machine due to excessive wear to the inside of the cutter head from the abrasive glacial conditions.

The surface equipment requirements of both machines must also be considered. EPB machines typically have higher power requirements resulting in a larger power drop at the surface and increased electrical cost. Additionally, the limited amount of surface space may also prevent the use of a conveyor for muck removal requiring the use of muck cars for an EPB machine. However, the slurry separation plant for a slurry machine must be sufficiently large enough to handle the flow rate for the expected advancement rate. Likely requiring a larger footprint at the launch shaft site than an EPB machine equipped with muck cars.

Both types of machines can perform the required excavation for the Suffolk Outfall Replacement Tunnel. Often there is less perceived risk among designers with Slurry compared to EPB, because the entire system is automated and less likely to be affected by the operator experience. To address the issue of operator experience, experience requirements were added to the contract documents. Additionally, regardless of the TBM selection, daily meetings between the tunnel superintendent and inspection team are required to discuss the anticipated ground conditions for each daily advance and the operators plan for maintaining control of the tunnel face.

CONTRACT REQUIREMENTS

In preparing the Contract documents for bidding, many of the risks associated with this project were considered along with the site constraints, environmental require-ments, ground conditions and other factors. Specifically, the Geotechnical Baseline Report, Technical Specifications and Drawings were all prepared recognizing that cer-tain issues needed to be specified very tightly and other items were left more open to allow the Contractors bidding the work flexibility in determining the best use of the means and methods of construction based on their experience, equipment and knowl-edge. This was done mainly through the GBR, but clearly is evident in the rest of the Contract Documents. Within the GBR a section describing the project constraints and project flexibilities was included. Among the project constraints are included (but not limited to):

- Tunnel is to be excavated using a pressurized face machine,
- Tunnel Launch shaft must use ground freezing techniques for temporary earth support, and
- Tunnel Shaft site is restricted to available working space shown on the Drawings.

Project Flexibilities included in the GBR (but not limited to):

- Pressurized face TBM can be either EPB or Slurry,
- Final Outside Diameter of Tunnel,
- Excavation Support for receiving Shaft open to ground freezing or secant pile wall, and
- Shape of the Launching Shaft.

SUMMARY

Construction of the Suffolk County outfall replacement tunnel will face many chal-lenges. The contractor will be required to carefully evaluate the cost, power require-ments, site size, and their own experience to make the decision on the selection of Earth Pressure Balance vs. Slurry Pressure Balance Equipment. In recent years, the use of additives and new injection systems have allowed a broader range of soil condi-tions to be excavated by the two types of machines. Ultimately, it was decided that the selection of the TBM type was best left up to the contractor for this project. Although EPB machines appear to have an advantage, ultimately the most cost efficient solu-tion will be determined by factors such as the available work area at the launch shaft, contractor equipment inventory, and cost of conditioning agents, and contractor per-sonnel experience, which cannot be quantified during design.

REFERENCES

DAUB (2011) "Innovations and Limitations of Two Long-Standing Soft Ground TBM Designs," Tunnel, 5/11, 20–35.

Lovat, P.R. (2006), "TBM Design Considerations: Selection of Earth Pressure Balance or Slurrry Pressure Balance Tunnel Boring Machines," Int. Symp. On Utilization of underground space in urban areas, 6–7 November 2006, Sharm El-Sheikh, Egypt.

SR99 Bored Tunnel in Seattle: Performance and Challenges of "Bertha," the Largest TBM Ever

Roger Escoda ▪ Dragados USA, Inc.
Juan Luis Magro ▪ Dragados USA, Inc.
Jorge Vazquez ▪ Dragados USA, Inc.

ABSTRACT

"Bertha," the 17.48 m diameter Hitachi Zosen EPB TBM procured by Seattle Tunnel Partners to bore the SR 99 BT in Seattle finally resumed mining in January 4th, 2016. This paper discusses the technical challenges faced since then, such as tunneling right below downtown Seattle through challenging geology, below water table, with shallow cover and underneath old utilities and already damaged structures, like the Alaskan Way Viaduct that the tunnel will ultimately replace. It also describes its performance and provide comments and insight into the design and operation of future extra-large diameter TBMs.

INTRODUCTION

The SR99 Bored Tunnel Project is one of the most publicized and real-time broadcasted projects in tunneling history. It was from project inception just because of featuring the largest TBM ever built by that time and even more soon after when the world-record TBM was down for two years after encountering an abandoned steel well casing. During that period, the TBM went through one of the most challenging repairs ever performed in this industry, including sinking an 80 ft. wide and 180 ft. deep access shaft in extreme conditions, right between the Puget Sound and the already damaged Alaskan Way Viaduct; manipulation of TBM components up to 2,200 tons of weight and precision machining performed in place to TBM components of over 10 meters diameter.

"Bertha" finally resumed mining in January 4, 2016, so now that it is approaching the end of its tunnel drive, time has come to recap the challenges the TBM faced and some of the keys of its performance.

Figure 1. Aerial view of the SR99 viaduct in downtown Seattle, WA, in early 1950s (top) and actual (bottom)

Figure 2. SR99 bored tunnel alignment in downtown Seattle, WA

PROJECT DESCRIPTION

Built in the early 1950s, the Alaskan Way Viaduct (AWV) is part of the State Route 99 that crosses downtown Seattle from South to North. It helped to relieve congestion of trains, trucks and wagons carrying cargo to and from ships. By the end of the last century, it was among the state's busiest and most important sections of highway, carrying 110,000 cars each day.

In February 2001, a 6.8 magnitude earthquake named "Nisqually" damages the AWV, which was closed several months for inspection and limited repairs but had the quake lasted a few moments longer, the AWV would have collapsed, so the AWV project began.

In 2009 and after considering more than 90 different alternatives, Governor, King County Executive, Seattle Mayor and Port of Seattle CEO recommend replacing the viaduct's central waterfront section with a bored tunnel beneath downtown and State Legislature approved bored tunnel funding.

The SR99 Bored tunnel is a Design-Build contract worth $1.34 Billion led by Washington State Department of Transportation (WSDOT), in partnership with the Federal Highway Administration, King County, the City of Seattle and the Port of Seattle, which was awarded to Seattle Tunnel Partners, a joint venture of Dragados USA, Inc. as sponsor and Tutor Perini Corp. in January 6, 2011. Project kicked off after Notice to Proceed #1 (NTP1) was issued in February 2011 and NTP2 in August 2011, once Environmental Impact Statement (EIS) was finally approved.

The SR99 Bored Tunnel Project consists of 9,273 feet of segments lined tunnel and about 1 million cubic yards of muck bored by and EPB TBM nicknamed "Bertha" of 57.5 ft. diameter, manufactured by Hitachi Zosen in Osaka, Japan. It goes right underneath downtown Seattle from South to North and connects the Access area right beside the Stadiums to the North Portal and disassembly area of the TBM, adjacent to the Space Needle. Both portals comprise roughly 540,000 yd^3 of excavation between pile walls and connection ramps plus one operations building per portal to hold all the permanent systems of the tunnel.

The double-deck roadway is being built right behind the TBM as a combination of cast-in-place, shotcrete and precast structures, for a total of 75,000 cubic yards of concrete poured and 14 million lbs. of rebar placed. Permanent systems such as electrical, mechanical, ventilation, gas monitoring, drainage and pumping, fire suppression, security, communication and SCADA systems are built behind the concrete work.

TECHNICAL CHALLENGES

Geology and Hydrogeology

Geological conditions around Seattle area are complex as a result of glacial scouring and sub-glacial erosion occurred during the recent Pleistocene Epoch. This complex glacial stratigraphy has also a strong influence on the hydrogeological regime and the nature of the groundwater flow. The permeability of glacial deposits typically differ by orders of magnitude between adjacent stratigraphic units and locally within a single stratigraphic unit.

The soils encountered during the bore of the tunnel fell within three major categories: till and till-like deposits, cohesionless soils (sand, gravel and silts) and cohesive soils (silt and clay) with transitional and interspersed lenses, dikes, layers, cobbles, boulders, logs and man-made debris such as timber piles, refuse, and organic soils.

The final result of this complex geology and hydrogeology is such that the tunnel face had never been in homogeneous soil conditions, therefore had been very difficult to anticipate any particular ground condition in advance requiring frequent adjustments to tunneling operations and logistics such as soil conditioning, cutter head rotation speed and penetration, tool inspection, both in atmospheric and hyperbaric conditions, mucking out logistics, and maintenance regimes among others.

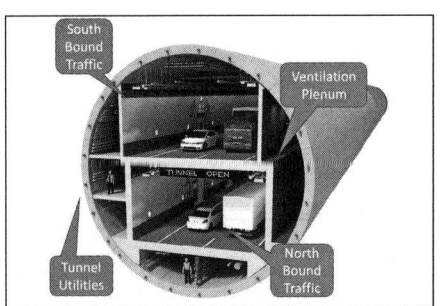 

Figure 3. SR99 bored tunnel cross section

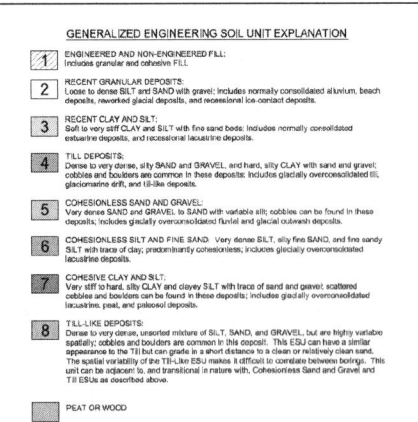 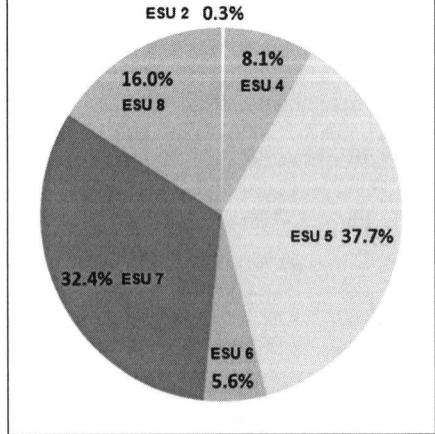

Figure 4. Engineering soil units characterization for the SR99 AWV replacement project

The maximum face pressure applied during the drive had been 60 psi at crown level or equivalent 100 psi at invert, given the size of the TBM, with a maximum water table elevation above the tunnel crown of 116 ft. The following are, among others, the major geological challenges Bertha had to overcome:

- **Sticky / Clogging clays.**
- **Cobbles and boulders** from igneous or metamorphic origin or sedimentary concretions of an anticipated unconfined compressive strength from 9,000 to 60,000 psi, and of sizes and quantities according to the GBR varying from,
  - <2 feet abundant—not measured
  - 2 to 8 feet 500
- **Abrasive granular soils** in glacial and interglacial sands, gravels, and till which had been classified from moderate to very high abrasive soils (% average quartz content between 34 and 77) and as such, with the potential to cause heavy wear on the TBM and ancillary equipment during excavation.
- **Shear zones and fractures** surfaces were present within several soil units in the over consolidated cohesive soils potentially causing wedge failures, block releases, spalling and sloughing of overhead and steeply inclined surfaces if not supported as well as transmitted hydrostatic pressures to the face and annular space along the outside of the shield increasing the likelihood of soil instability.

Several case studies, since early 20th century, on tunnel projects around Seattle area, prior to the AWV Replacement Project, have reported to have encountered very challenging ground conditions before which make it an even bigger challenge for such a large TBM.

Urban Environment

Downtown Seattle, where the tunnel boring runs from the south end of the waterfront until the north portal, nearby the Space Needle, is a congested urban area combination of old masonry buildings, some of them as old as late 19th century and modern constructions and skyscrapers, amongst the tallest in the west coast (more than 50 exceeding 400ft height).

Amongst the singular constructions Bertha had gone underneath are the elevated monorail, constructed in 1962 jointly with the Space Needle for the 21 World's Fair Exhibition, the Battery Tunnel, the Elliot Bay Interceptor, the Burlington Northern Santa Fe (BNSF) Tunnel, the Pike street adit and the Alaskan Way Viaduct which the tunnel will replace, without forgetting the innumerable utilities above and below ground to service the city.

To confirm ground movements at surface and subsurface were within specified criteria during the mining of the tunnel a thorough instrumentation and monitoring plan was developed and put in place in order to have real time data in all these adjacent buildings, utilities and singular structures. This included Near Surface Settlement Points (NSSPs), Automated Structure Monitor Points (ASMPs), Manual Structure Monitor Points (MSMPs) and Utility Settlement Points (USPs).

Moreover, the plan included inclinometers (INCL), piezometers (PZ) and Multiple Position Borehole Extensometers (MPBXs) with anchors located 5 and 10 feet above tunnel crown and 20 feet below grade, regularly distributed circa every 50 feet of tunnel advance. The aim was to have real time data of the potential disturbance of

 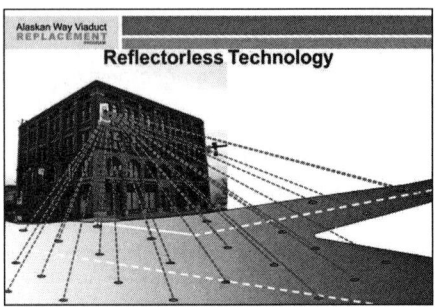

Figure 5. Automated Total Station being installed on top of a building

the ground induced by the TBM and have the ability to adjust tunnel operations and parameters as soon as possible to minimize the propagation of any potential ground loss to above structures and preserve them from any potential damage.

The data showed displacement of the MPBXs had been generally within criteria (<4 inches) achieving values normally at the bottom two anchors from −0.1 inches (heave) in shallow cohesionless ground from ESU5 and ESU6 to less than 1 inch settlement in over consolidated till and till like deposits from ESU4 and ESU8 and barely no movement in cohesive clay and silt from ESU7, whilst the top anchor has shown typically no discernable movement. Surface movement had also been within criteria (<1 inch) with movements typically ranging from 0 to 0.25 inches settlement which means a maximum volume loss of 0.2%.

The Alaskan Way Viaduct Crossing

Special attention deserves the crossing under the existing viaduct. This represented a major achievement for the tunnel construction as the world largest TBM was able to cross underneath the preexisting heavily damaged structure at a minimum distance of 5 feet from its bottom foundations without causing any settlement but slightly heave instead.

In order to protect the foundations of the piles a series of inclined micropiles were built, as far as reasonably possible depending on surface constraints and utilities location, to create a physical barrier to cut or mitigate the development of the potential settlement trough on each of the affected bents. The predicted settlement on the bents' piles was anticipated to be from −0.18 to −0.93 inches whilst the final settlement achieved was actually within +0.13 to +0.22 inches, thus heave.

Shallow South Portal Entrance and North Portal Exit

Since the tunnel replaces the existing elevated viaduct on Seattle's waterfront, means both ends connect with existing infrastructures (SR99 highway on the south and Aurora Avenue on the north). This makes the tunnel had to be very shallow in both ends in order to minimize the access and egress ramps, which implied a technical challenge for such a big machine to mine with less than a quarter diameter overburden.

In order to mitigate the potential surface impacts of such a shallow large tunnel in an urban environment two different approaches were taken. In the south portal were the tunnel run parallel to the existing viaduct, and a larger site surface area was available, a 1450 feet long secant pile wall was built at both sides of the tunnel alignment and was complemented with a jet grouting anti-buoyancy slab in order to encapsulate the

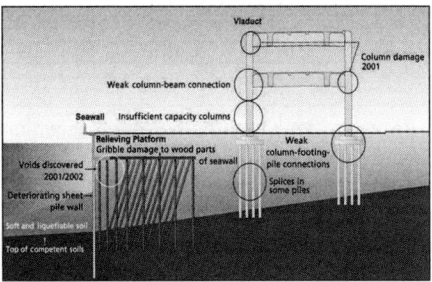 

Figure 6. Cross section and plan view of the AWV crossing

Figure 7. Tunnel profile alignment sections

tunnel, trying to minimize the settlement impact at surface level. In addition three mortar piles walls, called Safe Havens were built to allow for cutter head inspections. This ground improved stretch served also as a training section before going underneath the viaduct and downtown Seattle where to fine tune plans and procedures and validate settlement predicitons.

At the north portal where the space constraints were huge the mitigation measure was to build a 70 feet long canopy of treated ground along the crown of the TBM from the south secant pile wall in the extraction pit. The purpose of this canopy was to prevent ground loss during the final breakthrough in order to protect a significant number of utilities located at the intersection of 6th Avenue and Thomas Street where the final breakthrough took place. This was achieved via 21 horizontal sleeveport grout pipes and injection of more than 60,000 gallon of sodium silicate grout to treat a relatively loose and permeable ground above.

TUNNEL DRIVE—PRODUCTION

As commented before, geological conditions at SR-99 Project were very complex. As a result, underground stratigraphy and soils encountered during tunnel drive determined TBM performance and parameters values. In order to expose the project data regarding "mining production rates and consumables" the tunnel alignment was split on 5 stretches with similar geological characteristics, these stretches are stated below:

- SECTION 1, STA 194+50 to STA 198+00
- SECTION 2, STA 198+00 to STA 210+00
- SECTION 3, STA 210+00 to STA 237+00
- SECTION 4, STA 237+00 to STA 274+00
- SECTION 5, STA 274+00 to STA 287+20

Section 1, STA 194+50 to STA 198+00

Having performed the first 100 m of excavation as a "learning curve phase," TBM had to handle and overcome some engineering challenges as:

- Small coverage with recent granular deposits (less than half diameter), which implied:
  - Concrete slab at surface to avoid surface communication and TBM buoyancy, and
  - Ground treated with jet grouting, which make more complicated the muck flow throughout screw conveyors and TBM plenum. Also, some stoppages were needed to clean the TBM chamber.
- Mixed face (fill alluvium, till deposits and some clay)
- Mining under water level and by the west side of to the AWV (Alaskan Way Viaduct). So mining through saturated soil was constant in this stretch.

Main TBM parameters information at this stretch are shown in Table 1.

In relation to consumables, it is worth mentioning the following:

Table 1. Tunneling main parameters values, Section 1

Length Section 1	335 ft
Time to complete (working days, wd)	17 wd
Average per wd	25.26 ft/wd
Best day's production	45.93 ft

- Bentonite injection at the chamber was needed so that muck flow ability was adequate (increase of fine percentage).
- Foam agent specific for granular soils plus polymer was used due to saturated condition of the soil and to strength the foam (bubbles).
- Grout bi-component has been used during the whole tunnel drive, but a higher concentration of retarder was used in this stretch to allow 72 hours of stability.
- Cutting tools: disc cutters were distributed evenly at the cuter head face, since TBM had to mine through concrete walls at the beginning and end of this stretch. Apart from these disc cutter, rippers and cutting bits were used.

Section 2, STA 198+00 to STA 210+00

At this stretch the TBM had to face mainly a mixed face of cohesionless sands, silt and gravels with some clay at the beginning. In addition, same saturated conditions due to the water proximity and driving the TBM close to the AWV. The coverage was barely one diameter or less, and without the benefit of the concrete slab that there was in the previous section. In contrast, concrete piles were installed at both sides of the tunnel to either avoid excessive ingress of water from west side or influence any compromising structural behavior on the AWV.

At the end of this section, the TBM had a maintenance stop at the Safe Haven #3, just before mining underneath the AWV. Main parameters data at this section are represented in Table 2.

Regarding soil conditioners, recipes used explained below:

Table 2. Tunneling main parameters values, Section 2

Length Section 2	1,198 ft
TBM working pressure (average top sensors)	3.50 bar
TBM rate of advance (ROA) average	17 mm/min
Time to complete (working days, wd)	40 wd
Average per wd	29.92 ft/wd
Best day's production	58.40 ft
Hyperbaric interventions (#intervention/#dives)	2 int./ 168 dives

- Same recipe than previous stretch was used but consumption of polymer was higher due to both higher water presence and higher working pressure. Later on in this section, recipe was changed to anti-clay surfactant due to same clay at the cross section.
- Bentonite was injected mainly in the chamber to increase fine content (at the granular section).

Cutter head configuration was changed so that all cutting tools were rippers (no disc cutters), plus standard cutting bits at both sides of cutter arms.

Two hyperbaric intervention were performed at STA 204+70 and STA 210+00 respectively, with a total number of 168 dives; main information in this regard:

- Hyperbaric Intervention #1, STA204+70:
  - 43 dives total at 1.5 bar working pressure
  - Main activities: Cutter head cleaning, foam and water injection ports maintenance, and 18 cutting bits changed.
- Hyperbaric Intervention #2, STA210+00:
  - 125 dives total at 2.3 to 2.9 bar working pressure
  - Main activities: Cutter head cleaning, foam and water injection ports maintenance, and 9 cutting bits changed (out of 452 units)

Section 3, STA 210+00 to STA 235+00

The main challenge in this stretch was to mine underneath the AWV. During this stage, TBM worked 24/7 (total 14 days), while rest of the drive TBM was driven five day/week (24 hours each), performing maintenance trough the weekends.

Geologically, this section was predominantly composed by clay (ESU7) with some trace of sands. Due to these uniform characteristics, soil conditioning recipe used was "anti-clay foam agent" plus additional anti-clay polymer, with the objective to increase polymer percentage concentration. This was carried out based on the size of the cutter head, and the potential clogging of the spoke windows. As a consequence of this, using some extra anti-clay polymer exposed that it was most cost-effective than cleaning windows during the program maintenance. In addition, it contributes to reduce the necessity to perform additional hyperbaric interventions.

Another polymer implemented on this project was a "water absorber polymer" injected in the screw. It was used at this section when traces of sands, water ingress increased, were visible on the muck samples. Subsequently, a better muck control at the end of the screw was achieved using this additive. Main parameters data at this section are depicted in Table 3.

With respect to the cutter head maintenance jobs, another hyperbaric intervention was completed at STA 225+39, working pressure conditions were between 2.60 to 3.00 bar th 46 dives entirely performed, main activities accomplished on this intervention were: cutter head spokes cleaning, injection ports maintenance and cutting bits replacement (25 bits out of 452).

Table 3. Tunneling main parameters values, Section 3

Length Section 3	2,700 ft
TBM working pressure (average top sensors)	4.0 bar
TBM rate of advance (ROA) average	20 mm/min
Time to complete (working days, wd)	69 wd
Average per wd	39.14 ft/wd
Best day's production	66.6 ft
Hyperbaric interventions (#intervention/#dives)	1 int./ 46 dives

Section 4, STA 235+00 to STA 274+00

(Note: TBM has mined nearly 60% of this stretch at the time of this document section is being written.)

Geological conditions on the subsurface were mainly cohesionless sand and gravels, with a clay stratum at the top and some cohesionless silt and sand stratum at the bottom on the cutter head cross section. Therefore, a highly heterogeneous face condition was the main factor that influenced an unbalanced behavior at the cutter head force values. Additionally, boulders up to 3 ft. diameter were found as well. Maximum overburden was located at this stretch (64 meters), and TBM mined below water table for the whole length.

Respecting to soil conditioning, the recipe for no cohesive soils was used most of the time during the tunneling operation, plus adding water absorber polymer at the screw conveyor with the objective to dry the soil out. Also, Bentonite was added to the cutter face, providing fines to the ground due to its viscous and plastic properties (average of 35 m³). Table 4 exposed the main parameters information collected on this section.

Table 4. Tunneling main parameters values, Section 4

Length Section 3	3,900 ft
TBM working pressure (average top sensors)	3.5 bar
TBM rate of advance (ROA) average (section not completed)	30 mm/min
Time to complete (working days, wd) (section not completed)	48 wd (so far)
Average per wd (section not completed)	45.77 ft/wd (so far)
Best day's production (section not completed)	84.45 ft
Hyperbaric interventions (#intervention/#dives)	2 int./ 34 dives (so far)

Two programmed hyperbaric interventions were performed (36 dives at working pressure 2.90 bar carried out when abstract is being written), where the main maintenance activity aside from general inspection was cutting tools replacements (96 cutting bits out of 452 units).

Section 5, STA 274+00 to STA 287+20

(Note: TBM has not mined yet this stretch at the time this section is being written.)

Last section to be mined by the TBM has similar geological characteristics that previous one, except that in this one TBM will have to mine through ground above the water table.

It is worth mentioning that at the very end of the drive, tunnel coverage is less than half of the diameter, imply to be very cautious with TBM working pressure and communications with surface. A typical canopy has been design at the TBM breakout shaft.

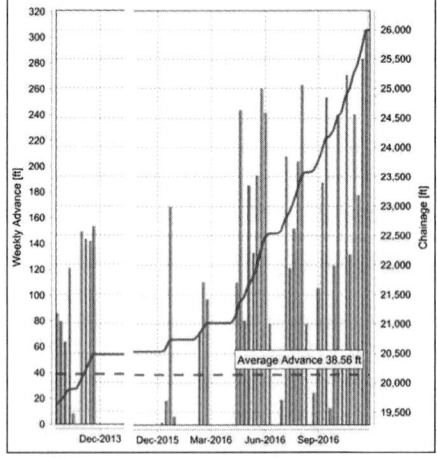

Figure 8. Average TBM performance to date

Due to similarity with section 4, soil conditioners to be used here are the same than previous section, but with exception of adjusting both the screw polymer (water absorber) and bentonite injection according to soil water content.

In regards to hyperbaric interventions, there are two ones scheduled in this sector, but potentially some of them can be skipped due to maintenance atmospheric jobs that are being regularly carried out on the weekends. This maintenance jobs is focused mainly on replacing the atmospheric cutting tools install at the cutter head arms (these ones are mean to be changed from inside of cutter head structure, so under atmospheric conditions, without performing hyperbaric interventions).

DESIGN AND OPERATION OF AN EXTRA-LARGE TBM

There are many factors to be considered to design and operate a mega TBM like "Bertha," especially to bore in extremely complex conditions like the ones depicted above. Some of them could be planned for because of previous experiences and good practice but some others were certainly unknown, a number of them related to the diameter of the tunnel section and, thus, the required size of the machine itself. Largest TBMs before "Bertha" ranked up to 50 ft. (15 m.) of diameter but despite the fact that Bertha meant just a 16% diameter increase, those extra 7.5 ft. opened a whole new world of challenges, but also of possibilities.

After a challenging break-in and a tough learning curve, interrupted by a two-year TBM shut down, tunnel boring effectively happened along 2016 and beginning of 2017 with the following aspects considered critical for its success:

Face Pressure Control

Earth Pressure Balance is always challenging to achieve, particularly in a mega TBM, where factors such variation of specific gravity along different areas and levels of the excavation plenum and heterogeneous mixture of soils inside it weight even more than in a "regular" size TBM. As a result, ground conditioning becomes even more relevant in order to homogenize the EPB soils mixture along the excavation face, so they all perform as a single viscous fluid, capable of transmitting the thrust of the TBM and supporting the pressure of the earth and water in the face.

In addition, face pressure fluctuations had to be carefully controlled since the large excavation surface at the face and around it increases the severity of any ground loss and volume loss event. Tight EPB operational thresholds were set accordingly (±5% of the working pressure established at any given location with maximum face pressure of 4 bar at spring-line) even during no-excavation periods.

The use of an automatic face and shield gap pressure control system ("keep" system) which automatically reestablishes pressure set by pumping bentonite (plus polymer if needed) into the excavation chamber and shield gap when pressure decreases below the set value has been proven very efficient to maintain face pressure within those narrow limits.

Last but not least, rate of filling of the excavation plenum had to be extremely high all the time and to do so, most of the air accumulated at the top of the excavation chamber as a subproduct of the degradation of the foam which conditions the muck (the so-called "air bubble") was efficiently eliminated by means of an automatic remotely controlled venting system.

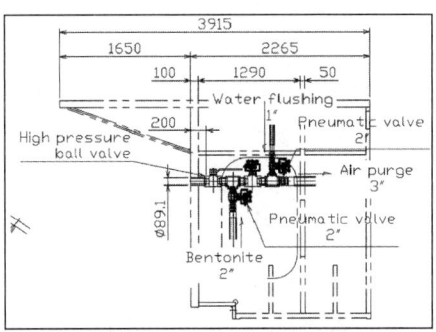

Figure 9. Automatic air venting system or "purge" at the top of the excavation plenum

Such venting system or "Purge" was operated from the operator's cabin, so the TBM operator could run it as part of the face pressure control process while mining.

Soil Conditioning

Performed primarily by means of foam and a given concentration of polymer, like in any other EPB machine. However, it was particularly important that the foam was injected right in the face by means of the all 22 ports in the cutterhead, as opposed to in the excavation plenum.

Cutterheads this size usually don't mix it very well, with relatively low rotation speeds designed up to 1.4–1.8 rpm but usually 0.8–1.2 rpm as working condition. Also large excavation plenums like that of Bertha, nominally contain more than one excavation shove in volume. Therefore, having the foam and bentonite directly added to the ground in the face was our best chance to achieve proper soil conditioning that had been impossible inside the plenum.

For instance, "Bertha" features a hydraulic center agitator with three rotating arms which spins up to 2 rpm in opposite direction to cutterhead rotation, covering 7.3 m of diameter (the maximum the cutterhead columns would allow). It represents almost half of the TBM diameter but when it comes to coverage, it only reaches 17% of the area of the plenum. Therefore, the remaining 83% had to be "mixed" by the 16 fixed arms in the back of the cutterhead and the four mixing arms attached to the bulkhead.

For similar reasons, as well as for having a large volume plenum where heavier grains and soils would settle at the bottom of the excavation while leaving liquids at the top, the addition of water absorber polymer to the plenum and eventually, to the screw conveyor was very helpful, as backup in case of insufficient ground mixing and when dealing with segregated soils with high content of not-mixed water.

Finally, soil conditioning needed large capacity but also flexibility to deal with changing ground and also, cutterhead geometry issues. Bertha's cutterhead design allows for certain cutting tools to be replaced from within the inside of the hollow cutterhead, at atmospheric condition, as it will be explained in more depth later. But at the same time, that feature made cutterhead arms wider than usual, so clogging was an issue to be paid even more attention than in any other EPB machine. Therefore, the main goal of achieving the best EPB paste possible would collide with the need of keeping the cutterhead spokes open and unplugged. In other words, conditioning system had to allow clay type configuration conditioning at the ports in the center of the cutterhead to prevent clogging from happening in the narrow cutterhead openings, in spite of having a sandy, gravelly and watery face, which demanded a more regular and totally opposite granular soils treatment.

Muck Volume and Weight Control and Reconciliation

Bertha has been particularly successfully controlling ground and volume loss and therefore, preventing and minimizing deformation above and around the tunnel. Key part of it has been a real time control of volume and weight of excavated muck and its continuous comparison with theoretical values, considering upper and lower limits with two level of alarms, yellow and red.

Firstly, every single fluid added to the excavation chamber while mining, such as foam, polymers, bentonite and water, was accounted for by means of individual flowmeters per line, sending readings to the PLC of the TBM, even those added while no mining periods.

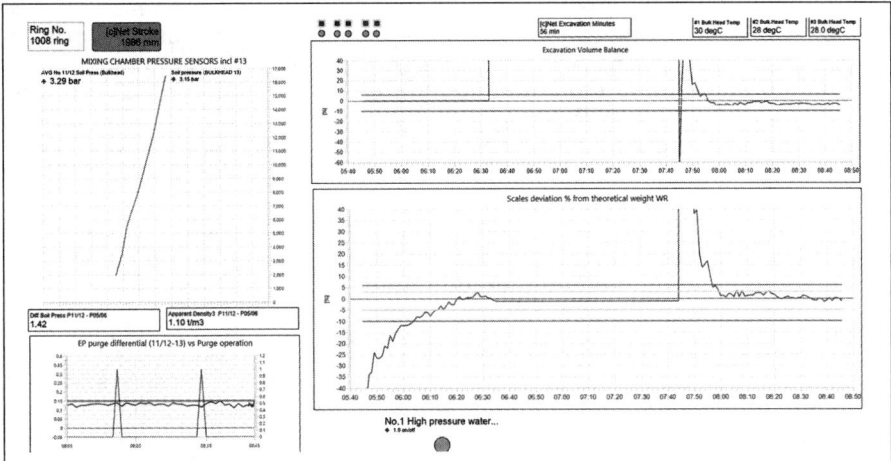

Figure 10. Muck weight reconciliation on real-time

Then, weight of the excavation muck coming out of the screw conveyor was recorded by two redundant scales, calibrated on a weekly basis and whose readings were double-checked by another two scales installed in the tunnel conveyor and outside in the overland conveyor. Thus, a continuous comparison between actual weight of excavation debris and theoretical weight based on baseline density values (as per GBR) was performed, in which every fluid added to the chamber was accounted for.

Real-time readings and comparison chart was displayed on the TBM cabins so TBM operators and QC engineers could follow trends and control muck volume and weight on time, without having to wait for the end of the push to do the after math, ignoring what happened throughout the excavation process.

Finally, QC engineer performed a final weight reconciliation per ring and also average per last three rings, to ensure perfect balance of theoretical versus actual excavated volume in the face.

It is worth to mention that the accuracy of scales, getting an average difference of readings between them of 1% or less and continuous adjustment per ring of theoretical specific gravity values have been proven critical to succeed in this task.

Annular Excavation Gap

The excavation gap of the TBM was designed to be 30 mm, the minimum possible, with a non-tapered shield TBM to minimize settlements while dealing with granular loosen soils with high water content, shallow conditions at both ends of the tunnel and mining under downtown Seattle, including live structures and utilities and tall buildings. This by itself was proven an effective measure.

That being said, the annular gap still represented a theoretical 5.2 m^3 per shove, so it was filled with bentonite and thickener polymer when needed thru six injection ports at the top of the shield, located in two rows of three ports, with one individual pump per line and port. Gap backfilling volume was controlled all the time, setting an alert upper limit of 130% of the theoretical amount, as a warning for potential voids or over excavated areas.

At the same time, gap injection was controlled by pressure in two ways, manual mode and "keep" or automatic mode. In manual mode, injection pressure was set comparing pressure in the gap by means of an EPB sensor aligned with the row of three injection ports, same type of those in the excavation bulkhead, with the requested working pressure set in the excavation chamber plus a calculated pressure loss because of the bentonite line. That way, bentonite pressure was sufficient to fill the excavation gap without bentonite diving into the excavation chamber.

In "keep" mode, bentonite pressure was directly interlocked to pressure sensors in the shield and alike the "keep" system in the excavation plenum, it would kick up when pressure dropped below the lower limit set, regardless of mining or shutdown periods.

Annular Ring Backfilling

Performed with bi-component mixture of cementitious mortar plus sodium silicate as accelerator and injected through nine backfilling ports at the end of the tail shield.

Once again, it was extremely relevant to minimize settlement and ground loss behind the shield, so annular backfilling was driven by both, pressure and volume. Backfilling volume was controlled by individual flowmeters per line (independent lines per port), with limits set for low or high volume over the theoretical 26 m^3 per push.

In addition and with such a large distance between ports, meaning up to 2 bar difference from bottom to top of the shield, the system was divided pressure wise in four zones: upper, middle upper, middle lower and bottom with 2, 2, 2 and 3 backfilling ports each and pressure sensors similar to those controlling EPB right beside the port. That way, backfilling pressure was properly controlled and set per zone, ensuring adequate filling of the gap while protecting the maximum pressure allowed tunnel liner and the metallic brushes at the end of the tail shield.

Atmospheric Cutter Changing Devices

TBM size growing significantly in the past years brings a number of new challenges that tunneling engineers are facing with not much experience from similar cases in the past, because there are not many. A clear example is an oversized cutterhead, which demands larger amount of cutting tools of different types which, combined with the usually difficult task to predict cutting tools wear and consumption, makes very difficult to choose the right cutting tool types and to anticipate the amount of replacements needed.

On the other hand, urban tunnels not always allow for ground improved areas or safe havens to be built along the alignment to perform cutterhead maintenance, especially when mining under downtown or at large depth. And both circumstances occurred in Seattle, concurrently with a mixed face conditions and potential for presence of boulders pretty much anywhere in the tunnel alignment.

Consequently, hyperbaric interventions to allow cutterhead inspections and cutting tools replacements are becoming more and more frequent in the past years, yet they involve some challenges, such as longer duration than performing work at improved or even "atmospheric" conditions inside a safe haven and the fact that conditions for a hyperbaric intervention cannot always be achieved, regardless of the supply of breathable air to the heading and caking and re-caking processes are very optimized these days.

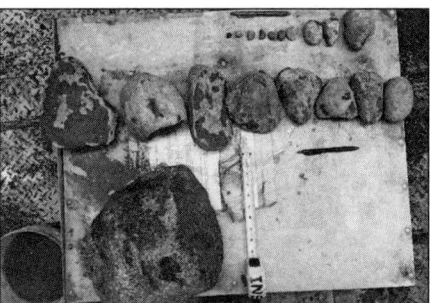

Figure 11. Boulder found at 40–50 ft depth during excavation at the North Portal, and cobbles found while mining

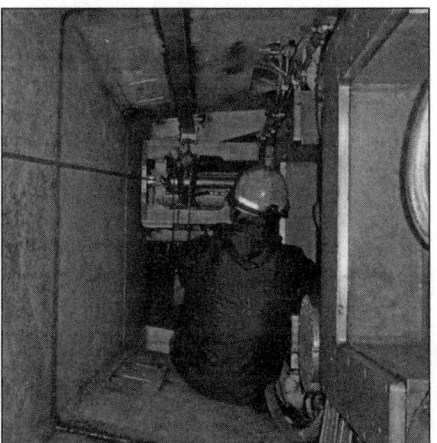

Figure 12. Atmospheric cutting tools replacement from within cutterhead arm

As a result, Hitachi Zosen as TBM supplier was tasked with designing and manufacturing a system to be able to inspect and replace a number of cutting tools from within the cutterhead hollow arms, taking advantage of its large size. Thus, Bertha features 106 cutting tools out of the 743 total that can be replaced this way. Those tools could be rippers or cutter discs and the philosophy implemented was replacing those more often to protect the ones that must be replaced at hyperbaric conditions.

In any event, atmospheric cutting tools were not meant to replace hyperbaric interventions but to supplement them, since direct observation of the rest of the cutting tools, injection ports and cutterhead structure is still necessary and recommended.

Continuous Monitoring of TBM Parameters

For this size record TBM, control of not only deformations around the TBM but TBM parameters itself was paramount. With more than 7,000 tags feeding the data logger with information every 3 seconds, being able to timely monitor them all was quite a challenge for everyone in the SR99 Team. The steps taken by the team to tackle this challenge were the following:

Clear definition of the TBM key parameters, such as Total Thrust, Cutterhead Torque, Speed of Mining, Hydraulic Pressure of Center Agitator, Bulkhead Temperatures and Screw Conveyors Torque, set in our Tunneling Work Plan (TWP). Three different levels were established per parameter (green=normal, yellow/amber=deviation within limits and red=limits exceeded) and actions set to correct deviations from stablished values being also part of the plans.

Clear definition of the TBM interlocks, set by the TBM manufacturer and listed in the TWP, where parameters such as Main Drive and Agitator Drive Lubrication and Seals Temperatures, Total and Differential Loads at the Cutterhead, Cutting Torque, Articulation Thrust and even EPB pressure were limited. Interlocks were adjusted as needed during execution of the tunnel and automatic actions coming out of them were set, going from a simple warning to total TBM shutdown.

Clear definition of the Ring Excavation Key Parameters, such as EPB face pressure, muck volume and weight and shield gap and concrete ring backfilling volume and pressure, as mentioned above, set in our Tunneling Monitoring Action Plan (TMAP), using as well the traffic lights system to characterize thresholds and status of any given parameter any time.

Integration of these three different set of parameters (TBM performance, interlocks and mining parameters), together with continuous readings of over 2,000 instruments distributed along the tunnel alignment and its area of influence in a clear and simple *communications and decision making scheme*, where flow of information and decisions were made based on the level of the alert trigger by any given parameter (green, yellow or red).

Continuous overview of every alarm triggered by means of the *Construction Monitoring Task Force* (CMTF), composite group of tunneling experts of the Design Builder and the Construction Manager and external experts and consultants when needed, meeting on a daily basis and more often when needed, available 24/7.

Logistics

TBM performance always relies on properly design and functioning logistics, even more for a mega TBM, with huge difference between production picks and planned long terms production rates, for planning purposes. In our case, it is worth to mention the following key logistic elements.

Mucking out by barges: with roughly 1,000 ton per excavation and picks of up to 14 rings per day, muck disposal was particularly challenging, especially when jobsite location in downtown Seattle and besides the congested Port of Seattle only permitted a temporary muck bin capable of holding a volume of muck equivalent to 6–7 rings, plus additional limitations to trucks traffic, not even allowed during rush hours and during events in the nearby Football and Baseball stadiums.

As a result, barging muck out of the tunnel was pretty much the only feasible alternative for muck disposal. However, that was both possible and efficient because of the selected disposal place had also direct access to the water, so it was a perfect fit for barging and the whole operation was planned starting at early stages of the project. Two deck barges carrying three-four rings a piece and one hopper barge carrying up to three rings plus two tug boats sufficed to carry most of the 1.5 million tons of excavation muck from the jobsite up to Mats-Mats quarry in Port Ludlow, 35 nautical miles north of the jobsite.

TBM supplies: excavation process demanded an average of 50 m^3 of bentonite and surfactants and 25 m^3 of bi-component grout per shove, so carrying totes and mixers wasn't really an option versus having those fluids coming to the stationary tanks at the TBM thru slick lines from the surface. It also required an array of hose reels in the TBM

Figure 13. Aerial view of Pier 46 in Seattle's waterfront where the mucking out by barge is taking place

back that took some extra room and demanded additional labor for adding and replenishing them but it reduced the need for unloading cranes and totes manipulation.

Finally, Bertha demanded a fairly large amount of bentonite for soil conditioning, gap injection and automatic face and gap pressure control of up to 2 tons per ring; so bentonite storage and pumping capabilities in the surface, as well as in the TBM had to be dimensioned accordingly. In turn, it saved plenty of time during caking and chamber refilling process as well as simplified pressure control operation during weekends by minimizing handling capabilities.

CONCLUSIONS

57.5 ft. or 17.5 meters is by no means the largest TBM diameter that can be built and operated to date and in fact, it is no longer the largest TBM ever built (Ø17.6 m Herrenknecht Mixshield launched in Hong Kong in 2015). However, the size is and still will be a key factor to be considered when designing and planning for one of this mega-size TBMs.

Every assembly, test, shipping, operation, maintenance and disassembly plan has to be developed ad-hoc and customized considering this circumstance, since extrapolation from previous experiences coming from smaller or more regular size TBMs may not be representative or applicable and could lead to dramatic failures.

That being said, the success of one of these mega TBMs still relies on the hard work and dedication of an open-minded and multidisciplinary team of engineers and the most expert crews, like those that made the dream of the SR99 Bored Tunnel possible.

REFERENCE

Parsons Brinckerhoff. SR99 Bored Tunnel Alternative Design-Build Project. Appendix G1 Revised SR99 Bored Tunnel Alternative Design-Build Project. Geotechnical Baseline Report. June 2010. Washington State Department of Transportation WSDOT.

Design and Implementation of a Large-Diameter, Dual-Mode "Crossover" TBM for the Akron Ohio Canal Interceptor Tunnel

E. Comis ▪ The Robbins Company
D. Chastka ▪ Kenny/Obayashi JV

ABSTRACT

The Ohio Canal Interceptor Tunnel (OCIT) Project involves construction of a conveyance and storage tunnel system to control combined sewer overflows for several regulators in the downtown Akron area.

The 27-foot (9.26m) finished inside diameter tunnel, approximately 6,200 feet long (1,890m), will pass through ground conditions that consist of soft ground, mixed face soft ground over bedrock, and bedrock. Depth to the invert of the OCIT will range from about 40 to 180 feet (12–55m). A Robbins dual mode type "Crossover" (XRE) Rock/EPB TBM, Ø9.26m bore in diameter, will be used to excavate the tunnel and install the precast segmental lining. Special design features will be implemented to overcome the changing geological conditions. The XRE TBM will feature characteristics of both Single Shield Hard Rock machines and EPBs for efficient excavation in mixed soils with rock, such as a flexible cutterhead design for proficient boring in both rock and soil conditions, adjustable main drive speed with an over-speed mode for operation in hard rock, and special screw conveyor wear protection measures. The paper will describe these design features, their manufacturing process and implementation in the field.

INTRODUCTION

The Ohio Canal Interceptor Tunnel (OCIT) came to be as a result of the City of Akron's Combined Sewer Overflow (CSO) Consent Decree with the USEPA and Ohio EPA. In accordance with the Consent Decree, the City of Akron developed a Long Term Control Plan (LTCP) Update Report to reduce CSOs and improve water quality in nearby rivers. The goal of the LTCP Update Report is that the projects associated with the report will decrease the annual volume of untreated combined sewer overflow being released directly to receiving streams.

The OCIT Project consists of the construction of a conveyance and storage tunnel system to control CSOs for several regulators referred to as Racks in the downtown Akron area. The major components of the project include a conveyance and storage tunnel, drop shafts, diversion structures, consolidation sewers, and related appurtenances (see Figure 1). The key component is the OCIT, which will be approximately 6,200 feet long (1,890m) with a finished inside diameter of 27 feet (9.26m). The tunnel will be excavated using a dual mode type "Crossover" (XRE) Rock/EPB TBM manufactured and supplied by The Robbins Company. The TBM will launch from a portal with a depth to invert of approximately 40 feet (12m) through a jet grout plug installed to provide a controlled launch environment. The TBM will excavate at a uniform slope of 0.15 percent through ground conditions that consist of soft ground, mixed face soft ground over bedrock, and bedrock. The OCIT will be lined with a steel fiber reinforced precast concrete segmental liner that will be installed concurrent with the mining

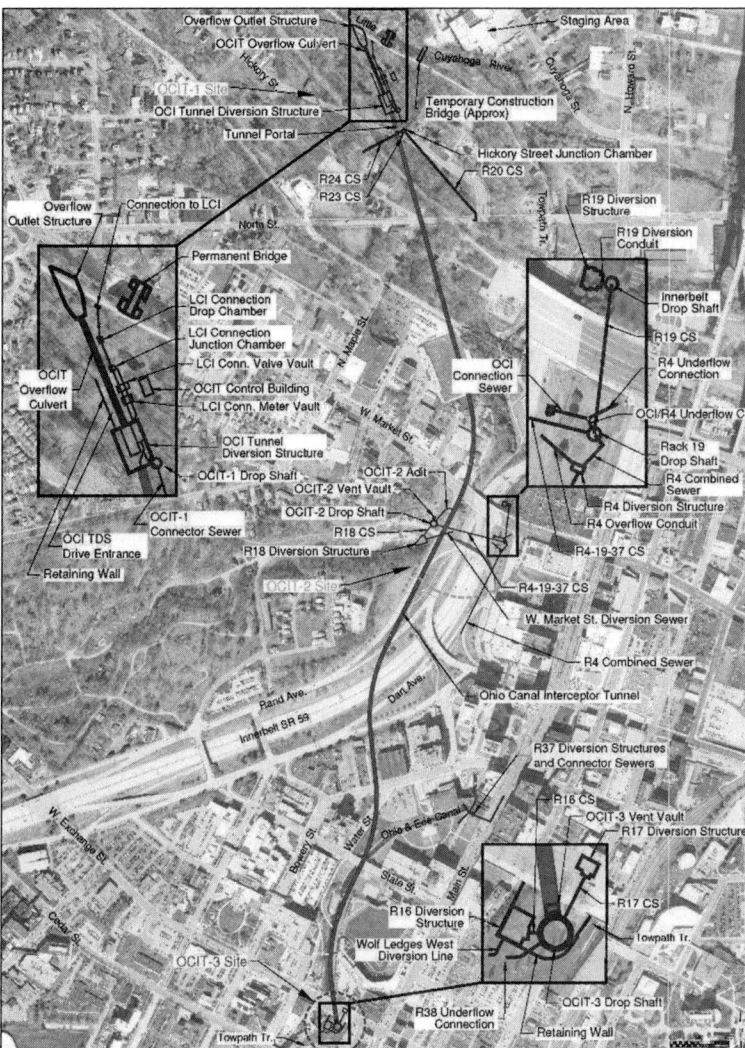

Figure 1. OCIT project layout

advance. The excavation will end at a retrieval shaft with approximately 180 feet (55m) from the surface to the invert of the tunnel.

The OCIT Project was awarded to Kenny/Obayashi V, a Joint Venture with a Notice to Proceed date of November 4, 2015. In order to meet the requirements of the Consent Decree the project must attain Achievement of Full Operation (AFO) by December 31, 2018. To be successful in realizing this schedule the TBM excavation is required to begin in March 2017 with a hole through in September 2017.

GEOLOGICAL PROFILE

The primary purpose of the Geotechnical Baseline Report (GBR) was to summarize the Geotechnical Data Report (GDR) in order to establish contractual baselines describing the anticipated ground conditions that will be encountered during construction of the OCIT Project. Data represented in these reports were as a result of geotechnical investigations including subsurface explorations, in situ testing, and laboratory testing programs. In addition to the geotechnical investigations, the GDR also incorporates data from the hydrological investigation performed.

The section of the GBR related to the OCIT identified three major reaches that were defined as distinctly different ground conditions. The TBM will launch from the portal into Reach 1 which primarily consists of soft ground. Excavation will continue into Reach 2 which is a transitionary zone with soft ground overlying bedrock. The final zone identified as Reach 3 is comprised of bedrock with two sections of low rock cover.

Reach 1—Soft Ground

This reach is 226 feet beginning at the construction portal located at Sta 11+24 to Sta 13+50. Surface elevations range from El 831 to El 885 with an approximate cover of 25 to 65 feet. A ground improvement zone was installed by Schnabel Foundation Company for the first 50 feet of the tunnel utilizing jet grout methods consisting of 140 each interlocking columns installed from the surface to facilitate launching of the TBM. Anticipated ground conditions in this reach consist primarily of silty sand containing interbedded layers of silt with a variable layer of glacial till deposits present following the top of the bedrock. Groundwater levels prior to the start of excavation were identified at El 812 approximately 4 feet from the crown of the tunnel.

Reach 2—Mixed Face

This reach is 600 feet extending from Sta 13+50 to Sta 19+50. Surface elevations range from El 868 to El 893 with an approximate cover of 45 to 70 feet. Excavation begins with a full face of soft ground with the bedrock rising from the invert throughout the reach until finally ending in a full face of bedrock. The top of the bedrock is depicted as having a consistent slope throughout the reach; however, the GBR identifies that configuration of bedrock may vary based on the jointing in the rock mass. The soft ground layer of Reach 2 emulates the profile anticipated in Reach 1 while the bedrock will consist primarily of shale with varying layers of siltstone. The top of the bedrock is identified as highly weathered at depths up to 10 feet. Groundwater levels will range from El 817 to El 820 which is approximately 4 feet from the crown of the tunnel at the start of the reach moving to at the crown of the tunnel at the end.

Reach 3—Bedrock

This reach is 5,398 feet extending from Sta 19+50 to 73+48. Surface elevations range from El 893 to El. 992 with an approximate cover of 70 to 165 feet. Bedrock will be entirely shale and siltstone with small amounts of sandstone present and the potential for clay seams. The top of the bedrock is identified as highly weathered at depths up to 15 feet. Groundwater levels will range from El 842 to El 946, which is approximately 15 feet from the crown of the tunnel at the start of the reach to 115 feet at the end. Due to the low permeability of the bedrock, infiltration is not anticipated to impact excavation operations with an anticipated average infiltration rate of 50 to 100 gpm per 1,000 feet. The GBR stated that there is a potential of flush flows up to 200 gpm but that they anticipate that these flows will be brief and will become less present as excavation progresses. The TBM will operate in open mode in Reach 3 with the exception of two sections that are identified as low bedrock cover from Sta 19+50 to Sta 27+50 and from Sta 51+00 to Sta 56+50. Additionally, it was documented that the tunnel will excavate below the existing St. Vincent-St Mary's landfill between Sta 26+00 and 33+00. Expected TBM operating conditions are summarized in Table 1.

Table 1. Akron OCIT project tunnel reaches and operating mode

Reach	Start Station	End Station	Footage	Ground Condition	Operation Condition
Reach 1	11 + 40	14 + 50	310	Full face soft ground, shale below invert	EPB Alpha 20
Reach 2	14 + 50	19 + 50	500	Mixed ground, partial face shale	EPB Alpha 25
Reach 3	19 + 50	27 + 50	800	Full face shale, low cover	EPB Alpha 25
	27 + 50	51 + 00	2350	Full face shale, good cover	Open Mode 40MPa Shale
	51 + 00	56 + 50	550	Full face shale, low cover	EPB Alpha 25
	56 + 50	73 + 40	1700	Full face shale, good cover	Open Mode 40–70MPa Shale
		Total	6210		
		Closed Mode	2160	35%	
		Open (Conditional)	4050	65%	

TBM DESIGN FOR MIXED GROUND GEOLOGY

A mixed ground condition consists of two or more geological formations, or the same rock formation with conspicuously different fracture intensity or weathering grades, with significantly different properties that will affect TBM operation. The mixed ground geology has been separated into 3 classes [1].

- Class 1: Layered or banded ground formed by rock beddings, dykes, faults or shear zones
- Class 2: Interface ground of soil and rock, or typically weathered materials above bedrock
- Class 3: Mixed face with locked cobblestones, rock blocks with soil materials, or isolated spheroidal weathering stone mixed with a soft formation.

Studies have been carried out to analyze factors that contribute to fast machine advance in such conditions [2]. Although a strong correlation between TBM performance and a quality ground conditioning regime has been identified, the probable geology, hydrology and face pressures of a mixed ground project have a heavy impact on the TBM design:

- Dress of cutterhead: disc cutters, scrapers, picks, bits, etc.
- Opening ratio of cutterhead
- Type of screw conveyors: ribbon or shafted
- Quantity and length of screw conveyors
- Abrasion-resistant cladding requirements: cutterhead, mixing chamber, mixing bars, screw conveyor flights and casing, etc.
- Face pressure related design: pressure bulkhead, thrust ram sizing, articulation ram sizing, tail shield seals, main bearing seals, man-lock and tool-lock, breathable air design, air compressors, etc.
- Ground conditioning foam, polymer and bentonite systems, air compressors, etc.

This paper will detail how these and other design features have been addressed on the Robbins TBM.

Cutterhead Design

Dress of Cutterhead

One of the main problems in mixed ground for TBM tunneling is related to abnormal flat and multi flat cutter wear, as the soft ground material in the mixed face cannot provide sufficient rolling force for cutters to overcome the pre-torque of cutter bearings [1]. The XRE TBM cutterhead is equipped with 56 housings that can eventually be dressed with either knives/rippers or 17" disc cutters (Figure 2). Although knives would be more suitable for the EPB section of the drive, the contractor and TBM manufacturer are taking the decision to dress the cutterhead with disc cutters.

Nevertheless the pre-torque of cutter bearings has been reduced by 25%, to require less rolling force for the cutter to rotate evenly even in soft ground. This could result in a shorter cutter life once in rock; however, the shale is not expected to be harder than 70 MPa so cutter ring wear will be very limited. Furthermore in order to avoid hyperbaric intervention in the first 1610 ft (490m), sacrificial rippers have been welded to intervene in case of a ring wear greater than 0.6 in (15 mm) (Figure 3). In the event a cutter gets blocked these rippers should be able to cut the face until the machine reaches an open mode condition, which won't require a hyperbaric intervention, even though the TBM is equipped with both a material and a man lock.

In consideration of the 65% drive in rock, the cutterhead has also been dressed with Hardox 450 faceplates and peripheral grill bars to reduce the risk of wear.

Cutterhead Opening

Cutterhead Opening Ratio is usually recommended to be 30–40% [3] for EPB application and 10–12% for rock TBMs. Being OCIT is a mixed ground project, the opening

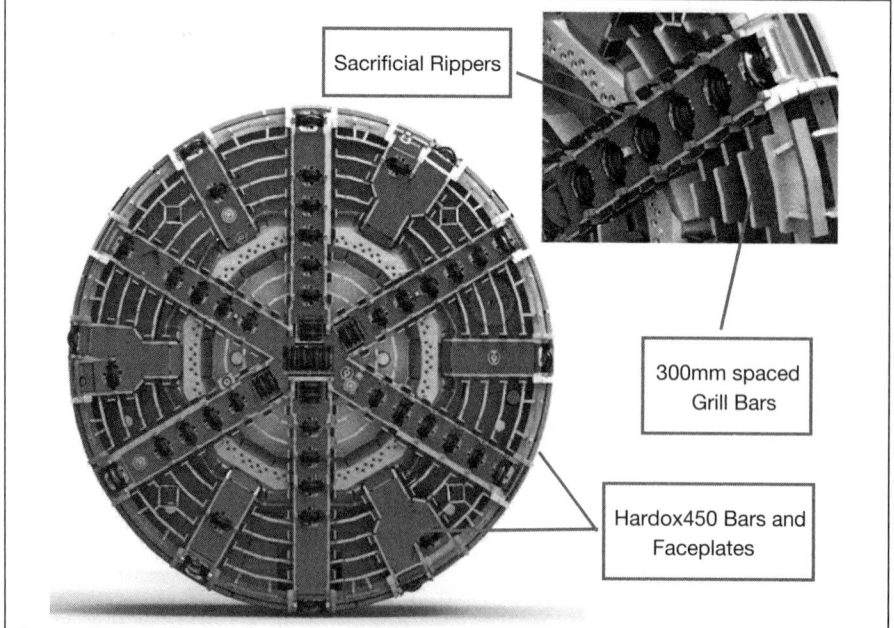

Figure 2. XRE TBM cutterhead for Akron OCIT

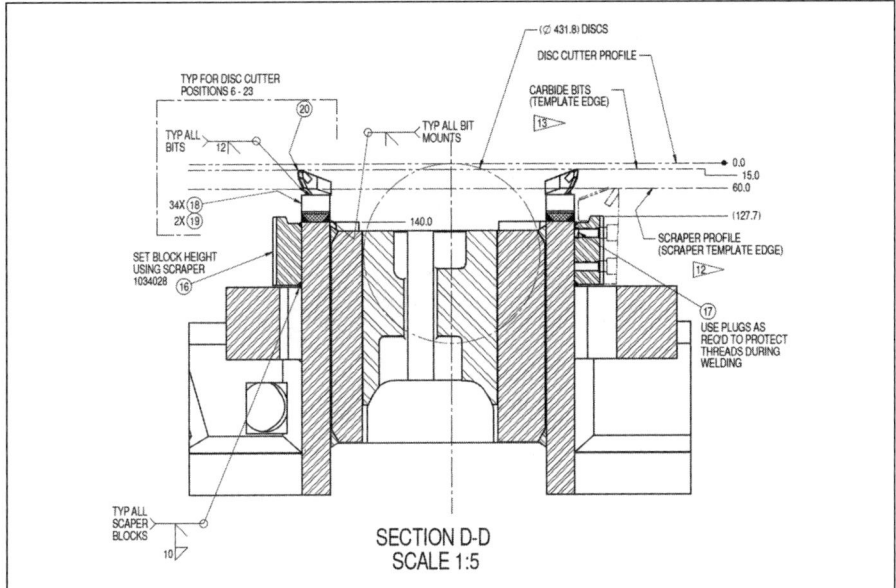

Figure 3. Sacrificial rippers installation

ratio of the XRE TBM has been set to 27%. This decision was made in consideration of the longer part of the drive in rock—during open mode operation the grill bars are required to reduce the size of rock that can get into the mixing chamber and eventually be conveyed through the 47 in (1200 mm) diameter screw conveyor. The maximum rock size has been set to 300mm (refer back to Figure 2).

Drive System Design

The torque equipped on a TBM is based off of the mechanical design of the cutter-head drive system. Power transmitted by the Variable Frequency Drive (VFD) controlled electric motors is based off of the motor properties, gearbox specifications, and the interaction of the gearbox output shaft and ring gear. The XRE TBM had 12×190 kW VFD electric drive units existing from a previous EPB project in Mexico City. EPB applications require high torque at limited speed whilst rock applications usually required higher rotation speed at lower torque. This EPB machine can operate in rock because the motors have been reworked to permit higher motor speed at reduced torque. A torque limit on the VFD has been implemented for the high speed rates, in order to avoid operating the motors in a region where the Breaking down Torque (BDT) would become smaller than the Running Torque (RT). Additionally, for operation in open mode an 80% efficiency has been considered for the losses due to the scraper design, seal friction and gear ratio (all typical for EPB machines). Total cutterhead torque vs speed is shown in Figure 4.

Screw Conveyor Design

The screw Conveyor is shaft type, 64.50 ft (19.66 m) long, Ø47 in (1200 mm) nominal diameter, with a tapered front nose to Ø30 in (762 mm). The system was previously employed in an EPB project in Mexico, hence it required the following modifications to operate in the OCIT geological conditions (especially considering the 65% drive in Shale bedrock).

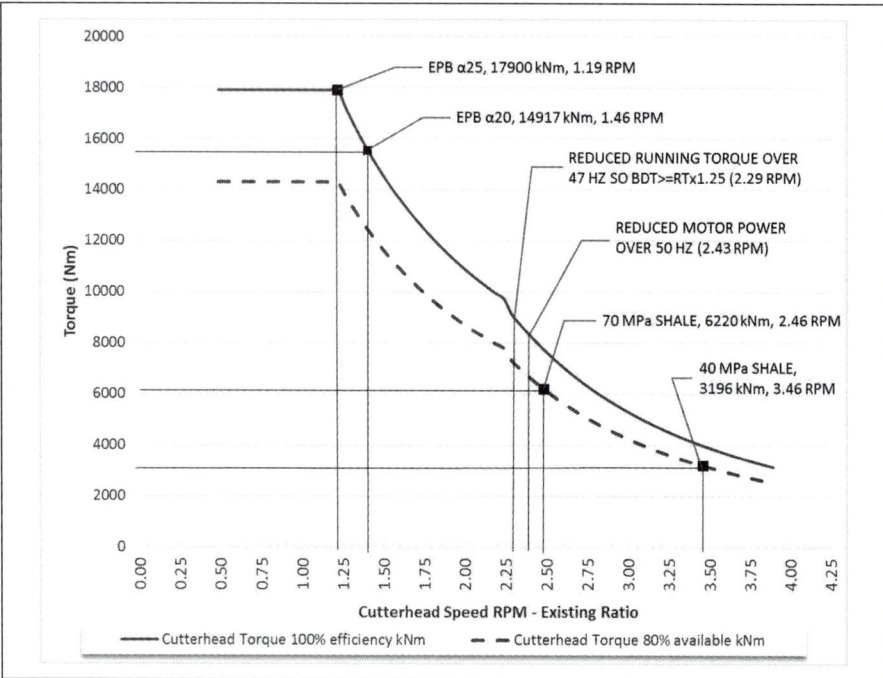

Figure 4. Drives torque-speed curve

Pump/Motor Drive Addition

The conveyor is driven by two hydraulic radial piston motors with large displacement and high power density (up to 240 kW). Originally the installed capacity on the hydraulic power unit was 2×110 kW motor/pump units to reach a max theoretical speed of the screw of 6 rpm. However this original screw conveyor was designed as part of a double screw system with a ribbon-type screw conveyor as the first stage and the shaft-type as second stage, and to be operated in EPB mode only.

In consideration for the OCIT geology and the necessity to muck out shale bedrock, the single shaft-type screw conveyor required a much higher speed. The hydraulic power unit has been consequently upgraded to 5×110 kW, which brings the max theoretical speed of the conveyor to 16 rpm at a limited torque of 232 kNm.

Wear Protection

In rock mode and mixed ground conditions, and to a lesser degree in EPB mode as well, the auger and the casings are in contact with abrasive material creating wear. The screw conveyor features the following characteristics to limit wear due to abrasion (refer to Figure 5):

1. The leading face of the front auger flight and the outside diameter is covered with welded-in wear plates, "inserts," made of Chromeweld 600™ and hardfacing in a crosshatch pattern

2. The auger shaft is covered in hardfacing in a crosshatch pattern

3. The inside diameter of the casing is lined with welded 0.4 in (10 mm) thick Chromeweld 600™ for the first ⅓ of the casing

Figure 5. Screw conveyor wear plates and inserts

4. The inside diameter of the remaining ⅔ of the casing is covered with hardfacing
5. No. 4 ports for lubricant material injection (foam, bentonite, etc…)

Chromeweld 600™ is a premium grade of chromium carbide wear plate, produced with a mild steel base plate and hardfaced/overlayed chromium carbide wire. Typical hardness ranges from 58–64 HRC based on weld deposit thickness.

A wear monitoring plan has also been prepared for the whole drive. Wear behavior is strictly related to material encountered. However the average HRC of the Chromeweld 600™ plate is 62 to 64 (much higher than AR type plate) and typically it has 5 to 10 times the life expectancy over AR400 type plate. Knowing the plate thickness and the overlay thickness of the wear plates installed as described along the screw conveyor, we will know when the overlay is getting close to wearing the base plate because the thicknesses of both at the start are known. With frequent measurements on material worn, wear rates can be estimated per every unit of measure (tonnage conveyed measured by belt scale).

In addition to regular visual inspection, thickness measurements of the wear plates accessible from the inspection ports will be taken using a handheld ultrasound machine. Ultrasonic analysis is one of the less complex and less expensive predictive maintenance technologies.

Thickness measurements will be taken on:

1. Two helical wear castings welded on the auger at each access location (front, mid and rear of the screw). Six points on each insert, three peripheral and three on the leading face (Figure 6).
2. Two rectangular wear plates welded on the casing in the front access location. Six points on each plate equally distributed.

3. Twelve random points at the hardfacing weld on the mid and rear locations.

In order to define a wear rate, the test period has to be adjusted depending on geology encountered. Referring to Table 1, thickness measurements will be carried out with the following frequency:

1. Up to station 27+50 (EPB Mode) measurements have to be taken every 250 feet bored

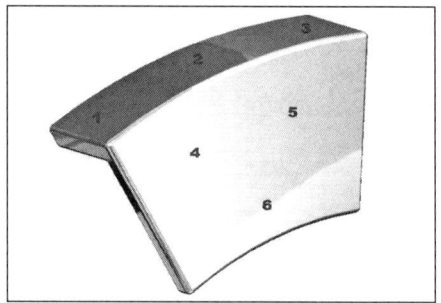

Figure 6. Insert, wear measurement points

2. From station 27+50 to 51+00 (Rock Mode) measurements have to be taken every 150 ft (45 m) bored

Based on results of these two first campaigns of data collection a wear rate will be defined for both EPB and Rock mode operation.

Probe Drilling

Crossover machines require probe drilling to determine what configuration the machine needs to be in, therefore the TBM has been equipped with two permanently installed drill rigs (Figure 7), located inside the shield for the face positions. Probing and drilling ahead of the

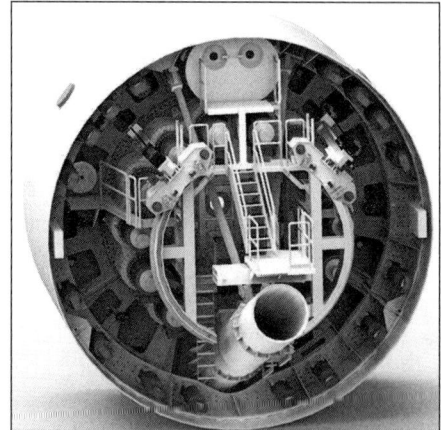

Figure 7. Probe drills inside the shield

face can be accomplished in open mode and also in closed mode conditions for the majority of the positions using blow-out preventer units.

CONCLUSIONS

The Ohio Canal Interceptor Tunnel (OCIT) Project involves construction of a 27-foot (9.26m) finished inside diameter tunnel, approximately 6,200 feet long (1,890m). The section of the GBR related to the OCIT identified three major reaches that were defined as distinctly different ground conditions. From the portal into Reach 1 geology will primarily consist of soft ground. Excavation will continue into Reach 2, which is a transitionary zone with soft ground overlying bedrock. The final zone identified as Reach 3 is comprised of bedrock with two sections of low rock cover. The tunnel will be excavated using a dual mode type "Crossover" (XRE) Rock/EPB TBM, with excavation scheduled to begin in spring 2017. The TBM has been designed for the mixed ground conditions in the following ways.

Cutterhead

The first dress of tools reduces pre-torque of cutter bearings, to require less rolling force for the cutter to rotate evenly even in soft ground. Sacrificial rippers have been installed in the event a cutter gets blocked. These rippers should be able to cut the face until the machine reaches an open mode condition, which won't require a hyperbaric intervention.

The opening ratio of the cutterhead has been set to 27% (whilst usually in EPBs the range is (30–40%), in consideration of the longer part of the drive in rock. In fact during open mode operation the grill bars are required to reduce the size of rock that can get into the mixing chamber.

Drive System

EPB applications usually require high torque at limited speed whilst rock applications usually required higher rotation speed at lower torque. This XRE machine can operate in rock because the motors have been reworked to permit higher motor speed at reduced torque for the open mode segments of the drive.

Screw Conveyor

In consideration for the OCIT geology and the necessity to muck out shale bedrock, the single shaft-type screw conveyor required a much higher speed than an EPB application. The hydraulic power unit has been consequently upgraded to bring the max theoretical speed of the conveyor to 16 rpm at a limited torque of 232 kNm.

Special Wear Protection measures have also been taken in consideration of the rock mode and mixed ground conditions. The leading face of the front auger flight and the outside diameter is covered with welded-in wear plates, and the auger shaft is covered in hardfacing in a crosshatch pattern. The inside diameter of the casing is lined with 0.4 in (10 mm) thick wear plates for the first $1/3$ of the casing length and the remaining $2/3$ of the casing is covered with hardfacing in a crosshatch pattern. A wear monitoring plan has also been prepared for the whole drive.

Probe Drilling

The TBM has been equipped with two permanently installed drill rigs, located inside the shield for the face positions. Probing and drilling ahead of the face will determine what configuration the machine needs to be in.

REFERENCES

[1] Hongsu, M., Lijun, Y., Qiuming, G. and Wang, J. 2015. TBM tunneling in mixed-face ground: Problems and solutions. *International Journal of Mining Science and Technology 25* (2015) 641–647.

[2] Roby, J. and Willis, D. 2014. Achieving Fast EPB Advance in Mixed Ground: A Study of Contributing Factors. *North American Tunneling: 2014 Proceedings* 182–194 Edited by Davidson, G., Howard, A., Jacobs, L., Pintabona, R. and Zernich, B.: SME.

[3] Japan Society of Civil Engineers, eds 2006. *Standard Specifications for Tunneling-2006.*

Cutterhead Protection in a Boulder Field Using Real-Time Vibration Monitoring

Jessica Buckley ▪ Jay Dee Contractors, Inc.
Ehsan Alavi ▪ Jay Dee Contractors, Inc.
Brian Hagan ▪ Jay Dee Contractors, Inc.
Michael A. DiPonio ▪ Jay Dee Contractors, Inc.
Mike Mooney ▪ Colorado School of Mines
Nathan Toohey ▪ Colorado School of Mines
Thomas Planes ▪ Colorado School of Mines

INTRODUCTION

Mechanized tunneling in boulder-laden glacial soils poses a significant risk to TBMs. Impacts with boulders and even cobbles can cause significant damage to cutting tools and the cutterhead body. Such damage can go unnoticed given the nature of closed face TBM tunneling, slowing productivity and increasing damage.

While boulders cannot be avoided, the TBM operator can reduce the risk of damage by reducing thrust and cutterhead rotation speed. However, the operator needs guidance on when to make these adjustments. To date, TBM operating data such as torque, thrust, advance rate, etc. are not sensitive to initial boulder impacts, unless the boulder is significant in size. And, the operator's cabin is now well back from the cutterhead where the operator cannot hear or feel boulder impacts.

This paper presents the implementation of a novel real-time vibration monitoring system installed on a Hitachi-Zosen (Hitz) EPB TBM during the Seattle N125 Northgate Link tunneling project in Seattle, Washington. A real time signal from cutterhead impacts with boulders was provided directly to the operator's cabin for immediate reaction, e.g., torque and thrust control. The system was used to record boulder/cobble activity along the alignment, and the information was used to help operate the TBM to avoid and minimize damage.

NORTH GATE LINK PROJECT BACKGROUND

The Northgate Link Extension tunnel project in Seattle, Washington will extend service north from the University of Washington to the University District, Roosevelt, and Northgate neighborhoods by 2021, and is expected to cost approximately $2.1 Billion. Most of this 6.9 km extension will be underground. The N125 (Northlink) contract includes the construction of 5.6 km of twin EPB tunnels. Also included are the excavations of the Maple Leaf Portal (MLP) where the light rail will transition from tunnels to elevated guide-way and two large underground station boxes, one for the University District Station (UDS) and one for the Roosevelt Station (RVS). JCM Northgate LLC was the tunnel and station contractor for the project.

The N125 tunnels are excavated through glacial and non-glacial sediments of the Puget Trough deposited during the Quaternary and Holocene periods. The Quaternary sediments are generally overconsolidated due to several glaciations, while the recent Holocene sediments are normally consolidated. The Engineering Soil Units (ESU) defined for this project are Engineering and Non-Engineered Fill (ENF), Recent

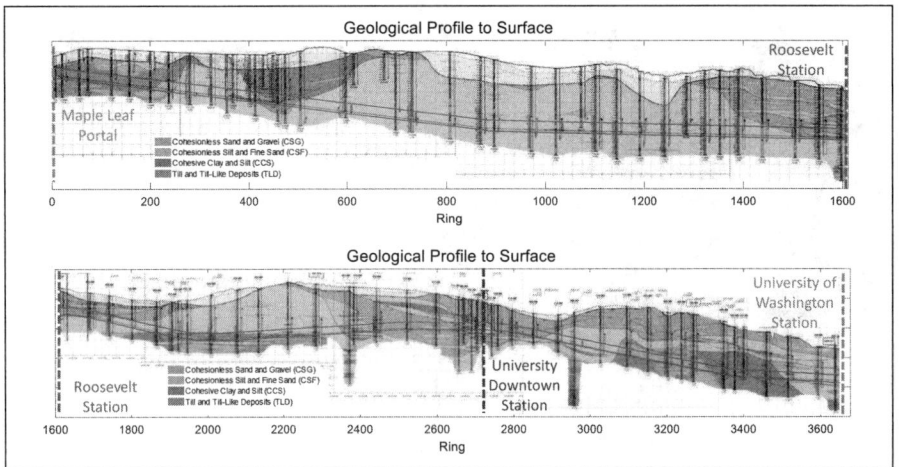

Figure 1. N125 (Northlink) geological profile and tunnel alignment (outlined in red)

Granular Deposits (RGD), Recent Clays and Silts (RCS), Till and Till-Like Deposits (TLD), Cohesionless Sand and Gravel (CSG), Cohesionless Silt and Fine Sand (CSF), and Cohesive Clay and Silt (CCS). ENF, RGD, and RCS are recent, normally consolidated sediments, whereas TLD, CSG, CSF, and CCS are glacial, overconsolidated sediments. The geology and tunnel alignment are shown in Figure 1.

TBM VIBRATION MONITORING SYSTEM

The TBM vibration monitoring system was composed of eight vibration sensors (accelerometers), a data acquisition system (DAQ) that collected (3 kHz sampling frequency), saved, and analyzed data, and an executable program that displayed summary results from the analyzed data. The sensors were very compact, which enabled them to be installed throughout the TBM without interfering with tunneling operation. The general layout of the Hitz 6.44 m EPB TBM, sensor placement, and basic functionality of the vibration monitoring system can be seen in Figure 2. The DAQ, including compact controller and associated power and sensor cabling, was housed in a vibration box that was mounted inside the TBM bulkhead and connected to an external 120 V power and Ethernet network connection. The controller was programmed to collect data when actively mining, intake certain operating parameters (such as ring number) from the TBM PLC network, and save the results in one minute files (that could be transferred over the network). Collected data was analyzed in 3 second windows and the resulting vibration amplitude, also referred to as Boulder Detection Variable (BDV) was summarized in terms of the average magnitude and number of impacts (values exceeding the specified threshold). The executable program could be installed on any computer and, if the computer was connected to the network, would display real time vibration data (for the most recent 3 second window of data) and plot results every 3 seconds for the current ring. All programming for the final version was completed in LabVIEW. See Buckley (2016) for further specifications and system development.

OPEN AIR TESTING (BEFORE LAUNCH)

Extensive testing of the system was carried out prior to launching the TBM. The goal of the pre-launch testing was to characterize measurable vibration amplitude and frequency content within the TBM due to the influence of the TBM electric motors on the recorded data and to establish a baseline for cutting tool impacts.

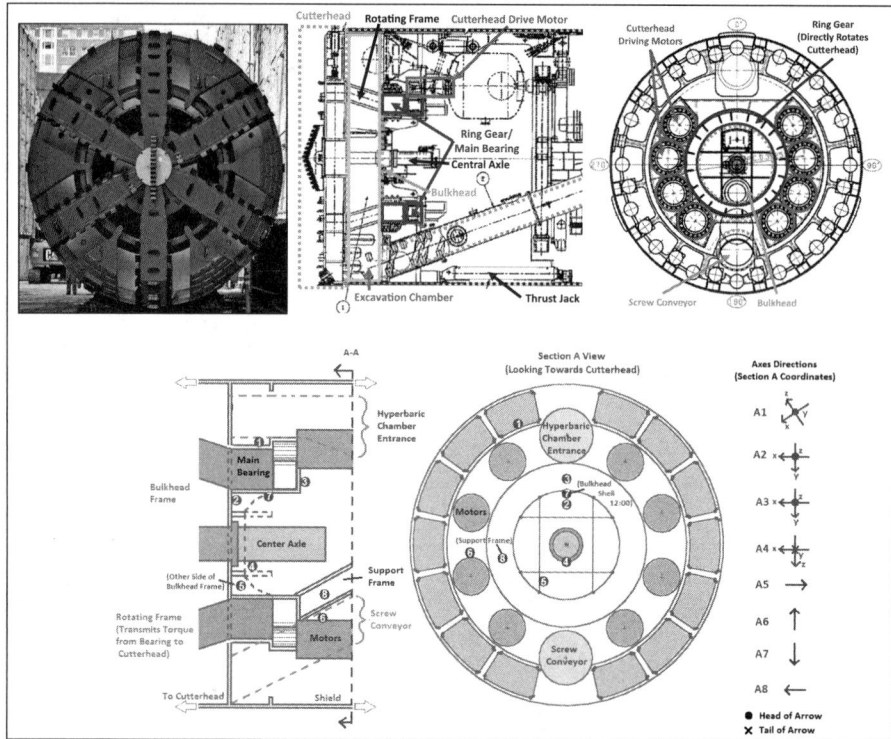

Figure 2. Vibration monitoring system installed on the Hitz EPB TBM. TBM cutterhead and general layout of the TBM are shown at the top, with sensor layout in the center and functionality of the vibration monitoring system at the bottom

Figure 3 shows ambient vibration levels on the order of 0.03g (g = 9.8 m/s^2) during 1.8 rpm cutterhead rotation (Figure 3b blue band). This signal was manifested across a number of frequencies consistent with drive motor planetary gear vibration (see Figure 3c). The amplitude and frequency response simply due to machine operation and cutterhead rotation was found to be rotation speed dependent (see Figure 4). This information was used in signal to noise ratio analysis. Figure 3 also shows the recorded signals when boulder impact was simulated via tangential hammer impact of various cutting tools. As shown by the arrows, some low level impacts are buried in the time domain ambient vibration signal while others are quite clear. In the frequency domain, the impacts are quite clear in their broadband response.

VIBRATION DURING EXCAVATION

Overall Response

Hitz TBM vibration was monitored consistently during the majority of the northbound tunnel construction. Figure 5 presents a summary of the number of vibration impacts, referred to as boulder detection value (BDV) impacts, above the threshold of 0.25g. Also shown in Figure 5 is the estimated geology within the excavated window, determined from the geological profile and tunnel alignment envelope in Figure 1. As shown in Figure 5, there was little vibration activity through the first 1400 rings while excavating through the cohesive clay and silt, and through the cohesionless sand and gravel.

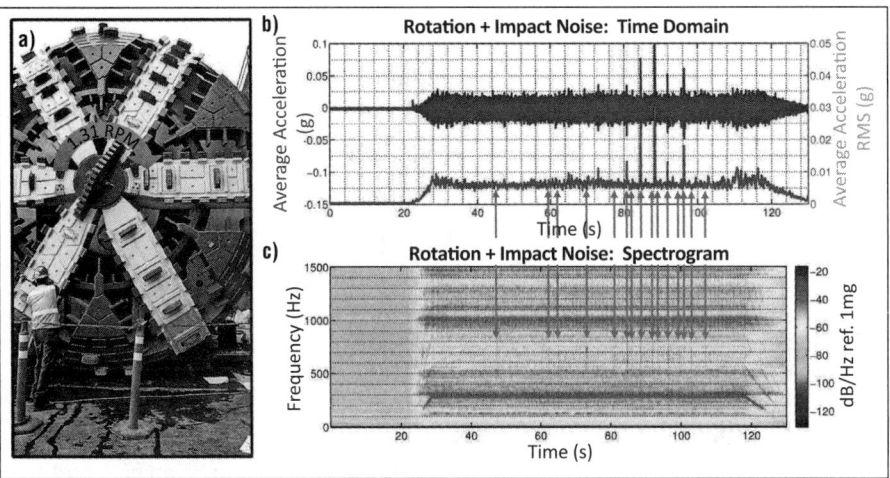

Figure 3. Vibration response during cutterhead rotation and during cutter impact testing: (a) picture of co-author impacting cutting tool tangentially; (b) time domain acceleration response inside the TBM (average of all sensors) due to 1.8 rpm rotation (blue band) and hammer impacts (shown via red arrows); (c) joint time—frequency domain response indicating dominant frequencies and their magnitudes

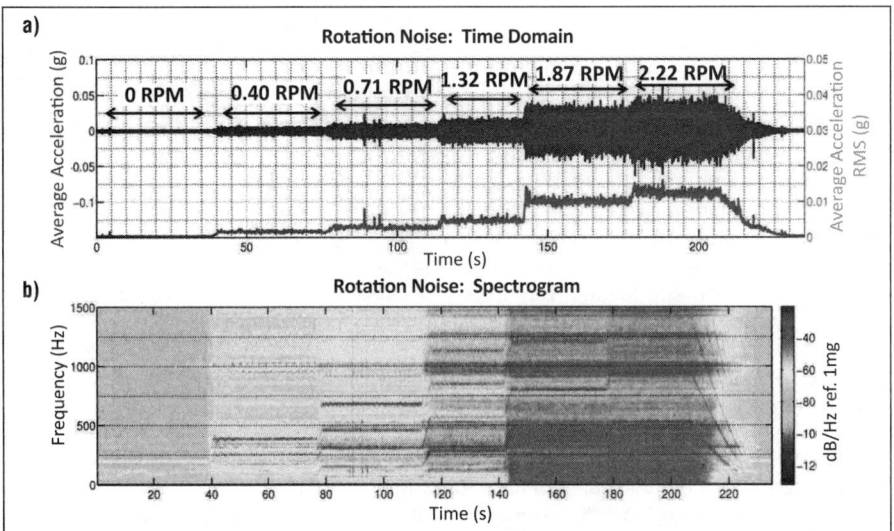

Figure 4. Ambient vibration (considered as "noise") recorded during cutterhead rotation prior to TBM launch. As shown, cutterhead rotation speed was varied from 0–2.22 rpm.

Vibration activity increased significantly when tunneling through the till deposits. These deposits are where one would expect to find cobbles and boulders based on the provided data in the Geotechnical Baseline Report for the project. Assuming correlation between vibration impacts and till deposits, an evaluation of Figure 5 shows that the original geological interpretation was correct in identifying when the till would first be encountered, but considerably incorrect in estimating the length of till deposits.

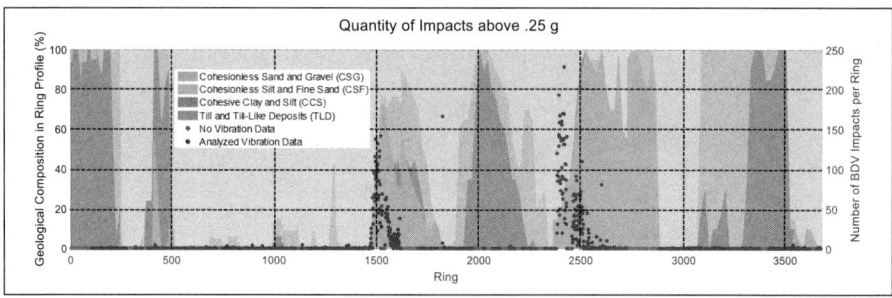

Figure 5. Summary of vibration (BDV) impacts exceeding a 0.25g threshold per ring during Hitz excavation of the northbound tunnel. The background shows the geology within the tunnel profile, estimated from Figure 1.

Relating Vibration Response to Boulders

A detailed analysis of recorded vibration impacts during the excavation of rings 1470–1611 was performed in an attempt to relate amplitude with physical meaning, e.g., with cobble/boulder size. One of the challenges with this study is that pressurized face TBM tunneling does not allow one to see the boulders being struck. The best that can be done is to sample the muck. This itself is challenging as each ring (subway size tunnel) excavation produced 50 m^3 of material. Furthermore, depending to the opening size of the cutterhead and the largest piece that the screw conveyor can handle, the large boulders will be disturbed and crushed prior to entering into the cutting chamber and screw conveyor of the TBM. To this end, our analysis had to be performed empirically and it was difficult to relate any given impact with a particular cobble or boulder. Nonetheless, muck sampling revealed significant number of cobbles, mostly rounded but some freshly fragmented suggesting they were affected by the cutting process. Figure 6 shows sample of cobbles with a mean diameter of 7 inches (175 mm) collected during the excavation.

Figure 6. Sample of cobbles collected from muck from rings 1470–1611. Cobble size varies, with a maximum diameter of 7 inches.

Figure 7a shows the number of vibration impacts above 0.25g through excavation of rings 1470–1611. Figure 7b shows for each ring a scatterplot of all impacts. The average value is presented in red. Figure 7c shows box and whisker plots of these same data per ring. These results show a greater number of impacts, presumably implying a greater number of cobbles/boulders, with higher concentration of till deposits in the excavation window. The mean vibration impact level was between 0.3 and 0.4g, and was reasonably constant through the 1470–1611 reach, indicating that perhaps the mean size of the cobble/boulder was not influenced by number of cobbles/boulders. Along a certain radius of the cutterhead, several scrapers and precutters were damaged along this portion of the excavation due to cobble/boulder impacts.

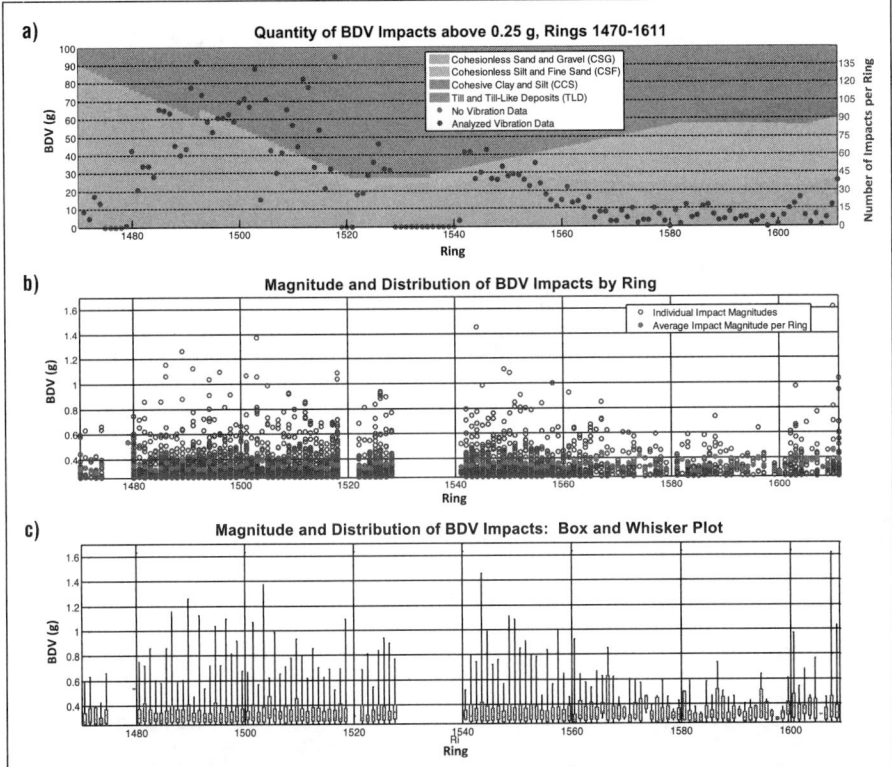

Figure 7. (a) Number of BDV impacts per ring for rings 1480–1700 and associated excavated geology; (b) BDV impacts per ring and average impact amplitude; (c) box and whisker plots showing the distribution of BDV impact magnitudes in each ring

Relating Vibration Response to Machine Operation and Damage

The second major region where vibration impact was significant was from rings 2384 to 2512. Here, recorded data is correlated with cutterhead damage that resulted in interventions. The damage zones were identified by recovering shavings from the cutters and cutterhead body by using a magnet installed on the conveyor belt beginning during the excavation of rings 2420 and 2500. Figure 8 summarizes the boulder detection value (BDV > 0.4g) during each advance together with key TBM operating parameters averaged over each ring advance.

As indicated in Figure 8, scrapers were first recovered in the muck conveyor during advance 2420, suggesting they were detached no later than during excavation of advances 2418–2419 (approx.), given that the chamber holds close to one ring of muck. A total of 13 scrapers were recovered from the belt conveyor between advances 2420 and 2430, prior to the first intervention during advance 2430 where a total of 41 missing or damaged cutters were replaced. Subsequently, four scrapers were recovered during excavation of advance 2500 and 17 scrapers were recovered during excavation of rings 2505–2506. A second intervention was performed at ring 2512 where 37 scrapers were replaced and 3 pre-cutters were replaced with disc cutters.

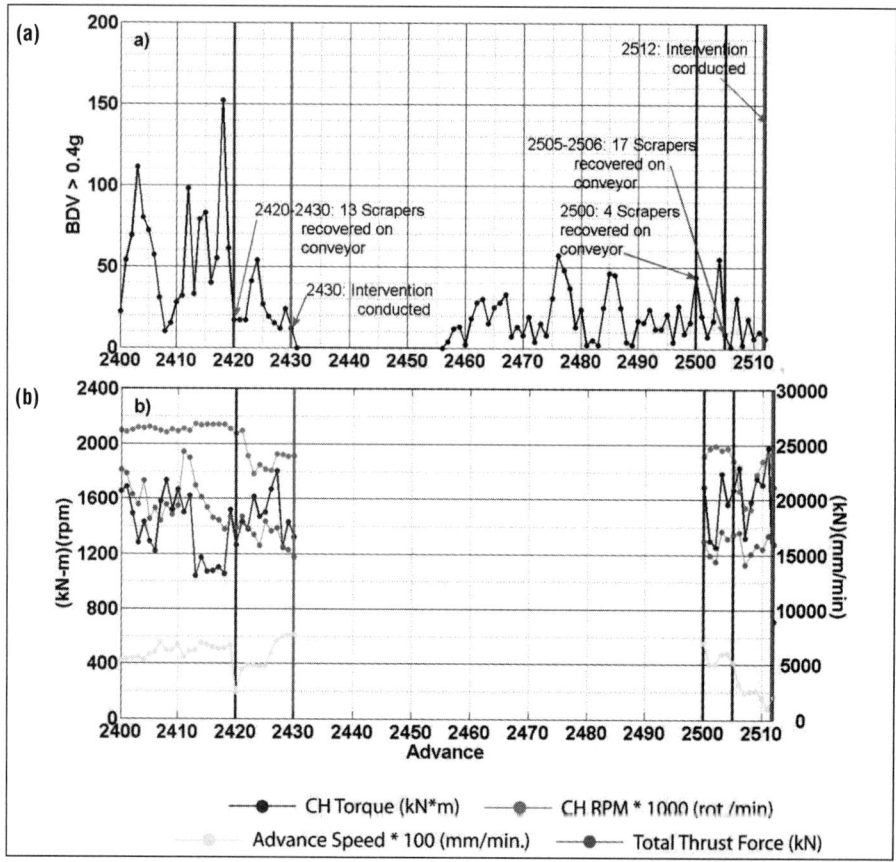

Figure 8. Rings 2400–2512: (a) number of impacts above threshold BDV > 0.4g; (b) TBM operational parameters (one average value per each advance)

Consistent with Figure 4, we expect vibration amplitude to be influenced by cutterhead rotation speed and torque, and advance speed and thrust. For example, vibration impacts were noticeably lower during excavation of rings 2500–2510, yet cutter damage similar to that achieved during excavation of rings 2400–2420 was observed. This decreased BDV activity is due to decreased cutterhead rotation speed and advance speed. This was done intentionally by TBM operators as a response to the feedback from the data.

We will focus here on advances 2418 to 2420 given that scrapers were first recovered on the conveyor during advance 2420. Vibration response is presented for each of these advances in Figures 9–11 together with TBM operating parameters. The number of vibration impacts > 0.4g was significant during advance 2418 (152 impacts) and decreased considerably during advances 2419 (61 impacts) and 2420 (17 impacts). The high BDV count observed during advance 2418 coincides with a time where scraper damage may have occurred, namely two advances prior to scraper recovery. The occurrence of the 152 impacts above 0.4g are spread throughout the 22 minutes of excavation during advance 2418. Very high amplitude impacts approaching or exceeding 1g occurred five times throughout the advance. There are also three time periods—at 5, 7 and 19 min—where impacts accumulate considerably. *The*

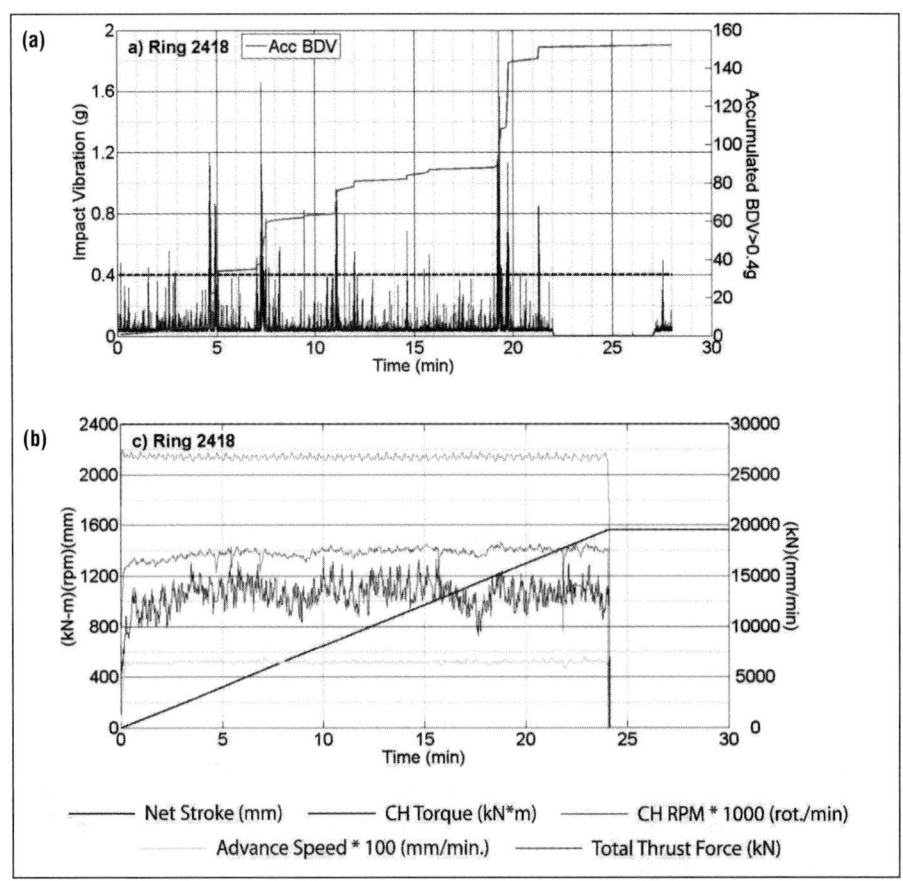

Figure 9. Ring 2418: (a) impact vibration and accumulated BDV>0.4g (152 impacts); and (b) TBM operational parameters

high amplitude impacts and the high number of lower amplitude impacts could have either caused or directly contributed (through fatiguing) to the scraper damage. This is speculation, however, as there is no actual evidence that scrapers were dislodged during excavation 2418. The TBM operating parameters are very constant during advance 2418. There is no evidence that the high impacts or high accumulation times influenced the TBM parameters.[*]

The number of impacts decreased considerably during advance 2419 (Figure 10), with only one area (around 6 min) where high amplitude vibration (>1 g) and high BDV rate was observed. TBM operation during advance 2419 remains mostly similar to advance 2418 in terms of advance speed (70 mm/min in 2419 vs. 65 mm/min in 2418) and cutterhead rotation rate (2.10 rpm for 2419 vs. 2.15 rpm for 2418). The corresponding penetration rates are 33 mm/rev and 30 mm/rev, respectively, for advances 2419 and 2418. Thrust force increased moderately (19 MN in 2419 vs. 17.5 MN in 2418) yet cutterhead torque increased considerably (1800 kN-m vs. 1100 kN-m in 2418). *Assuming the ground is the same, it is a stretch to tie the 60% increase in torque to a 10% increase in penetration rate.* The torque per penetration rate here has increased

---

[*]The vibration and operating parameter data are not time synchronized. They likely differ by 2–3 minutes.

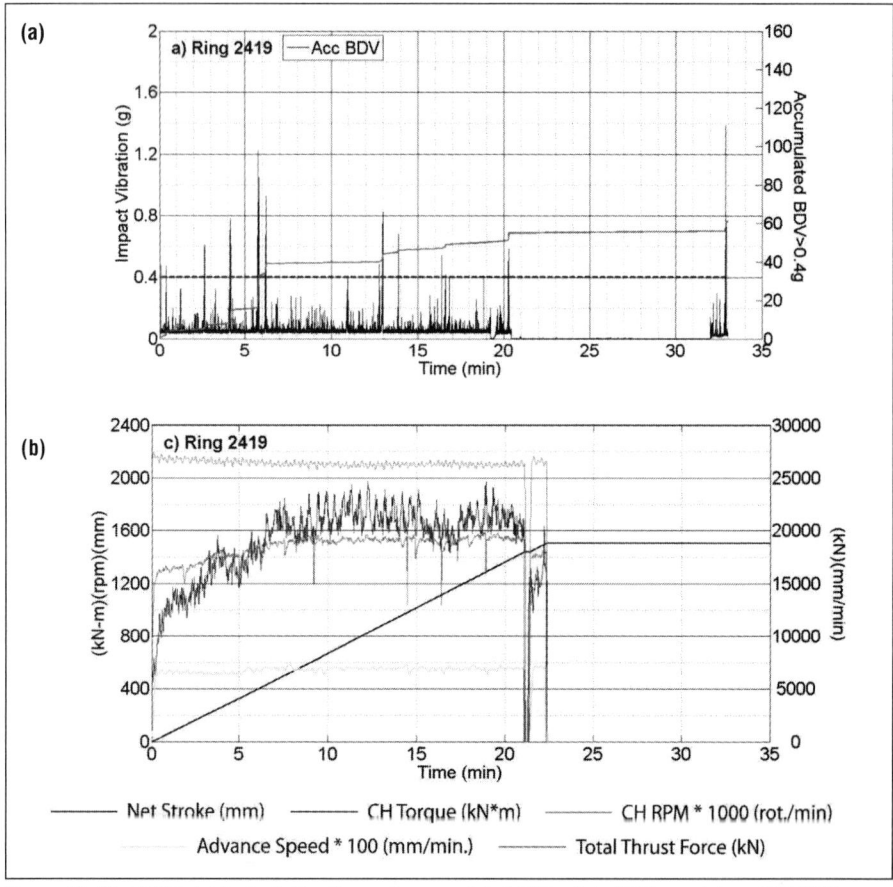

Figure 10. Ring 2419: (a) impact vibration and accumulated BDV > 0.4g (61 impacts); and (b) TBM operational parameters

from 37 to 52 kN-m per unit mm/rev. Interestingly, the torque increase from 1400 to 1800 kN/m initiates at the time (6 min) where there is high BDV rate. *It is reasonable to conclude that the reduction in measured impacts during advance 2419 is not attributed to changes in operating parameters, as these remained mostly the same as advance 2418.*

The change in advance speed and cutterhead rotation speed influence vibration. Specifically, an increase in penetration rate seems to decrease impact vibration amplitudes. *However, it is difficult to conclude that the significant reduction in BDV from advance 2418 to 2419 is due to a subtle increase in penetration rate. It is also possible that the noted decrease in BDV suggests less cutterhead impact with nested boulders/cobbles and perhaps some influence due to a loss in tools. Note: we believe that the flat cutterhead would not yield the same level of vibration impact that a tool would because a tool protrudes and strikes a boulder whereas the boulder would tend to slide past the cutterhead surface.*

Due to BDV activity, the TBM operator decreased the advance speed during mining of ring 2420. Initially, using similar advance speed and cutterhead rotation speed as 2419, the thrust force and torque increased sharply to 23–25 MN and 2200–2400 kN-m (at

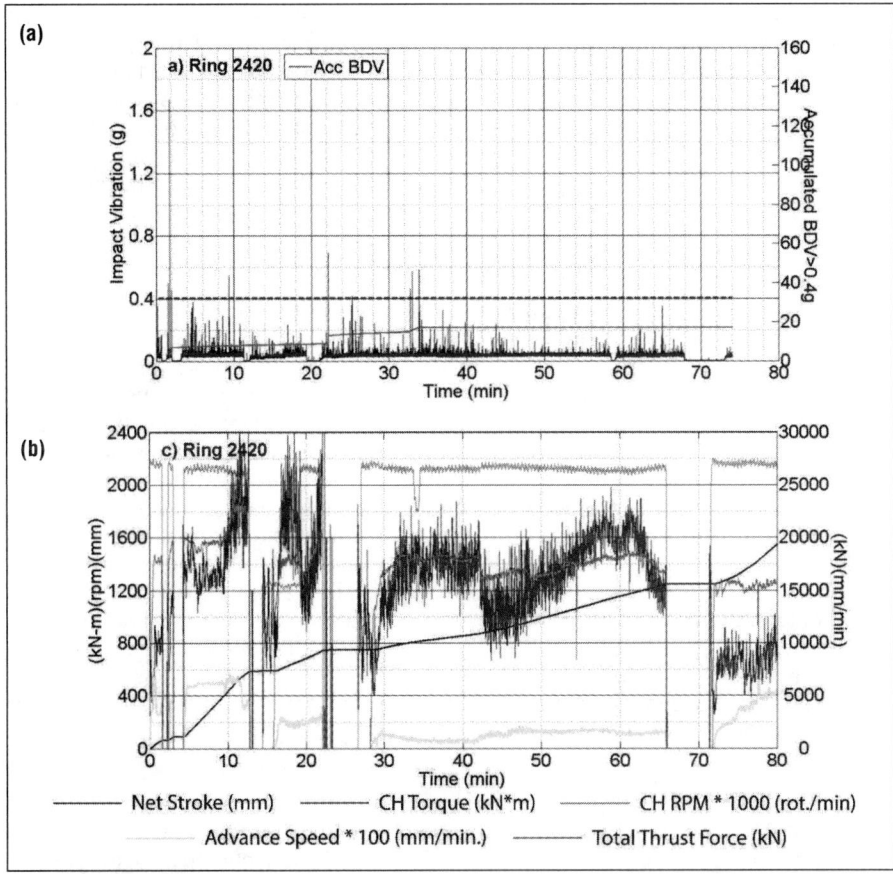

Figure 11. Ring 2420: (a) impact vibration and accumulated BDV > 0.4g (17 impacts); and (b) TBM operational parameters

10 min). The operator decreased the advance speed to 15–25 mm/min (penetration rate = 8–12 mm/rev) to maintain thrust at or below 18 MN and cutterhead torque at or below 1700 kN-m (30 min and greater). The resulting torque per penetration rate is comparatively high, 68 kN-m per unit mm/rev at 55–60 min). BDV activity is much lower during this time, *and very likely due to the lower advance rate.*

OPERATOR USE OF VIBRATION RESPONSE

After establishing the proof of concept by confirming the correlation between the data collected by the BDV system and the damage on the cutters and also collecting actual rock (cobbles and boulder) samples from the belt, the system was optimized and implemented to be used directly by the TBM operators. In order to address this, an executive program was installed on a laptop on the operator cabin of the TBM. The program was designed in such a way to provide both visual and sound alarms to the TBM operators. The operators were instructed to react immediately to the alarms by increasing the rotational speed of the cutterhead and reducing the advance rate (as a result reducing the penetration rate). By doing this, the amount of "bite" of the cutters engaged with the boulder was reduced significantly and as a result the amount of damage to the cutters was reduced. As an example, by using the recorded data from this system during the excavation of NB tunnel, a map of projected boulders and frequency was developed for the SB tunnel to be excavated next. During the excavation of the SB tunnel (NB and SB tunnels are parallel to each other and go through the same geology), the operators were instructed to operate with a slow advance rate (maximum advance rate of 25 mm/min with 2.1 rpm cutterhead rotation speed = 11.9 mm/rev) in the specified zone with boulders. As a result, the damage to the cutters was reduced significantly on the SB tunnel and the contractor was capable of passing the boulder zone without a need to perform hyperbaric intervention.

CONCLUSIONS

The Northlink Hitz EPB TBM was outfitted with a novel vibration monitoring system to detect, in real-time, vibration impacts between cutting tools and boulders. The monitoring system was used throughout northbound tunnel excavation. Measured vibration levels and number of vibration impacts were significant while mining through the till and till-like deposits, consistent geologically with where boulders and cobbles would be present and where they were discovered through muck sampling. Low vibration impacts were recorded while excavating through the clays, silts and gravels. High count levels of vibration amplitude measured by the monitoring system correlated well with observed cutter and cutterhead damage. Machine data such as torque, thrust, and penetration rate seem to be influenced when passing through the boulder/cobble fields. Using the system , the TBM operators modified the operational parameters of the TBM to excavate through boulder/cobble rich regions with no observed damage. During the subsequent SB tunnel excavation, the operator reduced the TBM advance rate in the regions identified as boulder-rich by the vibration monitoring system in the NB excavation. Consequently, there was significantly less cutterhead damage in the SB excavation.

REFERENCES

Buckley, J. Monitoring the vibration response of a tunnel boring machine: application to real time boulder detection. MSc Thesis, Colorado School of Mines, 2016, 112 pp.

Mooney, M.A., Walter, B.W., Steele, J.P.H., Cano, D. Influence of geological conditions on measured TBM vibration frequency, *Proc. North American Tunneling*, 2014.

Walter, B. Detecting changing geologic conditions with tunnel boring machines by using passive vibration measurements, PhD Dissertation, Colorado School of Mines, 2013, 163 pp.

EPB TBM Foam Generation

Mike Mooney ▪ Colorado School of Mines
Nils Tilton ▪ Colorado School of Mines
Dhrupad Parikh ▪ Colorado School of Mines
Yuanli Wu ▪ Colorado School of Mines

INTRODUCTION

Soil conditioning is often critical to high performance earth pressure balance (EPB) tunnel boring machine (TBM) tunneling. The injection of foam into the cutterhead tool gap and the subsequent mechanical mixing with scraped soil in the tool gap, cutterhead openings and in the excavation chamber, serves to transform the formation soil into a low permeability, low shear strength, compressible chamber material for steady face support and muck extraction. A significant body of research and literature exists on conditioned soil behavior, primarily based on laboratory studies (e.g., Vinai et al. 2007, Bezuijen et al. 2012, Budach and Thewes 2015, Williamson et al. 1999, Mori et al., 2016). However, very few if any published studies have focused on how foam is mechanically produced on board the TBM, what foam generation system characteristics are important, and how in-situ conditions such as ground pressure influence foam generation and system design. The lack of documented studies in the literature leaves many contractors 'foaming in the dark'. The first part of this paper presents an overview of on-board foam generating systems for EPB TBMs, discussing the desired foam properties and quantities, how they are estimated, and key system parameters that deliver and influence these properties. The second part of the paper presents the results of foam generation testing in a laboratory-simulated EPB TBM soil conditioning system, focusing on the influence of a number of key parameters such as pressure, air velocity, generator fillings, elbows, etc. on foam properties.

TBM FOAM GENERATOR

Plumbing and Geometry

While all TBMs are somewhat different in terms of size and configuration of systems, we present here a reasonable layout and geometry of a foam generation system for a 6.5–7.0 m diameter metro-size EPB TBM. Figure 1 presents a general layout of a foam generation system, including the main components and plumbing. Foam generation begins with compressed air tanks, surfactant and polymer containers and diluted surfactant and polymer tanks. Surfactants are the key ingredient to generating foam. Various polymers are sometimes used to bind or disperse soil and may be included directly in the foam liquid line prior to the foam generator or added to the foam transport line after the foam has been generated. Air and foaming solution flow separately and are metered at prescribed flow rates needed to achieve the desired foam injection ratio and foam expansion ratio, both of which are described below.

The compressed air and foaming solution flow lines merge immediately upstream of the foam generator unit (sometimes called the foam gun). Foam generators can vary in diameter and length, and often include fillings such as beads, tubes, or steel wool to create an environment that promotes foam bubble generation. The foam created in the generator exits and travels to a rotary fluid joint at the rear of the excavation chamber,

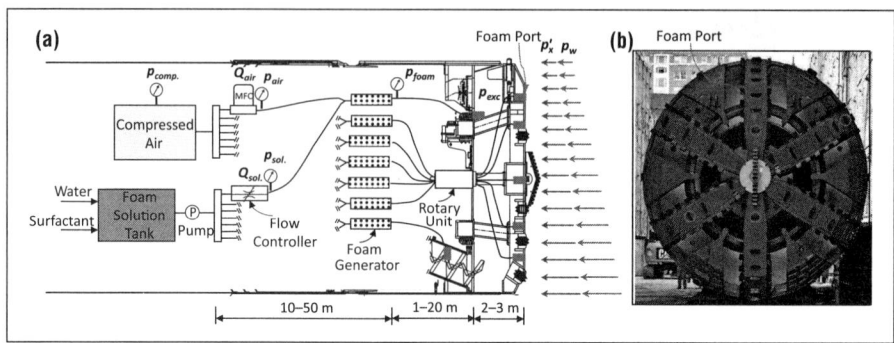

Figure 1. (a) Schematic of a generalized foam generation system in a TBM; and (b) photo of a metro-size EPB TBM cutterhead

i.e., the bulkhead wall. The foam flows through the rotary fluid joint in dedicated channels and then flows to dedicated cutterhead ports.

For a metro-size EPB TBM, there are generally 4–7 ports at the cutterhead face that deliver foam to the soil in the tool gap. The number of ports grows with TBM diameter. There are often additional ports in the excavation chamber and in the screw conveyor. Each cutterhead port has a one-way valve to allow foam flow when desired and to prevent pressurized groundwater and soil from flowing into the foam pipes. When foam is to be delivered through a desired port, the foam pressure exceeds that of the formation water. This pressure differential opens the valve and allows foam flow. There are a variety of valve configurations used to accomplish this.

Figure 1 is labeled with pressures p, volumetric flow rates Q, and velocities v at various critical points that will be discussed throughout the paper. An important aspect of foam generation is the pressure distribution along the system, bookended by the compressed air pressure $p_{comp.}$ and the formation groundwater pressure p_w. The pressure systematically decreases as the air, foaming solution and ultimately foam travel downstream.

Foam Delivery

Two governing foam parameters are the foam expansion ratio *FER* and the foam injection ratio *FIR*. The *FER* is the volumetric ratio and/or volumetric flow ratio of foam to foaming solution (Eq. 1). The *FIR* is the volumetric ratio and/or volumetric flow ratio of foam to excavated soil (Eq. 2). The volumetric flow rate of excavated soil can be represented by the product of advance rate (*AR*) and excavated area (A_{exc}). *FIR* is conventionally expressed as a percentage.

$$FER = FER_p = FER(p = p_w) = \frac{Q_{foam}}{Q_{sol.\,w}} \qquad (1)$$

$$FIR = FIR_p = FIR(p = p_w) = \frac{Q_{foam}}{Q_{exc.}} = \frac{Q_{foam}}{A_{exc.} \cdot AR} \times 100 \qquad (2)$$

The appropriate total *FIR*, i.e., the sum of foam delivered through all ports, can vary with soil and groundwater conditions. Generally, the *FIR* varies from 20–70% (EFNARC 2005). *FER* values can vary from 5–20 depending on soil types and groundwater conditions, e.g., the desire for a wet foam and less air introduced or a dry foam. While seldom clear during reporting, these *FIR* and *FER* values are generally calculated and reported at the pressure they are delivered into, e.g., p_w for cutterhead injection,

p_{exc} for excavation chamber injection, and p_{sc} for screw conveyor injection. For simplicity here, we will focus on the most critical and common foam injection—through the cutterhead. Because the groundwater pressure is not measured on the TBM, the excavation chamber pressure p_{exc}, most often at springline, is used as a proxy for p_w. A subscript p is sometimes used (FER_p, FIR_p) to clearly indicate that these values reflect the in-situ pressure.

The delivery of air and foaming solution typically begins with a pressure-corrected calculation. Assuming cutterhead injection and using p_{exc} as a proxy for p_w, the desired foam flow rate $\bar{Q}_{foam}$ (designated with a top line to reflect desired vs. actual Q_{foam}) is determined via equation (3) where D_{chd} is the cutterhead diameter. Note that the desired FIR remains constant and $\bar{Q}_{foam}$ varies with advance rate AR.

$$\bar{Q}_{foam} = \frac{FIR}{100} \cdot Q_{exc.} = \frac{FIR}{100} \cdot \frac{\pi}{4} D_{chd}^2 \cdot AR \qquad (3)$$

The corresponding desired $\bar{Q}_{air}$ and $\bar{Q}_{sol}$ are presented in equation (4). As shown in Figure 1, $\bar{Q}_{air}$ reflects the flow rate delivered by a servo-controlled air mass flow controller (MFC). Using the ideal gas law and assuming constant fluid temperature throughout the foam generation system, the MFC flow rate is pressure corrected to deliver the desired flow rate at the cutterhead. Because $p_{air} > p_{exc}$, Equation (4) shows that the desired MFC air flow rate is lower than that desired at the cutterhead. The foaming solution flow rate need not be corrected for pressure as it is incompressible.

$$\bar{Q}_{foam} = \bar{Q}_{air} \frac{p_{air}}{p_{exc}} + \bar{Q}_{sol.} \qquad (4)$$

Finally, the ratio of provided $\bar{Q}_{air}$ and $\bar{Q}_{sol}$ are determined by the desired FER per Equation 1.

Foam Flow Mechanics

Foam generation can be described as the vigorous mixing of two immiscible fluids (in this case air and liquid) to form a multiphase mixture in which tightly packed air bubbles are separated by thin liquid films called lamella. In TBMs, foam is generated by flowing air and a liquid water-surfactant solution through a pipe filled with porous media, such as randomly packed beads, steel wool, or perforated tubes. Two mechanisms of bubble creation within porous media are snap-off and lamella division (Ransohoff and Radke 1988, Radke et al., 1994) as shown in Figure 2. Snap-off (panel a) occurs when a bubble divides into one or more smaller bubbles as it is squeezed by a constriction. Lamella division occurs when a bubble ruptures as it flows through a branch point in the porous media, as in panel (b).

Our experiments also show that foam can be generated in the absence of porous media by simply flowing the liquid and gas through a pipe. In this case, bubbles likely form due to shear stresses and fluid mixing due to turbulence. These mechanisms likely play a role when foam is generated within porous media as well, particularly at high flow rates associated with turbulence. Consequently, one objective of our study is to vary the porous media and flow rates within the generator to investigate the competing roles of pore geometry versus fluid mixing. A better understanding of these mechanisms could allow one to tailor the porous media and flow rates to deliver foams with desired properties, and even vary foam properties dynamically to respond to varying soil conditions and operating pressures.

Considering the TBM foam generation system, the key pressures, volumetric flow rates and velocities along five key cross-sections (a - f) are shown in Figure 3. The desired

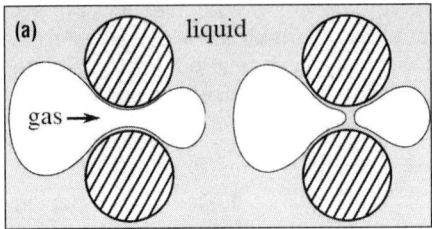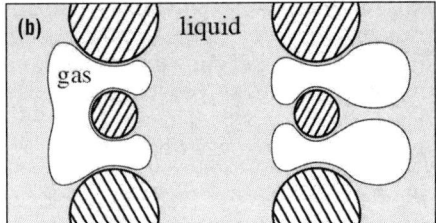

Figure 2. Foam bubble generation mechanisms in porous media: (a) snap-off and (b) lamella division (after Ransohoff and Radke, 1988)

and delivered foam flow rate is pressure driven, and significant pressure losses exist along the system. At the upstream end, the pressure losses within the pipes carrying single phase air and liquid are generally negligible compared to those incurred downstream, even for cases where the single phases must travel tens of meters from their source containers. The air and liquid merge at the generator or immediately upstream of the generator and flow through the porous media (beads, steel wool, tubes) in a multiphase fashion. Significant energy loss occurs, often greater than 2 bar, during multiphase flow through this porous media as foam is created. Upon exiting the generator, foam flow also induces significant pressure losses as it is delivered through the pipe joining the generator to the cutter head. When foam is generated in the rear of a TBM, such pipes are often on the order of ten meters long.

Foam is a complicated fluid that is both compressible and non-Newtonian with an effective viscosity that is typically orders-of-magnitude larger than that of water (Sibree 1933, Grove 1951, Wenzel 1967, David 1969, Wenzel 1970). Due to this large viscosity, foam flow through pipes is characterized by much larger pressure drops than for single phase flows or non-foam multiphase flows.

Understanding the source of pressure losses within foam generators is further complicated by the presence of the porous material. For single phase flows through porous media, such as pure water, pressure losses are well described by Darcy's law for low flow rates, and the Forchheimer equation for high flow rates (Joseph et al. 1982, Lage 1998). These equations show that pressure losses occur due to viscous effects in the small pore spaces, and additional inertial effects as the flow winds through tortuous pore paths. In the case of multiphase flow of air and water (without surfactant), additional pressure losses occur due to capillary effects and the fact that each phase must compete for the available pore space. Finally, in the case of foams, the restriction of the liquid phase to thin films lead to additional complicated flow phenomena that are not fully understood. As in pipes, foam flows through porous media are characterized by much larger pressure gradients than for comparable single-phase flows or non-foam multiphase flows. Considering these pressure losses, and that foam flow is pressure driven, it is important to design a foam generation system that has sufficient supply pressures – $p_{comp.}$ and p_{sol}. Naturally, with consideration to Figure 3, $p_{comp.} > p_{air}^a > p_{foam}^b > p_{foam}^d > p_{foam}^f > p_w$.

The pressure at a given location in a foam generation system is primarily set by the downstream conditions. Consequently, pressure variations within the system are best understood by beginning at the cutterhead and working upstream. The analysis begins at the cutterhead where the pressure p_w is determined by the depth below the groundwater table. Working upstream, the pressure at the outlet of the foam generator can be expressed as p_w plus the pressure drop required to deliver foam from the generator to the cutterhead. Similarly, the pressure at the inlet of the foam generator can

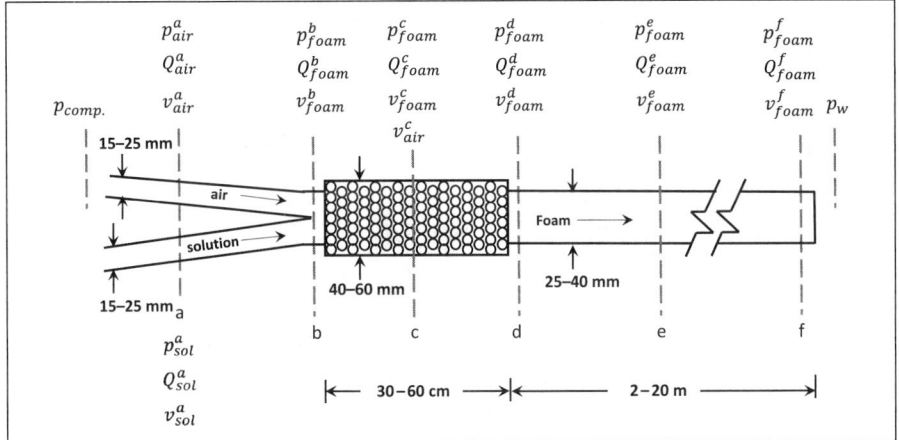

Figure 3. Flow parameters and plumbing dimensions of a foam generation system on a TBM

be determined by adding the pressure drop across the generator. Accordingly, with knowledge of the geometry of the foam generation system, e.g., pipe lengths, etc., one can estimate pressure losses and then determine the desired p_{air} and necessary p_{comp}. The primary difficulty is predicting the pressure drops across the generator and the pipe conveying foam to the cutterhead. A major objective of this study is to identify the various sources of pressure loss within TBM foam generation systems and develop simple practical relationships to predict pressure variations a-priori.

While the mass flow rates of air $\dot{m}_{air}$, solution $\dot{m}_{sol}$, and foam $\dot{m}_{air} + \dot{m}_{sol}$, are set by the mass flow controllers, quantifying the pressures p, volumetric flow rates Q, and flow velocities v, is complicated by the porous media, multiphase flow, and air compressibility. For the single-phase flows of air and solution upstream of the generator (location a), the velocities are computed per equation (5), where A is the cross-sectional area of the respective pipe, and ρ is the mass density (e.g., kg/m^3) of the air or solution.

$$v_{air}^a = \frac{Q_{air}^a}{A} = \frac{\dot{m}_{air}}{\rho_{air}^a A} \quad v_{sol.}^a = \frac{Q_{sol}^a}{A} = \frac{\dot{m}_{sol}}{\rho_{sol}^a A} \tag{5}$$

Within the foam generator (location c), the cross sectional area is reduced by the presence of fillings and there is uncertainty as to whether the liquid and air travel with different velocities and pressures. We take the pressure measured at the inlet and outlet of the foam generator to be characteristic of both the liquid and gas, e.g., $p_{sol}^b = p_{air}^b$, and we estimate the pressure at location c as $p^c = (p^b + p^d)/2$. Following the convention for multiphase flow through porous media (Bear 1972), we compute separate superficial velocities for the air v_{air}^c, and liquid v_{sol}^c, as in Equation 5, where A is taken to be the cross sectional area of the pipe containing the porous media. Though the liquid solution is considered incompressible, the local air density is computed from the local pressure using the ideal gas law. As a result, the volumetric flow rate Q_{air} and velocity v_{air} increase considerably across the generator due to the downstream pressure loss. With the formation of tightly packed bubbles separated by thin liquid lamella, there is uncertainty as to whether the two phases in a foam actually flow with different velocities. Consequently, we also compute an effective foam flow rate $Q_{foam}^c = Q_{air}^c + Q_{sol}^c$ and a superficial foam velocity $v_{foam}^c = Q_{foam}^c/A$. Finally, we characterize the volumetric flow rate and velocity at the inlet and outlet of the foam transport tube joining the foam

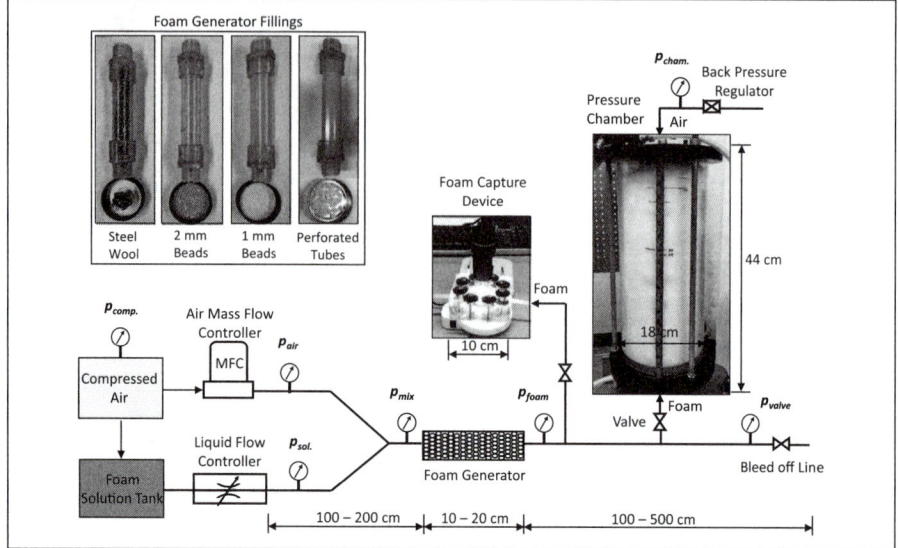

Figure 4. Schematic of laboratory foam generation system and foam testing devices (pressure chamber and foam capture device) and laboratory foam generators with different filling materials. Not to scale.

generator to the pressure chamber using an identical procedure, where *A* is then set to the cross sectional area of the tube.

EXAMINING THE CHARACTERISTICS OF FOAM

Foam generation principles were examined through careful laboratory testing with a foam generation system scaled appropriately from field conditions. The following sections describe the experimental setup, results and findings.

Experimental Setup

The scaled down foam generation system is shown in Figure 4. Compressed air and foaming solution are metered via mass flow controllers similar to a TBM system. Typical TBM supply pressures were duplicated, and in fact expanded to examine influence. Pipe diameters were scaled down because field volumetric flow rates are too high for laboratory testing. However, great care was taken to simulate the range of air and foam flow velocities observed on TBMs. Velocity is the important parameter because flow regime and energy loss is velocity-dependent and not volumetric flow dependent. We employ a variety of foam generator sizes in the experimental setup generator fillings, including 1 and 3 mm beads, steel wool, and perforated tubes.

A foam bubble capture device and digital microscope is used immediately downstream of the foam generator to capture and image the foam. This allows the detailed analysis of bubble size and bubble size distribution. A pressure chamber is used to mimic the pressurized environment that foam is deposited into in a tunneling environment, whether the formation water in front of the cutterhead or the excavation chamber. During foam generation in the laboratory, the chamber pressure is set and held constant at levels from 1–5 bar (absolute pressure).

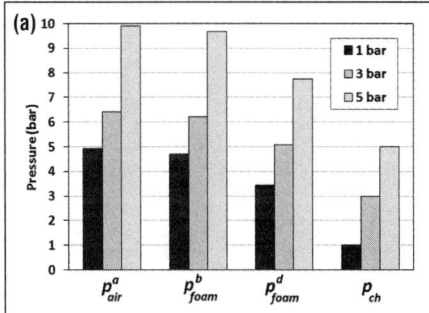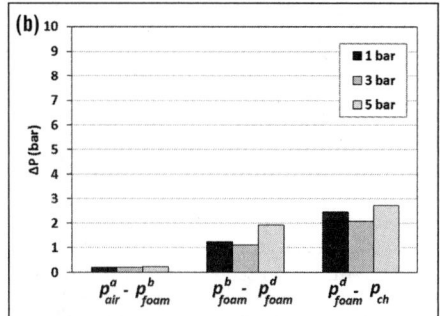

Figure 5. (a) Absolute pressure levels; and (b) pressure drops across the system for $v_{air}^c = 0.5$ m/s , 25 mm diameter, 20 cm long foam generator and 6 mm diameter, 1 m long foam transport pipe

Pressure and Velocity

A series of foam generation experiments was performed with the system shown in Figure 4. Three different excavation chamber pressures were simulated—1, 2 and 3 bar (absolute pressures, i.e., 1 bar = atmospheric pressure). At each chamber pressure, the air flow velocity v_{air}^a was varied over 4–5 different values for each p_{ch}. This created a matrix of 14 experiments. For this series of 14 experiments, the foam generator was 200 mm in length and 25 mm in diameter. 3 mm beads were used for filling material. The foam transport pipe length from the exit point of the generator to the chamber was 1.0 m long and 6 mm internal diameter, exactly the same diameter as the air and solution supply lines. The foaming solution included a commercial surfactant with 5% concentration. The *FER* was fixed at 15, and according with equation 1, was calculated using each chamber pressure.

Figure 5 shows the pressure regime throughout the system for three experiments with a consistent average generator superficial air velocity ($v_{air}^c = 0.5$ m/s) yet at different values of p_{ch}. The chamber pressure p_{ch} is prescribed and held constant through the use of a pressure relief valve. Figure 5 illustrates that the pressure drop through the 200 mm long generator ($p_{foam}^b - p_{foam}^d$) and through the 1 m long foam transport pipe ($p_{foam}^d - p_{ch}$) are significant. The pressure drop across the generator was on the order of 1–2 bar and the pressure drop along the exit pipe (from the generator to the chamber) was 2–2.8 bar/m of pipe.

Figure 6 shows the pressure loss across the generator and along the 1.0 m long, 6 mm diameter foam transport pipe for all 14 experiments as a function of v_{air}^e in Figure 6a and average foam transport velocity v_{foam}^e in Figure 6b. As expected, pressure loss increases with these respective velocities. The pressure loss also increases with chamber pressure.

The pressure loss in the foam transport pipe varies with foam transport pipe diameter. Three internal pipe diameters—6, 9 and 12 mm—were investigated (keeping all other system parameters the same). Figure 7a illustrates the pressure loss measured along 1 m of the transport pipe as a function of v_{foam}^e. What is evident from Figure 7a is that transport pipe pressure loss is strongly influenced by foam velocity and somewhat influence by chamber pressure. Figure 7b presents the same data in an interesting way. Each of the three pipe diameters are delivering generally the same Q_{foam}^e albeit using much different velocities. Figure 7b shows the benefit of larger diameter in reducing pressure loss.

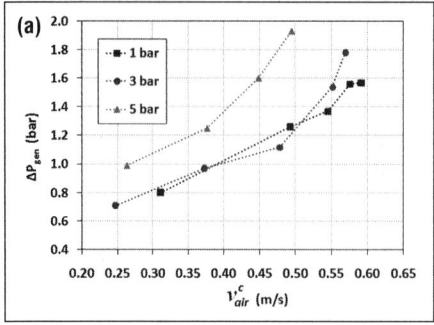 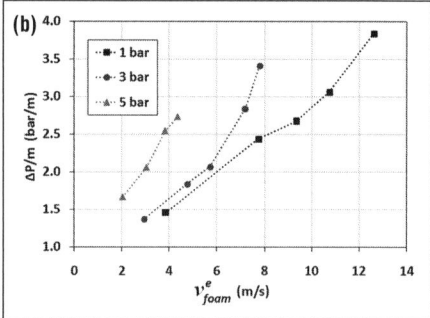

Figure 6. Pressure drops across the (a) foam generator and (b) foam transport pipe for 0.5 m/s average superficial generator air velocity, 25 mm diameter, 20 cm long foam generator and 6 mm diameter, 1 m long foam transport pipe

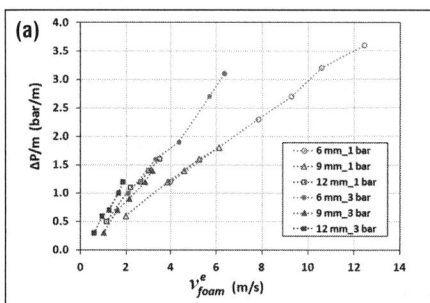 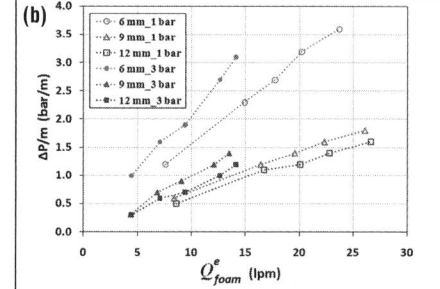

Figure 7. Pressure drop along the 1 m long foam transport pipe versus (a) v_{foam}^e and (b) Q_{foam}^e as a function of foam transport pipe diameter (6, 9, 12 mm) and chamber pressure (1, 3 bar). In all cases, foam generator is 25 mm diameter, 20 cm long, and beads are 3 mm.

Foam Behavior

For each of the 14 experiments described earlier, foam was captured, imaged and analyzed to characterize bubble size. The foam was captured immediately downstream of the foam generator as illustrated in Figure 4. A backpressure regulator was used to maintain the bubble capture device pressure equal to p_{foam}^d. This bubble capture process is described elsewhere (Mooney et al., 2016). While bubble characteristics are best reflected by a bubble size distribution histogram, a simplified mean bubble diameter is shown in Figure 8a as a function of p_{ch} and v_{air}^c. Figure 8a shows that both air velocity and chamber pressure have a significant influence on bubble size. Some additional results in Figure 8b show the influence of generator bead size on foam bubble size. Smaller beads have smaller pore sizes that leads to smaller bubbles through the snap-off and lamella mechanisms described earlier.

The engineering behavior of foam can be addressed by examining its stability and compressibility (or stiffness). Stability and compressibility were investigated using the pressurized chamber (Figure 4). Stability is typically measured by recording the drainage of liquid with time (EFNARC 2005). Compressibility is measured by recording the volume change due to a unit increase in pressure. Resilience in the context of foam behavior is measured by recording the recoverable deformation after five cycles of 1 bar pressure loading and unloading.

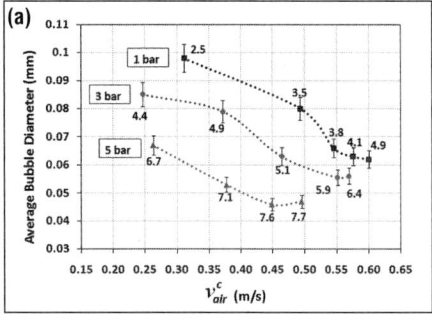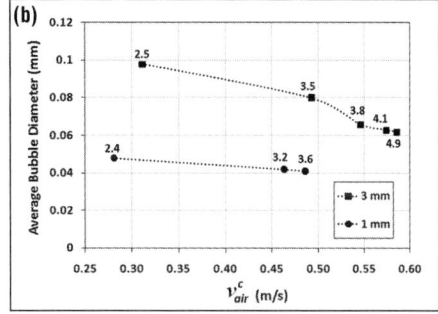

Figure 8. (a) Average bubble diameter measured at the generator outlet as a function of v_{air}^c for different chamber pressures (3mm beads, 20 cm length foam generator, 1m long, 6 mm diameter foam transport pipe); (b) average bubble diameter as a function of bead diameter (pch = 1 bar)

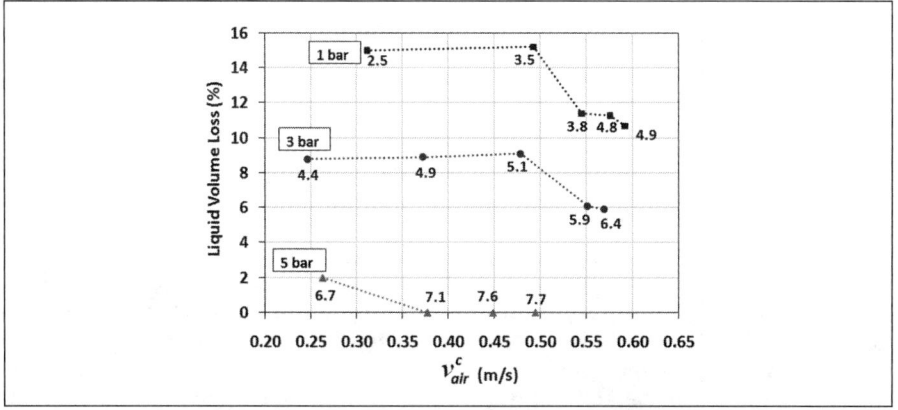

Figure 9. Foam liquid volume loss after 30 minutes as a function of v_{air}^c in three different chamber pressures

Figure 9 presents the liquid volume loss measured 30 minutes after foam generation at various air flow velocities and in the three chamber pressures. The air velocity has a small influence on stability in that higher velocities lead to more stable foam. The chamber pressure, on the other hand, has a significant influence on stability. Foams generated and deposited into higher pressure environments experience significantly greater stability. While not shown here due to paper length limitation, the reason for the increased stability has to do with smaller and more uniform bubble size at higher pressure. Smaller and more uniform bubbles are more stable and do not break down as quickly as larger and varied bubble sizes.

Figure 10 presents the compressibility measured by cycling the chamber pressure by 1 bar. The measured data points are accompanied by the theoretical compressibility of air for comparison. Figure 10 shows that the measured response matches theoretical air compressibility almost perfectly. The results also show that bubble size does not influence compressibility. The theory is that smaller bubbles have a greater internal pressure than larger bubbles, and therefore would be less compressible. This would manifest itself as a function of air velocity but these results show no influence of air velocity. It turns out that the difference in inner pressure between the bubble sizes we have observed is on the order of single kPa and therefore negligible compared to the 100, 300 and 500 kPa chamber pressures.

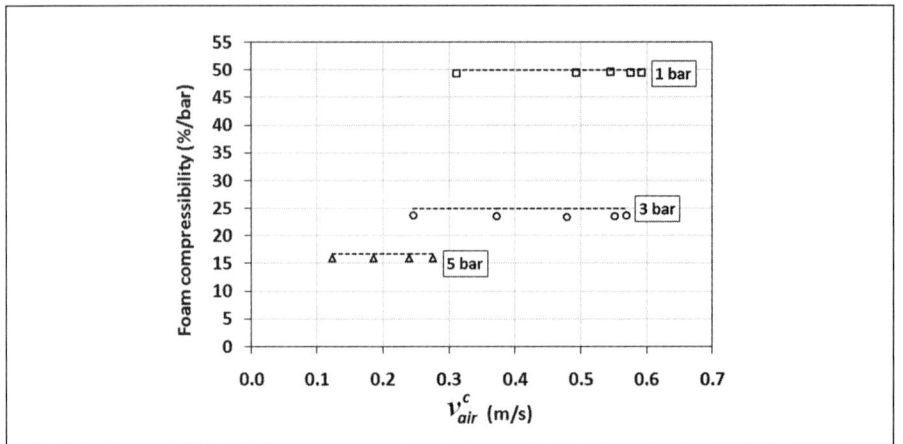

Figure 10. Foam compressibility at time = 5 mins for various range of flow rates at different chamber pressure using 3mm beads, 200mm length of foam gun and 1m-exit tube

Influence of Valves

Careful attention should be paid to cutterhead foam port valves and any valves where there is a significant reduction in cross-sectional area and corresponding increase in flow velocity. Such a rapid increase in velocity can cause cavitation that can destroy foam. The literature shows that foam can be severely degraded by flow through constrictions (Calvert 1988). The influence of the valve closing angle was investigated by physical simulation of a TBM valve in the laboratory. A ball valve is placed on the bleed-off line in the laboratory foam generation system (Figure 4). When the valve is completely open, foam flows through the bleed-off valve with good quality (i.e., uniform foam bubbles and consistent foam flow). However, when the valve is closed to some degree, considerably bigger bubbles are observed in the foam, the generated foam does not exhibit consistent flow and one can hear the air separation from the foam. To characterize this behavior, a series of tests was performed with four different valve opening angles, and the pressures along the foam generation system were monitored, particularly the pressure before the foam generator (p_{foam}^b), the foam pressure immediately after the foam generator (p_{foam}^g), and the pressure immediately upstream of the ball valve (p_{valve}).The test initiated with foam generation and flow using a fully opened valve (valve opening area is 100%). Pressures were measured and the ball valve was then closed gradually and in increments until poor foam quality appears as described above. The test results are shown in Figure 11. Foam exhibited good quality with the valve opening areas of 100%, 41%, and 30%. When the valve opening area reaches 22%, large foam bubbles were observed and foam flow was not consistent (air-liquid separation occurred). The valve pressure (p_{valve}) and foam pressure (p_{foam}^g) increase noticeably when the valve opening area decreases from 30% to 22% as shown in Figure 11. The most likely explanation for the poor foam quality is cavitation through the valve. Cavitation causes large vapor bubbles flowing downstream and air-liquid separation.

CONCLUSIONS

Foam generation on EPB TBMs is a complex process of compressible multiphase flow through a highly energy dissipative system. The pressure loss through typical foam generation lines is significant due to losses through the generator and through the foam transport line. Turbulent mixing and viscosity are key contributors to this.

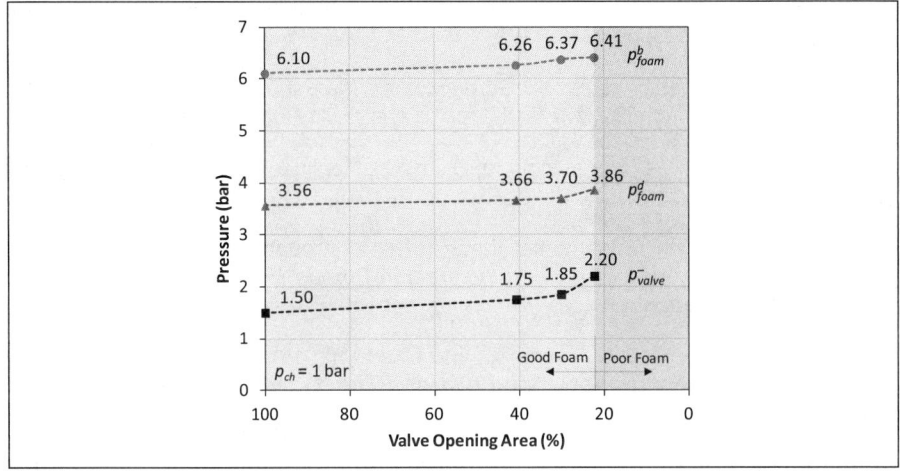

Figure 11. The influence of the exit valve opening area on pressures along the foam generation system

These pressure losses are velocity-dependent and foam transport pipe diameter-dependent. The air supply pressure required to overcome pressure losses and deliver foam into a formation water environment must be determined and delivered, otherwise foam velocity and flow rates will not meet the project needs. Foam bubble size and engineering properties (liquid loss with time, compressibility) are influenced considerably by pressure and flow velocity. Higher pressure and flow velocity produce foam with smaller, more uniform bubbles that have greater stability insofar as liquid drainage time.

ACKNOWLEDGMENTS

Financial support for this research was provided in part by BASF. We are very grateful to BASF for this support and for helping to make this research possible.

REFERENCES

Bear, J. (1972). Dynamics of fluids in porous media. American Elsevier Publications, New York.

Bezuijen, A. (2012). Foam used during EPB tunnelling in saturated sand, parameters determining foam consumption. In: *Proceedings WTC 2012*. 2012, Bangkok, Thailand, 267–269.

Budach, C., Thewes, M. (2015). Application ranges of EPB shields in coarse ground based on laboratory research. *Tunnelling and Underground Space Technology*. Vol. 50, No. 8, pp. 296–304.

Calvert, J.R. (1988). The flow of foam through constrictions. *International Journal of Heat and Fluid Flow*, 9(1): 69–73.

Calvert J.R. (1990). Pressure drop for foam flow through pipes. *International Journal of Heat and Fluid Flow*, Vol 11, No. 3, 1990, pp. 236–241.

David, A., S.S. Marsden. (1969). The Rheology of Foam, SPE paper 2544, *44th Annual Meeting of SPE AIME*. Denver, CO, 28 September–1 October 1969.

Grove C.S., Wise G.E., Marsh W.C., Gray J.B. (1951). Viscosity of Fire-Fighting Foam. *Ind. Eng. Chem.*, 43 (5), pp 1120–1122.

EFNARC (2005). *Specification and Guidelines for the use of specialist products for Mechanised Tunnelling (TBM) in Soft Ground and Hard Rock.* 44 (April). pp. 1–45.

Joseph, D.D., Nield, D.A., Papanicolaou, G. (1982). Nonlinear equation governing flow in a saturated porous medium. *Water Resour. Res.* 18, 1049–1052.

Lage, J.L. (1998). The fundamental theory of flow through permeable media from Darcy to turbulence. In *Transport Phenomena in Porous Media* (ed. D.B. Ingham & I. Pop), pp. 1–30. Pergamon.

Mooney MA., Wu Y., Mori L., Bearce R. and Cha M. (2016). Earth Pressure Balance TBM Soil Conditioning: It's About the Pressure. *Proc. World Tunnels Congress*, San Francisco, Apr. 22–28, 2016.

Mori, L., Mooney, M.A., Cha, M. (2017). Laboratory Tests to Determine the Relationship between Pressure and Foam Conditioned Sand Behavior in EPB Tunneling. *Tunneling and Underground Space Technology.* In press.

Radke C.J., Kovscek A.R. (1994). Fundamentals of Foam Transport in Porous Media, In: L.L. Schramm (Ed.), *Foams: Fundamentals and applications in the petroleum industry.* Washington DC: ACS Advances in Chemistry Series No. 242 (Am. Chem. Soc.).

Ransohoff T.C., Radke C.J. (1988). Mecahnism of foam generation in glass bead packs. *Society of Petroleum Engineering 1988*, pp. 573–587.

Sibree, J.O. (1934). The viscosity of froth. *Transactions of the Faraday Society*, 30(0): 325–331.

Thewes, M. & Budach, C. (2010). Soil conditioning with foam during EPB tunnelling/ Konditionierung von Lockergesteinen bei Erddruckschilden. *Geomechanics and Tunnelling.* 3 (3). pp. 256–267.

Vinai, R., Oggeri, C., and Peila, D. (2007). *Soil conditioning of sand for EPB applications: A laboratory research. Tunnelling and Underground Space Technology*, Vol. 23, No. 3, pp. 308–317.

Wenzel H.G., Stelson T.E., Brungraber, R.J. (1967). Flow of high expansion foam in pipes. *Journal of the Engineering Mechanics Division* 93: 153–166.

Wenzel, H.G., R.J. Brungraber, and T.E. Stelson. (1970). The Viscosity of High Expansion Foam. *J. Materials,* 5, 396–412.

Williamson, G.E., Traylor, M.T. & Higuchi, M. (1999). Soil Conditioning for EPB Shield Tunneling on the South Bay Ocean Outfall. In: *1999 RETC Proceedings.* 1999, pp. 897–925.

Challenges Encountered During Tunneling on the First Street Tunnel Project

Mina M. Shinouda ▪ Jay Dee Contractors Inc.
Thomas Costabile ▪ Skanska USA Inc.
Abdul-Ghani Mekkaoui ▪ Jay Dee Contractors Inc.
William P. Levy ▪ DC Water and Sewer Authority

ABSTRACT

The Clean Rivers Project in Washington, D.C., is a $2.7 billion program that will help improve water quality in the Potomac River and its tributaries when its massive tunnel program is completed by 2030. The project includes several large-scale tunnel projects, including the Blue Plains Tunnel ($330 million), the Anacostia River Tunnel ($253 million) and the First Street Tunnel ($157 million). The Northeast Boundary Tunnel—the program's largest in terms of footage—has yet to bid.

The FST is designed to mitigate sewer flooding in the District's historic and densely populated Bloomingdale neighborhood. Due to the proximity of the houses to the jobsite, special measures were taken to minimize the impact on residents which generated logistical challenges. An Earth-Pressure-Balance TBM was used to construct the 20ft diameter mainline tunnel. A slurry MTBM was used to construct a 7ft diameter connecting sewer as one of several adit connections. The MTBM was recovered thru the side of the mainline tunnel without a shaft; a first for this type of recovery. This paper will discuss challenges encountered during excavation of both tunnels through challenging ground conditions.

PROJECT DETAILS

In 2012, four major storm events swept through the District of Columbia and hit the Bloomingdale neighborhood particularly hard, resulting in the neighborhood experiencing severe flooding. The extreme nature of the flooding, the damage to private properties and the health risks associated with exposure to sewage led to a multi-agency response and community calls for answers and action. The Mayor assembled a Task Force, led by DC Water, to evaluate a range of potential actions to address combined sewer flooding including engineering, regulatory, code revision and management improvements. The Northeast Boundary Neighborhood Protection Project, the outcome of the Mayor's Task Force, initiated a three-step infrastructure initiative designed to offload the undersized sections of the existing collection system that serves the Northeast Boundary sewershed and Bloomingdale by implementing the following projects:

Step 1 (short-term). Construction of green infrastructure and storm sewers to help alleviate flooding.

Step 2 (medium-term). *McMillan Stormwater Storage Project*—Addition of 4 million gallons of stormwater storage within abandoned sand filtration cells; adding in-line storage to existing sewers and construction of bioretention areas to collect and infiltrate stormwater. Completed in spring 2014.

First Street Tunnel Project—Addition of 9 million gallons of combined sewer storage in a new tunnel to relieve undersized sewers that traverse the Bloomingdale neighborhood. The construction of this project has been accelerated to provide emergency flood relief. Substantial completion in October 2016.

Step 3 (long-term). *Northeast Boundary Tunnel Project*—Realigns a portion of the DC Clean Rivers Project tunnel system to mitigate flooding more effectively, reduce impacts to private properties and accelerate the construction schedule of the project by two years, from 2025 to 2023.

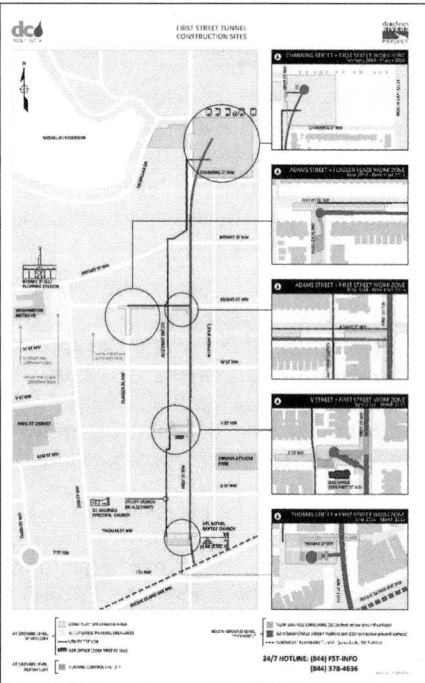

Figure 1. Overall project map

First Street Tunnel Project

The First Street Tunnel (FST) project was awarded to Skanska Jay Dee JV (SKJD) in October 2013. One of the main components of the project is a 2,700 foot long, 20 foot diameter precast concrete segmental lined tunnel that ranges in depth from 40 to 160 feet below ground (to the tunnel crown). This tunnel is to provide combined sewer storage during significant storm events and a temporary Pump Station at the downstream end was constructed to empty the tunnel into the existing interceptor system after the storm passes. The FST project includes four major construction areas; which are shown in Figure 1 and summarized as follows:

Channing Street. Tunnel construction site and includes a 65 foot diameter, 160 foot deep drop shaft.

Adams Street. Includes a 20 foot diameter drop shaft, diversion chamber and a 10 foot diameter de-aeration section and 7 foot diameter adit section.

V Street. Includes a 22 foot diameter drop shaft, three diversion chambers and a 16 foot diameter adit.

Pumping Station. Includes a 22 foot diameter shaft to house a temporary dewatering pumping station with a capacity of 6 mgd and an 8 foot diameter adit.

To minimize impact on residents, the three adits were constructed by means of underground construction methods. Two of the three adits (V-Street and Pump Station adits) were short and were built using ground freezing in conjunction with Sequential Excavation Method (SEM). The third adit (Adams Street adit) was longer and included the use of a slurry MTBM.

This project presented multiple significant challenges, ranging from technical to logistical to community concerns. Among these challenges are the TBM and MTBM mining into frozen ground, tight working areas in the center of a dense urban residential community, challenges in engaging craft labor with experience in this type of work, a schedule that reflected the owners desire to get the needed flood relief to this neighborhood as soon as possible, and the managing of community relationships and expectations during a multi-year period of heavy construction.

To provide construction impact mitigation to the residents, SKJD proposed ground freezing which helped with schedule, reduced material deliveries, and eliminated dewatering while providing stable, safe excavation support. Ground freezing was used extensively for shaft and SEM adit excavation support, and for the mining of the adit-to-main tunnel connections. It was also used for the excavation support of one of the approach channel walls at V Street, and even as a frozen strut at the bottom of cut for this excavation.

MAIN TUNNEL AND TBM

A TBM Named Lucy

To excavate the main tunnel, a 23-foot-diameter Herrenknecht TBM, named Lucy, was launched from the Channing Street mining shaft located at the northern end of the project. The TBM then headed south, curving from the shaft towards First Street before hitting a straight drive to the intersection of Rhode Island Avenue NW and First Street NW. Lucy was christened in a ceremony in April 2015, started excavation in the summer, took five months to mine the 2,700ft long tunnel alignment, and finished its journey on December 22nd 2015. Figure 2 shows the TBM and some details are listed below.

- TBM Manufacturer: Herrenknecht
- TBM shield type: EPB (Earth Pressure Balance)
- Shield diameter: 22.8 foot
- Shield length: 27.7 foot
- Shield & cutterhead weight: 974,442 lbs.
- Cutting diameter: 22.9 foot
- Trailing gear: 9 gantries + 1 bridge
- Total length of TBM: 351.75 foot.

Tunnel Eye

The TBM was launched from the Channing Street shaft which was 65 foot in diameter and 160 foot deep. The shaft was constructed using a 3.5 foot thick slurry wall as support of excavation and 4.0 foot thick cast in place concrete as final liner. The slurry wall panels were about 175 foot deep and were socketed in the bed rock which allowed

Figure 2. Tunnel boring machine—Lucy

for the dry excavation of the shaft. After pouring the base slab, the final liner was then installed in lifts from the bottom. A 23 foot diameter wooden form was utilized to create the tunnel eye and the slurry wall reinforcement at the eye location was fiber rebar.

In order to prevent any ground movement or material inflow at break-out from the shaft during the TBM launch, slurry walls were also utilized to create a 33 foot × 30 foot dry cell adjacent to the shaft along the tunnel alignment. This dry cell was dewatered before the TBM exited the shaft. In addition, an exit seal was installed in the shaft for added security.

The creation of the tunnel eye was challenging because it further restricted the already limited space at the shaft bottom for launching a machine of this size. Figure 3 shows the arrangement at the bottom of Channing Street shaft at the time of Launching the TBM.

TBM Launch with Umbilical Cords—First and Second Setup

The First St. Tunnel alignments was entirely in soft ground and under a tight delivery schedule with specified milestones which made constructing a tail tunnel infeasible. In addition, the size of the shaft provided a limited space at the bottom which was just enough to accommodate the TBM shields. These limitations prohibited the ability to assemble the gantries behind the machine in its normal configuration at launch time. Since these gantries housed necessary equipment for the operation of the machine, it was essential to find an alternative method to connect the gantries to the shields. Some of the concepts that were under consideration included staging the gantries at the bottom of the shaft beside the TBM shields, or hanging the gantries on the shaft walls. Both alternatives would have required building platforms to stack the gantries and multiple reconnections of the hydraulic and electrical systems. In addition to the cost and schedule impacts associated with building the platforms, the main concern about these two setups was the multiple reconnections of the hydraulic system. The hydraulic system would have had to be depressurized multiple times with a high possibility of system contamination and hydraulic oil leaks at the bottom of the shaft.

Ultimately, it was decided to assemble the entire gantry train at the surface adjacent to the shaft top and utilize long umbilical hoses/cables to allow the shield to advance forward enough distance for the gantries to be installed at one setup. This arrangement was especially challenging since the operators cab was located at the surface which mandated more reliance on communication/camera systems. Furthermore, this setup required an additional hydraulic power-pack to be integrated into the system at the bottom of the shaft to facilitate the transfer of the return oil back to the surface.

Shaft Setup

Due to the unique launch of the machine, the bottom of the shaft was initially set up differently than for production mining. After the TBM shield sections were lowered and assembled at the bottom of the shaft, a push frame was installed to provide the machine with sufficient reaction to move forward (Figure 4). The first ring was built in the tail section of the shield, pushed back to the frame, and fixed in place. The frame and shaft wall were used to hang the umbilical hoses/cables in a manner that would allow them to easily be pulled as the machine moves forward. As the machine moved forward, temporary rings were built across the shaft. With the TBM shields, temporary rings, and frame consuming almost half of the shaft, there was room for only one muck box to be used for the initial mining. Brackets and rollers were installed on the side of the tunnel as it was being built for the umbilical cords to ride on, this reduced the chance of over stretching the cables and hoses as the TBM advanced.

Figure 3. Arrangement at the launching shaft

Figure 4. Umbilical hoses and cables

A double drum "tugger" was used to move the box from the shaft to the heading and back (Figure 5), the tugger was installed at the back of the shaft and a snatch block was installed in the heading where the tugger cable was looped to allow for pulling the muckbox in both directions. This setup was used for about 300 feet at which time the top half of the temporary segments and push frame were removed and a locomotive was used to move three muck boxes at a time. Once the machine was about 600 ft away from the shaft, the mining was halted and the shaft was redressed for production mining.

Figure 5. Double drum tugger

It was prudent to have two muck trains to facilitate a more productive mining cycles. Each train comprised of a locomotive, six muck cars and two segment cars. One train would be at the heading being loaded with muck while the other would be at the shaft being emptied and loaded with segment stacks, this allowed for uninterrupted continuous mining operation. A rail switch and double track had to be installed to facilitate the usage of two trains. Since the shaft size did not allow for a full length of double track to accommodate a full size muck train, the double track was extended into the tunnel for 200 feet. This required the shaft bottom to be raised to an elevation where the platform in the tunnel would allow for the full width of double track.

Tunnel Mining

The entire tunnel alignment was located in the lower Potomac formation which was predominantly silty sand with very few inclusions of clay. Boulders were not antici- pated during mining. This led to the initial decision of dressing the entire cutterhead with ripper teeth and scrapers. However the slurry wall concrete strength was much higher than anticipated, reaching values in excess of 11,000 psi. This resulted in the cutting tools becoming completely worn out as the machine mined through the shaft wall. A free-air cutterhead intervention was conducted in the dry cell behind the shaft wall. While the ripper cutters were almost entirely worn, there was no measurable damage to the cutterhead structure itself. It was decided to redress the face with disc cutters which proved to be useful later on in the project as the machine mined through the frozen ground zones.

Tunnel-Adit Connections

The three adits were excavated from drop shafts offset from the main tunnel. The drop shafts receive the Combined Sewer Flows (CSO) from the local sewers and convey them to the main tunnel via the adits. V Street and Pump Station Adits were excavated using the SEM. The entire length of both Adits were frozen utilizing vertical freeze pipes that were installed from the surface. Once the frozen ground matrix attained its design strength and thickness, the Adits were excavated in 4 foot advances. Each advance of 4 feet along the alignment was followed by installing shotcrete as a tempo- rary support of excavation in the frozen ground prior to building the final lining.

Initially, the plan was to excavate each of these two Adits ahead of the main tunnel reaching the prospective location of their intersection. The Adit was to be extended into the envelop of the main tunnel and then the last section of the alignment would be backfilled with lean-mix concrete to allow for the future connection work to be done within a stable environment. However, as the project progressed it was realized that the main tunnel will reach the connection location ahead of the Adit excavation and thus the sequence of this work was changed to prevent significant schedule impacts to the tunneling operations. A decision was made to continue the main tunnel excavation and reevaluate the methodology of the connection work construction.

While this construction sequence eliminated the need for the lean-mix, it required that the ground freezing at the connection location be maintained longer until the connec- tion work is finished. The extended duration of the ground freeze produced a frozen block that was larger than intended which resulted in a full face of frozen ground for the TBM to pass through. This resulted in a slower rate of advance in the frozen zone. In general, while mining through the frozen zone was much slower, it was progressing well. However, an unanticipated extended stoppage resulted in the machine getting stuck at one of the frozen zones. This is discussed further below.

In addition, the need to maintain the stable frozen zone required some of the vertical freeze pipes to stay in place until the Adit connection is established. Some of these pipes were within the tunnel envelope and while the machine was capable of mining through them, a piece of pipe tore through the TBM's conveyor belt and resulted in an extended down time.

TBM Stuck at Freeze Location

A unique challenge of the tunnel alignment were the three zones where the ground was frozen in preparation for the adit connections. Extreme caution was used while tunneling through these three zones to ensure the cutterhead tooling remained effective for the duration of the alignment. Advance rates, thrust pressures, and ground conditioning were all adjusted and reevaluated based on progress through the zones. Most freeze pipes were positioned to avoid conflict with the tunnel alignment, but several were necessary to traverse the tunnel in order to establish an adequate freeze zone for safe excavation support to the adit connections. Those few pipes that remained were evacuated and sealed prior to the TBM arrival.

Figure 6. Damaged belt section

While tunneling through the second freeze zone at the V Street adit, the rubber belt conveyor system suffered a tear that effectively disabled the machine. A quick investigation determined that while tunneling through one of the abandoned freeze pipes, a small section of pipe became momentarily lodged in the discharge hopper. It became lodged between the pulley and the discharge chute housing causing a longitudinal tear in the conveyor which elongated to approximately 3.5 foot wide and 5 foot long after making several circuits before a crew member observed the damage. The size of the tear prevented any patches from being applied and required removal and replacement of the damaged section (Figure 6). It was decided to remove the damaged section of belt and splice in a new section utilizing two MLT Super-Screw field splices. This style of splice had the benefit of being flush with the surface of the belt, allowing the belt scraper system to still effectively clean the belt. This eliminated the need for a full vulcanization once the TBM cleared the freeze zone.

During the time between when the initial damage was observed to when the final repairs were made, the TBM was nearly frozen in place. The advance was force-stopped unexpectedly by the E-stop circuit, so no special provisions could be made to prepare the TBM for an extended stop in this critical zone. The total duration of this stop period extended to nearly 10 hours before the advance resumed. The temperature on the bulkhead dropped to nearly 10 degrees Fahrenheit during this time while the rest of the tail shield hovered just below freezing. Shortly after stopping, it was determined that it would be best to keep the hydraulics running and the cutterhead rotating at a very low rpm to ensure it did not get frozen in place. However, getting the cutterhead turning became a great challenge as it could not be rotated more than a few degrees before the torque load limiter shuts it down. The operator continued to work the cutterhead free by rotating a few degrees in either direction until we were finally able to achieve a full rotation. The cutterhead remained rotating at a low rpm while the hydraulic temperatures and EPB were closely monitored as repairs to the belt were made.

Once the trailing gear advanced beyond the adit location, additional freeze pipes were drilled and installed from within the tunnel to reestablish the freeze closer to the invert of the future adit opening.

TBM Disassembly and Removal

The TBM never officially "holed through" due to the fact that there was no retrieval shaft. The design called for the shields to be abandoned in place and bulkheaded off until the future Northeast Boundary Tunnel is constructed in 2023, and a connection made underground by SEM to the FST at which time, the temporary pump station will be decommissioned. CSOs would then be conveyed downstream by gravity to DC Water's Blue Plains Advanced Wastewater Treatment Plant approximately 12 miles away.

Therefore, after the approximately 2,700 foot, FST tunnel was completed, the majority of the TBM had to be removed through the original launch shaft at Channing St. In order to facilitate this work, the bottom of shaft elevation needed to be lowered, the dual rail track and switch removed. To save schedule, the support gantries were disconnected and reversed through the tunnel with as little work as possible being conducted in the tunnel itself. The goal was to perform most of the disassembly on the surface where access was much safer, while also allowing the follow-on work to occur in the tunnel. Reversing the support gantries through the tunnel was no easy task due to utility lines being in place, and freeze heads projecting from the tunnel liner at each of the adit locations. Non-essential utility lines were removed immediately after tunneling, but the majority were left in place to support the adit connection operation. Moreover, the additional freeze pipes installed from within the tunnel created a new set of projections into the tunnel envelop, further decreasing clearances.

Figure 7. Gantry section pulled through the tunnel

During mining, the support gantries were advanced forward on a jump rail system, but it was not practical to reinstall the system to move the gantries in reverse. Therefore, it was decided to pull the gantries to the shaft on rail cars. The gantries had to be rebalanced with strategically placed ballast in order to re-support and pull the gantries on the rail cars. The clearances between utility piping, freeze heads, and the gantry itself was measured in a few millimeters.

Following the removal of the support gantries, the majority of the tail shield systems were dismantled and removed. The major components for removal included the drive motors, screw conveyor, segment erector, and man-lock. To facilitate the removal of the screw conveyor, segment erector, carrier beam, and man-lock, a self-propelled hydraulic trailer was rented from a local crane rental company. The trailer was manufactured by Goldhofer and modified by W.O. Grubb to provide a means to drive the trailer into the heading, lift the components during disassembly, and drive them back out to the launch shaft. Clearances through the tunnel was a major challenge, nevertheless balancing and securing the assembly for the journey to the shaft required precise engineering calculations. In addition, the trailer needed modifications in order to operate on the curved invert of the tunnel.

Figure 8. Self-propelled platform used during TBM removal

In order to clear the obstacles at each adit location, a gravel ramp system was created. This allowed the Goldhofer to travel at a higher elevation, closer to springline, to ensure clearances. The slope of the ramp and the final elevation were critical to making sure the Goldhofer and the load it was carrying could safely traverse each adit location.

Termination Bulkhead

The FST terminus encompassed two separate bulkheads space about 100 foot apart. One was located at the downstream end of the tunnel, while the other was just south of the Pump Station adit. As previously mentioned, both bulkheads are temporary until the future Northeast Boundary Tunnel (NEBT) is constructed. The upstream bulkhead is a concrete mass with circumferential cast-in steel sections. There is also a cunette constructed abutting the upstream bulkhead, which diverts flow into the Pump Station Adit. The cunette is a reinforced concrete block anchored 4-inches into the FST liner with a bond-breaker in between. It is on the east side of the tunnel approximately 7-ft in height and begins to channel water 200 foot prior to the Pump Station Adit for controlled flow. It is also to be decommissioned once NEBT is constructed.

The downstream bulkhead encompassed the abandoned TBM cutterhead and shields. To construct this bulkhead, the TBM plenum was filled with a stable soil/grout mix. Grout was pumped into the plenum while the cutterhead is rotating to allow for it to mix with the soil within. Ultimately, the TBM was pushed forward a few inches to ensure the plenum is fully compact. Prior to dismantling the various TBM components, support brackets bearing against the leading edge of the last precast concrete ring were welded to the shield. Once completed, the thrust jacks, screw conveyor, segment erector, motors, air lock, and other components were removed.

ADAMS STREET ADIT

It is interesting to note that with all the complications of building a massive tunnel within a dense urban environment, that perhaps one of the most complicated portions of the FST project was an Adit. The Adams Street Adit involved a short run of approximately 280 foot of microtunnel completed under very challenging and unique circumstances. The depth (100 foot), excavation diameter (104 inches) and tightly constricted working area all contributed to the demands of the project.

The Adit began at the Adams Street drop shaft (AS-DS) at the corner of Adams Street NW and Flagler Place NW and paralleled the Adams Street right of way eastward under the residents front yards to the mainline tunnel under First Street. Due to the expected ground conditions, which included more silty sands with small amounts of clay below the groundwater table, no dewatering was allowed along the alignment. Also because of the adits's proximity to the residences, the use of a pressurized face

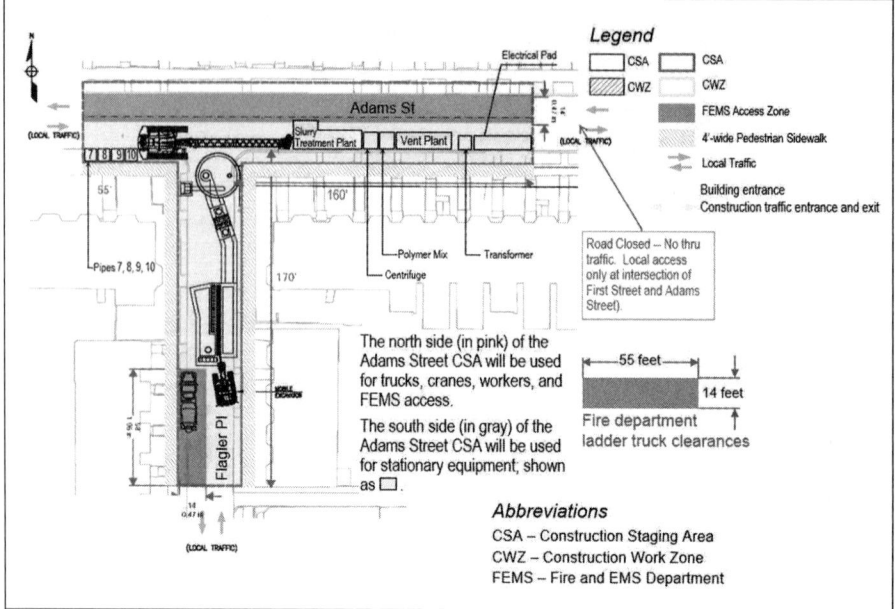

Figure 9. Adam Street construction site layout

machine was required by contract to minimize the risk of settlement. Figure 9 shows the construction work site proximity to the residences.

A MTS 2000M3 20C MTBM (nicknamed"Abigail"), was mobilized for this challenging work. The slurry MTS system included power pack, jacking unit, automated bentonite lubrication system and TACS guidance equipment. The machine, with a cutting diameter of 104 inches was used to install 84 inches internal diameter concrete jacking pipe. In total 45 pipes were jacked to complete this run, 10 of which were temporary pipes that were removed at the conclusion of the tunneling operation. Figure 10 shows the MTS slurry system components.

Due to the residential environment and the proximity of the houses to the jobsite, special measures were taken to minimize the impact on residents. The entire site was enclosed by sound-absorbing materials attached to the site chain link fencing. Equipment, including the slurry separation plant, was enclosed within noise-dampening structures. Additionally, restricted work hours, limited lay-down area for equipment and materials posed logistical challenges.

The Adam St. mining shaft (AS-DS) was constrained somewhat due to the already constructed permanent concrete liner for the de-aeration chamber in the adit and concrete vent riser pipe in the shaft. The shaft bottom was only a few inches longer than needed for the jacking frame. Lowering and setting the concrete pipe sections into the shaft required precise handling.

The de-aeration chamber was horseshoe-shaped and constituted the first 75 foot of the adit. It was excavated using a road-header through frozen ground from the AS-DS and lined with shotcrete. The dimensions of the chamber were larger than that of the MTBM and pipe, which required pre-blocking inside the chamber to restrain the string during mining and jacking operation. Crews set the MTBM in place and jacked

Figure 10. MTS slurry system

7 sections of pipe to push the MTBM though the chamber to the microtunneling break-out face. Once the MTBM was jacked through the chamber, excavation commenced.

Upon breaking out of the stabilized frozen zone outside of the de-aeriation chamber, the full hydrostatic groundwater head pushed the MTBM and pipe string back towards the shaft. While it is normal to have up to 6 inches rebound during pipe jacking, the movement in this case was much more significant due to lack of friction in the chamber around the jacking pipe. The string of pipe was almost following the jacking frame as it was pulling back. Typically this problem is resolved by installing a pipe clamp to arrest this movement but due to the shaft size restriction, this was not feasible. The jacking frame was modified to allow for holding the pipe string in place as it was pulled back to receive the next pipe. Ultimately, this movement gradually diminished as more pipe was installed into the ground.

Another challenge was waiting at the receiving end where the MTBM dead-ended into the mainline FST. The MTBM had to mine into the already-constructed segmentally lined tunnel and be recovered and extracted from within the tunnel. In preparation of receiving the MTBM, crews precut a reception ring in the mainline FST wall and cre-ated a grouted block outside the tunnel along the MTBM alignment. In the mainline tunnel, which has an inside diameter of 20 foot, crews constructed a concrete pad on which the MTBM would come to rest. Crews then encased the reception pad in the mainline tunnel within a block that was approximately 10 foot on either side of the MTBM alignment, and filled it with grout to create a sealed environment for the MTBM

Figure 11. Abigail sitting in the main tunnel at the end of its journey

entry into the mainline FST. Figure 11 shows the MTBM into the main tunnel as the reception chamber is being dismantled to recover the machine.

As the machine mined through the pre-cut tunnel wall into the grout block, the area outside of the mainline tunnel was also grouted to seal-off any ground water inflow. Then crews began to expose the machine by tearing down the reception block. Once exposed, the MTBM was lifted off the concrete pad, carefully rotated with only inches of clearance to be parallel to the tunnel, moved on to a self-propelled trailer, and transported to Channing street shaft where it was lifted out of the tunnel. With the variety of unique challenges facing the crew, planning and preparation were of the utmost importance. It took a tremendous effort and coordination between all parties involved.

The following statement by Andreas Thiele, an engineer for MTS who was onsite for the full duration of MTBM mining, sums up the challenges "the Adams Street Adit microtunnel was one of the most challenging jobs I have been involved with. What was so interesting about this project was all the unique and challenging aspects that came together. On a typical job, you might see one or two unique aspects, but here you had the combination of the constricted job site, the narrow shaft, the need to jack the MTBM through an existing structure, and a reception into a grout block within an existing mainline tunnel. It was a very interesting and very challenging piece of work."

CONCLUSION

The First Street Tunnel project presented many unique challenges and plenty of lessons learned. Through a collaborative effort between the neighborhood, Skanska/JayDee JV, and DC Water, the project team was able to overcome these challenges and finally provide the promised flood relief to this community years earlier than the original consent decree schedule.

ACKNOWLEDGMENTS

The authors wish to recognize the citizens of the Bloomingdale and LeDroit Park, including the Tunnel Forum members, ANC Commissioners, Ward 5 Councilmember and project's direct abutters, since without their understanding and cooperation this important flood mitigation project would not be built. The authors also wish to recognize the contributions of DC Water's Office of External Affairs, Greeley and Hansen, McMillan Jacobs Associates, EPC, and Parsons Brinckerhoff.

PART

Risk Management

Chairs

Greg Emslie
McMillen Jacobs Associates

Matt Trotter
Kiewit

Tunneling Risk Is Down, Uncertainty Is Up—
Fifty Years of Experience and Case Studies

Russell Clough ▪ Russell G. Clough Co.

ABSTRACT

Over the last fifty years I believe risks have decreased but uncertainties have increased causing a significant increase in costs and time for tunneling work. This paper will address our research regarding the cost and schedule challenges in our industry and possible remedies. How can we minimize tunneling uncertainties based on past histories? How can we better recognize and calibrate our biases creating some of these costly uncertainties? How do we help our young employees develop better risk judgment by giving them more trust, responsibility and reward?

BACKGROUND

I am a third-generation underground worker so this paper is based on biased opinions formed over fifty years in tunneling. I went into the Marine Corps out of high school in 1959 then worked summers as a laborer and equipment operator during college. After getting a civil engineering degree, I worked as a contractor's field engineer but changed back to work as a miner, walker, superintendent to learn the skills required of a competent project manager. I worked up to a project and area manager over the following twenty years. I then started my own contracting company doing small underground projects (mostly wine caves). I sold that company and for fifteen years taught at Stanford while doing mediation and dispute boards. I have been on over 70 DRB's which keeps me in touch with the business and provide background for this paper.

In 1968 I attended the Rapid Excavation Symposium in Sacramento (early RETC) and this provides additional basis for this paper. The attendee breakdown in 1968 and 2015 was about as follows:

Attendee Background	1968	2015	Difference
Contractor	30%	15%	−15%
Supplier	20%	10%	−10%
Owner (government)	20%	15%	−5%
Mining	5%	5%	0
Education & R&D	10%	10%	0
Consult/CM/Design	10%	40%	+30%
Miscellaneous	5%	5%	0

The Contractors attending the 1968 conference were Kiewit, Utah Construction, Atkinson, Lockheed, Al Johnson, Ball, Dravo, Brown and Root, Gates and Fox, Morrison-Knudsen, Redpath, Fenix and Scisson, Mullen, Kaiser, Pamco, Harrison, Emil Anderson, RA Wattson, and Oman.

1. Why are only two of the above nineteen contractors still in the tunnel business? I worked for four of the above contractors and worked in joint ventures with nine of the others so had direct contact with the principals from the contractors. Granted it was a smaller business in the sixties.

534

2. Notice the large increase in consultant, construction manager, designer attendees over the last fifty years. There has been an even greater decrease in papers presented by contractors and increase in consultant papers. What are the implications of those changes for our industry?

After the 1968 conference, the National Committee on Tunneling Technology was established and met various times to produce the 1974 report on Better Contracting for Underground Construction. I participated in those workshops and our suggestions have been implemented and repeated at conferences like this one for forty years. If implementing those solutions have not decreased our disputes and overruns, perhaps the above questions are key and/or we need to better define "risk vs. uncertainty." Our geotechnical investigations and means and methods have improved. Our analysis of risk is better quantified with probabilities. However, we have greater complexity and uncertainty coming from the hardwired instincts that move people in ways that we cannot explain, much less control. The public seems to be losing trust in our industry when it is believed that we do not take responsibility for our cost and time predictions.

RISK AND UNCERTAINTY

Our presentations on this topic explain the reasons behind collapsing tunnels to demonstrate these misunderstood uncertainties. College taught me about probabilities; working as a consultant taught me about support design; working for contractors taught me about losing other peoples' money; owning my own company taught me good lessons when I paid for my own mistakes; teaching helped me learn how to model these issues. However, learning how we might resolve our issues came from working as a shifter getting hurt when my breastboard collapsed on top of us.

We usually define risk as gaining or losing something of value. Uncertainty is better defined as the condition when there is insufficient knowledge to accurately estimate or quantify the impacts of an extraordinary event. Risk can be quantified using theoretical models but it is not possible to accurately predict uncertainty in quantitative terms. Risk can be controlled if proper measures are taken but uncertainty is mostly beyond our control. Risk analysis considers a set of mostly known circumstances but doing that for uncertainty is speculative if not impossible. We will also discuss the definition of systemic risk vs. non-systemic risk (uncertainty).

If you flip a fair coin 100 times, it will normally streak heads or tails six or more times in a row. There is a ten percent chance it will streak ten times in a row. These odds are not intuitive or commonly believed so we developed a contracting model based on streaks. *Systemic* risks come from conditions such as statistically predictable weather and ground conditions. On a well-run job, these risks are not changed by execution. On the other hand, non-systemic risks or uncertainties come from the unpredictable behaviors, actions, or inactions of people. This uncertainty does not lend itself to logical analysis using risk registers. Some people develop an intuition from experiences based on prior (often poor) judgments and learn not to depend on the advice of those who do not have to deal with the results of these uncertainties. These non-systemic risks are best handled by people with this "learned talent for judgment." As a sixteen-year-old tractor operator or an eighteen-year-old Marine, I was given more consequential responsibility than we now give young "managers." How can we better develop leaders who have "practiced" this judgment as medical doctors do in their internships and residencies?

RESEARCH

The following graphs are from our research on major infrastructure programs that included many tunnels constructed from the 1930s to the present. Our research indicates that there has been up to a tenfold increase in tunneling costs (inflation adjusted from project conception to operation).

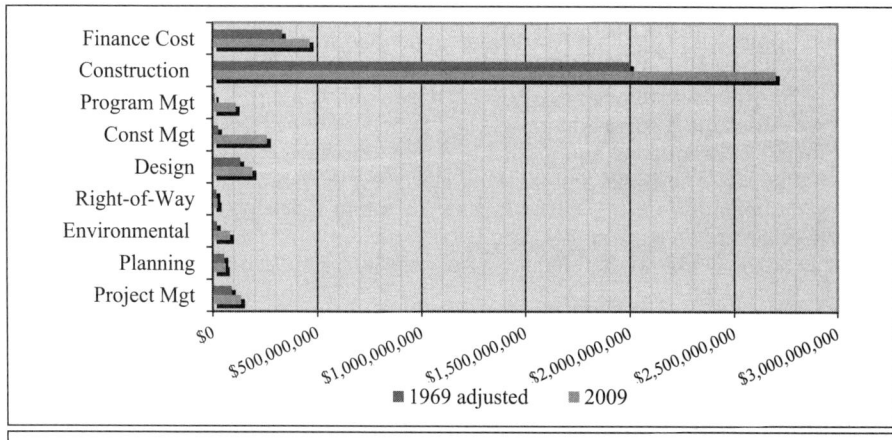

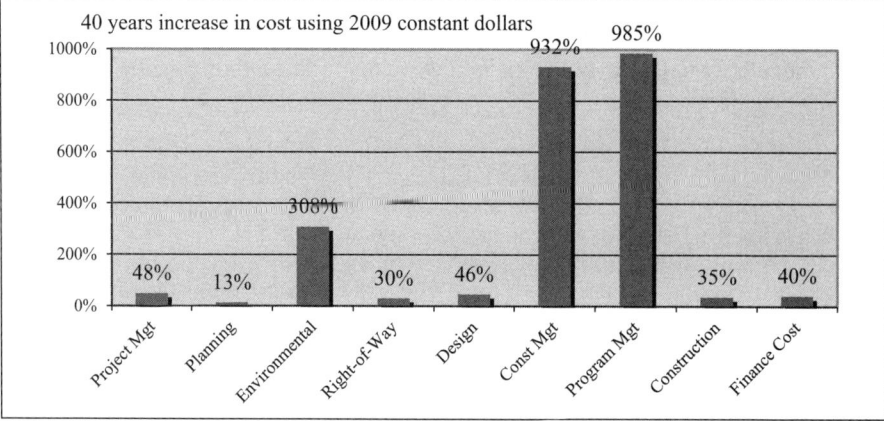

Are these increases the result of owners or designers inappropriately passing off responsibility for risk or is it because contractors bid and set up jobs without the skills to build the work? I believe a part of this issue began when we no longer allowed our young employees to learn from real world consequences. The Marine Corps trains their 23-year-old "strategic corporal" to be the key decision maker and gives them 100% backup. It is difficult for an engineer to design an efficient connection if he or she has not been a welder or ironworker putting together that kind of device. Can a geotech write a baseline if they never personally installed supports that failed when delayed by flawed equipment (not a ground or support problem)? Is it possible to be a qualified contractor's project manager if he or she has not been a laborer or equipment operator?

At these conferences, we need to discuss our failures as well as our successes so we clarify what is unacceptable design and construction. Because we know that our designs, procedures, people, etc. are not perfect, we load our systems with safety factors. Consultants who prescribe means and methods when that is the contractors responsibility create great uncertainty. I am asked to mediate disputes to determine which party exceeded the acceptable standard for errors. This is like deciding what is obscene, you may know it when you see it but it is impossible to define.

TUNNEL COST MODEL

We developed a Monte Carlo tunnel cost model which we demonstrate in our presentations. In this model an unbalanced coin flip which comes up nine times heads (profit) and one time tails (loss) represents a completed job. We run this model for thousands of jobs. Every fifty jobs, most of the retained profit is taken out to replace equipment or paid to shareholders. We know from years of actual cost reports that it is unusual for individual jobs to earn profits as great as potential losses since production greater than estimated will seldom balance the cost of an extraordinary event job that can easily double or triple the estimated time to complete. This model graphically demonstrates the major reason only two of the nineteen contractors from 1968 are here today.

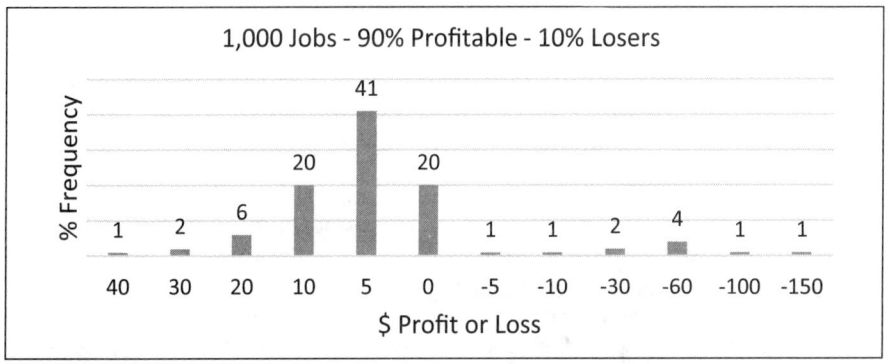

I am convinced that the costs of tunneling failures have increased since I started in the business and believe we can decrease these issues by addressing (along with other things) the following:

1. Risk allocation—System risks such as means and methods choices, weather, labor unrest, escalation, normal ground conditions, inexperienced personnel, etc. should be analyzed, priced, and assumed by contractors. Risks that are not inherent in the specific work or are new to the industry or out of the contractor's control such as political issues, permitting, environmental uncertainties, incomplete or inaccurate investigations, poor designs, funding issues, inexperienced owner or engineer personnel, etc. should be assumed by owners.

2. Owner involvement—More owner employees need on site experience so they can judge if their consultants, cm's, and designers understand how to build what is required. "Experts" marketing services without understanding constructability issues must be removed quickly.

3. Hands-on experience—Many of our contractors and consultants have not had experience in their early careers exercising poor judgment as an on-site worker. Our industry needs to develop a larger pool of leaders who have this experience and "talent for judgment."

Thank God—a panel of experts!

4. Inherent organizational failures—We know that around one hundred fifty people is the limit of our ability to control effectively. Individuals starting in larger companies sometimes do not get good opportunities to learn from command positions early in their careers. At DRB meetings I worry when we have representatives from five different organizations rather than just the owner and contractor as it was fifty years ago. Does this introduce too many personal and company agendas that are not healthy for our business?

CONCLUDING SOLUTIONS AND SUGGESTIONS

1. Replace conference papers with more panel discussions. Having four or five panelists from each industry segment provides a more productive and less biased discussion of our issues.

2. Put young employees to work in the field. Make sure they learn to hire and fire people. Train them well, support them fully, and make them responsible for the consequences of their decisions. Do not put them in the office processing paperwork.

3. Manage work by engaging and talking to people at all levels. Respect and give responsibility to craft people. Some crafts have lost the pride in their work that came from being engaged with means and methods decisions. I was often impressed with their creative solutions to problems we engineers did not discover. I think our costs will go down when more individuals at all levels are trusted and given responsibility for outcomes.

4. Get involved with your local high schools and colleges. Our students need people like you who understand what is necessary in the "real world." Endow university chairs for adjunct professors who have this kind of experience.

5. Let our team do a presentation for your company or project. We engage attendees in activities that demonstrate predictably irrational behavior. This practice helps your people better calibrate their ability to predict and deal with risk and uncertainties. We do this free of charge or through the associations committed to improving our industry such as the AGC and the Beavers.

6. Read the latest books about risk and uncertainty listed at the end of this paper. Each of them refer to real world research and case studies for their conclusions. Practice their exercises, take their tests, and go to the on-line sites they list that will help you better calibrate and improve your talent for judgment and uncertainty.

I owe much to the incredible people who trained me based on their experiences in the depression and WWII. We owe an industry that has been good to us. If we focus less on our personal and company agendas and more on our work's purpose, we make it better. Thank you.

REFERENCES

Apgar, David 2006. *Risk Intelligence*. Boston: Harvard Business School Press.

Ariely, Dan 2008. *Predictably Irrational*. New York: Harper Collins.

Evans, Dylan 2012. *Risk Intelligence*. New York: Free Press.

Gilovich, Thomas 1991. *How We Know What Isn't So*. New York: Free Press.

Hubbard, Douglas 2009. *Failure of Risk Management* Hoboken NJ: John Wiley.

Plous, Scott 1993. *The Psychology of Judgment and Decision Making*. New York: McGraw-Hill.

Taleb, Nassim Nicholas 2005. *Fooled by Randomness*. New York: Random House.

Vick, Stephen 2002. *Degrees of Belief*, Reston, VA: ASCE Press.

Contract Packaging and Formation—Risk Informed or Ignored?

Andy Thompson ▪ Mott MacDonald

ABSTRACT

Establishing the scope of an underground contract is relatively straightforward but translating that scope into a set of coherent contract documents that adequately frame the scope, provide a clear and suitable risk management approach to produce a constructible and biddable contract package that also satisfies the Owners goals can be a challenging process.

Examples of different contract formation approaches from completed projects will be examined to highlight the aspects of this project phase, which while lacking the glamour of the design plays a critical role in a projects outcome.

BACKGROUND

For underground construction, the focus of the discussion regarding risk and hence project outcome, is generally focused on the ground conditions that are expected to be encountered during excavation process. The Author believes however that there can be significant risks to project outcome that may be present but are unrelated to the ground conditions, the groundwater regime, the choice of excavation method and the design of the shaft and lining support systems. These risks are associated with the "environment" in which the project is being constructed and will vary in their potential impact depending on the factors that make up that environment, is the project part of a program or a stand-alone contract for example. Either way, these risks need to be recognized and managed to ensure that a biddable and coherent contract can be released to the Contractors.

PROJECT ENVIRONMENT

What is meant by the project environment? For the purposes of this paper this is the environment in which the project is being undertaken and is primarily governed by location. This includes the regulatory, legal, safety, insurance, project financing requirements as well as constraints imposed by the physical location of the project, is it in a mountain range under sacred lands, is it in an urban area and is it a stand-alone project or part of a program. Each of these factors will have differing impacts on the project outcome but all of them need to be assessed and evaluated to enable mitigation of their impacts on the project outcome to be incorporated into the contract documents.

CONTRACT INTERFACES

Although contract interfaces will likely have more impact on a contract that is part of a multi contract program there may be interfaces on a single contract project that need to be considered and addressed.

Major Program

The more complex the program the greater the number of interfaces that need to be considered and managed. For example, there is an ongoing program that will when its completed have awarded 35 third party construction contracts varying in size from $4m to $777m and issued 72 Railroad Direct Work Force Account contracts to complete the program over a 17-year period. Each interface between these contracts has the potential to impact project outcome as delay on one contract could potentially impact the critical path of an adjacent contract. Add in the wild card of the Direct Work Railroad contract over which there was only limited control as resources were supplied by the Railroads and this has the potential for cost and schedule impacts that can cause significant cost and schedule overruns.

From the start of this program a master schedule was developed that identified all the elements of the program and from this a contract packaging plan was developed. For this program a conscious decision was made to attempt to minimize the number of interfaces between contracts. Another early task was to identify what would be a minimum level of service that would be acceptable to the operator to enable service to commence. The overall program was designed to provide a new rail service of 24 commuter trains per hour into a new railroad terminal located in a major urban center. The minimum service requirements were determined to be 12 trains per hour. This information was considered when contract scope and sequence was determined, as the scope of the work needed to support 12 trains per hour was less than that required to provide 24 trains per hour. Obviously, the primary focus was on completing the entire project but as the project progressed decisions had to consider the provision of the minimum requirements.

The Integrated Program Schedule and Contract Packaging plan formed the basis on which the program then moved ahead. Both the schedule and packaging plan were considered dynamic and as the program progressed the IPS was updated monthly and the packaging plan was subject to regular scrutiny to ensure that it remained relevant.

For example, a decision was made to create a single very large +$1bn contract to excavate the new station caverns. To minimize the number of interfaces and push the risk of managing these interfaces to the Contractor this contract also included protection work of the structures located above the caverns, instrumentation installation, provision of offices for the Construction Manager, demolition of an existing rail storage depot to create space for the future station concourse and the demolition and construction of a new ventilation structure. For various reasons the contract was issued as an RFP to provide the owner with the opportunity to negotiate the final costs. During the negotiations, it became clear that the proposers were primarily interested in the excavation portion of the contract, and as such had proceed the other elements somewhat high. It was therefore decided that significant portions of the scope would be removed and repackaged into separate contracts. At that point the IPS and the Contract Packaging plan were then updated to enable the new interfaces to be established and the new Contracts created and procured.

Once these scope changes had been identified and incorporated into the IPS this then revealed new risks. It was recognized that the owner's procurement process would not support the schedule, so a decision was made to have the Program Management Consultant award contracts for certain specific elements of this work which included separate contracts for the provision of the CM's offices, instrumentation and protection works all of which had to be coordinated with the Railroad that operated in the structures above the cavern location and who would not permit any blasting and excavation work to proceed until all such elements were in place. As such the owner had to then identify and manage a rolling installation process to match the excavation contractors schedule to ensure baseline readings were taken and protection was in place prior to blasting occurring. As can be seen, solving the procurement problem for the large contract led to new interfaces that had to be identified, recognized and managed. Milestones, access restraints and allowance items were included in the new contracts to provide for impacts between the contracts.

Stand-Alone Contracts

A single stand-alone contract is unlikely to need to consider as many interfaces and be as constrained as a multi contract program. However, there may be other interfaces that need to be considered such as other owner works, third party works, third party restrictions and requirements etc.

For example, a non-related utility owner may have plans to undertake a service replacement that may impact access to the chosen work sites which will need considering, coordinating and some form of mitigation included in the construction contract. Absent this, the contract could be awarded and as soon as the utility blocks or restricts access to the work site there will likely be cost and schedule impacts that have not been considered in developing the budget and schedule.

Milestone and Access Restraints

Identification of and inclusion of provisions for the management of inter contract interfaces and third party interfaces is a critically important aspect that needs to be properly addressed during the development of the program/contract packaging plan. Once all the interfaces have been identified these need to be integrated into the contract during the contract formation period. These can be identified using contract milestones and access restraints.

A contract milestone is a tool that identifies a specific scope of work that needs to be completed within a pre-determined timeframe to typically enable a follow-on contractor, or the owner, to take the area over and undertake further work. Should milestones be used the scope of the milestone needs to be carefully developed to ensure that all elements needed for the follow-on contractor will be complete and available. In developing milestones, however there are several other elements that need to be considered. It is crucial that the contractual implications of taking work from a contractor prior to their contractual completion date be fully understood. For example, if the contractor is required to perform punch list work after substantial completion but before final completion, how is the punch list work going to be addressed for the scope of work defined in the milestone? Will the contractor be required to go back into this area and undertake punch list work when another contractor is working in there and may have blocked access to certain areas by the installation of electrical panels on the wall for example. Or will the follow-on contractor be required to remediate any "substandard" work elements left by the previous contractor? Obviously, the answer to this question will be partially dependent upon the conditions of contract that are in use and definitions of terms such as Substantial Completion, Beneficial Occupancy, Warranty Period, Defects Liability etc. What is not in question though is the need to rigorously identify such interfaces and consider the knock-on impacts of these interfaces thereby enabling the correct language to be incorporated into the contract documents. The author has frequently encountered such interfaces where limited consideration has been given to the contractual implications of the inclusion of such interfaces, which have inevitably led to quality, cost and schedule impacts.

Access restraints are somewhat easier to manage but it is very easy to overlook issues when establishing access restraints. An access restraint merely restricts a Contractors access to undertake work in a specific area until a specified time. For example, there may be an access restraint related to access to a completed element of a structure to install electrical equipment. On a multi contract program this could raise the following issues: where is the access route and who controls the access route to this location, who is responsible for provision of temporary utilities such as lights, ventilation, pumping etc. to this location. This can get quite complicated, for example if the contractor who is vacating the area is responsible for the temporary utilities, what happens when he leaves the project before the follow-on contractor? To the greatest extent possible these issues need to be properly fleshed out and provided for before contracts are issued. Obviously contract interfaces can change during the progress of a multi contract program but even then, allowance/contingency items can be included so that there is a recognition in the cost and schedule that there may be impacts.

THIRD-PARTY IMPACTS

In addition to the internal interfaces there is the potential for significant impacts to the project because of third party interfaces. Third parties range from building owners, regulatory bodies, utility owners, railroads, community and environmental activists, lending agencies and political impacts. Some examples are presented below.

On a recent single contract project the location of a railroad along the tunnel alignment had to be considered. The project was a relatively straightforward pressurized face TBM drive beneath a river between two shafts to replace an ageing force main. Located near the launch shaft is a railroad spur that is used by the Utility Authority for whom the tunnel was being built, for the delivery of oxygen to its treatment plant. The railroad spur was approximately 250 ft horizontally from the shaft and crossed the tunnel alignment at right angles. The spur is owned and operated by a national freight rail company and it was recognized that early discussions with the railroad would be necessary. Initial discussions with the railroad were not positive as despite the infrequent use of the spur the railroad refused to allow the tunnel to pass beneath the spur unless the tunnel was contained within a steel sleeve, which was their standard detail for utility crossing. Obviously, this was an impractical detail for a pressurized face segmentally lined tunnel some 100 ft. beneath the railroad. After much discussion, the following was included in the contract documents: the TBM had to mine continuously when beneath the railroad zone of influence which was defined using a 45 degree line from the edge of tie, instrumentation was needed on the rail to check whether there was any movement as the TBM passed beneath. These requirements were included in the contract documents giving the contractors the opportunity to price this risk, Given the proximity of the railroad to the shaft, the TBM would be mining in the railroad influence zone immediately after it had been fully launched, in the learning curve period. The successful contractor recognized these constraints and developed a work plan for the launch and initial operation phase that ensured the requirement was met and that there was no impact to the railroad. The early recognition and inclusion of mitigation measure for this issue was instrumental in managing this risk.

Railroad work can impact projects in other ways. Where work must occur on railroad territory there will be some form of railroad provided support necessary, the extent of which will be dependent on the specific railroads policy and the scope of the work being performed. For example, on a major program in the New York area the approach to the management of the railroad resources required to support the program has changed during the program. At the outset of the program master force account agreements were entered into between the project owner and the two main commuter railroads that would be impacted by the program and would be responsible for providing access and protection resources to enable the third-party contractors to undertake work within the Right of Way. These agreements somewhat optimistically as it turned out required that the project did not lead to any service disruptions to the railroads, identified the level of access and protection support envisaged but crucially did not bind the railroads into the provision of these resources. This meant that the needs of the program were not prioritized and that for example weekend access and protection requirements had to be "bid" against other work being undertaken by the railroads themselves. Given the fact that there was only a limited pool of resources from the railroad this meant that there was never any certainty that the required services would be provided. The lack of a dedicated pool of resources to support the program and the stance taken by the railroad and its unionized workforce has contributed to significant delay and cost to the program. Initially there was no provision for such delays in the individual contracts and this led to ongoing claims for delay and delay costs due to the non-provision of such protection services. For example, a $120m contract attracted some $80m in delay costs as well as becoming the program critical path. On later contacts language, has been inserted to ensure the contractors process for a stated amount of delay hours, up to 10,000 crew hours for example, and then a contingency is included for delays caused by non-provision of services beyond that. Despite this there are still significant cost and schedule overruns on these contracts due to the non-provision of protection resources from the railroads. Unfortunately, this particular issue is one that has proven rather intractable to manage, the issue has been recognized and is even addressed at the Chief Executive level but with little improvement as the main cause of the problem is the limited pool of resources from which the railroad draws.

Regulatory bodies may also have an impact on a project. For example, in New York the Metropolitan Transportation Authority is a State Agency and conducts its own Code Compliance duties. However, there are occasions where project work may impact a third party building and in that case the New York City Department of Buildings enters the equation. There are some slight differences between State and City Building Code requirements which need to be taken account of in developing the design. In addition a decision noodo to be taken with regard to the extent to which the DOB will be involved and typically it was determined that the Engineer of Record would perform Self Certification of designs to minimize the potential for delays that may occur waiting for a DOB Inspector to sign off work.

There are obviously similar impacts that need to be identified and resolved with utility companies, for example provision of temporary construction power, betterments demanded by utility companies would the project impact existing services. A comprehensive review of all items that impact construction and the new permanent facility must be undertaken to enable mitigations to be developed and where appropriate provisions incorporated into the contract documents.

PROJECT LOCATION

Project location will contribute significantly to the risk associated with the project. Excavating beneath a mountain range that may have sacred Native American water supplies on it, thereby requiring zero inflow to the tunnel, raises significantly different risks compared to excavating and working beneath a major urban area. However not all of the risks and issues will be obvious until detailed review of the project and its location are undertaken.

When working in an urban environment the location of work sites etc. raises issues of environmental impacts on the neighborhood. However there are other issues that can impact the contract interfaces and a couple of these are discussed below as an example of what needs to be considered.

On the East Side Access project in New York the original Environmental Impact Statement prepared for the project indicated that there would be minimal additional truck traffic added to Manhattan during the construction of the project. Obviously, this created certain challenges and had to be addressed in the construction contracts. What this meant was that the underground excavation in Manhattan was serviced through the tunnels that were constructed from Queens, and the station concourse work undertaken within the existing Grand Central Terminal was serviced from an existing MNR facility called BN Yard, some 9 miles north of Manhattan. These laydown locations also contributed to discussions regarding packaging as the underground contracts were essentially a 4-mile linear project with a single materials access point in Queens.

What this meant was that during the contract formation phase, the number of overlapping contracts in Manhattan was minimized as there was only one shaft, and three shifts available to supply the work. Ultimately this resulted in specific contracts being provided with specific shifts on which to utilize the supply route. This did not result in limitations on the shifts to be worked but more on the delivery routes for the underground work, meaning that the Contractors had to plan accordingly which resulted in palletization of loads and increased cost. From a schedule perspective, this meant that the schedule was constrained by the ability to feed material, which impacted the contract interfaces, milestones and access restraints also.

For the works in Grand Central Terminal there was the ability to feed this through the existing Metro North Service into GCT. The future concourse location was previously a mid-day storage yard for MNR and as such was connected to the MNR network. However, MNR had insufficient works trains to be able to provide the necessary capacity and this forced the project to purchase 20 flat cars and two locomotives for MNR. Understanding the capacity constraints resulting from this rail service was key to informing the contract schedules. Given that MTA was responsible for this capacity any under capacity would have resulted in a claim. MTA employed a separate subcontractor to manage this operation as at times there were up to 5 separate contracts utilizing the rail service and there were detailed requirements incorporated into the contract documents about the management of this operation. Requests for load placement on the rail service had to be made a week ahead for example.

While the commitment made in the EIS was understandable the impacts on the Program were not fully understood until contract formation was undertaken and the multiple access interfaces were identified.

CONCLUSION

The Contract formation period is critical to the success or otherwise of a project. Often the risks associated with such mundane items as site access, personnel access, support from third party entities etc. can be overlooked. It is therefore of critical importance that attention is paid to the environment within which the project is being undertaken to ensure that the risks are properly understood and managed. Ignoring these items can and will lead to adverse impacts on the project outcome.

DigIndy Tunnel System—Pleasant Run Deep Tunnel Optimization Yields Cost Savings and Improved Level of Service

Nick Maynard ▪ Citizens Energy Group
Leo Gentile ▪ Black & Veatch Corporation
Maceo Lewis IV ▪ Black & Veatch Corporation

ABSTRACT

The Pleasant Run Deep Tunnel is a 39,900-foot long component of the 28-mile long DigIndy deep rock tunnel system, the cornerstone of Citizens Energy Group's Long Term Control Plan (LTCP). Final planning optimized the conceptual design to achieve CSO capture objectives while eliminating expensive components and saving millions of dollars. The capacity of existing infrastructure was optimized by raising or adding weirs instead of installing new diversion structures and consolidation sewers. The number and length of required adits were minimized by optimization of the tunnel alignment and selection of drop shaft locations closer to the alignment. Several drop shafts originally planned were also eliminated. The overall level of control is expected to exceed LTCP requirements.

PROJECT DESCRIPTION

Background

Each year there are 60–80 combined sewer overflow (CSO) events that contribute an average of 8 billion gallons of untreated waste into the waterways of Indianapolis, Indiana (City). In accordance with a Federal Consent Decree, Citizens Energy Group (Citizens), owner of the wastewater system, is implementing a Long Term Control Plan (LTCP) to address CSOs on behalf of the City (Indianapolis DPW, 2007). Citizens acquired water and sewer infrastructure from the City in August 2011 along with the responsibility of implementing the LTCP according to the terms of the Federal Consent Decree.

The LTCP includes using existing system capacity, expanding and upgrading the existing treatment facilities and a new storage and conveyance system. The storage and conveyance system, known as the DigIndy Tunnel System, is a 28-mile long network of six 18-foot diameter tunnel segments. The DigIndy Tunnel System begins on the south side of the City at the Southport Advanced Wastewater Treatment Plant (AWTP), extends along the City's waterways and ultimately will end on the north near the Indiana State Fairgrounds and on the east in the Irvington neighborhood of Indianapolis. The system will capture 97 percent of sewage overflows in the Fall Creek watershed and 95 percent of sewage overflows in the White River, Pleasant Run, Pogues Run, and Eagle Creek watersheds in a typical year, reducing overflow events to less than two and four per year respectively. DigIndy will be the largest public works project in the City's history at a cost of approximately $2 Billion.

The DigIndy Tunnel System (Figure 1) will extend along White River, Fall Creek, Pogues Run, Pleasant Run, and Eagle Creek to create an underground storage and transport facility with the capacity to store over 250 million gallons of CSO during rainfall events. After a storm event, the tunnel system will be dewatered using a 90-million

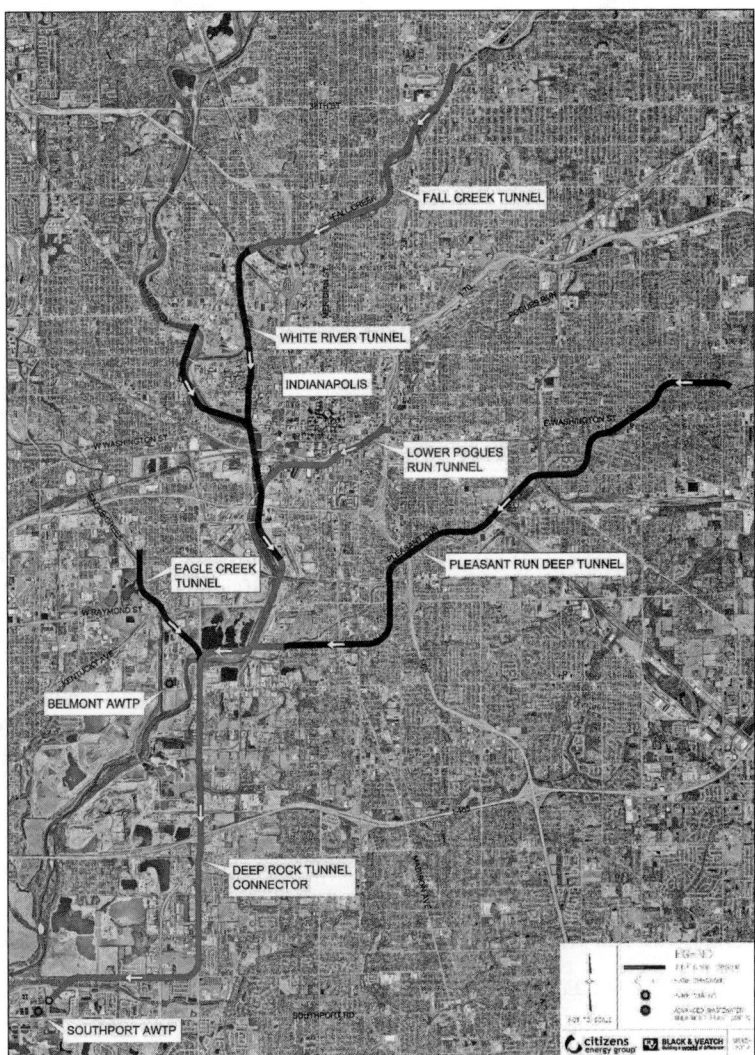

Figure 1. DigIndy tunnel system

gallon per day deep tunnel pump station located at the Southport AWTP. Table 1 summarizes the tunnel system components.

Under the Federal Consent Decree, a timeline to achieve full operation (AFO) of CSO compliance was established. By 2017 the Deep Rock Tunnel Connector (DRTC) must be on-line, which includes the deep tunnel pump station, followed by the Eagle Creek Tunnel (ECT) in 2018. Prior to December 31, 2021 Lower Pogues Run Tunnel (LPgRT) and White River Tunnel (WRT) must be complete and functional in the system. Finally, by the end of 2025 Fall Creek Tunnel (FCT) and Pleasant Run Deep Tunnel (PRDT) must also be on-line in the system.

Table 1. DigIndy system components

Component (AFO Year)	Tunnel Length (ft.)	No. Drop Shafts	Adit Length (ft.)	Total Volume (million gallons)
Deep Rock Tunnel Connector (2017)	39,000	3	800	75.2
Eagle Creek (2018)	9,200	1	0	17.5
White River (2021)	30,600	6	2,100	59.1
Lower Pogues Run (2021)	10,200	2	100	19.4
Fall Creek (2025)	20,100	14	6,100	40.9
Pleasant Run (2025)	39,900	8	1,500	74.7
Pleasant Run Extension (2025)	2,300	0	0	4.4

Advanced Facility Planning

Pleasant Run is an urban stream with very low base flows that becomes dominated by CSO during wet weather events. The Pleasant Run Interceptor and connecting sewers direct combined flows to the Citizens Belmont AWTP. There are 50 CSO outfalls located along the stream.

Regulators, an existing manhole or structure with a weir wall or elevated outfall pipe, maintain flow in the collection system and allow overflows to discharge into the CSO outfall pipes. Some regulators are directly connected to interceptor sewers, whereas other regulator structures have a small capture sewer that conveys the normal dry-weather flow to the interceptor sewer.

The initial concepts for Pleasant Run Deep Tunnel (PRDT) were described in the City's 2007 Raw Sewage Overflow LTCP and Water Quality Improvement Report. The 2011 AFP (Indianapolis DPW, 2011) refined the concept to capture CSOs along Pleasant Run. The major components of the system included:

- 42,600 feet of main tunnel.
- 4,000-foot long Bean Creek shallow tunnel.
- 10 drop shafts.
- 20,271 feet of consolidation sewer.

Through the AFP evaluation, the number of CSOs required to be directed to PRDT was reduced from 50 to 30. This was achieved by adjusting regulator weir heights in the existing Pleasant Run Interceptor and associated capture sewer diameter increases. Value Engineering (VE) during AFP also recommended alternatives to be investigated during final design. These included:

- Increasing tunnel slope to 0.2 percent to improve scour and reduce depth of drop shafts.
- Evaluating the need for several drop shafts by adjusting alignment and increasing consolidation sewer lengths.
- Partial tunnel lining.
- Eliminating a length of parallel sewer at two CSOs.

Design Optimization

Design concepts developed during AFP and its associated VE were reviewed and the design optimized to meet the following goals: help improve constructability, minimize disruption to the community, help increase level of control (LOC), and reduce cost.

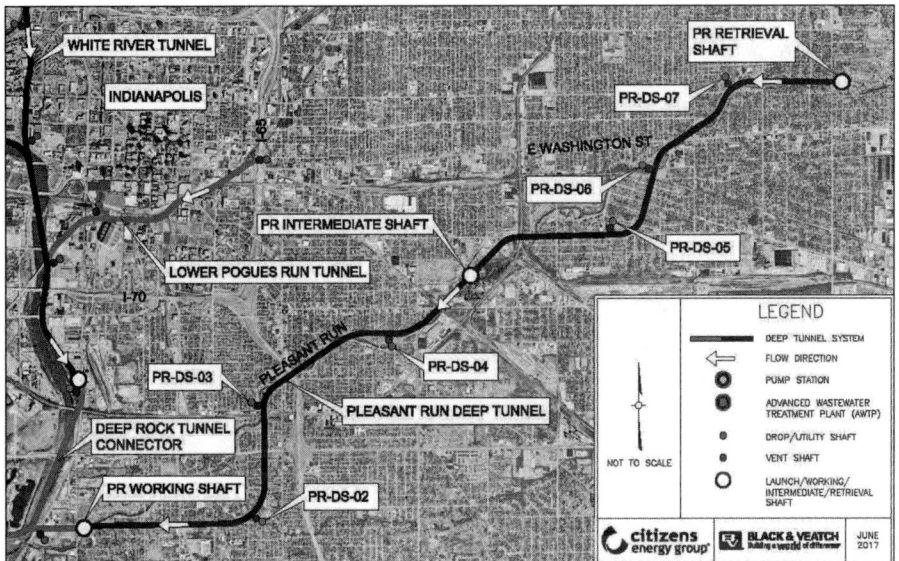

Figure 2. PRDT 50-percent design alignment

The following paragraphs provide the highlights of the salient design features adopted by Citizens to achieve these goals (Citizens, 2016).

PRDT is now in final design. The tunnel will extend approximately 7.5 miles east-northeast, beginning at the Pleasant Run Extension from DRTC near the intersection of Bluff Street and Pleasant Run Parkway. PRDT ends at Pleasant Run Golf Course near the intersection of Pleasant Run Park way and Arlington Avenue (Figure 2).

Alignment

Final design efforts included several changes to the initial horizontal and vertical AFP alignment, which generally parallel the Pleasant Run waterway and the associated CSO outfalls. The changes helped improve constructability, reduce length of consolidation sewers, and eliminate shallow tunnel segments. The selected alignment is shown on Figure 2.

The first change to horizontal alignment was to realign the tunnel from the working shaft eastward beneath W. Southern Avenue instead of beneath Pleasant Run. This change helps improve constructability, allowing the contractor to launch the tunnel boring machine (TBM) on an initial 6,000 foot straight run. The realignment also helps eliminate the proposed Bean Creek shallow tunnel by relocating drop shaft PR-DS-02 discussed further below.

The second modification to the AFP alignment includes a straight run beneath English Avenue upstream of the proposed Intermediate Shaft. This alignment reduced the number of subsurface easements, assisted with the elimination of the drop shaft and provides a more constructible segment for excavation and muck handling.

In addition, the alignment changes reduced the number of subsurface easements and impacted properties from 143 to 46. The alignment optimization lessens the impact to the community to a degree and helps Citizens avoid administrative cost of preparing and securing rights-of-entry and easements.

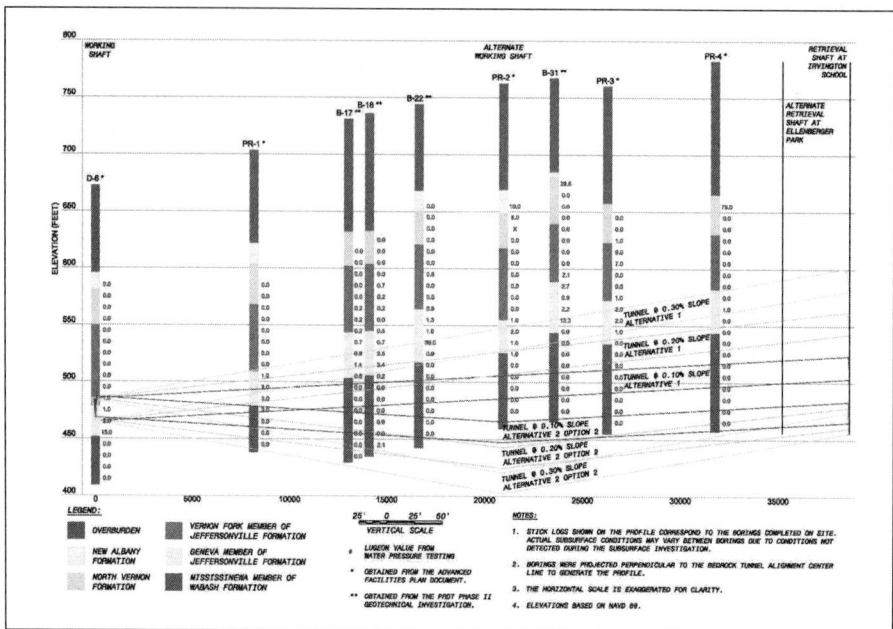

Figure 3. PRDT vertical alignment alternatives

The AFP VE recommended increasing tunnel slope to 0.2 percent to improve scour and reduce depth of drop shafts. Several vertical alignments were evaluated during the design and are illustrated on Figure 3.

The optimized slope of the tunnel is closely controlled by rock stratigraphy, constructability, and system hydraulics. Though 0.1 percent is the preferred slope for Citizens' DigIndy tunnels, alternative vertical alignments were evaluated as part of the design. The elevation and location of the downstream end of PRDT is fixed by the existing PRDT Extension that was excavated during construction of DRTC.

Tunnel slopes of 0.1, 0.2 and 0.3 percent were evaluated for constructability with respect to ground conditions identified in the geotechnical investigation. Tunneling considerations dictate a vertical alignment of the tunnel should be maintained within the most favorable rock conditions for tunneling efficiency, as well as to limit infiltration and exfiltration. The most favorable rock conditions for tunneling occur within the Mississinewa Member of the Wabash Formation, which is massive, has the least number of continuities, and is relatively the tightest formation based on geotechnical packer water pressure tests. Based on favorable construction experience from the recently completed DRTC, excavating in the Mississinewa Member at a 0.1-percent slope will help offset increased cost of deeper drop shafts.

An alternate vertical alignment with the working shaft located in the midpoint of the tunnel was also evaluated. The benefit of locating the working shaft in the midpoint is that the shaft and tunneling operations would be on an expansive property owned by Citizens. To allow groundwater to drain away from the working face, the tunnel would have sloped upward in both directions. The disadvantage is that this shaft would be the low point on the vertical alignment and require a remote pump station to completely dewater CSO from the tunnel during operation. This alternative was not selected. As stated previously, the 0.1-percent slope was selected.

Table 2. CSO design optimization strategies

Drop Shaft	CSO	Design Optimization
DS-01	019, 120	Raise regulator weir, eliminate 1,200 feet of 36-inch CCS, increase capture pipe to 24 inches.
DS-02	015, 016	Relocate drop shaft. Eliminate Bean Creek shallow tunnel.
DS-03	022, 149	Redirect flows to DS-04. Eliminate drop shaft.
DS-04	022, 025, 023, 149, 151, 119, 108, 127	Collect flows from CSOs, increase capture pipe size, remove diversion structures.
DS-05	028, 029, 072, 073	Raise regulator weirs to eliminate CSOs, increase capture pipe size, remove diversion structures.
DS-06	075, 076	Raise regulator weirs up to 5 feet, increase capture pipes up to 42 inches. Eliminate drop shaft.
DS-07	077, 078	New weir, regulator structure.
DS-08	080, 081, 224	Increase capture pipe up to 36 inches, weir height 4 feet.
DS-09	083, 084, 154	Additional consolidation sewer, raise weir up to 2.5 feet.
DS-10	088, 089, 090, 091, 092, 229	Increase capture pipe up to 18 inches, raise weir up to 1.5 feet.

Consolidation Sewers

Citizens' internal hydraulic modeling group performed a detailed evaluation of the combined sewer collection system in the Pleasant Run basin. The team characterized the Pleasant Run CSO system as very interconnected, lending it to optimization by raising regulator weir elevations so the capacity of existing infrastructure can be maximized. This allowed the design team to reduce the number of drop shafts and length of consolidation sewers and associated diversion structures. Table 2 summarizes the optimization strategies by drop shaft carried forward into the design.

The system hydraulic modeling helped develop weir height and regulator capture pipe diameter adjustments to eliminate 5,000 feet of CCS and two drop shafts.

- 27 of 50 CSO regulator weirs are raised from 1 to 4 inches.
- 17 of 50 regulator capture pipes will be increased by 9 to 30 inches in diameter.

Drop Shafts

Up to 10 drop shafts were envisioned during the AFP phase of the project. Some shafts were recommended for elimination during AFP Value Engineering. The hydraulic analysis conducted by Citizens during final design helped confirm that two shafts could be eliminated for final design through optimizing flows in the Pleasant Run Interceptor system. Further, all drop shafts will be tangential vortex-type (Table 3). Not only is this Citizens' preference, the reconfiguration of the existing interceptor system reduces flows such that direct-drop or baffle-type planned during the AFP are no longer required. The following drop shafts were reevaluated by Citizens during final design:

- PR-DS-02 was relocated to the southeastern side of Garfield Park, a historic and heavily used recreation area in the City. This alternative eliminated the extensive near surface consolidation sewer through Garfield Park and minimizes disruption to the park and the costly shallow ground tunnel previously referred to as the Bean Creek Branch Collection Sewer.
- The proposed site for drop shaft located near the intersection of East Pleasant Run Parkway South Drive and Napoleon Street, was eliminated. By raising the weir heights at CSO 022 and CSO 149, hydraulic modeling showed that

Table 3. PRDT drop shaft nomenclature

AFP	Type	Preliminary Design	Final Design	Type
DS-1	Direct Drop	PR-DS-01	PR-DS-01	Vortex
DS-2	Vortex	PR-DS-02	PR-DS-02	Vortex
DS-3	Vortex	PR-DS-03	—	—
DS-4	Vortex	PR-DS-04	PR-DS-03	Vortex
DS-5	Vortex	PR-DS-05	PR-DS-04	Vortex
DS-6	Vortex	PR-DS-06	—	—
DS-7	Vortex	PR-DS-07	PR-DS-05	Vortex
DS-8	Vortex	PR-DS-08	PR-DS-06	Vortex
DS-9	Vortex	PR-DS-09	PR-DS-07	Vortex
DS-10	Baffle	PR-DS-10	PR-DS-08	Vortex

drop shaft PR-DS-03 can be eliminated, while conveying these flows to drop shaft PR-DS-04. To accommodate these flows at drop shaft PR-DS-04, the weir height at CSO 151 will need to be raised and the size of the consolidation sewer leading to PR-DS-04 increased. Additionally, a new interceptor connection will be needed upstream of PR-DS-04, a capture pipe size will need to be increased, and a new regulator to the existing interceptor will be constructed.

- The proposed drop shaft located near the intersections of English Avenue and East Pleasant Run Parkway North Drive, was also eliminated. Flow will need to be diverted to drop shaft PR-DS-07, upstream of CSO 076, by constructing a regulator. By increasing the weir heights at CSO 075 and CSO 076 and upsizing their respective capture pipe diameters, the model shows that drop shaft PR-DS-06 can be eliminated. These flows will be conveyed downstream by the interceptor.

The capture pipe sizes will need to be increased and weir heights raised at CSO 074, CSO 077, and upstream of CSO 075 near the intersection of East Pleasant Run Parkway North Drive and Southeastern Avenue.

Intermediate Shaft

The PRDT Project includes four different types of tunnel shafts for construction and operation. These include the TBM working shaft, the intermediate shaft, the TBM retrieval shaft, and drop shafts. The TBM working shaft for PRDT will be situated near the east end of the Pleasant Run stub that was constructed as part of the DRTC project. The site is located at the southeast corner of the Bluff Road and West Pleasant Run Parkway North Drive.

For several reasons, an intermediate shaft is included. The intermediate shaft for PRDT is situated approximately at the midpoint of the alignment on the former Citizens Coke & Gas Plant site. The alignment optimization resulted in several turns that may affect muck handling, as was the case on DRTC. The intermediate shaft will allow a secondary access point to set up or maintain the TBM, transport materials, personnel, and equipment into the tunnel, and remove the tunnel spoil and waste material during excavation. The intermediate shaft will allow a central location for spoils removal along the tunnel which will more efficiently convey tunnel spoils.

The second reason is to allow the contractor the option of backing up the TBM after the full alignment is excavated. This technique is being successfully employed on other segments of the DigIndy Tunnel System. For example, the 9,000-foot long Eagle

Creek Tunnel was excavated as part of the DRTC contract. The TBM completed tunnel excavation, the utilities stripped from the tunnel and the TBM reversed to the main DRTC tunnel to complete its mining. While there is real cost to this effort, the cost is expected to be offset by eliminating a large-diameter retrieval shaft and a baffle drop shaft.

On PRDT, the intent is that the TBM will not be retrieved from the tunnel through a retrieval shaft at the upstream end, previously located at the Pleasant Run Golf Course in the Irvington neighborhood. This strategy reduces the shaft size of PR-DS-08 and thereby lowers impacts to the Irvington community and the golf course.

Material Repurposing

Citizens' past operations include industrial facilities that have been closed and are being considered for possible redevelopment. Tunnel muck from excavating various segments of DigIndy tunnels may be used to cover and regrade some of these closed facilities for future beneficial use. The PRDT Intermediate shaft is located at the former Coke & Gas Plant. Muck from excavating White River, Fall Creek and PRDT may be temporarily stored at the Coke & Gas Plant site and then graded to create a level site for future redevelopment. This beneficial reuse will help Citizens avoid the cost of purchasing tens of thousands of cubic yards of clean fill to achieve the same goal.

Level of Control

A Storm Water Management Model (SWMM) was been developed of the CSO outfalls, diversion structures, relief structures, consolidation sewers, drop shafts, adits, and the main PRDT tunnel to estimate design flows and hydraulic behavior to support the PRDT design as well as to verify the compliance with the defined levels of control (LOC) in the Long Term Control Plan (LTCP).

The modeling included the hydraulic analysis of the tunnel system but not hydrology. Some aspects of the PRDT design depend on other projects such as the Deep Rock Tunnel Connector (DRTC), Fall Creek Tunnel System (FCTS), and White River/Lower Pogues Run Tunnel (WRLPgRT) systems. Dewatering information for the tunnel system was provided by Citizens. All scenarios evaluated assumed that the Deep Tunnel Pump Station (DTPS) with a rated capacity of 90 million gallons per day (mgd) has 11.7 mgd capacity available to PRDT, and that the Southport AWTP has capacity to accept all the pumped flows. Additionally, all the scenarios considered that the entire volume of the PRDT was available for storage of the PRDT CSOs.

The PRDT SWMM model predicts 14 untreated overflow events along Pleasant Run and 98.1 percent capture during the 5-year design period. Both results comply with the defined LOC.

Construction Cost

The design optimization strategies described in this paper are expected to help reduce the overall project cost by millions of dollars. The strategies include:

- Eliminating two drop shafts.
- Utilizing the Intermediate shaft for TBM retrieval and a secondary working shaft and reducing the size of the upstream shaft DS-08 by eliminating the retrieval shaft from this location.
- Replacing a 30-foot diameter baffle drop structure with an 8-foot diameter vortex drop structure.

- Reducing the length of connecting sewers by 5,000 feet.
- Eliminating the shallow ground tunnel beneath Bean Creek through Garfield Park.
- Placing muck at Citizens facilities such as the Coke & Gas Plant site for cover regrading to allow redevelopment.

REFERENCES

Citizens Energy Group. 2016. Pleasant Run Deep Tunnel. Basis of Design Report. Citizens Project Number 92TU0534.

Indianapolis Department of Public Works. 2007. Raw Sewage Overflow Long Term Control Plan and Water Quality Improvement Report.

Indianapolis Department of Public Works. 2011. Advanced Facilities Plan for Pleasant Run Deep Tunnel. DPW Project Number CS-32-004A.

Reduce Urban Tunnel Utility Relocation Risk Through Early Relocation by Specialty Contractor

Gordon Evans ▪ DC Water
Carlton Ray ▪ DC Water
Tom DiLego ▪ Greeley & Hansen
Justin Carl ▪ Greeley & Hansen
Steven Bealby ▪ McKissack and McKissack
Aliuddin Mohammad ▪ Alfred Benesch & Company

ABSTRACT

The District of Columbia Water and Sewer Authority is implementing the $2.6 billion DC Clean Rivers Project (DCCR) to control combined sewer overflows to the Anacostia and Potomac Rivers and Rock Creek. To reduce risks of schedule delay and cost overrun on the $500–$600 million Northeast Boundary Tunnel (NEBT), DCCR will use an innovative approach for scheduling. By separately contracting the $18 million Northeast Boundary Tunnel Utility Relocations (NEBTUR), most of the utility relocation will be completed in advance of NTP for NEBT. Both NEBT and NEBTUR have similar issues of SOE installation, utility interruption, noise abatement and public outreach in a dense urban setting crowded with existing utilities.

INTRODUCTION

The District of Columbia Water and Sewer Authority (DC Water) is implementing the $2.6 billion DC Clean Rivers Project (DCCR) to control combined sewer overflows (CSOs) to the Anacostia and Potomac Rivers and Rock Creek and to provide flood relief to the Bloomingdale and LeDroit Park Neighborhoods in the District of Columbia. The Long Term Control Plan complies with the requirements of a Federal Consent Decree entered into by DC Water, the District of Columbia (the District), and the United States, as represented by the Environmental Protection Agency (EPA) and the Department of Justice. The Northeast Boundary Tunnel, a design-build project, will store combined sewage and deliver it to a system of downstream tunnels and ultimately to the Blue Plains Advanced Wastewater Treatment Plant (Blue Plains), where it will be treated. In addition, the NEBT is a major project associated with the Mayor's Task Force to alleviate extreme flood events in Bloomingdale and LeDroit Park. The Northeast Boundary Tunnel Utility Relocations (NEBTUR) design-bid-build (DBB) project is an accelerated component of the Anacostia River System, as shown in Figure 1.

During dry weather conditions the entirety of the sewage in the combined system is conveyed to Blue Plains. During storm events, when sewer pipe capacity is exceeded, the flow, which is a mixture of sewage and stormwater runoff, overflows into receiving waters through outfalls. There are a total of 47 active CSO outfalls along District waterways.

The influx of storm runoff into the combined sewer has an additional impact on the communities of Bloomingdale and LeDroit Park. The sewer system is outdated and undersized. In these neighborhoods, when the capacity of the sewer is exceeded, localized flooding occurs and backups into adjacent residences are prevalent. In any given year, there is a 50 percent chance that flooding will occur in these neighborhoods.

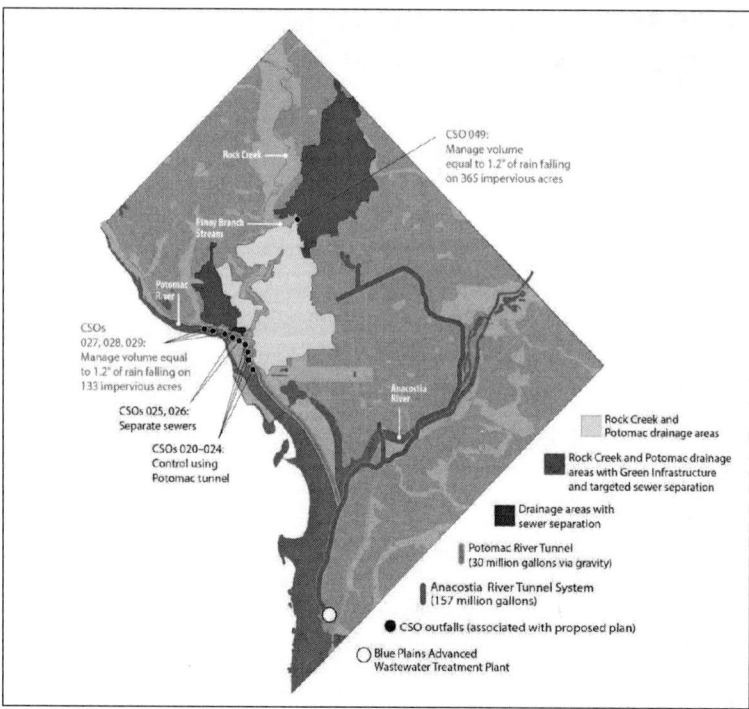

Figure 1. Extent of the DC Clean Rivers Project in the District of Columbia

In order to meet the CSO control objectives and flood relief requirements for the Anacostia River cowershed, DC Water is constructing a 13-mile-long tunnel system, roughly 100 feet below grade, to convey the combined sewage to Blue Plains. Diversion Structures and other Near Surface Structures are used on the combined sewers, to divert flow to adjacent Drop Shafts and down into the tunnel. All utility infrastructure in conflict with the Near Surface Structures, Diversion Structures and Drop Shafts must be relocated prior to construction of the tunnel facilities. As explained in the "Utility Protection While Tunneling in an Urban Environment for DC Clean Rivers Project," (Kottke, P. and Kantola, J.), DCCR is also concerned that tunneling under existing utilities in soft ground can lead to possible adverse impacts to these utilities. Thus, relocation of select utilities is required to prevent impacts from potential ground movement from tunneling activities.

Risk Reduction

The NEBTUR project was identified as a means to help navigate challenges and reduce risks associated with the NEBT. Previous projects nationwide revealed the problems with incorporating utility relocations into larger tunnel contracts. Common risks included the following:

- Tunnel Contractors not experienced with utility relocations
- Extended work in busy streets creating angry neighbors
- Prolonged work with public utility companies (electric, gas and communications)
- High probability of existing utilities failing due to age (e.g., 100 year cast iron water mains and brick sewers)

- Difficult maintenance of traffic (MOT)
- Major time delays associated with slower than anticipated utility relocations

Risks associated specifically with the NEBT project included the following:

- 20-inch prestressed concrete cylinder pipe (PCCP) water main
- 100 year old trolley tracks
- 100 year old 48-inch cast iron water main
- Coordinating shutdown for two different zones of a 48-inch water main
- High voltage (69 kV) transmission main
- Contaminated and potentially hazardous soil
- Electrical power transfer at Washington Metropolitan Area Transit Authority (WMATA) Red Line main tracking station
- Replacement of lead water services

A brainstorming team suggested the main risks experienced at other urban tunneling projects could be mitigated with the release of a separate DBB utility contract. Other risks such as coordinating shutdowns of 48-inch water mains required the design team planning and meeting at regular intervals with DC Water to prepare for major disruption to water facilities.

A collaborative bidding process with shortlisted contractors allowed for changes during the request for proposals (RFP) stage and even after the project was awarded. One post award change was discovered during the collaboration process with shortlisted NEBT DB Proposers. DB proposers warned about significant construction risk associated with the existing 20-inch PCCP water main lying adjacent to NEBT support of excavation (SOE). As explained in AWWARF/EPA 2008 paper, "Failure of Prestressed Concrete Cylinder Pipe," PCCP pipe installed between 1972–78 has historically shown a high incidence of failure—sometimes sudden failure—resulting in considerable consequential damage. Nearby NEBT construction would cause earth movement disturbances which could cause a normally stable and safe PCCP installation to be placed under new stresses that could lead to premature PCCP failure.

As the NEBT DB contractor would be responsible for protecting and maintaining adjacent PCCP pipe during NEBT construction, DC Water decided that NEBT risks of delay, cost overruns, and service interruptions to DC Water customers could be mitigated by scheduling the early removal and replacement of that PCCP pipe before the RFP stage of the NEBT. The task of removing the old PCCP and replacing it with new fully-restrained ductile iron water pipe was incorporated into the NEBTUR project. By early replacement of PCCP, DC Water eliminated risk of future breakage from its water distribution system. Thus, NEBT Proposers were able to lower their bid prices due to the lower construction risk.

Six key risk reduction elements are identified and discussed in the following sections.

Risk Reduction Measure 1—Schedule

From a cost perspective, utility relocations may only be a small percentage of a tunneling project but can cause significant delays in the major portion of the project. Experience from other projects, both locally and from around the country, has shown that utility relocations can delay complicated tunneling activities and lead to major cost and time claims. Thus, it may be better to separately schedule tunneling and utility relocation.

NEBTUR contract developed a complete schedule, which included both design and construction activities. The schedule accounted for utility design, coordination with private utility companies, the bidding process, and construction. The most important element was to design and relocate utilities in advance of the NEBT to prevent future delays with utility relocations and with scheduling private utility companies, such as Pepco, Washington Gas, and Verizon.

In September 2014, the decision was made to separate the utility relocation work into a separate DBB contract. Based on the schedule and anticipated construction duration, the goal is to have utility relocation completed by notice-to-proceed (NTP) for NEBT. The design for NEBTUR started three years prior to the anticipated NTP date for the NEBT. Key milestone dates were scheduled and adjusted as the design phase of the project progressed. The critical path utility work was identified at each site. Detailed site-by-site and overall schedules were developed based on the input from each Utility Agency. That schedule was further refined during the bidding phase. The allotted NEBTUR construction duration is 17 months.

Based on the anticipated construction schedule, concurrent work at multiple sites is critical for the completion of the utility relocation work prior to NEBT NTP. This would necessitate multiple construction crews, allowing concurrent work at multiple sites, while complying with the allowed restrictions in the NEBT Traffic Study. Due to the close proximity of some of the sites, the NEBT Traffic Study was used to determine sites where concurrent work would be permitted. Table 1 provides key milestone dates for the NEBTUR project.

Table 1. NEBTUR key milestone dates

Milestone	Date
Request for Qualifications (RFQ)	October 2015
90 Percent Design	August 2015
Final Design	January 2016
NTP	May 2016
Construction Completion	November 2017

The NTP was provided on May 19, 2016. The scheduled substantial completion date is November 3, 2017. The contract documents require the contractor to provide a "Baseline Schedule" and a detailed "Contractor's Construction Schedule" to plan, organize, and execute work. This schedule is critical path method (CPM) schedule and is required to be updated on a monthly basis to monitor and measure the progress of the project. The NEBTUR construction schedule revealed several potential problems in the early months of the contract. The most notable problem was that the contractor's schedule showed zero-float between critical path tasks. Any delay in one task would impact the completion.

Risk Reduction Measure 2—Specialty Contractor

Tunneling contractors are highly specialized in tunneling. Utility relocation may not be the strength, or the passion, of tunneling contractors. DC Water will award the NEBT contract to a DB with a specialized expertise in tunneling related construction to construct the largest tunneling project in the country in one of the busiest urban environments. Different construction skills and experience are required to perform the utility relocations in advance of the NEBT.

By separating the relocations from the tunneling work, DC Water was able to short-list potential bidders based on their expertise and experience with utility work in the District. This allowed DC Water to identify and select the best contractors for the utility relocation project.

Key factors in selection of a specialty utility contractor include ability to work on multiple types of utilities, large firms with enough bonding capacity to perform the work, multiple crews to work at multiple sites simultaneously and knowledge of working conditions in DC, especially working with key agencies such DDOT, PEPCO and Verizon.

Risk Reduction Measure 3—Completion Incentive

On NEBT and NEBTUR, there are nine sites where utilities are being relocated. Varying amounts of utility relocations are required at each site. The NEBTUR contract was divided into two parts with separate completion dates. Three sites were identified for early completion based on the complexity of the utility relocations and importance of the sites relative to work for the NEBT. Traditional construction contracts have a completion date or time period for completion. If the time period is not met, liquidated damages are triggered as monetary inducement for the contractor completing the work in a timely manner. The project completion time for six of the NEBTUR sites is 489 days for "Substantial Completion" and 534 days for "Final Completion." After 534 days, liquidated damages (LD) will be assessed at $5,000 per day.

An added incentive was proposed for the NEBTUR contractor to complete work early at three locations. The early completion sites are R Street NW, 4th Street NE, and Rhode Island Ave NE. The early completion time period for the incentive is 352 days. All work must be substantially completed at all three sites for the contractor to collect the incentive. This is a no excuse incentive and the contractor must complete all work in the 352 days. The incentive is $300,000 lump sum. The NEBTUR contractor is sharply focused on the early completion deadline. The NEBTUR contractor accelerated the submittals for shop drawings and schedules. The contractor is making attempts to resolve issues as quickly as possible. If the work is not completed within 381 days, LDs will be assessed. LDs commence a month after the early completion date.

Figure 2 shows the schedules of both the NEBTUR and NEBT projects relative to each other. The figure illustrates the NEBTUR (top bar) ending just in time (Sept. 2017) for commencement of the NEBT (bottom bar) contract.

The completion of the project will take into consideration Owner-caused delays, such as additional work added to the contract. The contractor will not be held responsible for delays considered "Acts of God," such as weather delays which are beyond normal weather delays. The contractor will also not be held responsible from delays caused by potential archaeological findings. Typical archaeological findings encountered include trolley tracks as shown in Figure 3.

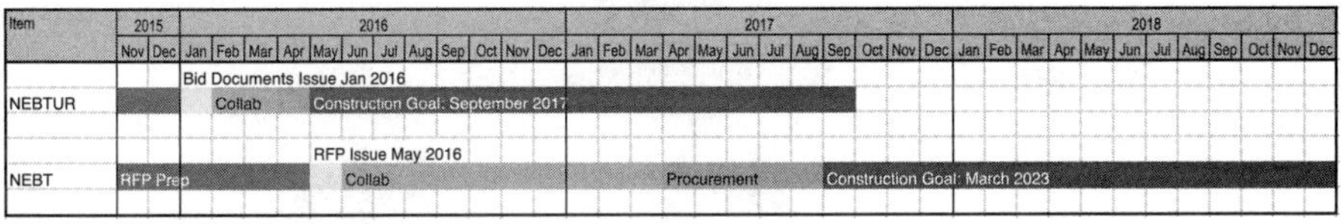

Figure 2. NEBT and NEBTUR schedules

Risk Reduction Measure 4—Hybrid Procurement Method

One of the biggest challenges for the NEBTUR is the selection of a qualified contractor to relocate multiple types of utilities, as well as managing subcontractors for the relocation of dry utilities. The contractor selection process involved utilizing both a traditional Design-Bid-Build (DBB) approach, as well as Design-Build (DB) approach. A typical utility project selection process is to advertise the project, accept bids, and select the low bidder without prequalifying or meeting with perspective contractors. This procurement method was not considered to be an effective method for the selection of the best contractor for the completion of the NEBTUR project. A "Hybrid Procurement Process" for selecting a qualified contractor capable of performing all the work in the time period allowed was utilized for the NEBTUR project.

The NEBTUR procurement process included the following steps; Request for Qualifications (RFQ), shortlist of contractors for bidding based on RFQ criteria, bid

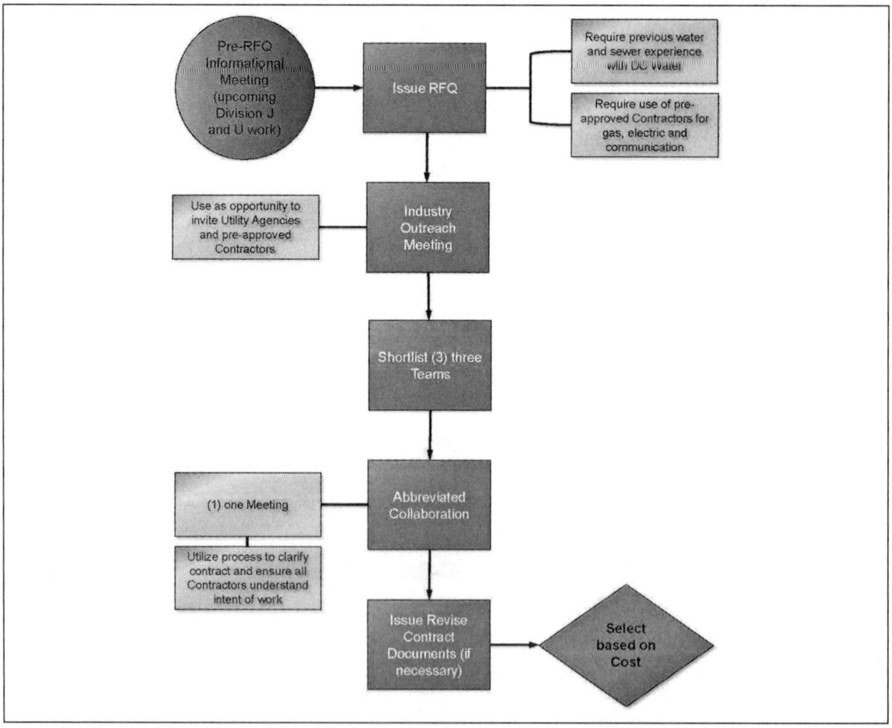

Figure 3. Trolley tracks

Figure 4. Contracting methodology with prospective contractors for "hybrid" option

documents given to selected contractors, collaboration meeting with each contractor after review of bid documents, bid opening, clarification and final contract agreement. Figure 4 shows a flow chart of the selection methodology.

The first RFQ for the NEBTUR project yielded only one response. A second RFQ was issued with some minor edits clarifying the scope of work required for the contract. After the second RFQ, five responses were submitted. DCCR formed an expert committee to review the responses to the RFQ packages and select contractors to bid on the project. The committee selected three Shortlisted contractors to bid on the project.

Selected contractors were given the construction documents to review. Each contractor met with DCCR for a collaboration meeting. Non-disclosure forms were signed by the contractors before meeting to assure that the information discussed was not shared with the other firms. Based on the collaboration meetings, several changes to the bid documents were made as part of addenda to the original bid documents. Addenda items included changes to the contract time, adjustment in work to provide more defined items for bidding, and final maintenance of traffic (MOT) plans. This process was used to provide more defined documents and reduce overall risk.

The apparent NEBTUR lowest responsible responsive bidder was determined. Before the final NEBTUR contract was awarded, a clarification meeting was held to discuss any issues the low bidder and DCCR felt were outstanding. Based on the clarification meeting, a few items were added to the contract documents to provide a clear understanding for both DCCR and the contractor.

Risk Reduction Measure 5—Permit Assistance

DC Water identified all permits required for the NEBTUR project, with each permit given a designated party responsible for obtaining that permit. DCCR identified long lead permits that would have the greatest impacts to the project schedule and tried to obtain those permits before the project NTP was issued. Permit responsibilities are shown in Table 2.

The longest lead permit item is the District Department of Transportation (DDOT) Construction Permit. Early in the NEBTUR design phase DCCR met with DDOT and prepared Maintenance of Traffic Plans (MOT plans) to assure that the DDOT Construction Permit would be issued in advance of construction. The permit would then be transferred to the Contractor. This would enable to the contractor to quickly obtain the DDOT Occupancy Permit. This permit requires the Contractor to re-submit monthly for as long as the Contractor is at each site.

DCCR specialist assisted and facilitated obtaining of permits for the Contractor. Permit specialist worked with the Contractor to assist and assure that all MBE/WBE goals are maintained and meet DC Water requirements.

Risk Reduction Measure 6—Public Utility Relocations

Public utility agencies, such as Gas (WGL), Electricity (Pepco), and Communications (Verizon), present special areas of concern when relocating utilities. Typically, these utilities require a purchase order to design relocations and a second purchase order to have crews relocate the utilities. The separate contracts can have long lead times for design and construction associated with them. The traditional practice of waiting for the contractor to set up contracts with these utility agencies, who in turn develop their own plans and perform construction, could lead to uncontrollable delays as experienced by other DCCR projects.

Table 2. Permits and approvals

| Agency | Approval or Permit | Responsibility | |
		Owner	Contractor
District Department of Consumer and Regulatory Affairs	Building Civil (BCIV) Permit	X	
	Miscellaneous Soil Boring Permit (helical pulldown micropiles)		X
	Support of Excavation Permit		X
	Building Permit for Site Trailers		X
	Trade Permit (MEP)		X
	After Hours Permit		X
	Environmental Intake Form	X	
District Department of Energy and Environment	Volume of Cut Fee	X	
	Generator Registration		X
	Asbestos Abatement		X
	Hazardous Waste Identification Number		X
District Department of Transportation	Construction Permit	X	
	Occupancy Permit		X
	Construction Permit for Support of Excavation		X
	Construction Permit for Soil Boring		X
	Tree Removal Permit	X	
	Steel Plate Permit		X
Environmental Protection Agency	Notice of Intent		X
	Stormwater Pollution Prevention		X
Washington Metropolitan Transit Authority	Letter of No Conflict	X	
District of Columbia Water and Sewer Authority	Plan Review	X	
	Availability Certificate	X	
	Inspections Invoice	X	
	Temporary Discharge Authorization		X
	Hydrant Use Permit		X
District Department of Public Works	W Street Facility Easement	X	
	W Street Facility Right of Entry Agreement	X	

In order to mitigate the contractual issues previously experienced with the purchase order methodology, DCCR met with the utility agencies between September and November 2014 to discuss potential alternatives. During the meetings, DCCR presented a contracting methodology in lieu of the traditional work order (W.O.) methodology. The alternative methodology provided contracting means for design and construction of utility relocations to alleviate the disadvantages previously encountered. The goal of this alternative was to provide DCCR with a direct contractual relationship with the contractor, to coordinate design and construction efforts, to manage scope creep, to quickly resolve field conflicts, to reduce neighborhood construction fatigue, to control messaging and public outreach and, potentially, to realize savings due to comprehensive regulatory efforts and elimination of iterative site restorations.

In order to ensure conformance of public utility design and construction activities, the standard W.O. agreements between DC Water and each utility agency were revised to document individual responsibilities, schedules, deliverables and coordination between utility agencies and DCCR. Furthermore, DC Water General and Supplementary Conditions, as well as insurance requirements, were used, as DC

Water would be holding the construction contract. The W.O. also addressed changes in design and construction, technical specifications, materials of construction, traffic control, permitting, restoration requirements and as-built information. Coordination of all utility work was transferred to the NEBTUR Contractor.

This contracting methodology allows DCCR to maintain control of construction schedule, costs and permitting efforts through a single prime contractor, contracted by DC Water. The prime contractor will contract with subcontractors acceptable to each utility agency, potentially hiring fewer subcontractors—each acceptable to multiple utility agencies—and thereby reducing coordination requirements, improving schedule and realizing saving. Additionally, the single NEBTUR prime contractor will allow the enforcement of mitigation measures within the community. Each utility agency (Pepco, WGL, and Verizon) has agreed on this methodology for implementation, completing the design either internally or through a third party consultant. The utility agencies will be required to revise their W.O. agreement with DC Water regarding design criteria, deliverables and interrelationships between utility agencies; each item considered paramount for project success.

All of the utility agencies, with the exception of Verizon, requested keeping the design in-house. DC Water was able to contract directly for the Verizon design. Where agreed-upon designs were incorporated into the NEBTUR bid documents, DC Water requested professional engineer sealed and signed plans from all the agencies. All agencies complied except Pepco which does not provide a PE seal and signature on plans they produce.

Pepco allowed for the construction of the buried concrete duct by a Pepco-approved contractor. Pepco will furnish, pull the cable and make termination connections back into their existing systems, thereby requiring a second W.O. All W.O.s for these services were agreed to and paid in advance by DC Water directly to Pepco. The NEBTUR Contractor was required to coordinate the work required for these utilities during construction.

Pepco was the largest and most important public utility requiring coordination. At one point, talks were elevated between the CEO of Pepco and DC Water to emphasize the importance of the project. Pepco design and construction generally has a long lead time. Many meetings with Pepco were held before a final design was agreed upon. Pepco did require that some of the work be performed in-house. During the construction process some unanticipated Pepco design changes were required, such as changing from 4-inch PVC conduit to 5inch fiberglass conduit. These changes were handled by change orders with the NEBTUR Contractor. It was also critical to look for interference with concurrent Pepco projects. Competing interests from larger projects, such as Buzzard Point Development and District of Columbia Power Line Undergrounding (DC PLUG), were analyzed to make certain that these future projects did not steal scarce subcontractor resources needed by NEBT or NEBTUR.

Verizon and Verizon Business (MCI) allowed for the construction of the duct through a pre-approved contractor. Both agencies will furnish, pull the cable and make termination connections back into their existing systems, thereby requiring a second W.O. All W.O.s for these services were agreed to and paid directly in advance by DC Water. The NEBTUR Contractor was required to coordinate the work required for these utilities during construction.

DCCR REPUTATION

The $2.6 billion DC Clean Rivers Project, though mostly 100 feet below ground, is a high-visibility project—the subject of many news interviews, tours, visitor tours, and ribbon-cuttings. Vice President Joe Biden attended one ribbon cutting, DC Mayor Muriel Bowser attended another. DCCR is a critical infrastructure asset that is designed to serve the community well for 100 years. DCCR managers strive to maintain a trusted reputation during the entire construction period through 2026.

Public outreach is a requirement for all contracts. Outreach programs include community notifications, attending ANC meetings, neighborhood meetings, and a dedicated phone line for complaints. The NEBTUR contract also requires noise and vibration monitoring for nighttime work, as well as strict traffic control requirements for vehicles, bikes and pedestrians. Advance notification for any public transportation interruptions or relocations are another important requirement.

Each contract has hiring goals that are tracked and enforced. DCCR strives to be a good neighbor assuring that local residents are able to contract for the work. DCCR projects have Minority Business Enterprise (MBE) and Women Owned Business Enterprise (WBE) goals of 32 percent and 6 percent, respectively. DC Water tracks these goals monthly on a project by project basis.

SAFETY ASSURANCE

Contributing to DCCR's position of public trust is the DCCR high safety standard on all projects. At a minimum this contract requires the Contractor to have a competent safety officer who shall be responsible for the supervision of the project safety requirements and be on the job at all times while work is in progress.

To ensure compliance with the requirements in the contract documents, the Contractor is required to have a Site Safety and Health Plan (SSHP) which shall be maintained at all times at the project site. The SSHP shall address site hazards related to environmental requirements, construction safety and occupational health. The SSHP also defines the protective measures that protect the environment, site personnel and the general public. Before the start of a specific activity, the Contractor has to prepare a Job Hazard Analysis (JHA) that details the activity steps, hazards associated with the activity, control measures, equipment that would be used, and any training requirements to perform the activity.

DC Water has its own safety officer monitoring the work to assure that the Contractor conforms to the safety standards. Safety meetings are held with the Contractor every other week, and any other project related meetings start with the safety discussion. Since the NEBTUR project requires the Contractor to work on multiple sites simultaneously, there is a need for the Contractor's safety officer to travel from site to site throughout the day, or sometimes the Contractor will have multiple safety officers to cover more than one site.

DC Water safety officers provide frequent jobsite safety inspections. Safety inspection and observation reports are issued to the Contractor as required to report unsafe conditions. The Contractor then must respond to each report with corrective measures that will be undertaken to assure that the condition is corrected and such incidents do not happen in future.

QUALITY ASSURANCE

Both the high public visibility of the DCCR and the 100-year design life of the DCCR Tunnel System require that quality construction be visible and assured during all stages of construction. DC Water requires full time Field Inspectors assigned to the project to assure that work conforms to the contract documents. Contractors are required to report their own quality missteps and correct them. Full time construction managers concurrently perform quality inspection and report variances from the contract documents. The inspectors preform review of the materials utilized on the job to assure the materials match approved submittals and the materials are installed to DC Water Standards.

Figure 5. DC Water approved manhole with DC Water stamp

Any unapproved materials and procedures used are rejected and Contractor is notified to perform corrective measures. In the crowded urban setting of the NEBTUR, all material (pipe bedding, valves, and all other material) is carefully reviewed to assure DC Water standards are being maintained. DC Water carefully inspects bedding and backfill material and compaction testing to assure proper backfill is achieved. Faulty construction may later result in leaking water lines or sinking roadways during the following NEBT project.

To assure the NEBTUR installed infrastructure, such as concrete manholes, is in top shape and strength for the following NEBT construction, DC Water approves all precast concrete material as it is manufactured at approved plants. A DC Water inspector reviews the manufacturing of the material and will stamp the material before it is shipped to each site. If a concrete pipe or manhole is shipped to a site without a stamp, the field inspector will not allow the material to be installed. Figure 5 shows a manhole with a DC Water approved stamp.

Other NEBTUR Inspectors' responsibilities include field measurements of construction work as it progresses, redlining Contract drawings to capture any field changes, and verifying that relocated utilities are outside the future NEBT support of excavation. Inspectors review the Contractor's monthly as-builts submittal updates to assure the information is correct. This is important since the information from NEBTUR will be relied upon by the following NEBT project. Both hard copies and AutoCAD drawings of the as-builts are submitted, for ease of use by the NEBT contractor.

REFERENCES

Kottke, P., Kantola, J, Ray, C. and Levy, W. 2014. Utility Protection While Tunneling in an Urban Environment for the DC Clean Rivers Project. *North American Tunneling 2014 Proceedings*, 424–431. Pub. Society for Mining, Metallurgy & Exploration.

Romer, A.E., Ellison, D., Bell, G.E.C., and Clark, B. 2008. *Failure of Prestressed Concrete Cylinder Pipe*. Denver: AWWA Research Foundation.

The Importance of Collective Safety Buy-In from Project Mobilization

Christina Lindstrom ▪ Obayashi Corporation / Kenaidan Contracting, Ltd.
Arthur Musisi ▪ Obayashi Corporation / Kenaidan Contracting, Ltd.
Crosstown Transit Constructors an Obayashi/Kenny/Kenaidan/Technicore Joint Venture
Eglinton Crosstown LRT Extension, Western Contract, Toronto, Ontario, Canada

ABSTRACT

Establishing a "Work Safe" culture is often met with resistance by upper management, especially after habits have been formed. Securing universal participation in safety at the onset of the Eglinton Crosstown LRT Extension was crucial in providing a solid foundation for an effective safety culture. Compliance with legislative requirements is equally important as everyone understanding why the investment in safety is beneficial. Lessons learned on this project encompass the effectiveness of multi-level hazard analysis, task planning, program management, incident response, accountability, and effective communication; all of which stem from unbridling project management support in all things related to safety.

INTRODUCTION

The Eglinton Crosstown LRT Extension West Contract is part of an overall transit project worth an estimated 9 billion dollars, making it the largest infrastructure project in all of Canada. The Obayashi Canada, Ltd., Kenny Construction Company, Kenaidan Contracting, Ltd., and Technicore Underground Joint Venture was tasked with mining 6.5 km of twin tunnels with eight cross passages, three shafts and included transferring both TBMs from one shaft to another. Utility relocation and installation of station headwalls posed independent risks and challenges. Given the dynamics of the project it was crucial to establishing an identity early in order to prevent lackluster safety performance.

Universal Commitment

Securing a universal commitment to safety prior to commencing a project seems like such an elementary concept at face value. Maintaining that high level of commitment throughout project evolution appears equally so. However, like most things in life that are worthwhile, preserving a robust safety culture amid the daily challenges of an ever-changing construction site can be challenging to say the least. The notion itself would be impossible without earnest support of upper project management and joint venture partners. Their influence spreads and inspires the entire team and delivers that first breath of life to the fledgling safety culture.

From top down, this project was supported in safety, risk management, engineering, development, and quality control. Regular audits were conducted by joint venture partners, the client, and insurance brokers. These reviews resulted in improvements within the core programs as well as highlighting successes. Throughout the 48 months of construction every member of the team was challenged to maintain that collective buy-in to safety that was established at the beginning of the project.

Time works in our favor in many ways when it comes to managing safety. Not only do we have generations of lessons and templates to draw from, we also are privy to technological advancements specifically designed to put safety first, protect workers and support a successful health & safety program. Even still the unexpected must always be expected; which is why clear expectations with consideration for accountability must be set. When these factors are combined with management support, the odds for accomplishment increase dramatically (Vincoli 1991).

Health & safety management and policy statements are the foundations of a safety culture. Nevertheless, it is the actions that are inspired by these documents that give them credence. Expectations related to safety were established early for all team members including:

- Communication
- Conduct
- Training
- Documentation
- Emergency Response
- Accountability
- Compliance

LESSONS LEARNED

A colleague recently commented upon his departure from the project, "If everything on this project had gone right, we would not have learned anything at all!" An accurate statement indeed. The simple fact is that we learn from our experiences including errors. Often times, it is events of crisis that teach us the most valuable lessons; especially when a philosophy of support and acceptance has been implemented (Carmeli, Schaubroeck 2008). Such was the case on this project with open communication and a willingness to adapt replacing antiquated philosophies.

Communication

Consistent and open communication with all levels of management is one of the most important components to maintaining a strong safety culture. Over the life of a project, the flow and topics of communication will evolve in concert with construction progress. During the initial phases of construction, dialog focuses on hazard identification, budgets, scheduling, training, and legislative compliance. As the project advances, programs, protocols, and approaches will require fine-tuning as a result of experiences, incidents, and changes in conditions. Job completion yields opportunity for reflection, evaluation, and consolidation. Within each discussion item, valuable insight can be gained by appreciating the diverse experience within the project staff.

Various approaches to communicate safety were implemented during the project and while some generated successful results, others were quietly abandoned knowing that there is always room for improvement. Methods of communication with the Project Manager that remained consistent include:

- Immediate notification of all incidents, accidents, and near-misses
- Immediate notification of Ministry of Labor visits
- Weekly, monthly, and quarterly review of all occurrences
- Direct involvement with hazard analysis

- Participation will all disciplinary action
- Follow-up on all outstanding inspection items

Such direct involvement from the Project Manager resulted in increased involvement from the Engineering and Superintendent staff. With communication lines open and trust developed between the staff and management, problems were generally solved in a timely manner.

Meetings

Conducting regular meetings and ensuring that safety is a present topic fosters the safety culture in an organization thereby maintaining safety in the forefront of everyone's mind. Throughout this project weekly staff meetings and regularly scheduled management safety meetings were successfully conducted. Each avenue allowed for open discussion of current events, upcoming challenges, and encouraged accountability throughout every department. Additional smaller meetings for specific site-related activities were also conducted with safety consistently leading the agenda.

While regular communication of safety performance is a high priority, statistical information and documentation also requires attention. Monthly and quarterly statistical data was delivered and maintained to the client and joint venture staff respectively. Within the quarterly statistical presentation, we had many opportunities to highlight new initiatives and successes as well as receive feedback or questions.

A risk register provides global insight to risk and hazards related to the construction process. The chaos of a construction site can sometimes overshadow outlying risks and hazards that might affect pedestrians, traffic, perception, and confidence. An in-depth risk register will take both internal and external risk sources into consideration and work towards mitigation to an acceptable level (Dunović, Radujković, and Vukomanović 2013). The risk register itself functions as a living document which evolves throughout the project providing valuable information to both direct users and those who might review the document from a distance whom might not be otherwise familiar with the terrain of the actual project.

Regular maintenance and reviews of the risk register were a significant undertaking which included various management members among the client and construction teams in addition to the Safety and Quality Managers. The bi-weekly meetings provided valuable insight and supported the flow of communication on relevant safety, risk, and quality topics. Detailed analysis was achieved for an expansive number of topics by focusing on one topic per meeting and only including those employees whose jobs are specifically relevant to the risk/hazard via either participation or expertise.

Tool box talks at CTC conducted on a daily basis at the start of every shift. The typical tool box talk informed the labor force of expected tasks for the day and what was done on the previous shift. We also encouraged a dialogue with our labor force that would notify management of any issues they encountered and concerns that they'd like to be addressed. By receiving feedback from the laborers who are actually doing the work, it made implementing new policies and developing procedures easier. The tool box meeting was usually the only opportunity that everyone working on that shift was together, thus it was a good time to communicate safety related topics including:

- Safety items from their previous shift
- Refresher training
- Discussing new initiatives

- New policies and procedures
- Review of mandatory requirements from all personnel
- Summarization of work related incidents/injuries on site
- Major construction related incidents/injuries on a different sites
- Overview new legislation and results of Ministry of Labor (MOL) inspections

The main objective of daily tool box talks was to create an inclusive relationship between labor and management and avoid having situations where the safety mandate came across as an authoritative dictation.

Documentation

Documentation is a contentious topic for many jobsites due to the redundant nature of forms, difficulties with file management, and ease of review or implementation. From the beginning of the project, all levels of management were willing to embrace technology that would allow for a near-paperless work environment. A comprehensive share drive with clearly defined organizational parameters paired with a mobile form application produced positive results.

Informal Job Safety Analysis Forms (JSAs) were developed in addition to tool box talks for various scopes of work to compliment formal analyses previously submitted to the client. They outlined work to be done and how it was going to be completed safely. A sudden increase in incidents and injuries led to this new approach and was instigated by the Project Manager. The TEAM (Total Efficiency Attainment Meeting) form is a document that was put in place to obligate Foremen and Lead Hands to plan their work; as well as to analyze and address any hazards associated with what they may be doing prior to starting a job. It had to be signed off by each member of the crew and handed in at the end of the shift, this created accountability.

The TEAM form (Figures 1 and 2) incorporates a stop and think method which is to stop and think about what the task is and think about how it is going to be done safely. The intent of the TEAM form initiative was to develop critical thinking from those directly involved in the tasks. Upon implementation, there was not only a decrease in incidents and injuries, but also a marked increase in participation in toolbox talks, task planning, and preparedness.

Formal Hazard Analyses were conducted for all major changes in conditions, contract requirements, or upon request. Each analysis involved direct coordination and signoff with the Project Manager. Upon completion, the form would be reviewed with site management as well as discussed in a documented toolbox talk. In some instances, emergency response would also be included in the form. In this case, the document would be laminated and posted in the immediate working area.

Site inspections provided the necessary input to those who may have not noticed hazards otherwise. Often times supervisors are so involved in a specific task that they overlook safety issues. Daily inspections allowed for feedback to the site management team and almost always highlighted areas that were positive or in accordance with regulations; as well as areas that needed improvement. By implementing this method, the goal was to not only zero in on safety infractions but to also compliment the team on maintaining compliance.

Initially, we conducted inspections on paper, scanned them to our internal hard drive, filed them in binders and then emailed the pdf to site management. This was a very

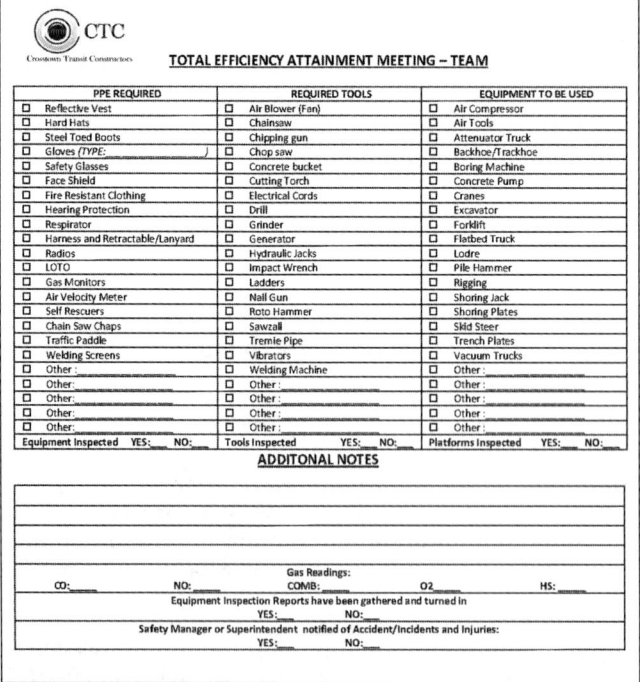

CTC
Crosstown Transit Constructors

TOTAL EFFICIENCY ATTAINMENT MEETING – TEAM

Foreman:_____   Date_____   Time_____
Site:_____   Task Location: _____

Review these items with the crew at the location of the task and check all items that apply

	Activity Hazards		Working tight areas/		Process Hazards
☐	Rigging	☐	Heavy Lifts/Awkward Body Position	☐ New Workers in crew
☐	Flying/Over head loads	☐	Repetitive motions/lifting	☐ Other crews in area
☐	Moving Equipment/Vehicles	☐	Noise	☐ Equipment Traffic
☐	Contact with energized equipment		Working at Heights (>6ft)	☐ New process/Change in process
☐	Pinch Points	☐	Guardrails in place	Access/Egress
☐	Hot Work-Cutting/Grinding/Welding	☐	Hole openings covered	☐ Ladders Secured
☐	Loose/Spalling material	☐	Protect from falling items	☐ Scaffolds have green tag
☐	Collapse of ground/walls	☐	Appropriate anchor points selected	☐ Pathway cleared of slip/trip hazards
☐	Dust/Flying Debris	☐	Other:	☐ Other:

TASK	HAZARD	CONTROL

DO NOT SIGN UNTIL YOU UNDERSTAND AND AGREE WITH THIS WORK ASSESSMENT. THIS DOCUMENT MUST BE REVIEWED AFTER EVERY ACCIDENT/INCIDENT/NEARMISS/CHANGE IN CREW/WORKING CONDITIONS/OR WORKING LOCATIONS.

Name:	Signature:	Name:	Signature:
	SINGATURE		SINGATURE
	SINGATURE		SINGATURE
	SINGATURE		SINGATURE
	SINGATURE		SINGATURE
	SINGATURE		SINGATURE
	SINGATURE		SINGATURE
	SINGATURE		SINGATURE
	SINGATURE		SINGATURE
	SINGATURE		SINGATURE

Figure 1.

CTC
Crosstown Transit Constructors

TOTAL EFFICIENCY ATTAINMENT MEETING – TEAM

	PPE REQUIRED		REQUIRED TOOLS		EQUIPMENT TO BE USED
☐	Reflective Vest	☐	Air Blower (Fan)	☐ Air Compressor
☐	Hard Hats	☐	Chainsaw	☐ Air Tools
☐	Steel Toed Boots	☐	Chipping gun	☐ Attenuator Truck
☐	Gloves (TYPE:)	☐	Chop saw	☐ Backhoe/Trackhoe
☐	Safety Glasses	☐	Concrete bucket	☐ Boring Machine
☐	Face Shield	☐	Cutting Torch	☐ Concrete Pump
☐	Fire Resistant Clothing	☐	Electrical Cords	☐ Cranes
☐	Hearing Protection	☐	Drill	☐ Excavator
☐	Respirator	☐	Grinder	☐ Forklift
☐	Harness and Retractable/Lanyard	☐	Generator	☐ Flatbed Truck
☐	Radios	☐	Hydraulic Jacks	☐ Lodre
☐	LOTO	☐	Impact Wrench	☐ Pile Hammer
☐	Gas Monitors	☐	Ladders	☐ Rigging
☐	Air Velocity Meter	☐	Nail Gun	☐ Shoring Jack
☐	Self Rescuers	☐	Roto Hammer	☐ Shoring Plates
☐	Chain Saw Chaps	☐	Sawzall	☐ Skid Steer
☐	Traffic Paddle	☐	Tremie Pipe	☐ Trench Plates
☐	Welding Screens	☐	Vibrators	☐ Vacuum Trucks
☐	Other :	☐	Welding Machine	☐ Other :
☐	Other:	☐	Other :	☐ Other :
☐	Other:	☐	Other :	☐ Other :
☐	Other:	☐	Other :	☐ Other :
☐	Other:	☐	Other :	☐ Other :
Equipment Inspected YES:___ NO:___		Tools Inspected YES:___ NO:___		Platforms Inspected YES:___ NO:___

ADDITONAL NOTES

Gas Readings:
CO:_____   NO:_____   COMB:_____   O2_____   HS:_____

Equipment Inspection Reports have been gathered and turned in
YES:____   NO:____

Safety Manager or Superintendent notified of Accident/Incidents and Injuries:
YES:____   NO:____

Figure 2.

time consuming process that often led to delayed notification emails. The result was conveying old information that may not have been relevant by the time it was sent out. Due to the fast pace and constant evolution of construction we needed something that was efficient and allowed for a quick turnaround. This deficiency led us to discover a mobile app that made it much easier to process safety inspections in real time. Suddenly, a process that used to take hours, essentially became instantaneous with the added value of imbedded pictures and visuals. Once the inspection was complete, the inspector would sign off and send it in for processing and an email notification would be sent out shortly thereafter. All site management personnel had smart phones with access to the app which meant they could view completed inspections on their phones and address relevant issues immediately.

Accident and incident investigations were also conducted using the mobile app platform. Upon notification of an event, members of management dispatched to the site location. Injured workers would be escorted by a member of management to ensure appropriate care and support. Investigations begin immediately by examining the scene, taking photos, securing witness statements, and analyzing contributing factors. Once the report is completed, it is reviewed by the Safety Manager and upon processing, sent to upper management, the client, and joint venture partners. All events are reviewed during progress meetings with the client as well as during audits and joint venture meetings.

Managing injured worker recovery included the management team as well. Regular communication is maintained with the injured worker and their progress communicated to the Project Manager. At the earliest opportunity, modified duty offers are presented with a commitment to ensuring the worker be accommodated regardless of limitations. When the worker returns to site, a meeting is held with the site Superintendent, Safety Manager to review limitations and establish a plan. If at any time the plan goes off schedule, assistance with the Insurance carrier is requested and another meeting is conducted and will include additional members of upper management.

The most common safety training in addition to orientations, included working from heights, scaffolding, first aid, tunnel rescue training and WHMIS training. Virtually everyone had gone through these trainings in one form or the other; and prior to working on site everyone had to have WHMIS and fall protection training in hand. As orientations were documented utilizing the new mobile app, maintenance of records was quite simple. However, it was the training that was tailored to the job that created be biggest impact. For instance:

- Loci training for all operators conducted by site mechanics was one way to avoid improper operation. Loci training also included an air brakes course to help drivers better understand how the loci's operated.
- We also used air monitors that required daily calibration. Specific people were trained on how to properly calibrate, maintain, and operate each device.
- Site management and even labor foremen completed Basics of Supervising training which defined their roles and responsibilities as a supervisor. This training proved to be very effective as it emphasized the legal liabilities of a supervisor as outlined in the OHS regulations.

An approach to progressive disciplinary action was also configured within the mobile app. Our internal policy at CTC detailed progressive enforcement in dealing with safety infractions. Courses of actions began with verbal warnings and progressed to suspensions or terminations. The difficulty in implementing the progressive discipline procedure came when determining who was responsible for issuing disciplinary action and what course of action should be taken for said infractions. Adjustments to the existing policy were made to include direct involvement and signoff by the Project Manager for each infraction in addition to the Safety Manager and other management members. As it turned out, discipline for safety infractions was one of the hardest points of emphasis on this project. The need to have a clear understanding of roles and responsibilities was also another reason for mandatory Basics of Supervising training. One of their many responsibilities includes making sure that employees work in compliance with the (Occupational Health & Safety Act) OHSA and its regulations (OHSA Sec. 27[1]).

CONCLUSION

If securing a universal commitment to safety is of the utmost importance, then certainly some contributing factors will outweigh others. One thing that must be acknowledged is the top down support sustained by the joint venture partners. They contributed professionals and tradesmen with diverse backgrounds and construction experience. Site management was specifically comprised of diverse backgrounds from around the world and included TBM maintenance, servicing, and assembly, shaft excavation, secant piling, tunnel surveying, and tunneling.

Contract structure and joint venture organization afforded multiple avenues for input when developing policies specifically for this project. This combined with previous experiences made for unique input on how to establish and maintain an effective safety plan. All of this would be moot without a collective buy-in to safety from upper management to apprentice level employees. Experience gained over the last four years has proven that easing the flow of communication is not enough. We must also provide each member of the team interactive tools to complete tasks as well as opportunities for interaction.

Of all the lessons learned on this project in terms of safety, it is clear that free flowing communication is critical to not only maintaining the collective buy-in, but also to fostering a positive safety culture.

REFERENCES

Vincoli, J.W. (1991). Total quality management and the safety and health professional. Professional Safety. 36(6): 27–32.

Dunović, I., Radujković, M., and Vukomanović, M. (2013). Risk register development and implementation for construction projects. Građevinar. 65(1): 23–35.

Carmeli A., and Schaubroeck, J. (2008). Organisational Crisis-Preparedness: The Importance of Learning from Failures. Long Range Planning, International Journal of Strategic Management. 41(2): 177–196.

Major Projects

Chairs

Frank Perrone
Mott MacDonald

David Smith
WSP | Parsons Brinckerhoff

Procurement and Delivery Strategies to Increase Competiveness on Tunnel Projects

Steven R. Kramer ▪ AECOM
Paul Nicholas ▪ AECOM

ABSTRACT

In recent years, as tunneling projects become more complex and funding sources are limited, owners are seeking to evaluate the best method(s) for delivering its projects. This paper explores the current trends in tunneling procurement being used across the Americas and the world. The expectations of owners and contractors are explored as well as how they impact the current atmosphere for bidding and project execution. This includes conventional and alternative delivery methods, such as design-build and public-private partnerships (P3s), but also covers a broader perspective to discuss fee at-risk, advance purchase of Tunnel Boring Machines (TBMs), pre-qualification and contract packaging.

INTRODUCTION

Across the world, owners, engineers and contractors are continuously seeking better and new methodologies for the safe and most cost-effective approach for delivering tunnel projects. This continues to be challenging in an environment where funding sources are frequently limited, yet project demands are high. Everyone wants to deliver projects safely, on time, within budget and with minimal impact to the communities where we live. These high demands have created an environment where we are continually seeking better procurement and delivery strategies that will create a competitive environment for delivering complex infrastructure projects, especially tunnels for transportation, water and energy. This paper explores various delivery options, ranging from conventional delivery to alternative delivery, with an expanded look at the variations and options that can be applied when using these methods, with particular emphasis on how to increase competiveness on tunnel projects.

CURRENT TRENDS IN PROCUREMENT AND BIDDING FOR TUNNEL PROJECTS

In the U.S., the underground construction and tunnel business remains highly competitive. There are several construction companies with excellent staff and equipment for successfully executing tunnel projects. For some owners and projects, this atmosphere is attractive and results in many bidders and pricing below the owner's or engineer's estimate. In an analysis of 10 recent water tunnel projects bid between 2013 and 2016, the winning bid for nine out of 10 of the projects was less than the owner's estimate. For these projects, the average was about 15% below the owner's estimate. The owner had short-listed three or four contractors for most of these projects. In most situations, the prices had significant ranges. This sample of bids demonstrates the competitive atmosphere for many tunnel projects.

It is estimated that approximately $5.25 billion was expended in 2015 on tunnel construction in the U.S. This is an approximation based on interviews with tunnel experts in both design and construction. This construction revenue includes both smaller-diameter utility tunnels and large transportation tunnels. In the last decade, there has

Table 1. Comparison of general construction cost drivers to tunnel cost drivers

General Construction Cost Drivers	Tunnel Cost Drivers
• Materials	• Differing site conditions
• Labor	• Design development & unforeseen design changes
• Equipment	• Linear nature of the work
• Design quality	• Scope changes & transfer
• Business climate/market	• Shaft construction
• Contract method	• Bonding & limitations
• Environment/location	• Limited contractor pool
• Indirects/insurance	• Specialized labor
• Inflation	• Ownership of risk

been an influx of international contractors into the U.S. marketplace who are bidding and winning projects. There are now both domestic and international contractors with the project resumes and qualified personnel to execute a broad range of tunnels, in both hard rock and soft ground. For many tunnel projects, there is a movement away from conventional delivery to alternative delivery methods, especially for large projects or where the methodology has not been decided. Alternative delivery methods may include design-build, progressive design-build, construction management at-risk and public-private partnerships (P3s). As alternative delivery methods are further utilized, the relationship between owners and consulting engineers changes. In the U.S., consulting engineers were traditionally working as owner's representatives and assisting them with minimizing design liability and financial risk. With alternative delivery methods, some engineers may be working for constructors as designers and other engineers may be serving as an owner's representative. This can create some possible conflicts for a consulting engineering firm, where one group of employees provides services to the owner on one project and another group of employees is working for a contractor an another project for the same owner.

GENERAL AND TUNNELING COST DRIVERS

In comparison to other types of infrastructure projects, tunnel projects are linear and can extend for many miles. Underground and tunnel projects are typically repetitious construction (e.g., mine the tunnel and construct the tunnel liner system). Due to the linear nature of tunnel works, there are limited items in the critical path and the constraints are complicated. Also for a tunnel project, there can be many third party impacts and there are limited work-arounds when a problem occurs. In general construction, both vertical and surface works (such as highways), the construction can be complicated but there may be many work-arounds when delays occur. It may be possible with general construction for the contractor to work on another part of the project to minimize delays. In contrast, on a tunnel project, obstructions to tunneling can have a critical impact on the schedule due to the project's linear nature.

Major drivers for general construction are materials, labor and equipment (Table 1). These factors are also important for a tunnel project, but changes in site conditions and underground conditions may play a large role. Also, the pool of contractors for tunnel construction is significantly more limited when compared to other types of heavy construction.

EXPECTATIONS OF OWNERS AND CONTRACTORS FOR TUNNELING PROJECTS

Table 2 lists some of the owner's needs on major capital projects. This is based on a limited sampling of owners who are conducting or plan to conduct water tunnel projects. Owners desire to create a competitive bidding environment, with a focus to deliver on time, within budget. They place a strong emphasis on candidates understanding their policies and procedures while providing technically astute, experienced people who also know how to deal with the public. Understandably, owners often prefer organizations that they know and trust.

Many owners now use several proactive approaches when implementing their tunnel projects. Owners frequently notify contractors and hold briefings many months before advertising the project to

Table 2. Owner's needs on major capital projects

- Competitive bidding atmosphere
- Delivered on time & within budget
- Safety culture
- Understand owner policies & procedures
- Willingness to partner & resolve issues
- Provide people with right experience & right equipment
- Be proactive before problems arise
- Understands public outreach & how to respond to public

bid. The owners are creating forums to foster relationships with the contracting community and to gain trust between the parties. When using minority and/or disadvantaged businesses, the owners are now identifying opportunities where these firms can be successfully used as part of the overall project.

Tunneling contractors are a relatively small, close-knit community. They have a good understanding of projects risks and are contract savvy. To spread the risk and potential liability, tunnel contractors frequently team on large projects. The contractors expect that the contract documents will be tailored to industry standards. To manage the risks, many contractors prefer to see owners implement Geotechnical Baseline Reports (GBR), Differing Site Conditions (DSC) clause and Dispute Review Boards (DRB). Due to the potential large investment of purchasing a Tunnel Boring Machine (TBM) for specific projects, the contractors prefer reasonable terms for payment of the TBM.

BIDDING, PROCUREMENT, AND DELIVERY STRATEGIES TO INCREASE COMPETITIVENESS

Owners now have many options for procuring their tunnel projects. This includes conventional and alternative delivery approaches. The preferred option will depend on many factors; it often is not a simple analysis to choose the best approach. There are advantages and disadvantages to each of the approaches and an owner must select the approach that best fits their particular situation. There is a growing interest in the use of alternative delivery approaches for tunnel projects. For water/wastewater tunnels at municipal agencies, the preferred delivery approach continues to be design-bid-build (DBB). In contrast, there seems to be a trend towards design-build (DB) for transportation projects—highway and transit tunnels. However, there are many other procurement approaches besides design-bid-build and design-build that have been implemented at both water and transportation agencies.

Table 3 outlines several of the approaches that owners have used for increasing the competitiveness of tunnel and underground projects. Each of the approaches listed in Table 3 has been previously used on tunnel projects in North America.

Some of the bidding and procurements strategies can also be used in conjunction with conventional design-bid-build delivery. There are several examples where owners have chosen alternative approaches for tunnel projects. This includes the following projects that have successfully implemented alternative approaches for constructing tunnel projects: City of Portland, OR Combined Sewer Overflow (CSO) Program – Modified Fixed Fee; Toronto Transit Commission – Pre-Purchase TBM; City of Atlanta Raw Water Tunnel – Construction Management at Risk (CMAR); Narragansett Bay Commission CSO Program, Providence, RI – Shared Savings Model; DC Water Clean Rivers Program, Washington, DC – Prequalification and Design-Build; Pentagon

Table 3. Bidding strategies for increasing competitiveness on tunnel projects

Bidding Strategy	Description
Advance purchase of TBM	Owner purchases the TBM in advance of the contract.
Best value	Contract is selected based on a combination of non-price and price attributes. The attributes can be weighted.
CM at-risk	Similar to a CMGC, the owner selects a contractor to serve as its construction manager. The final price may be submitted at the time of selection, or alternatively the CM may bid elements of the contract and then finalize its price.
CMGC (construction manager/general contractor)	Owner selects a contractor to serve as it construction manager and/or general contractor. This may occur before the design is complete. The CMGC will then bid the elements of the contract and then finalize its price.
Contract packaging	Owner elects to subdivide the project into several contract packages.
Design-build (DB)	Owner selects integrated design and construction team to execute the project. Owner conducts limited design, or issues performance- based specification.
Early contractor involvement (ECI)	Owner seeks input from contractors during the design process to take advantage of their experience. ECI can be formally or informally implemented.
Fee at-risk	Contractor is paid for work performed (including overhead) without profit. Profit is at-risk and based on performance during the contract.
Modified fix fee plus cost reimbursable	Contractor submits a fixed price for known items and unit prices for unknown items or cost +.
Pre-qualification of contractor (RFQ)	Owner defines criteria to qualify a limited number of contractors before the request for proposal is issued. Typically, the pre-qualification is based on a statement of qualifications submitted by the contractor.
Progressive/modified design-build	Owner selects a design-build team based upon qualifications. Design-build team then completes the design and finalizes its construction price.

Renovation Program Intake/Outfall Tunnel, Arlington, VA – Design-Build with Fee at Risk; Florida Department of Transportation, Port of Miami Tunnel, Miami, FL – P3 with Shared Contingency. Later in this paper, specific examples on where some of these approaches were implemented will be described.

Tunnel Methodology and Specifications

On projects where time to completion is critical, various strategies may be used in the procurement process to reduce the design and construction time. Frequently, design-build (DB) may be used to reduce the overall construction time, since design submissions may be staged, allowing early start-up of construction. In the DB process for tunnel construction, the longest delivery items are typically the tunnel boring machines (TBM). Also depending on the diameter, the location of the project site and manufacturing location, it will generally take from 10 to 14 months from equipment order to start of tunneling. The DB process allows the contractor greater flexibility to order the TBMs before the design process is complete.

The strategy for TBM supply may be critical to the success of the project and is an integral part of the risk assessment and construction process. The first issue is what level of involvement the owner and its team should have in the specification of the TBM. Table 4 outlines the advantages and disadvantages of two procurement strategies for the purchase of a new TBM.

Table 4. Procurement strategy for purchase of TBM (when project requires multiple TBMs)

Procurement by Contractor	Procurement by Owner and Supplied to the Contractor
Advantages • Little involvement or financial risk to the owner, as laid out in the contract documents • Contractor controls his own destiny • Negotiates on their own terms	**Advantages** • Likely to have single source of TBMs • Able to order in advance, negotiate better price, delivery and buy-back allowing earlier start to tunneling • One source for spare parts, service and maintenance
Disadvantages • Possible wrong choice of TBM and or tunnel methodology • Likely to have different TBMs and suppliers (when multiple contracts on a single project/program) • No common service or spare parts resources • Contractors will require finance/advances during the TBM manufacturing—Supplier will require Bank Guarantee	**Disadvantages** • Increased financial commitment to the owner • Potentially increased risk to the owner • Owner dictates Methods and Means • Complex three-way relationship for responsibility for TBM mechanical failures, service and maintenance • Owner and owner's engineer may not have the experience to specify and purchase TBMs

At one extreme, the owner may leave the method of tunnel construction totally up to the contractor by allowing all tunneling methods, from conventional to mechanized, and the contractor will make a choice based on the geotechnical data report (GDR) and geotechnical baseline report (GBR), if supplied as part of the bid documents and from his past experience. If the construction period and critical issues like settlement and traffic management are clearly identified and limits set, then in many cases leaving the tunnel method to the contractor is acceptable. However, contractors frequently will use the lowest cost method to ensure a low bid and in some cases this may be a methodology with higher risk to the client. Projects bid on this basis may suffer from cost overruns and delays from the increased risk.

At the other extreme, the owner and its team may carry out the ground analysis and specify the tunnel construction methodology by specifying the type of tunnel method based on the ground conditions and the final finish required e.g., a TBM may be specified and this may include the type of TBM: shielded or open gripper, closed or open-face, earth pressure balance (EPB or slurry), and may also specify the primary tunnel support and required final lining e.g., pre-cast tunnel segment rings or bolting and mesh with shotcrete. The specification may go on to include additional equipment and features to be included in the tunnel equipment such as on a TBM, to include man-locks, probing and grouting capability, minimum thrust and torque. It is generally recognized that the more detailed the specification for tunnel construction is made in the bid documents, the more risk and liability is carried by the owner and its team. If the equipment features specified are found to be unsuitable or detrimental to the tunneling process, it may result in claims and cost overruns. It is also recognized that a higher level of specification allows for a better cost estimate and may reduce overall risk to the project by allowing the risks to be better defined and limited. A higher level of specification also allows for a more accurate cost estimate and can result in closer construction bids.

In rare cases and generally on large projects with several tunnel contracts and where multiple TBMs will be required, the owner may pre-purchase the TBMs and provide them to the contractors for the tunnel construction. Initially this appears to be just an extension of the issue of what level of specification should be given. With a heavily specified tunnel requirement, the owner is already taking substantial risk. At this point it may be thought that if they are carrying the risk of the TBM choice they may as well gain some advantage and reduce the tunnel construction period by pre-ordering the TBMs early in the procurement process so that tunneling can start within a short period after the contract award and NTP. In the short term there are advantages; the owner may go out for competitive bids for the supply for all the TBMs from one source. This is likely to result in lower prices and better delivery. It also has the advantage that spare parts and maintenance will be shared in common on all the TBMs. The owner may make conditions as part of the purchase agreement, such as TBM buy-back and maintenance support.

However, there are also some less understood disadvantages to the TBMs being supplied to the contractors by the owner. One issue is the responsibility for operation and maintenance of the TBMs. Various formulas have been tried for handling the cost of maintenance, but where the operator of the TBM is not responsible for the cost of maintenance and downtime, there are likely to be contractual issues. This is amplified when a major breakdown occurs, such as a bearing failure, and raises the question of who is to carry the cost of the ensuing delays. In most cases, the owner does not have a thorough understanding of TBMs and when problems occur can tend to become confrontational, leading to claims.

Projects where TBMs have been pre-purchased include the Spadina Metro project in Toronto, Canada, where four identical EPB TBMs were purchased by the Toronto Transportation commission. There were some operation and maintenance issues where responsibility was shared. However, equipment failures led to substantial delays and cost overruns.

Another example was on the Doha Metro in Qatar, where 13 identical TBMs were pre-purchased by Doha Metro authority and increased later to 21 TBMs for completion of the project. In the construction contract, the contractor was required to assume full responsibility for TBM maintenance. However, the owner had a support and main-tenance team in Qatar from the TBM supplier, as part of the TBM supply contract. Overall, the 60 miles of rail tunnels were successfully completed.

On some other projects, notably Crossrail in London, various strategies for procuring the TBMs were evaluated, but in the end the TBMs were heavily specified towards a single source supplier and the contractors themselves saw the advantage to having a common supplier on the project. This resulted in eight TBMs being purchased from one supplier by the two contractors. In many ways, this is possibly the best outcome for the owner and the project overall.

Tunnel Linings

Another area of tunnel procurement which is open to various options is the design and supply of the final liner. This may be a cast in-situ secondary liner, a sliplined pipe or in many cases where TBMs are used for excavation, the primary liner of pre-cast segmental tunnel rings is also the final liner. In general, the owner wants to dictate the design of the final liner based on the tunnel usage and its operational requirements. These days the use of pre-cast reinforced concrete (RC) segmental tunnel rings is very common, as they act as the primary support system and the final liner. This will save a substantial amount of time when a cast in-situ RC liner is required, which is normally built after the tunneling is completed.

Once a decision is made to specify a pre-cast segmental liner, it is erected in the tail shield of a TBM in the excavation process. The design of the segmental rings is deter-mined by the ground loads and the final usage of the tunnel. Typically, the segments are designed in part or fully by the owner's engineer. The dimensions and specifica-tions are supplied to the contractors in the tender documents, so they can supply and be responsible for detail design and manufacturing, and delivery to the site. In this case, QA/QC is the responsibility of the contractor or it may be provided by an independent third party. In a large project with multiple tunnel contracts and contractors, the segments are likely to come from multiple casting yards with possible variations in detail design and quality.

In an effort to have a consistent tunnel segment design throughout a multi-contract project, the owner may complete the detail design including dimensions, reinforcement, concrete mix and ring type, and have the tunnel contractor only be responsible for casting. However, the design risk is carried by the owner, but the responsibility for molds and casting quality and erection in the TBM is carried by the contractor. In most cases of problems in segments, such as cracking and spalling, they are caused due to casting quality issues and mishandling in the casting yard, or by erection and thrust from the TBM. On large multi-contract tunnel projects, there are some advantages in having a single source of tunnel segments. It will be easier to monitor QA/QC, including consistency of materials and lab testing, and less manpower will be required to monitor multiple casting yards. In this case, there are some advantages and few disadvantages to the owner contracting directly for the supply of all the segments to the tunnel contractors. This was done on the Spadina Metro project in Toronto and the quality and supply of segments was a success, with very few quality issues at the casting yard or in erection in the TBM. The owner was able to subcontract QA/QC services at the casting yard to an independent provider and was able to use this verified data in any issues occurring in the tunnel. Overall, it is thought that the provision of design and supply of segments by the owner to the project was beneficial.

CASE STUDIES IN TUNNEL PROCUREMENT AND BIDDING

Design-Build Contract with Special Conditions: Tuen Mun–Chek Lap Kok Link, Hong Kong, China

This tunnel link is part of the new Pearl River crossing, connecting the NW New Territories and Macao to Lantau and the Hong Kong (HK) international airport, shortening the route to the airport by 13.7 miles. The project includes crossing the main navigation channel on the western side of HK, with a six-mile, two-lane trunk road, including three-mile 46-foot-diameter twin bore tunnels under the shipping channel. This tunnel is the deepest tunnel in the region, with 170 feet of water head. The design called for a third lane on the uphill sections to allow for slow moving heavy goods trucks.

Initially, in 2005, a new bridge was proposed for the crossing, but bridge closures due to bad weather and safety issues on the existing Tsing Ma Bridge eliminated a bridge scheme. This was replaced by an immersed tube design that has been used for all HK harbor crossings to date. After further study and objections to the immersed tube from the harbor authorities and the environmentalists, a bored tunnel approach was adopted in 2011. This new design was proven to better mitigate the construction and program risks.

The subsea bored tunnel had several technical challenges, including the type of TBM, the tunnel configuration, alignment, geology, interventions, TBM launch and retrieval shafts and the cross passages. As is common in Asia, the owner's engineer completed the detailed and reference design, tendering, contract management and construction supervision. As is standard in these types of contracts, a modified Design-Build procurement method was adopted.

Due to the active shipping channel, access for bore holes for geotechnical investigation was not possible. A typical geological profile was developed for the crossing and adopted using the information from the Tsing Ma Bridge and other construction projects in the area. This indicated the tunnel bore would excavate in saturated alluvium, clays, weathered granite and possibly unweathered granite. The DB tender document was fully developed in detail to ensure there would be minimal technical variation in the bids and to avoid predatory pricing. This approach allowed only the most experienced large-diameter tunnel contractors to bid. A "best value" criterion was used to determine the award. The tunnel segments were fully designed. The contract documents developed by the engineer specified the 46-foot sections of tunnel, the type of TBM was specified as a slurry mix shield, minimum thrust and torque, cutter wear sensor system and the requirement for saturation gas high pressure interventions. The DB tender also had special provisions for different "pay" rates for excavation in different ground types, so controlling possible differing site conditions (DSC) claims, as no GBR, was provided for the tunnel section.

The successful contractor after project award proposed an alternative to the cut-and-cover third lane. The contractor proposed a dual size TBM that would start at just less than 58 feet (the largest in the world) and complete the three-lane highway section and then be reduced in size to 46 feet to complete the two-lane undersea crossing section, refer to Figure 1 for TBM utilized.

Due to the flexibility in the DB procurement method, it was easy for the contractor to offer this alternative design and for the client to accept it. Meanwhile, a second TBM would complete the second bore at 46 feet for the full length of the required tunnel section. On these TBMs the contractor developed a robotic system to change cutters over about 60% of the face, but for the remaining cutter changes hyperbaric saturation gas interventions were used. The contractor also proposed an alternative method of constructing the 50+ cross passages using an innovative pipe jacking system, which removed the need for extensive ground treatment. Some lessons learned from this project are: 1. Flexibility in the Design-Build procurement method for this large complex tunnel allowed the owner to accept substantial changes in the construction methodology proposed by the contractor without changes to the contract or overall design; 2) Contract required the owner to provide an experienced tunnel team at the site to interact with the contractor, thereby ensuring a collaborative relationship between the contractor, the owner and owners' engineer; 3) Method of the owner having a single engineer (GC - General Consultant) to administer the program, geotechnical and environmental program, permitting, design, tendering and construction supervision, which is common in Asia, produced benefits for all parties.

Figure 1. Slurry 58.8 ft diameter tunnel boring machine (TBM)

Design-Build Contract with Shared Contingencies: Florida Department of Transportation/Port of Miami Tunnel, Miami, FL

There have been many technical papers written on the successful design and construction of the new Port of Miami Tunnel, which opened to traffic in 2015. For this paper, the commercial aspects of the contract are presented with particular emphasis on the shared contingency that was implemented to manage potential changes in the geotechnical conditions. The Port of Miami Tunnel was constructed to reduce car and truck traffic through downtown streets in Miami to the Port. The new tunnel connects I-395 to Watson Island and then the Port. Twin 4,200-foot-long, 39-foot ID tunnels were constructed between Watson and Dodge Islands, refer to Figure 2 for photo of TBM utilized. The geologic conditions indicated fill, plastic sandy silt /silty sand with zones of limestone. During the pre-liminary engineering and procurement, the owner was concerned about overruns and delays, due to the variable geologic conditions.

The Port of Miami was procured using a 30-year Design, Build, Finance, Operation & Maintenance (DBOM) concession. The total project cost was $1.062 billion, which included design, construction, bank debt, TIFIA loans and concession costs. Financial close occurred in October 2009 and the contract called for a 55-month construction period. The concession will operate and maintain the tunnel until October 2044.

Courtesy Florida Department of Transportation

Figure 2. Earth pressure balance (EPB) 43 ft diameter TBM

Many of the well-established risk allocation tools were used for this contract, including GBR, DRB, risk sharing and a defined contingency. The owner took a progressive and innovative approach with the contingency. The owner explained to potential bidders that a $180-million contingency would be created and potential bidders should not include a contingency for geotechnical risks in its bids. The contingency for unexpected geotechnical conditions was implemented as follows:

- Contractor responsible for initial $10 million of overruns for geotechnical conditions
- Owner responsible for next $150 million
- Contractor responsible for final $20 million
- Over $180 million; either party could terminate the contract

The Port of Miami was successfully completed and reports indicate that both FDOT and the contractor were pleased with the results. The total claims from the owner contingency paid out where in the range of $60 million. The authors believe that the success of the Port of the Miami Tunnel and the contingency approach used should encourage other similar innovations to be used for complex tunnel projects. Other important lessons learned are: 1. Bidders positively responded to the risk-sharing model proposed by FDOT, 2. Contingency models held by the owner can be an effective model for managing geological risks and other unknowns, 3. Risk-sharing approaches can be used to minimize confrontation between parties and help manage costs.

CONCLUSIONS

Owners, engineers and contractors have implemented a variety of delivery strategies to most cost-effectively deliver tunnel projects. These strategies have included many forms of alternative project delivery, and tools that can be implemented with conventional design-bid-build. There is not one specific solution for every project, due to the complex variables that can influence the design and construction of underground infrastructure. It is very encouraging to see the implementation of alternative approaches that can help reduce the risk, lead to lower costs and improve schedules without sacrificing quality, or and possibly even improve quality.

Design-Build Project Delivery: The Importance of Successful Coordination Between Designer and Contractor

Mark Johnson ▪ CH2M
Martin Ellis ▪ CH2M

ABSTRACT

The design-build procurement model is becoming increasingly common for large underground projects. It has proven to be a viable and effective delivery model, but the chances of successful implementation are greatly influenced by the ability and willingness of all parties to the Contract to communicate and coordinate effectively.

This paper draws upon the authors' experience of design-build tunnel contracts in the USA and around the world, and focuses on the critical relationship between the Contractor and Designer. The Paper discusses some of the key attributes required for an effective design-build delivery team, while recognizing the unique influence of constructability on design for underground structures, and the role of risk management to focus design and quality control. The author also highlights some of the approaches that can be taken to increase the chances of successful delivery through pro-active communication and coordination.

INTRODUCTION

The design-build procurement model is becoming increasingly common for large underground projects in the USA, and in recent years it has proven to be a viable and effective delivery model. Several major State and Local Authorities, such as LA Metro and DC Water, have adopted it as their primary delivery mechanism for large tunnel projects. One of the primary reasons for this is that design-build procurement enables the contractor and designer to work closely together to optimize the design approach and construction methodologies/sequencing to suit preferences of the contractor, while still meeting contract requirements. This can facilitate innovation and also potentially lead to cost and/or schedule efficiencies.

A good example of this is the design and construction of the dewatering and screening shafts for the Blue Plains Tunnel design-build contract, part of DC Water's Clean Rivers Program of work in Washington D.C., USA. The owner's reference design included these structures as two separate shafts, with an adit connecting between them. The successful DB contractors' proposal was based upon eliminating the adit (therefore eliminating the risk associated with constructing it) by placing the shafts in a figure-of-eight arrangement with a common wall between them. This proved to be more cost effective, reduced risk, and facilitated launching of the Tunnel Boring Machine from the shafts. During the final design phase following award of contract, the DB contractor (Traylor/Skanska/Jay Dee JV), their slurry wall subcontractor (Bencor) and the designer (CH2M HILL) worked very closely together to ensure that the design and construction of these shafts was as efficient as possible and remained viable. Due to the size (up to 139ft I.D.) and depth of these shafts (up to 196ft to the bottom of the base slab), the design and construction was extremely challenging. It was necessary to impose unusually tight tolerances for verticality of the slurry wall panels forming the support of excavation system for the shafts, in order to ensure that there would

Figure 1. Figure-of-eight layout of the dewatering and screening shafts for the Blue Plains Tunnel project in Washington D.C., USA

be sufficient contact area between adjacent panels. Without the ability of the designer, contractor and subcontractors to closely coordinate design, construction sequencing and monitoring requirements in real-time, it is unlikely that this approach would have been viable, or acceptable to the owner. See Figure 1 for a photo of the Blue Plains Tunnel dewatering and screening shafts.

As can be seen from this example, the ability to optimize the approaches to design and construction in this way is clearly an advantage of design-build procurement. However, the ability to capitalize on this potential advantage will only manifest if the designer and contractor can work as a coordinated team and in a transparent manner. Such coordination does not necessarily come automatically given that designers and contractors often have greater experience with more traditional design-bid-build procurement methods where the relationship between designer and contractor can sometimes be adversarial. It is also a fact that design and construction entities within a design-build team will have different drivers and needs influencing their behavior. For these reasons, the ability to coordinate effectively usually needs to be deliberately built-into the management and delivery approaches adopted by both parties throughout the lifetime of the project. This starts with the need to understand each other's goals and needs when working as part of a design-build team, recognizing that each has an important part to play in achieving the successful delivery of the project.

The following sections of this Paper examine some of the factors and mechanisms that can be implemented by a design-build team to maximize the potential benefits offered by a design-build procurement. The Paper is separated into the following sections:

- Building the team
- Coordination during the proposal phase
- Approach to the delivery phase

BUILDING THE TEAM

One of the key stepping-stones to a successful design-build project is the compatibility of the design and construction firms making up the design-build entity, which will affect their ability to work together in a collaborative and transparent manner. It is important that both parties understand the fundamental basis of design-build, which is that the approaches to design and construction are inter-dependent and therefore cannot be developed in isolation. Almost all (efficiently designed) underground structures will have some component of the loading on a structure that is dependent on the method of installation. The speed and stiffness of the method will affect the loading on a structure, generally with faster stiffer methods resulting in higher loads (but less movement). The Design-Build process is therefore iterative; in that each design decision or assumption can influence the construction method appropriate for that element of work. As the construction planning process is progressed and construction methods and sequences are confirmed, the design must be reviewed to verify that the selected construction methods are in agreement with the intent of, and approach to, the design.

As a result of this, the selection of design-build entity partners can have a significant influence over the outcome of the project, even though it normally occurs before the project is bid. Some of the factors that should be considered when evaluating teaming partners for a design-build project include whether the parties have a similar approach to innovation and risk, and whether both parties fully understand the differences between design-build and traditional design-bid-build procurement. The past working history of the designer and contractor can also be a factor, as it can be easier for both firms to collaborate effectively if they are familiar with each other and have personal relationships in place at multiple levels of their respective organizations. This can also facilitate resolution of issues throughout the project. For this reason, it can be advantageous for both designer and contractor to take a long-term view of their relationship.

It should also be recognized that not every person working for a designer or contractor has the ability to work well within a design-build team. Not everyone has the ability to collaborate at the level required for successful design-build delivery, or the requisite recognition that both the design and construction functions are important for the success of a design-build team. For this reason, the selection of key staff with the right personalities and approach to design-build is critical for both the design and construction teams. Where possible, it is advisable to have senior staff with significant past design-build experience who can act as an example to staff with less design-build experience. However, it may not always be possible to identify that a particular individual is not compatible with design-build delivery before the project starts. It can be very disruptive to the effective operation of a design-build team to have such an individual in their midst, and in such cases the senior managers on both sides should be prepared to remove and replace such individuals if it becomes apparent that they cannot adjust to such a collaborative environment.

There are also a number of roles within a design-build team that have more influence than others over the effectiveness of collaboration between the design and construction teams. These positions are particularly important and must be staffed with individuals with the appropriate attitude and understanding of design-build. Perhaps the most significant of these key positions is the Design Coordinator. The Design Coordinator's primary role is to facilitate the flow of information between design, construction and quality control teams, ensuring that the design and construction approaches are compatible, and that design and construction schedules are aligned. The Design Coordinator is the hub of the coordination process between the design team, construction team and also the construction quality management team, as indicated in Figure 2.

In addition to having the appropriate approach to, and understanding of, design-build delivery it is also beneficial if the design coordinator has an understanding of both

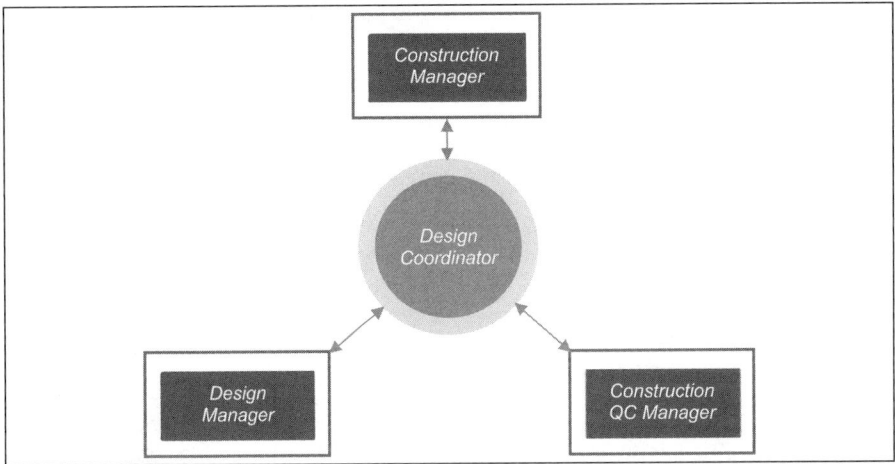

Figure 2. Key positions and relationships for coordination within a design-build team

design and construction processes so that they can better anticipate the needs of both teams and work pro-actively to facilitate smooth progress. The design coordinator typically comes from the contractor's organization, and needs to be given sufficient authority and independence to properly perform the role, which may require being able to deal with more senior members of their own organization on a peer-to-peer basis in the event of conflict between the needs of the design and construction teams. It is therefore recommended that the design coordinator should have direct access to the contractor's project manager, or even the project sponsor.

It is also recommended that the contractual terms and conditions between designer and contractor should be established prior to starting the proposal phase. These terms and conditions can influence behaviors of both parties, making one or other party more or less risk averse depending on how certain critical clauses (such as schedule or quantity risk) are worded. This, in-turn, can have a significant impact on the degree of conservatism that is built-in to the design or construction approach, and therefore influence the bid cost. A reasonable approach to risk sharing should be sought within the contract, with terms and conditions that encourage and reward collaboration, as far as possible.

COORDINATION DURING THE PROPOSAL PHASE

The construction team may or may not have prior experience of design-build procurement, and experienced traditional contractors are often concerned at the perceived lack of pace associated with the design process. Drawings and specifications are what they need to price the project, and incomplete drawings with errors can be an opportunity for a claim under a more traditional form of procurement. However, with design-build, changes to the design occurring between the bid phase and the delivery phase present a significant risk of abortive unrecoverable work and associated cost, most likely for the designer if caught in time, but potentially of even greater consequence for the contractor if caught later.

At the bid-stage, the design-build contractor will concentrate his efforts on understanding the quantum of the work and the schedule in order to properly price the project. But he will also pride himself on understanding the risks and making adequate contingency provision in the price and schedule. This is one of the key differences between design-bid-build and design-build that needs to be understood. It is important for the construction team to develop an appreciation of the designer's level of confidence and comfort with the adequacy of the bid design.

One method of achieving this appreciation is through the preparation of a bid-stage commercial risk register by the designer, as the bid-design progresses. This risk register can be used to record information on areas of design uncertainty that could have an impact on construction quantities or methodologies once the designer has a chance to progress the design further during final design. These design uncertainties can arise for a number of reasons; including the time and/or budget constraints that usually exist during the bid-phase, or sufficient information not being available during the bid-phase (such as condition surveys of utilities and structures potentially impacted by construction) which requires the designer to build assumptions into his design that may prove not to be entirely accurate at a later date.

Another potential problem area that could result in unanticipated cost increases for the designer and/or the contractor include the definition of the designer's scope of work. It is unusual for the lead design firm to have every single aspect of design encompassed by the project within their scope of work. There are usually areas of design scope that need to be discussed between designer and contractor before it can be agreed who will be responsible for them. An example of this could be the design of contractors' temporary works; some contractors will do temporary works design in-house, or have preferred sub consultants that they like to go to for that sort of work. It is therefore beneficial for both designer and contractor to find the time during the bid-phase to agree precisely what the lead design firms' scope of work will be, and ensure that it is accurately reflected in the designers fee proposal for final design services. This could even go to the level of agreeing which firm has primary responsibility for developing each specification. If this is not done with sufficient thoroughness, it is likely that there will be arguments over scope during the delivery phase, with potentially significant cost consequences for one party or the other.

It is obviously beneficial for a design-build team to work together to minimize changes between the bid-phase and delivery phase. One method of doing this is to ensure that both design and construction firms utilize staff in their bid teams who will also be part of the delivery teams. This not only provides continuity from one project phase into the next, with associated benefits such as retaining an understanding of the risks inherent in the bid-phase design or construction approaches, but also encourages those personnel to take ownership of their work during the bid-phase.

APPROACH TO THE DELIVERY PHASE

Alignment of Goals

One of the primary methods of facilitating successful delivery of a design-build project is to work to align the goals of the design and construction teams to the greatest extent possible. Fundamentally these should include safety, quality and profit.

Safety should not be considered entirely in the control of the construction team. It is the responsibility of a designer to consider the construction safety implications of the design. Indeed in some countries it is a requirement to demonstrate that the design is safe to build and to identify any unusual or significant hazards.

It is also important for both designer and contractor to understand the quality requirements of both teams so that quality control efforts can concentrate on the areas that count. During the design phase the control process should be geared to ensuring proper coordination of design and construction approaches during the initial design phases, and then more towards checking/ verifying at the later detailed design stages.

One of the key methods by which a design-build team can promote coordination and alignment of goals is by holding a chartering workshop immediately after Notice-to-Proceed is received from the project owner. The chartering workshop should include members of all parts of the design-build team, including design, construction, QA/QC, senior management and key sub consultant and subcontractors. The aim of the workshop should be to establish a common understanding throughout the team, including:

- Overall goals and scope of the project (both owner and design-build team)
- Organizational structure, roles and lines of communication between the parties
- Key risks and challenges that need to be mutually understood
- Management processes and procedures that will be implemented throughout the team, including the Design Quality Plan, Risk Management Plan, Construction Quality Management Plan, and document control process.
- Key contractual review and submittal processes

It is recommended that the project owner should also be invited to participate in parts of the chartering workshop. In addition to achieving a common understanding of the project, the workshop will also give team members a chance to build relationships, which will be valuable later in the project as challenges are inevitably encountered and everyone has to work together to overcome them.

It is also critical to make sure that approach to management of design, construction and quality assurance and control is integrated and aligned. This can only be achieved if the primary documents governing the management processes and approach for all parts of the design-build team are written as a coordinated set, with reference to each other and complementing each other, as indicated by Figure 3.

Once the design process has commenced, it is important to ensure that the right construction people review the early stage design submissions. This will help to guide the design far more effectively than waiting until all the calculations are complete and checked; which is a sure way to upset the designer, who will be well prepared with deeply technical and complex reasons why the contractor cannot change that particular aspect of the design so late in the design process!

Providing inter-discipline and construction comments is a vital part of the design-build process and should be properly accounted for in the design schedule, which should also be thoroughly integrated with the construction schedule. In fact, it is essential that

Figure 3. integration of the primary documents detailing the management approach for all parts of the design-build team will promote a coordinated approach to delivery

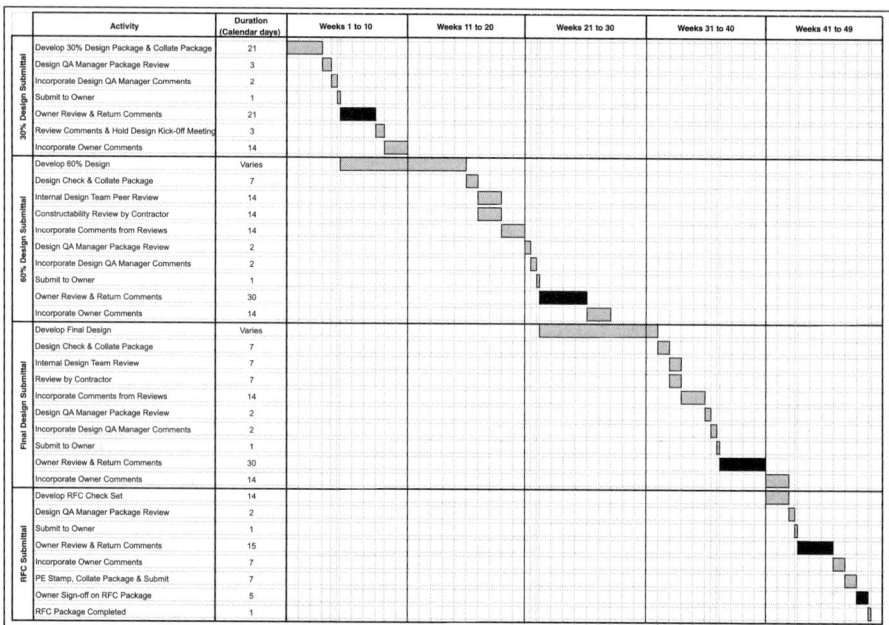

Figure 4. Design schedule for a typical element of design, incorporating detailed activities and durations for quality assurance

all stages of the quality assurance process are included within the schedule, with adequate time allocated to each step of the process. Figure 4 illustrates a typical schedule for a single generic design element, incorporating all phases of the QA process. It is common for an integrated design-build schedule to have some elements of the design on the critical path for construction. This can result in a highly pressurized design environment, particularly as the design is approaching released for construction status. In such an environment, it can be tempting to "squeeze" the durations allocated for quality assurance activities, but this invariably ends up being a mistake as it increases the likelihood of rework being required at the last minute, or the chances of the contractor needing to request additional information or design changes immediately before construction is about to commence. By incorporating every step of the quality process into the schedule, it makes it easier to resist such misguided temptation. A strong-arm design QA manager is also extremely important in this situation, someone without a vested interest in meeting deadlines and prepared to deal with the pressures from all sides of the team.

The uncertainties associated with underground design and construction also need to be properly managed in order to get the best overall solution for all parties. A designer will always have his eye on the permanent design scenario, which in many cases will be far beyond the duration of the design-build contract. For a designer to take on design responsibility for an underground structure, he needs to have confidence that his assumptions are reasonable. Since so much of underground design is dependent on the construction method, it is necessary that the designer is familiar with, and understands, the proposed method. That familiarity and understanding only truly comes from first-hand experience which means he needs to get his boots dirty and go and see construction in action. This does not happen enough, and when it does happen it is usually associated with things that have gone wrong. Extrapolation of this shortcoming to future generations does not bode well for designers' ability to produce practical and constructible designs!

During the construction phase the responsibility for the quality of the work is passed to a construction QC manager. In the absence of the design engineer on site the QC manager is the main custodian of the design intent. He should be charged with ensuring that the construction team is building the work in accordance with the designer's intent which means the drawings and specifications need to be followed so that all the assumptions made in the process of design remain applicable. The QC manager should be independent of the construction team, reporting directly to the design-build team managers and Project Sponsor. Maintaining the quality of the work should be seen as a cost saving exercise, with savings achieved through a right-first-time approach. To ensure that the construction QC manager and the inspection team has an adequate understanding of the critical elements and assumptions inherent within the design, it is beneficial for the design manager and construction QC manager to meet prior to issue of the released for construction drawings. Important aspects of the design that need to be encapsulated within the inspection and test plans can be discussed and understood by all parties.

Alignment of financial goals is a more difficult proposition. The design-build entity will be focused on the bottom line, which is often heavily skewed toward saving construction cost. However it could be argued that for a design-build project, the design process offers significant potential to achieve such savings, where refinement of design assumptions and confidence from additional site investigation can result in lower principle quantities. It should therefore be recognized that in some situations, more money spent on the design equals more cost saving for construction!

However, for underground projects equally big savings can derive from a well thought-out method of construction, which allows for more precise design assumptions closer to the start of the design process. Getting the appropriate decision makers from the construction team involved in the design details is therefore very important. One thing is certain, and that is that the ideal amount of design cost will not be known at the start of the project so the method of payment should recognize this and place the risks in the hands of the party best qualified to manage them.

Managing Risk During Delivery

Developing a coordinated approach to the management of risk is another good way to improve overall coordination between designers and contractors. A standardized approach to risk can be used to address all type of risks including those occurring during the design, construction and operational phases. Sharing knowledge of the risks promotes coordination and an appreciation of the different perspectives within the design-build team. A robust risk management process incorporating the identification of risk owners and mitigation measures can also help to identify where design can be used to manage construction risk and vice versa.

Risk management also has the potential to remove barriers to the development of a design solution within a design-build environment. For example, when a designer is faced with a range of parameters that have a major impact on the structure he can decide to adopt the more optimistic value by getting joint party agreement that there is a risk and that additional measures may be required during construction if the actual conditions turn out to be different.

Table 1. Summary of coordination mechanisms that should be employed by design-build teams

Project chartering session	Shortly after Notice-to-Proceed, the design-build team should hold a multi-day chartering session aimed at baselining expectations, project processes, and building relationships.
Co-location of key personnel of the design-build team	Key members of the design and construction teams should be co-located during design to facilitate communication. It is even more beneficial if some of the Owner's staff are also co-located.
Tailor design packages to suit the construction schedule	The design must be progressed in a way that supports the overall construction schedule. The design manager and design coordinator should work together to break the design into packages that can be progressed with relative independence to release for construction (RFC) level, enabling the design of the packages to be scheduled to align with the demands of the construction schedule. The deliverables of each design package include items necessary to build that job element (i.e., drawings, specifications, reports, design calculations, information on residual risks, etc.).
Kick-off meetings at the start of each design package	A kick-off meeting should be arranged within the first couple of weeks of starting a new design package, in order to ensure that the design team, construction team, owner's engineers and construction quality personnel all have a common understanding of the way in which a design package will be progressed, and the construction methods and key assumptions/criteria that will be used.
Constructability review by contractor's personnel before the end of the 60% design phase	The construction team should perform a detailed constructability review of each 60 percent design submittal prior to the design team commencing the final phase of design. The review will enable the contractor to confirm that the design is progressing in-line with their expectations for means and methods.
Coordination meeting between design and construction quality teams at 100 percent design submittal	The purpose of this meeting would be to enable the design team to high-light critical aspects of the design that require specific inspection or testing techniques, as well as to highlight design assumptions that will need validation by the construction quality control team.
Transfer of residual risk information from design to construction	The design team should utilize design hazard logs prepared for each design package that incorporate a requirement to record information on residual risks. These hazard logs should form part of the documenta-tion issued to the contractor with an RFC package.

The transfer of residual risk information from the design phase to the construction phase is critical for any type of underground project delivered via any type of procure-ment methodology. However, design-build delivery should make this easier to achieve if the design and construction teams are working together well. One tool that can be used to encourage this is the preparation of design hazard logs by the design team. Design hazard logs should be prepared for each design package; recording the design, construction and operational phase risks identified by the design team, the mitigations used to reduce the risk and, most importantly, information on any residual risk that remains and must be managed by the contractor during the construction phase. The design hazard logs should be included within each design submittal package.

Summary of Key Coordination Mechanisms

The above discussion has covered a broad range of coordination mechanisms and tools that can be used to enhance the chances of successful delivery of a design-build project. Table 1 provides a concise summary of some of these primary methods of coordination.

CONCLUSION

It is the author's belief that transparent communication and coordination between the design and construction teams is a key component in the success of a design-build project. Historic barriers to this relationship need to be removed and this will release the potential for ever greater benefits associated with the joining of the designer and contractor in a single design-build team.

The following list of recommendations embodies this philosophy and should be given due consideration for the mode of operation when delivering design-build projects.

- Appoint and identify key decision makers and make sure they attend meetings.
- Embrace co-location, and employ electronic document management systems to enhance the potential benefits from this.
- Establish rules of communication within a design-build team, and promote a mutual respect for the role of both designer and contractor.
- Develop, and then respect, a robust QA/QC process. Appoint a strong design QA manager who will not bow to pressure.
- Appoint a strong design coordinator who understands both design and construction processes.
- Find a way to get young designers on site to improve their knowledge of construction, and thereby improve the constructability of their designs.
- Use the risk register and risk management process to support the adoption of less conservative and more realistic design assumptions, potentially saving construction cost.

Construction of the Longest Road Tunnel in Mexico

Hector Canseco Aragon ▪ MHFCC Construcciones
Miguel Angel Banuet Rodriguez ▪ Ingenieros Civiles Asociados

ABSTRACT

Construction of the longest road tunnel in Mexico, as part of the project "Acapulco's Alternate Roadway to the Scenic Roadway," which is already included in the records of tunnel projects worldwide. Brief summary regarding details of the construction since the beginning of the work, done by skilled Engineers and Mexican workers. Outstanding execution of a work of such magnitude within an urban and tourist area. This project will certainly have a strong impact on the steady development of the great port of Acapulco.

PROJECT LOCATION

"Acapulco's Alternate Roadway Project" is currently under construction in the city of Acapulco, located in the Guerrero state of Mexico. It is an 8-km long roadway which starts at the Icacos neighborhood, heading towards Acapulco's Airport, crossing a drill-and-blast tunnel that is 3,160 meters long under Veladero Park, a mountainous area. The tunnel overburden reaches a depth of 380 m approximately at the central point of the horizontal alignment, and continues with a 4-km-long elevated road to connect to the existing "Viaducto Diamante Toll Road" (see Figure 1).

Tunnel construction has already been finished using two portals for drilling and blasting operations; Brisamar Portal on the west side and Cayaco Portal on the east side (see Figure 2).

GEOMETRIC DATA

Only two main excavation cross-sections were considered in the Executive Design. The first one corresponds to the running tunnel for a 3-lane vehicle section, with

Figure 1. Plant location of Acapulco's Alternate Roadway Project in Acapulco City, México

594

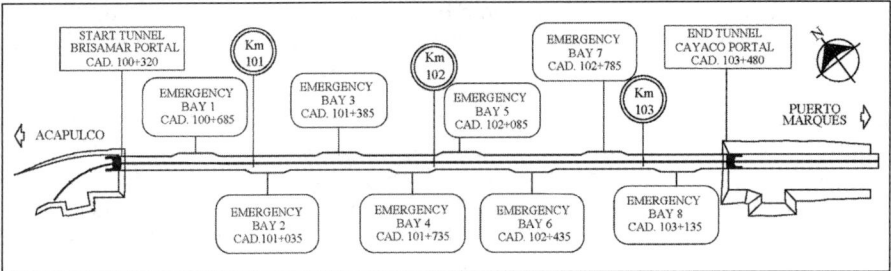

Figure 2. Schematic plan location showing Brisamar and Cayaco portals, and emergency bays along tunnel alignment

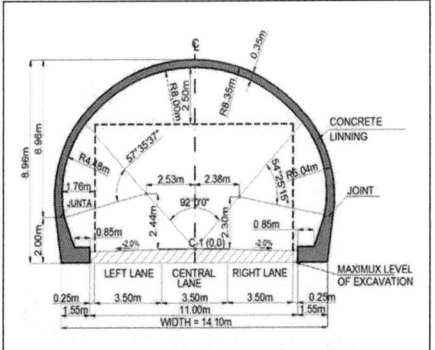

Figure 3. Cross section of the running tunnel   Figure 4. Cross section at bay tunnels

cross-section areas between 120 m² to 130 m², taking into consideration different geological and geotechnical conditions to be encountered (see Figure 3).

The second one corresponds to 8 emergency bays, each one 50 m long, laid down within the tunnel outline, located at 400 m from each one with excavation cross-sections between 156 m² to 169 m, in accordance to geological and geotechnical conditions to be encountered (see Figure 4).

The tunnel's horizontal alignment is completely straight, having a small curve within the vertical alignment on Cayaco portal, because there were some differences in level when tunnel excavation started.

Tunnel slopes were designed in such a way that rainwater does not run down into the tunnel at any time.

CONSTRUCTION PROCESS OF THE RUNNING TUNNEL AND BAYS, PRIMARY LINING OF TUNNEL, AND EQUIPMENT USED

Drill and blast construction was planned to be used for the entire tunnel excavation, adjusting the excavation sequence as a function of the geological-geotechnical conditions, resulting from the exploration carried out during the design stage; only 7 direct borings for sample recovery and 6 transient electromagnetic borings at distances between 370 to 530 m were executed. Taking into consideration that most of the direct exploratory borings were near the portals, 70% of the tunnel distance was not explored properly, and rock formation parameters were inferred in large measure. Because of these reasons, the executive design presented several constructive sequences to cover different mechanical conditions of the rock formations, which only were indicative of measures to be applied by contractor.

Because of the uncertainty due to the lack of information about geological and geotechnical conditions of site's rock formations, tunnel excavation was planned to be done in 3 stages; top-heading section first, keeping a 4- m bench for the lower section, which would be alternatively excavated after having advanced 500 m in the upper middle section. Drilling patterns and explosives were consequently adjusted accordingly to found rock-formation conditions.

Tunnel excavation at bays was done in two stages for the top-heading section, due to the larger dimensions. Bench excavation was also done in two stages, allowing access to the face at all times.

Primary support was in direct relationship to rock quality. Combination of steel fiber shotcrete in thickness from 5 to 20 centimeters, IPR profile steel arches for poor to very poor rock conditions, in accordance to criteria established by Bieniawski for RMR (Rock Mass Rating) determination. For regular rock conditions, it was recommended a 5-cm-thick layer of steel-fiber reinforced shotcrete, plus 6 m long friction rock bolts at the top covering a 160-degree area.

Micropiles Umbrellas and Rock Bolts

For very bad to bad rock conditions where underground water was leaking or flowing throughout joints and cracks, a systematic array of micropile umbrellas were installed first to prevent cave-in formations after each blasting, making easier the process of advancing safely under the micropile umbrella protection. 37 micropiles were originally considered for installation along the running tunnel arch, distributed at a 40-cm distance from each other. Micropiles were made of special steel pipe (4" inner diameter) and were installed simultaneously as boring drilling was being done (see Figure 5).

Drilling length for each micropile umbrellas was 12 meters, allowing 3 m length overlap between each umbrella, giving 9 m of reinforcing length to each one. The number of micropiles installed was directly related to cross section stability for rock conditions and water inflow, and thus this number was decreased or increased to cover tunnel arches and walls of the top heading as required. Steel arches and lagging installation were mandatory under these circumstances, as the micropile umbrellas needed a complimentary support as tunnel excavation progressed.

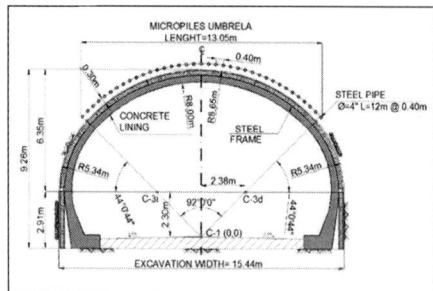

Figure 5. Micropile umbrella array for running tunnel section

Machinery Used for Excavation and Primary Support

Excavation of the Running tunnel has been done efficiently with great-performance machines, which reduced work cycle times and allowed for excavation production rates for the top heading section over 9 m long per 24 hours work day, when geological conditions were fair. This equipment is as follows:

- 3 boom electro–hydraulic jumbo for drilling
- Electro-hydraulic self-propelled mobile concrete sprayer
- Telehandlers 5 metric ton capacity and reach till 8 m height.
- Wheel loaders, 16.73 tons, 170 HP, 3.10 m^3 bucket
- Hydraulic excavator, 30.5 tons with hydraulic hammer
- Backhoe loaders of 6.79 tons, 74 HP, 0.73 m^3 bucket
- Extraction of excavated rock was carried out with many dump trucks, 14 to 16 m^3 capacity, owned by members of the local Union.

GEOLOGICAL AND GEOTECHNICAL CONDITIONS ENCOUNTERED DURING CONSTRUCTION

As mentioned above, the previous geological and geotechnical works consisting of direct and indirect exploration carried out for the executive project, were not enough for a proper evaluation of geological-geotechnical conditions on rock formations where the tunnel alignment was located.

Good practice in engineering recommends having sample recovery borings along the tunnel alignment at distances between 150 m to 300 m for a tunnel in rock with fair conditions, which means that in this case it was at least necessary to have 16 borings for direct exploration, complemented with geophysical borings for exploration to 300 m depth. Instead only 7 direct sample recovery borings that were made.

We are presenting the longitudinal tunnel profiles that show the differences between the real rock conditions encountered in comparison with the rock conditions predicted by the executive project (see Figures 6a and 6b). Figure 6b is the result of a very close follow up to geological-geotechnical reports which were taken every day as tunnel

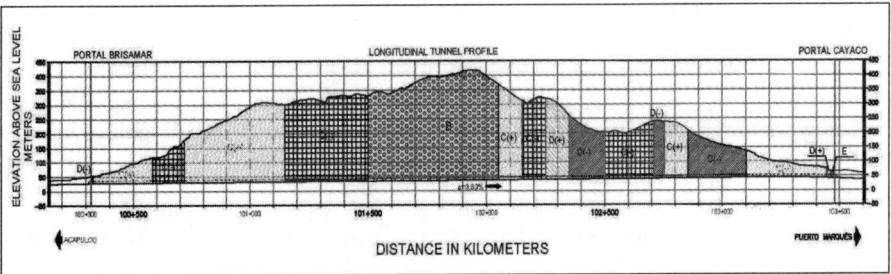

Figure 6a. Geotechnical conditions in accordance to the executive project "Acapulco's Alternate Roadway Project"

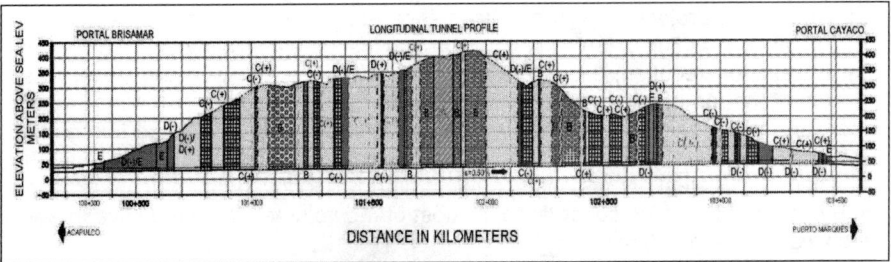

Figure 6b. Actual geotechnical conditions encountered

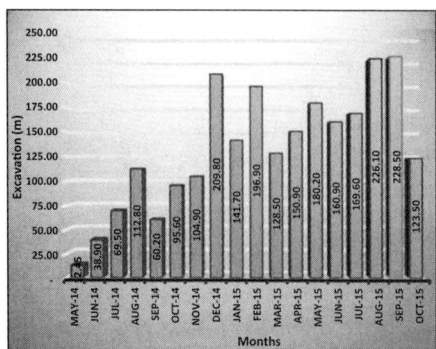

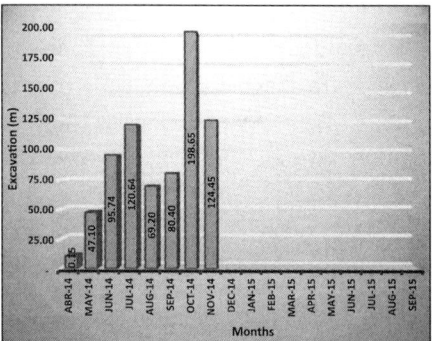

Figure 7a. Monthly tunnel excavation progress at the top heading section by Brisamar Portal

Figure 7b. Monthly tunnel excavation progress at the top heading section by Cayaco Portal

excavation progressed by both portals. The most remarkable difference is a very bad to bad rock condition encountered in the first 465 meters of the tunnel starting from Brisamar Portal, when the executive project indicated a regular-quality rock formation.

In this stretch some cave in happened when there was an omission to install micro-piles umbrella at places where rock fractures presented an unfavorable angle, allowing rock wedges to destabilize the tunnel section.

For the remaining length of the stretch, the real conditions of the rock formations in comparison to the ones indicated by the executive project, were also quite different, as it can be seen in the comparison chart.

It is suitable to mention that, since the beginning of the work, several approaches were made to complement the geological-geotechnical exploration that was missing. However, this attempt was not successful.

PROGRESS ACCOMPLISHED DURING TUNNEL CONSTRUCTION

Figures 7a and 7b show the monthly progress for the tunnel excavation at the top heading section by both portals, underlying a good production achievement for several months, to get 230 m maximum for a single heading, once most of the problems that arose during the start of the work were solved.

Description of details found during tunnel excavation, for each portal are the following.

Tunnel Excavation by Cayaco Portal

Tunnel excavation by Cayaco portal started on April 17, 2014, through a medium hardness metamorphic rock. 2 micropile umbrellas, 12 m long were installed as indicated by the executive project, with steel arches separated 1.0 m from each other. On June 23, 2014, after 114.45 m of tunnel excavation, the first cave-in event occurred suddenly after a blasting execution, due to the presence of unfavorably wedging formations in the arch of the top heading section. The blasted section length was 3.4 m. Seepage water ran throughout fractures of the wedging rock. The time taken to surpass this event was 7 days, after placing steel fiber shotcrete to stabilize sliding wedges, the installation of 13 steel arches and scaffolding formwork covering the entire area, and the pumping of hydraulic concrete to fill most of the volume left by the cave in event, which was close to 450 m^3.

After 120 meters of tunnel excavation, there was another cave in of minor proportions, due to similar rock wedging problems and water leakage. Steel fiber shotcrete and 7 steel arches were installed and formwork scaffolding and concrete were pumped for support of the area affected. As tunnel excavation progressed and regular to bad rock conditions were encountered, it was necessary to keep installing steel arches and steel fiber shotcrete, facing one more event of minor problems by fractured rock and water leakage. However, it is very important to say that production daily rates increased significantly as tunnel-worker skills and coordination improved.

Unfortunately, on November 19, 2014 tunnel works were suspended at Cayaco Portal, because of social problems between the state Government and former land owners, after a tunnel length of 746.3 m had been reached. Under these circumstances there was only one way to make the tunnel connection, which was through Brisamar portal.

Tunnel Excavation by Brisamar Portal

Tunnel excavation by Brisamar Portal began on May 17, 2014 through very weathered granite formations and water leakage. These rock formations were classified as very bad to bad quality, in accordance to the geological-geotechnical conditions found. The construction method for the first 465 m of tunnel required systematic installation of micropile umbrellas, steel arches separated between 1 to 1.5 m and a shotcrete layer 20 cm thick for primary support. Drill and blast operations were only partially used for excavation of the cross-section as mechanical excavation employing excavator and hydraulic hammer was necessary for tunnel stability reasons at the top heading.

Production rates achieved for tunnel excavation after the first 465 meters improved as the rock conditions upgraded to fair condition, although it several areas with bad rock were found, where some small cave-ins occurred in association with the presence of water leakage. The environmental conditions inside the tunnel began to be a problem, as ventilation calculations for the additional length of tunnel by Brisamar portal had not been adjusted. The problem was solved by replacing the whole ventilation system with more powerful and better-suited fans, and ducting to the actual conditions of the work.

It is important to mention that some important water inflows were found at this heading and the overall flow was close to 30 liters/second, Fortunately, rock conditions had improved and there was not a potential stability problem, as drilling and hoses were used to conduct all water inflow.

Tunnel excavation by Brisamar portal was successfully finished on October 27, 2015, making the breakthrough with the tunnel stretch that had been excavated by Cayaco Portal (see Figure 8).

Figure 8. Tunnel breakthrough on October 27, 2015 by Brisamar Portal. It concludes tunnel excavation at the top heading section.

SPECIAL CONSIDERATIONS FOR WORK AT THE BRISAMAR PORTAL

Taking into consideration that the Brisamar Portal is located at one side of "Joyas de Brisamar," a private high class residential development within an urban area, it was of high priority and importance to minimize the construction impact over buildings and especially the life quality of residents.

It was mandatory to take special measures to reduce all noise of drilling and blasting, as well as to buffer the effects of vibrations caused by blasting, a task that was not easy because residents reacted quickly to oppose the execution of the works, to the extent that legal judge order suspended the works. The relationship with representatives of the residents of "Joyas de Brisamar" was always full of friction and complaints. After several meetings, the following measures were taken:

1. Blasting was not allowed at night shifts.

2. Use only electronic detonators to reduce noise blasting. Do not use detonating cord.

3. Restrict the use of low explosive for tunnel excavation for the first 500 meters (below the zone of influence of the houses and buildings). Use only high explosive.

4. Keep neighbors constantly informed about blasting times through written notes delivered to the "Joyas de Brisamar" administration and use of a siren system to alert prior to the execution of any blasting.

Thanks to the measures taken, the buildings located within the radius of influence of blasting suffered only minor cracks and / or damage. Vibration and noise were satisfactorily controlled within allowable limits.

CONSTRUCTION PROCEDURE FOR TUNNEL LINING

Geometric sections of the final lining for the running tunnel section and Bay section are shown in Figures 1 and 2, appreciating the thicknesses of hydraulic concrete and reinforcing steel, as indicated by the executive project. These thicknesses correspond to theoretical sections of excavation, because actual thicknesses are based on geological rock conditions and measures taken to avoid over excavation and cave-ins during the stage of excavation.

Pouring the final lining of hydraulic concrete was originally planned using 2 monolithic and Collapsible steel forms, 15 m long each piece, supplemented with 2 steel form sections for the bays (see Figure 9), designed exclusively for the ceiling and walls; one for each portal, as well as several modular sections of metal formwork for curve, walkway, and a starting short wall section where the tunnel formwork overlaps.

Figure 9. Final lining construction activities at a bay area. The steel form has been assembled.

Having been suspended activities by Cayaco portal, the program related to the tunnel lining was fitted to the actual conditions of the work schedule, so both monolithic steel forms were armed and introduced by Brisamar Portal.

Activities for pouring Concrete in curves and walkways began on March 9, 2015 by the Brisamar portal, and activities for pouring concrete at the tunnel upper section and walls began until June 19, 2015. Due to logistical issues related to the activities for reinforcing steel bars installation, a geotextile liner combined with a water proofing geomembrane was fixed to the shotcrete primary lining.

Here it is important to emphasize the simplicity in the design of the steel formwork for concrete pouring in the bay areas, manufactured by a Mexican company recognized for its technology and ingenious design, making an easy fitting of both steel forms (running tunnel and bay).

CONCLUSIONS

Tunnel construction within an urban area has faced several special situations, due to the systematic use of explosives, however, the excavation was successfully completed and there is no doubt that it will bring great benefits to the people of Acapulco. National and international visitors will enjoy the benefits too.

The use of the electronic initiators for blasting operations was a successful measure for a substantial reduction of noise and vibrations.

The geological monitoring carried out in each blast was very useful for determining corrective actions required to be applied at subsequent blasts.

The work highlights the importance of adequate planning of prior geological-geotechnical studies, as well as of its magnitude and scope, in such a way that the executive project gets all the elements for the proper design of the primary tunnel support, and the best tools possible to avoid uncertainties during tunneling construction.

Micropile umbrellas have shown to be a good tool of support for safer and faster tunnel excavation when facing bad to very bad rock conditions.

All facts mentioned above have led to an excellent step further in training young Mexican engineers and skilled workers, to continue building the great tunneling projects that Mexico needs, with Mexican contractors specialized in tunneling and underground works.

The average advance rates in linear meters of excavated tunnel, made in combination with good coordination, fair to good rock conditions, high-production equipment and specialized machinery are very significant, close to 230 meters per month for a single tunnel heading,

Excellent ventilation and preservation of good environmental conditions inside the tunnel must be the starting point for the selection of the equipment that will be used in these tasks, thus ensuring the safety, health and efficiency of the engineers, technical workers and technical staff that work in underground projects.

PHOTOGRAPHIC DETAILS OF ACTIVITIES DURING TUNNEL CONSTRUCTION

With the purpose of offering a complete experience of the most important aspects of this project, details are illustrated through photographs that give the magic touch when an activity is executed. Photographs show tunnel excavation process through the portals, tunnel excavation at the top heading section, benching, concrete placement in curves, walkways, and tunnel vault.

REFERENCES

"Executive Project for Acapulco's Alternate Roadway Project to the Scenic Roadway." 2011. General Direction of Highways of The Secretary of Communications and Transportation. Commission of highway Infrastructure and airports of Guerrero State.

Technical Manual for Design and Construction of Road Tunnels—Civil Elements. December 2009. US Department of Transportation. Federal Highway Administration.

Figure 10. Cayaco Portal at the beginning of the excavation of the middle upper section of the tunnel. The ramp to access the tunnel is on first plan.

Figure 11. Aerial view at the beginning of excavation works at Brisamar Portal. Streets and houses of the residential area are shown.

Figure 12. Brisamar portal; shotcrete, tie beams, anchors and drains on the slopes are shown

Figure 13. Electrohydraulic 3 arms Jumbo during drilling oporotion. An excavated bay section over the left side. The average upper bay section was excavated in 2 stages.

Figure 14. Drilling the top heading using an electrohydraulic 3 arms drilling jumbo. Steel arches installed as excavation support seen in the photograph.

Figure 15. Electrohydraulic 3 arms jumbo drilling and installing micropiles umbrella at the top heading

Figure 16. Tunnel excavation with rock condition from fair to good. Steel fiber shotcrete and rock bolts are shown as primary support.

Figure 17. Mechanical excavation of the tunnel bench after blasting. Excavator and hydraulic hammer were used for scaling and rock breaking.

Figure 18. Tunnel lining operations by Brisamar Portal. Steel form is ready for the first concrete pouring as steel rebar reinforcement is installed.

Figure 19. Picture appreciating the Progress of tunnel final lining and the steel form along tunnel alignment

Semmering Base Tunnel: 17 Miles of SEM and TBM Tunneling Under Challenging Conditions in Austria

Michael Proprenter ▪ iC consulenten ZT GesmbH
Oliver K. Wagner ▪ OEBB – Austrian Federal Railways

ABSTRACT

The Semmering Base Tunnel is a major infrastructure project in Austria, Europe. This consists of two 17 miles of single-track railway tunnels, numerous cross passages and a complex underground emergency station including caverns, passages, tunnels and two 1,315 ft. deep ventilation shafts. Intermediate access for construction is provided by a 1.1 mile long, temporary tunnel with two 820 ft. deep sub-surface shafts at its end and two additional temporary shafts from the surface with large underground caverns at the shaft bases. The maximum overburden is around 2,850 ft. in difficult, frequently changing geological conditions with large fault zones and extensive water inflow of up to 300 l/s requiring special grouting and support measures. The underground construction works using SEM as well as TBM methods commenced in 2014 and will be completed in 2026.

PROJECT OUTLINE

General

After finalization the 17 miles Semmering Base Tunnel will form part of the Austrian section of the Baltic-Adriatic Railway Corridor stretching across Europe and connecting the Baltic Sea (Gdansk in Poland) and the Adriatic Sea (Bologna in Italy). The tunnels are located approximately 50 miles south of the Austrian capital Vienna connecting the towns of Gloggnitz in Lower Austria and Muerzzuschlag in Styria (see Figure 1).

The new tunnel system will replace the existing world heritage site Semmering-Railway, which includes numerous tunnels and viaducts built in the mid-19th century. The alignment of the existing railway with steep gradients and small radii does not fulfill the requirements for the operation of efficient and modern rail traffic. Heavy cargo trains require the use of an additional locomotive for the route. The newly built tunnels will improve the travel quality and time for passenger and freight trains crossing the mountainous landscape. The existing Semmering-Railway line will remain in place for maintenance purposes of the new tunnel, for local traffic and as a tourist attraction due to its breathtaking scenery.

The total investment costs for the project are around 3.3 bi. Euros (3.5 bi. USD), whereas 1.5 bi. Euros (1.6 bi. USD) are related to the civil construction works. Pre-construction works started in 2011 and underground excavation in 2014. The final opening for operation is scheduled for 2026.

Tunnel System

The final tunnel system includes two single track tubes connected by 56 cross passages with a maximum spacing of 1,640 ft. and a central underground emergency station with two 1,315 ft. deep ventilation shafts. For construction purposes four

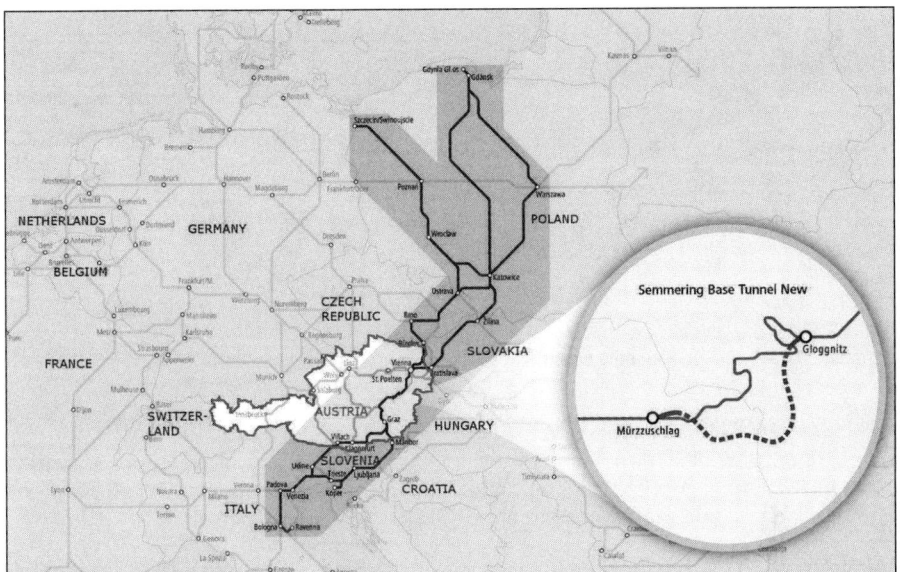

Source: ÖBB
Figure 1. General location overview

additional, temporary shafts are required. Two of the shafts will be excavated starting underground via a 1.1 miles long temporary access tunnel located around 660 ft.
above the main tunnel alignment.

The spacing of the mostly parallel running tunnel tubes is around 130 ft., which is
enlarged at the underground emergency station as well as in massive fault zones and
reduced in the portal areas. The alignment of the tunnel is based on a maximum train
velocity of 140 mph with a maximum inclination of 8.4%.

Due to the envisaged completion date and for logistic reasons, the main underground
works for the tunnels, shafts and caverns are divided into three separate construction
lots designated SBT 1.1, SBT 2.1 and SBT 3.1 (see Figure 2). An additional construction lot is foreseen at the southern portal including the cut and cover and open cut tunnel section with a total length of 0.4 miles as well as the modernization of the adjacent
railway station in Muerzzuschlag.

The excavation works for the underground structures are carried out by means of SEM
for a total length of 12 miles in the northern and southern parts of the tunnels. At the
peak of the excavation works a total of up to 12 SEM headings at various locations in
the main tunnel tubes will be running in parallel. Due to the geological conditions for
about 5 miles in the central section of the main tunnel tubes, TBM method is applied.

In the final tunnels a drained secondary lining will be installed. Due to legal reasons
in the northern section of the tunnels, mountain water has to be collected below the
drainage level and led of separately as to be available for the further use as drinking
water.

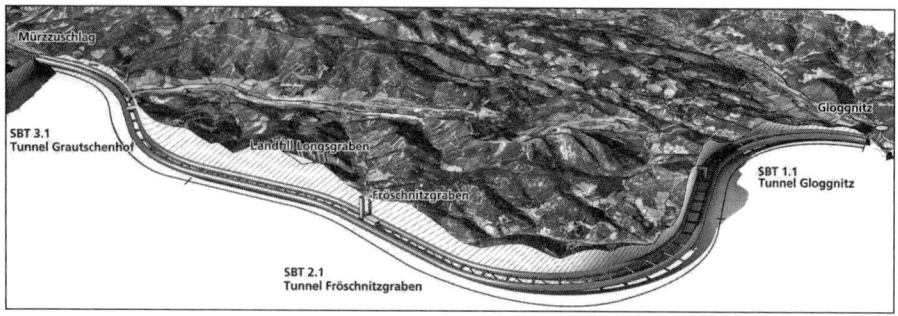

Source: OEBB
Figure 2. Tunnel scheme including construction lots

Geology, Hydrogeology, and Geotechnics

Overall the geological conditions of the tunnel system can be described as an area of intense tectonic imbrication. The main units are the so called Greywacke zone, the Semmering unit and the Wechsel unit.

Within frequently changing conditions rocks consisting of metasediments such as phyllite, schist, quartzite, locally sulphate rocks and metasandstone of various tectonic units as well as carbonate rocks and highly fractured metamorphic crystalline schist and gneiss will be encountered. The geological units are separated by distinct fault zones which formed during the orogenesis with extensive folding and nappe stacking, leading to an imbricate structure. Out of the numerous fault zones especially the ones consisting of cataclastic fault material with high overburden are challenging the construction procedure. Within the carbonate rock formations a considerable water inflow of up to 300 l/s is expected.

For the design and implementation, the tunnel was divided into 33 rock mass zones, of which more than 40 ground types were evaluated by the geotechnical design team. Rock mass zones can be characterized as tunnel sections, which show similar conditions in terms of geological structure or units, proportions of ground types and hydrogeological conditions. A ground type consists of a rock mass with similar in-situ properties, which refers to a volume of geotechnical relevance for the project.

MAIN STRUCTURES

Running Tunnels and Cross Passages

The main tunnels are single track tubes based on the clearance requirements of the Austrian Federal Railways for high capacity railway lines. The excavation area of the regular cross sections is around 810 ft² for both excavation methods. The inner contours of the tunnels include space for electrical and signaling installations and an emergency walkway. Certain sections of the tunnels are widened due to geotechnical or logistics reasons increasing the excavation area to a maximum of approximately 1950 ft².

The foreseen 56 cross passages between the main tunnels will act as emergency connections in case of an accident. In addition, the cross passages are equipped with technical rooms for electrical and communication purposes as well as maintenance installations. The excavation area of the cross passages is around 380 ft².

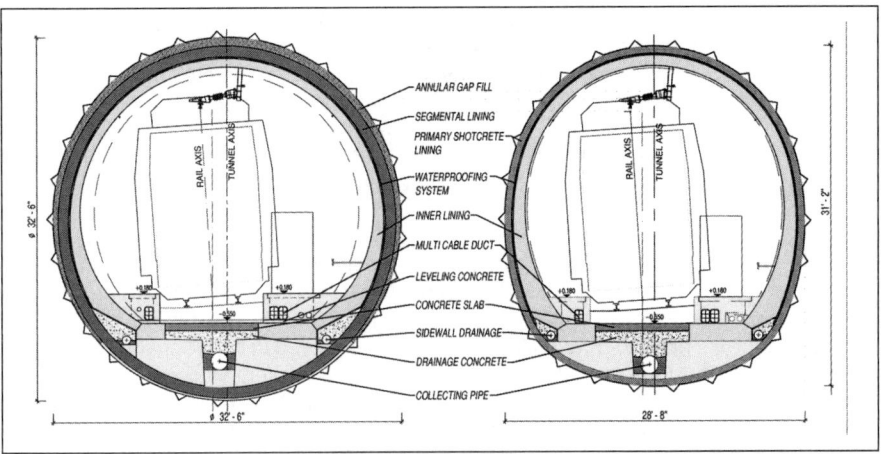

Source: PGST

Figure 3. Regular cross sections for TBM and SEM

The cross sections of the single track tubes, the cross passages and the caverns are designed with an outer and inner lining system (see Figure 3). Where the SEM method is applied, the outer lining system consists of reinforced shotcrete with additional support measures such as rockbolts and forpoling pipes. A mandatory 6+0 segmental lining system with a minimum thickness of 12 in. forms the primary lining system for the TBM method.

The inner lining for both methods is made of cast in-situ concrete, which is normally unreinforced. Exceptions with a reinforced inner lining are in areas of fault zone crossings or intersections with cross passages only. A fiber reinforced inner lining is applied in sections underneath residential areas or the main infrastructure where an improvement of the fire resistance properties is required.

The inner and the outer linings are separated by a waterproofing system consisting of a smoothening shotcrete layer, a waterproofing membrane and a geotextile. Groundwater will be drained off by lateral drainage pipes placed at abutment level in between the inner and outer linings.

In order to keep vibration and noise emissions below acceptable limits underneath residential areas the installation of a mass-spring track bed system is required for certain sections of the tunnel. For the main length of the tunnel a slab track system will be installed.

Underground Emergency Station and Ventilation Shafts

In accordance with Austrian and European laws, the overall rescue concept is based on self-rescue into a safe area, followed by the rescue by others (emergency team). In case of a fire on the train the main target is to reach the portals and leave the tunnel before stoppage. Due to the given length of the new tunnels and the expected available time before breakdown of the train, an underground emergency station is required within the tunnel.

The emergency station is located at the toe of the two permanent ventilation shafts in the central section of the tunnel and has a total length of about 0.6 miles. It comprises tunnels, caverns and passages of various sizes to accommodate all installations and areas required by the safety concept including a ventilation system.

In case of an accident trains shall stop in the emergency station and passengers shall be transferred via a safe and smoke free area to a rescue train approaching in the non-affected tunnel tube.

The central structure in the emergency station is the rescue tunnel, which is located in between the two running tunnels. Connections to the running tubes are provided by short escape passages with a maximum spacing of 165 ft., the central caverns and two regular cross passages. Ventilation tunnels are placed in between the rescue tunnel and the main tubes to provide a smoke free environment in connection with the ventilation shafts. Furthermore, four temporary logistic passages are located in the emergency station which are aligned for the use of conveyor belts for material transport during construction (see Figure 4).

The depth of the ventilation shafts is about 1,375 ft. with diameters of 40 ft. and 32 ft. The shafts will be lined with fiber reinforced concrete, which is separated from the primary shotcrete lining by a waterproofing membrane and geotextile. Reinforced concrete abutments in connection with deformation joints in the lining are foreseen at regular spacings of 200 ft. In the final configuration the shafts will be equipped with ventilation and inspection equipment. The top of the shafts will be crowned by a complex ventilation building housing the ventilators, electrical equipment and the control and telecommunication units.

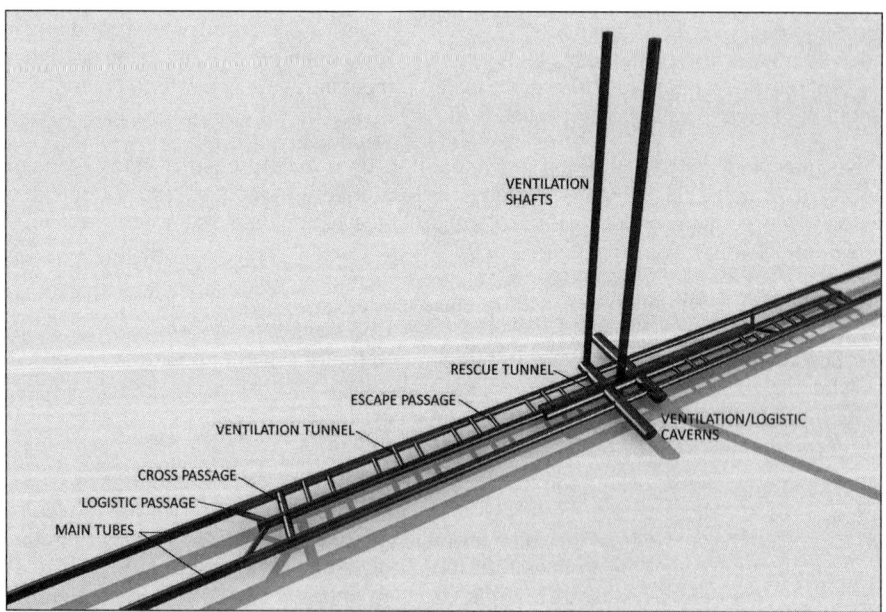

Source: OEBB

Figure 4. Scheme of underground emergency station

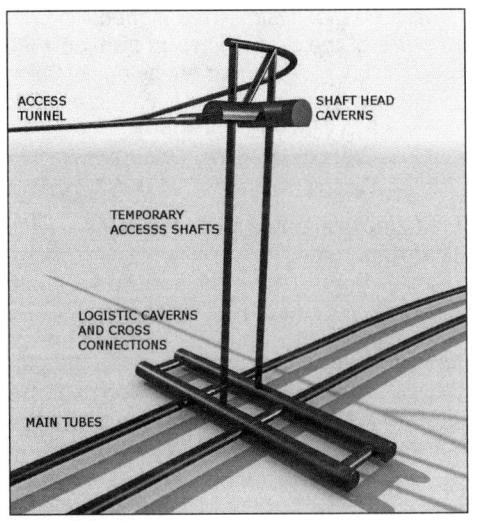
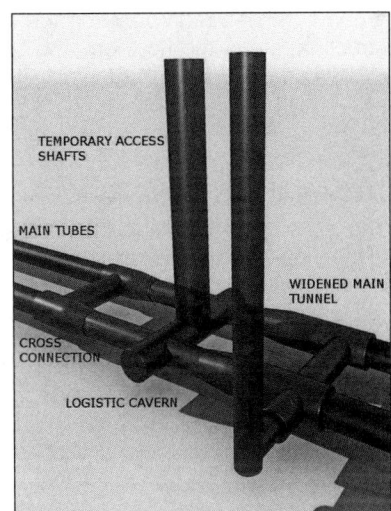

Source: OEBB
Figure 5. Scheme of cavern systems at temporary shafts

SELECTED DETAILS

Construction Logistics

One of the major challenges for the construction of the tunnel is to meet the tight time schedule for completion. Therefore, parallel excavation works are required in the three construction lots. Access to the tunnels is given at the north portal or via the permanent or temporary access shafts and tunnels.

Out of the total six shafts only two will remain as permanent structures for ventilation purposes (see Underground Emergency Station and Ventilation Shafts). The remaining four temporary shafts will be completely backfilled after the end of the construction period. Two shafts will be excavated from the surface and the others will be excavated underground via the access tunnel of 1.1 miles in length. The maximum depth of the shafts ranges between 330 ft. and 820 ft., with diameters between 23 ft. and 46 ft. At the top of the underground shafts a complex system of tunnels and caverns is foreseen to accommodate the installations for the hoisting operations during construction. In addition, at the shaft bases of the temporary shafts large underground caverns are located to provide sufficient site installation area for machinery assembly and material handling underground (see Figure 5). Apart from two caverns, which will house permanent cross passages, all other caverns will be backfilled after the end of the construction phase.

For logistic reasons the rescue tunnel in the emergency station is built as a large cavern over a certain length to accommodate site installations and supply including TBM assembly during construction. For the same reason two caverns are carried out perpendicular to the running tubes, which are also required for the construction process. The maximum dimensions of the large caverns are approximately 65 by 60 ft.

The deposit of the excavated material is foreseen in a newly planned landfill site approximately 1.4 miles from the site installation area of the central construction lot (see Figure 6). In order to provide a continuous discharge of the excavated material and for limitation of the intermediate deposit area in the central construction lot a conveyor belt runs between the shafts and the landfill site. The neighboring construction lots in the northern and southern parts of the tunnel have to transport the excavated material to this landfill site as well, which will be carried out by trucks using newly built construction roads as to avoid additional traffic in residential areas. The total volume of the landfill site is approximately 5.6 million yd^3.

Crossing of Major Fault Zones

There are numerous fault zones crossing the tunnel alignment along the entire stretch of the tunnel system. One of the most significant fault zones is the so called Grassberg-Schlagl fault system located approximately 3 miles from the northern portal. The length of the fault zone is about 0.5 miles with an average overburden of 1,640 ft.

The material can be characterized as tectonically heavily stressed, sheared and partially softened rock material consisting of sericite phyllites, quartzites, sulphate rocks and limestones. The fault material is sensitive to water, can be leached partially and contains large quantities of clay material prone to swelling phenomena. The occurrence of gases can be expected.

In order to provide an adequate design solution extensive investigation, testing and analyses procedures were carried out, which resulted in a specially designed cross section, heading sequence and support measures. A cross section with a rounded shotcrete invert was developed as to optimize the load transition between underground and lining.

The initial support consists of an initial shotcrete layer with a total thickness of 16 in., reinforced with two to three layers of wire mesh. Around 40 forpoling pipes consisting of self-drilling injection bolts with a length of 13 ft. will be installed as to provide sufficient overhead protection when opening the rounds. The expected radial deformations

Source: OEBB
Figure 6. Conveyor belts leading to landfill site

Source: PGST
Figure 7. SEM method—LSC elements

are in the range of 20 in. Therefore, up to six deformation gaps equipped with lining stress controllers (LSC elements) are foreseen (see Figure 7). The deformation slots are arranged equally distributed in the top heading. The LSC elements provide a certain support resistance when closing due to the deformations. In case of necessity an additional layer of 8 in. of reinforced shotcrete will be placed after the deformations are ceased.

Up to 20 self-drilling injection bolts with lengths between 20 and 30 ft. will be installed per round of heading. In order to increase the deformation capacity the bolts are installed in plastic hoses along a certain length at the bolt heads. The anchor plates are designed to be deformable accordingly.

The excavation cross section is divided into a top heading and bench/invert sequence, with a face opening in up to eight steps. The maximum round length is 3.3 ft. and the maximum invert ring closure distance is between 30 and 100 ft.

Grouting Measures

The hydrogeological conditions predict a water inflow in the range of 300 l/s for certain sections along the tunnel alignment in the carbonate rock formation. Furthermore, at the toe of the two temporary access shafts in the southern construction lot, crystalline gneiss with the potential to a complete loss of material strength and subsequent inflow of the water material mix will be encountered. Therefore, extensive grouting measures for sealing purposes and improvement of the ground conditions are foreseen. The target is to create a grouted ring surrounding the excavation boundaries with a minimum thickness of 9 ft.

Prior to the start of the actual grouting procedures extensive testing in differing underground conditions was carried out as to evaluate the most favorable method for optimum grouting results. Varying parameters for testing include admixture, pumping pressure and rate for cement and polymer based grout in differing depths and geological conditions. In case of necessity the grouting process will be carried out in a staggered pattern, starting with cement based grout followed by chemical grouting.

The grouting works in the tunnel are either carried out from logistic caverns at the shaft toes, definite grouting caverns or enlarged cross sections in the running tunnels.

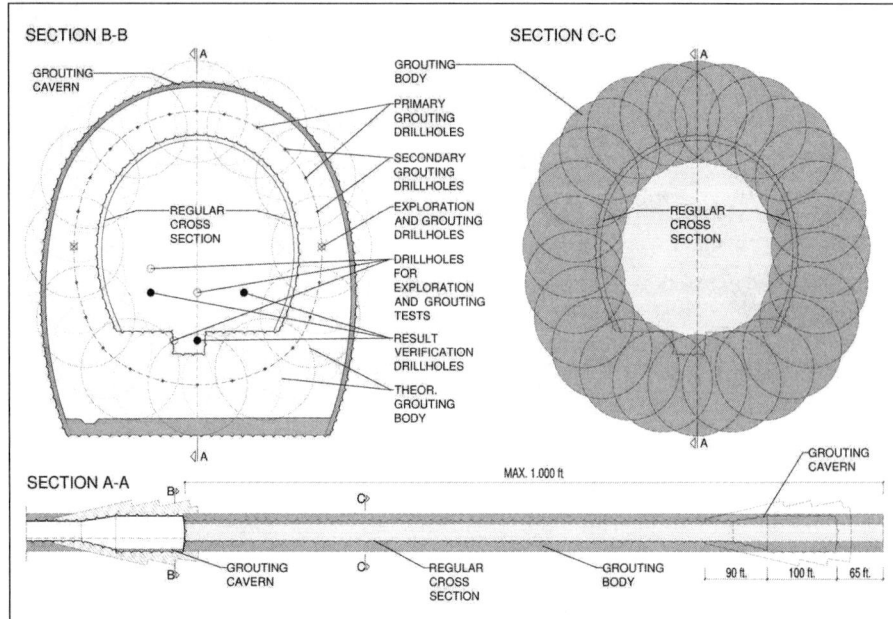

Source: PGST
Figure 8. Grouting measures in main tubes

These enlarged sections are required in order to provide sufficient space for the machinery and equipment handling for the grouting works.

For the testing and grouting procedures directional drillholes with lengths of up to 1000 ft. for a straight direction and 400 ft. for a curved line are to be carried out. In addition short drillholes with a length of 165 ft. are required in the caverns connecting the running tunnels. The drilling pattern around the excavation profiles is defined in detail in order to create the envisaged grouting body and volume (see Figure 8). Due to the existing ground water level all drillings are to be carried out under the protection of blowout preventers.

CURRENT STATUS AND OUTLOOK

At the moment all three underground construction lots of the Semmering Base Tunnel are in progress and the tender design for the western portal area is under preparation. The diversion of main roads in the portal areas, the construction of temporary construction roads and the preparation of the landfill site were tendered and finished ahead of the start of the excavation works.

The tunnel excavation at the northern construction lot which started at the end of 2015 has reached a length of about 1,0 miles for both tubes, with the cross passages following accordingly. The intermediate access tunnel is finished and the shaft head cavern excavation has already started. The subsequent shaft sinking shall be finished by the beginning of 2018 followed by the construction of the caverns at the toe of the shafts. The start of the SEM excavation works of the main tunnels is scheduled for 2019. The excavation works at this construction lot are to be completed by 2022.

Source: OEBB
Figure 9. Cavern excavation

The deep ventilation shafts in the central construction lot are finished and the excavation of the underground emergency station acting as logistic and TBM assembly area during the construction is under progress (see Figure 9). The two single shield TBM's are about to start the excavation of the running tunnels at the beginning of 2018, for about two years. The SEM works for the main tunnels in the opposite direction are planned between the years 2017 and 2021.

In the southern construction lot the two temporary shafts reached their final depths and the cavern excavation including the grouting measures started in spring of 2017 and shall be finished by 2018. The following excavation works for the main tunnel tubes are foreseen until 2022.

The excavation works in all three construction lots will be followed by the lining installation, which will be carried out separately in all three construction lots. The finishing works of tracks and electrical installations as well as signaling will be carried out in a combined lot for the entire tunnel system.

The overall schedule of the project with a partially overlapping construction and final installation phase with a planned final opening for operations at the end of 2026 is on target.

REFERENCES

Wagner, O.K., Gobiet, G., Druckfeuchter, H. 2016. Semmering Base Tunnel completely under construction. Geomechanics and Tunneling 9 (2016):382–390.

Wagner, O.K., Haas, D., Druckfeuchter, H., Schachinger, T. 2015. The challenges of contract SBT1.1 "Tunnel Gloggnitz." Geomechanics and Tunneling 8 (2015):554–567.

Klais, F., Wagner, O.K., Proprenter, M., Wolf, P. 2015. Particular aspects of the tendering contract SBT3.1 "Tunnel Grautschenhof." Geomechanics and Tunneling 8 (2015):568–580.

Gobiet, G., Wagner, O.K. 2013. The New Semmering Base Tunnel project. Geomechanics and Tunneling 6 (2013):551–558.

Boston's Central Artery/Tunnel Project—Lessons Learned

John Reilly ▪ John Reilly International
Fred Salvucci ▪ Massachusetts Institute of Technology
David Hatem ▪ Donovan Hatem LLP

INTRODUCTION

A decade has passed since substantial completion of Boston's Central Artery / Tunnel (CA/T) project, also called the "Big Dig." Despite the tremendous benefits and successes in planning, funding, designing and, having constructed this project with major innovations and a remarkable successful safety record, the project continues to serve nationally and internationally, both in perception and reality, as a "poster child" for much of what can go wrong with megaproject delivery.

In January 2016, the Boston Sunday Globe Magazine published an article entitled, "The Big Dig a Decade Later, Was It Worth It?" (*Boston Sunday Globe 2016*). In response to this question, on June 23 2016, a group of professionals held a seminar in Boston to discuss and review answers to that question. This paper summarizes key elements of that seminar including observations by panel members, related to benefits, lessons-learned, recommendations and improved procedures, which have been implemented for other megaprojects, as well as the future of megaproject delivery using P3s and design-build. It should be noted that Fred Salvucci, the former Massachusetts Secretary of Transportation who initiated the CA/T project, had previously outlined lessons-learned in 2003 (*Salvucci 2003*).

Project Background

The old Central Artery, a major elevated freeway dubbed the "Green Monster" by locals, divided the city, separated the North End from the West End, as well as dividing Government Center, Faneuil Hall and the Financial District from the Waterfront. It provided a never-ending source of smog and soot which covered surrounding buildings and engulfed the City in a wash of noise generated by the lumbering vehicular traffic. It plagued the City as an under-designed, ill-conceived means of traversing from the North to the South Shore in the I-95 corridor. The old Central Artery did not work and was dramatically obsolete—designed for 75,000 vehicles/day it was carrying around 180,000 in the late 1980s. It had 14 lanes converging onto 6 from the north, as well as a similar mismatch from the south and west. It was severely congested, strangling the city and hampering its development and urban quality of life.

The CA/T project resolved the capacity by planning for up to 245,000 vehicles a day, addressed access issues to the City and the Airport, and delivered 300 acres of parkland—the Rose Kennedy Greenway in Boston plus 100 acres on Spectacle Island, remediating a landfill. It spurred development in Boston and particularly South Boston, adding value and quality of life to the city and causing property values adjacent to the CA/T and the Greenway to appreciate substantially—in the order of +80% by 2004 for properties on the 1-mile strip next to the Greenway (*Boston Globe 2004*). Development in South Boston has been dramatic and has attracted offices, Government buildings, hotels and corporate headquarters, plus the new Silver Line for public transportation access that area and Logan airport. Access to the airport from the west, north and south was dramatically improved by the Ted Williams tunnel.

The CA/T project had the following key objectives:

1. Keep the city functioning during the construction process
2. Reunite the city—eliminate the dead center created by the existing elevated structure
3. Get traffic to work by a reliable 30 miles/hour travel time through the center and dramatically improve conductivity of Central Boston to the region
4. Expand the city—across the Fort Point Channel to South Boston to foster economic growth that is environmentally sustainable
5. Provide improved access to Logan Airport and the cities of Winthrop, Revere and East Boston

The CA/T Project's Inception

To address the problems and objectives noted above, a few key visionaries dreamed "Big" and determined that the best way to eliminate the soot, smog and noise, to reunite the City, to facilitate the development of adjacent areas in South Boston and to improve access to Logan Airport, was to plan, design and construct the CA/T Project by relocating the elevated Central Artery underground (in some areas as deep as 120 feet) and to construct an extension of I-90 to the Airport through an immersed tube tunnel (ITT) under Boston Harbor. As if the relocation of a 3.5-mile long interstate highway in a vibrant urban environment was not enough, it was necessary to underpin and maintain the existing elevated road in order to keep full traffic flowing above throughout construction. The project included constructing miles of slurry walls, an underwater immersed tube tunnel taking I-90 under the Fort Point Channel, within feet of the 100 year old Red Line transit tunnel, another major immersed tube tunnel under the harbor connecting I-90 to Logan International Airport, and construction of a signature gateway structure across the Charles River, the world's widest cable-stayed bridge.

Technical advances included construction of the largest circular cofferdam in North America; construction of miles of state-of-the-art slurry walls requiring specialized techniques, skilled individuals and equipment from around the globe; the deepest jet grouting used anywhere; and the use of newly developed ground freezing technology to enable full cross-section road sections to be jacked under active train tracks with minimal vertical clearance. Advanced design of elevated structures was developed, building on previous work with the Department's bridge section. In total, the project was one of the greatest engineering and construction feats and was the largest urban construction project in recent U.S. history.

Lessons learned from previously constructed Boston area megaprojects were put to good use on the CA/T, such as the major advanced utility relocation used for the Red Line North extension, reliable earth retaining slurry walls used by the Massachusetts Bay Transportation Authority (MBTA) during the Orange Line redevelopment project (Southwest Corridor) and integrated traffic management. Project management within budget was demonstrated by the above projects and the Boston Harbor Project, which all relied on experienced managers to successfully deliver those projects.

Given the positive cost outcomes of the Red Line North extension, Southwest Corridor and the Boston Harbor Projects, a major question is why the Big Dig had so many problems with cost. After decades of positive press during the project development and early construction phases in the 1980s, once the cost overruns became public in the late 1990s, the project received substantial criticism from the media and public (e.g., *Associated Press 2000; NCE 2000, Boston Globe "Spotlight" articles*). Fueled by allegations of a lack of oversight and cost overruns, the public demanded answers and "scalps." Ten years after substantial completion, because of the cost question, in the eyes of the public it remains questionable whether the public considers the project a success, even with the very substantial functional and value-added benefits noted previously. Other US megaprojects are often tagged with the "Big Dig" problem which makes it difficult for them to build public confidence.

THIS INITIATIVE, FOCUS OF THIS PAPER, PANEL QUESTIONS

To help move past the Big Dig's "stigma" ACEC Massachusetts co-hosted a conference with Donovan Hatem LLP on June 23, 2016 during which a panel of agency managers and consultants, who had direct involvement in various stages and facets of the Big Dig, reflected on the issues, the context and background at the time, and discussed improvements considered or implemented on other projects. Using the questions listed following, the panel reflected on the CA/T Project experience from different perspectives, addressed lessons learned and provided insights and analysis as to how the delivery, funding, and financing of present and future megaprojects has been, or will be, influenced by those lessons learned. The length limitation of this paper means that not all elements discussed in the seminar can be addressed—this paper gives a summary of key elements.

Questions Addressed

Looking Back:

- What were some of the significant challenges in the conception, planning, funding and realization of the CA/T Project?
- What are the most significant achievements and realized expectations of the CA/T Project?
- What are some of the more important "lessons learned" from the CA/T Project to improve the planning and delivery of other megaprojects?

Post CA/T Project to Present:

- Since the CA/T Project, what have project owners, project funders/grantors, and the design and construction industries understood as the important "lessons learned" from the CA/T Project and other megaprojects?
- What have been the major changes in megaproject planning and delivery since substantial completion of the CA/T project?
- Have those changes improved megaproject delivery in the United States?

Looking Ahead:

- What is the future and promise of megaprojects in New England and throughout the U.S.?
- How will megaprojects be planned, funded and financed in the next decades?
- How have (and will) design-build and public-private partnerships influenced megaproject planning and delivery?

Panel Members Included:

David J. Hatem, PC, Moderator; Fred Salvucci, Former MA Secretary of Transportation; Matt Amorello, Former Chairman and CEO of MA Turnpike Authority (2002–2006);

Bill Rizzo, Former President, Rizzo Associates; Jack Wright, Former Deputy Project Director, Central Artery Tunnel Project; Frank Leathers, Senior Principal, GEI Communications; Beth Larkin, PE, Operations & Management Consultant; John Reilly, Consultant, Logan Airport & Cross-Harbor Regional Transportation Program; Andrew Paven, Former Central Artery Tunnel Project Director of Community & Media Relations; Scott Bosworth, Chief Strategy Officer, Massachusetts Department of Transportation; Joseph Aiello, Partner, Meridiam Infrastructure; Tim McManus, VP of Infrastructure, McKinsey & Co.; Dan McNichol, Author *The Big Dig*; and Robert Rogers, Global Head of A&E Professional Liability, AIG.

LESSONS LEARNED, PROCESSES IMPLEMENTED, FUTURE CONSIDERATIONS

Background—Funding

At the time of the Big Dig's development, FHWA operated a program that provided 90% of funding for completion of interstate highway projects. Because Congress accepted Massachusetts' argument that a majority of the Big Dig's work was performed on I-93, a north-south interstate corridor that connects northern New England to the south shore of Massachusetts and the I-95 corridor, which stretches the length of the Eastern seaboard, the 90% funding policy for the depressed central artery (as was standard practice for completion of all interstate highway projects) was accepted in the 1976 reauthorization legislation. But the delay in finalizing the EIS until 1991, which was the necessary legal approval required for construction to start, not only added inflation costs, but also delayed implementation into a period when funding and the role of the Federal Highway administration changed. As part of the congressional reauthorization of 1991, Federal funding policy changed from 90% funding to one that required that all CA/T costs greater than $6 billion would be the responsibility of the State. This meant that the State of Massachusetts had to accept all cost risk above the 1991 estimate.

When the cost increases became public in 1997, their publication led to intense political and public pressure to rein in costs, which were now a major public focus. The media continued to define the project's price increase relative to "the number" of the early planning stages of the 1980s without explaining all the major utility relocations, inflationary costs, and required scope additions. The Boston Globe "Spotlight" articles further fueled public outcry and generated political resistance at the Federal level. It was noted that, if the public had been kept informed and shown progress throughout the project's life beneath the City's streets, the "tar and feather" mindset may not have reached such hysteria.

In all, the project wound up costing $14.5 billion[*], the greatest increase being due to project delays. The $8.5 billion increase has been attributed to the following elements:

- Inflation—55%;
- Environmental mitigation—15%;
- Scope growth—8%;
- Federal government accounting changes—7%;
- Traffic management and other contingencies—2%.
- Other—13%

---

[*]The federal "accounting" changes did not include the shifting of over $8 billion of cost from the federal government to the state.

The Cost Issue

The project cost estimate began with a published "number" of $2.6 billion in 1982 dollars included in the 1983 FEIS, which, in accordance with FHWA practice, was not inflated to the year of construction, did not include real estate acquisition or utility relocations, and was never intended to be the ultimate full cost of the project. FHWA did not require an inflation component in the initial estimates (normally this would be addressed every two years in the "cost to complete" report assessed by FHWA as the project advanced and budgets, including federal funds, were approved). Rejection by the Reagan administration of the 1983 final EIS, delayed federal environmental approval until 1991, which hugely increased cost because of inflation—but this was never recognized as the primary driver of cost increases that made the 1982 cost estimate moot.

Costs estimates increased from the $2.6 billion "number" to $5.8 billion in 1991 when the supplemental final EIS was accepted. As construction began, changes in cost were difficult to understand because inflation, scope additions, and federal fund proportions were all changing simultaneously, as well as added real estate and utility relocation costs not accounted for in the $5.9 billion estimate.

Panel Comments on Cost

In the lessons-learned seminar it was stated that a realistic cost, including many of the cost drivers associated with Federal requirements, schedule extensions, and additions to the project (environmental, functional, aesthetic) should have established the necessary cost baseline. See *National Academy of Sciences 2003* report for a commentary and summary of NAS recommendations in this regard.

The panel noted that, "We were not doing a good job of educating the public about what the (cost) increases were all about." This was a major problem, as tales of corruption and waste became rampant in the newspapers and nightly news programming, serving as a cautionary lesson for future megaprojects.

The panel members all agreed that the public's perception—that the cost overruns were due to mismanagement (*Associated Press 2000; NCE 2000*)—presented a significant challenge to the project team. As a result, a prime objective for those in charge of the project was to regain the public's support. One way that was accomplished was through increased public involvement.

Once the Zakim Bridge and I-93 tunnels were complete, the public was invited to walk the Bridge and the tunnel system to see first-hand what had been constructed, which is when many say the Big Dig won back some of the public's support. Those actions served as lessons for future megaprojects—it is essential to work with, involve and transparently communicate with those stakeholders directly affected by the project in order to gain their trust and support.

Implementation Example, Improved Process, Transparency, and Cost Discipline

The issues noted above make clear that cost and funding are two fundamental issues that must be adequately addressed during early megaproject development, and which have a fundamental impact on public trust in management (political and project-wise). An example of implementation of an improved process to better estimate and communicate the potential costs of megaprojects is the Washington State Department of Transportation's Cost Estimate Validation Process (CEVP) (*Reilly 2002*).

WSDOT Secretary MacDonald, who had managed the Boston Harbor project close to budget, needed to address the cost issue for a series of megaprojects (*MacDonald 2003*) and asked one of the authors (*Reilly*) to help develop a better and more transparent cost estimating process. CEVP was the result—the process reviews a project's base cost estimate (the cost that would result if all goes as planned), independently validates that cost and then factors in the potential cost of anticipated risk, resulting in a "range of potential cost" which probabilistically combines base cost and risk. This requires a comprehensive risk analysis, the result of which allows management to develop comprehensive risk management and risk mitigation plans. Such risks can be then addressed in the contractual process, e.g., some risks can be mitigated, some assigned to contractors, some can be insured, some shared and some accepted by the Agency (*Reilly 2008; 2013b*).

Roles and Responsibilities

Role of the Federal Government

The Panel expressed a strong opinion that the federal government must get back into the transportation business both as a monitor and contributor to public works projects. Ideally, the federal government would provide funding and a centralized point of expertise—since states construct major infrastructure projects relatively infrequently, leading them to retain outside consultants to assist with that process and in some measure "recreating the wheel" each time a new project is undertaken. To increase the probability of success for future megaprojects in the United States, the Panel suggested that the federal government should act as a repository of tunnel, bridge, environmental, traffic control, public relation consultants/experts and other specialty services which could be used throughout the nation on all large projects and megaprojects.

Local Considerations, Local Issues

Beyond the federal (macro) issues, there are also local (micro) considerations that must be addressed to gain public support and increase the likelihood of any project's success.

One such micro consideration encountered during the Big Dig involved East Boston which serves as the gateway to Logan Airport. It was incumbent upon the designers to create easier and smoother access to Logan Airport for the project to be considered a success. East Boston was struggling with congested traffic, and its residents feared that another access route to Logan would become the "Godzilla" that would swallow up East Boston. To solve this issue, the Ted Williams Tunnel (an immersed tube) was added to the project, providing direct access from I-90 to Logan Airport using airport land and avoiding impact to the residential community. The way in which this local consideration was addressed illustrates how engineering foresight and innovation, working with the local communities, can be used to better solve traffic, access, and satisfy political/local pressures and concerns.

Another concern raised during the Big Dig's design development was the potential for the improved roadway system to cause commuters to abandon public transportation in favor of their cars, leading to further congestion. State environmental officials sought to alleviate this concern by introducing: the new Silver Line bus LRT which would provide improved public transportation access to areas such as the Seaport's Innovation District, the South End and Logan Airport; the extension of the Green Line to the north; the expansion of the Blue Line to reach the Red Line and; replacement of aging transit rolling stock. Panel members expressed concern that some of these commitments were not fully honored, some were delayed, and not enough was done on the public transportation-side of the project to keep up with the improved automobile-commuter access. A public transportation-focused project is, perhaps, the next "megaproject" needed in and around Boston, one element of which is evidenced by the current development of the Green Line Extension (GLX) to augment public transportation vs. increased auto use.

Program Management

Projects such as the Big Dig depend on the an effective program management structure to "turn vision into reality." For the Big Dig in 1987 that management was provided by a core team of high-level Massachusetts Department of Public Works (MDPW)* managers with Bechtel/Parsons Brinckerhoff (B/PB) engaged in the role of Program Management Consultant.

Agency Owner/Operator Considerations

The project was constructed using the design-bid-build delivery method and all work was initially performed pursuant to Massachusetts Department of Public Works (MDPW) contracting procedures. However, a disconnect formed as the state delayed identifying the agency that would own and operate the project. In the Lazard Freres report of 1990 (*Lazard Freres 1990*) it was recommended that the eventual owner operator should be designated (and given a role in construction oversight) before construction began. But it was only in 1996 that the governor proposed, and the legislature approved, that the Massachusetts Turnpike Authority (MTA) would eventually own, operate and maintain the project. However the contracts to design and construct the CA/T project remained under MDPW management.

Change Orders, Accountability

This hugely complex project generated a multitude of change orders which were submitted to, reviewed and approved solely by MDPW. After the legislature directed that the Turnpike Authority be brought into the management process, in an attempt to increase transparency and accountability, a new system was established whereby all change orders were sent to a three-member Turnpike Authority Board for approval. This change made the Turnpike Authority Board fully aware of the increasing project costs (which later resulted in a toll increase), however the information came to the Turnpike Authority only after finalization of design and construction, when it was too late to make any meaningful changes.

Failure to give the Turnpike Authority major input to, if not full control over, the design, construction and cost from the project's outset led to problems with the system's configuration and operational requirements, as well as a lack of clarity of accountability. The lesson-learned is that maintaining continuity and control by one responsible entity, from inception through operation, provides better accountability, cost sensitivity and operational suitability.

Delivery Method Evolution and Impact on Future Megaprojects

Alternative project delivery methods have been developed and used on a regular basis in the US, encouraged by FHWA through the Special Experimental Project SEP-14 and 15 programs. SEP-14 was initiated by FHWA in 1990 to provide a means for evaluating project-specific recommendations when using federal funds. Measures included time plus cost (A+B) bidding, lane rental, warranties and design-build contracting. In

---

* Later renamed Massachusetts Highway Department (MHD)

2004, FHWA initiated SEP-15, an experimental program to allow contracting agencies to explore alternative and innovative approaches to the overall project development process, focusing primarily, but not exclusively, on applications with public-private partnerships. Use of such alternative contracting and delivery methods was not implemented for the CA/T project but may have been beneficial. Such methods are now routinely evaluated and implemented selectively on major transportation and civil infrastructure projects (*Hatem 2009; Reilly 2013c*).

Addressing Risk Through Contracting Method

Contract documents address risk, and each delivery system addresses risk in a different way. The traditional design-bid-build, low bid delivery system essentially seeks out the contractor that believes it can construct the project for the lowest cost, at the lowest acceptable quality. This procurement approach often leads to increased costs through change orders and claims, resulting in an untrustworthy baseline project cost estimate and award amount (*Reilly 2008, 2013b*).

The design-build delivery system is gaining increased use on megaprojects. Design-build projects address risk by placing responsibility on one entity for both design and construction. Design-build projects typically require the design-builder to construct the project within a guaranteed maximum price or for a lump sum. This reduces the owner's risk of shouldering cost overruns resulting from design or construction issues and claims. This system is intended to increase accountability and reduce risk to the owner by placing responsibility and control of all design development and construction in the hands of one entity. Even more significantly, it requires a bidder to place a price on a design when it is at an early enough stage that the owner can revise that design.

Public-private-partnerships (P3s) are an approach to megaproject delivery in which a private sector is responsible for providing financing and for the design and construction of the project. P3s are described as a means to align the long term interests of the government with those of the private sector. The government tends to falter as an asset manager—as evidenced by the estimated $3.6 trillion the United States needs to spend by 2020 to bring its transportation systems into a state of good repair—so the P3 delivery system places asset management responsibilities in the hands of the private sector, where it will consider everything from materials selection to potential financial repercussions for failing to maintain the asset per the contract requirements. All of these new delivery systems hope to build on and improve the design-bid-build system that was available at the time of the Big Dig (*Hatem 2009; 2017*).

These various project delivery methods each come with their own potential pitfalls (*Reilly 2013, 2016*). The risk for engineers on megaprojects is tied to effective management of client expectations which can influence the overall perception of the project's success. An improper analysis of risk factors, a lack of realism and strategic underestimation of the Big Dig's actual cost cast a cloud on its success for a decade. Although many believed the Big Dig was a success, it is only now starting to be seen as such.

CONCLUSION—PANEL COMMENTS

At the conclusion of the June 23rd seminar, the Panel agreed that the Big Dig was a success in terms of meeting its fundamental performance objects—while acknowledging the major cost increase and schedule extension concerns. The City of Boston has, and is, going through a resurgence led by the improved transportation access and reunification of the City that was made possible by depressing the Central Artery underground. Only time will tell if that innovation and "grit" that the City is known for will enable Boston to embark on and complete its next set of "megaprojects" including the GLX light rail project. As of this writing, a large potential cost increase from the GLX's preliminary, baseline cost estimate has marred the project and caused it to be put on hold, while cost containment measures are determined.

As Moderator, David J. Hatem, P.C. pointedly stated: "It is easy to say no and hard to say yes," especially when it comes to megaprojects. However, this "easy to say no" attitude can be countered if fear is substituted for real, balanced judgment and effective management processes.

REFERENCES

1. Associated Press 2000, news release, June 16.

2. Boston Globe, 2004. "For property owners, parks mean profits," Thomas C. Palmer Jr., June 14.

3. Boston Sunday Globe Magazine 2016, "The Big Dig a Decade Later, Was It Worth It?" January.

4. Hatem, David J., PC, Oct. 2009, "Megaproject Issues and Challenges: Some Informal Remarks." Design and Construction Management Professional Reporter, Donovan Hatem LLP.

5. Hatem David J., 2017: Co-editor and contributing author, Chapter 11, "Risk Allocation and Professional Liability Issues for Consulting Engineers on P3 and Design-Build Projects." Public-Private Partnerships and Design-Build: Opportunities and Risks for Consulting Engineers. American Council of Engineering Companies

6. Lazard Freres 1990 "State Financing Report in the 1990 FSEIR."

7. MacDonald, D.B, 2003 "Meeting the Perils of Early Cost-Estimating for Complex Construction Projects," Estimating Mega Construction Projects, TRB Conference, January 15, 2003.

8. National Academy of Sciences 2003, Completing the "Big Dig": Managing the Final Stages of Boston's Central Artery/Tunnel Project, http://www.nap.edu/catalog/10629.html.

9. NCE 2000, "Big Dig, Bigger Problems," New Civil Engineer, March 30 2000

10. Reilly, J J 2002 w. McBride, M, Dye, D & Mansfield, C - Guideline Procedure. 'Cost Estimate Validation Process (CEVP)' Washington State Department of Transportation, January.

11. Reilly, J.J 2008, prime author "Chapter 4, Risk Management" in "Recommended Contract Practices for Underground Construction," Society for Mining, Metallurgy, and Exploration, Inc. Denver.

12. Reilly, J.J. 2013c "Alternative Contracting and Procurement for Megaprojects," UCA/Tunneling Journal Cutting Edge Megaprojects Conference, Seattle, November.

13. Reilly, J.J. 2013b Author of the Foreword, co-author of the Chapters on Risk, Cost and Schedule management in "Managing Gigaprojects," ASCE press 2013, Edited by Galloway, Nielsen and Dignum.

14. Reilly, J.J. 2016b "Megaprojects—50 years, What Have We Learned?" ITA WTC2016 Proc. April.

15. Salvucci, F.P. 2003 "The 'Big Dig' of Boston, Massachusetts: Lessons to Learn," Proc. International Tunnelling Association Conference, Amsterdam, April.

Tunneling Challenges on the Auckland City Rail Link, New Zealand

Tom Ireland ▪ Aurecon
Bill Newns ▪ Aurecon
Shu Fan Chau ▪ Aurecon
Steve Hawkins ▪ Auckland Transport

ABSTRACT

The City Rail Link (CRL) is a $2.5B underground passenger railway, comprising 3.4km twin tunnels and two new underground stations that will double the capacity on Auckland's suburban passenger rail network by connecting the existing underground station at Britomart with the existing North Auckland Line adjacent to the existing Mt Eden Station.

This paper describes the underground engineering challenges including the mined platform tunnel station beneath heritage buildings at Karangahape Road, two large underground Y-junction caverns, and a narrow mixed ground pillar between running tunnels at the TBM arrival box at Aotea Station.

INTRODUCTION

Auckland's City Rail Link (CRL) comprises the construction, operation, and maintenance of 3.4km of underground passenger railway running between the existing Britomart Station and North Auckland Line (NAL). The CRL will provide more frequent trains with more direct services to the city center and allow a train every five to ten minutes from most Auckland stations. The alignment is shown in Figure 1.

The proposed works include cut-and-cover tunnels within Albert Street, cut-and-cover Aotea Station, mined Karangahape Station, twin TBM (Tunnel Boring Machine) tunnels from the southern portal to Aotea Station, mined tunnels/ caverns for the southern Y-junctions by sequential excavation method (SEM), cut-and-cover tunnels and trench excavations near Mt Eden Station where the underground tunnels rise to meet the NAL, and three cross passages with various lengths between the running tunnels.

Newns et al. (2015) has described the general information of the project, including the engineering challenges in relation to the alignment, interaction with existing and proposed buildings and developments, underpinning of heritage structures, utility infrastructure, ground conditions, fire and life safety, and management of traffic disruption. This paper will focus on the underground and tunneling challenges during planning and design stages, including the TBM launching obstructions at Mt Eden, a soft ground and close proximity pillar between running tunnels at the TBM arrival box at Aotea Station, two large underground Y-junction caverns, and the mined platform tunnels at Karangahape Station. The first two contracts commenced in July 2016 with procurement for the main underground works scheduled to commence in late 2017.

Figure 1. Plan alignment of Auckland City Rail Link

GEOLOGICAL AND HYDROGEOLOGICAL CONDITIONS

The bulk of excavation for the TBM and mined tunnels will be within the sedimentary East Coast Bays Formation (ECBF) of the Waitemata Group which underlies much of the Auckland urban area to an estimated depth of 500m or more (Kermode, 1992). The ECBF comprises a complex flysch sequence with interbedded turbidite sandstones and pelagic mudstones/siltstones found as shallow dipping alternating beds ranging in thickness from decimeter to several meters thick. In addition, there are occasional interbedded lenses of stronger channelized sandstone, found around Karangahape Station. The unconfined compressive strength (UCS) varies typically from 1 MPa to 10 MPa with the stronger beds of channelized sandstone 15–30 MPa.

The Waitemata Group sediments were deposited during the Miocene period in shallow seas and have been subsequently uplifted and eroded. During the Pleistocene period, material derived from the ECBF rocks were deposited in a shallow marine environment to form the Tauranga Group, a firm to stiff clay which is often found infilling paleo valleys above the ECBF. Intermittent volcanic activity has occurred throughout the region resulting in basalt cones and flows and several explosion craters that give rise to local depressions and tuff deposits. Recent alluvium comprising loose silty sand to very soft silty clay to silty sand, and older alluvium comprising firm to stiff silts and clays of medium to high plasticity and/or dense sands and gravels, have also been deposited in the low lying areas.

The geological and hydrogeological conditions along the alignments are summarized in Table 1.

Ground Investigations

Extensive existing borehole data available from Auckland Council records and a total of more than 130 specific CRL boreholes have confirmed the geological conditions along the CRL alignment. In-situ tests, including dilatometer and pressuremeter testing, and televiewer surveys of structure, were undertaken along with a comprehensive suite of laboratory tests to determine geotechnical design parameters.

Numerous piezometers have been installed along the route to establish pre-construction groundwater levels and given the potential for the Tauranga Group alluvium to consolidate, to monitor groundwater levels during construction. In addition pumping

Table 1. Geology and hydrogeology

Tunnel Section	Tunneling Method	Geology and Hydrogeology
Southern Y-junction	SEM tunnel/ cavern	Moderately Weathered to Unweathered ECBF with well-cemented sandstone through more typical very weak to weak interbedded sandstone and siltstone, to uncemented sand (EUs1). There are uncertainties in the extent of the uncemented sand (EUs1) layers. The ground water table at the Y-junction varies from 17m to 26m above tunnel crown.
Running Tunnels - Southern shaft to Karangahape Station	TBM tunnel	Un-weathered to Slightly Weathered East Coast Bays Formation with Well Cemented Sandstone and Conglomeratic Sandstone in approaches to the Karangahape Station. Regional groundwater table within ECBF is parallel to the topography and is about 8–10m above tunnel crown.
Karangahape Station	SEM tunnel	Thick layer of well-cemented, moderately strong sandstone (the unit is referred to as "channelized sandstone"). Bedding in this layer is horizontal or sub-horizontal. The unit drops in level from the north to the south of the station. A narrow fault zone is inferred to be intersected in the vicinity of the Mercury Lane Shaft. Fully saturated phreatic surface is above 20–23m above tunnel crown and a single perched water table at depths between 3–6m bGL (1–4m above saturated phreatic surface).
Running Tunnels - Karangahape Station to Aotea Station	TBM tunnel	Well Cemented Sandstone and Conglomeratic Sandstone in the vicinity of Karangahape, then Un-weathered to Slightly Weathered ECBF for the majority of the tunnel drive. At tunnel approaches to Aotea Station the geology is Highly to Completely Weathered ECBF with potential to encounter Tauranga Group Alluvial. Regional groundwater table within ECBF is parallel to the topography and is nearly at tunnel invert level.

tests were undertaken in the Karangahape Road precinct to characterize the channelized sandstone as a potential aquifer.

Another particular focus of the investigations has been the definition of the ECBF 'rockhead' (EU) and the differentiation of the Tauranga group alluvium (TA) from man-made fill materials (F) as these surfaces are significant from a design and construction risk perspective.

In common with other major investigations, the CRL investigations have found beneath the 'rockhead' level, phases of uncemented sands within the ECBF where the clay matrix is absent. These have potential to be problematic in large open face excavations.

The investigations have been progressively interpreted in 3D utilizing data from building records including piling records where these are available and the quality of the data supports their inclusion. The interpreted geological long section is shown in Figure 2.

TBM LAUNCHING AREA AT MT EDEN

The TBM tunnels are planned to be launched from the Mt Eden worksite where the CRL joins the existing North Auckland line rail network. From grade the rail lines will enter a series of cut and cover tunnels including the 16–26m deep, 80m long cut-and-cover crossover shaft. Within this shaft, the eastern connection box structure will cross underneath the main tunnel alignment structures. In addition, a large ventilation plenum, connecting to the four tunnel alignments will be constructed. Figure 3 shows the interface between tunnels and the shaft. Both Up and Down Mains have underground Y junctions to facilitate both the East and West facing connections.

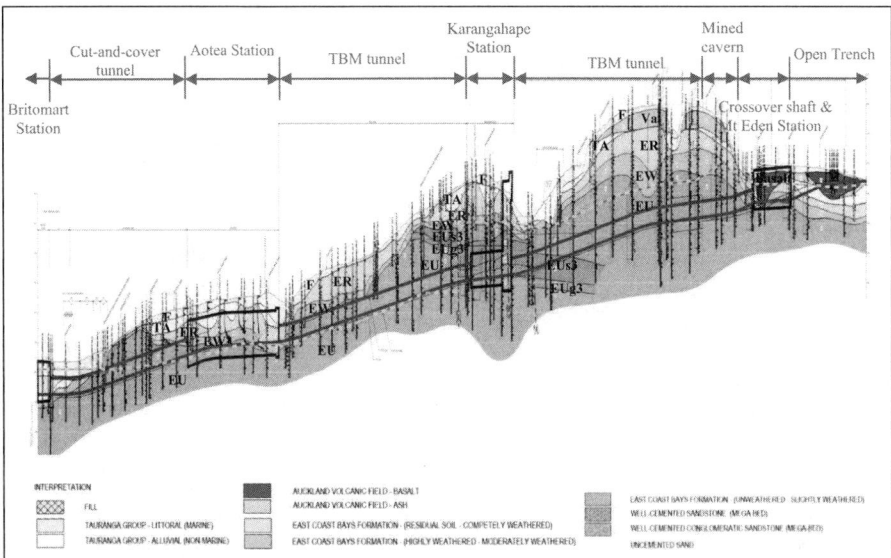

Figure 2. Interpreted geological long section for CRL tunnels

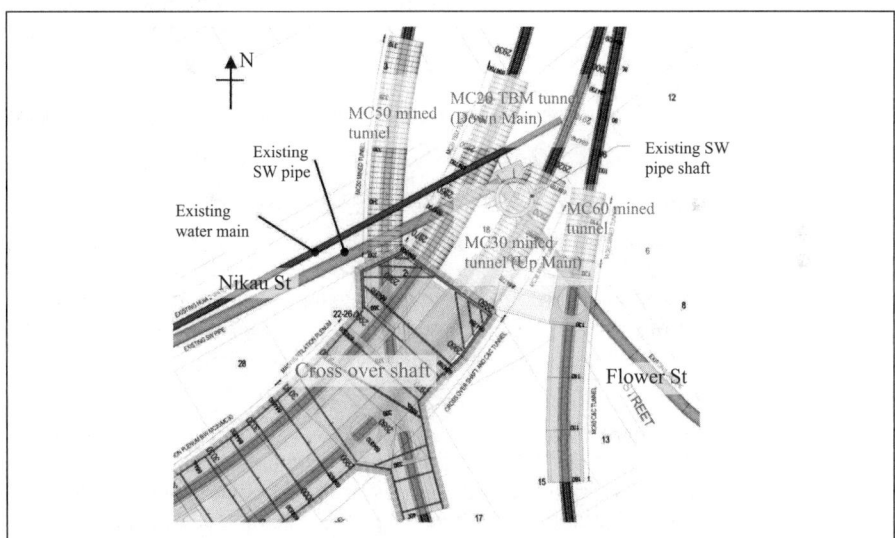

Figure 3. TBM launching interface at crossover shaft–plan

An existing 1950mm diameter stormwater pipe runs beneath Flower Street and Nikau Street and a pipejack shaft, previously associated with its construction, is located at the intersection of Flower Street and Nikau Street. The shaft, which is backfilled with sand and has an array of external rock bolts, clashes with the proposed alignment of the MC20 (TBM tunnel) and MC30 (SEM tunnel). The stormwater pipes and the existing shaft will be abandoned and diversion of the stormwater pipe between Water Street and Nikau Street will be carried out as an enabling contract.

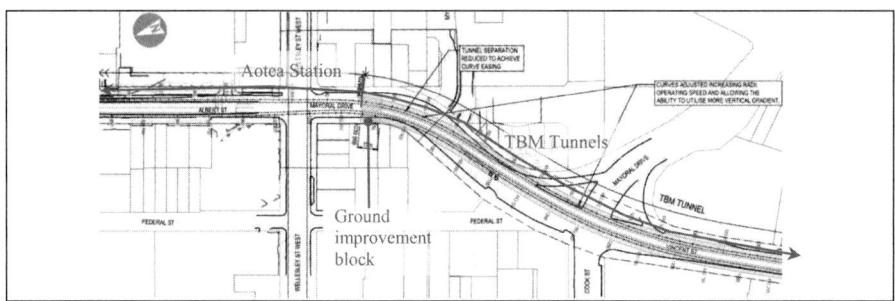

Figure 4. Layout of TBM tunnels and Aotea Station interface

In order to mine through the existing shaft along the Up Main tunnels, additional temporary ground support, e.g., canopy tubes and grouting, will need to be installed before excavation. For the Down Main tunnels, the rock bolts extending from the shaft are expected to interfere with the TBM tunnel drive. Two possible excavation methods were considered and evaluated. One was TBM excavation through a pre-excavated and backfilled pilot tunnel (3m(W) × 6m(H) × 30m(L)), which is to be constructed from the cross-over shaft to allow removal of the existing rock bolts. The other one was to mine a 30m long SEM tunnel from the cross-over shaft through the congested area before starting the TBM drive. Considering the lower risks to the construction pro-gramme and easier launch of TBM from the crossover shaft, the former method was preferable and documented for the reference design. It is expected that the future CRL contractor will design a temporary support system and select a method of excavation to pass through the existing shaft, based upon balancing resources, risks, and in con-sideration of the overall civil construction programme and interfaces with the overall fitout programme.

TBM INTERFACE WITH AOTEA STATION

The TBM drives will be retrieved at the Aotea Station, where a strutting system will be configured enable TBM breakthrough and removal to the surface. Due to the align-ment constraints, the tunnel separation between extrados for about 500m of the running tunnels southward from the Aotea Station, is less than one tunnel diameter. Alignment curves within this section are optimized for operating speed with the down main radius down to 150m adjacent to the station with a tunnel separation of 2.4m.

60m of TBM tunnel South of Aotea Station southward will pass through mixed ground including Residual Soil (RS) and Weathered (EW) /Unweathered ECBF (EU). Given the proximity of the tunnels and the relatively poor ground conditions, the interac-tion of the two tunnels and their construction sequence was assessed using PLAXIS. Ground relaxation of 60% was estimated based on 50mm tunnel crown convergence. The volume loss achieved based on the selected relaxation is less than 1% which is comparable to the previous back analyzed tunneling projects in Auckland with similar ground conditions.

Based on the analysis, the tunnel lining of the first tunnel is structurally capable to resist the additional forces and deformation induced by excavation of itself and the second tunnel bore. The determined movement and maximum segmental joint opening are small and within the design limit. The plastic zone induced from one single tunnel is within half of the pillar width. No ground treatment requirements were determined or lining enhancement of the first tunnel is required. However, an observation approach with continuous monitoring and a contingency plan shall be implemented during construction. This will extend to consider the specific risks of the TBM breakthrough.

Y-JUNCTION CAVERNS

Both Up and Down Mains have underground Y junction caverns to facilitate both the East and West facing connections. These caverns will be the largest underground spans constructed in the East Coast Bays Formation in Auckland. The layout of the Y-junction caverns is shown in Figure 5. Excavation of the mined tunnels and caverns are expected to be carried out by roadheader and other mechanical means. There are three different profiles of Y-junction caverns as summarized in Table 2.

Tunnel/Cavern Construction Sequence

As described above the main tunneling construction site is located at Mt Eden and will provide, TBM and SEM tunnels plant and machineries access and egress, spoil mucking out exit, housing of tunnel works facilities/equipment, segment storage, etc. The cut-and-cover tunnels and modification of Mt Eden Station will also be constructed within the construction site at a later stage.

Cavern Stability and Temporary Support Design

The Y-junction caverns have three different profiles, ranging in width from 11 to 18.5m and varying in height from 9–13m, as described in Table 2. The ground cover over the Down and Up Main Y-junctions range from 25–30m and 10–20m respectively, with only 1–2m of unweathered rock cover in the worst case location (i.e., Up Main Y-junction). The geological condition has been described in Table 1. Given the low rock cover relative to the span of the caverns, existing surface structures and impact of the ground settlements were key considerations in the design development of the tunnel excavation sequence and temporary initial support.

The primary tunnel linings were designed based on the geological model and parameters described above. The presence of uncemented sand were also considered in the lining design. The excavation sequences for the mined tunnels in terms of heading,

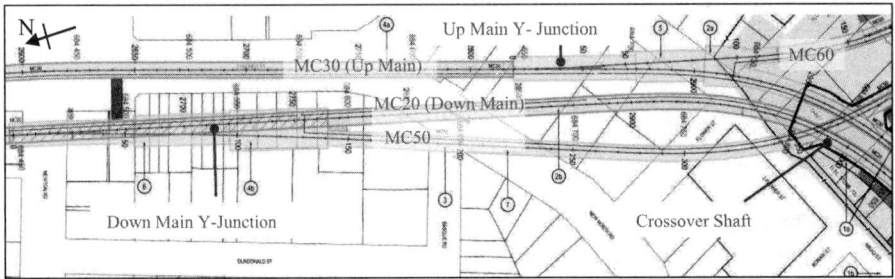

Figure 5. Southern Y-junction caverns layout

Table 2. Summary of dimensions of Y-junction caverns

	Down Main Y-junction			Up Main Y-junction		
	Small	Medium	Large	Small	Medium	Large
Length (m)	56	47	39	54	32	20
Span (m)	11	14	18.5	11	14	18.5
Height (m)	9	10.7	13	9	10.7	13

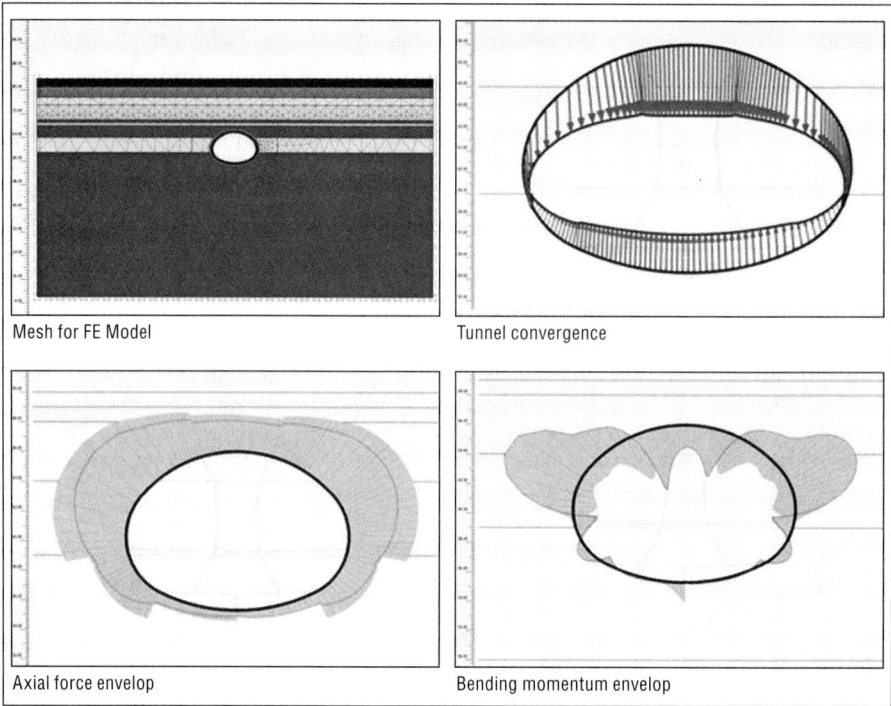

Mesh for FE Model	Tunnel convergence
Axial force envelop	Bending momentum envelop

Figure 6. Cavern stability FE model and outputs

bench and invert were developed to ensure stability of the ground during the excavation cycle. In the reference design, the primary linings were designed using the empirical approach and numerical modelling (i.e., FEM analysis), and determined with the overall assessment of the outcome from these approaches.

The global mechanical properties of a jointed rock mass were determined using the Geological Strength Index (GSI) to establish mechanical properties from Hoek–Brown with the equivalent Mohr–Coulomb strength parameters "c" and "angle of friction ø"; as well as elastic modulus (E) for numerical analyses.

The final design approach applied PLAXIS for simulating the construction sequences and the overall stability assessment. Samples of PLAXIS models for the primary lining of the largest section of the "Y" junctions are outlined in Figure 6.

Excavation Sequence and Support Types

The Y-junction caverns are designed to be excavated in multiple stages, to cater for the size of the Y-junction caverns and the expected variability, and in some cases poor ground conditions.

Based on the reference design, the temporary supports consist of steel fibre reinforced shotcrete (SFRS), rock bolt, lattice girder, etc. Lattice girder and rock bolt were compared in the temporary works design. Due to the ground conditions and the poor bond strength between the grouted rock bolts and the anticipated geological strata above the tunnel crown, lattice girder were adopted in the large and medium caverns. The design temporary support systems are shown in Table 3.

Table 3. Initial support system for Y-junction caverns

Cavern	Excavation stages	Initial Support
Large cavern – 18.5m span (both Y-junctions)	Heading: 3 stages	350mm SFRS + lattice girder @ 1m spacing
	Bench: 3 stages	350mm SFRS + lattice girder @ 1m spacing
	Invert: 1 stage	200mm SFRS
Medium cavern – 14m span (both Y-junctions)	Heading: 2 stages	250mm SFRS + lattice girder @ 1m spacing
	Bench: 1 stage	150mm SFRS
Small cavern – 11m span (MC20/MC50 Y-junctions)	Heading: 1 stage	200mm SFRS + lattice girder @ 1m spacing
	Bench: 1 stage	100mm SFRS
Small cavern – 11m span (MC30/MC60 Y-junctions)	Heading: 1 stage	200mm SFRS + 3m long rock bolt @ 1.2 c/c
	Bench: 1 stage	100mm SFRS

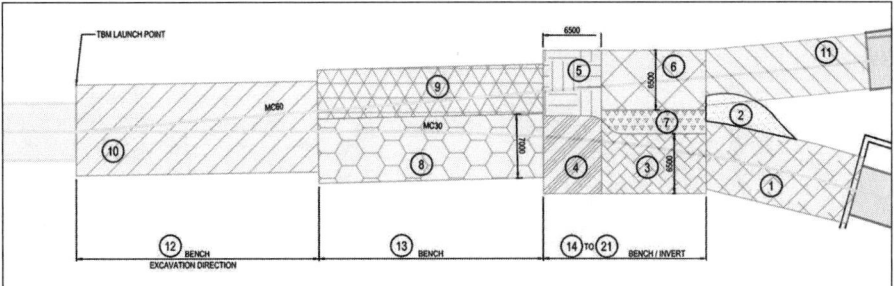

Figure 7. Excavation sequence for MC30/MC60 Y-junction

Cross Heading Excavation

Cross excavation of the top heading in the large cavern was considered.. Each leg of the Y-Junction tunnels cannot be excavated at the same time to form a full span heading excavation of the largest step plate. Cross excavation of heading at the end of the largest step plate is to ensure a safe and practical excavation method with lower risks. The sequence and initial supports for the cross heading excavation are shown in Figure 8. The cross heading excavation will be 7m wide and 7.3m high. The reference design assumes that the cross excavation will be initially supported by shotcrete (200mm thick) and rock bolts on roof and fiber glass bolts at tunnel eyes. The fiber glass bolts and shotcrete at both sides of tunnel eyes will be removed when excavation on both sides advance. A second layer of shotcrete (200mm thick) at a later stage will also be applied on the roof to form the full span temporary support of the large cavern.

In the Down Main Y-junction the excavation profiles are constrained by the temporary TBM tunnel linings. The reference design assumes spiling bars as pre-reinforcement extending from the edge of the first drift top heading will provide additional assurance for temporary support. The lining will be cut one segment at a time during cross cut advance. During the cutting of the lining, the lining ring is not maintained, and temporary fiber glass bolts are designed to hold the segments in place.

MINED PLATFORM TUNNELS AND SHAFT AT KARANGAHAPE STATION

Karangahape Station works include two platform tunnels, central passenger tunnel, escalator tunnel, four cross adits and two shafts (Mercury Lane Shaft and Pit Street Shaft), as shown in Figure 9. A historic heritage building, Mercury Theatre, is directly above the proposed Down Main platform tunnel and is 3 to 5m away from the Mercury Lane Shaft. The historic heritage building, with three stories of unreinforced brick

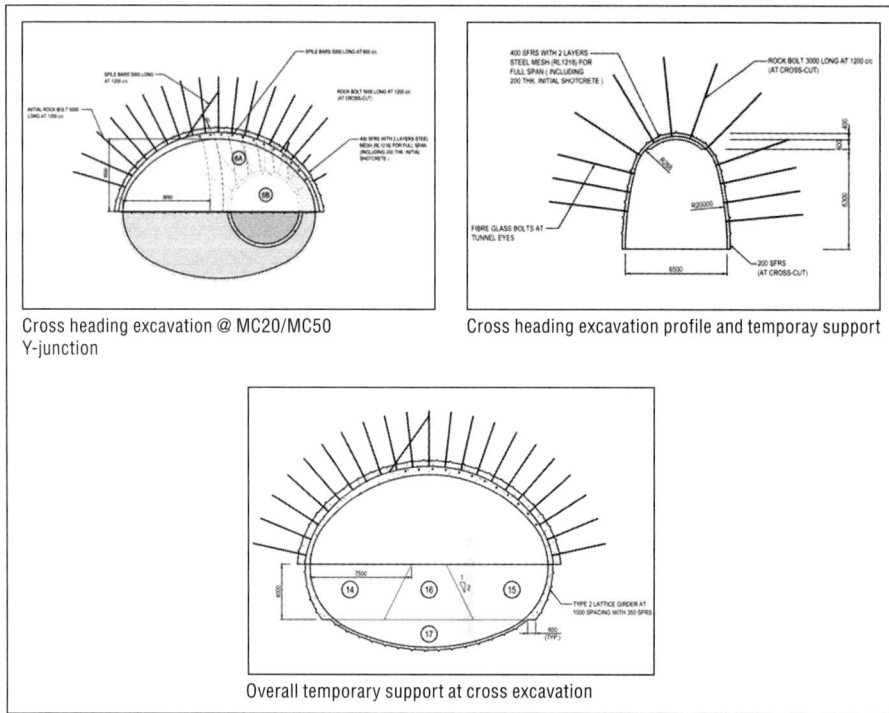

Cross heading excavation @ MC20/MC50 Y-junction

Cross heading excavation profile and temporay support

Overall temporary support at cross excavation

Figure 8. Cross heading excavation

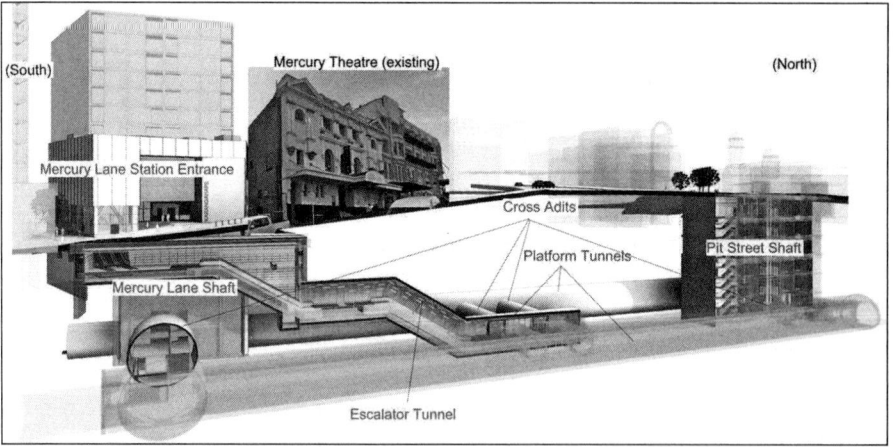

Figure 9. 3-Dimensional view of Karangahape Station

masonry walls, timber framed floors and concrete pad footings, is a building sensitive to ground movements. Control of ground movement and thereby impacts upon the historic heritage building are the most challenging factors to be considered in tunnel and shaft excavation design and are reflected in the temporary supports and staging of the reference design.

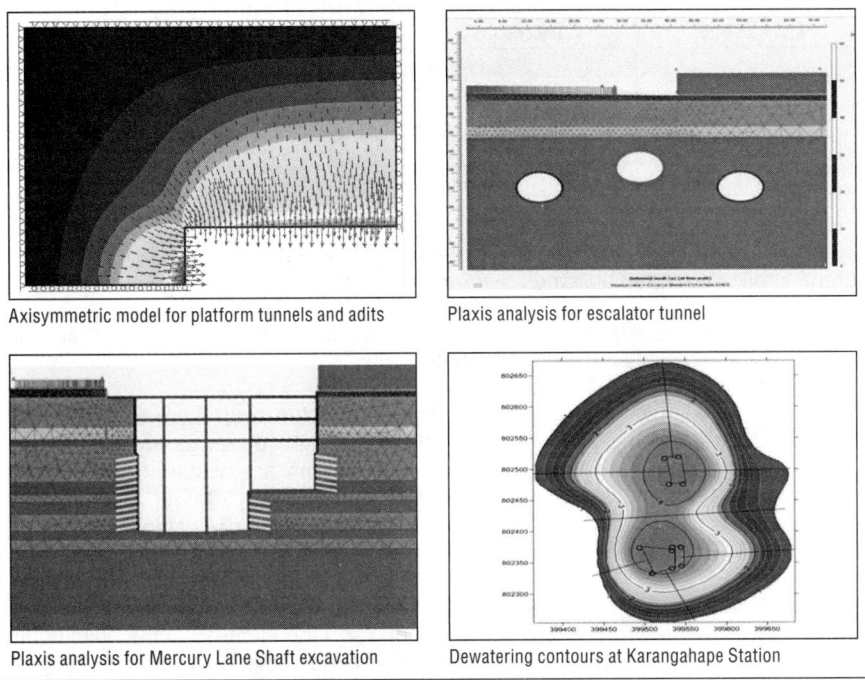

Axisymmetric model for platform tunnels and adits

Plaxis analysis for escalator tunnel

Plaxis analysis for Mercury Lane Shaft excavation

Dewatering contours at Karangahape Station

Figure 10. Settlement analysis of tunnels and shafts at Karangahape Station

To estimate the settlement effects from the platform tunnels and associated adits, an empirical method based on an inverted Gaussian distribution curve (O'Reilly and New (1982)) was initially adopted. Four axisymmetric models by using FE software Phase 2 were set up to determine the convergence at the perimeter of the tunnels. The approach was considered conservative, as these models provided an intrinsic response of the excavation and ignored the effects of the temporary support. The escalator tunnel in soft ground was modelled in PLAXIS. The determined convergence values were then converted into an equivalent volume loss as the input to the Gaussian model. The mechanical settlement effects from the Mercury Lane Shaft excavation were analyzed in PLAXIS, by modelling the staged excavations and installation of supports. Consolidation settlements induced by the groundwater drawdown were also assessed. The analysis models are shown in Figure 10. The total settlements at the ground surface are the result from a combination of the shaft mechanical settlement, the consolidation settlement and the estimated mined tunnel induced settlement.

For the purpose of building damage assessment, the vertical settlement extent was estimated and the settlement contour was plotted. The assessed settlement around the Mercury Lane Shaft and at the Mercury Theatre are 10 to 20mm. The building Damage Assessment Criteria proposed by Burland 1997 was referenced for the assessment. Burland et al (1977) demonstrated that the tensile strain is directly related to the degree of damage and that the specific limiting values of the tensile strain could be assigned to various categories of damage. Boscardin and Cording (1989) further developed these concepts and showed that the categories of damage proposed by Burland are related to the limiting tensile strength. The Burland Damage Classification is reproduced in Table 4. Buildings are considered to be at "low risk" if the predicted

Table 4. Building Damage Classification after Burland (1995) and Mair et al. (1996)

Risk Category	Description of Degree of Damage	Limiting Tensile Strain, %
0	Negligible	0–0.05
1	Very Slight	0.05–0.075
2	Slight	0.075–0.15
3	Moderate	–0.3
4 or 5	Severe to very severe	>0.3

degree of damage falls into the first three categories 0 to 2 (i.e., negligible to slight). For the shallow founded buildings, a three stage risk assessment was conducted:

Stage 1: Settlements less than 10mm and associated ground slopes of less than 1:500 have negligible damage potential.

Stage 2: For buildings within the area with the predicted settlement of greater than 10mm and/ or slope of greater than 1:500, the buildings are assumed as simply supported beams and the foundations are assumed to follow the settlement of the ground surface (i.e., greenfield settlement). The maximum tensile strains are calculated and the corresponding risk category of damage is obtained.

Stage 3: For the buildings with predicted damage levels greater than "Slight," detailed evaluations are required. The presence and the stiffness of the building and its foundations will reduce potential ground movements and thereby the building strains. The process involves reducing the horizontal and vertical movements that reduce the deflection ratio and horizontal strain, thus reducing the overall strain on the building (Potts and Addenbrook 1997).

The Mercury Theatre has been assessed for both a fully top-down excavation scheme and a bottom-up excavation scheme. The Stage 2 risk assessment indicated that the overall effects based on Burland's classification were "slight," for which a Stage 3 risk assessment is not required. However, given the building is particularly sensitive to ground movement, Stage 3 detailed evaluation, taking into account of the influence of the building configuration, has also been undertaken and a "very slight" category was suggested. Despite the above, the Mercury Theatre is recommended for specific preventive stabilization works for the southern wall, because of its historic heritage value and sensitivity to ground movement. Of course, detailed pre-construction structural survey and extensive building monitoring during construction are also proposed building on the inspections already undertaken for the purposes of analysis.

PROJECT UPDATE

The CRL project will be divided into eight works contracts, as summarized in Table 5. The first two contracts, C1: Britomart and C2: Lower Albert Street tunnels, have been awarded and the construction commenced in July 2016. The main civil works design and construction contract C3: Stations and Tunnels, including Aotea Station, Karangahape Road Station, running tunnels between Aotea Station and Mt Eden Station, and Newton junction to Western Line, is scheduled for procurement in 2017.

CONCLUSIONS

The CRL tunnels, consisting of TBM tunnels, mined SEM tunnels and deep shaft excavations, will be constructed within Auckland CBD, where mixed ground conditions, sensitive buildings/ structures, and clashes with existing utilities, etc. are anticipated.

Table 5. CRL works contract and procurement summary

Works Contract	Key Procurement Dates	
	Express of Interests (EOI)	Request for Proposals (RFP)
C1: Britomart	N/A	N/A
C2: Lower Albert Street	N/A	N/A
C3: Station and Tunnels	Prequal – Jan 2017/ EOI – Apr 2017	Q4 2017
C5: Western Line	Q2 2017	Q1 2018
C6: Mt Eden Stormwater Diversion	N/A	TBA
C7: Systems Integration, Testing & Commissioning	Jan 2017	Apr 2017
C8: Wider Network Improvements	TBA	TBA
C9: Britomart East	TBA: post construction of main works	

Source: Auckland Transport website https://at.govt.nz/projects-roadworks/city-rail-link/crl-procurement/

A robust tunnel design and construction arrangement to minimize the tunneling risks have been considered and documented in the reference design stage to provide a sound technical basis for procurement of the detailed design and construction Contract 3.

ACKNOWLEDGMENT

The authors acknowledge the expertise and commitment shown by the integrated team of engineers, architects and planners that have developed the CRL designs to date comprising principally Auckland Transport, Aurecon, Mott Macdonald, Grimshaw, Jasmax, and Arup.

REFERENCES

A History of the Auckland Railway Stations 1858–2004 Walker, G. (2004) *Auckland City Library* (Edited and Abridged by JF Webley.).

Auckland Rapid Transit Report to Government (1974) Unpublished.

Adhikary, T. (2001) Weathering Profiles and Characteristics of Waitemata Rocks in Auckland Region. *New Zealand Geomechanics News*, pp. 70–77.

Auckland Transport website (https://at.govt.nz/).

Britomart Transport Interchange, Auckland: Alterations and strengthening of the former Chief Post Office E.T. Sainsbury and M.L. Gibbs 2004 *NZSEE Conference.*

Boscardin, M.D. & Cording, E.g., (1989). Building response to excavation-induced settlement. *Jnl Geo Engrg, ASCE*, 115;1;1–21.

Burland J B (1997), Assessment of risk of damage to buildings due to tunnelling and excavation, *Earthquake Geotechnical Engineering, Ishihara (ed), Balkema, Rotterdam*, pp.1189–1201.

Burland J B and Wroth C P (1974), Settlement of buildings and associated damage. State of the Art Review, *Proceedings, Conference on Settlement of Structures*, Cambridge, Pentech Press, London, pp. 611–654.

Burland J B , Broms J B and de Mello VFB (1977), Behaviour of Foundations and Structures, SOA Report Session 2, *Proceedings of 9th International Conference*, SMFE, Tokyo, 2:495–546.

Ireland, T and Dowler, M (2011). Rosedale Outfall Project—Overcoming Extreme Conditions. *14th Australasian Tunnelling Conference*/ Auckland, New Zealand, 8–10 March 2011. pp511–527.

Ireland, T., Bishop, R., and Sheffield, M. (2011). Project Hobson—New Technology in Auckland. *14th Australasian Tunnelling Conference*/ Auckland, New Zealand, 8–10 March 2011. pp99–110.

Ireland, T., Spies, P., Eratne, S., and Newns, W. Risk Management and Use of a Geotechnical Baseline Report for a Competitive Alliance Procurement Process. *15th Australasian Tunnelling Conference*/ Sydney, Australia, 16–18 August 2015. pp99–110.

Kenny, J.A., Lindsay, J.M., Howe, T.M. (2012). Post-Miocene Faults in Auckland; Insights from Borehole and Topographic Analysis. *New Zealand Journal of Geology and Geophysics*.

Kermode, L.O. (1992) Geology of the Auckland Urban Area Sheet R11 Scale 1:50,000, IGNS Geological Map 2.

Namjou, P and Pattle, A (2000) Post-Audit of a Numerical Groundwater Flow Model developed for Britomart Transport Centre, Auckland, New Zealand.

New Zealand Geotechnical Society Inc. (2005). Field Description of Soil and Rock— Guideline for the Field Classification and Description of Soil and Rock for Engineering Purposes.

Newns, B., Hawkins, S. and Ireland, T. (2015). Auckland City Rail Link—The First 100 Year. *Proceeding of Rapid Excavation and Tunneling Conference 2015*, Pp 868–880.

Resource Management Act (1991) New Zealand Government.

Searle E.J. (1981). City of Volcanoes: A Geology of Auckland.

Innovation and Technology

Chairs

Kurt Braun
L-7 Services LLC

Peter Procter
Mott MacDonald

Game-Changing Technology for Overhead Ventilation Duct Construction in Large-Diameter Railway Tunnel

Tse-Hung Lee ▪ Ove Arup & Partners (Hong Kong) Limited
Li-Ling Chen ▪ Ove Arup & Partners (Hong Kong) Limited

ABSTRACT

The cast in-situ concrete is the most commonly-used system to construct the overhead ventilation duct in tunnels. The use of precast method is precluded for consideration due to the excessive self-weight of precast units, and insufficient headroom for lifting and erection operation. On a new railway project in Hong Kong, a game-changing technology utilizing a mix of cast in-situ and factory to produce precast units has been used to construct the overhead ventilation duct for a large diameter tunnel. This hybrid system which combines the in-situ and precast concrete has optimized the benefits of both forms of concrete construction, improving the buildability, accelerating the work cycle and reducing the cost. This paper summarizes the implementation of the hybrid concrete slab system and includes an overview of the key design aspects for this alternative composite structure in a confined building environment.

INTRODUCTION

In contrast with the traditional concrete structure where concrete is poured into a site-specific form and cured on site until it attained the required strength before the form removal, precast concrete is a structural element produced by casting concrete in a reusable mould, which is then cured in the casting yard under a controlled environment, transported to the worksite and lifted into place. Using the precast concrete system offers many advantages over the traditional cast in-situ concrete. The precast method enables a greater control over material quality and workmanship in a fabrication yard compared to a worksite. The structural steel moulds that used in the yard can be reused hundreds or thousands of times before they have to be replaced. This method is therefore more economical. The production of precast unit is performed on the ground level, which ensures safety throughout a project.

Although the precast concrete system is well-established as it is used in all types of building, and it provides solutions for a great variety and complexity of layouts, shapes and facade treatments, the method is generally precluded from consideration on the civil sector to construct structures inside a tunnel due to the excessive self-weight of precast units, and insufficient headroom for lifting and erection operation. In order to remove the essential constraint on the size of precast concrete slab to be lifted by crane, a hybrid concrete technology utilizing a mix of cast in-situ and factory to produce precast units has been implemented to construct the overhead ventilation duct for a large diameter railway tunnel in Hong Kong. This paper summarizes the hybrid concrete slab system, and includes an overview of the design principles underlying this composite structure built in confined environment. They should enable the engineers to form a realistic assessment of the application of precast concrete construction in a tunnel against some conventional biases.

638

SEMI-PRECAST CONCRETE SLAB SYSTEMS

General Considerations

With the emphasis on speed of construction, optimum economy and productivity of manpower, the use of hybrid concrete technology in suspended slab construction has drawn much attention among Clients and Designers. It is important to have a clear understanding of the various types of the semi-precast slab systems available with structural toppings. This makes it possible to apply the best suitable method to overcome a given set of conditions and constraints. The choice of a semi-precast slab system and the method of hybrid construction are largely dictated by the following considerations:

- Appropriateness of use
- Cost effectiveness
- Constructability
- Speed
- Health and safety

Semi-Precast Slab Options

Different types of precast slab systems may require different types of plant and equipment for their installation. They may also necessitate different minimum operational clearances from the nearest obstructions. If such clearances are not available, the use of certain types of plant and equipment and, failing that even certain types of precast slab systems themselves, may have to be ruled out. This becomes more critical for construction within tunnel, as the precast construction of internal structures can present a problem of headroom for certain types of plant and equipment, and impose additional constraints resulting from the confined geometry of the tunnel. Alternatively, system of hybrid construction can be considered. The main types of semi-precast slab systems that used to form the hybrid concrete slab are as follows:

1. Hollow core slabs—Hollow core slabs were developed in the 1950s when the dual techniques of long line prestressing and concrete production through machines were being developed by companies in United States and Europe. A hollow core slab is a precast, prestressed concrete member with tubular voids extending the full length of the slab. This makes the self-weight of slab about one half of a solid section of the same thickness without significant reduction on its structural stiffness. The reduced weight not only lowers the material and transportation costs, but also facilitates the installation. Figure 1 shows the hollow core slabs that used in construction of a multi-storey building. Hollow core units typically range in depth from 150mm to 400mm with a nominal width of 1,200mm. These hollow core slabs work most economically when the span to depth ratio is around 40. Continuity of hollow core slab units over supports can be achieved by placing reinforcing

Courtesy of PCI Design Handbook, 2004

Figure 1. Precast hollow core slabs in multi-storey building construction

steel in the grouted keyways in a composite structural topping, or by concreting bars into cores.

2. Ribbed slabs—A precast ribbed slab is made up of equally spaced reinforced or prestressed ribs spanning between supports. The precast ribbed slabs with in-situ concrete topping offer greater structural capacity at longer spans. Precast double-tee slabs, as shown in Figure 2, are the most common type of ribbed slab. These units originate from the United States, where they have been used extensively since the 1940s. The double-tee precast unit is typically manufactured in width of 2,400mm. The ribs can vary in depth from 400mm to 900mm. The top flange is usually 50mm or 60mm deep and the ribs taper from a minimum of 200mm at the bottom, widening upwards towards the underside of the top flange. The taper on each side of the rib allows easy lifting out of precast unit from a fixed mould. Because of the shallow flange depth, an in-situ concrete topping is required to ensure vertical shear transfer between adjacent units and develop horizontal diaphragm action in the slab plate. Although the ribbed sectional profile of slab reduces the quantity of concrete and reinforcement, and also the weight of slab, extra headroom between slabs is needed to accommodate the deepened ribs.

3. Lattice girder slabs—Lattice girder slabs are shallow precast slab, which have been used extensively since the early 1980s. The precast slab is generally 50mm to 100mm thick depending on the design span, and the depth of the complete slab is between 150mm and 200mm. The slab contains the bottom steel bars and the lattice girders partially embedded in the precast unit, which are necessary for structural purposes and acting as permanent formwork for the in-situ concrete topping. Void formers can be introduced in the form of polystyrene blocks or spheres for which to reduce the quantity of concrete used and also the self-weight of the slab. The lattice girders are fabricated from steel bars and installed at spacing of 300mm to 600mm centres. The protruding lattice girder ensures that composite action is to be developed between the precast unit and the in-situ concrete in the permanent condition. The slabs are produced in standard width of 2,400mm, and are generally propped during construction until the in-situ concrete topping has gained sufficient strength. A composite slab using precast lattice girder units with spherical void formers is shown in Figure 3.

Courtesy of PCI Design Handbook, 2004

Figure 2. Precast double-tee slab production in casting yard

Figure 3. Precast lattice girder slab under construction

OVERHEAD VENTILATION DUCT CONSTRUCTION

The railway tunnel construction in densely populated urban areas was constrained by a tight schedule. In practice, building internal structures within the tunnel could not start until the excavation of tunnel has been completed. On a new railway project in Hong Kong, the Contractor with support by Arup Hong Kong succeeded in replacing the conforming cast in-situ and "full" precast concrete overhead ventilation duct in tunnel through the use of a semi-precast concrete slab system and a hybrid construction method. This enables the construction of overhead ventilation duct to be carried out concurrently with the excavation of tunnel.

The tunnel is a twin-track, single tube tunnel with a central partition wall separating the uptrack and downtrack running tunnels. The horseshoe shaped tunnel is approximately 2,500m in length and have a finished internal span of 12.6m, except at the bifurcation cavern where the internal span is increased to 16.6m. The tunnel has an overhead ventilation duct system. It used to be created by suspending an in-situ and a "full" precast concrete slab along the tunnel lining on corbels. The clearance within the ventilation duct varies associated with the arched profile for a maximum height of 2.8m, and it extends from the mid-length position in the tunnel to the ventilation building. The tunnel was excavated by using the Drill-and-Blast method, which was the preferred method for tunnelling in hard rock condition in Hong Kong. A typical cross-section through tunnel showing the overhead ventilation duct can be seen in Figure 4.

After the contract was awarded, the Contractor was looking for an alternative method to construct the overhead ventilation duct that would deliver a quicker and more buildable project to the Client. Various semi-precast and precast prestressed slab options were considered for the ventilation duct. The selected method was based on the precast lattice girder slab with in-situ structural topping spanning at maximum of 7.2m between the in-situ concrete corbel and the central partition wall. The decision to use the hybrid concrete construction followed a value engineering exercise. By combining the speed advantage of precast installation and the cost advantage of in-situ concrete, the Contractor concluded that the hybrid construction method could give a high-quality and robust structure without compromising the performance requirements of

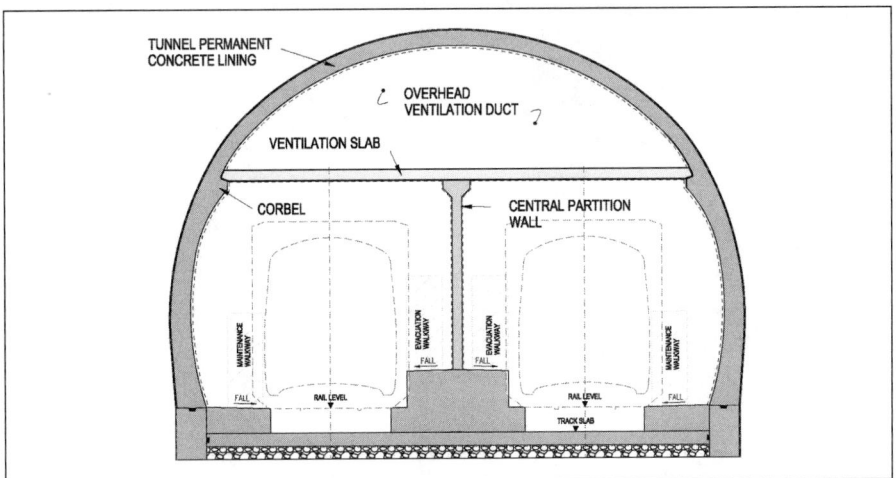

Figure 4. Typical tunnel cross-section with ventilation duct

ventilation duct in the conforming design. This led to a complete new design for the ventilation duct. The design principles and key considerations of this alternative composite structure are discussed in the following sections.

COMPOSITE CONCRETE SLAB DESIGN

Lattice Girder and Composite Slab Details

The main feature of this hybrid concrete slab system was the use of precast lattice girder panels to act as a permanent formwork for the slab. This reduced the weight of the slab for only a small reduction in flexural strength and structural stiffness. The precast panel was designed as one-way spanning slab between the corbel and the central partition wall. Lattice girders were partially embedded within concrete panel containing reinforcement in two directions, providing a precast panel for in-situ concrete topping. At the bifurcation cavern where the span between supports exceeded 6m, temporary propping at mid-span of the slab was specified to increase the load carrying capacity of the precast slab for the wet in-situ concrete.

The lattice girder was designed as a three-chord lattice with an isosceles triangular sectional profile. The lattice girder was manufactured with a 16mm diameter steel bar at the apex and two 16mm diameter steel bars at the base corners. The apex bar was separated from the side bars by a pair of 10mm diameter sinusoidally bent steel bars that were welded at each node to the main bars. No connection was required between the basal bars, as they were restrained horizontally by the concrete panel. Four types of precast slab were designed for the typical tunnel profile and to enable the transition to the bifurcation cavern. The precast panels were between 140mm and 160mm deep, and weighing about 4.3 and 5.4 tons respectively. A 90mm to 160mm thick in-situ concrete topping was added to the top of the precast panel, making the total construction span to depth ratio of about 22 and 24 for the typical tunnel profile

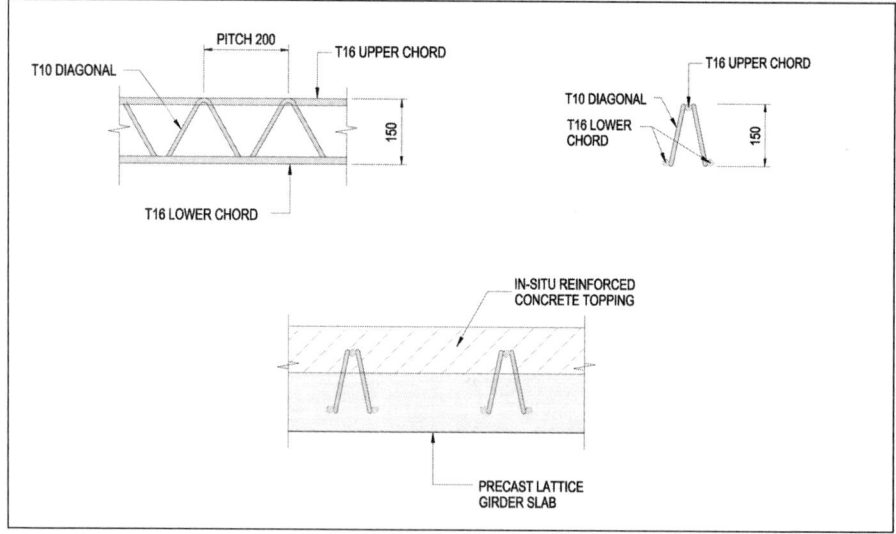

Figure 5. Lattice girder and composite slab details

and the bifurcation cavern respectively. Details of the lattice girder and composite slab are shown in Figure 5.

Design Principles and Criteria

The alternative ventilation slab was formed by two structural components: the precast lattice girder slab and the in-situ concrete topping. The use of a structural topping was aimed at limiting the self-weight of precast panel not to exceed the lifting capacity of plant and equipment. The precast slab was designed to act compositely with the in-situ structural topping, although the precast unit was capable to carry temporary loads without reliance on topping. The lattice girder allowed the in-situ concrete topping with light reinforcing bars and the precast panels to achieve composite action. This increased the spanning capabilities of the slab in permanent condition. The structural behavior of the proposed slab relied implicitly on composite action. The following criteria were considered in design of this composite structure:

- Flexural and shear resistance, and deflection
- Transfer of horizontal interface shear
- Differential shrinkage
- Joints and connections

A Two-Stage Computation

In order to assess the effect of topping concrete in the structural element, two distinct situations were considered in the composite slab design, before and after the structural in-situ concrete topping has reached its design strength. Despite the demoulding, transportation and stacking stages, the critical load case to the precast slab design was when the wet in-situ concrete was placed on the precast unit. During the temporary condition, the precast lattice girder slab was checked against the flexural and shear stresses, and deflection induced by its self-weight and the weight of wet in-situ concrete top-

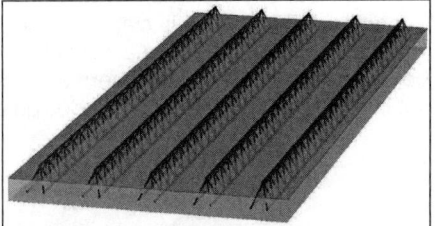

Figure 6. Oasys GSA analysis for precast concrete slab design

ping taking into account of a construction stage loading of 1.5kN/m². The strength of the precast slab was determined by adjusting the size of bars and the spacing between lattice girders. This procedure sought to provide sufficient shear stiffness for the wet in-situ concrete. The structural analysis of the precast lattice girder slab was performed by using Oasys GSA, as shown in Figure 6. At the bifurcation cavern, immediate propping was introduced to the precast slabs for which to reduce the construction stage stresses and deflection, even though this has posed some time and cost penalties on site, and induced negative "hogging" moments over props which was catered for in the detailed design.

In the permanent condition, the hardened in-situ concrete provided the compressive resistance. The overall section properties of the composite slab was used in the calculation. The composite slab was designed based upon the total weight of precast and in-situ concretes, and the most onerous combination of the characteristic superimposed live load of ±3kN/m² and ±2.5kN/m² resulting from the air ventilation and train aerodynamic respectively. The positive "sagging" moment at mid-span of the slab was carried by the bottom reinforcement, as specified in the ordinary precast panel.

The negative "hogging" moment was resisted by placing reinforcement in the in-situ concrete over supports. In order for the precast slab with the in-situ concrete topping to act compositely, full horizontal shear force transfer must be attained at the interface between the precast slab and in-situ topping. Since the interface shear stress due to bending exceeded the permissible value at joints given in the Hong Kong Code of Practice for Precast Concrete Construction (BD, 2003), shear resistance through dowel action provided by the diagonal bars in lattice girders was utilized to enable effective transfer of in-plane shear within the slab. Structural compatibility of the composite slab was also evaluated for the effect of differential shrinkage to ensure static equilibrium being achieved between the concretes.

Joints and Connections

Design of the joints between precast slabs, and at supports on corbel and wall must be capable of transmitting the worst anticipated combination of loads and catering to any movement of the structure during the erection of the precast unit and after the construction of the composite structure. The jointing system of the slab was designed to ease the pressure of field installation and connection of precast slabs that might cause to the structure. Particular attention was paid to the joint and connection details of the composite slab in order to ensure that adequate air tightness and fire sealing were provided to achieve a robust structure. Joints between precast slabs and at supports were detailed so that air passage during the tunnel operation, as well as flame and hot gases during the incident of fire is prevented, and that any heat transmission will not exceed the limits specified in the contract.

Precast slabs expand when heated and the joints tend to close during fire exposure. Therefore, non-combustible and flexible insulating materials in conjunction with caulking material, were designed at joints between the precast slabs and at support on corbel to provide the necessary thermal, flame and smoke barrier while permitting normal volume change movements. The design oonnection joint detail at corbel would assure

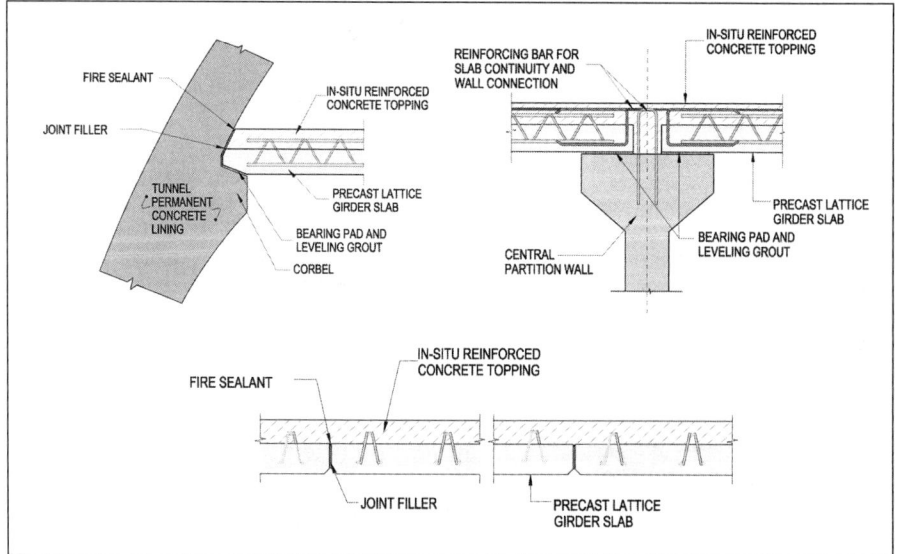

Figure 7. Design of typical connection joint between slabs and at supports

Stage 1: Precast lattice girder slab produced in casting yard

Stage 2: Precast slabs stacked at the storage area on site using timbers to prevent damage

Stage 3: Precast slab placed between corbel and central wall using gantry crane

Stage 4: Slabs topped with cast in-situ concrete to achieve a monolithic structure

Figure 8. Sequence of construction for composite concrete slab

the permanent tunnel lining and the ventilation slab be performed independently without any load transfer between the structures due to deformation of the tunnel lining. Structural connection between precast slabs and central partition wall relied on in-situ concrete topping detail. Continuity of reinforcement was provided by the lapping with projecting reinforcement in the central partition wall. This was an effective detail in structures where slab continuity was required as in construction and was inherently robust. Connection details between precast slabs, and at supports on corbel and central partition wall are shown in Figure 7.

COMPOSITE CONCRETE SLAB CONSTRUCTION

The precast slabs were installed by using gantry crane. This custom-made gantry crane was designed to allow the precast slabs to be rigged for balancing, remaining vertical and in line with their centres of gravity to prevent any undesirable and excessive movement, which could induce additional stresses on the precast units and lifting devices due to dynamic loading. The use of gantry crane has standardized the lifting devices and the procedures of lifting operation taking into account of the constraint of tunnel geometry and the confined working environment in tunnel. This procedure prevented the frequent changes of lifting method. The sequence of construction for the overhead ventilation duct using hybrid construction method can be seen in Figure 8.

CONCLUSION

The precast concrete system is generally precluded from consideration among civil engineers to construct structures inside a tunnel because of the excessive self-weight of precast unit and insufficient headroom for lifting and erection operation. In order to address this technical issue and remove the essential constraint on size of pre-cast slab, a game-changing technology that utilizes a mix of cast in-situ and factory-produced precast units has been adopted to build the overhead ventilation duct for a large diameter railway tunnel in Hong Kong, With the use of advanced prefabrication techniques and equipment for field installation, the hybrid concrete technology which combined the in-situ and precast concrete has optimized the benefits of concrete construction. The system not only improved the buildability of this project but also accelerated the work cycle and reduced the cost for construction of overhead ventilation duct inside a tunnel. The positive outcome of this case study has challenged engineers to form a realistic assessment of the application of precast concrete construction in any confined environment against some conventional biases.

REFERENCES

BD. 2003. *Hong Kong Code of Practice for Precast Concrete Construction.* Buildings Department, The Government of the Hong Kong Special Administrative Region. p. 61.

PCI. 2004. *Design Handbook on Precast and Prestressed Concrete*, 6th ed. U.S.A.: Precast/Prestressed Concrete Institute. p. 736.

The Use of Saturation Diving Techniques in Support of Pressurized Tunnels

Justin Costello ▪ Ballard Marine Construction, Inc.

INTRODUCTION

Hyperbaric conditions allow for increased access to the components of tunnel boring machines which are subjected to pressure from the surrounding soils and groundwater. This paper will briefly discuss some of the options and considerations when considering or being prepared to use Compressed Air Workers.

BOUNCE VS. SATURATION

Two distinct types of hyperbaric exposures are bounce and saturation. Bounce mode refers to a hyperbaric exposure followed immediately by decompression. Decompression maybe either "constant slide" or "staged." Constant slide is the continuous reduction of pressure until the exposed worker is at atmospheric pressure. Staged decompression is a reduction in pressure to various "stops." Each stop is timed and typically involves the inhalation of pure oxygen.

As an alternative to this, saturation may be used. In saturation, workers are compressed in a hyperbaric chamber and their body tissue can equilibrate with the pressurized gas in the system. At the point of equilibrium, the decompression duration is fixed. Any additional time spent at pressure will not require additional decompression time. This allows for work shifts which are not determined based on pressure. Cutter head maintenance or repair can be done around the clock with two or three crews.

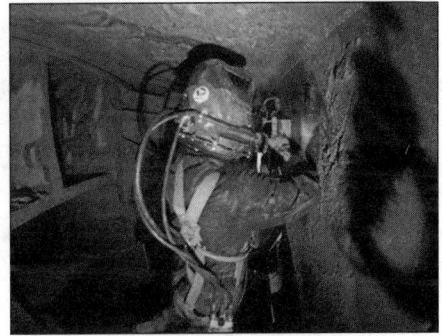

Workers live in a pressurized chamber, travel to and from the TBM in a pressurized shuttle, and work on the TBM itself at the same pressure. Work cycles of up to 28 days are possible with this technique. Decompression is typically a constant slide and takes place at the end of the work cycle. Decompression from saturation typically takes several days but is usually above ground and restarting of the TBM will not interfere with it.

Photo Courtesy of Ballard Marine Construction
Figure 1. Compressed air worker

NITROGEN VS. HELIUM

The composition of air is 20.9% Oxygen, 79% Nitrogen, and 0.1% other trace gasses. Nitrogen has a density of 1.251 g/L at standard temperature and pressure. Helium has a density of 0.08988 g/L at standard temperature and pressure. When pressure increases, as in the case of compressed air work, the density of the Nitrogen can make breathing difficult as the breathing medium feels viscous like breathing "syrup." When Helium is used in place of Nitrogen, this effect is reduced.

When pressurized, Nitrogen starts to have a narcotic effect. This increases with pressure. At high pressure Nitrogen narcosis becomes both a hindrance to safety and productivity. To avoid the effect, Helium can be used as an inert gas in the place of Nitrogen. Helium has the advantage of providing clear heads for the workers but there are also several drawbacks to the gas.

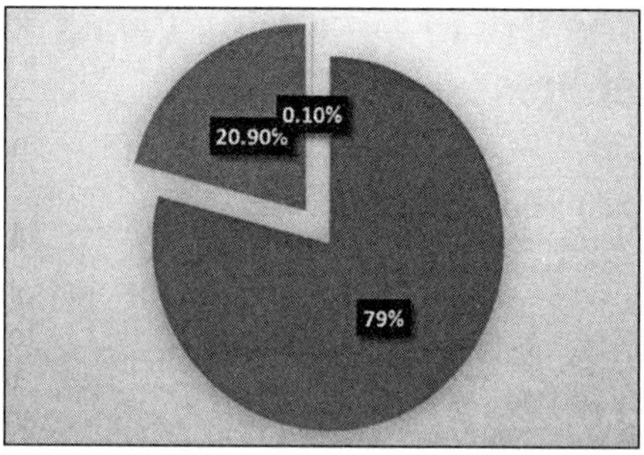

Figure 2. Air composition (% by volume)— 79% Nitrogen / 20.9% Oxygen / 0.1% Other

It can be difficult to find sufficient quantities of Helium and the cost is typically high. Long lead time in ordering the gas can delay an unexpected intervention if the necessary volume is not acquired ahead of time.

The thermal conductivity of Helium is high as well. Workers' lungs will quickly heat all inhaled Helium to body temperature then exhale the warmed gas. This can cause hypothermia if the gas is relatively cool. The living quarters are kept very warm for this reason.

Helium affects the voice as well. Workers' voices will become high pitched and difficult to understand. Specialized communication devices which lower the pitch of sounds coming through it are employed.

OXYGEN PARTIAL PRESSURE

Much like Nitrogen, Oxygen can cause problems in the body when concentrations get too high. The partial pressure of oxygen is the percentage times the atmospheres of pressure (absolute). Oxygen at partial pressures of 1.6 ATA and higher can cause central nervous system toxicity. Symptoms of this include blurred vision, tinnitus, tingling sensation, numbness, irritability, dizziness, and seizures. For this reason, the percentage of Oxygen is strictly controlled in saturation operations. One example of partial pressure of Oxygen used in saturation is 0.44 atmospheres.

COMPRESSING THE SYSTEM

Helium-Oxygen blends (heliox) must be brought on site for the initial compression of the habitat and shuttle.

Because the targeted partial pressure of oxygen at the "storage depth" is around 0.44 atmospheres, and partial pressure is a function of percentage and pressure, the oxygen percentage is quite low. With a storage depth of 7 bar, the oxygen percent is 5.5. The workers at this pressure can thrive on this concentration due to the pressure. Physiologically, the workers feel as though they were breathing 50% oxygen. A problem arises when compressing workers in the system, however, as 5.5% oxygen does not support life at atmospheric pressure.

This problem is overcome with the use of two blends of gasses. Workers enter the chamber system with normal air at atmospheric pressure. A "high mix" of heliox which still supports life at atmospheric pressure is used to pressurize the system to a predetermined depth. A "low mix," usually 2% oxygen and 98% Helium is then used to complete the process. Upon arrival at storage depth, the oxygen from the air, high mix, and low mix should combine to form the final percentage desired. This calculation is done at the time of each compression by a Life Support Supervisor. The quantities and blends of gasses, system volume, and final storage depth are taken into account.

SAFE BREATHING MIXES

As a safety measure, there must be a volume of safe breathing gasses on site which is adequate to accommodate a complete decompression. The breathing gasses will change their concentrations of oxygen as decompression takes place. As the pressure is reduced, the percentage of oxygen must increase to support life. This will allow for the safe decompression of a work crew even if the atmosphere in the system has been contaminated or otherwise compromised. Therapeutic gas must also be available for use in the event of decompression sickness during saturation. Like the safe breathing gas, several blends may be needed to assure the correct partial pressure of oxygen is always available at any pressures.

A drawback to this requirement is several banks of gasses must be kept in reserve with no plan for their use. This can be minimized through on-site gas blending capabilities; however, the increased cost of the equipment needed for blending may be prohibitive unless it is used elsewhere on the system as well.

Hypoxia is a condition characterized by deficiency in the amount of oxygen reaching the tissues, generally caused by inadequate supply of oxygen in the blood due to reduction of partial pressure of oxygen.

METABOLIC OXYGEN MAKEUP

Humans typically consume around half a liter of oxygen per minute. The atmosphere in the living quarters must be analyzed and monitored 24 hours per day. Because of the 50% oxygen equivalent, there is some room to allow the concentration to drop as the workers breathe. Periodically, however, the Life Support Supervisor must "dose" the atmosphere with pure oxygen to make up for the metabolic consumption. The oxygen is injected at a fan inside the system to allow for rapid mixing.

WORKING MIX

The largest quantity of heliox gas used on a saturation project is in the form of working mix. This mix is delivered at the TBM through hoses directly to the worker. An umbilical connects the worker's helmet to a distribution manifold. Heliox is inhaled through the umbilical then exhaled into the work space. During a typical shift, each worker can use 100 cubic meters of gas.

There are systems in use on commercial diving operations which capture the exhaled breath of a worker underwater and return it to the vessel for processing. This reclaim system includes equipment and operators to return the exhaled gas to a usable state. A technician must remove carbon dioxide from the captured gas then dehumidify it. The dehumidification process cools the mixture so reheating must then occur. A semi-permeable membrane may be incorporated to remove any Nitrogen which may have built up in the mixture through the incorporation of air during the locking-on and locking-off of chambers. The gas is then held in a tank with mixing paddles and Oxygen is added until the correct blend is achieved. In a batch by batch process, the gas is recompressed to be held in high pressure tanks for future use.

CARBON DIOXIDE

Humans exhale around half a liter per minute of Carbon Dioxide. The atmosphere is analyzed, and CO_2 is not allowed to accumulate in the system.

Carbon Dioxide is removed chemically through canisters filled with 4-8 mesh granulated soda lime. Fans circulate the gas in the chamber through the canisters either within the habitat or a standalone system. The chemical must be replaced often. Chemicals for odor removal may be mixed in with the soda lime as well. Soda lime is used at a rate of approximately 8 kg per person per day, or 250 kg per person per month. Problems can arise when this chemical must be imported through customs, as opening the containers will expose the chemical to carbon dioxide in the atmosphere and start the chemical reaction. If containers are not properly re-sealed, the product will become useless. There are on the market versions of this chemical which turn colors when the chemical becomes used. The use of this indicating version can improve efficiency and reduce unnecessary over use.

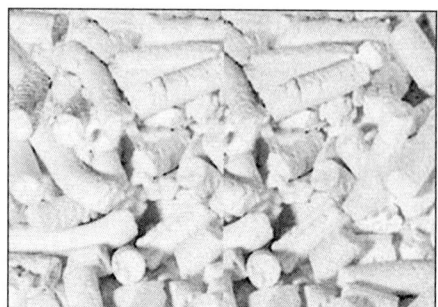

Figure 3. Example of color changing commercially available soda lime

TEMPERATURE

Due to the thermal conductivity of Helium, workers will heat inspired gas immediately then exhale the heated gas into the environment. Due to this heat transfer directly from the body's core, hypothermia can occur at higher than expected temperatures. The living quarters must be kept at a higher than normal temperature. Workers will feel comfortable at much higher temperatures than if they were at atmospheric pressure. Water or glycol systems with heaters and chillers are typically employed to maintain temperatures. A failure in the thermal control system is very serious and is cause to begin decompression.

HUMIDITY CONTROL

As pressure increases so does the capacity for gas to retain water vapor. With the warm environment in the saturation living quarters, humidity must be strictly controlled to avoid bacterial growth. All showers should take place in a separate chamber when possible. Shower water should be bilged out of the system immediately as well. When the shower is in use, doors should remain closed to reduce the spread of humid gas into the rest of the system. Saturation habitats with the ability to dehumidify are becoming increasingly more common.

SANITATION

The combination of high humidity and temperature creates an environment in the saturation living quarters which can support a high level of bacterial growth. For this reason, surfaces must be sanitized often. Deck plates must be removed by the workers and surfaces wiped down with a mild bleach solution, benzalkonium chloride, betadine solution, or other sanitizer. Workers should use prophylactic antibiotic ear drops daily as well to avoid ear infections.

CAMERAS

Due to the wide range of maladies which can occur in a hyperbaric environment, workers must be continuously monitored. A chamber operator has monitors to watch every action of the occupants. If an incident should occur, the operator is then able to react quickly to mitigate the damage. While this can be an uncomfortable invasion of privacy, the benefits outweigh the inconvenience.

FIRE SUPPRESSION SYSTEMS

There is a "fire zone" where combustibles ignite much more readily in a hyperbaric environment. When the oxygen percentage gets below this zone due to increased pressure (and thus, increased partial pressure), fire is much less of a risk. Nevertheless, hand held fire extinguishers which are specifically designed for use in hyperbaric facilities are employed. There is also often a water hose with nozzle and overhead sprinkler system inside some saturation habitats.

EXCURSIONS

In commercial diving, the pressure at which the workers live (storage depth) is often slightly different than the pressure of the water depth in which they work. This may be done due to the scope of work or to save resources. When lowered into the water in a pressurized bell, a hatch opens in the bottom at the depth of water equal to the pressure inside. Divers have a limited distance they can travel above or below this depth. This is called their excursion distance. In tunneling, excursions could be used in this manner as well. For example, workers could be stored at 6.0 bar and conduct work on the TBM at 6.4 bar. This would equate to a 0.4 bar excursion. A small amount of decompression would be needed to bring the workers back to their storage depth. This is only likely when resources are very limited, as the additional processes would likely slow production rates considerably.

CONTRABAND

There are many types of items which are not allowed in the saturation habitat. Sealed containers of any kind are prohibited. The change in pressure will crush nearly any object. Glass vessels are also not allowed. Electronics may or may not be permitted, based on the voltage and company policy. Fuels of any kind are to be kept out. Lighters, strikers, matches, perfumes/colognes, oil, grease, etc. are all closely monitored for. Personal bags may be checked before workers are allowed to enter the chambers for saturation.

NUTRITION

Due to the increased work time on the TBM, increased effort of breathing gas of a higher density, and the thermal conductivity of Helium, workers will typically require a higher caloric intake each day. A nutrition plan should be made in advance of work in saturation. "Bell lunches" are typically brought with the workers while they are on the TBM. Coffee, tea, and snacks should be available at all times to the workers. A pantry may be established on site to accommodate this. All

Figure 4. Carbonated beverages are frowned upon as well, as they introduce carbon dioxide into the pressurized environment

meals should be transferred into the saturation system just prior to consumption and all dishes should be removed immediately after. Meals may be brought in from local restaurants, prepared by a nutritionist, or directly on site. Careful attention to food allergies must be paid.

FOOD AND EQUIPMENT TRANSFER

In order to get food, beverages, equipment, and clothing in to the workers, transfer locks must be employed. A chamber in the sidewall of the saturation habitat serves as a food transfer lock. Each meal, pot of tea, and snack must travel through the lock to get to the workers and back out. When the CO_2 absorbent material is used up, canisters must also be sent out to be refilled. For larger items, such as laundry bags, a larger equipment lock may be used. It is also common for workers to leave their dirty laundry in the trunking during a Transfer Under Pressure (TUP) and allow support personnel to recover the bag once the shuttle locks off. Support workers may also load cleaned laundry into the trunking prior to the return of the shuttle so hyperbaric workers have clean clothes upon the completion of their shift on the TBM.

PERSONNEL TRANSFER

To service the TBM, workers must travel from the habitat, usually on the surface, to the man locks on the machine. This is accomplished using a hyperbaric shuttle. The shuttle must be large enough to house all workers plus a medic if necessary. All utilities to maintain life must be onboard as well. Power, heating and cooling, gas analysis, Heliox, metabolic Oxygen, audio and video communications, and fire suppression must all be present. The pressurized shuttle connects to the habitat via a flanged coupler. When the coupler is fastened, the space between the habitat and shuttle is compressed to the storage depth. Workers may then pass through the space and enter the shuttle. The process is then reversed to disconnect the shuttle from the habitat. The shuttle must travel to the TBM and connect to the man locks. Lifts, rails, pressurized tubes, or some combination of these may be used to establish the connection. Another flanged connection is made, and workers again pass through. The pressurized spaces of the TBM must either be filled with Heliox, or workers must don helmets which supply them with safe breathing gas. The latter is much more cost effective. Helmets provide not only safe breathing gas, but audio communication as well. Positive-pressure helmets provide protection against air ingress should the neck seal fail.

MEDICAL SURVEILLANCE

All workers entering a saturation system should undergo a physical examination. A medical surveillance program should be in place throughout the saturation process. Due to the length of decompression required, medical team members should be ready and able to enter the hyperbaric system should an injury occur. Once pressed in, medics must commit to the complete desaturation process which may take several days. Any mechanical injuries sustained in saturation may require specialized attention at the completion of decompression (broken bones re-set, e.g.). Upon completion of the 28-day saturation cycle, all workers should undergo an additional medical examination and remain under observation by the medical staff for 24 hours prior to leaving. Under no circumstances should workers fly on commercial airlines until at least 24 hours after exiting a saturation system.

REGENERATION

The saturation habitat must be dehumidified, heated, and scrubbed of carbon dioxide. This may be accomplished through internal mechanisms, or with the use of a regeneration system. The system is a set of external equipment which moves gas from the habitat through a series of processes to accomplish the goals stated above. Gas is passed through a large container of soda lime, chilled to remove moisture, and heated before returning to the habitat. A regeneration system allows for additional space in the habitat as well as the ability to use larger and more powerful equipment. An external CO_2 scrubber allows for the changing of soda lime less frequently and without disturbing or waking the hyperbaric workers. Additional space is required, however, and a regeneration system adds expense.

MALADIES

It is possible that workers can suffer from a mild form of decompression sickness while working in the pressurized air spaces of the TBM. If skin is exposed to air—70.8% of which is Nitrogen, and the inert gas which saturates their tissues is entirely Helium, a "switching" of the two may occur. This effect is called isobaric counter diffusion. The skin can react with pain, rash, itching, or mild swelling. This may be prevented using sealed suits which are filled with heliox. A small amount of Nitrogen in the breathing mix may also reduce the effect. Most saturation operations begin with the habitat and shuttle filled with air at atmospheric pressure and Heliox used to compress to the storage depth which will introduce a small fraction of Nitrogen. Each shuttle transfer will do the same.

High pressure nervous syndrome is a possible malady in saturation operations as well. The syndrome causes twitching of the hands and is typically a result of very fast compression to high pressure. HPNS is avoided by compressing workers into a system at a steady rate over the course of several hours when the pressure is high.

Pulmonary oxygen toxicity can occur if the partial pressure of oxygen is kept at an elevated level for prolonged periods of time. The resulting substernal burning sensation can be treated by lowering the oxygen concentration.

TOILETS

The flushing of toilets in saturation is a multi-step process. The pressure in the system, not gravity nor water pressure, is used. An interlock must be in place which prevents anyone from sitting on the toilet while the flush occurs. This is typically accomplished by incorporating an armature on the valve which is used during the flush. When a worker needs to flush the toilet, they will fill the toilet bowl to a predetermined level with water. They will then open a valve near the toilet. As the valve is opened, the armature impairs the ability to sit on the seat. When the internal valve is opened, the worker calls the Life Support Technician (LST) and asks for a flush. The LST will send an Assistant Life Support Technician (ALST) to the outside of the chamber in which the toilet is located. The ALST will confirm via phone that a flush is needed. The ALST will then

Figure 5. Pressurized facilities in Ballard 10-man saturation facility

open a spring-loaded valve which opens the piping from the toilet to a closed tank of a fixed volume. The contents of the toilet move from the habitat to the tank until the pressure is equalized. If the toilet bowl was filled to the proper level with water prior to flushing, no atmospheric gasses are lost. The ALST then shuts the spring-loaded valve connecting the habitat to the closed tank and opens a new valve which dumps the tank to a holding tank or portable restroom facility. The tank is left at atmospheric pressure and ready for another flush. At no time is a direct route from the habitat to atmospheric pressure created.

DECOMPRESSION

Decompression from the storage depth must occur well enough in advance to ensure the workers will reach atmospheric pressure within the 28-day window. The partial pressure of Oxygen is often increased and pressure is released slowly over the course of several days. It is common for "holds" to occur periodically as well during the night. Operators must add Oxygen more frequently as the pressure is reduced to maintain the correct partial pressure.

CONCLUSION

There are a few key decisions to make that will increase the chances of having a successful project. Selecting the right hyperbaric contractor for the project, and early contractor involvement are the keys to keeping a tunneling project moving. Equipment procurement and setup, planning, and permits can take many months and on some projects, years. Saturation techniques can allow for very high pressure work and long shift lengths on tunnel boring machines. There are many similarities between compressed air work and commercial diving but it takes an experienced hyperbaric services provider to produce the safe and efficient results needed.

Use of Infrared Technology to Detect Backfill Voids Behind Steel Lining in Tunnels

Alexander MacKinnon ▪ Hatch Corporation
Bruce Harland ▪ Hatch Corporation

ABSTRACT

The traditional method for discovering voids in the backfill behind a steel liner is by systematic tapping the liner interior with a hammer.

This method was specified for the Seymour Capilano Twin Tunnels Project, but issues were recognized and alternative methods were sought.

The hammer test was deemed to be subjective in determining 'pass or fail' criteria. Also hammer blows could potentially damage the pipe polyurethane lining. The paper explains how infra-red technology was employed successfully as an innovation and an improvement on the hammer test for detecting voids.

INTRODUCTION

At the Seymour Capilano Twin Tunnels (SCTT) project an infrared (IR) camera was used to inspect and direct annular and contact grouting operations behind steel pipe. Research could not find a record of infrared camera technology being used for this application. Methodologies for inspection using this technology on SCTT were developed and implemented from 2013 to 2015, to confirm the quality of the backfill throughout the steel lined segments of the tunnel.

Project Background

The SCTT consists of two 3.8 m diameter, 7.1 km long, TBM-bored tunnels in North Vancouver, British Columbia, Canada. The tunnels connect the Capilano Reservoir to the new Seymour Capilano Filtration Plant (SCFP). The Raw Water Tunnel carries untreated water from the Capilano Reservoir to the SCFP while the Treated Water Tunnel returns the treated water to the Capilano Valley for distribution. The tunnels have a capacity of 1.25 billion liters per day.

The unlined portion of the tunnel largely consists of granite and granodiorites reinforced with shotcrete, rock bolts and steel sets as ground conditions required. The tunnel was excavated entirely within bedrock, with cover ranging from 180 m at the Seymour Shaft to over 600 m. Approximately 9 lineal km of the two tunnels are in solid rock or lined with shotcrete.

In addition to the shafts, 4.8 km of steel pipe lining was installed in the tunnel. The steel liner consists of a 3.0m diameter

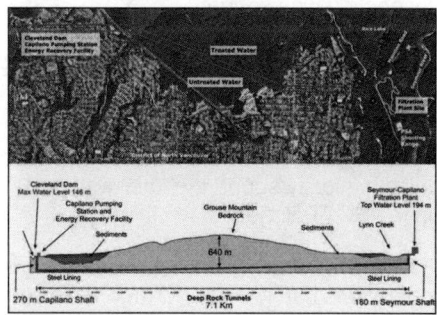

Figure 1. Plan and profile of the Seymour Capilano twin tunnels

polyurethane-lined steel pipe, with steel thicknesses varying from 12 mm to 34 mm depending on the expected external hydrostatic pressure. The liner prevents groundwater pressure increases in the lower elevation areas above the tunnels. After steel pipe installation, the annulus between the steel and excavated tunnel perimeter was backfilled with concrete, inspected, and then contact and void grouted.

Annular Backfill

The process of annular backfilling was performed through several injection ports located in each steel pipe. In each 12 m pipe, a minimum of 3 injection ports were located in the invert, and 5 injection ports in the crown. Additional injection ports were drilled in areas previously identified as likely to form an airlock during annular backfill.

The annulus was filled in six lifts to limit the backfill induced hydrostatic pressure placed on the unconfined steel pipe. Backfill lifts were placed in lengths of up to a kilometer prior to the succeeding lift. Backfill concrete was typically pumped 100 m into an injection port and required to flow 25 m in both the downstream and upstream directions from the port. The nearby injection ports were fitted with ball valves and standpipes to provide confirmation of backfill progress and depth.

The annular backfill was batched on the surface, remixed in the chambers and pumped through a steel slickline for up to 1.4 km to the point of injection. On several occasions plugs in the slickline, pump problems or problems with the batching system prevented backfill lifts from being fully placed prior to hydration of the backfill. This resulted in uneven backfill placement, irregular cold joints and increased potential for airlocks on upper annular lifts.

In several areas, groundwater inflows had the potential to wash away the annular backfill prior to curing. In these areas, groundwater was diverted around the annular backfill using geotextile fabric and plastic panning into drains installed in the tunnel invert. The drains were backfilled following the completion of the top annular lift.

To ensure the structural stability of the steel lining and reduce hydraulic conductivity along the tunnel alignment, all voids in the annular backfill had to be filled. Due to the port spacing and lack of ports near the springline of the pipe it was difficult to verify the successful installation of backfill using visual methods, such as borescope inspection. To confirm that any voids in the annular backfill were filled, an in situ method of backfill inspection was sought.

TESTING METHODS

Traditional Inspection Method

The traditional method of discovering voids in the backfill behind a steel liner is to systematically sound the liner with a 5lb hammer. This method was specified for use in the SCTT Project.

The SCTT has over 4.8 km of steel lined tunnel where contact and void grouting operations had to be performed. This corresponds to a surface area of 45,500 m². The scale of the sounding by hammer operation presents several significant challenges. On a 300 mm by 300 mm pattern, over 500,000 soundings would be required to test the entirety of the horizontal steel lined tunnel. Any increases in inspection resolution would also increase labour intensity to the square of the resolution change, discouraging more precise inspection. A significant effort would be required to produce results with a high degree of confidence of successfully testing the entire backfill volume.

As the pass/fail criterion in a hammer test is somewhat subjective, great care would be required to achieve consistency between hammer tests. Many factors could potentially affect test results such as variations in perception between crew members, variation in crew strengths, crew fatigue and variation in hammering blow intensity, or changes in working angle. Steel pipe up to 34 mm thick adds to the difficulty of reliably distinguishing between areas of competent backfill and areas which contain a small void.

For a 3.0 m diameter pipe scaffolds or a moveable platform would be required to access the crown, requiring a crew to remain with the inspectors for the duration of the hammer tests. Additional cleaning would be required to remove all water, dirt and foreign material from the invert prior to hammer testing to prevent irregular readings.

Use of hammer testing also could damage the polyurethane lining, possibly impacting the project critical path and creating additional expense.

Infrared Equipment Test

It was hypothesized that infrared cameras could detect voids behind the steel pipe by measuring changes in steel lining surface temperature. Hatch set up a mock void test and used an IR camera to test this theory.

The initial tests were done on pipes in the Owner's storage yard. The pipes were stored on a bed of sand. When viewed through the camera a temperature gradient was observed where the sand was in contact with the steel pipe.

Several small areas of sand removed from underneath the pipe and a PVC pipe was inserted under the pipe to expose portions of the pipe to the air. After a short interval the areas where the sand had been excavated became visible and later the area where the PVC pipe had been inserted became visible.

Once the surface tests were complete, the IR camera was tested underground. Cold water injected into the annulus was clearly observed with the IR camera through the steel pipe and polyurethane lining.

Infrared Inspection Technique

With the successful testing of the infrared camera, an inspection procedure was developed to detect voids in annular backfill. The IR camera procedure included providing a visual record of every installed backfill lift and details of any abnormalities.

Field tests of the IR camera on the initial backfill lifts were performed to determine what length of pipe could be tested from a single station. The camera used had a sensor resolution of 320 x 240 pixels. Given the 25° viewing angle of the camera lens, an effective resolution of 300mm by 300 mm could be achieved at a distance of three pipe lengths (36 m). In practical use some pipe details were found to wash out at a distance of approximately 30 m depending on the angle of the surface being viewed.

The inspection technique used required the inspector to begin at one end of the installed pipe and walk towards the opposite end of the pipe, checking the infrared camera viewfinder at every pipe joint (every 12 m). At every second or third pipe joint a photo was taken to provide visual records of each installed backfill lift. Photos were taken of any abnormalities encountered by the inspector. Additional photos and notes required to identify the exact location of the abnormality were also taken.

The IR camera had the ability to enter notes as text or record audio notes and attach to the IR photo. As the pipes are largely indistinguishable from one another in photos, this feature was used to ensure that the location, perspective and subject of every photo could be determined upon review. Once recorded on the camera, notes were bundled with the photo in a PDF photo report.

Capture of a single infrared photo, recording of notes and report generation took an experienced inspector an average of approximately two minutes to produce. No additional crew or supporting equipment was required for the inspections. A 1.5 km section of tunnel would typically take 2–4 hours to fully inspect depending on the number of detailed IR photos required.

In any section of steel lined tunnel, as many as seven sets of infrared inspections took place. Inspections were scheduled to occur from 24 to 48 hours after annular backfill was placed to ensure that the backfill heat of hydration was visible.

Once properly configured, the use of an IR camera required minimal training beyond familiarity with operation of a typical point-and-shoot camera. The IR camera used had an optional manual focus and contrast adjustment. The software included with the camera also allows for contrast adjustment and postprocessing after the photos have been downloaded to a computer to ensure that the appropriate features are highlighted.

INFRARED INSPECTION PRINCIPALS

Camera Operations

Most commonly available infrared cameras use an array of sensors called Microbolometers to form an image. Microbolometers measure minuscule temperature changes on the sensor surface to accurately quantify the amount of radiation striking the surface of the sensor. While microbolometers can be used to measure most types of electromagnetic radiation, the lenses used in thermal cameras only allow infrared radiation to pass through, while reflecting or absorbing other wavelengths.

A thermal camera turns the radiation intensities into a temperature by using an assumed infrared emissivity for a material, this appears as a given pixel in a photograph. The FLIR® T420bx camera that was used at the SCTT was able to measure surface temperature variations down to 0.045°C. In field use, the camera displays surficial temperature changes in as small as 0.1°C depending on the settings used.

Thermal Gradients

Three mechanisms were identified during the infrared inspection which enabled the thermal camera to detect possible void locations in the annular backfill.

First, the annular backfill being pumped had a temperature that was often considerably higher than that of its surroundings in the tunnel. Typical temperatures of the backfill prior to injection ranged from 18°C to 27°C. This thermal energy quickly conducts through the steel pipe lining providing a clear outline of where the annular backfill is flowing. Any areas where the backfill was unable to reach appeared colder than the surrounding areas.

Second, as the annular backfill hydrates, a considerable amount of heat is produced by the exothermic reactions of curing concrete. The energy released increases the temperature of the pipe lining for several days following backfill placement. Layers of hydrating annular backfill form warm layers at a consistent location around the pipe. Any areas without annular backfill curing would appear colder than the surrounding areas where hydration was ongoing.

Third, as the groundwater entering the tunnels was usually a different temperature than the tunnel air and/or hydrating annular backfill, a temperature differential develops between the groundwater and the tunnel air. Areas where the steel lining was backed with competent backfill would be well insulated from the groundwater while areas of the steel liner which were adjacent to, or near flowing groundwater were less insulated from this same temperature differential. The groundwater flows were generally of a lower temperature than the ambient air temperature and the hydrating backfill, therefore any flows would locally draw down the temperature of steel liner and annular backfill. In areas where the volume of flowing water was high, the reduction in temperature became visible even if the water was not coming in direct contact with the steel pipe.

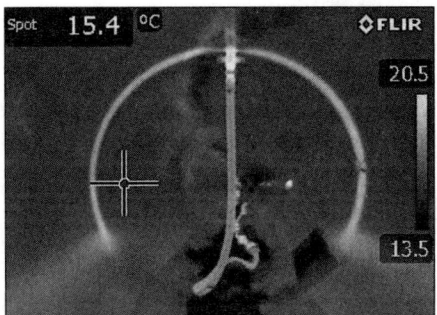

Figure 2. Warm backfill injected through crown port

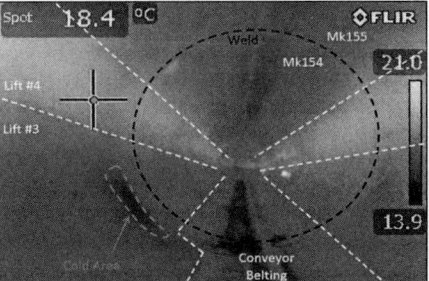

Figure 3. Hydrating concrete in upper lifts with a cold area in lower lifts

LESSONS LEARNED

While using an infrared camera is a relatively simple process, there are several factors which need to be accounted for to produce results which avoid false positives in looking for voids behinds the steel. Reflections, particularly bright sources of infrared, surface water and surficial differences between materials were found to have a considerable impact on the thermal camera readings.

Reflections

Infrared wavelengths can respond differently than visible wavelengths when refracted or reflected off a given material. Material such as glass which appears transparent in the visible spectrum may appear to be mirror-like when viewed at infrared wavelengths. Similar responses have been noted to occur on the polyurethane pipe lining used on the SCTT. Areas of pipe lining which appeared to have a matte or dull gloss finish to the naked eye often exhibited significant reflections in infrared photographs.

During an inspection, reflections were identified by changing the camera point of view and observing changes in the reported temperature relative to specific points in the photo. As the incident angle and reflection angle remain equal, the reflection would be seen to move at half the rate at which the camera was moved. In some cases where the reflection was only minimally distorted or the object had a distinct shape; the reflection could be identified without changing the point of view. Reflections which

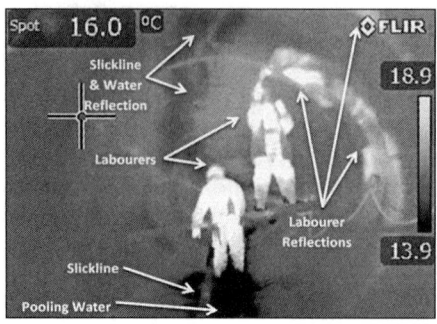

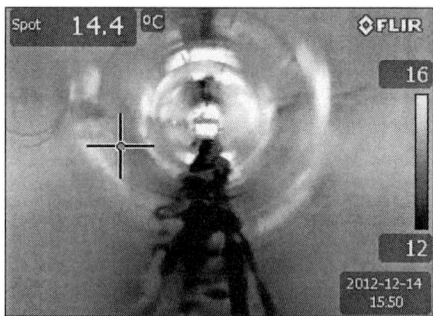

Figure 4. Labourers and reflections on tunnel lining　　Figure 5. Hot truck obscuring details

were not easily discerned from a single photo were noted during inspection to reduce the chance of being mis-identified during later photo analysis.

Contrast

Inspection of areas containing a particularly emissive source of infrared radiation, such as a crew of warm workers or a running vehicle required considerable attention to ensure that any voids would be visible. A large warm object has the same effect on an infrared photo as taking a visible spectrum photo with the sun visible in the background causing loss of foreground detail. Bright infrared sources resulted in photo details being easily hidden, shapes being washed out, and often rendered it impossible to identify voids when framing a photo from an angle parallel to the tunnel. To mitigate this, infrared photos were usually taken facing away from any intense infrared source, or by taking several photos at near perpendicular angles to the pipe. In photos taken at near perpendicular angles to the pipe, some reflections from the inspector would often be noticeable but generally did not obscure the presence of voids.

Water

Water is not transparent to infrared radiation. During hydration any ambient temperature water flowing on the surface of the pipe would show up as a cold area due to the lower temperature of the water and a reduced infrared emissivity relative to most materials in the tunnel. Groundwater flows encountered in the SCTT were generally between 8°C and 10°C.

Damp areas of the steel pipe were also found to undergo significant evaporative cooling, locally drawing down the surface temperature of the pipe lining. To ensure

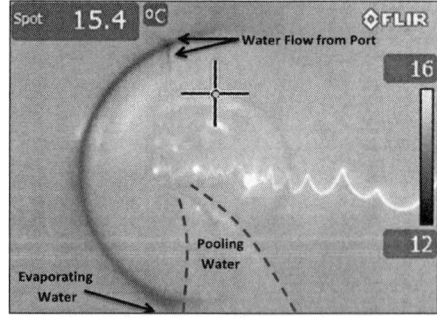

Figure 6. Water falling from injection port evaporation and pooling water

that water flows were not mistakenly identified as backfill irregularities during the photo analysis, inspectors had to note any significant water flows or dampness present during the infrared inspection of the steel pipe.

Emissivity

The IR camera uses the intensity of infrared emissions to determine the temperature of an object. Differing materials or surface finishes can emit different amounts of infrared radiation at the same temperature. The amount of infrared radiation that a material emits at a given temperature is called the materials emissivity or ε. The IR camera corrects for the material emissivity using an ε value as set by the inspector.

The emissivity of a specific material can be determined by using a thermocouple as a calibration point and by manually changing the emissivity until the camera temperature matches the thermocouple. Alternatively, emissivity can be found by applying a thin material of known ε to the surface of a material and changing the emissivity settings on the camera until the temperature of the unknown material is reported as the initial temperature of the tape. Electrical tape has a known ε between 0.95 and 0.97 and was used at SCTT to calibrate the camera. In areas where materials of multiple emissivities are to be measured, black electrical tape can also be applied across material boundaries to ensure a consistent ε.

Using electrical tape it was determined that all materials commonly encountered in the steel lining had an emissivity approximately equal to 0.95. Lists of typical ε values for various materials, surfaces and finishes can be easily found on the internet.

RESULTS

Infrared inspection was able to successfully identify and distinguish several types of defects in the annular backfill following installation in the SCTT. The inspection records have exceeded our expectations; they have been able to identify void formation both earlier (during annular backfill installation) and later (after annular backfill curing) than was initially thought practical.

After void identification, drilling locations were drawn on the pipe near the bottom of the void for injection and the top of the void for venting. All the locations drilled connected to a visible void, the groundwater aquifer or identified other type of defect. In several circumstances it was found that large voids in the annular backfill could be identified though upwards of 100mm of solid concrete. Installation fixtures in the annulus such as the presence of screen bulkheads or drainage panning were also identifiable from photos and compared to installation maps.

Several distinct types of backfill deficiencies were noted to have occurred in several locations. The recurring deficiencies were as follows.

Water Flows

Narrow groups of cold areas on a horizontal or sub-horizontal angle were noted in several areas of the tunnels. When drilled, these areas often contained a consistent water flow with no apparent void. When contact grout operations commenced, these areas were frequently found to have direct connections to the tunnel drain system, indicating that the water had bypassed the panning and geotextile used to divert the water flows around the annular backfill. These water flows washed away the annular backfill prior to the concrete curing.

The water flows were not immediately visible during installation or the early stages of curing, as the warm concrete contained enough energy to warm or obscure the water inflows. As the concrete hydrated, distinct cold areas would gradually become colder than the surrounding backfill. Water flows immediately adjacent to the steel liner were often visible in great detail, and visibly warmed as contact grout was pumped into the cracks. It was also noted that the water flows remained visible for several weeks after contact grout had finished curing. The small temperature differential between the tunnel air and the ambient temperature of the flowing groundwater was easily seen if the camera set to the correct temperature ranges.

During contact grout installation these types of features were typified by their small volume and rapid pressurization as grout was injected. Many of the grout injection ports also vented contact grout that was injected in nearby invert ports.

Cold Bands

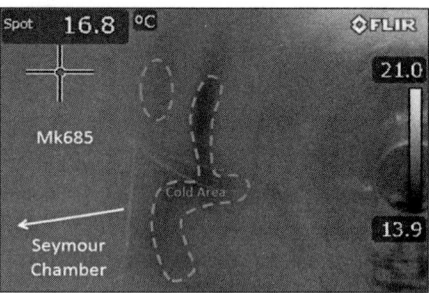

Figure 7. Water flow behind the steel liner

After the top two lifts of annular back-fill installation a number of similar band shaped features became visible from the springline to the crown. These Cold Bands were often 150–200mm wide and located immediately adjacent to a pipe port, extending through several layers of backfill installation. Upon checking records it was confirmed that majority of these locations were used for backfill on several distinct occasions. As the back-fill flowed down from the injection port, a thin layer of backfill would be left behind to harden after each injection. After several backfill lift installations a thick layer

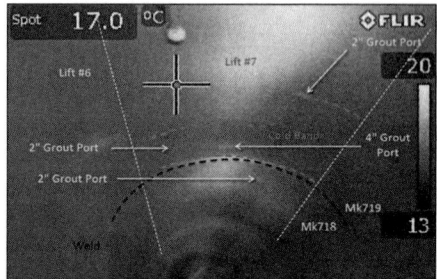

Figure 8. Cold band at 100mm grout port

of backfill would build up around each of these ports, insulating it from the hydrating concrete of the later backfill lifts.

These locations were identified as locations of possible blockage on upper annular backfill lifts due to the reduced clearance between the accumulated material and the tunnel walls. All cold bands were intersected as part of the prescribed contact grouting operations on the nearest grout injection port; no additional drill holes were required.

During contact grout installation these types of features were typified by a lack of grout take unless other unrelated features were present.

Cold Areas

Following the installation of the top two lifts of annular backfill, several areas of the pipe showed no signs of hydration occurring. Once hydration of the lower annular backfill lifts completed, these cold areas were often found to have completely disap-peared, matching the ambient temperature of the tunnel. These cold areas in the crown were often difficult to identify due to the presence of reflections from the com-pressed air and water lines in the tunnel invert.

The cold areas generally intersected a large number of grout injection ports, and were subsequently drilled as part of the planned contact grout operations. When drilled, the cold areas were found to reveal large voids where the annular backfill was unable to fill. These segments were often found to extend as much as 25m along the length of the tunnel and 450mm above the crown of the pipe.

Borescope inspection revealed that most cold areas were caused by clearance restric-tions between the steel pipe and the tunnel wall. Tunnel features such as steel set reinforcement or loose rock behind mesh caused restrictions to the annular backfill flow behind the steel liner. The annular backfill injection pressures were closely monitored to ensure that hydrostatic forces on the pipe from uncured backfill were controlled. When the restrictions to backfill flow rapidly increased, the back-pressure on the annular backfill injection system and the injection location was advanced down the tunnel, causing a segment to remain partially unfilled.

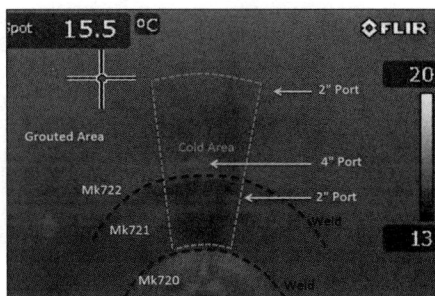

Figure 9. Cold area in Lifts 5 and 6

The cold area voids often consumed significant volumes of grout during the void grouting process. In order to ensure that any airlocks were removed from the annulus, ventilation lines were installed to reach the top of the void areas. Areas which were suspected of containing an airlock or cured before void grout completion were re-drilled for additional remediation. Several of the cold area voids intersected water bearing features in the tunnel. These areas had to be drained and vented prior to void grouting to ensure that non-diluted grout had completely filled the void.

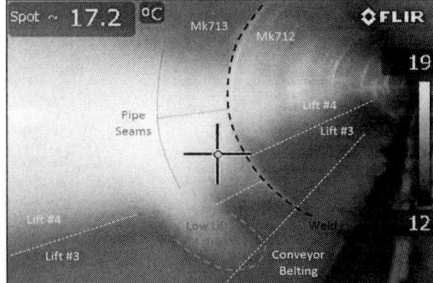

Figure 10. Delayed filling of Lift #3

Irregular Shapes

In several locations, geometric shapes were noted through several lifts of the annular backfill. In each instance the geometric shape was either colder or warmer than its surroundings immediately following backfill installation. Following initial detection, the irregular shape was marked on the pipe lining and monitored in all following lifts of annular backfill. The irregularities were drilled at the top and bottom to allow for contact grout injection and return, as well as dissipation of potential airlocks. The drilling of the injection and return ports often confirmed the likely causes of the irregularities.

The presence of panning plastic was found in the cuttings of several of the additional drill holes. The panning was often found immediately adjacent to the steel lining in the holes, confirming that exterior features such as loose panning resting on the steel pipe were the cause of several of the annular backfill irregularities.

CONCLUSION

Infrared inspection was highly effective at assuring backfill quality in the SCTT and provided the basis for all void remediation work in the steel lined section of tunnel.

The IR camera proved to be highly capable at detecting annular voids in a wide variety of situations. Approximately 1,800 sets of infrared spectrum photos were taken during the course of the annular backfill inspections in the twin tunnels. The photos of the irregularities provided a greater level of detail than would have been provided with the planned systematic hammering tests. The photos taken were then used to create records of any backfill deficiencies, providing illustrations to the contractor for use during the remediation, with minimal potential for misinterpretation or dispute. Irregularities were found regardless of the steel liner thickness, and in several cases were found behind both the steel liner and a small insulating layer of concrete.

Inspection of the lining was completed entirely off the critical path schedule and in the days immediately following backfill installation. No damage to the polyurethane liner was inflicted by the inspection process. Several months of labour intensive testing and lining repairs were successfully avoided. Elimination of hammer testing also greatly reduced the exposure to work at heights, climbing on slippery surfaces and repetitive physical motions, providing a significant safety advantage to the inspection team.

ACKNOWLEDGMENTS

We gratefully thank Metro Vancouver and Frontier Kemper Constructors for providing the opportunity and support needed to develop and implement this technique.

Use of "Command Chair" Simulator Technology to Optimize Modern TBM Performance

Steve Chorley ▪ The Robbins Company

TBM operator cabs and controls are often a long way down the list of priorities when it comes to the overall design of a Tunnel Boring Machine. However, their proper design and inclusion of the latest technology can mean the difference between a successful project and an unsuccessful one.

On soft ground machines the industry often puts operators in control of machines with little or no practical experience of that particular machine or the control systems. This practice can, in some cases, lead to major incidents such as ground heave or sinkholes in densely populated urban zones causing major disruption, downtime and untold damages and cost.

This paper will discuss the advances and developments in TBM operator controls and the use of "command chair" technology as both a training tool in the form of a boring machine simulator and as a way of controlling modern-day TBMs.

INTRODUCTION

For a long time now personnel entering construction sites have been required to carry out safety induction courses. These courses can range from a simple one-day course explaining basic safety principles to induction courses that can last a full week and cover a wide range of topics.

For personnel operating machinery or driving vehicles on site there are additional courses to be undertaken that include theoretical and practical training. In a lot of cases specialist training companies are hired to certify operatives of machinery, and personnel are issued with licenses or given approval to operate the said machine.

Boring in and around major world cities is often done under sensitive structures, roads and rivers. The potential for an incident caused by sinkholes or ground heave cannot be underestimated. Damage to structures or disruption to daily life can lead to multi-million dollar claims against contractors and untold damage to the reputation of the client and contractor.

It seems remarkable therefore that in today's world with tunnel boring machines operating in highly populated urban environments, TBM operators or pilots as they are sometimes known are hired based on, in the main, experience alone. In a lot of cases these operators are trained by personnel from the tunnel boring machine suppliers or by their peers whom, it can be said, have little or no formal training themselves.

Whilst this approach to TBM operator selection has served the industry well enough until now, albeit with numerous catastrophic failures, it would appear that with the technology available to us, the number of perceived higher risk projects being undertaken and the requirements of clients to adhere to much stricter operating parameters of machines, the time has come to introduce a more formal TBM operator training program.

OPERATOR TRAINING TOOLS

Currently there aren't many tools available for TBM operators to learn about the functions of the machine they are going to take control of. In the first instance the operator gets access to the tunnel boring machine manual. In most cases they are then put in an operator room or cab. The manual might explain the function of each control and the interlocks in any particular system on the machine, but it doesn't really tell the operator how to operate the machine to achieve safe, optimum production or how to overcome situations relating to changing ground conditions they encounter. A lot of the understanding of how the machine responds to operator inputs can only be gained during actual boring and this is where the fundamental problems lie.

Operator cabs in the modern day can, upon entering them for the first time, appear to be very sophisticated (they often are). They can also be daunting and personnel can in some cases be completely overwhelmed with the number of controls and information being presented to them. Couple this with the information being given to the operator in the form of verbal communication and it is easy to see how this can lead to simple mistakes that if not caught in time can escalate very quickly and lead to serious conditions and in the worst cases fatalities.

So how to improve the training? Commercial pilots are given classroom theoretical training and put in simulators for months before taking control of a modern airplane. The simulators give them the chance to not only learn the controls of a plane but also respond to most situations they may come across in real life situations.

Would it be possible therefore to come up with a tunnel boring machine operator training program that included theoretical classroom training and practical training on a TBM simulator?

ITA-CET Endorsed Training Program for Certified TBM Operators

Since the year 2000, the International Tunneling Association committee on education and training (ITA-CET) has identified education and training as one of the most important challenges and needs of the association and officially established the ITA-CET Committee during the ITA General Assembly in Prague, in May 2007.[1] ITA-CET's role is to promote education and training throughout the tunnelling and underground space association and assist in its coordination.

In regards to theoretical training for tunnel boring machine operators at the time of writing the ITA-CET is currently working in conjunction with one of the major tunnel boring machine manufacturers to develop a classroom training course that will lead to TBM operators being given some formal classroom training.

The course is foreseen to consist of 80 hours of theoretical training over a 10 day period covering a wide range of topics related to tunnel boring machine operations.

EPB SIMULATOR

As an enhancement to the theoretical course an earth pressure balance TBM (EPBM) simulator is also being developed by the Colorado School of mines (CSU) with input from a US tunnel boring machine manufacturer to serve as an additional valuable tool in the training and certification of EPBM operators. The EPBM simulator may also be used to train all tunnel engineers on TBM / ground interaction issues. The simulator may be implemented in multiple ways, i.e., as a stand-alone system operating on a PC or by linking the system with an actual boring machine operator cab. At the present time the simulator is simply being used as a stand-alone system operating on a personal computer.

There are an endless number of scenarios that can be created for EPBM operator training, depending on any particular project a machine is deployed on with known ground conditions. At the moment five (5) conditions / scenarios have been developed. These scenarios focus on EPBM operations during excavation and standstill (ring building) within each scenario. Geological parameters and tunnel depth can be modified (see Figure 1). Ring building will not be simulated. The current scenarios are as follows.

Scenario 1. Boring in Sand

At the commencement of a stroke the operator initiates and completes excavation of one 1.5 m ring advance. Excavation is straight (no curvature) and operation throughout ring build & waiting time (standstill for ring building for example) is simulated with regards to control of the chamber pressure for the ground conditions encountered.

Scenario 2. Boring in Clay

Again the operator initiates and completes excavation of one 1.5 m ring advance. Excavation is straight (no curvature) and operation throughout ring build & waiting time or standstill is simulated related to control of the chamber pressure for the ground conditions encountered.

The simulation administrator and/or user can adjust the following parameters:

- Geometry: (tunnel depth)
- The ground water table depth
- Soil parameters.

Some preset values will be provided that span the ranges of behavior, e.g., normally consolidated soft clay to heavily over consolidated stiff clay.

Scenario 3. Boring Through Transitional Geology

The operator initiates and completes excavation of one 1.5 m ring advance that transitions from sand to clay or clay to sand (both conditions programmable). Excavation is straight (no curvature) and operation throughout ring build & waiting time or standstill is simulated related to control of the chamber pressure in relation to the ground conditions encountered.

The simulation administrator and / or user can adjust the parameters of the clay and the sand (as described in scenarios 1 and 2), and prescribe at what distance into the ring the transition occurs.

Scenario 4. Boring in Mixed Face Conditions

The operator initiates and completes excavation of up to five 1.5 m ring advances in mixed face conditions—clay over sand or sand over clay (both conditions programmable). Excavation is straight (no curvature) and operation throughout ring build &

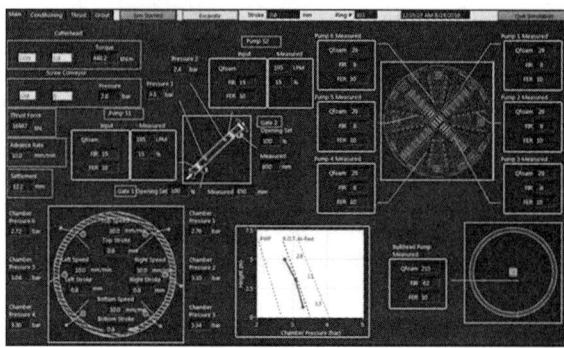

Figure 1. Screenshot of the EPB simulator operating on a PC

Figure 2. Students working with the simulator

waiting time or standstill is simulated, as well as control of the chamber pressure in relation to the ground conditions encountered.

The interface between the two soils changes with each advance hence the simulation of five excavation cycles.

The simulation administrator and / or user can adjust the parameters of the clay and the sand (as described in scenarios 1 and 2), prescribe whether sand over clay or clay over sand, and prescribe the depth of the interface with respect to the face.

Scenario 5. Boring in Homogeneous Conditions

Beginning from standstill (during ring build), the operator initiates and completes excavation of up to five 1.5 m ring advances in homogeneous clay or sand (both conditions programmable). Excavation is through vertical and horizontal curves.

The simulation administrator and/or user can adjust the geometry and soil parameters as indicated above, and can prescribe the horizontal and vertical curvature.[2]

The heart of the simulator is the suite of input-output relationships, algorithms and scenarios that relates actual operator inputs to produce 'reasonable' outputs. The simulator is still in the early stages of trials, but has recently been used in an industry course offered by CSU in June 2016 (see Figure 2).

INTERFACING THE SIMULATOR WITH THE COMMAND CHAIR

The next and perhaps most important phase of this program is implementing these simulator algorithms into any particular TBM training operator cabin. This requires the involvement of TBM and software suppliers. Further, the rendering of inputs and

Figure 3. An example of a simulator for operator training[3]

outputs can be done with any existing TBM supplier software, albeit with some work required to ensure both programs align correctly (see Figure 3—an example of a simulator from the excavator industry).

COMMAND CHAIR TECHNOLOGY

Operator cabs have evolved considerably from the early days of tunnel boring machines. On hard rock, open gripper style machines it was not uncommon for manufacturers to simply put a soft cushion on top of the lubrication tank. The operator had a simple bank of hydraulic controls and stop / start stations for controlling the necessary systems to allow the machine to operate.

Over the years there have been lots of improvements to operator control rooms. On most modern machines operator control rooms comparisons have been described as being like 'spaceships' amongst other things.

It is a fact however that operators of tunnel boring machines now have a large amount of systems to control and are receiving so much data from systems that we now consider data acquisition systems as the norm.

Operators are not only controlling the boring machine, they are sometimes responsible for grouting systems, foam systems and other additional systems that may be installed on the machine. This has led to operator cabins becoming much larger than they might need to be. Larger operator cabins are always the central meeting station on a boring machine and it is not uncommon to see 5 or 6 people crammed into a cab. This can lead to many distractions for the operator that as previously mentioned could cause the operator to make mistakes more easily than if he was left alone to concentrate on his work.

COMMAND CHAIR DESIGN & LOCATION

Working hours on boring machines are often long and carried out in hot, humid environments. More often than not operators are now controlling machines from sound proofed, air conditioned control cabs but one of the biggest complaints from TBM operators is the comfort of the chair they have to sit in. Many will have witnessed operators sat on stools, plastic chairs or been forced to stand up for long periods to control the machine. No particular thought went into the ergonomics of the chair for the operator either.

Furthermore, as previously stated operator stations are becoming so big that they are often placed on the boring machine back-up with little or no thought being given to what the operator sees visually. We rely on CCTV systems when in some cases it would be better for the operator to have a first-hand view of the situation. We could argue that as per a ship or other large machinery the operator should be sat in a location that gives him an overview of some of the more critical functions or areas of the boring machine. Should he for example be able to see the material on the conveyor belt or be able to see the ring build area (particularly the top 120 degrees of the machine)? In effect the operator should be located on the 'bridge' of the boring machine for the best view.

With this in mind, and with consideration to providing training to operators on the controls, a more compactly designed operator station or command chair needed to be developed. The chair essentially needs to be removable from the cab to allow on or offsite training and needs to be a self-contained unit with all systems necessary to control the boring machine included within the design of the chair.

Taking all the above into consideration, designers with input from onshore and offshore command chair manufacturers, and field service personnel (those who actually operate the machine), came up with an innovative design based around a helmsman's chair (see Figure 4).

The two panels on either side of the operator have been designed with modular control panels that are interchangeable based on the type of machine the chair will be installed on (see Figure 5).

A detailed explanation of all the functions on each of the modular control panels is beyond the scope of this paper, but suffice it to say that each modular panel is specific to that type of machine and is designed to be instinctive for the operator to start and control the machine. Functions on the control panels have been reduced to the absolute minimum with all other functions to control, monitor or stop and start systems being transferred to two touchscreen panels also incorporated in the command chair (see Figures 6–8).

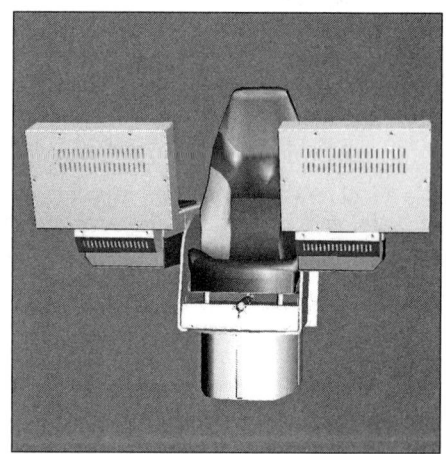

Figure 4. Preliminary 3D design concept

Figure 5. Overview of the modular control panel concept for an EPB machine

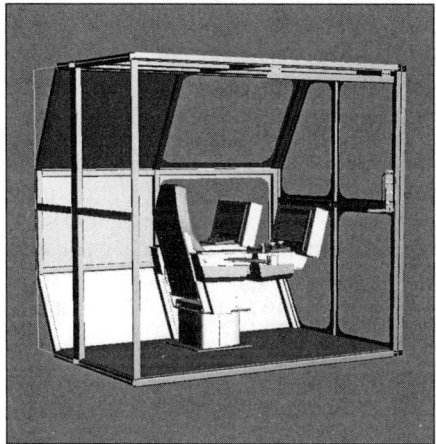

Figure 6. Command chair incorporated into a cabin

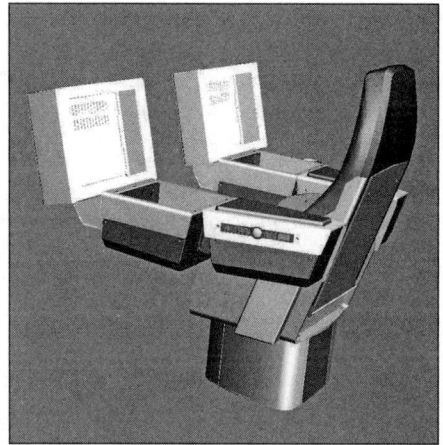

Figure 7. Side view of command chair concept[4]

SIMULATORS FOR HARD ROCK BORING MACHINES

At this stage the simulator has only been developed for soft ground (EPB) applications. In the future programs and scenarios will be added for Slurry and Hard Rock applications.

We may actually question the need for simulators for hard rock boring conditions, but there are scenarios related to ground collapses and squeezing ground that operators would benefit from by being put through their paces on a simulator. Hard rock ground conditions and incidents related to them can lead to months of downtime on a machine whilst bypass tunnels are constructed to free blocked machines.

Recent projects in Turkey and Austria where machines were down for 6 months and 3 months, respectively, have shown

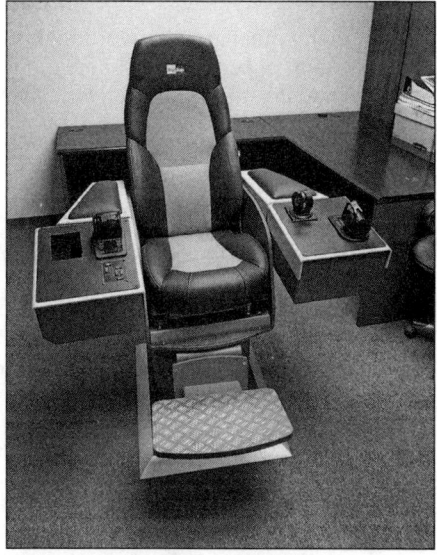

Figure 8. Actual commmand chair[5]

that simulator training would have been an invaluable training aid and may well have given the operator the ability to recognize what was happening with changing ground conditions. Familiarity with how to act could have given them options to stop and treat the ground or modify the machine (for example go into an overcutting mode) before getting the machine into a very difficult situation.

FUTURE USE OF THE COMMAND CHAIR

As the concept of the command chair is relatively new it is currently being introduced on several machines. These include projects in Nepal on a hard rock double shield machine, in India on EPB machines, and in Japan for slurry machines. These projects will come online in late 2017.

Developments Going Forward

There are several ideas currently being considered on how to best utilize the command chair and simulator technology in future. With the ever improving connectivity of machines via high speed internet there may in some circumstances be a valid case to argue for remote operation of boring machines outside a tunnel. Whilst this has been done in the past, most control cabins due to their size and cost have not been replicated outside a tunnel. Command chairs do not require operator cabs and can be easily installed in an office environment. Having the ability to operate a machine utilizing identical controls as those on the machine rather than from a keyboard can only give boring machine pilots more confidence in understanding that a remotely operated machine will respond to their inputs just as if they were actually on the machine themselves.

Additionally, due to the incorporation of highly sophisticated data acquisition systems it is only a matter of time before a real time database of every project and all the ground conditions encountered can be stored on the simulator to give operators firsthand experience of actual conditions on any particular machine encountered.

CONCLUSIONS

Tunnel boring machine operator training and formal recognition of training by means of certification or licensing will become the norm in the not too distant future and is a goal the industry should be working towards.

Providing operator instructors and experienced or new boring machine pilots with sufficient tools in the form of courses, lectures, operating manuals and practical training on simulators that replicate the controls on the machine (i.e., by use of the command chair technology) can only serve to improve industry standards and hopefully reduce unwanted incidents of sinkholes, ground heave or machines being out of action for longthy periods of time.

Having pilots familiar with machines and the control thereof before a project starts will be invaluable in terms of commencing boring and reducing learning curves on machine functions.

By incorporating information from data acquisition systems from all types of machines in any ground encountered into the simulator it is only a matter of time before a comprehensive database of rock or soil conditions can be drawn from to further enhance operator training.

REFERENCES

[1] https://www.ita-aites.org/en/wg-committees/committees/ita-cet.

[2] Mike Mooney, Director: Center for Underground Tunneling and SmartGeo, Colorado School of Mines.

[3] http://www.catsimulators.com/.

[4] Courtesy of The Robbins Company, Solon, OH, USA.

[5] http://norsap.com/.

Metro Doha—Continuous Tunnel Belt Application for One of the World's Largest Infrastructural Projects (A Challenge)

Marco Sonnenschein ▪ H+E Logistik GmbH
Georg Butsch ▪ H+E Logistik GmbH

No matter if infrastructural projects for utility purposes such as sewer, cable, hydro power tunnels or for traffic purposes such as road, railway or metro tunnels, first choice would be a mechanized tunnel drive with a TBM (Tunnel Boring Machine). Continuously extendable belt conveyors belong to the key equipment which lead to success in tunneling. Separating the ways of logistic of mucking out by conveyors and supplying the TBM on Rails or wheels is the first step to high performance and a safe operation This has been proven and is going to be continued, meaning continuously extendable tunnel belt conveyors are state of the art for mechanized tunneling.

This paper will introduce to one of the world largest infrastructural projects of time and explain the implementation of continuously extendable tunnel belt conveyor systems for 64km of tunnel. The distinctive climatic conditions were an additional challenge to overcome.

INTRODUCTION

In 2022 the FIFA World Cup is scheduled to take place in Qatar. There will be six stadiums in the capital of Doha, six more stadia are spread out in cities in the vicinity. In consideration of this global event the government invests more than 50 Billion US Dollar for the realization of the world championship. Especially in projects for the expansion of the infrastructure.

For the building up of a new and modern metrosystem in Doha the Herrenknecht AG from Schwanau, Germany was instructed to design, deliver and install 21 TBM's (tunnel boring machines) for the three main lines (Gold line, Green line and Red line), as shown on the overview in Figure 1, most of them with the opportunity to install continuously extendable belt conveyor systems, which is the state of the art of muck removal.

In cooperation with H+E Logistik GmbH from Bochum, Germany, a subsidiary of Herrenknecht and a specialized supplier of tunnel belt conveyor systems and back-up conveyors for tunnel boring machines, one of the largest and most advanced infrastructure projects in the world was realized.

A total of 111km of new and high quality tunnel have been created beneath the capital of Doha in incredible 26 months. This is a new Guinness world record. In August 2015 20 TBM's work simultaneously and so 2.5 km of new tunnel have been created per week.

This impressive drive was supported by H+E Logistik's high engineered conveyor belt technology. A total of 64 km of conveyor systems have been installed which extracted the material with capacities about 1080 t/h. This is construction and logistic at highest level.

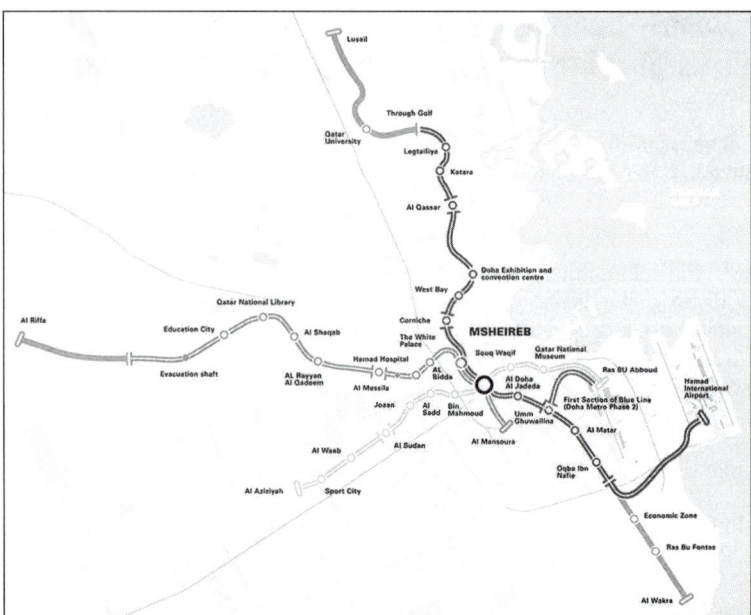

Figure 1. Overview Doha Metro

The extreme climatic and geologic conditions also with assembly, the dense urban development and tunnel alignment beneath densely populated areas were tricky challenges to overcome. With extensive experiences from previous projects and excellent engineering Know How of H+E Logistik the project was implemented satisfactorily with good prospects for further projects.

Figure 2. EPB shield

TBM

Acceleration, mobility and environmental protection are the driving forces behind this mega project. Katar's vision is to connect every corner of the country with public transport whether in the urban centers or in the countryside.

From August 2014 to September 2016, four international joint ventures realized this project in cooperation with Herrenknecht and subsidiaries like VMT, H+E Logistik, Techni-Metal-Systems, MSD and GTE professionals. Up to 125 service staff and specialists from Herrenknecht from 19 different countries worked 24/7 wherever contractors, machine technology and extraordinary events required them.

A total of 76 breakthroughs, including 42 intermediate breaktroughs, numerous TBM transfers and 31 final breakthroughs, are just one of many outstanding successes. Every TBM was designed and equipped specifically for the project. The geology, which mainly consists of Simsima limestone, and the high temperatures are requirements that had to be taken into account when designing the machines. The machine data are listed below.

6x EPB Shield:

- Shield diameter: 7,050 mm
- Drive power: 1,440 kW
- Torque: 4,769 kNm

9x EPB Shield:

- Shield diameter: 7,050 mm
- Drive power: 1,280 kW
- Torque: 4,239 kNm

6x EPB Shield:

- Shield diameter: 7,110 mm
- Drive power: 1,440 kW
- Torque: 4,769 kNm

As mentioned before the tunnel alignment (Red line south: 32,6 km; Red line north: 22,8 km; Green line: 23,3 km; Gold line: 33,4 km) proceeds beneath densely populated areas, like the high-rise neighborhoods of Doha City and the tourist centers with their attractive hotel facilities. Settlement-free tunneling was therefore one of the core requirements during construction.

CONVEYOR SYSTEMS

The de-mucking system being capable handling the advance rates of the tunnel boring machine is a conveyor system. The conveyor system comprises of a continuously extendable tunnel belt conveyor, following the TBM to the final length of the tunnel. With on surface conveyors the muck is piled up by a stationary stacker before transported away by trucks. The stacker is equipped with a telescopic chute as shown on Figure 3, which prevents big dust developments.

Figure 3. Telescopic chute (Green line)

Figure 4. Two conveyors fed the material on surface conveyor passing the switch chute (Green line)

The special thing about this project is the redundant system. A redundant system is a safety-related system that is designed in parallel so that in the event of a component failure, the others ensure the service. In this case, two parallel tunnel belt conveyor fed the material onto one surface conveyor with assistance of a switch chute as shown in Figure 4. This switch chute switches in the event of a failure of one conveyor system to the parallel adjacent conveyor system and thus guarantees a continuous service.

The tunnel belt conveyors are capable to transport 540t/h and are therefore equipped with 800 mm wide EP 800/4 5+3 UTS belts. The surface conveyors are capable to transport 1080 t/h and are equipped with 1400 mm wide EP 630/4 4+3 Y belts because of the double loading. The belts and the drives were covered with a thickness of 5 mm Top Cover and 3 mm Bottom Cover, as shown on Figure 4, to protect them from dust and the climatic conditions.

The tunnel belt is driven by a 200 kW main drive. In addition to the main drive of the tunnel belt conveyor three top strand and two bottom strand booster drives were installed during the tunneling operation. The booster drives are required due to the winding alignment of the tunnel with several curves of radii between 250 m and 5000 m as the belt tension, which constantly rises with the increasing tunnel length during the TBM drive, has to be reduced in order to track the belt through the curves. Each top strand booster is equipped with a 200 kW drive and each bottom strand booster with a 75 kW drive, in total 968,5kW (including Winch 18,5 kW) will keep the muck being conveyed from the TBM to the shaft. The shaft is the logistical bottle neck as the de-mucking equipment as well as everything to supply the TBM has to go through it. Figure 5 shows the main drive at the discharge. The surface conveyor is driven by 2x200 kW (Figure 3).

Figure 5. Main drive with covering (Green Line)

Characteristic for this project is that four machines starts from one shaft. This is logistic and construction on high level.

All drive units of the conveyor system and the belt tensioning winch are controlled by a Siemens PLC and communicate via Profibus. An interface links the belt PLC with the TBM main PLC to facilitate belt operation from the TBM control cabin. The Profibus signal was amplified every 1,000 m throughout the tunnel length, which required a power supply in the tunnel installed by the customer. Touch panel were installed in TBM control cabin for belt operation with complete emergency stop switches installed every 250 m along the tunnel. Including pre-start warning (sounder beacon and flash light) as well as misalignment switches. To ensure full electrical functionality Herrenknecht installed special air condition systems in the control cabins because of the climatic conditions.

A horizontal belt storage system, as shown on Figure 4, permits 250 m TBM advance before a new 500 m belt extension is vulcanized in the tunnel next to the portal. At the head station the belt is transferred to a horizontal position so the gravity could be used to clear the belt from the soil (Figure 4). To support the gravity is the most effective technical solution to clean such belts. Accordingly, a knocking pulley is installed at the horizontal discharge section of the head station.

The conveyor systems end with the stacker shown in Figure 3 with a discharge height of 19,065 m and a troughing angle of 37,5 degrees which is enough to pile up around 9,825 m^3.

PART  **13**

Tunnel Lining I

Chairs

Joe Clare
Mott MacDonald

Rick Gomez
Gomez International, Inc.

Load-Bearing Capacity of Fiber-Reinforced Concrete Tunnel Linings Under Combined Moment-Normal Force Loading Conditions

Axel G. Nitschke ▪ Shannon & Wilson
Erik S. Bernard ▪ Technologies in Structural Engineering

ABSTRACT

Fiber Reinforced Concrete (FRC) tunnel linings in soft-ground tunnels are typically designed as a two-dimensional beam structure under combined moment-normal force loading conditions. Structural design concepts for unreinforced or bar reinforced concrete are used by amending the stress-strain-relationship on the tension side, using data obtained from beam tests. By computing an equilibrium between the resulting forces from the stress distribution over the height of the beam with external forces, moment-normal force interaction diagrams for FRC are developed. The moment-normal force pairs from the tunnel lining are then compared with the bearing capacity of the material using the interaction diagram to determine if capacity is adequate.

This paper will describe this approach in detail, highlight some specific failure modes and highlight its weakness, namely that the full potential of FRC remains under-utilized. In addition, the concept is amended by introducing a concept of "plastic hinges" in the tunnel lining that allows the designer to take full advantage of the elasto-plastic failure characteristics of FRC.

INTRODUCTION

The use of Fiber Reinforced Concrete (FRC)—with steel or macro-synthetic fibers—has technical and economic advantages which stem from the fact that fibers transform the postcracking behavior from a brittle failure mode typical of unreinforced concrete into an elastoplastic behavior. Numerous codes and guidelines provide qualitative or quantitative design approaches [8,9,10,11,12,13]. However, modeling of the load-bearing behavior based on a Stress-Strain-Relationship (SSR) for tunneling applications appear to under utilize the structural and economic potential of FRC. This article discusses the modeling process and some typical results of a parameter study. It also identifies the weakness of the current concept and suggests a path to more fully utilize the structural and economic potential of FRC. The concept discussed herein is theoretical in nature and applicable for both steel and synthetic FRC. To limit the scope of this paper, the discussion is focused on the load bearing capacity under cracked conditions. Therefore, design concepts that do not utilize the toughness potential of FRC (i.e., by limiting it to uncracked conditions) are not discussed herein.

Different international codes and guidelines for FRC provide testing procedures based on simply-supported beam tests that are utilized to define a SSR by basically amending the known trapezoidal or parabolic SSR for concrete on the compression side with assumptions for an SSR on the tension side. The latter are the primary subject of this paper. For this discussion, it is irrelevant which type of macro fibers—steel or synthetic—is used, since the SSR models a homogeneous, composite material behavior and not discrete fibers. In general, the SSR design approach follows the concept to adapt existing concrete design concepts and procedures and simply extend the SSR on the tension side to account for the effect of the properties of the composite material.

This article is focused solely on combined moment-thrust or moment-normal force (M/N) loading of tunnel linings in which bearing capacity relies on a tunnel arch. This is typical for soft ground tunnel linings and rock tunnels with soft-ground-like behavior. Nonetheless, the ideas and concepts can be adapted in typical rock tunneling applications as well. However, they are not useful in tunnels with no arching effect, which is typical for tunnels with relatively thin linings or with an irregular shape. For these types of tunnel, (i.e., typical initial linings in classical rock tunneling) qualitative and empirical design concepts (i.e., Barton chart [14,15]) are available, which are not discussed in this paper.

The use of an SSR is typically evaluated on the basis of beam test data. Under elastic (uncracked) conditions the beam theory and the classical mechanics for materials apply. However, after the initial cracking of the FRC the material is no longer homogeneous and the theoretical conditions for beam-theory no longer apply. The bearing behavior of FRC in beam tests in a cracked state are better described using a stress-crackwidth relationship rather than a stress-strain relationship. It is important to understand that for the above reason a SSR cannot be measured directly in standard FRC beam tests. The codes and guidelines are therefore describing testing procedures that measure external forces and deformations, which are then transformed into stresses and "equivalent" strains via an equivalence model which implies several assumptions. Research by Nitschke [1] has discussed flaws in some of these models by back-calculation of tests using the SSRs. It was shown that these flaws can be significant under loading conditions of combined moment and thrust, typical for tunneling. The same work also provided modified models to provide more useable and accurate procedures [1].

STRUCTURAL BASICS OF FRC DESIGN

The biggest difference between the sectional strength of unreinforced or rebar reinforced concrete and FRC is that the concrete in unreinforced or rebar reinforced concrete has no bearing capacity in tension. In the modeling of conventionally reinforced concrete sections, all tension is supported by the rebar. Due to the fact that the location of the rebar is known, the location of the resulting tensile force is also known, and this simplifies the calculation of the equilibrium compared to FRC sections. The computation of axial equilibrium in FRC sections is much more challenging because the location of the resulting tension force is an unknown during the computation and moves if the external load and the distribution of the strain over the cross section changes.

The design assumptions for the calculation of the sectional strength for FRC based on a SSR can be summarized as follows ([4], ACI 318-14, 22.2 [6], ACI Design Handbook, 7.4[7]):

1. Equilibrium shall be satisfied at each section.
2. Strain in the cross section of the member shall be assumed directly proportional to the distance from the neutral axis (Bernoulli's theorem). The cross section also remains plane during loading.
3. The stress-strain-relationship for the FRC in compression is defined, thus the stress for a given strain is known within defined limits.
4. The stress-strain-relationship of the FRC under tension is defined, thus the stress for a given strain is known within defined limits.

A comparison of the essential design assumptions for moment and axial strength at sections for rebar reinforced concrete design in ACI 318 shows that the first and second assumptions—equilibrium (ACI 318-14, 22.2.1.1 [6]) and Bernoulli (ACI 318-14, 22.2.1.2 [6])—are adapted for FRC. However, by citing two additional design assumptions from ACI 318, two major differences between FRC and classical rebar reinforced designs assumptions can be highlighted as well. According to ACI 318-14, 22.2.2.2 [6]:

> "Tensile strength of concrete shall be neglected in flexural and axial strength calculations."

For sectional strength calculation of FRC the tensile strength under uncracked, as well as cracked conditions, is utilized. This is one of the major differences between the modeling of FRC in comparison with unreinforced or rebar reinforced concrete. According to ACI 318-14, 22.2.1.2 [6]:

> "Strain in concrete and non-prestressed reinforcement shall be assumed proportional to the distance from the neutral axis."

This design assumption is based on the hypothesis of perfect bonding between steel and concrete. While rebar reinforced concrete is modeled as a composite of concrete and steel, where each component has its own material properties (see i.e., ACI 318-14, section 22.2.2 for concrete and section 22.2.3 for non-prestressed reinforcement [6]), FRC is assumed to be a macroscopically homogeneous and isotropic material [3]. The material properties of a single fiber in the model becomes irrelevant. Therefore the fibers and the concrete are modelled using a single SSR relationship and not two, (i.e., as for steel rebar and concrete.)

After the cracking of the material under tension, the material properties in the model are based on strains rather than a discrete crack. In the model the cracked material is also viewed as homogeneous and isotropic. Since this is in the area around the crack, it is obviously not the case. This circumstance is very important to realize and understand when evaluating the sectional strength of FRC using a SSR. During the evaluation of material testing data based on beam tests (and subsequently the design of the structure), it is assumed that the crack is "smeared" over a certain length into an "equivalent strain," which is also referred to as "integral approach" [4].

Fibers influence the bearing behavior in multiple ways. However, three properties are most relevant for application in tunnels [3]. (1) They slightly increase the flexural tensile strength, which is mostly needed if improved properties under uncracked conditions are desired, (i.e., to design for serviceability.) However, for the case of ultimate bearing capacity of tunnel linings, the residual flexural tensile strength under cracked conditions (2) and the increase of the toughness (3), are the major benefits. The focus of this paper is on the performance improvements attributable to (2) and (3).

The provision of a reliable and usable post cracking tensile strength transforms the brittle failure mechanism of plain concrete into a ductile failure mode (see Figure 10). This is a material property that provides major engineering and economic advantages, especially if utilized to facilitate system failure of a tunnel lining rather than a cross section failure at one, presumably most critical location. A concept for the design of a system failure will be presented later in this article.

According to Dietrich [4] the load bearing response of FRC under bending can be subdivided into three phases. The first 'uncracked' phase is based on the behavior of the concrete matrix alone. The concrete matrix and fibers are assumed to be in "perfect bond" and the ratio of load supported by the concrete compared to the fibers is dependent on the moduli of elasticity of the materials. Due to the relatively small volume of fibers compared to concrete, the load bearing share of the fibers is relatively small.

Microcracks develop in the matrix during the second phase of load response. The development of cracks is hampered by the fibers and leads, according to Dietrich [4], to a more stable "strain softening" with a restricted expansion of cracks and less brittle material behavior. Phase two ends with crack widths of approximately 0.1 mm [4]. In the third phase the concrete matrix no longer provides significant bearing capacity at the crack. The opening cracks are bridged by the fibers and the load transfer is effectively provided by the fibers alone.

SECTION DESIGN OF FRC USING A STRESS-STRAIN-RELATIONSHIP

The three phases of crack development are also reflected in SSRs found in different codes and guidelines. Studies by Nitschke [1] have shown that by using all three phases in computer simulations, test results obtained using beam tests can be simulated very accurately. Typically all SSRs in codes and guidelines incorporate Phase I (elastic) and Phase III (macro crack) behavior. However, since the distinction between uncracked and micro cracking in Phase II is not clearly defined, Phase I and Phase II are often times lumped together or Phase II is completely neglected [1]. It is important to note, that for the modeling of ultimate load bearing capacity in the macro cracked phase, a detailed evaluation of the micro cracking phase II is irrelevant. However, it might be significant for serviceability design.

A generic SSR and nomenclature of the variables used throughout this paper is shown in Figure 1. The tension side is represented by the three sections discussed above. The compression side uses a classical parabolic-constant shape.

Nitschke has conducted numerous simulations of beam test results under pure bending as well as combined M/N-loading. The three load bearing phases observed during the experimental studies could also be reflected with the simulation of the load bearing behavior, using the SSR above. In general, it is possible to identify "typical" SSRs based upon typical load-deflectioncurves from either tests or the simulation of results. By adhering to certain boundary conditions, it is almost possible to look at each of the three phases separately [1].

The pure elastic (un-cracked) behavior is related to the first part of the stress-strain-relationship and conforms to the principles of elastic bending. The flexural strength f_{t1} results from the maximum elastic moment divided by the section modules. The range of the related strain ε_{t1} is very limited and can either be measured during the test or—based upon the used SSR—be calculated using the original modulus of elasticity. Alternatively, and if the major focus of the interest is the bearing capacity under cracked conditions, a generic value between $0,1\ ‰ \leq \varepsilon_{t1} \leq 0,15\ ‰$ (100–150 microstrain) will yield sufficiently accurate results, because the overall influence of the elastic section on the bearing capacity under cracked conditions is diminished [1].

The interim section of micro-cracking is reflected by the second section of the SSR on the tension side. In general, two different types of curves are used between ε_{t1} and ε_{t2}, (1) a plateau or (2) a linearly decreasing curve (trapezoid). By using a plateau the stress in the second section is constant ($f_{t2a} = f_{t1}$) (see Figure 1). In general, the plateau creates load-deformation curves with a distinct maximum and a "hard" decline of the moment bearing capacity in pure bending conditions. On the other hand

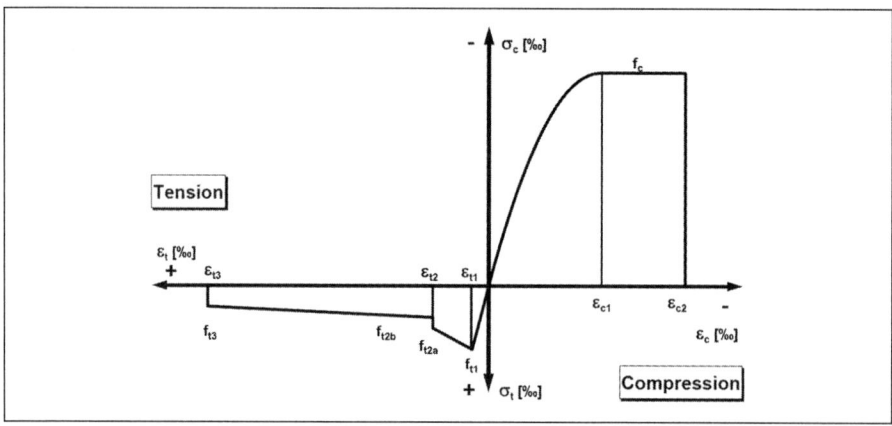

Figure 1. Generic stress-strain-relationship for fiber reinforced concrete

a declining curve in the second section ($f_{t2a} \geq f_{t2b}$) (see Figure 1) "softens" this area of the moment-deflection curve [1].

More complex curves can be used in the second section, however, the two selected types may encompass many other cases. As parameter studies have shown [1], the overall influence of the second section of the SSR controls the shape of a specific area of the simulation of the bearing capacity, but has only a small influence on the overall bearing capacity. It was also shown that more important than the value of the stress f_{t2b} is the specific strain ε_{t2}, which controls the shape of a moment-deflection curve in this area [1].

However, by far the biggest influence on the load bearing behavior under cracked conditions is the third section of the SSR. The tensile stress under cracked conditions is typically referred to as the "residual strength." Under consideration of the conducted beam tests with a maximum deflection of 3.5mm, SSRs up to a strain of $\varepsilon_{t3} = 25$ ‰ were investigated.

The load bearing capacity of a cross section based on the SSR is calculated by finding the equilibrium between internal and external forces. Only a discussion of the basic principle is covered in this paper. A complete solution for the calculation of the inner forces resulting from a specific strain scenario is provided by Nitschke [1]. For the calculation of equilibrium between internal and external forces acting on a cross section under typical tunneling conditions there are two equations (see Figure 2):

$$\sum N = 0 \quad \Leftrightarrow \quad C - T + N = 0$$

$$\sum M = 0 \quad \Leftrightarrow \quad -C \cdot z_C - T \cdot z_T + M = 0$$

The internal lever z, as well as the height of the compression zone x and height of the tension zone y are calculated as follows:

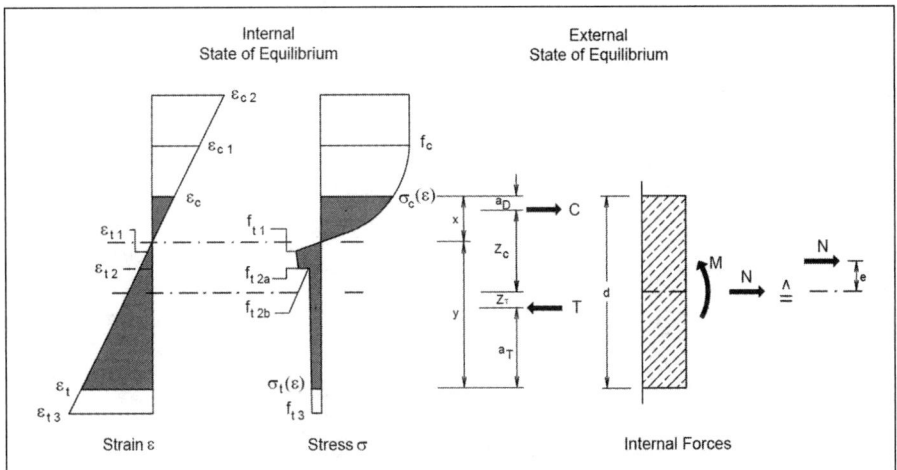

Figure 2. Calculation of equilibrium between internal and external forces

$$z_C = \frac{d}{2} - a_C \qquad z_T = \frac{d}{2} - a_T$$

$$x = \frac{|\varepsilon_c|}{|\varepsilon_c| + |\varepsilon_t|} \cdot d \qquad y = \frac{|\varepsilon_t|}{|\varepsilon_c| + |\varepsilon_t|} \cdot d \qquad for \quad \varepsilon_c \le 0 \ and \quad \varepsilon_t \ge 0$$

Since only two equations for the equilibrium are available, all but two variables must be known to compute a unique solution. However, at first there are four unknowns—the resulting thrust (C), the resulting tensile force (T), and their respective levers (z_C, z_T). All four unknowns are directly related to the existing strain condition. By selecting a specific strain condition, the moment capacity, as well as the normal force capacity, can be calculated and the result is unique.

Theoretically, the reversed approach—selection of the external forces followed by the calculation of the corresponding strain condition—is possible. However, this solution is practically not achievable, because typically a SSR of FRC is discontinuous and depends upon numerous parameters. In addition, the solution often provides multiple equilibriums and is therefore not unique [1]. As a result, an iterative process is necessary to solve the equations, which requires a lot of computation effort [2].

TESTING OF LOAD-BEARING BEHAVIOR UNDER TUNNELING CONDITIONS

The basis for the theoretical studies of described above include pure bending tests on beams and compression test on prisms, but are primarily based on bending in combination with a compressive normal force. Parts of the tests were conducted with the Moment-Normal force (M/N) test rig of the Ruhr-University Bochum in Germany reported by Nitschke (Figure 3) [5] or the Moment-Normal

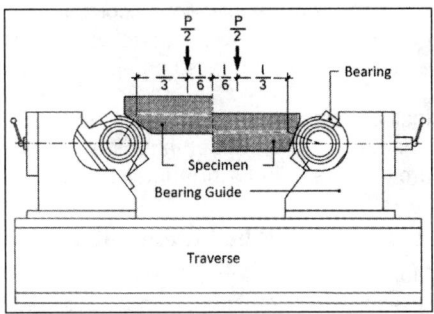

Figure 3. M/N-testing rig used by Nitschke at the Ruhr-University Bochum, Germany

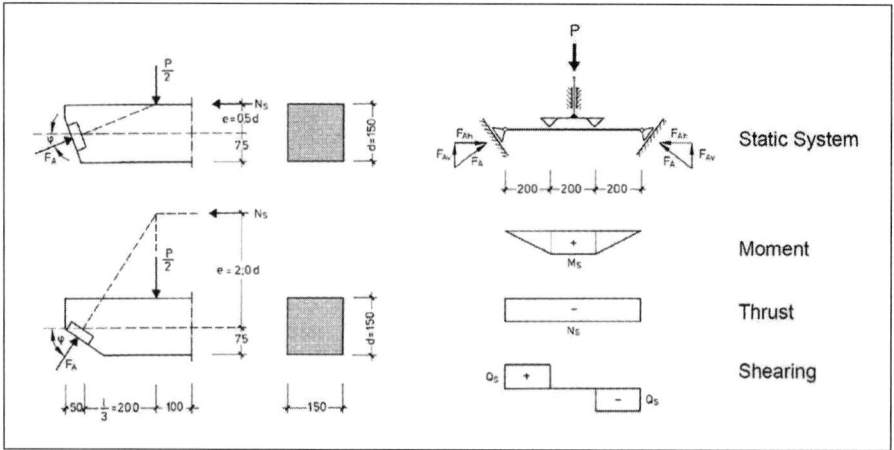

Figure 4. Geometrical and static basics of the M/N-testing rig used by Nitschke

force test rig at Technologies in Structural Engineering (TSE) in Australia by Bernard (Figure 5). In addition to classical four point bending, the beams in both test programs were loaded by an additional compression normal force.

The tests by Nitschke held the dimensionless eccentricity ratio e/d constant during the test. The eccentricity "e" over beam height "d" (e/d-ratio with e = M/N) is hereby controlled by the beams geometry (chamfered edges) and kept inherently constant (Figure 3). The testing concept allows evaluation of the entire spectrum of loading conditions and e/d-ratios that that are typical for tunneling.

The constant factor in Bernard's M/N-tests was the normal force, creating a compressive stress of either 0 (pure bending), 2.0, 4.0, or 6.0 MPa for each set of tests. In this way, the effect of a progressively increasing axial compressive stress on cracking and post-cracking moment bearing characteristics could be investigated. Standard EN14651 simply-supported notched beam tests were performed to characterize the flexural capacity of the FRC in pure bending. The test apparatus developed for this investigation is shown in Figure 5. Load was imposed on each beam-column specimen using two servo-hydraulic actuators and lateral deformation was measured using displacement transducers. Axial load was applied operating in load-control. The lateral load was imposed at two pivoted point loads 250 mm on either side of the centerpoint of the beam-column operating in stroke-control. A region of nominally uniform bending moment therefore occurred in the central 500 mm portion of the specimen, and a hinge was observed to occur in this region for all the test specimens. All the equipment was mounted on a slotted steel work plate.

Including test results from both testing assemblies is beyond the scope of this paper. Therefore typical observations of these tests, that can also be related to simulations, are discussed in further detail below.

LOAD-BEARING BEHAVIOR AND DESIGN OF FRC UNDER COMBINED M/N LOADING

Moment-Normal force Interaction Diagrams (MNID) are typically used during the design of tunnel linings (and columns under combined M/N loading in general) for rebar reinforced linings as well as FRC. However, while generic MNID are available for rebar reinforced members, a SSR-specific MNID has to be developed for FRC.

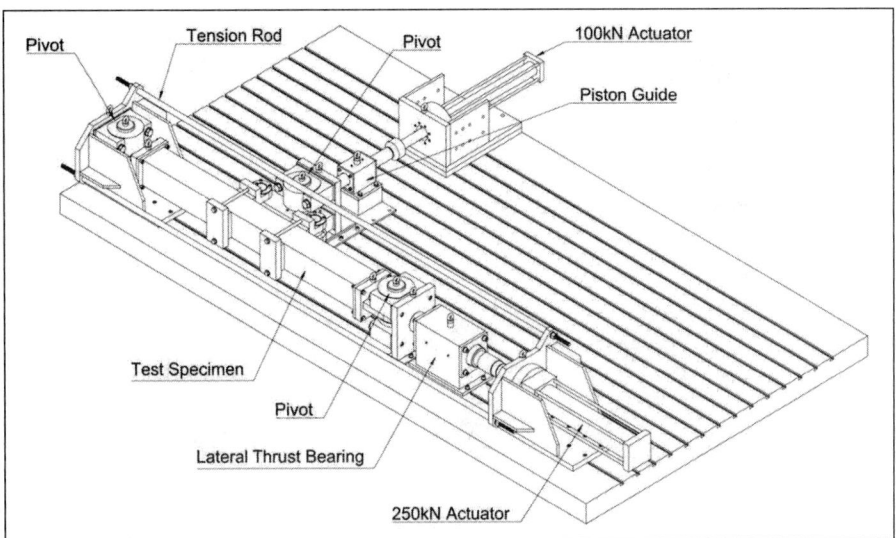

Figure 5. Test apparatus with twin-point central loading of beam-column by Bernard at TSE

Table 1. Stress-strain-relationship used in the parameter study

Stress	Tension				Compression	
	f_{t3}	f_{t2b}	f_{t2a}	f_{t1}	f_{c1}	f_{c2}
[N/mm²]	0.5	1.0	4.0	4.0	−40.0	−40.0
[% of f_c]	1.25%	2.5%	10%	10%	100%	100%
Strain	ε_{t3}	ε_{2b}	ε_{t2a}	ε_{t1}	ε_{c1}	ε_{c2}
[‰]	10.0	0.16	0.16	0.12	−2.0	−3.5

Generic MNIDs for FRC can be developed in a similar fashion to rebar reinforced members if the diagrams are dimensionless and all strength values are used (i.e., relative to the compressive strength f_c.) The dimensionless factor $n = N / (f_c \times b \times d)$ can hereby be interpreted as the utilization towards the maximum thrust under pure compression. The SSR used for the following parameter studies is defined in Table 1 and represents typical values for FRC (i.e., an initial tunnel lining.) For the nomenclature and shape see the SSR in Figure 1. The thickness of the lining was assumed to be 0.25m and a 1.0m wide tunnel lining section was assumed.

Figure 6 shows the complete MNID for the SSR presently used; while Figure 7 shows an enlarged section of the same MNID. Figure 8 shows the results of a parameter study with varying e/d-ratios; Figure 9 shows a parameter study for varying normal-forces. It is important to highlight that all figures are basically different ways of displaying the bearing capacity of the same material, defined in Table 1. The results represented, for example along the e/d = 0.5 - line in Figure 7, are the same as shown in Figure 8 for the identical case. The results represented in Figure 7 on a line with a constant normalforce, parallel to the x-axis, ie. N=1,000 kN is the same as shown in Figure 9 for the same thrust. The bearing behavior under pure bending (N=0 ; e/d = ∞) is represented on the x-axis in the MNID.

The example shown could also be transferred into a generic dimensionless MNID by using the following equations for the normal force and the moment:

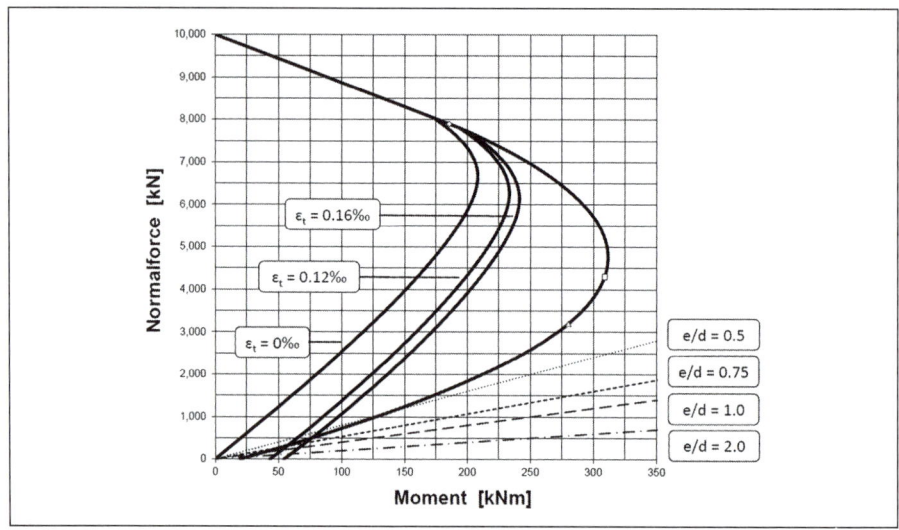

Figure 6. Moment normal force interaction diagram (MNID)

$$n = \frac{N}{f_c \times b \times d} \qquad m = \frac{M}{f_c \times b \times d^2}$$

The residual strength f_{ti} can be expressed as a percentage of the compressive strength (see Table 1). The dimensionless MNID would be valid for all cases where the ratios between the tensile strengths and the compressive strength are kept the same. The different lines in the MNID show cross section equilibriums for specific constant tension strains. Nitschke's tests, with a constant e/d ratio, would be represented on one of the inclined lines going through the origin (Figure 6), while Bernard's test, with a constant normal force, would be represented on a horizontal line, parallel to the x-axis.

A good rule of thumb is, that tunnel linings are typically using between 5% to 30% of the compression capacity of a member [1]. So for example, in the MNID in Figure 6, a typical utilization in a soft ground tunnel would be between 0.5 to 3 MN and only the lower third of the diagram would be relevant for the design; Figure 7 shows this area of the MNID enlarged. Tunnel linings in this part of an MNID generally fail under tension by reaching the maximum allowable tensile strain ε_{t3}.

The different lines in the MNIDs represent lines of specific strains (Figure 6, Figure 7). Left of the line marked with "ε_t =0‰" – tensile strain zero – all members are under full compression. The "ε_t =0.12‰" respectively "ε_t =0.16‰"-lines represent the end of the elastic behavior (phase I/II) respectively the beginning of macro-cracking (phase II/III). The "ε_t =10.0‰" line represents equilibriums reaching the maximum tensile strain defined for this example. Following this line into the area between N=0kN to N=500 kN (Figure 7) shows the area of pure bending in which axial thrust has a very small influence. In this area, the bearing capacity in the elastic state (ε_t =0.12‰) and the micro-cracked state (ε_t =0.16‰) is larger than the load in the cracked phase (0.16‰ < $\varepsilon_t \leq$ 10.0‰). That means the ultimate load is less than the peak load reached around the elastic / uncracked state (see also Figure 8, e/d = ∞ , 2.0, 1,0, 0.75 and Figure 9, N = 0. 250, 500 kN). This shows the typical strain-softening behavior of FRC in bending tests (see also Figure 10). The intersection of the ε_t =0.16‰ – line and the ε_t =0.16‰ – line in the MNID marks the point of quasi ideal elasto-plastic behavior; meaning the maximum load level reached under uncracked or microcracked conditions can be

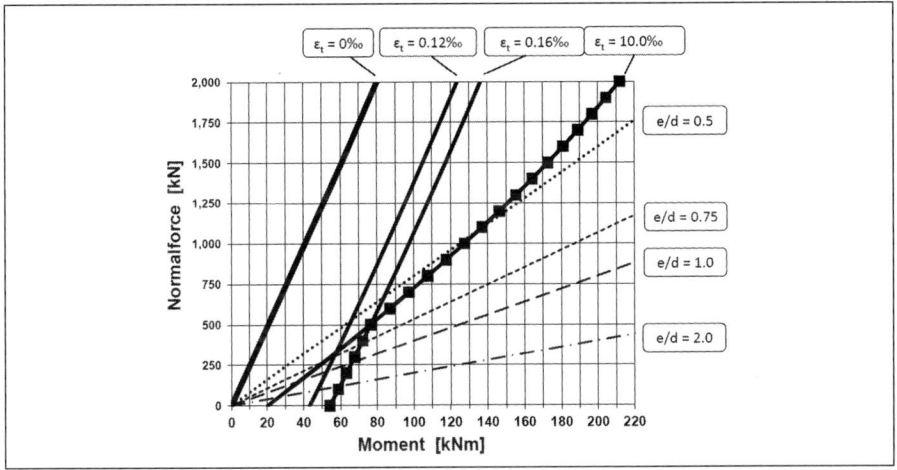

Figure 7. Moment normal force interaction diagram (MNID)—enlarged section relevant for tunneling

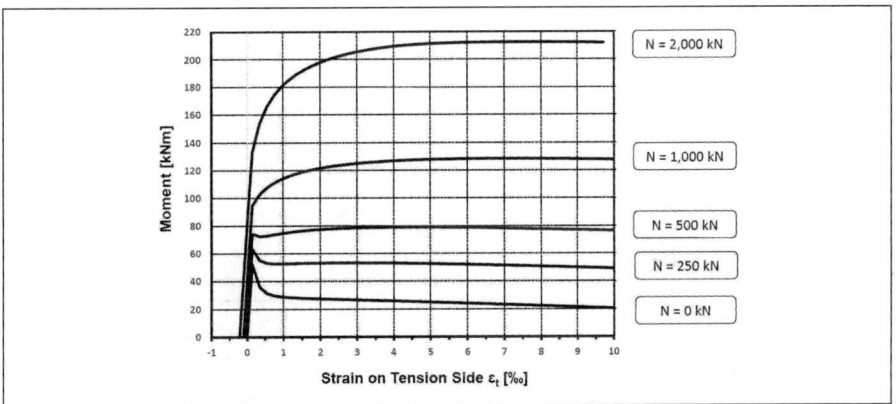

Figure 8. Moment-strain-diagram, parameter study e/d-ratio

maintained, which is reached in this example at roughly N=500 kN (Figure 7, Figure 9, Figure 10).

The simulated examples, which are representative of behavior observed in tests, also show that the moment bearing capacity in the elastic, the microcracking, as well as the cracked phase, are all increased under the influence of an increasing normal compressive force. While typical FRC simply supported bending tests show a strain-softening behavior, it can be observed that an increased normal force leads to a quasi elasto-plastic and a quasi strain-hardening effect. The term "quasi" is used because the bending behavior is a characteristic of the structural system, material properties do not actually change. For the same material the bearing behavior changes with an increased normal force influence (Figure 7, Figure 8, Figure 9, Figure 10).

Figure 10 shows the difference between elastic-brittle, elasto-plastic and strain-hardening behavior, and strain-softening in a simplified manner. Strain-softening behavior is typical for pure bending; the moment bearing capacity decreases after the peak load. An increased normal force influence leads to nearly elasto-plastic system

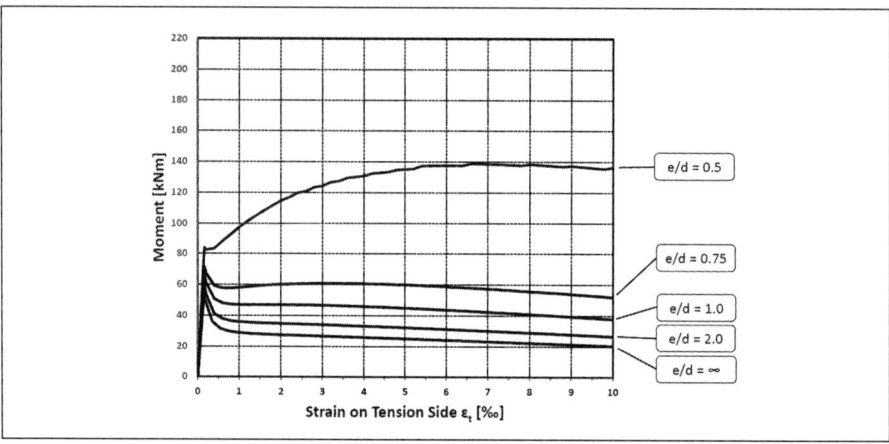

Figure 9. Moment-strain-diagram, parameter study normal force

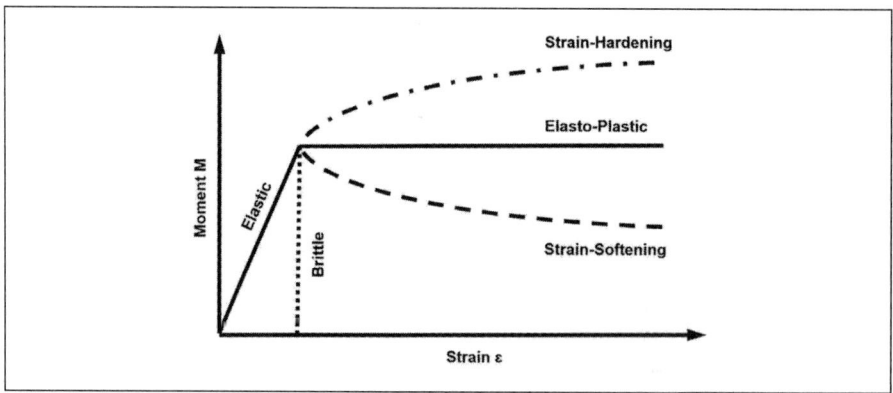

Figure 10. Schematic post peak response of fiber-reinforced concrete

performance (in our example for N ≈ 500 kN (see Figure 9) respectively 0.5 < e/d < 0.75 (see Figure 8)) and under a further increased normal force-influence the behavior progresses to a quasi strain-hardening effect. In addition to a change in failure mode, represented by the shape of the curve, there is also an increase in the peak moment capacity. While in the example the maximum bearable moment at 10‰ tensile strain is around 20kNm; every 100kN additional normal force increases the moment bearing capacity by roughly 10 kNm in this example.

FRC SPECIFICATION BASED ON RANGE OF NORMAL FORCE

What do these results mean for a tunnel lining design and FRC specifiation? The general desire from a structural perspective for a tunnel lining design under cracked conditions requires that the bearing capacity under cracked conditions shall be equal to or higher than the bearing capacity in the elastic state. Referring to Figure 10, the behavior shall be at least "elasto-plastic" or display "strain hardening." In the previous section it was shown that these conditions are highly dependent on the amount of normal force in the system. However, current tunnel designers do not take the range of expected normal force into consideration when specifying the material properties of FRC. Therefore a lot of structural potential of FRC remains under-utilized.

If elasto-plastic behavior or strain-hardening behavior is desired, material specifications should take the expected range of normal force, represented by the mean compressive stress in the lining, into consideration. The range of expected normalforce respectively the compressive strength in a lining can be easily evaluated based on preliminary lining designs. As shown in the parameter study herein, a project specific SSR could be developed that meets or exceeds the requirements. Subsequently, pure bending or M/N tests could used to prove that the SSR requirements could be met. Rather than the absolute values for the residual strength, it is suggested to specify an SSR with strength values relative to the compressive strength of the material.

INELASTIC STRUCTURAL ANALYSIS USING PLASTIC HINGES

ACI 318 and other international codes provide several options for a structural analysis of reinforced concrete structures. In a typical tunnel design the forces of the lining are determined in a linear-elastic model. Representative pairs of moments and normal forces from this analysis are then transferred into a MNID to ensure that the load combinations can be born by the FRC lining.

As discussed above, the inclusion of fibers increases the moment bearing capacity compared to unreinforced concrete when a section is subject to a large compressive normal force. However, typically even light rebar reinforcement can do the same or even exceed the bearing capacity of FRC. Where then is the benefit of FRC in the structural design? The benefit of FRC lies in the added toughness of the material, which allows—under elasto-plastic or strain-hardening conditions—to "hold" a moment in a lining even under severe deformation of the lining. However, these benefits are not utilized in a standard linear-elastic structural analysis. The structural and economic potential can be activated in an in-elastic structural analysis using, for example, a concept typically used for a simplified method for an in-elastic design of steel frames.

Structurally a cracked FRC lining acts like a "plastic hinge," which still transfers a moment while rotating. In a classical elastic analysis the capacity of the plastic hinges could be used as follows for a quasi in-elastic procedure: While increasing the load on a tunnel lining the peak elastic moment will be reached at a specific point. The elasto-plastic or strain-hardening behavior would allow for the introduction of a hinge at this location and altering the overall static (elastic) system of the lining. In a next step, the external load would be increased further until the peak elastic moment would be reached at another location and another hinge would be introduced at this location, and so forth.

At the end of this process the moments of each step would be superimposed and added. While increasing the external loads during this process, the values of the moments at the hinges could not be increased beyond the plastic moment, but the rotation could increase, making the overall system "softer." Following this approach the lining would be locally weakened to induce a load redistribution and eventually show a system failure rather than failure at a specific cross section. The ultimate stage would be reached either as a result of system failure or by reaching rotation thresholds at the hinges or some other pre-defined limits. Given the properties of a tunnel lining as a embedded beam, a system failure would basically mean formation of hinge next to hinge in close proximity. For this reason, other meaningful structural thresholds or definition of a maximum number of hinges seem to be a viable option. The procedure described above would allow full utilization of the properties and benefits of FRC in a structural analysis, while still using elastic structural analysis tools.

CONCLUSION

The article has presented and discussed the basics of FRC tunnel lining design using a selected Stress-Strain Relationship. The impact of normal force within the lining and the impact of a change of post-cracking behavior from strain-softening to strain-hardening was discussed in detail by means of a parametric study.

Current tunnel lining design does not fully utilize the potential of FRC because it disregards the positive benefits of the compressive force, which are not related to the material properties itself. Future material specifications for tunnels should consider the expected range of compressive stress in the lining and its beneficial influence on the ductility of FRC. The most advantageous property of FRC is its toughness when the tunnel lining has cracked. The potential of the toughness is currently not typically utilized when evaluating moment resistance in the cracked state under a simultaneous axial force. A procedure to introduce plastic hinges has been suggested that would allow utilization of the benefits of FRC using classical structural analysis tools and thereby realize the full structural and economic potential of FRC.

REFERENCES

[1] Nitschke, A.: Tragverhalten von Stahlfaserbeton für den Tunnelbau. Dissertation. (in German. Load Bearing Behavior of Steel Fiber Reinforced Concrete for Tunneling. Doctor Thesis.) Technisch-wissenschaftliche Mitteilungen des Instituts für konstruktiven Ingenieurbau der Ruhr-Universität Bochum, TWM 98-5, 1998.

[2] Ruhr-Universität Bochum, Lehrstuhl Prof. Maidl, Nitschke, A., Ortu, M.: Bemessung von Stahlfaserbeton im Tunnelbau. Abschlußbericht. (in German: Design of Steel Fiber Reinforced Concrete for Tunneling. Final Report) Research Project funded by the Deutscher Beton-Verein E.V. (DBV-Nr. 211) and the Arbeitsgemeinschaft industrieller Forschungsvereinigungen (AiF-Nr. 11427 N), Fraunhofer IRB Verlag, 1999. ISBN 3-81675465 4.

[3] Maidl, B.: Steel Fibre Reinforced Concrete. Ernst & Sohn, Berlin, 1995. ISBN 3-433-012881.

[4] Dietrich, Jörg: Zur Qualitätsprüfung von Stahlfaserbeton für Tunnelschalen mit Biegezugbeanspruchung. Dissertation. (in German: About Quality Testing of Steel Fiber Reinforced Concrete for Tunnel Lining under Flexural Tension. Doctor Thesis) Technischwissenschaftliche Mitteilungen des Instituts für konstruktiven Ingenieurbau der RuhrUniversität Bochum, TWM 92-4, 1992.

[5] Maidl, B.; Nitschke, A. ; Ortu, M.: Bemessung von Tunnelschalen mit dem M/NPrüfkonzept. (in German: Design of Tunnel Linings with the M/N Testing concept) In: Taschenbuch für den Tunnelbau 1999. Glückauf Verlag, Essen, 1998.

[6] American Concrete Institue Committee 318: Building Code Requirements for Structural Concrete (ACI 318-14). 2015.

[7] American Concrete Institute: The Reinforced Concrete Design Handbook. Volume 1: Member Design SP-17(14). Building Code Requirements for Structural Concrete (ACI 31814). 2015.

[8] American Concrete Institute: Report on Indirect Methods to Obtain Stress-Strain Response of Fiber-Reinforced Concrete (FRC) (ACI 544.8R-16)). 2016.

[9] American Concrete Institute: Design Considerations for Steel Fiber Reinforced Concrete (ACI 544.8R-16)). Reapproved. 2009.

[10] American Concrete Institute: Report on Design and Construction of Fiber Reinforced Precast Concrete Tunnel Segements (ACI 544.7R-16)). 2016.

[11] American Concrete Institute: Guide to Fiber-Reinforced Shotcrete (ACI 506.1R-08) First Printing. 2008.

[12] German Society for Concrete and Construction Technology (DBV): Guide to Good Practice—Steel Fibre Concrete. 2001.

[13] Deutscher Ausschuss fuer Stahlbeton (German Committee for Reinforced Concrete) (DAfStb): DAfStb Richtlinie fuer Stahfaserbeton (DAfStb Guideline for Steel Fiber Reinforced Concrte). 2010.

[14] Barton, N., Lien, R., and Lunde, J.: Engineering Classification of Rock Masses for the Design of Tunnel Support" Rock Mechanics, Vol. 6, No. 4.. 1974.

[15] Grimstad, E. & Barton, N.: Updating of the Q-System for NMT. In Kompen, Opsahl & Berg (eds), Proc. Of the International Symposium on Sprayed Concrete—Modern Use of Wet Mix Sprayed Concrete for Underground Support, Fagernes. 1993.

Engineered and Safe Approach to Tunnel Segment Lining Installation with Dowelled-In Connectors on the First TBM Tunnel in Qatar

Francois G. Bernardeau ▪ CDM Smith
Jacek B. Stypulkowski ▪ CDM Smith

ABSTRACT

Abu Hamour (Musaimeer) Surface & Ground Water Drainage Tunnel – Phase I (AHSO) is a 9.5k m long, 3.7 m ID storm water tunnel which was successfully and safely completed using circumferential dowelled-in connectors and guidance rods. The approach assured precise, safely sequenced and engineered radial segment fastening. The installation was designed with safety in mind. However, precision and quality assurance were the key requirements to assure that no newly or previously installed segments were not free standing during segment and ring installation process.

INTRODUCTION AND BACKGROUND

With annual rainfall of less than 100 mm, Qatar is one of the driest landscapes on earth. But individual driving rain showers lead to extensive flooding in the densely populated capital of Doha. The Abu Hamour Southern Outfall will form a critical element in overall storm water drainage network for greater Doha area in State of Qatar. Abu Hamour (Musaimeer) Surface & Ground Water Drainage Tunnel—Phase I (AHSO) is 9.5 km long with 3.7 m ID storm water tunnel about 30 m below ground surface. Completion of the project is scheduled for 2017. The Owner (or Employer), PWA-ASHGHAL, is the designated Engineer, who is represented by the Engineer's Representative CDM Smith. CDM Smith design team provided engineering design review, verification and approval services for all the design produced by the Contractor's team. CDM Smith is also the Construction Manager on this project. The bolt-less solution for segmental liner was pursued early by the contractor since they had very good experience on other projects and trained staff to handle it. To achieve this, all elements of tunneling had to be considered. The TBM was equipped with two component grouting system since the grout strength must be sufficient to restrain the ring before the hydraulic rams holding the segments in place are withdrawn to build one segment at a time. Each ring is being made out of 6+1 segments. Average time to mine one stroke was about half an hour. Contractor had to retain well train staff familiar with installation procedures from recently completed project. All elements which has contributed to successful tunneling with dowelled in connectors are describe herein.

GEOLOGY ENCOUNTERED

Integral part of supervision of the field activities was observation of rock as is exposed during excavation. Use of the shielded TBMs with segmental lining prevented continuous geological mapping. Rock mass therefore was properly surveyed only at all shaft locations spaced about 500 m apart. The wall mapping program of large scale rock mass provided geotechnical observations (Q, RMR) allowing assessment of the exposed rock face. For detailed geology description, methods used and mapping results refer to Pathak, 2015, Stypulkowski, 2016. The complete geological profile along the tunnel alignment as mapped is on Figure 1 where blue is limestone, yellow

694

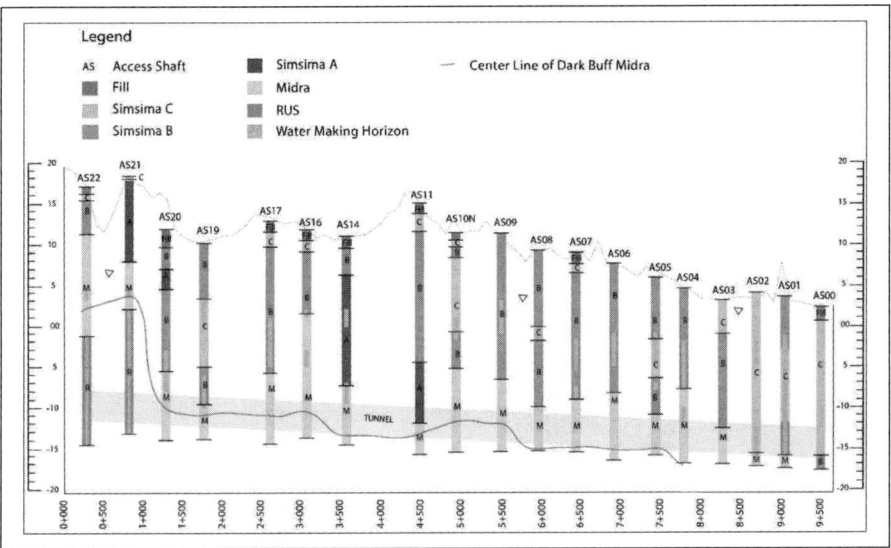

Figure 1. Interpreted geology for tunnel and access shaft locations for Phase 1 AHSO

is siltstone to claystone, and below the red line in the west of the alignment there are layers of shale, dolomitic limestone, clay, marl and claystone. Watermarking horizons observed in the shafts are marked in green. This database was compared with TBM performance data and projected back to the geology. Groundwater permeability were largely controlled by presence of connected solution cavities not the permeability of the matrix and they were very hard to predict. Temporary support application in shafts in 76 percent required use of shotcrete protective layer.

TBM OVERVIEW

The excavation was carried out by two identical TBMs with rotating cutter-head which was fitted with cutting wheels and cutting tools (buckets and slab cutters). The excavated material was collected by the "buckets" and transported through the openings provided between the cutters into the excavation chamber. A screw conveyor extracted the spoil from the chamber and discharged it onto the belt conveyor installed on the TBM back-up which in turn offloaded the material into muck skips (total 26). See Table 1 for additional details. The skips were operated by locomotives (total 7) which traveled on the tunnel rail track. The permanent lining of the tunnel consisted of precast concrete segments reinforced with steel fibers (Figures 2 and 3).

LINER

The Abu Hamour Tunnel is in a hot humid environment with high concentrations of chlorides and sulphates which are present in the ground and groundwater, as well as in the storm water and dewatering water that will flow through the tunnel. To avoid the deterioration of tunnel lining the mixes that were used are shown on Tables 2 and 3.

FINAL LINING CONCEPT

The tunnel is supported by precast unbolted/dowelled steel fiber reinforced concrete (SFRC) segmental lining (Figures 2, 3 and 4) consisting of 6 segments and a key 1.3 m (4.3 feet) wide and 250 mm (9.8 inch) thick with the concrete characteristic cylinder

Table 1. TBM specifications

Design Parameters	Curve radius (horizontal)	300 m
	Gradient	0.05%
	UCS	2–65 MPa
	Hydrostatic Pressure	2 bar
Segmental lining	Number of segments	6+1 key (7198 rings installed)
	Segment width	1,300 mm
	Segment thickness	250 mm
	Segment Backfill	Bi-component cement grout
Diameter	Bore Diameter	4.52 m
Cutterhead Style	Cutterhead	Mixed ground
	14" Cutters	17
	Scrapers	75
	Buckets	8
Cutterhead Drive	Cutterhead Speed	0–4.5 min^{-1}
	Breakout Torque	2,167 kNm
	Maximum Thrust	20,891 kN
	Cutterhead Drive	Electrical
	Cutterhead Power	110 kW × 6 motors
Thrust	Trust Jack Stroke	2 m
	Maximum Thrust	20981 kN @ 350 bar
TBM Conveyor	Screw Conveyor Diameter	600 mm
	Speed	0–46 min^{-1}
	Torque	59–74 kNm
	Back-Up Conveyor Belt Width	650 mm
Exploration/Ground Support	Probe Drill/Grout	8 peripheral ports; 1 drill
Protection	Methane Monitors	2 in cabin 1 in gantry # 7
	FPB Sensors	4 in chamber
Weights and Dimensions	Total length	124.55 m
	TBM weight	330 tonnes
	TBM core weight	172 tonnes
	Back-up system weight	130 tonnes

Figure 2. Segment and the key

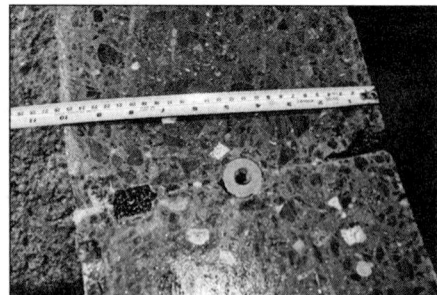

Figure 3. Saw cut liner

strength f_{ck} 45MPa (concrete grade at 28 days C55) and special mix consisting of a triple blend concrete suitable for harsh ground and groundwater conditions. The concrete rings were designed with a universal taper to follow curves in the alignment. This type of rings avoids cruciform joints, thus providing good leakage control and ring build safety. The ring segmentation was based on 6 - 56.842° segments and a

Table 2. Concrete mix

Structures	Materials	Other
Typical Tunnel Segments	Triple Blend Concrete: (OPC + GGBS 66-80% +MS) or (OPC+ FA 25-30% + MS) or (OPC min 50%+ FA + GGBS)	Steel Fiber Reinforced Concrete (SFRC)
Special Tunnel Segments (Close to the Adits)		SFRC + Stainless Steel Bar Reinforcement
Shafts & Adits		Carbon Steel Bar Reinforcement + PVC & Geotextile on Outer Surface

Table 3. Concrete mix requirements

Structural Element	Cement/ Binder Type	Maximum Equiv. Water/ Cement Ratio	Minimum Cement/ Binder Content (kg/m^3)	Maximum Cement/ Binder Content (kg/m^3)	Maximum Chloride Migration Coefficient[5] (m^2/s)
Typical tunnel segment SFRC C55	OPC + GGBS[1]+MS[4]	0.4	370	450	No requirement[6]
	OPC + FA[2] + MS[4]				
	OPC[3] + FA + GGBS				
Special tunnel segments close to the adit SFRC+Stainless Steel C55	OPC + GGBS[1] + MS[4]	0.4	370	450	No requirement[6]
	OPC + FA[2] + MS[4]				
	OPC[3] + FA + GGBS				
Adits & Shafts C37 Carbon Steel Rebar	OPC + GGBS[1] + MS[4]	0.4	370	450	2.2×10^{-12} @ 28 days
	OPC + FA[2] + MS[4]				4.1×10^{-12} @ 28 days
	OPC[3] + FA + GGBS				3.1×10^{-12} @ 28 days
Special Shaft Section	OPC + GGBS[1] + MS[4]	0.4	370	450	No requirement

Notes:
(1) GGBS = Ground granulated blast furnace slag, slag content of powder 66–80%
(2) FA = Fly ash, fly ash content of powder 25–30%
(3) OPC = Ordinary Portland Cement, cement content is minimum of 50%
(4) The addition of micro silica (MS) is limited to between 5–7% of the total binder
(5) Chloride migration test in accordance with NT Build 492 on laboratory samples (cast cubes/cylinders)
(6) Chloride migration testing is only for production monitoring purpose. No value larger than 10×10^{-12} m^2/s @ 28 maturity days. Testing performed on concrete without steel fibers.
(7) For all binder combinations, sulphate content (SO$_3$) of the concrete shall not exceed 3.6% of the binder content

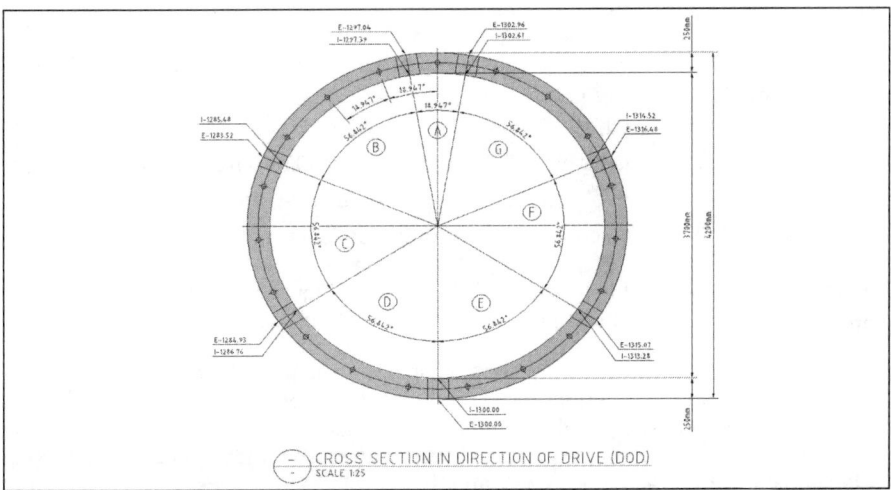

CROSS SECTION IN DIRECTION OF DRIVE (DOD)
SCALE 1:25

Figure 4. General arrangement of segment lining

Figure 5. Lifting socket FIP T142-005

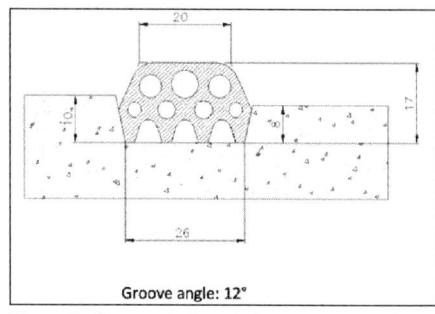

Groove angle: 12°

Figure 6. Geometric description of EPDM gasket (FIP T143 hardness 70 Sh A)

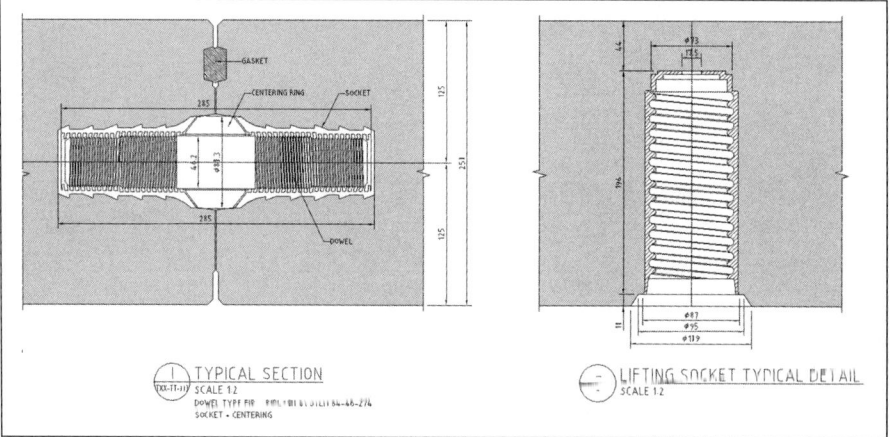

Figure 7. Typical section of the dowel and lifting socket

wedge shaped 18.948° key segment. The lining close to the adit junction (7 rings) was designed as stainless steel reinforced concrete and steel fiber reinforced concrete (SFRC).

Sealing was provided by EPDM gasket (FIP T143 hardness 70 Sh A). The mechanical seal was meeting the following requirements:

1. Design life more than 50 years,
2. Suitable for use in highly saline conditions (Cl-30,000mg/L, SO_4^2- 5000mg/l),
3. Water resistance to 4 bar external water pressure, and
4. Gasket properties certified by performance tests.

Lifting was facilitated by the socket (FIP T142-005—Figure 5) which can be and was used for grouting as well. FIP T142-005 (option 2 without the screwed cap).

The EPDM gaskets (Figure 6) were supplied in frames to fit the 7 types of segments— A, B, C, D, E, F and G. Each frame had 4 vulcanized corners and were installed on site just before lowering to the shaft bottom. The concrete surface of the EPDM gasket groove and the gasket itself was perfectly dry, free of any traces of contaminating or detaching agent before adhesive was applied.

Guidance rods (FIP Ø30 L=700 mm option A—Figure 8) were used to aid in segment fastening in the radial direction. The joint verifications considered the presence of 32 mm diameter guide rod hole for segment fastening in the longitudinal direction. The guide rods were installed prior to lowering the segments to shaft bottom.

Ring to ring fastening was achieved using circumferential connectors—dowels (FIP – BIBLOCK SYSTEM 84-46-274—Figures 7, 9 and 10). FIP Industriale has designed a mechanical system for the longitudinal connection of rings which they called Biblock system which is manufactured with special high-resistance, unbreakable and dielectric plastic materials. Each set consisted of the following elements:

- Two sockets casted in the segments concrete; and
- One connecting pin inserted in the sockets during installation of the segment.

The properties of FIP—BIBLOCK SYSTEM 84-46-274 dowel were verified considering the gasket force of 31kN/m (at 1 mm gap due to the presence of the packer). In select few instances around adit excavation reinforced FIP system was installed (84-46-274-100/150).

Bituminous 1 mm thick packers (Elastafip B (V3)—Figure 11) were used in between the rings to protect the segments from shove rams and avoid damages to the concrete. The packers were installed on the leading circumferential joint of the ring. A total of six packers were installed on segments B, C, D, E, F, and G whereas one only on segment A (key).

Figure 8. Guidance rod (FIP Ø30 L=700 mm Option A) during installation

Figure 9. Dowel (FIP–Biblock system 84-46-274) during installation

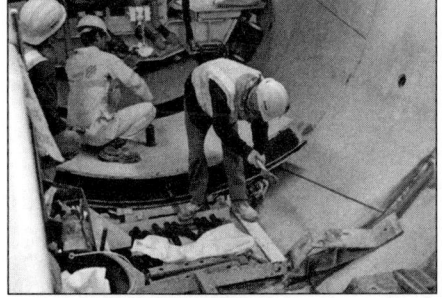

Figure 10. Dowel installation

Figure 11. Bituminous packers

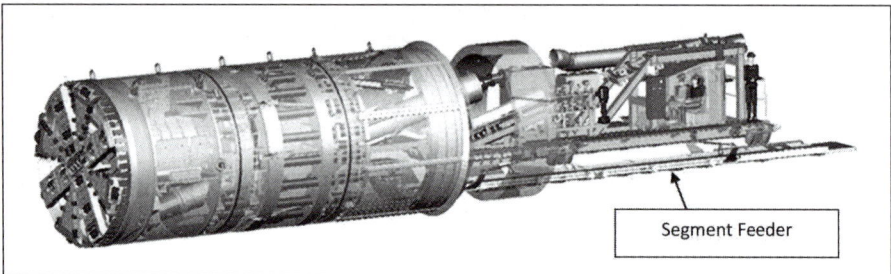

Figure 12. TBM machine, gantry no. 1 and segment feeder

INSTALLATION

The tunnel was lined with segments immediately after the TBM advance under the protection of the shield. The segments were moved into position by an erector, a remote-controlled crane arm, which picked them up mechanically and then put in place.

The segments were shifted from the storage yard area at the location of the main operation shaft from surface by a fork-lift and temporarily stacked on the concrete slab cast around the shaft within the reach of the gantry crane. From there the 50T gantry crane allocated for the supply of construction materials was used to transfer the segments from surface to shaft bottom (Figure 13). At the shaft bottom the segments were positioned on segment cars and then transported along the tunnel to the location of the TBM machine by a locomotive running on rail track. The train stopped in proximity of the TBM segment erection crane which grabbed the pre-cast segments from the segment car and delivered them on the segment feeder (Figure 12) to the TBM Segment Erector. The rings were erected according to a building sequence provided by Herrenknecht (VMT) with segment type "A" (key stone) being installed as last. The ring combination was governed by the VMT System. The following are key elements of the installation:

- Ring to ring fastening was achieved using circumferential connectors—dowels and guidance rod were used to aid in segment fastening in the radial direction.
- Bituminous packers were used in between the rings to protect the segments from shove rams and avoid damage to the concrete.
- The segment erector installed at the back side of the middle shield was used for the erection of the segmental lining in the tail skin portion of the TBM machine.
- The segments were lifted from the feeder by a claw bolt. The erector head was driven by hydraulic cylinders which was capable to move/rotate the segment along 3 axes.
- During the erection, the segment was held by the load handling equipment which includes a pulling cylinder plus erector head gripper and bolt inserted into the lifting socket already embedded in the segment.
- The erector was guided over the segment whilst the gripper grabbed the bolt and pull it against the erector head.
- The gripper held the segment in place by the friction force between the segment and the erector head.
- Only the required thrust rams were retracted, while the others were extended to be in full contact with the segments installed previously.

- The previously retracted thrust rams were extended to leading edge of segment before releasing the erector gripper from segment in place and retracting erector.
- The erector operator was in control of the erector and the thrust rams.
- Other two operatives were assigned to control the segment feeder and segment hoist.
- The Tunnel Filed Engineer was responsible for the whole ring building operation.
- The erector operator had to ensure that the erector is safe to lift, rotate, extend and retract.

Prior to starting the ring erection, the following steps were completed:

- The excavation was stopped and the machine did not move forward, TBM jacks fully extended supporting previously installed ring.
- The "ring building" mode was selected in the control cabin.
- The ring segments was ready on the segment feeder for the erector to pick up.

For all the rings the segments the same sequence (D–E–C–F–B–G–A) of erection was used.

Detailed step by step procedure implemented during segment erection was as follows:

1. Pre-install the FIP Biblock lock-up dowels + Centering Biblock on the trailing edge of segment D;
2. Retract the required thrust rams only, maintaining the others extended to be in full contact with the segments installed previously;
3. Transport the segment into the tail skin;
4. Pick up the segment with the erector gripper;
5. Slide segment D against the previous ring with the Biblock lock-up dowel + Centering Biblock entering the previous ring;
6. Check step/lip with adjacent segment within tolerance and adjust as necessary;
7. Extend thrust rams to leading edge of segment D;
8. Release the erector gripper from segment D and retract the telescopic cylinder of the erector completely;
9. Repeat steps 1. through 5. for segments E, C, F, B and G;
10. Apply grease or soap to radial joint of segments A;

Figure 13. Lowering of the segments

11. Pre-install FIP Biblock Pins + Centering Biblock to trailing edge of segment A;

12. Slide carefully segment A between B and G;

13. Check step/lip of segment A within tolerance and adjust as necessary;

14. Extend the TBM trust ram until touching the leading edge of segment A;

15. Release the segment erector from segment A;

16. Ring build complete and ring build records to be checked and filled; and

17. Resume the TBM excavation.

Figure 14. Segment installation

During the erection of the ring and retraction of the thrust cylinder, attention was paid to the condition of the gasket of the previous segment especially on the leading edge (Figure 14). If the gasket was found to be loose or not bonding to the segment groove, additional adhesive was applied to the segment groove and the gasket was re-installed back to segment groove following the same procedure as described above.

Each TBM was equipped with 19 thrust rams for segment erection and TBM advancing. A minimum of 3 thrust cylinders were retracted to enable the installation of one standard segment (i.e., all segments except the key stone) and care was taken to ensure that the radial joint alignment was not displaced when abutting segments together. The rams were re-extended to be in contact against the segment after it was placed in position, not just to hold it in position, but also to maintain face pressure.

HEALTH AND SAFETY RISK ASSESSMENT OF SEGMENT ERECTION

The following precautions were implemented during erection of segment by mechanical type segment erector:

1. Do not enter under suspended segment;

2. Operate the erector only inside the visible working area;

3. Do not stay in the danger zone of the mechanical erector system;

4. Secure loads only with suitable load handling equipment;

5. Keep eye contact between the erector operator and the assistant personnel;

6. Implement frequent maintenance/inspection of vacuum erector/control panel;

7. Maintain communication between operator and workers;

8. Never erect damaged segment;

9. Never use erect system with a damaged screw or a defective grabbing system;

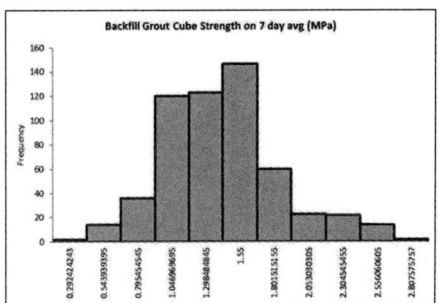

Figure 15. Steel joint plates around adit   Figure 16. Backfill grout cube strength on 7th day

10. Check segment, sealing and grabbing system before each lifting and erection;

11. Keep working place and working environment always clean and tidy;

12. Always keep escape route clear;

13. Always ensure adequate lighting; and

14. Protect erector/control panel from grease/grout/water leaking.

It should be noted that contractor's staff represented 18 countries with most coming from Asia and several languages were spoken on-site.

PERFORMANCE

Capacity of doweled connectors was tested once when the contractor installed one segment incorrectly. An unsuccessful attempt was made to pull it out of the socket using erection arm.

Several temporary steel props/arches and steel plates (Figure 15) were installed inside the main tunnel to secure the segmental lining against any potential unbalanced ground load deriving from the opening of the adit. The purpose of the steel plates was to hold the segments together to avoid potential opening of the radial joints. Since the opening of the radial joints was avoided by the application of the plates and the ground pressure acting on the extrados (through the annular grout) the axial forces and bending moments were transferred through the joints and the plates. No failure mechanism leading to instability of the ring developed. Plates were bolted with stainless steel requiring removal after completion. Backfill grout was tested for cube strength on 7 day (Figure 16) showing median value of 1.4 MPa (203 psi). Post-construction inventory of visually observed abnormalities was also conducted. Excessive step or lip (more than 5 mm) between adjacent segment or ring was noted in eastern tunnel in 96 rings out of 3834, which is 2.50 percent and in western tunnel 42 rings out of 3364, which is 1.25 percent.

TBM JACKING LOADS

During tunneling the advance cylinders were recorded for 4 groups as shown on Figure 17. Percentage of maximum designed force per thrust ram group was checked during tunneling for both drives. In general, measured forces rarely exceeded 60 percent of designed force (1100 kN-9.9 tons per single ram). Statistical summary is shown in Table 4. Jacking force group C – invert – was consistently higher than other three for both drives. Group A was 29 percent and 41 percent of the Group C for western & eastern drives respectively. Group C was smaller than Group D & B only in the few

Table 4. Statistical summary of jacking forces for individual thrust rams

	Western Drive			Eastern Drive		
	Mean	Mode	Std. Dev.	Mean	Mode	Std. Dev.
Jacking Force per Thrust Ram in Group A - Max - (kN)	280	194	85	229	232	77
Jacking Force per Thrust Ram in Group B - Max- (kN)	358	315	89	327	302	82
Jacking Force per Thrust Ram in Group C - Max- (kN)	511	551	79	474	474	96
Jacking Force per Thrust Ram in Group D - Max- (kN)	371	376	87	329	344	83

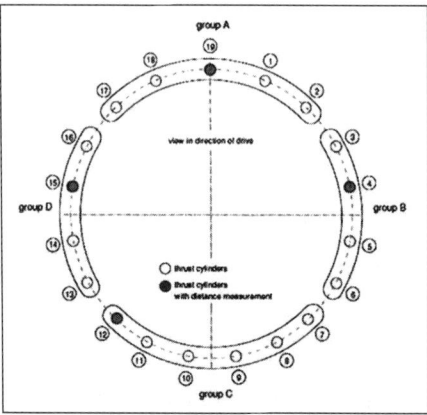

Figure 17. Advance cylinders   Figure 18. Finished tunnel

cases. Looking at distribution of Groups A, B & D they appear similar and always smaller than Group C.

PROJECT STATUS

All structures are completed and final project handover is planned for December 2016 one month ahead of schedule (Figure 18).

CONCLUSIONS

Two component continuous tail shield grouting has been implemented and grout had sufficient strength to restrain rings. Dowels were provided on the circumferential joints. Guide rods were also used on radial joints. Good control of the ring plane was maintained. There were no accidental withdrawals of the TBM rams for partially built rings.

ACKNOWLEDGMENT

The writers acknowledge Salini-Impregilo S.p.A who are successfully excavated the shafts and tunnels.

REFERENCES

Jafari, M.R., Bernardeau, F.G., Stypulkowski, J.B., Siyam, A.F.M., "Abu Hamour Outfall Tunnel Project In Doha, Qatar," in RETC—Rapid Excavation and Tunneling Conference and Exhibit, June 7–10, 2015, New Orleans, LA, pp 401–411.

Pathak, A.K., Stypulkowski, J.B, Bernardeau, F.G., "Supervision of Engineering Geological Activities during Construction of Abu Hamour Surface and Ground Water Drainage Tunnel Phase-1 Doha, Qatar," in International Conference on Engineering Geology in New Millennium at IIT, New Delhi 27-29. Special Publication, J of EG pp. 467–486, Oct 2015.

Siyam, A.F.M., Bernardeau, F.G., "Structural and Construction Challenges for Shafts on Abu Hamour Surface and Groundwater Drainage Tunnel" Arabian Tunneling Conference and Exhibition, Dubai, UAE, 23–25th November 2015.

Siyam, A.F.M., Stypulkowski, J.B., Bernardeau, F.G., "Improvement to Longevity of Tunnels in Aggressive Ground Conditions in the Middle East," Arabian Tunneling Conference and Exhibition Abu Dhabi, UAE, pp 163–185, 9–10th December 2014.

Stypulkowski, J.B., Pathak, A.K., Bernardeau, F.G.," Abu Hamour, TBM Launch Shaft, "A Rock Mass Classification Attempt for a Deep Shaft in Doha, Qatar." in EUROCK14, ISRM International Symposium, May 27–29, Vigo, Spain, CRC Press, Taylor and Francis Group, pp 117, 2014.

Stypulkowski, JB, Pathak, AK, Bernardeau, FG, "Engineering geology for weak rocks of Abu Hamour surface and ground water drainage tunnel Phase-1 Doha, Qatar," International Conference on Recent Advances in Rock Engineering, Specialized Conference of ISRM in Bengaluru, India, 16–18 November 2016, RARE2016, pp. 85–90.

Stypulkowski, J.B., Siyam, A.A.F.M., Bernardeau, F.G. and Al Kuwari, N.G., "Abu Hamour Drainage Tunnel, First TBM Mined Tunnel in Doha, Qatar." The First Arabian Tunneling Conference and Exhibition, Dubai, pp. 300–314, 2013.

Design of Steel Fiber-Reinforced Concrete Segmental Lining for the South Hartford CSO Tunnel

Mehdi Bakhshi ▪ AECOM
Verya Nasri ▪ AECOM

ABSTRACT

The South Hartford Conveyance and Storage Tunnel (SHCST) as a major component of the Hartford Metropolitan District's Clean Water Project is intended to capture and store Combined Sewer Overflows (CSO) from the southern portion of Hartford and Sanitary Sewer Overflows (SSO) from West Hartford and Newington. The 20 ft-diameter, 21,800 ft-long bored tunnel will be excavated in shale, siltstone and basalt through several fault zones with high groundwater pressures up to 9.6 bars. This paper discusses the implementation of the results of the geotechnical discontinuum analyses on the structural design of the tunnel segments for final service stage design as well as other critical load cases occurring during the segment production and transient stages. For the first time in North America, double hooked-end steel fibers (Dramix® 4D) were designed for reinforcement. This new type of steel fiber satisfies the serviceability requirements by limiting time-dependent effects of creep on crack opening and more significantly satisfies ductility requirements by providing an ultimate bending moment higher than the cracking bending moment.

INTRODUCTION

The South Hartford Conveyance and Storage Tunnel (SHCST) project is a significant component of the Hartford Metropolitan District's (MDC) Long Term Control Plan (LTCP) which is overseen by the Connecticut Department of Energy and Environmental Protection (CTDEEP). This project will address a portion of the MDC's Clean Water Project (CWP), which will reduce combined sewer overflows (CSOs); eliminate sanitary sewer overflows (SSOs); and reduce nitrogen released into the Connecticut River. The purpose of the SHCST project is to eliminate West Hartford and Newington SSOs, eliminate Franklin Area CSOs discharging to Wethersfield Cove and to minimize CSO discharges to the South Branch Park River. In 2010, MDC prepared a Preliminary Design Report (PDR) for the SHCST project. However, subsequent to the PDR, the objectives of the SHCST have slightly shifted with no need for relief of the Folly Brook Trunk Sewer, and new relief points proposed within the Franklin area to be diverted to the SHCST. Figure 1 shows the current recommended tunnel route (Alignment F). Alignment F was identified as the preferred alignment and recommended to advance to the final design as a result of an alignment study conducted on Seven (7) conceptual rock tunnel alignments and associated consolidation conduit options (Nasri et al., 2015). This alignment provides the maximum reduction in consolidation conduit length which reduces the associated cost, business impacts and construction risk. SHCST construction, as shown in Figures 1 and 2, includes a main conveyance and storage tunnel; tunnel boring machine (TBM) launch and retrieval shafts; several hydraulic drop shafts with associated adits, deaeration chambers, ventilation shafts, CSO connection structures, approach channels, odor control structures; consolidation conduits and manholes; and a tunnel pump station (TPS) at the downstream (eastern) end of the main tunnel. The main tunnel will be a deep rock tunnel 21,750 ft long with an invert located at El. -170 at its downstream end, and a finished diameter of 18 feet. The

Figure 1. Selected alignment (Alternative F)

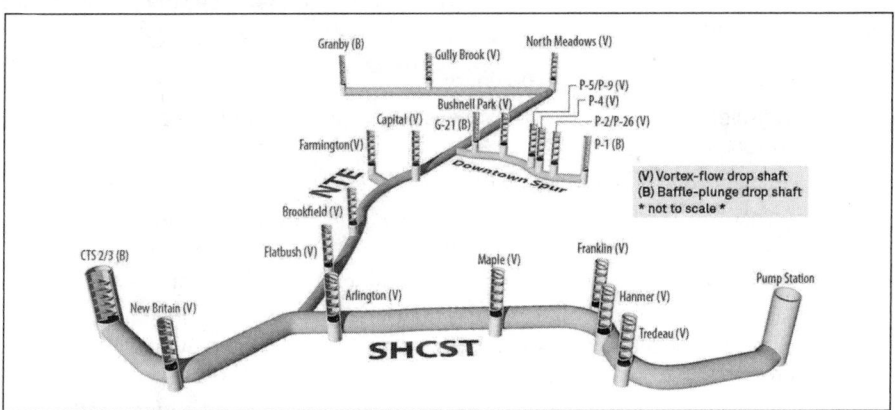

Figure 2. SHCST and North Tunnel integration

tunnel rises in grade at a proposed slope of 0.1 percent to its upstream end. At some time in the future, an additional deep rock tunnel (NTE) will be constructed that connects to and convey flow into the SHCST.

This paper discusses the implementation of the results of the geotechnical discontinuum analyses on the structural design of the tunnel segments for final service stage design as well as other critical load cases occurring during the segment production, transient stages, and construction. As an innovative solution, double hooked-end steel fibers (Dramix® 4D) were designed for reinforcement. This new type of steel fiber satisfies the cracking serviceability requirements and guarantees ductility requirements by providing an ultimate bending moment higher than the cracking bending moment.

Table 1. Tributary overflows to the SHCST

Contribution	Design Storm	Peak Flow (MGD)	Volume (Mgal)
West Hartford/Newington SSO	25-yr	27	17
South Branch Park River CSO	1-yr	68	6
Franklin Area Relief	18-yr	313	39
Total			62

DESIGN CRITERIA

The sizing of the tunnel was based on the volumes from the 1-year, 18-year and 25-year design storm per the LTCP and updated collection system modeling from the MDC's Program Management Consultant. The LTCP specified a different level of control for each tributary area. Table 1 shows the peak flows and volumes to be stored in the SHCST for each major source and respective design storm. Sediment deposition can present an ongoing maintenance burden if not controlled. Based upon the initial sediment deposition analysis and modeling, a slope of 0.1% appears adequate for the deep rock tunnel to cost-effectively minimize sediment deposition issues. This slope is consistent with the state of practice for other large diameter CSO tunnels as steeper slopes will increase project cost. The tunnel will still require periodic maintenance to remove sediment build-up over the life of the facility. The capacity of the tunnel dewatering pump station has been established by the MDC as 50 MGD. At this rate, the 25-yr design volume for both the North and South tunnels will be dewatered in 4.2 days. The 1-yr design volume would be dewatered in approximately 2.5 days. A typical CSO tunnel is dewatered in 24 to 48 hours. The tunnel is designed to meet the following performance criteria:

- A service life of at least 100 years
- Self-cleaning ability to prevent solid depositions, blockage, odor, and corrosion
- A finished diameter of 18 feet
- An allowable infiltration rate of 100 gallons per day per inch diameter per mile of tunnel
- Resistance to corrosion and chemical attack in CSO environment
- Sufficient strength to withstand ground loads, external hydrostatic pressure, internal operating pressure, forces due to temperature variation, seismic ground motion, etc.
- Limited deformation of underground opening and ground surface settlement

GEOTECHNICAL SETTINGS

The site area lies in the Central Lowlands physiographic province that extends in a north-south direction in the middle of the state of Connecticut. The central lowland area consists mainly of the sedimentary rocks and the associated igneous basalts of Triassic and Jurassic age. The Hartford Basin of Connecticut and southern Massachusetts is a half graben in structure, 90 miles long, and filled with approximately 13,000 ft of sedimentary rocks, and basaltic lavas and intrusions (Hubert et al, 1978). The source area for the sedimentary rocks was mainly the metamorphic rocks of the Eastern Highlands. Volcanic flows separated the deposition of the lacustrine and fluvial deposits, which were derived from the erosion of the highlands to the east. Displacements along the faults continued throughout the depositional period. Following the deposition of most of the sediments, the tectonic activity during Triassic/Jurassic period continued along the east edge of the basin resulting in displacements

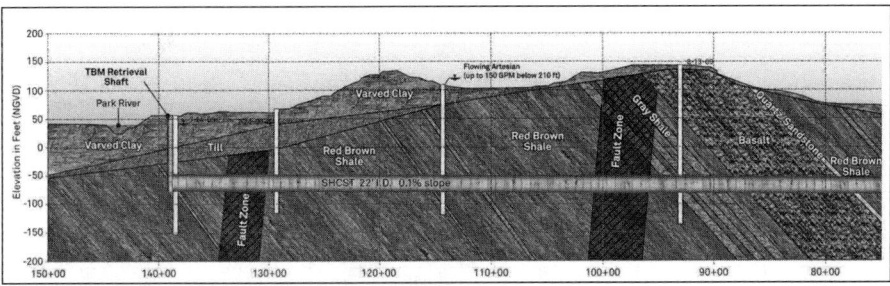

Figure 3. Expected geological formations along SHCST

along the eastern border fault which rotated the basin downward to the east. This resulted in the easterly dipping beds. Last activity along the faults is dated approximately 175 million years ago, therefore all faults in the project area are considered to be inactive. The region has undergone a period of glaciations that ground down the area's peaks, scraping away any weak or weathered rock and laying down a heterogeneous layer of ground-up rock. More recent alluvial deposits are common along the Connecticut and Park Rivers and their tributaries. In the site area, the following soils are present overlying the bedrock, in general order of sequence from ground surface downwards: Artificial Fill, Alluvium, Beach Deposits of Lake Hitchcock, Glaciolacustrine Deposits, Glaciofluvial Deposits, and Glacial Till. The formations that potentially could be encountered along the proposed tunnel are the Portland Arkose, the Hampden Basalt, and the East Berlin Formation. These units consist of shale and basalt with fractured and fault zones (Figure 3).

For the purposes of characterization, the tunnel alignment has been divided into three reaches summarized below based on differences in lithology, bedrock structure, and anticipated ground behavior during tunneling.

- Reach 1 (Station -0+80 to Station 93+78) – Portland Formation East
- Reach 2A (Station 93+78 to Station 98+83) – Hampden Basalt
- Reach 2B (Station 98+83 to Station 124+66) – East Berlin Formation
- Reach 2C (Station 124+66 to Station 143+00) – Holyoke Basalt, Hampden Basalt & Portland Formation – Regional Fault Zone
- Reach 3A (Station 143+00 to Station 185+00) – Portland Formation West
- Reach 2B (Station 185+00 to Station 217+54) – Portland FM West – Bedrock Depression

Recommended engineering properties for these formations including properties of intact rock and discontinuities are presented in Table 2. There were several geotechnical challenges which needed to be investigated and addressed during the design. The main geotechnical challenges were:

- Consolidation settlement of varved clay as a result of groundwater lowering due to shaft and tunnel construction and the settlement impact to the existing facilities
- Faults and fracture zones along the tunnel drive and tunnel construction impact
- Artesian groundwater inflow with pressures as high as 30 ft above the ground surface in the central portion of the tunnel drive and design and construction

Table 2. Design geotechnical properties for rock formations expected along different reaches of SHCST

| | | Intact Rock Properties | | | | | Properties of Rock Discontinuities | | | | | |
| | | | | | | | Joint Set 1 - Bedding | | | Joint Set 2 | | |
Reach		Unit Weight (pcf)	Modulus of Elasticity ($\times10^6$ psi)	UCS (ksi)	Internal Friction Angle (deg)	Tensile Strength (psi)	Joint Spac. (mm)	Joint App. Dip Dir.	Joint App. Dip	Joint Spac. (mm)	Joint App. Dip Dir.	Joint App. Dip
1	Max	170	9	20	N.A	2600		60	30		315	85
	Min	160	1.6	5	N.A	40		135	5		245	65
	Average	165	4.7	12.5	45	940	250	97	15	1500	305	75
2A	Max	181	11.5	44	N.A	3000		270	25		285	90
	Min	173	3.6	19	N.A	1200		90	0		270	65
	Average	178	9.6	30	50	2100	360	165	7	730	277	80
2B	Max	172	7.3	36	N.A	2700		120	30		285	90
	Min	164	1.4	19	N.A	900		60	0		270	65
	Average	167	5.2	26.5	45	1700	300	90	14	780	277	80
2C	Max	185	12.9	50	N.A	3000		270	25		285	90
	Min	177	4.5	22	N.A	1500		90	0		270	65
	Average	180	13	34	50	2200	280	165	7	340	277	80
3A	Max	170	9	20	N.A	2600		30	35		255	79
	Min	160	1.6	5	N.A	40		120	5		290	50
	Average	165	4.7	12.5	45	940	320	75	22	380	275	62
3B	Max	170	9	20	N.A	2600		30	28		255	73
	Min	160	1.6	5	N.A	40		120	5		290	45
	Average	165	4.7	12.5	45	940	260	75	20	580	277	59

| | | Properties of Rock Discontinuities | | | | | | | | | | | |
| | | Joint Set 3 | | | Joint Properties | | | | | | | | |
Reach		Joint Spac. (mm)	Joint App. Dip Dir.	Joint App. Dip	Peak Joint Fric. Angle (deg)	Peak Joint Coh. (psi)	Resid. Joint Fric. Angle (deg)	Resid. Joint Coh. (psi)	Joint Normal Stiffness (kci)	Joint Shear Stiffness (kci)	$E_{int.}/E_{mass}$	GSI	M_i Hoek Brown
1	Max		270	90	36	70	31	60					5
	Min		255	75	17	0	15	0					9
	Average	1500	263	82	30	35	26	16	232	91	3.1	60	7
2A	Max		90	90	38	70	32	50					30
	Min		105	80	22	0	15	0					20
	Average	730	97	85	34	35	29	10	391	163	3.7	70	25
2B	Max		90	90	36	70	31	60					15
	Min		105	80	17	0	15	0					5
	Average	780	97	85	30	35	26	16	1100	413	1.8	72	10
2C	Max		90	90	38	70	32	50					30
	Min		105	80	22	0	15	0					20
	Average	340	97	85	34	35	29	10	545	218	4.8	71	25
3A	Max		275	50	36	70	31	60					9
	Min		295	30	17	0	15	0					5
	Average	380	285	35	30	35	26	16	329	138	3.6	58	7
3B	Max		275	42	36	70	31	60					9
	Min		295	28	17	0	15	0					5
	Average	580	285	33	30	35	26	16	407	171	2.8	63	7

ANALYSIS AND DESIGN

Design Methodology

Based on the above geotechnical challenges and evaluation of cost estimate and risk mitigation, it was concluded that shielded TBM with one-pass lining system is the preferred alternative for the deep rock tunnel. Precast segmental lining will be installed using a double-shield TBM as the initial and final lining for the main tunnel. As shown in Figure 4, a 5+1 rhomboidal system assembled ring by ring is selected for SHCST tunnel lining. It consists of five full-size parallelogrammic and trapezoidal segments and one small key segment in a ring. Advantages include staggered longitudinal joints, continuous ring building and compatibility with a dowel type connection in circumferential joints, which results in a faster ring assembly process comparing to rectangular systems. Universal rings are selected for this project assembled from rings with circumferential joints inclined to the tunnel axis on both sides. One of the main advantages of this ring system over other systems (e.g., left/right rings) is using only one set of forms for segment production. With 18' as the internal diameter of the tunnel, 12" was selected as the thickness of segments which is the common value used in practice for this size of tunnel. Lining thickness was verified during the design procedure. Longitudinal and circumferential joints were designed as completely flat joints which are advantageous to other types of joints for transfer of loads between segments and rings, and also have proven to have a superior sealing performance. Bolt connection was designed for longitudinal joints and dowels were chosen for connecting rings in circumferential joints as they require less work for the construction of the segment form and less manpower in the tunnel as the insertion is automatically performed by the erector when the segment is positioned. The gasket profile was designed as anchored gasket (DATWYLER M 389 33 "type Doha") providing watertightness under the maximum expected groundwater pressure of 140 psi (9.6 bar). This gasket profile guarantees watertightness for 1.5 times maximum working water pressure considering a combination of gasket differential gap of 0.2" (5 mm) and bearing surface offset of 0.4" (10 mm).

The analysis and design of the segmental lining complies with the latest guideline (ACI 544.7R, 2016) in order to satisfy the intended objectives of the project during construction and service life of the tunnel. The design is carried out using the load and resistance factor design (LRFD) design method. During segment production and construction, concrete segments must be capable of withstanding stripping, storage and handling loads, thrust force of jacks needed to drive the TBM forward during excavation, and contact/backfill grouting. After installation, segments must resist loads imposed by the surrounding rocks and hydrostatic pressure. Segments were designed for final service stage and verified for production, transient and construction stages loadings.

Materials and Concrete Requirements

Precast concrete tunnel segments for SHCST can be made of reinforced concrete (RC) or fiber-reinforced concrete (FRC). However, only design of FRC segments is presented in this paper as FRC segments are superior in the crack control and more cost effective, and therefore, more likely to be used in this project. Cement type is Portland cement type II per ASTM C150. Type II Portland cement is sulfate-resistant cement, recommended for concrete linings exposed to the sulfate attack. In wastewater tunnels, the lining must also be designed to withstand corrosion attack from the influents which contains sulfate, as well as the groundwater that may contain corrosive chemicals. A quantitative approach, adopted by EPA and ASCE, was used to

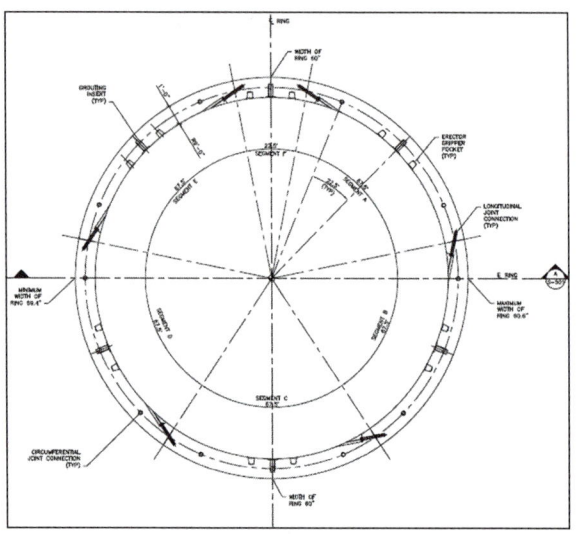

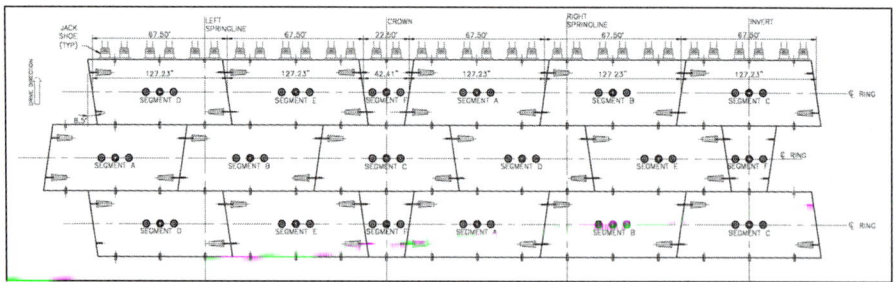

Figure 4. SHCST segmental ring geometry

assess the corrosion of the final lining. Loss of concrete material as a function of time, concrete properties and CSO characteristics was estimated and considered in this project. Minimum compressive strength of concrete is recommended to be 6,500 psi. To achieve a high early-age strength segment and high performance and durable concrete lining, maximum water cement ratio was designed as 0.40 with the addition of silica fume in the level of 5–7 percent of cementitious materials. Maximum aggregate size was specified as ¾" and chloride ion penetrability to less than 1,000 coulombs following ASTM C1202.

Cold-drawn wire ASTM A820 type I steel fibers, with a minimum tensile strength of 145,000 psi was specified for segment reinforcement. Other specified characteristics of fibers include double hooked ends, a minimum length of 2-inches, a maximum diameter of 0.035 inch and an aspect ratio (length/diameter) ranging from 65 to 85. Minimum required fiber content for segments reinforced with steel fibers was 67 lb/yd³, in order to obtain a minimum first-peak flexural strength (f_1) and residual flexural strength (f^D_{150}) of 340 and 360 psi, respectively, at the time of segment stripping according to ASTM C1609. At 28 days, minimum first-peak flexural strength (f_1) and residual flexural strength (f^D_{150}) were considered 540 and 580 psi, respectively. FRC segments were designed following ACI 544.7R (2016) using these parameters and verified versus results of different analysis in the next section. Note that for the first time in North America, double hooked-end steel fibers (Dramix® 4D) were designed for reinforcement. This new type

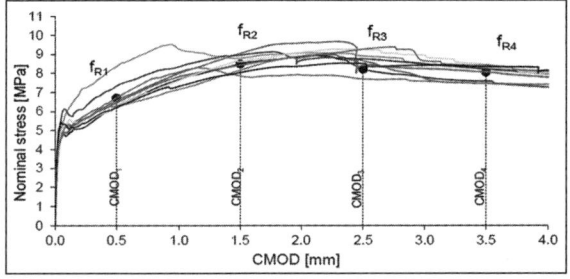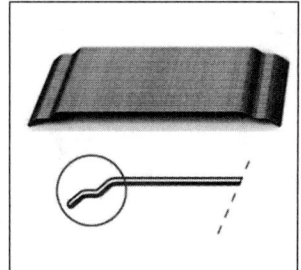

Figure 5. Results of standard beam bending tests (Caratelli et al., 2015)

of steel fiber satisfies the serviceability requirements by limiting time-dependent effects of creep on crack opening and more significantly guarantees ductility requirements in conventional fiber dosage rates by providing an ultimate bending moment higher than the cracking bending moment. This is shown in Figure 5 as a result of bending tests on standard beams reinforced with 67 lb/yd^3 of Dramix® 4D fibers. This is especially important for the loads applied on segments during production, storage, handling and transportation when segment is subject to pure bending loads.

Structural Analysis and Design Approach

Design for Final Service Stage

The discrete element method (DEM) was selected as the most suitable analysis method to calculate member forces in the segmental lining due to loads imposed by the surrounding fractured rock. The geometry of the excavation and the joint pattern—one bedding set and two orthogonal sub-vertical sets—justify the use of two-dimensional analysis. This method which is implemented in the numerical code UDEC (Universal Distinct Element Code) was used for analysis of tunnel lining in SHCST. Using UDEC, the domain is simulated by a group of discrete blocks and joints which are simulated as boundary conditions between the blocks, thus simulating explicitly a fractured rock mass behavior. The Mohr-Coulomb elastic-perfectly plastic failure criterion was used for the intact rock material and rock joints with the friction angle equal to the peak friction angle. Mohr Coulomb Slip Joint model with dilation option was used for the joint planes. Coefficient of lateral earth pressure and other rock discontinuities properties were applied in the models using the recommended values in Table 2. Effect of hydrostatic pressure on the tunnel lining was added separately including the artesian pressure in the central portion of alignment. In order to properly take into account the 3D face effect, a relaxation factor of 50% of internal stress in 5 intervals is applied between tunnel excavation and installation of precast concrete segments. Results of DEM analysis for the load case of ground and groundwater pressure for all different reaches including Reach 1, 2A, 2B, 2C, 3A and 3B are shown in Figure 6. This figure presents analysis results in comparison with the axial force-bending moment interaction diagram of FRC section. This interaction diagram was developed following recommendations of ACI 544.7R (2016), Appendix A, and using 28-day strength parameters presented in previous section for FRC materials. Note that based on ACI 544.7R (2016) recommendations, load factors of 1.35 and 1.25 were considered for the ground and groundwater pressure, respectively. Results indicate than precast concrete tunnel segments reinforced with steel fibers with residual flexural strength of 580 psi (4 MPa) can withstand final service stage loads induced by surrounding ground and hydrostatic pressure.

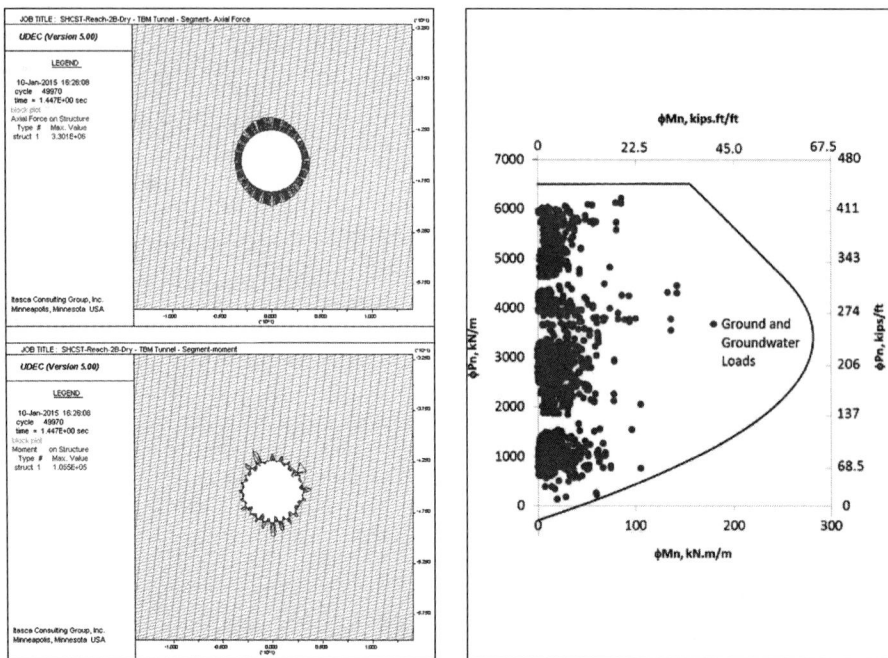

Figure 6. Results of DEM analysis for the load case of ground and groundwater pressure in comparison with axial force-bending moment interaction diagram of FRC segment

Design Verification for Production and Transient Loads

The production and transient loading includes all the loading stages starting from the time of the segment casting up to the time of the segment erection inside TBM shield. Main load cases include segment stripping, segment storage, segment transportation and segment handling.

Load case of segment stripping represents the effect of lifting systems on stripping precast concrete segments from the forms in the segment manufacturing plant. Figure 7a shows the stripping phase which is modeled by two cantilever beams loaded under their own self-weights (w).

The design is performed with regard to the specified strength when segments are stripped (i.e., 3–4 hr after casting). As shown in Figure 7b, the self-weight (w) is the only force acting on the segment, and therefore, the applied load factor in ULS is 1.4 per ACI 318 (2014).

Segment stripping is followed by segment storage phase in the stack yard where segments are stacked to gain specified strength before transportation to the construction site. As shown in Figure 8a, all segments comprising a full ring are piled up within one stack. Designers provide the distance between the stack supports considering an eccentricity of e = 4" (0.1 m) between the locations of the stacks support for the bottom segment and the supports of above segments. A simply supported beam loaded as in Figures 8b and 8c represent this load case. As shown in the figure, dead weight of segments positioned above (F) is acting on designed segment in addition to its self-weight (w). Therefore, corresponding load combination is 1.4 w + 1.4 F per ACI 318 (2014).

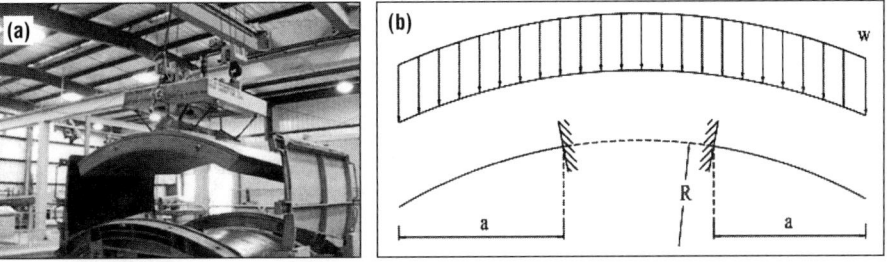

Figure 7. (a) Stripping segments from the forms in manufacturing plant; (b) forces acting on segments

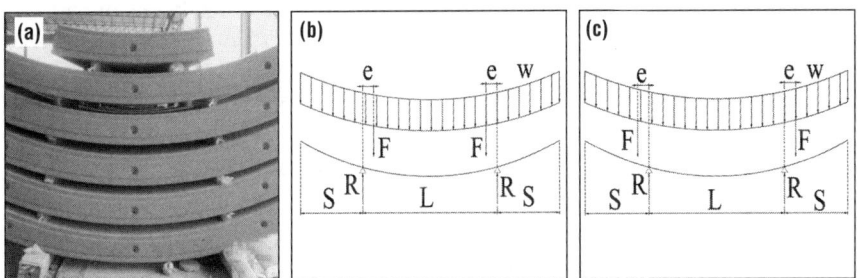

Figure 8. Segments stacking for storage and schematics of forces acting on bottom segment

Table 3. Segment design checks for production and transient stages

Phase	f'_{D150}, psi (MPa)	M_u, kipf-ft/ft (kNm/m)	ϕM_n, kipf-ft/ft (kNm/m)
Demolding	360 (2.5)	1.13 (5.04)	5.91 (26.25)
Storage	360 (2.5)	4.05 (18.01)	5.91 (26.25)
Transportation	580 (4.0)	4.68 (20.80)	9.44 (42.00)
Handling	580 (4.0)	2.26 (10.08)	9.44 (42.00)

During the segment transportation phase, precast segments which are stored in the stack yard are transported to construction site and TBM trailing gear. Segments may encounter dynamic shock loads during this phase and usually half of the segments of each ring are transported in one car. Wood blocks provide supports for the segments. An eccentricity of 4" (0.1 m) is recommended for design. Similar to segment storage phase, simply supported beams represent the load case of transportation with dead weight of segments positioned above (F) and self-weight (w) as the acting loads on designed segment. In addition to load combination of 1.4 w + 1.4 F per ACI 318 (2014), a dynamic shock factor of 2.0 is applied to the forces for the transportation phase.

Segment handling from stack yard to trucks or rail cars are carried out by specially designed lifting devices or vacuum lifters. Inside the TBM, segment handling is usually carried out using vacuum lifters while other methods may be used occasionally. This load case is simulated similar to segment stripping shown in Figure 7. Self-weight (w) is the only force acting on segments and therefore, a dead load factor of 1.4 in ULS per ACI 318 (2014) and a dynamic shock factor of 2.0 are recommended for design. Maximum bending moment and shear forces developed during above-mentioned stages are used for design checks.

Design checks for the production and transitional loads are shown in Table 3. As one can see, the values of ultimate bending moments (M_u) developed under the designated load cases are less than the factored strength of the segments (ϕM_n), and the design requirements are satisfied.

Design Verification for Production and Transient Loads

The construction loading stages include loading cases starting from the time of segment erection inside the TBM shield up to the time when installation of a ring is completed. Main load cases include TBM thrust jack forces and tail void and localized backfill grouting pressures. In this paper, only design checks for critical load case of TBM thrust jack forces is presented. After assembly of a ring, TBM moves forward, as shown in Figure 9a, by pushing its jacks on bearing pads placed on the circumferential joints of the newest assembled ring. This action results in development of high compression stresses under the pads, as well as bursting tensile stresses deep in the segment and spalling tensile forces between the pads. Maximum thrust force for each jack pair (J) is obtained by dividing maximum total thrust of TBM, if known, over the number of jack pairs. It is recommended by ACI 544.7R (2016) to apply a load factor of 1.2 on jack forces applied on each pad. A TBM with maximum total thrust of 5,620 kips (25,000 kN) applied on 16 jack pairs was considered for this project. Maximum thrust forces on each pair is therefore 351 kips (1.562 MN). The length and width of the contact area between the jack pads and segments, considering a maximum eccentricity of e = 1 in (0.025 m), are a_l = 34 in (0.87 m) and h_{anc} = 8 in (0.2 m), respectively. Dimensions of fully spread stresses are a_t = 11.1 ft /3 = 3.7 ft (1.13) m and h = 12 in (0.3) m in tangential and radial directions, respectively. Conforming to simplified equations of ACI 318 (2014), bursting force (T_{burst}) and its centroidal distance from the face of section (d_{burst}) in radial and tangential directions are:

Tangential direction: $d_{burst} = 0.5(a_t - 2e) = 0.5(3.7 - 2 \times 1/12) = 1.77 \text{ft} (0.54\text{m})$

$$T_{burst} = 0.25 P_{pu}\left(1 - \frac{a_l}{a_t - 2e}\right) = 0.25 \times 351 \times (1 - \frac{34}{3.53 \times 12}) = 17.32 \text{kipf} (0.077 \text{ MN})$$

Radial direction: $d_{burst} = 0.5(h - 2e_{anc}) = 0.5(12 - 2 \times 1) = 5 \text{in} (0.125\text{m})$

$$T_{burst} = 0.25 P_{pu}\left(1 - \frac{h_{anc}}{h - 2e_{anc}}\right) = 0.25 \times 351 \times (1 - \frac{8}{12 - 2 \times 1}) = 17.55 \text{kipf} (0.078 \text{ MN})$$

Using this method of analysis, the maximum bursting stresses developed in radial and transverse directions considering a load factor of 1.2 are determined as

Tangential direction: $\sigma_p = \dfrac{1.2 T_{burst}}{\phi h_{anc} d_{burst}} = \dfrac{1.2 \times 17.32 \times 1000}{0.7 \times 8 \times 1.77 \times 12} = 174 \text{psi} (1.2\text{MPa})$

Radial direction: $\sigma_p = \dfrac{1.2 T_{burst}}{\phi a_l d_{burst}} = \dfrac{1.2 \times 17.55 \times 1000}{0.7 \times 34 \times 5} = 177 \text{psi} (1.22\text{MPa})$

These stresses are less than 28-day specified residual tensile strength of FRC segment as $\sigma_p = 0.34 \times f'D_{150} = 0.34(580) = 197$ psi (1.36 MPa), and the design is valid for load case of TBM thrust jack forces.

CONCLUSIONS

With more than 170 ft of cover, the SHCST is considered as a deep rock tunnel. With exception of small zones of fault impacted rock, the entire length of the tunnel is expected to be constructed in competent rock. The concrete lining of the tunnel will have an inside diameter of 18 feet with a design thickness of 12 inches. Precast FRC segments were designed as initial and final lining for the main tunnel. The geotechnical and structural analyses indicate that precast concrete segments reinforced with steel bars or short steel fibers provide sufficient strength and serviceability requirements for the one-pass TBM-bored tunnel lininxg. Double hooked-end steel fibers (Dramix® 4D) designed for reinforcement in this project is the first use of this type of fiber in North

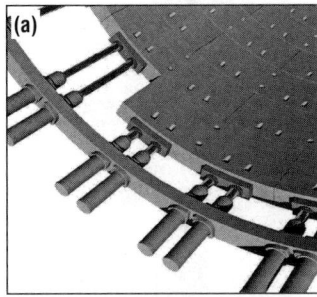
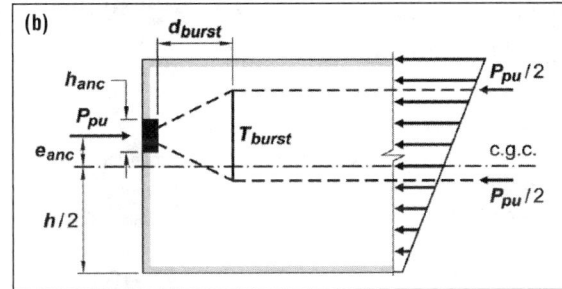

Figure 9. (a) Thrust jacks pushing on circumferential joints; (b) bursting tensile forces and corresponding parameters per ACI 318 (2014)

America. Special anchorage of this type of fiber can satisfies ductility requirements by providing an ultimate bending moment higher than the cracking bending moment.

REFERENCES

ACI 318. 2014. Building Code Requirements for Structural Concrete and Commentary. American Concrete Institute Committee 318, Farmington Hill, MI.

ACI 544.7R. 2016. Report on Design and Construction of Fiber-Reinforced Precast Concrete Tunnel Segments. American Concrete Institute Committee 544, Farmington Hills, MI.

ASTM A820-16. 2016. Standard Specification for Steel Fibers for Fiber-Reinforced Concrete. American Society for Testing and Materials, West Conshohocken.

ASTM C1202-12. 2012. Standard Test Method for Electrical Indication of Concrete's Ability to Resist Chloride Ion Penetration. American Society for Testing and Materials, West Conshohocken.

ASTM C1609-10. 2010. Standard Test Method for Flexural Performance of Fiber-Reinforced Concrete (Using Beam with Third-Point Loading). American Society for Testing and Materials, West Conshohocken.

Caratelli, A., De Rivaz, B., Meda, A., and Rinaldi, Z. 2015. Full-Scale Tests on Precast Tunnel Segments in Fiber Reinforced Concrete, ITA World Tunnel Congress 2015, Dubrovnik, Croatia, May 22–28, 2015.

Hubert, J., Reed, F., Dowdall, W., and Gilchrist, J. 1978. Guide to the Mesozoic Redbeds of Central Connecticut, Guidebook No. 4, Connecticut Geology and Natural History Survey, Hartford, CT.

Nasri, V.; Bent, W., and Hogan, W. 2015. South Hartford CSO Tunnel and Pump Station, Rapid Excavation Tunneling Conference 2015, New Orleans, LA, June 7–10, 2015, pp. 140–151.

Final Lining Design of the Ohio River Bridges East End Crossing Tunnel

Wern-ping Chen ▪ Jacobs Engineering
Mohammad Tughral Shaikh ▪ Jacobs Engineering
Sharma Narasimharajan ▪ Jacobs Engineering
Clement Uhring ▪ Walsh Vinci Construction Joint Venture

ABSTRACT

The Ohio River Bridges East End Crossing Tunnel is the largest twin tube highway tunnel in the US, 55-ft wide by 30-ft tall, constructed by sequential excavation method. Geologic units along the tunnel profile include karstic limestone, shale, and dolomite. Tunnel final lining support types were predetermined during early design phase and confirmed or adjusted/redesigned based on actual ground condition mapped in the field. This paper presents rationales in selecting the tunnel final lining design approaches, its associated rock load and groundwater load derivations, design details to meet service life requirement, and lining support type selection process during construction.

INTRODUCTION

The Ohio River Bridge East End Crossing (ORBEE) Tunnel is part of a Public Private Partnership (PPP) project, which connects transportation network between Louisville, Kentucky and southern Indiana. Located on State Highway KY841 in Louisville, Kentucky, the twin-tube highway tunnel provides northbound and southbound lanes of the Kentucky approach to a new bridge across the Ohio River. Tunnel lengths are approximately 1,675-ft and 1,685-ft for the northbound tube and the southbound tube, respectively. Connected by two cross-passages, both tubes have a roadway width of 40-ft, for three (3) traffic lanes and one (1) 4-ft shoulder for unidirectional traffic, and a vertical clearance of 26.5-ft, as shown on Figure 1.

The purpose of this paper is to present the methodologies of the tunnel final lining designs, the tunnel waterproofing and drainage systems, the durability and service life requirements of the permanent tunnel components, and the fire design approach.

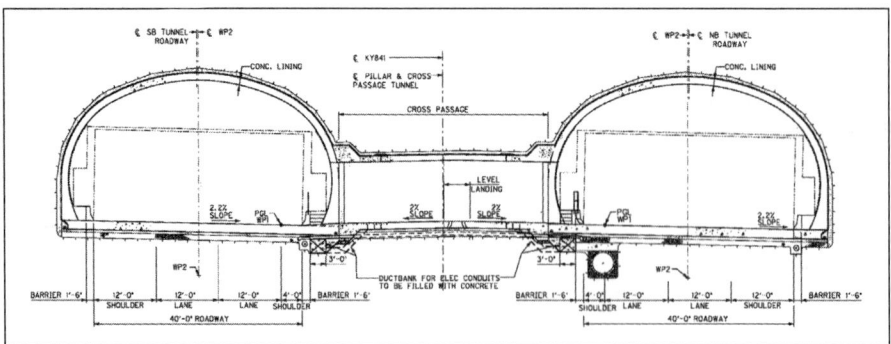

Figure 1. Typical tunnel cross section with cross passage

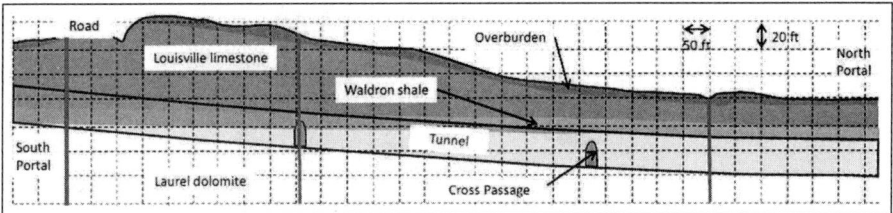

Figure 2. Schematic tunnel and geologic profiles

Challenges in the designs and constructions of the tunnel include the potential of encountering karst formation, slaking behavior of the Waldron Shale when exposed to groundwater and air, tunnel excavation in mix-ground condition, tunnel excavation with large tunnel diameter in shallow cover condition, derivation of geotechnical parameters for tunnel design, waterproofing and drainage between cross passage tunnel and main tunnels, and fire design of the tunnel final lining.

GEOLOGICAL PROFILE

The project site locates near the western edge of the Cincinnati Arch, a rift basin that was uplifted during the Cambrian Period. It is underlain by a relatively thin layer of residual soil derived from in-place weathering of rock. Rock below these soils, and within the proposed tunnel horizon, consists of Middle Silurian limestone, shale, and dolomite. The geologic units, arranged in increasing depth below the residual soil are Louisville Limestone, Waldron Shale, and Laurel Dolomite, as shown on Figure 2.

GROUNDWATER CONDITIONS

Hydrogeological investigations include surficial geologic mapping, surface-based geophysical investigations, observation wells, down-hole geophysics, and in-situ packer tests. The permeability of the intact rock is fairly low, on the order of 1×10^{-6} cm/sec or less. Groundwater primarily moves through solution joints and channels rather through pores in the rock, with hydraulic conductivity on the order of 1×10^{-3} cm/sec or more, in accordance with Geotechnical Baseline Report (GBR) (Golder Associates, 2012). Groundwater monitoring wells were installed at the soil-rock interface, the Louisville Limestone, the interface between Louisville Limestone and Waldron Shale, and the Laurel Dolomite. Based on readings from observation wells as described in the Geotechnical Design Memorandum or the tunnel (GDM) (WVC, 2013), Figure 3 summarizes interpreted groundwater levels for tunnel final lining design.

The GBR anticipates a flush flow of 300 gallons per minute (GPM) per incident at the excavation heading and decreased to 50 GPM within a week. It also anticipates a sustained total flow of less than 200 GPM at each tunnel portal. These data are the primary baselines for tunnel groundwater drainage designs.

TUNNEL EXCAVATION AND INITIAL SUPPORT

Because of the large tunnel cross section, the tunnel was determined to be constructed by Sequential Excavation Method (SEM), with three (3) top heading drifts and one (1) bench excavation. Figure 4 shows tunnel excavation sequences in cross section and plan views, respectively. Due to the high Unconfined Compressive Strength (UCS) of the Louisville Limestone, drill and blast excavation was selected as a means of tunnel excavation, instead of using mechanical excavation equipment, such as roadheaders.

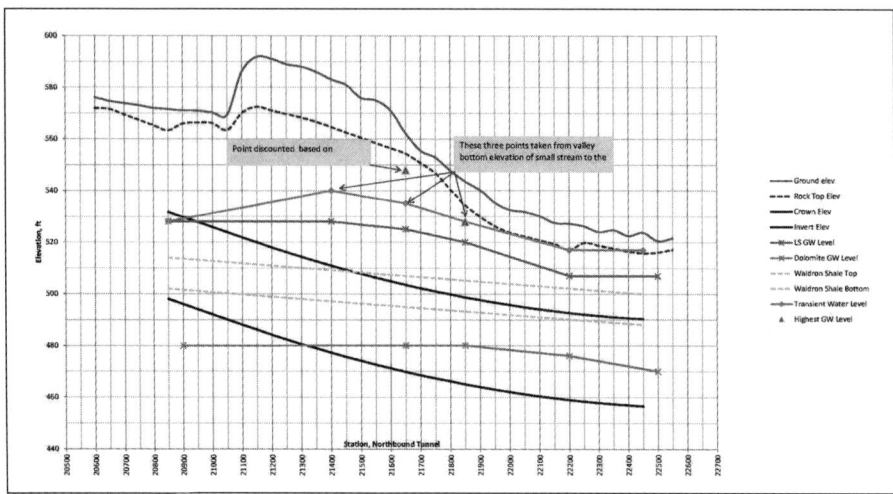

Figure 3. Tunnel final lining design groundwater levels

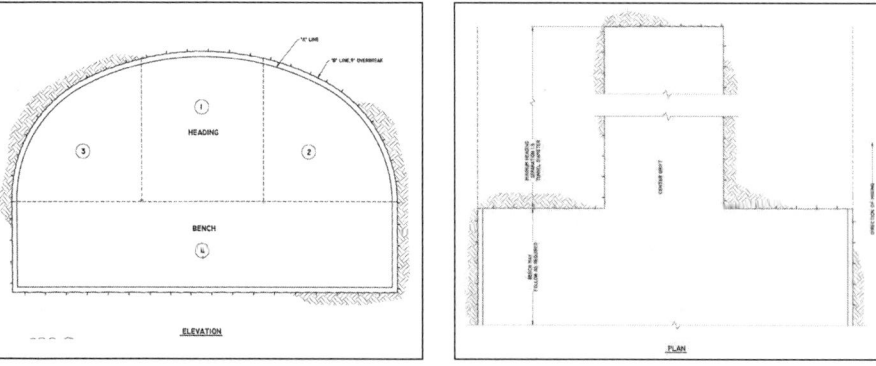

Figure 4. Tunnel excavation sequences—cross section and plan

In accordance with the Technical Provisions (TP), the initial supports during tunnel constructions are not permitted to be counted as part of the tunnel final lining system. During design stage, ground support classes were designed, as a tool box, based on Ground Classes identified. Actual ground support to be installed in the field is based on actual ground condition mapped during tunnel excavation. Typical ground initial supports include spot rock dowels, pattern rock dowels with shotcrete, and pattern rock dowels with shotcrete and lattice girders. Steel sets with timber lagging were installed at and near the north portal areas with least rock cover.

Geotechnical instrumentations include structural monitoring points, multiple position piezometer, multiple position borehole extensometer, and subsurface settlement point. Probeholes are specified in the TP. Depending upon tunnel Station, two (2) or four (4) probeholes are required ahead of tunnel excavation advancement to investigate geologic anomaly, such as solution cavities, and the potential of encountering excessive groundwater inflows.

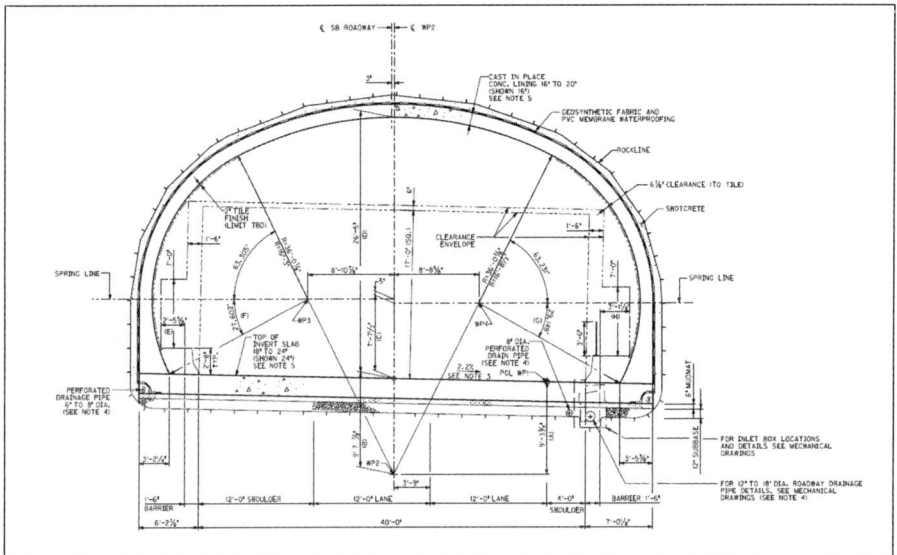

Figure 5. Typical tunnel cross section with waterproofing membrane and drainage systems

TUNNEL FINAL LINING

TP stipulates the tunnel final lining design to be in accordance with AASHTO (loads, load factors, and loading combinations) and AASHTO (2010) Technical Manual for Design and Construction of Road Tunnel—Civil elements, wherever is applicable.

Geometry

During the initial stage of the design, the tunnel geometry was developed so that the same formwork can be used for both tunnels with minor adjustments. Also, several types of tunnel cross sections with different tunnel lining and invert slab thicknesses were investigated in order to minimize tunnel excavation and concrete volume for both tubes.

Drainage

Groundwater. TP requires a waterproofing membrane system around tunnel perimeter such that seepage through tunnel final lining will not develop. Two (2) side drainage pipes, at either side of the tunnel final lining, and one (1) drainage pipe at the low point of the tunnel invert were designed to collect groundwater around the entire perimeter of the tunnel, as shown on Figure 5. Cleanouts were also designed, at 80-ft on center to facilitate future tunnel drainage pipe maintenance.

Roadway Surface Water. Two additional drainage systems were also provided. They are the tunnel roadway drainage inlets and the portal transverse roadway trench drain. The tunnel roadway drainage inlet, spaced at about 90-ft on center, was designed to collect tunnel wash/fuel spill water and tunnel emergency fire-suppression water. The portal tunnel trench drain was located at the south portal to prevent any surface run-off water entering into tunnel. Water from all drainage system will be collected and floatable separated prior to releasing into the Wellhead protection area.

Material Properties

Steel fiber reinforced concrete alternative was evaluated; however, with the size the tunnel, it was determined infeasible from flexural capacity perspective; therefore, traditional rebar reinforced concrete was recommended. Based on our experience, plan concrete alternative was not considered since tunnel lies within Seismic Zone 2 which requires reinforcement. The f_c', 28-day concrete strength, proposed for the final lining is 4,400 psi, with water cement ratio (W/C) of 0.41 and 20% fly ash added as supplementary cementitious material. All reinforcing bars comply with A615, with yield strength, f_y, of 60 ksi.

Unit weight of the concrete, with the inclusion of steel reinforcement, is 150 pcf; with Poisson's ratio of 0.21. Its long term modulus of elasticity is assumed to be $28,500 \cdot (f_c')^{1/2}$, which is equal to 50% of its short term modulus, in order to consider concrete long term creep phenomenon (Kuesel, 1987).

Design Loads

Groundwater Load and Its Distribution. Groundwater water level for final design is based on observation well readings as shown on Figure 3. Its loading distribution from groundwater level to tunnel invert level is based on AASHTO (2010) recommendation, which is based on empirical observations from previous projects and assuming tunnel groundwater drainage systems are provided as described in the previous section. The water pressure at the invert level was assumed to be 10% of its full hydrostatic pressure at the invert level for the design of the ORBEE Tunnel. This approach is to consider the potential long-term partial blockages of the drainage systems. Theoretically, the tunnel drainage systems are designed to drain all groundwater around and below tunnel invert and the hydrostatic pressure exerting on the tunnel lining shall be minimum.

The tunnel is divided into four (4) hydro geologic sections based on four different groundwater levels as shown in Table 1. No 200-year flood level was considered for lining structure design. The lowest roadway elevation at the centerline of KY 841 is 460.47' at STA 127+11.52 (220' north of North Portal), which is higher than Ohio River 500-YR and 100-YR flood elevations i.e., 457' and 452' respectively. Transient groundwater loads were also considered for the tunnel final lining.

Dead Load and Surcharge Load. Dead load includes the self-weight of structural components and non-structural attachments such as signs, lighting fixtures, signals, jet fan (5 tons/EA), architectural finishes, waterproofing, etc. Unit weight of concrete, including the weight of steel reinforcement, of 150 pcf was used in the design and analysis. No surcharge load to the tunnel level is anticipated, since there is no future development on the top of the tunnel due to historic preservation.

Horizontal Earth Pressure. The horizontal earth pressure exerting to the finite element models, as described in the Vertical Rock Load section, was based on constrained boundary condition, i.e., by Poisson's effect only. Its corresponding K_o is about $v/(1-v)$, which is generally smaller than 0.5 and is a conservative assumption.

Vertical Rock Load. The TP stipulates "Rock loading of a *loosened slab* of thickness *not less than 10 feet* up to the *full width of the Tunnel* and shear

Table 1. Based on GDM, the groundwater level along the proposed tunnel alignment

Section	Tunnel Station (Along Centerline of KY841)	Static Groundwater Level Elevation, ft.
1	108+10 to 114+00	528
2	114+00 to 118+00	525
3	118+00 to 120+25	520
4	120+25 to 124+90	507

deformations along bedding planes and joints in Reach 1 (from Station 50+30 to Station 55+00) and Reach 2 (Station 55+00 to Station 61+00) and load of *not less than 30 feet* up to the full width of the tunnel or maximum overburden (whichever is applicable) in Reach 3 (Station 61+00 to Station 69+85)." The TP also requires the tunnel lining system to be modeled using an appropriate Finite Element Method (FEM) and or Finite Difference Method (FDM) acceptable by IFA; closed-form solutions are not acceptable.

It is believed that the "no less than 10-ft" and "no less than 30-ft" statements above were derived from empirical rock mass classification methods. To investigate these parameters numerically, parametric numerical FEM models, based on Midas GTS, were performed. Two typical tunnel stations were selected for this purpose, one with the highest overburden and in Class III Ground condition and one with the shallowest ground cover at the north portal area, referred to as Scenario 1 and Scenario 2, respectively. Rock was simulated by plane strain elements, with Mohr-Column constitutive model and elastic-plastic failure criterion. Tunnel lining was simulated by elastic beam elements. Symmetric boundary condition was applied to the FEM model, i.e., only one tube was simulated with roller boundary condition at the plane of symmetry.

For rock mass properties, Jacobs performed an independent rock mass properties study, which includes entire boring database for consideration. Concrete material properties were adopted as those defined in the Material Property section.

Based on these rock mass properties and assuming 30% ground relaxation prior to the lining installation, the derived rock loads are about 10-ft for Class I (10<Q<40) and Class II (1<Q≤10) Grounds and 15-ft for Class III (Q≤1) Ground; rock Classes are as defined in the GDM. These rock loads extend to the full width of the tunnel (Note— the equivalent rock load was derived from the total reaction at the base of the beam elements divided by the width of the tunnel and the unit weight of the rock.). These vertical rock values were adopted in the development of the Release for Construction (RFC) drawings.

Seismic Load. Based the TP, the structural capacity check shall comply with FHWA Technical Manual (2009). From the review of the U.S. Geological Survey (U.S.G.S) seismic risk maps, the Section 4 alignment lies within Zone 2, corresponding to an area exposed to potential earthquakes of magnitudes exceeding 7.0 on the Richter Scale, which corresponds to a Peak Ground Acceleration (PGA) of 0.1g for Louisville, KY area (Wang, 2008).

Underground structures are subjected to three primary modes of deformation during a seismic event: ovaling/racking, axial, and curvature deformations. Specifically, ovaling/racking refer to the deformation acting in the plane of the tunnel cross section caused by seismic waves propagating perpendicular to the tunnel longitudinal axis. Vertically propagating shear waves are typically the most critical type of waves for this mode of deformation. The axial and curvature deformations are caused by components of seismic waves that propagate in the longitudinal direction.

Since the ORB tunnel is situated in hard rock, seismic effects on tunnel lining are expected to be minimum. A practical seismic analysis was performed in this case is by the simplified and conservative closed form solutions based on formulations recommended by Hashash et al. (2001), which is consistent with the approach taken by the FHWA.

Temperature. Temperature loading was applied at the tunnel portal areas (approximately 60-ft into portals) only, since temperature variation decreased rapidly from tunnel portals. The design temperature variation was plus or minus 25 degrees Fahrenheit (JSCE, 2006); corresponding coefficient of thermal expansion is 0.000006 inch/inch/degree F. For the purpose of calculating stresses induced by temperature variation, the modulus of elasticity, E_c, of the concrete is only 50% of its associated value determined using AASHTO.

Shrinkage. Shrinkage reinforcement is based on AASHTO requirement.

Swelling. Swelling pressure exerting on the tunnel final lining was implicitly included in the rock load verification after tunnel excavation and geologic mapping are completed. Since NGI–Q index was used in the field mapping, its corresponding rock load was derived based on field mapping. The swelling factor was considered in the Q index from the interpretation of the Stress Reduction Factor (SRF).

Durability

TP requires the service life of the tunnel permanent structures to be 75-year, including the tunnel final lining and all drainage components. Concrete material (strength, W/C, and SCM), concrete reinforcement cover, and environment exposure conditions (pH, sulfate, chloride, and CO_2 concentration levels) will impact the service life requirement. Service life prediction model was conducted prior to the final design of the tunnel final lining (Chen, 2016a). It concludes that the proposed concrete mix in associated with the following concrete reinforcement covers for each of the tunnel components will meet the project service life requirement:

- Tunnel lining face against waterproofing membrane—2 inches
- Tunnel lining face exposed to traffic—3 inches
- Tunnel invert, top face—4 inches
- Tunnel invert, bottom face—3 inches

Crack Width Control. AASHTO limits crack width by reinforcement distribution. In addition to satisfy strength requirements, the amount of reinforcement shall also satisfy the crack width limit requirements. In accordance with AASHTO, Class 1 exposure condition (with 0.017" crack width limit) applies when cracks can be tolerated due to reduced concerns of appearance and/or corrosion; Class 2 exposure condition (with 0.0085" crack width limit) applies to transverse design of concrete lining for any loads applied prior to attaining full nominal concrete strength and when there is increased concern of appearance and/or corrosion. Since the tunnel lining has a waterproofing membrane on its exterior face, corrosion concerns from the environment are reduced; therefore, a Class 1 exposure condition was applied to the tunnel lining design. The top face of the tunnel invert will experience de-icing salt/agent exposure in winter seasons, which increases corrosion concerns; therefore, a Class 2 exposure condition was applied to the tunnel invert design.

DESIGN APPROACH AND METHODOLOGIES

As to the design of tunnel initial supports, tunnel final lining support type can only be determined after tunnel excavation and ground classes are mapped in the field; therefore, during initial design phase, certain tunnel final lining types were designed based on anticipated ground conditions to be encountered as depicted in the GDM and GBR. If the actual ground condition encountered in the field was worse than those anticipated in the design phase, then final tunnel lining type was redesigned and revised.

In addition to portal beam design, eight (8) different final tunnel lining support types were analyzed for both tunnels, including:

- Type A (SL) Support—Final lining designed for Class I and Class II Ground conditions, with 10-ft rock load, symmetrical loading condition (applying vertical rock load on full tunnel width).
- Type A (USL) Support—Final lining designed for Class I and Class II Ground conditions, with 10-ft rock load, unsymmetrical loading condition (applying vertical rock load on half tunnel width).
- Type B (SL) Support—Final lining designed for Class III Ground condition, with 15-ft rock load, symmetrical loading condition (applying vertical rock load on full tunnel width).
- Type B (USL) Support—Final lining designed for Class III Ground condition, with 15-ft rock load, unsymmetrical loading condition (applying vertical rock load on half tunnel width).
- Type C (S) Support—South portal section, about 60-ft long, with 10-ft rock load and including temperature load.
- Type C (N) Support—North portal section, about 60-ft long, with 15-ft rock load and including temperature load.
- Type D (S) Support—Tunnel lining designed for the south cross passage area, with 10-ft rock load.
- Type D (N) Support—Tunnel lining designed for the north cross passage area, with 15-ft rock load.

Each tunnel final lining support type has a panel length of approximately 40-ft long, except the most southern and the most northern panels of Type C Support, which have panel lengths of approximately 21-ft to 25-ft long. Figure 6 shows the tunnel final lining reinforcement panel layout. Each panel has a designated number, and its associated final tunnel lining support type is assigned after tunnel excavation and field geological mapping is confirmed.

Design Approach. The entire tunnel length was divided into four (4) "Sections" based on ground water levels and rock mass classes. Groundwater pressure was applied to design models based on Figure 3 readings for each tunnel section and its associated groundwater distribution was based on the descriptions in the Groundwater Distribution section. Loading combinations analyzed consider lower bound and upper bound values for each type of load. Concrete reinforcement cover thicknesses were based on the results discussed in the Durability section. Each tunnel "section" is subdivided in to two (2) separate STAAD models based on two types of rock loadings, (Type A and Type B), except Section-1 which has only Type A rock Loading.

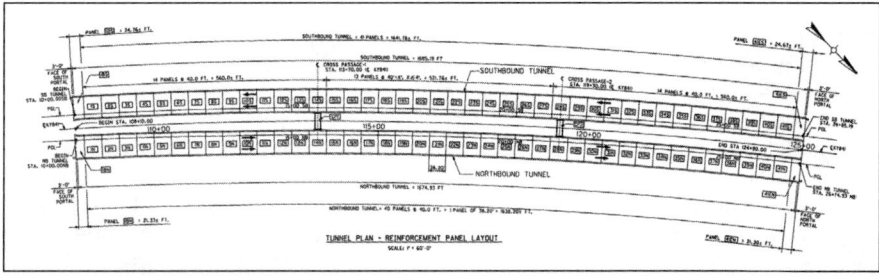

Figure 6. Tunnel reinforcement panel layout—Plan

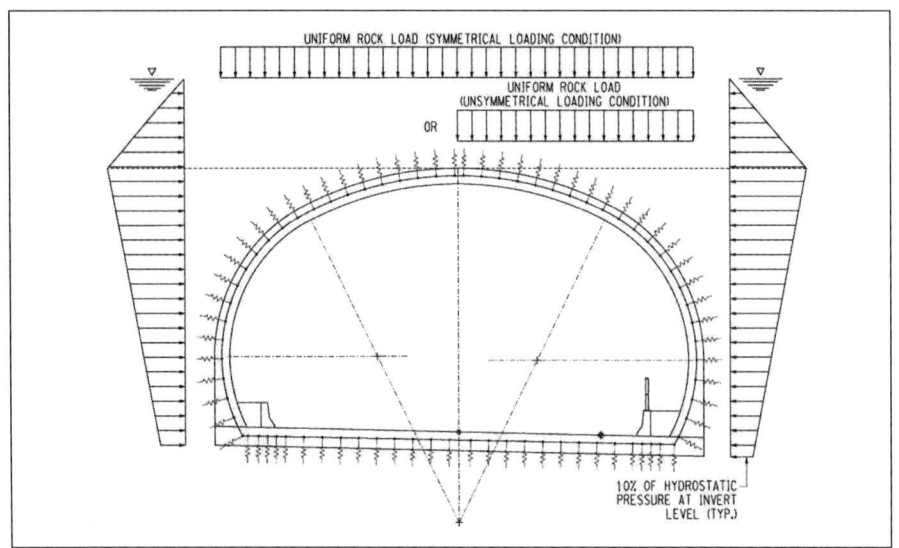

Figure 7. Typical STAAD model of tunnel cross section with rock loads, water loads, and spring support

In addition, more detail groundwater distribution within each tunnel section was analyzed at three locations, i.e., at the beginning station, the mid station, and the end station of the tunnel section.

Loads and loading combinations considered for the permanent lining design were as per "Structural Design Criteria Report," which was based on the AASHTO 2012, Load and Resistance Factor Design (LRFD) method.

Methodologies. The TP requires that the design to be performed with acceptable numerical models and with associated "rock loading of a loosened slab." A two-dimensional (2D) beam-spring soil-structural interaction model, by STAAD-Pro (2013) software, was adopted. The "loosened slab" is either the 10-ft or the 15-ft vertical rock load described in the Vertical Rock Load section. The tunnel final lining is simulated by a series of "beam" elements (each with three degree of freedoms), and the surrounding rock is simulated by normal springs (K_n) in compression only, oriented perpendicular (normal or radial) to the liner and closely spaced at regular intervals. Since waterproofing membrane is to be installed around the tunnel, lining-ground interaction from the tangential springs (K_t) is small; therefore, the tangential springs were ignored in the 2D model. The contribution of PVC membrane on the normal spring constant was not considered as a normal practice. The K_n and K_t can be derived from Winkler theory. USACE (1997) suggests: $K_n = E_m/(R_{eq} \cdot (1 + \nu))$, and $K_t = K_n/3$.

Where E_m is the rock mass modulus of elasticity; R_{eq}, the equivalent tunnel radius; and ν, the rock Poisson's ratio. The springs, simulating the rock below the tunnel invert, were derived from Winkler beam on elastic foundation theory; they can also be derived from FEM models by back calculations. For invert slab, the sub-base was also modeled using the spring supports in compression only in vertical direction. The equivalent stiffness of the spring support was calculated using composite sub-base modulus of subgrade reaction so that they closely model the behavior of the sub-base under load. Figure 7 demonstrates a similar discretized 2D beam-spring model.

Temperature Reinforcement. In accordance with AASHTO requirements, reinforcement for temperature induced stress was provided in the *near surface of the tunnel lining* that is exposed to daily temperature changes only.

Minimum Reinforcement. AASHTO's minimum reinforcement provisions are intended to reduce the possibility of brittle failure by requiring the flexural resistance greater than the cracking moment of a flexural member. The flexural resistance moment, M_r, shall be at least equal to the lesser of:

- 1.33·M_u, where M_u is the factored moment
- M_{cr}, where M_{cr} is the cracking moment

Construction Joints. Construction joints for the lining were provided approximately at every 40-ft, matching the panel reinforcement width. Reinforcements were discontinued at these joints, to avoid the potential of reinforcement corrosion causing concrete spalling in the future.

Contact Grouting. Contact grouting is required to fill the gap between the final tunnel lining and the waterproofing membrane. The grouting pressure is about 30 psi higher than the hydrostatic pressure at the point of injection. Three (3) contact grouting hoses were designed for the main tunnel and one (1) contact grouting hose was designed for the cross passage. Figure 8 shows typical contact grout details.

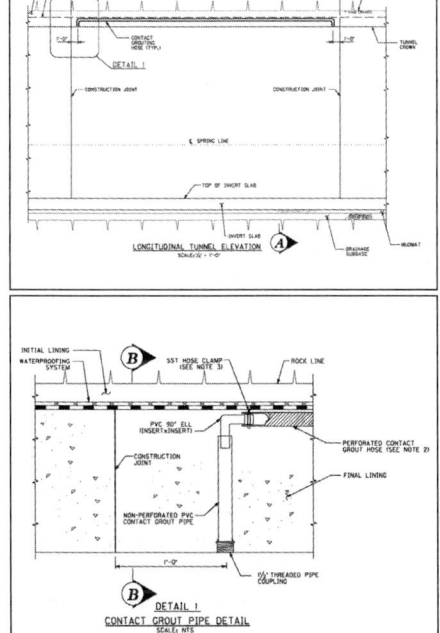
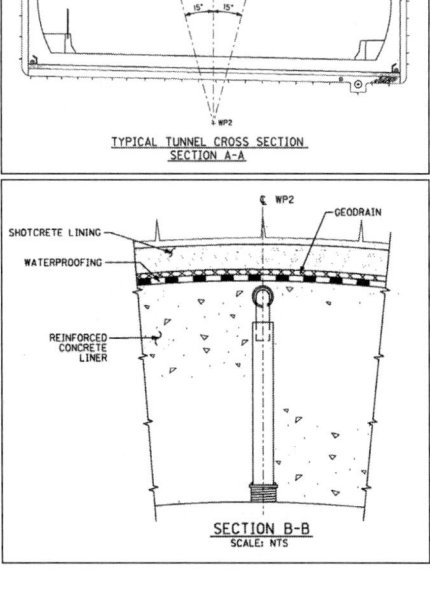

Figure 8. Typical main tunnel contact grout systems

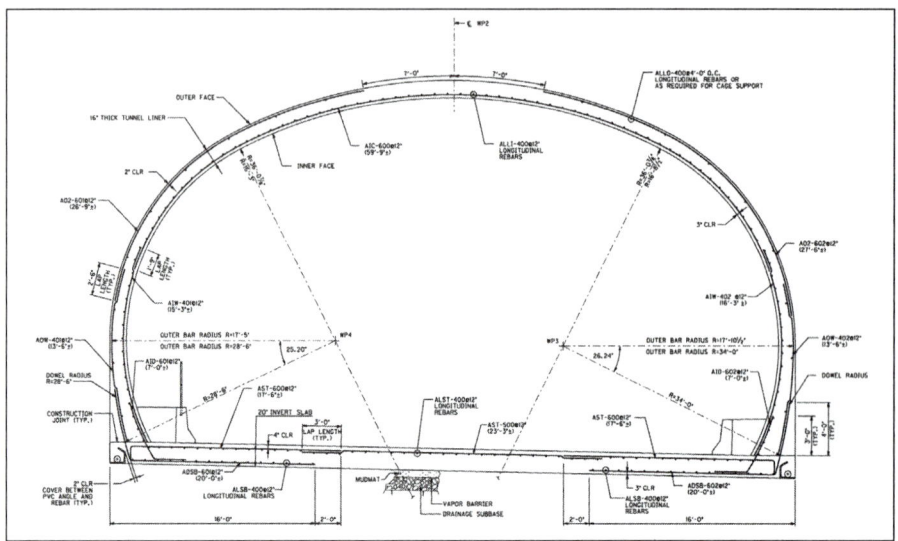

Figure 9. Typical Type A(SL) tunnel support

Fire Design. Fire design for tunnel final lining was conducted, based on the explosive spalling fire test results (Chen, 2016b), it was concluded that a passive fire protection system is not needed for the ORBEE Tunnel.

Results

Tunnel lining final design was performed in conjunction with the tunnel excavation. It was concluded that the thicknesses of the tunnel final lining and the invert are 16-in and 20-in, respectively. The first RFC drawing present the eight (8) typical tunnel support types based on the loading combinations, the durability requirements, and the functional space requirements; however, it does not provide the exact locations of the typical tunnel support types. After the tunnel top headings were excavated and the associated geological mappings were completed, the tunnel reinforcement panel type was mapped for each tunnel panel shown on Figure 6. In addition, some tunnel reinforcement types were revised and redesigned based on geological mappings and NGI-Q rock load derived:

- Type C (N)—rock load was revised from 15-ft to 20-ft.
- Type D (S)—rock load was revised from 10-ft to 15-ft.
- Typed D (N)—rock load was revised from 15-ft to 20-ft.

The reason for these revisions is primarily from the field observations and interpretations of the RQD and the following Q indices: the joint water reduction factor (J_w), the joint roughness number (J_r), and the Stress Reduction Factor (SRF) (Jacobs, 2015). Typical tunnel lining support types are shown on Figure 9 and Figure 10, for Type A symmetric (SL) and non-symmetrical (USL) loading conditions, respectively. Please note the symmetrical and non-symmetrical loading conditions were derived from both the NGI-Q index, from actual geological mapping, and kinematic analysis using UNWEDGE software developed by Rock Science. The un-symmetrical loading conditions generally occur in tunnel reach with sub-vertical joint set that is parallel to the tunnel axis.

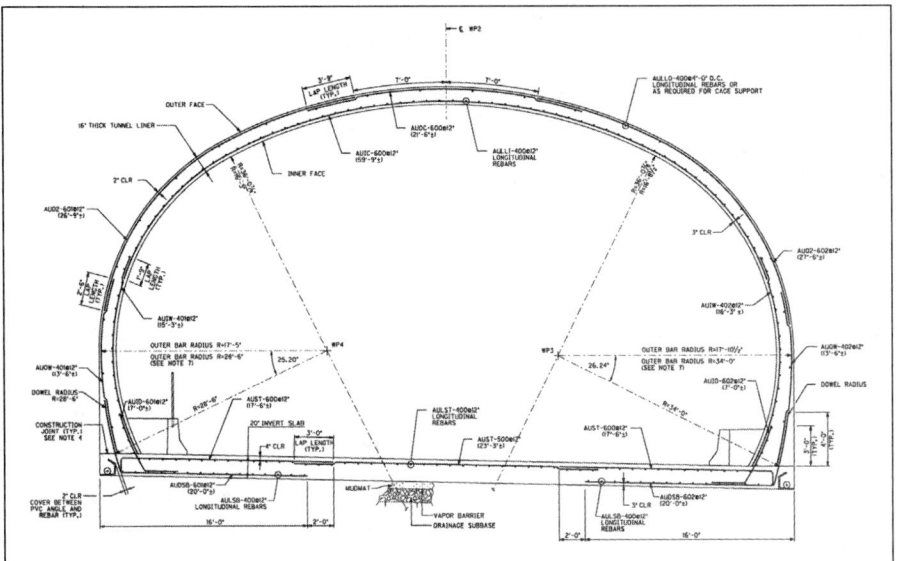

Figure 10. Typical Type A(USL) tunnel support

As can be seen from Figures 9 and 10, the primary difference between the symmetrical and un-symmetrical loading designs is the reinforcement pattern at tunnel crown area. The un-symmetrical loading case requires double layers of reinforcement and the symmetrical loading case does not.

CONCLUSION

The final lining design of the US largest highway tunnel was successfully executed following the principles of the observation method. Tunnel final linings were installed based on actual ground conditions encountered and observed in the field.

The application of the equivalent "loosened slab" rock loading with beam-spring model was proved prudent. This is especially critical for tunnel excavated in mix-ground condition with the slake-prone shale formation rock, such as the ORBEE Tunnel, where the overlying limestone with potential solution zones and karst features. The conventional continuum FEM models will not be able to capture the behavior of the shale formation and its interfaces with limestone and dolomite during tunnel excavation, especially when under groundwater level condition.

State-of-the-art fire design approaches, including the first concrete fire explosive spalling test methodology, will set the norm for tunnel fire design principles in the tunnel industry worldwide.

REFERENCES

AASHTO, 2012. AASHTO LRFD Bridge Design Specifications.

Barton, N., Lien, R. & Lunde, J., 1974. Engineering classification of rock masses for the design of tunnel support. Rock Mechanics. 6: 4: 189-236.

Bieniawski, Z.T., 1989. RMR Classification Guide for Excavation and Support in Rock Tunnels.

Chen, W. 2016a. Service Life Prediction for the Ohio River Bridge East End Crossing Tunnel. World Tunneling Congress 2016, San Francisco.

Chen, W. 2016b. Fire Design for the Concrete Lining of the Ohio River Bridges East End Crossing Tunnel. World Tunneling Congress 2016, San Francisco.

FHWA-NHI-10-034, 2009. Technical Manual for Design and Construction of Road Tunnels, Civil Elements.

FLAC Version 5, 2013. Itasca.

Galera, J.M., Alvares, M., and Bienawski, Z.T., 2007. Evaluation of the Deformation Modulus of Rock Mass using RMR. ISRM Conference "Underground Works under Special Conditions," Madrid.

Golder Associates, 2012. Tunnel Geotechnical Baseline Report, The East End Crossing, Public-Private Agreement, Book 2, Technical Provisions.

Grimstad, E., K. Kankes, R. Bhasin, A.W. Magnussen and A . Kaynia, 2002. Rock Mass Q used in designing Reinforced Ribs of Sprayed Concrete and Energy Absorption, 4th Int. Symp. on Sprayed Concrete, Davos, Switzerland

Hashash, Y.M.A. Hooka, J.J. Schmidtb, B. and Yaoa, J.I. 2001. Seismic design and analysis of underground structures, Tunnelling and Underground Space Technology.

Jacobs Engineering. 2015. Ohio River Bridges—East End Crossing, Section 4 Tunnel Segment.

Japan Society of Civil Engineers (JSCE). 2006. Standard Specifications for Tunneling.

Kuesel, T.R. 1987. Principles of tunnel lining design, Tunnels & Tunnelling, April 1987.

RWS curve. 1979. Rijkswaterstaat, Ministry of Transport in the Netherlands.

S&ME, 2012. RID Document GE-4.04, Tunnel Vertical Borings.

STAAD Pro 2 Structural Analysis and Design. 2013. Bentley Systems.

Technical Provisions (TP), Book 2. 2012. The East End Crossing (Louisville-Southern Indiana Ohio River Bridges Project) Public-Private Agreement.

US Army Corps of Engineering (USACE). 1997. Tunnels and shafts in rock. Engineering Manual 1110-2-2901.

Walsh Vinci Construction Joint Venture (WVC), 2013. Geotechnical Design Memorandum - Tunnels.

Wang, Z. 2008. A Technical Note on Seismic Microzonation in the Central United States, Journal of Earth System Science, March.

Construction Logistics for East Side Access CM006 Manhattan North Structures—A Study from Queens to Manhattan

Sam Lo Grasso ▪ Mott MacDonald
Roberto Adames ▪ Frontier-Kemper Constructors, Inc.
Lonnie Jacobs ▪ Frontier-Kemper Constructors, Inc.

ABSTRACT

This paper describes the logistical challenges that were encountered during the construction of East Side Access (ESA) CM006 (Manhattan North Structures). While the daily logistics of transporting materials, personnel and concrete throughout the construction site may seem mundane (given the sheer enormity of ESA), this is a critical operation that directly impacts the project schedule. Coordinating the delivery of the necessary materials is a complex endeavor, given the high demand, the dense urban setting and the large geographical scope of the project. The specific delivery system for concrete and other materials to the tunnel site will be discussed in detail in this paper.

INTRODUCTION

At $10.4bn, ESA is the largest federally funded transportation project in the United States (US) and when complete will provide 160,000 passengers each day a one-seat ride from Queens and Long Island into a new terminal being constructed 120 feet beneath the iconic 100-year-old Grand Central Terminal (GCT) in midtown Manhattan. ESA is the first expansion of the Long Island Rail Road (LIRR) in over 100 years, and will also relieve overcrowding at Penn Station, which is currently the terminal for all LIRR services into New York City as well as for the local subways that connect Penn Station to the east side of Manhattan.

The $294m CM006 Manhattan North Structures Contract which is being built by Frontier Kemper Constructors, Inc. (FKC) with Construction Management services provided by Mott MacDonald, is comprised of two distinct areas: The first area being the construction of new tunnels within Manhattan and the second being the rehabilitation of the existing 63rd Street tunnel which traverses the East River. In Manhattan, there are two levels of new tunnels being constructed, the lower level tunnels start below 37th Street and join the existing 63rd Street system below 1st Avenue and running above these tunnels are two additional tunnels which join the lower level tunnels below 59th Street.

The construction of the caverns and the rehabilitation of 63rd Street Tunnel will allow the movement of trains from Sunnyside Yard in Queens to two new station caverns built under Grand Central Terminal in Manhattan. Figure 1 shows the ESA alignment and the extent of the CM006 work. The project spans a total linear distance of 3 miles, from Grand Central Station in Manhattan to Northern Boulevard in Queens.

Scope of CM006 North Structures

The project consists of over 1500 individual concrete and Pneumatically Applied Concrete (PAC) placements and includes the construction of the permanent lining for 11,000 ft. of hard rock tunnels excavated under an earlier contract together

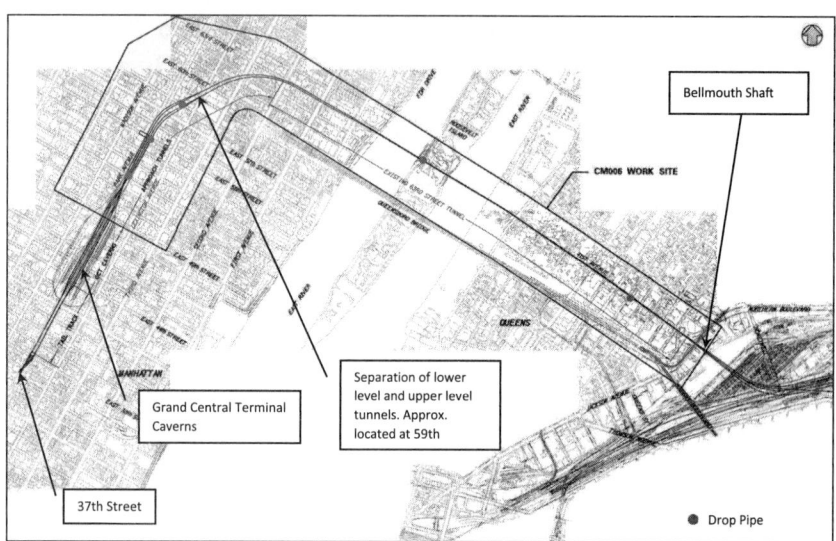

Figure 1. Alignment of ESA showing the extent of the CM006 contract

with 3 caverns including various wyes and cross overs, 2 multi-level underground ventilation structures, 24,000 ft. of ductbench and approximately 1,000,000 sq.ft. of waterproofing.

In addition to the construction of the new underground structures, the project also involves the rehabilitation of the existing 63rd Street Tunnel which was completed in the late 1980s. The scope of the rehabilitation includes asbestos and lead abatement, concrete spall repair, leak remediation, drainage repairs and running tunnel invert modifications.

At any one time there can be multiple activities going on within the worksite, including

- PAC,
- Cast in Place (CIP) Tunnel Lining for fiber reinforced sections and conventional reinforced sections,
- Reinforcement installation,
- Substrate preparation involving dry shotcrete,
- PVC waterproof installation,
- Asbestos and lead abatement,
- Electrical conduit installation, and
- Remediation of the existing 63rd Street Structure including leak remediation and concrete spall repair.

This varied scope of work puts a significant strain on the supply of materials and manpower; there is a constant demand for materials across all areas within the work site. Each of these activities need personnel with different skills sets, different auxiliary construction support and construction materials. To meet these demands, CM006 has access to six concrete drop pipes across Manhattan, all other materials are transported by locomotive from the Bellmouth Shaft (located in Queens). Figure 2 shows a selection of the construction activities that occur on a daily basis.

As mentioned above CM006 has access to six drop pipes, as shown in Figure 1. For such a long tunnel alignment, access to only 6 drop pipes presents significant challenges. FKCI, due to intense surface development, chose to not use the 23rd Street drop pipe and opted to instead use a concrete boom truck located at the Bellmouth Shaft to convey concrete to the tunnel. Table 1 shows the distribution of the drop pipes and the concrete pumping distances that are required to service the site. The large distances inbetween drop pipes and the required distances that the concrete needs to be pumped means that special attention needs to be given to the timing of delivery of the concrete, the traffic conditions within Manhattan, the ambient temperature, the

Figure 2. Typical construction activities: (a) cavern reinforcement installation, (b) back-of-house structure, (c) cast in place tunnel lining, (d) ductbench installation

Table 1. Drop pipe locations showing maximum pumping distance

Drop Pipe #	Drop Pipe Location	Terms of Use	Relative Separation of Drop Pipes (ft.)
1	50th Street & Madison Avenue, Manhattan.	7 days a week. 24 hours	1 to 2 = 800
2	52nd Street & Park Avenue, Manhattan.	7 days a week. 24 hours	2 to 3 = 700
3	55th Street and Park Avenue, Manhattan.	5 days a week. 7am to 10pm	3 to 4 = 1100
4	58th Street and Park Avenue, Manhattan.	5 days a week. 7am to 7pm	4 to 5 = 5800
5	Roosevelt Island, Queens.	5 days a week. 24 hours	5 to 6 = 4300
6	23rd Street, Queens.	7 days a week. 24 hours	6 to end of site = 1500

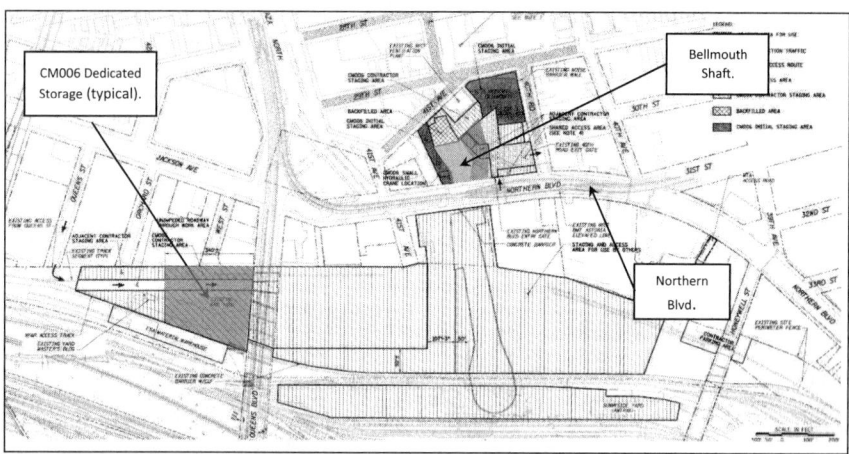

Figure 3. Available laydown areas (orange) and Bellmouth shaft (green)

mix design and the rate of progress of the concrete pour. Often the concrete placement needs to be slowed down to ensure that there is constant delivery of the concrete
to the pump to ensure that no cold joints are formed within the structure. All of these
factors need to be considered and balanced on a daily basis to ensure consistent and
efficient progress during the construction phase.

LOGISTICAL CHALLENGES FOR CM006

Storage and Site Layout

While the dense urban setting of the project presents very obvious challenges for the
supply of personnel and materials there are other less obvious challenges that needed
to be overcome during the construction of the project. With the exception of concrete
and PAC all materials had to be transported from the Bellmouth Structure in Queens.
To do this, the contractor had a limited area for their shared crane and laydown operations as shown in Figure 3. The areas highlighted in orange were the only dedicated
areas available for the storage and laydown of materials and this limited available
space meant that there was typically no surplus of material on site.

The separation of the dedicated storage area and the Bellmouth Shaft by Northern
Boulevard presented a physical barrier for the transportation of materials into and out
of the tunnel. Northern Boulevard (one of the main arterial routes to the 59th Street
Bridge which provides vehicular access across the East River to Manhattan) carries
approximately 174,000 vehicles per day.* Traffic incidents, rush hour traffic and road
works regularly create bottlenecks with vehicles blocking Northern Boulevard, thus
delaying intersite deliveries. Transportation of materials from the dedicated storage
area to the Bellmouth Shaft may take upwards of 30 minutes. In addition, most of the
material that arrives at the site must be double handled; once at the dedicated storage
area and then again at the Bellmouth Shaft.

Most materials had to be palletized and then lowered by crane at the Bellmouth Shaft
to the locomotive crew, who would then transport the materials along the worksite.
Once the materials entered the tunnel, their distribution to the required locations
along the alignment presented the next challenge. Due to the length of the worksite,

---

*New York State Department of Transportation. 2014. New York City Bridge Traffic Volumes Report.

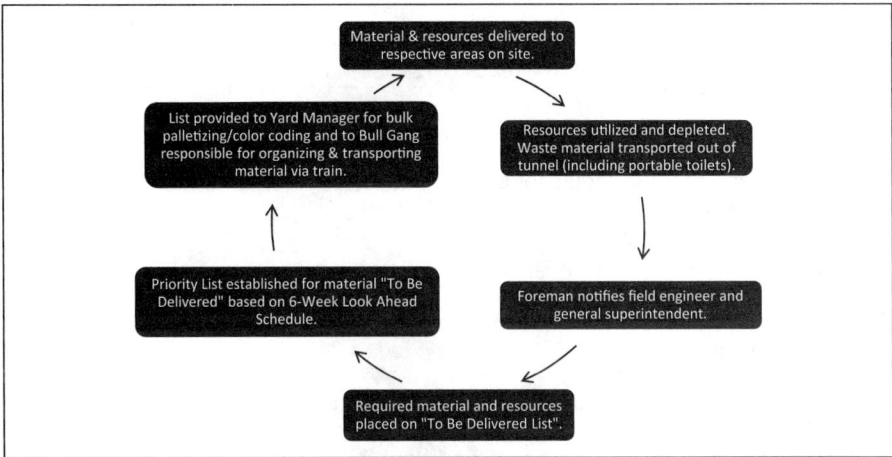

Figure 4. Cycle and supply of materials and waste materials

the delivery of materials was often prone to disruption. Reducing congestion of work crews was also a challenge; as crews got closer to each other along the alignment, the competition for materials, clear work space and storage space increased and tended to slow the progress of the work. Materials were often appropriated by other crews who were closer to the Bellmouth Shaft. To reduce the appropriation of materials the contractor used color identification for each work zone along the alignment. Materials were tagged with the designated location and the unique identifying color code in the event the tags fell off. The color coding made it simpler to distribute materials to all areas of the site and regulated the unproductive competition over materials between adjacent work crews. Figure 4 demonstrates the typical resupply cycle of materials and disposal of waste materials for CM006.

Use of Construction Locomotives for Delivery of Materials and Personnel

The supply of materials and personnel via locomotive was a critical component to CM006. At the height of production, approximately 500 people had to be transported in and out of the tunnel over 3 shifts. A number of locomotives and rolling stock, as outlined in Table 2, were utilized to transport personnel, materials and equipment. In order to service the multiple work headings and due to the overall length of the underground work, the fleet of 4 locomotives and 17 flat cars and 5 mantrips covered approximately 1300 miles per month.*

The rail access route was not only shared with the adjacent contractors but also had to be used for the transportation of the workforce. There was one construction rail track which had to service multiple locomotives. The locomotives had to travel up to 3 miles (one way), unload their cargo, get re-loaded and then continue to the next destination. Multiple switch points and spurs were added along the alignment to create passing opportunities for the locomotives, but locomotives still had to wait significant amounts of time to pass each other, further delaying the delivery of materials to the intended destination.

For single, linear tunnels, the tunnel is typically simultaneously excavated while the tunnel lining or rock bolts are installed behind the Tunnel Boring Machine (TBM). This

*Based on 3 shifts per day and each locomotive making 2 trips for material or personnel delivery per day.

was not possible during the construction of CM006 due to a number of factors including that the excavation had already been completed under a previous contract, the geometry of the tunnel system being built, the contractual requirements to maintain a rail route for adjacent contractors and the varied nature of the work being undertaken. To overcome these constraints, one tunnel was dedicated to provide a construction rail route while the concrete lining in the other tunnel was being constructed.

Table 2. Locomotive and rolling stock fleet

Type	Quantity	Comments
Locomotives		
Brookville Locomotive	2	193 hp—20 ton diesel engine.
Goodman Locomotive	2	150 hp - 20 ton diesel engine.
Flatcars		
Hudson Flatcar	17	7.5 ton flatcar with 20 ton payload.
Mantrip		
Man trip	5	23-person capacity.

The single rail route and access point required congestion within the worksite to be constantly monitored. Construction activities produce a tremendous amount of debris and waste materials that must be removed from the tunnel to maintain accessibility, minimize fire hazards and maintain a good level of housekeeping. Productivity, along with quality and safety considerations, are key in all projects but the access constraints on CM006 make waste removal essential. Insufficient waste removal would have quickly led to significantly decreased production. The most efficient way to manage the build-up of waste materials within the tunnel was to ensure that no trains returned to the portal empty. After delivering construction material, trains were re-loaded with construction debris. However, this constant loading and unloading cycle meant that round trip times to the Bellmouth Shaft were even further extended.

Once the materials were transported into the tunnel, they were a significant constraint to the delivery of subsequent material. There was little available storage space in the tunnels; for example, reinforcement that had been previously transported into the tunnel frequently has to be installed before any additional material was brought past that point. This challenge was typically addressed by skipping over the areas of tunnel lining requiring reinforcement and first building sections of tunnel lining reinforced with fibers. This allowed the tunnel lining could be advanced independent of the reinforcement installation. Once all sections of the fiber reinforced tunnel were built from south to north, the direction of operations was reversed and the reinforced sections of the tunnel lining would be placed. The installation of the reinforcement and concrete operations were decoupled, meaning the overall progress of the tunnel lining was not delayed by deliveries of reinforcement steel from the Bellmouth Shaft in Queens.

Delivery, Placement, and Performance of Concrete Mixes

Since the rail route was consumed with the general material, equipment delivery and personnel transportation, all concrete and PAC had to be pumped via a series of drop pipes in Manhattan and Queens (refer to Figure 1). The location of these drop pipes, which are situated next to prominent business and residential buildings presented a challenge in itself.

All of the concrete drop pipes located along Park and Madison Avenues are surrounded by 'high end' retail and residential buildings. Community boards and city agencies remain on high alert for work sites producing excessive noise or contain material and equipment that is not aesthetically pleasing. The use of decorative wraps for MPT, sound-mitigating blankets and electric concrete pumps assist in keeping noise to a minimum and reducing the visual impact of the drop pipe locations so as not to disturb the surrounding businesses and residents. The sensitive nature of the sites meant that

Figure 5. Road closures along concrete truck delivery routes for the papal and presidential visits

concrete placements had to be timed so as not to affect the surrounding buildings. There was a strict limit to the number of idling concrete trucks that were allowed at any one drop pipe, which meant that the timing of the deliveries had to be evenly spaced so as not to crowd the pumping locations. Additionally, deliveries of emergency supplies to the drop pipe locations had to be coordinated with concrete truck traffic to ensure successful concrete operations.

The maximum distance between any two drop pipes was approximately 5800 ft. meaning that the longest theoretical distance to any one placement could be 2900 ft. The long pumping distances between drop sites, the traffic conditions on the surface, the vicinity of the drop pipes to areas of political and religious significance (most notably the United Nations Building and Saint Patrick's Cathedral in Manhattan) all posed potential issues for the reliable delivery of concrete. In the event of visits by Papal, Presidential or UN Officials, concrete deliveries had to be coordinated with private security teams and local law enforcement officials including the Secret Service. Partial road closures, were sudden often covered the drop pipe locations and their duration was dependent on which officials were being transported across the city. Trucks were re-routed and redirected, with the assistance of law enforcement officials to ensure the delivery of concrete did not affect the visiting political or religious figures.

Table 3. Typical fiber mix design

Component		Dried Weights per Cubic Yard	
	Unit	Amount	
Cement	lb.	529	
GGBF*	lb.	176	
Sand	lb.	1279	
Stone	lb.	1685	
Air entrainment admixture	oz.	2	
% Air	—	6	
Water reducing admixture	oz.	24.6	
Hydration controlling admixture	oz.	35	
Fibers	lb.	50	
Water	Gals.	35	
Initial slump	In.	9.5	
W/C	—	0.41	

*Ground granulated blast furnace slag

To overcome the obstacles from concrete batching to point of placement, a concrete mix for the tunnel lining was designed with emphasis on an extended set time. The concrete mix had to retain workability and pumpability, but also meet the requirements for early stripping strength within a reasonable time. A plain mix and a fiber mix as shown in Table 3 was used for the casting of the tunnel linings. The addition of the hydration controlling admixtures extends its workability, upwards floatation of the form, due to the liquid head of the concrete, needs to considered and factored into the rate of the concrete pour. Strict controls were placed on the amount of the hydration controlling admixture that was used. The additional of too much admixture would require the pour to be slowed down, to counter the effects of floatation, which would in turn would delay the discharge of the concrete trucks which would lead to overcrowding of the drop pipe locations and would delay the completion of the pour.

At peak production, CM006 was placing and stripping tunnel lining on a daily cycle. A typical cycle would include,

- Placement of the tunnel lining during day shift,
- Concrete strength gain to the required 1000 psi stripping strength and commencement of form stripping during swing shift, and
- Complete stripping and re-setting of the tunnel form during grave shift.

To limit potential delay to the cycle, early strength concrete compressive strength results were estimated using the Maturity Meter Method. A maturity meter curve was established for the typical fiber mix design. The curve was established in accordance with ASTM C1074-98 "Standard Practice for Estimating Concrete Strength by the Maturity Method." The test involves casting at least 15 specimens; temperature sensors, connected to the maturity meter, are inserted into each specimen. Once the cylinders have reached sufficient strength for a compression test, standard compression tests are recorded hourly up to 1 day and then tested at 3, 7, 14 and 28 days. At each compressive strength test, the average maturity meter reading is recorded. From this data, a curve and function can be derived for the relationship between the Time Temperature Factor (TTF) of the mix and the compressive strength of the mix. Figure 6 shows the predicted strength curve, the corresponding compressive strength in accordance with ASTM-C39 and the field measured corresponding compressive strengths from the maturity meter. There was good correlation between the predicted strength, the 7-day compressive strength tests and the field recorded strength data, even up to an extended elapsed time while the maturity meter was continuously recording. The advantage of the maturity meter over compressive strength tests is that continuous data is obtained. The stripping process does not have to be delayed while cylinders are tested and processed by the testing laboratory.

RATE OF PRODUCTION

The varied nature of the CM006 project makes it hard to quantify true production rates. However, to date, 1,500,000 lbs. of reinforcement, 1,000,000, sf. of PVC waterproofing and over 75,000 cy. of concrete and PAC for tunnel linings, caverns linings, crash walls, underground facility structures and ductbenches have been placed and installed as part of CM006. The typical running tunnel diameter for CM006 is approximately 22ft as shown in Figure 7.

Figure 8 quantifies the total concrete and PAC placements for every structure, and have been converted to monthly installation rates of a typical running tunnel arch. From this perspective, CM006 contractor has constructed 33,000 lf. of equivalent running tunnel lining to date (excavation was completed under a previous contract) at a monthly average rate of 1200 ft.

CONCLUSION

This paper has outlined some of the logistical challenges that were faced during the construction of the CM006 Manhattan North Structures as part of East Side Access. As in any large urban project a balance between community needs and construction and schedule requirements must be struck. CM006s presence along the alignment was minimal, 6 discrete drop pipes for concrete delivery were installed for the

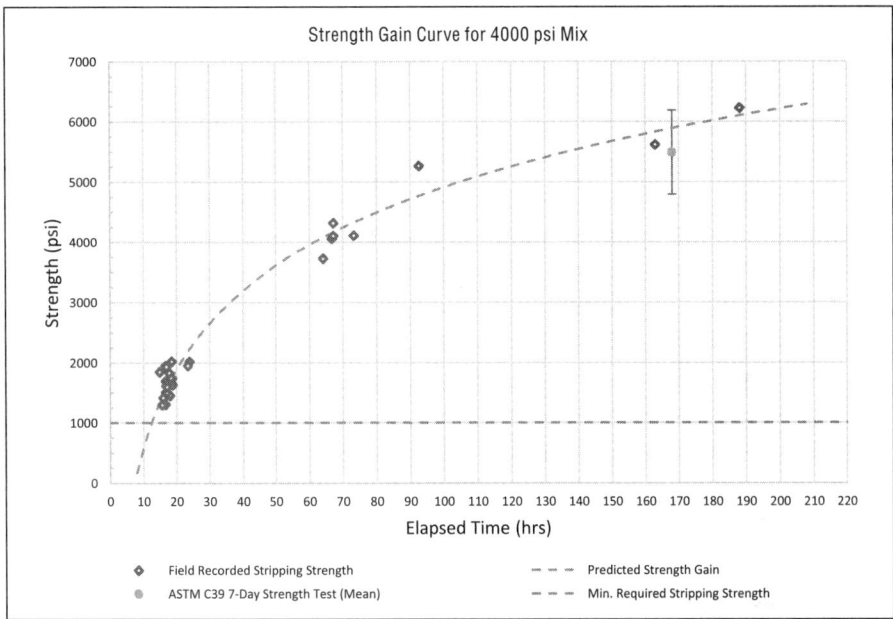

Figure 6. Predicted strength and recorded strength results for typical concrete tunnel lining mix

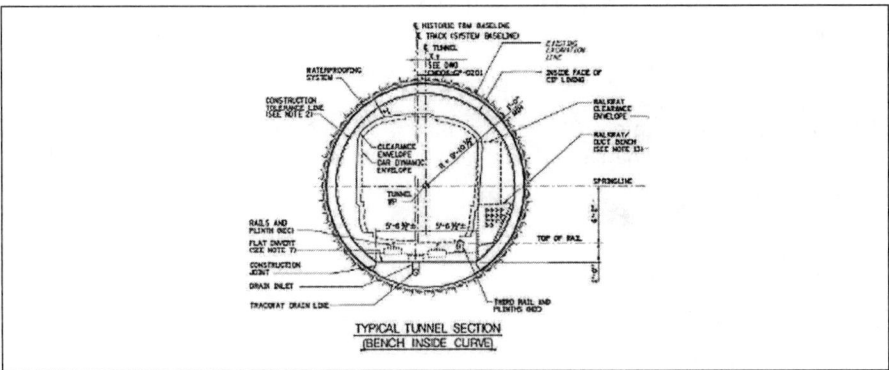

Figure 7. Typical running tunnel

contractor's use with all other material having to be delivered from the Bellmouth Shaft in Queens.

The limited storage area meant that the contractor had no room to stockpile materials, extra attention had to be given to the logistics cycle to ensure that work crews did not run out of resources. A color coded system was implemented along the worksite that allowed for easy identification and distribution of materials.

Special consideration was given to the mix design used for the tunnel arch lining. The design mix placed special emphasis on an extended work time to account for any delays in delivering the concrete to the drop pipes in Manhattan and Queens.

This paper has attempted to illustrate some of the less obvious considerations that that the project faced on a daily basis. The transportation of construction materials

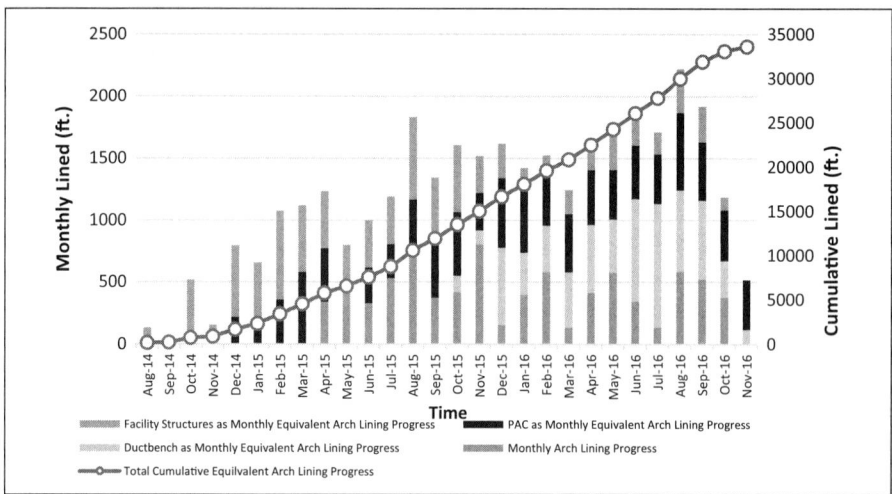

Figure 8. Monthly and cumulative equivalent tunnel lining progress

into the tunnel and waste materials out of the tunnel was a daily logistical challenge that had to be overcome as was the balance between concrete mix workability and set time and the traffic issues associated with the Manhattan drop pipes. These engineering challenges combined with the residential, retail and community environment of the worksite made the construction of CM006 North Structures a unique experience.

REFERENCE

New York City Department of Transportation. 2014 New York City Bridge Traffic Volumes. www.nyc.gov/html/dot/downloads/pdf/nyc-bridge-traffic-report-2014.pdf.

Cost-Effective Seismic Station-Tunnel Connections on Westside Subway Extension Project Section 1

Anthony Harding ▪ CH2M
Hisham Nofal ▪ CH2M

ABSTRACT

When designing metro systems for strong seismic motions there is a wealth of information on the design of both conventional bored tunnels and cut and cover structures like stations. However, there is very little literature treating the connection between the two. This paper describes how the closed form solutions often used to approximate the longitudinal soil-structure behavior can underestimate the movements at connections, and how these shortcomings were overcome in the Westside Subway Phase 1 project in Los Angeles. With reference to previous projects in the LA area, some useful conclusions for future designs are drawn, including how larger displacements can be allowed for in commonly used details with negligible impact on their cost.

INTRODUCTION

The effects of strong seismic motions on underground stations can be very serious for underground structures in highly seismically active areas. Structures must be designed with sufficient strength and ductility that they can accommodate the forces and deformations imposed by an earthquake. In order to address the risk of loss of watertightness, serious structural damage, or even collapse of underground structures in severe earthquakes, LA Metro have developed detailed guidance on the development of ground motions and the verification the associated structural behavior. This guidance is similar to that provided by other authorities on the west coast of North America, and in the literature. It prescribes methods for the verification of both the tunnels and the underground box structures typically employed for stations and crossovers.

However, the behavior of the connections between tunnels and stations is highly complex, subject to the geometry of both tunnels and station, the ground, and the detailing of the connection between the two structures. For this reason the criteria do not prescribe methods for determining the behavior of the connections, but simply outline the aspects that need to be considered and analysis methods that could be used.

In order to provide more material to help designers of such structures, this paper describes the analysis and design of the connections for the Westside Subway Extension Phase 1 project, explaining of the behavior of station/tunnel connection and lessons that have been learned from both the analysis and the design. In particular the results of a full soil structure interaction analysis of a shear wave generating longitudinal and transverse movements at a station is presented to illustrate the benefits that may be derived from more complex approaches. The paper presents an economical seismic joint that was designed to accommodate the movements, and reflects on how the design has benefited from the more detailed analysis.

Table 1. MRDC earthquake criteria

Event	Probability of Occurring over Design Life	Return Period	Criterion
ODE	50%	150 years	Asset can be put back into service after a simple inspection
MDE	4%	2,500 years	Asset should suffer only repairable damage

It is hoped that these lessons will assist the designer in identifying the behaviors that must be quantified, the level of analysis that might be required to quantify them, and some design solutions that may be adopted.

BACKGROUND

Westside Subway Phase 1 is a 6.2km (3.9 mile) long extension to the alignment of the existing Purple Line subway in Los Angeles area. The maximum depth of the subway system is about 90 feet, kept almost parallel to the ground surface over much of its length. It includes new underground stations, a twin-tunnel reaches for transportation between the stations, and cross passages to connect between the tunnels. The design was awarded to a Skanska-Traylor-Shea Joint Venture (STS) in 2014. CH2M are a subconsultant providing design services—primarily tunnels and fire life safety—to lead designer Parsons. The design of the seismic connection between the station and the tunnels is within CH2M's scope.

The design of special seismic joints—joints designed for the effects of strong earthquake motions—have been employed the United States for some time, notably in the construction of the Bay Area Rapid Transit (BART) and the Los Angeles Metro systems. During the 1989 Loma Prieta Earthquake, the ventilation buildings for the immersed tube tunnels in the BART facilities sustained no damage because special seismic joint design (Bickel and Tanner, 1982) was used to accommodate differential movements while maintaining watertight connections (Hashash, 2001). However, during the 1995 Kobe Earthquake, major separation in construction joints were observed in the collapse occurred for the Daikai subway station (Nakamura et al., 1996), because no special seismic provisions had been used in the design of the subway station, and longitudinal cracks up to 1-inch wide were observed after the seismic shaking (Yoshida, 1999). Successful design of seismic joints is therefore vital to the performance of critical infrastructure during an earthquake.

Performance of underground assets is specified by the Metro Rail Design Criteria (MRDC)(Metro, 2013) at two levels: Ordinary Design Earthquake (ODE) and Maximum Design Earthquake (MDE). These are defined in Table 1.

For both station and tunnel structures prescriptive methods are provided that are suitable for most design situations, along with guidance for situations where those prescriptive methods should not be used and the alternatives that might be deployed. For connections between stations and tunnels the design guidance refers to the more complex methods for determining tunnel displacements.

A comprehensive discussion of the literature on the design of the connections between stations and tunnels is beyond the scope of this paper, but two notable papers are mentioned. Hashash et al.(2001) makes the following observations based on previous research:

- A rigid (monolithic) connection will attract very large loads that may be difficult to design for.

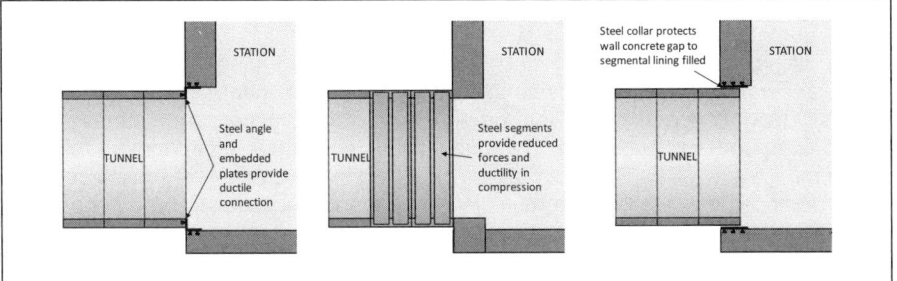

Figure 1. Sample connection types for a seismic joint

- Longitudinal displacements are typically larger than transverse.
- Design can be based on separate determinations of displacement of the two structures from which differential displacements can be derived.

The determination of displacements or forces can be made using a variety of methods varying from a simple pseudo-static beam on springs analysis to full three dimensional soil structure interaction analysis. An analysis of this type was discussed in a more recent paper by Gregor and Shobayry (2011) which summarizes a 3D finite element model that included a station and two running tunnels. The model applied a shear and pressure wave to determine maximum transverse longitudinal displacements between the two structures. It showed relative transverse displacements at the connection to be small relative to longitudinal displacements.

DESIGN OF SEISMIC CONNECTIONS

A number of different seismic joints have been applied to station/tunnel connections in the Los Angeles area, a sample of which are provided in Figure 1. While there are a variety of different solutions, each of which suit differing site conditions and structural configurations, they generally fall into two categories: rigid (or monolithic) connections that resist the loads, and flexible connections that accommodate the deflections.

For Westside Subway Phase 1 the most adverse transverse and longitudinal deflections arose from the passage of shear waves. Nevertheless, many of the observations made during the design would apply to pressure waves or Rayleigh waves for situations where those wave types dominate.

The passage of shear waves generates two types of motion: axial curvature (shear and bending) and longitudinal. Both of these movements had been determined for the tunnel lining design using closed form solutions that accounted for the stiffness of the tunnel. In order to determine whether a fixed or flexible joint would be adopted, the impact of these loads on the station structure were evaluated. This analysis showed that if a fully monolithic connection were to be employed, the longitudinal loads would be an order of magnitude larger than could be economically designed for. Therefore, a flexible joint was selected.

Relative Movements

The design of a flexible joint requires the determination of both transverse and longitudinal displacements in relation to the tunnel, as shown in Figure 2. The transverse displacements were considered a small concern relative to the longitudinal displacements. Previous literature suggested that longitudinal displacements would dominate, including Gregor and Shobayry (2011), whose analysis showed the relative transvers displacements to be 1/30 the maximum longitudinal displacements. Furthermore, segmental linings can easily accommodate slip at every ring joint, particularly close to a free end where there is no axial load and hence no friction across the joint. The simple closed form analysis of the 'green-field' scenario (where the stiffness of the tunnel is ignored) the maximum ring to ring displacement would be 2mm for the ODE and 10mm for the MDE. Therefore, it was expected that the transverse displacements could be accommodated by the multiple ring joints in the vicinity of the connection, rather than the connection itself.

The longitudinal displacements could not be dismissed. The maximum displacements associated with the 'green-field' analysis of the tunnels alone were 24mm and 143mm for the ODE and MDE respectively. While the station was significantly longer than the wavelength of the shear wave, and hence not expected to move as much, it could result in higher still differential movements. If a factor of safety was added to these figures to account for the uncertainties in the interactions between the two structures, then a design gap of 200mm could easily be required. It was likely that a large omega seal would be required to accommodate movements of this level, and there was a desire to design something more cost effective. Therefore, it was decided that a more refined analysis would be required.

It was hoped that the longitudinal restraint offered by the tunnel would help reduce the displacements. The restraint offered by the tunnel can be determined using closed form solutions provided in the MRDC, which are identical to those in Hashash (2001). The stiffening effects of a tunnel in a longitudinal wave are illustrated in Figure 3.

When the tunnel is present it is too stiff to deform to the same extent as the soil and deforms less. The soil next to the tunnel is deformed in shear between the reduced movements of the tunnel and the larger movements of the far field. The shear forces generated on the tunnel create tension and compression, with associated deflections that are reduced, (significantly so in the case of longitudinal movement).

Using this closed form analysis, the displacements were reduced to 4.6mm and 14.6mm for the ODE and MDE respectively. While these numbers were

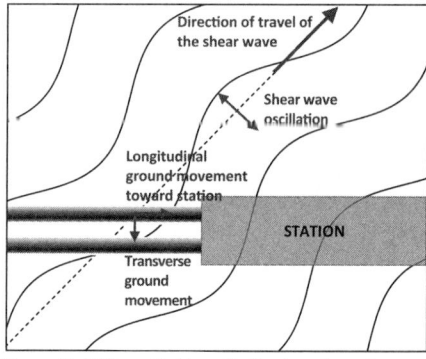

Figure 2. Differential longitudinal motion of a station and tunnel

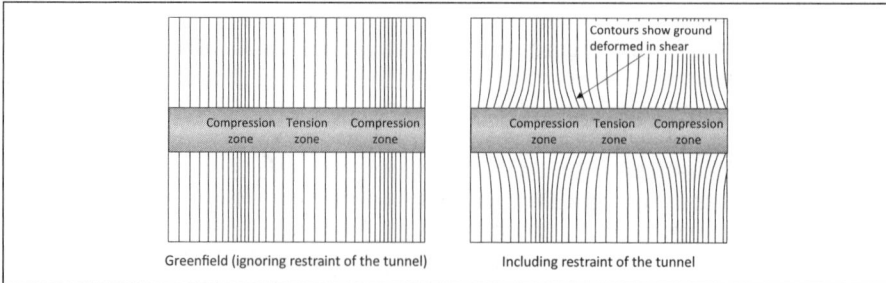

Figure 3. Longitudinal motions without and with tunnel restraint

acceptable for the design of the tunnel lining in compression (as they yielded maximum compressive stresses) there was a concern for the joint that the analysis was based on an infinite length of tunnel. At any point on an infinite length of tunnel displacements are restricted by compressive (or tensile) forces on the other side. If the restraint of a continuous tunnel is removed then the displacement of the tunnel will change, and likely increase. This is illustrated in Figure 4, which shows to ground deformations in the conventional continuous tunnel as solid lines and the effects of the free end

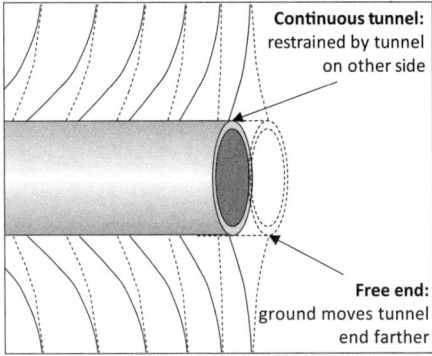

Continuous tunnel: restrained by tunnel on other side

Free end: ground moves tunnel end farther

Figure 4. Amplification of motion at a tunnel end

as dashed lines. With a free end there is no restraint at the end of the tunnel resisting deformation, so the deformation increases. Closed form solutions do not allow for the analysis of this effect.

Furthermore, several other factors control the stress distribution in the longitudinal and the transverse directions, such as the geometric shape of the cross-sections, the depth, the nature and direction of the imposed loads, soil-structure interface characteristics, as well as the flexibility characteristics of the system. Underground structures with different stress distributions are expected to behave differently during the shaking from the same seismic event. Although underground structures have the tendency to conform to the free field ground deformation, it is not always intuitive to predict the response of structures with different stress distributions combined in one model from the combined responses of each individual structure in separate model. To address these concerns and those arising from the simple analyses undertaken initially, a one-step approach was developed. In this approach, the underground structures and the soil are analyzed in one model taking into account the site-response and the wave passage effects from the ground motions.

DESCRIPTION OF THE 3D ANALYSIS

The actual site selected for the analysis is at the West end of the alignment, at La Cienega Station. This was the location of the strongest shaking motions for both MDE and ODE, due to its proximity to the seismic source and the increased depth to bedrock compared to the other station locations. The cross-section at the station-tunnel connection consists of a reinforced concrete underground station box structure with an arch roof module connected to a twin circular precast concrete segmentally lines tunnels. The connection between the stations and the tunnel reaches are flexible and designed to be watertight. The tunnel alignment is horizontal at the connection with the station structure, gradually transitioning to a longitudinal inclination less than 2%, as illustrated in Figure 5. The subway structure is completely submerged under a hydrostatic head from ground water at depth of 10 feet below the ground surface. The station box structure will be cut-and-cover construction, whereas the tunnels are bored by tunnel boring machine (TBM).

Site Ground Profile

The project is located within a deep sedimentary basin with ground profile near the surface consisting of thin, alluvial deposits underlain by marine deposits of the San Pedro Formation. At La Cienega the thickness of the alluvium is fairly uniform with an

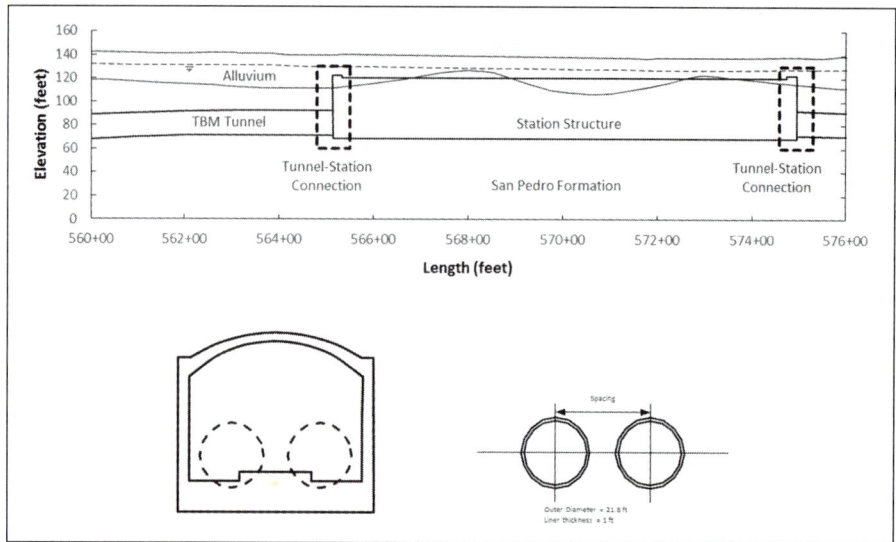

Figure 5. Longitudinal profile and cross sections of station analyzed

average of about 27 feet at the location of the reference subway structure. Both the alluvium deposits and San Pedro Formation are heterogeneous in nature and mainly composed of (typically overconsolidated) cohesive soil with occasional cemented, granular soil layers. The cohesive soil is typically overconsolidated, and most of the time, classified as lean clay and clay of high plasticity according to the Unified Soil Classification System. The deeper sedimentary rock of the San Fernando Formation which underlies San Pedro Formation was not encountered during the field investigation and primarily consists of siltstones with fragments of micaceous minerals. Almost the entire excavation for the subway structure in the study area is within San Pedro formation with the deepest section at the station structure extending to a depth of about 79 feet below the ground surface.

For the chosen site, the variation of the shear wave velocity with depth has been developed based on field measurement from seismic cone penetration tests (SCPT). As indicated in the project report (AMEC, 2013), data from field measurements were highly variable and not consistent between the SCPT locations, and average values were adopted for the shear wave velocity profile at the site. The stiffness and damping ratio were determined using the Darendeli (2001) soil model for fine grain cohesive soil; and the Seed and Idriss (1970) soil model for coarse gain granular soil. Specific details for the basis of selecting these soil models can be found elsewhere (GDC, 2015).

Seismic Hazard and Input Ground Motions

For the site location, the United States Geological Survey (USGS) web-based application was used to generate the ground motions for seismic hazard corresponding to 2% probability of exceedance in 50 years to target the bedrock outcrop motion for an average shear wave velocity of 560 m/sec. at a depth of 45 m below the ground surface. Table 2 shows seed ground motion records from the three major earthquakes used in the analysis. The set of ground motions correspond to a reverse and strike-slip faulting mechanism.

Table 2. Strong motion seed records

Event	Station	Component	Moment Magnitude
Cape Mendocino, 1992	Cape Mendocino	090	7.01
Northridge, 1994	Jensen Filter Plant Generator	292	6.69
Kobe, 1995	Nishi-Akashi	090	6.90

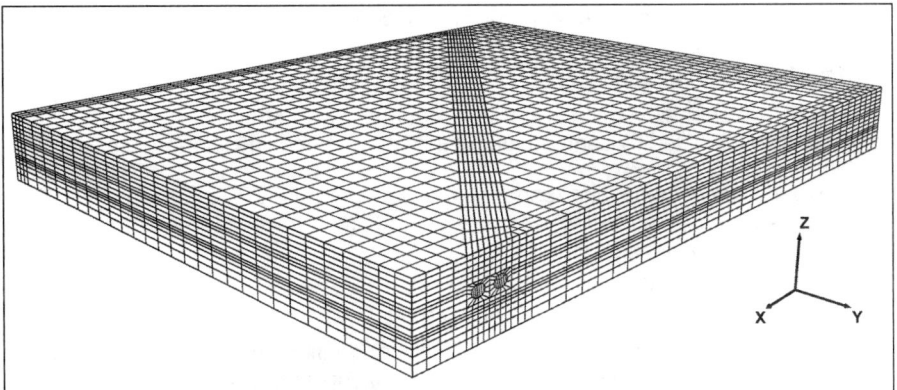

Figure 6. The finite difference grid

Model Description

The seismic response of the structures was studied with the explicit finite difference (FD) code FLAC3D (Itasca, 2010). The overall geometry is shown in Figure 6. The key feature of the grid is the fact that the tunnel and station are located at an angle of 45° from the vertical boundaries of the model. This geometry was adopted to facilitate the passage of shear waves at an angle of 45° to the structure, after attempts to apply the shear wave at an angle of 45° to the boundary proved problematic.

Some other important aspects of the model are noted below:

1. The finite end segments of the tunnels and the station box at the exterior boundaries, i.e., the first and the last segment of the subway structure, were assumed to be rigidly connected to the surrounding ground and follows their motions at those locations; since having structural elements at the exterior boundary of the model was unavoidable, and to allow for an effective transmission of the outgoing waves without introducing complexities and uncertainties into the solution.

2. The effects of the soil non-linearity on the soil-structure interaction is not significant and it is sufficient to capture these effects using hysteretic damping.

3. The interface behavior between the subway and the soil is of Mohr-Coulomb type characterized by friction coefficient and adhesion. Three values were selected: a soil value of 30°, a 'worst credible' value of 17° and 0°.

4. The base of the model represents an elastic-half space which allows for the passage of vertically propagating shear waves.

5. The alignment has been assumed to be perfectly level, ignoring the longitudinal inclination.

6. Three different station masses were assumed: 50%, 100% and 200% of actual expected mass of the station perimeter walls and slabs.

The grid utilized homogenous and linearly elastic three-noded flat shell elements to discretize the tunnels and the station box. These elements have membrane and bending capabilities suitable for modeling thin shells with transverse shear deformations associated with Kirchhoff plate theory. The hysteretic damping model in FLAC3D was used for the soil zones to capture the non-linear soil effects. The model allows for stiffness degradation using strain-dependent damping ratios and secant modulus function in a manner similar to the "equivalent-linear" method commonly used in earthquake engineering to calculate wave propagation and SSI effects.

Verification and Validation

As well as substantial verification of soil, boundary and structure behaviors, model behavior was validated by comparison against a number of known problems:

- One-dimensional site response analyses from DEEPSOIL (Hashash, 2008) were compared to those from the FLAC model without the structures provided with identical inputs

- Comparison of the closed form solution for a single tunnel with a shear wave propagating at 45° to the FLAC3D model with a single tunnel

- Application of a sinusoidal shear wave identical to the at assumed in the closed form solutions, propagating horizontally in a model with the tunnel with a free end.

RESULTS AND COMPARISON OF METHODS

The results of the displacements for the different runs of the final model are shown in Figure 7.

The results presented in Figure 7 illustrates the effect of the interface friction angle from the three ground motions on the maximum relative joint displacement for three different station wall thicknesses. The results for the full slip ($\psi - 0°$) case provide the maximum displacement, which has decreased to around a third at $\phi = 17°$. The behvaiour between between 0° and 17° was considered of academic interest only and therefore not studied, becuase $\phi = 17°$ was a worst credible value for a grouted segmental tunnel lining. Increasing the mass of the station structure will slightly increase the maximum relative joint displacement. This is important because the mass of the station in an earthquake will be greater than simply the external walls, and so this should be taken into account. For the actual station analyzed, the 200% case was more representative of the actual mass of the station.

The maximum relative joint displacement in all examined cases was nearly equal to 95 mm (3.75 inches) occurring during Cape Mendocino ground motions, but the

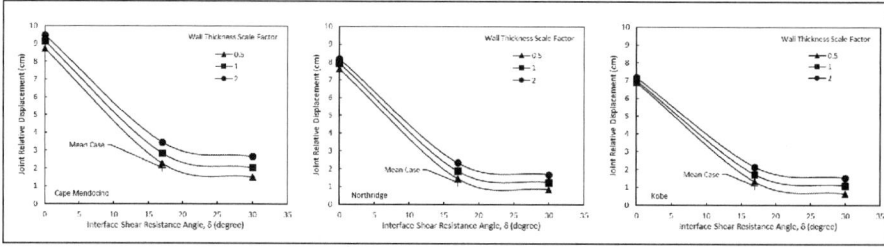

Figure 7. Joint relative displacement results summary

maximum under credible inputs was 35 mm (1.5 inches) for the same motions. This is compared to the results of the other analyses in Table 3.

The results in Table 3 suggests that the closed form green-field displacements are likely to overestimate the ground movements. The closed form soil-structure interaction displacements are closer to the actual results, but still less than half the actual results. This means that using closed form soil-structure interaction displacements will significantly underestimate the movement for the design of a seismic joint.

SEISMIC CONNECTION DESIGN

The seismic connection is shown in Figure 8. It uses a nitrile gasket clamped to the inside of the two structures to provide a waterproof, flexible bridge across a 76mm (3 inch) gap between the two structures. In the event of seismic movement towards the structure the gap will close somewhat, but the gasket will remain sealed. Longitudinal tensile movements are ignored as they are expected to be small: significant tensions cannot develop in the lining due to the relative flexibility of the dowels that bridge the circumferential joints in the tunnel lining. Therefore, the lining will be restrained against large tensile movements due to friction between the ground and the lining.

The compressive movement is allowed for by a 76mm (3 inch) gap, filled with compressible material. The 76mm adopted offers factor of safety of around 2.0 against the expected movements. However, it should be noted that 76mm is only around a third of the total length of the gasket between the clamps—the remainder being a function of the edge distance required for the capacity of the clamp bolts. Therefore, the cost of providing such a factor of safety is small. Nevertheless, the distance between clamps is approaching the practical limit of this detail, so if the green-filed longitudinal movements were used instead then a more expensive (omega seal) solution would be required. This shows that the additional analysis has provided the required justification for a more cost-effective detail.

Table 3. Predicted relative displacements from different methods (MDE)

Method	Longitudinal Displacement
Closed form, green-field	143mm
Closed form, soil-structure interaction	15mm
FLAC3D, actual ground motions, $\phi = 17°$	35mm

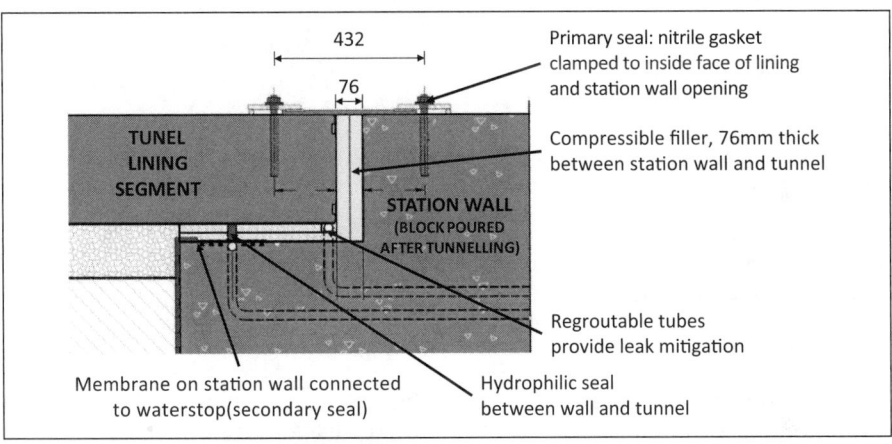

Figure 8. Adopted seismic connection design

Secondary sealing systems are provided as required by the design criteria and good engineering practice. These take the form of a waterstop and gasket that seals between the station wall membrane and the back of the segments. The secondary seal does carry a risk of damage in the event of an MDE earthquake, but the presence of the regroutable tubes provide the facility to seal gaps and control leakage after such high movements. Meanwhile, every component of the nitrile gasket primary seal can be removed and replaced: the gasket itself, the clamps and the holding bolts. This provides a robust joint that can easily be repaired by replacement of damaged components, even after an MDE event.

LESSONS LEARNED

Through the process of analysis and design the following lessons have been learned that may be of use to other designers.

Rigid Connections Between Structures Attract Significant Load During Seismic Events. Resisting the seismic loads can be prohibitively expensive in many scenarios. Therefore, it is likely that most connections can be designed more cost effectively as flexible joints.

Closed for Solutions for Soil Structure Interaction (SSI) Will Significantly Under Predict the Displacements at a Flexible Connection. The closed form solutions assume an infinite length of tunnel that provides stiff restraint at every point along the tunnel, reducing displacements. The flexible joint is a free end condition where this restraint is not present, and displacements are higher.

An Order of Magnitude Check Should Be Undertaken to Help Guide the Design and the Level of Analysis Required. The expectation early in the design was that the true differential displacements would be somewhere between the green-field displacements and the soil-structure interaction closed form solution. This was borne out in practice. However, before the results were available the design was based on an estimate. In the absence of any data to the contrary an initial value of 45mm was chosen. The basis for this was that 45mm was three times the results of the SSI closed form, while the green field displacements were three times the 45mm initial estimate. Therefore the initial estimate was of the form:

$$n = \sqrt{\frac{\delta_{green\_field}}{\delta_{closed\_form\_SSI}}}$$

$$\delta_{estimate} = \frac{\delta_{green\_field}}{n} = n \cdot \delta_{closed\_form\_SSI}$$

where $\delta_{estimate}$, $\delta_{closed\_form\_SSI}$, and δ_{green_field} are the estimated displacement and the closed form SSI and green-field displacements respectively.

This formulation was little more than an intuitive guess and has no theoretical basis. However, this guess was only 20% higher than the results of the actual analysis, suggesting it might be useful as a rule of thumb to obtain in initial estimate. This could be investigated further in other designs that employ these methods.

Inertial Effects Within the Stations Can Have a Significant Effect on the Movements. A twofold increase in the station mass from the mass of the walls alone to a more representative mass of the station resulted in approximately 30%

increase in relative longitudinal deflections. Therefore, a representative mass of the whole station should be selected.

Transverse Movements Are Usually Less of a Concern Than Longitudinal for a Segmental Lining with a Flexible Joint. Where a flexible joint is employed the lining is free to follow the ground in shear and shear displacements typically result in small ring to ring displacements rather than one large displacement at a particular location.

Friction Between Lining and Ground Can Be a Concern for Some Mined Tunnels. The ($\phi = 0°$) case shows a doubling of the displacements for the analysis. This is of only academic interest to the design for Westside subway, due to the use of a grouted segmentally lined tunnel. However, the fact that the displacements have the potential to rise significantly at low friction levels may be a concern for mined tunnels with a permanent lining comprising a fleece and membrane, as the friction coefficients of such systems can be low. Therefore, careful attention should be paid to the friction coefficient when designing such systems for seismic effects.

Cost-Effectiveness of Design Solutions. One final lesson learned (or rather an old lesson reconfirmed) is that 3D dynamic finite difference or finite element analyses are complex and require substantial verification and validation. The time required is high and can run into the hundreds of thousands of dollars in design fees. Therefore, it is important that the designer is clear what the likely benefits of the analysis will be. Simpler solutions will yield greater displacements as they tend to be more conservative, and those increased displacements will likely translate into an increase in construction cost. However, if that increase in construction cost is small it may well be less than the cost of the additional analysis, making it more cost effective to construct for the higher displacements based on simple analyses.

In saying this, it is important to distinguish two separate scenarios: one where we can be confident that a simple analysis yields a conservative result, and one where the uncertainties in behavior mean that we cannot. In the latter case more complex analyses are required to address the uncertainty. However, in the former a cost-benefit analysis of the additional work should be undertaken.

CONCLUSIONS

The design of connections between stations and tunnels subject to strong earthquake motions requires an understanding of the relative movements of two structures that significantly restrain the ground movements and move significantly one relative to the other. Design of a connection requires an understanding of these relative displacements—whether they are to be resisted by a rigid connection, or accommodated with a flexible one. The closed form or simple 2D solutions often adopted for design of the stations and tunnels themselves are not usually adequate to describe this problem, and so more complex analyses are required.

To address this a dynamic three-dimensional finite element or finite difference model approach was adopted on the Westside Subway Extension Phase 1 project. The outcomes of the analysis show that the longitudinal displacements are the most difficult to accommodate, and that true relative displacement is somewhere between the 'green-field' predictions, and those arising calculated by the simplest closed form soil-structure interaction solutions.

Reviewing the lessons learned suggest that the considerable expense of such complex modelling will sometimes only result in small changes to the joint detail, and the additional design effort is not worth the construction savings. However, this was not the case on the Westside Subway Phase 1 project, where the use of the 3D dynamic analysis has justified displacements low enough to permit a cost effective seismic joint detail.

REFERENCES

AMEC. (2013). Geotechnical Data Report—Wilshire/La Cienega Station, Westside Subway Extension Project, Section 1.

Bickel, J.O., Tanner, D.N., (1982). Sunken tube tunnels. In: Bickel, J.O., Keusel, T.R. (Eds.), Tunnel Engineering Handbook, chapter 13. Van Nostrand Reinhold, New York, pp. 354–394.

Darendeli, M.B. (2001). "Development of a new family of normalized modulus reduction and material damping curves." PhD dissertation, Univ. of Texas at Austin, Tex.

GDC. (2015). Geotechnical Report. Site response analyses at structures and four locations along tunnel reaches, Westside Subway Extension—Section-1, Los Angeles, Calif.

Gregor, T., Shobayry, R. 2011. Seismic analysis of large underground structures using FLAC3D, FLAC /DEM Symposium 2011, February 14–16, 2011, Melbourne, Australia

Hashash, M.A.Y., Hook, J.J., Schmidt B., Yao, I.J., (2001). "Seismic design and analysis of underground structures." Tunnelling and Underground Space Technology, 16, pp. 247–293.

Hashash, M.A.Y. (2009). "DEEPSOIL-1D site response analysis program." Univ. of Illinois at Urbana-Champaign, Urbana, IL. (http://www.uiuc.edu/~deepsoil).

Itasca, (2012). FLAC3D—Fast Lagrangian Analysis of Continua. ITASCA Consulting Group, Inc., Minneapolis, Minn.

LA Metro, 2013. Metro Rail Design Criteria (MRDC)—Section 5—Structural/Geotechnical.

Nakamura, S., Yoshida, N., Iwatate, T., (1996). Damage to Daikai Subway Station During the 1995 Hyogoken-Nambu Earthquake and Its Investigation. Japan Society of Civil Engineers, Committee of Earthquake Engineering, pp. 287–295.

Seed, H.B., and Idriss, I.M. (1970). "Soil moduli and damping factors for dynamic response analysis." Rep. No. EERC 70-10, Earthquake Engineering Research Center, Berkeley, Calif.

Yoshida, N. (1999). Underground and Buried Structures, Earthquake Geotechnical Engineering, Seco Pinto (ed.) Balkema, Rotterdam, ISBN 90 5.

Stations and Cross Passages

Chairs

Nate Long
Jay Dee Contractors

Steve Maggipinto
Schiavone Construction Co. LLC

Cross-Passage Mining Using Different Supports in Different Grounds

Satoshi Akai ▪ Obayashi USA
Kenji Yamauchi ▪ Obayashi USA
Hiroyoshi Kawasaki ▪ Obayashi USA
Darrell Liebno ▪ Obayashi USA
Guido Venturini ▪ SWS Canada

ABSTRACT

Cross Passage (CP) mining is a unique form of excavation and there are many ways to skin a cat. In conjunction with the ground improvement such as dewatering, jet grouting, ground freezing and compressed air, different types of initial support can be applied to maintain safe mining. Particularly, CP mining heavily relies on ground conditions and ground improvements therefore the initial support has to be designed accordingly. Through the course of the Project, eight cross passages were excavated along the 6.4 km of twin bored tunnel. Each cross passage had unique ground conditions and different ground support systems were applied at each CP. This paper summarizes our approach on a case by case basis and includes our lessons learned.

PROJECT OVERVIEW

The Greater Toronto and Hamilton Area is Canada's largest urban region and one of the most rapidly growing regions in North America. The Eglinton Crosstown Light Rail Transit (LRT) is a new public transportation project being delivered in the city of Toronto, Ontario by Metrolinx, a provincial transportation agency. The Crosstown project is a 19 km long LRT system including a 10 km of underground section. In October 2012, the West Tunnels project was awarded to Crosstown Transit Constructors (CTC), a joint venture between Obayashi Canada Ltd., Kenny Construction Company, Kenaidan Contracting Ltd. and Technicore Underground Inc. The West Tunnels project scope consists of approximately 6.4 km of TBM-bored twin tunnels, eight cross passages, three shafts and associated surface works. The twin tunnels consist of four drives. The first two drives (Drive 1 and 2) were mined 3.5km from Keel to Allen (Figure 1). After TBM transportation, the second two drives (Drive 3 and 4) were mined 2.9km from Allen to Yonge. The TBMs were abandoned at end of Drive 3 and 4.

GROUND CONDITION

The Geotechnical Baseline Report (GBR) was developed by Metrolinx to describe anticipated subsurface conditions for the West Tunnels project. The anticipated ground conditions were comprised of plastic/non-plastic till and interstadial sand to gravel along the tunnel alignment. Glacier till ground provides irregular strata along the tunnel alignment and each CP had unique ground conditions. The GBR also describes the existence of a regional confined aquifer at the cross passage locations with a maximum head of 6 m above the crown of the CP. The project stipulated that the Contractor is responsible for the temporary design including dewatering and initial support. Ground conditions, ground improvement and propping used at each cross passage are summarized in Table 1. Remarkably, jet grouting was specified for CP-7 and CP-8. This is because the city sewer was not capable of handling the amount of

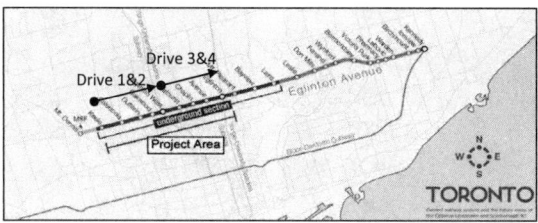

Figure 1. Project alignment

Table 1. Summary of cross passage No. 1–No. 8

CP No.	Location	Soil Description at CP Face	Ground Improvement	Propping
1	Drive 1&2	Full Face Clay	Dewatering, Grouted Pipe Spilling & Chemical Grouting	Full Compression Ring (Specified)
2		Full Face Sand/Silt		
3		Silt above Springline(SL), Sand below SL		
4		Silt above SL, Sand below SL		
5		Full Face Silt/Sand		
6	Drive 3&4	Clay above SL, Silt below SL		
7		Clay above SL, Sand below SL (Full JG Face)	Jet Grouting (Specified)	Frame (VE)
8		Full Face Silty & Sand (Full JG Face)		

groundwater that would have been generated if dewatering was used. Hence, jet grout was used to avoid dewatering rather than for structural purposes.

PROPPING

The first step in cross passage plan was to consider the cross cut opening in the tunnel lining. It is one of the most sensitive tasks for both designers and contractors as it can have a significant impact on the cross passage construction. While the designer tends to pick the more robust design in attempt to limit the lining deformation as much as possible, these designs often place least priority on the ease of installation, less attention to the amount of encroachment into the tunnel section that often results in restricted space for tunnel access and the tunnel utilities. Furthermore, because the propping design may affect the permanent lining during the process of cross passage excavation, the designer tends to remain conservative and grant limited freedom to the Contractor to optimize the design. As a result, contractors are often left with no choice but to use the most conservative propping design that poses schedule and cost challenges from the contractor's standpoint.

Full ring beam steel propping is still one of the most commonly used design approaches to transfer hoop load in the segmental lining where openings are created. In recent years, with the aid of modern 3D design methods, a greater variety of load transfer approaches are being developed and becoming available which maintain the integrity of permanent lining at the cross cut similar to more established approaches.

Under this contract, conventional ring beam steel propping was specified and implemented for CP-1 to CP-5 which was constructed in the first drives (Drive 1 & 2). Implementation of the full ring beam propping installation, however, resulted in considerable schedule delay due to both access and constructability issues.

One of the unique features of the second two drives (Drive 3 & 4) was both TBM were abandoned in place following excavation. This resulted in dead-end tunnel for Drive 3 and Drive 4 which limited access from only the Launch Shaft. A Value Engineering

Figure 2. Full compression ring propping (left) versus frame propping (right)

Table 2. Comparison of different propping system

	Full Compression Ring	Frame
Advantages	• Ease of reuse	• Ease of installation • Minimum interference tunnel utilities • Possible concurrent CP construction during TBM excavation
Disadvantages	• Installation challenge due to its 360 degree shape • High interference with rolling stock & tunnel utilities	• Robust and precise Shear connector installation required • Limited application for the large opening and/ or under high pressure

(VE) was proposed by the contractor to change propping design for CP-7 and CP-8 in an attempt to overcome this adverse impact of one way access tunnel during cross passage construction and tunnel demobilization.

Under the VE proposal process, the contractor supported by his consultant SWS Canada engineered a solution utilizing a Frame Design Propping that takes into account the composite behavior between steel members and concrete lining using 36 mm diameter threaded bar as shear connectors. Since the composite reaction of the lining and the steel frame is the critical part of the frame propping design, design of robust and precise shear connectors was carefully evaluated taking into account both the designer's intent and installation feedback from the contractor in an effort to precisely replicate the calculated model in the field.

To validate the model and ensure the integrity of the permanent lining was not affected six sets of vibrating wire strain gauges were installed at the steel beam flange which were closely monitored during all excavation phases (Figure 3). Analysis of the strain values showed close correlations to the predicted strain values therefore it can be concluded that the frame design propping developed and utilized in this project was successfully implemented in the jet grouted ground (Figure 4).

Aside from ease of installation, an additional benefit of implementing this frame propping was the compatibility with other facilities located inside the tunnel. In combination with a customized platform built to fit several components in the tunnel, these facilities included: Frame propping, Tunnel conveyor, ventilation bag line and Rolling stock, the propping system allowed versatile flexibility which provided the opportunity to work at multiple locations within the tunnel without removing the existing tunnel facilities that may become vital, such as tunnel conveyors, for concurrent operations in the tunnel.

Despite the fact that excavation by both TBMs was completed prior to CP construction in this particular project, excavation setup for the cross passage using these frame

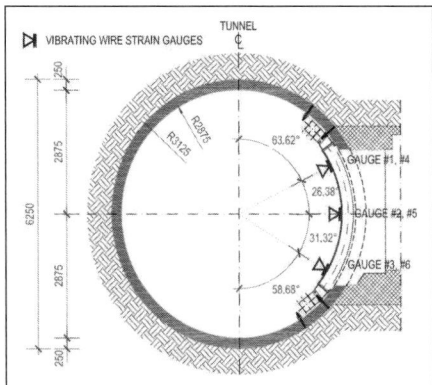

Figure 3. Strain gauge layout

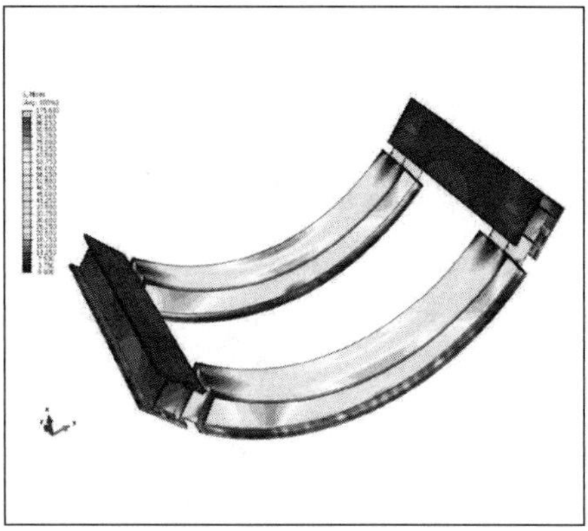

Figure 4. Predicted tensile stress on propping

propping and customized platform systems make cross passage construction during the TBM excavation phase possible which could result in substantial schedule savings.

INITIAL SUPPORT

The contract drawings specified steel liner plate as the initial support for CP excavation with the option for the Contractor to propose an alternative support system. In the last decade, Sequential Excavation Method (SEM) using shotcrete and lattice girder has become more popular for CP excavation and wellrecognized in the US industry. As the changing from steel liner plate to shotcrete lining is a fundamental design change, it was expected that obtaining approval from the Owner would be a lengthy process. Also SEM method requires a lot of logistics which may create disruptions in the tunnel. Hence, the suggested initial support, steel liner plate lining was adopted.

There are two types of steel liner plate (four-flange and two-flange) in the market. The contract drawings specified four-flange steel liner plate but two-flange steel liner plate was substituted in the project. There are several advantages to using the two-flange steel liner plate. The lapped longitudinal joint of the two-flange plate makes continuous circumferential corrugation and provides stronger connection. The deeper corrugation makes a bigger moment of inertia and provides greater stiffness. It also makes more room behind the liner plate therefore the annular grout can be pumped more easily. The steel liner plate lining is one of the classic initial support systems but it was recognized that there are advantages and disadvantages compared to shotcrete lining. One of the biggest advantages is that steel liner plate does not require special skills and equipment. This simplifies mining operation underground. The rigid steel segment also protects the crew from minor raveling. One of the disadvantages is that steel liner plate is not structurally solid until it is fully closed and grouted. It relies on the ground standup time until the steel liner plate is erected and grouted. The same condition is applicable to shotcrete lining. The steel liner plate, however, requires longer standup time than the shotcrete lining. Because soft ground often has shorter standup time, the steel liner plate may not be fully grouted within the standup time. Thus, we proposed customized steel liner plate for the soft ground application (Table 3).

The contract drawings show a 1.83m diameter pilot tunnel prior to fully opening the face. The purpose of the pilot tunnel is to explore the ground and have a better understanding of the ground conditions before opening the large heading. It also reduces

Table 3. Original and customized profile

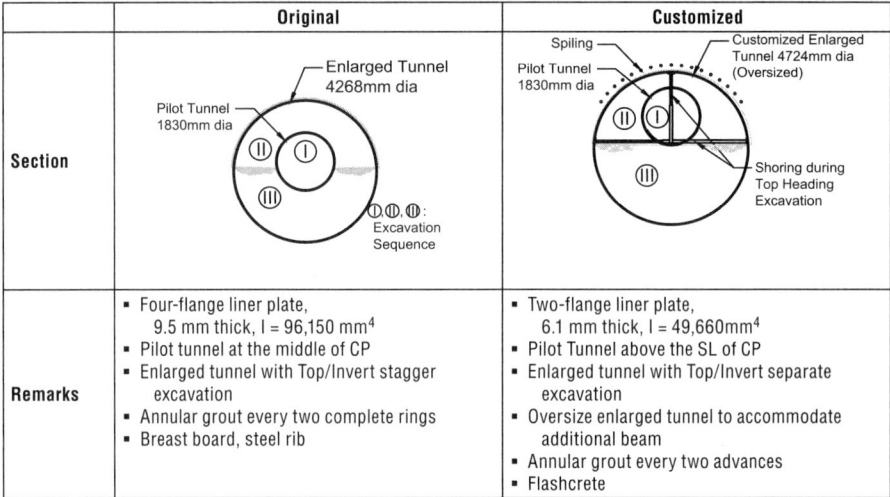

	Original	Customized
Section	Enlarged Tunnel 4268mm dia / Pilot Tunnel 1830mm dia / (I)(II)(III): Excavation Sequence	Spiling / Customized Enlarged Tunnel 4724mm dia (Oversized) / Pilot Tunnel 1830mm dia / Shoring during Top Heading Excavation
Remarks	• Four-flange liner plate, 9.5 mm thick, I = 96,150 mm[4] • Pilot tunnel at the middle of CP • Enlarged tunnel with Top/Invert stagger excavation • Annular grout every two complete rings • Breast board, steel rib	• Two-flange liner plate, 6.1 mm thick, I = 49,660mm[4] • Pilot Tunnel above the SL of CP • Enlarged tunnel with Top/Invert separate excavation • Oversize enlarged tunnel to accommodate additional beam • Annular grout every two advances • Flashcrete

the excavation volume in enlarging the face and shortens the excavation time before erecting the initial support. It works like the face bolt contributing to the ground stability. Thus, we followed the pilot tunnel method at all cross passages. The only adjustment was to raise the pilot tunnel completely above the springline so that the top heading excavation volume was minimized.

The excavation sequence was also changed by completely separating the top heading and invert excavation so as to avoid a high vertical face at the heading during full face excavation. The annulus behind the liner plate was filled every two top heading advances and post shoring was erected to reinforce the unclosed steel liner plate as shown in Table 3. This grouting method speeds up the grouting cycle and shortens the period the ground is unsupported. In addition, sodium silicate accelerator was added to the annulus grout to get an earlier set time to support the ground.

In addition, flashcrete (25–50 mm thick) was applied to the exposed ground before erecting the steel liner plate. A dry shotcrete pot was set at the heading and bag dry shotcrete with powder accelerator was supplied. The purpose of flashcrete is to maintain the integrity of the ground so that raveling can be reduced therefore increasing the standup time. Also flashcrete was utilized to control difficult ground but also to close any gaps in the steel liner plate before grouting. Moreover, flashcrete was applied to the face instead of the breast board and shot inside the steel liner plate instead of steel rib. It became a great tool to supplement the steel liner plate lining in the project.

Case 1: Clay Till Ground Mining

Clay till has high strength ($Cu= 130–300kPa$), low permeability ($k= 2\times10^{-5}$ cm/sec) and a long standup time. If clay till is present throughout the full face, it could be favorable ground for open face mining but ground squeezing needs to be taken into account. Although no squeezing condition was predicted by Peck's stability factor, it was found that the steel liner plate squeezed in the clay ground after studying a similar case near the Project. Hence, steel liner plate was oversized so that the steel rib (W150×22.5) could be erected or shotcrete could be applied inside the liner plate in case excess deformation is detected (Table 3). Throughout the project, we never encountered these squeezing conditions or even excess deformation. It was thought that the customized support system contributed to prevented excess deformation.

Moreover, there was concern about the clay roof condition. Clay thickness varies due to the features of the glacier till. Even though clay till may be present at the CP heading, there may be a sand layer close to the CP profile. Importantly, the clay roof thickness is critical as there could be disastrous consequences if the clay roof collapses and flowing sand runs into the CP heading. Some ground improvement measures were carried out in order to mitigate the catastrophic risk.

Firstly, deepwells were installed in the aquifer above the clay roof and the groundwater was depressurized. By installing the deepwells, the risk was minimized but still existed. The geological profile of the glacier till is complex and the boring result can't capture local dips. Thus, spiling was employed to reinforce the clay roof. Self-drilling anchors (32mm diameter, 3m long) were installed every 30cm over the crown and cement grout was injected through the self-drilling anchor (Figure 5). The spiling served not only to reinforce the clay roof but also to probe the local ground. Additional measures such as cutoff grouting or additional dewatering were implemented based on the probing results. As shown in the Table 4, dewatering and roof support were applied depending on the type of clay roof and the upper aquifer. If the clay roof was in a stable condition, spiling could be eliminated. However, clay roofs sometime fissure or slickenside and therefore the ground condition were carefully examined before making a decision. Flashcrete was a good provision to maintain the integrity of the clay roof regardless of spiling.

Case 2: Sandy Ground Mining

For safe mining purposes, CPs shall be designed in clay ground as much as possible. However, NFPA regulation or the Owner's specification defines maximum distance between station and CPs in the tunnel. Thus, it was unavoidable to place the CP in sandy ground. Sandy ground under the water table tends to flow during CP mining. It is essential to make ground improvements before CP mining. The dewatering approach was applied to sandy ground at CP-2 through CP-6. Since the interstadial silt/sand/gravel structure was so complex, it took time to lower the groundwater level. The detail of dewatering was published as "WellPoint installation under Hydraulic Pressure" in Tunnel Associate of Canada Conference 2016.

Figure 5. Spiling to clay roof

Table 4. Provisions for clay roof

Location	CP-1	CP-3	CP-4	CP-6
Upper aquifer above clay roof	Water-bearing Silty sand	Non-plastic till with waterbearing sand lens	Non-plastic till Dry	Fill
Type of clay roof	Plastic glacio-lacustrine 3 m thick	Plastic till 4 m thick	Plastic till 2 m thick	Plastic till 7 m thick
Dewatering above clay roof	Deepwells	Deepwells & Eductor wells	None	None
Roof support	Grout Spiling	Grout Spiling	Partially Spiling	None

After dewatering was successfully achieved, saturated sand becomes dry and its behavior changes from flowing to raveling. Now the sandy ground tends to ravel. In fact, the sand never became totally dry after dewatering. Dewatering left residual moisture and capillary tension that resulted in firm to slow raveling behavior. The sandy ground has some standup time. As long as the initial support is erected within the standup time, ground stability can be maintained. However, it is hard to estimate the standup time. Thus, the in situ ground behavior must to be observed carefully and erection time of the initial support needs to be adjusted accordingly.

Figure 6. Bench cut during invert excavation

CP-3, CP-4 and CP-6 had sandy ground below the springline. In order to avoid a high vertical face and minimize the risk of face raveling, the section below the springline was divided to bench and invert (Figure 6). The steel liner plate was erected every two or three advances. The struts were placed inbetween the bench and annulus grouting was performed to support the lateral earth pressure. When localized seepage or flowing was observed, the excavation advance was shortened and the bench excavation was staggered or advanced on only one side. This field adjusted excavation method ensured the difficult ground conditions were controlled and maintained a safe mining environment at all times.

CP-2 and CP-5 had sandy ground covering the full face. This presented a significant risk to the crew who had to mine the sandy ground against gravity at the top heading. To reinforce the sandy ground, sodium silicate grouting was attempted but the effect was minimal. Chemical grout has to be injected in static groundwater conditions. If groundwater flow exists, the chemical grout runs out of the CP area. Moreover, chemical grouting may lower hydraulic conductivity which could affect the dewatering system. Instead, it was recognized that utilizing a pilot tunnel and spiling approach was much more effective and reliable to control the sandy ground at the top heading.

Case 3: Jet Grouted Ground Mining

Two of the cross passages (CP-7 and CP-8) were located in high permeable sand and silt. A jet grout block was the specified ground improvement method for CP-7 and CP-8 in lieu of dewatering. The GBR also presented a baseline nominal leakage of 20 m3 /day (equivalent to 14 Liter/min) through the jet grout block. As is often the case, the actual leakage rate would depend not only on the workmanship of the jet grouting but also on the jet grout suitability for the in situ ground. The specified jet grout quality control criteria for jet grout blocks in non-cohesive soils (present at CP-7 and CP-8) are summarized in Table 5.

As the ground stability strongly relies on the water tightness of the jet grout block, some uncertainty remained in the definition of what "nominal leakage" actually meant in the GBR. In actuality, not only did the flow rate of leakage matter, but also the nature of the leakage mattered. The critical part of the quality, in author's opinion, was whether or not the leakage has direct hydraulic connection to the untreated ground outside of the jet grout block for potential washout of untreated material. In this project, jet grout columns were installed by the triple fluid jet method prior to main TBM excavation. The required quality of the jet grout columns was satisfied in accordance

Table 5. Quality control criteria for jet grout block test program criteria

Test Program	Criteria	
	Single Test	Average/Aggregated
Minimum Core Recovery in a test borehole	85%	
Minimum 28 days compressive strength by in situ core	0.5 MPa	2.0 MPa
Maximum Hydraulic Conductivity	1×10^{-5} cm/sec	2×10^{-5} cm/sec (90% of tests)
Maximum untreated length in a single borehole	300mm	1,000mm
Minimum designed jet grout column overlap	⅛ of diameter or 150mm	

with the testing program set by the specifications. Despite the strength and quality of the jet grout being assured per specification, the contractor decided to take a careful approach to minimize the chance of ground loss during the cross passage excavation as follows:

- Pre-excavation grout with microfine cement from the segmental lining around the cross passage opening was performed in order to cut-off potential water passing through the annulus of the TBM cut and/or possible joint through the jet grout block near the TBM cut.

- Advanced 1830mm diameter pilot tunnel above springline to fully investigate the quality of the jet grout through the excavation prior to top heading and bench excavation. This is the same approach used in the untreated and dewatered ground in CP-2 to CP-6.

- While the strength and the thickness of the jet grout was proven to have necessary strength by itself to perform as initial ground support, a combination of the shotcrete and liner plate installation were fully implemented to provide additional safety measures (with the exception of contact grouting being done only after the full length liner plate installation accounting for the longer stand up time expected for the treated ground).

In essence, until the in situ quality of the jet grout columns had been fully investigated, the jet grout was only deemed advantageous for the purpose of controlling water and had no significant advantage to the structural strength of the ground to allow for aggressive advance of excavation.

During the first jet grouted cross passage excavation at CP-8, a notable amount of sand inclusions were observed in the pilot tunnel. The sand inclusions initially observed in the pilot tunnel at this point appeared to be fully depressurized from the protected columns above and from the sides.

Unfortunately, this was not the case when the excavation proceeded in the bench near the springline (Figure 7 Left). Initially, just a few drips were observed from the sand inclusion located near springline, then shortly after these drips developed into a steady flow. Dry shotcrete was immediately applied to prevent further washout of the sand inclusion and in an attempt to control the inflow, only to see these inflow ultimately start to bring in some material (Figure 7 Right). More shotcrete were immediately applied to build a bulkhead with a few water relief pipes. In addition, weep holes were drilled from the segmental lining side for the purpose of pressure relief to protect the on-going excavation inside the CP. Once the inflow inside the cross passage was all diverted to the weep holes and secured, a microfine cement cut-off grouting program was executed in several stages to create a grout curtain around the breached jet grout column. Following microfine cement grouting, chemical grouting

 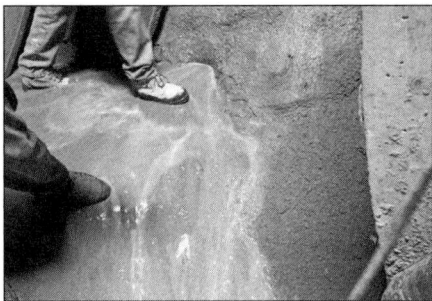

Figure 7. Observation of "wet" sand inclusion (left) and initiated inflow (right) at bench excavation

were performed in several stages to account for some ground loss observed from the inflow event.

Using lessons learnt, prior to excavating CP-7, a full length core investigation were performed on springline level where the column cover over the excavation envelope becomes the least. Full length horizontal coring at the most critical part of the excavation (area of least treated ground cover) prior to excavation provided information on stand up time of the in-situ sand inclusion, if any existed. This investigation method was successfully implemented to the second jet grouted CP excavation at CP-7 in this project.

A few lessons obtained from this incident are summarized below:

- The significances of how to set the quality assurance criteria under the jet grout testing program are realized. Untreated zones remaining inside the jet grout column have considerable consequences when the jet grout column is meant to be used as a water barrier, therefore, careful attention should be made in conjunction with the range of treated ground versus the size of the excavation envelope.
- The location(s) of the investigation cores have to be carefully chosen to obtain information from the most critical part of the ground/water control system. Vertical in situ core may have limited information compared to the information that can be obtained from the horizontal core.
- If untreated jet grout zones are unavoidable by its nature, the risk of these untreated zones shall be clearly acknowledged in advance and proper measures, such as having more than one column on both side of excavation, shall be taken into account at the earliest stage, ideally at the design phase, so as not to risk the workers and the public involved.

CONCLUSION

Cross passage excavation in urban soft ground can often pose complex and sometime serious challenges even with careful design and planning. The Eglinton Crosstown LRT West Tunnels Project was no exceptional and the project team faced some serious challenges due to the complex and unique geological conditions that differed from one cross passage to another.

The lessons learned in this project are summarized below.

- Significant benefit can be gained from investigating the wide variety of the propping systems available to select the best suited system while balancing between the lining deformation risk and constructability;
- Successful application of steel liner plate can be achieved by customizing the system for soft ground;
- Evaluate the clay roof condition thoroughly and have some provisions to avoid the worst case scenario;
- Carefully observe in situ ground condition and adjust the excavation method and the support system; and
- Inclusions created in the jet grout columns, whether it is avoidable or not, have significant effects on cross passage excavation and shall be fully investigated and risk shall be acknowledged and properly managed at the earliest stage, ideally at the design phase.

The paper provides some insight into cross passage mining from our experience. Hopefully it inspires the designer and the contractor and the lessons learned shared in this paper can be taken into consideration in future projects.

REFERENCES

Hatch Mott MacDonald (2012), Keele to Yonge Geotechnical Baseline Report, Eglinton-Scarborough Crosstown Twin Tunnels Contract #ECLC1-15.

SWS Canada (2016), Calculation note for Design of Propping System.

Earth Tunnel with Steel Support, Proctor & White.

Satoshi Akai, Garett Urban, "Wellpoint installation under hydraulic pressure," Tunnel Associate of Canada Conference 2016 Proceedings, Ottawa, Oct 17–18.

Systematic Cross-Passage Design and Construction Planning for Transit Tunnels

Peter Chou ▪ Parsons
Yue Shi ▪ Parsons
Matthew Burdick ▪ Skanska Taylor Shea Joint Venture
Patrick Nicholson ▪ Skanska Taylor Shea Joint Venture

ABSTRACT

Long transit tunnels are becoming popular in California's metropolitan areas. Due to their lengths, many cross passages (CP) are required between twin bored tunnels through urban areas in gassy, tar-rich, and seismically active ground. The considerable number of CP, along with varying operating functions, heterogeneous ground conditions (including mixed face) and limited space create a unique design and construction planning challenge. This paper presents a practical, systematic design and planning approach for long tunnels. The paper covers critical aspects of CP design: geological setting, constructability, functionality, system layout and life-safety. This paper discusses key features of the design and construction planning necessary for long transit tunnel CPs, and presents recommendations for an effective CP planning solution including the use of Building Information Modeling (BIM).

INTRODUCTION

Long transit tunnels have become a popular transportation solution in California's urban areas typically utilizing twin bored tunnels between which cross passages (CP) serve as emergency egress, ventilation and house critical electrical/systems/firefighting equipment. The location, spacing, geometry, clearance envelope, equipment layout and structural design of the CPs are driven by various factors, starting with the mandatory safety requirements specified in National Fire Protection Association (NFPA) standards. However, each project has unique aspects that must be considered. Even for the same project, individual CPs may have different functionalities, sizes, and could be constructed in varying ground which would require different construction techniques.

This paper describes a practical and systematic approach used on a recent California transit tunnel. With more than 20 CPs to design, each CP was planned such that similar templates were utilized during construction. The effective use of Building Information Modeling (BIM) for space proofing is discussed.

SYSTEMATIC DESIGN AND CONSTRUCTION

When planning twin bored transit tunnels requiring many CPs, a systematic design and planning approach is necessary to provide a practical and economical solution for tunnel construction. Once the tunnel alignment and profile are fixed, the following systematic planning approach is used as illustrated in Figure 1. Each stage is also discussed below.

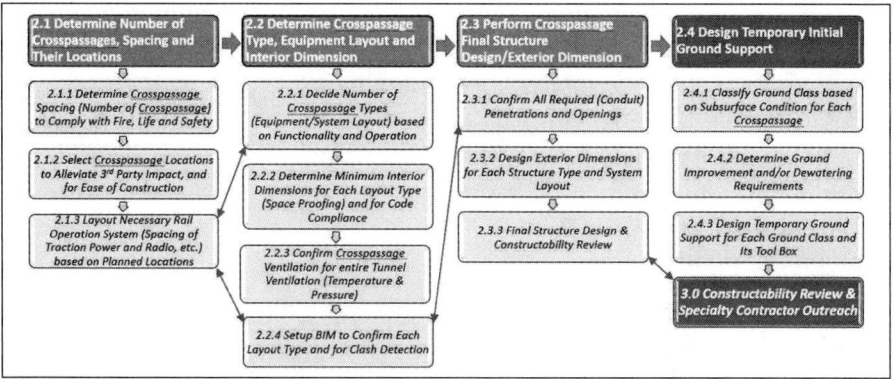

Figure 1. Systematic cross-passage planning flowchart

Cross Passage Number, Spacing, and Locations

Design. At an early planning stage, preliminary tunnel ventilation, drainage and discharge should be first identified. Additionally, any mandatory requirement needs to be considered as follows:

- Ventilation: Determine if a ventilation CP is required to control critical flow velocities. Determine if CP needs to be pressurized to stop ingress of smoke. This is done by evaluating tunnel emergency ventilation and examining the different incident/accident scenarios that will affect the ventilation and emergency egress of the whole system.
- Drainage: Locate CP to be coincident with the tunnel low points where sump structures will be required. The sump pit will be incorporated into CP structures, resulting in a deeper excavation to satisfy the required storage volumes. The sump pump can discharge upward toward the surface storm drain or pump laterally back to the adjacent station drainage system.

In addition, NFPA 130 specifies the required number of CPs along the tunnels as follows:

- CP spacing shall not exceed 800 feet, to allow the passengers to take refuge in the safe tunnel in the case of fire. CPs shall be divided by a 2 hour-rated fire wall, with doors having a 1.5 hours fire protection rating.
- The maximum 800 feet spacing could possibly be optimized if a performance-based design is performed to demonstrate a shorter evacuation time.
- CP egress path shall be separated from the air plenums and sump pits.

Once the maximum spacing is defined and strategic locations are identified, selection of the remaining CP locations should focus on ease of construction. The following critical considerations should also be considered (see Figure 2).

- Avoid locations relative to surface cross streets and intersections if dewatering and/or ground improvement will be required for the following reasons:
    - Traffic control becomes much more difficult when close to an intersection
    - Overhead restrictions from traffic signals, and bi-directional underground utilities

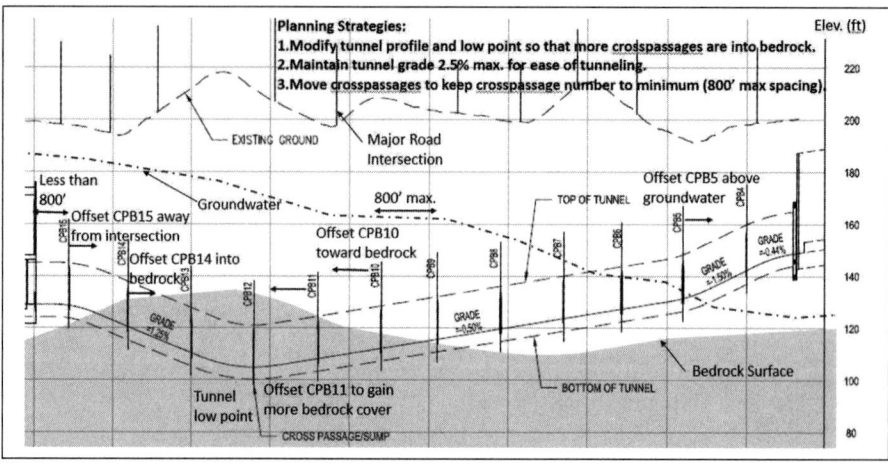

Figure 2. Selecting cross-passage locations based on ground/surface conditions

- Median location and sidewalks:
  - Attempt to not remove more medians than absolutely necessary.
  - Saving trees. There is a limit on the number of trees that can be removed to facilitate the project.
- Underground utilities:
  - Layout the utilities to evaluate the impact to drill vertically for ground improvement
- Avoid sensitive noise receptors such as residential buildings (if possible):
  - Certain properties are deemed as sensitive noise receptors with environmental restrictions on noise and vibration levels and restricted work hours
- Ground conditions for Sequential Excavation Method (SEM).
  - Prefer cohesive and stable material above groundwater that does not require full ground improvement. Adjust tunnel alignment profile if possible (making sure the rail profile grade stays within acceptable limits), and move the CPs to the locations with stable ground.
  - It is preferred to have a full face of one material to avoid possible difficulties with mixed-face conditions (unless it can be fully in bedrock)
- Sump locations for discharge to surface storm drain option:
  - Overhead drill clearance
  - Ability to stage concrete trucks for subsequent tunnel and CP's concrete delivery
  - Proximity to storm and sanitary sewer lines with adequate capacity for permanent discharge

Finally, from an operating system perspective, identifying all system equipment space requirements (i.e., traction power, radio equipment), and if they should be inside, embedded, or crossed through the CPs.

Cross Passage Type, Functionality, and Equipment Layout

Cross Passage Type

Typical CP provides egress passage and equipment housing. Ventilation (or pressurized) CPs may be required at strategic locations to maintain escape routes free from

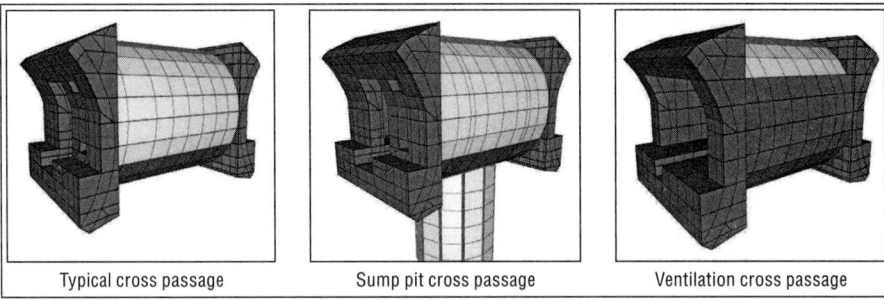

| Typical cross passage | Sump pit cross passage | Ventilation cross passage |

Figure 3. Different cross-passage structures 3D view

Table 1. Overview of typical equipment inside transit cross passages
(bold equipment could affect layout arrangement)

Ventilation	Electrical	Communications	Tunnel Drainage
• **Fans & ducts** • Air intake/exhaust • Damper Switch • Control Panel • Fire smoke damper • Vents	• Light and exit sign • **Control power panel** • Transformers • Enclosed circuit breaker • Grounding bus • **Local control station**	• Radio cabinets • **Power supply and attery cabinet** • **Interface cabinet** • Blue light station	• **Sump pumps & discharge pipes** • Sump control • Level control • Lifting guiderail • Ladder • Discharge piping • Totalizer
Fire Wet Standpipe	**Safety**	**Track Operation**	
• **Cross connections & fire hose control values** • Pressurized piping	• Gas detectors • Smoke detectors • Camera • Emergency phone	• Corrosion control test box • Traction power cross-bonding	

hazardous smoke and at sufficient flow velocities. Sump pit CPs need to be located at tunnel low points. These three types (typical, ventilation and sump pit) of CPs require different geometric configurations as shown in Figure 3, and will likely require different equipment layouts.

Pumps, fans and duct work are not the only large pieces of equipment that must fit inside specifically functioned CPs. Communication, electrical and fire suppression systems are also space-consuming and unless specially designed underground vaults are planned along the tunnels, all will have to be housed inside the CPs. Table 1 shows a list of typical equipment inside a transit tunnel CP.

Minimum Interior Dimensions

Code-required minimum dimensions for egress path, as well as specific operational clearance for electrical panels, dictate CP inside dimensions and equipment layouts. Important code related criteria include:

- Width: The CP shall have a minimum clear unobstructed width of 44 inches (6 feet 6 inches is preferred). Space for ventilation or drainage equipment located within the CP shall be provided in addition to the unobstructed width.
- Height: The CP shall have a desirable height of 8 feet and a minimum height of 7 feet.
- Access Doors
  - Minimum dimensions of the door opening shall be 46 inches (three feet eight inches) wide and 80 inches (six feet eight inches) high.

- Doors shall be provided at each end and arranged to open into the cross passage.
- The doors shall be offset to the same side of the cross passage.

Based on past experience with equipment arrangements within tight spaces, the following are recommended:

- Fire standpipe cross connections are typically required at 1,600 ft spacing (every other CP), similar to communication radio cabinet arrangement. Therefore, they can be staggered in different CPs. However, depending on the authority having jurisdiction, standpipe cross connection could be required at each CP (800 feet apart). If both are required in the same CP, they will be placed on opposite walls of the CPs.

- For below-floor sump pit (see Figure 3) where hatch doors are required for pit access, radio cabinets with their power supplies could take up too much space, and thus should be avoided in sump pit CP.

- Traction power cross bonding, fire standpipe cross connection and sump pit drainage pipes are typically embedded in the floor/wall (along with system and electrical conduits). Thus, the CP layout should avoid including all three systems together in the same CP.

- For operational safety of electrical panels/cabinets, layout should avoid placing the power supply cabinet and the control power panel on the same side of the CP where pressurized water piping (sump pump discharge and fire wet standpipe valves) are present.

- Keep duct work hanging from the roof so that seven feet tall safety egress and 3ft × 3ft × 6ft electrical clearance in front of electrical control panels can be satisfied.

In urban areas, transit tunnels may be situated below the existing street within the City's right-of-way limits to minimize 3rd party impacts. Thus, the available length of CPs are defined by "pillar width" between twin bored tunnels. Sometimes this length could be as short as 15 feet or less. Even though CP "width/diameter" can be increased to fit all planned equipment, enlarging the width (or height) of the CP poses several concerns as follows:

- Increased volume of excavation and time to construct the CP,
- Increased risk of SEM excavation safety, and
- Increased complexity of segment design and ring support when multiple segment ring removal is required at the CP break-in and break-out.

Therefore, an effort is made to keep CPs as small as possible to benefit both design and construction. Generally, code-required minimum unobstructed width plus system cabinet depth and required space for wet standpipe (or sump pump discharge) should provide sufficient width for placing all operational and security equipment. This width can provide code-compliant doors located at both ends and allow the evacuation of passengers within the acceptable elapsed time.

Cross-Passage Ventilation and Temperature

In addition to tunnel emergency ventilation demands, the CP itself (and tunnel) have a maximum operating temperature. Due to heat dissipation from the equipment, additional ventilation (or cooling as necessary) may be required to alleviate overheating concerns inside a relatively small CP. An internal height complying with the minimum

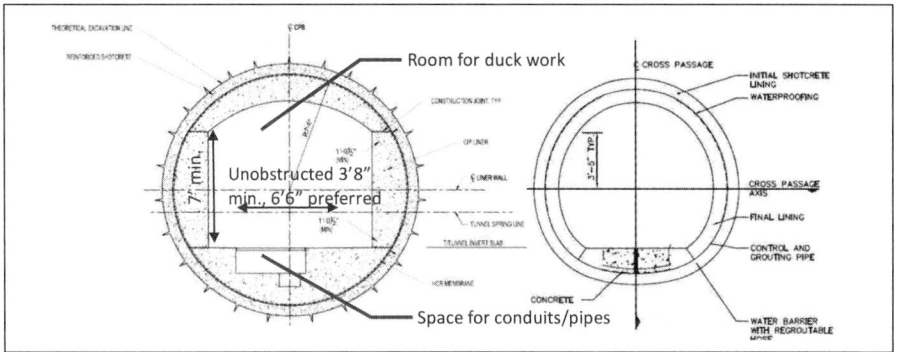

Figure 4. Efficient cross passage section examples

height plus space for air ducts (or discharge piping) is typically sufficient to meet ventilation and temperature requirements. Generally speaking, an internal horse shoe (or sometimes egg shape) section (see Figure 4) can provide a most efficient CP interior space for design and construction.

BIM and Space Proofing

CPs are usually the most crowded space within a tunnel system. To ensure the proper fit of equipment, conduit and code/operational required clearance, three-dimensional fully integrated BIM model should be prepared for each type of cross passage (including its transition to the tunnels). These BIM models define accurate dimensions and physical boundaries, and are essential to the success of space proofing as well as system and conduit management. In addition, a comprehensive model allows for cutting a section through any part of the model, rotated 3D views and plans requested by the designer and contractor. An example of a CP BIM model for space proofing is shown in Figure 5.

BIM delineates architecture hardware, and simplifies material quantity assessment and hardware schedule for each contract item. A comprehensive BIM model also includes equipment clearance, structural reinforcement, waterproofing, construction/expansion joints and temporary ground supports for the ease of constructability review.

Structural Design and Exterior Dimensions

Once interior dimensions (space for necessary equipment layout) are set up, the permanent CP lining design can be finalized in compliance with the transit agency standard criteria. A typical structure consists of CP room and bulkhead walls at each end. Lining design should consider 1) earth loads, 2) surcharges, 3) structural support of the permanent segmental lining at break-in and break-out, 4) fire resistance/rating and 5) durability of the permanent structure. Optimization of the tunnel shape (usually circular or oval shape) is critical to minimize the amount of reinforcement required to meet the design life (usually 75 years).

In seismic zones, development of prefabricated reinforcement details for the junctions between the main tunnels and the CPs is also important to enable faster construction and minimize risks of damaging the waterproofing membrane. Structural design of the permanent cast-in-place concrete CP could involve 3D finite element modeling (integrated with BIM if possible). Structurally CPs are also designed to withstand seismic forces and deformations in order to remain functional during earthquakes. Three typical approaches are available for the seismic analysis of CPs, each with different assumptions and level of computational effort. The first method uses closed-form solutions with consideration of soil-structure interaction (SSI). A pseudo-static approach is the second method and employs free-field racking of soil to induce the desired structural deformations. Lastly, a dynamic time history analysis is recommended using an SSI model to obtain the deformation history. Based on previous past experience, dynamic SSI approach as shown in Figure 6 can provide more accurate structure behavior during a seismic event.

The direction of shear wave propagation in 3D space (i.e., longitudinal forces from tunnel segments) should be investigated to determine critical load cases for bulkhead wall design. Additionally, bulkhead walls must be detailed to withstand the design train fire and other project-specific fire requirements (time-temperature curve for transit tunnels) without compromising overall tunnel structure stability and passenger safety. Design fire for transit tunnels should be clearly defined, as structural fire protection criteria may govern bulkhead design in compliance with NFPA 502. Once concrete

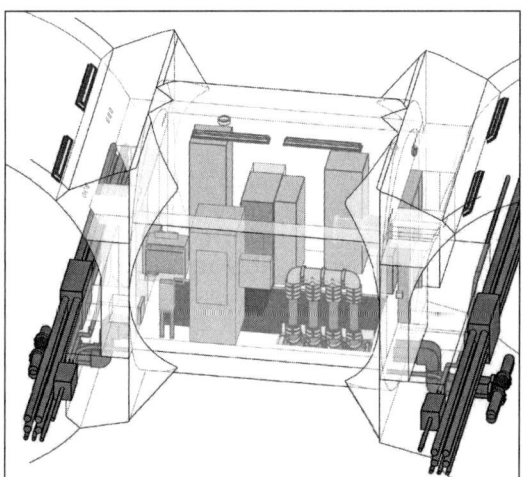
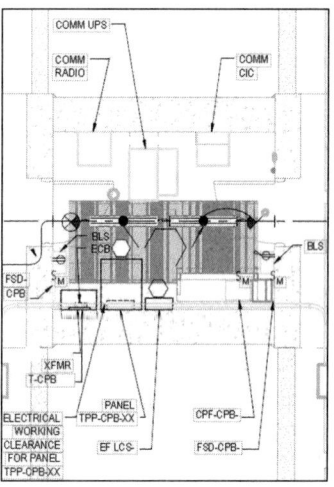

Figure 5. Example 3D cross-passage Revit model in BIM

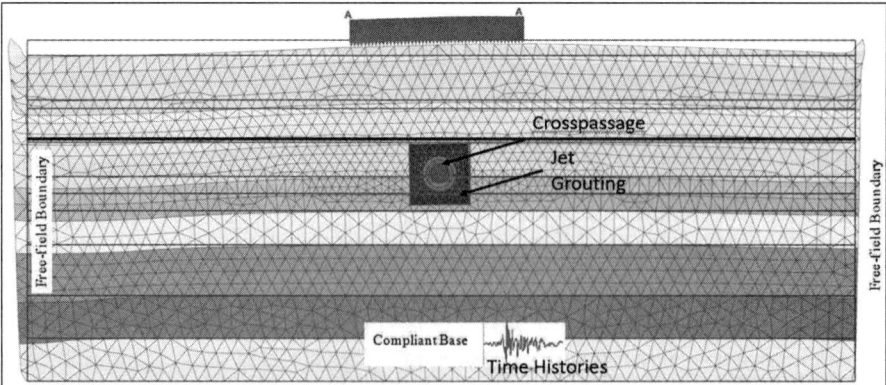

Figure 6. Dynamic SSI analysis using time histories

thickness (and reinforcement) are determined, the outer edge of the CP geometry is defined.

Temporary Ground Support for Excavation

CPs are typically excavated by a mechanized SEM approach. Temporary ground support system is used to ensure a stable, safe opening and to limit groundwater seepage. Depending on ground conditions, temporary initial ground support may be accompanied with ground improvement, dewatering and/or depressurization. The sequence of construction (and initial ground support) will be determined based on ground condition classification and/or methods of ground improvement. Therefore, when more than 20 CPs are considered, systematically categorizing/classifying ground classes for temporary ground support is crucial. From the contractor's viewpoint, a systematic approach (involving a limited number of construction sequences and support techniques) would greatly simplify work flow and equipment used by the construction crew.

An example (shown in Table 2) illustrates four temporary support ground classes for a California tunnel project that accounts for the variability in ground conditions expected at CP locations. The temporary construction support requirements also include construction probing ahead of CP excavation.

CONSTRUCTABILITY AND TECHNICAL CHALLENGES

Revision to the tunnel profile to optimize ground conditions for CP excavation may shift tunnel low point, which impacts the CP sump pit location. Further adjustments to CP location to avoid potentially problematic ground, maintain an offset from major intersections, and avoid top side utilities or specific business/residential properties may change the number of CPs required along the tunnel alignment. To mitigate potential delays to the design process and gain full optimization for design and construction, it is recommended to undertake these planning activities concurrently with constructability input from ground improvement and SEM specialists. The following are other technical challenges that may be encountered.

Optimization of Egress Path and Ventilation

The maximum CP interval of 800 ft is an NFPA 130 requirement to minimize passenger evacuation time. However, an estimate of actual evacuation time could further reduce the number of CPs, but this is a function of CP spacing, walkway/door width, design fire scenarios, etc. In addition, CP ventilation involves overall tunnel ventilation, which would require complicated Computational Fluid Dynamics (CFD) analyses. These project-specific design elements cannot be determined at an early planning stage. Thus, past experience becomes an important asset to advance the planning, while emergency egress and ventilation design are still under evaluation.

Penetrations

Ideally placing all conduits within planned cable trough/tray could simplify interdisciplinary coordination. However, high voltage electrical cables must maintain a minimum offset from system signal cables (due to electrical interference), and "wet conduits" should not be placed in the same trough/tray with electrical conduits. Also, block outs and construction joints always happen to encroach each other.

Some of the penetrations listed in Figure 7 need to be considered when laying out the structural reinforcement. Space proofing coordination with mechanical, electrical and

Table 2. Proposed ground classes for cross-passage temporary supports

Ground Type	Ground Type 1		Ground Type 2		Ground Type 3	Ground Type 4
Number of Cross Passages	2		11		4	6
Des. #	CPB-4, CPB-5		CPB-7, CPB-14, CPB-15, CPB-16, CPB-17, CPB-21, CPB-22, CPB-23, CPB-24, CPB-25, CPB-26		CPB-6, CPB-8, CPB-9, CPB-10	CPB-11, CPB-12, CPB-13, CPB-18, CPB-19, CPB-20
Soil/Bedrock Condition	Shallow in San Pedro (less than 50 ft)		Deep in San Pedro (more than 50 ft)		Mixed Face or Shallow in Fernando	Deep(10+ feet) in Fernando
Ground Improvement Required	Permeation Grouting	Jet Grouting (300 PSI)	Permeation Grouting	Jet Grouting (300 PSI)	Permeation Grouting within San Pedro Zone	Not Required (water control required)
Excavation Method	Heading/Bench	Full Face	Heading/Bench	Full Face	Heading/Bench	Full Face
Pipe Canopy	Pipe Umbrella 3.5" dia. @ 18"	Not Required	Pipe Umbrella 3.5" dia. @ 18"	Not Required	Pipe Umbrella 3.5" dia. @ 18"	#8 Grout Bar/Bolt @ 12"
Initial Ground Support (Shotcrete thickness with lattice girder)	5,000 PSI Shotcrete (In lieu of WWM, 70 lbs/CY Steel fiber can be considered).					
	Double 4x4-W4.0xW4.0	Double 6x6-W4.0xW4.0	Double 4x4-W4.0xW4.0	Double 6x6-W4.0xW4.0	Double 4x4-W4.0xW4.0	Double 6x6-W4.0xW4.0
	10"	8"	12"	8"	10"	8"
	3-bar #70/#6/#8 @3ft	3-bar #70/#6/#8 @4ft	3-bar #95/#6/#10 @3ft	3 bar #70/#6/#8 @4ft	3 bar #70/#6/#8 @3ft	3-bar #70/#6/#8 @4ft
Legend:	Alluvium	Fine San Pedro	Coarse San Pedro	Fernando		
Ground Improvement Required	Jet Grouting (300 PSI)		Jet Grouting (300 PSI)		Jet Grouting to Fit into Fernando Formation	Not Required (Only Water Control Required)
Excavation	Heading and Bench					

structural design disciplines is crucial and requires several iterations to achieve an optimum solution.

Waterproofing. One important aspect for CP design and construction is waterproofing the entire structure. This is especially true when the CPs are located in water bearing and gassy ground. Soft ground SEM typically features a dual lining system by which waterproofing is inserted between the initial shotcrete lining and the final cast-in-place lining. Due to complex excavation geometry (i.e., oval shape), inconsistent surface roughness, and difference in structure stiffness (i.e., precast segment v.s. cast-in-place bulkhead walls), it is difficult to find a simple solution (or even material) that can provide resistance to expansion and contraction of the structure. Design considerations include:

- Satisfactory elongation, (i.e., high tensile strength) to accommodate tunnel movements (i.e., seismic).

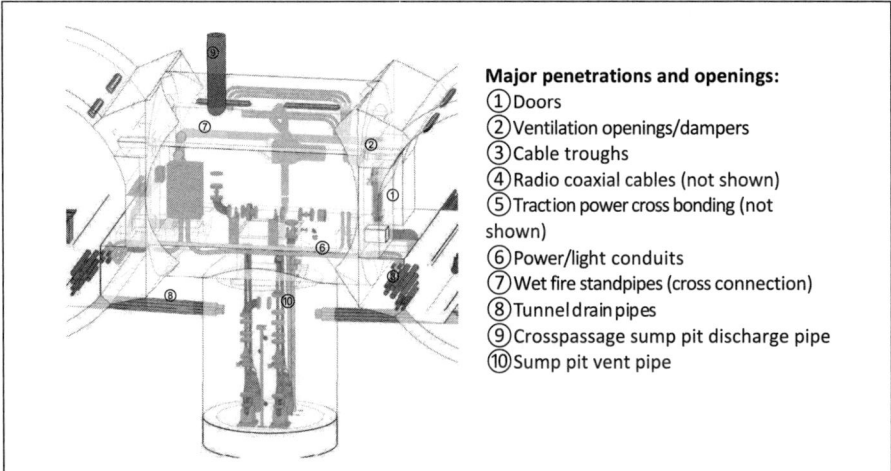

Major penetrations and openings:
① Doors
② Ventilation openings/dampers
③ Cable troughs
④ Radio coaxial cables (not shown)
⑤ Traction power cross bonding (not shown)
⑥ Power/light conduits
⑦ Wet fire standpipes (cross connection)
⑧ Tunnel drain pipes
⑨ Crosspassage sump pit discharge pipe
⑩ Sump pit vent pipe

Figure 7. Sump pit cross-passage 3D view with multiple penetrations

- Satisfactory flexibility to overcome concrete surface irregularities, potholes, undulations and unevenness

In addition, the waterproofing membrane should provide sufficient adhesion/bonding to the structural concrete so that if any area of the membrane fails, for whatever reason, the adhesion enables the membrane to compartmentalize the failure to localized areas. Therefore, if failure of the membrane occurs, the extent of the damage is limited, the location is known, and the repair time is minimized, thus avoiding unnecessary interruptions to the operating transit system.

Sprayed on membrane is another advantageous alternative to prefabricated waterproofing membrane under typical tunneling conditions, especially in geometrically complex areas such as CPs where installation of sheet based waterproofing membranes is inherently difficult and locating leaks is challenging. Additional advantages to owners, contractors and designers from the use of spray-on waterproofing include flexibility in the design, increased productivity and overall flexibility, as well as improved logistics. This leads to significant schedule and cost savings and reduced long-term maintenance costs.

CONCLUSION

800 feet maximum CP spacing is typically used for transit tunnel projects. If optimization of CP layout and number is desired for a long transit tunnel, the following priority topics need to be evaluated during planning and systematic design:

Ventilation

Crosspasage interior dimensions are governed by safety codes (evacuation time) and also by specific equipment operating criteria. Tunnel ventilation can also become a decisive factor in determining CP size and the number of CPs required. Thus, ventilation design and evacuation simulation contribute to design decisions and identify areas of concern, and should be reviewed as early as practical in the planning stage.

3rd Party Impact Mitigation

The construction of CPs may require pre-construction ground improvement or dewatering, and these activities may trigger adverse impacts to the nearby structures. Thus sometimes adding an additional CP to avoid CP location in the vicinity of sensitive structures or road intersections could become the better choice after all.

Seismicity and Structure Movements

Steel reinforcement congestion is always a critical constructability concern for bulkhead wall design. In California, seismic events will likely govern the structural design. During earthquakes, tunnel structures will move with the surrounding ground. Thus, the structures should be designed to accommodate the deformation imposed by the ground, particularly for structures with considerable structural discontinuities (such as joints between tunnel segments and CPs, and joints between tunnel segments). CPs are relatively short and are the most rigid structures within the whole tunnel system; thus they will attract more load if deformation/relaxation is not allowed. Detailed deformation analyses should consider the effects of these discontinuities on the earthquake resistance of CPs, in order to optimize structural performance.

BIM Implementation

There is no doubt that intelligent 3D BIM has become one of the greatest innovations for clash detection and planning collaboration in the tunneling industry. To maximize its usefulness for CP planning, it is important to establish project specific standards at the early stages, and then enforce periodic unified model updates across all disciplines for early coordination and clash detection. Large-scale tunnel projects require data management through one centralized model. Linked databases facilitate the visualization of planned works. This makes planning significantly more effective, error-free, and improves collaboration. BIM-integrated structural analysis significantly reduces design effort. BIM management should be in-place as early as possible when the tunnel project begins.

ACKNOWLEDGMENTS

The authors wish to thank the Los Angeles Metropolitan Transportation Authority Engineering team for their valuable comments, and also acknowledge the contributions of the tunnel contractor STS, design partner CH2M Hill, and Parsons colleagues Jon Kaneshiro and Sean-Philip Bolduc.

REFERENCES

"NFPA 130 Standard for Fixed Guideway Transit and Passenger Rail Systems," 2003 Edition, Published by the National Fire Protection Association, 1 Batterymarch Park, Quincy, MA 02269-9101.

"NFPA 502 Standard for Road Tunnels, Bridges and Other Limited Access Highways," 2014 Edition, Published by the National Fire Protection Association.

Station Excavation and TBM Tunnel on Los Angeles Crenshaw Project

Ran Chen ▪ J.F. Shea Construction
Jesse Salai ▪ J.F. Shea Construction
Ben Schatz ▪ J.F. Shea Construction

ABSTRACT

The LA Metro Crenshaw/LAX Corridor project is 8 miles of light rail which includes (3) underground cut and cover stations and one mile of twin tunnels. Support of Excavation (SOE) for the underground stations included various methods, such as drilled soldier piles, Cutter Soil Mix (CSM) wall, and jet grout. Tunnels were excavated by an earth pressure balance (EPB) tunnel boring machine (TBM). Precast sixsegment universal rings with 5.47-m (18.83-ft) inside diameter were erected after each 1.52-m (5-ft) of TBM advance. Efforts made by Owner and Design-built team turns the project into a big success, although there were many of challenges during construction.

INTRODUCTION

The LA Metro Crenshaw/LAX Corridor project (Crenshaw project) is a part of a Los Angeles County Metropolitan Transportation Authority's (Metro) major capital expansion program aimed at improving public transit service and mobility in Los Angeles County. This project, a design-built job, was awarded to Walsh/Shea Corridor Constructions (WSCC) for $1.27-billion teamed with HNTB as the Designer. The Crenshaw project runs through South LA and will service Los Angeles, Inglewood, and El Segundo along with portions of unincorporated Los Angeles County. The project's guideway connects the existing Metro Exposition Line at Expo/ Crenshaw Station to the existing Metro Green Line at Aviation/ LAX Station. Overall, the project has 8 stations, two of which were bid options that have been exercised. The guideway includes at grade trainway, aerial structures, cut and cover box structures, and bored tunnel, as shown in Figure 1. As a part of a separate contract, Los Angeles World Airport (LAWA) will connect the Crenshaw Line to airport terminals via a "people mover."

Excavation of three underground stations and bored tunnel sections is the main focus of this paper. The general construction scheme is predicated on constructing stations and tunnels simultaneously. First the stations were decked over to

Figure 1. LA Metro Crenshaw/LAX Corridor Project alignment

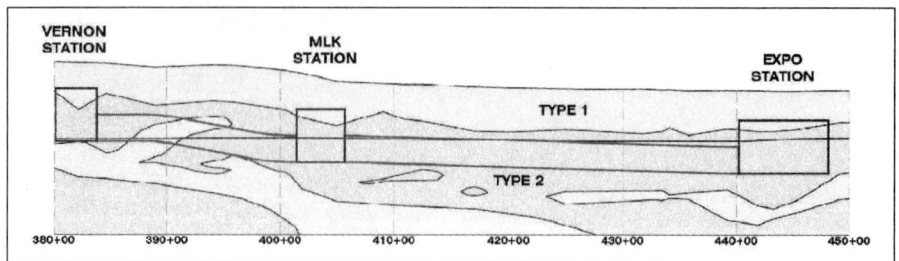

Figure 2. Geological profile for underground stations and tunnel (reproduced from GBR)

restore the street traffic on Crenshaw Boulevard while station work continued under the deck. Utilities, such as water main, storm drain, sewer line, duct bank, gas line, etc. were supported from the deck while the excavation and bracing continued to the bottom invert, approximately 20 m (65 ft) in depth. After the station was excavated to the designed elevation, the structural invert was placed and lowest level of bracing was removed. Once the lowest SOE bracing was removed at Expo Station and overhead clearance was obtained, TBM operations began, including positioning the machine to the portal, setting up thrust frame, etc. The TBM was assembled in the main construction site, Expo yard, positioned to Expo station head wall, and used to excavate the southbound tunnel from the Expo station to MLK station, and then to Vernon station. The machine was disassembled into three components and shipped back to the main site after the hole-through at Vernon station, and the whole process was repeated for the northbound tunnel. Post-tunneling operations will not be stated in this paper, such as invert and walkway concrete, cross passages construction, etc.

GEOLOGICAL INFORMATION

This project is located in the northern part of the Los Angeles Basin, which is directly underlain by unconsolidated Quatornary-age sandy sediments [1]. These generally are subdivided into unconsolidated Holocene-age sediments (Young Alluvium) and late-Pleistocene materials (Old Alluvium). Young Alluvium typically consists of surficial sediments including clay, sand, and gravel. Old Alluvium typically consists of sediments including pebble-gravel, sand, silt, and clay. The northern part of the project alignment along Crenshaw Boulevard is directly underlain by Young Alluvium over Old Alluvium. Cohesionless soil with blow counts consistently below 30 blows per foot and cohesive soil with blow counts consistently below 15 blows per foot are considered Young Alluvium. Sub-surface soil in the excavation zone of the underground station and tunnel is classified for excavation purposes as Fill, Type 1 Soil, and Type 2 Soil based on the engineering parameters derived from laboratory testing. Type 1 and 2 Soils are anticipated within both Young Alluvium and Old Alluvium. A generalized sub-surface profile for the underground excavation is shown in Figure 2.

The estimated volume percentages of the Fill, Type 1 Soil, and Type 2 Soil anticipated to be encountered during excavating the underground stations and bored tunnel are summarized in Table 1. Type 1 Soil consists predominantly of fine-grained silt, clay, and organic soil. Type 2 consists predominantly of a mixture of fine- to coarse-grained sands and gravels and includes some cobbles and boulders. The engineering properties of these two type of soils are shown in Table 2.

STATION EXCAVATION

Support of Excavation

Due to a contaminated plume left from a dry cleaning business in a nearby lot, the Contract dictated that Expo Station be constructed without dewatering to prevent spread of the plume. Since the designed excavation was approximately 4.6 m (15 ft) below the ground water table, a cut off wall was required for SOE. A 1-m (3-foot) thick wall with the designed UCS of 2 MPa (300 psi) was created along the perimeter of Expo and MLK station using cutter soil mix (CSM) methodology. W24 solider piles were installed in the CSM wall at the typical spacing of 1.67 m (5'6") before the soil-cement mixture is hardened. Where there was a gap along the CSM wall because of utility interruption, jet grout columns were installed to fill the gap and achieve a continuous sealed SOE wall. Both the CSM panels and jet grout columns were extended to a depth of 30 m (98 ft) to be toed into a clay lens below the station invert to create an impermeable "bathtub" and minimize the groundwater recharge into the station from invert slab. An identical CSM wall with jet grout columns were selected for MLK station SOE due to similar excavation depth below the groundwater table, presence of a clay layer below station invert, and cost savings on dewatering system, even though it was not the contractual requirement. However, Vernon station presented a dry excavation, and traditional drilled soldier piles and wood lagging was chosen as the station SOE. W24 piles at the typical spacing of 1.8 m (6 ft) were set in 0.9 m (3 ft) diameter drilled holes. Solider piles were toed into 17 MPa (2,500 psi) concrete and embedded in the lean mix concrete backfill.

Once the SOE wall was installed, cap beams and cap wales were set on top of the piles to support the decking. Deck beams were typical W40 beams at the typical spacing of 3.68 m (12'1") to receive 3.66 m × 1.52 m × 254 mm (12'-0" × 5'-0" × 10") precast concrete deck mats. Deck beams had a stub beam to transfer shoring load from piles and make deck beams act as top level strut. Once the decking was completed, street utilities were supported from it using typical back to back channel as stringers and coil rod or clevis style pipe hangers. Expo and MLK stations had 3 levels of bracing below

Table 1. Estimated volume percentages of fill, Type 1, and Type 2 Soil

Soil Type	Exposition Sta.	Bored Tunnel	MLK Sta.	Bored Tunnel	Vernon Sta.
Fill	1%	0%	1%	0%	1%
Type 1	42%	0%	30%	15%	66%
Type 2	57%	100%	69%	85%	33%

Table 2. Soil engineering properties (from GDR)

Soil Type	Approximate Blow Count Range (blows/ft) (blows/30 cm)	Total Unit Weight (pcf) ($\times 10^3$ kg/m^3)	Effective Stress Strength Parameters		Undrained Shear Strength (psf) (Mpa)
			c' (psf) (Mpa)	ϕ' (deg)	
Type 1 Soil Young Alluvium	2–42	125 2.0	250 5	20	500 10
Type 2 Soil Young Alluvium	5–29	125 2.0	250 5	30	
Type 1 Soil Old Alluvium	12–100	125 2.0	600 13	28	3600 75
Type 2 Soil Old Alluvium	13–100	125 2.0	0	35	

Figure 3. Decking operation at night time on fully closed Crenshaw Blvd.

the deck and Vernon had 2 levels. Wale sections were typically W36X395 with welded connections making the beams continuous around the station perimeter. The wales were connected to the soldier piles with WT12 sections split from the pile cut-offs and supported by lookout beams. Pipe struts were 914 mm (36") diameter by 25 mm (1") wall thickness with 25 mm (1") horizontal and vertical knife plates at each end to prevent crushing. The selection of heavy sections for wales and struts was based on the shoring loads accounting for over consolidated soil loading, which was theorized post Metro Gold Line Project in East LA. This over consolidated condition postulated that seismic activity in the LA basin area caused high soil loads, which caused bracing buckling at the Gold Line excavations.

Decking Operation and Wale/Strut Installation

The decking operations at each station were performed with a full closure of Crenshaw Boulevard for several weeks at a time. Street excavation was performed by Caterpillar 345 excavators and deck beams were set by rental truck cranes, typically 110 ton in size. The crane set 3 to 4 beams before picking up outriggers and advancing onto the matted deck. Infill bracing beams were erected with bolted connections. Once the deck was set, bracing work and utility hanging continued alongside the excavation below. The picture in Figure 3 shows the decking operation at night time on the fully-closed Crenshaw Boulevard. Crews set the wale packing and lookout beams, then used a string line to cut the packing to allow proper layout of wales for the future rebar and waterproofing construction. The wales were lowered into the excavation and set with Caterpillar Extreme forklifts after the packings and lookouts were ready. Pipe struts with a maximum length of 33.5 m (110 ft) were lowered into the stations and were dually handled with Cat 973C track loaders and the Extreme forklifts. Great care was taken to rig, handle, and install the struts which weighed up to 133 kN (30 kips) and were precarious to handle underground.

Ground Monitoring

Public safety always brings Owner, Designer and Contractor's attention to the ground movement on any underground related project, especially in urban areas. This project is located at one of the most crowded areas in the US. The geotechnical instrumentation system, which adequately monitored buildings, utilities, ground movement, and ground water behavior within zone of influence of station SOE and tunnel excavations, was set to nearly the most strict settlement criteria of the industry. This system

Table 3. Manual monitoring points

	Settlement	Settlement (Feno type)	Building	Utility
Quantities	626	170	531	34
Limit	15 mm(0.6")	15 mm(0.6")	13 mm(0.5") or L/600	15 mm(0.6")

Table 4. Automatical monitoring instruments

	Excavation Support	Inclinometers	Multipoint Borehole Extensometers	Observation Wells	Tilt-meters
Quantities	101	35	56	16	56
Limit	18 mm(0.72") H or V	0.25 mm(0.01") × depth	38 mm(1.5")	0.9 m(3') (±)	0.1°

consists of manually and automatically monitored instruments, the quantities and limits of which are listed in Tables 3 and 4. Instrument data was logged into a cloud-based Geographic Information System (GIS) and made available to different levels of project management team. With CSM technology being utilized, ground surface, utilities, and buildings around Expo station and MLK station were held stable, and settlements were below maximum allowable values. Where solider piles and lagging were used for support of excavation in lieu of CSM (Vernon station), there was overall greater settlement than the other two stations, with some max level exceedances. In addition to settlement monitoring, the water levels in the shallow and deep aquifer and the movement of the clay aquitard were closely monitored. Vibrating wire piezometers and fixed point extensometers were used to verify the effectiveness of the CSM wall seal and excavation base stability while excavating the stations, which have base of excavations below the groundwater table.

Challenges on Station Excavation

1.68 m (66") Diameter North Outfall Sewer Bypass and Replacement

Some of the most challenging and tedious work on this project was related to the locating, protecting, supporting, and relocating of utilities. Most utilities were supported and protected in place, relocated into stacked clusters passing though the SOE, or replaced by the equivalent material temporarily. The largest undertaking of this work was the replacement of a 27.4-m (90-foot) section of 1.68 m (66") North Outfall Sewer pipe located at the intersection of Crenshaw Blvd and Rodeo Rd. This section of sewer line, made of fiberglass reinforced polymer mortar pipe, was grouted in a half of RCP and casted in a reinforced concrete collar 30 cm (1 ft) above the crown of the pipe during an emergency repair two decades ago. Because joints of the encased pipe section was in poor condition due to the emergency repair, and the section was too heavy and fragile to be hung from the SOE decking, the bypass and replacement option was proposed and proceeded. To bypass the designed peak flow rate of 2180 l/s (34,560 gpm) for the pipe replacement, two 6.1 m × 4.6 m (20' × 15') pits were established outside of the station box over the pipe for suction and discharge. Five 505 l/s (8,000 gpm) pumps connected the two pits by five 500 mm (20") HDPE pipes on each side of the pumps, with five identical stand-by pumps for City agencies' 100% redundancy requirement. After the bypass system was operational, two inflatable plugs were installed to isolate the middle section, the concrete collar with encased pipe was wire sawn into several sections for easier demolition, and new pipe sections were installed on cradles which were supported from SOE decking. After the new pipe passed the test for leakage, the bypass plugs were removed and the pumps and HDPE pipes were demobilized within planned schedule. Figure 4 shows the bypass and replacement operation in different stages.

Figure 4. 66" sewer pipe bypass and replacement

CSM Wall Leakage Repair

One of the biggest challenges during the station excavation was due to leakage in the CSM wall, which was intended to be impermeable below groundwater table. Though many utilities were relocated before the CSM wall installation, many large clusters of utilities were not allowed to be relocated, making it impossible to install a continuous sealed CSM wall. Large efforts were made to side shift or rotate the CSM panels to minimize the gap between panels. In attempt to seal off the remaining gaps completely, jet grout columns were installed. As the excavation progressed below the water table in Expo station, the monitoring wells began to show significant drops in ground water levels indicating leakage through the SOE wall and the invert. Sump wells were installed below the base of excavation to draw down ground water within the SOE until leaks could be identified and sealed. These leaks were caused by large clusters of nested cobbles that caused shadowing and webs of voids in the jet grout columns. This unforeseen concentration of cobbles was later recognized as a differing site condition by the Owner. The high inflow of up to 18.9 l/s (300 gpm) through this type of ground posed a big challenge for the final 4.5 m (15 feet) of station excavation. Various grouting methods were implemented to stabilize the groundwater inflow.

TUNNEL CONSTRUCTION BY TBM

General Description of TBM Operation

Twin tunnels with the centerline separation of 11.93 m (39.13 ft) were excavated by an earth pressure balanced tunnel boring machine (EPB TBM) between three underground stations. Total excavation length was approximately 3.22 km (2 miles). Tunnels were bored to the diameter of 6,550 mm (21.49 ft). To support LRT vehicle operation, gasketed precast segment rings with 5,740 mm (18.83 ft) I.D. were erected in the tail shield of the TBM after each mining cycle as a one-pass final lining. Like typical TBM tunnel construction, this TBM was advanced 1.5 m(5 ft) by jacking against existing precast segment ring assembled previously and rotating the cutterhead to shave the ground into the excavation chamber. During TBM advancing, the following activities were happening simultaneously:

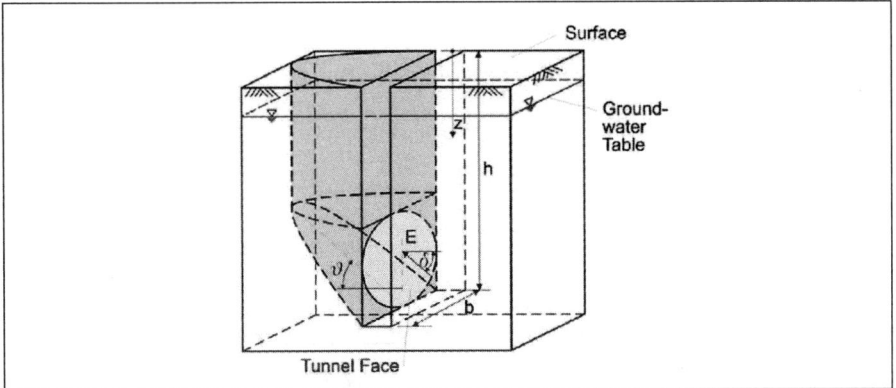

Figure 5. Failure mechanism of Piaskowski & Kowalewski model

    a. soil conditioning (foam) was injected in front of the cutterhead to transform the in-situ material into a plastically flowable mixture;

    b. bentonite slurry was also injected around the shield to lubricate the contact area between the TBM shields and ground and stabilize the excavated ground temporarily;

    c. the screw auger and conveyor belt were discharging conditioned muck from the TBM muck chamber to muck boxes which travelled along the excavated tunnel by a locomotive;

    d. A/B grout was injected into the annular area between the segments and ground where the tail shield has just moved forward.

After each five-foot advance,

    a. the screw auger and conveyor belt were stopped;

    b. discharge gate of the screw auger was shut to hold the face pressure in the excavation chamber and at the cutterhead;

    c. the muck was delivered to the opening at Expo station for the crane to lift and dump; the train also delivered segments and tunnel utilities when it travelled back to the TBM with empty muck boxes;

    d. assemble a precast ring which includes six segments between the existing ring and jack cylinders using the erector.

TBM Face Support Pressure and Soil Conditioning

As a systematic operation, tunnel construction by EPB TBM includes excavation, mucking, segment erection, ground support, ventilation, etc. The key parameters to keep excavation effective and the ground safe are face support pressure and foam injection parameters. For the face support pressure prediction, DIN 4085 model [2] was adopted in this project. In this model, three-dimensional active earth pressure is calculated based on the failure mechanism theory of Piaskowski & Kowalewski, as shown in Figure 5. The tunnel face is divided into multiple horizontal strips to calculate the horizontal earth pressure at each depth. Due to the relatively small width of each strip in comparison to its depth, there will be an arching effect in the horizontal plane, and it will rearrange the pressure distribution in the ground. The three-dimensional active earth pressure acting on each strip is calculated with the two-dimensional active earth pressure method, adjusted by reduction factors, as shown in Eq. (a).

$$p'_{i,e} = \mu_i \cdot K_{a,i} \cdot \sigma_{V,i} - 2 \cdot \sqrt{K_{a,i}} \cdot c'_i \qquad\qquad (a)$$

where, $\sigma_{V,i}$ is the effective vertical earth pressure at the center of soil strip i, which includes the pressure from permanent and temporary loads on the ground surface and soil pressure above the soil strip i considered;

$K_{a,i}$ is active earth pressure coefficient, $K_{a,i} = \dfrac{1 - \sin\varphi'_i}{1 + \sin\varphi'_i} = \tan^2\left(45° - \dfrac{\varphi'_i}{2}\right)$;

φ_i is the effective internal friction angle of this soil strip;

c'_i is the effective cohesion at the center of soil strip i;

μ_i is the effective horizontal earth pressure reduction factor, which is

$\qquad \mu_i = -0.005\alpha_i^3 + 0.0155\alpha_i^2 - 0.178\alpha_i + 0.9947$, if $\alpha_i < 10$;

$\qquad$ or $\mu_i = 0.25$, if $\alpha_i \geq 10$;

$\qquad \alpha_i$ is the ratio of strip depth to tunnel diameter, d_i/D;

$\qquad d_i$ is the center depth of each strip i; and

$\qquad D$ is the tunnel diameter.

To ensure stability of the tunnel face, it is necessary to counterbalance the total of effective horizontal earth pressure and water pressure with face support pressure from the support medium (i.e., foam-soil mixture in the plenum) at any location of the tunnel face. For a typical ground condition, the target operational pressure is governed by the pressure at the tunnel crown, which can be calculated as:

$$p_{cr} = FS_e \cdot p'_{cr,e} + FS_w \cdot p_{cr,w} \qquad\qquad (b)$$

where, FS_e is the safety factor for earth pressure, FS_w is the safety factor for water pressure, $p'_{cr,e}$ is the effective horizontal earth pressure at the tunnel crown, computed using Eq. (a), and $p_{cr,w}$ is hydrostatic water pressure at the tunnel crown. Some additional safety checks need to be performed, such as invert stability and crown blowout stability. Due to the material removal by the screw auger, the chamber pressure distribution was typically disturbed around the auger. Therefore, the operational face support pressure was monitored by transducers at or near spring line. To consider the operational tolerance, the lower limit of face support pressure can be calculated using Eq.(b) with reduced safety factors of $(0.2 \cdot FS + 0.8)$, and the warning lower limit can be calculated using Eq.(b) with reduced safety factors of $(0.6 \cdot FS + 0.4)$. Similarly, the error upper bound of the face support pressure can be calculated using Eq.(b) with increased safety factors of $(2.5 \cdot FS - 1.5)$, and the warning lower bound can be calculated using Eq.(b) with increased safety factors of $(1.8 \cdot FS - 0.8)$. Here FS is FS_e or FS_w.

Foam injection is one of the most commonly used soil conditioning methods for transforming in-situ material into a plastically flowable mixture. Foam injection application is necessary for this project to modify soil behavior to enhance the fluidity of the spoil in the plenum, reduce abrasion, and reduce cutterhead torque. In addition, foam injection was able to minimize face support pressure variations while excavating, control groundwater ingress, and minimize the spoil treatment. Soil conditioning foam consists of the foam solution (a mixture of surfactant agent and water) and air. Three parameters, foam Solution Concentration (C_F), Foam Expansion Ratio (FER), and Foam Injection Ratio (FIR) control the foam quality and, to a certain degree, the efficiency of excavation and cutterhead torque reduction. The concentration of foam solution (C_F) and the foam expansion ratio (FER) typically follow the manufacturer's recommendations. They can be kept constant through the whole project after they are adjusted to get the best conditioning results at the beginning of the project. Different conditioned soils could partially achieve the same quantifications concerning the workability just by varying the FIR, maintaining stable foam parameters. Laboratory testing is still the best way to determine FIR. However, it is not practical to carry out FIR tests at relatively short spacing along the tunnel alignment. Therefore, the formula below, developed by Rheological Foam Shield Tunnel Association of Japan [3], was used to derive FIR values for the project to start with. This formula builds an intrinsic relationship between soil particle size distribution and FIR needed to turn soil into a workable material.

$$FIR = \frac{a}{2} \cdot (X + Y + Z)(\%) \qquad\qquad (c)$$

where,　FIR is the foam injection ratio, FIR = 20% if FIR < 20%;
$X = 60 - 4.0 \cdot x^{0.8}$, $X = 0$ if $X < 0$;
$Y = 80 - 3.3 \cdot y^{0.8}$, $Y = 0$ if $Y < 0$;
$Z = 90 - 2.7 \cdot z^{0.8}$, $Z = 0$ if $Z < 0$;
x is percentage by weight passing a #200 (0.075 mm) sieve;
y is percentage by weight passing a #40 (0.42 mm) sieve;
z is percentage by weight passing a #10 (2 mm) sieve;
a is a coefficient based on the Uniformity Coefficient, $C_U = \dfrac{D_{60}}{D_{10}}$

$$a = \begin{cases} 1.6, \text{if } C_U < 4 \\ 1.2, \text{all others}, \\ 1.0, \text{if } C_U > 5 \end{cases}$$

D_{60} is the diameter for which 60% of material is finer than D_{60};
D_{10} is the diameter for which 10% of material is finer than D_{10}.

For clays or cohesive silts, a minimum *FIR* of 30% was applied even when the calculated *FIR* results in values close to zero to prevent adhesion and minimize wear. In this case, an anti-clay agent can be added into the foam to break clay bonds and reduce swelling effects. However, if the fines content (*x* value in the formula above) are below 20%, like coarse, clean sands and gravels (poorly-graded), addition of polymer was added to increase the soil cohesion and strengthen the viscosity effects [4].

The water content of in-situ material plays an important role in getting a soil-foam mixture which has the best physical and mechanical properties for the mining and mucking system. If the material is too dry, some water in the pre-generated foam will be absorbed by soil particles. Foam solution concentration and FER will increase and the stability of foam will reduce. If the soil is too wet, there will be some free water in the soil-foam mixture. Foam solution concentration will decrease and the stability of foam will reduce. There is not a formula available to calculate the optimal water content for a certain type of soil. The optimal water content of two typical soil samples for this project was determined to be around 13.25% by laboratory tests [5]. Along the tunnel alignment, the average water content of the tunnel face is between 6% and 17%. When the soil is a dense and coarse material and the groundwater table is in the tunnel range or below, the water content is typically below the optimal. Since the calculated FIR is for a saturated material, the additional water is needed to get a well-conditioned mixture for this type of soil if the calculated FIR is used. However, it is not practical to pre-inject the additional water into the tunnel face to let in-situ material reach the optimal water content. The easiest, but not the most economical approach is to increase FIR. It can be as high as 60% in this project.

Main Challenges on TBM Activities

One challenge of the construction of this tunnel was the high cutterhead torque in the dry sandy ground. At the beginning of the first tunnel excavation, cutterhead motor overheating with a low advance rate was an issue. After reviewing the excavation parameters recorded by data collecting system, including advance rate, cutterhead RPM and torque, total thrust force, contact force at the face, chamber EPB pressure, FER and FIR for conditioning, and screw auger RMP and torque, it was concluded that the EPB pressure in the mixing chamber was not able to accurately represent the face support pressure happening at the face. This was confirmed by the fact that the contact force at the face was much higher than required. That was because the ground was relatively dry and dense, and material was not fully mixed at the tool gap to turn into a uniform mixture, which caused a large pressure gradient between the face and chamber bulkhead where pressure transducers were mounted. The mixture will be more uniform at the tool gap and this gradient will be small if the tunnel is below the groundwater table, and the soil conditioner has a longer life mixing with wetter material. Since ground can't be prewetted practically, in order to minimize this adverse effect from the dry dense sandy ground, lowering contact force at the face by safely lowering EPB pressure, lowering FER and increasing FIR, and adding a torque reducing agent were some effective ways to keep cutterhead from over-torquing and improve the advance rate.

Settlement control is always a big challenge for any soft ground tunnel, especially for Metro subway project. Any EPB machine is designed to minimize the ground settlement by counterbalancing the total of effective horizontal earth pressure and water pressure with face support pressure from the support medium in the pressurized chamber. The long shield also prevents ground from caving in before tunnel segments are installed and back grouted annularly. However, the ground has to be over-excavated radially 25 mm (1 inch) for this machine to allow machine to steer. A typical EPB machine is not able to control the convergence when the ground is excavated but not yet supported and back grouted. A shield gap bentonite slurry injection system was installed to counter soil and water pressure to prevent collapse of the annular void outside the shield until this void can be filled with two component grout. The shield gap injection was volumetrically controlled to match the advance rate, and also had low and high pressure limits to trigger pump operation. These values accounted for bleed even when not advancing the TBM. The injection pressure was set close to the calculated face support pressure at the tunnel crown. The effectiveness of this settlement control approach was confirmed by ground surface settlement monitoring. The cloud-based GIS collecting settlement information was linked to the TBM data management system to make all real-time settlement information available to TBM operators. In early stage of TBM operation, there were some ground settlement monitoring points approaching the warning limit after the TBM had passed. It was found to be in correlation to periods of malfunction of TBM shield bentonite system. Ground movement was stable with the approximately 6 mm (¼") of settlement after each tunnel drive in average along the tunnel alignment, as long as the TBM shield bentonite system was properly functioning.

Another challenge is that a tight project schedule required the station invert concrete placement to happen much earlier than TBM launch and very close to the tunnel portal. There was very limited room available for TBM launch. A new methodology, "move-up" thrust frame, was used to launch the machine. Different from the typical stationary thrust frame, in which rear braces are blocked and columns are anchored into invert to provide enough horizontal shear and overturning moment resistances for TBM advance, the "move-up" thrust frame was initially braced towards the front to minimize the room required, and eventually braced towards back when TBM was advanced.

Additionally, different from the typical thrust frame which requires several "dummy" segments erected next to the thrust frame which also need to be disassembled later on, the "move-up" thrust frame was moved up after each TBM advance to eliminate any "dummy" ring erection in the station. A 4-segment thrust ring bolted to thrust frame was moved up with the thrust frame by a winch and was rapidly disassembled for the other three-time reuse. Instead of blockouts and anchorages in the traditional case, a thicker concrete invert and two embedded beams were installed precisely. A bolt hole pattern on the top flange of embedded beams matches the one on the end plates of braces and columns of the "move-up" thrust frame.

CONCLUSION

The underground stations and bored tunnel of the LA Metro Crenshaw/LAX Corridor project are being constructed with soft ground techniques refined for use in urban environments. Major work items included permeable and impermeable shoring, station decking, heavy bracing installation, ground monitoring and settlement control, mass station excavation, utility support and/or relocation, and EPB TBM tunneling. As with all construction endeavors, each item whether innovative or tried and true presented its own set of obstacles. Overcoming the obstacles gives the project management team and Design-built team experience to carry out future work and provides the city of LA with knowledge to justify continued underground infrastructure.

ACKNOWLEDGMENT

The authors acknowledge Los Angeles County Metropolitan Transportation Authority (MTA) for guiding and supporting this project.

REFERENCES

[1] Hatch Mott MacDonald, Geotechnical Baseline Report of Crenshaw/LAX Transit Corridor Project, 2012.

[2] DIN 4085 - Calculation of earth-pressure, 2011.

[3] RFSTAJ, Rheological Foam Shield Tunneling Method, The Rheological Foam Shield Tunneling Method Association, 1993.

[4] Ball, R.P.A., D.J. Yong, J. Isaacson, J. Champa, Research in Soil Conditioning for EPB Tunneling Through Difficult Soils, *2009 RETC Proceedings*.

[5] BASF, Soil conditioning Tests (Type 1 & 2 Soil) for Crenshaw/LAX Project, Los Angeles, CA, 2011.

Third Street Light Rail Phase 2, Central Subway Stations, San Francisco, CA—Utilization of Multiple Foundation Techniques

Albert Neumann ▪ Bencor Global Inc.
Kevin Bolton ▪ Bencor Global Inc.
Jeffrey J. Bean ▪ Bencor Global Inc.

ABSTRACT

In October 2013, Bencor Global (a Keller Company) was awarded a subcontract by the successful prime contractor, Tutor-Perini Corporation, to perform key elements of the construction of the support of excavation system to facilitate the installation of underground subway stations for the project owned by the San Francisco Municipal Transportation Agency. The Central Subway (Contract 1300) is the second phase of San Francisco's Third Street Light Rail Program. It is a 2.74 kilometer extension that will expand light rail service within the city, improve regional connections to Caltrain, BART, Muni Metro, and provide an alternate transportation option to congested vehicular surface traffic.

The key elements of this project included the utilization of three different advanced foundation techniques using diaphragm walls, compensation grouting and jet grouting at the new Chinatown Station (CTS), Yerba Buena/Moscone (YBM) and the Union Market Street Station (UMS). This paper will discuss the complexities associated with the planning and performance of geotechnical construction work in the heart of downtown San Francisco within densely populated neighborhoods and complex underground utility systems. It will also highlight the logistics associated with coordinating key elements of a complex foundation system and the benefits of utilizing one geotechnical contractor to perform all three elements on the project.

INTRODUCTION

The Central Subway (Contract 1300) is the second phase of San Francisco's Third Street Light Rail Program. The project is being majority funded by the Federal Transit Administration's (FTA) New Starts program. This contract was for the construction of four new subway stations—4th and Brannan on the surface and three underground stations, Chinatown Station, Union Square/Market Street Station and the Yerba Buena/Moscone Station as well as surface, track and system construction (see Figure 1). The stations and head houses were constructed atop and adjacent to the twin tube tunnels previously installed under separate contract. Each station utilized a different combination of construction techniques and had unique logistical challenges to overcome.

Chinatown Station (CTS-1254R). This is the northernmost station along the alignment of this contract. It is located at the intersection of Stockton and Washington Streets on a small corner lot. The site has a rectangular shape 56 meters long by 22 meters wide (see Figure 2). The bottom of excavation is approximately 39 meters below ground surface. There are existing buildings adjoining on the north, west and south sides and borders Stockton Street on the east side with no open space available for staging and layout (see Figure 3). This station required the use of two foundation techniques:

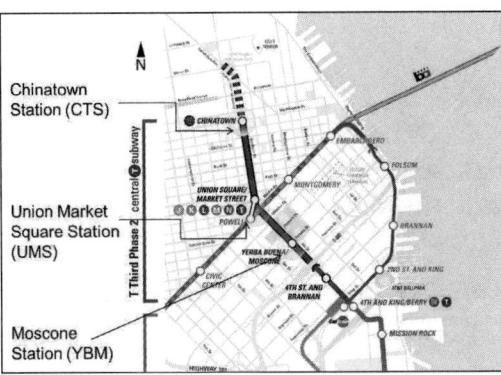

Figure 1. Project overview

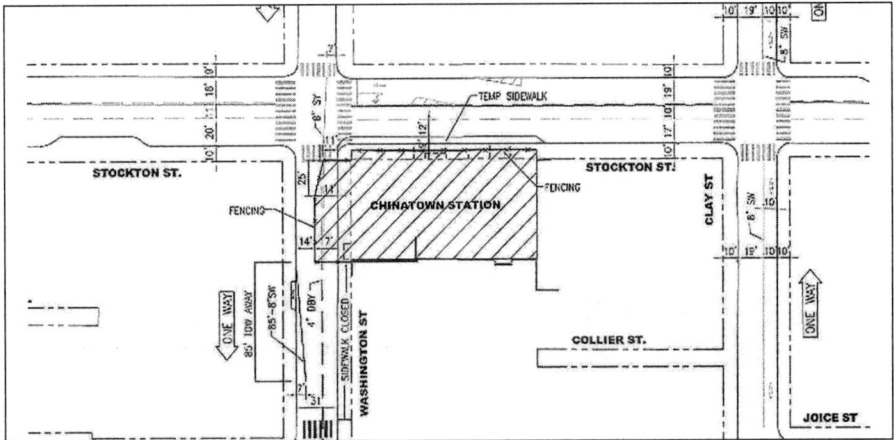

Figure 2. Chinatown Station plan

diaphragm walls and compensation grouting. The diaphragm walls were designed by the authority to provide rigid temporary support of excavation for the head house, which provides access to the station, as well as become the permanent structural walls for the head house. The diaphragm walls also serve as structural support for the station floors and roof slabs. Since this project is in San Francisco near active faults, the structure was designed up to the latest building codes to resist seismic forces.

The small project site along with heavy pedestrian and vehicular traffic complicated the construction operations requiring close logistical coordination. Daily delivery of reinforcing steel cages as well as staging many concrete trucks—along the streets had to be closely coordinated to maximize the process while still maintaining a safe project site for the work crews and pedestrian traffic.

The 1.1 meter thick diaphragm wall was installed along the 154 meter perimeter of the subway station head house, utilizing 48 individual panels totaling 6,780 square meters in area. The completed diaphragm wall panels were installed to depths of up to 46 meters and founded in the Franciscan formation (see Figures 4 and 5). The Franciscan formation consists mainly of soft friable rock however large boulders of extremely hard "green stone' were encountered during excavation. Each of the

Figure 3. Existing buildings surround the congested Chinatown Station site

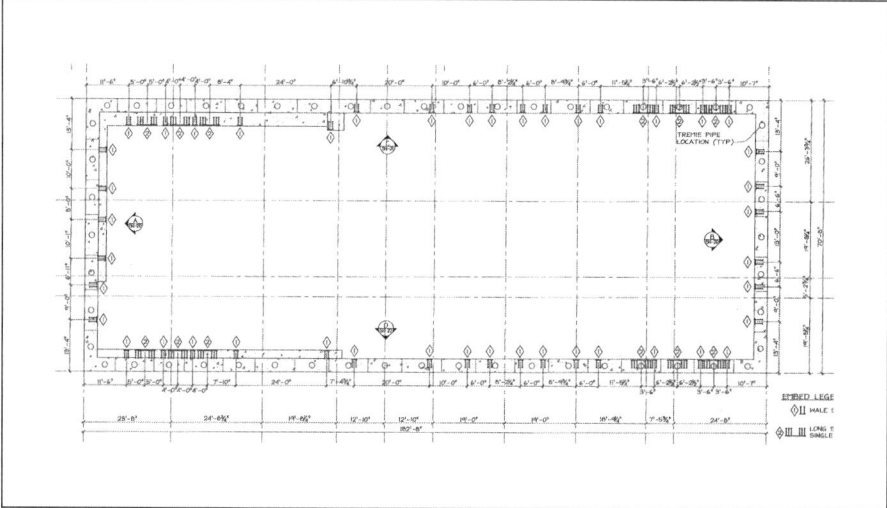

Figure 4. Chinatown Station diaphragm wall plans

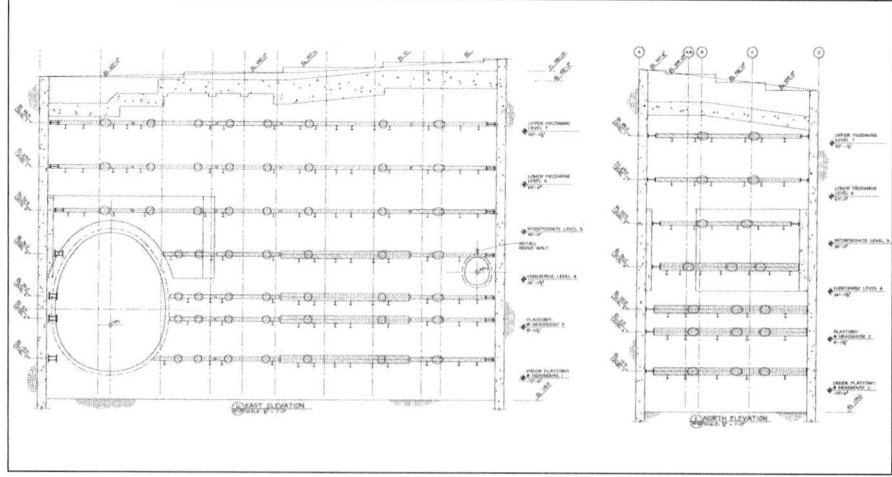

Figure 5. Chinatown Station diaphragm wall elevations

48 reinforcing cages was unique. Due to the complexity of the reinforcing arrangement, Building Information Modeling was utilized to aid in the construction of the cages. The reinforcing cages consisted of thousands of form savers and couplers for connections to the reinforcing in the station floor slabs as well as up to two layers of #36 @ 152 mm on both sides along with #16 ties @305 mm (see Figure 6). The high density of reinforcing steel along with the 46 meters depth produced very heavy cages weighing up to 37 metric tons which required the use of large cranes on the small site. Reinforcing cages were constructed at an offsite location approximately 100 km from the job site and delivered to the job site in 2 pieces. Oversized deliveries were restricted to early morning hours.

The small site was a logistical challenge for construction operations. As a result, diaphragm wall construction operations were conducted in two phases with the stationary equipment relocated on the site between phases (see Figure 7). The diaphragm wall excavation required the use of large and heavy equipment including a slurry desanding plant. Since there was no off-site laydown or staging area available, the plant had to be located on the site and relocated to enable the construction of the diaphragm wall. With the jobsite located on a hill side, a level platform could not be constructed, further complicating the excavation and concreting of the panels.

After completion of the diaphragm walls, the excavation was supported utilizing up to 6 levels of heavy internal bracing that was installed top down as the station excavation progressed down to subgrade.

In addition to the diaphragm wall construction at the head house, compensation grouting was required to mitigate potential settlement of the surrounding structures that may result from the station cavern excavation operations (see

Figure 6. Reinforcing cages weighing up to 37 metric tons

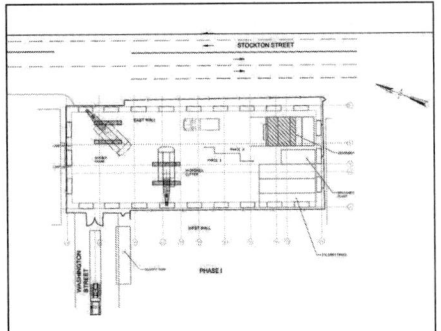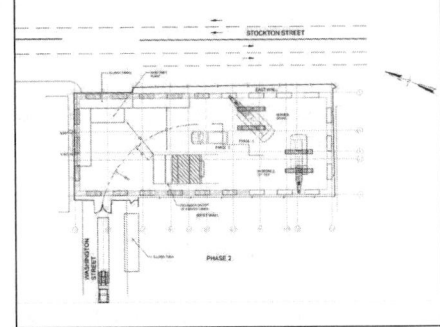

Figure 7. Diaphragm wall operations were conducted in two phases

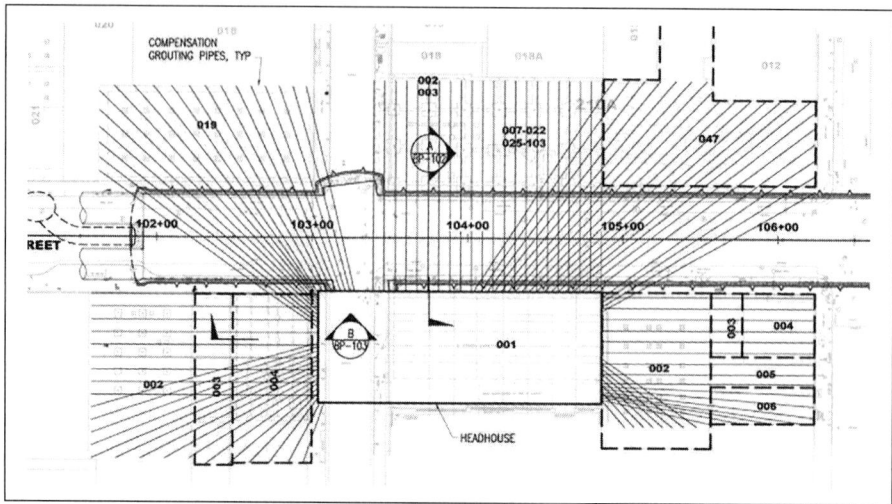

Figure 8. Chinatown Station compensation grouting site plan

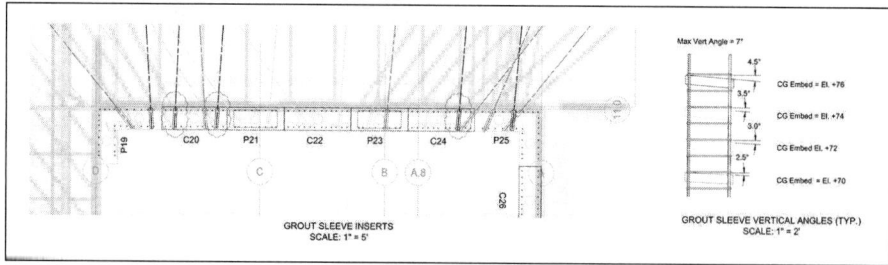

Figure 9. Chinatown Station compensation grouting details

Figure 8). The compensation grouting program was designed as a two-step process; Pre-conditioning to induce initial lift and compensation grouting to recover any settlement induced by station construction. CTS station cavern, unlike YBM and UMS is to be construction utilizing sequential excavation methods.

As the excavation advanced to a depth of roughly 12 meters below grade, drill rigs were brought in to drill and install 94 tube-a-manchetes (TAM) through PVC sleeves installed in the reinforcing cages for the diaphragm wall panels to distances of up to 56 meters (see Figure 9). A compensation grouting program was performed that included the injection of 325,500 liters of grout pumped in at pressures up to 2 MPa. Since both of the diaphragm wall and compensation grouting were performed by Bencor, the installation details were able to be closely coordinated to maximize the efficiency for the project as opposed to having two separate contractors trying to work this out, such as hole locations and sleeves.

Compensation Grouting Statistics:

- Total Grout Holes = 94
- TAM Installed = 4060 meters
- Grout Injected = 325,550 liters
- Treatment = 10 Buildings – 3,820 sq. m.

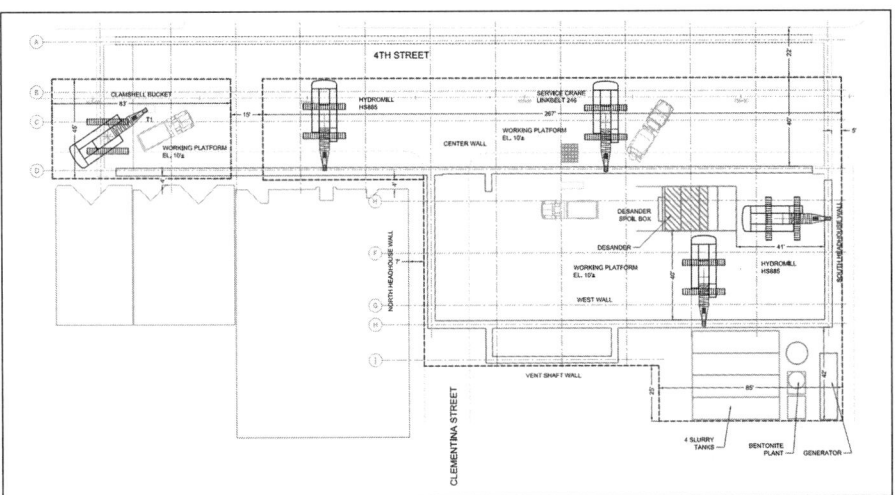

Figure 10. Yerba Buena/Moscone Station site plan

Yerba Buena/Moscone Station (YBM-1255). This station is located along the center of the contract alignment. It is located at the intersection of 4th Street and Folsom Streets on a small corner lot as well as under 4th Street. The site is 98 meters long by 40 meters wide. The bottom of excavation is roughly 90 meters below ground surface. Similar to the Chinatown Station, there are existing buildings adjoining on all sides with 4th Street splitting the site. There was minimal open space available for staging and layout (see Figure 10). 4th Street could not be completely closed during construction, so the diaphragm walls were installed in two phases with limited lane closures on 4th Street. Stationary equipment was relocated at the site between phases. This station required the installation of diaphragm walls for both the head house and the station box. These diaphragm walls were also designed by the authority to provide rigid temporary support of excavation as well as become the permanent structural station walls. The diaphragm walls also serve as structural support for the station floors and roof slabs. This structure is also heavily designed to the latest codes to resist seismic forces.

Due to its location this station required the installation of three parallel rows of diaphragm walls. Two parallel walls to provide support for the station excavation above the existing tube tunnels which run beneath 4th Street and the head house with a common diaphragm wall in the center for structural support of both the head house and the station box. The east and center walls were installed within 254 mm of the existing tunnels. Frequent verticality checks were required during installation of panels at these critical walls to insure that the tunnels were not contacted or damaged.

Prior to construction of the diaphragm walls, two test panels were installed and load testing was conducted using Osterberg cells. The toe of one test panel was post-grouted and one was not. Based on the results of the load testing, it was determined by the owner's engineers that the post-grouting was beneficial and would be required for each load bearing panel.

The diaphragm walls at this station were 0.9 meter and 1.1 meters thick, installed along the 325 meter perimeter of the subway station, utilizing 95 individual panels totaling 10,405 square meters in area (Figures 11 and 12). The finished diaphragm

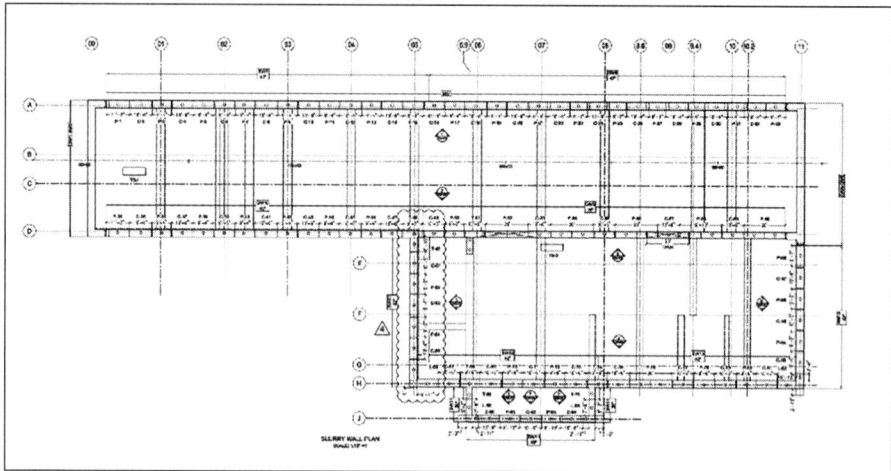

Figure 11. Yerba Buena/Moscone Station diaphragm wall plan

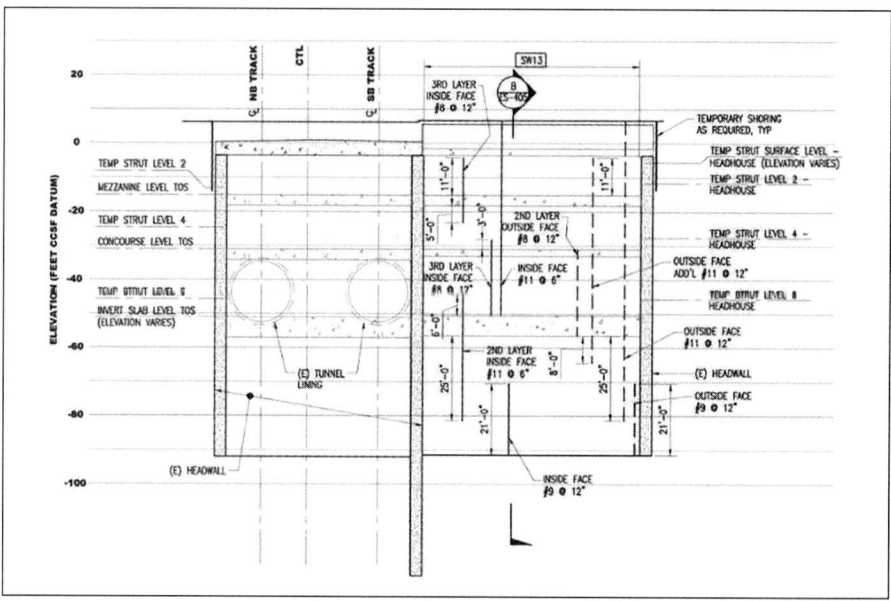

Figure 12. Yerba Buena/Moscone Station diaphragm wall section

wall panels were up to 42 meters deep. The reinforcing cages consisted of #36 @ 152 mm on both sides along with #16 ties @304 mm. In addition to the basic rebar, each cage contained thousands of form savers and couplers to connect the walls to the station floor slabs. The complexity of the reinforcing cages was such that BIM was used to determine the configuration of each unique cage. The high density of reinforcing steel along with the large depth produced very heavy cages at this station as well, which required the use of large cranes to place them into the panels. The cages were constructed at an offsite location approximately 100 km from the site and shipped to the job site in two pieces. Due to traffic considerations, oversized loads could only be delivered to the site in the early morning hours.

As the diaphragm walls were founded in the soft material of the Bay Mud, the toe of each panel was post-grouted through steel sleeve pipes installed with the reinforcing cages. After completion of the diaphragm walls, the excavation was supported utilizing 3 levels of heavy internal bracing that was installed in lifts as the station excavation progressed down to subgrade.

Union Square/Market Street Station (UMS-1253). This station is located along the southern end of the contract alignment. It is located on Stockton Street and between Geary and Ellis Streets. The site is 256 meters long by 15 meters wide. The bottom of excavation is approximately 30 meters below ground surface. This station was designed to be supported utilizing a battered tangent pile system that needed to be angled away from the centerline of the station to accommodate the excavation below which is wider than the space that was available at street level in order to not make contact with the existing tube tunnels below ground. Tangent walls, by design, are permeable so a jet grouting system was installed along the back (unexcavated) side of the wall to provide a water cutoff. As sections of this tangent wall were completed, the jet grouting operations began.

Four distinct test programs were performed to verify the installation process as well as to confirm the effectiveness of the parameters of the jet grouting to meet the project requirements for strength, column diameters and permeability.

A total of 307 jet grout columns having diameters of 1.1 and 1.5 meters were installed. The total volume of treated material was 15,780 cubic meters and column depths were 43 meters. Jet grouting was constructed using a two phase system due to project requirements. A pilot hole was initially installed via traditional rotary drilling techniques and jetting was subsequently conducted through the previously drilled pilot hole. Pilot hole drilling operations were inefficient due to difficulties drilling through abandoned basement walls and through concrete over pours from the tangent pile installation. The jet grout cut-off was initially designed as a two row system to insure quality, however, due to schedule constraints, a single row system with larger columns was implemented for the completion of the cut-off (see Figure 13).

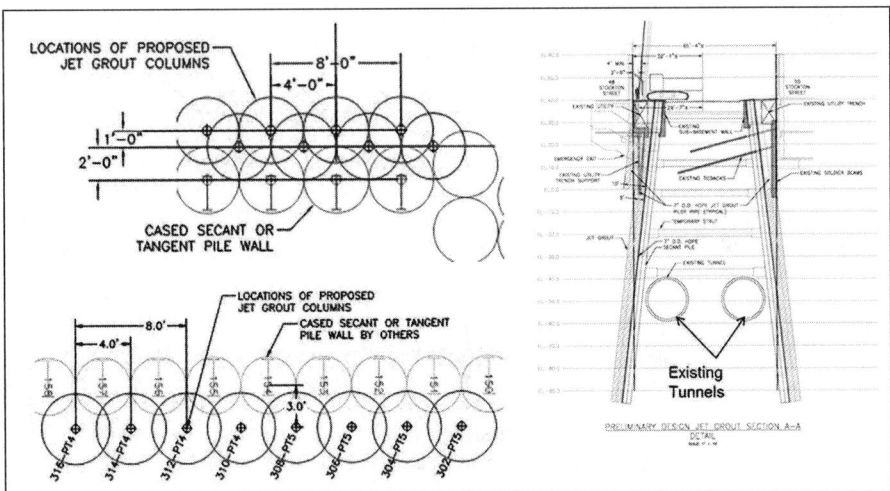

Figure 13. Union Square/Market Street Station column arrangements

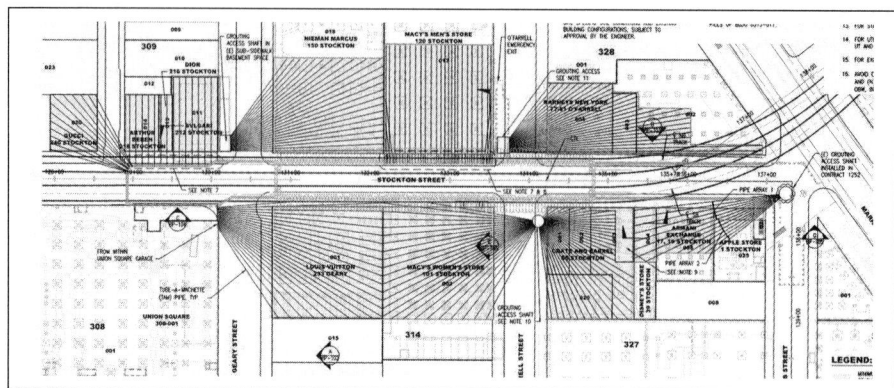

Figure 14. Union Square/Market Street Station original compensation grouting plan

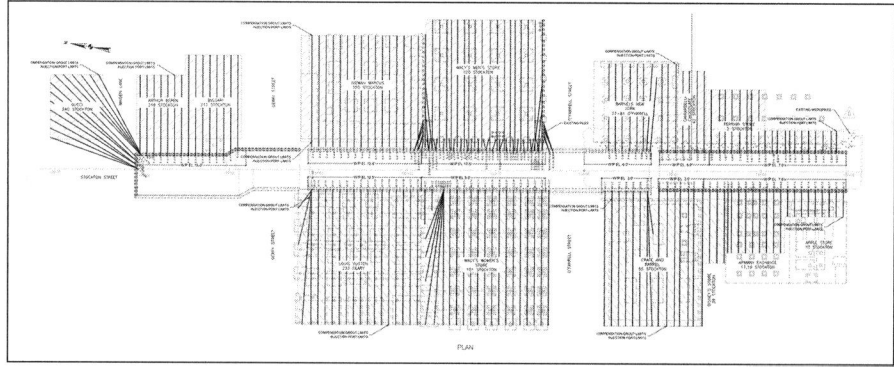

Figure 15. Union Square/Market Street Station actual compensation grouting plan

In addition to the jet grouting cutoff wall for the tangent pile wall, compensation grouting was also required to mitigate potential settlement of surface structures from the station excavation operations. The compensation grouting was originally designed to be constructed from circular shafts as in Figure 14. After award, Bencor proposed an alternate approach and after some discussions with the authority, the decision was made to conduct the compensation grouting operations from inside the station excavation that provided time and cost savings as well as eliminating the need to install multiple access shafts that would also have to be backfilled. Careful consideration was taken in the layout of the boreholes to avoid the reinforcing steel in the tangent and secant piles installed as a part of the station and existing utility and building support structures as shown in Figure 15.

Compensation Grouting Statistics:

- Total Grout Holes = 209
- TAM Installed = 5,975 meters
- Grout Injected = 772,224 liters
- Treatment = 15 Buildings – 11,781 square meters

CONCLUSION

Phase 2 of the Third Street Light Rail presented difficult geotechnical conditions, complex site logistics and required close coordination between the phases of work. Projects that require multiple types of specialty geotechnical construction techniques that are required to be performed by several subcontractors during the same time-frame typically lead to problems for all involved. Competition for space in the same work areas usually leads to schedule impacts, resulting in delays and claims. On this particular project, the limited availability of staging and laydown areas along with small sites in a densely populated downtown setting would have only exacerbated such conditions. The diaphragm walls, jet grouting and compensation grouting work were all conducted by a single specialty geotechnical contractor on this project. This was beneficial to the project on multiple fronts:

- Location and angle of entry for compensation grouting holes through the diaphragm wall at Chinatown station was based both on the wall reinforcement design and compensation grouting requirements. Several iterations of hole locations were performed to provide adequate coverage for the compensation grouting while maintaining the constructability of the diaphragm wall.

- Jet grouting and compensation grouting were completed simultaneously on Union Square/Market Square Station. In some cases, due to urban congestion, crews were shared between the two operations to maximize efficiency. Both operations shared the same space and the benefit of in-house coordination minimized productivity impacts and space conflicts.

- A continuous project management team provided a consistent form of contact and communication with the Prime Contractor and Owner during all the construction.

Risk Reduction, Management, and Mitigation from Experience-Based Learning During Construction of Cross Passages, Seattle, Washington

Sandeep Pyakurel ▪ Gall Zeidler Consultants
Walter Klary ▪ Gall Zeidler Consultants
Vojtech Gall ▪ Gall Zeidler Consultants
Nate Long ▪ Jay Dee Contractors, Inc.
Anthony Pooley ▪ Jacobs Engineering Group

ABSTRACT

Cross passages are critical elements in transit and highway tunnels, providing a means of safe emergency egress between adjacent running tunnels. Although usually short in length, they are often technically challenging and can pose significant construction risks. Two recent projects in Seattle—Sound Transit's University Link Light Rail Contract U230 and Northgate Link Extension Contract N125—involved cross-passage construction between twin single-track Metrorail tunnels. This paper describes risk reduction, mitigation and management experience gained from the U230 and N125 contracts during construction of cross passages. Specific emphasis is given to challenges associated with excavation in glacial deposits under high ground water pressure and the ground improvement measures implemented, which included dewatering, grouting and ground freezing.

INTRODUCTION

Cross passages are critical safety elements in transit and highway tunnels, providing refuges or a means of egress between adjacent tunnels during emergencies such as fire in a tunnel or any incident which results in the closure of a section of a tunnel. For this reason, placement of cross passages along the tunnel alignment has important safety implications and must be carefully considered. National Fire Protection Association Standard "NFPA 130: Standard for fixed Guideway Transit and Passenger Rail Systems" requires cross passages to be constructed between the main tunnels for safety and evacuation. For twin bore tunnels, cross passages may be used in lieu of emergency exit stairways to the surface, at a maximum spacing of 244m (800ft.). They require minimum internal dimensions of 1120mm (44in.) in clear width and 2100mm (7ft.) in height.

Two recent projects—Sound Transit's University Link Light Rail Contract U230 and Northgate Link Extension Contract N125 involved construction of cross passages between twin single-track Metrorail tunnels in Seattle (Figure 1). This paper describes the challenges encountered during the construction and methods used to mitigate and minimize the construction risk and experience gained to manage risk during the cross passage construction. Specific emphasis is given to difficulties associated with excavation in glacial deposits under high ground water pressure, the methods used to control ground movement during excavation and implementation of the ground improvement program, which included dewatering, grouting and ground freezing. U230 and N125 are major tunnel construction contracts, forming part of Sound Transit's University Link and Northgate Link projects respectively. Both projects are

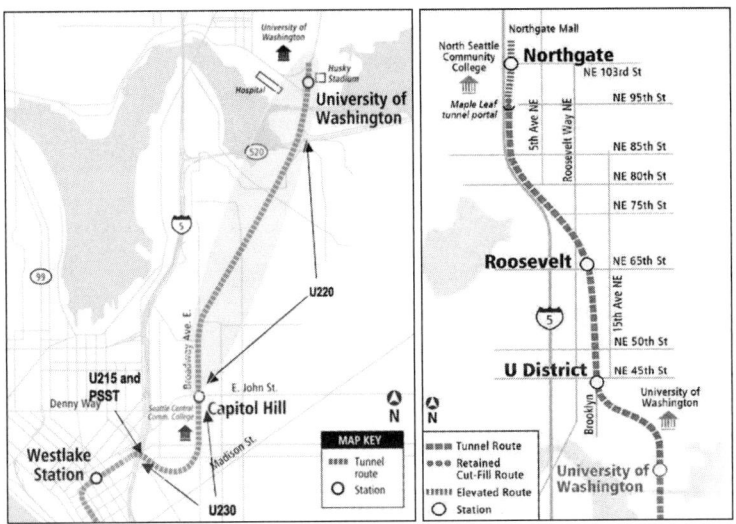

Courtesy: Sound Transit
Figure 1. Alignment Locations for U230 and N125

part of a large-scale expansion of the Seattle area's light rail system. U230 which was completed in 2013 included one mile long twin bore tunnels running between Downtown Seattle and the Capitol Hill neighborhood to the north with five cross passages between the TBM tunnels. N125 is expected to be completed in 2018 and includes approximately 3.4 miles of twin bore running tunnels and 23 cross passages running from University of Washington to the Maple Leaf Portal in north Seattle. The main running tunnels were bored using Earth Pressure Balance TBMs, and lined with a single pass, gasketed segmental lining, 10 inches thick. The finished internal diameter of each tunnel is 18ft 10in.

Geology and Ground Conditions

The geology of the area consists of soft ground deposits comprising both glacial and non-glacial deposits of the quaternary period overlying tertiary volcanic and sedimentary bedrock. The area was subjected to several glaciations and at the project area, thickness of the advancing ice sheets exceeded 3000ft. leading to the soil deposits being over-consolidated from very high overburden.

Both glacial and non-glacial deposits consist of clays, silts, sands and gravels in various proportions, combinations and densities. The distinction between the glacial and non-glacial deposits is made based on observation of sediment type, textures, sedimentary structures, amount of organics present, and identification of old soil horizons and other geologic indicators. Boulders are also present in both glacial and non-glacial deposits with higher amount in tills and diamicts of glacial deposits and along erosional contacts between different soil units. The groundwater system mostly comprises of aquifers and aquitards and there are changes in hydrologic heads when transitioning from one hydrologic regime to another.

Due to the considerable variability in the soil units, the ground conditions were described in terms of soil groups (SG) that exhibit similar behavior and characteristics. Each soil group was comprised of several geologic units which were based on soil index properties, particle size, Atterberg limits along with strength and deformation

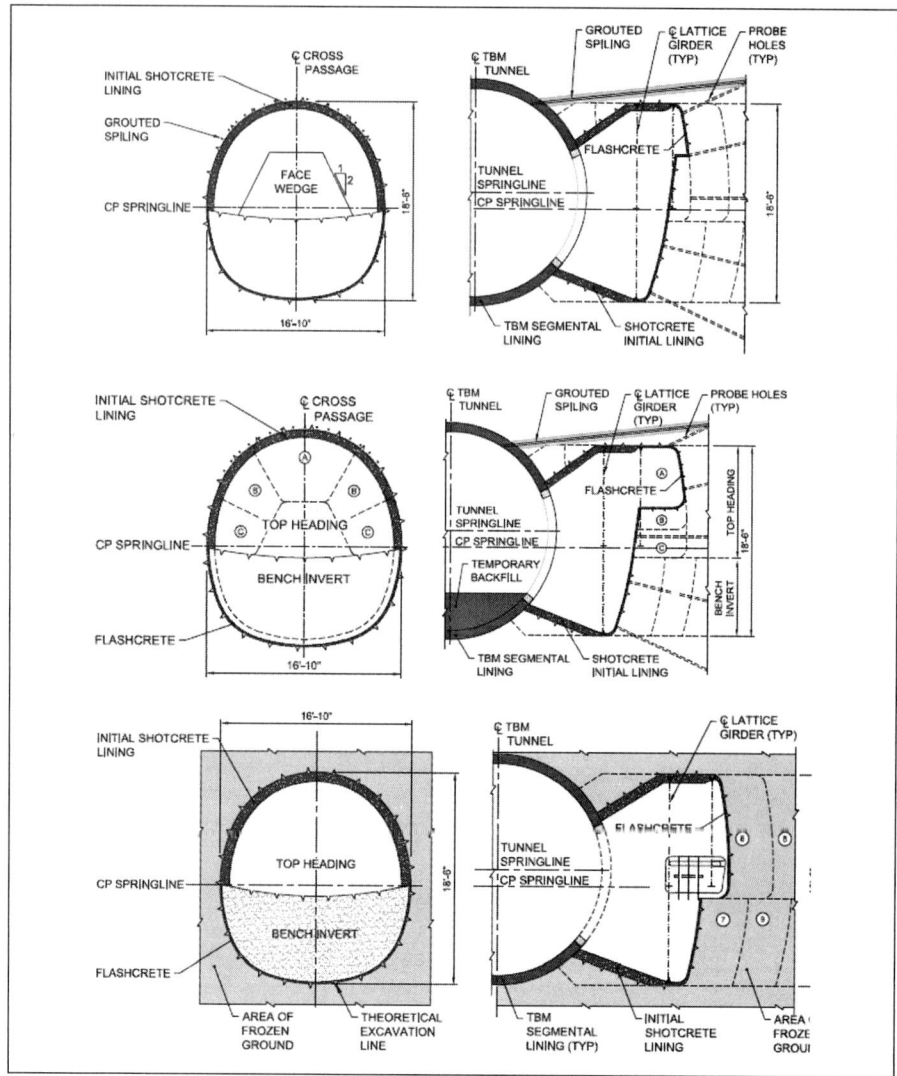

Figure 2. Ground Support Categories 1, 2 and 3 (from top to bottom)

properties. Most of the cross passages were excavated entirely below the groundwater table in the glacial deposits which were grouped into Engineering Soils Units comprising of Till and Till-Like deposits (TLD), Cohesionless Sand and Gravel (CSG), Cohesionless Silt and Fine Sand (CSF) and Cohesive Clays and Silts (CCS).

Ground Support Categories. The initial design of the cross passages was based on data from borehole logging, pumping tests including permeability of the ground and interpretation from the geotechnical baseline report. Based on these findings, the cross passages were categorized into three different support categories (Figure 2) to reflect the soil and groundwater conditions and anticipated behavior for the corresponding ground classes.

Ground Support Category 1 comprised of systematic pre-support and within a competent ground that did not require any extra support measures.

Ground Support Category 2 comprised of systematic pre-support accompanied with pocket excavation and dewatering prior to excavation.

Ground Support Category 3 comprised of excavation in treated ground using jet grouting or ground freezing methods. However, during construction ground freezing was chosen as a preferred method along with elimination of spiles and pre-support.

Support Category 1 did not require any ground improvement due to the soil competency, Support Category 2 required dewatering of the CSG, CSF and TLD when ever encountered. The anticipated soil types in most of the Ground Support Category 2 cross passages had the inherent potential for fast raveling to flowing when under pressurized groundwater and exposed in a free face. The more cohesive clay and clayey materials were expected to display better stand-up properties in the tunnel face. For the Support Category 3 cross passages, which were constructed across or in close proximity to soil contacts between fine and coarse grained deposits or in large aquifers where significant drawdown was required relative to the saturated thickness, ground freezing was implemented to improve the ground around the cross passage. Support Category 3 required ground improvement through ground freezing because of a presumed inability to effectively dewater the soil due to the high flow rates and localized boundary conditions.

The U230 and N125 cross passages have similar structural support systems, comprising shotcrete and lattice girders as the initial lining and cast-in-place concrete final linings. However, their geometries differ slightly (Figure 3). The U230 cross passages have a "dog-bone" longitudinal profile, with larger cross-sections at each end, adjacent to the running tunnels and a smaller cross-section in the middle. The section varied in height from 13ft 2in to 15ft 6in and in width from 12ft 10in to 14ft 10in. The N125 cross passages had a uniform, slightly larger cross section, being 18ft 10in high and 17ft 2in wide. This larger cross-section is close in size to the running tunnels and makes the N125 cross passages some of the largest in North America, relative to the running tunnel size. The larger, uniform cross section provided more internal space for permanent equipment, and made the installation of waterproofing and reinforcement for final lining more straightforward.

The initial lining comprised of two inches of a steel fiber reinforced sealing shotcrete layer (flashcrete) and six inch sprayed fiber reinforced shotcrete with exception of four inches of flashcrete for the Support Category 3 cross passages. The additional two inch thickness accounts for sacrificial shotcrete near the contact with the frozen ground; which is unable to develop full strength due to the cold temperature from freezing. The final lining comprised of 10 inch thick cast in place concrete lining. A waterproofing membrane was used between the initial and final lining system. The excavation profile was maintained using pipe lasers and lattice girders.

Cross Passage Excavation and Support

Probe Drilling. Probe holes were drilled prior to installation of pre-support to verify geology before excavation of the cross passages (Figure 4). The probe holes were arranged in a systematic manner to cover the entire excavation face area and a zone around the excavation perimeter, and inclined to ensure that as many granular layers/lenses as possible could be intercepted in one boring.

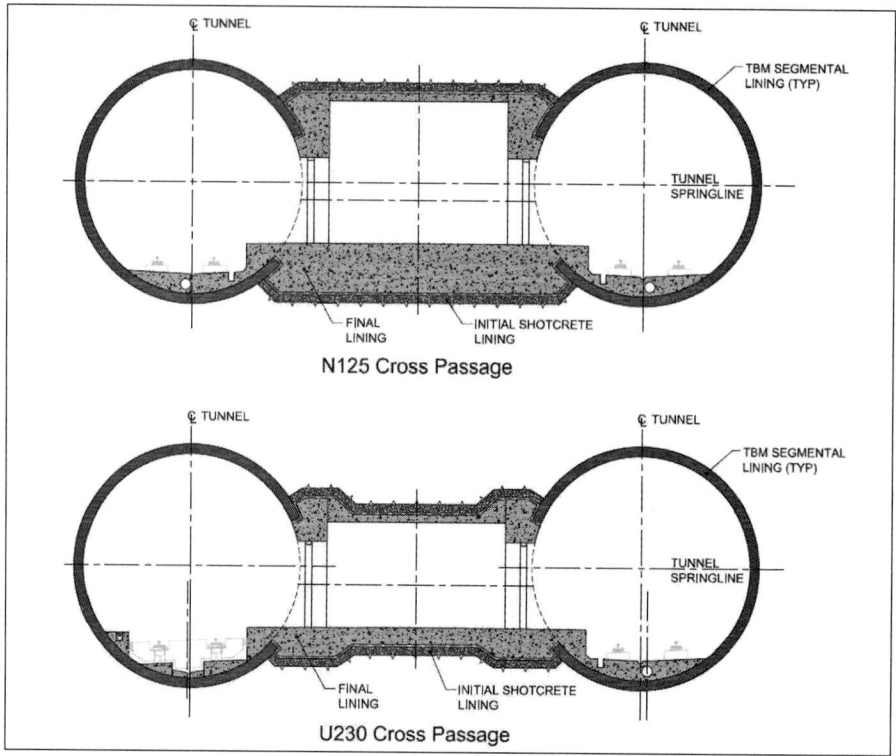

Figure 3. Typical longitudinal section along the cross passages

For Support Categories 1 and 2, a minimum of nine probe holes (6 in the top heading and 3 in the bench/invert) were drilled with the final number depending on the actual conditions found during probing. These probe holes were converted to drainage holes whenever groundwater inflow was encountered and used to dewater the soil. In some cases, they were attached to a vacuum system for dewatering.

For coarser materials with low flow rates gravity drainage was sufficient. For fine-grained material (silt, fine sand) or soils with more clay content, vacuum depressurization of the soil was used to extract groundwater from these pipes. During drilling, probe holes were continuously logged to assess the actual ground and groundwater conditions encountered. The logged data were used to assess the need for application of vacuum dewatering or the installation of screened pipes or additional probe/depressurization holes. These dewatering holes were used to depressurize a seven foot zone around the tunnel opening. In case of Support Category 3, only short, small diameter probe holes were drilled to verify the temperature of the frozen soil to ensure competency of the frozen ground.

Pre-Support. Prior to the commencement of break-out from the TBM tunnel into the cross passage, pre-support was installed using a drill rig and consisted of a steel self-drilling hollow bolt (IBO) grouted spiles to allow for grouting through the tube in a systematic manner. Systematic pre-support was installed only for the cross passages in Support Categories 1 and 2. For Support Category 3 cross passages, the extent of ground freezing was sufficient to stabilize the soil and provide pre-support and therefore grouted pipe spiling was not required. However, the effectiveness of ground freezing was carefully verified through thermal couplings and evaluated by the ground freezing engineer.

Cross Passage Breakout Support. Shear Bicone Dowels were installed between the running tunnel lining segments at cross passage breakout locations and the segments were propped using vertical steel propping. The Shear Bicones were used to transfer loads from cross passage opening to the running tunnel lining. The TBM lining segments were saw-cut to form the required opening (10'x10') for the break-out (Figure 5). Segments were broken with a hydraulic hammer and removed. The edges of the segmental lining were protected with temporary wooden protection blocks to avoid damaging the segments. Upon removal of the segments (or parts thereof), steel fiber reinforced shotcrete was immediately applied over the exposed ground surface to stabilize the face.

Excavation and Support. The cross passages were excavated with a staggered heading based on the Sequential Excavation Method (SEM). This method involved the development of a top heading with a face wedge, where required, and a bench / invert to ensure safe tunneling conditions and control the development of any instabilities. At the break-in a smaller temporary opening was excavated, which was enlarged to a full cross passage size after the first two

Figure 4. Ground probing from inside of the TBM tunnel

Removing rebar and bicone dowels

Breaking upper left segment

Exposed grout behind segment

Excavating after removal of segments

Figure 5. Break-in to the cross passage

Figure 6. Lattice girder and shotcrete installation at the top heading

rounds of excavation and support. The full ground support comprising of lattice gird-ers and steel fiber reinforced shotcrete was installed immediately after completion of each excavation round. A two inch thick flashcrete layer was installed at the face and covered all other exposed ground surface immediately after completion of each exca-vation round. Following the flashcrete application, a pre-fabricated lattice girder was installed, followed by application of an additional six inch fiber reinforced shotcrete layer (Figure 6). The flashcrete was an integral part of the eight inch thick shotcrete initial lining. All cross passages were over-excavated by three inches to account for construction tolerances and anticipated deformations. The general excavation and support sequence is:

1. Installation of temporary support for TBM tunnel at break-out location
2. Ground treatment (ground freezing or dewatering), if required
3. Probe drilling with in tunnel dewatering installation
4. Installation of systematic pre-support (grouted pipe spiling) from within the TBM tunnel at the break-out location
5. Removal of TBM tunnel lining segments
6. Excavation and support of the cross passage in a sequential manner includ-ing enlargement

Risks and Challenges Identified During Construction

Both U230 and N125 were constructed in highly over consolidated glacial deposits consisting of clays, silts, sands and gravels in various proportions, combinations, and densities and very high water head. Such heterogeneity and variability in ground con-ditions resulted in considerable and frequent changes in the soil behavior. One of the biggest challenges was uncertainty in the ground conditions. Geology changes are very frequent and can occur over very short distances. For example, at a given cross passage elevation, the boring at one end showed the soil type as gravel while other end showed sand. This leads to a very complex and non-uniform geology along the cross passage alignment (Figure 7). Sometimes face stability would also be a problem due to short standup time during excavation, again due to the varying soil behavior. In some cases, running or flowing ground was experienced in the cohesionless soils or raveling in more cohesive soils.

To mitigate risks associated with uncertain ground conditions, a thorough ground prob-ing program was conducted to identify geology ahead of the tunneling face. Ground probing comprised of both horizontal and inclined probe holes. The probe holes were installed outside of excavation profile of the tunnel to reduce the risk of encountering

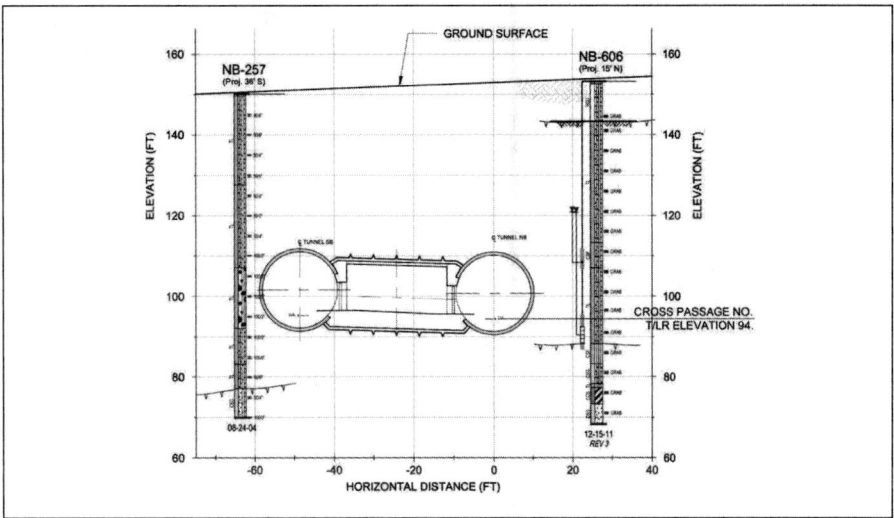

Figure 7. Cross-passage showing variability in geology between two tunnels

unanticipated soil or groundwater conditions. Probing was designed to investigate the soil conditions a minimum of five feet above the crown of the cross passages.

Hydrogeological conditions were also an important factor, posing challenges during excavation. The excavation area showed a varied hydrogeological regime, including a number of aquifers, aquitards and hydraulic connections. The heterogeneity in soil composition led to variations in the ground permeability which was compounded by high ground water pressure. Groundwater heads have been identified up to 74ft above the tunnel. One area may show ground as dry while another, nearby area has very high flow. The presence of groundwater at the face compounds the unstable conditions during excavation. Therefore, managing groundwater flow was a significant challenge.

The risk associated with instability from groundwater inflow was minimized using systematic dewatering. Surface dewatering was used to dewater ground associated with coarser soil deposits particularly in areas where larger flows were expected. In the case of fine grained soil deposits, dewatering was implemented from inside the TBM tunnels using well points with gravity drainage. If the gravity drainage was not sufficient, the drainage system was connected to a vacuum pump to suck the water from narrow pores of the fine grained soil. Water encountered during excavation was collected using drain mats and pipes. Typically, nine well points were drilled from the tunnel and additional well points were installed as needed, based on the results of the probe drilling and observed water inflow. The ground was depressurized over a minimum of seven feet zone outside the cross passage excavation boundary prior to starting open face excavation. Dewatering was successfully achieved using this combination of surface wells, gravity drainage and a vacuum dewatering system. It was important to run the dewatering system until completion of the final lining.

Although, dewatering works well in areas with lower permeability, it becomes expensive and risky to operate in soils with high permeability that have hydraulic connections with high groundwater recharge zones. In such scenarios, dewatering requires very long pumping times and will produce large quantities of pumped water which needs to be disposed of properly. Managing and removing such a high volume of water becomes a costly operation, adding to the project cost since there are high discharge fees to the local combined sewer system.

As a means to minimize risk, pumping tests were conducted before each cross passage excavation to assess the ground permeability and possible water inflow. If the pumping test results showed groundwater flow to be very high, making dewatering impractical; ground freezing was implemented in lieu of dewatering. This change reduced risks associated with managing a high volume of groundwater and eliminated costs for its treatment and disposal.

Therefore, as a mitigation measure, thorough testing and a detailed study of hydro-geological nature of the associated ground water including flow rate, permeability and transmissivity has to be conducted. Detailed investigation and planning is required if high permeability is suspected in certain areas to assess the viability of dewatering and the recharge rate of the pumped aquifer. For example, the top of the permeable layer during a U230 excavation was encountered above the bottom of the excavation contrary to the anticipated 15 ft. to 20 ft. below the bottom of excavation. This led to installation of more dewatering wells. Such scenarios were expected also during N125 and to avoid this, probe drilling and pumping tests were carried out to assess the ground for further ground treatment in terms of dewatering or ground freezing. A backup power systems were also arranged to provide an uninterrupted power supply in case of power outage or failure, to ensure continuous operation of the dewatering pumps.

Ground Freezing. Ground freezing is a method of ground treatment where the ground is frozen to provide stability during excavation. Freezing converts the in-situ pore water into ice which binds the soil particles together and makes the ground stronger and impermeable. Freezing increases both strength and stiffness of the ground. Ground freezing is a proven technology which was originally developed and used in Germany in 1883 for a shaft sinking project in a coal mining application (Schultz, 2008).

Ground freezing was used for N125 cross passages where pumping test results indicated very high flow rates and where layered geology made dewatering impractical. Ground freezing was implemented at eleven locations using two different methods: ground freezing from the surface and ground freezing from inside the TBM tunnel. Ground freezing from the surface was performed by installing vertical zone freeze pipes from the ground surface and short angled haunch freeze pipes through the tunnel liners. The short pipes were to maintain the freeze adjacent to the tunnel during excavation where the freezing is most susceptible to the warmth from the TBM tunnels. Temperature monitoring pipes were installed to actively monitor the frozen zone. The chilled brine for the vertical zone freeze pipes was supplied by chillers located on the ground surface whereas chilled brine for the haunch freeze pipes inside the tunnel was supplied by small chillers located in each tunnel. The annulus around the freeze pipes was grouted over the freeze depth range to create effective contact with the surrounding soil. Such method is particularly useful if cross passage construction is in critical path, however it has challenges with logistics and surface restoration. The surface installation had to remain in place for the duration of the freeze, requiring temporary road closures and other traffic restrictions. Other third party considerations which complicated this method included the need to obtain power drops from the local utility provider, community concerns and the extensive street restoration required at each location once the freeze was decommissioned. Five cross passages on the N125 contract utilized ground freezing from the surface as a primary means of temporary ground stabilization. The freeze design was intended to freeze the soil between the

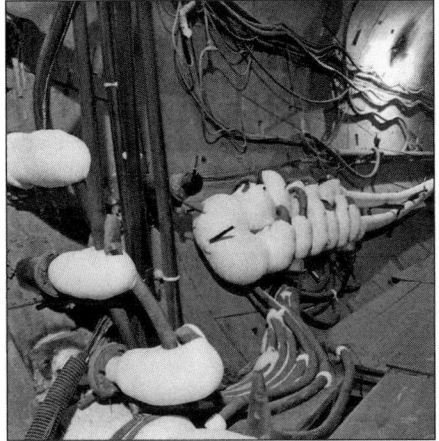

Figure 8. Ground freezing from inside the TBM tunnel

two running tunnels to 20ft above and below the tunnel springline at a minimum distance of 13.5 feet either side from the cross passage center line.

Ground freezing from inside the TBM tunnel is implemented by installing horizontal freeze pipes around the periphery of the cross passages (Figure 8). This process employs primary refrigeration plants to chill a secondary coolant which is continuously circulated through a closed-loop distribution manifold and refrigeration pipes installed within the tunnels. The entire system is a closed circuit with no materials injected into the ground. Horizontal freeze pipes were drilled from the southbound tunnel and short inclined pipes drilled from the northbound tunnel for each cross passage. Ground freezing from inside the tunnel is particularly useful when cross-passage construction schedule is not on the critical path, and it also eliminates significant third party tasks and surface restoration at the ground surface. This method also simplifies excavation since freeze pipes are outside the excavation profile and do not interfere during excavation as in the case for the surface ground freeze method. Six cross passages on the N125 contract utilized in-tunnel freezing as a primary means of temporary ground stabilization.

As of December 2016, eight of the 11 cross passages have been excavated using ground freezing method, with three remaining to be excavated. Even though ground freezing provides stable ground for excavation, there are challenges and risks associated with the ground freezing operation.

The biggest challenge with the ground freezing operation is the coordination between the ground freeze contractor and SEM crew to prevent damages to the freeze pipes. The freeze pipes could easily be damaged by the excavator releasing the brine inside the pipe. This brine could thaw the neighboring frozen soil mass triggering tunnel instability. Care should be also given while reconnecting freeze pipes from the top heading to the invert during excavations that utilize the surface freezing method.

Such risks can be managed with proper coordination between the SEM crew and ground freeze contractor. Using hand excavation around the vicinity of the pipes and careful supervision with the SEM engineers on site helps to prevent damage to the pipes. Possible damage to the freeze pipes and leaking of the freeze brine can also be avoided by conducting an as-built survey of the freeze pipe installation prior to excavation. Experience showed this was critical especially at locations where only two to three feet of frozen ground remained.

Additionally, ground heave from the soil freezing induced deformation in the TBM tunnel lining. In one instance, the TBM tunnel lining underwent a maximum movement of 2.75 inches. In an effort to minimize tunnel deformation horizontal struts were installed at this location in the TBM tunnel to restrain tunnel movement. Further, thawing of the frozen ground may led to additional movement but these impacts are unknown at this time. Ground heaving from the surface freeze also posed a significant challenges since the heaving had potential to damage sensitive surface installations, particularly utilities. At surface freeze locations, near-surface frost heave of 0.5 to 2.0 inches was observed during the freeze down process. The heave was observed via surface and near-surface settlement monitoring, and utility settlement points. It was also observed as an apparent trend in extensometer readings; the surface monument of the extensometer moved upwards causing an apparent downward movement in all of the subsurface extensometers. The rate and amount of heave showed correlation to the brine temperature at any given time. Generally, the lower the brine temperature, the more rapid the trend in heave.

To manage risk associated with ground freezing to control ground movement, sufficient monitoring mechanisms have to be in place with pre-defined trigger levels. The trigger levels dictate further action to be taken once set levels are exceeded such as adjusting the temperature of the brine and selectively turning off parts of the freeze temporarily or permanently, as allowed by the ongoing cross passage construction. Attempts to mitigate the heave were also made by installing heat trace tape and circulating warm air into the annulus of freeze pipes, but these measures showed no significant beneficial effect.

Monitoring development of the frozen soil around the vicinity of the cross passage are necessary to ensure ground achieves the required stability A drainage test from inside the tunnel has to be conducted to ensure sufficient tightness of the frozen ground mass. Attention should also be given to the groundwater flow velocity since high groundwater velocity retards the rate of ground freeze. In such circumstances grouting could be adopted in addition to the ground freezing but this was not required on N125.

Before break-in, the extent of the frozen zone was evaluated and discussed with the ground freezing contractor, the Designer of Record and the CM Team to ensure the frozen ground had achieved its design requirements. Additionally, a second redundant power supply system for ground freezing is needed to ensure uninterrupted operations of the chiller plant and freezing system. There were a few instances of power outages and backup systems were utilized to run the ground freezing smoothly which could otherwise comprise the integrity and structural support of the frozen ground mass.

The size of the N125 cross passages in relation to the main TBM also posed construction challenges during installation of the pre-support, since the sizes of the cross passage and TBM tunnel were very similar. The original design employed horizontal spiles as a pre support measure which was impossible to install near the cross passage crown during break-out since drill rig could not be positioned at such high levels within the tunnel. As a solution, those spiles were installed at an angle going slightly upward from the cross passage crown. Spiles were also designed to be installed at each heading through the lattice girders. Due to the geometry of the cross passage opening, this installation is basically impossible. In this instance it was decided to install full length spiles from the running tunnel, to ensure a complete canopy across the entire length of the cross passage.

One important aspect of addressing challenges and risk mitigation during cross passage construction is communication and coordination between various stake holders including contractor, SEM crew and owner's representative. Daily site meetings were held between SEM crew, contractor, design teams and owner's representative to ensure efficient communications and planning for each day of operation including discussion of construction progress, encountered difficulties during construction and remedial measures. These meetings greatly helped to allow different crew members working synchronously during construction to avoid conflicts in schedule and efficiently utilize logistics.

CONCLUSIONS AND LESSONS LEARNED

Cross passages can be constructed safely in very challenging ground, provided the construction is commenced with careful planning and implemented after proper knowledge of the ground is obtained from ground probing and test results that supplement the geotechnical baseline information. Experience gained from the U230 showed a planned approach needs to be implemented before proceeding with excavation, especially in terms of ground probing and ground treatment. Several challenges were encountered during excavation which were successfully addressed with modification of the ground support systems and ground treatment. Overall, the cross passages at U230 (5) and most of the cross passages at N125 (23) have been successfully constructed without any major delays or issues despite significant construction challenges posed by difficult ground conditions. The following list summarizes important lessons learned during construction of the cross passages.

- Daily site meetings between the SEM crew, contractor, design teams and owner's representative are considered to be very important to ensure efficient communications and planning for each day of operation including discussion of construction progress, encountered difficulties during construction and their remedial measures.

- Probe drilling is very important to verify ground conditions ahead of the face, especially where ground conditions can vary significantly between nearby locations. Probing aided to confirm the ground support type for particular cross passages. Further, probe drilling becomes most valuable if done as early as possible, to give time to react if anything unexpected is found.

- Pocket excavation is recommended to limit over-breaks in difficult ground. As many as 23 pockets were excavated in one cross passage to ensure stable excavation face.

- In cohesionless glacial deposits, such as the CSG, the ground has to be completely depressurized or frozen to achieve stable excavation.

- If dewatering is planned, a pumping test should be carried to verify groundwater flow as it will confirm the risks and likely success of dewatering. At certain cross passage locations, results from pumping tests showed very high ground water flow which led to reclassification of five cross passages from Ground Support Category 2 to Category 3.

- Excavation of frozen cross passages requires close coordination between the Ground Freeze Contractor, SEM Superintendent and Excavation Crew.

- Before opening of the segments, probing at the center of the frozen ground is recommended to verify the frozen ground conditions. There were instances where the frozen ground mass was very solid around the periphery of the tunnel but was relatively soft at the center of the face. This shows that even though temperature monitoring showed frozen ground, the center core was not completely frozen which resulted in flow of water requiring additional depressurization. A probe hole at the center of the frozen mass, about a week in advance before the breaking of the segments helps to avoid such situations.

- Backup power system—A second redundant system is required. There were multiple instances of power outages and the backup system had to be used which is critical for both dewatering and ground freezing operation.

- The potential for heaving or bulk expansion of the frozen ground shall not be underestimated since it may exert significant pressures on the lining. In one case the frozen ground moved the lining as much as 2.75 inches and deformed the circular TBM lining into an oval shape.

- Experienced SEM Consultants, Superintendents and Contractors are required for safe, high quality and on time completion of large and complicated projects, such as U230 and N125.

REFERENCE

Schultz, M.S., M. Gilbert, and H. Hab. 2008. *Ground Freezing: Principles, Applications and Contracting Practices*. Tunnels and Tunneling International. September 2008.

Closing the Gap for Bogotá River Sanitation System Tunnels (Colombia)

Michael B. Gilbert ▪ CDM Smith
Harlem Suarez ▪ CDM Smith
Mahmood Khwaja ▪ CDM Smith

ABSTRACT

Located within challenging terrain, the Interceptor Tunjuelo-Canoas (ITC) (4.2 m) and the Emergency (3.2 m) tunnels form 11 km of underground conduits of the Bogotá River sanitation system. Vertically offset, the tunnels were to be connected via a pump station; logistical and contractual challenges resulted in a two-year construction delay. CDM Smith, subsequently retained by the owner, Empresa de Acueducto de Bogota (Owner), designed the interconnection as a vortex drop-shaft structure. This paper documents project and geotechnical synopsis; novel approach to vortex chamber construction through layered claystone and sandstone geology, without utilizing working shafts; and, modifications to the segmentally lined tunnels.

INTRODUCTION

Located within challenging terrain in Bogota, Colombia, the ITC is a 8.9 km long, 4200 mm ID tunnel with 250 mm thick precast segmental lining. Almost aligned horizontally along the centerline of the ITC is the underlying 2.2 km long Emergency Tunnel (ET) that has 3200 mm ID with 200 mm thick precast segmental lining. Combined, they form almost 10 kilometers of underground conduits of the Bogotá River sanitation system. Initially, it was intended that the two tunnels would be connected via a pump station. The springline to springline distance between the tunnels is 7.21 m. Considering the lining thicknesses the clear distance between the outside of the linings is 2.31 m. Due to problems between Owner and Contractor the tunnels were not connected. The resulting logistical and contractual challenges resulted in a two-year construction delay.

Subsequently retained by the Owner, CDM Smith initially considered an inter-tunnel connection as a vortex drop structure; our computational fluid dynamics model, however, indicated that the vortex structure was not adequately sized to dissipate energy and that the hydraulic design was limited by the 2.3 m of drop that was inadequate for the design flow velocity. As an alternative, CDM Smith proposed a new hydraulic structure that used a straight alignment with a 2:1 (H:V) slope to connect the tunnels. Along this slope energy dissipaters were designed. To provide space for the released air the sloped conduit kept the inverts of both tunnels as initially designed and create additional volume by designing the tunnel as a chamber.

This paper describes the design development of the project, focusing on geotechnical and hydraulic considerations; novel approach to chamber structure construction through layered claystone-sandstone geology, without utilizing working shafts; and, modifications to the segmentally lined tunnels.

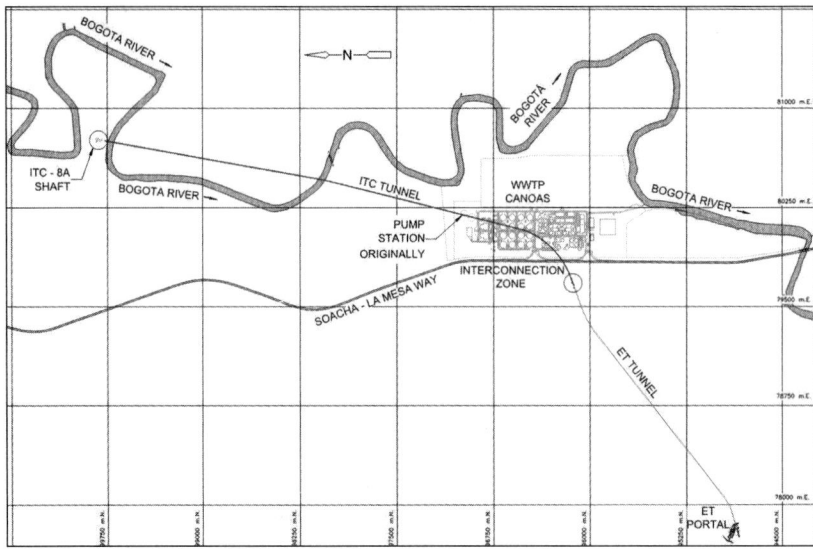

Figure 1. ITC alignment passing under the future Canoas WWTP

ORIGINAL PLANNED FUNCTION OF THESE TUNNELS

The ITC tunnel was constructed to carry the flow from Tunjuelo River and Soacha's municipal waste water system, to Wastewater Treatment Plant (WWTP) Canoas. The ET was constructed as an emergency system in the event that WWTP pump station or the plant itself could not operate.

The initial idea by the owner was that a shaft would be constructed between the two tunnels. Figure 1 shows the alignment of the two tunnels. The present termination of these tunnels is shown in the circled area just to the west of the proposed WWTP. Figure 2 is the existing hillside terrain at this termination point of the two tunnels. The tunnels would then advance into the shaft and the Tunnel Boring Machines (TBMs) would be recovered. The end use of the shaft would be to function as a pump station. This shaft was designed and the contractor was ready to start construction. The Owner, however, encountered significant problems in during land purchase negotiations and, eventually, elected not to purchase the property for shaft construction. Because there was now no receiving shaft the ET was stopped. And the ITC tunnel contractor was instructed to extend that tunnel to a new zone located approximately 600 m to west which place it just above the ET.

Once the two tunnels were at locations established by Owner, the Contractor put forward an alternative for a recovery shaft and future pump station. The horizontal overlap of the two tunnels is approximately 7.6 m as indicated in Figure 3. This overlap meant that the shaft would have to be built to enclose both tunnels, significantly increasing constructability risks and risk to project budget. The Contractor's estimated cost for their

Figure 2. Present site condition at tunnel termination point

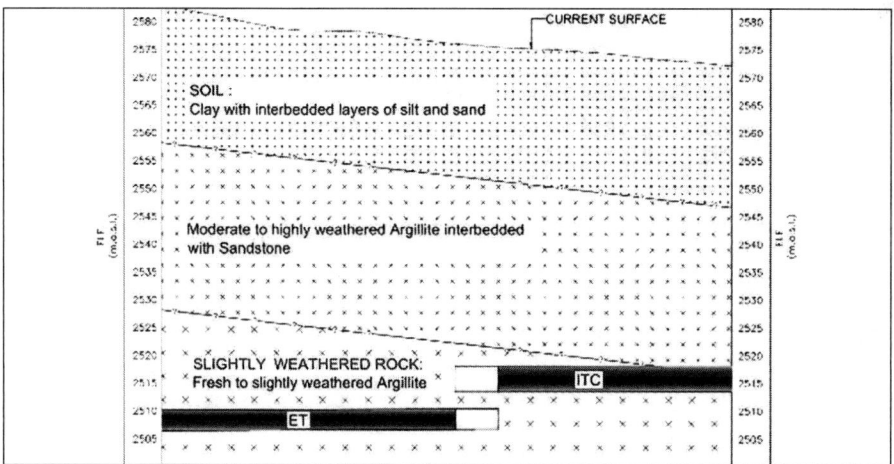

Figure 3. Surveyed position of the two tunnels and subsurface profile

proposed plan confirmed significant increase in project cost without mitigating any of the construction risks; the plan was abandoned and the two parties, through the court system, argued over the project issues including the contract, and retrieval of the TBM and trailing equipment from both tunnels.

Subsequently, Owner requested CDM Smith to develop and design a solution to interconnect both tunnels without the shaft. Initially, the solution had to be designed by abandoning the machines in place; later, the Owner requested that solution needed to be designed with the assumption that the TBMs were to be recovered. CDM Smith supported Owner to find an economical means of retrieving the two TBMs: to meet the legal requirements, both machines would have to be disassembled and the working parts removed back through each of the respective tunnels.

As a first step, CDM Smith mobilized a survey crew to determine the exact location of the two tunnel headings. Starting December 2015, the survey sub-consultant on the design team started survey of both tunnels. This work, consisting of about 6.5 km of tunnel alignments took approximately three weeks to complete, and it served as the base plan for the design.

The survey confirmed vertical separation between ITC and ET tunnels to be 2.31 m, and in-plan overlap of 7.6 m. This is shown in Figure 3. Access to ET is through a portal, located in challenging terrain and accessible only through a single lane road. The tunnel heading is 2.2 km from the portal Access to ITC is through an elliptical shaped, 14m diameter shaft, ITC-8A; the shaft depth is approximately 20 m, with the excavation supported using a system of secant piles.

As part of the inspection the ITC tunnel, the TBM and the trailing gear in both tunnels was tested and found to be functional, even after almost three years of downtime. As the legal proceedings continued, the Contractor maintained ventilation, and lighting within the tunnels; the cutter heads of each TBM were also maintained to ensure working order.

During the downtime involving legal issues, the groundwater flows into the tunnels was observed to be minimal. An estimated inflow quantity of each tunnel was less than

2 liters per minute (lpm) based on a visual inspection of the tunnels within the last year. Seepage inflow from ITC-8A shaft was estimated at 4 lpm.

Also at the maintenance work area of the ITC shaft site there are several pieces of transportation equipment including: locomotive, muck and flat cars and rail tracks of two machines. Based on the appearance of these items the assumption is that they are all functional or could be made functional with minimal maintenance.

SUBSURFACE CONDITIONS

A subsurface investigation consisting of two (2) test borings was conducted at the proposed underground connection site as a supplement to the existing data available from the original tunnel design. In general, the site consists of clayey gravel or sandy clay alluvium to, approximately, El. 2553. This represents below ground surface (bgs) of approximately 22 m— see Figure 3. Bedrock was encountered at, approximately, El. 2553 in in both borings and generally consists of alternating layers of Argillite and Sandstone. The top 32 m of bedrock were highly-weathered, before transitioning to more competent rock quality approximately at El. 2521.

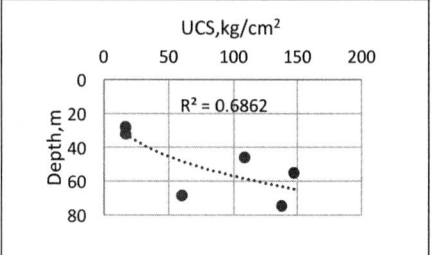

Figure 4. UCS vs. depth

The results of the subsurface investigation indicate that the Argillite layers with the overall profile are up to 6 m thick, and the Sandstone layers are up to 16 m thick. Groundwater was not observed in the test borings.

LABORATORY TESTING

Based on the subsurface exploration and laboratory testing results, three distinct strata with unique strength characteristics were identified.

Geotechnical laboratory testing was conducted to evaluate the engineering properties of the rock encountered at the project site. The laboratory testing program performed supplemented the existing test data. The testing performed on the new borings at the connection site of the two tunnels included Uniaxial Compression Strength and Brazilian Indirect Tensile. These results were added to the database and used to model rock behavior.

DESIGN REQUIREMENTS

The design requirements based on both the court ordered requirements and of the Owner are as follows.

Court-Mandated Requirements

- Inspect the machines, to determine their operating condition;
- Disengage them from the tunnel (i.e., remove all utility connections and support from the tunnel);
- Remove all equipment from the tunnel; and
- Reassemble all equipment on the ground surface under same conditions that was established during the inspection.

Owner Requirements

- Non-turbulent air flow;
- Operating system with minimal long term maintenance;
- Capable of handling a hydraulic flow of 36 m^3/sec in a non-steady condition;
- No shafts at the site; and
- Preference for vortex drop shaft between the two tunnels.

EVOLUTION OF THE DESIGN PROCESS

The design process required very close coordination among hydraulics, structural, geotechnical and tunnel engineers to develop a functional design that would allow construction to proceed and machinery to be retrieved—all to be done without the benefit of a working shaft.

Simple Drop Connection

From a geotechnical and tunneling perspective the simplest solution was to remove the equipment and then excavate a drop hole from one tunnel to the other. However, from a hydraulics consideration this solution was flawed because of limited capacity to handle the flows and volume of air created at the connection point; this option was rejected on those bases.

By-Pass Tunnel

A second alternative was considered to serve as a by-pass tunnel; the tunnel would extend around the tunnel equipment in both tunnels. This by-pass tunnel would start 10 m upstream of the 4.2 m diameter TBM with two curves 12 m radius and end 10 m downstream of the 3.2 m diameter TBM (bottom tunnel) with two curves 12 m radius. The energy dissipation design for this connection consisted of a steep chute, following a procedure proposed by Ohtsu et al in 2004. The principal characteristics are, inclination angle 18 degrees, length 20 m, step height 0.65 m, number of steps 10. At the end of the steeped chute a basin was designed to contain the hydraulic jump before the inlet to the bottom tunnel.

This was only a consideration during the period when a lower court requirement was to abandon the construction equipment in place. When this decision was overruled in favor of removing the equipment, this by-pass alternative was rejected.

Vortex Drop

With this reversed court ruling the Owner's preference to use a vortex drop shaft was then analyzed. A design following the Hager procedure for subcritical flow using a 3.5 m diameter drop shaft with a length of 2.80 m was developed. The dimensions used were a result of survey showing the relative locations of the two tunnels.

Because of time constraints on completing the design the decision was made to advance both the hydraulics design and tunnel excavation designs at the same time. The vortex drop chamber resulted in very complex geometry.

A 3-dimensional (3-D) model of the proposed connection structure was created using FLAC^{3D} 5.01 and details of the model is discussed below.

Table 1. Conservative rock strength values

Unit	Density (kg/m³)	Elastic Modulus, E (MPa)	Poisson's Ratio, ν	Bulk Modulus, K (MPa)	Shear Modulus, G (MPa)	σ_{ci} (MPa)	mb	s	a
1	1700	1900	0.30	1583	731	--	--	--	--
2	2300	4400	0.25	2933	1760	10.6	1.80	0.0067	0.504
3	2300	4700	0.25	3133	1880	12.3	1.80	0.0067	0.504
Concrete	2300	17000	0.20	9444	7083	--	--	--	--
Shotcrete	--	10500	0.25	7000	4200	--	--	--	--
Conservative Rock Strength Values									
2	2300	3080	0.25	2053	1232	7.5	1.80	0.0067	0.504
3	2300	3290	0.25	2193	1316	9.0	1.80	0.0067	0.504
				D=0.1 [1]					
3	2300	1678	0.25	1100	670	9.0	1.80	0.0067	0.504
				D=0.5 [1]					
3	2300	946	0.25	630	380	9.0	1.80	0.0067	0.504
				D=0.9 [1]					
3	2300	497	0.25	330	200	9.0	1.80	0.0067	0.504

Notes:

1. Values calculated after Hoek, E., Diederichs, M.S., 2006, Empirical Estimation of Rock Mass Modulus, International Journal of Rock Mechanics and Mining Sciences Vol 43, Issue 2, pp. 203-215.

DETAILED MODELING METHODS

A total of six (6) design cases were evaluated during the modeling of the proposed connection structure. The cases were selected to evaluate the effect of variations in rock strength, excavation methods and workmanship, and the addition of an initial support system. Each of the design cases is described below. While the stratigraphy and model geometry unchanged between each of the cases, the material properties used in each analysis were modified for each of the six cases to study the ground behavior during excavation. These cases are detailed below, and the material properties used in the various cases are included in Table 1.

In addition, the following assumptions were made as part of the modeling effort:

- The Hoek-Brown failure criterion (Hoek and Brown, 1988) using properties of the intact rock, and accounting for a reduction in rock mass strength by applying factors that describe the characteristics of joints and discontinuities in the rock mass. However, the model assumes a homogenous behavior for each rock unit, and as such, local anomalies due to the presence of discontinuities are not represented. It was decided that this model would reasonably approximate the behavior of the bedrock at the project site.

- It is assumed that the rock mass surrounding the excavation will be fully drained (Hoek and Brown, 1997); however, a scenario, Case 5, was also run using effective stress parameters to evaluate the effect of groundwater taken at the soil – rock interface, approximately El. 2553.

- Bedrock within 1–2 m of the excavation is anticipated to have a strength reduction due to disturbance from blasting or mechanical excavation. The sensitivity of this parameter was modeled to evaluate its influence of the modeling results. It was found to be noticeable but relatively minor with regards to rock behavior.

Case 1

Used "best estimate" value for each rock strength parameter was selected based on the results of laboratory testing.

Case 2

A conservative reduction in anticipated rock strength was used. Appropriate parameters were reduced by approximately 30 percent, approximately one standard deviation of the laboratory testing data set.

Case 3

A disturbance factor of D=0.1 was applied to the rock surrounding the excavation to a distance between 1 and 2 meters from the excavation wall. This disturbance factor is representative of high-quality mechanical excavation or closely controlled blasting in poor quality rock. Conservative strength values were used for this case and also for Cases 4, 5 and 6.

Case 4

A disturbance factor of D=0.5 was applied to the rock surrounding the excavation to a distance between 1 m and 2 m from the excavation wall. This disturbance factor is representative of lower quality blasting.

Case 5

A disturbance factor of D=0.9 was applied to the rock surrounding the excavation to a distance between 1 and 2 meters from the excavation wall. This disturbance factor is representative of very poor quality or production blasting accompanied by significant stress relief.

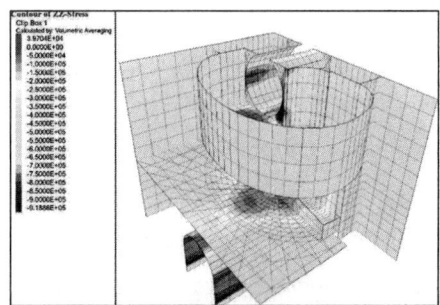

Figure 5. Case 1 stress concentrations

Case 6

Similar to Case 5, except a 100 mm thick shotcrete layer is applied to the excavation walls immediately following the excavation. Assumed strength properties of the shotcrete are shown in Table 1.

The model was used to evaluate the stresses and displacements that are anticipated to develop as part of the construction of the structure as shown in Figure 5 that shows the results of Case 1 and Figure 6 that presents the stresses developed from the conditions imposed in Case 5.

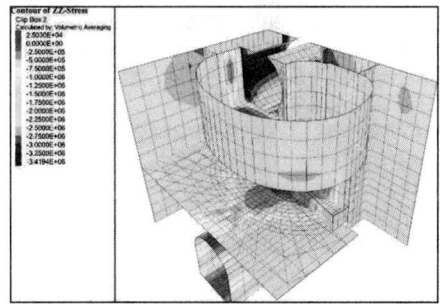

Figure 6. Case 5 stress concentrations

GENERAL MODELING SEQUENCE

These modeling steps were based on a anticipated construction sequence that consisted of:

1. Initial in-situ stresses were established based on our understanding of the stratigraphy and soil and rock properties at the project site. These stresses were used to accurately represent the response of the rock to excavation and construction of the connection structure.

2. The existing ITC was assumed to be constructed instantaneously.We considered this assumption as valid since the ITC was constructed several years ago and any stress change due to the construction of the tunnel has had adequate time to revert to long-term conditions.

3. The connecting adit cross-section was assumed to be constructed with a horseshoe shape in stages using 2 m blasting rounds.

4. The vortex structure was excavated in 5 stages, working around the chamber area in a pie shape excavation following completion of the connecting adit.

5. The drop shaft was constructed in 2 m lifts to connect the vortex structure to the horseshoe connecting drift.

EVALUATION OF VORTEX EXCAVATION MODELING

Based on the geometry of the proposed structure and the strength of the surrounding rock, we anticipated that the excavation can be successfully conducted in general accordance with the sequence outlined. The stress redistribution that is anticipated to occur during and after excavation of the structure is not anticipated to cause major failures; based on numerical modeling results, the deformation within the excavation is limited to negligible amounts. Stresses are not generally anticipated to exceed the failure envelope using the Hoek-Brown failure criterion. Stress contours at the completion of excavation for Cases 1 and 5 are shown in Figures 5 and 6.

In general, maximum displacements are expected to range between 1 mm (Case 1) and 10 mm (Case 5), depending on the actual quality of the rock encountered, and the excavation methods employed. Displacement contours at the completion of excavation these same cases 1 and 5 are shown in Figures 7 and 8.

As shown in the figures, localized zones of generally higher displacements are anticipated to develop at critical areas, including the crown of the ET connection tunnel, where the drop shaft intersects the CT and at select locations along the wall of the vortex structure.This modeling showed that the excavation would be stable and displacements were within an acceptable range. Some local zones of higher displacement may need additional primary support and such support would fall under the contractor means and methods approach.

While, from a structural and geo-mechanics perspective, the vortex option was a viable alternative, results from the concurrent hydraulics modeling, however, did not satisfy design intent. Using CFD, FLUENT program (ANSYS), hydraulic flow within the conveyance system was simulated; the results indicated that the size of the de-aeration chamber was insufficient to carry the peak flow of 36.2 m^3/sec; subsequent

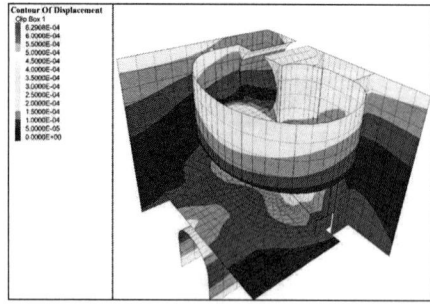

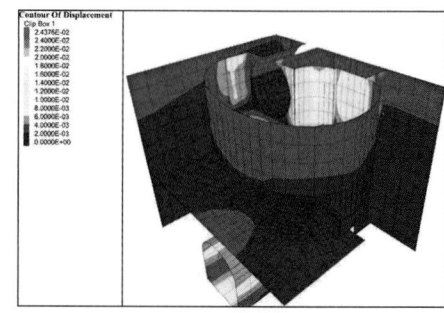

Figure 7. Case 1 displacement   Figure 8. Case 5 displacement

analysis that modelled an increase in the vortex structure size were unsuccessful in satisfying design requirements.

CHAMBER DESIGN

The hydraulic simulations results suggested that the vertical drop of 2.3 m was the bottleneck in dissipating the energy; the vertical drop could not be increased due to relative positions of the existing ITC and ET tunnels. An alternative approach was needed. As a result, a new alternative to dissipate the energy was conceived. The Energy Dissipation Blocks (EDB) alternative is a structure with a baffled chute that dissipates the energy similar to the concept used in previously rejected by-pass tunnel alternative. It differs from the vortex drop shaft by blocking the flow with baffles. The hydraulic requirements to obtain a non-turbulent airflow concluded that a chamber with a clear width of 6.5 m, and a transition in height up to 12.10 m was needed.

As shown in Figure 9 the designed structure is 6.5 m wide with 8 rows of concrete blocks of 1.20 m height spaced horizontally at 1.75 m on center; longitudinally, the blocks are spaced 2.34 m from face to face of blocks. From a geo-mechanics perspective, the construction approach for this alternative was very simple: Remove the tunneling equipment; drill-and-blast a linear ramped connection tunnel from the face of the upper ITC to a desired connection point along the alignment of the lower ET. As the hydraulics design was developed and advanced, modifications to the excavation

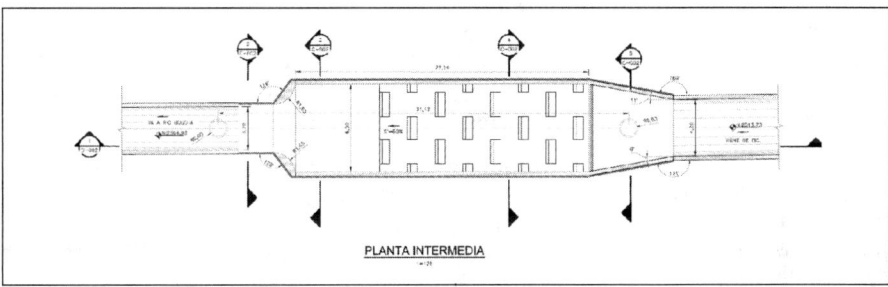

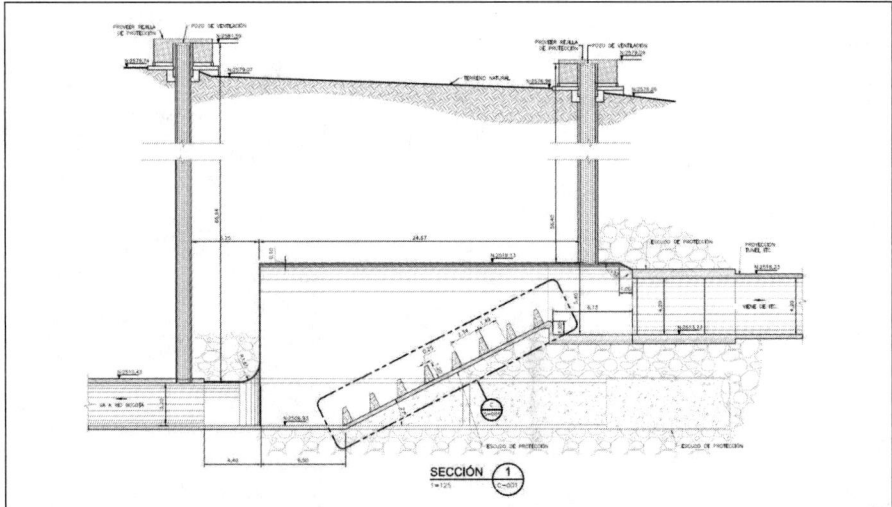

Figure 9. Final design configuration of excavation and gap closure

envelope were made to reflect the hydraulic design requirements. The transition structure needed two vent shafts, one of the two tunnels (ITC and ET); an increase in vertical excavation height. 2-D FE modeling showed that this underground chamber, with a sloped floor, could be excavated by drill and blast, provided the excavation is sequenced. The rock conditions expected for the connection tunnel in the previous design were similar to the roof of the chamber. Initial tunnel conditions and resulting support system were based on RMR value of 52 and Q rating of 6; excavation support required a rock bolt pattern (2 m × 2 m) with a final shotcrete lining (100 mm) with WWF. The same roof support design was utilized.

The specifications require the excavated length of the ET beyond the limits of the chamber to be filled with a low strength flowable fill or grout. This cavity filling will be required to be completed prior to excavation activities for the chamber. The purpose of this sequencing is to mitigate issues with blast waves affecting the integrity/stability of the excavation. The major issues that were considered are as described in the following text.

Starting from ITC easier access for the workers to drill and blast from the higher tunnel down to the lower tunnel. This would involve starting from the existing tunnel face of the ITC. The overall chamber length is 33 m. Tunnel muck would be removed from the tunnel to the ITC-8A shaft, a haul length of more than 4 km.

If the excavation was to commence from the ET the heading, then the haul length is significantly less at approximately 2 km. However, excavation will require tighter control as the chamber invert will be inclined upward. Relatively minor amount of groundwater inflow is expected and can be handled with relative ease, particularly when face advance is inclined upwards. Finally, the excavated material can be hauled through a portal opening, instead of a shaft as would be necessary for the ITC.

For the Chamber excavation, a sequential excavation method is required for stability of the exposed rock walls. The height of these chamber walls was required to control the air flow. Our analyses showed that the excavation is stable with exposed wall height of 4m. Three stages of benching are required. The result of this SEM approach is the bolting and stress field as shown in Figure 10.

Due to uncertainty about the in-situ condition of the rock prior to blasting and after blasting the structural designer decided not to anchor each bump into rock. Instead, it was concluded that it was best to anchor the blocks into the slab. This made for an easier and more economical construction. Because of the limited maintenance that is expected to help mitigate damage to the structure due to the energy in the flow and the potential solids in the waste water, a final lining on the walls and concrete slab with thickness 0.30 m was also designed.

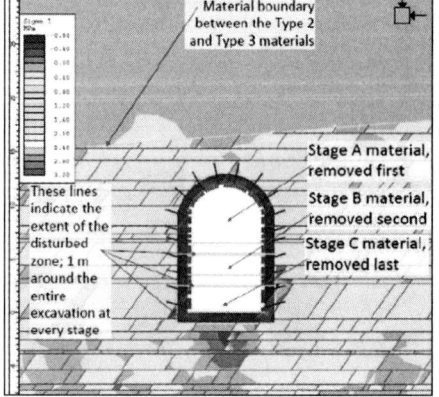

Figure 10. Initial support of chamber

CONCLUSION

Design required very close coordination of geotechnical/structural and hydraulic input to develop the design.

The major challenge was to select the optimum hydraulic structure that was constructible for the given ground conditions. With limited time to advance the design and address the identified technical issues, international conference calls were held every other day to present progress and evaluate modeling output. The result was a design that shows great promise to meet all the court ordered and owner requirements and provide sufficient information for bidders to take to the next step which is a physical hydraulic to confirm the computer modeling. Some minor modifications to the dimensions are expected.

PART  **15**

Grouting and Ground Modification II

Chairs

Adam Curry
Moretrench

Jack Nakagawa
Tutor Perini

Leak Mitigation Grouting for New York Subway Tunnels

Paul M. Gancarz ▪ Sovereign-Thyssen L.P.
John E. Minturn ▪ Sovereign-Thyssen L.P.
Nico J. Grobler ▪ Sovereign-Thyssen L.P.
Deon Van Dyk ▪ Sovereign-Thyssen L.P.

ABSTRACT

Groundwater intrusion into tunnels is a severe challenge encountered by contractors and owners. Significant cost over the tunnel's lifetime illustrates the pivotal role of dewatering and leak mitigation. Polymer-based emulsion grouts have provided permanent cutoff barriers in dams, mines and tunnels for over forty years. Our study addresses the failure of membrane waterproofing systems in mitigating groundwater infiltration, illustrates the technology, performance and material characteristics of polymer emulsions, and how its application to water cutoff grouting compares with available cementitious and resinous systems. Case histories into the use of polymer emulsions for tunnel rehabilitation in New York will be provided.

INTRODUCTION

Preliminary Investigations

Given the tremendous burden tunnel leakage can pose during construction (both as a cost and a scheduling impediment), it is prudent to set up some form of leak monitoring once the tunnel's concrete liner is placed and has had time to cure. This can be done by assigning one or several field inspectors to regularly walk the site searching for leaks—paying special attention to areas which permanently house sensitive equipment, behavior during cold seasons and after a significant rainfall event. In the event that unacceptable tunnel leakage is detected, a significant lead time is gained to address these leaks.

Once the determination is made to address any leaks, project background is valuable in treating them quickly and effectively. Understanding the hydrogeology of the surrounding ground medium—variations in the groundwater table, nearby bodies of water, geotechnical properties of the surrounding rock or soil—all play a role in how water infiltrates the structure. Equally important is the tunnel structure itself. Of particular interest is the initial waterproofing system and the nature of how it was installed, as tunnel leakage is the result of deficiencies to the waterproofing. Knowing concrete thickness, distribution of rebar, stress distributions (and therefore where there is likely to be a concentration of structural cracks)—all of this information is valuable, and can make a difference in the duration and effectiveness of a leak mitigation program.

Polymer-Based Emulsion Grout

Polymer-based emulsion (PBE) was developed in the early 1970s as a means of sealing water with high velocity washout rate as well as very high hydrostatic pressure, typically applied to deep shaft mines in South Africa. In its fluid state, PBE exhibits waterlike properties including a low viscosity, a specific gravity comparable to water and miscibility in water. The particle size of PBE is less than one micron, which allows

PBE to permeate through very small cracks. Due to its waterlike properties PBE is capable of spreading out long distances, greatly reducing the required number of grout holes (Sovereign, 2011).

PBE can be activated in three ways—mechanically, chemically, or environmentally (De Bruin, 1991). Friction due to injection can cause PBE to adhere and flocculate along the crack or annulus being treated. Alternatively, PBE grout set can be chemically accelerated as fast as four seconds with the use of an activator (conversely, set time can be prolonged several days with the use of an inhibitor). When PBE sets, it adheres to surfaces as a durable, rubberlike grout membrane which is highly elastic. Capable of elongating up to 350%, PBE has been documented sealing inflows as high as 200 liters per second under hydrostatic pressures of 2,900 PSI.

Leak Mitigation Grouting in Underground Structures

Once a tunnel lining is placed and excessive leaks are discovered, there are two post-construction grouting methods which can be used to remedy the problem: crack-injection and curtain grouting. Crack-injection, or "negative-side grouting" (Warner, 2004) is used to correct individual leaks. By performing targeted drilling (usually at an angle, stopping before reaching the exterior of the concrete) the individual crack is intercepted. A mechanical packer is fixed to the hole and grout is injected directly into the plane of the crack, sealing the leak. By contrast, curtain grouting aims to treat an entire area which is leaking (usually an earth-bearing wall or a roof slab in contact with the ground) by drilling a gridlike pattern of holes through the exterior of the concrete (Karol, 2003). Grout is then injected into the annular space between the tunnel's structural liner and the surrounding ground medium, creating an impermeable membrane "curtain" for that area. Unlike crack-injection which is a localized solution, curtain grouting is a "coarse" technique which aims to prevent all leaks in a given area. While crack-injection is a less involved process, it is not recommended beginning treatment with this technique as the water will simply migrate to an adjacent crack or the "second-least path of resistance." This paper will explore the use of curtain grouting with polymer-based emulsion for leak mitigation.

Curtain grouting operations are sequential, one following after another in a deliberate fashion in order to push the water in a specific direction. Mini drill rigs used to create grout holes typically consist of an electric drill with a diamond core bit (usually ¾-in. or 1-in.) on a guided drill stand mounted to the wall. All grouting is performed from a central pump room, where all pumps and supply of grout is stored. Air-powered double-acting plunger pumps are used, with discharge pressures from 800 to 2,000 PSI. When curtain grouting with PBE, three pumps are employed (one for grout, one for activator, one for inhibitor). High-capacity 5,000 PSI-rated hydraulic hoses are used, with three separate delivery lines run from the pump room to the treatment area, where grouting is performed through a valved injection manifold.

Prior to grouting, a series of injection tests are performed using dye water to simulate the injection process. This is done to observe which holes will take grout, gauge injection pressures and observe for surface reports. Actual grouting with PBE is carried out sequentially for all holes in the treatment area. For application in tunnel structures, injection pressures are typically kept in the range of 300 and 700 PSI.

Specifications and Contracts

It is important to understand what the desired end state (dryness performance criteria) of a treatment program is. The major challenge faced by designers in laying out specifications for leak mitigation grouting work is that, unlike geotechnical grouting work, it is not guided by an explicitly laid out procedural flow chart or refusal criteria. This issue is exacerbated by a general misunderstanding of how the work is actually carried out. Execution sections of contract specifications should be focused on a particular end state and written flexibly, allowing the grouting specialist to inspect the site and submit his/her own work procedure for approval. Open dialogue should be maintained between the grouting specialist and the designer to address any concerns prior to treatment and to ensure all parties understand the objective of the program. Measurement should be based on a set monitoring period after treatment. Once an area is treated, it is then inspected regularly for any leaks over the monitoring period and payment based upon completing successful milestones.

Contracts are typically unit price; based on area the grouter is responsible for treating. A warranty period is usually outlined during which the grouter agrees to treat any leaks which may recur in the area covered in the contract.

CASE HISTORIES

South Ferry Terminal Complex Rehabilitation

The original South Ferry Loop Station was constructed in 1905 as part of the original IRT subway line in New York City, see Figure 1. This station was decommissioned in 2009, when a new South Ferry platform was constructed beneath the original loop. In October 2012, the new South Ferry station was severely impacted when Lower Manhattan suffered extensive flooding due to Hurricane Sandy. Service was suspended, and the old loop was reopened in 2013 until such time that the new South Ferry could be rehabilitated. Initial demolition of flood damaged finishes and transit systems was finished in 2014, at which time rehabilitation and flood resiliency became the Transit Authority's top priority.

Preliminary investigations were conducted by the owner's engineer to assess the extent of leakage in the station and 700-ft of tunnel running north of South Ferry in late 2014. After all available options were considered, they concluded that polymer-based emulsion was the best option for ensuring a dry station without risk of delay to project schedule. In January 2015 Sovereign-Thyssen L.P. was tasked with performing the extensive leak mitigation grouting program of South Ferry, and the polymer-based

Figure 1. Lower Manhattan–South Ferry Station

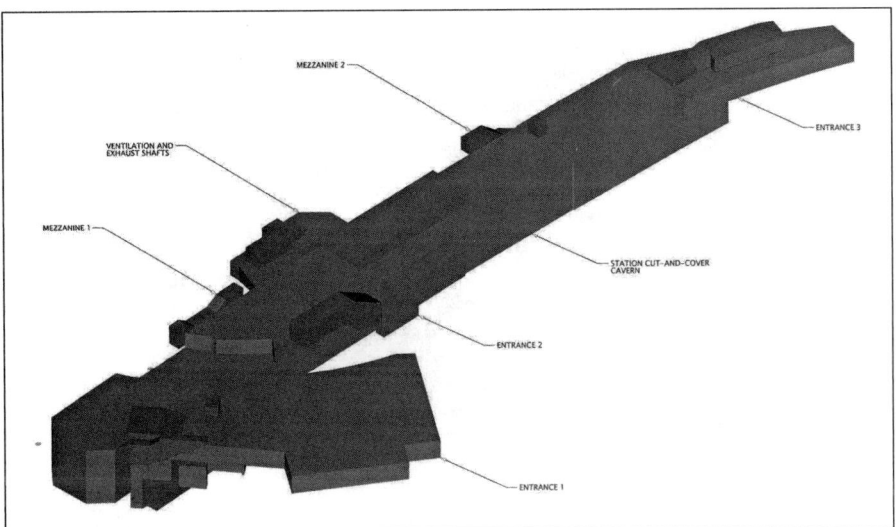

Figure 2. South Ferry terminal complex and station

NOH2O emulsion grout would be used to treat tunnel walls, station walls, roof slabs, exterior walls and slabs for mezzanine and concourse levels, and entrance adits.

At the beginning of treatment, the water infiltration at South Ferry was mostly observed as damp surfaces throughout the tunnel and station. There were many areas where previous attempts at grouting with resins had been unsuccessful, and those resins were being forced out of station wall and roof cracks. The first areas treated were the exterior tunnel walls immediately north of the station portal, so as to form a gasket preventing water from traveling between the tunnel and station. The goal was to push the water south and out of the system. The main station cavern is a cut-and-cover reinforced concrete box roughly 40-ft tall, 55-ft wide and 600-ft long, see Figures 2 and 3. The east and west walls were trea-ted sequentially, working south in 200' × 40' treatment areas at a time. Altogether 470 holes were drilled through a 2'-9" reinforced concrete liner and 36,370 liters of NOH2O were injected over a two month period, treating 41,378 square feet.

Following the successful sealing of the station box, the remaining mezzanine and concourse levels were grouted, as well as all entrances and rail tunnels running north of the station to the project limits. In total, 1,231 holes were drilled throughout the course of production grouting and 60,372 liters of NOH2O were pumped in five months from February to June 2015, treating 91,753 square feet. Additional response work covered by warranty has been performed in ancillary locations throughout the facility since completion of primary treatment. In the months that followed, monitoring of treated areas for leak recurrence was carried out on a periodic basis. Treatment has successfully mitigated groundwater inflow, allowing the general contractor to advance installation of finishes and station systems.

Second Avenue Subway—G3/G4 Tunnel

Upon completion in 2016, Phase 1 of Second Avenue Subway will connect riders in an historic way with Manhattan's Upper East Side—one of the most densely populated and heavily developed urban centers in the world. Although construction for Phase 1 broke ground in 2007, much of the planning and preliminary work for a Second Avenue

Figure 3. Core drilling at South Ferry—station cut-and-cover box

line was done over the last century. The G3/G4 Tunnel (BMT-63rd Street Line) was originally constructed in the late 1980s as part of Route 131-A and runs north from 57th St. and 7th Ave. beneath Seventh Avenue, Central Park and 63rd Street before connecting with Lexington Av/63rd St. Station as shown in Figure 4. This tunnel was designed to bring trains from the 7th Avenue Line to the eventual 2nd Avenue line. Since their construction the G3/G4 tracks have not been put into regular commercial service. Part of the work scope for Phase 1 of the Second Avenue Subway Project was the structural rehabilitation and restoration of rail systems in the G3/G4 Tunnel.

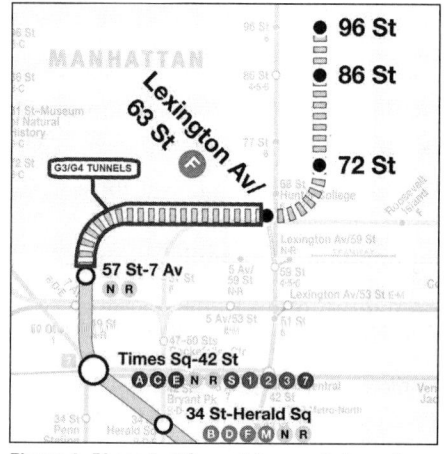

Figure 4. Phase 1 of Second Avenue Subway & G3/G4 Tunnel

The G3/G4 Tunnel runs in very close proximity to the Central Park Pond. Over time the high groundwater table resulted in excessive leaking in this unused tunnel, as seen in Figure 5. In September 2015 Sovereign-Thyssen L.P. was invited by New York City Transit to assess the extent of these leaks and propose a plan for remedying them. After conducting several site investigations and referring to record drawings for the tunnel, it was decided that the sealing of 1,850 linear feet of the G3 Track and 900 linear feet of the G4 Track would be sufficient to prevent out-of-tolerance groundwater intrusions into the tunnel throughout the remainder of construction and revenue service. The proposed solution would be to drill a series of holes through the reinforced concrete tunnel liner and inject NOH2O polymer emulsion grout in order to create an impervious grout membrane between the tunnel and the surrounding ground medium.

Work began in February 2016 in the G3 track area with a crew of eight. It was decided that for the duration of grouting work, all other track and structural work would be done in the G4 track area so as to not interfere. A work train was used to bring all drills, pumps, grout lines, scaffolds and a sufficient supply of NOH2O and its additives to the G3 track. A crossover passage at the center of the 1,850 linear foot work limit would serve as a central pump room for grouting operations. For the G3 track, two to three drill locations were laid out along arch—one hole 2-ft from the crown, one hole 2-ft from the springline and (where applicable) one hole on the sidewall. This drillhole pattern would be spaced every 10 feet along the 1,850 feet of G3 track as shown in Figure 6.

Holes were drilled off of a work scaffold which was rolled down the existing G3 tracks. Due to the need for expedited work and the high water pressures anticipated, a pneumatically-powered Mid-Western 83 Jackleg drill with a 1-inch carbide bit was used to drill through the concrete liner. The jackleg drill with carbide bit can drill around six times faster than electric drills with diamond core bits through reinforced concrete. Furthermore the jackleg will not short out if it gets wet, unlike electric core drills. This was significant, as some holes drilled through the liner discharged water at 20 gallons per minute. A separate scaffold would be used to install 1-inch mechanical packers in drillholes and perform grouting work. All grout and pumps would remain in the cross-over, and grout lines would run out to the scaffold. This sequential method of work was performed along the entire alignment. Once the first round of grouting was complete, attention was given to treating leaks in the G3 crossover passage, where the leaks and cracks were most extensive. Working in this fashion, we were able to drill and grout 416 holes and inject 20,774 liters of NOH2O polymer emulsion, providing a grout curtain for 40,700 square feet of tunnel along the G3 track in 24 shifts and less than four weeks.

Figure 5. Leakage in G3/G4 Tunnel—G3 track area

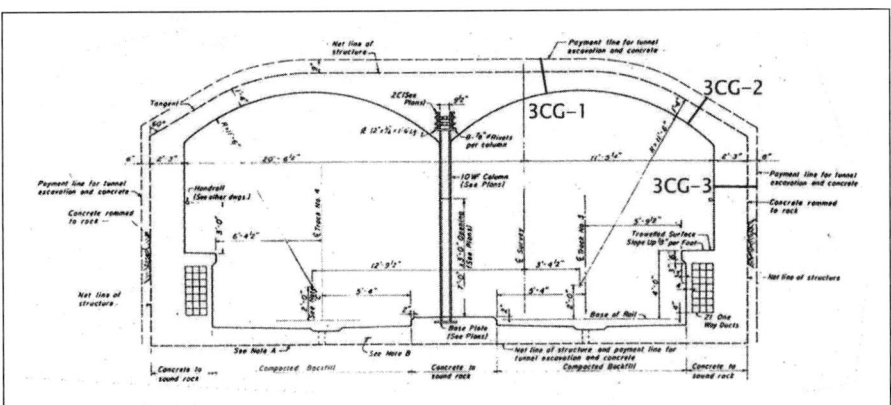

Figure 6. G3/G4 double-arch tunnel cross-section—G3 three-hole layout

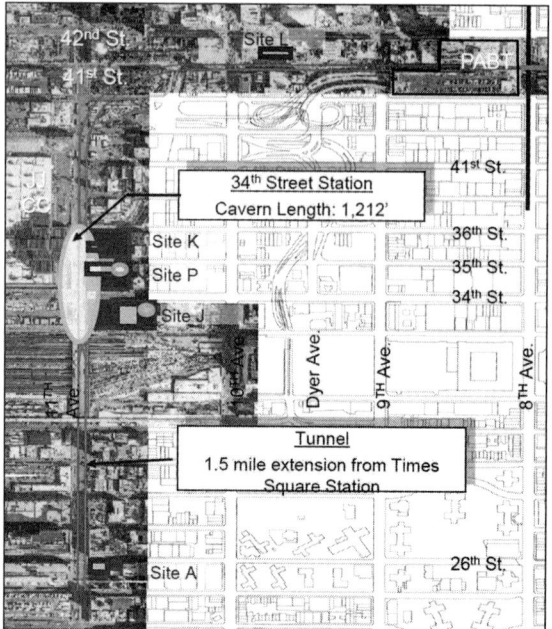

Figure 7. Midtown West Manhattan—34th St. Hudson Yards

The second phase of treatment began in August 2016, when the G4 track became available to conduct grouting operations. The G4 track transitions from running adjacent to G3 as a double-arch tunnel (as shown in Figure 6) to running below G3 as a stacked tube. The 900 linear foot work limit addressed the double arch section of the G4 track. A work train transported equipment and a fresh supply of grout to a new pump location in the G4 track area, and a crew of four proceeded in a similar fashion to G3. By mid-September, we had drilled and grouted 132 holes and injected 7,030 liters of NOH2O polymer emulsion, providing a grout curtain for 13,500 square feet of tunnel adjoining the G4 track in 5 weeks.

In this application, the primary constraint was schedule. The client needed a reliable product and an equally reliable solution which could be implemented flexibly and expeditiously. This sequential, hard-and-fast approach to positive-side curtain grouting with polymer-based emulsion yielded a production rate of 250 square feet per man-hour, including time required for mobilization, second pass and follow-up work. Providing a dry tunnel in this short span of time greatly helped the general contractor to push forward with track and systems work undeterred. Replicating these results using conventional crack-injection techniques would require dozens of workers and an extensive inspection and monitoring team be set up, and even then results would not be guaranteed as the water would retain its potential to migrate.

34th Street Hudson Yards—Entrance at Site J

One of MTA Capital Construction's major programs is the No. 7 line extension. This expansion of New York City Transit's IRT line into Midtown Manhattan West (Figure 7), which includes 1.3 miles of twin bored tunnel and the new 34th St.—Hudson Yards Station, began in October 2007 and was opened for revenue service in September 2015. After a few months of operation, commuters expressed concerns about water dripping through ceiling panels over the entrance escalators.

At the time, only one of the two station entrances (Site J) was operational. This entrance features two reinforced concrete arches which are inclined to span from Mezzanine to Concourse level (roughly 60-ft change in elevation). The south tunnel incline (E1) features one escalator and two inclined elevators, see Figure 8. The north tunnel incline (E2) features four escalators. Unable to shut down either incline, MTA chose to address these drips at night during off-peak hours.

Sovereign-Thyssen L.P. was brought in to address ongoing leaks throughout the entrance at Site J. Of primary importance was to stop the water dripping through the crown of the E1 and E2 inclined tunnels. Additional leaks in non-public areas on the concourse level and in personnel staircases were also specified for treatment. Polymer-based emulsion was requested and NOH2O grout was selected for production grouting.

Primary grouting operations started in late March 2016. Initial treatment began with the installation of a grout curtain barrier at the vertical endwall at the top of both incline tunnels, so as to prevent displacing water up the incline and into the station entrance. Once complete, a series of holes were drilled through the reinforced concrete arch along the incline tunnels and grouted sequentially—starting with the north tunnel (E2) and finishing with the south tunnel (E1). This first effort to seal leaks along the inclined tunnels above the escalators required four six-day weeks, with grouting work taking place at night and periodic monitoring during the daytime for possible reports of water leakage. Altogether 68 holes were drilled (24 in the endwall, 22 each in both inclined tunnels) and 23,250 liters of NOH2O were pumped in order to seal leaks along these two tunnels and endwall—treating a total of 21,450 square feet.

An interesting observation was made during the grouting of tunnel E2; prior to the selection of PBE, attempts were made by the general contractor to grout using several hydrophobic polyurethanes and a multi-component methacrylic resin. While injecting NOH2O positively into the tunnel annulus, a report was observed to a nearby construction joint. The material communicating from the joint was a murky white liquid

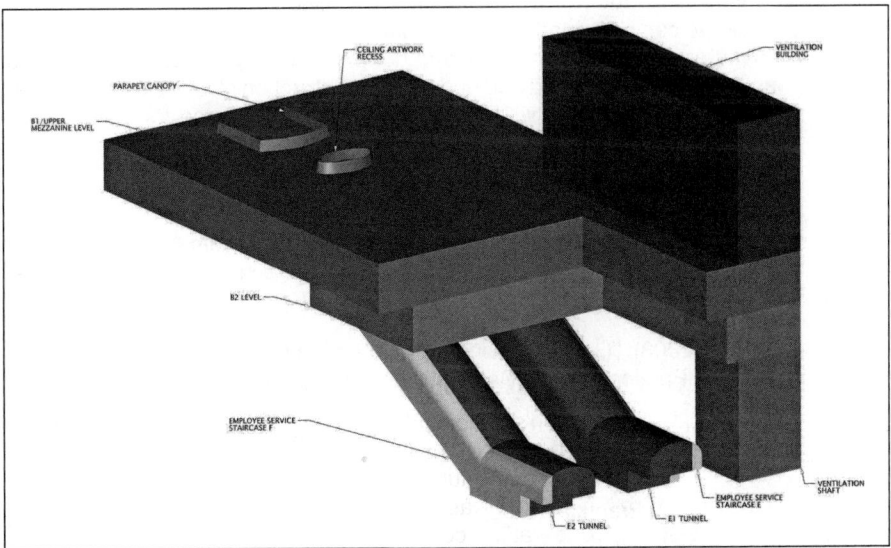

Figure 8. 34th St. Hudson Yards—Entrance at Site J

which was later identified as the B-component of the methacrylic resin, which had been injected unreacted and remained dormant in voids adjacent to the tunnel annulus until it was pushed out due to grouting pressures.

Once the inclines were finished the remaining non-public areas for station personnel were addressed. In the aftermath of grouting the inclines, some of the water had migrated to the landing of E1 tunnel. Ceiling panels were removed and holes were drilled through the liner. It was here that several holes drilled struck a void behind the liner which was channeling much of the groundwater to this area. These holes were allowed to drain overnight and were grouted soon after. In total, 225 holes were drilled throughout the course of production grouting and 28,964 liters of NOH2O were pumped in three months from end of March to mid-June 2016, treating 27,780 square feet. The leaks above both escalator inclines were eliminated, while station service continued uninterrupted throughout the duration of treatment.

CONCLUSION

Economic and punctual delivery of a tunnel depends on the reliability and effectiveness of the construction operations required to build it. Tunnel leakage has become a preeminent obstacle in the timely delivery of a finished tunnel, due to the inadequacy of usual countermeasures. Prolonged treatment in the latter phases of construction can have a detrimental effect on the project schedule—dragging out furnishing of finishes, installation of sensitive equipment and diagnostic testing of station facility systems. Conventional waterproofing systems like PVC or HDPE sheet membranes are effective at mollifying the bulk of water inflows, but have proven ineffective at total water cutoff. Crack injection with resinous grouts does not correct the problem at its source, instead diverting the water to nearby cracks. Many times an inappropriate grout is selected for the task. For example, hydrophobic polyurethanes exhibit poor bond strength to leaking cracks in concrete and are frequently pushed out.

There is a need for a fundamental change in the approach and underlying assumptions on how to treat tunnel leakage in a systematic, predictable and reliable manner. Understanding that tunnels located beneath the groundwater table are always susceptible to water inflows is crucial. The case histories outlined in this paper demonstrate how tunnel leakage can be remedied expediently and soundly by curtain grouting and taking full advantage of the penetrability and durability of polymer-based emulsions. With emphasis placed on the rapid and economic construction of tunnels in the United States, polymer-based emulsion will continue to be a driving force for innovation in waterproofing applications.

REFERENCES

De Bruin, N.J.H., Grobler, J., and Pollard, C.A. (1991). "The Scem 66 Water Control System." Institute of Shaft Drilling Technology Annual Technical Conference, Las Vegas, NV.

Karol, Reuben H. (2003). "Grout Curtains," Chapter 17 in Chemical Grouting and Soil Stabilization. Third Printing. Marcel Dekker Inc., New York, NY.

Sovereign International Inc. (2011). Technical Data Sheet: NOH2O/SCEM66, Sovereign International, Kansas City, MO

Warner, James (2004). Practical Handbook of Grouting. Second Printing. John Wiley & Sons Inc., Hoboken, NJ.

Complex Inner-City Tunnel Excavation by Means of the New Austrian Tunnel Method in Combination with a Hyperbaric Atmosphere

Thomas Wechner ▪ BeMo Tunnelling GmbH

ABSTRACT

The Tunnel Karl-Friedrich-Strasse is located directly in the city of Karlsruhe (Germany) and is being excavated using the New Austrian Tunneling Method. The total length of the tunnel is 250 meters. 40 meters will be a highly complex tie-in structure with a cross section of 180m^2. Due to the inner-city location and the very high water table the chosen construction method was to excavate the tunnel in a hyperbaric atmosphere of circa 1.0 bar. This document describes the extraordinary challenges of NATM Tunnel in an urban area in combination with hyperbaric atmosphere.

PROJECT OVERVIEW

This paper will focus on the special challenges of this extraordinary tunnel project in Germany. The so-called Tunnel "Tunnel Karl-Friedrich-Strasse" abbreviated called "TKF" is a nearly straight twin-track Metro tunnel and runs along the so-called "Karl-Friedrich-Street." On average, the tunnel has an overburden of 6.0 up to 7.0 meters, depending on the excavation profile and the tunnel gradient. Two unique public structures must be highlighted on this 250-meter-long tunnel stretch. A nearly two-hundred-year-old obelisk which had to be crossed with a vertical distance of less than 6.5 meters and a 125 year old sewer at the final tie-in section of the tunnel.

The entire tunnel excavation is located under the natural groundwater table. Due to the above mentioned structures and the close proximity of several public buildings it was not feasible to drop the water-level. In addition, the sewer above the tunnel had to be maintained during construction. Therefore cut and cover methods such as slurry wall or steel piling were not possible.

To facilitate a conventional tunnel excavation, the unrivaled method of a hyperbaric atmosphere (compressed air) had been chosen. This method allows NATM in dry ground condition due to the positive effect of a high pressure atmosphere, which pushes the water from the tunnel. For this purpose, the ground conditions had to be adapted previously with a systematic permeation grouting program from the surface. Aim of those injections was one hand, to improve the structural ground condition by grout cement suspension into the soil and on the other hand to seal the ground by injection of gel.

HYPERBARIC ATMOSPHERE

Tunnel excavation in a hyperbaric atmosphere is very uncommon because of the high complexity of this method in combination with NATM. Key parts of this technology are the compressors, which produce the pressurized air, and the different locks, which separate the hyperbaric air zones from the atmospheric conditions. The required air pressure depends on the current groundwater level, which has to been seen in relation with the deepest level of tunnel excavation for each individual phase.

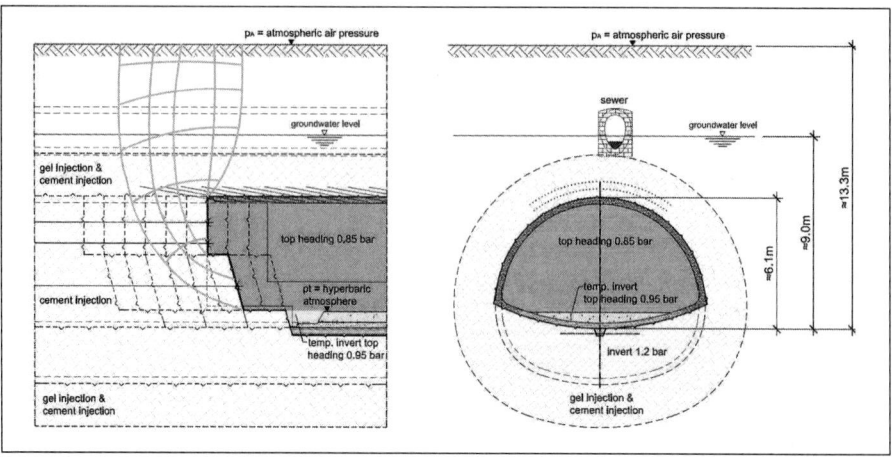

Figure 1. Air pressure levels

In Figure 1, the three different excavation phases are shown including the required air pressure to maintain a dry face. The individual locks as well as the entire lock building are designed for a maximum pressure of 1.3 bar. Depending on the local groundwater level, the tunnel gradient and the size of cross section, the required air pressure can vary slightly.

Air-Locks and Challenges

To be able to transfer material and personal inward our outward of the hyperbaric zones, locks must be implemented into the overall tunnel concept. In total, three different air locks need to been designed:

1. Manlock for staff and labor
2. Equipment- and materials lock
3. Muck/Spoil removal lock

The people air lock is approved for 20 individuals and is made up of three chambers. In each chamber oxygen mask are available for the discharge procedure. Compression for inward transfer only takes a few minutes. For the outward decompression transfer a timeframe from 5 minutes up to 1.5 hour has to be considered. The discharge times are exactly dictated in the German Compressed Air Regulation and depending on the air pressure and the individual exposure time. In terms of site organization and construction sequencing it is an extraordinary challenge to manage the tunnel operation without any interruptions. Quick transfers are not possible. Everything must be planed accordingly to avoid any down times. The same applies for the equipment- and material lock. Small management errors can lead to a total works stop.

To be able to transfer the equipment without further assembly, the equipment lock made from reinforced concert and two massive steel doors is 4.0 meters wide, 4.5 meters high and 16.0 meters long. For the spoil removal, a separate lock has been created especially for this project. The muck disposal lock consists of an 8m³ muck bin, a vertical inclined rail system and a transfer chamber. The operational capacity of this structure is around 70 m³/ hour.

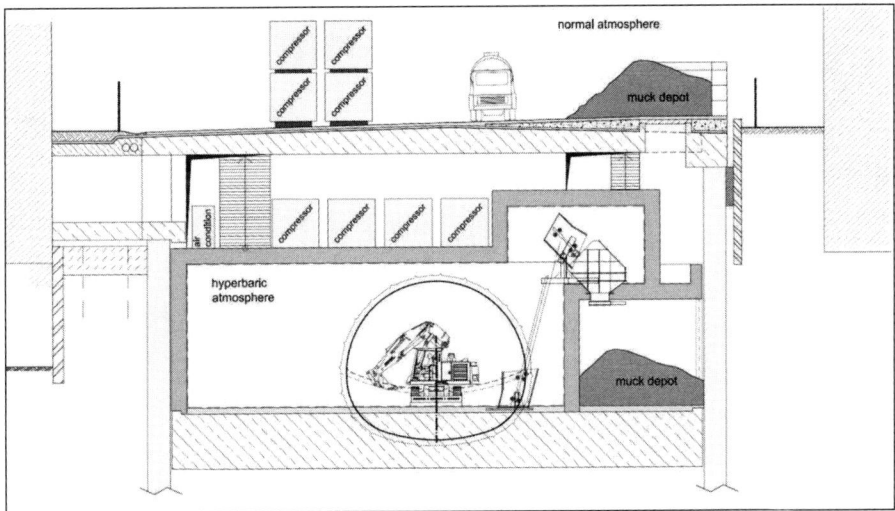

Figure 2. Cross section lock arrangement

Compressors and Air Consumption

The required compressor capacity depends on the calculated air requirements in the tunnel which is required to keep the water outside the excavation zone. The latter dictates the air pressure, which must be supplied at all time. The air loss can vary significantly depending on a multitude of factors such as:

- Air pressure in the hyperbaric zone
- Overburden of the tunnel
- Nearby utilities
- Adjacent geology
- Speed of excavation
- Excavation phase
- Size of the open tunnel face
- Number of compressions/decompressions

Because of the number of parameters it is nearly impossible to foresee the exact air demand in advance. Compressed air calculations assist by choosing the right number of compressors but uncertainties remain. The calculations are based on the air loss direct through the face, the air loss through the tunnel lining and the operating air loss from the locks and pipelines. In addition, in Germany the Compressed Air Regulation, dictates the basic guidance for designing, organizing and operating a pressurized tunnel excavation. This document dictates that $2/3$ of all compressor capacity must deliver the total required air usage. [1]

The requirements for air volumes had to undergo a relatively quick adjustment At the beginning the compressor facility was designed for an air loss of maximum 100 m^3/minute. In Figure 3 the air consumption during the excavation period is illustrated. The different graphs indicate the different target air pressures. Blue stands for 0,85 bar, orange 0.95, green for 0.75 bar and red for 1.25 bar.

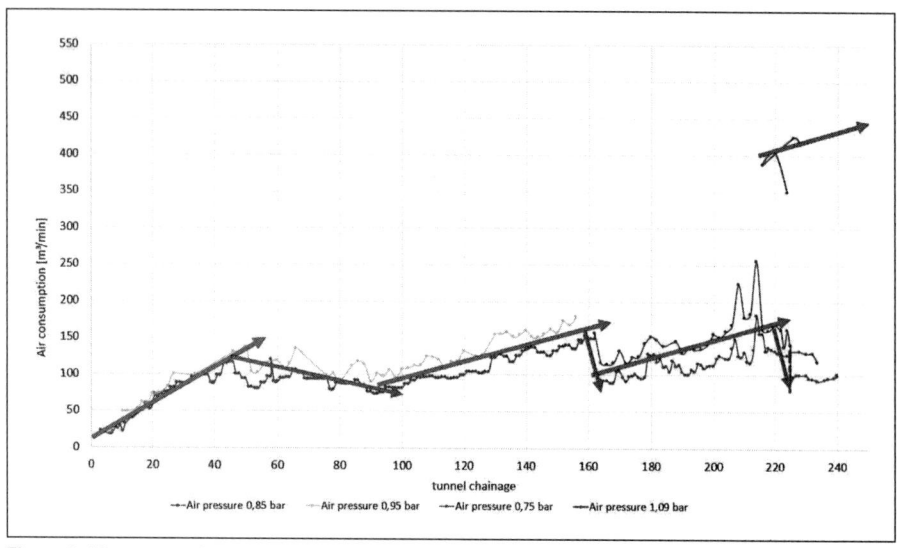

Figure 3. Air consumption

Already after the first 30 meters of tunnel excavation the predicted maximum air demand for the whole tunnel was reached. Because of this realization in the first stage, the compressor capacity was upgraded significantly. As the first upgrade was finished, the air consumption decreased suddenly without any traceable reason. After passing the obelisk at chainage 100, the air loss started to increase relatively steadily. Due to the fact, that the groundwater level drops towards the end of the tunnel a slightly air pressure reduction of 0.1 bar became feasible around chainage 165. This decrease on the pressure range reduced the air consumption by 50 m³/ minute. Nonetheless the increasing tendency continued toward the tie-in section. During the pipe arch works, sealing activities had been applied in the tunnel, which reduced the consumption by 35 m³/min. Air losses around the tie-in section were stable. No significant changes could be recorded. For the invert excavations, the air pressure had to be raised up to 1.20 bar. This increase of 0.5 bar moved the air loss up to 450 m³/minute.

To keep up with the frequently changing air demand the compressor capacities had to be upgraded 3 times. From originally 4 units up to 12 units.

Table 1 shows the timeline. In total a maximum volume of 490 m³/minute pressured air was available with a buffer of 280 m³ remaining as required by the ⅔ rule in by the German Compressed Air Regulation. Another factor to consider in hyperbaric atmosphere is the requirement to maintain a certain air temperature in the tunnel. The German Compressed Air Regulation dictates a temperature in the tunnel between a minimum 10° C and

Table 1. Air compressor setup

Compressor Units	Performance	Creditable
1	50 m³/min	50 m³/min
2	50 m³/min	50 m³/min
3	50 m³/min	50 m³/min
4	60 m³/min	60 m³/min
5	70 m³/min	70 m³/min
6	70 m³/min	70 m³/min
7	70 m³/min	70 m³/min
8	70 m³/min	70 m³/min
9	70 m³/min	buffer
10	70 m³/min	buffer
11	70 m³/min	buffer
12	70 m³/min	buffer
	770 m³/min	**490 m³/min**

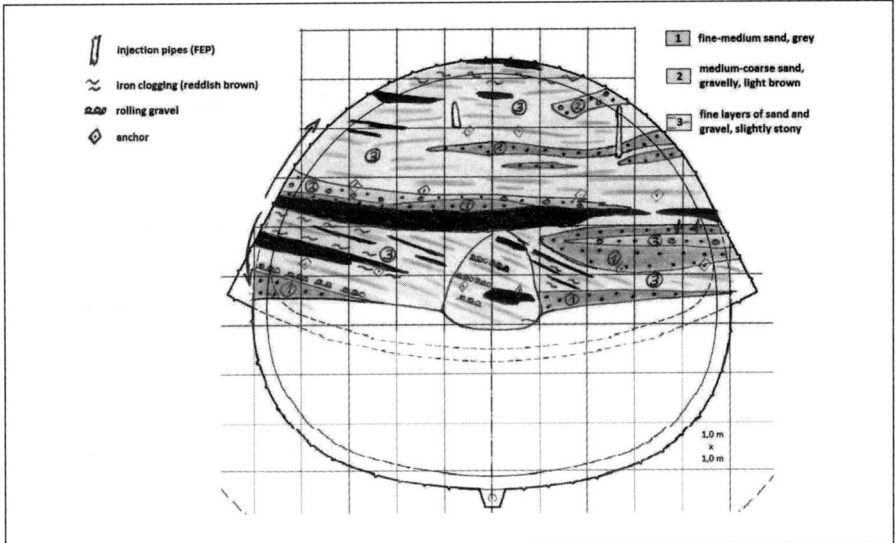

Figure 4. Geological face mapping

a maximum 25° C. To maintain this temperature range, a powerful air condition system was installed in the tunnel.

TUNNEL EXCAVATION AND TIE-IN STRUCTURE

One of the most challenging parts of this project was the tunnel excavation of the "Tunnel Karl-Friedrich-Strasse." A very shallow overburden in combination with an extremely variable geology made the tunnel excavation works very difficult. A mixture of gravel, sand and fine sand made up the general picture of the geology. In addition, layers of rolling gravel where detected as well as chunks of injected cement grout. Overall the geology was very mixed. Due to the hyperbaric tunnel excavation method, the sand and fine sand lost their original properties. Figure 4 shows a typical mapping of the tunnel face. Because of these complex circumstances a very elaborate excavation class had to be developed together with the designers to address those conditions. In the standard sections a sequencing of top heading/temporary invert and a late subsequent bench/invert had been designed. The excavation of the top heading was the most critical stage. To control the tunnel face, very small excavation windows with immediate support had been implemented. In addition, the face was supported by face dowels, shotcrete and lattice girder, and a massive support core to address the structural requirements. A special challenge was to apply the wet shotcrete onto the very dry ground. To make this possible the adhesion properties of the shotcrete had to be adapted by developing a special mix with elastic plastic fibers and a microsilica. The individual advance length was limited to maximum 1.0 meter. Large amounts of forepoling with more than 60 steel spiles was mandatory to ensure a safe tunnel excavation environment. After two single top heading advances the temporary invert had to be established to obtain a quick ring closure.

The Tie-in Structure

The 40-meter-long tie-in section was the biggest challenge on this project. The tunnel cross section increased from 75 m² up a maximum of 180 m². At the same time the 125 year old sewer had to be crossed. This historical listed structure is 17 m² large

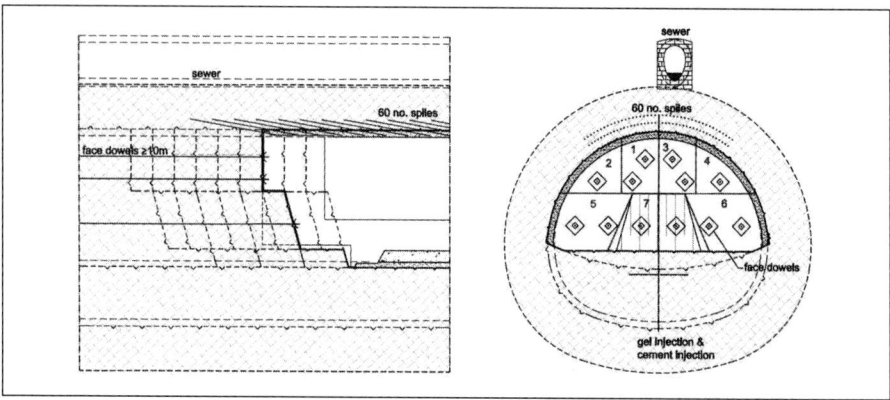

Figure 5. Tunnel cross section

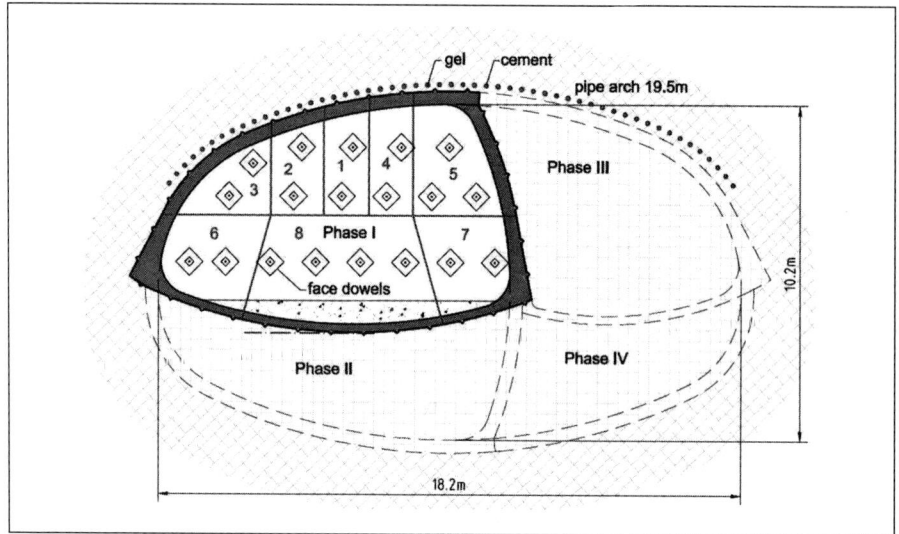

Figure 6. Cross section tie-in section

and crosses above the excavation works with a distance of less den 750 mm. "Tunnel excavation with large cross-sections in shallow overburden requires that the size of the excavation face be reduced to limit surface settlements." [2]

In our case a single sidewall drift was designed to meet those criteria's. 25 meter of this tie-in section were excavated in separate drifts. The following sequencing was established.

1. Excavation: Left top-heading side wall drift
2. Excavation: Left invert/bench side wall drift
3. Excavation: Right top-heading side wall drift
4. Excavation: Right invert/bench side wall drift

Before the removal of the massive 10-meter-high sidewall was allowed, the entire cross section had to be finished. In addition, two massive pipe arch were established

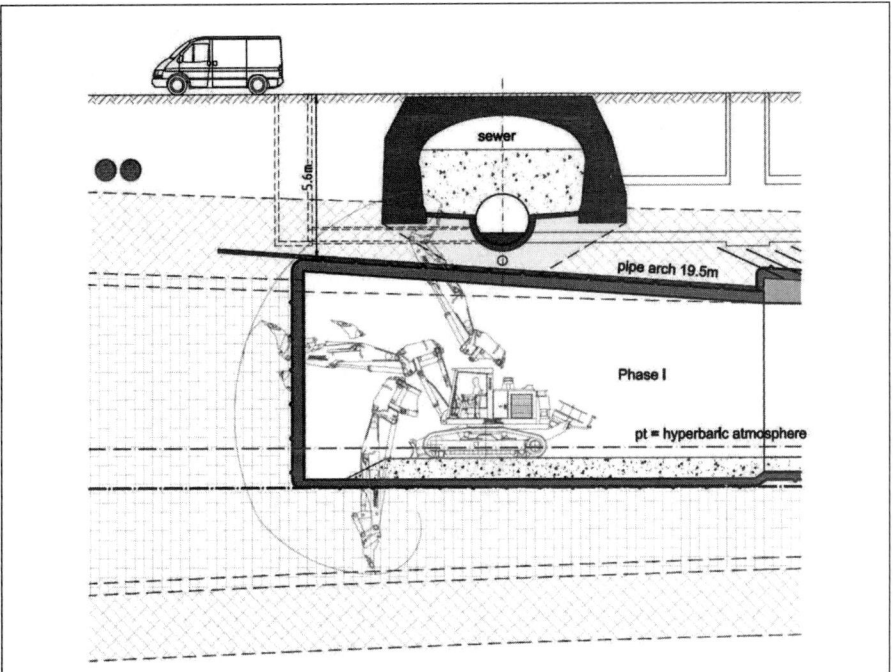

Figure 7. Long section tie-in structure

to distribute the loads on the one hand and to create a safe work environment on the other hand. The first pipe arch consists of 60 no pipes with a single length of 19.5 meter and 139 mm thick steel pipes. For the second pipe arch 74 no pipes with a length of 9.5 meter had been used. In particular, the close proximity of the sewer in combination with hyperbaric atmosphere requires a grouted pipe arch.

The permanent danger of a hyperbaric tunnel excavation is the risk of a blowout. A blow out is a radical air loss through a channel or opening between the pressurized tunnel zone and the atmospheric air environment. If the sudden air loss cannot be compensated by the buffer capacity of the compressors or stopped by means of special actions, the air pressure will drop in the tunnel. If so, the water pressure will act on the tunnel face and the underground structure becomes unstable. If this situation cannot be controlled immediately, a collapse of the tunnel and in last consequence a day light event is not avoidable anymore.

Because of this circumstances a very safe and robust design for the tunnel excavation works is required. Furthermore, a detailed emergency-management-plan must be in place. To decrease this risk to an absolute minimum a special pipe arch has been adapted.

Every 330 mm four openings had been implemented on each of the pipes. Immediately after a single pipe was drilled, gout was injected at each of the openings. The injection was done either with cement grout or with a chemical gel. In total, more than 3,500 injections were made just for the first pipe arch. The main purpose of this very time consuming activity was to fill potential hollow spots in the soil and to back grout the space between the single pipes. Due to this additional measure, a safe tunnel

Figure 8. Side wall drift excavation, tie-in section   Figure 9. Temporary road closure; surface tie-in structure

excavation was consistently possible. Figure 6 shows the delicate tunnel excavation situation with the historical sewer as described above.

INNER-CITY CHALLENGES

Constructing a tunnel in an urban area is always a very demanding task. In our special case the hyperbaric atmosphere severely increased those challenges. Air loss in adjacent basements, elevated surfaces, an extraordinarily laydown area, temporary roads closure, maintaining a real-time ground movement monitoring system, are just a few of the problems that occurred.

SUMMARY

NATM is one of the best methods to construct complex tunnel structures. The unique hyperbaric atmosphere, however, adds high complexity to this method. Without an adequate design, a well-organized site organization, high quality materials and excellent equipment a successful hyperbaric tunnel excavation would not be possible. Furthermore, it must be mentioned that all the equipment used in the hyperbaric atmosphere must be operated fully electrical. Due to the high levels of variables to be controlled, hyperbaric excavation forms a high risk construction method among regular NATM excavations. Several contingency measures and quick reaction to problems is required to ensure safe excavation. Program and cost considerations must certainly be secondary to the above considerations and a fixed price contract, for such excavations, is not recommended.

REFERENCES

[1] ArbSch 2.2.02;2013, *Druckluftverordnung (Compressed Air Regulation)*. p. 13.

[2] Austria Society for Geomechanics Divison "Tunnelling" Working Group "Conventional Tunnelling," 2010, *NATM The Austrian Practice of conventional Tunnelling. Salzburg.* p. 15.

Geologically Targeted Pre-Excavation Grouting Along the WestConnex M5 Tunnel, Sydney, Australia

Ulrike Pelz ▪ CDS Joint Venture
Joan Casado ▪ CDS Joint Venture
Harry Asche ▪ Aurecon Jacobs New M5 Joint Venture
Jack Raymer ▪ Aurecon Jacobs New M5 Joint Venture
David Crouthamel ▪ McMillen Jacobs Associates
Scott Fidler ▪ Golder Associates Pty Ltd

ABSTRACT

The WestConnex M5 motorway includes deep bedrock tunnels 70 m deep to pass under a 40 m deep paleochannel. Beneath the paleochannel, the rock contains large, systematic fractures caused by valleyfloor thrusting and a regional wrench-fault system. Pre-excavation grouting from the surface and underground will be used to control groundwater inflows to the tunnel. The surface phase uses an acoustic televiewer to geologically target each grout stage and find the optimal place to set the packer. This reduces the potential for bypass and allows higher pressures and volumes to be used, thus reducing cost by reducing the number of grout holes. The surface program creates the foundation for the underground program, which uses high-pressures and ultrafine cement to provide the final sealing of rock mass.

INTRODUCTION—THE WESTCONNEX NEW M5 PROJECT

WestConnex Stage 2 is a new road tunnel project in Sydney. It will comprise two new motorway tunnels, each around 9 km in length, connecting the existing M5 East Motorway at Kingsgrove to an interchange in the east at St Peters. The tunnels will pass underneath the suburbs of Kingsgrove, Bexley North, Bardwell Park, Bardwell Valley and Arncliffe, before crossing beneath the Cooks River then continuing under Tempe and St Peters. Each tunnel will be built with long term capacity for three or five lanes, but will initially operate with two lanes (see Figure 1).

Like other similar tunnels in Sydney, this will be a drained tunnel. The total allowable inflow rate is 1 liter per second per kilometer (25 gpm per mile). Due to the wide and varying spans of the tunnels and the high hydrostatic head at Arncliffe, it is not practical to build a tanked tunnel. This leaves preexcavation grouting as the principal means for controlling excessive inflows.

TUNNELS AT ARNCLIFFE

One of the major sites on the project is located at Arncliffe. A worksite, located in the former grounds of the Kogarah Golf Club, provides construction access to the tunneling activities via two drives. A shaft and adit is located on the south-west corner of the worksite, while a decline is located to the north east corner of the worksite. Permanent works within and adjacent to the Arncliffe worksite include twin 4 lane tunnels (wide enough for future expansion to 5 lanes) heading towards St Peters, twin 2 lane tunnels (wide enough for future expansion to 3 lanes) heading towards Kingsgrove, and associated twin Y-junction caverns to accommodate a future Southern Connector project. Within this are included two major ventilation shafts and connecting tunnels,

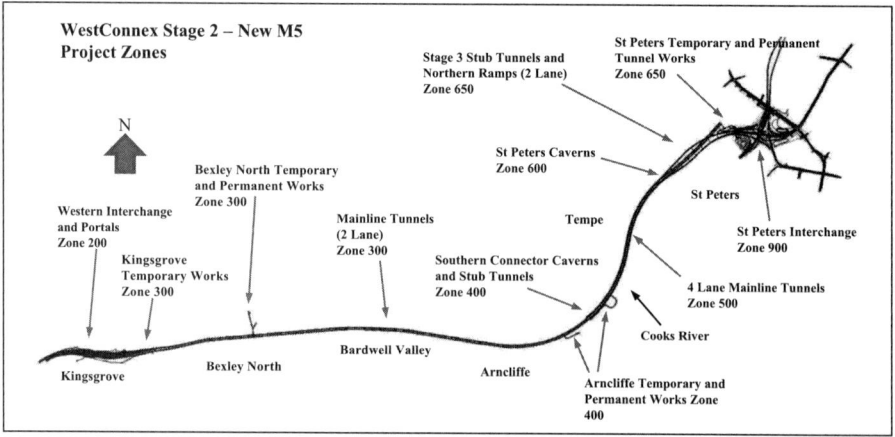

Figure 1. Schematic showing WestConnex Stage 2 project zones

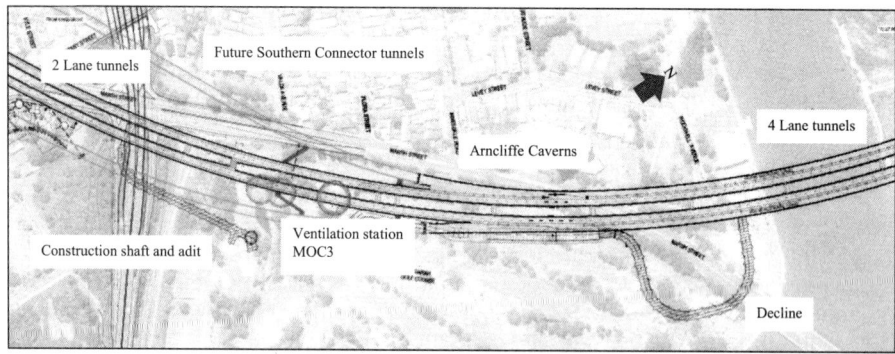

Figure 2. Subsurface structures at Arncliffe

a vehicular cross-over and the low point sump facility. Figure 2 shows the significant components in the area.

GEOLOGY AND HYDROGEOLOGY

Geology

The Hawkesbury Sandstone formation is the predominant bedrock in Sydney. The Hawkesbury Sandstone is typically comprised of fine to coarse grained quartz sandstone with very minor shale and laminate lenses. Hawkesbury Sandstone is exposed near the surface and commonly occurs in clifflines and outcrops along water courses.

The Hawkesbury Sandstone formation occurs within the geological unit called the Sydney Basin and has been subjected to bending and associated faulting. Major named faults, minor faults, and various shear zones occur along the alignment. Within the Sydney Basin, a prevailing tectonic horizontal stress is believed to run with principal stress direction in the NW-SE direction. This stress is magnified by local features and can reach significant values.

Current and ancient watercourses have cut valleys into the Hawkesbury Sandstone. In many places these watercourses (paleochannels) have been infilled by Pleistocene and Holocene alluvial and estuarine sediments.

At Arncliffe, a major paleochannel exists which crosses the tunnel alignment. The tunnel is believed to cross the main channel within the worksite, and also to cross an inferred tributary under the Cooks River. Figure 3 shows the surface geology and the inferred locations of some of the major structural features.

Hydrogeology

Over the Sydney area, there is a general topographic induced groundwater flow from recharge areas in higher areas towards the valleys which eventually head towards the coast. Thus in Figure 3 it is expected that the general groundwater flow is to the south east into Botany Bay with localized flows in different directions caused by local topography and drainage features.

The conceptual hydrogeological model for the Arncliffe and Cooks River area is shown in Figure 4. The Arncliffe Cavern and Cooks River crossing will be constructed within the Hawkesbury Sandstone and may intersect the Woolloomooloo Fault and various fracture networks associated with the paleochannel which has been scoured into the sandstone by the prehistoric Cooks River at the worksite. The hydraulic conductivity of Hawkesbury Sandstone is generally dependent on the jointing and bedding partings. The large scale average hydraulic conductivity of the Hawkesbury Sandstone is interpreted to be low (approx. 1×10^{-7} m/s) but is much higher in areas of faulting or open shearing. Along the alignment of paleochannels, the hydraulic conductivity of the Hawkesbury Sandstone may be enhanced as a result of the development of stress relieved bedding and joints predominantly within the upper approximately 15 m

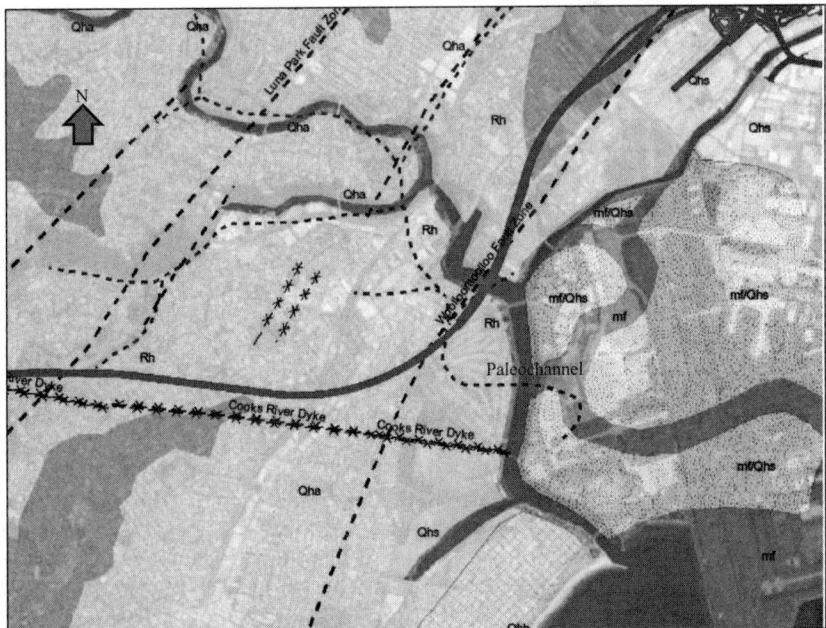

Figure 3. Surface geology at Arncliffe

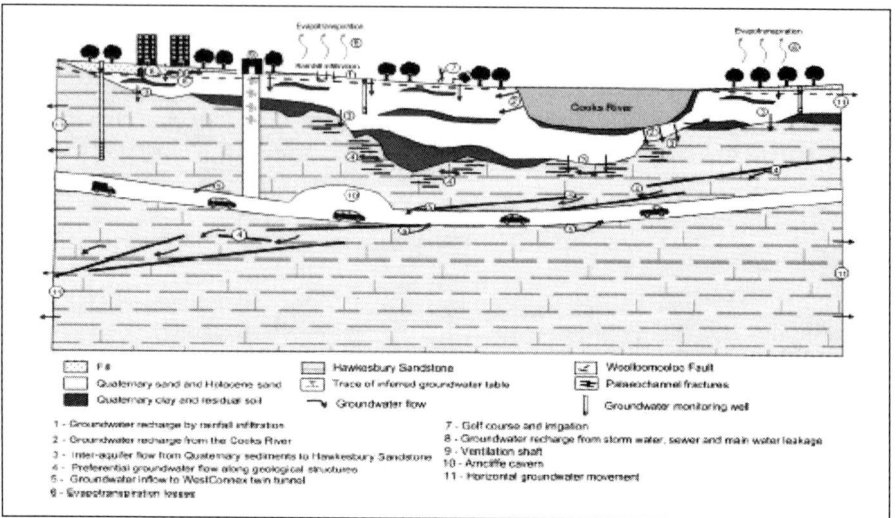

Figure 4. Conceptual hydrogeological model at Arncliffe, noting that to show the main features, a "cartoon" like representation is used

of the bedrock. Preferential flow paths may therefore exist in the sandstone along the alignment of paleochannels.

Holocene and Pleistocene sediments overlie the Hawkesbury Sandstone at the work-site. The hydraulic conductivity of these sediments is variable, reflecting the presence of interbedded sands and clays. Groundwater levels in the Holocene sediments at this location are close to the water level in the nearby Cooks River. Groundwater levels in the Hawkesbury Sandstone are similar, although in some locations a very small upwards gradient is evident, interpreted to be indicative of discharge to the Cooks River. Pumping tests carried out for the project (see below) indicate a high degree of connection between the Hawkesbury Sandstone aquifer and the overlying sediments.

Under the influence of drainage to the proposed tunnels, groundwater flow is antici-pated to occur preferentially along the Woolloomooloo Fault and through stress-relief fractures associated with the paleochannel, which is likely to increase groundwater inflow to the tunnel relative to the inflow rates that will be experienced elsewhere. Recharge to groundwater, which will act to limit the drawdown in alluvial sediments overlying the rock, is anticipated from the Cooks River, irrigation of a Golf Course close to the tunnel alignment, leakage from water mains, sewers and storm water, in addition to natural processes.

Groundwater Drawdown and Associated Induced Settlement

The paleochannels consist of layers of interbedded sand and clay. Holocene sedi-ments are close to normally consolidated, and Pleistocene sediments at the base of alluvial sequence are overconsolidated. When the tunnel is driven groundwater flow to the permanent tunnel drainage system will underdrain the soil, and potentially lead to depressurization and consolidation of compressible soil layers. The extent to which pore pressures in the alluvium is reduced and to which the water table is lowered will depend on relative vertical permeability of the rock and different lithological units within the alluvium, and also on recharge. If the conductivity of the rock or of clay lay-ers at the base of the alluvium are low enough, the majority of head loss will occur

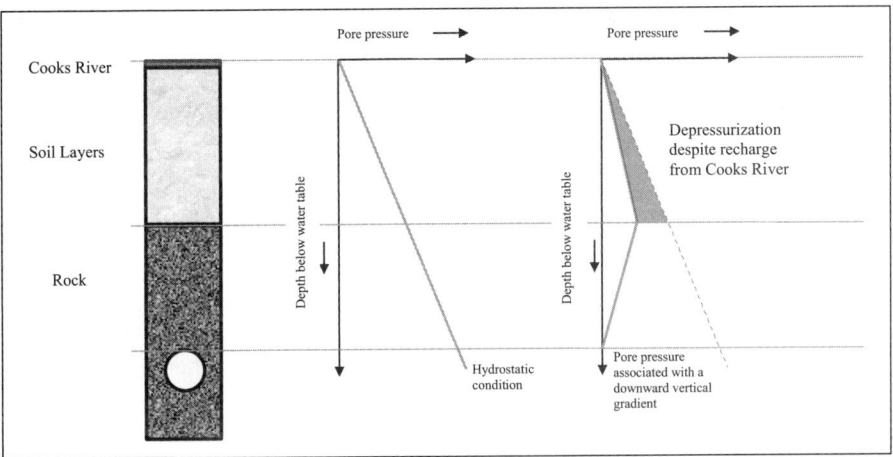

Figure 5. Depressurization associated with under-drainage

through these materials, and groundwater pressure in the overlying layers will be relatively unaffected. The change in pore pressure that will result from tunnel drainage are illustrated conceptually in Figure 5.

Pumping Tests

To refine the conceptual hydrogeological model for the area, and to establish input parameters for numerical models, pumping tests were carried out in the Arncliffe area. Three pumping wells screened in fracture zones in the Hawkesbury Sandstone were installed. Monitoring of the pumping tests included multiple monitoring locations in the sandstone and at multiple levels in the overlying alluvium. The pumping tests have shown that:

- Drawdowns of up to 16 m were observed at monitoring locations in the Hawkesbury Sandstone, in response to combined pumping. Very rapid responses to changes in pumping were observed at some monitoring locations, and somewhat slower responses were observed at other locations. This difference is interpreted to relate to the proximity of the vibrating wire piezometer sensors to higher permeability features.

- Drawdowns of up to approximately 4 m developed in the alluvium, at locations of up to 195 m from the pumping wells. It is noted that at some locations in close proximity to one of the pump wells, relatively little drawdown was observed in the alluvium, compared to the drawdown in the underlying Hawkesbury Sandstone. The response in the alluvium appears to be variable, which is interpreted to reflect the interbedding of sands and clays within the alluvium.

- Settlement of up to 10 mm resulted from the drawdowns associated with the pumping test. Rebound of settlement after the termination of pumping was observed at some locations.

The magnitude of response observed at monitoring locations in the alluvium in the pumping tests within a relatively short period of time indicates a high degree of hydraulic connection between the Hawkesbury Sandstone and the alluvium. This is consistent with the stratigraphy encountered in investigations in the area, which indicates that the alluvium comprises interbedded sands and clays with discontinuous, inter-fingered lenses of sand and clay.

Numerical Models

A 3D numerical groundwater model was developed for the Arncliffe area in order to estimate groundwater inflows to the tunnel, and changes in groundwater pressure in the vicinity of the tunnel as a results of permanent tunnel drainage. The model was developed using the 3D finite difference modelling package MODFLOW, developed by the United States Geological Survey (USGS).

A relatively fine model grid was used to allow individual tunnel features to be repre-sented, and to allow geological structures such as sub-horizontal shears and sub-vertical faults to be represented explicitly in the model. A higher than normal vertical discretization was also used in the model to allow accurate representation of the steep vertical gradients in this area. The local scale model was calibrated to match the pump test results over the period of the pumping tests.

Base Case Groundwater Model Results

The so-called base case model scenario is that in which the tunnels are drained, and no measures are taken to reduce the permeability of high permeability features that are expected to be intersected by the tunnels. For this case, the estimated average inflow rate over the full length of the tunnels is close to 1 L/s/km/tunnel at both the start of operation and at quasi-steady state. While the average over the total length of tun-nel is calculated to be close to 1 L/s/km/tunnel, some sections of tunnel are calculated to have inflows that will exceed 1 L/s/km/tunnel locally. This is particularly the case in the Arncliffe area where the tunnels/caverns are in close proximity to or intersect high permeability structures. The calculated drawdowns in the Arncliffe area were between 38 m and 49 m in the Hawkesbury Sandstone and up to 16 m in the overlying soils. Drawdown was greatest in the vicinity of the Woolloomooloo Fault.

Settlements were calculated from the change in calculated groundwater pressure, and applying this as a change in effective stress. The calculations are based on the layers identified from boreholes and cone penetrometer tests, and using compress-ibility parameters assessed from these. Settlement of cohesive soils was assessed using consolidation theory, taking into account preconsolidation stresses. Settlement of cohesionless soils was assessed elastic theory.

An assessment of the building impacts following from this settlement has been carried out using the methodology of Boscardin and Cording (1989), as later expanded by Burland et al (2001). Although the predicted settlements are relatively large, the pre-dicted effects on the buildings is not high for small buildings with shallow foundations. This is due to the small gradients and curvature of the settlement profiles predicted.

Prediction of Grout Treatment Effect

Numerical model runs were also performed to simulate the effect of grouting to reduce the permeability of high permeability structures at Arncliffe. The approach to grouting is discussed below. The cases considered in the groundwater modelling were:

- Base Case—as described above with no grouting
- Case 1—hydraulic conductivity values for grouted geological structures in the Hawkesbury Sandstone reduced to 5×10^{-7} m/s within the areas to be grouted

Table 1. Results of settlement calculations with respect to grout treatment

Grouting Case	Settlement Maximum Value Calculated	Comments
Base Case – no grouting	155mm	Some risk of building damage
Case 3	50mm	Risk level of building damage negligible
Case 4	35mm	Risk level of building damage negligible

- Case 2—hydraulic conductivity values for grouted geological structures in the Hawkesbury Sandstone reduced to 3×10^{-7} m/s within the areas to be grouted
- Case 3—hydraulic conductivity values for grouted geological structures in the Hawkesbury Sandstone reduced to 1×10^{-7} m/s within the areas to be grouted
- Case 4—hydraulic conductivity values for grouted geological structures in the Hawkesbury Sandstone reduced to 1×10^{-8} m/s within the areas to be grouted.

It is noted that in practical terms, it may not be possible to achieve a reduction in permeability to 1×10^{-8} m/s over large areas.

Again, settlements were calculated from the change in calculated groundwater pressure. Table 1 shows the results of settlement calculations. Negligible risk of damage to small buildings with shallow foundations is predicted for groundwater depressurization associated with Cases 3 and 4.

Groundwater Inflow Predictions

A significant reduction in groundwater inflows is expected due to grouting, in addition to the reduction settlements induced by depressurisation.

The project requirements are for inflows less than 1 L/s per any given kilometer length of tunnel. Generally, inflows are predicted to be less than this, however at Arncliffe the inflows are predicted to be locally greater than this. Figure 6 shows the predicted inflow as a moving average over 1 km long intervals, along the tunnel length.

GROUTING PROPOSED

Grouting is proposed to take place over the area shown in Figure 2. The intention is to reduce the conductivity in the rockmass around the tunnel to an average level of 1×10^{-7} m/s, the intention being to meet the case simulated in the numerical model Case 3.

This program is not intended to create a dry tunnel. Water inflows are expected to occur and where they do, they will be treated by the water resisting strategy, involving stripdrain or membrane to divert the water to the sides of the tunnels.

The work includes both grouting from the surface and grouting from underground. Each method has its own advantages and disadvantages. By using both, we hope to realize the advantages of each in the most economical and efficient manner. The surface grouting is being performed first and will be followed by the underground grouting.

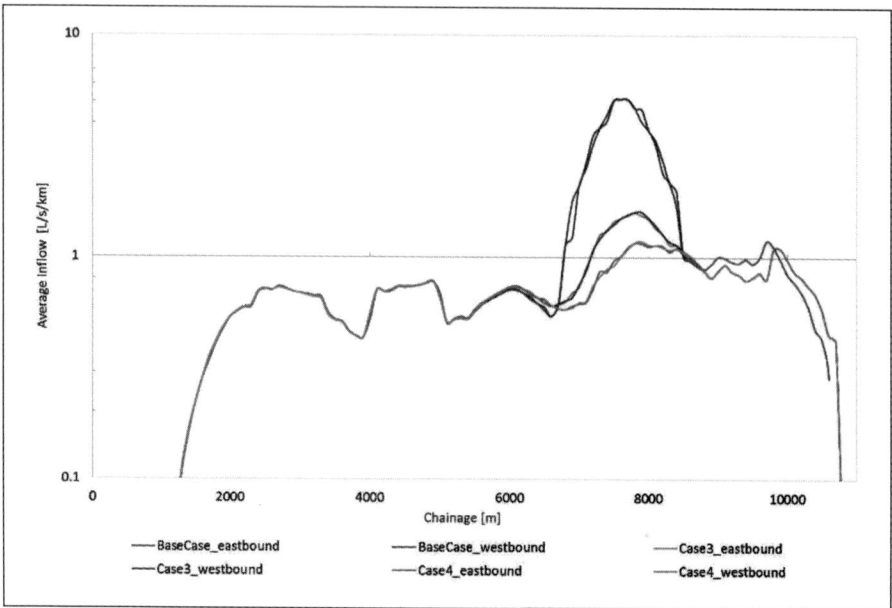

Figure 6. Predicted inflows

Surface Grouting

Philosophy

The goal of the surface grouting program is to reduce the rock mass permeability over a broad zone around the tunnel. This broad zone of low permeability rock mass will become the foundation for the underground program. The underground program will inject grout under high pressures into the finer permeability zones that remain following the surface grouting.

There are two reasons we are using a combined program. The first is that the underground program can proceed with high-pressure injections without having to first fill the larger fractures. This reduces effective overall work effort by reducing the number of holes and the pumping time needed for underground grouting. The second reason is that the surface holes and underground holes intercept the fractures at different angles. Horizontal fractures, in particular, as well as larger fractures, are grouted more efficiently from the surface.

The boreholes for the surface program penetrate below the tunnel excavation limits (approximately 15 m) and therefore are deep and individually expensive. We are treating the maximum amount of ground possible per borehole. This is done by specifically targeting the open fractures using a borehole acoustic televiewer and then pumping as much cement as possible into each stage. Field engineers from AJJV, with a high level of experience in geology and grouting, are providing advice on site to specifically target grout stages.

The grout is allowed to travel as far as it can be pushed, within reason. There is no evidence of runaway takes occurring following the recommended mix schedule.

In general, the main targets of the surface grouting program are intervals of rock with permeability greater than about 5×10^{-7} m/s.

Drilling

The surface grout holes are being drilled using reverse-circulation water-rotary methods. The pattern includes primary holes, secondary holes and tertiary holes. The primary pattern includes both vertical and inclined holes. The vertical holes are on an approximate 20 m grid. The inclined holes are used to reach areas that are outside the surface work zone.

In the early stages of the grouting program, it has been found desirable to keep at least 30 m between a hole being grouted and an adjacent open borehole that has not yet been grouted. This is so that grout from the first hole does not fill the second borehole before it can be properly grouted under pressure. The primary holes are drilled to a vertical depth of 85 m, which is approximately 10 m below the tunnel. The inclined holes are drilled longer in order to reach a vertical depth of 85 m.

Secondary holes are being drilled in areas where the primary holes encountered broken rock or that took at least 3x the theoretical hole volume of grout for the respective grouting stage, regardless of the mix. Secondary holes do not have to be drilled if all the adjacent primary holes had intact rock or took no grout.

Tertiary holes are being drilled in areas where the secondary holes encountered broken rock or that took at least 3x the theoretical hole volume (approximately 1 T of cement) for the respective grouting stage, regardless of the mix. Additional tertiary holes can be drilled inside the pattern shown above if it is judged that they are needed for particular circumstances.

Picking the Stages

An acoustic televiewer was run down the hole after drilling was complete. The acoustic televiewer showed the individual fractures that intersected the borehole; it showed their exact depth, apperture and orientation in three dimensions. The grouting stages were picked based on this data. The televiewer was run during the float period between when the hole is drilled and grouting begins, so it did not add additional time to the schedule.

Using the televiewer saved time and money. The grouted zone was 45 to 50 meters long in each hole. If standard 6-m stages had been used, then it would have taken 8 stages to grout each hole. Even if a stage takes little or no grout, it still requires at least 30 minutes to move and set the packers, set up the stage and attempt the process. With the televiewer, the holes typically needed only three or four stages, with each stage specially focused on a specific zone of fractured rock. This saved at least two hours of wasted time per hole.

The televiewer eliminated the guesswork that comes from trying to interpret driller's logs, and it eliminated the need to closely inspect each hole while it was being drilled. The televiewer showed whatever was encountered in the hole in precise, depth-controlled detail.

The televiewer data was used to update the fracture model for the site as the work progressed. This updated model was used to predict where additional fractured rock was likely to be located. This allowed the program to be expanded sensibly rather than randomly and wastefully.

Grouting

The grouting is performed upstage using a single grout packer. (Upstage grouting is where the hole is drilled to total depth and then grouted in stages from the bottom up.)

The grout is mixed in a grout plant and then pumped down the hole through a drop pipe. Well head pressure is measured using a compound pressure gauge at the hole collar. Bottom-hole pressure can be calculated from the well-head pressure. Because the stages are deep and the fractures large, the weight of the grout column was often sufficient to suck the grout down the hole, which typically resulted in negative pressure at the well-head. In some holes this condition persisted for hours until pressure began to build.

Grouting continues until the stage refuses (grout take becomes less than 2 L per minute) or until the designated work day ended (the site permit does not allow work past 6:00 pm). The grout is not thickened in order to induce refusal. The grout is thickened in order to reduce bleed if it is thought that the hole could take the richer grout without inducing artificial refusal.

Each hole is grouted initially with a 1:1 W/C ratio (by weight). If the hole takes well, then the mix is thickened gradually. It is important to increase the thickness slowly so as not to shock the flow system, which can lead to the thicker grout seizing.

Automated batching equipment is used to mix bulk cement at any specified ratio. A superplasticizer, (BASF Rheobuild 1000) is added to reduce the viscosity of the grout. The dosage rate is 800 to 1600 mL per 100 kg of cement, depending upon the air temperature.

Results

In the primary holes, significant interconnection has been noted even over a distance of 30 m in a few cases. This has been confirmed from the televiewer logs which have shown open, horizontal fissures of 100 mm size. These large horizontal fissures are interpreted to be from stress relief due to the removal of the overburden pressure when the paleochannel cut down 40 m into the rock.

Underground Grouting

Philosophy

The goals for the underground grouting is to treat water bearing fissures which allow groundwater to flow into the tunnels, such that the amount of the inflow is permanently minimized, and more importantly, such that the depressurization of the aquifer in the softer sediments above the tunnels is minimized. The grouting methodology proposed is to identify the water bearing fissures by probing/verify ahead of the tunnel face and fill them with grout under pressure as they are identified. The following describes some of the assumptions of the grouting program:

- As much as possible, the water bearing fissures should be grouted while minimal water flow is occurring into the excavation. Grouting features with uncontrolled flowing water, in the open excavation, is rarely successful because the water bears the unset grout away. Furthermore, water can migrate along open seams in the already open excavation. For this reason, the probing and grouting is ahead of the face, and the trigger for grouting, based on probe hole flows is set low to facilitate through rock mass treatment and suitably identify areas requiring treatment. Post-excavation grouting, or pre-excavation grouting with a high probe trigger is not likely to be successful to meet the project design goals.

- The base approach is to target a perimeter of the excavation, ahead of the advancing face. To this end, the probe holes and the grout holes are drilled in a fan from the tunnel face, far enough ahead that the water bearing features can be treated in static water conditions as above or controlled drainage condition to maintain treatment within the excavation perimeter. The ability to control drilling accuracy limits the length that the fan is drilled ahead of the face. As the length of drilling increases, the danger is that deviated holes miss an important feature or treatment is not sufficiently placed in the correct location.

- The assumption is that with a grouted perimeter, just outside the full circumferential limits of the excavations, flow around the grouted perimeter and up through the face is prevented. Therefore it is not necessary to grout the rock that will be excavated. If however water is flowing through a feature through the face inside the grouted perimeter, which can threaten to wash the grout out, the geometry of the grouted perimeter would have to be modified and the feature treated.

- The approach also assumes that grouting an individual high flow feature in the perimeter zone will control water flow. This assumption needs to be tested and it may be that if we can identify discrete high flow features, targeting those features with specific holes may be advantageous to restricting the work effort to the identified water bearing locations.

- Identification of significant inflows ahead of the face will require staged treatment through shorter holes intercepting the significant features. This will increase the effectiveness of treatment of high flow discrete features when present.

- It is important that the grout is permanent. For this reason, cementitious grout is the primary material proposed. Chemical grouts would only be used if leakage through the face is occurring, where wash out may not be controllable, and the effects temporary.

In the area of the surface grouting, given that the surface grouting has treated some of the larger waterbearing fissures, the underground program can proceed with high-pressure injections without having to first fill the larger fractures. This will reduce the number of holes and the pumping/placement time needed for underground grouting. The surface holes and underground holes will approach the fractures at different angles. Horizontal fractures, in particular, can be grouted much more efficiently from the surface, while sub-vertical features, such as major and secondary fault zones can be treated from the underground program.

The design goal of the underground grouting ahead of the advancing face, is to 1) Identify the location and relevance of water bearing features, 2) Reduce the hydraulic conductivity of these features to a design level approaching 1×10^{-7} m/s, recognizing that the final result will vary locally.

The proposed procedures described for underground grouting are proven methods and applications for intensive ground treatment and moderately high head conditions anticipated for the project. The program has elements which strive to maintain effectiveness in identifying water bearing zones and their treatment while maintaining a suitable level of overall efficiency. To do this, it is envisioned to use high grout pressure, large volume grout plants, multiboom drill jumbos and the use of ultrafine cements, as needed for low inflow conditions, to enhance penetrability of the rock mass. Positioning of the probe and grout holes will be drilled from the face where it is most efficient and lookouts to position the holes primarily outside the limits of the excavation to create a zone of treated ground at and outside the limits of the excavation, where long term water inflows and ground treatment are relevant. The initial probe drilling must be uniform around the full excavation perimeter to identify water bearing features for subsequent treatment. The key to the success of this program is the use of these variable procedures and materials correctly and collectively to ensure the program carries a balance between efficiency and effectiveness.

Grouting Intensity

Additional modelling has included a discrete fracture model which assists with the selection of the inflow triggers for probing and grouting as discussed below. This model suggests that in the more permeable fault and shear zones, treatment in part with ultrafine grout is necessary. The model simulations show that an effective grouting program requires a probe hole flow criteria of 0.25 L/min/m of probe hole length. The grouting efficiency at this inflow criteria suggests that the grouted zones will have an average hydraulic conductivity around 1×10^{-7} m/s, which is the minimum that can be achieved using high early strength (HES) Portland cement. Because this will not be always achieved for practical reasons, we will also be attempting to reduce some of the effective rock mass hydraulic conductivity to 5×10^{-8} m/s, the minimum assumed that can be achieved with ultrafine cement. Using both HES and ultrafine cement will provide an average conductivity of about 1×10^{-7} m/s. As indicated by the simulation results, effectively grouting the zones where fault zones communicate to the top of rock has the greatest effect at reducing tunnel inflow.

Execution

Holes will be drilled up to a nominal length of 30 m using multiboom drill jumbos ahead of the excavation face. Drilling will be performed with a percussive drill and water flushing. All holes will be flushed with water after the rods have been tripped out, to ensure that all holes have the opportunity to accurately report steady-state flows and treatment of the holes is as through as possible.

The holes will be drilled from within 0.5 m of the supported excavation limits with lookout angles of 10 to 15 degrees depending on the distance of features found ahead of the face. The holes will be terminated a maximum of 7 m past the limits of excavation which is anticipated to be just outside the limits of the initial support.

Probe holes, which become the primary grout holes, have a uniform circumferential spacing of 8 m or less. All of the probe holes will be fully drilled. Typically for the four lane spans (nominal span 19 m) a 13 hole pattern uniformly spaced around the circumference would be used. It is critical that the entire mandatory probe hole pattern be drilled to identify the location of water bearing features ahead of the face and their orientation and distance relative to the advancing face. This can allow the decision after the holes are drilled and treated, how far to advance the face. Water inflows, for each hole, will be measured individually for each hole with immediately adjacent holes closed off unless sustained inflows are readily recognized to be measurably greater than the recommended inflow criteria. The inflow criteria for any single hole is a sustained inflow of 0.25 Liters/min/m of probe or verification hole for any single hole, with all other holes closed off. Inflow measurement will be staged to attempt to measure inflows while opposing holes are drilled. If the above criteria can reliably be measure and is met or exceeded, all of the drilled probe/verification hole shall be grouted which produce measureable flow. Probe holes will be considered primary grout holes. Verification holes will be drilled where holes have shown inflow beyond the designated inflow criteria and will be split spaced adjacent to the exceeding hole. Any hole which does not meet the designated inflow criteria will be fully backfilled with a non-shrink blocker grout.

The verification split spaced holes on either side of the exceeding hole are drilled after the previous adjacent holes are grouted and have gone through initial set. This will be repeated for all patterns past the mandatory probe holes as long as the inflow criteria is exceeded in any given hole.

Unless the inflows exceed 4.0 Liters/min/m hole, grouting will commence with micro-fine cement, otherwise use OPC cement.

Maximum grout pressures are 50 kPa per m of rock cover depth or up to a maximum 50 bar unless grout leakage at the face or downstage of the face occurs.

Grout should be thickened if pressure does not increase within 30 or more minutes. Typical thickening of the mixes should not cause significant viscosity changes to avoid shocking the hole and causing premature refusal. The maximum amount of cement placed in any hole should be targeted to not exceed 15,000 kg/30 m of hole.

CONCLUSIONS

The level of investigation carried out for this tunnel is now considered to provide a high level of understanding of the actual situation. The modelling carried out is to a high standard and calibration and is expected to predict the situation well.

The drawdown and settlement anticipated after the grouting is carried out are accept-able and involve small impacts only to the surrounding environment. Grouting to meet the situation described by Case 3 is sufficient to achieve the environmental require-ments. There is only a slight improvement when Case 4 is reached. The grouting is specified to be performed to a high level of execution and suitable durability.

REFERENCES

Boscardin, M.D. and Cording, E.J., 1989. Building response to excavation-induced settlement. ASCE Journal of Geotechnical Engineering, Vol. 115, No. 1, 1–21.

Burland, J.B.; Standing, J.R.; Jardine, F.M. 2001. Building response to tunnelling. Case studies from the Jubilee Line Extension, London. CIRIA Special Publication 200. Thomas Telford Publishing, London, 2001.

Doherty, J., 1994. PEST: Model-Independent Parameter Estimation. Watermark Numerical Computing. Third edition.

ESI, 2016. Environmental Simulations Incorporated, www.groundwatermodels.com.

Harbaugh, A.W., 1992. A generalized finite-difference formulation for the U.S. Geological Survey modular three-dimensional finite-difference ground-water flow model: U.S. Geological Survey OpenFile Report 91-494, 60 p.

Houlsby, A.C., 1990, Construction and Design of Cement Grouting—A Guide to Grouting in Rock Foundations. John Wiley & Sons.

Stille H, 2015. Rock Grouting—Theories and Applications. BeFo Stockholm.

A Proactive Approach to Tieback Anchor De-tensioning

Sean Peterfreund ▪ McMillen Jacobs Associates
Grant Finn ▪ McMillen Jacobs Associates

ABSTRACT

For urban underground development involving tieback anchor excavation support walls, authorities having jurisdiction usually require tiebacks to be de-tensioned before project completion. De-tensioning typically involves the construction of block outs in the permanent concrete basement walls, which are labor intensive to install and back-fill, and tend to interrupt the continuity of the basement's waterproofing system.

This paper presents a tieback de-tensioning approach that is integrated into the design of the excavation support system: one that maintains the integrity of the waterproofing system and eliminates the need to sequence de-tensioning activities with basement build-out.

STATE OF PRACTICE

Tiebacks are a popular method of excavation support for underground development when clear and unobstructed access to the excavation is essential for construction activities. In urban areas, tiebacks present a number of unique challenges, especially when dealing with zero-lot-line development. Some of these include:

- Limited feasibility due to basements of existing structures on adjacent properties.
- Careful coordination to avoid existing utilities
- Requirement to obtain temporary construction easements under adjacent parcels
- Approval from local jurisdictions (especially local departments of transportation) to install tiebacks under publically owned rights-of-way (ROWs)
- Near universal requirement to de-tension tiebacks under adjacent property or public ROWs upon completion of project

The ideas expressed in this paper attempt to deal with the last item in particular.

Figure 1 shows a typical soldier pile and lagging support of excavation with tiebacks. The project shown in this figure is Capitol Hill Station, part of Sound Transit's University Link light rail extension in Seattle, completed in 2016. Because of a high groundwater table, that station employed a sheet waterproofing membrane with temporary penetrations required at each tieback head.

Current state of practice provides two alternatives for de-tensioning tiebacks. The more common of the two employs the following sequence in a fully waterproofed (tanked) structure:

1. Install excavation support using vertical elements restrained by tiebacks at a certain vertical and horizontal spacing (a typical spacing may be 8-foot horizontal and 12-foot vertical).

2. Cast base slab (invert), providing a "prop" at the base of the shoring wall.

3. De-tension tiebacks between invert and lowest elevated floor slab.

4. Install waterproofing system.

5. Cast wall lift and floor slab.

6. Repeat steps 3 through 5 until permanent basement structure is in place and temporary support of excavation is no longer required.

In the other alternative, tiebacks may be left tensioned until the permanent basement structure is in place, as long as openings (or "windows") are cast into the walls to de-tension tiebacks at a later date. Tieback windows accommodate taller spaces between floors and reduce construction sequencing logistics. However, they are expensive and complicated to install; are aesthetically problematic when concrete is the finished surface; and when membrane waterproofing is required, involve patches and seams in areas where proper quality control measures are extremely difficult.

Capitol Hill Station used both of these alternatives in different areas due to a variety of sequencing requirements and floor configurations. Figure 2 shows formwork for a tieback window in progress for 3-foot-thick transit station walls and membrane waterproofing. Note that the worker is installing couplers in order to reinforce the window pour-back adequately.

Given the extreme loads resisted by tieback heads, de-tensioning using acetylene torches is also a safety concern, whether or not windows are utilized. Figure 3 shows a welder torching the tieback heads in a confined space, preparing for the moment when the stress is released.

In their final condition, tieback windows can have a significant aesthetic. Figure 4 illustrates the typical appearance when

Figure 1. Soldier pile and tieback support for of excavation for the Capitol Hill Station box, with membrane waterproofing

Figure 2. Installing formwork and couplers for a temporary tieback window at Capitol Hill Station

Figure 3. Using an acetylene torch to de-tension tiebacks at Capitol Hill Station

exposed concrete is the finished surface. Note the seepage around the backfilled window area (squares), indicating reduced effectiveness of the waterproofing.

IMPROVED APPROACH FOR SOLDIER PILE AND TIEBACK SYSTEMS

De-tensioning of tiebacks is seldom planned for in design documents, and is usually left up to the contractor to coordinate. For a typical commercial development with a basement garage with close floor spacing and minimal waterproofing requirements, this approach may be adequate. However, construction of other underground development such as transit stations, cut-and-cover tunnels, laboratories, or other facilities with finished below-grade spaces stand to benefit greatly from designing with tieback de-tensioning in mind.

Figure 4. Final condition of tieback windows at Capitol Hill Station

The lessons learned at Capitol Hill Station, completed in 2016, allowed the design team to proactively address the issue at subsequent cut-and-cover stations. U District Station, part of Sound Transit's Northgate Link light rail extension in Seattle, is currently under construction and scheduled to open in 2021. Support of excavation for the station box was designed in such a way as to minimize the effects de-tensioning has on schedule, reliability of waterproofing, safety, and finish quality by using twin wide flange soldier piles, stitch-welded together at their flanges. The following methods were employed to reduce construction impacts of de-tensioning.

Vertical Spacing of Tiebacks

Vertical spacing of tiebacks was carefully chosen to avoid floor slabs. Further, spacing was chosen to allow for a floor to be cast; tiebacks in the next wall lift to be de-tensioned; construction of wall lift; floor casting; etc. This method works well when floor slabs are spaced sufficiently close to limit pile deflection during each step, but is of limited use with tall underground spaces such as transit platforms. When membrane waterproofing is required, using windows as the only means of coordination still requires the waterproofing subcontractor to be on-site for a long period of time, coordinating with the concrete subcontractor at each level.

Permanent or Semipermanent Construction Easements

For deep excavations, the lowest row of tiebacks may not be of any long-term consequence to owners of adjacent properties. An example of this would be properties that have already been developed to their highest and best use. When such a situation arises, permanent easements may be obtained for the deeper tiebacks, eliminating the need to de-tension at the deeper rows, as was the case at some of the properties adjacent to U District Station. A second approach was also used at U District Station for underdeveloped properties: semipermanent easements were obtained for deeper rows of tiebacks. This arrangement allowed a permanent tieback easement (that is, for tiebacks that are no longer needed but are still tensioned) to be considered null and void at the adjacent property owner's request, as well as the easement owner's approval for removing the tiebacks using extraction equipment.

Twin Wide Flange Soldier Piles

U District Station soldier piles were a unique design of twin wide flanges, welded together at the flange corners, shown in Figure 5. This design was originally intended simply to allow for the tiebacks to be installed with no stiffeners and a smaller drilled shaft diameter. After installation, it was realized that one could use the space between the two webs filled with lean mix concrete as a conduit from the surface. A 6-inch core drill can be used to cut each of the tieback strands from the surface, without any risk of wandering due to the thick steel webs and flanges.

Figure 5. Installing twin wide flange soldier piles at U District Station

In order to make the most use of this method, the reinforcing of the outside face of the wall was bundled, as shown in Figure 6. This allowed for a cavity in the wall thickness at each pile. Adding a continuous vertical smoothing layer over the entire height of each pile covering the tieback heads allowed for the waterproofing to run continuously over each pile with no additional seams. (This smoothing layer can be visualized by imagining a speedbump over each soldier pile, running vertically up the face of the wall from the deepest tieback to the shallowest.) The result is a holistic system providing several key benefits:

- All tiebacks can be quickly de-tensioned from surface after permanent structure is in place.
- Installation of waterproofing membrane can occur at any point before concrete is placed, minimizing coordination issues with concrete subcontractor.
- All windows, and the significant costs associated with them, can be eliminated, maximizing reliability of waterproofing, and minimizing aesthetic concerns of exposed concrete.
- Remote de-tensioning of tiebacks from the surface is significantly safer than using acetylene torches with a welder's face extremely close to the stressed anchor.

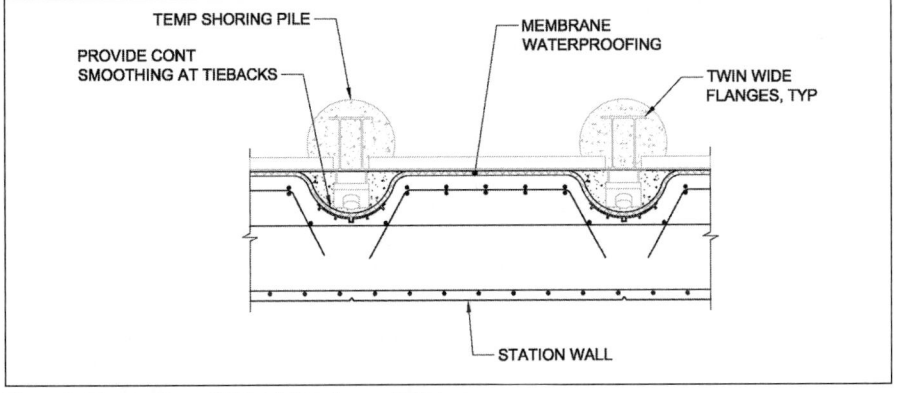

Figure 6. Plan section of U District Station wall design

As of December 2016, U District Station excavation was complete, and construction of the station box itself was scheduled to begin mid-2017. Therefore, de-tensioning is not scheduled to occur until 2019.

NEW TIEBACK DE-TENSIONING DEVICE FOR DIAPHRAGM WALLS

The concept of drilling from the surface to de-tension tiebacks led the authors to develop an insert for contiguous concrete shoring walls with water-cutoff properties (including slurry (diaphragm) walls, secant pile walls, and cutter soil mixing (CSM) walls). This patent-pending device consists of two small wide flange sections welded together at the flange tips. At each desired tieback location, a standoff and anchor bearing plate is welded to the flange face of the wide flanges. The assembly consists of a rectangular HSS, sized to align with the wide flange web spacing, and a bearing plate with a hole welded to the end. Figure 7 shows the assembled device, which may extend up and down the full height of the excavation with standoffs corresponding to tieback row spacing.

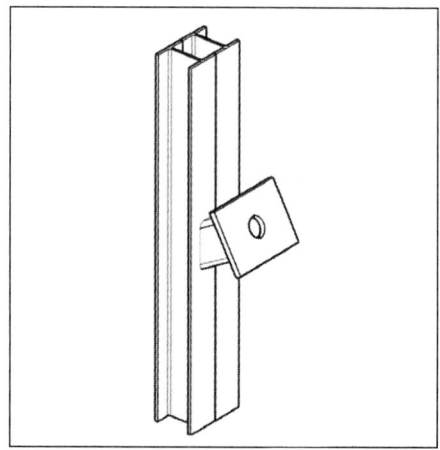

Figure 7. Isometric view of prefabricated tieback de-tensioning device

This device is shown installed in a generic concrete wall section in Figure 8. Figure 9 is a plan view of the device installed specifically in a slurry wall, showing how it may interact with a typical rebar layout. The insert is expected to maintain all the benefits of the soldier pile and lagging arrangement discussed above, plus:

- The vertical member, when tied into a rebar cage, provides additional rigidity to the cage which aides in picking.
- Tiebacks are arranged precisely in the vertical and horizontal direction, without the need to tie in pipes, or "trumpets," typically used for tiebacks in contiguous walls.

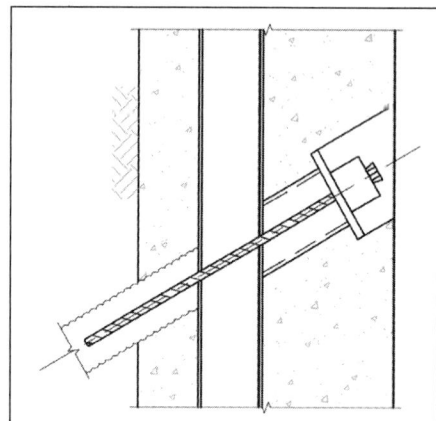

Figure 8. Section view of tieback de-tensioning device as installed in diaphragm wall (rebar not shown for clarity)

- The small size of the device allows for the bearing plate to be recessed into the face of the wall. Providing a foam block-out on the bearing plate creates a pocket in the wall, allowing for tieback install. This pocket can then be grouted, creating a smooth wall with almost no evidence of tieback heads.
- This arrangement seals out the majority of seepage typically evident at tieback heads in contiguous shoring walls.

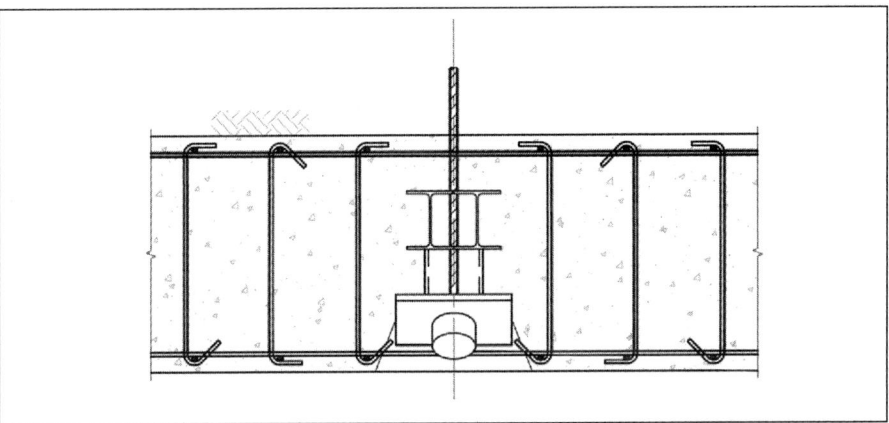

Figure 9. Plan view of tieback de-tensioning device as installed in diaphragm wall

- The resulting surface is ready for sheet or spray-on waterproofing membrane with no need for cutouts at each tieback.

CONCLUSIONS

Tieback de-tensioning can be a costly and complicated process in many forms of underground development. However, through careful discussion and planning among the design team, owner, contractor, adjacent property owners, and authorities having jurisdiction, a plan may be developed to minimize cost and scheduling impacts while maximizing quality of the final structure. Two underground transit stations were examined: one used the typical approach of delay in dealing with tieback heads and de-tensioning until the construction phase (Capitol Hill Station); the other created a plan for each tieback's final condition, how it related to the permanent structure, and how tiebacks interacted with the structure's waterproofing system (U District Station). Finally, a patent-pending insert device was discussed that the authors developed for contiguous diaphragm walls to arrange tiebacks and provide a means of de-tensioning from the surface. For an increased investment in the shoring system, the insert results in a smooth wall face, more rigid rebar cages, and allows for de-tensioning to occur quickly at any point in time after the permanent structure is in place.

ACKNOWLEDGMENTS

Figures 1 through 6 are the property of Sound Transit and used with permission. The authors would like to thank Sound Transit for the use of these figures and for providing a project environment where innovation and process improvement are encouraged and welcomed.

Ground Freezing for Tunnel, Shafts, and Adits

Joseph A. Sopko ▪ Moretrench American Corporation
Adam Curry ▪ Moretrench American Corporation
Bianca Messina ▪ Skanska USA Civil
Stephen Njoloma ▪ McMillen Jacobs Associates

ABSTRACT

Construction of a 0.81km long, 6.1 m diameter sewer tunnel, a component of D.C. Water's Clean River Project required construction of three drop shafts and connecting adits. The location of these three shafts and adits was complicated by their locations in a populated suburban neighborhood. Ground freezing from the surface was selected as the method to provide temporary earth support and groundwater control during construction. The ground freezing option was not only the most technically appropriate method, it provided significant advantages to accommodating the citizens in the neighborhood. Not only did the drilling and installation process provide a minimal equipment footprint at each site and reduce cuttings and spoils, but the refrigeration plants were located at a remote location reducing noise levels during the freezing process and construction. This paper discusses the design and implementation process of this complex freezing system that included drilling refrigeration pipes at complex angles in order to minimize traffic interruption and avoid utilities. Additionally, horizontal freeze pipes were drilled under 3 bars of pressure from the EPB tunnel in order to ensure that the adit connections were sufficiently frozen. The intricate cooling system extended over half a mile on both the surface and underground tunnel with instrumentation and monitoring along all points.

INTRODUCTION

The relatively short (0.81km) First Street Tunnel had three individual drop shafts designed to transport combined sewage from higher elevation sewers to the tunnel via small diameter (2.4-5 m) adits. The three shafts and adits are shown in section in Figure 1.

Construction of these shafts and adits was complicated by their locations within a residential neighborhood. In selecting the support of excavation for the shafts and the required ground improvement for the adits, consideration was given to minimizing the impact on the residents as well as minimizing noise. Secant piles and slurry diaphragm walls were considered for the shafts while jet grouting was a technical feasible method of ground improvement for the Sequential Excavation Method (SEM) for the adits.

While both methods were technically and economically feasible, there were some significant disadvantages related to their compatibility with working in a residential neighborhood. Specifically there was concern that space was limited for slurry diaphragm wall construction due to the reinforcing steel on-site storage and placement. Jet grouting would result in a significant amount of cuttings and spoils that would not only create a less than pristine site, but would also result in significant traffic for disposal. Furthermore, jet grouting would require significant utility relocations to avoid angled drill holes.

858

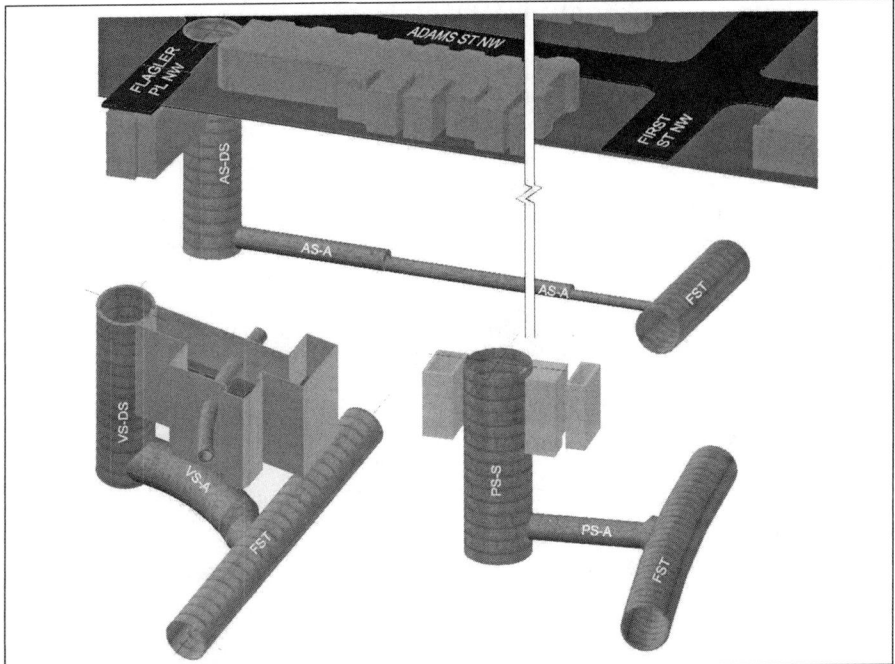

Source: Rob Chamberland, MTAC
Figure 1. First Street Tunnel shafts and adits

Ground freezing was originally considered, however immediately ruled out due to the constant operations of the refrigeration plants at each of the three shaft and adit sites. Ground freezing, however, had several technical advantages to the methods. Specifically:

- Ground freezing could be used for groundwater control and temporary earth support for the shafts and adits requiring one equipment mobilization.
- Cuttings and spoils would be minimized.
- Freezing could be used to ensure a water-tight connection from the adits to First Street Tunnel.
- Vehicle traffic near the residential structures would be minimized.
- Construction noise and vibration would be minimized

These advantages were too significant to totally discard the freezing concept. After reviewing the site plan in detail, an alternative approach was considered. The Channing Street Shaft site where the Tunnel Boring Machine (TBM) would be launched was not only relatively far from the residences, but had ample laydown room to place the refrigeration plants, pumping skid, and instrumentation trailer. This unprecedented approach was to use the Channing Street site as the lay-down area and transfer the refrigerated coolant to each of the three sites using a subsurface distribution manifold. Using this concept, the ground could be freezing at each of the sites literally unnoticeable to the residents of the community.

COOLANT DISTRIBUTION MANIFOLD

The initial design computations concluded that three electrically powered mobile refrigeration units would be required to provide sufficient cooling capacity to the shafts and adits at the required times as proposed by the original schedules. As shown in Figure 2, the plants and pumping system would be located near the Channing Street Shaft. The refrigerated calcium chloride brine would be pumped to each of the sites through a subsurface manifold system as shown in Figure 3. The Pumping Station Site was located 670 m from the plants. In addition to the hydraulic requirements as related to the pumping capacity and pipe size, heat loss through the system was also evaluated.

The manifold was buried and essentially undetected during the process. The heat loss was mitigated by backfilling the trench with polyurethane insulation. This insulation resulted in a heat gain of less than 2 °C at the Pumping Station, consistent with the thermal design requirements of providing 25 °C brine or colder through each subsurface refrigeration pipe. It should be noted that in addition to maintaining the cold brine temperature within the manifold pipes, the insulation also served to reduce or eliminate any frost effects to the sidewalks and pavements due to the freezing temperatures.

While the design of the frozen earth structures stipulated a brine temperature of –25 °C or colder, sufficient flow through each individual freeze pipe or circuit of freeze pipes

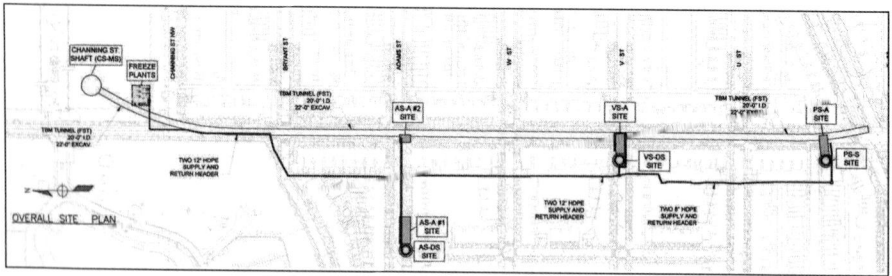

Source: Rob Chamberland, MTAC
Figure 2. Freeze plant and pumping system

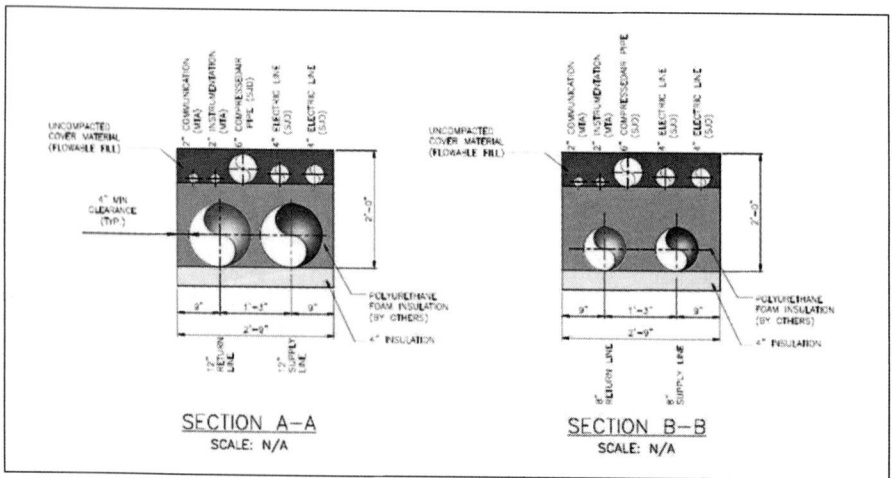

Source: Rob Chamberland, MTAC
Figure 3. Subsurface manifold system

Table 1. Frozen shaft dimensions (m)

Shaft	Shaft Inside Diameter	Excavated Depth	Depth to Rock
Adams Street	6.0	29.9	42.8
V Street	7.0	27.3	44.0
Pumping Station	7.0	26.8	40.0

is just as important to the timely and successful formation. To ensure both temperature and flow requirements, an instrumentation system specifically designed for this project was designed and fabricated. Figure 3 shows the placement of conduits, one to be used to transmit information for the shaft sites back to a centralized computer and SCADA system.

The performance of the distribution system was monitored with flow meters and also the supply and return temperatures of the circulating brine. Flow meters were used to evaluate the overall mass flow in and out of the pumping system. The brine supply and return temperature was measured at each pipe or circuits of pipes. By evaluating the difference between the supply and return temperatures it was possible to not only confirm sufficient flow, but also to balance the system across each individual structure.

FROZEN EARTH SHAFTS

The frozen earth for the shafts provided both temporary earth support and groundwater control. The three frozen shafts are described in Table 1.

Figure 4 shows the typical system for this project. To provide bottom stability of the excavation, the frozen wall was terminated in competent rock below the excavation invert. A structural evaluation of the frozen shafts prior to design was completed following a frozen soil test program completed on samples retrieved following the contract award. The analyses concluded that at 2.5 to 3.0 m thick freeze wall would provide the required resistance to lateral earth and hydrostatic pressures and the time and temperature dependent creep properties of the frozen earth. A thermal analysis estimated a freezing time of approximately eight weeks, using refrigeration pipes at 1.0 m spacing.

The freeze pipes were drilled with mud rotary methods and verified for verticality with an orientable inclinometer. Once the freezing process was initiated, the ground temperature data indicated freezing times compatible with the thermal model. As the frozen wall was formed, pore water within the shaft was forced to the center as a result of the expansion created as the water was converted to

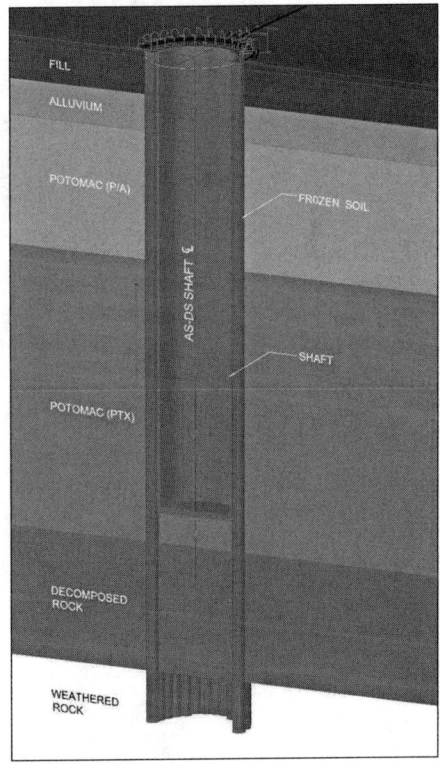

Source: Rob Chamberland, MTAC

Figure 4. Typical shaft freeze system

ice. This expansion was observed through pressure measurements in a piezometer drilled and installed in the interior of each shaft.

Following the indication of the closure of the frozen wall, an additional two to three weeks of freezing was required to achieve the required 2.5 to 3.0 m thickness. Once this thickness was achieved excavation commenced.

Prior to achieving the 3.0 m thickness excavation was able to begin within the upper fill area, which contained no pore water. Standard soldier piles and lagging were used in this non-frozen area to ensure ground stability. Once the 3.0 m thickness of frozen wall was achieved, excavation through the frozen soil began using a Brokk hammer along with mini excavator.

As excavation progressed a 10cm thick layer of foam insulation was sprayed onto the exposed earth. Due to site noise and work hour restrictions, excavation, form, and concrete works had to be performed during weekdays and daytime hours only. Excavation of each of the three frozen shafts took approximately 2.5 months each.

Immediately after bottoming out the shaft excavation, the base slab was poured followed by the formwork for the walls of the shaft. The formwork was installed in 4.9 m lifts followed by pouring of the concrete liner. Once poured the formwork was stripped and raised for the next lift. Cycle time per lift averaged approximately 1 week. Wooden block-outs within the liner were also installed at the adit connection locations.

After the entire height of shaft line was poured and cured, the freeze pipes for the shaft were shut off.

FROZEN ADITS AND TUNNEL CONNECTIONS

The frozen soil for the adit connections provided both a temporary earth support system as well as a ground water cutoff. The adits were designed as solid frozen soil masses. The dimensions of the frozen masses are outlined in Table 2.

At a minimum, 1.8 m. of frozen soil was required by design around the designed excavated area.

Unlike the shafts, the adits were considered sufficiently frozen when temperature monitors combined with thermal modeling indicated a complete frozen mass. As a precaution, probe drilling was performed ahead of the excavation to verify anticipated frozen soil conditions.

Excavation of the adits was performed utilizing Sequential Excavation Methods (SEM). A Brokk utilizing a roadheader type mining attachment was used to excavate the adits. Upon completion 1.2 m of excavation, a lattice and girder support system was installed and 28 cm of shotcrete, installed in 2 passes, were used as additional excavation support. This was necessary due to the fact that freeze pipes were removed during excavation and thawing would occur prior to the installation of the final lining.

Table 2. Frozen soil mass dimensions (m)

Adit Location	Length	Width	Height
Adams Street Adit	26.5	8.2	8.2
Adams Channing Tunnel Connection	8.5	7.9	8.2
V Street Adit	24.7	9.7	11
Pump Station Adit	25.9	9.1	10

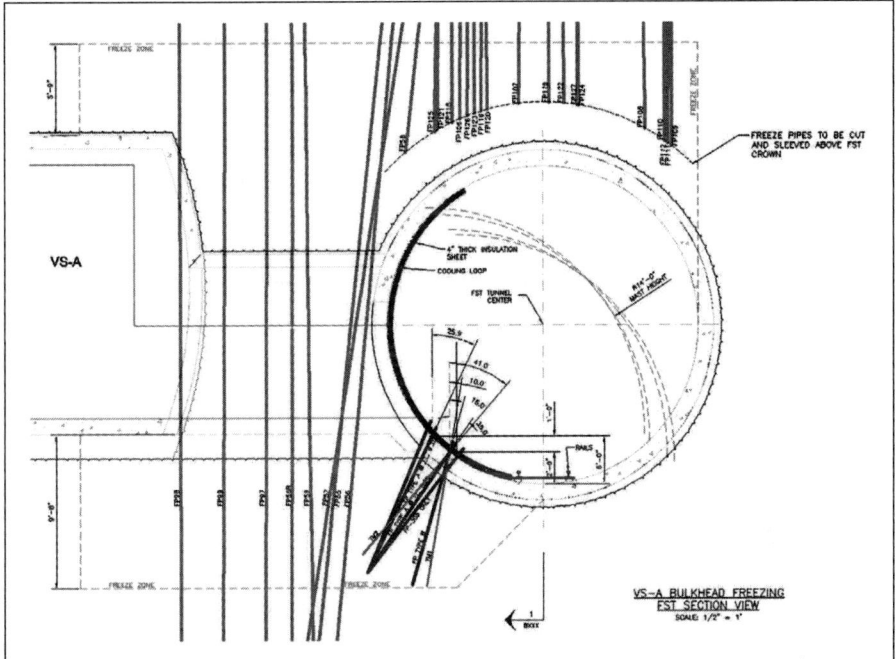

Source: Rob Chamberland, MTAC
Figure 5. Tunnel connection drawing

Schedule required the passing of the TBM through the adit connection areas prior to completion of the adits. As such, a latter connection between the frozen adit and the newly mined tunnel was necessary. Freeze pipes that were installed prior to the TBM mining had to be decommissioned once the main tunnel passed. To facilitate such connection, with the concerns of thawing, additional freeze pipes were installed from within the tunnel through the invert to ensure a sufficient frozen mass beneath the connection.

Trumpets were mounted on the tunnel liner and core holes were drilled though a series of valves and wiper seals to prevent excess soil loss in the event that unfrozen soil was encountered. Once the freeze pipes were installed, urethane grout was used to seal the protrusion through the tunnel liner. Additionally, cooling loops were secured to the tunnel lining along the excavation line to ensure that a sufficient seal was maintained between the concrete lining and the frozen soil.

HEAVE AND MITIGATION

Because of the difficulty in accurately estimating the freeze related heave and/or settlement, the approach for the First Street Tunnel was to go beyond evaluation of freeze related heave through modeling and to incorporate freeze heave mitigation as part of the project-wide protection of structures process. The protection of structures process required performing analyses to determine allowable and maximum movements or deformation limits and, if required, mitigations for any impacts during construction. During construction, information from instrumentation and monitoring of the existing utilities and structures was used in the selection and timely deployment of mitigation measures for excessive heaving. Typical mitigations that were used included

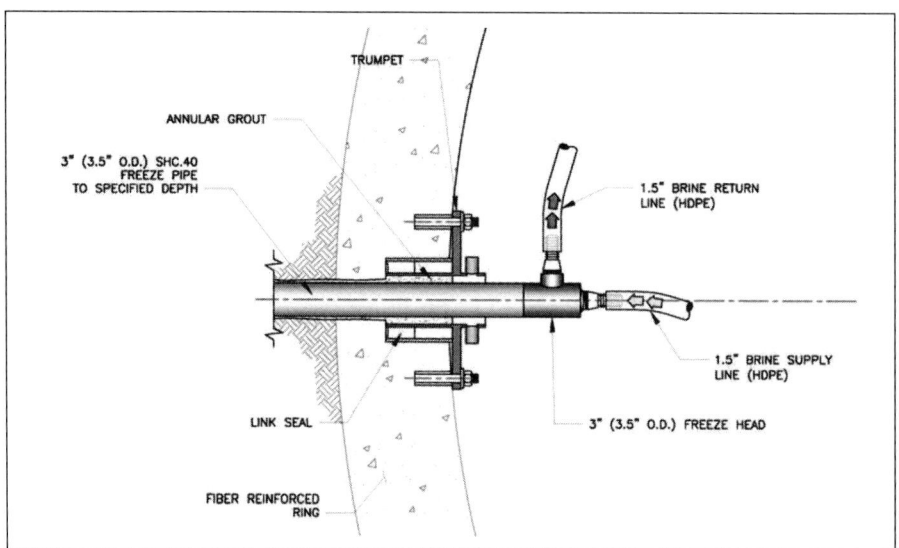

Source: Rob Chamberland, MTAC
Figure 6. Trumpet and freeze pipe detail

installation of heat trace pipes to cut off or slow down freeze growth and modifications to the brine flow in the freeze system. Other mitigation measures included systematic isolation or shut-down of portions of the freeze system to control freeze related ground deformations and protect existing utilities and structures.

CONCLUSION

In order to accommodate the specific needs of this project all methods of work had to be evaluated. Due to the rigorous constraints relating to available area, noise, structural stability, and environmental concerns, ground freezing was proven to be the necessary method of support of excavation.

With the unique structure and layout of the First Street Tunnel project, alternate ideas and concepts of supply and install of brine were critical to the success of the project. Implementation of innovative concepts within the ground freeze technology allowed for lessened impact to the surrounding community. Accurate monitoring and survey throughout the entire process proved critical to ensuring real-time data updates for schedule progression along with mitigation of any issues within the system.

Exposing Young Engineers to Multiple Facets of Tunneling

Jeff Brandt ▪ Traylor Bros., Inc.
Eren Kusdogan ▪ Traylor Bros., Inc.
Nick Tabor ▪ Traylor Bros., Inc.
Dillon Tew ▪ Traylor Bros., Inc.

ABSTRACT

Firsthand experience is best suited to expose and train new engineers in unfamiliar tunneling methods. This paper illustrates the advantage of how hands-on training through the collaboration of different, non-competing companies can be advantageous in preparing young engineers for the challenges of technical leadership positions. To further broaden their engineering background, four engineers were temporarily placed internationally with a partner company on two existing projects, the Neubau Zierenberg Tunnel and Tunnel Karl-Friedrich-Strasse. The experiences gained have been indispensable in implementing the Sequential Excavation Method and the New Austrian Tunneling Method on their projects, increased exposure to the variety of toolbox selection methods, as well as shown the differences with foreign client-contractor relationships.

INTRODUCTION

Underground projects across North America have been growing larger in scope and complexity. As these jobs are built, they often involve several different tunneling methods and require a large number of engineers and superintendents to plan the work. With several projects on the horizon for many contractors, it is a necessity to have enough personnel with specialized technical knowledge in a variety of tunneling methods. The opportunity to get this experience from within one organization can be difficult, especially as some companies favor a "niche" project set and are therefore limited in their exposure to different tunneling methods.

A distinct focus of Traylor Brothers is to develop their young engineers in as many different methods of tunneling as possible early in their careers to give them the skill set that they need to succeed in any tunneling environment. However, as many professionals can attest to, the best way to gain this experience is through hands-on experience. Three of the projects that Traylor Brothers has recently been awarded involve both mechanized and conventional tunneling in difficult ground requiring staff that were competent with a variety of tunneling methods. Without any similar work being performed among active projects, there were few opportunities to learn the technicalities of difficult conventional excavation methods within the Traylor Brothers organization. Therefore, Traylor Brothers looked outside of their company to a familiar partner, BeMo Tunneling.

BeMo has been a foundation in tunneling construction for the last 50 years in the European market. Traylor Brothers and BeMo have a history together since the end of the 20th century working as a team to complete challenging jobs in the underground industry, particularly when projects call for expertise in NATM. The experience brought by BeMo to many Traylor-led projects has allowed Traylor Brothers to pursue work when the personnel requirements do not match the availability within the company. BeMo's specialties made them the perfect partner for this unique learning opportunity.

Traylor Brothers needed to place their next generation of engineers inside a NATM environment without having any ongoing NATM work. In order to get their hands dirty, BeMo and Traylor Brothers implemented a "knowledge-share" program where four engineers were placed on various NATM jobs throughout Germany. For three months they integrated with the BeMo project teams and tried to learn as much as possible. The projects that were chosen had similar characteristics to those that the Traylor engineers would be directly responsible for managing in the future. These projects included large caverns with low overburden in both urban and rural environments, mixed face excavations with drill and blast typical mechanized excavation, smaller diameter tunnels with difficult access, and even a soft ground tunnel with a high water table where compressed air was utilized to balance the earth pressure. With this diverse project set, there were many experiences and lessons-learned that could be shared with the visiting engineers who in turn could bring them back to their projects in the states.

EXPOSURE PROBLEM

Providing the necessary training required to plan and manage large difficult work is a challenge common throughout the underground industry. Young engineers and personnel that are new to this kind of work need training outside of the classroom to get a full understanding of what is at hand. A problem encountered within the industry with large project scales and durations is the amount of time that an individual will spend on a single job or specifically focused on a single method of tunnel excavation. While the employee excels in their field, they may become quite specialized in a single method of excavation without much knowledge in other styles. When the opportunity presents itself to manage a different project, the employee may lack the proper skills to plan and execute the work as efficiently as possible. This sense of specialization can seem limiting in comparison to peers who have a larger variety of work experience but at a much smaller level of responsibility. This creates the dilemma where an exceptional engineer that may have the necessary aptitude to manage the work can get overlooked for an individual that has more experience with the method being implemented.

This limited firsthand experience can also promote a confirmation bias within a company. As firsthand experience becomes limited to fewer past projects under a consistent set of factors, guidance for the upcoming generation can become limited. Fewer individuals have seen an application or method performed under different circumstances or can offer objective comparisons of the different factors that were at play. "This was how it was done in the past, it worked for that application," but all the contributing factors are not evident without seeing comparisons. With larger projects, there are larger consequences to not catching some of these contributing factors.

Smaller companies still need to expose the upcoming generation of engineers to a large range of differing job sites. In order to diversify the experience within the company, training opportunities must be sought outside the normal scope of work. Great exposure can come from placing employees in a completely foreign work environment with a non-competing company that is committed to exposing that engineer to their full range of capabilities. This is a win-win for both companies as the host can not only advertise their capabilities and develop their relationship, but spur their own employees to examine their practices and single out what works and what needs improvement. The exchange engineers in the program were exposed to multiple generations worth of practical field experience and were able to bring back a significant amount of information that otherwise would have gone unrealized. This had immediate impacts on the upcoming work and was a contributing factor to many of the decisions made during the preplanning activities.

REALIZED BENEFITS

Equipment Selection

One of the most advantageous takeaways from the experience was the exposure to a large array of different mechanical excavation and related hauling equipment. The introduction to new types of equipment not commonly used in North America was beneficial to look at the processes in a different way. The dialogue and conversation with the operators, foremen, and engineers that purposefully selected that piece or pieces of equipment for the intended application had immeasurable benefits. While a certain setup was preferred for the job at hand, there were discussions as to when and why they would prefer something else. There were years of experience and knowledge on hand that within BeMo was considered common knowledge. They had all been exposed to the same information and experiences and had collectively disseminated those experiences among themselves. To them it was nothing new but to the uninitiated, it was a wealth of information that had immediate impacts. To be specific, there were opportunities on the Maryland Purple Line to revisit the equipment selection and ultimately see huge cost savings.

ITC vs. Tunnel Excavator

On the Maryland Purple Line the project team was deciding between using an ITC or a Tunnel Excavator to excavate the Plymouth Tunnel. Based on Traylor's past projects and other successful jobs in the area, the decision at bid time was to pursue the ITC. While on the BeMo projects, there were opportunities to speak with the operators and superintendents that had seen both setups in multiple ground conditions and varying heading sizes. They focused on what made each setup successful rather than the equipment itself and were able to get past the general sales pitch and cut sheet comparison. Their decision-making put less focus on brand recognition or specs and instead emphasized the specific needs of each individual project. The past projects were not successful because they used the ITC, they were successful due to the ITC being the right selection for those projects. This type of knowledge was only achievable through real hands-on experience and having lived through the learning curve. Placing engineers under the influence of outside professionals enables them to pick up knowledge that broaden the discussion within an organization. Ultimately for the Purple Line, further discussions with the teams back home led to making the switch to the tunnel excavator and revising the mucking scheme accordingly.

Scoop Trams vs. Loaders

The original proposal for Purple Line was to use haul trucks for the ITC and scoop trams with the bottom bench. Based on the revised excavation equipment and consequently the larger top heading, it became possible to go with articulated haul trucks and a conventional loader. This was the preferred method for the BeMo teams whenever possible and it looked like a good fit based on their factors. Further discussions back in the US brought out other points that needed to be considered, mainly limited NATM experience in local labor force, different safety culture & expectations, and how union labor rules would impact each. Ultimately the decision was made to not pursue the haul trucks due to the limited size and the safety concerns with inexperienced personnel loading the haul trucks in such a confined environment. For the traveling experienced BeMo crews, this had become second nature but for the untrained local work force in the US, this was a large safety concern. Since the scoop trams were planned for the bottom bench phase anyway, it was possible to tram the material out the length of the tunnel and no additional equipment was needed. Again, while each point may or may not be correct, at least the questions are being asked and the status quo is being reaffirmed.

Grout Truck and Equipment Skids

Challenging the status quo also brought up changes to the means and methods that could have huge impacts on productivity. Previous methods to handle a simple operation, grouting for example, had always been done the way they'd historically been done. This was the case within both companies however each had chosen to optimize the operation differently. While Traylor had come up with setups when necessary, the model was always around the same equipment and a locomotive transportation method. This covered the majority of instances so there just wasn't the need to develop a method to handle the one off grouting situations on the surface or after demobilization. Due to the nature of NATM work, BeMo had been forced to create a mobile grout truck that was self-sufficient and provided everything they needed for their operation. While in itself a simple item, it was completely new to those unfamiliar with their workflow. This was another example of an item that was deemed common knowledge by one company yet is completely knew and foreign to another. While many senior personnel may have seen something similar or used a different setup that works equally as well, these items are so small that they don't stand out as being essential at first. There's only a limited time to pass along certain concepts during passing discussions and many of the smaller details are left up to the younger generation to figure out or preplan. It's only after such an item is missed that the true importance is realized but by then we're already playing catchup. While in itself a very simple setup, it would have been a huge stumbling block and probably still is for a few companies. The act of grouting went from a major activity requiring preplanning, mobilization, dedicated crew time, and storage space in the heading to just another mobile piece of equipment that is self-sufficient, can work independently, and can integrate into the work cycle without interruption.

Similar ideas were evident in the equipment skids. Supplies and materials were kept on skids rather than the invert and moved along with the face. This minimized time spent moving stockpiles multiple times and cut down on travel time from the face. Not only were the skids used to store material, but years of experience had led to further measures being in place. For example, the transformer skid also provided secure storage for the superintendent as well as the most expensive surveying equipment. Why? Because the one thing the operators are most likely to not hit is the largest, brightest, lethal box that is powering the machines. Outlet and drop plugs are mounted on the front of the wire mesh skid. Why? Because the wire mesh is best stored in the large sheets and then cut to fit based on the round length for the day. This requires grinders, requiring power. Hooks are mounted to keep just such tools at the stations they are needed. Lights are mounted, and permanently wired to a common plug, so they provide light to the areas of each station. Simple steps can be taken in terms of lighting, power drops, charging stations, personal storage locations, utility hangars, spare lights, wedges, bolts, etc. All of these steps while they may be common knowledge to an employee of company A, they are only evident to employee of company B when they experience it in the field. These small details can be easily missed during the initial planning phases and would only come to realization after the operation was underway. For a company such as BeMo who is often brought in as a consultant after work has commenced, being able to have all of these preplanning measures taken care of ahead of time can greatly increase their productivity from the start.

Shotcrete

As those familiar with NATM know, shotcrete is one of the most important but ever changing factors in the quality and safety of the tunnel. The protective support that the shotcrete creates is wholly important and can be disastrously weakened by small errors in the placement procedure. When the entire system works as designed, the shotcrete shell is a consistent supportive measure that creates a quality finish with minimal waste, both of product and time. If small issues are not investigated thoroughly or necessary adjustments are not incorporated into the system, the results can be disastrous or even deadly. A quality program that tracks the right factors and limits variations in the system is key to establishing a standard for success. One of the most important ways to create a quality product in a cost-effective manner is to streamline the operation as much as possible, eliminate waste, identify and correct mistakes early, and repeat the process as safely as possible. The opportunity to watch several strong tunneling operations with different shotcrete systems and varying application environments exposed the Traylor engineers to first hand field experiences that will be indispensable to their future success.

At BeMO, the project teams considered quality testing procedures and documentation to be as integral to the tunnel as the excavation itself. Implementing methods to check the whole shotcrete system in real time were the paramount duties of the heading engineers. Shotcrete test panels were produced on a daily basis for internal testing, ensuring the exact method of placement for the tunnel was being used to make the test panels. In this manner, the actual batched concrete, equipment, nozzleman and placement orientation could be verified. Testing included pull out strength, temperature, and constituent reactivity properties. All of this data was compiled on a daily basis and compared to the temperatures of the batched concrete and accelerator, placement rates, and life of the concrete. Using this extensive data, small adjustments could be made to the shotcrete system to maximize the properties of the tunnel lining. In addition, the same testing would be done by outside third party testing laboratories on a weekly (or if required, more frequent) schedule. The results were compared and any discrepancies immediately addressed before advancing the tunnel further. Further, core samples were taken intermittently when there was cause for concern or to verify that proper layering of the liner was sufficient.

Early strength results were compared with latent strength and poorly performing placement was not accepted. The early strength criteria are delicately balanced. Enough early strength is obviously necessary to support the immediate deformation of the surrounding ground and allow the continuous tunnel work. However, when the shotcrete achieves too much strength early in its life, it may not reach its full design strength to support the final deformation of the ground for the tunnel's design criteria. This is a critical item for tunnel liners that need to be watertight or are subjected to a very high loading, and must be explored in depth when determining the mix design. Shotcrete that achieves a very high early strength generally has a lower final strength. When the load from the deforming ground passes to a weak initial liner, cracks can form between the bonded shotcrete layers and create leaks or a potential mode for liner failure.

The balance of designing a shotcrete mix that reaches the proper final design strength while meeting the necessary early strength that does not limit desired advancement rates can be difficult. When the solution is not determined by changes in the mix design alone, other options were useful. The lattice girder in the tunnel advancement is often used only as a form for the tunnel lining, providing a guide for the profile and thickness of the tunnel. When the design of the lattice girder can be considered with the initial support system, the lattice girder will see some of the loading from the deforming ground and reduce the load that the initial shotcrete lining must support. This would then require a lower early strength requirement and provide time for the shotcrete mix to reach a higher latent strength.

An additional method observed to help increase both the initial and final lining strengths were the incorporation of fibers into the shotcrete matrix. This requires some additional cost for batching and definitely requires further quality control, but the results can be significant. Synthetic and steel fibers are commonly used in the shotcrete lining to provide additional strength to the system, often resulting in a more ductile lining that can yield more to deformation before failure and still allow a higher final strength. The addition of steel fiber can be investigated in comparison to traditional steel reinforcement, and in some cases, replace the traditional steel reinforcement in the liner entirely. However, the addition of the fiber to a shotcrete mix can have a great impact on the equipment used, the placement method, and the overall quality control measures that must be taken to keep the system operating. The most common issues that were observed when placing shotcrete were plugs in the system due to "balls" of tangled fibers that did not allow the concrete to be pumped properly and ultimately do not become part of the homogenous concrete mix. The impact to the cost of materials and installation rates must be considered before committing to the addition of fiber to the shotcrete mix, however it was noted that it was a preferred system by many of the engineers.

Similar to testing the properties of the shotcrete product, the equipment was cleaned and calibrated every shift. After every use, all the shotcrete systems were thoroughly cleaned and inspected. The accelerator and water lines were metered, and the dosage and the placement quantity was sampled and calibrated on a weekly, if not daily, basis. This allowed for consistent placement between nozzlemon and ensured that any orrors or adjustments were well-documented and communicated. It is much easier to troubleshoot a problem when only one or two variables are changing, as opposed to the whole system. The crews became so efficient at testing and calibrating their equipment prior to use that they could perform their tests without having to waste any time or material unless there was a flaw in their system. The integration of quality control beginning with the operators and their equipment definitely lends its strength to the quality of the product provided to the tunnel lining system. Traditionally, the amount of money wasted in shotcrete has a direct correlation to the amount of rebound present. Instead of accepting rebound and waste as a necessary evil of a shotcrete operation, exposure to different methods and techniques to manage the constantly changing variables showed what was possible. By committing to owning the shotcrete production QC rather than just for compliance, it is possible to target the desired characteristics and slightly alter the mix design to ensure high quality shotcrete. Not only will this produce a safer environment, it will save money through minimal rebound and waste. An example of this was experienced directly in the compressed air tunnel at Karlsruhe as there was excessive rebound being observed.

During a short stretch of the tunnel, shotcrete rebound began to build up at an excessive rate and the tunnel lining was leaking air to the surface at a volume of roughly 50 L^3/min. This required the compressed air system to be increased in production and power, resulting in a much costlier operation. All of the variables to the shotcrete system were diagnosed and there did not seem to be any issue with the equipment, mix, or application method. The mix was the same as previously provided in the tunnel and

Figure 1. Dillon Tew looks on as tunnel crews install spiles at the Neubau Zierenberg Tunnel

the issue could not be determined. The temperature of all materials were well documented, the strength of the mix was acceptable, the placement rates were unchanged from any previous settings and there had been no change in the supply system. However, after further investigation with the batch plant, it was discovered that there was a change in the gradation and shape of the aggregate in the mix. The change in aggregate still passed the initial criteria for the mix, but it affected the application of the shotcrete from the nozzle. It was discovered that the larger aggregate was not being powered through the air far enough to sufficiently stick to the paste of the shotcrete on the surface of the excavated tunnel. The larger aggregate would fall to the earth before reaching the wall, while the lighter aggregate and paste would still form the liner. This thin, poor matrix of material would dry and form tiny cracks that allowed the compressed air to pass through to the surface. The mix design had to then be changed and the gradation had to be more closely inspected. The results were positive, but the final adjustments had to be made at the nozzle itself. The distance that the shotcrete was being applied was decreased to provide a better quality final product. Being able to play a part in diagnosing this problem and working it through with tunneling crews that had so much experience in similar situations made this a short problem in the life of the tunnel construction, but a valuable lesson learned.

CONCLUSION

Traylor Brothers and BeMo Tunneling have had an ever growing relationship over the years. Through multiple projects, BeMo has provided the technical expertise and helpful guidance needed to round out the Traylor Brothers skill set. For the first time, Traylor engineers made the leap to the BeMo projects overseas and were able to experience the jobs firsthand. This not only provided additional training to the next generation of Traylor engineers but further strengthened the relationship between BeMo and Traylor Brothers. Both parties gained from this unique training opportunity and hopefully both will continue to reap the benefits for years to come.

Trenchless Tunneling

Chairs

Justin McCain
Tutor Perini

Richard Taylor
Traylor Bros., Inc.

Paradise Raw Water Intake: Fighting the Green River

Nicholas Joens ▪ Kiewit Infrastructure Co.
Matt Roberts ▪ Kiewit Infrastructure Co.

ABSTRACT

The TVA Paradise Combined Cycle Project included a Raw Water Intake structure designed to pull water from the nearby Green River in Paradise, Kentucky. The original plan was to sink a 100-foot-deep shaft into rock, and drive a microtunnel into the bottom of a cofferdam located 335 feet away in the river. Heavy rainfalls and fluctuating river elevations delayed progress of the cofferdam, forcing a change to the design. It was decided to eliminate the need for the cofferdam by raising the elevation of the tunnel into soft ground, and performing a wet retrieval in the river. This paper will walk through the decisions that led to the original design, the change and the outcome.

INTRODUCTION

Originally called Stum's Landing, the town of Paradise, Kentucky was established as a supply post located in Muhlenberg County along the Green River. Positioned within the Western Kentucky Coal Field region, the area and its underlying coal was sought after and ultimately mined by the Peabody Coal Company. The subject of John Prine's song "Paradise," coal mining was the biggest source of funding for the county, but arguably took its toll on the town and its surrounding area.

In the late 1960s, the Tennessee Valley Authority ("TVA") built a coal-fueled steam generating plant in Paradise, and in 1967 the town's last residents moved away for good. After mining operations concluded in the 1970s, the area adjacent to the coal plant was reclaimed by placing the excavated overburden material back over the mine site. This reclaimed area became the future site of a new, cleaner, gas-powered combined cycle plant contracted by TVA to Kiewit Power Constructors Co. ("Kiewit") in 2014.

PROJECT DESCRIPTION

The Paradise Combined Cycle Project was an Engineering Procurement and Construction ("EPC") contract with the TVA to construct a new gas powered production facility on the Paradise Reservation near the town of Drakesboro in Muhlenberg County, Kentucky. The new plant includes three Series Five combustion turbine generators, three heat recovery steam generators, and one steam turbine generator. Kiewit executed a subcontract with Layne Heavy Civil, Inc. ("Layne") to design and construct a Raw Water Intake system to supply the new power plant with fresh water from the nearby Green River. Working together with Kiewit, and their designer Bell Engineering ("Bell") out of Lexington Kentucky, Layne developed a plan to install a new concrete wet well shaft and a tunnel into the Green River.

Geological Conditions

LE Gregg Associates was contracted by Bell to complete a Geotechnical Report, while S&ME Inc. was asked to perform the drilling on site. Borings were taken in December of 2014, but were later abandoned due to alignment issues concerning the future intake's proximity to coal barge operations in the river. Starting in January of 2015, a

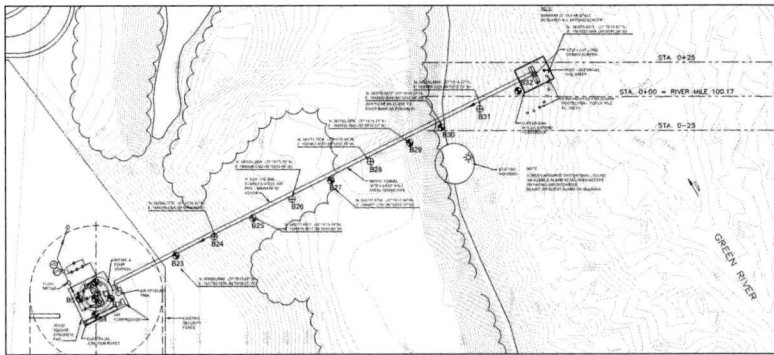

Figure 1. Tunnel alignment with borehole locations by Bell Engineering

new alignment was selected and six new borings were taken at roughly 50ft intervals along the tunnel alignment terminating at the river's edge (see Figure 1). The plan called for three borings to be taken out in the Green River, but a winter storm raised the river level, flooding it over the banks, and delaying any in-river work from occurring. A possible foreshadowing of things to come, it wasn't until May of 2015 that the three river borings were completed (see Figure 2).

Figure 2. S&ME barge drilling in Green River

Results from the borings determined the bed rock to be approximately 55ft below grade; diving deeper as it got closer to the river. However, the in-river borings showed a higher elevation indicating that the rock bottomed out in-land rather than beneath the river; likely a result of the river shifting over time. It's important to note, during the initial drilling, furthest away from the river, an obstruction was encountered that was believed at the time to have been a boulder. Previous geotechnical drilling performed elsewhere on the site indicated several zones of sandstone boulders upwards of ten feet thick. Later, while excavating elsewhere on the site, civil crews encountered a boulder the size of a small car. As such, the job team believed encountering a boulder was a significant risk to the future tunneling.

LE Gregg's lab report indicated subsurface materials categorized as alluvium; expressed primarily as lean silty clays and sand with flood plain deposits. Beneath the alluvium was the Carbondale Formation, consisting of sandstone, shale, limestone and coal. Compressive strengths in the rock were generally around 7 to 10ksi, with localized peaks ranging upwards of 30ksi when encountering a very hard bioclastic limestone material. Luckily the layers encountering the bioclastic material were very thin, less than 18 inches and could be tracked over the course of the alignment. Conversely the soil material had "N" values in the range of 0–10 exhibiting very soft, zero blow-count material at times. With this information in-hand it was decided to place the future tunnel alignment in rock to eliminate the risk of boulders and provide for a more consistent, and conservative approach to the tunneling.

DESIGN

In order to supply the new plant with sufficient cooling water to meet the demand, 3ea—200hp pumps were deemed necessary. These pumps would be supported by a 3ft thick concrete slab, overtop of a 100ft deep, 20ft inside diameter shaft that would serve as the wet well. The bottom third of the shaft would be positioned in bed rock with a reduced diameter of 18ft, while the bottom 6ft would be slopped towards the pumps to prevent any silt from building up. Although the design was performed by Bell Engineering, the shaft was primarily designed by Shirk & O'Donovan because of their experience designing sunken caisson shafts for Layne's collector well division.

At the bottom of the shaft, a standard wall 36in uncoated steel interlocking pipe would connect the shaft bottom, through the rock, to a roughly 13ft long stainless steel intake screen positioned inside the Green River 335ft away (see Figure 3). Steel protection piles would surround the intake protecting it from debris and a pneumatic air-blast system would be attached to the screen to clear it of any smaller debris that may obstruct water from entering the shaft.

Layne subcontracted Huxted Tunneling ("Huxted") to perform the microtunneling and install the interlock pipe. The microtunnel machine was planned to be received inside of a cofferdam located just off shore in the river. This cofferdam was planned to be a roughly 25ft-by-25ft sheet pile box with four internal walers and was sized to provide adequate room to remove the machine and install the remaining pipe upon completion. The cofferdam would sit on rock and a "slot" would be blasted out of the rock to allow for machine retrieval. This entire process was planned to be performed in dry conditions. At the time, a dry retrieval inside of the cofferdam was planned to reduce the need for underwater divers, and the risk of contamination that may result from a wet retrieval.

The Paradise site is situated so close to the Green River that environmental issues were always a top priority and concern. Additionally, since the intake would require work to be performed in the river, the cofferdam and microtunneling were considered the riskiest elements of work performed on the entire project from an environmental perspective. Even a small issue would result in a substantial fine.

Figure 3. Raw water intake original design model

CONSTRUCTION

Construction began in July of 2015 by sinking the new shaft in overburden using the sunken caisson technique. The shaft consisted of six different concrete lifts. The first lift was 7ft tall and contained a steel cutting shoe. This first lift also included steel fibers within the concrete to help reduce any damage from the subsequent blasting operations. In addition, grout packers were placed horizontally through this first lift, in the event they would be needed later on to help seal the soil-rock interface. The remaining concrete lifts were each 12ft tall and reinforced with No.6 rebar throughout.

After each concrete lift was placed on the surface, the inside of the shaft was excavated using a crane and clamshell bucket while hydraulic cylinders attached to a concrete deadman pulled the new concrete into the ground. The shaft was excavated and sunk until it was roughly 2ft above grade, where the process of placing a new lift would begin again. This cycle continued until the cutting shoe of the first lift was resting on rock.

Once the shaft was bearing on rock, and sufficient cure was achieved, work began on sinking the remaining 32ft in rock. Ring beams and liner plates were used to support the in-rock portion of the shaft, while the annular void was backfilled with grout. The rock in the bottom of the shaft ended up being soft enough to excavate using a hydraulic breaking hammer for all but three lifts of limestone where the use of explosives was required. Upon reaching the bottom, a working slab was placed in preparation for microtunneling.

Changes in the River

While the shaft was being excavated, workers roughly 300ft away began installing the cofferdam in the river. However, as pile driving operations commenced it was quickly discovered that the anticipated depth of rock in the river was not matching the data obtained from the in-river borings. In early September 2015, after several weeks of driving test piles, it was discovered that the rock formed a 10ft cliff, dropping off at the midpoint of the proposed cofferdam location (see Figure 4). This cliff effectively rendered the original four waler cofferdam design insufficient. A more level surface was needed.

Figure 4. Three-dimension image of river bottom

The test pile operation continued for another few weeks until a flat surface could be found in the river. Eventually, an area was located approximately 15ft away (diagonally) from the original location, extending the length of the microtunnel by about 10ft. However, this new location was deeper than the original and making matters worse, geologist-performed lab results indicated the soil in this new area was softer than anticipated with a soil friction angle (Phi) of only 15. Phi angle is a property used to measure shear strength in soils, essentially the ability of soil to stand on its own.

Re-Designing the Cofferdam

Coupled with the increased depth, this new Phi angle was enough of a change to force the job team to attain a new cofferdam design. After several more weeks of working

Figure 5. Green River flooded above elevation 375.5

with the designer, it wasn't until November 2015 that a new more robust five-waler design was completed, with one external tension brace. In addition to the extra waler and brace cost, the lower-than-expected Phi angle provided less passive resistance "toe" to the cofferdam, effectively increasing the amount of sheet pile steel required, and forcing the need to buy entirely new sheets.

With a new location, design and revised permit in hand the job team was again free to start installing the cofferdam into river. However, at this time, in early December 2015, the Green River was flooded over its banks due to an abundance of precipitation in the area. With only a handful of productive work days in December, and even less in January the next year, the job team met in early February 2016 to discuss the impact the river was having on the schedule, and what, if-any, changes could be made to get back on track.

Eliminating the Cofferdam

Late 2015 and early 2016 saw considerably high water levels in the Green River during construction of the early phases of the cofferdam as shown in Figure 5. The initial design assumed a maximum water level of elevation 375.5. There were five periods and 57 total calendar days where the water level exceeded elevation 375.5 due to heavy rain fall. The high water levels left the completion of the cofferdam construction unattainable without a complete redesign which would add a considerable amount of cost to the Project.

With river levels too high to work and with the wet season approaching, the team including TVA, Kiewit and Layne, met in February 2016 to discuss options to keep the construction moving forward and mitigate the loss of schedule associated with the high river levels. If something didn't change soon, commissioning, and overall completion of the plant would be delayed. Redesigning the cofferdam to accommodate higher river levels was analyzed as the first option to salvage and include items of work that had already been completed. Initially, this was thought to be the best way to mitigate the schedule loss and eliminate the risk of additional high-river delays.

It soon became apparent that re-designing the cofferdam for the higher river levels encountered would require the cofferdam construction to essentially start from the beginning, and require more and much larger steel members, in order to provide a dry and safe work environment for the microtunnel machine recovery.

At the same time the job team was evaluating the re-design of the cofferdam, an option to eliminate the cofferdam altogether was also being entertained. Eliminating the cofferdam would also eliminate the possibility of a dry retrieval of the microtunnel machine. Layne brought their microtunnel subcontractor, Huxted, in to discuss the feasibility of performing the microtunnel drive with a wet retrieval directly into the river. Collaboration meetings were held between TVA, Kiewit, Layne and Huxted to fully understand the process and associated risks with this type of change. Finally, after weighing all options, it was jointly decided that eliminating the cofferdam, raising the microtunnel into soft ground, and performing a wet retrieval afforded the job its best opportunity to save valuable time and money.

Microtunneling

As the partially-built cofferdam was being removed from the river, work began to facilitate the higher microtunnel alignment. This higher alignment was now located in a different geologic condition. As previously mentioned, the lower alignment was initially chosen in rock to reduce the risk of obstructions in the soft ground and to encounter a more consistent condition for the microtunnel machine. After the alignment was raised the ground expected to be encountered was obviously softer and the potential to encounter obstructions like boulders increased. This change of anticipated ground conditions warranted Huxted to revise the cutterhead configuration of the machine from rock to soft ground as shown in Figure 6. Fortunately, Huxted was able to utilize the same Iseki Unclemole machine that was planned to be used in rock, by adapting the cutterhead to the softer ground prior to mobilizing on site.

Figure 6. Huxted's Iseki Unclemole MTBM used at Paradise

With the potential risk of encountering an obstruction (boulders) now greater in the higher soft ground alignment, the team met to discuss mitigation measures if an obstruction was encountered. In addition, the team also discussed the potential to drill new bore holes or to drill a pilot hole to identify any obstructions along the new alignment. Additional vertical and/or horizontal bore holes would have been done prior to the drive, however, it was ultimately decided not to perform the bores. Rather, if an obstruction was encountered, a small drilled shaft would be utilized to remove the obstruction along the alignment as a contingency.

In preparation to launch the machine, the completed shaft that was excavated to the lower microtunnel alignment was filled with road base gravel, and a concrete mud slab was placed on top. Huxted mobilized in the end of March to setup their slurry tank farm, power, utilities, and control container. The machine was launched on April 5, 2016 and went substantially better than originally scheduled, taking only seven days to complete the 345ft drive with an average production of 50ft per day.

One of the biggest challenges and highly-focused tasks of the microtunnel operation was the wet retrieval of the machine at the river. Prior to penetrating the river the bulkhead door at the tail end of the machine was also closed and sealed to eliminate flooding the machine. Once the machined holed through into the river, it was temporarily supported by pre-driven piles with cross braces. Furthermore, a barge crane and chain fall system provided additional support until the interlock pipe was pushed into its final position. Before the machine could be retrieved from the water, a 36in butterfly valve

Figure 7. Removing the MTBM from the river

was installed in the shaft before completely filling the pipe with water. This was done to eliminate the risk of flooding the shaft and to reduce the risk of injury to the divers while cutting the machine free. Once the last section of pipe was cut free the machine was then lifted to the barge as shown in Figure 7, and the intake screen was installed.

CONCLUSIONS

The Paradise Raw Water Intake was a prime example of collaboration between Owner, Contractor and Subcontractor to solve a problem and achieve a common goal. Risks were identified, discussed and given ownership early in the design process between all parties. It's because of this, that when things started going poorly in the river, focus could be aimed towards developing a solution quickly to save schedule timo and money.

Changing a plan you've spent countless hours developing, is not easy and pride of authorship can often impede one's ability to react appropriately. However, in the case at Paradise, careful study of all factors and making decisions jointly as a team was what prevented the problem from escalating too far. According to Socrates, "The secret of change is to focus all of your energy, not on fighting the old, but on building the new."

ACKNOWLEDGMENTS

The authors would like to thank Adam Ralph of Layne Heavy Civil, Jamie Cook with the Tennessee Valley Authority and Huxted Tunneling for their help in supplying information, and for permission to publish this paper.

REFERENCES

Roberson, Cleo and Anderson, Jan. 2008. *Images of American Muhlenberg County.* South Carolina: Arcadia

LE Gregg Associates. 2015. *Geotechnical Engineering Exploration*—TVA Paradise Combined Cycle Power Plant Raw Water Intake—Drakesboro, Kentucky. Kentucky

S&ME. 2013. *Final Report of Geotechnical Exploration*—Combined Cycle Plant TVA Paradise Operations—Drakesboro, Kentucky. Kentucky

Dugway South CSO Relief and Conveyance Sewer, A Critical Connector

David Mast ▪ AECOM
Karrie Buxton ▪ Northeast Ohio Regional Sewer District
Amanda Foote ▪ AECOM
Irwan Halim ▪ AECOM
Alison Schreiber ▪ Northeast Ohio Regional Sewer District

ABSTRACT

The Dugway South Relief and Consolidation Sewer is a critical component of the NEORSD Easterly CSO control system in Cleveland, Ohio, connecting four (4) other CSO tunnels in a dense urban neighborhood. Design and construction challenges included: previously fixed pipe elevations necessitated both soft ground and rock tunneling on the same alignment, underground hydraulic structures in poor ground conditions, tunneling beneath a critical unreinforced concrete box culvert built in the 1910s, contract constraints necessary for coordination with adjacent CSO contracts, and potentially gassy ground. The flexible bid documents allowed for a variety of construction means and methods, improving bids.

INTRODUCTION

The Northeast Ohio Regional Sewer District entered into a consent decree with the United States and Ohio Environmental Protection Agencies (OEPA/USEPA) and the Department of Justice (USDOJ) in 2011 to implement, within a 25-year time limit, a long term control plan to control combined sewer overflows (CSO) currently impacting Lake Erie and its tributary streams. The consent decree is a public document that can be accessed through the NEORSD Project Clean Lake website. The plan includes a series of control measures comprised of wet weather plant upgrades, collection system improvements, remote storage tanks, and implementation of green infrastructure technologies limiting storm inflow. The Dugway South Relief and Consolidation Sewer (DSRCS) is part of Control Measure #6, Euclid Creek/Dugway Storage Tunnel System, identified in Appendix 1 of the consent decree.

In 1998, the Northeast Ohio Regional Sewer District (NEORSD) undertook the Easterly CSO Phase II Facilities Plan (P2FP) to identify causes and develop solutions for the activation of combined sewer overflows (CSOs) in the Easterly Service District. Under the P2FP, diversion structures, consolidation systems, and drop structures were required to convey the second highest peak flow in a typical year, and the tunnels were required to capture and convey all but the four highest volumes produced by typical year storms. As a result of the consent decree negotiations, the conveyance structures must be sized for the highest typical year peak flow (equal to typical year Storm 60), and the tunnel system must capture all volume which is less than the two highest typical year storm volumes.

The Dugway East Interceptor Relief Sewer (DEIRS) and Dugway West Interceptor Relief Sewer (DWIRS) projects were developed and presented as Advanced Facilities Plans in 2004 and 2005, respectively. Because of property acquisition issues that arose during the design of DWIRS and impacted the southern portion of DWIRS, the

Dugway Storage Tunnel, the Doan Valley Storage Tunnel, and the interfaces between these projects, the configuration of these systems was revised. The DWIRS scope was reduced to exclude the realigned sewers and structures that are now the DSRCS.

CRITICAL CONNECTING SEWER SYSTEM

The DSRCS system will be connected to adjacent CSO control points that have been designed by others. Figure 1 shows the entire Easterly CSO Control system and how the DSRCS project is interrelated. At the upstream end, DSRCS will receive both dry and wet weather flow from the Doan Valley Relief and Consolidation Sewer (DVRCS), providing surcharging relief and capacity for the DVRCS system. The DSRCS will also receive up to 50 million gallons per day (MGD) of flow from the Doan Valley Tunnel (DVT) during storm events, and will be used after each rain event to dewater the DVT system. Two DSRCS control structures (DSRCS-1/2 and DSRCS-3) will regulate flow at the downstream end into one of the following three structures: Dugway West Interceptor Relief Sewer (DWIRS), Dugway Storage Tunnel (DST), and the existing Dugway West Culvert. Specifically, DSRCS flows up to 56 MGD will be passed through DSRCS-3 to the DSRCS-1/2 structure for conveyance to DWIRS. DSRCS flows between 56 MGD and 90 MGD will be conveyed to DST via the DSRCS-3 structure until DST reaches storage capacity. Once the DST is full, a gate structure adjacent to DSRCS-3, and designed by others, will close and stop all flow to DST, resulting in the remaining flow being conveyed to DSRCS-1/2. Flow entering DSRCS-1/2 at a rate of 92 MGD or less will be directed to DWIRS. If DWIRS is unable to accept the flows (i.e., pipe capacity has been reached) or if flow rates into DSRCS-1/2 exceed 92 MGD, excess will be diverted directly to the Dugway West Culvert.

The DSRCS system is comprised of the mainline alignment with six hydraulic diversion or junction structures, the Primrose Avenue Consolidation Sewer (PACS), and the Phillips Avenue Relief Sewer (PARS) and shown in Figure 2.

The Mainline Alignment is a 90-inch internal diameter tunnel approximately 2,700 linear feet in length. The six hydraulic structures, numbered to increase from downstream to upstream, serve as junction structures or regulator structures. At the downstream end, DSRCS-1/2 diverts flow to the DWIRS system or Dugway West Culvert. DSRCS-1/2 also joins flow from the PACS into the mainline alignment. DSRCS-3 is a regulator structure to divert flow to the DST during wet weather events. DSRCS-4 is a baffle drop structure to drop flow from the PARS into the mainline alignment. DSRCS-5 and DSRCS-5A are junction structure and contribute flow from two local regulators into the mainline alignment. DSRCS-6 is a junction structure that will combine flow from the DVT and DVRCS and send to the DSRCS mainline alignment.

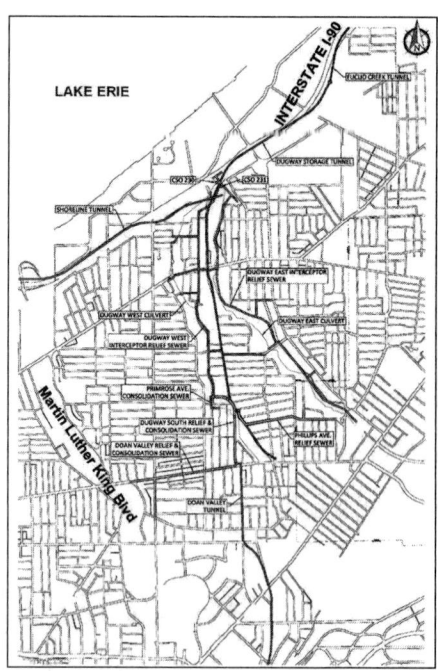

Figure 1. NEORSD Easterly CSO control system

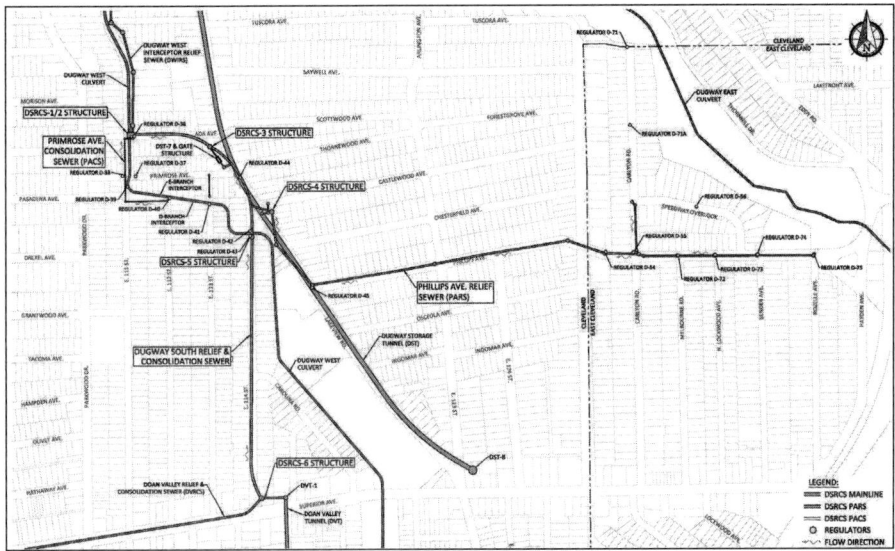

Figure 2. Dugway South relief and consolidation sewer project layout

The Primrose Avenue Consolidation Sewer is approximately 690 LF of new open-cut consolidation sewer ranging in ID from 12 to 30-inches. This sewer collects flow from five local existing regulators and conveys the flow to the DSRCS-1/2.

The Phillips Avenue Relief Sewer is approximately 4,100 LF of new relief sewer ranging in ID from 24 to 42-inches. Approximately 2,500 LF of the PARS was designed as a microtunnel and the rest of the sewer was designed as open-cut. The PARS is collecting flow from eight regulators for conveyance to the mainline alignment via DSRCS-4.

The DSRCS system is a critical connector to four surrounding NEORSD projects. During design of the DSRCS, two of the four projects were either well into construction or just beginning the bidding phase (DWIRS and DST, respectively). The DWIRS contractor constructed a shaft near the DSRCS-1/2 structure for the DSRCS contractor to make a connection between the two projects. Connection to the DWIRS project fixed the downstream invert elevation for DSRCS. At the upstream end, the invert was bound by the highest feasible invert for the DVRCS project and by dry weather flow velocity minimum requirements required to facilitate self-cleaning.

TUNNELING DESIGN

Geotechnical Conditions for Tunneling

Geotechnical conditions along Dugway South Relief Consolidation Sewer (DSRCS) tunnel alignments dictated the selection of tunneling method. These conditions are depicted on subsurface profile shown in Figures 3 to 5. The mainline tunnel (MT) alignment between DSRCS-1/2 and DSRCS-3 shaft structures (approximately 500 LF) is expected to be constructed in very stiff to hard cohesive glacial till near DSRCS-1/2 and into a mixed ground with highly weathered shale and competent shale toward DSCRC-3. The MT alignment between DSRCS-3 and DSRCS-6 (approximately 2,200 LF) is expected to be mostly in fair to good quality shale with occasional poor shale zones along the alignment. Highly weathered top of shale zone encroaches toward or near the tunnel crown in several locations.

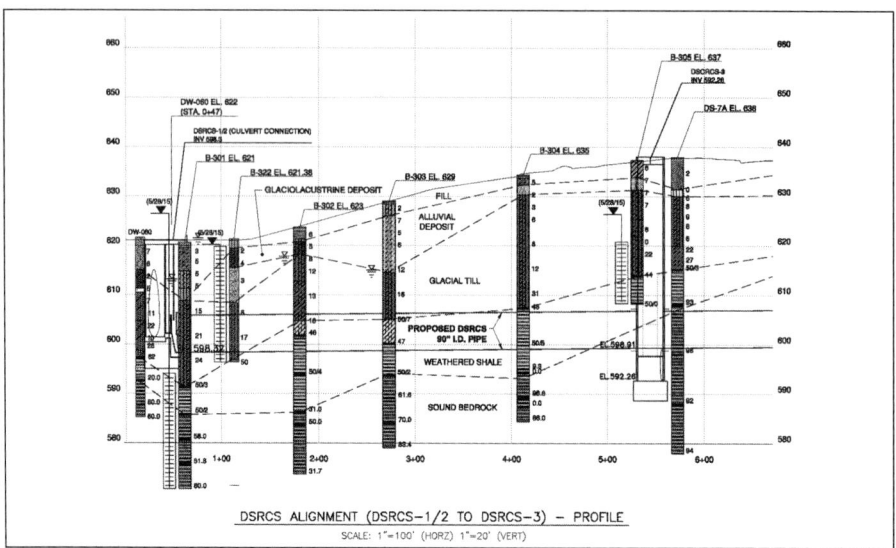

DSRCS ALIGNMENT (DSRCS-1/2 TO DSRCS-3) — PROFILE
SCALE: 1"=100' (HORZ) 1"=20' (VERT)

Figure 3. Subsurface conditions for mainline tunnel (MT)

The Phillips Avenue relief sewer (PARS) segment, west of Carlyon Road, must be constructed between 20 and 25 feet below ground surface to achieve hydraulic connections, with a significant portion of this sewer will be in competent shale rock. Open cut excavation in this material would be difficult using hydraulic hammer and rock ripper methods. As a result, this segment is being constructed using trenchless construction or tunneling methods, whereas the remaining shallower portion of PARS will be constructed in open cut.

Bedrock along the project alignment consists of relatively weak Chagrin Shale which is frequently interbedded with stronger siltstone and sandstone layers. When present, seams of siltstone and sandstone may constitute up to 22% of excavation face within the tunnel horizon which will impact the tunneling equipment design. Furthermore, methane gas is not uncommon in the Chagrin Shale formation. Gas was encountered on the west side of the project site near DSRCS-1/2 in previously drilled borings, as well as in observation well installed as part of this project geotechnical exploration. Therefore the DSRCS tunnel is considered potentially gassy during construction.

Existing Man-Made Structures

The City of Cleveland was established in 1796, and the adjacent City of East Cleveland was incorporated as a village in 1895. Rapid residential growth occurred in both cities in the late 1800s and early 1900s. The Dugway Brook originally flowed in a natural open stream channel. During residential development pre-1900, the Dugway Brook channel was partially realigned. The Dugway Brook likely originally extended across low, wet ground which was unfavorable to development. In approximately 1913, the Dugway Brook was realigned again and an unreinforced concrete culvert structure was built to enclose the Dugway Brook and its flows. The new culvert was constructed and backfilled with unknown materials and filled over to create additional buildable land. The Dugway West Culvert (DWC) alignment is shown on Figure 2. The new culvert cross section varied along the total length. Two of the structural cross sections present in the DSRCS Project area are illustrated in Figure 5.

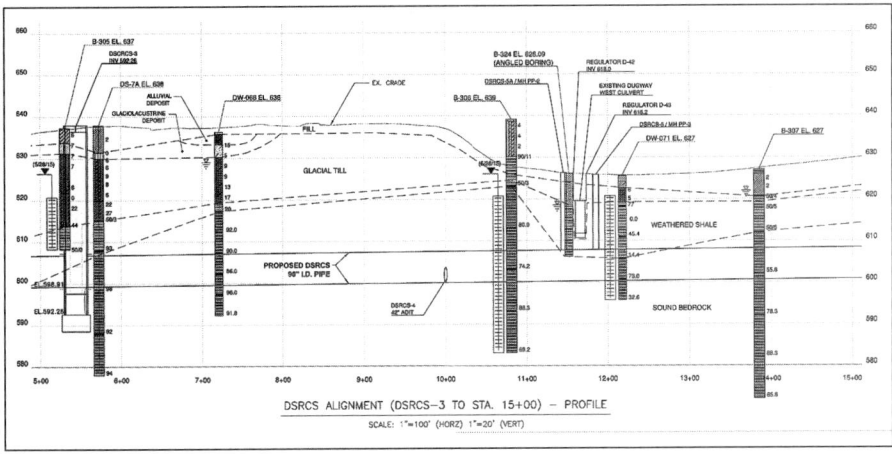

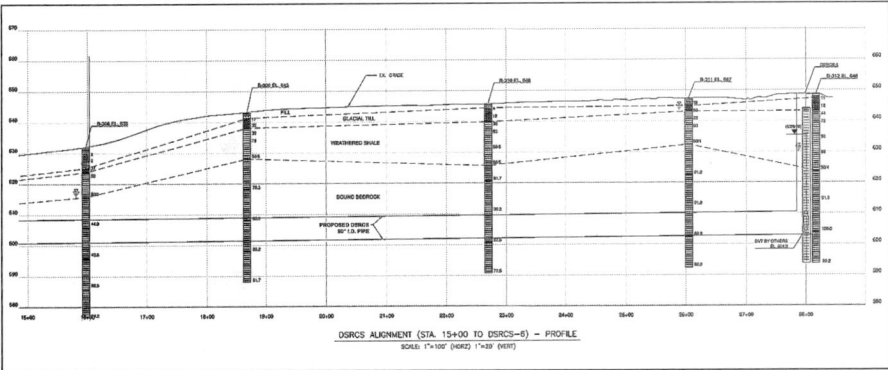

Figure 4. Subsurface conditions for mainline tunnel (MT)

Structure DSRS-1/2 connects the DSRCS mainline sewer to the existing DWC. Construction of the structure will require excavation of and removal of a portion of the DWC. During previous construction, it was found that the zone around the culvert was backfilled with loose random materials with groundwater inflow potential that must be considered during the new structure connection. Furthermore, the DSRCS will cross the existing DWC where the mainline tunnel must be excavated and a final lining installed directly beneath the existing DWC. Based on site investigation in this area, the culvert structure is founded directly on highly weathered natural shale bedrock.

Tunnel Design Considerations

During the design evaluation, geotechnical conditions, present risks, local contractor capabilities, schedule impacts, and costs were evaluated for each potential excavation method. The principal design approach was to open up alternative construction methods with similar underground risk profiles to foster a competitive bid environment. The most suitable tunneling method for the MT from DSRCS-3 to DSRCS-6 is anticipated to be a single-shield open-faced rock TBM with a 2-pass lining system consisting of tunnel initial support and final grout-in pipe liner. Initial support consisting of patterned rock dowels in fair to good quality shale and steel ribs in poor quality shale can be used in conjunction with a gripper TBM. As an alternative, a fully lagged steel ribs support that can also be used for TBM propulsion may be installed along the entire tunnel

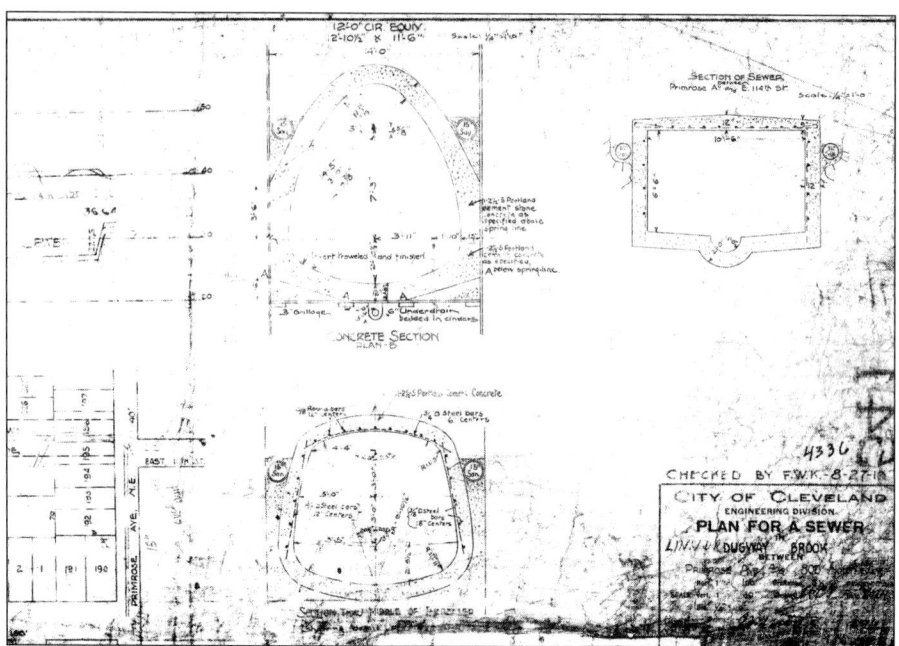

Figure 5. Portion of historical design drawings for Dugway West culvert structure

length. When crossing underneath the existing Dugway West Culvert, steel ribs support with much closer spacing and full-circumferential lagging is required to prevent damage to the existing culvert structure.

For the remainder of the MT alignment from DSRCS-1/2 to DSRCS-3, the tunnel was assumed to be built using a combination of dewatering and hand mining methods due to the mixed ground condition. The anticipated tunneling method in this portion was hand mining with mechanical excavator or digger shield with a 2-pass lining system consisting of steel ribs initial support and final grout-in pipe liner. However, the Contractor was not precluded from using a TBM outfitted with a cutter head to suit the mixed ground condition.

Along PARS, the tunneling ground condition is anticipated to be mostly in fair to good quality shale with occasional poor quality shale. The anticipated tunneling method was microtunneling construction with MTBM and either RC (Reinforced Concrete) or CCFRPM (Centrifugally-Cast Fiber Reinforced Polymer Mortar) jacking pipe, which would serve as the final carrier pipe (1-pass method). The MTBM will be launched and retrieved from working shafts, which are installed to construct the future access manholes along the pipe alignment. Alternatively, the Contractor was allowed to use an oversized single-shield open-face rock TBM with an initial support and final grout-in carrier pipe (2-pass method). Due to the limited size tunnel, rock dowel installation was assumed to not be efficient, and therefore a single type of initial support consisting of steel ribs with full-circumferential timber lagging was assumed by the designers.

FLEXIBILITY AND CONTRACT SETUP

Bidding Approach

The District and AECOM recognized that building flexibility into the bid documents would result in the most competitive bids by allowing each contractor to use its preferred equipment and practices, while at the same time not excluding potential bidders. For each unique sewer section, design details were included in the bid documents for at least one assumed construction method, primary lining system, and secondary liner. Bidders were permitted to select alternative methods within stated parameters and base their bid prices on the method chosen. Either new or refurbished tunnel or microtunnel boring machines were acceptable, provided that the machines could meet the requirements of the specifications. Contractors were required to disclose their selected methods in response to the bidders' qualification questionnaire. Figure 6 was included in the geotechnical baseline report (GBR) as a guideline of acceptable deviations from the detailed plans.

Excavation dimensions were provided in the contract drawings to accommodate the most probable space requirements. The contractor was assigned responsibility for determining the necessary excavation dimensions to accommodate the chosen means and methods, excavation equipment, initial supports, and final liner. Furthermore, the expectations outlined in the GBR were based on the most probable construction methods and the details provided in the contract documents.

Alternative Construction Methods and Materials

Shaft DSRCS-1/2 to DSRCS-3

The sewer reach from Shaft DSRSC-1/2 to DSRCS-3 comprises a 310-linear-foot straight section out of Shaft DSRCS-1/2 and a 174-linear-foot curved section with a 350-foot radius into Shaft DSRCS-3. The required, finished inside diameter of this sewer reach is 90 inches. Anticipated subsurface conditions included soft ground, mixed face, weathered shale and sound Chagrin Shale bedrock. Because of the short total length of this reach, it was assumed that mobilization of a separate TBM to handle the various conditions to be encountered would not be cost-effective. Consequently, the design was based on mechanically excavated hand mining, which would be feasible for all bidders regardless of the type of mining equipment selected for other reaches. The main excavation detail consisted of a horseshoe-shaped excavation supported by W4x13 steel ribs and timber lagging. Alternatively, bidders were permitted to use a TBM bored or shield driven tunnel with a primary support systems of steel ribs and full circumferential lagging or steel liner plates.

Shaft DSRCS-3 to DSRCS-6

From Shaft DSRCS-3 to DSCRS-6, the excavation zone was expected to be fully within Chagrin Shale ranging in quality from highly weathered to sound. The finished inside diameter of the secondary liner is 90 inches. TBM bored or shield driven tunnel was considered the most practical excavation method. The primary support detail, referred to in the bid documents as Type I, consisted of four equally spaced rock dowels and welded wire fabric in the upper 120 degrees of the tunnel excavation. The longitudinal spacing of the rock dowels was specified to be no more than 5 feet. A more robust primary support detail, Type II, was required in areas where poor quality shale was expected within the tunnel zone, where weathered shale was anticipated within one tunnel diameter or less above the tunnel crown, and under the existing unreinforced concrete box culvert. The Type II detail required ribs and welded wire

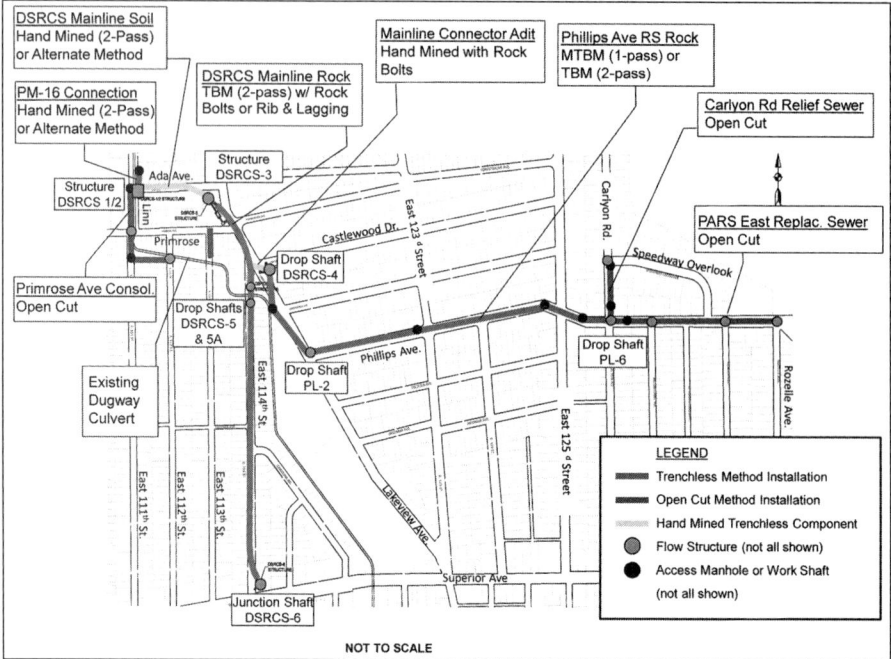

Figure 6. GBR figure providing guidelines for acceptable tunnel methods

fabric (WWF) lagging in the upper 220 degrees of the excavation. Full circumferential timber lagging was permitted In lieu of WWF. Contractors were also given the option of installing steel liner plates rather than ribs and lagging for the entire tunnel length.

Phillips Avenue Relief Sewer

The sewer reaches along Phillips Avenue and Lakeview Avenue, referred to collectively as the Phillips Avenue Relief Sewer (PARS), afforded the greatest challenges and the maximum opportunity for contract flexibility. Initially, open-cut trenching was considered a viable option for construction of the entire PARS. The sewer reaches on Phillips Avenue east of Carlyon Rd. were shallow enough to be cost-effectively constructed via open-cut methods. West of Phillips Ave., the depths necessary to meet hydraulic performance requirements and geometric constraints were in the 30-foot range. The required 40-inch inside diameter of the pipe would necessitate a wide trench. Furthermore, Phillips Ave. is a residential street in a densely populated urban area; and therefore, open-cut construction would be result in excessive disruption to the community. These factors, coupled with the cost of pavement restoration, increased the risk and estimated cost associated with open-cut construction. Consequently, AECOM and NEORSD determined that trenchless methods would be required for construction of the 42-inch sewer from Shaft DSRCS-4 to Manhole PL-6.

For the trenchless portion of the PARS, microtunneling appeared practical due to the relatively small sewer diameter. However, numerous directional changes made curves or intermediate shafts necessary. Although a curved microtunnel was recently constructed as a contractor-proposed modification on a NEORSD project, with risk accepted by the contractor, NEORSD had not yet bid a project with a curved microtunnel. Therefore, for DVRCS, Intermediate shaft locations were chosen to accommodate straight-line microtunneling. TBM bored or shield driven excavation could navigate curves and would therefore allow for reduction of the number of shafts. The annular space between the primary support system would be filled with low density cellular grout. In order to maximize competition, the bid documents included both microtunneling details and initial tunnel excavation support details, and contractors were permitted to base bid prices on either option. Additionally, bidders could elect to eliminate intermediate shafts that would not be finished as flow structures.

Bid Results

The engineer's opinion of probably construction cost (EOPCC) was $36,800,000. Eight bids were received, and the lowest bid price was $26,088,400. The second and third lowest bid prices were also below $30,000,000. Seven of the bid prices were less than the EOPCC. All but the two highest bids were based on using a TBM for the trenchless portion of the PARS. Most of the cost difference between the EOPCC and the lowest bid could be attributed to five bid items: 90-inch diameter trenchless sewer, tunnel boring machine for 90-inch diameter trenchless sewer, 42-inch diameter trenchless sewer, TBM/MTBM for 42-inch diameter trenchless sewer, and manholes. A comparison of the bid prices from the lowest bidder, second bidder, seventh bidder, and the EOPCC for those key bid items is provided in Table 1 along with information regarding chosen methods and assumptions. The bid results indicate that the flexibility provided in the bid documents led to significant cost savings for the NEORSD

Final Construction Approach

After the bids were received the District and AECOM were able to review the contractor's chosen construction methods based on the bid flexibility. The lowest and best bidder's construction method, primary lining, and secondary lining methods are described in the following paragraphs.

DSRCS Mainline Sewer Modifications

The contractor proposed to adjust the alignment of the DSRCS mainline tunnel by increasing the radii of the curve between DSRCS-1/2 and DSRCS-5 and the curve entering DSRCS-6. In lieu of the original 350 ft. radius curve from DSRCS-1/2 to DSRCS-3 and the 500 ft. radius curve from DSRCS 3 to DSRCS 5, the contractor proposed a continuous 770 ft. foot curve with a radius of 536 ft. beginning at Station 2+43.21 that would mine through DSRCS-3 and then to DSRCS-5. The contractor also proposed a radius change from DSRCS-5 to DSRCS-6 from the planned 248 ft. long 400 ft. radius curve to utilize a 289 ft. long 600 ft. radius curve to ease mining operations. The centerline of the TBM's entrance and exit points at the shaft interface at DSRCS-3 would remain the same; however there is a five (5) degree difference to the north when entering DSRCS-6. These proposed alignment changes were found by the District and AECOM to not cause issue with the overall concept to the project and were accepted. A Lovat 112" TBM was selected to perform the DSRCS mainline sewer run from DSRCS-1/2 to DSRCS-6. DSRCS-1/2 will be utilized as the TBM launch shaft and DSRCS-6 will serve as the TBM's retrieval shaft, with the TBM mining through the DSRCS-3 site while the shaft floor elevation is at springline of the tunnel. After the TBM has mined the soft ground conditions from DSRCS-1/2 through DSRCS-3 and mining excavations are expected to be in competent shale, the operations will be moved from DSRCS-1/2 to DSRCS-3. With these proposed alignment changes, the tail and starter tunnels at DSRCS-1/2 and DSRCS-3 are unnecessary and the work will not be performed. The remaining mainline tunnel run from DSRCS-3 to DSRCS-6 will be mined uphill at a grade of 0.11%, with water allowed to flow to DSRCS-1/2 and DSRCS-3 by gravity flow. The tunnel rib sets for the portion between DSRCS-1/2 and

Table 1. Comparison of bid prices for key items

	Bid Price ($ million)			
	Lowest Bidder	2nd Lowest Bidder	7th Lowest Bidder	EOPCC
Total	26.09	28.78	35.93	36.81
2,666 lf of 90-inch-dia. trenchless sewer	5.93	5.32	8.56	10.42
TBM for 90-inch-dia. trenchless sewer	0.3	0.84	0.25	1.02
Selected method for DSRCS-1/2 to DSRCS-3	112-in. TBM with tooling configured for soft ground, ribs and full lagging	Tunnel back hoe shield equipped with a breaker attachment for rock and bucket attachment for till; ribs and wood lagging	110-in. TBM, ribs and boards	Tunnel back hoe shield equipped with a breaker attachment for rock and bucket attachment for till; horseshoe-shaped ribs and wood lagging
Selected method and TBM for DSRCS-3 to DSRCS-6	112-in. conventional TBM, already owned by the contractor; ribs and lagging	130-in. TBM, already owned by contractor; ribs and wire mesh/lagging	Existing 102-in. TBM, to be upsized to 110-in. for the project	New 108-in. TBM
2,520 lf of 42-inch-dia. trenchless sewer	3.40	4.21	6.30	4.20
TBM/MTBM for 42-inch-dia. trench-less sewer	0.25	0.39	0.05	0.51
Selected method and machine	100-in. conventional TBM, already owned by the contractor; ribs and lagging; two-pass system	96-in. TBM, already owned by contractor; steel ribs and wood lagging; two-pass system	63-in.-OD MTBM, 48-in. jacking pipe in lieu of 42-in. pipe	New 48-in. MTBM
Manholes	1.30	0.78	3.00	2.33
Eliminate intermediate shafts?	Yes	Not stated	No	No

DSRCS-3 will be at 4 ft. spacing while the section from DSRCS-3 to DSRCS-6 will be 4.5 ft. spacing. All sections will be fully lagged throughout the tunnel.

Phillis Avenue Relief Sewer Modifications

The Contractor's proposed changes for the Phillips Avenue Relief Sewer were in lieu of the microtunneling approach as originally outlined in the contract documents. The Contractor elected to use a Lovat 100" open face TBM with a conventional two-pass mining operation to perform the portion of the project. As a result the Contractor was able to eliminate shafts PL-1, PL-4 and PL-5 with minor adjustments to the alignment to allow for the curved radii. The Contractor has elected to utilize both DSRCS-4 and PL-6 as TBM launch shafts and PL-2 as the TBM double retrieval shaft. This will give the contractor the opportunity to mine uphill from DSRCS-4 to PL-2, allowing the water to run by gravity DSRCS-4 and be controlled from within the shaft. The mining operation and TBM will then be moved to PL-6; where mining operations will continue downhill from PL-6 to PL-2. This mining will require the contractor to pump all accumulated water from the tunnel heading back to the PL-6 launch location. Tunnel rib sets in both runs will utilize a 4.5 ft. spacing and will be fully lagged.

ACKNOWLEDGMENTS

The authors would like to thank the Northeast Ohio Regional Sewer District for approval to publish the paper.

Long-Distance Microtunnelling at Toronto Pearson International Airport

Robert Ofori ▪ Hatch Corporation
Jordan Schreiner ▪ Hatch Corporation
Marc Gelinas ▪ Hatch Corporation
Ajay Puri ▪ The Regional Municipality of Peel
Walter Trisi ▪ CRS Tunnelling Inc.
Joe Mulville ▪ CRS Tunnelling Inc.

ABSTRACT

Identified as the last link in the twinning of Peel Region's East Trunk Sewer, the Etobicoke Creek Trunk Sanitary Sewer Twinning project was initiated to service future development in the sewershed. The project involved long-distance microtunnelling below operating surfaces that included a runway, two taxiways and access roadways at Canada's busiest airport, Toronto Pearson International Airport (TPIA). This paper discusses the challenges overcome in completing this project including working within TPIA's restricted "airside" environment, long-distance pipe jacking in mixed-ground conditions, providing for tunnel face access under hyperbaric conditions and minimizing tunnelling impacts to airport operations during construction.

INTRODUCTION

The Regional Municipality of Peel (Peel) operates a lake-based wastewater conveyance system which services the City of Mississauga, City of Brampton and parts of the Town of Caledon. This wastewater conveyance system includes the West Trunk (Credit Valley Sanitary Trunk Sewer) and East Trunk (Etobicoke Creek Sanitary Trunk Sewer). Over the past decade, Peel has twinned the entire length of the East Trunk Sewer with the exception of an approximately 600m (1,970') long section underneath Runway 23 at Toronto Pearson International Airport (TPIA). At this location, the trunk sewer consists of a single, 1,650mm (66") ID pipeline.

Peel's Water and Wastewater Servicing Master Plan identified the need to complete twinning of this final portion of the East Trunk Sewer to service projected demands in the sewershed. The Etobicoke Creek Sanitary Trunk Sewer Twinning project (referred to hereafter as the 'project') was undertaken to complete this last link.

SITE AND PROJECT DESCRIPTION

The project involved the construction of approximately 600m (1,970') of new 1,800mm (72") ID trunk sanitary sewer beneath the busiest runway (Runway 23) at TPIA, as well as four deep access/flow diversion maintenance hole structures and sewer interconnections. The tunnel alignment runs parallel and east of a culvert carrying Spring Creek, a tributary of Etobicoke Creek. Between the project tie-in points, the tunnel alignment runs beneath Runway 23, two taxiways (Taxi J and H/H2), a high-speed taxiway exit (Taxi H2) and two airside service roads (North Perimeter Road and Hotel Service Road) all of which had to remain in operation during construction. Figure 1 shows the plan alignment for the trunk sewer, as well as the location of the tunnel and shaft construction staging areas.

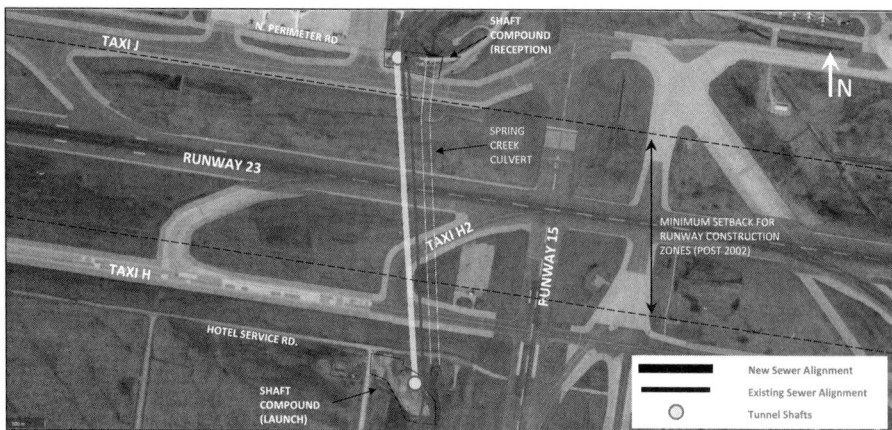

Aerial photograph 2015 (City of Mississauga Achives)
Figure 1. Etobicoke Creek sanitary trunk sewer twinning project alignment

DESIGN AND CONSTRUCTION TEAMS

Peel conducted a competitive proposal process to select a design team. The process was closed, with only four design firms considered as having the appropriate experience and technical background to complete the assignment. Hatch Corporation (Hatch) was the successful proponent and was retained by Peel to undertake detailed design and contract administration services for the Project.

Given the challenge of completing a major runway crossing by tunneling methods in both soft ground and hard rock conditions, obtaining an experienced Contractor was considered critical to the success of the project. Peel completed a prequalification process for teams consisting of a general contractor, microtunnel subcontractor and compressed air intervention specialist to bid on this project. Contractor teams were evaluated on a number of criteria including financial strength and stability, corporate experience and proposed project staff experience. In addition to these criteria, the microtunnelling subcontractor was also required to demonstrate that they either owned or could lease a Microtunnel Boring Machine (MTBM) and an airlock unit for the tunnelling operation.

In April 2015, the project was awarded to Dibco-CRS, a joint venture of CRS Tunnelling Inc. and Dibco Underground Ltd (hereafter referred to as Dibco-CRS) for a bid of CAD 16.7 million. Dibco-CRS acted as both the general contractor and microtunnelling subcontractor, with ASI Marine as their compressed air intervention specialist.

SUBSURFACE CONDITIONS

Subsurface conditions were identified as one of the most significant risks to the project. Given that the project involved construction of a gravity sewer pipeline, the tunnel alignment was constrained by the existing trunk sewer inverts at the tie-in locations. Available records showed that the tie-in points were generally situated at the bedrock contact. A review of historic aerial photographs revealed that the historic channel for Spring Creek also traversed the tunnel alignment in at least two locations.

A total of five geotechnical borings were drilled along the tunnel alignment. The planned investigation had proposed a total of six boreholes, spaced at roughly 100m

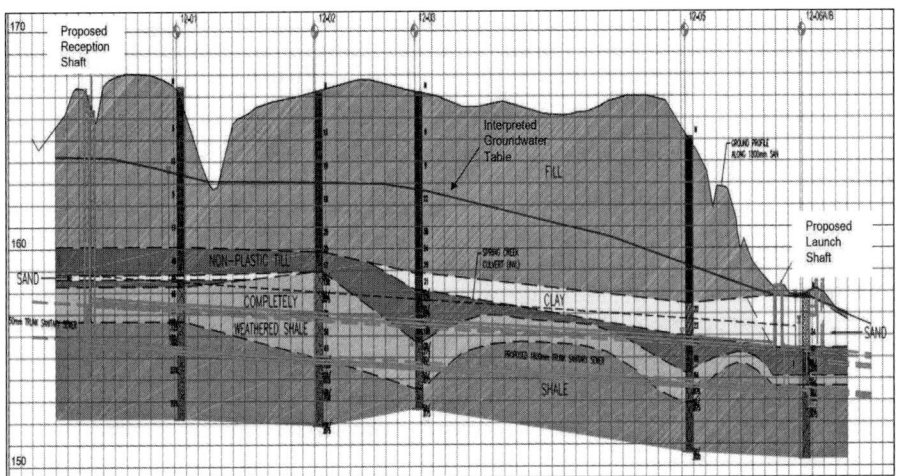

Geotechnical Baseline Report, 2015

Figure 2. Stratigraphic profile along the Etobicoke Creek sanitary trunk sewer alignment

(330') intervals. Unfortunately, one of the proposed borings had to be removed from the program in the field due to utility conflicts.

As part of the field investigations, a geophysical survey was completed along the tunnel alignment to map the depth to bedrock and provide additional information regarding the overburden material. The geophysical survey included Testing and Imaging using Seismic Acoustic Resonance (TISAR, also known as Seismic Resonance), seismic refraction and Multi-channel Analysis of Surface Waves (MASW) methods to collect data along the alignment.

The subsurface soil and bedrock conditions along the tunnel alignment, as interpreted from boring logs and geophysical surveys, generally consisted of fill materials associated with airport development overlying till deposits, which were underlain by highly weathered to fresh Georgian Bay shale bedrock (Figure 2). Clay and sand deposits believed to be related to Spring Creek were found locally overlying the till deposit in the creek valley (launch shaft area). Clay seams and stronger limestone interbeds were observed in the shale bedrock cores. The intact strength of the shale bedrock material ranged from 8 MPa to 30 MPa (1,160 psi to 4,350 psi).

Subsurface groundwater conditions were established based on the monitoring of three standpipe piezometers constructed within project borings. The entire tunnel was identified as being below the groundwater table with water head ranging from 3m to 7m (10' to 23') above the tunnel crown.

DESIGN CHALLENGES

In the following sections, several of the design challenges will be discussed and details regarding how they were addressed will be provided.

Disruptions to Airport Operations

Of primary concern for the project team was ensuring minimal disruptions to airside operations and preserving the integrity of operating surfaces at TPIA. TPIA is the largest and busiest airport in Canada with over 41 million passengers and 443,000 flights in 2015 (GTAA n.d.). Runway 23 is a principal runway at TPIA and is currently the most used runway in Canada. As such, full closures of the runway for any significant period of time (i.e., beyond a few hours between midnight and 6 am, or a few minutes between 6 am and midnight) would not be allowed.

In order to minimize disruption to airport operations and preserve the integrity of operating surfaces at the airport, pipe jacking using a Slurry Pressure Balance MTBM was specified for the project. To further reduce the impacts of tunnelling, shaft compounds were strategically located where they would have the lowest impact to airport operations (Figure 2). The initial plan proposed a reception shaft at an existing sewer stub, located north of the proposed alignment, between Taxi J and Runway 23. This sewer stub was installed in 2001/2002 to facilitate future twinning underneath the Runway 23. In light of recent (post 2002) revisions to the minimum setback requirements for runway construction zones, a reception shaft at the stub location was not feasible due to its close proximity to the runway. As a result, the reception shaft was relocated to a new location, midway between North Perimeter Road and Taxi J.

Subsurface Obstructions and Abrasive/Resistant Bedrock

One of the key geologic risks identified for the project was encountering an obstruction in either the fill or native soils while tunnelling. Furthermore, the presence of resistant and abrasive limestone interbeds in the shale bedrock was identified as posing an increased risk for cutter tool wear and/or damage.

Owing to airside infrastructure above the tunnel alignment, the use of mid-drive rescue shafts for obstruction removal and/or tooling changes was considered impractical. In order to mitigate the risk of obstructions or tooling wear/damage, the contract called for a new or newly refurbished MTBM with new cutting tools and hard facing. Additionally, the MTBM was specified to include face access and back-loaded tooling to facilitate obstruction removal or tooling changes as required. Given the potential for encountering unstable ground conditions when accessing the tunnel face, such as within historic creek bed deposits, the machine was specified to include the ability to perform compressed air intervention at the face.

Pipe and MTBM Sizing

In order to accommodate anticipated future flows, the minimum pipe internal diameter for the twin sanitary sewer was calculated to be 1,650mm (66"). However, for constructability reasons, 1,800mm (72") ID pipe was adopted as the design basis. The main reasons for selecting a slightly larger diameter pipe included (1) local availability (a 1,800mm (72") diameter pipe was considered a "stock" item while a 1,650mm (66") diameter pipe was considered a "custom" item) and (2) availability of airlock units (discussion with MTBM manufacturers indicated a clear preference for a minimum 1,800mm (72") MTBM if an airlock was to be included). In addition, the larger 1,800mm (72") MTBM offered more power (torque-thrust-speed) capabilities and greater accessibility compared to a 1,650 (66") MTBM (which is typically an up-skinned 1,500mm (60") MTBM).

CONSTRUCTION CHALLENGES

Tunnel construction commenced on Oct. 27, 2015. An Akkerman SL86P peripheral drive MTBM with a cut diameter of 2,275mm (89.5") was used to install the 1,800mm (72") ID reinforced concrete microtunnelling pipes. A photograph of the MTBM is shown in Figure 3. The MTBM was equipped with a mixed ground cutter head with back-loading disc cutters, carbide face scrapers and bucket scrapers. As required by

Figure 3. Peripheral drive MTBM SL86P manufactured by Akkerman Inc.

the contract documents, the MTBM included a central access door for accessing the face of the tunnel. The microtunnelling pipe used on the project was manufactured by Decast Ltd. (formerly Munro Ltd). The pipes were noted to have a maximum jacking load capability of 1,035 tonnes (1,141 tons).

In the following sections, several of the construction challenges will be discussed and details regarding how they were addressed will be provided.

Working Within TPIA's Restricted Airside Environment

As the work was to be completed within restricted-access airport lands and regulated wetland areas, the project required extensive coordination efforts in securing permits/ approvals from the Greater Toronto Airports Authority (GTAA, the airport operator and primary stakeholder) and other government agencies including:

- Facility Alteration Permit (FAP) issued by the GTAA upon completion of a comprehensive review process confirming including building code compliance and contractor's safety documentation have been reviewed.
- Land-use clearance required by Navigation Canada (Canada's civil air navigation service) for all construction equipment used within the airside areas that could interfere with aeronautical navigation
- Toronto and Region Conservation Authority (TRCA) required for the temporary construction activities in regulated wetland areas and adjacent shorelines.

It required several weeks and numerous internal and external coordination meetings to obtain all the necessary approvals for work to begin on site.

In addition to the minimum 150m (492') setback from runway centerlines for construction zones, all contractor's operations within airside areas were required to remain below the obstacle limitation surface (OLS) in accordance with Aerodromes Standards and Recommended Practices (Transport Canada 2015). Figure 4 shows an illustration of the OLS along the tunnel alignment. As the tunnel launch shaft and reception shafts were in close proximity to active runways and taxiways, special consideration had to be given to the methodology for construction of the shafts as well as all crane elevations for insertion and removal of the MTBM. This resulted in reconsideration of the type of equipment used to construct the tunnel launch shaft (use of a drilling rig with a lower mast height), redesigning the tunnel reception shaft from interlocking secant pile to a system of steel ribs with timber lagging, and finally obtaining a temporary night-time runway shutdown in order to lower the MTBM into the tunnel launch shaft.

Slurry Microtunnelling in Clay-Rich Ground

Soils with a high clay content (30%) including clayey silt till and highly weathered shale were expected to be encountered in the tunnel. To effectively remove the clay fines from the return slurry, the contractor initially operated two and later three Derrick DE-7200 centrifuges (3–5μm cut-size) during tunnelling. Figure 5 presents a summary of the planned and observed tunnel mining production. The first 28 shifts (260m (850')) of the tunnel drive) represents the period when polymer usage was largely ineffective, increasing the need for slurry off-hauling/replenishment to support the tunnelling operation. During this period, the mining production was low, averaging about 9.5m/day (30ft/day) for two 12 hour shifts. Following the successful application of polymers to the separation equipment (centrifuge), it was observed that tunnel mining production increased to an average of 11.6m/day (38ft/day) and the volume of slurry off-hauling/ replenishment was reduced considerably.

Long-Distance Pipe Jacking

Special attention was paid to lubrication during tunnelling. Bentonite lubrication was continuously pumped into the annular space between soil and jacked pipe by means of an automatic pipeline lubrication system. The volume of lubricant injected per

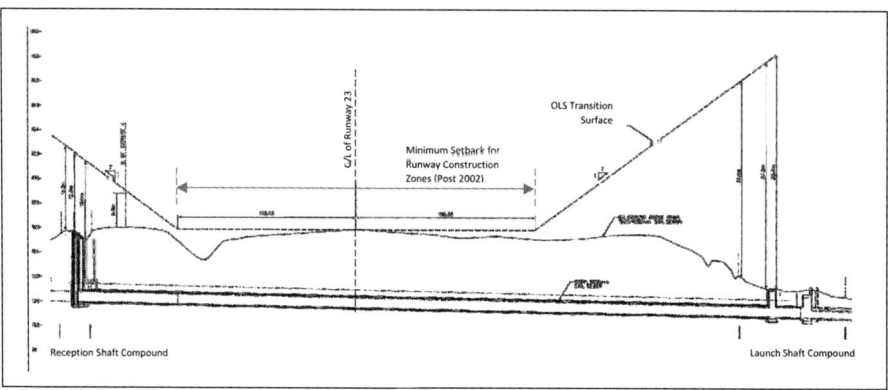

Figure 4. Obstruction limitation surface (OLS) transition details

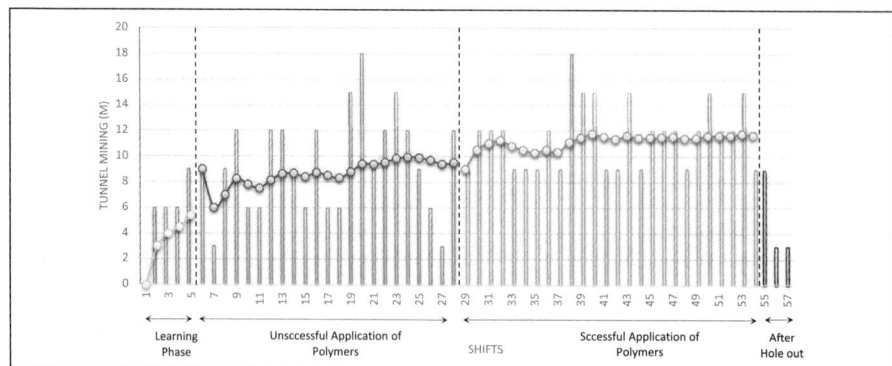

Figure 5. Tunnel mining production

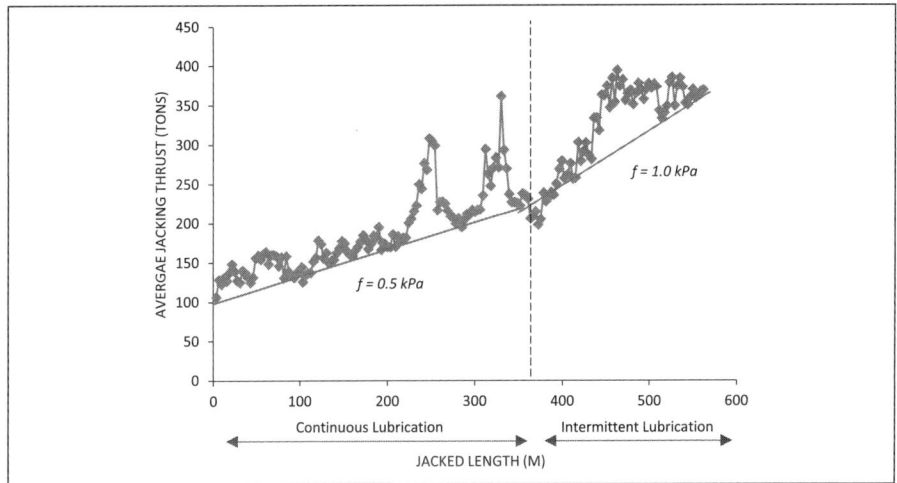

Figure 6. Variation in average jacking thrust

jacked pipe ranged from 189 liters (50 gal) to a maximum of 11,356 liters (3,000 gal). Figure 6 presents a plot of the variation in the average jacking thrust for each pipe for the project. The average jacking thrust per pipe generally varied between 97 tonnes and 358 tonnes (107 tons and 395 tons), representing less than 35% of the allowable pipe jacking load. Owing to such low jacking loads, none of the installed intermediate jacking stations were utilized.

No mining-related impacts to operating surfaces were observed. However, one instance of frac-out from bentonite lubrication required a temporary shutdown of Taxi H2 as a precaution. The MTBM was approximately 370m (1,215') into the drive (roughly 200m (655') past the frac-out location) when the incident occurred. The frac-out occurred shortly after injecting large quantities of bentonite lubrication in excess of 7,570 liters (2,000 gal). While not positively confirmed, it appears that the frac-out was more likely an issue of an existing flow pathway being exploited as opposed to lubrication pressures exceeding earth pressures. As a contingency measure, the lubrication was pumped intermittently via lubrication injection ports outside of the affected zone for the remainder of the drive.

The total jacking thrust data presented in Figure 6 were subsequently used to determine the soilpipe unit friction value (f) using the minimums of the jacking thrust curve (FSTT 2006). As evident in Figure 6, the unit friction increased from 0.5kPa to 1.0kPa (0.07psi to 0.15psi) after the jacking pipes were subjected to intermittent lubrication. The increased unit friction was attributed to the reduced annular space between the jacking pipe OD and the tunnel bore cut diameter, leading to tightening of the ground around the pipe. The results show that maintaining a constant and ample quantity of lubricant during pipe jacking can reduce the soil-pipe friction and significantly lower the required jacking force.

Ground Displacements Above Tunnelling Works

Monitoring stations (in-ground and surface points) were installed to monitor ground displacements above the tunnel. In order to minimize impacts on airport operations, surveys were typically scheduled at night-time on a limited-time basis. In general, surveys were completed for one monitoring station in front of the tunnel face and

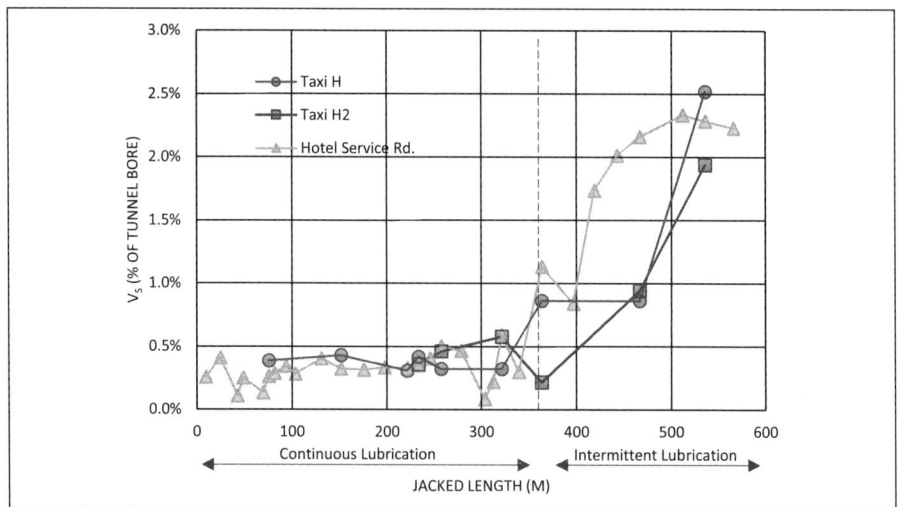

Figure 7. Variation in ground loss for select monitoring stations along the tunnel route

all monitoring stations behind the tunnel face. The results were compared to baseline data and were used to estimate the ground loss per unit length of tunnel (Vs), expressed as a percentage of the tunnel bore volume. Figure 7 presents the variation in Vs with jacked length for select monitoring stations along the tunnel route.

Overall, the observed ground loss was generally less than 2.5% of the tunnel volume. The results compare well with the reported range of 1% to 2.5% for shield tunnelling in glacial deposits (O'Reilly and New 1982). As evidenced in Figure 7, roughly 70% increase of the observed settlements, while still within acceptable limits, occurred shortly after the jacking pipes were subjected to intermittent lubrication. The increase in tunnelling related settlements can be attributed to the collapse of the annular space between the installed pipe and the tunnel excavation due to reduced lubrication quantities. The results underscore the importance of maintaining a stable annular void by continuous lubrication during microtunnelling operations. Another factor to be noted is that the tunnel line and grade were maintained as per design and hence helped minimized settlement impacts; versus a tunnel drive that required continuous correction to keep line and grade, increases both friction and settlement.

MTBM Face Access Under Hyperbaric Conditions

For this project the hyperbaric chamber was designed to mate with the pipe string via a mechanical bulkhead at the launch shaft. If face access, under pressure was required, the entire tunnel would be pressurized, equalizing pressure to the tunnel face pressure. With pressure equalized, the MTBM face could be safely accessed by skilled divers. This is thought to be the first known application of airlock technology to a microtunnelling project in Canada. While being available onsite for the duration of the tunneling operation, the hyperbaric chamber was not required for the tunnel drive, a testament to both the robust design of the MTBM and the experience of the contractor's MTBM operators.

LESSONS LEARNED

The project demonstrates how the use of innovative technologies such as microtunnelling can significantly reduce the impact of construction. Microtunnelling equipment was capable of working in the extremely small compound sizes available on site, reducing impacts to airport operation. Microtunnelling also allowed for the direct installation of the product pipe, minimizing the tunnel diameter and thereby limiting the potential for settlement.

In clay-rich ground, slurry testing and the proactive use of polymer additives (i.e., advance testing and identification of required polymer type and dose rates) to remove the fines content below the cutsize for the separation equipment (centrifuges) is critical to efficient tunnel mining and to minimizing the need for slurry off-hauling/replacement.

In long-distance pipe jacking, maintaining a constant and sufficient quantity of lubricant around the jacking pipes can reduce the soil-pipe skin friction and significantly lower jacking forces. For longdistance pipe jacking alignments, the authors strongly recommend the use of automated pipeline lubrication systems be mandated in the contract documents to provide continuous lubrication of the pipeline, reduce jacking forces on the installed pipe, maintain bore stability, and reduce the risk of surface settlement.

Lastly, the project demonstrates how a collaborative effort between owner, engineer, contactor and project stakeholders can lead to a successful outcome. It is often said that even the best of plans seldom survive their first contact with reality. This is especially true in construction. For this project, several plans had to be changed during construction. These changes ranged from complete redesigns of shaft shoring systems to changes in polymer additives (type and dose rates) for the efficient removal of fines in the separation process. In all cases, the changes were handled in a collaborative way, with all parties maintaining the completion of the project as their primary priority.

REFERENCES

French Society for Trenchless Technology (FSTT). 2006. Microtunneling and Horizontal Drilling: Recommendations. Wiley-ISTE. p. 342.

Greater Toronto Airports Authority (GTAA). N.d. About the GTAA.N.p. Web. 12 Dec. 2016.

O'Reilly, M.P. and New, B.M. (1982). "Settlements above tunnels in the United Kingdom—their magnitude and prediction." Tunnelling 82, London: IMM. pp173–181.

Transport Canada. 2015. Aerodromes Standards and Recommended Practices (TP 312), 5th ed. N.p.

Upper Limit Microtunneling Application to Meet Dam Safety and Operational Longevity

Babs Marquis ▪ McMillen Jacobs Associates
Everette Knight ▪ Gannett Fleming Hazen Joint Venture
Emory Chase ▪ NYC Environmental Protection
John Vickers ▪ NYC Environmental Protection
John Arciszewski ▪ Southland Renda Joint Venture

ABSTRACT

Gilboa Dam in upstate New York impounds Schoharie Creek, forming Schoharie Reservoir—a part of NYC's water supply system. To ensure the dam's long-term performance and reliability, a Low-Level Outlet (LLO) is being constructed to facilitate reservoir draining to meet conservation releases. It consists of two 9-foot-diameter tunnels (water leg and land leg) totaling 2,160 linear feet, terminating at a new chamber releasing water to the Creek downstream. Excavation is by microtunnel boring machine (MTBM). The land leg tunnel terminates downstream while the water leg tunnel upstream, and penetrates into a coffer dam at the reservoir bottom elevation 977 (-153 feet). This paper presents LLO microtunneling design considerations and construction methods.

INTRODUCTION

The Gilboa Dam, located in Schoharie County, New York, was constructed between 1919 and 1927 to create the Schoharie Reservoir, a key component of New York's water supply system. The original dam is a classic gravity dam design by New York City's Department of Environmental Protection (NYCDFP). It consists of a 160-foot-high by 1,326-foot-long spillway overflow structure. This structure is constructed of mass Cyclopean concrete with a 3- to 5-foot-thick Ashlar masonry façade of mortared quarried stone on the entire downstream face, and on a portion of the upstream face. The dam is abutted on the west by a 160-foot-high by 700-foot-long earth-filled embankment section consisting of homogenously rolled earth fill with a concrete core wall. The stepped overflow structure, also constructed of Cyclopean concrete with stone veneer facing, cascades water from the dam spillway into a lower side channel spillway, which varies from 80 to 270 feet in width. The stepped façade is intended to dissipate energy as water overflows into the side channel spillway.

Based on the age of the dam and as part of NYSDEC's forward looking program, an evaluation was undertaken of the dam and appurtenances after their having being in service for over eight decades. The evaluation revealed evidence of deterioration and the need for upgrade rehabilitation. The reconstruction and upgrade are intended to extend the dam's service life in line with current New York State Department of Environmental Conservation (NYSDEC) dam safety guidelines and operational standards (Gannett Fleming/Hazen and Sawyer Joint Venture July 2008).

A series of construction projects are being executed as an integral component of the NYCDEP Gilboa Dam Reconstruction Program to extend and ensure continued reliable operation of the dam over the next 100 years. This includes the 108-inch LLO Tunnels, which are to be excavated by microtunneling techniques. The LLO Tunnel

900

project is scheduled to be completed in phases, as described in the following sections. Upon completion, the Low Level Outlet and associated facilities will provide:

- The means to manage Schoharie Reservoir levels and minimize flooding during snowpack offset management;
- The ability to provide environmental low flow releases to support critical downstream habitat;
- And most importantly, a Low Level Outlet that meets current NYSDEC dam safety standards and provides a means to maintain the Gilboa Dam and to minimize risk during critical dam safety operation needs.

GILBOA DAM RECONSTRUCTION PROGRAM AND PHASING PLAN

To expedite initiation of construction work of the Dam Reconstruction Program, NYCDEP elected to perform the work under multiple construction contracts, with construction beginning in approximately 2008. The contracts were phased and prioritized based upon facility risk and critical need for improvement as well as permitting and design lead time. The major work phases were preceded by a significant site preparation phase that provided for the access roads, staging areas, and space for construction phase support offices. The components of work completed as part of the reconstruction and upgrade activities include:

1. Installation of anchors and tiedowns (2008–2009)
2. Installation of crest gates and notch (2009–2011)
3. Clearing and preparation of the project site for heavy construction (2009–2011)
4. Reconstruction to improve dam stability—spillway façade and side channels (2011–2015)
5. Schoharie Reservoir Low Level Outlet—tunnels, discharge valve chamber, and intake structure (2015–2021)

The LLO Construction Contract (item 5 above), the subject of this paper, was awarded in 2015 with Notice to Proceed in June 2015, and the LLO project is currently under construction. This construction requires a deep jacking shaft, two tunnels (108 inches in diameter), and two receiving sites to retrieve the MTBM. The tunnels will be constructed using microtunneling techniques to tap into the bottom of the reservoir for the LLO water intake, where the MTBM will be recovered by wet retrieval. The wet retrieval approach is a method not commonly used for tunnels of this size, making the installation unique for today's microtunneling standards.

SUBSURFACE CONDITIONS FOR THE LLO ALIGNMENT

The subsurface conditions along the tunnel alignment are summarized herein. The summary is limited to the tunnel horizon from the intake structure location at the bottom of the Schoharie Reservoir to the tunnel portal at the downstream end of the alignment, where the tunnel terminates.

The LLO Tunnels are expected to be mined in bedrock on either side of the jacking shaft for approximately 1,725 linear feet, about 80 percent of the entire alignment. As such, the discussion regarding the subsurface conditions, for the scope of this paper, is limited to the area from the location of the jacking shaft to the receiving sites at the intake and portal locations (that is, the reaches of the interface/mixed ground conditions and the bedrock). This information is extracted from the Conformed Contract Documents and is summarized below and illustrated in the geologic profile in Figure 1.

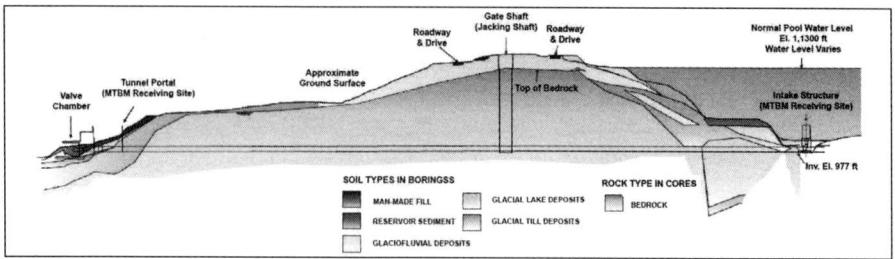

Figure 1. Geologic profile for the LLO shaft, and land and water leg tunnels

Gate Shaft (MTBM Jacking Shaft)

Subsurface conditions at the jacking shaft location consist of a 27- to 29-foot-thick overburden that is predominantly glaciofluvial deposits consisting of stiff to hard silty/clay with traces of sand and gravel. These soil materials have low permeability, with the groundwater level varying between 6 and 10 feet below the ground level. The bedrock lies directly below the overburden strata and consists of fine grained sandstone with occasional layers of siltstone and clay seams. The rock is generally moderately fractured to massive with high recovery and Rock Quality Designation (RQD) percentages ranging from 88 to 100 percent, with an average recovery of 99 percent. The bedrock that will be encountered during excavation of the Gate Shaft is expected to be strong and abrasive. Laboratory test results indicate that the Unconfined Compressive Strength (UCS) varies from 17,870 psi to 28,990 psi, with an average of strength of 23,200 psi. The Cerchar Abrasivity index ranges from 4.4 to 5.1.

LLO Tunnels: Land and Water Leg

The subsurface condition in the tunnel horizon for both the land and water leg alignments is similar to the rock condition described for the Gate Shaft rock, except for the short reach of mixed ground conditions at the terminal ends of the tunnel. The profile of the bedrock along the tunnel alignment varies, with the high point located at the Gate Shaft where top of bedrock is recorded at 29 feet below grade. From the Gate Shaft, the top of bedrock dips downward on both sides of the tunnel alignment.

On the land leg side, the top of bedrock slopes towards the portal wall of the Valve Chamber, where it appears to strike the crown of the tunnel towards the end of the tunnel drive, which is considered to be overburden fill with cobbles and boulders, as the topography slopes down toward Schoharie Creek downstream of the Gilboa Dam.

Similarly for the water leg, the top of bedrock dips downward from the Gate Shaft towards the intake structure location in the reservoir and appears to strike the crown of the tunnel, where it seems to transition from bedrock to soft ground. This soft ground is characterized as very fractured and weathered rock with an RQD varying between 0 and 37 percent. The soft ground anticipated in the tunnel horizon comprises very compact granular glacial till soil materials predominantly consisting of coarse to fine sand with moderate amounts of silt and varying amounts of gravel. Occasional cobbles and boulders are anticipated from the rock interface to the location of the receiving site. The bottom of the reservoir consists of silt and miscellaneous debris deposited over the years.

Groundwater

The maximum water level within the Schoharie Reservoir is approximately El. 1,130 feet. Groundwater appears to follow the land topography, varying from 6 to 10 feet in the overburden. In situ packer permeability test results conducted at the Gate Shaft location indicate permeability ranging from 1.75×10^{-5} cm/sec to a high of 3.62×10^{-4} cm/sec. Groundwater inflow is likely to occur during Gate Shaft excavation.

LLO TUNNEL DESIGN CONSIDERATIONS

The LLO Tunnel project entails construction of the 2,400 cfs Low Level Outlet while maintaining normal service of the 19.5 billion gallon reservoir.

During preliminary design there were four options investigated for the LLO configuration. In some cases more than one alternative or alignment for an option was given consideration. Based upon the geotechnical studies, consideration of risk, and cost, Option 1 was defined as the preferred alignment alternative. Option 1 originally consisted of two alternatives described as Options 1A and 1B. These two alternatives both consisted of a 120-inch-diameter tunnel that comprised a water leg that extended from the east side of the lake northeast approximately 950 feet to a gate shaft located near the east abutment of the Gilboa Dam. A land leg then extended approximately 1,250 feet to the northwest to near Schoharie Creek. Both Options 1A and 1B had a portion of the tunnel sections (either the land leg or the water leg) passing under the dam, which raised concerns for conflicts with the anchor system installed during early phases of dam reconstruction.

As planning and design advanced, there was a preference to have all tunnel work occur outside of the existing dam footprint to minimize risk to existing facilities. With that understanding, Option 1C was developed, as presented in Figure 2.

Option 1C presents the water leg of the tunnel beginning at approximately 650 feet south of Station 7+50 along the dam and extending 950 feet northeast to a Gate Shaft that would be located on the east side of State Route 990V. From the Gate Shaft, the land leg would extend approximately 1,250 feet to the northwest to an area near the Schoharie Creek.

Groundwater Control

Because of the close proximity of the Schoharie Reservoir, with an average water elevation of 1,130 feet and relatively high rock permeability, a pre-excavation grouting

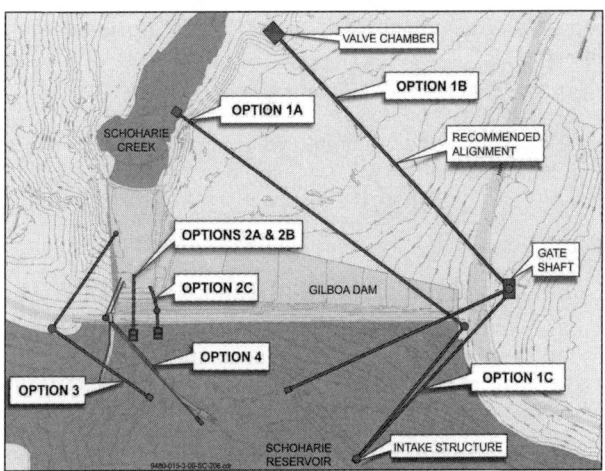

Figure 2. Preliminary tunnel alignment considerations

program was required around the Gate Shaft's perimeter. The pregrouting extends the full depth of the shaft to seal potential pathways for groundwater inflows during shaft construction and launch of the MTBM.

MTBM Receiving Site at Tunnel Portal for Land Leg Tunnel

The overburden in the portal area where the MTBM will be received consists of random fill material (which includes cobles and boulders). The means, method, and equipment selected to mine through the mixed face condition are provisioned for the ground condition described within the Geotechnical Baseline Report (GBR).

Intake Structure and MTBM Receiving Site for the Water Leg Tunnel

Because of the soft silt/clay deposits at the bottom of the reservoir, dredging will occur to remove the soft sediments to competent depth for the intake structure foundation. Ground modification will be performed to replace the dredged area with tremie concrete to stabilize the cofferdam and improve the soils for microtunneling operation. When the MTBM breaks into the cofferdam, it will be rigged and hoisted to the surface via the wet retrieval method with assistance from a specialty diving operation. Multiple bulkheads are required to be installed prior to decoupling of the MTBM for wet retrieval and to prevent reservoir loss.

U.S. Fish and Wildlife Permit Restrictions

With the presence of bald eagles (an endangered species) nesting at the site, construction activities within 660 feet and all blasting activities within 0.5 mile of the eagle nests are prohibited between January 1 and July 31. This restriction leaves a limited window to commence and complete drill and blast for the jacking shaft construction and support of excavation.

Tunnel Construction Requirements

It is stipulated that the land leg tunnel will be constructed first so that lessons learned with MTBM performance and ground behavior can be carried into the water leg operation. Also, as a requirement for the water leg construction, three bulkheads are to be installed, as follows:

- Bulkhead No. 1: On the inside face of the land leg in the Gate Shaft.
- Bulkhead No. 2: Directly behind the MTBM to provide a seal at the front of the lead pipe.
- Bulkhead No. 3: At the rear of the lead pipe.

The two lead bulkheads are to be fabricated watertight with power cables, slurry lines, and access hatch to the face of the MTBM in the event of obstruction intervention and disk cutter maintenance. In addition, the contract stipulates an air lock compatible with the MTBM.

MICROTUNNELING CONSTRUCTION FOR LLO TUNNELS

This section discusses the major components of the trenchless construction features associated with the LLO construction activities, construction methods and sequencing, and challenges that must be overcome to successfully complete the LLO Tunnel project and achieve its objective(s). A brief description of the associated trenchless structures and anticipated construction methods is also included.

Low-Level Outlet Microtunneling Alignment

Water leg Tunnel. The microtunneling alignment originates at the intake structure located at the bottom of Schoharie Reservoir, upstream side of the Gilboa Dam, and under approximately 153 feet of water. The water leg of the microtunneling alignment projects in the northeasterly direction outside of the dam footprint to the Gate Shaft, at the right abutment, which also serves as the jacking shaft for both the water and land legs of the microtunneling operations. From the jacking shaft, the land leg of the LLO Tunnel projects in the northwesterly direction towards Schoharie Creek downstream of the dam, and terminates at the face of the proposed portal, where the MTBM will be received (see Figure 3).

Gate Shaft (Microtunneling Jacking Shaft). The Gate Shaft is 43 feet in excavated diameter and extends from the ground surface at approximately El. 1,156± feet to the top of bedrock at El. 1,127± feet. The shaft then extends from top of bedrock at 40.5 feet excavated diameter down to the bottom of the shaft, at approximately El. 969.5 feet. The geologic profile at the Gate Shaft includes a 30-foot overburden stratum atop the bedrock, with a total shaft excavation depth of approximately 187 feet. The shaft excavation sequence and construction involve perimeter pregrouting in the overburden to about 25 feet into bedrock to control groundwater inflow during shaft excavation and launch of the MTBM for microtunneling operations. Pregrouting in the bedrock section of the jacking shaft is planned to be completed in a series of 40-foot zones through battered grout holes circumferentially drilled from within the shaft.

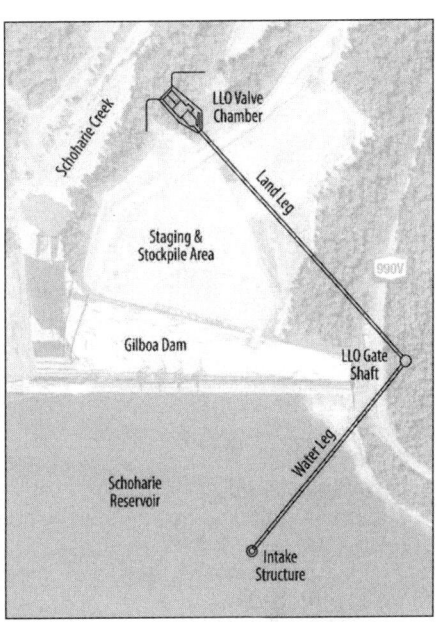

Figure 3. Low-level outlet tunnel alignment

As drill and blast excavation is advanced into the grouted zone, four 3-inch probe holes will be drilled to verify the effectiveness of the grouting. If the groundwater inflow exceeds 5 gallons per minute (GPM) per hole drilled, the zone in question would receive additional grout treatment until excavation is advanced to the bottom of the shaft. The face of the excavated shaft is designed to be secured with 10-foot rock bolts and treated with 3-inch-thick shotcrete reinforced with welded wire fabric.

After tunnel construction is complete, two wheel-mounted gates (roller gates) will be installed within the lower part of the Gate Shaft. These will be situated at the outlet of the upstream tunnel and the inlet of the downstream tunnel. The roller gates will provide a means for dewatering the tunnels as well as the Gate Shaft for future inspection and maintenance. The roller gates will also provide a means for additional security and a source of redundancy to maintain reservoir levels if there are problems with the downstream control valves. The shaft will be concrete lined to an approximate finished inside diameter of 36 feet.

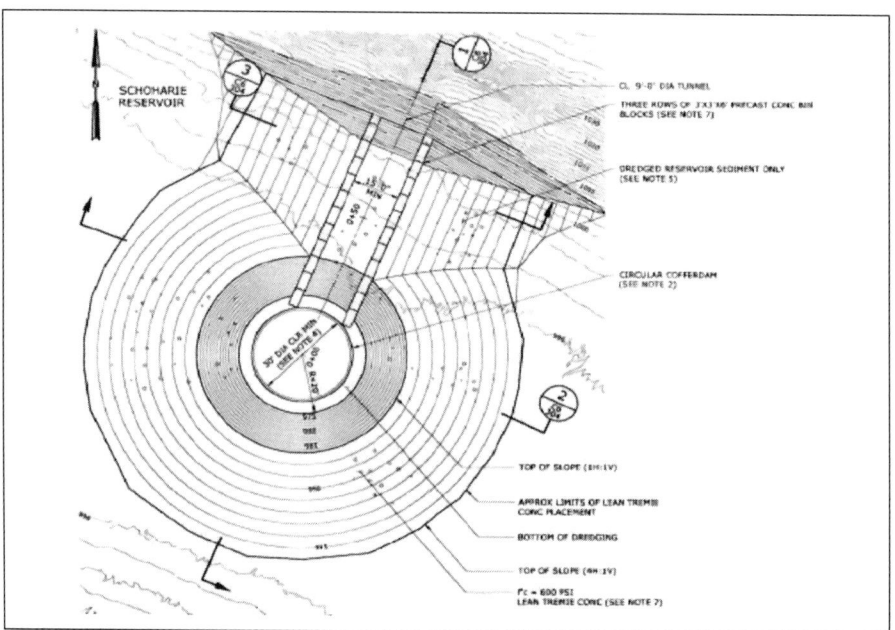

Figure 4. LLO intake receiving site dredge plan

Receiving Site at Intake Structure Location. The initial phase of construction for the intake structure will be preparation of a suitable receiving pit for the wet retrieval of the MTBM. This portion of the work involves dredging and underwater grading, predominantly accomplished from the reservoir surface from a working barge platform. The finished grades will be verified by sonar. The full pool elevation of the reservoir is at CL. 1,130 feet, but the actual level varies considerably and can potentially rise 30 feet in one day. The sediment layer at the location of the intake and MTBM receiving site is recorded at approximately El. 996 feet. As part of the receiving site and intake structure foundation preparation, it is required that the limits of the receiving site be dredged with hydraulic dredging techniques to minimize the creation of turbidity within water supply reservoir (see Figure 4).

Land Leg Tunnel and Receiving Site. The receiving site for the land leg tunnel excavation is located downstream of the Gilboa Dam, where microtunneling terminates at the portal, along the bank of Schoharie Creek. This portal provides for the open-cut excavation where the bifurcation piping will be constructed between the tunnel outlet and the Valve Chamber Building. The guard valve, control valves, and primary flow control are managed within the Valve Chamber Building. As part of the preparation required to receive the MTBM, a 10-foot grout plug cube will be installed at the portal face to stabilize the ground for MTBM break-out at completion of the land leg tunnel. After the land leg tunnel construction is complete, the contractor will continue with other work necessary for installation of the valve chamber structure, connection with the tunnel, and finishing for the LLO water release into Schoharie Creek.

The two tunnel segments as described above are scheduled to be constructed utilizing a 108-inch-diameter MTBM with slurry pressurized face (AVN2200AB) manufactured by Herrenknecht. Per contract requirements, the new MTBM is specifically designed and fabricated for the LLO Tunnel project (Figure 5a). The MTBM design requirements include the ability to mine mixed ground conditions containing cobbles

Figure 5a. Microtunnel boring machine   Figure 5b. Hyperbaric chamber

and boulders in sizes of up to 30 percent of the MTBM outside diameter, as presented by the GBR (Gannett Fleming / Hazen and Sawyer 2014). The MTBM is designed to operate under an external hydrostatic pressure equivalent to 153 feet of water head and allow for underwater retrieval. The machine is required to be equipped with an air lock hyperbaric chamber to allow for safe personnel access to the tunnel face in the event of obstruction intervention or cutter maintenance (Figure 5b). An extensive list of additional functionalities for the MTBM is also included in the design.

The jacking pipe scheduled for use as the initial and final ground support is a 108-inch-diameter, 1.375-inch-thick Permalok® steel pipe with T-7 gasketed joints for both the land and water leg tunnels. Per the contract, two intermediate jacking stations are required for each of the tunnel legs. Utilization of butt welded steel jacking pipe joints are not allowed for this project.

PROJECT CONSTRUCTION STATUS

The construction contract was awarded to the low cost responsive bidder, Southland Renda Joint Venture, with a construction bid price of $142.6 million with a Notice to Proceed (NTP) date of June 29, 2015. The contractor has mobilized to the site, working with the construction manager in providing critical submittals for long lead equipment, materials, and permits. Preparatory work has begun, and includes site clearing, temporary access road construction, and geotechnical borings in the valve chamber area for wing wall foundations.

Phase 1 grading is complete site wide, including the dredge spoils pond. All marine equipment has been mobilized for the intake construction including the 230-ton crawler crane on the barge. Hydraulic dredging is nearly 75 percent complete and the 35-foot liner plate and rib cofferdam is fully assembled. Once the hydraulic and mechanical dredging is complete the cofferdam will be set on a leveling slab under 160-feet of water and the tremie concrete cast around it; this will serve as the underwater receiving pit for the MTBM. The manufacturing of the Permalok pipe has begun and is in full production; manufacturing is inspected by NYCDEP at the plant, and the pipe is being stored there until it is ready to be installed.

The pre-excavation grouting of the Gate Shaft is complete as well as the soft ground excavation and support. Blasting of the shaft started in August and will likely continue into the beginning of 2017. The 40-foot-diameter shaft is being excavated in 10-foot lifts through hard sandstone and supported by bolts, mesh, and shotcrete. Groundwater infiltration has been minimal to this point thanks to the extensive grouting program installed by SRJV. The upper 80 feet of the rock excavation were blocky with multiple sand and clay seams, but has tightened up considerably since then. The local power company made the temporary power drop and the power setup for the MTBM is complete.

MTBM Selection

The bedrock will account for approximately 93 percent of the land leg tunnel excavation and approximately 58 percent of the water leg tunnel excavation. The only excavation method that is allowed by the project specifications is an MTBM pressurized slurry machine in which excavation, muck removal, and pipe jacking occur simultaneously. In order to successfully excavate the bedrock and meet acceptable production rates, the MTBM was configured to excavate this bedrock. Items considered in the MTBM selection include cutterhead torque and rotation speed, configuration of the cutting tools, and thrust. The diameter of the tunnels allows for the machine to be equipped with significant cutterhead torque and rotation to cut the anticipated bedrock. Intermediate jacking stations will be required in order to provide the necessary thrust to cut the rock as well as jack the steel pipe for the distances required.

Typically, 17-inch-diameter (or larger) disc cutters are used on hard rock tunnel boring machines in rock of significant strength. However, with the smaller diameter MTBM required on the LLO Tunnel project, the size and placement of the discs is limited. Harder rock, such as will be encountered on this project, tends to be more brittle, and can allow for a larger cutter spacing provided the supplied thrust will allow for efficient breaking of the rock. If the cutter spacing is too far apart, the necessary crack propagation between cutters may not occur and, thus, the rock may not break in an efficient manner. Conversely, if the disc cutters are spaced too close, excessive pressure will be required for the discs to engage the rock, which will necessitate excessive torque to turn the cutterhead. Additionally, if the disc cutters are spaced too close, the rock will be ground into a fine powder, instead of chips. This can strain the separation plant and potentially seize the MTBM if the cuttings are allowed to migrate to the outside of the machine. Smaller diameter disc cutters tend to wear more rapidly than larger diameter disc cutters, and it is anticipated that regular maintenance and/or changing of the disc cutters will be required. The MTBM is equipped with back loading cutting tools in anticipation of the abrasivity and high water pressures expected.

The MTBM manufacturing is complete, and the machine has been delivered to the site. This machine represents the upper limit in terms of size for a microtunnel boring machine. At 113.75 inches in diameter, it is believed to be the largest in North America and one of the largest in the world. Commissioning and testing is planned for the first part of 2017, followed by excavation of the land leg tunnel. The approximately 1,200-foot-long land leg tunnel is expected to be complete by the middle of 2017, at which time the MTBM will be set up for the water leg tunnel.

REFERENCES

Gannett Fleming / Hazen and Sawyer. 2014. CAT-212C Schoharie Reservoir Low Level Outlet Conformed Contract Documents. NYCDEP.

Gannett Fleming/Hazen and Sawyer Joint Venture. July 2008. Gilboa Dam Reconstruction Environmental Assessment and Project Description. NYCDEP.

Microtunneling in Georgian Bay Shale: Rebecca Trunk Wastewater Main, Oakville, Ontario

Paul Headland ▪ Aldea Services LLC
Guadalupe Monge Fabian ▪ Aldea Services LLC
Rajab Ali ▪ Aldea Engineering Services Ltd.
Kanchan Mohammed ▪ Cole Engineering Group Ltd.
Mark Bajor ▪ Halton Region

ABSTRACT

The Rebecca Trunk Wastewater Main Construction project comprises 4.05km (13,287 ft.) of microtunnel (MTBM) construction including a total of 14 microtunnel drives (12 rock drives & 2 mixed face drives) varying in length from 165 m (540 ft.) to 403 m (1,325 ft.) located in Oakville, Ontario. Twelve of the drives are curved in part based upon requirements to limit impacts to existing utilities, protected trees, and existing structures, minimize easement requirements, maximize clearance beneath environmentally sensitive creeks, and connection locations required to decommission existing pumping stations. The rock drives are located primarily within the Georgian Bay Formation (Upper Ordovician) which comprise shales with interbedded Siltstone and Limestone ("hard layers"). The mixed face drives comprised of Georgian Bay Formation and Glacial Till. All drives are below the water table. Design considerations include provisions for swelling Shale, interbedded Limestone & Siltstone ("hard layers") within the softer Shale, high horizontal stress, areas of limited ground cover (less than 0.4 MTBM diameters), curved and composite curved alignments (minimum radius of 430 m/1,410 ft.), and pipe requirements including provision for grout ports (for lubrication during jacking and grouting after jacking), and watertight gaskets. Construction requirements include a continuous lubrication system, an automated guidance system; and an MTBM equipped with rear loading cutters, a crushing chamber, and a minimum required overcut. The paper presents the design approach taken and the requirements placed within the Contract Documents for the project. Construction commenced in December 2015.

INTRODUCTION

In order to balance the needs of growing the Region of Halton with the protection and preservation of its natural, environmental, and heritage resources, Halton Region initiated Sustainable Halton which developed goals that serve as a blueprint for building a sustainable and healthy community. The need to upgrade the existing wastewater network to service future intensification in Oakville's Urban Growth Centre was identified as part of Sustainable Halton.

The Regional Municipality of Halton Pumping Station Master Plan was prepared to identify different upgrading alternatives. The preferred alternative selected in the Master Plan was the construction of a deep trunk wastewater main from the Urban Growth Centre to the Oakville Southwest Waste Water Treatment Plant. The trunk wastewater main will relieve the existing constraints along the existing Rebecca Trunk Sewer (900mm/35.5 inch) and allow for the decommissioning of a number of existing wastewater pumping stations by redirecting flows from the pumping stations sub-catchments to the new trunk sewer.

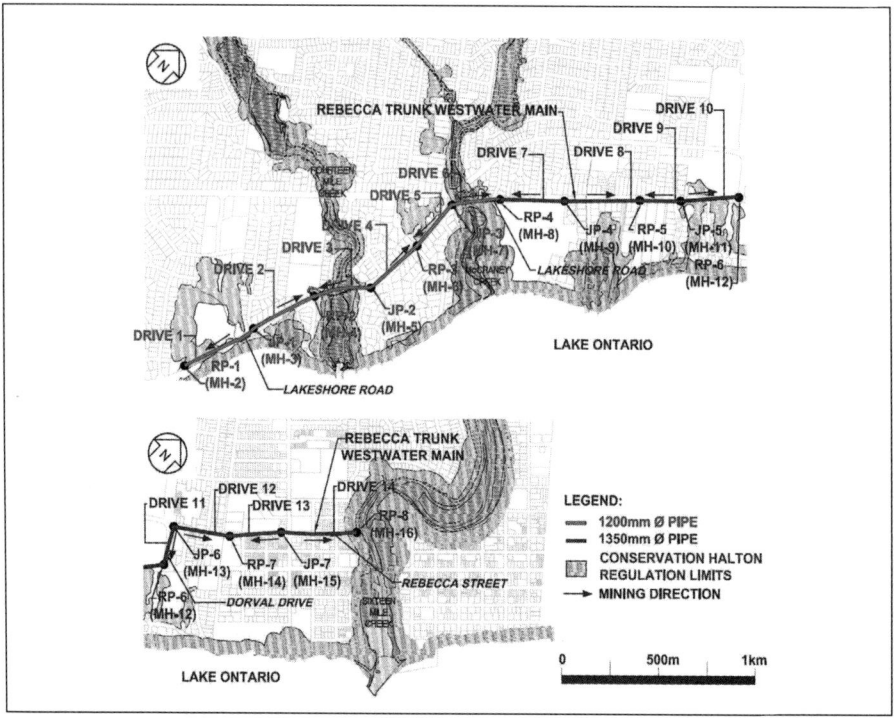

Figure 1. Project location plan

The Rebecca Trunk Wastewater Main Construction Project consists of 4.13 km (13,550 ft.) of 1200mm/47.25 inch (ID) and 1350mm/53.15 inch (ID) pipe by trenchless construction (microtunneling) comprising seven (7) jacking pits (JP1 to JP7), eight (8) receiving pits (RP1 to RP8), and fourteen (14) microtunnel drives (Drive 1 to 14) between the Oakville SW Wasterwater Treatment Plant (WWTP) and the West Abutment of Sixteen Mile Creek at Forsythe Street. The project location including microtunnel drive, jacking pit, and receiving pit are presented on Figure 1.

The microtunnel drives are located at depths up to approximately 14 m (46 ft.) below the ground surface and are to be excavated using a slurry microtunnel boring machine (MTBM) on a grade of approximately 0.15%. With the exception of two (2) of the microtunnel drives (mixed face), microtunneling shall be through rock. The microtunnel drives comprise straight, and curved drive alignments as follows:

- Straight (3 Drives)—Drive 1, Drive 4, Drive 5,
- Curved (11 Drives)—Drive 2, Drive 3, Drive 6, Drive 7, Drive 8, Drive 9, Drive 10, Drive 11, Drive 12, Drive 13, & Drive 14

All the curved drives include an initial straight section of at least 10m (33 ft.) prior to commencing a curve section. Many of the curved drives also incorporate multiple curves of varying radii (composite curves).

The microtunnels pass beneath creeks at the following locations:

- *Drive 1 (1350mm/53.15 inch)*—beneath Unnamed Tributary with approximately 854mm (33.65 inch) of cover. Creek is outside of the O.Reg. 162/06 regulated area).

- *Drive 3 (1350mm/53.15 inch)*—beneath Fourteen Mile Creek with approximately 600mm (23.65 inch) of cover (less than 0.45D). The watercourse crossing falls within the O.Reg. 162/06 regulated area. Fourteen Mile Creek is a habitat for Redside Dace which are designated an endangered species in 2009 under the Ontario's Endangered Species Act, 2007.

- *Drive 6 (1350mm/53.15 inch)*—beneath McCraney Creek with approximately 1627mm (64 inch) of cover. The watercourse crossing falls within the O.Reg. 162/06 regulated area.

- *Drive 8 (1200mm/47.25 inch)*—passes beneath an Unnamed Tributary at Chainage 2+370 with approximately 3317mm (130.60 inch) of cover. The creek crossing is outside of the O.Reg. 162/06 regulated area.

Microtunnel access to the ground surface shall be via jacking pits and receiving pits. The jacking pit (JP1 to JP7) work areas shall be used as the base for microtunnel construction operations and the receiving pits (RP1 to RP8) shall be used for retrieval of the microtunnel boring machine (MTBM) following completion of each microtunnel drive. Jacking pits and receiving pits are anticipated to be excavated using a combination of mechanical equipment above and below the top of the bedrock surface (below top of rock). No blasting is permitted during construction of the work. A section of open cut pipeline shall be constructed to facilitate tie-ins to the Oakville Southwest WWTP located at the western end of the alignment.

GROUND CONDITIONS

The Halton Region is located within the Niagara Escarpment physiographic region. The Escarpment bisects the Region on a north-south axis and represents an erosional landform created by glaciers and water differentially eroding the rock strata comprising the escarpment. The rock formations in the region include the shallow-dipping Ordovician age Queenston and Georgian Bay Formations. The Queenston Formation comprises predominantly red, hematitic, fissile, micaceous calcareous shale. The Georgian Bay Formation comprises grey-green shales with interbeds of sandstone, grey-green siltstone, and grey argillaceous limestone (Ontario Geological Survey, 1976). The surficial geology is represented by glacier and glacial lake sediments/soils although more recent alluvial (stream) and organic deposits are also common. Below the escarpment, the rolling plains overlying the Queenston Formation consist primarily of thick silty to sandy clay till referred to as the Halton Till.

The hydrology and hydrogeology of the Region is closely linked to the landform characteristics. Many streams originate on or immediately below the Niagara Escarpment and flow southerly or south-easterly towards Lake Ontario. These include, from west to east, Grindstone, Bronte (also known as Twelve Mile Creek) and Oakville (Sixteen Mile Creek). Below ground, it is anticipated that the till material is generally not water bearing. However, there will likely be sand and gravel seams with perched water.

A detailed desk study was performed for the project and included review of available published geotechnical information including Ontario Geological Survey Borehole Database and historical geotechnical reports provided by the Region.

The geotechnical investigation was completed in two phases. The investigation comprised the following components:

- Phase 1 Geotechnical Investigation (20 test borings).
- Pilot Geophysical Survey (500 m/1,640 ft.)
- Phase 2 Geotechnical Investigation (6 test borings).
- Full Alignment Geophysical Survey (3,600 m/11,812 ft.)

Representative soil and rock samples were selected for geotechnical and environmental laboratory testing. Soil testing included natural moisture content, grain size analysis, and atterberg limits. Rock testing included point load index, unconfined compressive strength, brazilian (splitting) tensile strength, punch penetration, thin section analysis, cerchar abrasivity index, slake durability, bulk density, and moisture content. Swell tests was performed on select rock samples including free-swell, semi-confined swell, and null-swell tests in addition to calcite content, water content and salinity tests. Environmental testing was also performed on soil, rock, and groundwater samples.

The geophysical investigation was performed using seismic refraction tomography and MASW (Multi-channel Analysis of Surface Waves) methods. The primary purpose of the geophysical survey was to determine the variation in the top of rock elevation (soil/rock interface) along the alignment. The results of the pilot geophysical survey were evaluated (ground truthed) using the Phase 1 test boring results to confirm proof of method before commencing work of the full alignment Geophysical Survey. The results of the geotechnical investigation were presented in the Geotechnical Data Report and the Geophysical Investigation Report.

The ground conditions along the alignment based upon the results of the geotechnical investigation comprise the following:

- Fill
- Glacio-Lacustrine Deposits (*Sand to silty Sand/sandy Silt*)
- Cohesive Glacial Till (*silty Clay to clayey Silt*)
- Till/Shale Complex (*very dense sandy Silt to sandy Gravel Till/very stiff to hard clayey Silt Till matrix containing extensive broken bedrock slabs and fragments*)
- Queenston Shale (*dominantly red, hematitic, fissile and micaceous calcareous shale. Reduction zones having a green coloration occur parallel and discordant to bedding*)
- Georgian Bay Formation (*grey-green shales with interbeds of sandstone, grey-green siltstone, and grey argillaceous limestone*)

The Till/Shale Complex is only locally encountered (in four test borings) underlying the cohesive Glacial Till and overlying bedrock. The bedrock formations, consisting of the Queenston Formation underlain by the older Georgian Bay Formation, were encountered in all borings along the alignment at relatively shallow depths ranging between approximately 1m and 8m below the ground surface. The Georgian Bay Formation is the predominant bedrock formation that will be encountered during construction. In most areas the Queenston Formation has been completely eroded. A weathered layer of rock is present at the soil/rock interface (top of rock) which is more permeable than the overlying Till and underlying intact (unweathered) Shale bedrock. The bedrock is characterized by well-developed horizontal or sub-horizontal bedding and occasional vertical or near-vertical joints. Groundwater inflow during pit and tunnel construction will occur where the soil/rock interface and any Lacustrine & Outwash Sands are encountered.

Table 1. Soil baseline properties

Parameter	Units	Fill	Glacio-Lacustrine Deposits	Cohesive Glacial Till	Till/Shale Complex
Unit weight	kN/m³	21	22	22	22
Effective friction angle	degrees	28	30	32	32
Effective cohesion	kPa	0	0	10	5
Undrained shear strength	kPa	0	0	150	100
Permeability	m/sec	1.0×10^{-6}	1.0×10^{-6}	1.0×10^{-7}	1.0×10^{-5}
Behavior above water table*	—	Firm to Slow Raveling	Slow Raveling	Firm to Slow Raveling	Firm to Slow Raveling
Behavior below water table*	—	Slow Raveling to Flowing	Flowing	Slow Raveling	Slow Raveling

* Based on Tunnelman's Ground Classification.

Groundwater levels measured in monitoring wells range between approximately 2m (6.5 ft.) and 6m (20 ft.) below ground surface and indicate that the bedrock is hydraulically connected to the overlying soil. Long term groundwater levels are expected to fluctuate seasonally by up to 1.5 m (5 ft).

Soil Properties

A Geotechnical Baseline Report was prepared for the project. The baseline soil properties are presented in Table 1. Included in this table is the anticipated ground behavior of the soil units based upon the Tunnelman's Ground Classification (Terzaghi, 1950 and Heuer, 1974), which describes behavior of completely unsupported material unmodified by grouting or other processes.

The frequent grinding of augers and the high SPT N-values recorded during drilling indicate that cobbles and boulders as well as rock slabs and fragments, derived from the weathering of the underlying rock, are expected to be present in the soil units. The maximum dimension of the boulders may be up to 1000mm (39.35 inch) with a maximum strength of 200 MPa (29,000 psi). The Boulder Volume Ratio (the ratio of the total volume of all boulders to the total volume of excavated soil) is 0.50%. All of the soils are expected to have medium to low clogging potential.

Intact Rock Properties

Rock characteristics and properties vary significantly by rock type. The Shale exhibits low to medium durability and medium abrasiveness and has an average Moh's hardness of approximately 3. The Siltstone/Limestone ("hard layer") exhibits medium to medium high durability and medium to high abrasiveness and has a hardness of between approximately 3.5 and 4. The baseline intact rock properties are presented in Table 2.

A comparison was made between "*all of the compressive strength data*" and "*compressive strength data within the microtunnel zone only*" (the microtunnel zone is from 1.5m above the tunnel crown to 1.5m below the tunnel invert) and there was no significant difference between the results.

Punch penetration test results for the Shale and Limestone/Siltstone indicate that the peak slope ranges from 7 kips/inch to 248 kips/inch indicating ductile to very high brittle rock (Yagiz, 2015).

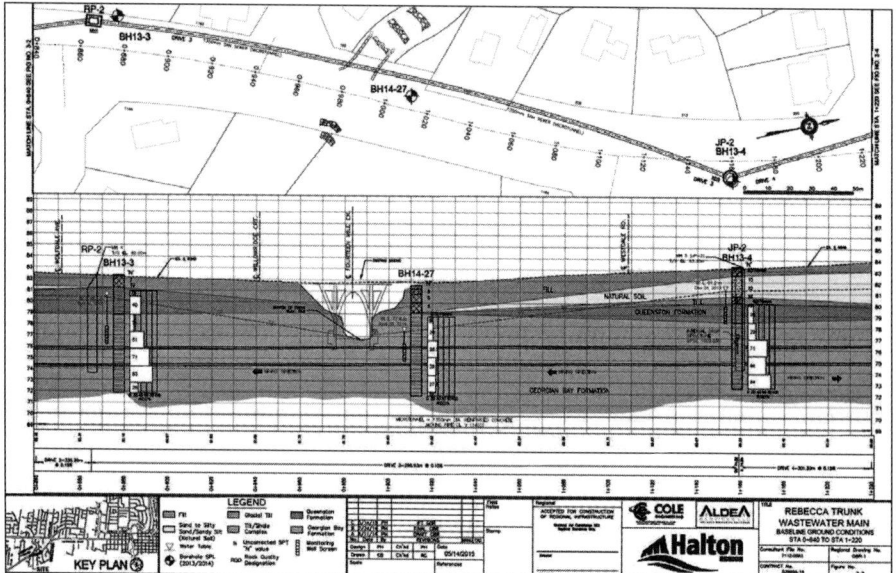

Figure 2. Baseline ground condition profile

Table 2. Intact rock properties

Parameter	Units	Shale		Limestone & Siltstone ("hard layers")	
		Average	90th Percentile	Average	90th Percentile
Unconfined Compressive Strength	MPa	18	34	53	95
Point Load Index	MPa	18	36	114	174
Brazilian Tensile Strength	MPa	2.2	2.8	8.02	10.55
Cerchar Abrasivity Index	—	1.80	1.94	2.77	3.31
Punch Penetration Index	kips/inch	46	148	127	189

Rock Mass Characteristics

The upper surface of the bedrock is moderately to highly weathered but below this layer weathering generally occurs along the surfaces of discontinuities. Rock core recoveries (total and solid) within the Shale and Limestone/Siltstone ("hard layers") were generally good with an average total core recovery of 95% and solid core recovery of approximately 70%. Rock Quality Designation (RQD) values ranged between 0% and 84% with an average of approximately 50%, indicative of poor to fair rock quality (Deere, 1988). Rock core recoveries were typically poorer near the rock surface and generally increased with depth. The percentage of Siltstone/Limestone ("hard layers") layers per core run was approximately 26%. The layers were typically less than 100 mm (4 inch) thick, however, the Georgian Bay Formation is known to contain very strong limestone or siltstone layers up to 600mm (23.65 inch) in thickness and thin, closely spaced groupings of layers may collectively be up to 1m in thickness. Discontinuities within the rock mass include generally bedding planes and planes of fissility with occasional oblique and subvertical joints.

Equivalent hydraulic conductivity values for the rock mass were estimated based on water pressure tests (Packer Tests) performed during the geotechnical investigation.

The measured equivalent hydraulic conductivity ranges from $<1\times10^{-5}$ to 6×10^{-4} cm/s. In some intact zones there were no water intakes and a couple of the higher values ($>1\times10^{-3}$ cm/s) are considered to be from a single discontinuity or zone of broken rock.

Gas

The Queenston and Georgian Bay Formations are known to contain pockets of combustible gas and the shale is classified as "potentially gassy" according to the Occupational Safety and Health Administration (OSHA Standard 1926.800, 2015).

DETAILED DESIGN

The detailed design included Preliminary Design, 50% Design, 90% Design, and 100% Design stages. The design phases each had schedule milestones to address project constraints including crossing beneath regulated waterways (14 Mile Creek & McCraney Creek), tree protection ordinance, noise ordinance, ongoing and future adjacent construction, and permitting and interaction with permitting agencies (Conservation Halton & Ministry of the Environment). The detail design comprised the following key items:

- *Alignment Layout*—location and optimization of pipeline alignment, micro-tunnel drive lengths, jacking and receiving pits, construction laydown areas, services connections (at manholes or structures), existing pump station connections (to facilitate removal offline), and temporary and permanent easement requirements.

- *Environmental Concerns*—identify creek crossing geometry, fish impacts (redside dace in-water windows), and impacts to trees.

- *Existing Utilities & Adjacent Structures*—locate and document all existing utilities and adjacent structures to minimize impacts and identify utilities requiring relocation. See Figure 3 for utility congestion on Drive 14.

- *Ground Conditions & Ground Behavior*—determine jacking pit, receiving pit, and microtunnel drive ground conditions and ground behavior.

- *Tunnel Method Selection*—evaluate tunnel vs trenchless, and select *"the lowest risk, best technical construction method available"* based upon ground conditions, ground behavior, and project constraints. Particular issues of concern include swelling shales, varying rock strengths in tunnel face (Shale vs "Hard Layers"), and low ground cover at creek crossings (14 Mile Creek—less than 0.45D).

- *Microtunnel Equipment & Systems Requirements*—determine specific equipment requirements including rear loading cutters, automated lubrication system, continuous guidance system, intermediate jacking stations, and swell suppressing slurry additives (e.g., CETCO HYDRAUL-EZ®). Contact grouting following completion of each drive will require a low strength elastic grout (less than 10 MPa/1,450 psi) to mitigate long term transfer of swelling pressures onto the pipe.

- *Drive Length & Horizontal Alignment Evaluation*—based upon required hydraulic grade, evaluate drive lengths and need/benefits of curved alignments. Calculation of jacking forces, required overcut (30mm/1.18 inch), and minimum drive radius (430m/1,410 ft.) and maximum pipe length for curve sections (2.5m/8.2 ft. for 1200mm & 3.0m/9.85 ft. for 1350mm).

- *Pipe Requirements*—determine pipe internal diameter (based upon hydraulic analysis), pipe wall thickness, and overcut requirements. Evaluate need, location, and long term benefits of corrosion protection measures (additives or cast in place linings). Calculate joint deflection on curved alignments to confirm watertightness. Requirements for pipe manufacturer testing including D load and hydrostatic pressure testing (based upon test pressure of 21.35 psi or 15m/49 ft. groundwater head).

- *Impacts on Adjacent Structures*—determine effects of microtunneling and pit construction upon utilities and structures. Perform calculations to evaluate settlement (horizontal & vertical), tensile strain, and joint slip and rotation.

- *Geotechnical Baseline Report*—GBR prepared and workshops performed with Owner (including O&M, Legal, & Procurement staff), third party stakeholders, and designer (all engineering disciplines) to discuss and agree upon baseline statements and understanding of project geotechnical risk apportionment between Owner and Contractor.

- *Geotechnical Instrumentation & Monitoring Plan*—develop geotechnical instrumentation plan to monitor ground movements during construction with respect to specified action levels and maximum levels as required to protect adjacent structures.

- *Local Ordinance Compliance*—determine local Town Ordinance requirements including allowable working hours and noise levels and process for obtaining exemptions.

- *Permitting*—clear understanding of all permits required, application requirements, and permit approval process (MOE Permit to Take Water [PTTW], MOE Environmental Compliance Approval [ECA], & Conservation Halton). Permits placed on Project Schedule and keys milestone reviewed at regular intervals.

- *Easement Acquisition*—identify and start easement acquisition process early in detailed design process.

- *Public Outreach*—perform regular public outreach to inform local businesses and residents of project and planned construction related impacts.

- *Risk Register*—preparation of Risk Register and Risk Workshops during detailed design process to identify and manage major project risks as identified by the project design team, Owner, and third party stakeholders.

Key Design Challenges

Curved Alignments

Due to the limited availability of jacking and receiving pit locations, presence of overhead utilities, limited potential for lane closures, areas of highly congested underground utilities, and required alignments at creek crossings curved microtunnel drive alignments were incorporated into the project design to limit the number of pits, reduce the size and number of easements required, and limit impacts to businesses and residents. The curved alignment design required special consideration of eccentric/non uniform jacking forces (calculated in accordance with ASCE 27-00—Standard Practice for Direct Design of Precast Concrete Pipe for Jacking in Trenchless Construction), minimum allowable radius (for 1200mm/47.25 inch & 1350mm/53.15 inch pipe), maximum allowable pipe length (for 1200mm/47.25 inch & 1350mm/53.15 inch pipe), minimum required overcut, and maximum allowable joint deflection on curved to maintain required watertightness.

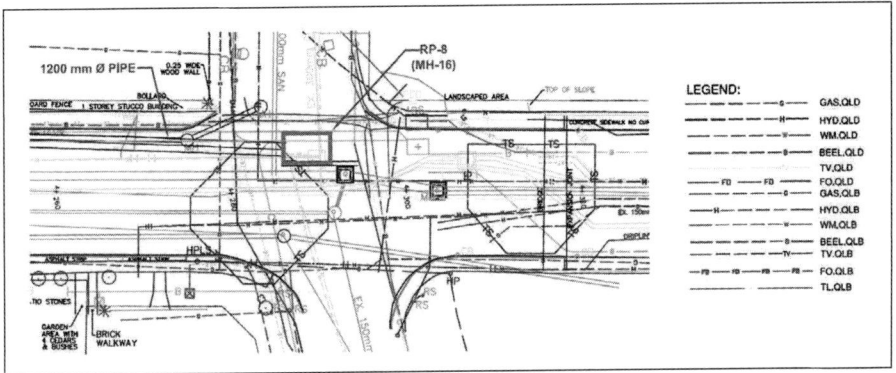

Figure 3. Utility congestion at Drive 14 and Receiving Pit 8

Time-Dependent Deformation

The Georgian Bay Formation (Shale) is known to possess high horizontal stresses and time-dependent deformation. Upon relief of the high horizontal stresses in the rock, time dependent creep-like deformations are expected to persist beyond the initial elastic deformations and will typically exceed the magnitude of the elastic movements. During microtunneling, the excavated rock envelope shall be sealed using bentonite and swell suppressing additives (e.g., CETCO HYDRAUL-EZ®) injected at the tunnel face and via grout ports in the microtunnel pipe during pipeline installation in order to fill the annular space between the outside of the pipe and the excavated rock surface. The filling of the annular space will limit the availability of water and thereby limit the ability for swelling to occur. The injection of bentonite and swell suppressing additives into the annulus during mining operations, and performance of contact grouting using low strength grout following completion of mining operations will limit swelling of the rock, reduction in overcut and loading on the pipe.

Hard Rock Layers

Limestone/Siltstone ("hard layers") layers within the Shale will be encountered during microtunneling and will increase cutter tool wear and tool cutter damage and reduce penetration rates and the broken nature of these layers may cause localized face instability and behave like cobbles/boulders. Adequate slurry line capacity, properly sized muck openings, crushers and access to the face if needed shall be provided. Gauge cutter wear and rate of jacking force shall be monitored during microtunneling as excessive wear of the gauge cutters will result in reduction of the overcut and potential for an increase in jacking forces.

14 Mile Creek Crossing

Due to the limited ground cover at 14 Mile Creek and the permitting requirements placed upon the design team for long term protection of the creek bed from scouring, a pre-engineered solution to mitigate long term impacts to the shale bedrock in the creekbed was evaluated The pre-engineered solution, as detailed in Figure 4, included localized excavation of the creek bed, installation of a box culvert, backfilling around and over the box culvert with concrete to reinstate creek bed profile, and infilling of box culvert with low strength concrete. This box culvert will be tunneled through by the MTBM during mining for Drive 4. The slurry rheology following mining through the low strength concrete inside the box culvert shall be tested during construction and the slurry properties shall be modified as necessary. When concrete is ingested by the slurry system, the calcium in the concrete "kills" the hydration effect of the bentonite in the slurry water and viscosity is significantly reduced potentially leading to destabilization of the face in poor ground at some distance beyond the concrete fill. The entire slurry system should be replenished with new water after passing the concrete to improve bentonite hydration and viscocity.

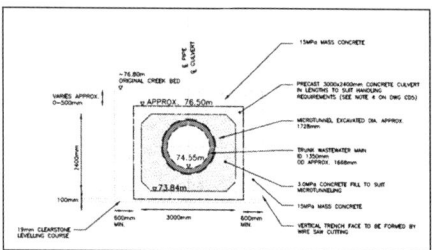

Figure 4. Details of 14 Mile Creek mitigation pre-engineering

PROCUREMENT

The project procurement comprised a microtunneling contractor Request for Qualification (RFQ) and a Request for Proposal (RFP). A total of eight (8) microtunnel contractors submitted an RFQ package and based upon review of financial strength, corporate experience, corporate references, staff experience, staff references, and Health & Safety/Training a total of six (6) microtunnel contractor were prequalified. Due to the number of local microtunnel subcontractors with experience of long curved microtunneling (>400 m/1,300 ft.) in Georgian Bay Shale

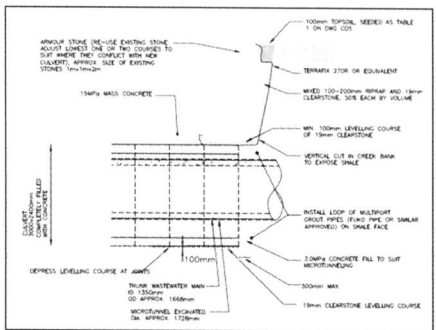

Figure 5. Details of 14 Mile Creek mitigation pre-engineering

the drives were designed with a maximum length of 400m (1,300 ft.). A total of four (4) contractor teams submitted bids for the project of $42,58M, $42,62M, $46,9M, and $64,03M. The project was awarded to the low bidder. Based upon the length of microtunnel being approximately 4,050m (13,287 ft.) the bid price cost including all jacking pits, receiving pits, jacking pipe, and all appurtenant project work is $10,513/meter ($3,205/foot).

POST-AWARD VALUE ENGINEERING

Following award of the Contract the selected microtunneling subcontractor, who has significant experience with curved drives and local experience in the Greater Toronto Area (GTA) mining through the Queenston Shale and Georgian Bay Formation, proposed extending five of the microtunnel drives thereby removing five receiving pits (RP-2/MH-4, RP-3/MH-6, RP-4/MH-8, RP-5/MH-10, & RP-7/MH-14). The proposal was submitted by the microtunneling subcontractor in accordance with Section 1036 (Value Added Change Proposals). The modified proposed drive lengths including proposed mining duration and rates are presented in Table 3. The use of additional intermediate jacking stations was included in the proposal. The design change was approved and the benefits include reduced impact and disruption to businesses and residents, fewer expensive utility relocations, and fewer tree removals.

CONSTRUCTION

Notice to commence work was provided to the Contractor from Halton Region in October 2015. Construction activity commenced in December 2015 with preparation of works areas, traffic control, construction of jacking pits and receiving pits using reinforced concrete caisson shaft construction methods, and installation of geotechnical

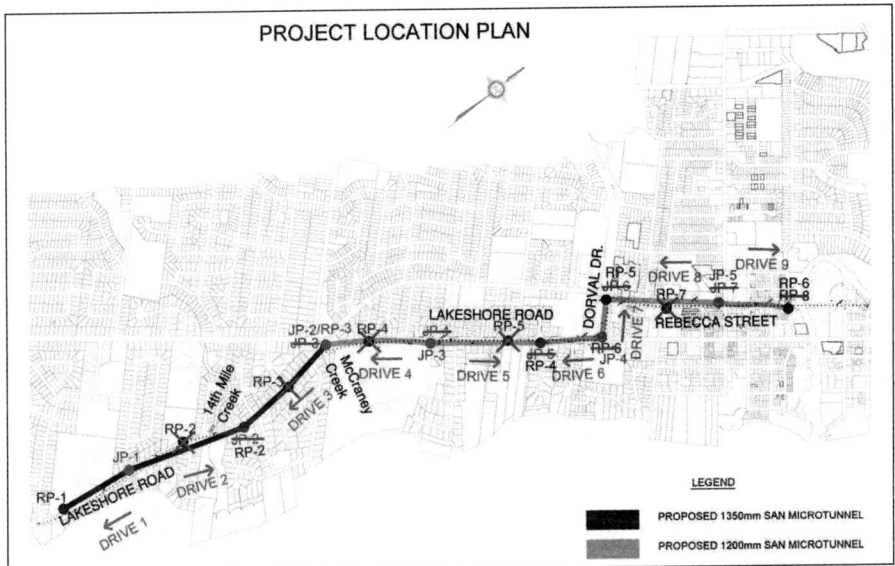

PROJECT LOCATION PLAN

LEGEND

PROPOSED 1350mm SAN MICROTUNNEL

PROPOSED 1200mm SAN MICROTUNNEL

Figure 6. Modified design

Table 3. Summary of as-built microtunnel drive mining duration and mining rates

Drive	Pipe Internal Diameter (mm)	Length (m)	Ground Conditions	Alignment Type	Duration (days)	Mining Rate (m/day)
Drive 1	1350	382	Mixed Face	Straight	23	16.58
Drive 2	1350	626	Rock	Composite curve	44	14.26
Drive 3	1350	576	Rock	Composite curve	64	9.00
Drive 4	1200	523	Rock	Composite curve	42	12.44
Drive 5	1200	564	Rock	Composite curve	54	10.42
Drive 6	1200	311	Rock	Composite curve	25	12.49
Drive 7	1200	180	Rock	Composite curve	10	18.11
Drive 8	1200	587	Rock	Composite curve	52	11.31
Drive 9	1200	328	Mixed Face	Composite curve	33	9.95

Drive 1—approximately 13% Overburden, 75% mixed face and 12% rock
Drive 9—approximately 77% mixed face and 23% rock.

instrumentation. All microtunnel drives were completed on 25 October 2016 ahead of the scheduled construction completion date of 30 December 2016. The as-built drive lengths, mining duration and mining rates are presented in Table 3. In this section we discuss predicted vs actual jacking forces for straight and curved drives in the Georgian Bay Shale, back calculated friction factors, and issues encountered during construction.

The total as-built mined length was 4082.02m. The total as-built mining duration was 347 days (Proposed mining duration = 263 days). Construction was performed 24 hours per day (two 12 hour shifts) for 5.5 days per week between Monday 7:00am through Saturday 7:00pm.

As-Built Jacking Forces

The jacking forces during construction were monitored constantly and data collected at regular intervals by the Contractor and verified by regular spot checks by the Owner's Engineer. The entire data set was provided to the Owner at the end of each drive. A summary of the as-built jacking forces for each drive is presented in Figure 7.

Intermediate jacking stations were installed on all drives but were not used at any time on any of the drives.

A summary of the actual jacking forces versus predicted jacking forces, and the actual back-calculated friction factor per drive are presented in Figure 8.

Based upon published literature the recommended friction factor for rock is between 0.3 and 0.4 psi (ASCE 27-00).

Curved Drives

All drives are curved with the exception of Drive 1 (MH3 to MH2) have a proportion of the drive constructed on a curve. Of the eight (8) curved drives Drive 3 had the least amount of curve (20%) and Drive 6 had the greatest amount of curve (91%). The percent of total curve vs straight length per drive are presented in Figure 9.

Longest Curve and Tightest Radius Drives

An evaluation of the longest curved drives and the tightest radius curves has been performed to determine the impact upon jacking forces. A summary of the jacking forces and friction factors for the longest curve in each drive are presented in Table 5 and on Figure 10.

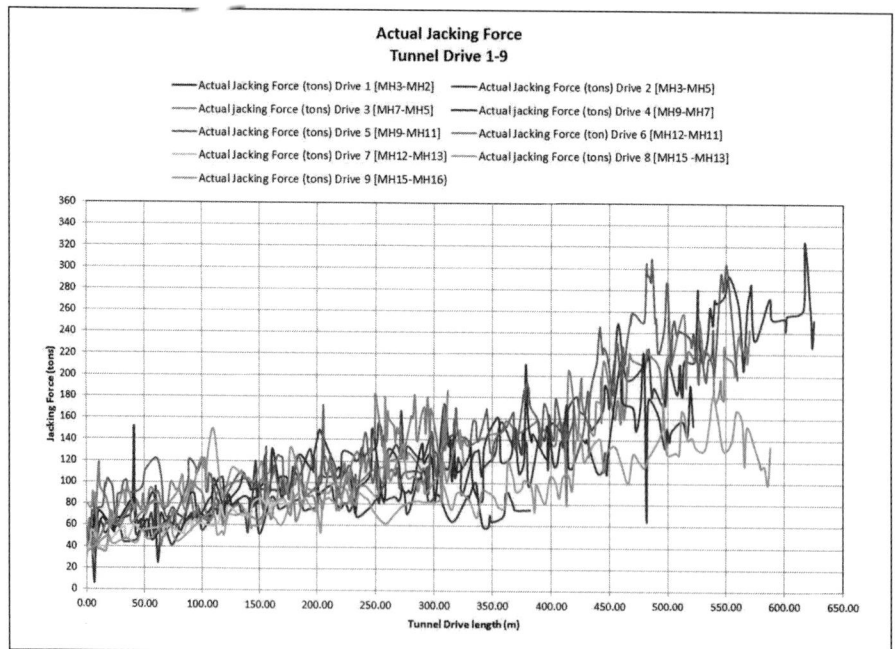

Figure 7. As-built jacking forces

Table 4. Summary of as-built jacking forces by drive (full drive length)

Drive	Inside Pipe Diameter (mm)	Length (m)	Curved Length* (m)	Jacking Force Range (tons)	Jacking Force† (tons)	Face Pressure‡ (tons)	Skin Friction Force§ (tons)
Drive 1	1350	382	0	25–113	82	46	36
Drive 2	1350	626	433	6–325	265	30	235
Drive 3	1350	576	115	32–258	227	50	177
Drive 4	1200	523	466	43–218	165	43	122
Drive 5	1200	564	352	41–308	270	41	229
Drive 6	1200	311	284	35–186	158	35	123
Drive 7	1200	180	100	36–94	85	36	49
Drive 8	1200	587	420	30–182	117	30	87
Drive 9	1200	328	102	36–150	129	80	49

* Curved Length = total distance of drive constructed as a curve
† Jacking Force = average of the last five (5) readings
‡ Face pressure estimated at start of drive (generally within first pipe length).
§ Skin Friction Force = jacking force – face pressure

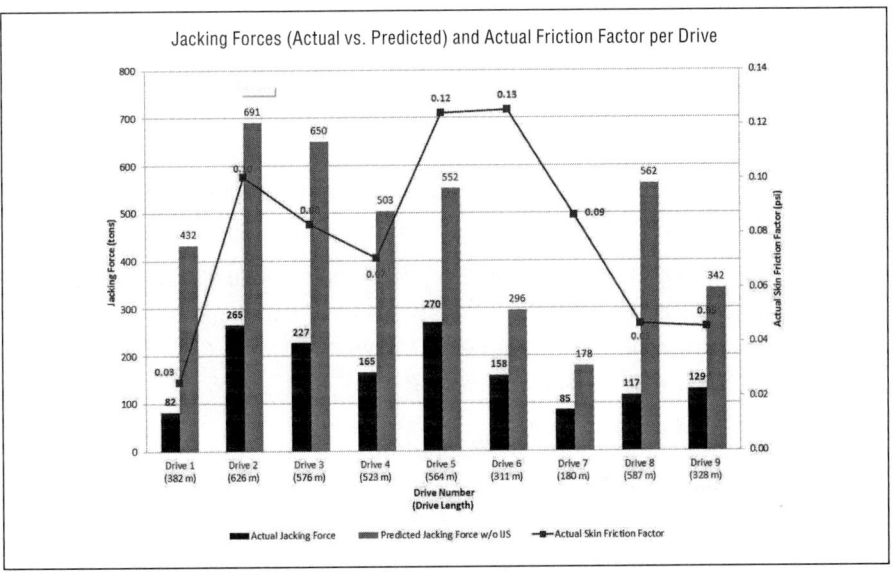

Figure 8. Jacking forces and skin friction factor per drive (total drive length)

A summary of the jacking forces and friction factors for the curves less than 500 m radius are presented in Table 6 and Figure 11.

General observations for as-built mining data (all data):

- Actual jacking force as percent of the predicted jacking force = 19% to 54%
- Actual Jacking Force in Rock Drives (1200mm) = 30 to 308 tons
- Actual Jacking Force in Rock Drives (1350mm) = 6 to 325 tons
- Friction Factor in All Rock Drives = 0.05 psi to 0.13 psi

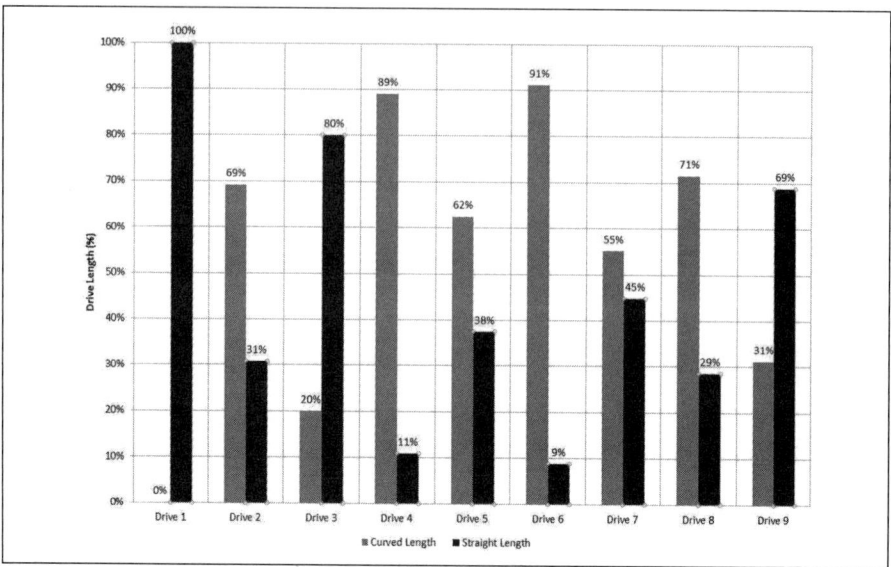

Figure 9. Percent of total curve vs. straight length per drive

General observations for longest curves:

- Actual Jacking Force in Rock Drives (1200mm) = 46 to 172 tons
- Actual Jacking Force in Rock Drives (1350mm) = 80 to 191 tons
- Friction Factor in Rock = 0.02 psi to 0.14 psi

General observations for the tightest curves:

- Actual Jacking Force in Rock Drives (1200mm) = 36 to 150 tons
- Actual Jacking Force in Rock Drives (1350mm) = 6 to 120 tons
- Actual Jacking Force in Mixed Face Drives (1200mm) = 36 to 150 tons
- Friction Factor in Rock = 0.06 psi to 0.12 psi

CONCLUSIONS

The following general conclusions can be made with respect to the construction:

- *Friction Factor.* The friction factors experienced during construction (0.05 to 0.14 psi) in Georgian Bay Shale were approximately 15 to 35 percent of the commonly published values for rock (0.3 to 0.4 psi)
- *Time-Dependent Deformation.* Mining with a minimum overcut of 25mm and an annulus filled with bentonite lubricant (with swell suppressing additives) no significant increase in jacking forces were observed, even after week-end shutdowns, in some cases for a week or more. Design and construc-tion means and methods were developed to prevent development of TDD/swelling and resulting locking of pipe in jacking place.
- *Jacking Forces.* The rate of development of jacking forces on properly designed and constructed curved drives in Georgian Bay Shale is not

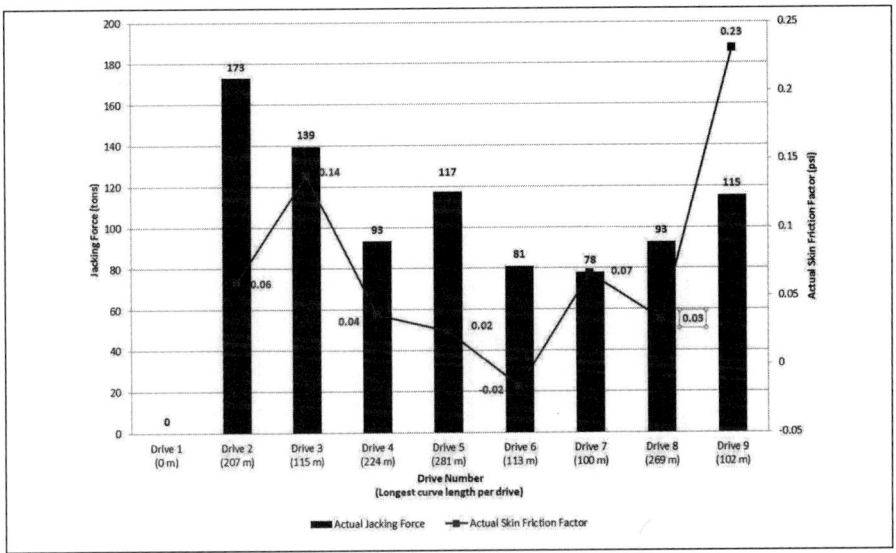

Figure 10. Jacking forces and skin friction factor per drive (longest curve)

Table 5. Summary of as-built jacking forces for longest single curve per drive

Drive	Pipe Diameter (mm)	Longest Curved (m)	Curve Radius (m)	Jacking Force Range (tons)	Jacking Force* (tons)	Face Pressure† (tons)	Skin Friction‡ (tons)
Drive 1	1350	0	—	—	—	—	—
Drive 2	1350	207	600	89–191	173	126	47
Drive 3	1350	115	1000	80–147	139	80	59
Drive 4	1200	224	1710	46–152	93	67	26
Drive 5	1200	281	1200	61–172	117	95	22
Drive 6	1200	113	3382	60–90	87	81	6
Drive 7	1200	100	1531	52–93	78	57	21
Drive 8	1200	269	800	53–120	93	65	28
Drive 9	1200	102	500	36–150	115	37	78

* Jacking Force = average of the last five (5) readings.
† Face pressure estimated at start of drive (generally within first pipe length).
‡ Skin Friction Force = jacking force – face pressure.

significantly different from straight drives assuming that the pipe is designed to take into account the radius of curvature, drilling mud is cleaned/separated from spoil and mud rheology controlled, and automated in-pipe lubrication systems are used.

- Intermediate Jacking Stations. IJS's were installed within the pipe string on all drives but not used at any time during mining due in large part to the in-pipe lubrication system installed/operated and the careful control/cleaning of the mud to limit the amount of spoil entering the annulus (between excavated ground and pipe outside diameter) which has an impact on the development of higher jacking forces.

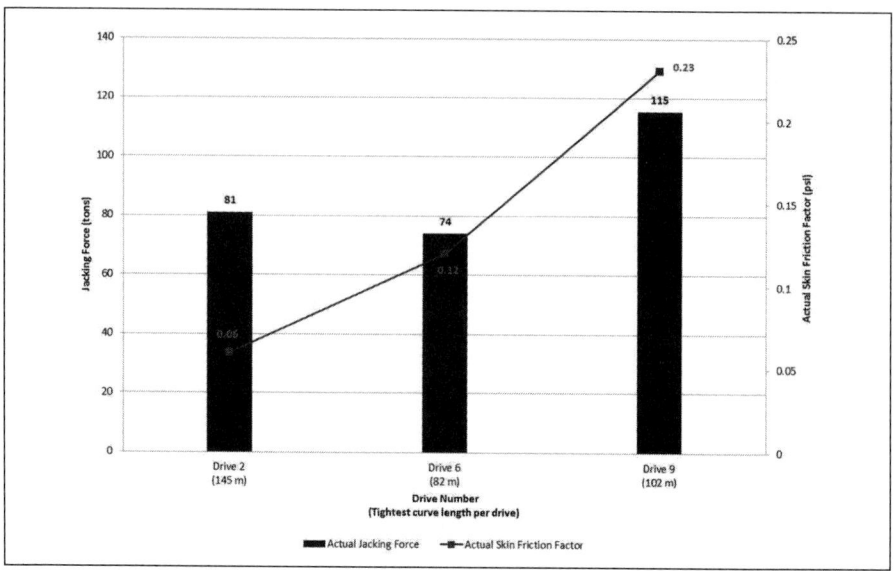

Figure 11. Jacking forces and skin friction factor per drive (tightest curve)

Table 6. Summary of as-built jacking forces for tightest curve per drive

Drive	Pipe Diameter (mm)	Tightest Curve Length (m)	Curve Radius (m)	Jacking Force Range (tons)	Jacking Force* (tons)	Face Pressure† (tons)	Skin Friction‡ (tons)
Drive 2	1350	145	430	6–120	81	48	33
Drive 6	1200	82	450	41–87	74	43	31
Drive 9	1200	102	500	36–150	115	37	78

* Jacking Force = average of the last five (5) readings.
† Face pressure estimated at start of drive (generally within first pipe length).
‡ Skin Friction Force = jacking force – face pressure.

REFERENCES

Canadian Standards Association (CSA), 2014, *Reinforced Concrete Culvert, Storm Drain and Sewer Pipe* Standard CAN/CSA-A257.2-14.

Deere, D.U. and Deere, D.W., 1988. *The Rock Quality Designation (RQD) Index in Practice*. Rock Classification Systems for Engineering Purposes, ASTM STP 984, ASTM, Philadelphia, 1988, p 91–101.

Heuer, R.E., 1974, *Important Ground Parameters in Soft Ground Tunneling*. Subsurface Exploration for Underground Excavation and Heavy Construction, ASCEW Specialty Conference.

Ontario Regulation 162/06, 2013, *Halton Region Conservation Authority: Regulation of Development, Interference with Wetlands and Alterations to Shorelines and Watercourses*, February 2013.

Ontario Geological Survey, 1976, *Paleozoic Geology, Hamilton, Southern Ontario*, 1:50,000, OGS Map 2336, 1976.

Terzaghi, K, 1950. *Geologic Aspects of Soft Ground Tunneling.* Chapter 11 in Applied Sedimentation, ed. P. Trask, John Wiley and Sons, New York.

Yagiz, S., 2015, *The Punch Penetration Test for Estimating Machine Performance,* Paper Ref #41, International No-Dig Istanbul 2015, 33rd International Conference and Exhibition.

United States Department of Labor, Occupational Safety and Health Administration (OSHA), *Safety and Health Regulations for Construction,* Standard Number 29 CFR 1926.800(h)(1)(ii), 2015.

Environmental Protection Act, 2014. Ontario Regulation 153/04, *Records of Site Condition,* Part XV.1 Of The Act, under Environmental Protection Act R.S.O. 1990, c. E.19, January 2014.

Canadian Standards Association (CSA), 2014, Standard A23.1-09/A23.2-09 (R2014) - *Concrete Materials and Methods of Concrete Construction/Test Methods and Standard Practices for Concrete,* 2014.

PART

SEM/NATM

Chairs

Bianca Messina
Skanska

Josh Jonasen
Traylor Bros., Inc.

Steep Inclined SEM Excavation—The "Uphill Machine"—at London Crossrail: Development and Application of a Safe Excavation System in Soft Ground

Rainer Antretter ▪ BeMo Tunnelling GmbH

ABSTRACT

Escalator tunnels for passenger station access are usually built applying declined excavation method in order to avoid excavated material from falling down along the excavated section putting individuals and plant in danger. If this method cannot be executed due to geometric or contractual restrictions a challenging situation is encountered. This was the case at the Whitechapel and Liverpool station sites at London's Crossrail, contract C510.

This paper describes development of the basic idea, initial layout, final design and practical application of a mechanised uphill excavation and support system. The machine was used for excavation of multiple inclined tunnels with a 30-degree gradient.

INTRODUCTION—SITE CONDITIONS AND BASIC TASK

During tendering Contract 510 the challenging situation of excavating a connection from the Crossrail platform tunnel upwards to the existing Northern Line was encountered. From the tender documents, it was obvious that no access from the existing Northern Line tunnel system could be provided to excavate the link to the new Whitechapel station in declined manner. A methodology had to be developed to excavate safely from the station tunnel inclined to the slurry wall of an existing shaft box ending blind in front of the box - the birth of the Uphill Machine or the Uphill excavator.

How Could a Steep Inclined Excavation System Look Like?

A rough idea from existing mining techniques and in house experience of a self-moving suspended drilling system for spiles and blasting holes in the crown of tunnels formed the description of a suitable system to be inquired at special suppliers on the market which needed to fulfill all steps of SEM excavation, see Figure 1.

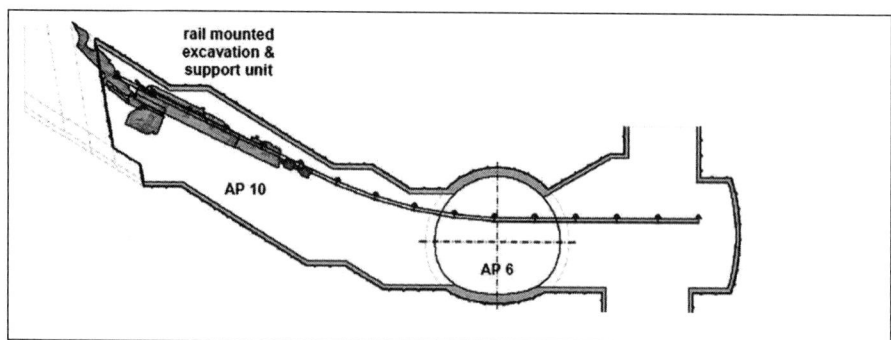

Figure 1. Early stage of development the Uphill Machine

The link passage stairway ES1 requires excavation from the lower level (AP6) up to the Northern Line Platform Tunnel level. This is not a standard technique for the construction of passages with this degree of inclination (up to 30°) and introduces additional hazards relating to the increased face size and excavating below the face. In addition, the passage is to be constructed closely underneath an existing escalator tunnel.

Minimize health & safety risks when tunnelling in an upward direction, requires a highly mechanized excavation & support machine. The system must be capable of excavation, mucking, applying shotcrete, and operating a working basket. The machine could be railbound (e.g., rack/pinion) or use the shotcrete side walls to climb up/down. The unit must be able to move in the tunnel by itself.

When the machine reaches its working position, it must somehow be braced or locked firmly in place.

The operator must be able to control all the operations of the machine from within the safe environment (e.g., elevated position) of a cab providing good sight and overview of the working area.

The cross section of the tunnel in combination with the tunnelling unit must be big enough to allow falling material to pass under the machine without posing any threat to the operator or the machine.

The Godfather

As one of the key construction parts of the Ermenek hydroelectric plant in Southern Turkey, an inclined shaft with a slope of 27° and a length of appr. 620m had to be executed by NATM in upward direction. The tunnel had a cross section of 42m², which was mined in drill and blast full face excavation.

Due to the gradient, special equipment was necessary to ensure safe working conditions and efficient tunnelling. The principle of excavation and drilling machines running on a monorail in the tunnel soffit has been used extensively in German mines. Alpine-BeMo (now BeMo Tunnelling GmbH) together with GTA Maschinensysteme GmbH, Haminkeln Germany developed a proto-type *monorail* drill rig based on this principle, adapted to the project. It consisted of a telescopic boom with a hydraulic rock drill and a second boom with a man basket for safe changing or extending the monorail in the top of the tunnel. The man basket was also used for safe installation of all required support measures. The inclination presented challenges in respect to safe upward advance and fail-safe braking. During drilling, reaction was provided by grippers acting against sprayed concrete lining, see Figure 2.

Figure 2. The Ermenek unit in the drill-and-blast penstock tunnel

Execution

Several suppliers were asked to design and propose a suitable machine. Most of them have refused to make a technically and economically elaborated proposal or replied with a budget figure only. The German supplier GTA was that offered exactly what was asked for since this company designed and delivered similar "crown suspended" excavation machines before.

In addition, entirely different systems of solution have been born within the JV team on site. Either of these with the serious disadvantage of not offering a suitable and effective solution against falling lumps from the face towards to the excavation machine and personnel, see Figure 3 as an example. Therefore, for the sake of safety reasons—which was the core concern—the team responsible

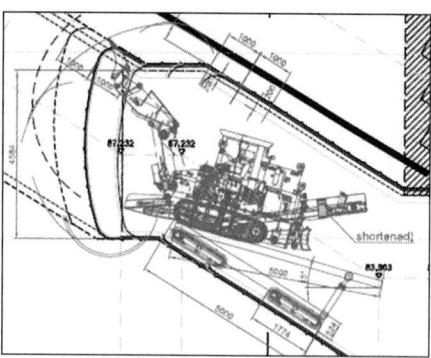

Figure 3. Alternative idea, tunnel excavator on tracked and tiltable platform

for development of the uphill excavation system finally decided to abandon alternative systems and proceeded with the crown suspended excavating and shotcreting machine.

GENERAL ASPECTS AND PRESENTATION OF DEVELOPMENT

The excavation and spraying unit, together with the operator's man basket, was intended to hang on rails from the tunnel crown. The proposed method was selected in accordance with the requirements of the British Standard for Tunnelling (BS6164).

The time frame for development of details was not very tight at the beginning. During autumn 2011 the major features were developed and proposed mainly by the manufacturer after discussions of geometrical and technical requirements. In that stage the main sizes, movement velocities and abilities and the reach of the Uphill Excavation Machine were determined and agreed upon.

Presentation of Main Parameters for All Stages of Excavation

A presentation of the designed top hanging excavation system in which all features had to be reported to the client was scheduled at the end of January 2012.

The already well-engineered design of the Uphill Machine itself had been prepared and all steps of an excavation cycle were determined beforehand and presented within this meeting, see Figure 4.

Main Parameters

- Reach of excavator and spraying boom in dependence of early strength of shotcrete
- The envelop around the Uphill Machine—the pilot tunnel of the escalator tunnel—was flexible to a certain degree in size but was not supposed to be too large since the extreme gradient created an enlarged vertical face area
- Surrounding equipment as ventilation duct, walkable stair as access to close the face and mucking assistance equipment needed to be accommodated within the pilot tunnel as well
- Measures for rail placing and probe drilling were designed and determined

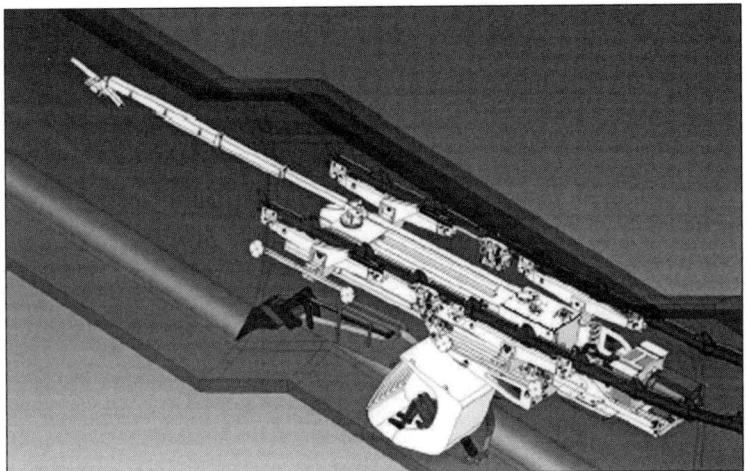

Figure 4. General arrangement of mainframe, excavator boom, and spraying boom on two rails

Equipment Set-Up

- Mainframe hanging on two rails braced with four grippers to the lining during operation
- Excavator boom mounted on and able to travel along main frame fitted with a quick hitch for additional tools beside the excavation bucket: Man basket and hydraulic hammer
- Spraying boom mounted on and able to travel along main frame; a light weight telescopic type boom which is located between the tracks on the top-side of the mainframe
- Operator's cabin mounted on the main frame
- Man-riding basket attached to the excavator boom; Basket to be operated from the basket

Gripper Function

In working position the mainframe is braced by four gripers to the tunnel lining fixing the whole equipment during excavation and spraying operations. This is necessary to prevent uncontrolled movement of the unit while excavating and spraying and avoids dynamic loads into the track support, see Figure 5 and Figure 6.

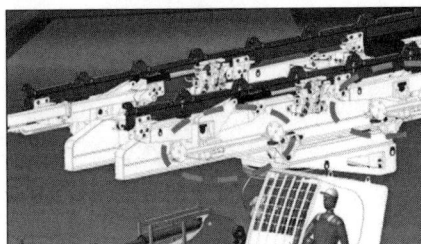

Figure 5. Mainframe in driving position (grippers retracted position)

Figure 6. Mainframe in working position (grippers extended position)

Driving Mechanism

The whole unit moves along the tracks using three pairs of clamps with one set having a pair of hydraulic cylinders attached. The carriage is alternately fixed with the two sets of fixed clamps (also for excavation mode) or the one set of the clamps with cylinders, see Figure 7. When the cylinder is retracted the clamps on the mainframe are closed. Then the cylinder extends and the clamps on the carriage close. Now the clamps on the mainframe are opened and the cylinders extend, pushing the unit forward. For reversing the process is reversed.

Figure 7. Driving mechanism

The unit covers about 0,5m with every "step" and can travel roughly 20m/min.

Excavation Boom Unit (Including Man Basket)

The excavator boom is a custom-made design to provide the kinematics for the required reach and flexibility to excavate the profile of the Pilot Tunnel. It consists of three boom sections and can rotate left and right on a swing bearing between the mainframe and the first boom section. The whole unit can travel over a distance of 2,5m on guiding rails along the mainframe. This movement is powered by a hydraulic cylinder, allowing the excavator boom to operate in any position along its 2,5m range, see Figure 8.

The boom is equipped with a quick hitch system allowing for different buckets as well as a basket which will be utilised to mount the lattice girder used for holding the support bolts in place as well as to extend the rail track in the same fashion, see Figure 9.

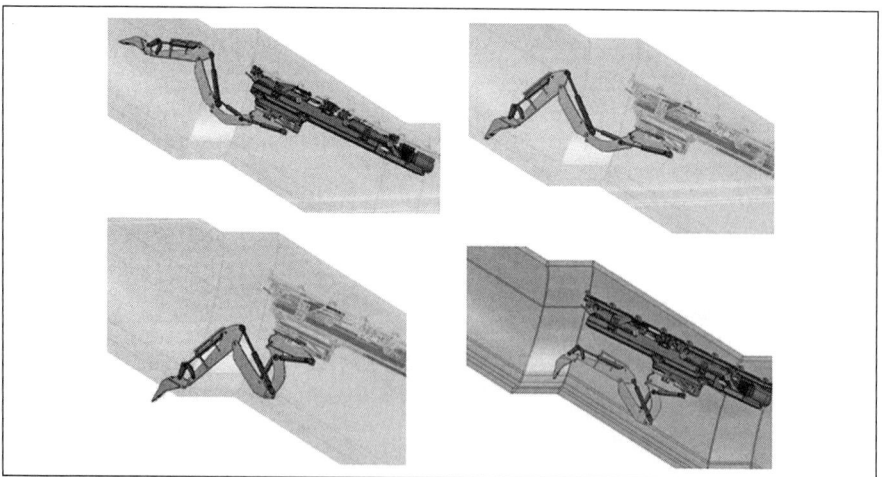

Figure 8. Kinematics of excavator boom

Suspension, Rail Tracks, and Installation

Track extensions, as shown in Figure 10, are *1m long* and have *32kg each*. They are standardized I-shaped profiles (I140E) installed from man riding basket. To avoid heavy manual lifting activities the excavation unit will be equipped with a man basket with an attached bracket that supports the track extension while it is moved into the correct position using the basket.

Rails are installed on the shotcrete lining after approx. 3 days using two possible installation systems (both methods were used in combination), suspended on brackets and a chain system, see Figure 11.

Time Frame

The presentation meeting held with the client at the end of January 2012 went well and the responsible persons of Crossrail could be convinced of the system and its safety advantages compared to an invert based system. Of course, the discussion also created some additional technical items to be solved.

The further time frame was continued by solving of the additional items followed by submitting the final Uphill Excavation Concept in summer 2012. After technical clarification, the general decision by the client to order the system took place in autumn 2012. Finally, the order to the manufacturer was placed in January 2013. Delivery was

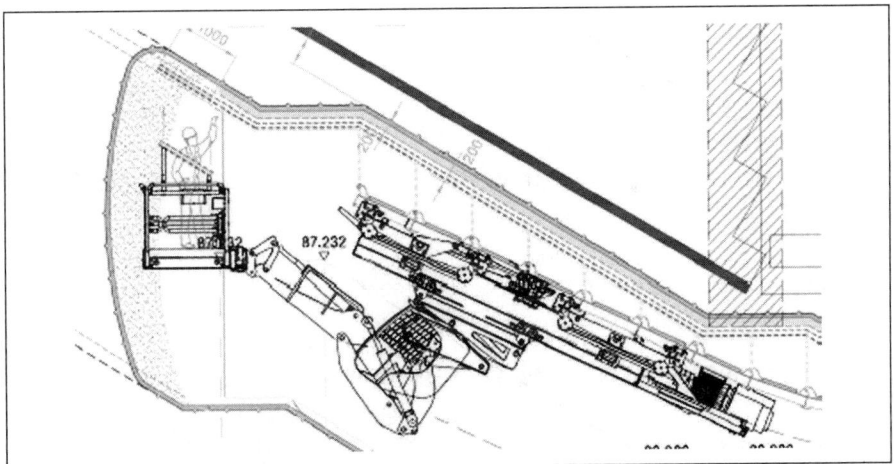

Figure 9. Placement of lattice girders for support bolts

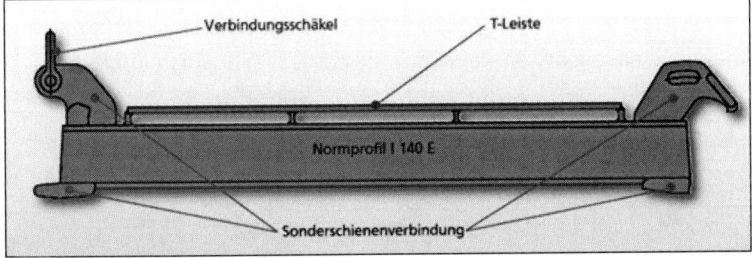

Figure 10. General layout of standard rails

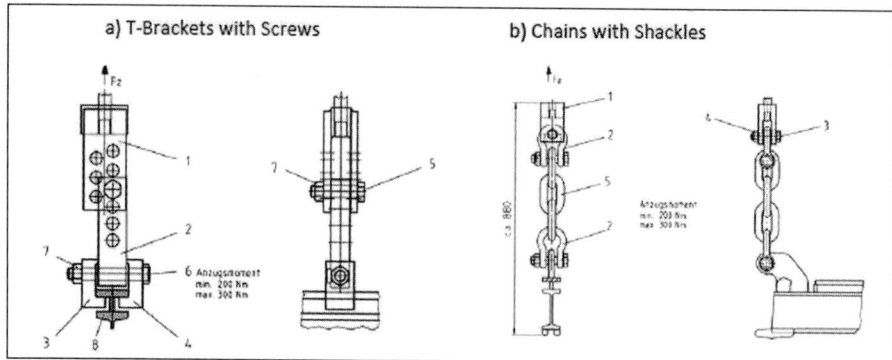

Figure 11. General layout of standard rails

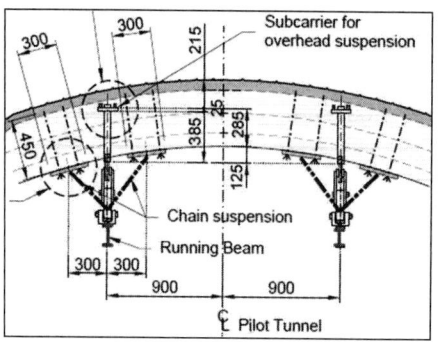

Figure 12. Bracing system—cross section in pilot

Figure 13. Bracing system—longitud. sect. in pilot

scheduled eight month after order and the first utilization was intended in October 2013.

The real date of the first application was summer 2014 after some additional preparations due to design issues of the tunnel. The most important preparation was a pipe arch at the starter tube of ES1.

DEVELOPMENT OF BRACING SYSTEM—APPLICATION IN ES1 TO ES3

The selected *twin-rail system* required adequate suspension at the shotcrete surface in the pilot tunnel crown. Typical practice in mining sections would be chaining and counterbracing the rails by brackets from and against the steel arches of the section. In a SEM tunnel in soft ground steel arches would normally be sprayed in shotcrete and therefore not available for brackets and in the case of C510 no arches and mesh had been used at all. All shotcrete was steel fiber reinforced using Lasershell™.

It was therefore necessary to develop a carrier system which could be sprayed in shotcrete and able to bear the 18,5tons weight of the Machine and a chaining/counterbracing system to keep it stable during operation with the maximal load of more than 50kN specified digging force, see Figures 12 and 13.

Within the wide station sections the chaining system was designed to consist just of vertical suspenders and horizontal bracing, see Figures 14 and 15.

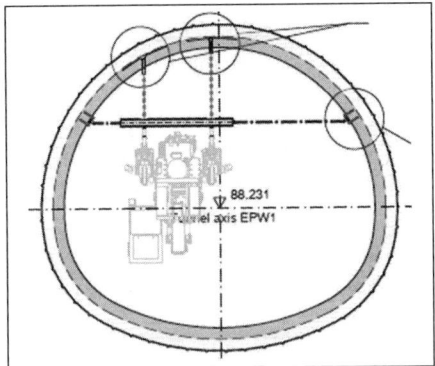

Figure 14. General layout of standard rails

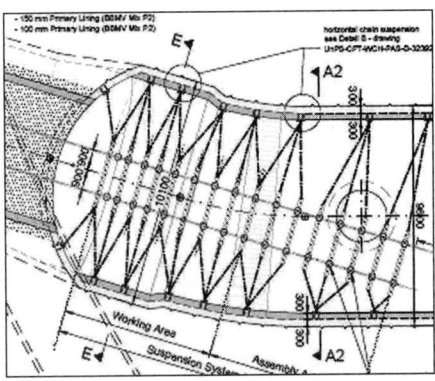

Figure 15. General layout of standard rails

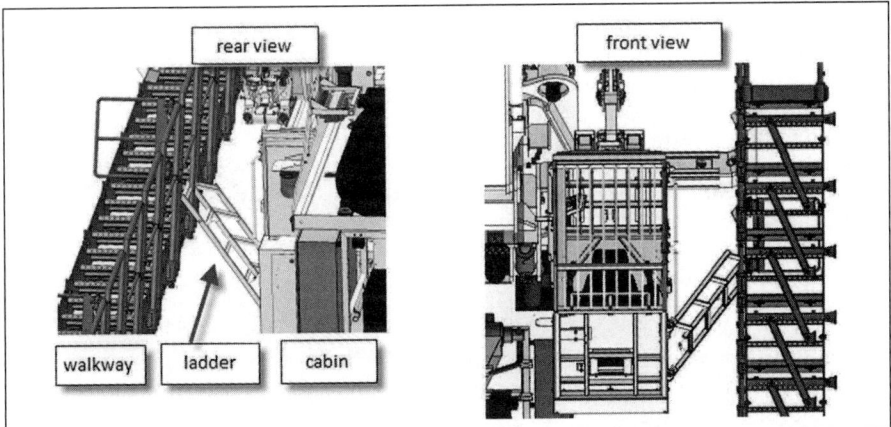

Figure 16. Bracing system—cross section in pilot tunnel

An additional measure was the introduction of an elevated walkway on the left-hand side of the tunnel, allowing the operator of the Uphill Machine to leave the cabin via a short emergency ladder which could be folded off the cabin and laid against the walkway. This ladder also served as a safety door on the cabin, see Figure 16.

Application in ES1, ES2, and ES3

First application of the Uphill Machine took place in ES1 which was a tunnel with 14 inclined advances with 1m each, second and third application were tunnels ES2 with 28 and ES3 with 35 inclined advances. ES2 and ES3 even had the specialty of seven (ES3) and 12 (ES2) horizontal advances at their upper ends. At each of the tunnels the first 5 advances were excavated with the standard tunnel excavator Liebherr R944 and shotcreted with a Meyco Potenza spray robot from the station floor level.

All the single pilot tunnel drives were excavated safely and without serious incidents. The elevated walkway separated personnel well from excavation operation. The machine system with its boom reach provided shelter for personnel within the hardened shotcrete shell sufficiently far behind open sections at the face during operation, see Figure 17.

EXPERIENCES—CONCLUSION

Excavation speed was accepted to be limited by the shotcrete strength; it was intended to excavate one meter per day with the load from the Machine being introduced to the shotcrete shell four meters behind the face. Together with an additional layer of shotcrete this distance ensured sufficient strength of the shotcrete shell at the front load points. A strict regime of bolts testing via a test device prior to this load transfer was carried out on each advance.

Figure 17. Uphill Machine during digging ES1— walkway left-hand side

In practical use the intended excavation speed has only partly been reached but the purpose of the innovative excavation system as such has fully been achieved. In particular, from a safety prospective but also from a clearly separated flow of materials with attack and supply from a suspended system in the cross sections upper part and discharge and disposal of muck in the lower part, the Uphill Machine operated without additional measures in logistics.

Figure 18. Mucking down ES3 tunnel

Shotcrete was pumped from a pump situated in the horizontal station tunnel via a hose to the robot arm and applied by the operator from the cabin by using remote control.

Mucking down the pilot tunnel was supported by a chain conveyor in the long ES3 tunnel, see Figure 18, in the short tunnels the digger boom mucked the remainder which did not slide down itself.

Benefits of the Innovation

- Removing the escalator excavation from the projects critical path
- Allowing to work almost completely independently from others, without interference and/or constraint
- Reduced the time frame required in the station shafts and helped to optimization of the overall Crossrail program
- Significant Cost Savings for all involved
- Fully electrical operation reduced demand on ventilation and influence to the workers
- Safe system of work for constructing incline tunnels from the bottom up

Downtown Bellevue Tunnel—Concept Optimization Through Team Collaboration

Derek Penrice ▪ Mott MacDonald
Hong Yang ▪ Mott MacDonald
Chad Frederick ▪ Sound Transit
Jacob Coibion ▪ Atkinson Construction

ABSTRACT

The Downtown Bellevue Tunnel (DBT) is a key component of Sound Transit's East Link Project—a $3.7 billion, 14-mile light rail extension connecting Seattle with the cities of Bellevue and Redmond. The DBT construction contract was awarded to Atkinson Construction in the Fall of 2015. Subsequently, at the request of Sound Transit, the Contractor was asked to provide recommendations for reducing the DBT construction schedule by 3 months, to mitigate a delay in right-of-way acquisition and provide flexibility on an adjacent contract. This paper discusses the collaborative process involving Owner, Contractor and Engineer that supported the evaluation and implementation of the Contractor-generated proposals—including revisions to the prescriptively designed SEM excavation sequence, and elimination of pipe canopy, to accommodate the revised project schedule demands.

INTRODUCTION

The East Link Project is a $3.7-billion (2015 $), 14-mile extension of the existing Sound Transit (ST) light rail transit (LRT) system from downtown Seattle, across Lake Washington via I-90, to the cities of Bellevue and Redmond on the east side of Lake Washington. Final design of the South Bellevue to Overlake Transit Center Section of the Project was awarded to an H-J-H joint venture comprising HNTB Corporation, Jacobs Engineering and Hatch Mott MacDonald now operating as Mott MacDonald (MM) in February of 2012. MM retained design responsibility for DBT. The DBT extends approximately one half mile through downtown Bellevue under 110th Avenue NE between Main Street at East Main Street Station and NE 6th Street at the Downtown Bellevue Transit Center (BTC) Station. The location and extent of the DBT is indicated in Figure 1. A fuller description of the project setting and the DBT itself is provided in Sound Transit East Link—Downtown Bellevue Tunnel (Penrice et al., 2016).

DBT

The DBT comprises approximately 2,237 feet of tunnel structure involving both cut-and-cover and sequential excavation (SEM) tunneling methods. At the DBT's southern end, a 250 ft long cut-and-cover structure provides a transition from an at-grade track section to the SEM tunnel in an area of limited ground cover where an enlarged structure cross section is also required to accommodate emergency ventilation fans.

The 1,987-foot SEM tunnel portion of the DBT extends between the south cut-and-cover structure and a similar structure to the north, to be constructed as part of a separate contract. The typical SEM tunnel cross-section, as measured to the outside of the initial support is a 36.7 feet wide by 30.5 feet high ovoid. The final lining is comprised

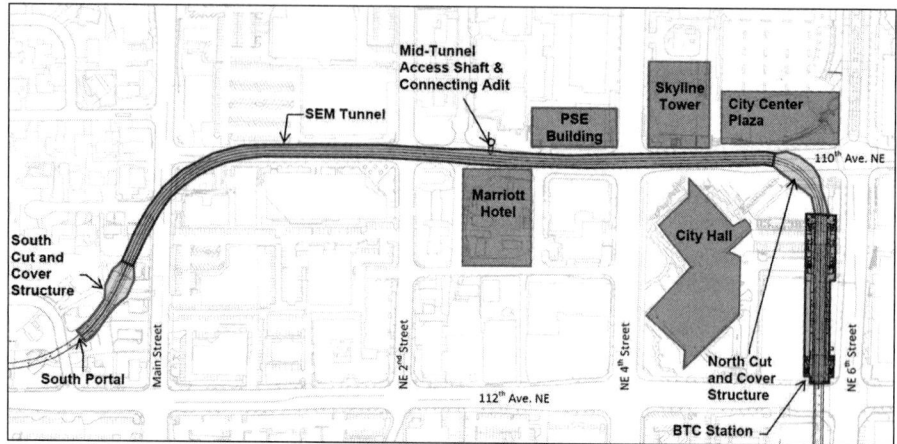

Figure 1. DBT alignment

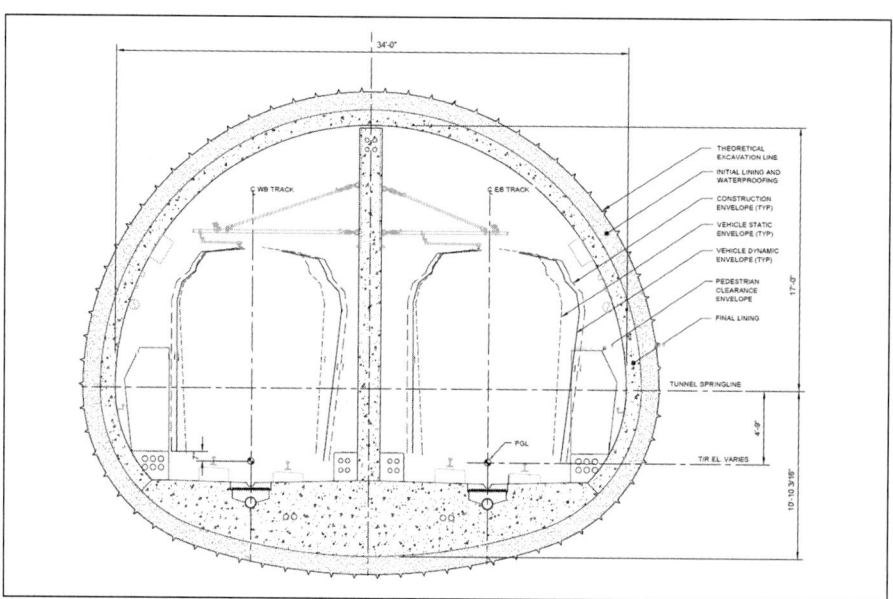

Figure 2. Typical SEM tunnel cross-section

of 12-inches of 4,000 psi reinforced cast in place concrete, with varying invert thickness and a central fire separation wall, as shown in Figure 2.

Procurement

The DBT was advertised for bid in the summer of 2015, with an Engineer's Estimate of approximately $157M and an estimated range of costs between $130 million and $180 million.

A total of 7 bids were submitted for the project. Bids were opened in October of 2015 and ranged from a low of approximately $121M to a high of $170M. The mean of the bid estimates was $154M and the median was $163M, both of which compared favorably with the Engineer's Estimate, which fell near the midpoint of the construction bid amounts. The highest bid was within the maximum population standard deviation of $172M. This suggested that the Contract Documents were clear, and that the contractual mechanisms for risk sharing were understood and acceptable. After review of bids and Supplemental Criteria Submittals Sound Transit issued Atkinson Construction (Atkinson), the lowest responsive and responsible bidder, with Notice of Intent to award the construction contract in November of 2015.

Value Engineering

After the project award, in December of 2015, ST and Atkinson met to identify scopes of work that could potentially be modified to improve the Project's duration and reduce the impact of a delayed Notice to Proceed (NTP) caused by delays in right of way acquisition. As the construction of the SEM tunnel is the longest single activity on the Project's critical path, recommendations for scope modification were focused on the tunnel. Atkinson's recommendations included the following:

- Replacement of north portal presupport (pipe canopy) with controlled low strength material (CLSM)
- Proposed modified excavation sequence
- Substitution of Synthetic Fiber for Steel Fiber in the Shotcrete
- Elimination of lattice girders
- Spiling Modification
- Use of a thinner shotcrete lining

All parties agreed that there was merit in further evaluating these recommendations. The processes and outcomes of these evaluations are presented herein.

NORTH PORTAL CLSM REPLACEMENT

At the northern limit of the SEM tunnel, the cover over the tunnel crown is approximately 13 feet, which is shallow relative to the tunnel excavation size. The original expectation was that the material overlying the tunnel crown was principally Vashon Till, overlain by a limited 1–2 foot depth of fill and roadway pavement. A pipe canopy, comprising a double row of 6-inch diameter pipes was proposed to be installed over a length of approximately 70 feet. These conditions, with the inclusion of the canopy were demonstrated to be sufficient to maintain a stable excavation. However, greater scrutiny of existing borings during the final design phase, and removal of borings more remote from the tunnel alignment suggested that the fill was far deeper than originally anticipated, presenting significant risk to the tunneling.

A series of mitigation options were studied including grouting the fill—using either jet grouting or permeation grouting; removing and replacing the fill with CLSM, and reducing the extent of the SEM tunnel by extending the adjacent north cut-and-cover structure approximately 125 feet to the south. Due to the shallow depth and proximity to utilities and thin walled basements jet grouting was rejected, as was permeation grouting based on the expected 30–40% of fines within the fill material. It was originally anticipated that extending the cut-and-cover tunnel would be the least expensive and most practical option, however, independent estimates of the cut-and-cover and CLSM in conjunction with the SEM tunneling demonstrated otherwise. Correspondingly, the Contract Documents required that fill material be removed over the width of the tunnel and over a length of 125 feet from the north SEM portal.

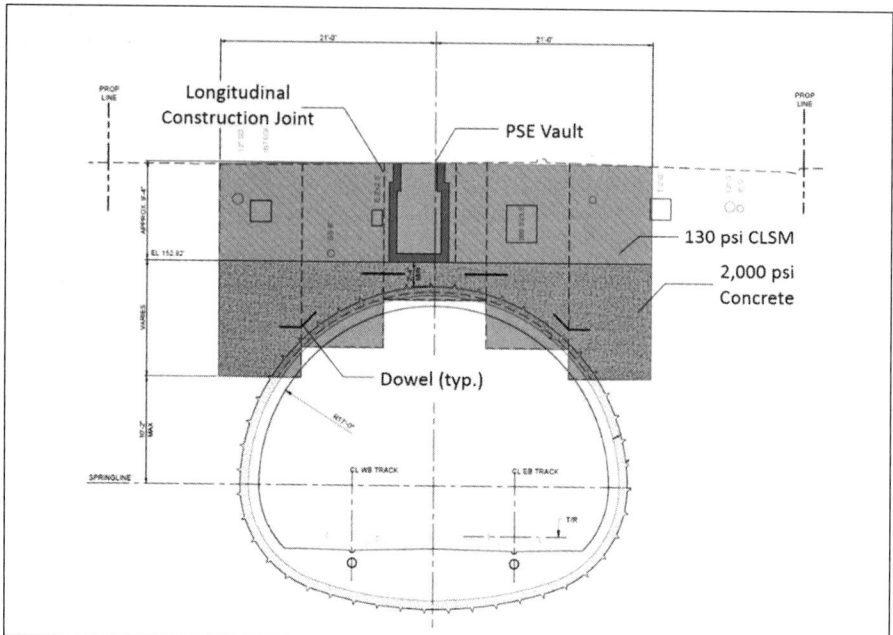

Figure 3. Proposed CLSM/concrete configuration at PSE vault

The installation of the pipe canopy at the north SEM portal would require Atkinson to coordinate possession of the north cut-and-cover excavation site with an adjacent contract, introducing risk of delays and conflicts for both contracts. As mitigation, Atkinson proposed deepening the existing CLSM to create a 'structural' arch over the crown of the tunnel, as indicated in Figure 3, and thereby avoid the need to install the pipe canopy.

A further complication in the north portal area was the realization that a large electrical utility vault was deeper than previously identified by PSE and could conflict with the pipe canopy installation. As part of the evaluation of the deepened CLSM, a separate structural concrete arch solution capable of spanning the tunnel excavation was developed for the vault.

To maintain surface traffic, Atkinson further proposed to install the structural concrete and CLSM in a series of 5 longitudinal trenches, as indicated in Figure 3. As part of a no cost change to Sound Transit, Atkinson engaged MM to evaluate the concept, and received reimbursement for this from Atkinson. The details of the extended CLSM replacement, CLSM construction joints, CLSM/concrete properties, and required analyses and modeling inputs and parameters were agreed with ST and Atkinson staff at a kick-off workshop. Progress was thereafter reported to Atkinson on a weekly basis over the short duration of the assignment.

The size and strength of the CLSM and concrete arches at 130 psi and 2,000 psi were confirmed from a series of 2D geomechanical models using the Rocscience RS2 software. For the structural concrete arch dowels were required at each longitudinal joint to accommodate shear transfer. For the CLSM due to the extent of the joint surface area, only roughening and cleaning of the vertical cold joints was necessary.

The numerical modeling demonstrated the feasibility of the deepened CLSM solution. As the solution eliminated a contract interface and reduced costs on two separate construction contracts this change was implemented with the agreement of all parties.

Once the change was implemented, Atkinson and ST worked with the City of Bellevue and Microsoft, the most impacted neighbor, to allow for a full road closure of 110th Street to execute this work even more efficiently. These changes made it possible to complete the CLSM placement in 10 weeks—6 weeks with a full road closure and 4 weeks with a partial road closure. Despite uncovering several unidentified utilities, Milestone #1, Acceptance of the CLSM, was finished 3 months ahead of the approved baseline schedule. With the elimination of the north portal pipe canopy, Milestone #2, the north portal canopy installation was functionally complete as well.

EXCAVATION SEQUENCE

Of the identified value engineering recommendations, the most significant was in modifying the SEM Tunnel excavation sequence, which suggested a potential 2 to 3-month schedule improvement.

The geologic profile along the DBT alignment generally consists of a thin layer of fill material overlying overconsolidated glacially overridden soils, including Vashon Till, Advanced Outwash Deposits, and Lacustrine Deposits. Based upon local experience, the Vashon Till and Lacustrine Deposits are expected to be excellent tunneling media, with good standup times. The nearby Lincoln Square Tunnel was excavated in the Vashon till, with reported standup times in excess of one week (Breeds et al., 2010). The southern 50% of the SEM tunnel alignment is expected to be entirely within the till, and groundwater, apart from water bearing sand lenses, is anticipated to be below the excavation invert as shown in the soil profile in Figure 4.

A combination of design and construction issues including varying tunnel geometry, relatively shallow cover—particularly in the northern half of the alignment, potential for randomly occurring water bearing sand lenses, overlying utilities, an anomalous zone, and proximity to adjacent building basements resulted in the development of a total of 6 excavation and support sequences as a means of appropriately managing construction risk. The excavation and support sequences maintain the same drift and heading

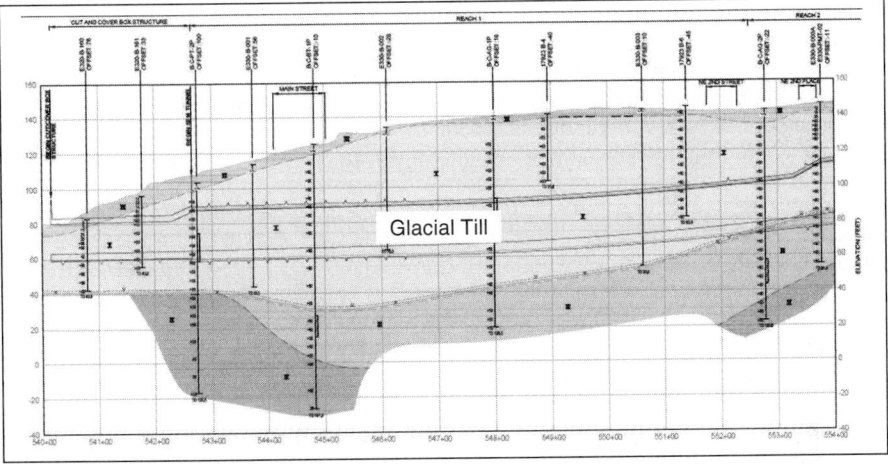

Figure 4. DBT geologic profile, southern 50% of tunnel alignment

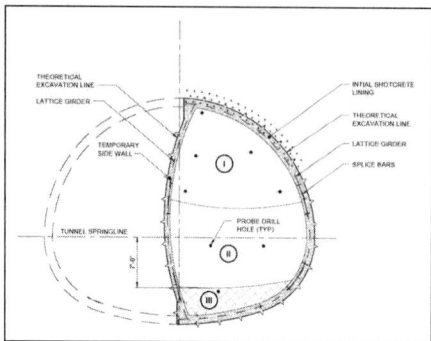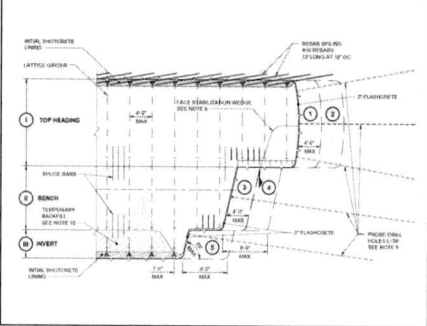

Figure 5. Typical 6-heading SEM tunnel excavation sequence

configuration for simplicity of construction, but presupport and round length are varied. The most common excavation and support sequence is indicated in Figure 5.

The SEM tunnel was designed to be progressively excavated and supported with a 6-heading sequence consisting of left and right drifts, each of which is further subdivided into top heading, bench and invert excavations, as shown in Figure 5. Drifts/headings will typically be advanced in 4-foot round lengths, in a top-top-bench-bench sequence followed by removal of 8-feet of invert. With sloped bench and invert ring closure of the leading drift occurs approximately 30 feet behind the face. A minimum distance of 24 feet is required between the two sidewall drifts is required.

Standard support measures for all excavations include presupport comprising either pipe canopy, rebar spiling or grouted pipe spiling, in conjunction with a 10-inches of fiber reinforced shotcrete lining, also containing lattice girders. In addition to the prescribed support, supplemental SEM toolbox items including face bolts, grouting, metal sheets, vacuum dewatering, additional spiling and shotcrete will be utilized as required by the prevailing field conditions.

Recognizing that the southern 50% of the tunnel alignment was almost entirely with the Vashon Till, that groundwater was below invert and that many other identified construction issues such as potential obstructions and adjacent tall buildings were concentrated in the north of the alignment, Atkinson submitted a proposal requesting that an alternative excavation sequence comprising full top heading, bench and invert excavations be adopted over this extent of the tunnel alignment. The proposal excavation sequence included an advanced top heading approach as used for the Caldecott 4th Bore Tunnel and elsewhere. The proposal suggested based on 2D analyses, that ground movements and settlements would be similar to the as-designed excavation sequence, and that associated risks would be comparable.

The project team met in January of 2016 to review the proposal, and determine whether it should be evaluated in detail. As an introduction to the meeting ST requested that all parties keep an open mind throughout the evaluation. This approach was important in fostering cooperation and team spirit. Over the following weeks the project team held regular discussions on the proposal, the excavation sequence, and relative risks. Concerns were raised over the extended time that the ground would be exposed, with potential for increased ground movements, surface settlements and infrastructure damage. A primary concern with the advanced top heading concept was the ability to quickly revert to the as-designed 6-heading sequence in the event of unexpected ground behavior. Atkinson subsequently committed to a proposed alternate SEM excavation sequence comprising full top heading, bench and invert excavation with early ring closure, which satisfied all parties. In April of 2016 ST and Atkinson signed a term sheet agreeing to a no-cost change order. One of the most important conditions of the term agreement was that the Engineer of Record would retain the right to direct Atkinson to revert to the original 6-heading excavation sequence if deemed necessary for the good of the project.

At the request of ST, MM submitted a proposal to undertake the necessary analysis to demonstrate the feasibility of the revised 3-heading concept, and to undertake the necessary additions and revisions to the Contract Documents. As part of the no-cost change order, Atkinson was required to compensate ST for the design costs and were dutifully concerned about the expense. It was therefore agreed to phase the design effort such that once the numerical modeling and analysis necessary to verify the performance of the excavation sequence was completed, the project team would reconvene and make a go/no-go decision on whether to continue with the preparation of contract documents to represent the 3-heading concept. The proposed evaluation allowed for 2 months to perform numerical modelling followed by an estimated period of 4–5 months for calculations, contract document development—including revisions to the existing documents as well as preparation of new drawings to reflect the 3-heading concept, and quality assurance.

At this point the modified sequence design took on more of a design build flavor, with Atkinson and MM coordinating regularly on excavation sequence, cycle times and shotcrete rate of gain of strength to be used in the numerical modeling. As an example, instead of mirroring the design excavation sequence of top-top-bench-bench-invert Atkinson requested that the sequence be modified to a top-bench-top-bench-invert sequence to allow greater flexibility with equipment selection. This modified excavation sequence is indicated in Figure 6. Weekly meetings were held during which progress was measured against and agreed schedule. With construction submittals, preconstruction testing and tunnel construction itself looming, it was critical that the evaluation remained on schedule.

To verify the feasibility of the proposed 3-heading sequence, numerical modeling of ground-structure interaction and structural analyses for initial shotcrete lining have been performed. The numerical analyses include one 3-dimensional (3D) model and three 2-dimensional (2D) models at three representative locations. The 3D model calibrated the 2D models in terms of ground relaxation and model performance. The 2D models were also used for parametric analyses to account for the range of several key input ground parameters.

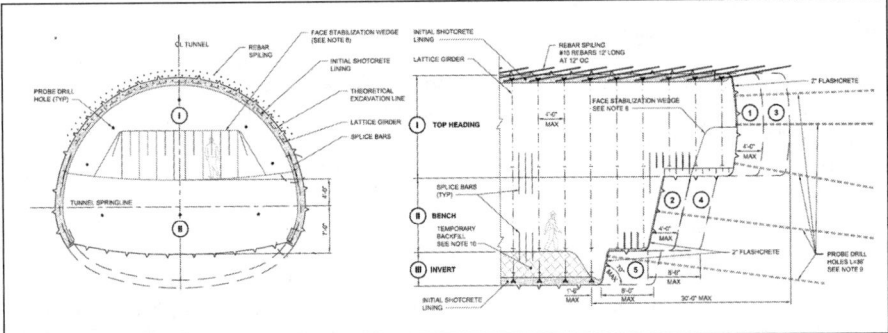

Figure 6. Modified 3-heading excavation sequence

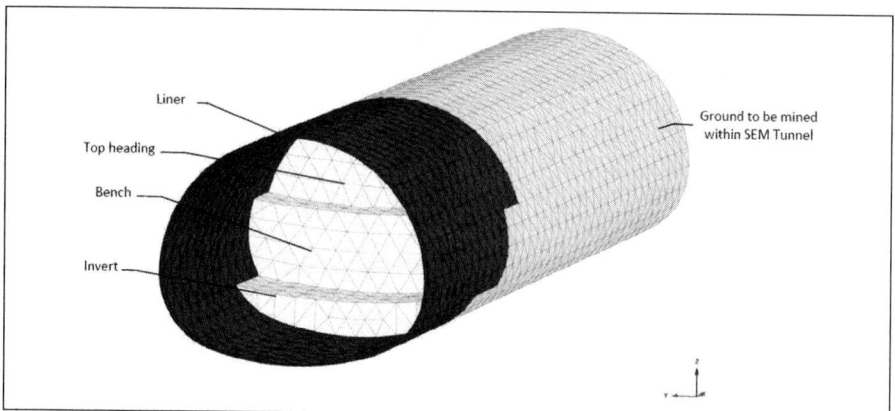

Figure 7. 3D numerical model for SEM tunnel using 3-heading sequence

The soil stratigraphy, soil parameters and 28-day shotcrete strength remained the same as the analyses for the originally prescribed design. The shotcrete strength at each sequence/stage in the 3D modeling was based upon a construction timeline prepared by Atkinson. The rate of gain of shotcrete strength was initially proposed per the project specification. However, this was modified based on discussions with Atkinson, to reflect values that were thought to be achievable based upon the proposed shotcrete mix design. Pre-support measures for the design, i.e., rebar spiling and grouted pipe canopy, were not modeled in the analyses, which was consistent with the modeling approach for the original design. Based on the modeling results, tunnel and ground responses have been obtained, ground volume losses have been estimated, and the liner demands were used as input for the structural analyses to confirm the liner structural capacity.

In tho 3D model (Figure 7), the anticipated construction duration for each heading advance and excavation cycle, and an anticipated shotcrete strength gain (curve a in Figure 8) as provided by Atkinson were modeled.

The initial 3-heading construction sequence proposed by Atkinson, referred to as Sequence A, involves applying 6" shotcrete in 2 layers comprising 2" flashcrete and 4" shotcrete, followed by excavation of next advance, then applying another 4" shotcrete at the first advance location. However, structural analyses indicated that the shotcrete for Sequence A would have insufficient capacity during this construction. Therefore, Atkinson proposed a modified sequence, referred to as Sequence B, in which the full initial lining thickness comprising 10" shotcrete is applied in layers in one operation before moving to the subsequent excavation.

Based on the modeling and analyses performed for both Sequences A and B, it was found that: 1) The proposed 3-heading SEM tunnel construction Sequence B is feasible, as demonstrated by the ground and face stability and structural capacity of shotcrete liner at each stage of construction; 2) The volume loss due to the proposed 3-heading tunneling will be generally small, ranging from 0.08% to 0.17%; the predicted maximum ground surface settlement is 0.3 inch. The predicated volume loss for the 3-heading sequence is also generally similar to the 6-heading sequence, which is mainly attributed to elastic response of the ground and relatively large stiffness of the ground; 3) The settlement impact on existing buildings and utilities from 3-heading sequence will remain the same as those from the 6-heading sequence; 4) The face

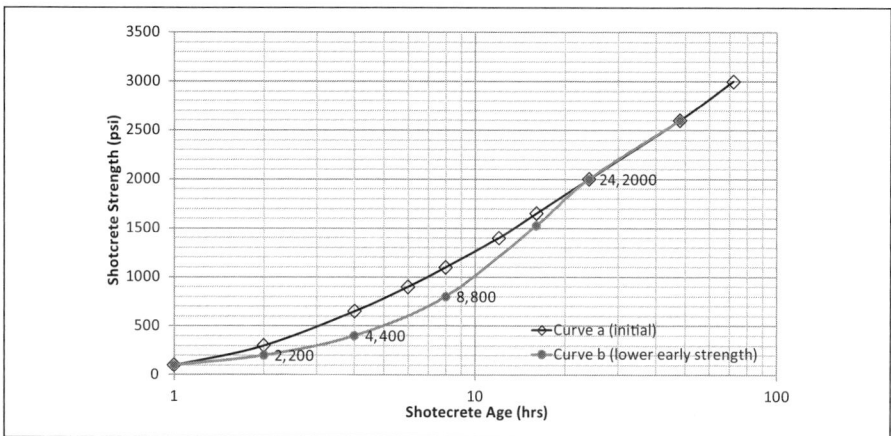

Figure 8. Early age strength of initial shotcrete liner used in 3D numerical model

of tunnel excavation at each stage will be stable when mean ground conditions (soil cohesion) are encountered; and will be marginally stable when lower bound ground condition is encountered. This is demonstrated by the results of both 3D modeling and empirical evaluation of face stability, which conservatively consider a full face of lower bound ground conditions, which is unlikely. However, it is anticipated that the application of flashcrete on the excavation face in addition to the toolbox items will keep the face stable when the lower bound soil conditions are encountered.

The feasibility of the 3-heading sequence was demonstrated by the results of the Sequence B analysis. Several critical conditions in the construction cycle were identified in which the shotcrete had barely sufficient capacity to meet code requirements. In these cases, the strength of the lattice girder was included to provide additional capacity. The project team convened in June of 2016, to discuss the modeling outcomes and the path forward. At this point, the 3-heading sequence received a tentative 'go' decision. This was based upon the perception that the 3-heading sequence inherently retained higher risk than the 6-heading sequence due to the much larger heading size. Correspondingly specific criteria, that would result in excavation reverting to the original 6-heading sequence were also made contingent upon a 'go' decision. These criteria included: excessive ground movement beyond current specification thresholds; excessive convergence of initial lining that is beyond current specification thresholds; time of ring closure as compared with durations used to develop the prescribed modeling sequence (no maximum duration if design distances are maintained); face mapping and geological conditions encountered; incident of face instability; and ST direction. In conjunction with these criteria requirements for additional instrumentation and convergence monitoring were introduced, as were requirements for increased monitoring of instrumentation and soil sampling. As these resulted in additional costs, Atkinson had to consider whether they wanted to continue with the proposal. Ultimately the revert criteria and requirements were reconciled between all parties, and a 'go' decision was agreed.

Following the Go decision for the 3-heading sequence, Atkinson performed shotcrete mix design testing and field trials. Atkinson had planned to use a shotcrete mix with a large dose of hydration stabilizer to put the shotcrete to "sleep" for up to 12 hours. Initial field test results indicated that with the hydration stabilizer, the shotcrete early

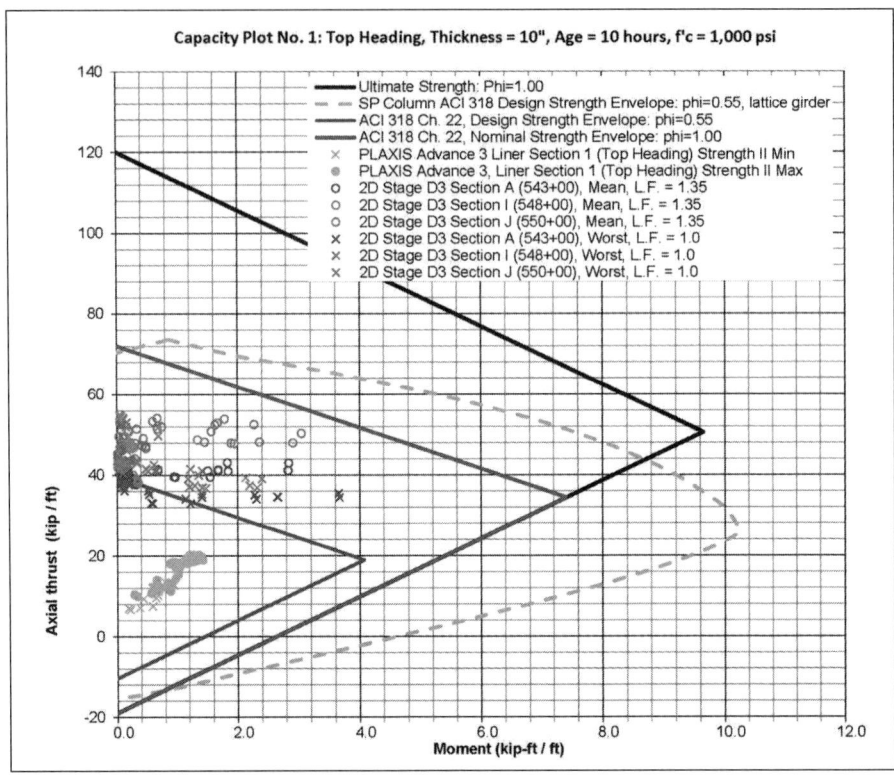

Figure 9. Example of liner capacity plot for Sequence B of 3-heading sequence

ctrength gain (0 to 12 hours after application and "waking up" the shotcrete) could not meet the strengths initially used in the model (see curve 'a' in Figure 8).

As such, additional 3D modeling and structural analyses were performed based on the reduced shotcrete strengths (curve 'b' in Figure 8), and the results indicated that the proposed 3-heading Sequence B is still feasible with the originally proposed cycle times and the initial liner will still have adequate capacity. An example of typical capacity plot is shown in Figure 9. Thus it is confirmed that, during the construction, the proposed excavation advance rate for the 3-heading sequence can be maintained as long as the shotcrete can achieve the reduced strength (curve 'b', Figure 8), and wire mesh will not be required. However, the reduction in shotcrete strength poses higher risk due to reduced safety factor and increased reliance on the lattice girder as reinforcing. It is therefore critical that initial lining quality, including shotcrete performance, thickness control, and lattice girder assembly is maintained during construction.

FIBER REPLACEMENT

The Contract Documents stipulated that the initial shotcrete lining should be reinforced with steel fiber, with a minimum specified dosage of 50 pounds per cubic yard. As part of their value engineering review Atkinson proposed that the steel fiber be replaced by macrosynthetic fibers, for several reasons including safety, quality, and construction schedule. Worker safety is improved through elimination of hazards of handling and application of the steel fibers and contending with the steel fibers protruding from the tunnel wall. Project quality is enhanced by removing the potential for protruding steel fibers damaging the overlying waterproofing membrane.

Potential construction schedule and cost benefits were also suggested by reducing the thickness requirement of a 1.5" thick unreinforced 'smoothing' layer of shotcrete between the initial lining and the waterproofing membrane. However, the shotcrete in this layer is comprised of 0.16" aggregate in lieu of the normal ⅜" requirement, as the smoothing layer serves the function of providing a more uniform surface to install the waterproofing membrane against, as well as encapsulating protruding fibers. This is particularly important for a spray applied membrane, as allowed by contract, for which a rougher as shot surface with ⅜" aggregate can result in the need for additional passes of spray application and thicker membrane to overcome the surface finish.

As the substitution request affected only the initial support, which is purely temporary in nature, there were no concerns over longer term performance of the fiber rein-forced lining with respect to creep deformation. Correspondingly, the proposal to sub-stitute the steel fiber with macrosynthetic fiber was accepted. It was also agreed that requirements for a prescriptive fiber dosage should be removed, with the fiber dosage determined from preconstruction testing to meet mix performance requirements for compressive strength and flexural toughness. The testing method was changed from the beam tests used for steel fiber to the circular panel method.

OTHER RECOMMENDATIONS

The proposal to eliminate the lattice girders, which was originally provided for profile control and to act as a support for spiling installation, was not implemented as a result of the outcome of the analysis of the 3-heading excavation sequence. The results of the numerical modeling and analysis showed that based upon Atkinson's proposed excavation sequence and cycle times, the lattice girder was temporarily required to provide additional structural capacity beyond that provided by the early age shotcrete.

The proposed spiling modification would replace the as-designed 12-foot long spile installed with every advance to a 16-foot spile installed with every second advance. This would enable Atkinson to save time by eliminating 50% of the stoppages required to install the spiling, estimated at 4 hours per round. Over the course of the tunnel length this could save approximately 30–40 days on the construction schedule. However, this proposal has yet to be formally addressed.

The final proposal for the use of a thinner shotcrete lining has not been evaluated and likely will not be submitted. During the 3-heading SEM modeling it was determined that the full 10-inches of shotcrete meeting early strength requirements was neces-sary. Based on these model results, Atkinson will forego the idea of a thinner initial liner system.

SUMMARY AND CONCLUSIONS

To mitigate project construction schedule delays arising from delays in right-of-way acquisition ST asked Atkinson to provide time-saving recommendations to reduce the DBT construction duration. The recommendations included significant revisions to the project design, including removal of pipe canopy and revisions to the SEM tunnel excavation sequence, which can often be contentious to both Owner and Engineer of Record. However, in maintaining an open mind, concentrating on the good of the project, and through regular and frequent communication, the project team has worked collaboratively in a design-build type framework to implement several of these recommendations. This team approach has led to the early achievement of the first two project milestones, and has continued with the evaluation and finalization of the design of revised 3-heading excavation sequence, the results of which will be known in approximately 2-years.

ACKNOWLEDGMENTS

The authors would like to thank Sound Transit for the permission to publish this paper. In addition, the contributions of the H-J-H Team—HNTB Corporation, Jacobs Engineering, and Golder Associates, the Construction Management Team of HDR and McMillen-Jacobs Associates, and the MM SEM Resident Staff—Michael Murray and Mike Wongkaew are gratefully acknowledged.

REFERENCES

Penrice, D., Schutt, J., Dorn J., and Hachey, J., 2016 "Sound Transit East Link—Downtown Bellevue Tunnel," Proceedings of the World Tunnel Congress 2016.

Breeds, C., Leone, L. and Gonzalez, D. (2010), "The Lincoln Square Tunnel: Tunneling Between Two Parking Garages Using Sequential Excavation Mining," Proceedings of the North American Tunneling Conference 2010.

Comparative Application of NATM, TBM, and RBM Technologies

Pedro Pino Véliz ▪ PEK Teknep Overseas Engenharia S.A.
Patricia Kong Diaz ▪ PEK Teknep Overseas Engenharia S.A.

ABSTRACT

Last expansion of Rio de Janeiro subway began in 2010 aiming the 2016 Olympics. The project included seven underground stations and 17 km tunnels. Biotite-gneiss, augen gneiss, weathered rock, marine sands, soils and urban material were excavated. NATM, TBM and RBM technologies were applied to each ground type and urban condition. Innovative methods of excavation associated to last generation electronic blasting, ground control, roof support, instrumentation in surface and underground environments were some of the most challenging tasks. Almost all the current underground construction methods were applied resulting in a very valuable experience. This paper outlines strategic guidelines for assessment in subway construction in densely populated areas and variable rock conditions. It also describes the main elements of design with respect to project requirements.

LINE 4 EXPANSION

Rio de Janeiro Metro, also referred to as Metro Rio, is a partially underground railway network that initiated operations in 1979. The network has two lines and serves half a million passengers each day. The Metro Rio network consists of 33 stations on two lines. Line 1 (Orange line) that runs from Saens Pena to Ipanema is completely underground. It covers 18 stations. Line 2 (Red line) that extends from Pavuna to Sao Cristovao is a diagonal line connecting 15 stations is mostly surface in 11 stations. Both lines are served by old metro rolling stock of A Type. Line two includes A Type and the new B Type rolling stock. As line 2 is a conversion from light rail to metro, its rolling stock includes trains that have been converted from light rail to metro. The

Source: Consórcio Linha 4 Sul - http://www.metrolinha4.com.br/o-que-e-o-projeto/
Figure 1. Overview of Line 4 (Barra da Tijuca) and connection to Line 1 subway (Ipanema)

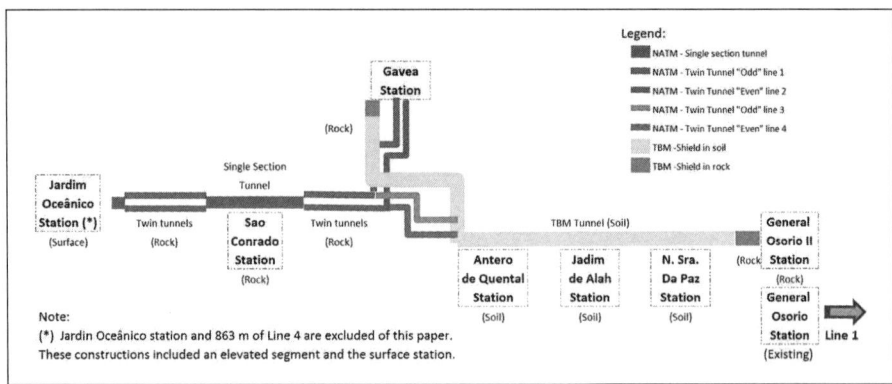

Figure 2. Diagram of Line 4 expansion project

trains are powered by a 750 V cc third-rail system. Entire rolling stock is fully air-conditioned and the trains comprise six passenger cars.

Following the plan to expand the necessary infrastructure to support the 2104 World Cup and the Olympics of 2016 the Metro Rio developed engineering design to build the Line 4 based on NATM and TBM excavation methods which connected to Line 1. First move was the purchase order of the tunnel boring machine to a German company in 2009. Construction of NATM tunnels started in 2010 from the south edge of Line 4 (Barra da Tijuca) and a central point located at Sao Conrado that allowed to excavate the tunnels in both directions south and north. In 2012 the excavation of the connecting station General Osorio II in edge north was initiated. This station also used NATM method. The TBM arrived to Rio de Janeiro in March 2013 and started operation in December 2013.

To ooocoo the main line from the south edge an auxiliary tunnel of 260 m (~82 m² avg. section) was excavated to reach the initial segment of 646 m (~85 m²) of the main line as a single tunnel containing paired rail lines ("Even" and "Odd"). From that point and along 4030 m a twin tunnel (33.6 m² section each) connected to 370 m segment in single section (~85 m²) then Sao Conrado station. Another segment of 263 m length as single section (~85 m²) after the station connected again to 936 m of twin tunnels. From this point onwards Line 1 and Line 2 as twin tunnels ran along ~1270 m (33.6 m² section each) then each one of these lines split in two additional "Eve" and "Odd" lines. Line 2 continues as Line 2 to Gavea station and divert to Line 4 ("Even") to connect to the TBM tunnel. In the other side Line 1 Continues to Gavea station and deflect to Line 3 ("Odd") that also connects to TBM tunnel near General Osorio II station.

GEOLOGIC ASSESSMENT

The morphology of the coastal region of Rio de Janeiro principally the several steep rocky domes that emerge close to the coast is varied in soil and rock types. Such scene resulted from a combination of differential weathering controlled by the presence of a variety of Neoproterozoic gneisses, Cambrian granites and strong valley incision along subvertical faults and fractures. Main geological structures resulted from tectonic processes and regional metamorphism.

Geology of the rock along the subway line include biotite-gneiss, augen gneiss and kinzigite gneiss with pegmatite intrusions and some faults NW-SE /85°SW dip. Predominant foliation is N55°E /10°SW dip.

Table 1. Rock properties in the Line 4 expansion

Property	Biotite*	Pegmatite (Wet)	Pegmatite (Dry)	Biotite[†]	Augen Gneiss[‡]	Augen Gneiss[§]
Uniaxial compressive stress	70	80	102	108	122	146
Deformation modulus	50	35	41	47	48	54
Poisson's ratio	0.157	0.097	0.162	0.095	0.118	0.139
Dry density	2694	2600	2600	2694	2651	2600
Empty space	0.6	0.6	0.6	0.6	0.7	0.6
Porosity	1.2	1.3	1.3	1.2	1.2	1.2
Sample depth	30	135	135	30	30	50

Notes:
* Parallel to foliation
† Transversal to foliation
‡ Predominant rock in Line 4
§ Predominant rock in Line 1

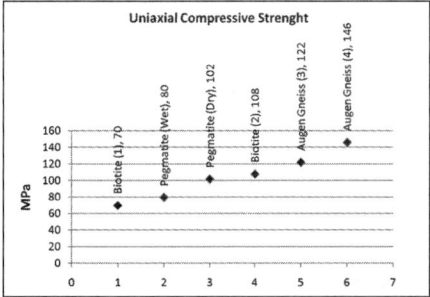 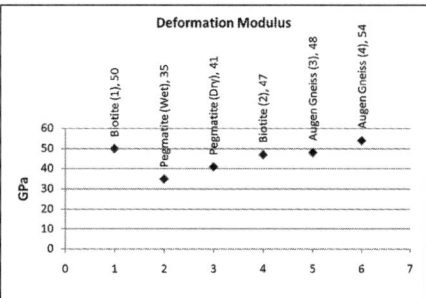

Figure 3. Comparative charts of main geotechnical properties of the Rio de Janeiro rocks

ROCK CHARACTERIZATION

Augen gnaisses are the predominant rocks excavated in the Line 4 project. These rocks were selected because of its geotechnical properties. Pegmatite intrusions are the weaker rocks because of its aptitude for degradation and loss of mechanical properties in presence of water. Table 1 presents a summary of the properties of the main rocks found in the Line 4 expansion.

ESTABLISHING CONSTRUCTION PARAMETERS

The main part of the team involved in the Line 4 project included engineers and designers experienced in the previous phases of the Metro Rio as well as experienced contractors and consultants to oversee every aspect of the design and construction. This distinctive feature conditioned the design parameters and resulting construction elements.

Project criteria:

1. Vibration limits defined for the Rio de Janeiro Metro project: Vp max: 3 mm/s measured in the vertical direction and 4.2 mm/s as result of the integration of three directions (X,Y,Z) of the seismograph. This standard is recommended by CETESB of the Sao Paulo State reference D7.013 of 1992, which is more strict that any other standard in Brazil in regards of vibrations. The Brazilian standard NBR 9653:2005, which is applied to open cats mine in the surroundings of the urban areas was not applied to this project. This standard define Particle Velocity (Vp) with frequency of 15 Hz increasing to 40 Hz

varying from 15 mm/s to 20 mm/s. Above 40 Hz Vp should increase from 20 mm/s to 50 mm/s.

2. Use of shotcrete lining minimum fck:25 MPa combined to steel bars of 1" diameter (500 MPa minimum yield strength) × variable length from 1.5 m to 6 m and welded steel mesh of 100 × 100 mm space and 3.3 mm diameter. Thickness of shotcrete varying from 100 mm to 300 mm.

3. Application of 100 mm thickness lining to rocks Class I or Class II RMR (Bieniawski, 1990).

4. Application of 150 mm thickness lining to rocks Class II or Class III.

5. Application of 200 mm thickness lining to rocks classified as Class IV.

6. Application of 200 to 300 mm thickness lining with use of steel structure to rocks Class IV or Class V.

7. Rate of drilling and advance according to the rock quality and area (m^2) of the tunnel.

8. Use of explosives and accessories of first class companies. Not allowed electrical initiation.

9. TBM—The first 300 meters in open mode, then for about 3,000 meters in closed EPB mode.

10. TBM—Finally, around 1,300 meters will again be tunneled in open mode.

11. TBM ED 10,460 mm and lining in 5 sets of moulds ID 10,330 mm, carrousel system, handling equipment, evacuation line and others.

COMPUTER TOOLS USED

All the series of the modules of the Rocscience software for design and modeling. Eventual use of FEM with use of Solidswork adapted to mechanical properties of rock, steel and concrete structures. This tool was used for structural dynamic analysis and simulation of seismic offooto from blasting. The figures below illustrate some applications for modeling of shotcrete structures with varies thickness in response to rock loads in the lining.

NATM APPLICATION

NATM method was applied to all types of sections from 36 m^2 to almost 400 m^2 area. In general the average rate of advance was of order of 3 m/day in each face.

Typical operations include: Geological mapping, drilling with two or three boom jumbos, blasting with use of aluminized ANFO based slurries and electronic initiation (with

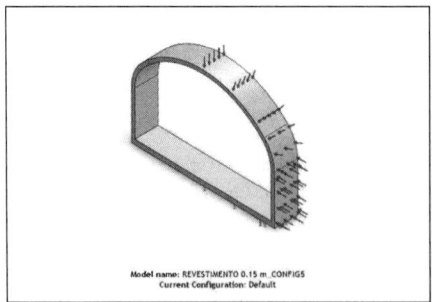
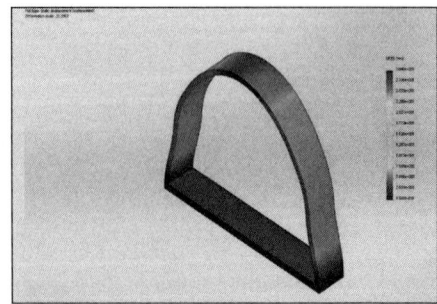

Figure 4. Modeling of shotcrete lining 150 mm thickness: deformation

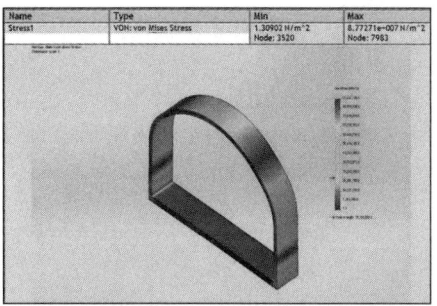

Figure 5. Modeling of shotcrete lining 150 mm thickness: Von Misses and factor of safety

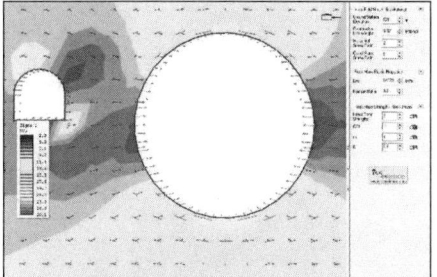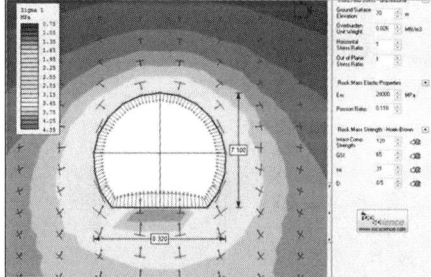

Figure 6. Stress analysis (Sigma 1, Sigma 2, Differential Stress, Von Misses, Horizontal Displacement, Vertical Displacement, and others in tunnels)

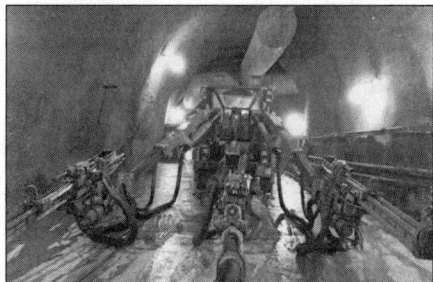

Figure 7. Excavation of subway station 18 m wide and three boom jumbo

seismometric monitoring), cleaning with backhoes or FEL, transport of blasted rock to waste disposal areas some 20–30 km distance, roof and wall inspection with removal of rock fragments, geological face mapping and confirmation of rock RMR Class, roof bolting associated to steel mesh then shotcreting according to the project.

TBM APPLICATION

The tunnel boring machine used in the Rio de Janeiro subway was an Earth Pressure Balance Shields (EPB) shield (Ø 11.46 meters) for the 4.7 kilometer long section of the Line 4 between the stations Gavea and General Osorio as shown in Figure 2. To ensure the machine should cope with the different ground conditions along the tunnel route, the TBM was designed to be converted in the tunnel. The first 300 meters in open mode, then be converted for about 3,000 meters in closed EPB mode. Finally, around 1,300 meters will again be tunneled in open mode. In open mode, the firm hard

Figure 8. Excavation of Class IV rock and lining with steel structure 1" diameter and 300 mm thickness shotcrete

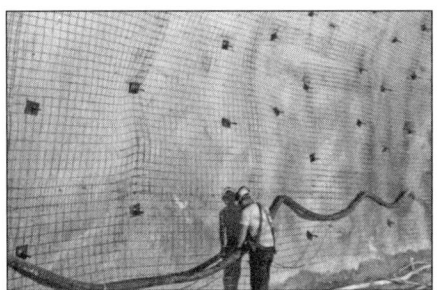

Figure 9. Application of shotcrete and welded mesh associated to steel bars 6 m length

Figure 10. TBM was used to excavate almost 30% of the Line 4 West sector. The picture of the right side shows the preparation to initiate TBM operation from the General Osorio station (after excavation using NATM method).

Figure 11. Preparation of the first TBM lining from the General Osorio station and tunnel excavated

Figure 12. Infrastructure required to operate the TBM; conveyor belting and slurry pond

rock (gneiss) is transported away on a conveyor belt. In closed EPB the cohesive soils are removal by a screw conveyor whose conveying speed simultaneously safely and precisely regulates the face support pressure during excavation in the predominantly sandy soil. Another special feature of the machine for Rio de Janeiro's Metro Line 4 was the articulation in a diameter range of more than eleven meters. This design allowed tight bends with a radius of just 250 meters in some cases.

TBM was a natural response to very complex conditions encountered in very dense populated areas and soil or marine sediments. NATM method was studied and found unfeasible to accomplish with expected construction schedule to arrive to the Olympics in 2016. Its application justified the investment of $ 50 million made.

RBM APPLICATION

Raise boring machine (RBM) normally applied in mining development was success-fully applied in a very complex situation that required a creative solution to provide fresh air to the General Osorio II and main access to the Lagoa exit (a 650 m length tunnel). Two shafts of Ø 4.5 meters × 100 m height and one shaft Ø 5.5 meters × 100 m height were excavated in hard gneiss in the center of one of the most populated urban areas of the city with minimum interference and minimal vibration.

Figure 13. Excavation of three exhaust shafts with use of Raise Boring Machine normally used in mining operations. Picture of right side show the pilot hole 12 ¾" diameter to connect the reamer.

 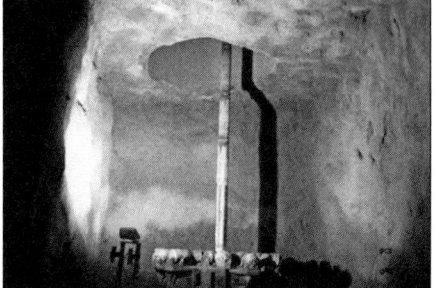

Figure 14. Reamer prepared to excavate the shaft and shaft already completed

Figure 15. Protection in the lower part of the shaft and second picture shows the protection in the upper part of the shaft

DIAMOND WIRE CUTTING EXCAVATION

The connection between General Osorio II station (Line 4) and General Osorio in operation (Line 1) required the cut of a vertical pillar 8 m width between both stations then hydraulic diamond wire cutting was the solution as shown in Figure 15.

CONNECTING NATM TO TBM TUNNELS

Challenging operation was the connection of NATM tunnel with the TBM in open mode with rigorous ground stability control.

COMPARATIVE PERFORMANCE DATA

The Line 4 expansion project permitted to compare in almost the same conditions (terrain, schedule, contractors and designers) two methods of excavation that are generally analyzed separately. In this case we had NATM applied to single tunnels, NATM applied as twin tunnels, TBM in open mode applied to hard rock and TBM in closed mode excavating soils.

CONCLUSIONS

The application of TBM when compared to NATM in soil in the Metro Rio's Line 4 was without discussion the best method. The possibility of small diameter TBM to excavate

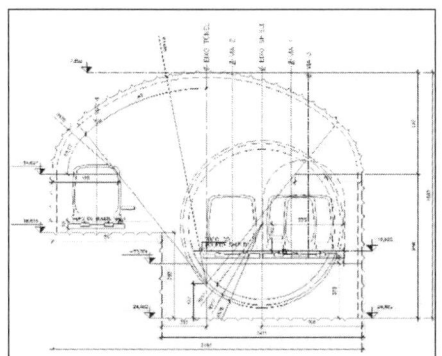

Figure 16. Connection of NATM tunnel with TBM in open mode operation

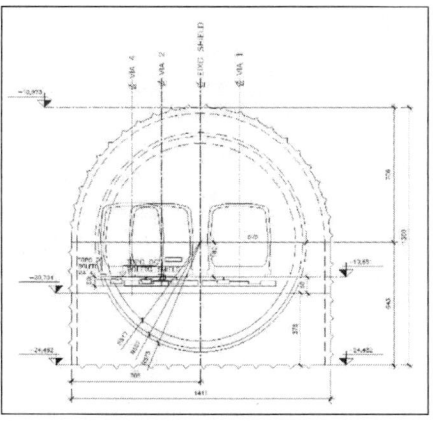

Figure 17. TBM full face inside the NATM tunnel

Table 2. Comparative section (m²) × rate of advance in the Line 4 and total excavated

Description	m²	m/ day	Total (m)
Single tunnel Section - Hard rock	88.90	2.34	1,800
Twin tunnel Section (only one) - Hard rock	33.60	10.78	14,940
Twin tunnel Section - pair*	67.20	5.39	7,470
TBM tunnel open mode	111.22	1.90	880
TBM tunnel closed mode	103.15	8.30	3,834

* Compares the required advance to equal twin tunnels to single section containing two rails ("even" and "odd")

Table 3. Summary of experience in the Line 4 expansion

Description	NATM Single Tunnel	NATM Twin tunnel	TBM Open Mode (hard rock)	TBM Close Mode (soil)
Time required to basic engineering	One to two years	One to two years	One to two years	One to two years
Equipment Spec and Purchase Order	Search market for contractor	Search market for contractor	1 to 2 years to order tailor made TBM supported on basic engineering	1 to 2 years to order tailor made TBM supported on basic engineering
Detailed engineering required	6 to 12 months before excavation	6 to 12 months before excavation	2 years before excavation during fabrication period	2 years before excavation during fabrication period
Fabrication	Subcontracted	Subcontracted	2 years	2 years
Flexibility to change route and geometry	Adaptable	Adaptable	Rigid	Rigid
Time required to start up excavation	3 months	3 months	6 to 9 months	6 to 9 months
Tunnel fronts excavated at the same time	Should be used multiple fronts	Should be used multiple fronts	Only one front	Only one front
Infrastructure required to operate	Very few only to accommodate equipment fleet	Very few only to accommodate equipment fleet	Large infrastructure (internally in the tunnel area) and in surface to construct lining segments	Large infrastructure (internally in the tunnel area) and in surface to construct lining segments
Rate of advance (meters excavated per day in equivalent conditions)	Lower rate than in the twin tunnel system	Higher rate when operating the two tunnels at the same time with more efficient application of equipment	Lower speed than all the other methods due the rock hardness and large section	High speed with some peaks with rate of around 15 m/day
Volume of rock or soil required to accommodate the section required for two rails ("even" and "odd")	Less material than TBM	Less material than TBM	A little more material than in closed mode	More material than NATM and TBM open mode
Level of vibration transmitted to the surface and surrounding areas	Higher that TBM but controllable according to the regulations	Higher that TBM but controllable according to the regulations	Lower level of vibration than NATM	Lower level of vibration than NATM
Equipment after completion of the subway expansion	Return equipment to the contractors, clear the area and abandon	Return equipment to the contractors, clear the area and abandon	If no other expansion is expected then equipment should be discarded . In case of new project this new expansion must be in the continuation of the previous rout	If no other expansion is expected then equipment should be discarded. In case of new project this new expansion must be in the continuation of the previous rout
Does the selected method of excavation the best option due the geology and urban constraints?	In some cases the single tunnel option was justified, mainly as auxiliary entries to access the main subway line	Twin tunnels shown several advantages as better use and efficiency of equipment, better geological control and tunnel spa reduced with better roof control	Probably no other option existed to develop the connection between different materials (rock × soil)	No doubt that the TBM option wa the best solution to excavate in very dense populated areas and soil. NATM is not feasible in this terrain and cut & cover should n be viable for the required route
Costs after completion of the subway expansion	No costs	No costs	Maintenance of TBM to prevent corrosion and protection to sensitive electronics. Depending how long will be the time to resume operation the technology of the equipment should be obsolete.	Maintenance of TBM to prevent corrosion and protection to sensitive electronics. Depending how long will be the time to resume operation the technology of the equipment should be obsolete.

twin tunnels in soil probably would be another alternative. One of the problems identified when shield is used refers to secondary accesses, stations and other auxiliary opening that needed other methods of excavation (NATM or Cut & Cover). In the case of Line 4 those secondary underground spaces totaled almost 5–7% of the volume excavated. One of the main problems that require an appropriate answer is: "what to do with the TBM after the present expansion?" if no new contracts are expected in the mid-term (say two to three years).

NATM shown its flexibility, short-term start up and relatively low initial investment because a good part of the required equipment (jumbos, FEL, shovels, trucks) are available in the country and should be contracted to several companies that work in tunneling for roads, hydropowers and underground mining. NATM in the twin tunnel mode demonstrated to be flexible, efficient and relatively cheaper than TBM in this case, but not appropriate to soils.

REFERENCES

Bieniawski, Z.T., Tunnel design by Rock Mass Classifications, US Corps of Engineers, Update of Technical Report GL-79-19, Jan 1990.

Efron, Nathaniel, Read, Megan, Analysing International Tunnel Costs, An interactive Qualifying Costs, Worcester Polytechnic Institute, Feb 2012.

Januario, Marcelo, Unprecedent machine, M&T Manutencao e Tecnologia Magazine, Dec 2012 www.revistamt.com.br/index.php?option=com_content&view=article&id=140.

Nelson F. Fernandes, Miguel Tupinambá, Claudio L. Mello, Maria Naíse O. de Peixoto, Rio de Janeiro: A Metropole Between Granite-Gneiss Massifs, Chapter Geomorphological Landscapes of the World pp 89–100, 07 November 2009.

Nobrega, P., 2012. Metro Line 4, Rio de Janeiro / Brazil. Tunnel Magazine 6/2012.

Oliveira, D.G.G. Dias, C.C., Silva, A.B., Kuwajima F.M., Infra7 Consulting and Engineering, São Paulo University & Technological Institute of Aeronautics (ITA), São Paulo, Brazil. Monteiro, M.D. São Paulo, Brazil, Cyrillo, G.R.& Pierri, J.C.D.D., Consortium of Line 4 South, Rio de Janeiro, Brazil, Castro, A.R., Promon Engineering, Rio de Janeiro, Brazil, Geological and Geotechnical Aspects of Line 4S RJ - Stretch Morro do Cantagalo-Gávea, Proceedings of the World Tunnel Congress 2014.

Prompt, Halcrow Engenharia, Study on Unit Price for Civil Construction, TGV High-speed Train Project Between Campinas and Sao Paulo-Rio de Janeiro, 2008.

Queiroz Galvao Engenharia, Tatuzao da Linha 4 do Rio falta escavar apenas 300 m, Queiroz Galvao website news, Feb 2016, http://www.grupoqueirozgalvao.com.br/tatuzao-da-linha-4-do-rio-falta-escavar-apenas-300-metros/.

Saliba, F.S.A., Holanda M.T. and Gropello I.J., Odebrecht, Rio de Janeiro, Brazil, Kuwajima, F.M. Infra 7, São Paulo, Brazil. Rio de Janeiro Metro—Vibrations due to blasting in urban environment, World Tunnel Congress 2014.

Tunnel Tech, Cost benefits of large-diameter bored tunnels, Arup North America Ltd, Washington, US, Apr 2015.

Vieira, Rodrigo, Secretary of Transports of the Rio de Janeiro State, Presentation Mobilidade Regiao Metropolitana do Rio de Janeiro, 2012.

Sequential Excavation Method with Ground Freezing for DC Water's First Street Tunnel

Ivan Hee ▪ Skanska USA Civil
William Bracken ▪ Skanska USA Civil
Harald Cordes ▪ WSP | Parsons Brinckerhoff
Stephen Njoloma ▪ McMillen Jacobs Associates

ABSTRACT

Mined tunnel excavation for DC Water's First Street Tunnel project in an urban residential neighborhood with complex ground conditions; required the utilization of ground improvement with ground freezing. A total of three connection tunnels (adits) of different size and geometry have been constructed between deep drop shafts and the main tunnel, which was excavated by Tunnel Boring Machine. This paper is focused on the adit design challenges and contractor's means and methods to successfully deliver this project. The application of vertical ground freezing required innovative construction sequencing, unique solutions to maintain the freeze during excavation, and deliberate equipment selection. Lessons learned from the design through construction are provided.

INTRODUCTION

DC Water's First Street Tunnel (FST) is part of the $2.6 billion Clean Rivers Project designed to reduce the occurrence of combined sewer overflows into local waterways. Due to flood events in the Bloomingdale and LeDroit Park neighborhoods, the FST Project was accelerated by DC Water to mitigate future flooding by boosting storage capacity and thus relieving the undersized combined sewers.

The FST Project was finalized in 2016 as a collaborative effort between DC Water, their consulting team consisting of Greeley & Hansen and McMillen Jacobs Associates, and the design-build team of Skanska, Jay Dee Contractors (SKJD) with WSP Parsons Brinckerhoff as their designer.

Scope of Project

The FST Project (Figure 1) included four (4) primary shaft sites, a large 20 ft internal diameter, 2,700 ft long bored tunnel, and three (3) adit connections from the off-line shafts to the large diameter tunnel. The adits which have variety in size and length are the subject of this paper. The Channing Street site (Figure 2) was established as the main site for an Earth Pressure Balance TBM and included a 160 ft deep shaft with slurry wall support of excavation and 65-ft internal diameter final cast-in-place permanent concrete liner. The construction sites at Adams Street, V Street, and Thomas Street each consisted of off-line shaft structures tied to near-surface sewer diversion chambers, and ventilation facilities. These three off-line shafts were connected to the large bored TBM tunnel via adit tunnels. Each of the three (3) adits had a portion or the entire length excavated by Sequential Excavation Method (SEM) in frozen ground.

The Adams Street adit (Figure 3) was constructed in two phases. The first phase is the 75 ft long and 16 ft diameter section with immediately outside the drop shaft which also serves as the de-aeration chamber and was excavated by SEM. The second phase which is 300 ft long and 10 ft in diameter was excavated by Micro TBM launched from within the permanently lined de-aeration chamber. MTBM excavation commenced towards a reception chamber that was excavated by SEM from the previously bored FST TBM tunnel. This SEM excavation had a 17 ft diameter and a length of 15 ft.

The V Street adit was the largest SEM excavation of the FST Project and included a 20 ft diameter, 70ft long SEM tunnel. This SEM tunnel was excavated with an 80 ft radius curve from the 110 ft deep drop shaft towards the previously bored TBM tunnel. The reference design by DC Water initially included a straight alignment that was revised by the design-build team to a constant curve to avoid passing beneath a residential property and to allow the utilization of vertical ground freezing for both the SEM excavation and the deep Near Surface Structure excavation above.

The Pumping Station adit at Thomas Street had an excavation diameter of 13 ft. The 50 ft long SEM tunnel was excavated from a 90 ft deep shaft that will function as the temporary Pumping Station until the adjacent Northeast Boundary Tunnel connects the FST to the Clean Rivers Project.

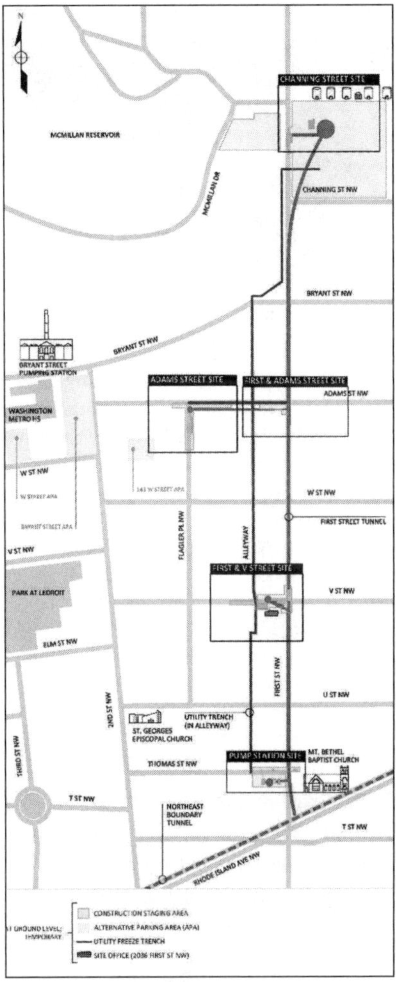

Figure 1. Project overview

Ground Conditions

Ground conditions along the FST alignment consisted of an upper layer of recent Fill, followed by Quaternary Alluvium, Cretaceous Potomac Group above Bedrock.

The Potomac Group was previously overlain by several hundred feet of soil deposits that were later eroded away an d fine-grained cohesive soils are hard and over-consolidated. The coarse-grained cohesion-less soils are dense to very dense. The upper portion of the Potomac Group consists of Patapsco/Arundel Formation with transitional layers of fine-grained soils with high plasticity (G1) and lower plasticity (G2). SEM excavation was fully located in the lower portion of the Potomac Group, and consisted of the Patuxent Formation which included predominantly non-plastic silty or clayed sand (G3A) or non-plastic silty or clayed gravel (G3B). These G3 sub-groups are distinguished from each other solely on the basis of whether the sand or the gravel fraction has the higher percentage in the particle size analysis. The G3 soil group within the Patuxent Formation is transitional interlayered with G1 and G2 clay, fine to coarse sand with traces of gravel and fines (G4) and fine to coarse gravel with traces of sand and fines (G5).

Figure 2. Channing Street site overview

Figure 3. Adams Street site (note freeze pipes)

A shallow groundwater aquifer exists, predominantly in the Fill and Quaternary Alluvium. This unconfined alluvial aquifer is generally perched on the confining clay units of the upper Potomac Group of the Patapsco/Arundel Formation. Groundwater within the confined water bearing layers and lenses of the lower Potomac Group of the Patuxent Formation exhibit artesian conditions.

All three adits have been excavated predominantly through G3A soils with various amounts and thicknesses of G1 and G2 layers in the face. When unsupported in the excavation, these G3A soils will exhibit fast-raveling to flowing behavior. The G1/G2 clay is prone to slide or fall as discrete blocks or wedges along fissured and slicken-sided fractures and therefore ground freezing was utilized to stabilize these soils for all three SEM excavations.

DESIGN AND CONSTRUCTIONS

SEM Designs

Due to desired low impact on the urban and historic residential neighborhood and to mitigate risks of SEM excavation in unstable ground under high groundwater pressure, ground freezing to temporarily improve the ground was utilized for all three (3) adit excavations.

In a first step, the properties of the freeze body around the excavation were established based on empirical correlation of similar soils with a target temperature of −10°C. With these pre-established frozen ground properties, the required limits of frozen ground around the SEM excavation were developed with the help of numerical modeling. Initial support for the adits included shotcrete with lattice girder and welded wire mesh. The final liner for the V-Street and Adams Street adits included cast-in-place reinforced concrete and the Pumping Station adit was furnished with a Hobas pipe.

In a second step, in order to verify the frozen ground properties derived by empirical correlation, frozen soil testing was performed. Representative split spoon samples from the ground investigation program were selected and re-compacted to simulate undisturbed field conditions. Samples were saturated and tested in the laboratory under design target temperatures. A total of three (3) unconfined compression tests and three (3) pulse velocity tests on frozen soil test specimens were performed in accordance with ASTM standards. The test results confirmed previous empirical assumptions with a Young's Modulus of 2,280ksf to 4,100ksf, thus finalizing the SEM design.

A freeze pipe layout with vertical and slightly inclined freeze pipes drilled from the surface was generated to provide a required freeze boundary with a target temperature of

−10°C. However, since the SEM excavation was cutting through vertical freeze pipes, the numerical model assumed a reduction of frozen ground properties (thawing) while the shotcrete liner was gaining strength at the same time. Information of shotcrete strength gain curves, sprayed on frozen ground, was utilized from previous projects and a trial test was performed prior to and during construction to support and verify the required shotcrete strength.

Brine, as the freeze medium, was circulated via a utility trench to all three (3) sites from the central freeze plant located within the Channing Street site. The brine had an initial temperature of −29°C going out and typically returning around −27°C, with a warming of about 2°C. Freeze formation of the three (3) adits, in order to generate the required freeze boundary with pre-defined properties, took between 40 and 60 days and was confirmed via temperature monitoring pipes with thermocouples at various depth intervals that were read automatically.

Convergence monitoring in the SEM excavation was installed to verify proposed excavation length and initial support measures to allow possible adjustments during excavation.

V Street Adit Construction

The V Street adit (VS-A), which was the largest of the headings to be excavated, was the first adit to be constructed. The drop shaft size limited the equipment selection due to size constraints. The physical constraints required that the adit be excavated with a heading and bench excavation (Figure 4). The top heading cross section was excavated at a 14 ft height for the full length of the adit. Once the shotcrete initial liner was installed, the bench was excavated to the full 20 ft cross section.

Vertical freeze pipes were installed to accelerate the overall schedule of adit construction, as well as facilitating ground support for the North Capital Street Diversion

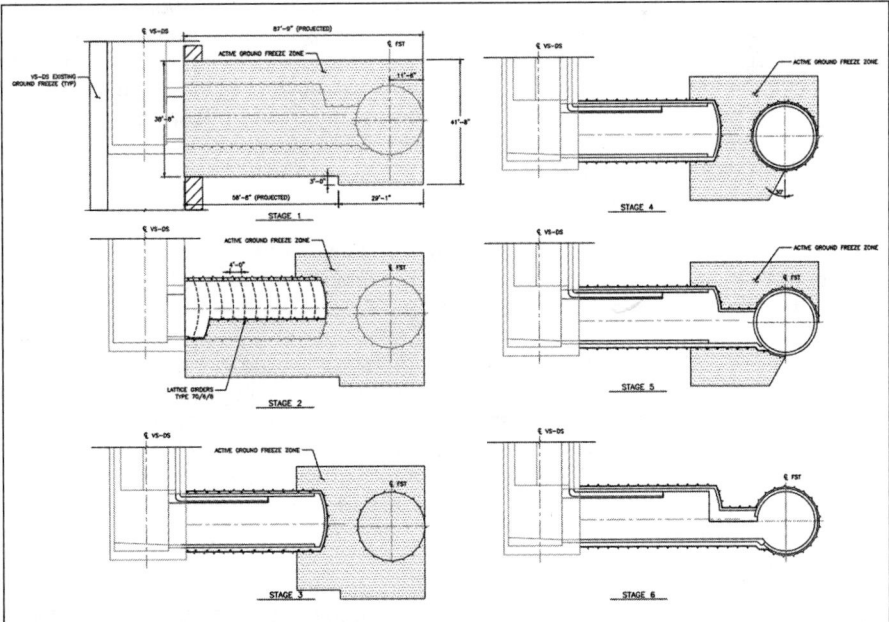

Figure 4. V Street adit construction sequence

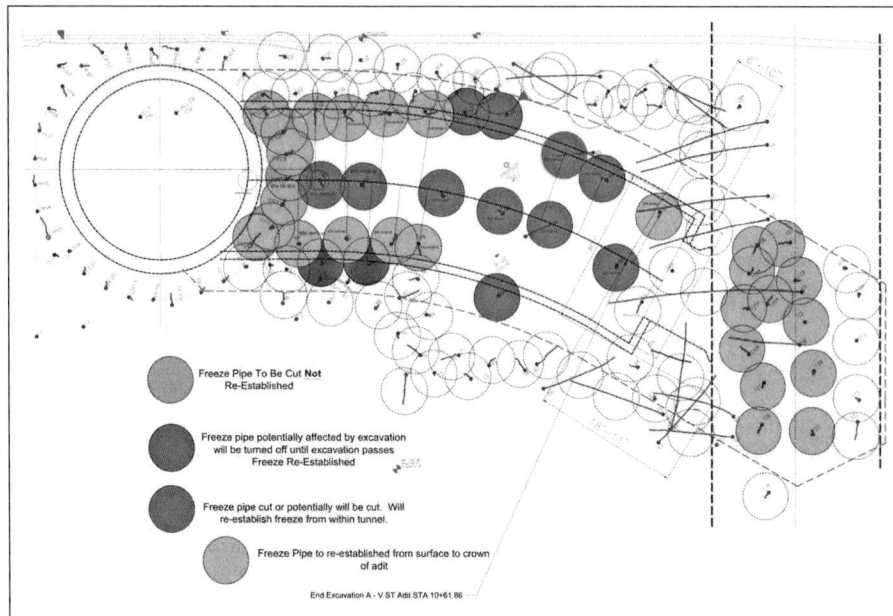

Figure 5. V Street adit freeze pipe layout

Chamber, directly above the adit, that was being excavated concurrently with adit construction. The vertical freeze pipes were activated as the final liner of the drop shaft was being constructed. This saved over 6 weeks of schedule as the freeze was able to be developed during the lining process. Excavation of the adit began as soon as the cast-in-place liner of the 23 ft diameter V Street Drop Shaft was complete. The SEM was developed for the first 7.5 ft using a Brokk 400 electric demolition robot with an Atlas Copco SB552 hydraulic hammer. The first two lattice girder arch sections were installed and shotcreted to full lining thickness. The excavation duration in the top heading dictated that the freeze pipes be re-established both on the crown and in the bench excavation to maintain the freeze temperatures (Figure 5). A separate glycol freeze unit was setup onsite and a header was installed down the shaft into the adit. Each invert freeze pipe was hooked into this header once it was cut out of the top heading cross section (Figure 6).

Once the initial heading was developed with the Brokk 400, an Antraquip AQM 150 roadheader was selected for excavation of the remainder of the adit. A skidsteer was used to tram the muck from the conveyor to the shaft. Excavation progressed by excavating 4 ft of total cross section, installing the lattice girder, wire mesh and shotcrete. Both dry and wet shotcrete were tested and approved for use in the adits, but due to low sprayed shotcrete quantities and advantages in timely supply, dry shotcrete was selected for the initial support. The small size of the site and concurrent excavations on the site dictated that the shotcrete setup had to be setup and torn down each shotcrete cycle. Quickcrete 5,000 psi shotcrete was the supplied mix with 3% dry accelerator premixed into the super sacks. A silo and Putzmeister GM-060 were used for the shotcrete plant.

Construction of the adit's top heading excavation used a Meyco Oruga shotcrete robot that was modified with a dry shotcrete nozzle. Due to the surface area and height of the heading spraying with the robot was more safe and efficient as the nozzleman did

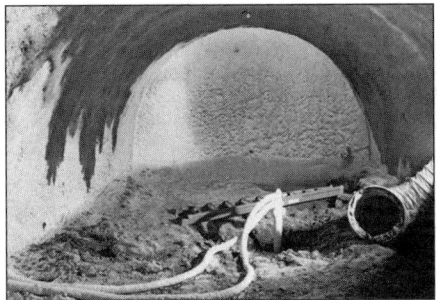

Figure 6. VS-A Re-establishment of invert freeze   Figure 7. Completed excavation of VS-A

not have to stand under the shotcrete as it was being sprayed. A temporary concrete invert was placed with the skidsteer to protect the bench frozen ground. This was completed before the shotcrete was applied. The adit was excavated to within 5 ft of the First Street Tunnel (FST). The tunnel was not constructed by the time the excavation of the adit was complete. A shotcrete bulkhead with wire mesh was installed until the FST tie in was made. The bench was excavated in 8 ft advances with a Brokk 400 diesel demolition robot. As the re-established invert freeze pipes were encountered, each pipe was removed and backfilled. At each 8 ft advance, the lattice girders from the top heading were completed through the invert. Wire mesh and shotcrete were applied to complete the initial support (Figure 7).

Adams Street Adit Construction

The Adams Street adit (AS-A) was excavated concurrently with the bench excavation for the V Street adit and the Pumping Station adit SEM excavation. The Brokk 400 electric demolition robot was used to develop the first 7.5 ft of heading similar to the V Street adit. The cross section of the Adams Street adit allowed for full face SEM excavation. The Antraquip AQM 150 Roadheader was used to excavate the de-areation section of the adit. Mucking was done by a 2 cy muckbox on light gage rail. Excavation was advanced in 4 ft sets. Each set would allow for installation of lattice girder, wire mesh and shotcrete. Due to limited space, a shotcrete robot could not be utilized and shotcrete was applied by hand in this excavation. Once the adit was excavated to the end station of the de-areation chamber a shotcrete bulkhead with wire mesh was installed to allow for construction of the cast in place lining (Figure 8).

Pumping Station Adit Construction

The Pumping Station adit (PS-A) was excavated in full face with a Brokk 400 electric demolition robot with hydraulic breaker. Muck was removed by skidsteer from the heading to the shaft. The 12 ft excavation was excavated in 4 ft advances. Each advance allowed for lattice girder, mesh and shotcrete installation. Shotcrete was applied by hand for this heading (Figure 9). Similar to the other headings the adit excavation was completed before the FST tunnel was mined. A shotcrete bulkhead with wire mesh was installed 5 ft from the FST. The adit was then lined with a 96 inch Hobas pipe and grouted.

Adit Connections to First Street Tunnel

As the TBM excavated the FST, each location of the adit tie in was blanketed with a surface freeze loop on the concrete segments and a series of invert freeze pipes were drilled and installed through the pre-cast segmental liner from within the FST to allow for continuance of freeze and water cutoff at the invert of the adit tie in. This was work done on non-production shifts during mining of FST. Additionally, the aluminum freeze pipes that were cut while the TBM advanced were sleeved to the surface and re-established to maintain the freeze block at the tie in location at the crown and springline of the FST. Once the TBM was decommissioned and buried in place at the tunnel termination, a steel propping ring assembly was installed at each adit tie in location. The segments were then saw cut and removed in the cross section of the excavation. The SEM excavation at the Pumping Station adit (Figure 10) and V Street adit were tied into the adits that were constructed from the drop shafts. Each were 4 ft advances with lattice girders, wire mesh and shotcrete. The Adams Street adit connection was excavated by the same method but was terminated 15 ft from the FST springline to provide enough space to receive the MTBM within the reception chamber (Figure 11). A shotcrete bulkhead was constructed for the MTBM reception chamber. Once the cast in place tie-ins were constructed, the invert and surface freeze pipes were removed from service.

LESSONS LEARNED

Overall, the freezing process and SEM excavation worked out very well, with stable ground for both shaft and adit excavations (Figure 12). Maintenance of freeze was achieved with re-sleeving cut freeze pipes (either by TBM or SEM excavation) from the top with a smaller diameter pipe and re-establishment of brine flow. A couple of freeze pipes were punctured by excavation equipment, but shut off valves enable system isolation which enable freezing to be maintained in other sites while repairs were made. Overall, aluminum pipes were easily mined through by the TBM. Due to concerns of warming in the invert zone once the TBM passed through, additional freeze pipes

Figure 8. AS-A permanent liner reinforcement

Figure 9. Shotcrete application by hand

Figure 10. Breaking out from FST with Brokk

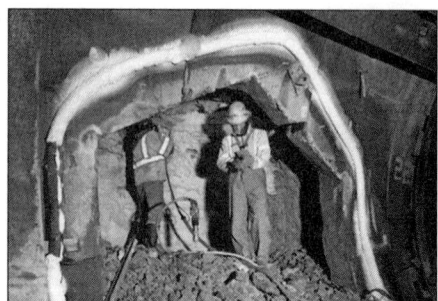

Figure 11. MTBM reception chamber

Figure 12. Stable face of frozen ground   Figure 13. Antraquip roadheader

were successfully installed by drilling through the precast segmental lining to maintain the invert freeze which was not able to be reached by the re-sleeving from the top.

There was a learning curve in shotcrete operations. The Pumping Station shaft SOE was selected as a field trial location to spray the shotcrete on exposed frozen ground. The shotcrete was then cored at required time after application in order to determine and verify actual shotcrete strength grain on frozen ground under field conditions. Nozzleman testing and certification was done in the frozen shaft, enabling the nozzleman to practice shooting on frozen ground. Utilization of premixed dry shotcrete that already included 3% accelerator was very beneficial to eliminate an additional step during shotcrete application and to avoid any over dosage. During placement, the heat of hydration of shotcrete coupled with frozen ground caused thick fog to develop and limited visibility of the shotcrete nozzleman, thus causing high rebound. Increased ventilation and blow-in exhaust proved to be sufficient in clearing the fog. The design for the initial support of shotcrete allowed for 2 inches of sacrificial shotcrete that was not required for the full design loads. This sacrificial layer insulated the remainder of the shotcrete layer from contact with the frozen ground and maintained the heat of hydration for the initial 6 inch pass of shotcrete.

The sizes of the drop shafts were the main factor in the selection of equipment for SEM operations. Due to the multiple heading sizes, choosing a single piece of equipment that could satisfy the size constraints of the shaft as well as heading size took collaboration with the equipment suppliers. Antrquip Corporation was selected to provide the roadheader (Figure 13) for excavation of the headings. A modified Antraquip AQM 150 was utilized which used a AQM 150 turret, boom, and transmission assembly mounted to an AQM 100 body which allowed for installation in the smaller diameter shafts and allowed full reach of the machine for the SEM headings. Other modifications were made to accommodate the unusual conditions caused by the freeze in the roadheader's cooling system, electrical, and hydraulic system.

Due to scheduling constraints, all three headings were excavated concurrently at a certain phase of the schedule. The Brokk 400 demolition robot was used due to its nimbleness and small footprint from within the shafts. The unit was able to reach the full height of the headings and muck during the initial development of the portal in each heading. The Brokk 400 was used to completely excavate the Pumping Station adit which had the least effect on the schedule, due to removal for mucking this was the most inefficient of the mucking cycles. The V Street adit bench excavation utilized a Brokk 400 diesel demolition robot which sat on the bench and chipped material into the bucket of a skidsteer at the toe of the bench. In general, the headings cycled roughly 2 to 2.5 cycles/week which included mobilization and demobilization of the shotcrete plant for each cycle.

The adit tie-ins were excavated with the Brokk 400 Diesel demolition robot due to the necessity to move within each adit location in the tunnel. The machine was able to tram itself to the subsequent adit tie in once excavation was complete. Both the roadheader and Brokk 400 proofed to be good and suitable equipment in the abrasive sand and gravel frozen ground. Roadheader teeth were checked after every round and worn teeth were replaced. Selected equipment was suitable for the sizes of SEM excavation and confined space.

Convergence Monitoring in the adit excavations showed very little to no deformation in the excavation and stayed well below the normal predicted convergences of 0.4 inches in the design. This verified that the design parameters for the frozen ground were prudently selected. Excavation at the face and within unsupported section prior to shotcrete application was stable at all times and allowed application of shotcrete initial support the following day.

All adits were constructed at locations with existing structures and utilities in close proximity. These structures had to be monitored for potential movements due to both ground deformations due to the adit excavation and the freeze related heave and/or settlement. Because of the difficulty in accurately estimating freeze related heave and/or settlements it is important to have a robust and proactive instrumentation and monitoring program. There is need for close coordination between the freeze designers and the instrumentation team to allow for early mobilization of mitigation actions when excessive freeze related movements are observed. There were several successful mitigations to counter impacts of heave on adjacent utilities that included installation of heat trace rows to counter the growth of the ground freeze body. The success of these types of measures is dependent on proactive monitoring. Furthermore, the selection and locations of ground monitoring equipment should consider the frozen mass. For example, an inclinometer will not be responsive to excavation related ground deformation if it is installed within a mass that eventually freezes. Our recommendation for adit instrumentation is to have beyond SEM monitoring a plan that addresses the effects of ground freezing and installation of heat pipes at critical utilities such that freeze mitigation measures can be implemented immediately as necessary.

SUMMARY

Ground freezing proved to be a good solution for ground improvement of SEM excavation for the FST project. This low impact and less intrusive method of ground improvement enable SKJD to adhere with the strict contract working hours of 7am to 7pm. The stable ground created by ground freezing enabled efficient use of the limited working hours. A significant benefit that ground freezing provides during the SEM mining cycle is the flexibility in stand-up time. For example, during the week work days, partially excavated ground could be left by covering with blankets at the end of the day, without the need to for temporary support, such as a flash coat of shotcrete. This is unlike other ground improvement techniques, where temporary support needed to be installed before leaving for the day or in some instances, the requirement of round-the-clock construction. Re-establishment of freeze pipes was considered a more critical SEM construction factor than partially completing the initial support and ground freezing allowed for this flexibility in the sequence of construction.

PART **18**

Difficult Ground

Chairs

Gregory Hauser
Dragados USA

Robert Marshall
Frontier-Kemper Constructors

Design of the Fort Wayne CSO Tunnel Through Complex/Wet Rock

Aswathy Sivaram ▪ Black & Veatch
Mark H. Bradford ▪ Black & Veatch
T.J. Short ▪ Fort Wayne Utilities

ABSTRACT

The Three Rivers Protection and Overflow Reduction Tunnel (3RPORT) is a critical component of Fort Wayne's Long Term Control Plan to reduce the volume of combined sewer overflow (CSO) entering the waterways of Fort Wayne, Indiana. The project consists of 24,580 ft (7,500 m) of 16 ft (4.9 m) finished diameter deep rock tunnel, 9 drop structures, Tunnel Boring Machine (TBM) launch and retrieval shafts, a 30 MGD (113,500 m³/day) deep dewatering pump station (DDPS), and associated deaeration chambers, adits, and near surface infrastructure. The local geology consists of horizontally bedded dolomite interbedded with chert and fractured dolomitic bank reef formations with artesian groundwater. Key considerations in the tunnel design will be presented including planned groundwater mitigation strategies for construction, which is scheduled to start in early 2017.

INTRODUCTION

The City of Fort Wayne, Indiana has entered into a Consent Decree with the U.S. Environmental Protection Agency, U.S. Department of Justice, and the Indiana Department of Environmental Management to implement a combined sewer overflow (CSO) Long Term Control Plan (LTCP) to reduce the volume of combined sanitary and storm sewage that is discharged into the city's waterways. The Three Rivers Protection and Overflow Reduction Tunnel (3RPORT) is a critical component of this LTCP. The tunnel is anticipated to be fully operational by the end of 2022. The tunnel will receive flows from existing CSOs to reduce overflows to the St. Marys and Maumee Rivers to no more than four overflow events within a typical year. The tunnel will then convey the flow to the Wet Weather Pumping Station (WWPS) for transfer to and storage in the Wet Weather Ponds or directly to the Water Pollution Control Plant (WPCP) for treatment.

The tunnel begins at the western limits of the WPCP that is located east of downtown Fort Wayne, IN along the southern bank of the Maumee River. The tunnel alignment extends west from the WPCP generally following the alignment of the Maumee River, crosses beneath a residential neighborhood, and follows the St Marys River. The alignment then passes through the heart of downtown, beneath Headwaters Park, continues west and then turns southwest following the St. Marys River. The tunnel alignment continues south beneath the St. Marys River until it runs beneath Vesey Road in a southerly direction, crosses beneath Indian Village Park, the St. Marys River again, and terminates at the north end of Foster Park at the retrieval shaft. The invert of the tunnel ranges from 175 to 225 ft (53 to 70 m) below the ground surface. The major components of the 3RPORT Project–Tunnel and Shafts package include:

- 24,580 ft (7,500 m) of 16 ft (4.9 m) inside diameter (ID), hard rock tunnel, at 0.15% slope

- 30 ft (9.1 m) diameter working shaft
- 60 ft (18.3 m) diameter deep dewatering pump station (DDPS) shaft
- 21 ft (6.4 m) diameter retrieval shaft, with an integral drop shaft
- 11 drop and vent shafts, finished ID ranging from 2 to 7 ft (0.6 to 2.1 m)
- 740 ft (225 m) of 7 ft (2.1 m) finished ID adit

An alternate bid was also solicited to extend the tunnel approximately 4,000 ft (1,200 m) in length, locating the retrieval shaft at the south end of Foster Park. The tunnel alignment, along with the shaft locations and associated adits for both the base and alternate bids, are shown on Figure 1. The DDPS shaft and the working shaft will be converted into a 30 MGD (113,500 m³/day) pump station and screening/upflow shaft, respectively, in a later design package.

SUBSURFACE CONDITIONS

Geologic Setting

The City of Fort Wayne is the largest city in Allen County. The county is located between two regional bedrock structures: the Cincinnati Arch to the south and the Michigan Basin to the north. Relief varies throughout the county as a result of pre-glacial and glacial stream erosion on rocks of varying hardness. There is no exposed bedrock in Allen County, and the depth to top of rock varies from as little as 50 ft (15 m) to more than 300 ft (91 m) depending on the location within the county. The thickness of the unconsolidated deposits along the tunnel alignment and shaft sites varies between approximately 60 and 100 ft (18.3 and 30.5 m). The unconsolidated deposits

Figure 1. 3RPORT tunnel alignment

are the results of the most recent Wisconsin glaciation period, and are associated with an end moraine glacial environment.

Devonian and Silurian carbonate rock underlies most of Fort Wayne and the valleys of the rivers, which includes the project area. The local rock unit formations include, in descending order: Devonian age Antrim Shale; and Traverse and Detroit River Formations, both of the Muscatatuck group. The underlying Silurian age formations include the Wabash, Louisville Limestone, and Waldron Shale. The Fort Wayne rock geology is similar in age and structure to the rock geology in which most of the Milwaukee CSO tunnels were constructed, and lie along roughly the same Michigan Basin "contour."

The Antrim Shale is calcareous shale and is not present along most of the alignment. The Traverse Formation is a medium grained argillaceous limestone with scattered chert nodules. The Detroit River Formation is a gray and dark brown to tan dolomite. The Detroit River Formation is comprised of thinly bedded, alternating bands of conglomeritic dolomite and banded coarse grained dolomite. A bentonite shale marker bed is located approximately 6 to 9 ft (1.8 to 2.7 meters) above the base of the Detroit River formation. Thin, continuous bands of chert, 2 to 8 inches (50 to 200 mm) in thickness are interbedded within the Detroit Formation. The Wabash Formation consists of a fine-grained, slightly dolomitic limestone and dense to fine-grained, vuggy, dolomite. The Louisville Limestone is comprised of the light colored, thickly bedded, argillaceous dolomitic limestone. Major portions of the Wabash Formation and Louisville Limestone are comprised of bank reef facies with scattered vugs and solution channels of various sizes. The 3RPORT will primarily be constructed within the Silurian Wabash formation. The shafts will be constructed primarily through the Traverse, Detroit River, and the Wabash formations.

While no faults or shear zones have been identified along the project alignment, it is likely that minor faults and shear zones do occur in the bedrock. The predominant joint set trends approximately N60E to due east and is near vertical. A secondary less prevalent near vertical joint set generally trends approximately N20W to due north, and is nearly perpendicular to the primary set. The near vertical primary and secondary joint sets are based on observations at the Hanson Ardmore Quarry located approximately three miles southwest of the 3RPORT retrieval shaft location. The northern highwall of the Hanson Ardmore Quarry showing the rock formations is shown on Figure 2.

Geotechnical Investigation Program

Geotechnical investigations were conducted over three major phases to obtain project specific geologic, hydrogeologic, and geotechnical data to support design and construction of the 3RPORT project. The earliest investigation started in 2012 and the latest phase was complete in the summer of 2016. The latest phases of the geotechnical investigation were performed to obtain additional hydrogeological information to support vertical tunnel alignment, lining design, and tunnel boring machine (TBM) selection.

The field investigation program involved drilling 27 vertical and inclined deep rock borings up to 320 ft (97.5 m) deep, and 28 shallow vertical borings depths varying from 50 ft (15.2 m) to top of rock. Observation wells were installed in all the vertical borings, with the wells in the deep borings screened at or near the anticipated tunnel zone. Water pressure testing of the full rock column was performed in the deep borings, as well as slug testing in all observation wells. Laboratory testing on soil and rock was also performed to obtain information critical to the design of the shafts and

Figure 2. Quarry highwall

tunnel for the project. The rock properties of the Devonian and Silurian Age rocks are similar to those encountered in other tunnel projects in the Midwestern cities of Chicago, Milwaukee, and Indianapolis. Tests confirmed that rock was of moderate to high strength with good borability and low abrasivity properties.

Four large diameter wells were installed in the bedrock during the Phase III investigation in which aquifer pumping tests were performed. The first phase of aquifer testing included 72-hour pumping tests in two wells located near borings in which water pressure testing indicated zones of high hydraulic conductivity. Wells W-1 and W-2 were cased through the overburden with 12 in (305 mm) ID steel casing seated into the top of rock, and then the rock was drilled to a depth of approximately 260 ft (79.2 m) below grade. The intent of the aquifer testing of the first two wells was to pump test the entire rock column in each well, and then isolate the bottom 80 ft (24.4 m) of the 12 in (300 mm) diameter bedrock well by installing and grouting in a 8 in (200 mm) inner casing to a depth of 180 ft (55 m), and test the bottom portion of the well. The second phase of aquifer testing included drilling wells W-3 and W-4 near borings where water pressure testing indicated the rock had relatively lower hydraulic conductivity than the first two well sites. The last two wells were designed such that only the bedrock at the tunnel zone, 170 to 223 ft (51.8 to 68.0 m) depth, was to be tested by utilizing telescopic steel casing cement grouted in-place to a depth of 170 ft (51.8 m). The final bedrock wells, W-3 and W-4, were approximately 8 inches (200 mm) in diameter over the depth tested, and were pumped and monitored for a period of 7 days. Groundwater levels were also monitored utilizing downhole transducers and hand measurements in the pumping wells and nearby observation wells, screened in both the bedrock and overburden, to determine the connectivity of the deep and surficial aquifers.

A series of downhole geophysical tests were also performed at five locations (three boreholes and two pumping wells) during Phase III investigation to obtain information on rock lithology, fractures, permeability, and porosity. Geophysical testing including 3-arm caliper, optical televiewer, resistivity, gamma, and heat-pulse flowmeter.

Subsurface Profile

The overburden along the 3RPORT alignment consists of manmade land or fill at a number of isolated locations overlying the natural alluvium, glacial till, and outwash deposits. The thickness of the overburden varies from approximately 60 to 100 ft (18.3 to 30.5 m) depending on the location along the alignment. The near surface soils consist of normally consolidated fine grain soils with zones of silty sands and sandy silts. The surficial soils are the result of a post glacial alluvial environment and overlie thick, discontinuous, glacial till sheets that are heavily over consolidated and primarily cohesive in nature with varying amounts of sand and gravel. Deeper outwash sand and gravel soils are prevalent along the project corridor and vary from fine silty sands to coarse sand and gravels with cobbles. Hydraulic conductivity of the outwash soils is very high, in the range of 1 to 1×10^{-1} centimeter per second (cm/sec). At many locations, the outwash soils are not continuous with depth and are interbedded with lenses of fine-grained glacial till. Igneous and sedimentary boulders are present within the outwash soils as well as within the glacial till deposits. Some test borings encountered isolated zones of nested cobbles and boulders at the soil-rock interface at proposed shaft locations. The groundwater level along the alignment is controlled by the nearby rivers and monitoring indicates that the near surface aquifer responses quickly to changing river levels and precipitation.

Antrim Shale was only encountered in three core borings near the northern end of the project alignment in thicknesses ranging from 2 to 4 ft (0.6 to 1.2 m). Limestone bedrock was encountered in 24 of the 27 deep borings below the overburden. This Traverse formation is thinly to thickly bedded, ranging from 2 in to 3 ft (50 to 915 mm) thick, with near horizontal bedding and consists of light to medium gray, medium grained, fossiliferous argillaceous limestone with chert nodules. The chert nodules vary from widely spaced and sporadic to tightly spaced within a nearly continuous horizontal zone that comprises up to 10% of the unit thickness. Fractured limestone varies in thickness from 20 to 60 ft (6.1 to 18.3 m) and makes up the bedrock-overburden interface surface in the majority of the borings.

The Detroit River formation is thinly to thickly bedded, ranging from 4 in to 3 ft (100 to 915 mm), with near horizontal bedding, and is blocky in nature. The upper portion of Detroit River consists of light to medium gray, fine to medium grained, thickly bedded to massive, conglomeritic dolomite. Argillaceous layers of varying thickness are present in this upper zone. The lower portion consists of gray and dark brown to tan dolomitic limestone or dolomite, banded with white, crystalline calcite and shale beddings, with some coarse, sandy zones. Pinhead sized vugs are present in the lower zone in some borings. Multiple discrete lenses of chert layers are noted within this unit and vary in thickness from 2 to 8 inches (50 to 200 mm). The thickness of this formation ranges from 35 to 50 ft (10.7 to 15.2 m) in the project area.

The Wabash formation is thickly bedded to massive and free of chert. It consists of medium to dark gray, dense, vuggy, massive argillaceous dolomite with greenish-gray shale partings, and near vertical features. Pressure dissolution features such as stylolites, evidence of bioturbation, and reef facies are prevalent in this formation. This unit is vuggy and reef facies comprise 75% of the rock. Bedding within this unit is not distinct; however, based on visual observation of the reef facies structure within the Hanson Ardmore Quarry, the observed apparent bedding based on fractures and coloring is inclined between 15 and 25 degrees from horizontal. The remaining 25% is materially identical, but bedding is near horizontal with distinct bedding planes and fewer vugs. The Wabash formation ranges in thickness from 50 to 110 ft (15.2 to 33.5 m) in the project area. The Louisville limestone is materially and visually similar to the Wabash formation with only a slight color change and fewer vugs. The Louisville limestone was typically encountered below the final design invert of the 3RPORT. The subsurface profile as interpreted from boring information and the final design tunnel profile are shown on Figure 3.

Groundwater Infiltration

Hydraulic conductivities were calculated using water pressure packer testing, with a double packer assembly. The hydraulic conductivity distribution which includes the data from 530 packer tests conducted is shown on Figure 4. During packer testing, there were 39 instances where in the desired pressure could not be achieved due to limitations of the equipment, and free flow conditions were encountered. The pressure and flow developed in these instances were used to calculate the hydraulic conductivity (k), for those intervals. The intervals for the distribution correspond to the intervals used when estimating groundwater inflow to tunnels.

The hydraulic conductivity distribution with data from packer tests conducted only in the Wabash formation is shown on Figure 5. Out of the 39 unrestricted flow instances,

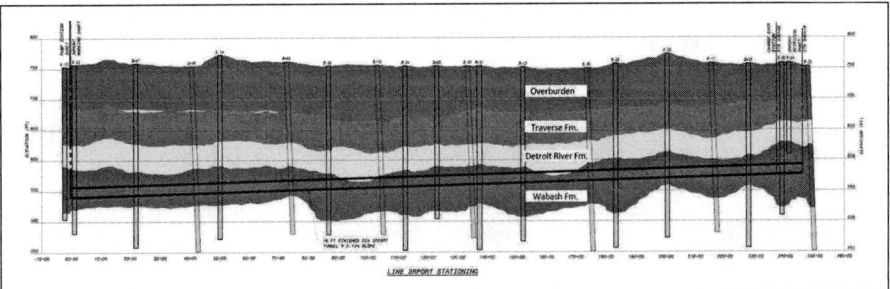

Figure 3. Subsurface profile

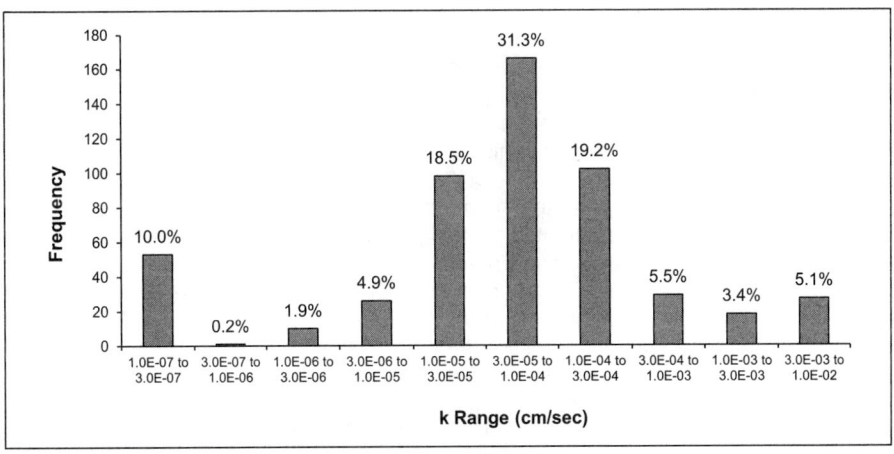

Figure 4. 3RPORT hydraulic conductivity distribution—all test data

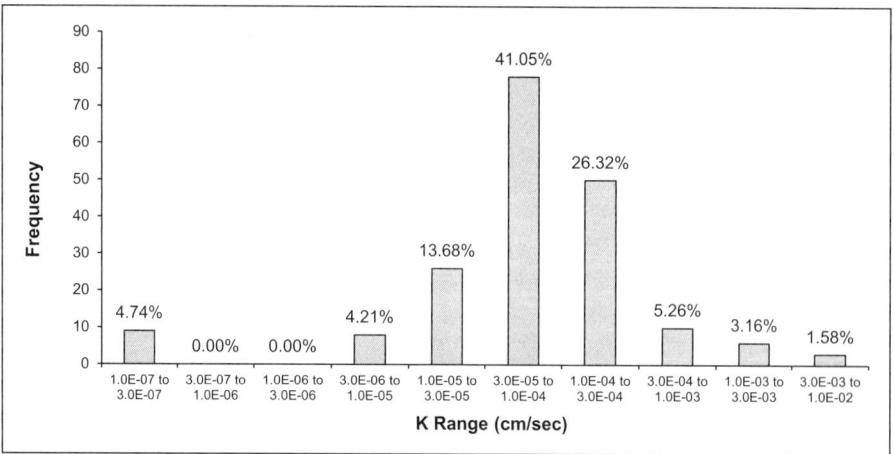

Figure 5. 3RPORT hydraulic conductivity distribution—Wabash formation

seven occurred in this formation. The tunnel alignment will be constructed in the Wabash formation.

The estimated steady state infiltration rate for an unlined tunnel is approximately 15,000 gpm (950 liters/second) using Heuer's method (Heuer 1995 and 2005) and the calculated hydraulic conductivity values for the 39 instances where free flow was observed in the water pressure testing. Steady state infiltration will increase as the tunnel length increases to a maximum of approximately 15,000 gpm (950 liters/second) that will likely be reached when the tunnel excavation is 60 to 80 percent complete. Initial heading flow, also known as flush flow, over a maximum duration of 36 hours is estimated to be approximately 1,250 gpm (80 liters/second). However, flush flows in excess of 5,000 gpm (320 liters/second) are possible.

The permeability calculated from packer tests is not a true permeability that can be translated for use in standard hydrology equations (Heuer 1995). Hence the packer hydraulic conductivities cannot be compared to those computed from the aquifer tests. While the hydraulic conductivity from packer tests are representative of near hole conditions along a 10 ft (3 m) zone, the values from aquifer tests are representative of the overall aquifer conditions.

Pumping rates in test wells installed during the geotechnical investigation along the alignment varied from 97 to 777 gpm (6.1 to 49 liters/second). The aquifer storativity coefficients varied by up to 18 orders of magnitude. The calculated hydraulic conductivity of the rock mass tested, which is controlled by joints, bedding planes, and discrete discontinuities, ranged from 2×10^{-3} to 3×10^{-2} cm/sec, only varying by one order of magnitude. The aquifer testing performed in wells W-3 and W-4 indicate that the wells were essentially connected to near infinite sources of recharge based on the quick rate at which the drawdown curve reached steady state conditions. Wells W-1 and W-2 also reached steady-state conditions, but at a much slower rate, and none of the wells indicated a distinct boundary condition indicating that the well production would dissipate with time. Slug testing that was performed in the observation wells screened in the Wabash formation indicate similar hydraulic conductivity values to those calculated from the aquifer testing. Some of the observation wells screened in the deep bedrock aquifer along the tunnel alignment showed changes in the groundwater level almost immediately to corresponding changes in river levels.

DESIGN AND CONSTRUCTION CONSIDERATIONS

The design team sought out the opinions and lessons learned from previous projects of the prequalified contractors during the final design phase of the 3RPORT Tunnel and Shafts package. Multiple meetings between Fort Wayne Utilities, individual contractors, the designer, and the program manager were held after providing the contractors preliminary plans and geotechnical data to discuss key design and construction issues. The details of each meeting were kept confidential between the design team and each respective contractor. The insight and contractors' lessons learned were reviewed, considered, and incorporated into the final contract documents where the design team found value for the project.

Shaft Construction

Dewatering for construction of the tunnel shafts and the drop and vent shafts is not allowed on the 3RPORT project. Dewatering could lead to potential environmental concerns at some sites. The increase in effective stress at some sites due to the amount of time required for deep shaft construction could cause consolidation settlement of cohesive alluvial soils. This settlement would likely cause damage to structures and infrastructure near these shaft sites. Watertight ground support through the overburden and socketed into the top of rock is required. The initial support of excavation for the larger tunnel shafts (i.e., Pump Station Shaft, Working Shaft, Retrieval Shaft, and Alternate Retrieval Shaft) through the overburden is to be either slurry diaphragm walls or secant piles socketed into rock. Drilled shafts through the overburden utilizing wet drilling methods and initial support of continuous welded steel casing socketed into the top of rock will likely be the method of installing the initial support for the drop and vent shafts through the overburden.

As evident from the aquifer testing, groundwater inflow to the shafts through the bedrock will be high. Pre-excavation grouting of the rock around and below is required for the large shaft excavations. A performance specification approach was taken in the contract documents rather than a prescriptive approach to allow the contractor to select the preferred method of pre-excavation grouting. The contractor is allowed to select the method of excavating the rock at the drop and vent shaft excavations, and based on the selected method, pre-excavation grouting may or may not be needed for those shafts.

Tunnel Alignment Evaluation

A desire to have uniform rock conditions drove the final invert elevation for the tunnel within the Wabash formation. The Wabash formation is a relatively consistent, mostly massive, competent rock suitable for tunneling as evident from the rock core test borings, laboratory testing, and visual observation at the nearby quarry. The Detroit River formation, while comprised of competent rock, is thinly and nearly horizontally bedded, has alternating bands of rock with varying strength and isolated shale layers. Test borings drilled along the last 4,700 ft (1,400 m) of the base bid tunnel alignment indicate that a 1 to 3 in (25 to 75 mm) thick layer of shale has been washed out due to pumping from bedrock wells near the retrieval shaft. Distinct, continuous layers of chert are also prevalent within the Detroit River formation. Past experience has shown that the varying strength and hardness properties of the rock in the Detroit River formation will give rise to blocky ground conditions during mining, causing high impact loading on the TBM disc cutters. This will result in more frequent cutter changes and interventions. The nearly horizontal bedding of the rock also makes it difficult to intersect the bedding planes that are the primary water producing features. The Traverse formation is more consistent in regards to rock strength and hardness, but the near horizontal water producing features and presence of chert would negatively impact the effectiveness of pre-excavation grouting and disc cutter life.

Tunnel Construction

The anticipated groundwater inflow to the tunnel during excavation of the 3RPORT necessitates the use of a one-pass, gasketed and bolted, precast concrete segmental lining system to be installed as the tunnel is advanced through the ground. Monitoring well data, along with results from the aquifer pumping tests, indicate that the deep bedrock aquifer is hydraulically connected to the overburden aquifer, and is in communication with the surface waters along the alignment. This connection is through a complex and extensive system of joints and discontinuities, including near horizontal bedding planes of varying aperture thicknesses, connected to near vertical joints that extend to the overburden, which provides an infinite source of recharge. Regions of high hydraulic conductivity are not confined to any particular formation, and are spread throughout the rock mass and along the tunnel alignment.

Horizontal features in the rock along with sub-vertical features contribute to the groundwater inflows that need to be managed in the tunnel. Groundwater inflow rates for the tunnel have also been estimated using the numerical model program known as *MODFLOW*. The modeling results are presented in Table 1. The maximum tunnel heading flush inflow anticipated ranges from 3,000 gpm to 10,000 gpm (190 to 630 liters/second).

The effectiveness of pre-excavation grouting is highly dependent on the probe holes encountering water producing features, the grout mix, contractor workmanship, grouting pressure, the interconnectivity of the features as well as other issues not listed here. The amount and location of groundwater inflow at the TBM face and around

Table 1. Summary of *MODFLOW* results

Tunnel Length (Model Cell Size) (ft)	Horizontal Rock Hydraulic Conductivity, k_h (cm/sec)	Horizontal to Vertical Rock Hydraulic Conductivity Ratio, k_h/k_v	Rock Hydraulic Conductivity After Pre-excavation Grouting, k (cm/sec)	Transient Water Inflow, Q	
				After 6 hours (gpm)	After 12 hours (gpm)
50×50	$5.3×10^{-2}$	2	No pre-excavation grout	10,920	10,880
50×50	$3.5×10^{-2}$	1	No pre-excavation grout	9790	9750
50×50	$3.5×10^{-2}$	2	No pre-excavation grout	7,700	7,670
50×50	$3.5×10^{-2}$ in and around excavated tunnel, $1.8×10^{-2}$ elsewhere	2	No pre-excavation grout	7,440	7,390
50×50	$3.5×10^{-2}$ for excavated tunnel, $1.8×10^{-2}$ elsewhere	2	No pre-excavation grout	5,690	5,610
30×30	$3.5×10^{-2}$	2	No pre-excavation grout	5,440	5,430
50×50	$1.8×10^{-2}$	2	No pre-excavation grout	3520	3410
50×50	$3.5×10^{-2}$	2	$1×10^{-3}$	540	540
50×50	$3.5×10^{-2}$	2	$3.5×10^{-4}$	200	200

the shield is anticipated to change as the TBM advances. Water inflow through the annular space in between the segments and excavated rock can potentially be detrimental to the grouting operations behind the segments and must be planned for during construction. The experience the design team gained from the construction of the OSIS Augmentation Relief Sewer (OARS) project in Columbus, Ohio, the Lake Mead Intake No. 3 project outside of Las Vegas, Nevada, and other tunnels constructed in ground where high groundwater inflow was encountered; as well as the meetings with prequalified contractors, helped shape the final contract documents.

The Contractor was allowed the option to bid either a shielded TBM, operating at atmospheric pressure within a continuously grouted tunnel envelop (Open Mode TBM), or a shielded TBM, operating continuously with a closed slurry muck circuit under normal groundwater head conditions (Slurry Mode TBM). The Open Mode TBM must be designed and equipped with probing and pre-excavation grouting capabilities, maintain the ability to close off the excavation chamber in the event of an emergency, and adequately retard groundwater travel to allow grouting of the annular space between the segmental lining and excavated rock. The Slurry Mode TBM must be designed and equipped to minimize disruptions due to interventions, operate without affecting normal groundwater head conditions, and adequately retard groundwater travel to allow complete grouting of the annular space between the segmental lining and excavated rock. Excavation chamber interventions for the Slurry Mode TBM are required to be performed at atmospheric conditions. It is anticipated that certain locations along the tunnel alignment will require little pumping or pre-excavation grouting, allowing for interventions under atmospheric conditions.

The adits and the deaeration chambers are anticipated to be excavated by drill and blast methods after pre-excavation grouting. The horizontal alignment of the tunnel was developed to minimize the length of adits connecting drop shaft deaeration chambers to the main tunnel. The short length of the adits will allow for most or all of the pre-excavation grouting of the deaeration chamber and adits to be performed from the ground surface at drop shaft sites if the contractor so chooses.

CONCLUSIONS

"In tunneling, groundwater is enemy No. 1!" as James McKelvey used to say. On this project the enemy is prevalent. The plan selected was to allow the contractors to choose the specific TBM and operational characteristics, and provide operational baseline quantities in the bid documents associated with the contractor's preferred TBM configuration. The operational baseline quantities include hours of pre-excavation grouting ahead of the TBM and an allowance pay item for contractor-selected grouting materials for the Open Mode TBM. The bidding documents include a baseline quantity of hours to prepare for atmospheric interventions on the excavation chamber for a Slurry Mode TBM. Additional allowances were also included in the bid documents. This approach was developed through an extensive partnering with the pre-qualified contractors, pre-bid interviews and information sharing, and several collaborative design team workshops between the Program Management team of CH2M, Arcadis, and Aldea Services, LLC, Black & Veatch, and Fort Wayne Utilities.

REFERENCES

Heuer, R.E. 1995. Estimating rock-tunnel water inflow. In *Proceedings of the Rapid Excavation and Tunneling Conference,* June 18–21, 1995, p 41.

Heuer, R.E. 1995. Estimating rock-tunnel water inflow–II. *Proceedings of the Rapid Excavation and Tunneling Conference,* June 18–21, 2005, p. 394.

Rondout West Branch Bypass Tunnel—TBM Boring in Hard Rock Against High Water Pressure and High Water Inflows Beneath the Hudson River in New York

David Terbovic ▪ The Robbins Company
Martino Scialpi ▪ The Robbins Company

ABSTRACT

A single shield hard rock tunnel boring machine is set to bore in hard rock, high water inflows and high water pressure in New York State. To overcome the difficult conditions the TBM is designed to handle 2500 gpm water inflows and seal against 30 bar of pressure. The TBM will bore a tunnel to replace a damaged portion of the Delaware Aqueduct that supplies half the raw water to New York City. The 2.5 mile bypass tunnel passes beneath the Hudson River with geology consisting of shale and limestone. Due to the high water pressure and inflows the TBM was designed with new sealing systems for the main bearing and to close the TBM off if high water inflows are encountered. The TBM is to be equipped with two dewatering systems and multiple drilling and grouting systems for pre excavation grouting and segmental lining backfill. Systematic drilling and grouting procedures specific to the project were developed and incorporated into the TBM and backup design to ensure that the TBM can handle the extremely difficult ground conditions of the project.

INTRODUCTION

The construction of the 85 mile long Delaware Aqueduct commenced in 1939 and operations began in 1944. It conveys about 500 million gallons per day of water to the city of New York, about 50% of the city's drinking water supply.

Since the 1990s the New York City Department of Environmental Protection (DEP) has been monitoring leaks in a portion of the aqueduct that connects the Rondout Reservoir to the West Branch Reservoir, finding specific areas of the Rondout West Branch Tunnel (RWBT) leaking approximately 35 million gallons of water per day into the Hudson River. As a percentage of the capacity of the tunnel, the leaks are not excessive, but the location and nature of the leaks, together with the fact that they are concentrated in one location, is a cause for substantial concern. In 2009 the DEP (which owns and operates the tunnel) determined that a repair was needed and the investigation and design process was initiated.

In addition to historical construction records and pictures of the interior of the tunnel taken by an autonomous underwater vehicle in 2003 and 2009, in 2014 a remotely operated vehicle was put into the RWBT. Cracks in the lining were observed and mapped. The conclusion was that leaks could not be repaired from within the tunnel. Thus, it was decided to proceed with the construction of a by-pass tunnel right under the river and build two shafts at each end between the towns of Wappinger and Newburgh, New York.

Construction of each shaft provides access points for construction of the by-pass tunnel and enables the by-pass to connect with the existing tunnel. The tunnel will be bored by a single shield TBM.

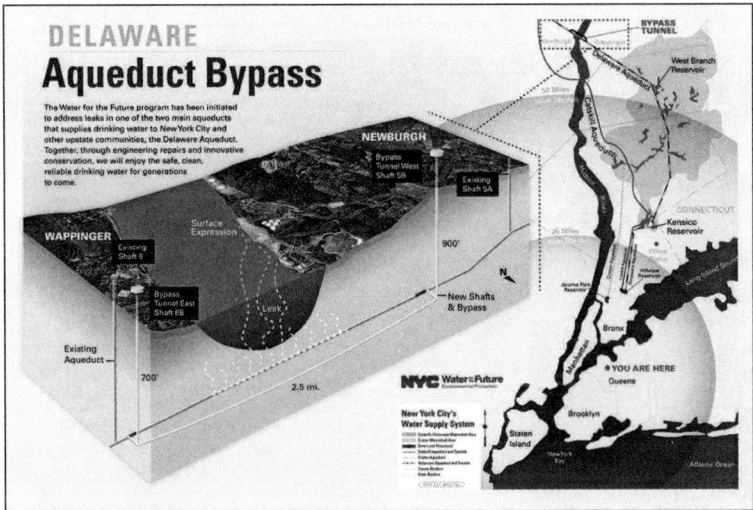

Figure 1. Delaware aqueduct bypass infographic

Once the shafts and by-pass tunnel are constructed, the aqueduct will be shut down and dewatered to repair the leaks and connect the by-pass tunnel to the existing tunnel. The city will implement a number of additional projects to supplement DEP's water supply during the shutdown period. Project location, summary and infographic are shown in Figure 1.

THE BYPASS TUNNEL

The Rondout bypass tunnel construction was separated into two contracts:

- BT-1 containing Shaft 5B and Shaft 6B, with the 5B site able to support tunneling operations (room for tunnel segments, concrete plants, water treatment facilities, etc.);
- BT-2 for excavation and lining of the entire by-pass tunnel, including the intersections with the shafts and connection tunnels to the RWBT

Locations of shafts 5B and 6B are shown in Figure 2.

The Shaft 5B (completed in late 2016 by drill and blast method) is located in the west connection site and extends to a depth of approximately 900ft (274m) below local grade. Support equipment and a crane are located adjacent to the shaft, to provide access for material and equipment.

The finished interior diameter of the shaft is 28 feet (8.5m), with a 300-foot starter tunnel at the bottom for TBM assembly and launching.

Figure 2. Rondout WBT bypass project shaft locations

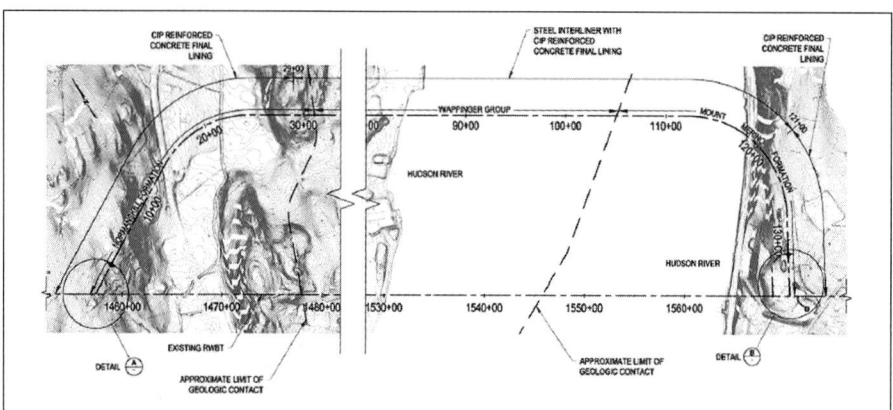

Figure 3. Bypass tunnel details

The shielded TBM, assembled at the Robbins facility in Solon, Ohio, will be delivered in pieces to site in the spring of 2017. The machine components will be lowered down the 28ft diameter shaft and assembled at depth.

The 21.4ft (6.5m) diameter TBM will operate on the 12,500ft (3,800m) drive by installing pre-cast reinforced concrete segments as tunnel lining concurrently with the excavation. Such supports will be the by-pass tunnel interior walls. The voids between the lining and the rock will be sealed by injecting cement grout, creating an effective barrier against the seepage of water into the by-pass tunnel.

The by-pass tunnel will be constructed in a constant minor downhill slope, with a total elevation change of 5ft (1.5m) over its length.

At completion of the by-pass tunnel excavation, the TBM will be disassembled and removed through the 700ft (213m) deep shaft 6B, at the east connection site.

After TBM removal, an additional steel liner (made up of steel segments) will then be installed to reinforce the tunnel in certain sections surrounded by weaker rock, for a total length of 9,200ft (2,800m).

Following the installation of the interliner pipe, a cast-in-place concrete liner will be installed throughout the full length of the by-pass tunnel, as well as the starter tunnel and connection tunnels. The locations of the CIP concrete and steel interliner pipes are shown in Figure 3.

PROJECT CHALLENGES: GEOLOGY, WATER PRESSURE, AND WATER INFLOW

The Rondout by-pass tunnel crosses through various geologic formations. The Wappinger Group consists of dolomite, dolomitic limestone and limestone rock types. The Normanskill and Mount Merino Formations consist of slatey shale, argillite and sandstone rock types. Figure 4 shows the tunnel alignment overlaid onto the geologic formations below. Very high UCS values up to 372MPa with an average of 241MPa and low to medium boreability are anticipated for the Wappinger Formation. The expected strong and tight dolomitic limestone have influenced the TBM selection and its design along with other anticipated ground conditions such as spalling behavior, raveling blocky ground, squeezing ground conditions.

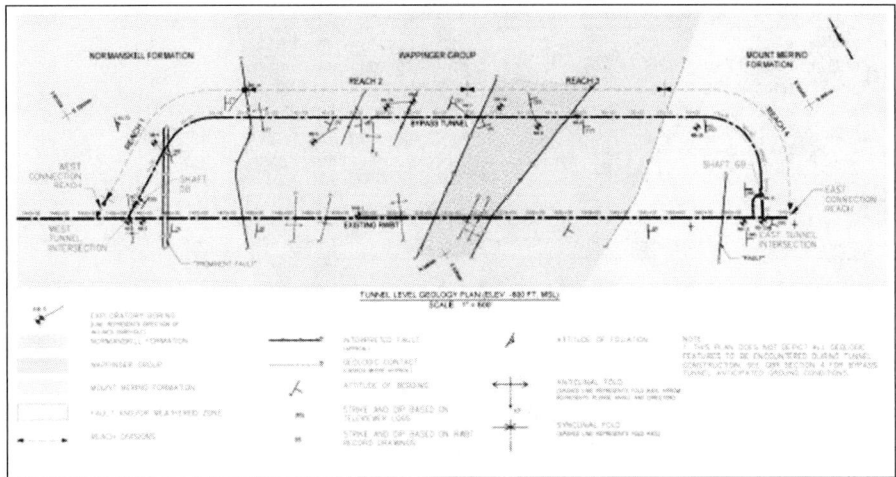

Figure 4. Tunnel level geology plan

The bypass tunnel alignment crosses under the Hudson River for 3,451ft (1,052m). The groundwater head along the tunnel ranges from a minimum of 600ft (183m) under the river to a maximum of 875ft (267m) and 700ft (213m) respectively on the west and the east side. The anticipated high conductivity of the Wappinger formation with the fault zones in the Normanskill Formation represent the most challenging tunneling conditions, especially if and when combined with high head conditions.

A groundwater model was developed to predict water inflows along the alignment and determine the excavation methodology (drill and blast vs TBM).

In favor of the TBM option, the gasketed segmental lining installed behind the machine would beable to minimize the water inflow. However in the Wappinger formation higher inflows up to 1,300gpm (4,920 l/min) were predicted from the full face TBM heading, compared to D&B multi-stage excavation methodology. In order to handle these potential inflows, mandatory probing ahead and intensive pre-excavation grouting were introduced into the requirements (where the pre-excavation grouting is intended to lower the groundwater inflow potential of the rock mass, not to increase its strength or improve its stability).

With the tunnel designed to be bored downhill, another challenge is represented by the water accumulating at the heading. This will need to be pumped to Shaft 5B for discharge to the ground surface.

To manage geological adversities in conjunction with exceptionally high water pressure (over 20bar) and water inflow (over 1300gpm), unique features have had to be implemented in the design of the machine, together with ad hoc procedures for pre-grouting and dewatering.

TBM DESIGN

TBM General Specifications

The NYDEQ and Kiewit/Shea were heavily involved in the specification and design of the TBM. This approach was critical to ensure the best chance of project success.

Table 1. Robbins single shield TBM general specifications and features

• Bore Diameter 21' 7" (6,583 mm) • Cutterhead Speed 0–8.8 rpm • Bolted Cutterhead Design • 9 × 330 kW Variable Freq. Drive Motors • Pressure Compensated 19" Disc cutters • Sealable TBM design – 30 bar • Stabilizers shoes on forward shield • Gripper shoes on rear shield • Skewing ring for roll correction • Large dewatering system capacity	• Mucking via muck train • Rapid segment unloaders • Stepped Shield Design • Bolted CHD design – no welding at site • Drill in ports in CHD, and Forward shield • Dedicated dewatering ports to dewater from the heading • Water handling at conveyor discharge

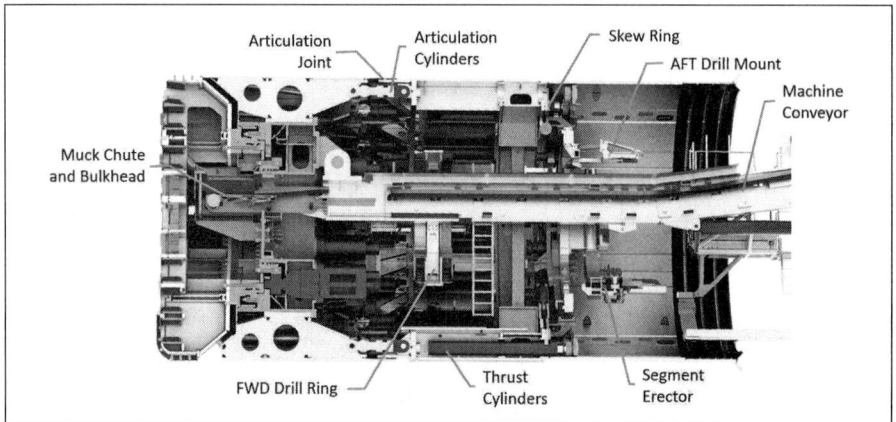

Figure 5. TBM general layout

Specifications and features of the Robbins Single Shield TBM are listed in Table 1 and the TBM general layout is shown in Figure 5.

TBM Special Features

The contractor added additional specifications for the TBM and backup systems in order to optimize the TBM with respect to drilling and grouting operations, to handle muck-laden water and to optimize operational efficiency of the backup system. De-mucking of the cutterhead was studied with a priority placed on how to seal the TBM against water inflows.

The project specifications required that the TBM be capable of withstanding 20 bar of hydrostatic water pressure with a safety factor of 1.5. Pressure compensated disc cutters were selected for the project. The special cutters are equipped with a pressure equalization system to ensure protection of the cutter bearings at high water pressure.

High Water Pressure Design Considerations

A new main bearing and sealing system design was required for the project due to the 20 bar static water pressure expected on the project. The system is comprised of normal TBM lip seals and emergency inflatable seals. The advantage of the inflatable seals is that they are not in running contact with moving parts of the sealing system during boring and can be activated when needed for additional pressure protection of the main bearing of the TBM. The seals are flushed and lubricated with grease. Grease flushing was selected because it provides better protection of the seals when exposed to the water with fines that is expected on the Rondout project.

High Water Inflow Design Considerations

In order to protect the TBM and personnel from sudden inrushes of water the TBM was designed to be sealed quickly in the event of a sudden inrush of water. The below steps are required for sealing of the TBM against inflows and are shown in Figure 6:

1. Close knife gates over muck chute
2. Retract conveyor frame
3. Retract belting out of cutting chamber
4. Retract bulkhead sealing plate
5. Close Stabilizer doors

Due to the water inflows measured during the geotechnical investigation the contract specification required a robust dewatering system for the tunnel. Two dewatering capacities were required on the TBM.

- Continuous Pumping—800 gpm during boring with no impacts on boring or ring building
- Emergency Pumping—2500 gpm emergency capacity for water from heading and construction water

The continuous dewatering system is capable of collecting water from the CHD chamber, TBM shields/ring build area and also at the transfer point between TBM conveyor and Transfer conveyors. The dewatering system is designed to transfer fines up to a ¼ inch in size from boring through the piping and tanks to the tunnel dewatering system. The TBM is equipped with two 10 cubic meter dewatering tanks with mixing pumps inside the tanks to prevent fine settlement, thus reducing shutdowns for maintenance and cleaning.

The emergency pumping system bypasses the dewatering tanks and transfers water directly from the backup system to the tunnel dewatering system via a telescoping pipe extender on the backup. A general system schematic of the dewatering system is shown in Figure 7.

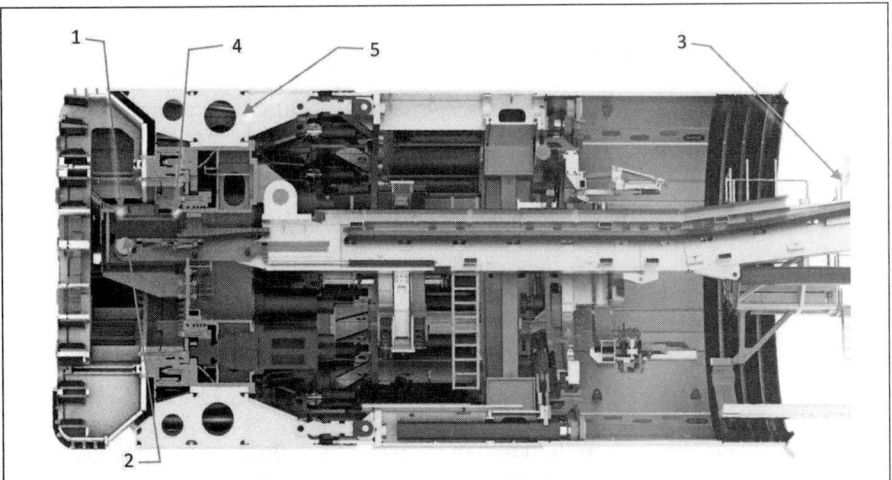

Figure 6. TBM sealing sequence

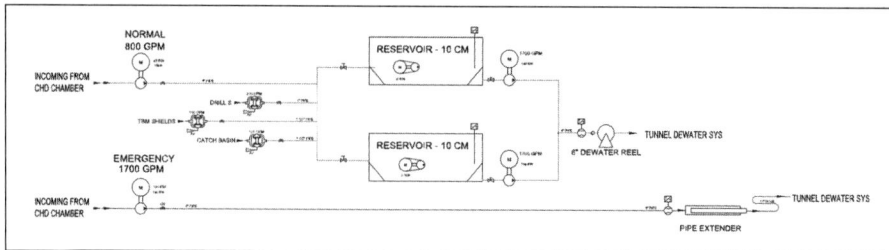

Figure 7. Dewatering system line diagram

Drilling and Grouting Methodology

The project specification requires a mandatory probe drilling program for the entire tunnel alignment that includes water inflow measurements at the probe hole locations. The TBM is required to drill 4 probe holes every 200 feet and to measure water inflows. When water inflows exceed contract allowable values grouting will be required to reduce water inflows to acceptable levels. The TBM can then advance inside the grouted area of the alignment. The TBM is equipped with two types of grouting systems. The pre-excavation grouting (PEG) system is a mono-component grout system used to grout ahead of the TBM. The two-component (A+B) grout system is used to backfill the annular gap between the segmental lining and the bored tunnel.

Drilling Systems

The TBM is equipped with two drill systems for probing and grouting operations. The forward drill system is used for drilling operations through the tunnel heading at angles of 0 degrees and up to 5 degrees measured relative to the tunnel alignment. The drill system consists of two independent drill positioners mounted on a fixed ring that can position each drill 360 degrees radially to drill and grout through 16 cutterhead drill ports. This system is the primary drilling and grouting system used on the TBM and will be used to probe and grout along the tunnel alignment.

The aft drill system is a single drill permanently mounted to the segment erector. The segment erector is used to position the drill for drilling and grouting operations through 14 peripheral shield ports. The ports are at 7 degrees to the tunnel alignment and are used for umbrella drilling and grouting to form a grout curtain to cut off water inflows surrounding the tunnel alignment.

The drills selected by the contractor are down-the-hole water hammers manufactured by Wassara. The water hammer is ideally suited to this project for the following reasons:

- Down-the-hole drilling reduces size of drill equipment inside TBM
- Water used to power drill also provides flushing of cuttings
- Drills longer straighter holes compared to top hammer type drills
- Water used to power drill will not erode borehole
- Core drill units are relatively short and compact, making them better suited to fit inside TBM
- API drill rods are used in drill string to power drill with high pressure water

Drill testing completed with water hammer drills near the jobsite area verified the drill performance and suitability for the project.

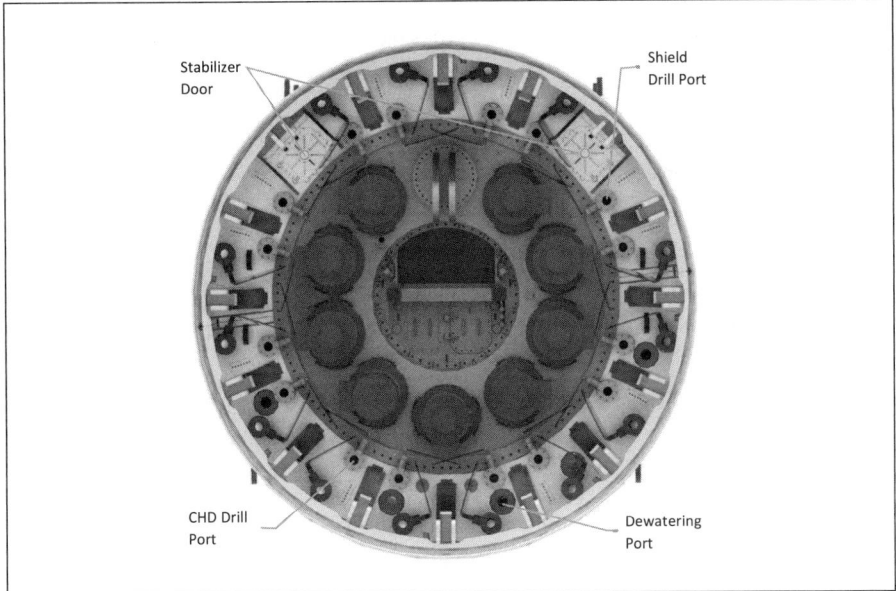

Figure 8. Forward shield drilling and dewatering ports

The bridge area of the TBM was also designed to allow for radial drilling with a portable drilling platform in order to verify backfill grouting. Forward shield drilling/grouting and dewatering ports are shown in Figure 8.

Grouting Systems

The TBM is equipped with two PEG mixing and grouting plants for grouting ahead of the TBM. In addition the PEG system can be used for proof grouting through the segmental lining to counteract high water pressure and to mix and inject bentonite around the TBM shields to reduce friction in squeezing ground conditions.

The A+B grout system is supplied from a batch mixing plant at the surface and pumped directly to TBM. A+B Storage tanks on the TBM backup are equipped with level sensors that start and stop pumps at the contractor-supplied surface grout batch plant. This ensures grout is available when required for grouting operations on the TBM.

Due to high static water pressure the contractor required the TBM to have the ability to backfill grout through the TBM tail skin and not through the segmental lining in order to increase worker safety. On previous projects grout penetration plugs have been dislodged at high velocities that endanger worker safety. Grouting through the tail skin reduces the segmental lining penetrations and improves safety by eliminating these penetrations.

Project Update and TBM Assembly Challenges

The project site presented several challenges to the assembly and launching of the TBM. Due to limited space at the launch shaft the shipping of TBM assemblies was carefully planned to coincide with the assembly sequence. The TBM is to be assembled at the bottom of a 900 foot vertical shaft. To facilitate the assembly and launch of the TBM the weight and lifting of major parts of the TBM was carefully coordinated with the contractor to ensure the machine design worked well with the site constraints.

The contractor muck haulage crane has a maximum lifting capacity of 100 US tons. The TBM components were designed so that all lifts are less than 100 US tons and will fit down the limited shaft window designated by the contractor. The TBM is to be assembled on a moving cradle at the bottom of the shaft that can then be moved to the tunnel face.

The TBM is currently being assembled at the Robbins Facility in Solon, OH and is tentatively scheduled to begin boring in the spring of 2017.

CONCLUSIONS

The Rondout West Branch Tunnel Bypass Project presents many challenges that required innovative solutions. The difficult geology and large water inflows required the TBM to have features typical to a normal rock TBM's. The high static water pressure required the TBM to have many features that are typically associated with pressurized TBMs. The TBM is to be equipped with robust drilling, grouting and dewatering systems to overcome the expected challenges of the project. In addition, custom sealing systems have been developed so the TBM can bore at atmospheric conditions and then be sealed in a short time to withstand 30 bar water pressure. The innovations developed for the Rondout project show that TBM technology continues to improve to allow hard rock TBMs to tackle more difficult geology than ever before.

REFERENCES

Dowey, T., Sozer, Z., and Brion, P. 2015. Rapid Excavation and Tunneling Conference 2015 Proceedings. Edited by M.C. Johnson and S. Oginski: Society of Mining and Engineering

Dowey, T., Kharivala, B., and O'Connor, P. 2014. Rondout West Branch Bypass Tunnel Construction and Wawarsing Repairs, Delaware Aqueduct Bid Set. New York City Environmental Protection

Water for the Future: Delaware Aqueduct Rondout-West Branch Tunnel Repair, www.nyc.gov/html/dep/html/environmental_reviews/rwb_tunnel_repair_project.shtml

Innovations on West Trunk Sewer Contract 2

Jon Hurt ▪ Arup
Jörg Riechers ▪ Herrenknecht Formwork Technology GmbH
Mike Ghasemi ▪ Technicore Underground
Tony DiMillo ▪ Technicore Underground
Vanessa DiMillo ▪ Ewing Fabricators
Ajay Puri ▪ Region of Peel

INTRODUCTION

Construction of Contract 2 for the twinning of the West Trunk sewer tunnel for the Region of Peel in the City of Mississauga consists of approximately 3,800 m (12,500ft) of 3.0 m (10ft) diameter trunk sanitary sewer. The new sewer is being constructed by tunneling using two EPB machines and installing precast concrete segments as the final liner. Two significant innovations are included in the construction of the tunnel. The first is due to the tunnel being constructed in the Georgian Bay Shale, which is known to exhibit time dependent deformation (TDD) after excavation. The forces generated by TDD are significant and are typically dealt within tunnels by delaying the installation of the final lining to allow unrestrained swelling to occur, or by use of a compressible material outside the tunnel lining. For the precast concrete segments, which are installed immediately behind the TBM a compressible grout was developed to accommodate the TDD. The second innovation is the use of Herrenknecht Combisegments® over a 250 m length of the tunnel where a corrosion resistant lining was specified in the contract. This paper discusses the development and use of these innovations.

PROJECT DESCRIPTION

The West Trunk Sewer Contract 2 tunnel project is located at in the City of Mississauga, Ontario, Canada. The project consists of approximately 3.8km (12,500ft) of 3.0 m (10ft) diameter trunk sanitary sewer to be constructed by tunneling. Contract 1 was previously completed and followed the alignment of Erin Mills Parkway from north to south, terminating at Manhole 5 (MH5), located just south of Highway 403. The Contract 2 tunnel alignment also primarily runs under Erin Mills Parkway, heading southeast from MH5 for 3.15km before turning at MH3 through a 90 degree bend, heading 0.65km northeast towards MH2 and MH1. At MH4, there is a drop shaft, with the tunnel from MH5 entering MH4 with an invert elevation of 122 m, and the tunnel to MH3 leaving MH4A at an invert elevation of 94 m, a drop of just under 30 m. MH4 and MH4A are two adjacent 3.6 mID shafts that will be constructed within the same shaft excavation with a 3.75 m connection between them. The Contract 2 alignment and profile are shown in Figures 1 and 2.

The ground conditions along the alignment consist of soil deposits overlying Georgian Bay Formation. The soil deposits are interbedded cohesive and granular layers, ranging from clayey silt to sand and gravel, and including deformation till, a deposit that includes slabs of limestone and shale at various stages of fracturing and weathering. The Georgian Bay Formation is characterized by blue-gray shale interbedded with siltstone and limestone layers.

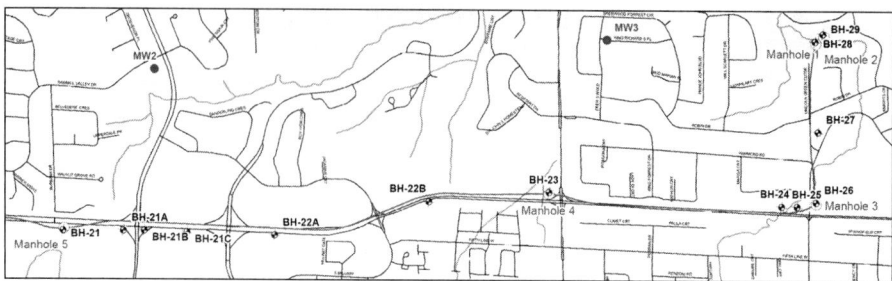

Figure 1. West Trunk Sewer Contract 2—tunnel alignment

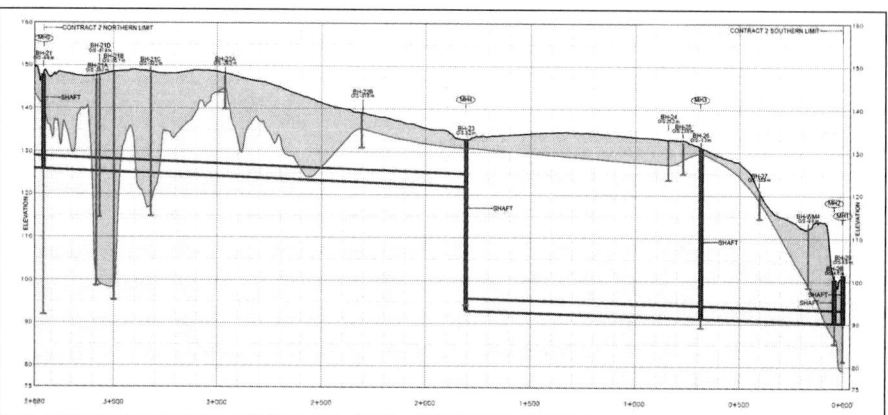

Figure 2. West Trunk Sewer Contract 2—tunnel profile (figure from GBR by Golder Associates)

The tunnel (see Figure 3) is being excavated with two drives, commencing at the two ends of the alignment and heading towards MH4/MH4A. The drive from MH5 to MH4 runs along the interface between the overburden soils and the shale, and the drive from MH½ to MH 4A is within the shale except for the initial section. While the mixed face section from MH5 was always specified to be constructed with an EPBM, the contract envisaged that the shale tunnel would be constructed with an open TBM with a cast in place lining. Given the alignment, the availability of equipment and the schedule, Technicore opted to construct both reaches in the same manner, with an EPBM and a steel fiber reinforced concrete segmental lining. Segment production was undertaken by Technicore subsidiary Ewing Fabricators (see Figure 4) with design by Arup and component supply and coordination from David R. Klug and Associates.

The use of a segmental lining, which was essential in the mixed face reach, introduced two challenges to the project—accommodating time dependent displacement in the shale without overloading the lining, and providing HDPE concrete protective liners in some sections of the tunnel. The project team developed innovative solutions to these two challenges—compressible grout and an integrated corrosion resistant lining respectively—which are described in the following sections.

COMPRESSIBLE GROUT

Approximately 3.3km of the tunnel is being constructed in the Georgian Bay Shale, which is known to exhibit time dependent deformation (TDD) after excavation. The Georgian Bay Shale unit consists of typically moderately weathered to fresh, grey to

Figure 3. Tunnel under construction

dark grey, fine to very fine grained fissile shale interbedded with slightly weathered to fresh grey, fine grained calcareous silt-stone and limestone interbeds. There are two distinctive features of the shale in the greater Toronto area. One is a high horizontal stress regime, and the second is long-term time dependent swelling behavior. The swelling is a consequence of the reduction in confined stress in the rock that occurs upon excavation in combination with a differential gradient in salinity between the saline rock pore-water and freshwater or even humid air. Osmotic and diffusive processes result in

Figure 4. Tunnel segment manufacture at Ewing Fabricators

a decrease in the salinity of the rock porewater achieved by an overall increase in the water content, resulting in volumetric expansion of the shale rock over time. The development of this time dependent deformation (TDD) relative to the time of installa-tion of the permanent lining has a direct impact on the long-term moments and forces induced on the lining.

Geotechnical investigation of the site was carried out by Golder Associates in 2009. Based on the geotechnical investigation report, time dependent deformation (swelling) of the Georgian Bay shale should be expected during and after excavation in the tun-nel. For baseline purposes, the time dependent horizontal and vertical swell rates were defined to range from 0.05% to 0.5% and 0.1% to 2.5% per log cycle of time, respec-tively. These rates are based on the swell testing that was conducted as part of the geotechnical investigation and have been adjusted upwards based on published data.

As noted in Cushing (2016), TDD movements are dependent on the excavation shape and the ratio between in-situ stresses. In particular, a high horizontal stress in shallow overburden will give a high potential for swelling. Greater investment in stress mea-surements on projects would be useful in identifying sections with high potential for damaging TDD. The in-situ stress were not measured in for this project, though it is known from tests in the area that the in-situ stress can be very high near the bed rock surface. For baseline purposes, the baseline maximum major and minor horizontal stresses were defined as between 6 MPa and 12 MPa and between 2 MPa and 9 MPa, respectively, which are at the high end of published data.

Technicore hired Arup to design the precast concrete segmental liners and evaluate the forces resulting from TDD acting on the precast concrete segments. If the shale around the tunnel was highly restrained, which would be the case if a conventional annular grout was used, the TDD would lead to high stresses in the ground. Consequently, it was decided to use a compressible grout. Based on numerical modeling of the TDD, Arup evaluated the anticipated movements and applied pressures specified the grout requirements for this project. The grout needed to have sufficient stiffness at lower stress levels to provide sufficient support to the tunnel lining, but allow sufficient deformation at higher stress levels to avoid the build-up of high loads on the segmental lining. The grout performance was specified to provide a compression capacity up to twice the anticipated time-dependent rock deformation, which was a strain level of around 8% in the annular grout in this case. Consequently, a compressible grout mix capable of supporting up to 16% compression was specified.

The roles of compressible grouting behind precast concrete segments in rock with TDD are as follows:

- Providing a stable backfill for the concrete segments
- Accommodating the deformation up to 16% of the annular space (the percentage of compressibility could be increased by changing the ratio of particles in the mix design)
- Avoiding overstress on the liners due to TDD
- Minimizing water leakage into the tunnel (as a backup to the gasket)
- Minimizing penetration of gases to into the tunnel (methane etc., if present, as a backup to the gasket)

Technicore hired BASF to prepare a design mix, using materials supplied by Technicore. The compressible grout mix was comprised of a hydraulic binding agent, bentonite clay, foam particles, water-reducing admixture, water and air. The design mix ratio was determined by BASF. The proportions of the ingredients were selected based upon the desired compressibility of the compressible grout mix in the hardened state that is in turn based on the expected time dependent deformation of the rock through which the tunnel is excavated. Six grout specimens were prepared and shipped to the Queen's University laboratory for physical testing. A test procedure was developed using a steel cylinder to provide lateral confinement to the cylinders to allow confined compression strength testing utilizing 14 and 28-day cured specimens.

For all of the testing, each sample was cut to prepare cylindrical samples having nearly parallel end faces. The samples were subjected to failure within a servo-controlled compression frame. All tests were performed under axial strain control at rates approximating $2\times10^{-4}s^{-1}$ (equivalent to an axial deformation rate of 0.033 mm/s) and, for these tests, simultaneous recording of axial force and axial deformation was performed from which determination of standard failure parameters (Young's modulus and peak compressive strength) were made. Each unconfined compression test was permitted to undergo axial deformation equivalent to 20% axial strain prior to completion of testing.

The pre- and post-test views of a confined compression test of a 28 day cured specimen of compressible grout of the present invention is illustrated in Figure 5. The strain-stress chart is shown in Figure 6.

Placing the Compressible Grout in the Tunnel

The compressible grout is being made by batching plant at the job site and transported into the tunnel in a grout car, equipped with an agitator and pump. The compressible grout is injected behind the precast concrete segments through the grout sockets. The pressure of the grout is monitored by a pressure gauge and is limited so as to not exceed of ground water pressure plus half a bar.

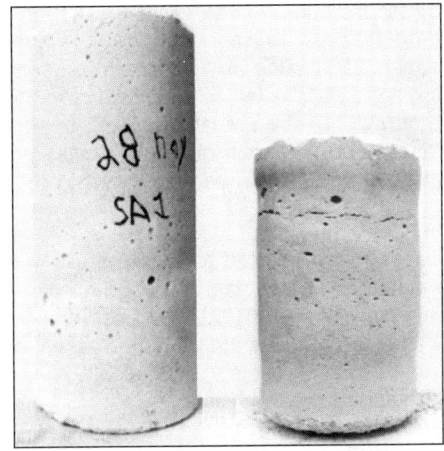

Figure 5. Pre-test (left) and post-test (right) samples of compressible grout (28 day confined compression test)

Some concerns during preparation and installation of grout in the tunnel included:

- Some percentage of the foam particles may float to the top of the grout, and for this reason mixing the elements should be done by an automated system
- A special grouting pump is required to pump out the grout to avoid clogging of the pump
- The filter of regular pumps should be cleaned frequently
- The density and viscosity of the grout should be checked at each pour.

INTEGRATED CORROSION-RESISTANT LINING

The requirement for the corrosion resistant lining was determined by the Region of Peel and their designer, WSP, during the design period. The contract documents contained detailed requirements for the location and properties of the HDPE concrete

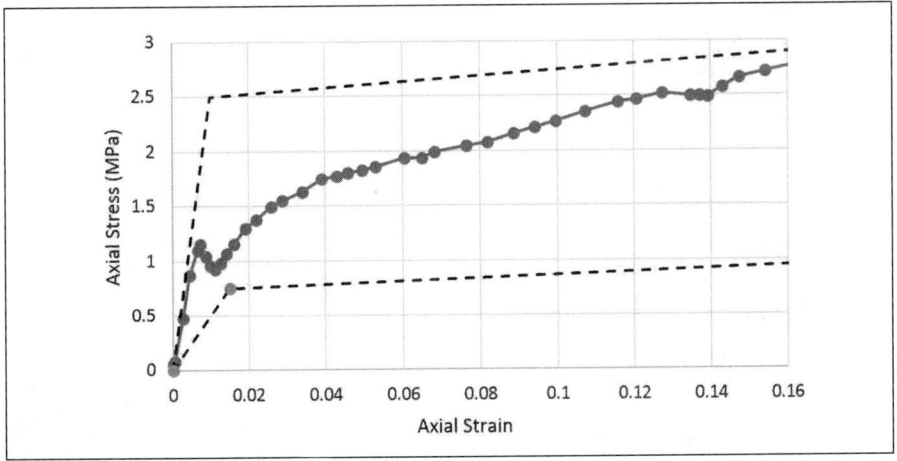

Figure 6. Typical stress/strain plot of sample failure test at 28 days cure

protective liners (CPL). In the stretch of tunnel between MH4 and MH5, the 150 m of tunnel closest to HM4A required protection. This length was always envisaged to be constructed using an EPBM with a segmental lining, but the exact method of installing the CPL was left to the contractor. Immediately downstream of MH4, in the length of tunnel between MH2 and MH4, the contract required a 100 m length of CPL. This length of tunnel was originally envisaged to be constructed with an open TBM with initial support and a secondary lining with the corrosion resistant lining incorporated.

With the decision to use an EPBM with segments along the entire alignment, Technicore reviewed the options to install this 250 m length of CPL. Since the segmental lining (based on available molds) being used had a larger diameter than the minimum specified (3.0 m ID vs 2.4 mID), one option was to form an internal secondary lining that includes the CPL as a cast in element. The other element was to use a CPL integrated into the segmental lining. On evaluation of the cost and schedule, it was decided to proceed with the integrated option based on the use of the Herrenknecht Combisegments® Type II HDPE, hereinafter referred to as "Liner System," was chosen by Technicore. This system provides compliance with design life requirements and allows an accelerated tunnel construction in a single pass. In addition to meeting durability characteristics, the Liner System provides application benefits, i.e., accelerated segment casting, handling with vacuum erectors and the feasibility to construct the tunnel in one step with a continuous protective membrane.

The CPL was specified to be formed with AGRUSAFE Sure Grip® Type 560 HDPE by AGRU Kunststofftechnik GmbH with a minimum thickness of 5 mm. The CPL system was required to be continuous and impenetrable to water and gases. All HDPE CPL liner had to be extruded with 13 mm high anchoring studs, a minimum of 420/m^2, manufactured during the extrusion process in one piece (integral) with the sheet so there is no welding and no mechanical finishing work to attach the studs to the sheet. The liner shall have a pull-out of minimum 800Newtons per anchoring stud at 23°C. The 5 mm liner shall have a "signal layer" composing of 2 mm top yellow and 3 mm bottom white. The CPL was also required to have the following properties:

- All AGRUSAFE Sure Grip® HDPE CPL and welding rods had to be manufactured from the same resins
- Density of 0.945g/cm^3 as defined by ASTM D792–86;
- Melt Flow Rate (MFR) of 0.7 -1.2g/10 min. (Cond. 190/5) as defined by ASTM 1238–88;
- Heat Reversion (Dimensional Stability) of less than 2 per cent as defined by ASTM D1204 (212°F/1h);
- Yield Stress of equal to or greater than 2,320 psi as defined by ASTM D638-89;
- Elongation of yield of equal to or greater than 10 per cent as defined by ASTM D638-89;
- Elongation at break greater than 400 per cent as defined by ASTM D638-89;
- Minimum puncture resistance 120lbs by ASTM D4833;
- Maximum working temperature of 60°C; and
- Fire Classification of HB as defined by UL - 94(DIN 4102-B2).
- checked when completed by visual checking and by spark testing all welded joints and surfaces using an approved electrical holiday detector. Spark testing shall be performed in accordance with the test equipment manufacturer's recommendations.

Components and Functions

The Liner System consists of a High Density Poly-Ethylene (HDPE) membrane for covering the main segment surface and a pDCPD frame for edge protection with an overmolded anchored gasket. A force-locking connection between material sections HDPE and pDCPD is achieved during plastic processing by means of chemical welding in the interface. During the reaction injection molding (RIM) process the gasket is integrated without additional agents or adhesives. Plastic inserts for handling and fastening purposes are integrated in the HDPE membrane, specifically adjusted to the TBM in use. Each Liner System is exactly adjusted to necessary segment dimensions. Labelling on each Liner with indication of part number, segment designation and

Figure 7. 3D ring model Combisegments® Type II HDPE

Direction of Drive (DOD) enables a confusion-free handling during casting. On the basis of a successful factory acceptance including measurements of segment molds, each Liner System and casted trial segments, a continuous quality control during manufacturing ensures dimensional stability of the Liner System.

For the amount of 210 rings of Combisegments® necessary, the Liner System consists of the following project related build-in components:

1. AGRU Sure Grip HDPE CPL, t=5 mm, coextruded (3 mm top yellow, 2 mm bottom white), 420 anchor studs/sqm, stud height 13 mm
2. Telene pDCPD frame for edge protection, with integrated
3. Dätwyler EPDM Profile M 38920
4. Telene pDCPD frame anchors
5. 2 × Optimas Type III bayonet connection adapter
6. 1 × Optimas Type III Socket, short
7. 1 × Optimas Type III Socket, long, for secondary grouting
8. 2 × Optimas Type III Screw-Caps, yellow.

Components (1) to (4): On the basis of the chosen EPDM gasket profile, the necessary pDCPD frame of the Lining System was designed to ensure a precise and long-lasting integration and to cover all segment edges. By means of this key function sealing of all circumferential and longitudinal joints is achieved through the ring build itself.

Available frame anchors are produced in the same step during plastic processing. Stabilization of the gasket zone is achieved by trapezoidal frame anchors, considering stress loads applied to segments during tunneling, such as loads induced by trust cylinders of shield tunnel boring machines, as well as during handling and transport, (Figure 8).

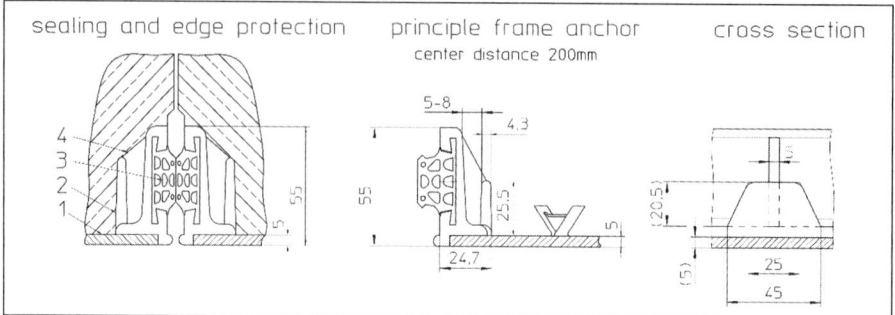

Figure 8. Detailed lining system: sealed and protected joints (left); frame cross section (right)

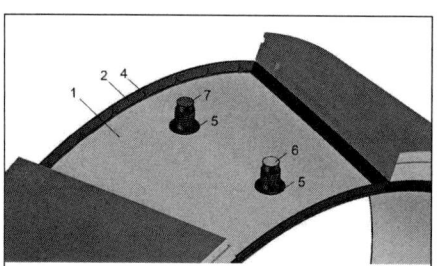

Figure 9. Overview: 3D design detail (left); liner system in segment mold (right)

Components (5) to (8): The refurbished TBM being used for excavation is equipped with a vacuum system for segment handling and erection. The vacuum plates use two shear cones that require voids in the segments at specific locations for pre-positioning of the segment and provision of additional safety during rotation and holding. There was also a need to provide a resealable grout hole for primary and secondary grouting. In conventional segments the necessary recesses for shear cones are shaped in concrete by forming parts on the bottom of the segment mold. Furthermore, plastic sockets with a tube extension are cast in for grout injection, if necessary. If this approach was used with a conventional protective liner, additional patchwork and associated welding and inspection would become necessary at each void or insert. Consequently, there was a need to integrate inserts with watertight seams and openings that are sealable without welding into the Liner System which (a) provided voids for the shear cones and (b) provided access through the segments for secondary injections.

The selected Liner System fulfills specification requirements, construction demands and the overarching objective on durability by integration of two Optimas Type III Bayonet Connection Adapters. Adapters are integrated into the tailored HDPE CPL prior to the RIM process. As a result of the bayonet connection, matching Type III Sockets can be installed on the job-site prior to concreting. Before using the vacuum plate, a metal insert is mounted in the socket for thread protection. The remaining cavities provide room for shear cones of the erector plate. After ring build, the thread facilitates water- and gastight sealing by means of a screw-cap without additional welding (Figure 9).

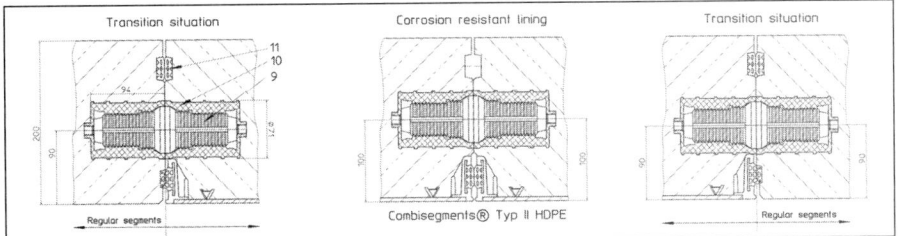

Figure 10. Ring connection—alignment of dowels in different tunnel sections

Ring Connection

Precast concrete segments are installed within the tail skin of the TBM. Positioning and connection of segments in the longitudinal joints is achieved by means of a bolt-less system using guiding rods. The solid hard rubber rods are mounted on a semi-circular notch along the longitudinal segment surface. Circumferential ring connection is achieved using the thermoplastic dowel system SOF-FIX. The system consists of a socket (10) which is cast in the segment and a matching dowel (9), installed shortly before segment erection.

In order to enable a continuous and homogeneous corrosion resistant lining, the Liner System incorporates a gasket sealing in its liner-frame. The resulting gasket position (intrados) is in contrast to the tunnel sections with regular segments (gasket extrados (11)). However, a seamless transition between the regular segments and the Combisegments® section is achieved by installation of a transition ring which is equipped with two gaskets—one on the intrados to seal against the Combisegments® and one on the extrados to seal against the regular segments. In addition, the transition ring serves to adjust the vertical alignment of the dowel system, a specific project requirement which is attributable to the reuse of existing segment molds with a different segment design for production of regular segments and Combisegments® requiring a minimum height for its liner-frame (Figure 10).

A flexible height adjustment is achieved by means of a conversion kit for Herrenknecht Formwork molds, enabling casting of both Combisegments® and adjusted Combisegments® for connection to regular segments.

CONCLUSION

Use of a compressible grout allows the placement of a relatively rigid precast concrete segmental lining in ground subject to time dependent displacement, without the lining suffering damage due to the imposed stresses. The grout mix can be tailored to provide a specified higher stiffness during initial loading, and to then allow additional deformation at a relatively low stiffness. The grout has successfully been placed in the initial sections of the West Trunk Sewer Contract 2 tunnels. Tunneling is currently ongoing.

Tunneling using the Liner System Combisegments® Type II HDPE prevents any concrete contact of aggressive gases, fluids and dissolved agents. Furthermore, penetration of substances in deeper concrete layers is hindered without any rework necessary. Structural integrity of the Liner system and reinforced concrete body is provided by integrated anchor studs and circumferential frame anchors. By means of a suitable mix design of Steel Fiber Reinforced Concrete (SFRC) the widths and spacing of any possible cracks are reduced and thus the concrete durability is improved (Siyam et al., 2014). The Liner System is currently in production after successful injection trials and subsequent trial ring approval. One set of special molds from Herrenknecht Formwork were installed in December 2016 followed by first Liners, supplied in January 2017. Use of prefabricated Combisegments® for the relevant tunnel section will commence in February 2017.

REFERENCES

DiMillo, T., Marsland, D., Ghassemi, M., Marinov, N., Harold, J., Hurt, J., Monroe, C., Hernandez, C., and Archibald, J. 2016. Development of a grout mix to be used as annular fill behind pre-cast concrete segments installed in tunnel with time dependent deformation character. Tunnelling Association of Canada 2016 Annual Conference, Ottawa.

Cushing, A., Hurt, J. and Carvalho, J. 2016. Time-Dependent Deformations of Excavations and Tunnels in the Greater Toronto Area. World Tunnel Congress 2016, San Francisco.

Siyam, A.A.F.M.; Stypulkowski, J.B. and Bernardeau, F.G. Improvements to longevity in aggressive ground conditions in the Middle East. In: Proceedings of the ITA-AITES Arabian Tunneling Conference and Exhibition 2014. Abu Dhabi, United Arab Emirates: Society of Engineers of UAE, SOE UAE Tunneling Chapter. p.163–184.

Tunneling Through a Fault Zone at West Trunk

Behzad Khorshidi ▪ McNally Construction Inc.
Alireza Ramezani ▪ McNally Construction Inc.
Nik Crawford ▪ McNally Construction Inc.

PROJECT BACKGROUND

The majority of the West Trunk sewer twinning project is aligned along Erin Mills Parkway in Mississauga, Ontario. The tunnel was expected to be driven through the Georgian Bay Formation's blue-grey shale and so the primary tunnel rock support was limited to steel channel arches, timber lagging, and rock bolts. This is overlain by the Queenston Formation's typically weaker brownish-red shale with soil above that which consists of top soil, fill material, and cohesive and granular deposits depending on the borehole. The final lining is Ø2.4m ID cast in place concrete.

The Robbins Ø3.023m main-beam tunnel boring machine (TBM) was launched from a 22m deep shaft, S5, approximately 1km south of Mississauga's highway 403. Within approximately 575m the TBM encountered the first fault zone. Dewatering was commenced and full steel sets were employed behind the TBM's limited roof shield to support the ground. Probe drilling and grouting were used to stabilize the roof ahead of the TBM. The overall length of the fault zone was approximately 20m.

A second fault zone was found approximately 1.5km from the launch shaft and 800m short of the next 36m deep shaft S6. Again, grouting and steel sets were used to control and support the ground. The overall length of the second fault zone was approximately 60m.

VALLEY DISCOVERY AND GEOTECHNICAL INVESTIGATION

What was originally assumed to be a third fault zone was encountered approximately 5.16km from the launch shaft and 350m short of the next 48m deep shaft S8. The project team attempted to continue forward in the same manner used through previous fault zones and initially succeeded in making short gains, however significant water inflow quickly stymied any further progress. The geotechnical baseline and data reports suggested significant rock cover throughout this section of the tunnel drive, though the boreholes immediately adjacent to the TBM's location were approximately 1.7km apart. The information gleaned from these borehole logs can be found in Table 1.

To complement the widely spaced boreholes in the initial geotechnical investigation a series of three additional boreholes drilled from the surface to the TBM invert elevation were commissioned to determine what lay ahead of the TBM. A summary of these boreholes' findings can be found in Table 2. Meanwhile the Ø3.023m main-beam TBM was prepared for a long-duration shutdown while water was pumped from the tunnel through an alignment hole approximately 100m back from the face.

The first three boreholes confirmed the existence of an extensive subterranean valley rather than the relatively short fault zones encountered prior. They also revealed that cobbles and boulders infilled with sand, silt, and clay lay within the tunnel alignment. This is further overlain by dense to very dense sand. Building a jet grout block was

Table 1. Rock data from boreholes immediately ahead of and behind the TBM face

Borehole	Distance from TBM Face (m)	Rock Depth (m)	Termination Depth (m)
BH12 (Shaft S7)	1340	3.3 (Queenston Formation) 29.8 (Georgian Bay Formation)	35.3
BH12A (Shaft S7)	1340	3.7 (Queenston Formation) 29.6 (Georgian Bay Formation)	70.1
BH11 (Shaft S8)	350	6.8 (Queenston Formation) 34.8 (Georgian Bay Formation)	70.1
BH10 (Near Shaft S8)	590	5.0 (Queenston Formation)	6.2

Table 2. Rock data from additional boreholes ahead of the TBM face

Borehole	Distance from TBM Face (m)	Soil at Tunnel CL	Rock Depth (m)	Termination Depth (m)
BH1	7.5	Infilled cobbles & boulders infilled with clay,silt,sand	N/A	47.4
BH2	23	Infilled cobbles & boulders	Weathered @ 47.8	49.4
BH3	40	Infilled cobbles & boulders	Weathered @ 49.0	49.1
BH4	50	Infilled cobbles & boulders	N/A	49.7
BH5	100	Infilled cobbles & boulders	Weathered @ 46.8	48.8
BH6 (Recovery Shaft)	240	Georgian Bay shale good quality	28.5	43.1

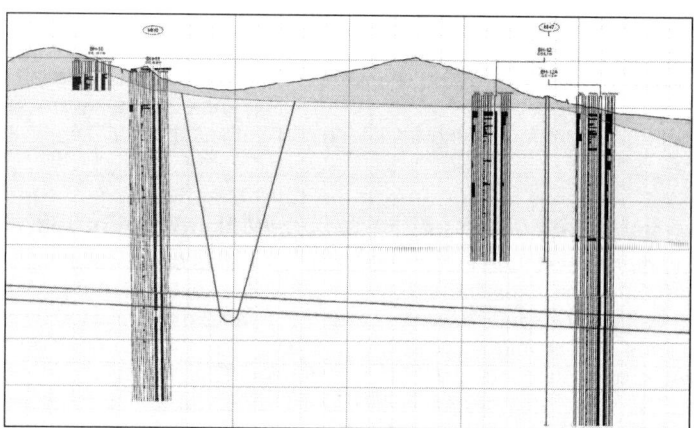

Figure 1. Original GBR soil profile showing S7 and S8 BH findings. Subterranean valley approximation added based on additional BH findings.

considered as a means to get the TBM through the valley because it would provide a competent tunneling medium, both for thrusting against (grippers) and for initial roof support (shield-less main-beam TBM). Unfortunately, additional investigative bore-holes indicated that the valley is more than 100m in width. This length of a jet grout block was deemed economically infeasible. This second set of additional boreholes revealed a similar geological profile to the prior three.

A simplified graphical representation of the valley is shown applied to the original ground profile in Figure 1.

TBM INTERCHANGE

It was decided that a shielded soft-ground TBM would be required for the short 350m drive from the next shaft, S8, back towards the northbound tunnel face. The main-beam TBM was backed down the tunnel and removed to be relaunched northbound

Figure 2. General overview of the construction area about the subterranean valley

from within the competent rock of S8. The nearest open shaft behind the TBM, S6, was approximately 2.85km back and so a recovery shaft was excavated approximately 220m from the face. It was kept back to ensure it would be excavated into competent rock as well as to minimize the effect on surface traffic because the TBM was stopped beneath the intersection of Erin Mills Parkway and Britannia Road—two major arterial roads.

The face and any voids ahead of and above the cutting head were grouted prior to pulling the TBM back from the rock face. As the TBM was pulled backward, a bulkhead was erected against the face. The recovery compound was laid out within the confines of the two left-turn lanes and median along northbound Erin Mills just south of Britannia. To accommodate this narrow compound a 6.0m wide by 8.0m long elliptical shaft was constructed to a depth of approximately 50m using secant-piles for overburden support and rock bolts with wire mesh for rock support. The approximate layout around the intersection can be seen in Figure 2.

The main-beam TBM cutting head was removed and the machine was extracted using two lifting eyes installed atop the side-supports just behind the cutting head mount. A second crane was connected to the gripper assembly to rotate the TBM to level once out of the shaft. A diagram and photo of the recovery shaft and the TBM lift can be found in Figures 3 and 4, respectively. Similar hoists were completed for each piece of the TBM's trailing gear. While out of the ground for the duration of the southbound tunneling operation, the main-beam TBM was subject to a complete servicing to prepare it for the remaining 4.26km of its northbound drive. This included the fabrication of a completely new cutting head to improve the machine's cutting and excavating efficiency.

One of the major challenges in tunneling back towards the recovery shaft was that S8, while designed for use as a mucking shaft, was not intended for TBM assembly and launch, and thus had a diameter of only 5.8m at its tightest point. To

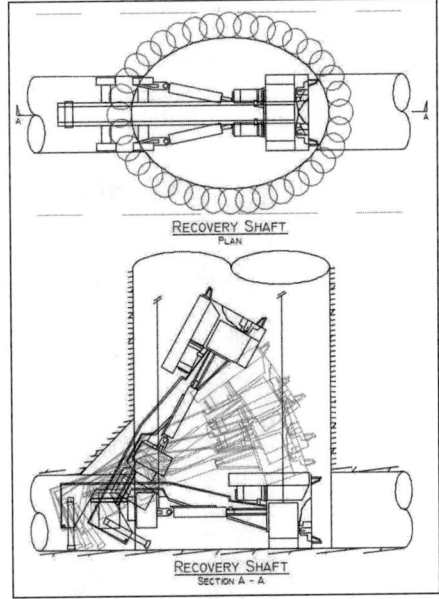

Figure 3. Main-beam TBM extraction from recovery shaft diagram

overcome this, a cavern was carved out of the competent rock at the base of the shaft complete with a 48.5m long northbound starting tunnel and a 9.2m long southbound starting tunnel. A photo of the cavern can be found in Figure 5.

A Lovat Ø3.277m earth pressure balance (EPB) TBM was already being prepared for a short 390m drive near the northern extent of the project and was thus selected as the TBM for the drive from S8 back to the recovery shaft. Although the EPB TBM was originally designed to erect and thrust against 2.744m ID concrete segments, it was being fitted with a thrust ring for pushing against steel ribs and timber lagging—a more cost-effective primary support for a final cast-in-place tunnel lining. The TBM cutting head was dressed mainly with cutters, however rippers were assigned to the gauge-cutting positions.

With no overhead crane access in the cavern, a stationary base-mounted launch frame was selected. McNally was already in possession of such a frame however it was previously designed for and used with the Ø6.2m Caterpillar EPB TBMs on

a Subway Extension project. The frame members were cut down to accommodate the smaller TBM while the overall original design and connections were preserved. Micro-piles required to tie the frame down on the previously used project were eliminated owing to the lower thrust loads and shorter overturning moment arm.

The TBM backup gantries were lined up in the northbound starting tunnel and then the TBM itself was lowered shell-by-shell on to the support cradle and pulled forward into the southbound starting tunnel. Assembly for a short-mode, umbilical launch was completed within the cavern and southbound starting tunnel. The launch frame was assembled behind the machine and a 2.4m long steel 'dummy'

Figure 4. Main-beam TBM extraction from recovery shaft photo

Figure 5. Photo from northbound starting tunnel into cavern; EPB TBM support cradle is visible

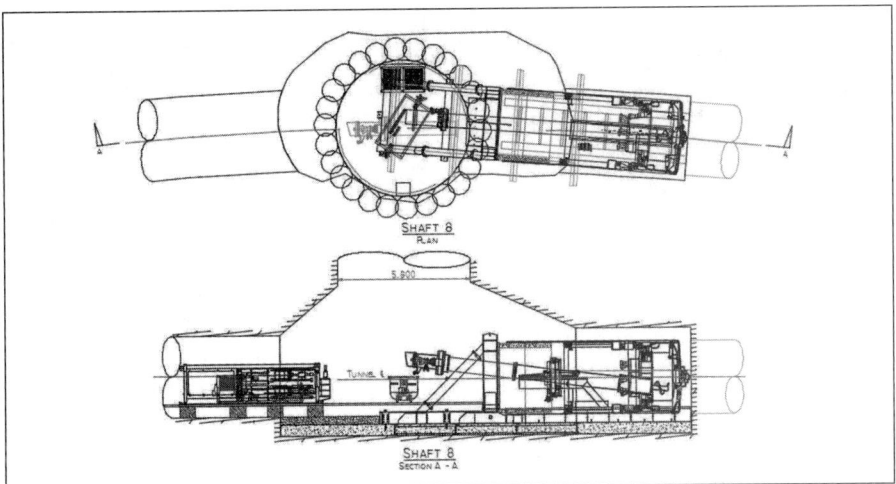

Figure 6. Diagram of the EPB TBM launch from within the S8 cavern

ring was erected within the trailing shield to begin thrusting against the launch frame. This arrangement can be seen in the diagram in Figure 6.

Because steel and timber sets are not as stable to thrust against as concrete rings are when erected outside of the bored tunnel, a limited number of concrete rings were used to launch the TBM. Eight concrete rings were built after the dummy ring in order to bury the entirety of the TBM's shields within the competent rock. Five of these concrete rings were assembled outside the bored tunnel and had to be tied down to the support cradle with synthetic slings (Figure 7) to provide the external load necessary for preventing excessive deformation and/or collapse. With severely limited access and no means to construct a bulkhead at the bored tunnel eye, the three rings within

Figure 7. Steel dummy ring and tied down concrete rings

the bored tunnel were backfilled with pea-gravel blown through the segments' grout ports. A photo of the launch frame can be seen in Figure 8.

In an effort to reduce the tunnel lining costs steel ribs and timber lagging were used as the primary lining through the competent rock. An adaptor ring was anchored to the leading edge of the eighth segmental ring which accepted the timber lagging from the first steel rib in order to adequately transfer the TBM thrust back to the launch frame. The thrust pads were removed and replaced by the thrust ring while no tail brushes had been installed prior to the launch. Two steel and timber sets were erected within the trailing shield and the machine began to push forward against the ribs. This transition can be viewed in Figure 9.

Figure 8. EPB TBM launch frame as viewed from the base of S8

Figure 9. Concrete ring to steel rib transition and TBM operating in short/umbilical mode

The competent rock was excavated in open mode with an average daily advancement rate of 7.7m. The ribs were designed for a tight clearance within the trailing shield to minimize distortion when expanded against the rock. Steel bar stock was welded at locations around the entire inner circumference of the trailing shield to act as longitudinal runners preventing the ribs from catching on the brush blocks which were left in place for easy brush installation later. With enough bored tunnel behind the TBM, the transport beam and unloading stations were installed, and the backup gantries were pulled across the cavern into the tunnel and connected to the TBM. Muck-cars were hoisted through the open tops of the gantry cars until all the gantry cars were in the tunnel.

After 158 steel and timber sets—approximately 193m—the TBM was converted back to a segment-erecting EPB machine in order to traverse the valley. The transition between support systems occurred prior to tunneling beyond the competent rock and required another simple steel adaptor ring to be installed between the ribs and segments. This adaptor ring included a built-in steel bulkhead to allow for grouting behind the segmental rings and is shown in Figure 10. The conversion also involved the removal of the thrust ring and the installation of thrust pads and tail brushes. To accomplish this, two extra steel and timber sets were excavated and then dismantled, exposing the rear edge of the trailing shield.

A compressible cellular grout was used as annular backfill through the remaining length of competent rock to allow adequate release for any in situ stresses. Typical backfill grout was used through the valley for improved ring support in the soft ground and to better protect against wash-out.

The immense wear on the cutting tools due to the initial rock excavation has resulted in very slow soft ground excavation, averaging a little over one concrete ring per shift. The average EPB pressure over the length of the submerged drive so far has been 1.5 bar. Additionally, steering the EPB TBM through the boulder-filled valley has become noticeably difficult due in-part to the worn cutting tools.

Figure 10. Transition back to concrete segments from steel ribs and timber lagging

Throughout the setup and current progress of the southbound tunneling operation, an alpine miner was employed to widen the existing 220m length of Ø3.023m tunnel north of the recovery shaft. This will allow the EPB TBM to push itself toward to the recovery shaft without the need to cut any of the rock in the previously completed tunnel.

At the time of writing—with respect to the TBM interchange operation—the EPB TBM continues to tunnel southbound whilst the main-beam TBM refurbishments were completed previously. Preparations are underway to scale back the EPB TBM operation and relaunch the main-beam TBM northbound.

RECOVERY SHAFT ROCK SUPPORT

Based on the borehole information, 2-dimensional finite element method was employed to model two different critical cross sections of the recovery shaft using RocScience Phase2 V9.0. Proper rock parameters were selected to model the rock behavior at the shaft location based on the GBR information.

For the weak rock zone, 300mm of 30MPa shotcrete complete with two layers of 102mm × 102mm × 4mm welded wire mesh was specified. For the strong rock zone, a 2.0m × 2.0m pattern of Ø22 mm × 1,800mm long rock bolts complete with 150mm × 150mm × 6mm welded wire mesh was specified. The finite elements models based on these rock support designs can be found in Figure 11. Their results can be found in Figures 12 through 15.

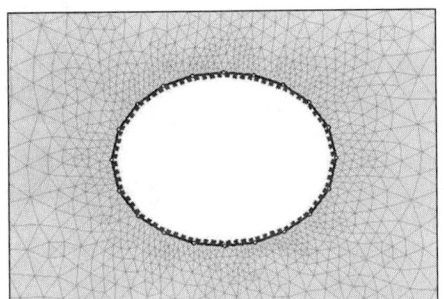 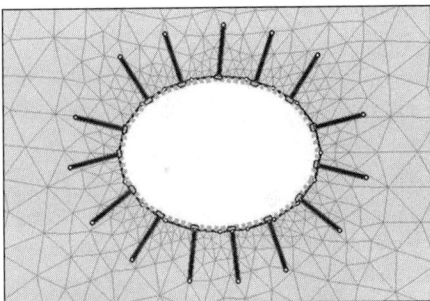

Figure 11. Weak and strong rock finite element models, respectively

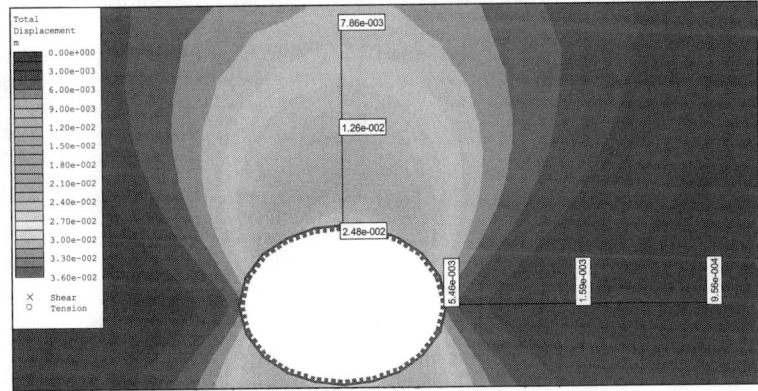

Figure 12. Weak rock displacement contours

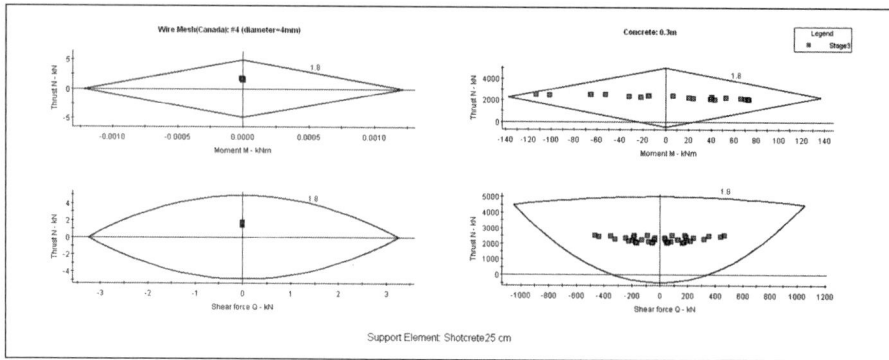

Figure 13.Weak rock shotcrete and wire mesh capacity diagrams

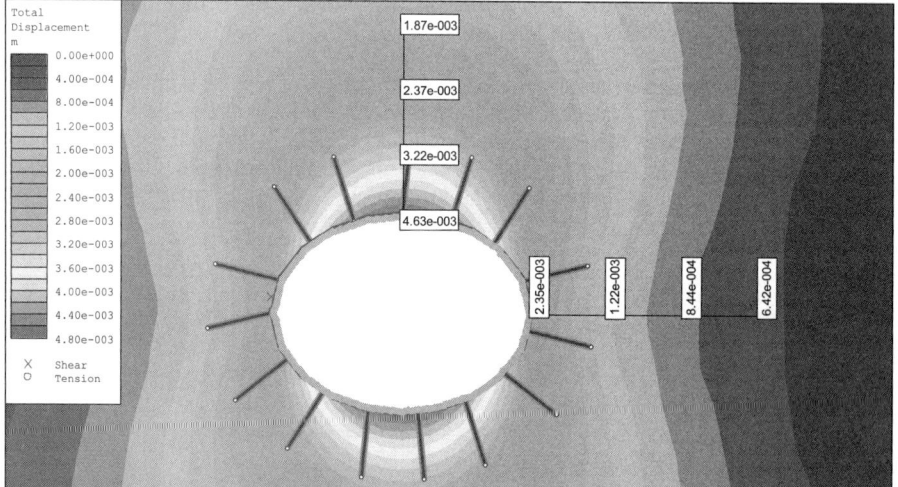

Figure 14. Strong rock displacement contours

SWELLING PRESSURE

The concrete rings erected within the in Georgian Bay formation are anticipated to be subject to the formation's well documented swelling pressure. The design of the rings needed to consider every load case including swelling pressure. The swelling potential of the Georgian Bay formation is stress dependent. Based on the data from laboratory testing and a review of published information (Zalucki et al., 2012), the concrete rings should be designed for an additional external pressure of approximately 0.3 MPa to 1.8 MPa in the horizontal and vertical directions, respectively.

CONCRETE RING DESIGN

Manual calculations employing the Muir-Wood method as well as finite element modelling were used to determine the internal forces in the concrete rings. These were then compared with capacity of the rings to determine the suitability of the concrete rings for use within the competent rock. In the series of calculations, two different loading scenarios are chosen: first it is assumed that the swelling pressure is carried entirely by ring and second a compressible grout is introduced between the rock and the ring to release the swelling pressure causing little to no load transfer to the ring itself. The results show, predictably, that the first loading scenario leads to excessive

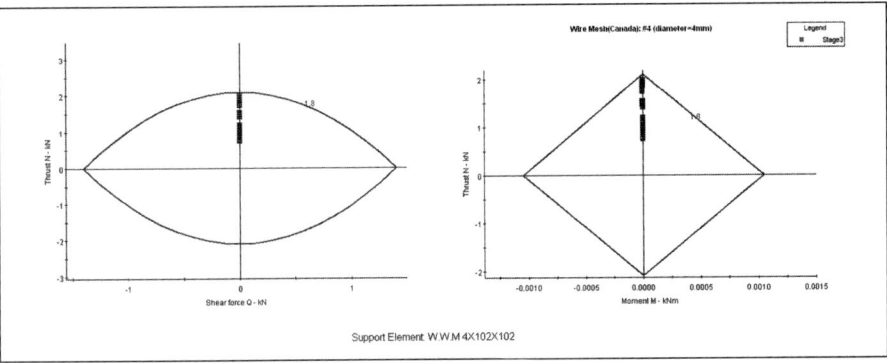

Figure 15. Strong rock wire mesh capacity diagram

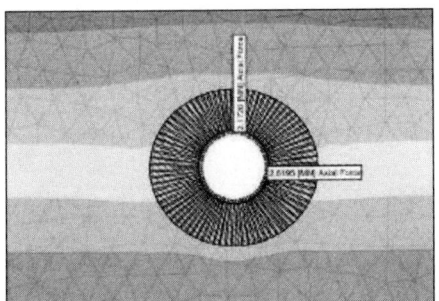

Figure 16. Axial load distribution based on finite element analysis

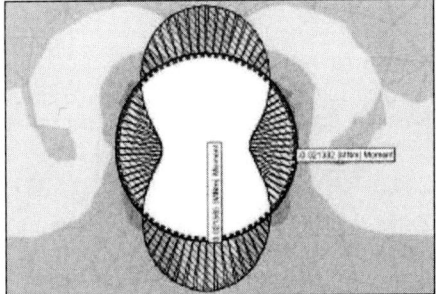

Figure 17. Bending moment distribution based on finite element analysis

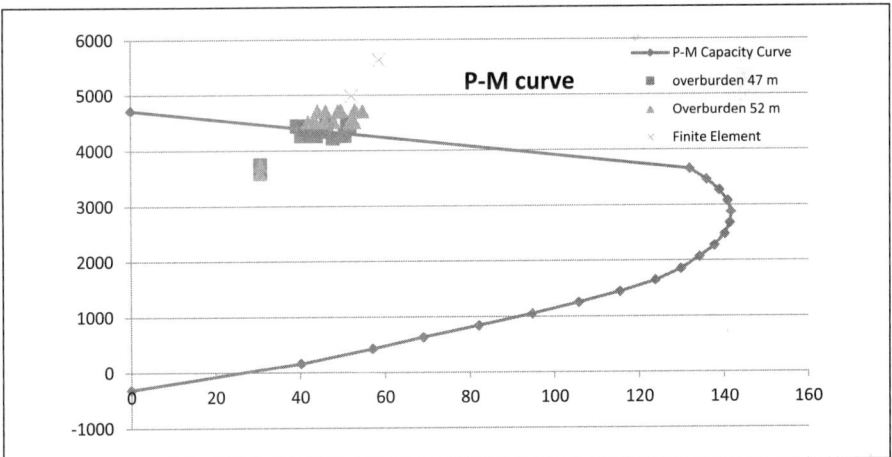

Figure 18. Concrete ring compression-moment interaction curve versus the first loading scenario

internal forces as a result of swelling pressure whereas the second loading scenario is within the safe limits of the precast concrete ring. This confirms that the first loading scenario should be mitigated through the use of compressible grout. The axial and bending moment finite element results are shown in Figures 16 and 17, respectively. Compression-moment interaction curves for the concrete rings for each loading scenario can be seen in Figures 18 and 19.

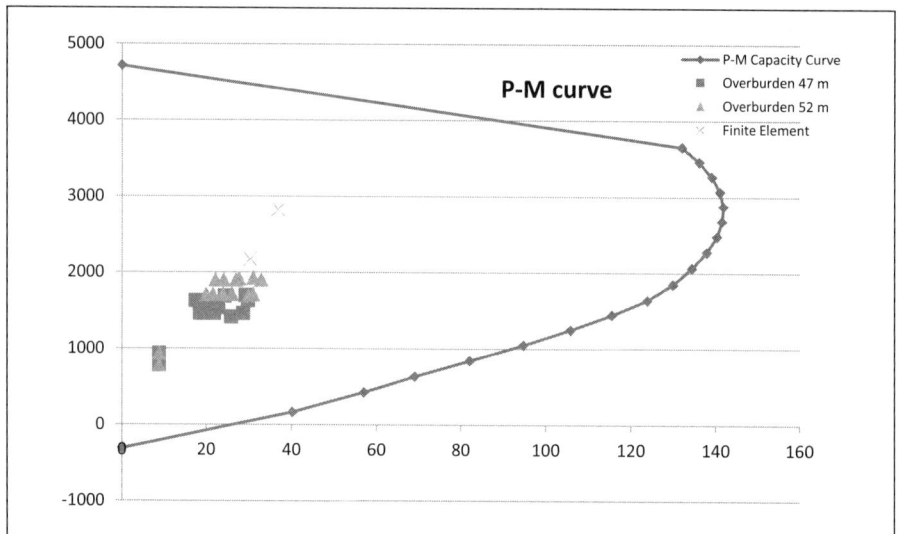

Figure 19. Concrete ring compression-moment interaction curve versus the second loading scenario

REFERENCE

Zalucki, T., Horwitz, A., Poschmann, A.S., and Telesnicki, M.J. 2012. *Design Recommendations: Detail Design and Contract Administration for the West Trunk Sewer, Project 08-2205, Region of Peel, Ontario*. Technical memorandum, project no. 09-1111-6069. Golder Associates Ltd.

Tunnel Rehabilitation

Chairs

Pierre Ciuffarin
Frontier-Kemper Constructors

Shaun Firth
CH2M Hill

Sumner Tunnel Rehabilitation

S.C. Quinn ▪ WSP | Parsons Brinckerhoff
J. Rigney ▪ MassDOT Highway Division

ABSTRACT

First opened to traffic in 1934, the Sumner Tunnel currently carries traffic travelling from East Boston to Boston under the Boston Harbor. It was constructed using cut and cover construction techniques near the portals. Between the cut and cover sections the tunnel was bored/shield-driven to form a 31-ft diameter tube using steel liner plates to support the ground and later reinforced with concrete. Given the age and constant use, the Sumner Tunnel concrete liner, invert slab, ventilation system, electrical system and drainage system are in dire need of repair and upgrading. MassDOT contracted with WSP | Parsons Brinckerhoff to prepare construction documents for a complete "gut renovation" and upgrade.

INTRODUCTION

The Sumner Tunnel was designed by the Boston Transit Commission and first opened to traffic in 1934 as a two-lane tunnel with bidirectional traffic. In 1961 the parallel Callahan Tunnel was opened to provide two lanes of traffic from Downtown Boston to East Boston and Logan Airport and the Sumner Tunnel was reconfigured to provide two lanes of unidirectional traffic from East Boston to Downtown (Figure 1). The tunnel consists of an open approach section leading to a cast-in-place concrete box tunnel to a 27 foot long construction shaft. The circular section was initiated in this construction shaft in East Boston and driven 4,800-ft to Downtown Boston and terminated into an approx. 400 foot long cast-in-place concrete box section.

With the exception of where the tunnel crosses a vent building or the construction shaft, the tunnel is comprised of either a 31-ft diameter circular or 13-ft × 26-ft box section. Figure 2 shows the typical cross sections from the original 1930s drawings.

The existing ventilation system of the tunnel consists of full transverse ventilation. It contains a hung ceiling that forms an exhaust duct above the roadway and a roadway

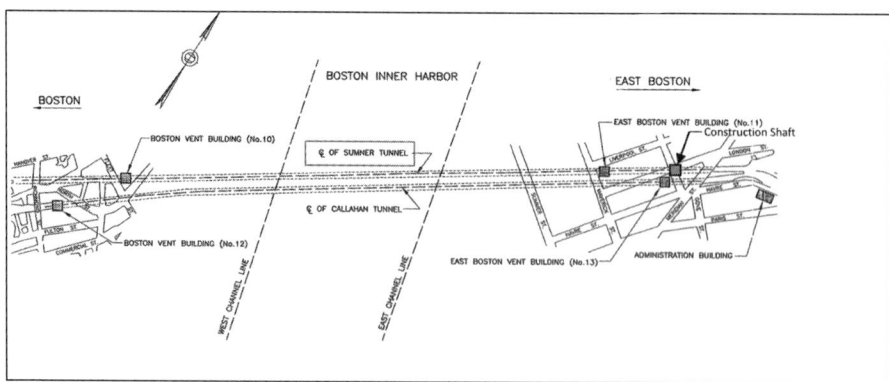

Figure 1.

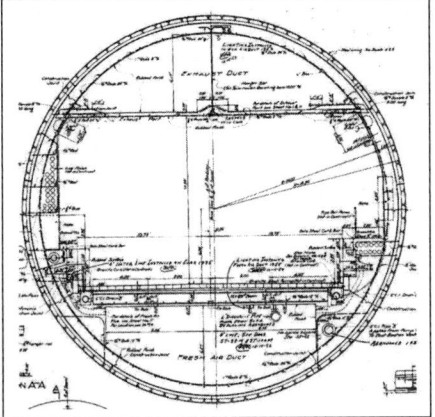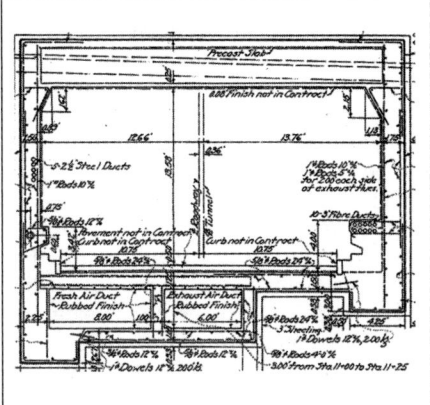

Figure 2.

deck that acts as a bridge over a fresh air supply duct. Fresh air is supplied to the roadway through vent holes in the barriers and exhaust air is drawn from the roadway through small openings in the ceiling slab. A ventilation building on each end of the tunnel contains two supply and three exhaust centrifugal fans that control the exhaust and fresh air flow. Ventilation buildings also house all the main electrical equipment that powers the tunnel fans, lights and pump stations.

Tunnel drainage lines are installed at each curb with inlets spaced along the tunnel and catch basins at the interfaces and portals. Three pump stations are associated with the tunnel (one storm water pump station at each portal and a mid-river/low point pump station). The mid-river pump station was built in the fresh air duct at the low point collecting the tunnel drainage water and pumping it to the portal and out to the surface. The Boston side pump station seized functioning throughout the tunnel's service life but tunnel's serviceability was not impacted.

Fire/life-safety systems within the tunnel include CCTV cameras, a fire stand pipe system, two way radio, roadway telephones, overhead vehicle detectors, portal closure traffic signals and AM/FM radio.

Since the opening of the tunnel in 1934, the tunnel underwent various rehabilitation projects that include but are not limited to:

- Modernization and reconfiguration in the early 1960s that included removing the original cobblestone wearing surface and replacing it with bituminous pavement and replacing the 4 inches thick concrete ceiling;
- Tunnel wall tile repairs and the installation of polymer panels on the barrier faces in 1975;
- In 1994, CCTV and an emergency telephone communication system were installed and the fire protection standpipe system was replaced;
- In 1996, the wall panels and hung ceiling was replaced with fire rated materials with an architectural finish. During this 1996 upgrade, some concrete repairs were also initiated, however not completed. Deteriorated concrete was removed in the arch but not replaced leaving much of the reinforcement steel exposed and corroding. In 2000, ventilation and electrical upgrades were completed. The upgrade consisted of a total replacement of the tunnel fans.

As part of the work performed by Parsons Brinckerhoff, to design the renovations to this tunnel, PB started with a detailed structural, architectural and systems inspection. This inspection demonstrated that approximately 30% of the concrete liner within the exhaust plenum was delaminated and or spalled with the transverse reinforcement being exposed and attacked by corrosion. Due to its deterioration, the most important portion of the rehabilitation project will consist of the repair of this liner. Since the ceiling creating the exhaust air plenum will be removed to perform this repair a complete closure of the tunnel will be required. To limit distribution to the public, it was determined that the complete closure of the tunnel had to be limited to four months. Therefore the design was developed around repairing this concrete with structural shotcrete instead of the more conventional form and pour method. It should be noted that an evaluation of the tunnel steel liner plates (used as a support of excavation for the tunnel construction) and the secondary 18" thick concrete liner can each support the tunnel design loading. Thus, shallow depth demolition prior to shotcreting of the tunnel liner will not compromise the structural integrity of the tunnel.

A roadway rating analysis also demonstrated that the existing roadway slab and floor beams were not structurally adequate for supporting HL-93 loading. Therefore the construction scope will also contain upgrading the roadway slab.

Other scope items include replacing the wall panels, rebuilding the sidewalks and lining them with polymer concrete panels, installing a new tunnel lighting system, CCTV cameras, drainage inlets, rebuilding the mid-river pump station, installing a new fire alarm and fire protection system, upgrading vent building electrical systems and investigation of removing the hung ceiling and installing ten jet fans to provide emergency ventilation.

The current design was brought to 100% including plans, specifications, a construction cost estimate and schedule. Upon reviewing the complete construction cost estimate the client requested that a value engineering or cost saving scheme be developed to cut 30% of the costs while still achieving the major results originally scoped in the contract. A cost saving scheme was therefore developed with the following work tasks modified from the original design:

- Repair of the concrete arch only in areas that are showing delamination or spalling—(30% of the area in the exhaust duct)
- Reinstall a new ceiling to protect the travelling public from future spalling in areas that were not repaired. Leave in ceiling under the ventilation buildings.
- Do not install jet fans and the decorative metal screen
- Replace the detailed installation of the south wall chase with a concrete duct bank in that area
- Replace new wall panels including costly stainless steel unistruts and anchors with painted spray on fire proofing
- Remove precast panel replacement work on the walls in the approaches

Currently both the original design and the savings option are being contemplated as options. No decision has been made on the resolution of this cost savings scheme and both repairs and upgrades will be described below.

REPAIRS

Structural

Arch of Circular Tunnel

The first item in the construction sequence will be the removal of the suspended ceiling and the existing systems that are used to support the lighting and other electrical mechanical equipment. As part of the removal of the suspended ceiling the existing anchorages used for the permanent and temporary ceiling supports will be removed, and disposed offsite. The suspended ceiling removal should be reasonably quick since the ceiling was assembled with a minimum of bolted connections and yoke and pin connections used for the ceiling hangers. The epoxy anchors will be cut flush and embedded into the repair material. The arch repair will begin with the removal of the unsound concrete and repairing of existing reinforcement steel. Depending on the option chosen, this repair will either be performed on the entire arch (from 10 am to 2 pm) or only on the spalled and delaminated portions. This will be done using the hydro-demolition process where high pressure water is applied to the concrete surface removing any unsound materials and cleaning the reinforcement to a white metal condition (Figure 3).

Exposed reinforcement will be protected with a zinc rich coating. New galvanized welded wire mesh will then be installed in the areas that are to be repaired using "J" hooks at 18" spacing. Also depth pins will be installed to guide the shotcrete applicator to obtaining a uniform depth of the repair. Following the final saturation of the surface to be repaired to SSD, shotcrete will be installed using a pre-packaged fiber reinforced Portland cement mortar to a depth of 3" to 4". If the savings option is chosen, a new ceiling will be installed following completion of the repairs to the ceiling. Currently a structural beam system is envisioned spanning from wall to wall and resting on corbels which will be integral with the tunnel liner. Precast concrete panels will then be installed on top of the beam system creating a new exhaust plenum. No hangers are required for this design creating a system that is supporting the ceiling in shear and bearing.

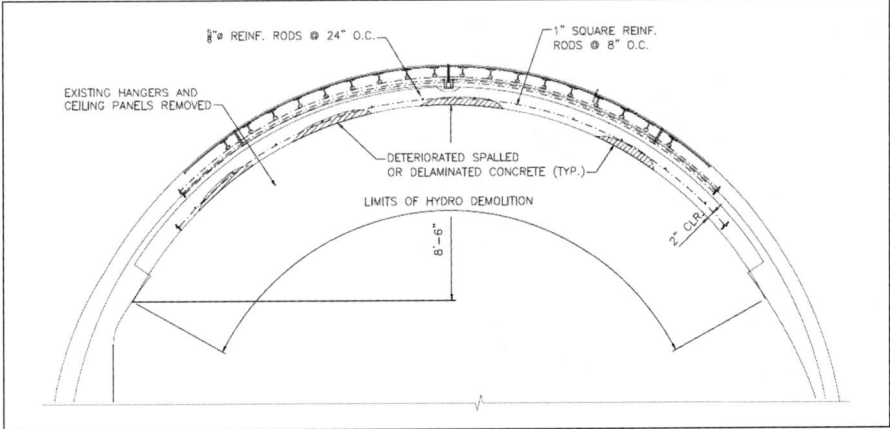

Figure 3. Existing arch conditions

The arch will be protected from heat during a fire event either through the use of another hung ceiling or by use of polypropylene fibers introduced into the shotcrete mix (which will melt under a fire event and allow the concrete to expand—acting similarly to air entrainment which alleviates stress in concrete pores under extreme cold temperatures).

Roadway Slab

The existing roadway slab typically spans 18'-0" in length and is simply supported. It is comprised of concrete encased W10×89 structural steel sections spaced at 5'-0" intervals along the longitudinal ring section of the tunnel. The concrete encasement contains additional reinforcement bars and is 14.88" thick with a typical clear cover 2" above and below the top and bottom flanges. See Figure 4 for typical cross section from the original 1930s drawings.

The existing floor beams appear to have been originally designed for H10 highway loading and are not structurally adequate for supporting HL-93 loading. Therefore the following upgrade is proposed. After removal of the existing roadway pavement the concrete overlaying the floor beams will be removed using hydro- demolition leaving the top of the concrete between the beams with a roughness of ¼" amplitude. Then a 4 ½" inch thick layer of high performance concrete reinforced with galvanized welded wire fabric will be cast. The top of the new slab will be waterproofed and paved using HMA. See Figure 5 for proposed detail.

Sequencing

Sequencing of these two major work tasks was coordinated to be performed one lane at a time and half a tunnel at a time to ensure continuous access for supply of the work to the jobsite and access for emergency responders. The south side of the roadway and ceiling will undergo reconstruction first with a staggered start to the north side.

Jet Fan Installation

The installation of jet fans is being proposed as part of the original project design. The removal of the existing ceiling panels will change the full-transverse ventilation system to a modified transverse and longitudinal system. The purpose of the jet fans is to compensate for

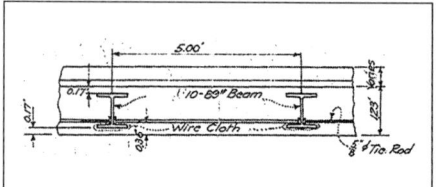

Figure 4. Section

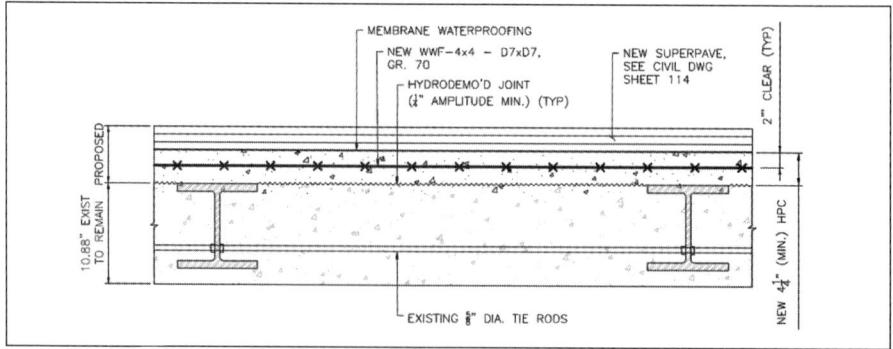

Figure 5.

any reduction in the required performance of the tunnel ventilation system in emergency conditions. This will be discussed further in the ventilation section below.

The proposed location of the new jet fans will be in the arch of the ceiling above the centerline of the tunnel; one jet fan per location. The current design consists of removing two 2 foot wide strips of the concrete liner at 11 foot spacing and installing new concrete beams/columns in its place. Steel embedments will be centered in the arch of the tunnel to support the steel jet fan support beams. Should the decision be made to reinstall a new ceiling, then these fans will not be installed.

Electrical

Two ventilation building electrical systems are supplied at 13.8 kV by two separate power feeders at either end of the tunnel. These feeders provide power to original Westinghouse metal-clad 15kV switchgears from the 1930s. Electrical upgrades were designed to handle the ten new jet fans and new tunnel lighting. Four new motor control centers are proposed to be added to serve the new proposed 50 HP jet fans, two in each ventilation building. Some shifting of loads from one switch board to another is also planned to keep the existing exhaust and supply fans operational in an emergency condition. All new installations were designed with future growth in mind and will be per the NEC code. These upgrades will be greatly diminished should the savings option scheme be chosen.

Communications

As part of the communications upgrades, this contract will replace the existing fixed closed circuit TV cameras with 15 new pan tilt zoom cameras mounted either in the ceiling or on the side walls of the tunnel. They will be used for automatic incident detection and traffic surveillance. A new SCADA system will be installed to control tunnel lighting, low point pumps, jet fans, air quality sensors, portal closure traffic signals and vehicle detectors. A new AM/FM rebroadcast system which will employ an in-tunnel commercial AM/FM Radio Rebroadcast System with emergency override capability to allow the Highway Operations Control Center (HOC) to broadcast in-tunnel emergency messages to the commuters over the in-car commercial radio receiver for traffic advisories and alerts in case of emergency. An In-tunnel Broadcast Monitoring System (IBMS) will also be installed in order to allow the tunnel operator personnel at the Highway Operations Control Center to listen and verify the content and quality of the AM/FM rebroadcast transmission within the tunnel facilities. The existing two way radio will be upgraded to allow interagency two-way radio communications among all agencies responding to an emergency, to communicate with their individual dispatch facilities, and between their mobile response units, from all points within the tunnel. A new fire alarm system will be installed consisting of 30 pull stations located on the high walk and connected to fire alarm control panels in both ventilation buildings alternately.

Lighting

The new tunnel lighting design was performed by following the standard practices defined by the American National Standards Institute and Illuminating Engineering Society, AAHTO and the National Fire Protection Association. The purpose of a tunnel lighting system is to provide an approaching motorist adequate contrast to perceive hazards within the tunnel, as well as sufficient illumination to maintain a consistent traffic flow throughout the tunnel. During daylight hours, a driver's adapted state is such that the tunnel interior can appear like a "black hole," and hence, obstructions within the portal become less visible. Lighting must be divided into zones to account for this occurrence with the lighting at the entrance portal to mimic daylight and to then reduce with distance from the portal. The tunnel environment is such that lighting equipment needs to be durable having a longevity of 20 years and to withstand vibration, abrupt temperature shifts, dirt and particulate build-up, direct water spray, and galvanic degradation of the materials. The fixture components are designed to be replaceable without tools, to reduce the duration of exposure a maintenance crew will have to traffic. Light Emitting Diodes (LED) sources meet these requirements and have the ability to be adaptive to the lighting requirements through modulation of the light output. LED tunnel luminaires manufactured by HOLOPHANE or similar are currently specified for use in the tunnel to be mounted on the sidewalls of the tunnel at the haunch area to remain out of the striking zone. HOLOPHANE uses a glass lens and an internal reflector system with LED arrays. Wired lighting control system panels will be mounted in both ventilation buildings that will allow the operators to dim light levels as lighting levels outside the tunnel change throughout the year.

Mechanical

Both the fire protection system and tunnel drainage system will be replaced. The majority of the main standpipe currently consists of a 6" pipe that is located in the supply air duct below the roadway however there are sections of the standpipe that are 4" pipe located on top the safety barrier along the left travel lane in the cut and cover tunnel sections at the tunnel entrance and exit. The standpipe main has 2-½" branch pipes that serve the hose valves in the tunnel. The branch pipes from the 6" pipe section located in the supply air duct are routed through existing supply air flues up to hose valves within the tunnel roadway area. The branch pipes to the hose valves from the 4" pipe on the bench wall are fairly short runs as they are not routed through the supply air flues. The new fire protection system will consist of a 6" main standpipe with hose connections along the south wall of the tunnel and fire department connections outside at the ventilation buildings.

The tunnel drainage system consists of drainage inlet structures located at intervals along the roadway that feed into a collection pipe, embedded drainage collection pipes that serve as the conduit from the drainage inlets to the low point pump station, low point pump station consisting of pumps and related controls, duct sump drainage system that collects water in the supply duct and discharges it to the low point pump station and a discharge force main pipe from the low point pump station to the East Boston Ventilation Building. The entire system will be replaced in kind. Drainage inlets and catch basins will all be replaced during this renovation while the roadway slab is being upgraded. The mid river pump station, located at the low point of the tunnel will be completely gutted and rebuilt with new pumps and controls.

Ventilation

The purpose of the full-transverse tunnel ventilation system is to dilute the vehicle exhaust emissions to safe levels for motorists using the tunnel during normal tunnel operations and to manage the spread of smoke and heat in the event of vehicle fire (i.e., emergency operations).

During normal tunnel operations, the ventilation system is operated in a "balanced" mode; that is, equal quantities of air are supplied to and exhausted from the tunnel. The required dilution is provided by the outside air supplied to the tunnel. This supply air is routed to the tunnel through evenly spaced flues which supply the air from the supply plenum below the roadway. Four supply fans are located in the ventilation buildings. The function of the exhaust air duct is to remove the vehicle exhaust-laden air from the tunnel roadway thereby reducing the quantity of tunnel air discharged through the Boston portal (i.e., the exit portal). However with free-flowing traffic, there is always a significant amount of longitudinal airflow through the tunnel that is generated by the "piston action" of the moving vehicles.

In the event of a vehicle fire, the ventilation system is operated in a mode that creates a longitudinal airflow through the tunnel in the direction of traffic movement while exhausting smoke via the exhaust air duct. The actual fan operating mode depends on the location of the vehicle fire.

In the original design scheme the aging ceiling system is being removed which will remove the exhaust air plenum from the ventilation system. The proposed design scheme allows the existing supply and exhaust fans to continue to handle the ventilation requirements during normal tunnel operations and will keep the supply air functioning and the exhaust fans running but without a plenum. The removal of the ceiling panels will alter where the exhaust air is extracted from the tunnel, but it will not affect the dilution capability of the existing fans. The dilution of the vehicle emissions is provided by the supply air system which will remain unchanged.

To remove smoke from the tunnel in an emergency—fire event, ten jet fans will be installed at a spacing of about 350 feet. These jet fans along with the existing exhaust fans will be controlled to direct the direction of smoke removal. Jet fans are efficient for generating longitudinal airflow through a tunnel. In addition, the airflow and air velocity along length of the circular cross-section tunnel will be constant. By contrast with the existing ceiling panels in place, the longitudinal airflow and velocity vary from point to point along the roadway, since tunnel air is being removed continuously from the tunnel. Therefore, the addition of jet fans will enhance the ventilation system's smoke control capability over the existing capability.

In the design of the savings option a new ceiling is being installed thus making the jet fans impossible and unnecessary to be installed. With a new exhaust plenum the current ventilation scheme will remain as is.

Finishes

The tunnel finishes consist of the majority of items visible to the motorist, and therefore carry special significance for appearance, ease of cleaning, and durability. All finishes items are subject to the provisions of NFPA-502, Standard for Road Tunnels, Bridges, and Other Limited Access Highways. Many of the existing finish materials date from the tunnel rehab effort in 1996, and are nearing the end of their functional lifetime. The two major finishes systems in the tunnel are the porcelain enameled ceiling panels and wall panels. The ceiling system along with its hangers and stringers will be removed in the original design scheme and replaced with a precast panel system in the savings scheme. The wall panels which are supported from the concrete liner along the walls of the tunnel will be removed and replaced with new panels and a stainless steel support system in the original design scheme.

A fire proof chase is envisioned along the south wall which is created using fire proofing board materials and the new wall panels on a stainless steel support structure. This chase would house a new electrical duct bank to replace conduits that are currently housed in the exhaust and supply plenums and to supply power and controls to the new jet fans and communication system. In the savings scheme option this chase has been replaced with a concrete encased duct bank along the south wall housed in the infill area above the safety barrier. Wall panels and fire proofing boards have been

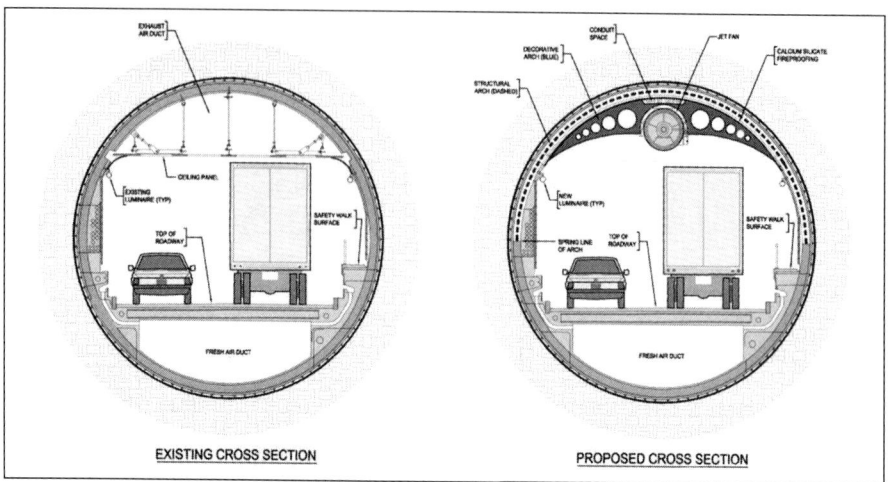

Figure 6.

replaced with spray on fire proofing and ceramic paint. There are existing 1-½" thick polymer concrete facing panels applied to the concrete traffic barriers on either side of the roadway. These will be replaced by new polymer concrete panels shaped in a safety barrier shape on both sides of the tunnel roadway. The bottom of the barrier will be tied into the upgraded roadway slab.

The stainless steel handrails mounted on the high walk on the north wall of the tunnel will simply be removed, cleaned and re installed as they are in good condition. Utility cabinets will all be constructed using stainless steel materials to ensure a durable and long life in the harsh tunnel environment.

CONCLUSION

PB will prepare the final bid package when it is decided whether to perform the complete gut renovation repair or instead the cost savings scheme (Figure 6). Construction is anticipated to begin in 2018.

The Arlberg Tunnel Project—A Milestone in the Austrian Efforts to Increase Safety of the Road Tunnel Network

Michael Hoellrigl ▪ BeMo Tunnelling GmbH
Norbert Fuegenschuh ▪ BeMo Tunnelling GmbH
Christoph Wanker ▪ ASFINAG Bau Management GmbH

The Arlberg Road Tunnel with a length of almost 14 kilometers is both, the longest road tunnel in Austria as well as the most important east to west connection between Tyrol and Vorarlberg. Built between 1974 and 1978 and after being in service for almost 40 years the tunnel is currently undergoing a major renovation program. This includes construction of additional emergency bays, emergency galleries, caverns for new electrical equipment and a complete replacement of the electromechanical equipment.

INTRODUCTION

The Arlberg Road Tunnel is the only all-year-connection between Austria's federal states of Tyrol and Vorarlberg. A road across the 1,800m high Arlberg pass can be used during summer months but gets closed in winters due to sometimes extreme snow and avalanche conditions.

To get in compliance with Austria's "Tunnel Safety Law" (STSG) requires a general renovation with completion no later than 2019. Current traffic frequency and forecast values make it possible that the Arlberg Road Tunnel will also in the future be operated as a two-lane, single tube traffic tunnel. Up to 8,000 vehicles pass the tunnel daily, forecasts for 2025 indicate that this will increase slightly to 9,500 vehicles daily. After the current general renovation, the Arlberg Road Tunnel will be the safest of its sort.

A special feature of the Arlberg Tunnel scheme is the existence of the operational Arlberg Railway Tunnel which was built in the 1880s running parallel to the road tunnel. Already between 2004 and 2007 a total of seven escape tunnels have been built between road and railway tunnel with an average spacing of approximately 1,500 meters. Latest EU regulations require additional escape routes for the road tunnel with a maximum spacing of 400 meters.

PROJECT OVERVIEW

The concept for the escape passageways includes 37 new walkable and barrier-free escape tunnels, which mainly run from to the traffic lane level up to the supply air duct on the intermediate ceiling. From there the escape routes either lead to the so called collection caverns or directly to the tunnel portals.

Eight new emergency bays (see Figures 5 and 6) on the northern side wall of the tunnel have to be constructed during the ongoing renovation, in addition to the 16 existing emergency bays on the Southern side wall. According to current safety standards new fire extinguishing units (niches) will be built in both, the new as well as the existing emergency bays. With the renewal of the drainage gutter the previous combined-water

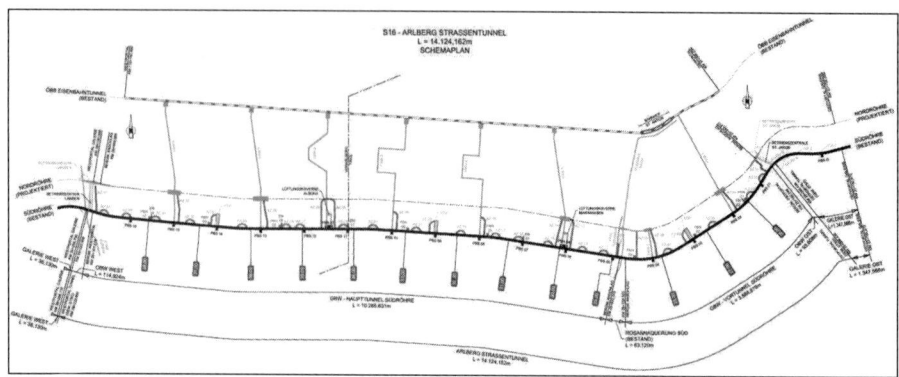

Source: Antretter, WTC 2016
Figure 1. Arlberg tunnel scheme and project overview

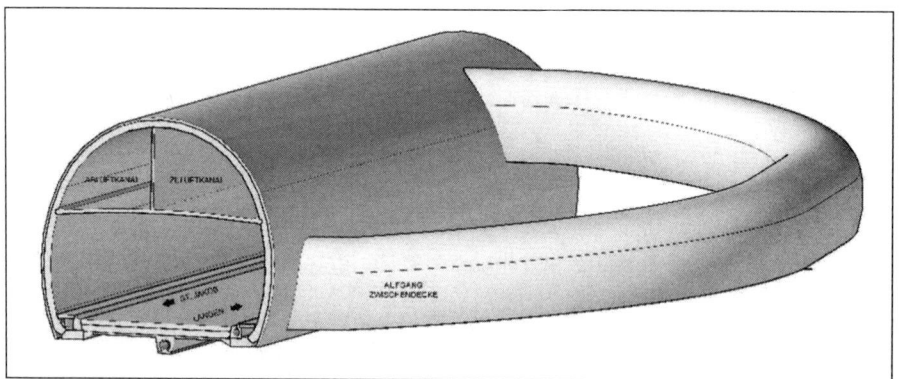

Source: Antretter, WTC 2016
Figure 2. Walkable and barrier-free escape tunnels

system will be modified and adapted to a state of the art separating system. Moreover, the whole fire emergency system will be replaced including construction of 167 new niches for improved maintenance of the tunnel drainage system.

Furthermore, three operating tunnels, each about 50 meters in length, have to be set up for medium-high voltage equipment and for transformers. For both galleries, in the east and the west side of the tunnel, with a total of 1600 m, extensive rehabilitation measures have to be realized. Thermal scanners are set at both tunnel portals, in order to prevent the entering of overheated trucks.

A water-mist spray system will be assembled above the traffic lanes to improve the chances for a safe rescue and to support fire extinguishing efforts in case of a tunnel fire. Besides the adaption and renewal of the ventilation facility, even the lighting and the whole electromechanical equipment will be replaced (video, radio, emergency and traffic control). The tunnel will also get a new reflective coat.

PREPARATION WORK

Design works for the project engineers started in November, 2010. Very soon it turned out that the complexity and size of the project would make longer periods of traffic closure necessary. Construction of a new parallel second tube (as it was done for many other Austrian road tunnels) was considered but put aside because of financial restrictions. Still it was very important to find a doable alternative that allowed to keep traffic unrestricted rolling during the winter months when a safe ride over the 1,800m high mountain pass cannot be guaranteed. A complete closing of the tunnel was finally limited to two May–November periods (2015 and 2017). During all winter months (December to April) construction was limited to night hours.

The project was divided in different parts:

- Works without traffic disturbance—for example works outside the drive lanes
- Works compromising traffic which had to been done during nights or in the periods of the complete closings.

Because of the importance of the Arlberg road tunnel and due to the complex connections between structural and electro-technical requirements, the client, ASFINAG, considered a main contractor model—a contractual set-up that has been successfully applied in similar, though smaller, tunnel renovation projects before.

A contractual model, in German called "Kreatives Bauen im Bestand," (roughly translated into "Creative Construction in Existing Structures")—got developed and allows the contractor to introduce optimizations after project award. As an additional incentive, a monetary bonus for cost savings is stipulated. The contractor receives a period of time between project award and start of construction to elaborate on optimization potential. The idea is to utilize the practical experience of the contractor and by doing so to utilize innovative ideas not only for the current project but also for future design work and similar projects. Increased efficiency as a result of this process should also result in cost reductions. Proposals get then evaluated and incorporated into the construction contract.

The Arlberg road tunnel project became a pilot project for this "Creative Construction in Existing Structures"—Model. Despite the complexity of the construction and the very special conditions on this project several innovative ideas could be implemented to the advantage of both contracting parties.

REALIZATION

In June 2014 a joint venture consisting of four companies (PKE Verkehrstechnik, Jäger Bau, G. Hinteregger & Söhne and BeMo Tunnelling) was awarded the contract. As a general contractor, this joint venture is responsible for both, planning and construction of the contracted works. As part of the vertically split joint venture PKE Verkehrstechnik GmbH is responsible for the electromechanical part of the works, whereas the actual construction is done by a joint venture consisting of the other three companies.

As a result of the technical planning a series of optimization proposals could be implemented. Examples are the video surveillance system, the stainless steel suspension of the intermediate ceiling and the construction sequence in the emergency bays which could be optimised with regard to cost reduction.

The contractual commencement of construction on September 5th, 2014, also marked the start of the first construction phase during which nine out of in total 37 walkable escape tunnels were excavated by implementing alternating night time traffic stops at the tunnel portals.

Phase I—Works During Night Hours

Without any restrictions of day traffic the construction works in Building Phase I (winter months from November 2014 to April 2015) were realized in the night hours from 8 pm to 5 am. The special challenges in this phase were a strict compliance with safety standards and regulations as wells as the sophisticated transport logistics. Particularly the transport of shotcrete and other supporting materials for excavation works and the removal of muck material had to be coordinated with the traffic flow in order to prevent long waiting times and, as a consequence, inefficiencies. Due to these circumstances the shotcrete concept for the rescue tunnels was adapted. The small quantities and the need for short-term availability of shotcrete led to the use of the so called „Schretter Halbnass-System". This system is especially developed for small tunnel profiles and combines a series of advantages, namely the immediate availability of dry shotcrete and a significant reduction of dust and rebound by premoistening the dry material. The premoistening process adds up to 60 percent of the total water needed and takes place before the dry material enters the shotcrete machine. The containers with a capacity of 6 to 10 cubicmeters get filled by silo trucks and with the dry shotcrete mixture required.

Source: Rainer Antretter

Figure 3. Concrete silo with premoistening unit and shotcrete machine

This concept got also used during summer of 2015 when the tunnel was completely closed for regular road traffic. During this time the remaining escape tunnels were built.

Phase II—Works During Full Closure

The main phase of the construction works started with the beginning of the full closure of the tunnel on April 21st, 2015. More than 400 different construction sites of various sizes had to be completed in about 6 months and were distributed over a length of approximately 14 kilometres. Besides, over 2,300 meters of drill and blast excavation with cross sections from 15 square meters (walkable rescue tunnels) up to 50 square meters (operating tunnels for energy installations) as well as 8 new emergency bays and more than 200 new fire extinguishing and dewatering niches had to be excavated and completed. Furthermore, over 17,000 m of fire extinguishing—and sewage pipes as well as sidewalks had to be renewed. To be able to fulfil this enormous variety of construction requirements the number of blue-collar workers had to be quadrupled to more than 400 at the start of this full tunnel closure. Altogether, over 650 workmen, working on 3 shifts and 24/7 were in action during this main phase (including the working crews for installation of electromechanical equipment and works ongoing outside the tunnel).

The excavation works for the rescue tunnels were done in a three-shift operation. The challenge was the one hand to drive about 30 separate tunnels with lengths between

Source: Rainer Antretter
Figure 4. Excavation works in walkable, barrier-free escape tunnels (cross section 15 m^2)

50 and 130 meters (total length around 1,500 m) within a period of only 4 months and on the other hand to optimize the utilization of equipment and working personnel.

Workgroups consisting of 7 men were formed and put in charge for construction of 2 to 3 excavation sites with one set of key equipment. By doing so it was possible to construct 9 of these emergency tunnels concurrently with 3 such workgroups (in a 24/7 operation).

In addition to the 30+ emergency rescue tunnels also the 3 new energy stations with single lengths of about 50m and the 8 new emergency bays were excavated during the time period of this first full closure.

The finishing works of the rescue tunnels and the energy stations were done subsequent to the excavation works using reinforced concrete for the invert and reinforced shotcrete for the tunnel lining.

Already with the beginning of the technical planning and optimization process the special focus was to develop an efficient and optimized process for the time-critical realisation of the 8 new emergency bays. A series of contractor proposals got implemented in the early design stage of the final design resulting in an optimized construction process:

- The actual concrete thickness of the existing lining was examined and used as a basis for further calculations.
- Instead of the planned subdivision of the tunnel cross-section into top heading and bench excavation in a full-face operation was realised. This reduced the number of construction steps significantly and led to a shorter construction duration.
- The installation of the bolts in the tunnel roof for fixing the tunnel lining above the break out of the bays could already be done in phase I. The use of a special lightweight drilling system made it possible to work from the intermediate ceiling. By doing so tunnel traffic was not compromised in any respect.

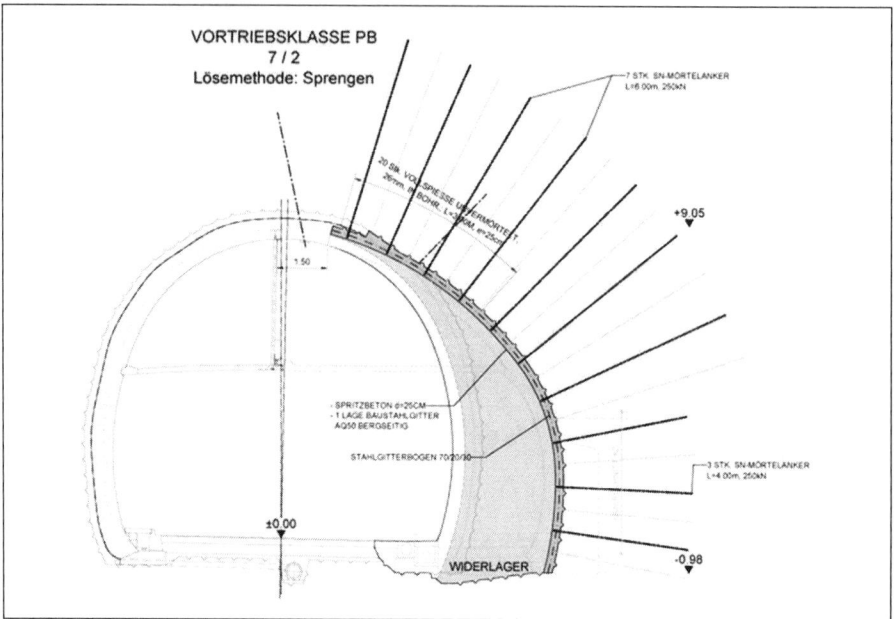

Source: Michael Hoellrigl, Eurock 2015
Figure 5. Design emergency bay excavation

The fact that the existing inner lining could already be prepared in phase I allowed for a quick and efficient start of the actual excavation works right at the beginning of the full closure period.

The existing inner lining including the intermediate ceiling had to be sawcut in predetermined sections. To increase the work safety and to minimise dust, the ceiling was dismantled in 2 meter sections. A specially prepared wheel loader was used to handle the ceiling parts.

The excavation and support works were started in the middle of each of the future emergency bays after radial perforation of the existing inner lining. The first portion of the existing inner lining was taken out applying a hydraulic hammer mounted on a tunnel excavator. This slot had to be big enough to allow for drilling of blastholes parallel to the existing tunnel and by doing so establishing the basis for an efficient drill and blast operation.

Source: Rainer Antretter
Figure 6. Excavation works in emergency bays

The blasting operation could be optimised with the goal to prevent damages of the remaining parts of intermediate ceiling and vertical partition walls. Due to geological

Source: Hoellrigl
Figure 7. Spraying of shotcrete inner lining against waterproofing membrane

and geotechnical conditions, all emergency bay excavation steps were done with a length of 1.50 meters.

Light lattice girders type 70/20/30, rebar spiles with lengths of 6 and 4 meters and 25 centimetres of reinforced shotcrete (1 layer of WWF on the outside) were installed as support materials.

Waterproofing membrane and CIP tunnel inner lining got installed at all emergency bay locations before the end of the tunnel closure, i.e., before mid of November 2015. Existing PVC waterproofing, got in some areas connected to the new waterproofing membrane in the new emergency bays. Where no old membrane was found or where it was in bad condition expandable waterproofing strips in combination with grouthoses were used to achieve the goal of a watertight construction.

The final lining below the intermediate ceiling was done with reinforced CIP-concrete, whereas in areas above the intermediate ceiling reinforced shotcrete was applied directly on the waterproofing membrane—a construction method that has been used by BeMo Tunnelling at several projects before (see for example Russia Wharf Tunnel, Boston, Massachusetts/ Weehawken Tunnel, Weehawken, New Jersey/ Tunnel Bad Wildbad, Germany/ Pfändertunnel, Austria/ and other projects).

It's remarkable that the permanent shotcrete inner lining (in areas not open to public traffic) was already part of the original project—certainly a result of good experiences from other sites. Wire mesh was used in 2 layers and kept in the right distance to the membrane and to each other by a series of "spiders" which were drilled and anchored through the membrane (BA anchors). The spiders (see figure 7) prevent the wire mesh from vibrating and allow for exact positioning of the wire mesh in a certain predetermined distance from the membrane. The actual shotcrete application has to be done in layers and by experienced nozzlemen.

The new intermediate ceiling has a thickness of 20 centimetres and is suspended from the tunnel roof with high-grade steel bars.

After reopening the tunnel for traffic a new reflective tunnel coating on the side walls as well as final touches in the escape tunnels have been done. In addition, the major part of the electromechanical equipment had to be installed or renewed.

In April, 2017, the tunnel will be closed again for 6 months to finalize all remaining works as well as the intense testing of the new equipment. The new and improved Arlberg Road Tunnel will be opened for traffic on September 26, 2017.

CONCLUSION

Execution of such a complicated and comprehensive construction site was only possible with intense preparation and by utilizing the joint capability and experience of contractor and owner to optimize construction sequences and details.

Complicated logistics, safety requirements, ventilation issues, handling of human resources and many other problems proved to be challenges in addition to the "normal" problems of a tunnelling site.

REFERENCES

Hoellrigl, M., Tschofen, J., and Wanker, Ch. 2015. Tunneling under challenging conditions—general renovation escape and safety passageways via the supply air duct at the Arlberg Road Tunnel. Eurock 2015–7.

Antretter, R. 2016. Ventilation Challenges during Consruction of Emergency Galleries. WTC 2016. San Francisco, CA.

Fuegenschuh, N. 2006. Spritzbetonanwendung unter schwierigen Verhältnissen beim Projekt "Russia Wharf Tunnel" in Boston, USA. Paper at Spritzbetontechnologie Alpbach, Austria.

Fuegenschuh, N. 2008. Användning av sprutbetong som en del av förstärkningsåtgärder i tunnelbyggnader med dåliga bergförhållanden. Paper at BK 2008, Stockholm, Sweden.

Large-Diameter Sliplining Under Extreme Conditions: Rehabilitating the Oakland-Macomb Interceptor While Maintaining Service to 830,000 Customers

Curtis Rozelle ▪ Jay Dee Contractors, Inc.
Abdul-Ghani Mekkaoui ▪ Jay Dee Contractors, Inc.
Fritz Klingler ▪ FK Engineering Assoc.
Saju Sachidanandan ▪ NTH Consultants, Ltd.
Sid Lockhart ▪ Office of the Oakland County Water Resources Commissioner

ABSTRACT

The Oakland Macomb Interceptor Drain (OMID) serves 830,000 residents in Macomb and Oakland Counties, providing sanitary sewer service to large portions of the northern suburbs of Detroit. The sewer has had multiple failures since its construction in the 1970s and was in need of major repairs. The contracts preceding Contract 4 involved constructing flow control facilities and interceptor repairs for preparation of sliplining. OMID Contract 4 was comprised of sliplining four runs of large diameter interceptor over 11 miles. The sliplining lengths in PCI-5, PCI-6, PCI-7 and PCI-8 were approximately 14,760 linear feet (LF), 3,300 LF, 3200 LF, and 4,200 LF, respectively, and the finish pipe diameter ranged from 8-foot to 10-foot internal diameter (ID). The annulus was filled with over 30,000 cubic yards (CY) of cellular grout. Jay Dee Contractors (JDC) had to overcome many challenges during this project which included: flow control management, long travel distances with pipe, difficult tunnel geometry, and logistics of working in a live sewer. This paper presents a brief history of the interceptor and the design process, followed by a review of the methods JDC used to overcome the unique challenges.

INTRODUCTION

In December 2016, almost eight years after design was initiated, rehabilitation of the Oakland Macomb Interceptor was completed. This project was unprecedented in scale, complexity, and innovation.

Contract 4 of the rehabilitation was almost certainly the most complex and difficult part of the overall program. Because of the non-redundancy of the OMID, a series of flow control gates (constructed under Contracts 1 and 2) were utilized, allowing for in-system storage and simultaneous repair without ever interrupting service to the 830,000 upstream users. Most of the nine major access points used under Contract 4 were directly under 480 KV power transmission lines, making construction and access to the sewer extremely challenging. Sewer rehabilitation was accomplished within discrete areas of the sewer that were drained each day, while simultaneously storing sewage within other sections; then releasing flow at night. Tunnel rehabilitation was accomplished using a number of methods, including surface and in-tunnel grouting, shotcrete lining, cured-in-place lining, and over 26,000 LF of complete sliplining.

HISTORY OF THE OMID SYSTEM

The OMID is approximately 20 miles in overall length, and generally flows from north to south, terminating at the Detroit Water and Sewerage Department's (DWSD) Northeast

Sewage Pumping Station (NESPS) just south of 8-Mile Road. The diameter of the sewer ranges from 42 inches at one of the upstream extremities, to 12.75 feet in the southern half of the system.

During the original construction of the OMID in the early 1970s, a tributary sewer failed during its construction, as a result of soil and groundwater seeping through relatively small cracks in the unreinforced tunnel lining near a manhole penetration. In 1980, a section of the OMID itself failed, due to a similar failure mechanism. Specifically, the failure mechanism was determined to involve cracks in the unreinforced liner, combined with a high-water table and silt/fine sand surrounding the tunnel liner, which allowed excessive amounts of surrounding fine soils to be piped through the lining. This created voids and removed outside support on the lining, resulting in collapse. The failed section of the interceptor together with upstream and downstream sections also displaying distress were repaired in 1980 through 1982 by sliplining approximately 2,300 LF with ASTM C-76 reinforced concrete pipe.

Following a catastrophic collapse in 2004 of a nearby section of interceptor that flows into the OMID, sewer service was seriously threatened for all the upstream users of the system. The repair in 2004 involved a 10-month 24/7 construction and emergency bypass effort that was completed in 2005, and ultimately cost rate payers over $56 million. More critically, and because it was known that the soil and groundwater conditions similar to

Figure 1. Aerial photograph from 2004 catastrophic collapse

Figure 2. Construction of large-diameter access shaft to accommodate GFRPMP pipe sliplining

those contributing to the previous failures were present throughout the OMID alignment, the 2004 collapse (together with the experiences of the previous collapses) alerted the engineers maintaining the OMID, that the entire system was vulnerable to similar future collapse in other locations.

Because of this potential danger to the OMID, the remainder of the system was inspected in the years following the 2004–2005 repair. The investigation indicated there were extensive areas in various portions of the OMID with cracking, fracturing, infiltrations, and eroding/corroding concrete liner. Geophysical investigations confirmed these conditions, and more specifically identified probable areas of loose ground (indicating possible soil loss). In 2009, the Detroit Water and Sewerage Department (DWSD) transferred the system to the newly created OMID Drainage

District (OMIDDD), and the OMIDDD then undertook a $170 million 7-year repair program, ultimately to be conducted under 6 separate contracts.

SUMMARY OF THE 2009–2016 REPAIRS

Contracts 1 and 2 (Segment 1) of the program involved construction and installation of flow control and access structures at six points along the sewer system in Sterling Heights and Warren. Contract 3 (Segment 2) of the program involved leak sealing and spot repairs in the lower 10 miles of the system (Reaches PCI-5, PCI-6, PCI-7, and PCI-8 as shown in Figure 1). This work was designed to prepare the deteriorated pipe in advance of the slip-lining that was accomplished under Contract 4.

Contract 4 (Segment 3) of program has involved lining about 26,000 LF of tunnel with glass fiber polymer mortar pipe, within four sections of the OMID system between Metro Parkway (16-Mile Road) and the Northeast Sewage Pump Station (just south of 8-Mile Road).

Simultaneously during the completion of Contract 4, Contracts 5 and 6 (Segment 4) of the OMID repair program were undertaken and completed. These repairs included intermittent lining and internal spot repairs to the northern half of the OMID sewer system (upstream of Contract 4). The need for coordination of flow control within the overall system during this simultaneous work further complicated the Contract 4 effort.

Figure 3. Grouting suspected voids under Contract 3 prior to GFRPMP pipe installation

EXISTING OMID CONSTRUCTION CONDITIONS

Based on the as-built information, the portions of the OMID that were repaired under Contract 4 were constructed in-tunnel, with a primary rib and lagging liner and a secondary unreinforced cast-in-place concrete liner. The sewer extents along a 480 KV power corridor, below various major roads and freeway, and (due to alignment issues during construction) below several large industrial buildings.

The original tunneling was completed through varying subsurface conditions, including granular soils found significantly below the groundwater elevation. The sewer was constructed with many grade changes due to the varying subsurface conditions and difficulties experienced during groundwater control.

The corrosive environment of the interceptor can generally be described as containing moderate to high concentrations of hydrogen sulfide, and smaller quantities of other potentially corrosive compounds. The high hydrogen sulfide readings tend to be localized downstream of connections, and downstream of flow control structures. There have been periodic odor complaints from property owners along the OMID alignment over many years, attesting to the high concentrations of hydrogen sulfide produced by the sewer, and the high potential for conditions leading to proliferation of Thiobacillus bacteria and related sulfuric acid production. During various investigations conducted between 2006 through 2008, pH readings as low as 1.0 have been measured on the sewer walls. Such conditions have resulted in section loss as great as 3 inches, with steel reinforcing (where it exists) exposed in some areas.

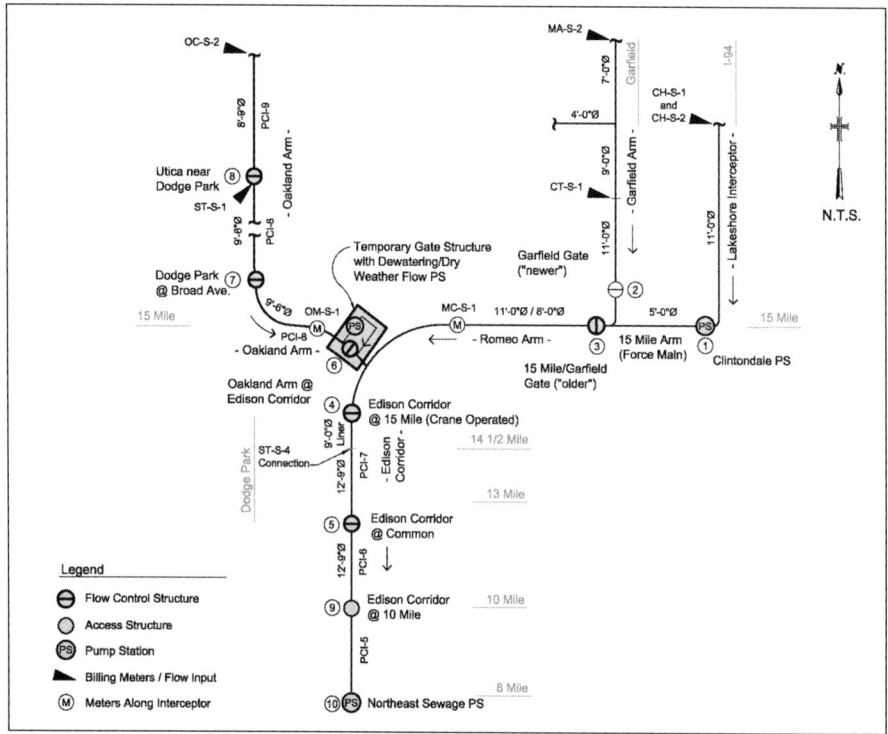

Figure 4. Layout of Oakland-Macomb interceptor drain

FLOW CONTROL AND MANAGEMENT SYSTEM

A series of flow control structures were constructed under Segment 1 to accommodate subsequent in-sewer repairs. In addition, the NESPS was modified under Segment 2, to facilitate the dewatering of the downstream portion of the interceptor (PCI-5) for the purposes of lining rehabilitation and other repairs. The design concept for the flow control allowed for sewer rehabilitation within discrete areas of the sewer that were drained each day, while simultaneously storing sewage within other sections; then releasing flow at night.

The flow control structures were designed such that the work could be performed during dry weather conditions, which was generally defined by the Contract Documents as any period excluding precipitation events greater than ½ inch over the service area. The design modeling predicted that wet weather events greater than ½ inch would result in elevated levels and flow rates in the OMID system along with the potential for surcharging. During such conditions (expected about 60 days per year), all work in the sewer was prohibited.

Flow storage was primarily accomplished by five remotely operated gates (CS-3, CS-5, CS-6, CS-7, and CS-8) and two remotely operated pumps (Pump CS6 and NESPS Pump 4). The flow control system operation was designed so that the Contractor's flow control operator could monitor and control all flows in the sewer remotely from any location, through a web-based portal. Each morning the incoming weather conditions together with sewer levels could be assessed; the gates closed to isolate sewer sections and begin the storage sequence; and pumping stations activated to dewater certain sections. Throughout the work day, the safety of the sewer could be assessed remotely for access of equipment and work crews; and if weather conditions warranted, the notification could be made for removal of personnel and equipment from the interceptor, then return of the remote gates to the open position.

DESIGN APPROACH

During the initial planning and basis-of-design stages of the project, it became clear that full scale relining of the pipe would be necessary for a major portion of the existing alignment. At the same time, a number of challenges were recognized, that together might severely limit the options for such relining. Major project challenges included:

- Sewer Depth and Access: The area of the sewer to be relined is 80 to 100 feet below ground surface, making entry and egress from existing shaft locations difficult. Further, the depth and ground conditions ensured that any new access shafts would be a major cost to the project. Because of this, the design had to balance the number of new shafts with the cost of long-haul sliplining.

- Ground Conditions: The sewer extends through several areas of predominantly silt and fine sands. In addition, the depth below groundwater table results in groundwater pressure head against the lining of 75 to 100 ft.—more than enough to drive silt and fine sand particles through very small cracks in the existing lining or any new lining. Therefore, the new lining needed to be essentially impermeable.

- Flow Control: The sewer dewatering necessary to provide for sliplining and other work could only be accomplished in 7 to 15 hour intervals, depending on the location; and work each day would be followed by release of flow through the work area. Because of this, all insertion of materials, men and equipment, all construction, and all removal of men and equipment had to be accomplished in the most efficient means possible.

- Initial Sewer Construction: Most of the sewer was constructed as a tunnel in deep soft ground with a steel rib primary liner and an unreinforced cast-in-place concrete secondary liner. The type of sewer wall combined with the evidenced deterioration and the ground conditions, resulted in a requirement for a full structural lining.

Initial efforts to identify potential lining options that would adequately address the challenges revealed a very limited number of products that might be adequate. In fact, no product could be identified that possessed a proven track record under similar conditions and project requirements. Further, research identified a number of examples where one or more of the conditions of the OMID project had resulted in major problems (and even catastrophic failures) with respect to lining installation and/or performance. In short, it became clear that given the depth, diameter, and system constraints involved in the proposed relining under Contract 4, the scope and scale of the proposed work was unprecedented.

Based on initial research, it appeared that certain products might be able to meet the challenges, although the available manufacturer information was inadequate to make a final determination. One thing that was certain, was that the pool of prospective product providers would be very limited, and price competition would be minimal, unless several options could be considered by the Contract 4 bidders. For these reasons, a two-stage approach to bidding was developed. Stage 1 consisted of a "Request for Technical Submission" (RFTS) phase where lining manufacturers were invited to submit detailed product information. After detailed review and analysis of manufacturer's submittals, seven products were "pre-approved." Stage 2 consisted of a final design and bidding stage, during which contractors were invited to bid on a base and alternate scope of work, using one (or a combination) of the pre-approved lining systems.

FINAL DESIGN AND BIDDING

The Contract 4 final design included lining all of PCI-5, sections of PCI-6, PCI-7 and PCI-8, and performing localized grouting in PCI-8. In addition, the design called for the Contractor to construct up to six access shafts, as necessary for their operation. Within PCI-5, 14,760 LF of sewer was designed to be re-lined to an inside diameter of 10 feet. Within PCI-6, about 3,300 feet of sewer below the I-696 expressway was designed to be re-lined with an inside diameter of 10 feet. Within the Northern Section of PCI-7, about 2,400 feet of 9-foot diameter sewer was designed to be re-lined to an inside diameter of 8 feet.; and 830 feet of the 12.75-foot diameter sewer was designed to be re-lined to an inside diameter of 10 feet. Within PCI-8, repairs consisted of slip-lining of about 4,100 feet of sewer to a finished 8-foot diameter, and grouting and leak sealing from within the sewer.

The Engineer's opinion of probable cost was $64.1 million. A total of seven base bids and two alternate bids were received. Base bids ranged from $46.4 million to $68.9 million, and alternate bids ranged from $43.7 million to $60.7 million. After evaluation, the low base bid, submitted by Jay Dee Contractors, Inc. (JDC), was determined to have the greatest value to the Owner, and Contract 4 was awarded to JDC for the low base bid amount.

FLOW CONTROL CHALLENGES

The primary challenge was the available number of working hours In the interceptor. The Contract Documents specified storage durations of 10 hours in PCI-5 & 6, 7 hours in PCI-7 and 6 hours in PCI-8. Further, if work was to take place in PCI-7 and PCI-5 or 6 concurrently, the storage time was 4.5 hours and 10 hours respectively. Due to contract schedule constraints, JDC determined it was necessary to install pipe at two locations concurrently. One crew would install pipe on the north end of the job in PCI-8 and PCI-7, while another crew would install pipe on the south end in PCI-6 and PCI-5. Working hours in the interceptor at the upstream end of PCI-8 were very limited. JDC coordinated with the Flow Control Manager to perform observational storage tests to improve working durations. During the first week of concurrent pipe installation in PCI-8 and PCI-6, gate closing times and crew start times were adjusted to determine the best combination of start time and gate closing times. After a week of testing, JDC determined the best way to maximize working hours was to start the PCI-8 crew 2 hours before the PCI-6 crew and shut the downstream gate (CS-5) 2 hours before the PCI-6 crew started. Crews in PCI-8 were able to get up to 8 hours working time in the interceptor while crews in PCI-6 were able to get up to 10 hours.

Once crews moved downstream, a different set of challenges were encountered. In PCI-5, crews had to work adjacent to the NESPS while crews in PCI-7 were working immediately downstream of a manual stop gate (CS-4) that had no remote monitoring. It quickly became apparent that the previously described flow control method would not work because leaving CS-5 gate open until the PCI-7/8 crew started, created too long of drainage time and the PCI-5 crew was delayed entering the interceptor. On the other hand, if the CS-5 gate was closed too early and the run between CS-4 and CS-5 was not sufficiently drained, the water would back up and flood out the CS-7 crew.

After another week of adjusting start times and gate closing times, JDC and the Flow Control Manager determined that CS-4 gate had to be closed 2 hours before the PCI-7 crews start time and CS-5 gate would be closed 1 hours later. The crew in PCI-5 was then able to start 2 hours after the CS-5 gate was closed and crews in PCI-7 were able to start 1 hour after CS-4 gate was closed. Crews at the NESPS were still impacted by the long drainage durations controlled by number and type of pumps available at the downstream pump station, but as they worked north the times significantly improved.

PCI-5 & PCI-6 CONSTRUCTION

JDC elected to slipline both PCI-5 and PCI-6 with one of the base-bid approved products, which was a centrifugally cast Glass Fiber Reinforced Concrete Mortar Pipe (GFRPMP), produced in the USA. JDC had considerable past experience using GFRPMP, and could attest to its reputation as a viable slipline material. JDC constructed an additional 27-foot ID tangent pile Access Structure just south of Toepfer Road, at PCI-5 station 59+55. The tangent piles were drilled from existing ground surface, 120 feet deep. Deep gravity and educator type dewatering wells were used to control ground water. JDC utilized this structure to install pipe from the midway point to CS-9 (STA 90+00) back, and from the NESPS at STA 0+00 back. In PCI-6 JDC utilized an existing structure, Control Structure 9 (CS-9), located approximately 5,400 linear feet (LF) from the northern most pipe installation zone and 5,600 LF from STA 90+00 in PCI-5. Since CS-6 was existing and required minor modifications, JDC's installed the northern most pipe first and used this as a field test.

Since flows needed to be released every night and the pipe installation in PCI-5 and PCI-6 required JDC to carry the pipe far distances, fast moving mobile equipment was developed and constructed specifically for this project. The pipe carrier was a "motorcycle" style rubber tire carrier that could carry 20 foot sections of pipe and set them in place. In order to install blocking, JDC added work decks on small utility vehicles to allow crews to access the crown of the host pipe and the polymer mortar pipe. The vehicles were sized to fit inside installed liner pipe and allow the pipe carrier to deliver and home additional polymer mortar pipe in one coordinated operation. This method allowed personnel to work at the heading while the carrier was picking up pipe from the shaft.

Figure 5. Lowering motorcycle style pipe carrier into access structure

Determining and implementing the proper blocking scheme was critical to successful installation and grouting of the pipe. JDC filled the annulus with cellular grout. JDC used blocking at the crown and boards in the invert to help maintain tunnel grade.

After record rain events in the summer of 2014 resulting in the rare event of the interceptor being surcharged, JDC discovered sections of installed but ungrouted pipe had been displaced. It was determined that during the filling and dewatering of the interceptor, water filled and drained behind the slip lined pipe at substantially slower rates than pipe interior, resulting in the pipe blocking being subject to much greater forces than anticipated. Following this event, JDC coordinated with the Engineer to increase the blocking for full hydrostatic uplift. The revised system included additional crown blocking and blocking at springline. The more robust blocking system proved to be adequate for all different load conditions subjected to the pipe during different phases of construction.

Grouting of the slip lined pipe was self-performed by JDC. Due to the uncertainty of working hours and pipe installation schedule, it was not feasible or practical to have a grouting subcontractor on site. The volume in the annulus that needed to be filled was approximately 1.52 CY of cellular concrete fill per LF. JDC pumped the ready-mix slurry and foam was injected inline to create cellular grout with

Figure 6. JDC's crew installing blocking for full hydrostatic uplift in PCI-5 and PCI-6

a minimum compressive strength of 150 psi. Since the pipe was blocked for 100% of hydrostatic uplift in the revised design, theoretically the pipe could be grouted in one lift, but this was not feasible based on the ready-mix supplier's ability to supply large quantities of slurry and working hour limitations. Grout was pumped through ports which were previously installed in the pipe. Heavier grout was used in the invert to displace any trapped water. Laser survey and diameter measurements of the liner were performed during the grouting operations to monitor to pipe to ensure no deformations were occurring that exceeded the manufactures recommendations. Bulkheads were installed halfway between manholes so the grout crew could grout the pipe while pipe installation progressed. Utilizing two crews for grouting and pipe installation allowed JDC to gain schedule time.

All in all, JDC's method of utilizing a high-speed pipe carrier and simultaneously grouting the pipe allowed PCI 5 and PCI-6 to be completed on time. Construction methods were modified during installation based on the crews' feedback, observations, and lessons learned during the course of the project. In PCI-5 and PCI-6, 14,600 LF and 3,430 LF, respectively, were successfully installed and grouted in place.

PCI-7 & PCI-8 CONSTRUCTION

Similar to PCI-5 and PCI-6, JDC used GFRPMP to line the PCI-7 and PCI-8 sections. The majority of the pipe was 8-foot ID pipe with a short 830 LF section in PCI-7 being 10-foot ID pipe. PCI-8 had an existing flow control structure (CS-6) in place that was utilized for installation of 4,200 LF of pipe and JDC excavated a 20-foot ID shaft and removed a 11.5-foot long section of tunnel for pipe installation in PCI-7. Both pipe runs presented a unique set of challenges.

PCI-8s as-built drawings depicted a 190-foot radius curve immediately upstream of CS-6 and one more 180-foot radius curve 2,700 LF upstream. Upon initial investigation, it was apparent the curves were not smooth but tangents of varying lengths. JDC mobilized a survey crew and measured the deflections of each tangent and determine pipe lengths needed. The surveyors determined a 20-foot section of pipe would be able to be carried around the curves. This was confirmed by walking a mandrel through the tunnel. A pipe lay schedule was developed utilizing varying pipe lengths, some with elbows, in order for the pipe length to match the alignment tangents.

In addition to the flow control challenges discussed above, JDC had to manage a 24-inch 3.0 CFS live connection (ST-S-2) 2,750 LF upstream of CS-6, which was in the second curve. During sewer investigation and planning it was apparent the flow would need to be stopped or bypassed for pipe installation. JDC explored bypass pumping, but running piping on the surface would be too obstructive to the public and since the connection connects near surface sewers, storage times were unknown. After further field investigation and reviewing old as-built drawings from the local municipality (Sterling Heights), an old existing sewer that appeared to connect ST-S-2 to the interceptor upstream of CS-7 gate was discovered. JDC coordinated field testing with Sterling Heights and confirmed the sewer could be plugged and flows would passively divert upstream of CS-7. The plug was installed and removed every work day, which allowed JDC to work in

Figure 7. JDC testing rail mounted pipe carrier

the interceptor with lowest flow conditions possible, and construct the polymer mortar pipe connection to ST-S-2 to the interceptor in the dry.

Since pipe needed to be installed at two locations simultaneously, JDC had to develop a second string of pipe installation equipment. Based on past projects JDC completed, it was determined a locomotive and rail based pipe carrier would be most cost effective. JDC utilized existing Control Structure CS-6 to install the lining pipe in PCI-8.

PCI-7 consisted of lining a previously collapsed and slip-lined zone with 2,400 LF of 8-foot diameter pipe and then installing a transition from 8' diameter to 10' diameter pipe followed by 830 LF of 10' diameter pipe.

JDC constructed an additional 18-foot ID Access Structure south of CS-4 (just north of the Redrun Drain). Steel ribs and wood boards were utilized to support the excavation. Deep gravity type dewatering wells were utilized to control ground water.

The PCI-7 as-built drawings showed a distorted invert, and upon investigation JDC found water was ponding up to 21 inches deep. To mitigate the water depth JDC used a small loader to push the water out of the dips upstream to CS-4 and then

the water was pumped behind CS-4. This allowed JDC crews to work in relatively dry conditions.

JDC used the same cellular grout as PCI-5 and PCI-6. After problems in PCI-6 with pipe displacement, JDC adjusted the blocking to allow for 100% uplift of hydrostatic pressure.

JDC also self performed the grouting of the pipe in PCI-6 and PCI-7. Since the grout volumes were only .8 CY/LF of tunnel, JDC coordinated with the ready-mix supplier to deliver short loads and JDC

Figure 8. JDC crew installing pipe blocking in PCI-8

pumped foam into the trucks to expand the volume of slurry. The foamed slurry was then pumped through a pump and injected through grout ports in the pipe. The grouting was done in several lifts, with the first lift utilizing heavier grout to expel the water from the invert, then the following lifts completely filled the annulus. This grouting method proved to be effective.

All in all, JDC's simple approach to installing pipe in the shorter runs on the north end of the project proved to be successful. Crews had to overcome short working hours and difficult geometric challenges to install 3,100 LF and 4,200 LF of pipe in PCI-7 and PCI-8, respectively.

LESSONS LEARNED

Although there were many lessons learned from this never-before-attempted set of challenges that comprised this project, several lessons stand out:

- The approach of advertising and prequalifying products, successfully allowed for product suppliers, contractors, and the designers, to better understand the challenges ahead and to plan, design and implement a successful project under very adverse conditions.

- The approach of allowing the Contractor to choose the number and locations of access shafts versus and balance that cost with the cost of long-hauling pipe, resulted in a very cost effective bid and finished project.

- Pipe blocking design is an extremely critical component to a successful slip-line pipe installation and grouting project. JDC originally designed the pipe blocking for the uplift forces from grout, but it after working in the interceptor it became apparent that a more robust blocking system would need to be used to accommodate the full uplift from water given all potential conditions. JDC worked closely with the Engineer and the manufacturer to determine a blocking systems that met the requirements and could feasibly be installed, which involved short term problems, but ultimately lead to a very successful project.

- Working in a live interceptor proved to be a very challenging task, but one that was successfully handled using the unique remote flow control system developed for this project. As previously discussed, JDC performed many flow control tests to determine the best working hours. Since the flows were controlled by gates, there were times when debris would obstruct gates and they could not be closed completely, thereby shortening work hours. Also, the pumps at the NESPS were in the process of being repaired, and therefore the tunnel was not always turned over the JDC in a fully dewatered condition. To help mitigate these problems, JDC was in constant communication with all entities to ensure work could proceed the following work day. JDC's crews were flexible with start and finish time to allow maximum working hours in the interceptor. Coordination and good communication between JDC, the Engineer/Owner and all third parties was critical to the completion of this project.

CONCLUSION

OMID Contract 4 involved installing over 21,200 LF of 10-foot diameter and 6,600 LF of 8-foot diameter pipe and backfilling the annulus with over 30,000 CY of cellular grout. Challenges faced during this precedent-setting repair program were immense. From the early schedule and funding challenges, to public involvement, to technical innovation necessary to maintain service to upstream users, to the technical innovation necessary to accomplish the construction, the overall project was unique in the industry.

At the completion of this project, the public is left with a completely rehabilitated sewer, a remote flow control system that will be useful into the future in the control and management of flow, and various new access shafts to allow for future maintenance and repair. Further, because of the methods used to repair and line the sewer, groundwater infiltration has been reduced by approximately 90 percent, resulting in an estimated savings of $10 million in sewage pumping and treatment costs.

The project owners, designers, and contractor have been privileged to have contributed to this unprecedented effort.

REFERENCES

[1] Klingler, F., Edberg, J., McMahon, M., Lockhart, S., and Groboske, K., "Balancing Innovation, Risk, and Cost: Innovative Procurement Approach for Sewer Tunnel Rehabilitation" Proceedings of the North American Tunneling Conference 2014, 22–25 June, 2014, Los Angeles, U.S.A., pp. 618–623. Society for Mining, Metallurgy, and Exploration, Inc., Englewood Colorado, 2014.

[2] Miller, M.S., Klingler, F.J., and Alberts, J.B., "Rehabilitation of the Oakland Macomb Interceptor Drain: Innovative Design of Multi-Tiered Shafts to Accommodate Low Overhead Clearance" Proceedings of the North American Tunneling Conference 2012, 25–27 June, 2012, Indianapolis Indiana, U.S.A., pp. 806–812. Society for Mining, Metallurgy, and Exploration, Inc., Englewood Colorado, 2012.

[3] Mercado, V., Fujita, G., Shukla, R., Price, H.R., Swaffar, K.M., Klingler, F.J., "Emergency Repair of Oakland-Macomb Sewer Collapse: A Case History" Proceedings of the North American Tunneling Conference 2006, 11–14 June, 2006, Chicago, Illinois, U.S.A., pp. 415–424. Taylor & Francis Group plc, London, UK, 2006.

Index